LIE GROUPS AND THEIR REPRESENTATIONS

LIE GROUPS AND THEIR REPRESENTATIONS

SUMMER SCHOOL OF THE BOLYAI JÁNOS MATHEMATICAL SOCIETY

EDITED BY

I. M. GELFAND

A HALSTED PRESS BOOK

JOHN WILEY & SONS

New York — Toronto

Proceedings of the Summer School on Group Representations,
Bolyai János Mathematical Society
(Budapest, 1971)

Technical editor
K. MÁLYUSZ

Published in the U.S.A. and Canada by Halsted Press,
a Division of John Wiley & Sons, Inc., New York.

Library of Congress Cataloging in Publication Data

Summer School on Group Representations, Budapest, 1971.
Lie groups and their representations.
"A Halsted Press book"
1. Lie groups — Congresses. 2. Representations of groups — Congresses.
I. Gelfand, Izrail Moiseevich.
II. Bolyai János Matematikai Társulat. III. Title.
QA387. S86 1971 512'.55 75—4811

ISBN 0—470—29600—3

Published as a co-edition of
ADAM HILGER LTD
Rank Precision Industries Ltd 29 King Street London WC2E 8JH
and
AKADÉMIAI KIADÓ, BUDAPEST

Printed in Hungary at the Szegedi Nyomda

CONTENTS

PREFACE

This volume is a collection of lectures given at a Summer School in Budapest 1971. The participants of the School express their thanks to the Bolyai János Mathematical Society for the organization of the conference and to the Hungarian Academy of Sciences for their assistance.

The reader will observe a strengthening connection between group representations and number theory in the titles already. The connections with quantum mechanics, field theory, harmonic analysis and number theory, and the role of deep ideas common with algebraic and differential topology as well as algebraic geometry, make representation theory an important part of mathematics. Some of these connections are reflected in the lectures of the School.

Besides the scientific reports presented in this volume a series of introductory lectures has been held. Since the readers of the two parts may not be the same, the introductory lectures will be published in a separate volume together with the special lectures held by Professor G. W. Mackey for physicists.

The following are the contents of the companion volume:

Andrianov, A. N: Local fields

Kirillov, A. A.: Introduction to the representation theory of finite and compact groups

Bernstein, I. N., Gelfand, S. I.: Finite-dimensional representations of semi-simple Lie algebras

Želobenko, D. P.: Gelfand—Cetlin formulas for representations of $GL(n, \mathbf{C})$

Andrianov, A. N.: Modular forms, zeta functions and elliptic curves

Tanaka, S.: Representations of $SL(2, \mathbf{F}_q)$

Mackey, G. W.: Induced representations and their applications to quantum mechanics

The introductory course on representation theory was a necessary part of the School and we hope it succeeded in raising interest among young participants, mathematicians, physicists, and other scientists, in this field of mathematics.

The participants of the School are obliged to Academician B. Szőkefalvi-Nagy for the support and help he has given in organizing the Summer School and in editing the present volume. The participants of the School fully acknowledge that without the activity of K. Mályusz this School would have been scarcely possible. His selfless work reached far beyond the usual responsibilities of the scientific secretary of a conference.

I. M. Gelfand

ZETA FUNCTIONS AND THE SIEGEL MODULAR FORMS

A. N. ANDRIANOV

Automorphic forms and zeta functions are probably the area of number theory where the influence of representation theory is most apparent. (Or conversely, should we speak about the influence of number theory in representation theory?) A convincing example is the connection between modular forms of one variable and the theory of infinite-dimensional representations of the group of second order matrices. Unfortunately, in the multidimensional case we are far from that kind of a solid and complete theory. One of the reasons is that in the majority of many-dimensional cases there was no success, so far, in constructing a satisfactory theory of zeta functions corresponding to automorphic forms, analogous to Hecke's theory in the one-dimensional case. In this lecture we set forth the bases of such a theory for the case of Siegel modular forms of genus 2. A detailed presentation of the results mentioned below, can be found in the paper [1] of the author.

§ 1. Modular forms

The problem of integer representations of the natural numbers by positive-definite integral quadratic forms naturally reduces to modular forms for subgroups of the modular group $SL_2(\mathbf{Z})$. In a similar way, the problem of representations of quadratic forms of n variables by positive definite integral quadratic forms of m variables, $m \geqq n$, reduces to modular forms for subgroups of the Siegel modular group $\mathrm{Sp}_n(\mathbf{Z})$ [2].

In this § we briefly recall the facts and definitions about modular forms for the full Siegel modular group $\Gamma_n = \mathrm{Sp}_n(\mathbf{Z})$, necessary for our purpose. For the general theory see [3], [4].

Notations. $\mathbf{Z}$, $\mathbf{Q}$, $\mathbf{R}$, $\mathbf{C}$ denote the ring of rational integers and the fields of rational, real and complex numbers, respectively. For any commutative ring $\mathbf{A}$ with unit element, write $\mathrm{M}_n(\mathbf{A})$, $GL_n(\mathbf{A})$, $SL_n(\mathbf{A})$ for the ring of quadratic matrices of order n, the full linear and the special linear group over $\mathbf{A}$ respectively. E_n stands for the unit element of the ring $\mathrm{M}_n(\mathbf{A})$. For $X \in \mathrm{M}_n(\mathbf{A})$, ${}^tX$ denotes the transposed matrix and $\sigma(X)$ the trace of X.

The upper Siegel half-space of genus n

$$\mathfrak{Z}_n = \{Z = X+iY \in \mathrm{M}_n(\mathbf{C}) | {}^tZ = Z, Y > 0\} \tag{1.1}$$

obviously has the structure of $\frac{n(n+1)}{2}$-dimensional complex manifold. The real symplectic group of genus n

$$\mathrm{Sp}_n(\mathbf{R}) = \left\{ M \in \mathrm{M}_{2n}(\mathbf{R}) | {}^tM \begin{pmatrix} 0 & E_n \\ -E_n & 0 \end{pmatrix} M = \begin{pmatrix} 0 & E_n \\ -E_n & 0 \end{pmatrix} \right\} \tag{1.2}$$

acts on $\mathfrak{Z}_n$ as a group of analytic automorphisms

$$M: Z \to M\langle Z \rangle = (AZ+B)(CZ+D)^{-1}, \quad \left(Z \in \mathfrak{Z}_n, M = \begin{pmatrix} A & B \\ C & D \end{pmatrix} \in \mathrm{Sp}_n(\mathbf{R}) \right). \tag{1.3}$$

Denote by $\Gamma = \Gamma_n = \mathrm{Sp}_n(\mathbf{Z}) = \mathrm{Sp}_n(\mathbf{R}) \cap \mathrm{M}_{2n}(\mathbf{Z})$ the Siegel modular group of genus n.

A function $f(Z)$, holomorphic on $\mathfrak{Z}_n$, is called a *modular form* of *genus n* and *weight k* ($k > 0$, integer) if it satisfies the following two conditions:

1) for every $M = \begin{pmatrix} A & B \\ C & D \end{pmatrix} \in \Gamma_n$,

$$f(M\langle Z \rangle) = \det(CZ+D)^k f(Z), \tag{1.4}$$

2) $f(Z)$ is bounded in every domain of the form

$$\{Z = X+iY \in \mathfrak{Z}_n | Y \geqq cE_n \quad (c > 0)\}.$$

In the case $n > 1$ any holomorphic f satisfying *1)* also satisfies *2)* (see [5], [4]). The set $\mathfrak{M}_k^n$ of all modular forms of genus n and weight k is a finite-dimensional vector space over $\mathbf{C}$. Every form $f \in \mathfrak{M}_k^n$ expands into a Fourier series

$$f(Z) = \sum_{N \in \mathcal{N}_n} a(N) \exp(2\pi i \sigma(NZ)), \tag{1.5}$$

where $N = (n_{ij})$ runs through the set $\mathcal{N}_n$ of all symmetric half-integer matrices (i.e. for which $n_{ii}, 2n_{ij} \in \mathbf{Z}$) of order n, satisfying $N \geqq 0$. The Fourier coefficients $a(N)$ of f depend, in fact, only on the class of the matrix N in the sense of proper equivalence of the corresponding quadratic forms, i.e. $a({}^tUNU) = a(N)$ for any $N \in \mathcal{N}_n$, $U \in SL_n(\mathbf{Z})$.

The spaces $\mathfrak{M}_k^n$ for different n are naturally connected by the Siegel operator Φ:

$$(\Phi f)(Z_1) = \lim_{\lambda \to \infty} f \begin{pmatrix} Z_1 & 0 \\ 0 & i\lambda \end{pmatrix} \quad (f \in \mathfrak{M}_k^n, \ Z_1 \in \mathfrak{Z}_{n-1}, \ n > 1).$$

It turns out that the limit exists and $\Phi f \in \mathfrak{M}_k^{n-1}$ [4]. As shown by Maass [6], for a sufficiently large integer k ($> 2n+1$) the homomorphism $\Phi: \mathfrak{M}_k^n \to \mathfrak{M}_k^{n-1}$ is an epimorphism.

Define Φ for $n=1$ by the condition $\Phi f = \lim_{\lambda \to \infty} f(i\lambda)$ $(f \in \mathfrak{M}_k^1)$, and put for all $n \geqq 1$

$$\mathfrak{N}_k^n = \{f \in \mathfrak{M}_k^n \mid \Phi f = 0\}.$$

The forms in $\mathfrak{N}_k^n$ are called *parabolic forms* of genus n and weight k, and can be characterized by the condition that in their Fourier expansion (1.5) all coefficients $a(N)$ with $\det N = 0$ equal zero. The following estimates are well known for parabolic forms and their Fourier coefficients [7]: if $a(N)$ are the Fourier coefficients of $f \in \mathfrak{N}_k^n$ $(n \geqq 1)$, then

$$|f(Z)| = O\big((\det Y)^{-k/2}\big), \quad |a(N)| = O\big((\det N)^{k/2}\big), \tag{1.6}$$

where $Z = X + iY \in \mathfrak{Z}_n$, and the constants contained in O depend only on f.

Using the mapping Φ we can reduce many problems about modular forms (in particular, those connected with zeta functions) to the case of parabolic forms, by purely technical tools. Therefore, we will speak in what follows, only about parabolic forms.

§ 2. The Hecke operators

Put

$$S_m = S_m^n = \left\{M \in M_{2n}(\mathbf{Z}) \,\middle|\, {}^tM \begin{pmatrix} 0 & E_n \\ -E_n & 0 \end{pmatrix} M = m \begin{pmatrix} 0 & E_n \\ -E_n & 0 \end{pmatrix}\right\},$$

$$S = S^n = \bigcup_{m \geqq 1} S_m^n. \tag{2.1}$$

Define, analogously to the one-dimensional case, the Hecke ring of the pair (Γ_n, S^n) [8]. Denote by $R = R(\Gamma_n, S^n)$ the free $\mathbf{Z}$-module spanned by the double cosets $(M) = \Gamma_n M \Gamma_n$ $(M \in S^n)$. Introduce in R a product by the rule

$$(M) \cdot (N) = \sum_{(T)} C_{M,N}^T (T), \tag{2.2}$$

where for $(M) = \bigcup_i \Gamma_n M_i$, $(N) = \bigcup_j \Gamma_n N_j$ (disjoint decompositions to left cosets) $C_{M,N}^T$ is equal to the number of pairs (i, j), for which $\Gamma_n M_i N_j = \Gamma_n T$ (every double coset $\Gamma_n M \Gamma_n$, $M \in S^n$ contains only finitely many different left or right cosets modulo Γ_n). Extending this product to R by linearity, we obtain an associative ring, the *Hecke ring* of the pair (Γ_n, S^n). It is well known [8] that R is a commutative integral domain.

The most important property of R is the existence of standard representations of this ring on the spaces of modular forms. For the sake of definiteness we restrict ourselves to parabolic forms, although this is not essential in most cases. For $M = \begin{pmatrix} A B \\ C D \end{pmatrix} \in S^n$ and a function f on $\mathfrak{Z}_n$ put

$$f|M = f|_k M = \det(CZ+D)^{-k} f\big((AZ+B)(CZ+D)^{-1}\big). \tag{2.3}$$

This is easily seen to be again a function on $\mathfrak{Z}_n$, and for any $M, M' \in S^n$ we easily get

$$f|M|M' = f|MM'.$$

Now let $f(Z) \in \mathfrak{N}_k^n$ (the space of parabolic forms of genus n and weight k); $\Gamma_n M \Gamma_n \in R$ $(M \in S^n)$ and let $\Gamma_n M \Gamma_n = \bigcup_i \Gamma_n M_i$ be a decomposition into disjoint left cosets. Consider the function

$$(M)f = \sum_i f|M_i. \tag{2.4}$$

Since $f|M' = f$ for $M' \in \Gamma_n$, the function (2.4) does not depend on the choice of the system of representatives M_i. Since for any $M' \in \Gamma_n$ the matrices $\{M_i M'\}$ also form a complete system of representatives of the different left cosets of $\Gamma_n M \Gamma_n$, we have $((M)f)|M' = (M)f$ $(M' \in \Gamma_n)$. It is easy to see, finally, that the function $(M)f$, as well as f, satisfies the purely analytic condition of the definition of parabolic forms, too (holomorphic on $\mathfrak{Z}_n$, convergence to zero as the boundary of $\mathfrak{Z}_n$ is approached). Thus, $(M)f \in \mathfrak{N}_k^n$. Extending this definition to $R = R(\Gamma_n, S^n)$ by linearity, we obtain a ring homomorphism $R \to \mathrm{End}_{\mathbf{C}}(\mathfrak{N}_k^n)$. The images of the elements of R under this homomorphism are called the *Hecke operators* on the space $\mathfrak{N}_k^n$.

The Hecke operators which correspond to the sum of all double cosets contained in S_m^n are especially important in zeta functions. It is appropriate to introduce it with the following normalization:

$$T(m)f = m^{nk - \frac{n(n+1)}{2}} \sum_{M_i \in V(m)} f|M_i, \quad (f \in \mathfrak{N}_k^n,\ m \geqq 1), \tag{2.5}$$

where $V(m)$ is a complete system of representatives of the different left cosets in S_m^n modulo Γ_n. For the case $n=1$ the operators (2.5) coincide with the operators $T(m)$ defined in our lecture [14] for the spaces $\mathfrak{N}_k(\Gamma_0(1))$.

Now we present a series of well-known properties of the operators $T(m)$. In the first place,

$$T(m)T(m_1) = T(m_1)T(m), \quad (m, m_1 \geqq 1);$$

$$T(m)T(m_1) = T(mm_1), \quad ((m, m_1) = 1) \tag{2.6}$$

(see [9]). For any prime number p the formal operator series $\sum_{\delta=0}^{\infty} T(p^\delta)t^\delta$ is formally equal to a rational function of t:

$$\sum_{\delta=0}^{\infty} T(p^\delta)t^\delta = \frac{P_p(t)}{Q_p(t)}, \tag{2.7}$$

where $P_p(t)$ and $Q_p(t)$ are polynomials in t with coefficients in the ring of Hecke operators and of degree 2^n-2 and 2^n, respectively [10]. In particular, for the space $\mathfrak{N}_k^2$ the following formula holds [8]:

$$\sum_{\delta=0}^{\infty} T(p^\delta)t^\delta = \tag{2.8}$$

$$= (1-p^{2k-4}t^2)\left(1-T(p)t+[T(p)^2-T(p^2)-p^{2k-4}]t^2-p^{2k-3}T(p)t^3+p^{4k-6}t^4\right)^{-1}.$$

Finally, just as in the one-dimensional case, in every space $\mathfrak{N}_k^n$ $(n\geqq 1, k\geqq 1)$ there is a basis consisting of eigenfunctions of all Hecke operators $T(m)$, $(m\geqq 1)$ [9].

§ 3. Zeta functions corresponding to Siegel modular forms of genus 2

We present in this § our basic results on zeta functions corresponding to parabolic forms of genus 2 [1]. For comparison, we first formulate the basic statements of the classical theory of Hecke on zeta functions corresponding to parabolic forms for the ordinary modular group $\Gamma_1=\mathrm{Sp}_1(\mathbf{Z})=SL_2(\mathbf{Z})$ (the proof of these assertions can be found in our lecture [14] or in [11]). Since in all cases there is a basis consisting of eigenfunctions of all Hecke operators, we will speak only about eigenfunctions.

$n=1$. According to Hecke's theory, to parabolic forms of one variable we can assign Dirichlet series having a decomposition as Euler product, admitting an analytic continuation to the whole plane, and satisfying a functional equation, as follows. Let

$$f(z) = \sum_{n=1}^{\infty} a(n)\exp(2\pi i n z) \in \mathfrak{N}_k^1 \quad (z\in \mathfrak{Z}_1)$$

be a parabolic form of weight k for the group $\Gamma_1=\mathrm{Sp}_1(\mathbf{Z})$ $(=SL_2(\mathbf{Z}))$. Assume that f is an eigenfunction of all Hecke operators $T(m)$ (see (2.5)):

$$T(m)f = \lambda(m)f, \quad (m\geqq 1).$$

Then the following assertions hold:

(I)$_1$ (Relation between eigenvalues and Fourier coefficients of the eigenfunctions.) Put

$$L_f(s) = \sum_{m=1}^{\infty} \frac{\lambda(m)}{m^s},$$

$$R_f(s) = \sum_{m=1}^{\infty} \frac{a(m)}{m^s}.$$

Then the series are absolutely convergent in the half plane $\mathrm{Re}\, s > \frac{k}{2}+1$, and we have the identity

$$R_f(s) = a(1)L_f(s).$$

(II)$_1$ (Multiplicativity of the Hecke operators.) The Dirichlet series $L_f(s)$ has the Euler decomposition

$$L_f(s) = \prod_p (1-\lambda(p)p^{-s}+p^{k-1-2s})^{-1},$$

where p runs over all prime numbers.

(III)$_1$ (Continuability and functional equation.) The Dirichlet series $L_f(s)$ has an analytic continuation to the whole s-plane;

$$\psi_f(s) = (2\pi)^{-s}\Gamma(s)L_f(s)$$

is an entire function and satisfies the functional equation

$$\psi_f(k-s) = i^k\psi_f(s).$$

Now we turn to the case of Siegel forms of genus 2. First we make some remarks. According to § 1, the Fourier coefficients of the parabolic forms in $\mathfrak{N}_k^2$ can be parametrized by half-integer positive-definite matrices of second order, i.e. matrices of the form $\begin{pmatrix} a & b/2 \\ b/2 & c \end{pmatrix}$ where $a, b, c \in \mathbf{Z}$, $a>0$, $b^2-4ac<0$. Every matrix of this form corresponds to a positive-definite integral binary quadratic form $aX^2+bXY+cY^2$, and conversely. We will identify these quadratic forms with the corresponding matrices. Two forms with matrices N and N_1 are said to be properly equivalent, if $N_1={}^tUNU$ for some $U\in SL_2(\mathbf{Z})$. As mentioned in § 1, the Fourier coefficients of modular forms of genus 2 depend only on the class of the parametrizing matrices N in the sense of proper equivalence. The number of classes of proper equivalent forms $N=\begin{pmatrix} a & b/2 \\ b/2 & c \end{pmatrix}$ of fixed discriminant $D=b^2-4ac$, is finite. Assume now that $d<0$ is a fundamental discriminant, i.e. it coincides with the discriminant of the imaginary quadratic field $\mathbf{Q}(\sqrt{d})$. Let $(M_1), \ldots, (M_n)$ be the different classes of proper equivalent positive-definite integral binary quadratic forms of discriminant d, M_i representatives of these classes. Associating to every form

$$M_i = \begin{pmatrix} a_i & b_i/2 \\ b_i/2 & c_i \end{pmatrix}$$

the integral ideal

$$\mathfrak{a}_i = \left(a_i, \frac{b_i-\sqrt{d}}{2}\right), \quad (\operatorname{Im}\sqrt{d}>0)$$

of the ring of integers of the field $\mathbf{Q}(\sqrt{d})$, we obtain a bijection between all considered classes of forms and all classes of ideals of the field $\mathbf{Q}(\sqrt{d})$. Thus we can introduce on the set of classes of integral positive-definite binary forms of discriminant d a group structure, isomorphic to the group of ideal classes of the field $\mathbf{Q}\ (\sqrt{d})$ (a group structure on the set of classes of forms can also be defined directly, using the Gaussian composition of the forms [12]). For the notions and facts used above see [13].

$n=2$. In this case the analogue of Hecke's theory goes as follows. Let

$$F(Z) = \sum a(N)\exp(2\pi i\sigma(NZ)) \in \mathfrak{N}_k^2 \quad (Z \in \mathfrak{Z}_2)$$

$$N = \begin{pmatrix} a & b/2 \\ b/2 & c \end{pmatrix} > 0, \quad a, b, c \in \mathbf{Z}$$

be a parabolic form of genus 2 and weight k (over the group $\Gamma_2 = \mathrm{Sp}_2(\mathbf{Z})$). Assume that F is the eigenfunction of all Hecke operators $T(m)$ (see (2.5)):

$$T(m)F = \Lambda(m)F.$$

Then the following assertions hold:

(I)$_2$ (Relation between eigenvalues and Fourier coefficients of the eigenfunctions.) Let $d<0$ coincide with the discriminant of the imaginary quadratic field $\mathbf{Q}(\sqrt{d})$; $M_1, \ldots, M_n$ a complete system of representatives of the classes of properly equivalent positive-definite integral binary quadratic forms of discriminant d (we identify a form with its matrix). For $m \geqq 1$ and an arbitrary character χ of the group of classes of forms of discriminant d (or, what is the same, of the group of ideal classes of the field $\mathbf{Q}(\sqrt{d})$), put

$$a(m;\chi) = \sum_{i=1}^{h} a(mM_i)\chi(M_i).$$

Then we have the following identity

$$L_{\mathbf{Q}(\sqrt{d})}(s-k+2;\chi)\sum_{m=1}^{\infty}\frac{a(m;\chi)}{m^s} = a(1;\chi)\zeta(2s-2k+4)\sum_{m=1}^{\infty}\frac{\Lambda(m)}{m^s},$$

where $L_{(\mathbf{Q}\sqrt{d})}(s;\chi)$ is the L-series of the field $\mathbf{Q}(\sqrt{d})$ with character χ, and $\zeta(s$ the Riemann zeta function. All Dirichlet series appearing in this identity are absolutely convergent in the half-plane $\mathrm{Re}\, s > k+1$.

(II)$_2$ (Multiplicativity of Hecke operators.) The following Euler decomposition holds:

$$L_F(s) = \zeta(2s-2k+4)\sum_{m=1}^{\infty}\frac{\Lambda(m)}{m^s} =$$

$$= \prod_p \left[1-\Lambda(p)p^{-s}+(\Lambda(p)^2-\Lambda(p^2)-p^{2k-4})p^{-2s}-\Lambda(p)p^{2k-3-3s}+p^{4k-6-4s}\right]^{-1},$$

where p runs over all prime numbers.

(III)$_2$ (Continuability and functional equation.) Assume $a\left(\begin{pmatrix}1&0\\0&1\end{pmatrix}\right) \neq 0$ (this assumption implies, in particular, that k is even). Then the Dirichlet series $L_F(s)$

(see $(\mathrm{II})_2$) can be continued to a meromorphic function on the whole s-plane, the function

$$\psi_F(s) = (2\pi)^{-2s}\Gamma(s)\Gamma(s-k+2)L_F(s)$$

is regular everywhere except possibly for simple poles at the points $s=k$, $s=k-2$, and satisfies the functional equation

$$\psi_F(2k-2-s) = \psi_F(s).$$

§ 4. About the proofs

Assertion $(\mathrm{II})_2$ is a corollary of relation (2.6) and Shimura's formula (2.8).

For the proof of assertion $(\mathrm{I})_2$ we shall first find a convenient expression for the Fourier coefficients of the images of the modular form

$$f(Z) = \sum_{N\in\mathcal{N}_2,\, N>0} a(N)\exp\left(2\pi i\sigma(NZ)\right)\in\mathfrak{N}_k^2 \tag{4.1}$$

under the Hecke operators. By virtue of their multiplicativity, it is sufficient to restrict ourselves to operators of the form $T(p^\delta)$, where p is a prime number. Put

$$T(p^\delta)f = \sum_{N\in\mathcal{N}_2,\, N>0} a(p^\delta, N)\exp\left(2\pi i\sigma(NZ)\right). \tag{4.2}$$

The success of computing the coefficients $a(p^\delta, N)$ essentially depends upon the choice of the system of representatives $V(p^\delta)=\Gamma_2\diagdown S^2_{p^\delta}$ (see (2.5)). The following system of representatives seems convenient:

$$V(p^\delta) = \left\{\begin{pmatrix} A & B\\ 0 & D\end{pmatrix}\middle|\, D\in\begin{pmatrix} p^\alpha & 0\\ 0 & p^{\alpha+\beta}\end{pmatrix}R(\beta),\right.$$
$$\left.\alpha\geqq 0,\ \beta\geqq 0,\ \alpha+\beta\leqq\delta,\ A=p^\delta\,{}^tD^{-1},\quad B \bmod D\right\}, \tag{4.3}$$

where $R(\beta)$ is the set of matrices $\begin{pmatrix} u_1 & u_2\\ u_3 & u_4\end{pmatrix}\in SL_2(\mathbf{Z})$, whose first rows $\{u_1, u_2\}$ run through a complete system of representatives modulo the equivalence relation

$$\{u_1, u_2\}\sim\{u_1', u_2'\}\pmod{p^\beta};$$

this means that there is a $v\in\mathbf{Z}$, $(v, p)=1$, such that $vu_1\equiv u_1'$, $vu_2\equiv u_2'$ $(\mathrm{mod}\ p^\beta)$, and B mod D in (4.3) means that B runs through a complete system of representatives of the classes of integral matrices of second order, satisfying the relation ${}^tBD={}^tDB$ modulo the equivalence relation

$$B\equiv B'\pmod{D}\Leftrightarrow (B-B')D^{-1}\in\mathrm{M}_2(\mathbf{Z}).$$

Standard computations, applying formula (2.5) and the above system of representatives result in a closed formula for $a(p^\delta; N)$. For writing out this formula, it is convenient to introduce some definitions. Denote by $\mathscr{H}$ the set of all complex-valued functions a on the set $\mathscr{N}=\mathscr{N}_2$ of all symmetric half-integer matrices of second order, constant on the classes of proper equivalent matrices (i.e. $a({}^tUNU)=$ $=a(N)$ for $U\in SL_2(\mathbf{Z})$). The Fourier coefficients of any modular form of genus 2 are functions in $\mathscr{H}$ (see § 1). Define two operators on $\mathscr{H}$:

$$(\varDelta(p^\alpha)a)(N) = a(p^\alpha N), \quad (a\in\mathscr{H},\ \alpha\in\mathbf{Z},\ N\in\mathscr{N});$$

$$(\varPi(p^\beta)a)(N) = \sum_{U\in R(\beta)} a\left(\begin{pmatrix} p^{-\beta} & 0 \\ 0 & 1\end{pmatrix} UN{}^tU \begin{pmatrix} 1 & 0 \\ 0 & p^\beta\end{pmatrix}\right), \quad (a\in\mathscr{H},\ \beta\geqq 0,\ N\in\mathscr{N}), \tag{4.4}$$

where outside the set $\mathscr{N}$ all functions in $\mathscr{H}$ are considered zero, and $R(\beta)$ is defined above (see (4.3) and after). In these terms the formula for $a(p^\delta; N)$ is of the form

$$a(p^\delta; N) = \sum_{\alpha+\beta+\gamma=\delta} p^{(k-2)\beta+(2k-3)\gamma}\big(\varDelta(p^{-\gamma})\varPi(p^\beta)\varDelta(p^\alpha)a\big)(N). \tag{4.5}$$

For a further transformation of formula (4.5) we should understand the meaning of the operators $\varPi(p^\beta)$. They turn out to be connected with the composition of binary quadratic forms (see § 3). Since the composition is simpler when the discriminants d of the forms considered are fundamental, i.e. coincide with the discriminants of the fields $\mathbf{Q}(\sqrt{d})$, we restrict ourselves to this case. Let us fix the matrix $M=\begin{pmatrix} a & b/2 \\ b/2 & c\end{pmatrix}$ of an arbitrary integral positive-definite binary quadratic form, whose discriminant $d=D(M)=b^2-4ac<0$ coincides with the discriminant of the field $K=\mathbf{Q}(\sqrt{d})$. Let $m\geqq 1$ and $(m, p)=1$. Then for an arbitrary function $a\in\mathscr{H}$ one of the following formulae holds, depending on the type of the factorization of the ideal (p) into prime ideals in the field K:

1) $(p) = \mathfrak{p}\bar{\mathfrak{p}} \quad (\mathfrak{p}\neq\bar{\mathfrak{p}})$,

$$(\varPi(p^\beta)a)(mM) = a(m\{M\cdot N(\mathfrak{p}^\beta)\}) + a(m\{M\cdot N(\bar{\mathfrak{p}}^\beta)\}) \quad (\beta\geqq 1)$$

2) $(p) = \mathfrak{p}^2$,

$$(\varPi(p^\beta)a)(mM) = \begin{cases} a(m\{M\cdot N(\mathfrak{p})\}) & \text{if } \beta=1 \\ 0 & \text{if } \beta>1,\end{cases}$$

3) $(p) = \mathfrak{p}$,

$$(\varPi(p^\beta)a)(mM) = 0 \quad (\beta\geqq 1),$$

where for two forms M and N we denote their composition by $\{M\cdot N\}$, and for integral ideals $\mathfrak{a}$ of the field K we write $N(\mathfrak{a})$ for the corresponding binary form (see § 3). The proof of these formulae, once they are written out, is not particularly hard. Assertion $(\mathrm{I})_2$ can easily be deduced from the above formulae by more or less standard computations with generating functions.

Now we turn to $(\mathrm{III})_2$. According to $(\mathrm{I})_2$, for the analysis of the series $L_F(s)$ we can start from any fundamental discriminant $d<0$, if only for some character χ the coefficient $a(1, \chi)=\sum a(M_i)\chi(M_i)$ is different from zero. For the sake of simplicity (single class in field $\mathbf{Q}(\sqrt{-4})$), we start from the discriminant $d=-4$, $\chi=1$. Therefore, we have to impose the restriction $a=a\left(\begin{pmatrix}1&0\\0&1\end{pmatrix}\right)\neq 0$. In this case $(\mathrm{I})_2$ yields

$$Z_{\mathbf{Q}(\sqrt{-4})}(s-k+2)\sum_{m=1}^{\infty}\frac{a\left(\begin{pmatrix}m&0\\0&m\end{pmatrix}\right)}{m^s}=aL_F(s), \tag{4.6}$$

where $Z_{\mathbf{Q}(\sqrt{-4})}(s)$ is the Dedekind zeta function of the field $\mathbf{Q}(\sqrt{-4})$. Consider the following subset of the Siegel half plane $\mathfrak{Z}_2$:

$$H=\left\{Z=\begin{pmatrix}x+iy & -t\\ -t & -x+iy\end{pmatrix}\middle|\, y>0,\ x,t\in\mathbf{R}\right\}\subset\mathfrak{Z}_2,$$

$$S=\left\{Z=\begin{pmatrix}x+iy & -t\\ -t & -x+iy\end{pmatrix}\in H\,\middle|\,|x|\leqq\frac{1}{2},\ |t|\leqq\frac{1}{2}\right\}\subset H.$$

For $Z=\begin{pmatrix}x+iy & -t\\ -t & -x+iy\end{pmatrix}\in H$ put $u=u(Z)=(y;w)$, where $w=x+it$, and denote by f the restriction of the form F to H:

$$f(u)=f(y;w)=F(Z)\quad(u=u(Z)=(y;w)).$$

Integrating term-by-term the Fourier expansion of f, we get

$$\int\limits_{|x|\leqq\frac{1}{2},\ |t|\leqq\frac{1}{2}}\int f(y;x+it)\,dx\,dt=\sum_{m=1}^{\infty}a\left(\begin{pmatrix}m&0\\0&m\end{pmatrix}\right)\exp(-4\pi my)\quad(y>0)$$

whence

$$\int\limits_{S}f(u)y^{s-1}du=(4\pi)^{-s}\Gamma(s)\sum_{m=1}^{\infty}\frac{a\left(\begin{pmatrix}m&0\\0&m\end{pmatrix}\right)}{m^s}, \tag{4.7}$$

where $du=dy\,dx\,dt$. Now we consider the set H. Denote by G the subgroup of the symplectic group $\mathrm{Sp}_2(\mathbf{R})$, consisting of all matrices of the form

$$M=\begin{pmatrix}a & a_1 & b & -b_1\\ -a_1 & a & -b_1 & -b\\ c & c_1 & d & -d_1\\ c_1 & -c_1 & d_1 & d\end{pmatrix}\in\mathrm{Sp}_2(\mathbf{R})$$

and by $G_{\mathbf{Z}}$ the intersection of G with $\Gamma_2=\mathrm{Sp}_2(\mathbf{Z})$. It is directly seen that for $M\in G$, by restricting to H the automorphism $Z\to M\langle Z\rangle$ of the domain $\mathfrak{Z}_2$ (see § 1), we obtain an automorphism of the domain H. To every matrix of the above type we associate the following complex matrix of the second order

$$\varphi(M)=\begin{pmatrix}a+ia_1 & b+ib_1\\ c+ic_1 & d+id_1\end{pmatrix}.$$

Direct computation shows that φ gives an isomorphism of the group G with $SL_2(\mathbf{C})$. This provides, at the same time, an isomorphism between $G_{\mathbf{Z}}\subset G$ and $SL_2(\mathfrak{O})$, where $\mathfrak{O}=\mathbf{Z}[\sqrt{-1}]$. With the matrices described the automorphism $Z\to M\langle Z\rangle$ of the domain H can be written in terms of $\varphi(M)$ and coordinates $u=u(Z)$ as follows:

$$u=(y;w)\to\sigma\langle u\rangle=\left(\frac{y}{\Delta_\sigma(u)};\ \frac{\alpha\bar{\gamma}y^2+(\alpha w+\beta)(\bar{\gamma}\bar{w}+\bar{\delta})}{\Delta_\sigma(u)}\right),$$

where $\sigma=\begin{pmatrix}\alpha&\beta\\\gamma&\delta\end{pmatrix}=\varphi(M)$, $\Delta_\sigma(u)=|\gamma|^2y^2+|\gamma w+\delta|^2$. Since $F(Z)$ is a modular form of weight k on $\mathfrak{Z}_2$ over Γ_2, its restriction to H satisfies for any $\sigma=\begin{pmatrix}\alpha&\beta\\\gamma&\delta\end{pmatrix}\in SL_2(\mathfrak{O})$ the relation

$$f(\sigma\langle u\rangle)=(\Delta_\sigma(u))^k f(u)\quad(u\in H). \tag{4.8}$$

Now put $\Gamma_0=SL_2(\mathfrak{O})/(\pm E_2)$ and $\Gamma_{0,\infty}=\left\{\begin{pmatrix}1&\beta\\0&1\end{pmatrix}\middle|\beta\in\mathfrak{O}\right\}$. Let D be a fundamental domain for Γ_0 on H and $\Gamma_0=\bigcup_i\Gamma_{0,\infty}\sigma_i$ the decomposition into left classes. Then $S'=\bigcup_i\sigma_i D$ will be a fundamental domain for $\Gamma_{0,\infty}$ on H. Since the set S, on which the integral in (4.7) is taken, will also be a fundamental domain for $\Gamma_{0,\infty}$ on H, and the integrand is invariant under the transformations in $\Gamma_{0,\infty}$, we obtain from (4.6):

$$(4\pi)^{-s}\Gamma(s)\sum_{n=1}^{\infty}\frac{a\left(\begin{pmatrix}m&0\\0&m\end{pmatrix}\right)}{m^s}=\int_{S'}f(u)y^{s-1}\,du=\sum_i\int_{\sigma_i D}f(u)y^{s-1}\,du=$$

$$=\sum_i\int_D\frac{y^{s-1}}{\Delta_{\sigma_i}(u)^{s-k+2}}f(u)\,du=\int_D E(u;s-k+2)f(u)y^{k-3}\,du \tag{4.9}$$

where $\operatorname{Re}s>k+1$ (see (1.6)), and

$$E(u,s)=\sum_{\sigma\in\Gamma_{0,\infty}\setminus\Gamma_0}\frac{y^s}{\Delta_\sigma(u)^s}$$

is called the Eisenstein series (we have made in every integral the change of variables $u=\sigma_i(u')$ and applied (4.8)). Assertion $(\mathrm{III})_2$ can be deduced from (4.6) and (4.9) by usual arguments, based on the representation of Eisenstein series as Mellin transforms of quaternary theta-series and the Poisson formula for the latter.

Remark. The restriction $a\left(\begin{pmatrix}1&0\\0&1\end{pmatrix}\right)\neq 0$ is a purely technical one, from which we can free ourselves by an analogous argument based on an arbitrary field $\mathbf{Q}(\sqrt{d})$ $(d<0)$ instead of $\mathbf{Q}(\sqrt{-4})$.

REFERENCES

[1] Andrianov, A. N. (Андрианов А. Н.), Ряды Дирихле с эйлеровским произведением в теории зигелевых модулярных форм рода 2, *Труды МИАН СССР*, **112** (1971), 73—94.
(Dirichlet series and Euler product in the theory of Siegel modular forms of genus 2.)

[2] Siegel, C. L., Über die analytische Theorie der quadratischen Formen, *Ann. of Math*, **36** (1935), 527—606.

[3] Siegel, C. L., Einführung in die Theorie der Modulfunktionen n-ten Grades, *Math. Ann.*, **116** (1939), 617—657.

[4] *Séminaire H. Cartan, 1957/1958, Fonctions automorphes,* Paris, 1958.

[5] Koecher, M., Zur Theorie der Modulformen n-ten Grades, I, II, *Math. Z.* **59** (1954), 399—416, **61** (1955), 455—466.

[6] Maass, H., Über Darstellung der Modulformen n-ten Grades durch Poincarésche Reihen, *Math. Ann.*, **123** (1951), 125—151.

[7] Raghavan, S., Modular forms of degree n and representations by quadratic forms, *Ann. of Math.*, **70** (1959), 446—477.

[8] Shimura, G., On modular correspondences for $\mathrm{Sp}(n, \mathbf{Z})$ and their congruence relations, *Proc. Nat. Acad. Sci. USA,* **49** (1963), 824—828.

[9] Maass, H., Die Primzahlen in der Theorie der Siegelschen Modulfunktionen, *Math. Ann.*, **124** (1951), 87—122.

[10] Andrianov, A. N. (Андрианов А. Н.), Сферические функции для GL_n над локальными полями и суммирование рядов Гекке, *Математический сборник*, **83**:3 (1970), 429—451.
(Spherical functions for GL_n over local fields and the summation of Hecke series.)

[11] Ogg, A., *Modular forms and Dirichlet series,* Benjamin, New York, 1969.

[12] Venkov, B. A. (Венков Б. А.) *Элетентарная теория чисел*, М.—Л., pp. 87—122.
(Elementary number theory.)

[13] Hecke, E., *Vorlesungen über die Theorie der algebraischen Zahlen,* Leipzig, 1923.

[14] Andrianov, A. N., Modular forms, zeta functions and elliptic curves, *see p. 7 of this volume!*

DIFFERENTIAL OPERATORS ON THE BASE AFFINE SPACE AND A STUDY OF $\mathfrak{g}$-MODULES

I. N. BERNSTEIN, I. M. GELFAND, S. I. GELFAND

The present work consists of two parts. In the first part we study the ring of regular differential operators on the base affine space of a complex semisimple group. By the *base affine space* of a group we mean the quotient space $A = N_+ \diagdown G$ of the group G by a maximal unipotent subgroup N_+. Experience in representation theory suggests that for many problems in representation theory the solution results from a careful study of the base affine space. In particular, the structure of the ring of regular differential operators on A seems to be closely connected with the representations of the real forms of the group G. In addition to the connections with representation theory, the study of this ring yields an instructive and rather advanced example for the study of the rings of regular differential operators on algebraic varieties, an area in which not much is known so far.

We approach the study of the differential operators on A by establishing a connection between the regular functions on the group G and the regular differential operators on the base affine space A. We would also like to draw the reader's attention to Conjecture II, where the notion of the generalized Segal—Bargmann space for a representation of a compact Lie group is introduced.

The second part of the work is formally independent of the first and is devoted to the algebraic study of modules over the Lie algebra $\mathfrak{g}$ of the group G. We restrict ourselves to a category of $\mathfrak{g}$-modules, which is closely connected with the theory of highest weight. We shall call this category of $\mathfrak{g}$-modules the category O. The category O contains in a natural way every finite-dimensional representation of the Lieal gebra $\mathfrak{g}$. The fundamental result of this part lies in constructing a resolution for finite-dimensional $\mathfrak{g}$-modules. The simplest objects of the category O are the modules M_χ and it seems important that the resolution consists of modules which are direct sums of these simplest modules. The description of the modules occurring in the composition series of the modules M_χ, which is given in the Appendix, is also useful. Unfortunately, the complete structure of these composition series is not known to us yet.

We think that the methods developed in the second part of this paper may turn out to be useful in the further study of questions considered in the first part.

The fundamental content of this work is concentrated in Theorem 6.3 and Conjectures I and II in the first part and Theorems 8.12 and 10.1 in the second.

§ 1. Notations and preliminaries

$\mathfrak{g}$ is a semisimple Lie algebra of rank r over $\mathbf{C}$, $\mathfrak{h}$ is a Cartan subalgebra of $\mathfrak{g}$. Δ denotes the root system of $\mathfrak{g}$ corresponding to $\mathfrak{h}$, with a fixed ordering, Δ_+ and Δ_- the system of the positive and negative roots, respectively, Σ the set of simple roots, and $\varrho=\frac{1}{2}\sum_{\gamma\in\Delta_+}\gamma$. $E_\gamma\in\mathfrak{g}$ is the root vector corresponding to the root $\gamma\in\Delta$. Here we have $\gamma([E_\gamma, E_{-\gamma}])=2$.

$\mathfrak{n}_+$ is the subalgebra of $\mathfrak{g}$ spanned by the vectors E_γ, $\gamma\in\Delta_+$, while $\mathfrak{n}_-$ is the subalgebra of $\mathfrak{g}$ spanned by E_γ, $\gamma\in\Delta_-$. $\mathfrak{b}=\mathfrak{h}\oplus\mathfrak{n}_+$. $U(\mathfrak{g})$, $U(\mathfrak{n}_+)$, $U(\mathfrak{n}_-)$ are the universal enveloping algebras of $\mathfrak{g}$, $\mathfrak{n}_+$, $\mathfrak{n}_-$, respectively; $Z(\mathfrak{g})$ is the centre of $U(\mathfrak{g})$.

$\mathfrak{h}^*$ is the dual space of $\mathfrak{h}$.

G is a complex semisimple Lie group with Lie algebra $\mathfrak{g}$; H, N_+, N_- and B are the subgroups of G corresponding to the subalgebras $\mathfrak{h}$, $\mathfrak{n}_+$, $\mathfrak{n}_-$ and $\mathfrak{b}$, respectively.

$A=N_+\diagdown G$ is the base affine space of the group G.

Additional notations, used in Part 2.

$\mathbf{Z}_+$ is the set of non-negative integers.

$\mathfrak{h}^*_{\mathbf{R}}$ denotes the real linear subspace of $\mathfrak{h}^*$ spanned by all roots $\gamma\in\Delta$.

$\langle\,,\rangle$ is the scalar product in $\mathfrak{h}^*$ constructed with the help of the Killing form of the algebra $\mathfrak{g}$; $\|\cdot\|$ is the corresponding norm in $\mathfrak{h}^*_{\mathbf{R}}$.

$\mathfrak{h}^*_{\mathbf{Z}}$ is the lattice in $\mathfrak{h}^*_{\mathbf{R}}$ consisting of those $\chi\in\mathfrak{h}^*$ for which $2\langle\chi,\gamma\rangle/\langle\gamma,\gamma\rangle\in\mathbf{Z}$ for all $\gamma\in\Delta$.

$$K=\{\chi\in\mathfrak{h}^*\,|\,\chi=\sum_{\alpha\in\Sigma}n_\alpha\cdot\alpha,\ n_\alpha\in\mathbf{Z}_+\};\quad K\subset\mathfrak{h}^*_{\mathbf{Z}}.$$

$\chi_1\geqq\chi_2$ means that $\chi_1-\chi_2\in K$ ($\chi_1, \chi_2\in\mathfrak{h}^*$).

W is the Weyl group of the algebra $\mathfrak{g}$, $\sigma_\gamma\in W$ is the reflexion corresponding to the root $\gamma\in\Delta$, i.e. $\sigma_\gamma\chi=\chi-2\langle\chi,\gamma\rangle\langle\gamma,\gamma\rangle^{-1}\gamma$. We note that $\sigma_\alpha\varrho=\varrho-\alpha$ for $\alpha\in\Sigma$.

$\chi_1\sim\chi_2$ for $\chi_1,\chi_2\in\mathfrak{h}^*$ means that there exists an element $w\in W$ such that $\chi_1=w\chi_2$.

$l(w)$ is the length of the element $w\in W$, i.e. the smallest possible number of factors in a decomposition $w=\sigma_{\alpha_1}\cdot\ldots\cdot\sigma_{\alpha_k}$, $\alpha_i\in\Sigma$.

$$W^{(i)}=\{w\in W\,|\,l(w)=i\}.$$

$\Xi_\gamma=\{\chi\in\mathfrak{h}^*_{\mathbf{R}}\,|\,\langle\chi,\gamma\rangle=0\}$; the connected components of $\mathfrak{h}^*_{\mathbf{R}}\diagdown(\bigcup_{\gamma\in\Delta}\Xi_\gamma)$ are called the *Weyl chambers;* $\bar{C}$ is the closure of the Weyl chamber C; C^+ is the Weyl chamber containing ϱ. The group W acts on the set of Weyl chambers simply transitively. Two Weyl chambers C_1 and C_2 are called *neighbouring* if $\dim(\bar{C}_1\cap\bar{C}_2)=\dim\mathfrak{h}^*_{\mathbf{R}}-1$. In this case there exists a unique element $\gamma\in\Delta_+$ such that $\sigma_\gamma C_1=C_2$ and the hyperplane Ξ_γ separates C_1 and C_2;

$$D=\mathfrak{h}^*_{\mathbf{Z}}\cap C^+.$$

An element $\chi \in \mathfrak{h}_{\mathbf{R}}^*$ is called *regular* if $\langle \chi, \gamma \rangle \neq 0$ for all $\gamma \in \Delta$.

Let M be a $\mathfrak{h}$-module, $\chi \in \mathfrak{h}^*$. Put

$$M^{(\chi)} = \{f \in M \mid xf = \chi(x) \cdot f \text{ for all } x \in \mathfrak{h}\};$$

$$P(M) = \{\chi \in \mathfrak{h}^* \mid M^{(\chi)} \neq 0\}.$$

Let M be a $\mathfrak{g}$-module and $0 = M_0 \subset M_1 \subset \ldots \subset M_k = M$ its Jordan—Hölder composition series, $L_i = M_i / M_{i-1}$ are simple $\mathfrak{g}$-modules. The collection of the modules L_i, with multiplicity, is called the Jordan—Hölder decomposition of M and is denoted by $JH(M)$.

PART I

DIFFERENTIAL OPERATORS

§ 2. Introduction

Let G be a connected complex semisimple Lie group of rank r, B a Borel subgroup, N_+ the unipotent radical of B, $H \subset B$ the Cartan subgroup of B. The quotient space $A = N_+ \setminus G$ plays a fundamental rôle in representation theory; it is called the *base affine space* of the group G.

The aim of this part is to study the ring $\mathscr{D}(A)$ of regular differential operators on A.

First of all we study the possible connections between the space of regular differential operators on A and the space $\mathscr{E}(G)$ of regular functions on G. More precisely, Conjecture I claims the possibility of embedding $\mathscr{E}(G)$ into $\mathscr{D}(A)$ (operation $f \to \tilde{f}$); here $\mathscr{D}(A) = U(\mathfrak{h}) \otimes_{\mathbf{C}} \mathscr{E}(G)$, where $U(\mathfrak{h})$ is the universal enveloping algebra of the Lie algebra $\mathfrak{h}$ of the group H. In this part we prove a result (Theorem 6.6) weaker than Conjecture I, namely we construct an isomorphism between the L-modules $L \otimes_{\mathbf{C}} \mathscr{E}(G)$ and $L \otimes_{U(\mathfrak{h})} \mathscr{D}(A)$, where L is the quotient field of the ring $U(\mathfrak{h})$.

Further on we construct a scalar product in the ring $\mathscr{E}(A)$ of regular functions on A which is invariant under the action of the maximal compact subgroup $K \subset G$. The completion of the space $\mathscr{E}(A)$ by this scalar product consists of analytic functions on the complex manifold A. This space is a generalization of the Segal—Bargmann space. Conjecture II states that an operator adjoint to a differential operator is again a differential operator, hence the ring $\mathscr{D}(A)$ is selfadjoint with respect to the introduced scalar product. It has to be noted that the introduced involution in $\mathscr{D}(A)$ does not preserve the order of a differential operator. For instance, in the case of the group of matrices of order h, the adjoint to the simplest operator of order zero will be an operator of order $h-1$.

§ 3. Regular differential operators

In this section the rings of regular differential operators on G and A are introduced. We shall consider G and A as algebraic varieties over $\mathbf{C}$. The projection $\pi: G \to A$ is a morphism of algebraic varieties. The rings of regular functions on G and A will be denoted by $\mathscr{E}(G)$ and $\mathscr{E}(A)$, respectively. Let $\pi^*: \mathscr{E}(G) \to \mathscr{E}(A)$

be the embedding induced by the mapping π. The image $\pi^* \mathscr{E}(A)$ consists of exactly those functions which are constant on the left cosets of the subgroup N.

The variety A is non-singular and quasi-affine. More precisely, let

$$\hat{A} = \text{Spec max } \mathscr{E}(A)$$

be the affine algebraic variety corresponding to $\mathscr{E}(A)$. Then there is a natural isomorphism between A and a dense open subset of $\hat{A}$.

Let us define in the space $\mathscr{E}(G)$ the left and right representations L^G and R^G of the group G by the usual formulas

$$(L^G_{g_0} f)(g) = f(g_0^{-1} g), \quad (R^G_{g_0} f)(g) = f(g g_0), \quad g, g_0 \in G.$$

It is a common property of both of these representations that every element $f \in \mathscr{E}(G)$ is contained in some finite-dimensional invariant subspace.

The group G acts naturally on the space A (right translations). In addition one can also define the left-hand action of the Cartan subgroup on A by associating to each element $h \in H$ the transformation $x \to hx$. The element $hx \in A$ is well defined because H normalizes N.

Let us define in the space $\mathscr{E}(A)$ the representations L^A of the group H and R^A of the group G by means of the formulas

$$(L^A_h f)(x) = f(h^{-1} x), \quad (R^A_g f)(x) = f(xg); \; x \in A, \; h \in H, \; g \in G.$$

Obviously, $L^G_h \pi^* = \pi^* L^A_h$ and $R^G_h \pi^* = \pi^* R^A_h$.

By differentiating R^A we obtain for each $X \in \mathfrak{g}$ an operator $R^A_X : \mathscr{E}(A) \to \mathscr{E}(A)$. These operators determine a representation of $\mathfrak{g}$ that extends to a representation of $U(\mathfrak{g})$; here the operator corresponding to an element $X \in U(\mathfrak{g})$ will also be denoted by R^A_X. We define the similar representations R^G_X and L^G_X of the algebra $U(\mathfrak{g})$ in $\mathscr{E}(G)$ and the representation L^A_Y of the algebra $U(\mathfrak{g})$ in $\mathscr{E}(A)$.

Definition 3.1. Let X be a quasi-affine variety and $\mathscr{E}(X)$ be the ring of regular functions on X. A linear mapping $D : \mathscr{E}(X) \to \mathscr{E}(X)$ is called a *regular differential operator of order* $\leqq k$ $(k \geqq 0)$ on X if it satisfies the condition

$$[f_{k+1}[f_k \cdots [f_1, D] \ldots]] = 0 \tag{3.1}$$

for any $f_1, f_2, \ldots, f_{k+1} \in \mathscr{E}(X)$. In (3.1) f_i denotes the operator of multiplication by f_i.

The differential operators on X form a ring which will be denoted by $\mathscr{D}(X)$.

Remark. The definition given here coincides with that of a differential operator on an arbitrary algebraic variety given in [1].

It is easy to see that a differential operator of order zero is an operator of multiplication by a function $f \in \mathscr{E}(X)$.

Any vector field over X determines a differential operator of order $\leqq 1$ on X. It can be verified that on composing an operator of order $\leqq k$ with an operator of order $\leqq l$ we obtain a differential operator of order $\leqq k+l$, moreover the commutator of two such operators is of order $\leqq k+l-1$.

We shall use the following properties of differential operators (see [1]).

Lemma 3.2. 1. *If $Y \subset X$ is a dense open subset (in the Zariski topology) of a quasi-affine variety X, than every regular differential operator can be restricted to Y. More precisely, there exists a unique differential operator $D': \mathscr{E}(Y) \to \mathscr{E}(Y)$ whose restriction to $\mathscr{E}(X) \hookrightarrow \mathscr{E}(Y)$ coincides with D.*

2. *Let X be a non-singular variety and $Z_1, \ldots, Z_n$ be a system of vector fields over X which defines a basis in the tangent space at each point $x \in X$. Then every differential operator D on X has a unique representation in the form*

$$D = \sum a_{i_1 i_2 \ldots i_n}(x) Z_1^{i_1} Z_2^{i_2} \ldots Z_n^{i_n},$$

where i_j are non-negative integers, $a_{i_1 \ldots i_n}$ are regular functions only a finite number of which are different from zero.

For any element $X \in U(\mathfrak{g})$ the operators R_X^G and L_X^G are differential operators on G, moreover L_Y^A, $Y \in U(\mathfrak{h})$ and R_X^A are differential operators on A. The description of the ring of differential operators on G yields the following proposition.

Proposition 3.3. The mapping

$$\vartheta: \mathscr{E}(G) \underset{\mathbf{C}}{\otimes} U(\mathfrak{g}) \to \mathscr{D}(G)$$

given by the formula

$$\vartheta\left(\sum f_i \otimes X_i\right) = \sum f_i L_{X_i}^G$$

is an isomorphism of left $\mathscr{E}(G)$-modules.

The proof of Proposition 3.3. follows easily from Lemma 3.2.

In what follows the element $\vartheta^{-1}(D) \in \mathscr{E}(G) \otimes U(\mathfrak{g})$ will be called the *standard form* of $D \in \mathscr{D}(G)$.

Now let us turn to the study of the ring $\mathscr{D}(A)$. In this ring we can define a representation of the group G. Indeed, we put

$$D^g = R_g^A D R_{g^{-1}}^A, \quad g \in G, \quad D \in \mathscr{D}(A).$$

Similarly, we put

$${}^h D = L_h^A D L_{h^{-1}}^A, \quad h \in H, \quad D \in \mathscr{D}(A).$$

We notice that the variety A is smooth, but it is not an affine variety and there may exist differential operators on A which cannot be expressed by operators of the first order. (See e.g. Example 2.)

We shall say that a differential operator D' on G is a *lifting* of the operator D on A, if

$$D' \pi^* f = \pi^* D f, \quad f \in \mathscr{E}(A).$$

Theorem 3.4. [2]. *Every differential operator on A can be lifted to G.*

In order to prove this theorem we need the following lemma.

Lemma 3.5. *Let $0=\pi(e)\in A$. There exists a mapping $\eta:\mathscr{D}(A)\to U(\mathfrak{g})/\mathfrak{n}_+ U(\mathfrak{g})$ such that for all $D\in\mathscr{D}(A)$ and all $f\in\mathscr{E}(A)$*

$$(R_X^A f)(0) = (Df)(0) \tag{3.2}$$

for every element $X\in U(\mathfrak{g})$ belonging to the coset $\eta(D)$.

Proof. It is easy to see that if $X\in\mathfrak{n}_+$ then $R_X^A f(0)=0$ for all $f\in\mathscr{E}(A)$. Therefore the validity of the equality (3.2) does not depend on the choice of the element X in a $\mathfrak{n}_+ U(\mathfrak{g})$-coset.

Now let $X_1, \ldots, X_N$ be a basis in $\mathfrak{h}\oplus\mathfrak{n}_-$. Then the vector fields $R_{X_i}^A$ form a basis in the tangent space at each point of a certain affine neighbourhood V of the point $0\in A$. According to Lemma 3.2, in this neighbourhood V the operator D can be expressed in the form

$$D = \sum a_{i_1 i_2 \ldots i_N}(R_{X_1}^A)^{i_1}\cdot \ldots \cdot(R_{X_N}^A)^{i_N}.$$

Put

$$X = \sum a_{i_1 i_2 \ldots i_N}(0) X_1^{i_1} \ldots X_N^{i_N} \in U(\mathfrak{g}).$$

It is easy to see that the image $\eta(D)$ of the element X in $U(\mathfrak{g})/\mathfrak{n}_+ U(\mathfrak{g})$ satisfies the condition of the lemma.

Note that the element $\eta(D)$ is uniquely determined by the equality (3.2) (see [2]).

Let $\tau: U(\mathfrak{g})\to U(\mathfrak{g})$ be an anti-automorphism such that $\tau(X)=-X$ for $X\in\mathfrak{g}$. Obviously, $\tau(\mathfrak{n}_+ U(\mathfrak{g}))=U(\mathfrak{g})\mathfrak{n}_+$, so τ determines an isomorphism

$$\bar{\tau}:\ U(\mathfrak{g})/\mathfrak{n}_+U(\mathfrak{g})\to U(\mathfrak{g})/U(\mathfrak{g})\mathfrak{n}_+.$$

Definition 3.6. Let $D\in\mathscr{D}(A)$. Define a function $\sigma_D(g)$ on G with values in $U(\mathfrak{g})/U(\mathfrak{g})\mathfrak{n}_+$ by the formula

$$\sigma_D(g) = \bar{\tau}\eta(D^g).$$

One can verify [2, § 8] that $\sigma_D(g)$ is a regular function on G, i.e.

$$\sigma_D(g)\in\mathscr{E}(G)\otimes U(\mathfrak{g})/U(\mathfrak{g})\mathfrak{n}_+.$$

Now we are ready to complete the proof of the theorem. Let $D\in\mathscr{D}(A)$ and consider an arbitrary element $\sigma_D'\in\mathscr{E}(G)\otimes U(\mathfrak{g})$ which is sent into $\sigma_D(g)$ by the natural projection

$$\mathscr{E}(G)\otimes U(\mathfrak{g})\to\mathscr{E}(G)\otimes U(\mathfrak{g})/U(\mathfrak{g})\mathfrak{n}_+.$$

Put

$$D' = \vartheta(\sigma_D'(g))\in\mathscr{D}(G).$$

We shall show that D' is a lifting of D to G, i.e. that

$$(D'\pi^* f)(g) = (\pi^* Df)(g),\quad f\in\mathscr{E}(A),\ g\in G \tag{3.3}$$

(3.3) can be rewritten in the form

$$(R_g^G D' R_{g^{-1}}^G \pi^* R_g^A f)(e) = (\pi^* D^g R_g^A f)(e).$$

It is easy to see that $R_g^G D' R_{g^{-1}}^G$ is a regular differential operator on G and that $\vartheta^{-1}(R_g^G D' R_{g^{-1}}^G)$ is projected into σ_{D^g} under the natural mapping

$$\mathscr{E}(G) \otimes U(\mathfrak{g}) \to \mathscr{E}(G) \otimes U(\mathfrak{g})/U(\mathfrak{g})\mathfrak{n}_+.$$

Therefore (replacing D by D^g and f by $R_g^A f$) it suffices to prove that $(D' \pi^* f)(e) = (\pi^* Df)(e)$.

Let $X = \sigma'_D(e) \in U(\mathfrak{g})$. Clearly, then $(L_X^G f)(e) = (D' f)(e)$ for $f \in \mathscr{E}(G)$. Moreover, the image of $\tau^{-1}(X)$ in $U(\mathfrak{g})/\mathfrak{n}_+ U(\mathfrak{g})$ is equal to $\eta(D)$, hence

$$(\pi^* Df)(e) = (Df)(0) = (R_{\tau^{-1}X}^A f)(0) = (R_{\tau^{-1}X}^G \pi^* f)(e), \quad f \in \mathscr{E}(A).$$

Therefore, the required equality is implied by the following lemma.

Lemma 3.7. *If* $Y \in U(\mathfrak{g})$, $f \in \mathscr{E}(G)$ *then*

$$(L_{\tau Y}^G f)(e) = (R_Y^G f)(e).$$

Proof. If $Y \in \mathfrak{g}$ then the lemma follows from the definitions of L^G and R^G. Assume now that the lemma is valid for $Y_1, Y_2 \in U(\mathfrak{g})$. Then we have

$$(R_{Y_1 Y_2}^G f)(e) = (R_{Y_1}^G R_{Y_2}^G f)(e) = (L_{\tau(Y_1)}^G R_{Y_2}^G f)(e) = (R_{Y_2}^G L_{\tau(Y_1)}^G f)(e) =$$
$$= (L_{\tau(Y_2)}^G L_{\tau(Y_1)}^G f)(e) = (L_{\tau(Y_1 Y_2)}^G f)(e).$$

(Here we use the fact that L_Y^G and $R_{Y'}^G$ commute for any $Y, Y' \in U(\mathfrak{g})$.) Hence we have the lemma for $Y = Y_1 \cdot Y_2$, and the proof is complete.

Proposition 3.8. Let us denote by I_+ the left ideal in the ring $\mathscr{D}(G)$ spanned by the operators L_X^G, $X \in \mathfrak{n}_+$.

1) Let $D \in \mathscr{D}(G)$. Then D gives rise to a differential operator on A (i.e. $D(\mathscr{E}(A)) \subset \mathscr{E}(A)$) if and only if

$$[L_X^G, D] \in I_+ \quad \text{for} \quad X \in \mathfrak{n}_+.$$

2) $D \in \mathscr{D}(G)$ gives rise to the zero operator on A if and only if $D \in I_+$.

The proof of this proposition is rather simple and is left to the reader.

We shall now describe how these conditions can be expressed in terms of the standard form of the operator D. We remark that

1) $D \in I_+$ if and only if

$$\vartheta^{-1}(D) \in \mathscr{E}(G) \otimes U(\mathfrak{g})\mathfrak{n}_+.$$

2) If $\vartheta^{-1}(D) = \sum f_i \otimes X_i \in \mathscr{E}(G) \otimes U(\mathfrak{g})$ and $X \in \mathfrak{g}$ then

$$\vartheta^{-1}([L_X^G, D]) = \sum L_X^G f_i \otimes X_i + \sum f_i \otimes [X, X_i].$$

Corollary 3.9. Let $D \in \mathcal{D}(A)$. Then it has a unique lifting D' such that

$$\vartheta^{-1}(D') \in \mathcal{E}(G) \otimes U(\mathfrak{h} \oplus \mathfrak{n}_-).$$

Definition 3.10. *1)* The element $\vartheta^{-1}(D') \in \mathcal{E}(G) \otimes U(\mathfrak{h} \oplus \mathfrak{n}_-)$ is called the *standard form* of the operator $D \in \mathcal{D}(A)$ and will be denoted by $s(D)$.

2) The *lowest term* $s_0(D)$ of an operator $D \in \mathcal{D}(A)$ is defined as the element of $\mathcal{E}(G) \otimes U(\mathfrak{h})$ which is equal to the projection of $s(D)$ under the decomposition

$$U(\mathfrak{h} \oplus \mathfrak{n}_-) = U(\mathfrak{h}) \oplus \mathfrak{n}_- U(\mathfrak{h} \oplus \mathfrak{n}_-)$$

Definition 3.11. Let us denote by Wu the subring of $\mathcal{D}(A)$ consisting of the operators

$$L_X^A, \; X \in U(\mathfrak{h}).$$

§ 4. Conjecture I; examples

Conjecture I. *There exists a mapping* $\mathcal{E}(G) \to \mathcal{D}(A)$, $f \to \tilde{f}$ *with the following properties:*

1) $\tilde{f} = f$ *for* $f \in \mathcal{E}(A) \subset \mathcal{E}(G)$.

2) The mapping $f \to \tilde{f}$ *commutes with the representations* R_g *and* L_h.

3) The mapping $\mathcal{E}(G) \otimes Wu \to \mathcal{D}(A)$ *given by the formula*

$$\sum f_i \otimes Z_i \to \sum \tilde{f}_i Z_i, \; f_i \in \mathcal{E}(G), \; Z_i \in Wu$$

is an isomorphism of Wu*-modules.*

4) Let $s_0(\tilde{f}) = \sum f_i \otimes X_i$. *Then* $f = \sum f_i \cdot X_i(-\varrho)$ (here $\varrho \in \mathfrak{h}^*$ is half-sum of the positive roots; $X_i \in U(\mathfrak{h})$ is considered as a polynomial function on $\mathfrak{h}^*$).

A weakened version of Conjecture I is

Conjecture I′. $\mathcal{D}(A)$ *is a free* Wu*-module.*

We remark that the mapping $f \to \tilde{f}$ (if it exists) is not uniquely determined by the properties *1)—4)*. We assume however that there exists a "natural" mapping $f \to \tilde{f}$. The following examples will perhaps illuminate to the reader what we have in mind.

We shall now present several examples illustrating the notions and facts expounded above.

Example 1. $G = SL(2, \mathbf{C})$ is the group of 2×2 matrices of determinant 1. We choose as $N_+ \subset G$ the subgroup of all matrices of the form

$$\begin{pmatrix} 1 & x \\ 0 & 1 \end{pmatrix}.$$

In this case $A=N_+\diagdown G$ can be identified with the punctured complex plane i.e. $A=\mathbf{C}^2\diagdown\{(0,0)\}$, while the mapping $\pi: G\to A$ is of the form

$$g = \begin{pmatrix} u_1 & u_2 \\ z_1 & z_2 \end{pmatrix} \to (z_1, z_2).$$

Let

$$E_- = \begin{pmatrix} 0 & 0 \\ 1 & 0 \end{pmatrix}, \quad H = \begin{pmatrix} 1 & 0 \\ 0 & -1 \end{pmatrix}, \quad E_+ = \begin{pmatrix} 0 & 1 \\ 0 & 0 \end{pmatrix}$$

be a basis of $\mathfrak{g}$. Then

$$L^G_{E_-} = -u_1\frac{\partial}{\partial z_1} - u_2\frac{\partial}{\partial z_2},$$

$$L^G_{E_+} = -z_1\frac{\partial}{\partial u_1} - z_2\frac{\partial}{\partial u_2},$$

$$L^G_H = z_1\frac{\partial}{\partial z_1} + z_2\frac{\partial}{\partial z_2} - u_1\frac{\partial}{\partial u_1} - u_2\frac{\partial}{\partial u_2}.$$

The algebra of regular functions $\mathscr{E}(G)$ is $\mathscr{E}(G)=\mathbf{C}(u_1, u_2, z_1, z_2)/(u_1z_2-u_2z_1-1)$ and $\mathscr{E}(A)\subset\mathscr{E}(G)$ consists of the functions $f\in\mathscr{E}(G)$ which satisfy

$$-L^G_{E_+}f = z_1\frac{\partial f}{\partial u_1} + z_2\frac{\partial f}{\partial u_2} = 0.$$

The ring $Wu\subset\mathscr{D}(A)$ coincides with the ring of polynomials of the single generator

$$L^A_H = z_1\frac{\partial}{\partial z_1} + z_2\frac{\partial}{\partial z_2}.$$

Now we show how to construct the mapping $f\to\tilde{f}$. Let T be an irreducible representation of $SL(2, \mathbf{C})$. We denote by $\mathscr{E}_T$ the largest subspace of $\mathscr{E}(G)$ such that the restriction of R^G to this subspace is a multiple of T. Furthermore, for each $n\in\mathbf{Z}$ we denote by $\mathscr{E}^n_T$ the subspace of $\mathscr{E}_T$ consisting of all functions $f\in\mathscr{E}_T$ for which $L^G_H f=nf$.

Lemma 4.1. *1)* $\mathscr{E}(G)=\bigoplus_{T,n}\mathscr{E}^n_T$.

2) Let $\dim T=l+1$. *Then* $\dim\mathscr{E}^n_T=l+1$ *for* $n=-l, -l+2, \dots, l-2, l$ *and* $\mathscr{E}^n_T=0$ *for the remaining values of n. Those n for which* $\mathscr{E}^n_T\neq 0$ *are weights of T.*

3) $\mathscr{E}^n_T$ *is invariant under* R^G *and the restriction of* R^G *to* $\mathscr{E}^n_T$ *is equivalent to T.*

4) $\mathscr{E}^l_T$ *consists of vectors of highest weight with respect to* L^G *(i.e.* $L^G_X f=0$ *for all* $f\in\mathscr{E}^l_T$ *and* $X\in\mathfrak{n}_+$*).*

The proof of this lemma follows easily from simple properties of the representations of $SL(2, \mathbf{C})$.

It suffices to construct the mapping $f\to\tilde{f}$ on each space $\mathscr{E}^n_T$ separately. We may assume that the restriction of L^G to the smallest invariant subspace of $\mathscr{E}(f)$

containing f is equivalent to T. Let us put $f_i=(L^G_{E_+})^i f$. Then obviously $f_i \in \mathscr{E}_T^{n+2i}$, and in particular $f_i=0$ for $i>\frac{1}{2}(l-n)$. In accordance with property 2 of the mapping $f\to\tilde{f}$ (see Conjecture I), we shall look for the operator $\tilde{f}$ in the form

$$\tilde{f} = \sum_{i=0}^{\frac{1}{2}(l-n)} f_i (L^G_{E_-})^i \alpha_i \tag{4.1}$$

where $\alpha_i \in Wu$.

Let us assume now that the right-hand side of (4.1) defines an operator on A. Applying the relations

$$[E_+, E_-^i] = E_-^{i-1}\cdot i(H-i+1)$$

we find easily that the elements $\alpha_i \in Wu$ satisfy the equations

$$\alpha_{i-1}+i(H-i+1)\alpha_i = 0.$$

Therefore, putting $\frac{1}{2}(l-n)=p$ we obtain

$$\alpha_i = (-1)^{p-i}\frac{p!}{i!}(H-p+1)\cdot\ldots\cdot(H-i)\alpha_p.$$

Consequently, we have the operator $\tilde{f}$ if we determine α_p. It is clear that property 2) requires $\alpha_p \in \mathbf{C}$. Moreover, $s_0(\tilde{f})=f\otimes\alpha_0$ and $H(\varrho)=1$. Therefore, to assure the validity of property 1) we have to take

$$\alpha_p = (p!)^{-2}.$$

Then all statements 1)—4) of Conjecture I will be valid.

Example 2. Let $G=SL(3, \mathbf{C})$ and N be the subgroup of upper triangular matrices with units on the diagonal.

Let T_i be the i-th fundamental representation of G, $i=1, 2$. Both representations T_i are three-dimensional and the spaces $\mathscr{E}_{T_i}$ (see Example 1) are of dimension 9. Here $\mathscr{E}_{T_1}$ consists of the linear combinations of the matrix elements g_{ij}, $1\leqq i, j\leqq 3$ of the matrix $g\in G$, and $\mathscr{E}_{T_2}$ consists of the linear combinations of the second order minors of g. Let us construct the mapping $f\to\tilde{f}$ for $f\in\mathscr{E}_{T_1}$. Put $f_i=\alpha_1 g_{i1}+\alpha_2 g_{i2}+\alpha_3 g_{i3}$, $i=1, 2, 3$. Then for arbitrary α_1, α_2, α_3 the elements f_i form a subspace of $\mathscr{E}_{T_1}$ such that the restriction of R^G to this subspace is equivalent to T_1. Here f_3 is a vector of highest weight, that is $\tilde{f}_3=f_3$. Let us now give formulas for $\tilde{f}_1$ and $\tilde{f}_2$. Let E_{ij} $(i\neq j)$, $E_{11}-E_{22}$ and $E_{22}-E_{33}$ be the basis in $\mathfrak{g}$. We put

$$L^G_{E_{ij}} = \hat{E}_{ij}, \quad L^G_{E_{ii}-E_{jj}} = \hat{Z}_{ij}.$$

These operators act on the functions f_i as follows

$$\hat{E}_{ij}f_k = -\delta_{jk}f_i,$$

$$\hat{Z}_{ij}f_k = (\delta_{jk}-\delta_{ik})f_k,$$

where δ_{ij} is the Kronecker symbol.

Let us consider in $\mathscr{D}(G)$ the following operators:

$$\tilde{f}_3 = f_3$$

$$\tilde{f}_2 = f_2\hat{Z}_{32} - f_3\hat{E}_{32}$$

$$\tilde{f}_1 = f_1\hat{Z}_{12}(\hat{Z}_{13}+1) + f_2\hat{E}_{21}(\hat{Z}_{13}+1) + f_3(\hat{E}_{32}\hat{E}_{21} + \hat{E}_{31}\hat{Z}_{12}).$$

It is easy to verify that the operators $\tilde{f}_1, \tilde{f}_2, \tilde{f}_3$ belong to $\mathscr{D}(A)$ (i.e. $\tilde{f}_i(\mathscr{E}(A)) \subset \mathscr{E}(A)$) and the mapping $f_i \to \tilde{f}_i$ satisfies all conditions of Conjecture I. Thus the mapping $f \to \tilde{f}$ has been constructed for $f \in \mathscr{E}_{T_1}$.

We also remark that the operator $\tilde{f}_1$ is an operator of the second order on A which cannot be expressed in term of first order operators on A.

The mapping $f \to \tilde{f}$ for $f \in \mathscr{E}_{T_2}$ can be constructed in the same way.

In Examples 1 and 2 we were dealing with representations T for which all weight subspaces were one-dimensional. It is a more difficult task to construct the operation $f \to \tilde{f}$ in the case when these subspaces are not one-dimensional. Now let us see a simple example of that kind.

Example 3. Let us put, as in Example 2, $G = SL(3, \mathbf{C})$ and T be the adjoint representation of G in its Lie algebra $\mathfrak{g}$. Let f be a vector of highest weight with respect to L^G in $\mathscr{E}_T$. We shall introduce the following notations

$$f_{13} = f, \quad f_{23} = \hat{E}_{21} f, \quad f_{12} = -\hat{E}_{32} f,$$

$$h_{12} = -\hat{E}_{21} f_{12}, \quad h_{23} = -\hat{E}_{32} f_{23},$$

$$f_{21} = -\hat{E}_{21} h_{23}, \quad f_{32} = -\hat{E}_{32} h_{12}, \quad f_{31} = \hat{E}_{32} f_{21}.$$

Then the restriction of L^G to the subspace spanned by f_{ij} and h_{ij} is equivalent to T. Here h_{12} and h_{23} generate a two-dimensional weight subspace of T. We define the operation $f \to \tilde{f}$ in the following way

$$\tilde{h}_{12} = h_{12}\left(\hat{Z}_{12}\hat{Z}_{23} + \frac{2}{3}\hat{Z}_{12} - \frac{2}{3}\hat{Z}_{23}\right) + h_{23}\left(-\hat{Z}_{12}\hat{Z}_{23} + \frac{1}{3}\hat{Z}_{12} - \frac{4}{3}\hat{Z}_{23}\right) +$$
$$+ f_{12}\hat{E}_{21}(3\hat{Z}_{23}+1) - f_{23}\hat{E}_{32}(3\hat{Z}_{12}+2) + f_{13}(3\hat{E}_{21}\hat{E}_{32} + \hat{E}_{31}).$$

$$\tilde{h}_{23} = h_{12}\left(-\hat{Z}_{12}\hat{Z}_{23} - \frac{4}{3}\hat{Z}_{12} + \frac{1}{3}\hat{Z}_{23}\right) + h_{23}\left(\hat{Z}_{12}\hat{Z}_{23} - \frac{2}{3}\hat{Z}_{12} + \frac{2}{3}\hat{Z}_{23}\right) -$$
$$- f_{12}\hat{E}_{21}(3\hat{Z}_{23}+2) + f_{23}\hat{E}_{32}(3\hat{Z}_{12}+1) - f_{13}(3\hat{E}_{21}\hat{E}_{32} + 2\hat{E}_{31}).$$

We remark that in the subspace spanned by $\tilde{h}_{12}$ and $\tilde{h}_{23}$, there is, up to multiplication, only one operator of first order namely $\tilde{h}_{12} - \tilde{h}_{23}$; other operators in this space are of order 2.

§ 5. The generalized Segal—Bargmann spaces

In this section we shall consider a generalization of Bargmann's construction [3] of representations of the group $SU(2)$.

Let K be a maximal compact subgroup of the group G, $\mathfrak{k}$ be its Lie algebra, and $\mathfrak{i}:\mathfrak{g}\to\mathfrak{g}$ be the corresponding Cartan involution. We assume that K is chosen so that $\mathfrak{i}(\mathfrak{n}_+)=\mathfrak{n}_-$. Let T_i, $1\leqq i\leqq r$ be the representations of G corresponding to the fundamental highest weights of $\mathfrak{g}$. Let f_i be a vector of highest weight in T_i. We define a function $H_i(g)$ on G by the formula

$$H_i(g) = \|T(g^{-1})f_i\|_i^2,$$

where $\|\cdot\|_i$ is a K-invariant norm on T_i such that $\|f_i\|_i=1$. Clearly, $H_i(ng)=H_i(g)$ for $n\in N_+$, hence H_i can also be considered as a function on A. Let $u(t)$ be a positive, rapidly decreasing function defined for $t>0$. Let ϱ_i $(1\leqq i\leqq r)$ be positive numbers. We define the weight function $\varrho(x)$ on A by the formula

$$\varrho(x) = u\left(\sum \varrho_i H_i(x)\right).$$

Definition 5.1. The *Segal—Bargmann space* of the group G is defined as the completion of $\mathscr{E}(A)$ with respect to the scalar product

$$\{f, g\} = \int_A f(x)\bar{g}(x)\varrho(x)\omega,$$

where ω denotes the G-invariant measure on A.

It is obvious that this scalar product in $\mathscr{E}(A)$ is invariant under R_k^A, $k\in K$.

Conjecture II. *There exists a function $u(t)$ such that for any $D\in\mathscr{D}(A)$ we have $D^*\in\mathscr{D}(A)$; here D^* denotes the adjoint operator of D with respect to $\{\,,\}$.*

Let us consider the simplest case $G=SL(2, \mathbf{C})$, $K=SU(2)$. In this case A is the plane $\mathbf{C}^2$ without the point $(0, 0)$, and $\mathscr{E}(A)$ is the space of polynomials of two variables z_1, z_2. The scalar product is introduced in $\mathscr{E}(A)$ by the formula

$$\{f, g\} = \int e^{-\varrho(|z_1|^2+|z_2|^2)} f\bar{g}\, dz_1\, dz_2\, d\bar{z}_1\, d\bar{z}_2.$$

Now the ring $\mathscr{D}(A)$ is generated by the operators z_i and $\frac{\partial}{\partial z_i}$, $i=1, 2$. It is easy to verify that $\left(\frac{\partial}{\partial z_i}\right)^* = \varrho z_i$ and therefore

$$(z_i)^* = \varrho^{-1}\frac{\partial}{\partial z_i}.$$

This shows that Conjecture II is true in the present case.

The above construction for $SU(2)$ was suggested by Bargmann [3].

Using the examples given in § 4 we can show that Conjecture II is also valid for $G=SL(3, \mathbf{C})$. As $u(t)$ we take the decreasing positive solution of the equation

$$t\frac{d^2u}{dt^2}+2\frac{du}{dt}+u = 0.$$

It seems that for $G=SL(n, \mathbf{C})$ the function $u(t)$ has to satisfy the equation

$$\frac{d^{n-1}}{dt^{n-1}}(ut^{n-2})+(-1)^n u = 0.$$

A more precise version of Conjecture II is the following

Conjecture II'. *Let* $g \to g^*$ *be the anti-automorphism of* G *corresponding to the anti-involution* $(-\mathfrak{i}):\mathfrak{g}\to\mathfrak{g}$. *The numbers* ϱ_i *can be chosen in such a way that for any function* $f\in\mathscr{E}(G)$ *the equality* $(\tilde{f})^*=(\widetilde{f^*})$ *is satisfied. Here* $f^*(g)=f(g^*)$.

§ 6. The mapping π_*

In this section we shall give a construction which yields a weaker version of Conjecture I. More precisely, for every function $f\in\mathscr{E}(G)$ a collection of regular differential operators on A will be constructed. In addition to this, we shall show that all differential operators on A can be obtained in this way. In the construction we apply an operation π_* which maps functions on G into functions on A and which, we believe, is of independent interest. This operation is an algebraic analogue of the averaging operation over a subgroup (which is unipotent in the present case). It is remarkable that π_* transfers the operation of multiplication by a function $f(g)$ into an "almost" differential operator on A. The exact formulation of these facts is given in Theorems 6.3 and 6.5.

Lemma and Definition 6.1. *There exists a unique mapping* $\pi_*:\mathscr{E}(G)\to\mathscr{E}(A)$ *such that*

1) $\pi_* R_g^G = R_g^A \pi_*$ *and* $\pi_* L_h^G = L_h^A \pi_*$

for all $g\in G$, $h\in H$,

2) $\pi_*\pi^*\varphi = \varphi$ *for all* $\varphi\in\mathscr{E}(A)$.

Proof. First we prove that $\pi_* f$ is uniquely determined by the conditions *1)* and *2)*. It is enough to consider the case when f lies in a subspace V irreducible and invariant with respect to L^G; further we can assume that f is a weight function of weight χ with respect to the restriction of L^G to H. If χ is a highest weight of the given irreducible representation, then $f\in\operatorname{Im}\pi^*$, i.e. $f=\pi^*\varphi$, hence in view of *2)*, $\pi_* f=\varphi$.

Assume now that χ is not a highest weight. Let us denote by f_0 a vector of highest weight in V and by χ_0 the corresponding highest weight. Then, under the action of R^G, f and f_0 are transformed by the same irreducible representation of G. From *1)* and the fact that every irreducible representation of G occurs in $\mathscr{E}(A)$ only once (see [2]) it follows that $\pi_* f$ and $\pi_* f_0$ belong to the same subspace, irreducible and invariant with respect to R^A. But then the weights of

$\pi_* f$ and $\pi_* f_0$ with respect to L^A coincide (see [2]). From *1)* it follows that the weight of $\pi_* f$ is equal to χ and the weight of $\pi_* f_0$ is equal to $\chi_0 \neq \chi$. Since $\pi_* f_0 \neq 0$ we have $\pi_* f = 0$.

From this proof we immediately obtain a construction for π_*. Indeed, if f is a weight vector, not of highest weight, lying in a subspace, irreducible and invariant under L^G, then we put $\pi_* f = 0$. On the other hand, if f is a vector of highest weight then $f = \pi^* \varphi$ and we put $\pi_* f = \varphi$. The lemma is proved.

Let us denote for each $\gamma \in \Delta$ the operator $L^G_{E_\gamma}$ by $\hat{E}_\gamma$. It follows from the construction of π_* that $\pi_*(\hat{E}_\gamma f) = 0$ for all $f \in \mathscr{E}(G)$ and $\gamma \in \Delta_-$.

By means of the mapping π_* one can construct differential operators on A in the following way.

Definition 6.2. Let $f \in \mathscr{E}(G)$. We define an operator $\hat{f}$ in the space $\mathscr{E}(A)$ by the formula $\hat{f}(\varphi) = \pi_*(f \cdot \pi^* \varphi)$, $\varphi \in \mathscr{E}(A)$.

Theorem 6.3. *There exists a non-zero element $Z \in Wu$ such that $Z\hat{f}$ is a regular differential operator on A.*

Proof. A differential operator D on G is called a *chain* if it can be expressed in the form $\hat{E}_{\gamma_1} \hat{E}_{\gamma_2} \dots \hat{E}_{\gamma_k}$, $\gamma_i \in \Delta_+$. The weight of this chain is defined to be $\gamma_1 + \gamma_2 + \dots + \gamma_k \in \mathfrak{h}^*$. Let us denote by Ξ the set of all chains D such that $Df \neq 0$ and by $\Xi_0 \subset \mathfrak{h}^*$ the set of all their weights. Obviously, Ξ and Ξ_0 are finite sets. For any function $\varphi \in \mathscr{E}(A)$ and any chain D we have $D(f\pi^*\varphi) = Df \cdot \pi^* \varphi$, since $\hat{E}_\gamma \pi^* \varphi = 0$ for $\gamma \in \Delta_+$. Therefore, if $D \notin \Xi$ then $D(f\pi^*\varphi) = 0$. Let us denote by U the subspace of $\mathscr{E}(G)$ consisting of all functions u such that $Du = 0$ for any chain $D \notin \Xi$.

Lemma 6.4. *There exist a regular differential operator T on G and an element $Z \in Wu$ such that $Tu = Z\pi^* \pi_* U$ for all $u \in U$.*

The theorem is an immediate consequence of this lemma, since for any function $\varphi \in \mathscr{E}(A)$ we have $f \cdot \pi^* \varphi \in U$, and so

$$T(f\pi^* \varphi) = Z\pi^* \pi_*(f \cdot \pi^* \varphi) = \pi^* Z\hat{f}(\varphi),$$

i.e. $T \circ f \circ \pi^* = \pi^* \circ Z \circ \hat{f}$. It follows from this equality that the differential operator $T \circ f$ preserves $\mathscr{E}(A) \subset \mathscr{E}(G)$, or in other words, that $Z\hat{f}$ is a differential operator on A.

Proof of the lemma. Let $H_1, \dots, H_r$ denote a basis in $\mathfrak{h}$. The elements of Wu are polynomials of $H_1, \dots, H_r$ and, as above, we may consider them as polynomial functions on $\mathfrak{h}^*$. Let Δ be the Laplace operator of the second order on G (constructed by means of the Killing form). Then there exists an element $P \in Wu$ such that for any vector $\varphi \in \mathscr{E}(A)$ the equality $\Delta\varphi = P\varphi$ is satisfied, or equivalently, $\Delta\varphi = P(\chi_0) \cdot \varphi$, where χ_0 is the weight of the vector φ.

Let B be the restriction of the Killing form of the algebra $\mathfrak{g}$ to $\mathfrak{h}$, and $Q(\chi)$ the dual quadratic form on $\mathfrak{h}^*$. It follows from results of Harish–Chandra [4] that $P(\chi)=Q(\chi+\varrho)-Q(\varrho)$, where ϱ is half-sum of the positive roots.

For an arbitrary weight β we denote by P_β and Z_β the elements of Wu corresponding to the polynomial functions

$$P_\beta(\chi) = P(\chi+\beta)$$

and

$$Z_\beta(\chi) = 2\langle\beta, \chi+\varrho\rangle-\langle\beta, \beta\rangle$$

respectively ($\langle , \rangle$ denotes the scalar product in $\mathfrak{h}^*$ corresponding to the quadratic form Q). Let $T=\prod_\beta (P_\beta-\Delta)$, $Z=\prod_\beta Z_\beta$, where β runs through $\Xi_0\setminus\{0\}$.

We shall show that for all $u\in U$ the equality $Tu=Z\pi^*\pi_* u$ is satisfied.

Clearly, it suffices to verify this equality when u is a weight vector lying in a subspace V which is irreducible and invariant under L^G. Let u_0 be a vector of highest weight in V, χ and χ_0 be the weights of u and u_0, respectively, with respect to $\mathfrak{h}$. It follows from the uniqueness of the vector of highest weight that $u_0=cDu$ where D is an appropriate chain and $c\in\mathbf{C}$, and therefore $\chi_0-\chi\in\Xi_0$.

Case 1, $\chi\neq\chi_0$. The restriction of Δ to V is multiplication by $P(\chi_0)$. Therefore

$$(P_\beta-\Delta)u = (P_\beta(\chi)-P(\chi_0))u = (P(\chi+\beta)-P(\chi_0))u = 0,$$

if $\beta=\chi_0-\chi\in\Xi_0\setminus\{0\}$. Thus $Tu=0$, and since $\pi_* u=0$ in this case, we have $Tu=$ $=Z\pi^*\pi_* u=0$.

Case 2, $\chi=\chi_0$. Then

$$(P_\beta-\Delta)u = (P(\chi_0+\beta)-P(\chi_0))u = (Q(\chi_0+\beta+\varrho)-Q(\chi_0+\varrho))u =$$

$$= (2\langle\chi_0+\varrho, \beta\rangle+\langle\beta, \beta\rangle)u = Z_\beta(\chi_0)u,$$

hence

$$Tu= (\pi(P_\beta-\Delta))u = \pi Z_\beta(\chi_0)u = Zu = Z\pi^*\pi_* u.$$

The proof of the lemma and of Theorem 6.3 is complete.

It can be shown that the order of the operator $Z\hat{f}$ is equal to card Ξ_0-1, i.e. the order of Z'.

Theorem 6.5. *Every regular differential operator* D *on* A *can be written in the form* $D=\sum Z_j \hat{f}_j$, *where* $Z_j\in Wu$ *and* $f_j\in\mathscr{E}(G)$.

Proof. Let D' be a lifting of D to G (Theorem 3.4). By means of simple transformations D' can be transformed to the form

$$D' = \sum Z_j f_j+\sum A_k\hat{E}_{\gamma_k}+\sum \hat{E}_{\delta_i}B_i,$$

where A_k and B_i are differential operators on G, $Z_j\in Wu$, $f_j\in\mathscr{E}(G)$, $\gamma_k\in\Delta_+$, $\delta_i\in\Delta_-$.

Since $\hat{E}_{\gamma_k}\pi^*\varphi = 0$ and $\pi_* E_{\delta_i} f = 0$, we have

$$D\varphi = \pi_*\pi^* D\varphi = \pi_* D'\pi^*\varphi = \sum Z_j \pi_*(f_j \pi^*\varphi) = \sum Z_j \hat{f}_j(\varphi),$$

which proves theorem 6.5.

Let L be the quotient field of the ring Wu. Then the operation $f \to \hat{f}$ extends to a mapping $\Omega: L \underset{\mathbf{C}}{\otimes} \mathscr{E}(G) \to L \underset{Wu}{\otimes} \mathscr{D}(A)$. Here Ω commutes with the actions of R_g and L_g.

Theorem 6.6. *Ω is an isomorphism of L-modules $L \otimes \mathscr{E}(G)$ and $L \underset{Wu}{\otimes} \mathscr{D}(A)$.*

Proof. It follows from Theorem 6.5 that Ω is a surjection.

In Theorem 6.3 we constructed for every function $f \in \mathscr{E}(G)$ an operator $Z_f \in Wu$ such that $Z_f \cdot \hat{f} \in \mathscr{D}(A)$.

Lemma 6.7.

$$s_0(Z_f \cdot \hat{f}) = f \otimes Z_f + \sum f_i \otimes Z_i,$$

where $\deg Z_i < \deg Z_f$.

Proof. The lifting of the operator $Z_f \cdot \hat{f}$ has the form $T \circ f$, where $T = \prod_\beta (P_\beta - \Delta)$ (see the proof of Theorem 6.3). It follows from [4] that

$$\Delta = P + \sum_{\gamma \in \Delta_+} c_\gamma \hat{E}_{-\gamma} \hat{E}_\gamma, \quad c_\gamma \in \mathbf{C}.$$

Consequently

$$P_\beta - \Delta = Z_\beta - \sum_{\gamma \in \Delta_+} c_\gamma \hat{E}_{-\gamma} \hat{E}_\gamma.$$

Let $K = \operatorname{card}(\Xi_0 \setminus \{0\})$. It is easy to show by induction on K that

$$T = \prod_\beta Z_\beta + \sum_\lambda c_\lambda H_\lambda X_\lambda Y_\lambda,$$

where $c_\lambda \in \mathbf{C}$, and H_λ, X_λ and Y_λ are products of suitably chosen operators $H_i \in Wu$, $\hat{E}_{-\gamma}$, $\gamma \in \Delta_+$, and $\hat{E}_\gamma$, $\gamma \in \Delta_+$, respectively. We also have $\deg H_\lambda + \deg X_\lambda \leq K$ for all λ, and $\deg H_\lambda < K$ if $X_\lambda = 1$. Moreover, $Z_f = \prod_\beta Z_\beta$. From this it follows that the operator $T \circ f$ is of the form $f \cdot Z_f + \sum f_\mu H_\mu X_\mu Y_\mu$, where $f_\mu \in \mathscr{E}(G)$, H_μ, X_μ and Y_μ satisfy the same conditions as H_λ, X_λ, Y_λ. Therefore, $s_0(Z_f \cdot \hat{f})$ (see Definition 3.10) is of the form $s_0(Z_f \cdot \hat{f}) = f \otimes Z_f + \sum f_i \otimes Z_i$, where $\deg Z_f = K$, $\deg Z_i < K$. The lemma is proved.

Let us now consider the element $D = \sum \tilde{Z}_i \otimes f_i \in \mathscr{E}(G)$, where $\tilde{Z}_i \in L$, $\tilde{Z}_i \neq 0$, and the f_i are linearly independent. We shall show that $\Omega(D) \neq 0$. Let us multiply D by an element $Z \in Wu$ such that $ZD = \sum Z_i \otimes f_i$, where $Z_i \in W_u$ and Z_i is divisible by Z_{f_i}. Let $l = \max \deg Z_i$. Then $\Omega(ZD) \in \mathscr{D}(A)$ and by Lemma 6.7

$$s_0(\Omega(ZD)) = \sum f_i \otimes Z_i + \sum f'_j \otimes Z'_j,$$

where $\deg Z'_j < l$. This implies that $s_0(\Omega(ZD)) \neq 0$, consequently $\Omega(D) \neq 0$.

We have now proved that Ω is an injection and therefore an isomorphism.

PART II

THE RESOLUTION OF A FINITE-DIMENSIONAL $\mathfrak{g}$-MODULE

§ 7. Introduction

Let $\mathfrak{g}$ be, as above, a complex semisimple Lie algebra. Let V be a finite-dimensional irreducible $\mathfrak{g}$-module. The cohomology groups $H^i(\mathfrak{n}_-, V)$ play an important rôle in the theory of representations (see [5] and [6]). They have the following properties expressed in Bott's theorem [5]. Let W be the Weyl group of the algebra $\mathfrak{g}$. Then $\dim H^i(\mathfrak{n}_-, V) = \operatorname{card} W^{(i)}$, where $W^{(i)} = \{w \in W : l(w) = i\}$ and $l(w)$ is the length of the element $w \in W$.

The fundamental result of this part is Theorem 10.1 which improves on Bott's theorem.

For any $\chi \in \mathfrak{h}^*$ we denote by M_χ the $U(\mathfrak{g})$-module generated by a vector of highest weight $\chi - \varrho$ (the exact definition is given below). In Theorem 10.1 we construct a resolution

$$0 \leftarrow V \leftarrow C_0 \leftarrow C_1 \leftarrow \ldots \leftarrow C_s \leftarrow 0$$

of the $\mathfrak{g}$-module V such that

$$C_i = \bigoplus_{w \in W^{(i)}} M_{w\chi},$$

where $\chi - \varrho$ is the highest weight of V. Bott's theorem follows from Theorem 10.1 since M_χ is a free $U(\mathfrak{n}_-)$-module with one generator.

In this part we shall make a systematic use of a certain category of $\mathfrak{g}$-modules which we call category O (see [12]). § 8 is devoted to the exposition of the properties of category O.

In § 9 several results concerning the cohomologies of Lie algebras are presented. In particular, in this section a purely algebraic proof of Bott's theorem is given which, it seems to us, is simpler than the proofs presented in [5] and [6]. This proof has several points of contact with Kostant's proof [6], but it does not make any use of the Hermitian structure. The observant reader will notice that the resolution constructed in the proof of Bott's theorem is dual to a part of the de Rham resolution well-known from the theory of formal differential forms.

In the Appendix we describe the modules occurring in the Jordan—Hölder decomposition of the modules M_χ. The study of the structure of the modules M_χ was initiated in Verma's work [7]. We remark that the works [7], [8] and [9] contain everything we know about the modules M_χ. All these facts are also contained in Theorems 8.7, 8.8, 8.12 and 10.1 of the present work.

§ 8. The category O

Definition 8.1. Category O is the full subcategory of the category of left $U(\mathfrak{g})$-modules consisting of all modules M such that

1) M is a finitely generated $U(\mathfrak{g})$-module,

2) M can be made $\mathfrak{h}$-diagonal, i.e. there exists a basis in M consisting of weight vectors,

3) M is $U(\mathfrak{n}_+)$-finite, i.e. for any $f \in M$ the space $U(\mathfrak{n}_+)f$ is finite-dimensional.

Definition 8.2. Let $\chi \in \mathfrak{h}^*$. We denote by J_χ the left ideal in $U(\mathfrak{g})$ generated by the elements E_γ, $\gamma \in \Delta_+$ and $H - \chi(H) + \varrho(H)$, $H \in \mathfrak{h}$. Let us put $M_\chi = U(\mathfrak{g})/J_\chi$. We shall denote by f_χ the image of $1 \in U(\mathfrak{g})$ in M_χ.

For the sake of convenience we formulate the elementary properties of the category O and the modules M_χ in Propositions 8.3, 8.5, 8.6. These propositions are simple consequences of Harish-Chandra's theorem about Laplace operators (see [4]).

Proposition 8.3. *1)* The category O is closed under taking submodules, factor modules and finite direct sums.

2) Let $M \in O$. Then all the spaces $M^{(\psi)}$ $\psi \in \mathfrak{h}^*$ are finite-dimensional and $M = \bigoplus_{\psi \in \mathfrak{h}^*} M^{(\psi)}$. In addition to this $P(M)$ is contained in a finite union of sets $\chi_i - K$, $\chi_i \in \mathfrak{h}^*$.

3) Every element $M \in O$ has a finite Jordan—Hölder composition series.

4) M_χ is a free $U(\mathfrak{n}_-)$-module with f_χ as a generator.

5) $M_\chi \in O$.

6) In M_χ there exists a maximal proper submodule. The corresponding irreducible factor module will be denoted by L_χ.

7) Every irreducible module in the category O is of the form L_χ, $\chi \in \mathfrak{h}^*$.

We denote by $Z(\mathfrak{g})$ the centre of the algebra $U(\mathfrak{g})$ and by Θ the set of all homomorphisms $\vartheta: Z(\mathfrak{g}) \to \mathbf{C}$.

Definition 8.4. Let M be an arbitrary $\mathfrak{g}$-module. To each element $f \in M$ which is an eigenvector with respect to all the operators $z \in Z(\mathfrak{g})$ we can assign a homomorphism $\vartheta_f \in \Theta$ such that $zf = \vartheta_f(z) \cdot f$ for all $z \in Z(\mathfrak{g})$. The set of all such homomorphisms ϑ will be denoted by $\Theta(M)$.

Proposition 8.5. *1)* $\Theta(M_\chi)$ consists of a single element which we shall denote by ϑ_χ.

2) $\vartheta_{\chi_1} = \vartheta_{\chi_2}$ if and only if $\chi_1 \sim \chi_2$.

Proposition 8.6. Let $M \in O$. Then

1) $\Theta(M)$ is finite.

2) For any $\vartheta \in \Theta$ we put $I_\vartheta = \operatorname{Ker} \vartheta \subset Z(\mathfrak{g})$. Let $M_\vartheta^{(n)} = \{f \in M \mid I_\vartheta^n f = 0\}$. Then $M_\vartheta^{(n)}$ stabilizes for large values of n. The obtained submodule of M will be denoted by M_ϑ.

3) $\Theta(M_\vartheta) = \{\vartheta\}$.

4) $M = \bigoplus_{\vartheta \in \Theta(M)} M_\vartheta$.

5) The mapping $M \to M_\vartheta$ is an exact functor in O.

Now we shall pass on to the study of the modules M_χ. The following two theorems give a complete description of the homomorphisms between modules M_χ.

Theorem 8.7 [7]. *Let* $\chi, \psi \in \mathfrak{h}^*$. *Then either*

1) $\mathrm{Hom}_{U(\mathfrak{g})}(M_\chi, M_\psi) = 0$.

or

2) $\mathrm{Hom}_{U(\mathfrak{g})}(M_\chi, M_\psi) = \mathbf{C}$, *and every non-trivial homomorphism* $M_\chi \to M_\psi$ *is an injection.*

Theorem 8.8. *Let* $\chi, \psi \in \mathfrak{h}^*$,

then

$$\mathrm{Hom}_{U(\mathfrak{g})}(M_\chi, M_\psi) = \mathbf{C}$$

if and only if there exists a sequence of roots $\gamma_1, \ldots, \gamma_k \in \Delta_+$ *satisfying the following condition (A) for the pair* (χ, ψ).

Condition (A).

1) $\chi = \sigma_{\gamma_k} \cdot \sigma_{\gamma_{k-1}} \cdot \ldots \cdot \sigma_{\gamma_1} \psi$.

2) Put $\chi_0 = \psi$, $\chi_i = \sigma_{\gamma_i} \cdot \ldots \cdot \sigma_{\gamma_1} \psi$. *Then* $\chi_{i-1} - \chi_i = n\gamma_i$, *where n is a non-negative integer.*

In particular $\mathrm{Hom}_{U(\mathfrak{g})}(M_\chi, M_\psi) \neq 0$ only if $\chi \sim \psi$ and $\chi \leqq \psi$.

Theorem 8.8 was formulated in [7] as a conjecture; a proof of the sufficiency of condition (A) was also given there. A complete proof of Theorem 8.8 was given in [9].

The structure of the submodules of the modules M_χ is most interesting when $\chi \in D$. We shall study this case in more detail. For this purpose we introduce the following partial ordering in the Weyl group W.

Definition 8.9. If $w_1, w_2 \in W$ and $\gamma \in \Delta_+$ then $w_1 \xrightarrow{\gamma} w_2$ means that $w_1 = \sigma_\gamma w_2$ and $l(w_1 = l(w_2) + 1$. (Sometimes we shall omit the symbol γ above the arrow.) We put $w < w'$ if there exists a sequence $w_1, w_2, \ldots, w_k$ of elements of W such that

$$w \to w_1 \to w_2 \to \ldots \to w_k \to w'.$$

Theorem 8.8′. *Let* $\chi \in D$, $w_1, w_2 \in W$. *Then* $\mathrm{Hom}_{U(\mathfrak{g})}(M_{w_1\chi}, M_{w_2\chi}) = \mathbf{C}$ *if and only if* $w_1 \leqq w_2$.

In what follows all modules $M_{w\chi}$, $w \in W$, $\chi \in D$ will be considered as submodules of M_χ.

Theorem 8.8′ is an immediate consequence of Theorem 8.8 and the following lemmas.

Lemma 8.10. *Let* $\chi \in D$, $\gamma \in \Delta$, $w \in W$. *Then* $w\chi - \sigma_\gamma w\chi = n\gamma$, *where* $n \in \mathbf{Z}$, $n \neq 0$, *while* $n > 0$ *if and only if* $l(\sigma_\gamma w) > l(w)$.

Lemma 8.11. *Let* $w \in W$ *and* $\gamma \in \Delta$ *be such that* $l(\sigma_\gamma w) < l(w)$. *Then* $w < \sigma_\gamma w$.

The proofs of these lemmas will be presented in § 11.

The complete structure of the submodules of the modules M_χ is not yet known. Some information is contained in the following theorem.

Theorem 8.12. *Let* $\psi, \chi \in \mathfrak{h}^*$. *Then* $L_\chi \in JH(M_\psi)$ *if and only if there exists a sequence* $\gamma_1, \gamma_2, \ldots, \gamma_k \in \Delta_+$ *satisfying condition* (*A*) *for the pair* (χ, ψ) (see Theorem 8.8).

Since we shall need this theorem in § 10, we shall give the proof in the Appendix.

Corollary. Let $\chi \in D$. Then the Jordan—Hölder decomposition of the module $M_{w\chi}$ consists of the modules $L_{w'\chi}$, where $w' \leqq w$ (possibly counted with multiplicities). $L_{w\chi}$ occurs in this decomposition exactly once. The example in [9] shows that the modules M_ψ may contain submodules M which are not generated by submodules $M_\chi \subset M$.

In fact, even the following proposition can be proved.

Proposition 8.13. Let $\mathfrak{g} = sl(4, \mathbf{C})$. Consider the module M_ψ corresponding to a weight $\psi \in D$. Then M_ψ contains a submodule M such that

$$M \neq \sum_{M_{w\psi} \subset M} M_{w\psi}.$$

This statement is equivalent to the fact that the number of elements in $JH(M_\psi)$ is greater than the number of elements of W.

§ 9. Cohomology of Lie algebras

In this section we shall recall a number of results concerning the cohomology of Lie algebras. Moreover, Bott's theorem will also be proved here.

Let $\mathfrak{a}$ be an arbitrary complex Lie algebra and M an $\mathfrak{a}$-module. The exact sequence of $\mathfrak{a}$-modules

$$0 \longleftarrow M \longleftarrow C_0 \xleftarrow{d_1} C_1 \xleftarrow{d_2} \ldots, \tag{9.1}$$

where each module C_i is free over $U(\mathfrak{a})$ is called a *free resolution* of M.

Let N be another $\mathfrak{a}$-module. Consider the complex

$$0 \longrightarrow \mathrm{Hom}_{\mathfrak{a}}(C_0, N) \xrightarrow{d_1'} \mathrm{Hom}_{\mathfrak{a}}(C_1, N) \xrightarrow{d_2'} \ldots$$

and put

$$\mathrm{Ext}^i(M, N) = \mathrm{Ker}\, d_{i+1}' / \mathrm{Im}\, d_i'.$$

Let $\tau: U(\mathfrak{a}) \to U(\mathfrak{a})$ be the anti-automorphism defined by the formula $\tau(X) = -X$ for $X \in \mathfrak{a}$. We denote by N^τ the right $U(\mathfrak{a})$-module whose underlying space coincides with that of N and on which the action of $U(\mathfrak{a})$ is defined by the formula $f \cdot X = \tau(X) f$ for $f \in N$, $X \in U(\mathfrak{a})$.

Now consider the complex

$$0 \longleftarrow N^\tau \underset{\mathfrak{a}}{\otimes} C_0 \xleftarrow{d_1''} N^\tau \underset{\mathfrak{a}}{\otimes} C_1 \xleftarrow{d_2''} \dots$$

and put

$$\mathrm{Tor}_i(N^\tau, M) = \mathrm{Ker}\, d_i'' / \mathrm{Im}\, d_{i+1}''.$$

The following standard facts hold (see [10]).

1) The groups $\mathrm{Tor}_i(N^\tau, M)$ and $\mathrm{Ext}^i(M, N)$ are independent of the choice of the resolution (9.1).

2) Let $\mathfrak{a}$ be a finite-dimensional Lie algebra and M, N be finite-dimensional $\mathfrak{a}$-modules. Then

a) $[\mathrm{Ext}^i(M, N)]^* = \mathrm{Tor}_i(N^*, M)$,

where $N^* = \mathrm{Hom}_{\mathbf{C}}(N, \mathbf{C})$ is a right $\mathfrak{a}$-module ([10], Chapter XI, § 3, Proposition 3.3),

b) $\mathrm{Tor}_i(N^*, M) = \mathrm{Tor}_i(M^\tau, (N^\tau)^*)$.

([10], Chapter VI, § 1).

The cohomology group of $\mathfrak{a}$ with coefficients in M is defined by the formula $H^i(\mathfrak{a}, M) = \mathrm{Ext}^i(\mathbf{C}, M)$, where $\mathbf{C}$ is the trivial one-dimensional $\mathfrak{a}$-module.

The computation of the cohomology groups is done by means of the standard resolution $V(\mathfrak{a})$ of the module $\mathbf{C}$, which is defined in the following way.

We put

$$C_k = U(\mathfrak{a}) \underset{\mathbf{C}}{\otimes} \Lambda^k(\mathfrak{a}), \quad k = 0, 1, \dots$$

Then we define a homomorphism $d_k: C_k \to C_{k-1}$ of $\mathfrak{a}$-modules by means of the formula

$$d_k(X \otimes X_1 \wedge \dots \wedge X_k) = \sum_{i=1}^{k} (-1)^{i+1} (X X_i \otimes X_1 \wedge \dots \wedge \hat{X}_i \wedge \dots \wedge X_k) +$$

$$+ \sum_{1 \leq i < j \leq k} (-1)^{i+j} (X \otimes [X_i, X_j] \wedge X_1 \wedge \dots \wedge \hat{X}_i \wedge \dots \wedge \hat{X}_j \wedge \dots \wedge X_k).$$

Here $X \in U(\mathfrak{a})$, $X_i \in \mathfrak{a}$ and the symbol ^ means that the corresponding element is to be omitted. Furthermore, we define $\varepsilon: C_0 \to \mathbf{C}$ by the formula $\varepsilon(X) =$ (the constant part of X), $X \in U(\mathfrak{a})$.

As was shown in [10], Chapter XIII, § 7, the sequence

$$0 \longleftarrow \mathbf{C} \xleftarrow{\varepsilon} C_0 \xleftarrow{d_1} C_0 \xleftarrow{d_2} \dots$$

is a resolution of the $\mathfrak{a}$-module $\mathbf{C}$.

Subsequently we shall need a generalization of the resolution $V(\mathfrak{a})$ for the case of relative cohomology.

Let $\mathfrak{a}$ be a complex Lie algebra and $\mathfrak{p}$ a subalgebra. The adjoint action of $\mathfrak{p}$ in $\mathfrak{a}$ yields a representation ϑ of the algebra $\mathfrak{p}$ in the linear space $\mathfrak{a}/\mathfrak{p}$. The corresponding representations of $\mathfrak{p}$ in the linear spaces $\Lambda^k(\mathfrak{a}/\mathfrak{p})$ will be denoted by the same symbol ϑ.

Let us consider for each k, $k=0, 1, 2, \ldots$, the module

$$D_k = U(\mathfrak{a}) \underset{U(\mathfrak{p})}{\otimes} \Lambda^k(\mathfrak{a}/\mathfrak{p}).$$

We define the operators $d_k : D_k \to D_{k-1}$ in the following way. Let $X_1, \ldots, X_k$ be elements of $\mathfrak{a}/\mathfrak{p}$. Let $Y_1, Y_2, \ldots, Y_k \in \mathfrak{a}$ be arbitrary representatives for $X_1, \ldots, X_k$, respectively, and put

$$d_k(X \otimes X_1 \wedge \ldots \wedge X_k) = \sum_{i=1}^{k} (-1)^{i+1}(XY_i \otimes X_1 \wedge \ldots \wedge \hat{X}_i \wedge \ldots \wedge X_k) +$$

$$+ \sum_{1 \leq i < j \leq k} (-1)^{i-j}(X \otimes \overline{[Y_i, Y_j]} \wedge X_1 \wedge \ldots \wedge \hat{X}_i \wedge \ldots \wedge \hat{X}_j \wedge \ldots \wedge X_k).$$

Here $X \in U(\mathfrak{a})$, and $\bar{Y}$ is the image of the element $Y \in \mathfrak{a}$ in $\mathfrak{a}/\mathfrak{p}$. It is easy to verify that the operator d_k is well defined, i.e. independent of the choice of the representatives Y_i.

In addition to this, we introduce the augmentation $\varepsilon : D_0 \to \mathbf{C}$ by putting $\varepsilon(X \otimes 1) =$ (the constant part of X). Thus we have constructed a sequence $V(\mathfrak{a}, \mathfrak{p})$ of $U(\mathfrak{a})$-modules

$$0 \longleftarrow \mathbf{C} \xleftarrow{\varepsilon} D_0 \xleftarrow{d_1} D_1 \longleftarrow \ldots .$$

Direct computation shows that this sequence is a complex, i.e. $d_{i-1}d_i = 0$, $\varepsilon d_0 = 0$. We shall call this complex $V(\mathfrak{a}, \mathfrak{p})$ the *relative chain complex* of the algebra $\mathfrak{a}$ with respect to the subalgebra $\mathfrak{p}$. Clearly, $V(\mathfrak{a}, 0) = V(\mathfrak{a})$.

Theorem 9.1. *The complex* $V(\mathfrak{a}, \mathfrak{p})$ *is exact.*

Proof. Our proof will be similar to the proof of exactness of the standard complex $V(\mathfrak{a})$, given in [10].

We define a filtration in $V(\mathfrak{a}, \mathfrak{p})$ by writing $A \in D_k^{(l)}$ if $A \in D_k$ can be written in the form

$$A = \sum c_i(X^{(i)} \otimes X_1^{(i)} \wedge \ldots \wedge X_k^{(i)}),$$

where

$$c_i \in \mathbf{C}, \quad X^{(i)} \in U(\mathfrak{a}), \quad X_j^{(i)} \in \mathfrak{a}/\mathfrak{p} \quad \text{and} \quad \deg X \leq l-k.$$

It is clear that $d_k(D_k^{(l)}) \subset D_{k-1}^{(l)}$. Therefore, to prove the theorem, it is enough to show that for every l the complex

$$0 \longleftarrow M^{(l)} \longleftarrow D_0^{(l)}/D_0^{(l-1)} \xleftarrow{d_1^{(l)}} D_1^{(l)}/D_1^{(l-1)} \xleftarrow{d_2^{(l)}} \ldots$$

is exact. Here $M^{(0)} = \mathbf{C}$ and $M^{(l)} = 0$ if $l > 0$. It follows from the Poincaré—Birkhoff—Witt theorem that $D_k^{(l)}/D_k^{(l-1)}=S_{l-k}(\mathfrak{a}/\mathfrak{p})\otimes\Lambda^k(\mathfrak{a}/\mathfrak{p})$, where $S_{l-k}(\mathfrak{a}/\mathfrak{p})$ denotes the set of all homogeneous elements of degree $l-k$ in the symmetric algebra of the space $\mathfrak{a}/\mathfrak{p}$. The operator

$$d_k^{(l)}: D_k^{(l)}/D_k^{(l-1)}\to D_{k-1}^{(l)}/D_{k-1}^{(l-1)}$$

is given by the formula

$$d_k^{(l)}(X\otimes X_1\wedge\ldots\wedge X_k) = \sum_{i=1}^{k}(-1)^{i+1}(XX_i\otimes X_1\wedge\ldots\wedge\hat{X}_i\wedge\ldots\wedge X_k).$$

Therefore, the complex Gr $V(\mathfrak{a}, \mathfrak{p})$:

$$0\leftarrow \mathbf{C}\leftarrow\bigoplus_l D_0^{(l)}/D_0^{(l-1)}\leftarrow\bigoplus_l D_1^{(l)}/D_1^{(l-1)}\leftarrow\ldots$$

coincides with the Koszul complex (see [10], Chapter VIII, § 4) of the space $\mathfrak{a}/\mathfrak{p}$. This means that the complex Gr $V(\mathfrak{a}/\mathfrak{p})$ and also the complex $V(\mathfrak{a}/\mathfrak{p})$ are exact. Theorem 9.1 is proved.

Proposition 9.2. Let $\mathfrak{p}$ and $\mathfrak{q}$ be subalgebras of $\mathfrak{a}$ such that $\mathfrak{a}=\mathfrak{p}\oplus\mathfrak{q}$ (as a linear space). Then $V(\mathfrak{a}, \mathfrak{p})\approx V(\mathfrak{q})$ as complexes of $U(\mathfrak{q})$-modules.

Proof. We define a mapping of complexes $\varphi: V(\mathfrak{q})\to V(\mathfrak{a}, \mathfrak{p})$ by the formula

$$\varphi_k(X\otimes X_1\wedge\ldots\wedge X_k) = X\otimes\bar{X}_1\wedge\ldots\wedge\bar{X}_k,$$

where $X\in U(\mathfrak{q})$, $X_i\in\mathfrak{q}$ and $\bar{X}_i$ is the image of X_i in $\mathfrak{a}/\mathfrak{p}$. The theorem of Poincaré—Birkhoff—Witt implies that φ is an isomorphism.

Remark 1. Proposition 9.2 owes its significance to the following fact. Assume that $\mathfrak{q}$ is a subalgebra of $\mathfrak{a}$ and that there exists in $\mathfrak{a}$ a subalgebra $\mathfrak{p}$ which is complementary to $\mathfrak{q}$. Then the action of the algebra $\mathfrak{q}$ on $V(\mathfrak{q})$ can be extended to the action of the whole algebra $\mathfrak{a}$. We remark that this extension depends essentially on the choice of $\mathfrak{p}$.

Remark 2. Let A be a complex Lie group, P a Lie subgroup of A, $\mathfrak{a}$ and $\mathfrak{p}$ the Lie algebras of A and P respectively. Let us consider the de Rham complex $\Omega=\{\Omega^k\}$ of formal analytic differential forms at the point e on the space A/P. More precisely, let $z_1, \ldots, z_n$ be a system of coordinates on the complex manifold A/P in a neighbourhood of the point e. Then Ω^k consists of the forms

$$\omega = \sum a_{i_1 i_2\ldots i_k}(z)\,dz_{i_1}\wedge\ldots\wedge dz_{i_k},$$

where $a_{i_1 i_2\ldots i_k}(z)$ are formal power series of the variables $z_1, \ldots, z_n$.

The group A acts on the space A/P. Of course, it cannot act on the complex Ω. However, the Lie algebra $\mathfrak{a}$ acts on Ω. It is easy to verify that the complex Ω is dual to the complex $V(\mathfrak{a}, \mathfrak{p})$ constructed above. Therefore the exactness of $V(\mathfrak{a}, \mathfrak{p})$ is also a consequence of the exactness of de Rham's complex Ω.

In what follows we shall be interested in the case when $\mathfrak{a}=\mathfrak{g}$ is a complex semisimple Lie algebra, $\mathfrak{p}=\mathfrak{b}=\mathfrak{h}\oplus\mathfrak{n}_+$ is a Borel subalgebra of $\mathfrak{g}$. We shall study the structure of the members D_k of the complex $V(\mathfrak{g}, \mathfrak{b})$.

Lemma 9.3. *Let V be a $\mathfrak{b}$-module and let $V^{\mathfrak{g}} = U(\mathfrak{g}) \underset{U(\mathfrak{b})}{\otimes} V$. The mapping $V \to V^{\mathfrak{g}}$ generates an exact functor from the category of $\mathfrak{b}$-modules to the category of $\mathfrak{g}$-modules. If in addition V is a one-dimensional module, $Hv = \chi(H)v$, $E_\gamma v = 0$ for $H \in \mathfrak{h}$, $\gamma \in \Delta_+$, $v \in V$ then $V^{\mathfrak{g}} = M_{\chi+\varrho}$.*

The proof follows easily from the fact that $U(\mathfrak{g})$ is a free $U(\mathfrak{b})$-module. The second statement follows from the definition of $M_{\chi+\varrho}$.

Lemma 9.3 enables us to present an easy description of the modules D_k occuring in the complex $V(\mathfrak{g}, \mathfrak{b})$.

Definition 9.4. Let Ψ be a finite collection of weights (we allow that some of them coincide). We shall say that the module M is of *type* Ψ if there exists a filtration $0 = M^{(0)} \subset M^{(1)} \subset \ldots \subset M^{(l)} = M$ such that $M^{(i)}/M^{(i-1)} = M_{\psi_i}$ and the collection of weights $\{\psi_i\}$ coincides with Ψ.

Lemma 9.5. *Let N be a finite-dimensional $\mathfrak{h}$-diagonalizable $\mathfrak{b}$-module, $\Psi(N) = \{\varphi + \varrho\}$, where φ runs through all weights of N (with multiplicities). Then the module $U(\mathfrak{g}) \underset{U(\mathfrak{b})}{\otimes} N$ is of type $\Psi(N)$.*

The proof is an immediate consequence of Lemma 9.3.

Corollary. The module D_k in the complex $V(\mathfrak{g}, \mathfrak{b})$ is of type $\Psi(\Lambda^k(\mathfrak{g}/\mathfrak{b}))$.

Since the Lie algebra $\mathfrak{g}$ acts on the complex $V(\mathfrak{g}, \mathfrak{b})$, we can distinguish in it a subcomplex corresponding to the "zero" eigenvalues of the elements of $Z(\mathfrak{g})$. More precisely, let $\vartheta = \vartheta_\varrho \in \Theta$. We consider the subcomplex $V_\vartheta(\mathfrak{g}, \mathfrak{b})$ of $V(\mathfrak{g}, \mathfrak{b})$ consisting of the submodules $(D_k)_\vartheta \subset D_k$ and the module $\mathbf{C}_\vartheta = \mathbf{C}$. It follows from Proposition 8.6, §5 that the complex $V_\vartheta(\mathfrak{g}, \mathfrak{b})$ is exact.

Proposition 9.6. Let $\Psi_k = \{w\varrho \mid w \in W^{(k)}\}$. Then $(D_k)_\vartheta$ is of type Ψ_k.

First we prove the following lemma.

Lemma 9.7. *Let M be a module of type Ψ and $\vartheta \in \Theta$. Then the module M_ϑ is of type Ψ_ϑ, where Ψ_ϑ is the collection of all weights $\psi \in \Psi$ such that $\vartheta_\psi = \vartheta$.*

Proof. Let $0 = M^{(0)} \subset M^{(1)} \subset \ldots \subset M^{(k)} = M$ be a filtration of M for which $M^{(i)}/M^{(i-1)} = M_{\psi_i}$, $\psi_i \in \Psi$. From the exactness of the functor $M \to M_\vartheta$ it follows that the modules $M_\vartheta^{(i)}$ form a filtration of M_ϑ and $M_\vartheta^{(i)}/M_\vartheta^{(i-1)} = (M_{\psi_i})_\vartheta$. It follows from Proposition 8.5 that $(M_{\psi_i})_\vartheta = M_{\psi_i}$ if $\vartheta_{\psi_i} = \vartheta$ and $(M_{\psi_i})_\vartheta = 0$ if $\vartheta \neq \vartheta_{\psi_i}$. This implies Lemma 9.7.

It follows from Lemma 9.7 that $(D_k)_\vartheta$ is of type $[\Psi(\Lambda^k(\mathfrak{g}/\mathfrak{b}))]_\vartheta$. Now we shall study this set.

Let Φ be a subset of Δ. Put $|\Phi| = \sum_{\gamma \in \Phi} \gamma$. Since the set of weights of $\mathfrak{g}/\mathfrak{b}$ coincides with Δ_-, the collection of weights of $\Lambda^k(\mathfrak{g}, \mathfrak{b})$ (with multiplicities) coincides with the collection of weights of the form $-|\Phi|$ for all $\Phi \subset \Delta_+$ such that card $\Phi = k$. Therefore,

$$[\Psi(\Lambda^k(\mathfrak{g}/\mathfrak{b}))]_\vartheta = \{\varrho - |\Phi| \,|\, \Phi \subset \Delta_+, \ \text{card}\, \Phi = k, \ (\varrho - |\Phi|) \sim \varrho\}.$$

For any element $w \in W$ we put $\Phi_w = \{\gamma \in \Delta_+ | w^{-1}\gamma \subset \Delta_-\}$. Then card $\Phi_w = k$, for $w \in W^{(k)}$ (Lemma 11.1). Thus Proposition 9.6 is a consequence of the following lemma.

Lemma 9.8. *Let* $w \in W$, $\Phi \subset \Delta_+$. *Then* $\varrho - w\varrho = |\Phi|$ *if and only if* $\Phi = \Phi_w$.
A proof of this lemma is presented in § 11, see also [6].
Now we can formulate the main theorem of this section.

Theorem 9.9. *Let* V *be an irreducible finite-dimensional* $\mathfrak{g}$*-module with highest weight* λ. *Then there exists an exact sequence of* $U(\mathfrak{g})$*-modules*

$$0 \leftarrow V \leftarrow B_0^V \leftarrow B_1^V \leftarrow \ldots \leftarrow B_s^V \leftarrow 0$$

where $s = \dim \mathfrak{n}_-$ *and* B_k *is a module of type* $\Psi_k(\lambda) = \{w(\lambda+\varrho) | w \in W^{(k)}\}$.

Proof. In Proposition 9.6 the required exact sequence of the $\mathfrak{g}$-modules $B_k^{\mathbf{C}}$ was constructed for the case $V = \mathbf{C}$ (i.e. $\lambda = 0$). In the general case we consider the exact sequence $B_*^{\mathbf{C}} \otimes_{\mathbf{C}} V$ and put

$$B_k^V = (B_k^{\mathbf{C}} \otimes V)_{\vartheta_{\lambda+\varrho}}.$$

Now we prove that the sequence

$$0 \leftarrow V \leftarrow B_0^V \leftarrow B_1^V \leftarrow \ldots$$

satisfies the conditions of the theorem. Its exactness follows from the fact that $M \to (M \otimes V)_\vartheta$ is an exact functor. Now we show that B_k^V is of type $\Psi_k(\lambda)$.

Lemma 9.10. *Let* $\chi \in \mathfrak{h}^*$, V *be a finite-dimensional* $\mathfrak{g}$*-module. Let us denote by* Ψ *the set* $\{\lambda + \chi\}$ *where* λ *runs through all weights of* V *with the corresponding multiplicities. Then* $M_\chi \otimes V$ *is of type* Ψ.

Proof. Let $e_1, \ldots, e_l$ be a basis in V consisting of weight vectors and $\lambda_1, \lambda_2, \ldots, \lambda_l$ be the corresponding weights. We choose an enumeration of the vectors e_i such that $\lambda_i < \lambda_j$ implies $i > j$. Let $a_i = f_\chi \otimes e_i \in M_\chi \otimes V$ and $M^{(k)} = U(\mathfrak{g})(a_1, \ldots, a_k)$. Then $0 = M^{(0)} \subset M^{(1)} \subset \ldots \subset M^{(l)}$.
To prove Lemma 9.10 it will suffice to show that

$$M^{(i)}/M^{(i-1)} = M_{\lambda_i+\chi} \quad \text{and} \quad M^{(l)} = M_\chi \otimes V.$$

Let $\bar{a}_k$ denote the image of a_k in $M^{(k)}/M^{(k-1)}$. It is obvious that $\bar{a}_k$ is a generator of $M^{(k)}/M^{(k-1)}$ of weight $\chi + \lambda_k - \varrho$ and $E_\gamma \bar{a}_k = 0$ for $\gamma \in \Delta_+$. Therefore $M^{(k)} = U(\mathfrak{n}_-)(a_1, \ldots, a_k)$. We shall show that $M^{(k)}$ is a free $U(\mathfrak{n}_-)$-module with generators $a_1, \ldots, a_k$. Let $X_i \in U(\mathfrak{n}_-)$, $1 \leqq i \leqq k$ and let p be the largest among the degrees of X_i (with respect to the natural filtration in $U(\mathfrak{n}_-)$). Then

$$\sum_{i=1}^{k} X_i a_i = \sum_{i=1}^{k} X_i f_\chi \otimes e_i + \sum_{j=1}^{l} Y_j f_\chi \otimes e_j \neq 0,$$

since the degrees of the elements Y_j are less than p and M_χ is a free $U(\mathfrak{n}_-)$-module.

This means that $M^{(k)}/M^{(k-1)}$ is a free $U(\mathfrak{n}_-)$-module with the single generator $\bar{a}_k$, i.e. $M^{(k)}/M^{(k-1)} \approx M_{\chi+\lambda_k}$.

Similar considerations will show that $M^{(l)} = M_\chi \otimes V$. This completes the proof of Lemma 9.10.

It follows from this lemma that $B_k^{\mathbf{C}} \otimes V$ is of type Ψ, where $\Psi = \{\lambda_i + w\varrho \mid \lambda_i$ are the weights of V (with multiplicities), $w \in W^{(k)}\}$. Therefore, Theorem 9.9 follows from Lemma 9.7 and the following lemma.

Lemma 9.11. *Let V be a finite-dimensional irreducible* $\mathfrak{g}$*-module with highest weight λ. Then for every $w \in W$ there exists exactly one weight $\mu \in P(V)$ such that $\mu + w\varrho \sim \lambda + \varrho$. Moreover the weight μ has multiplicity one in V.*

Proof. Let $\mu \in P(V)$, $w_1, w_2 \in W$ be such that $w_1(\mu + w\varrho) = \lambda + \varrho$; since $w_1\mu$ is a weight of V, we have $w_1\mu \leqq \lambda$.

In addition, $w_1 w\varrho \leqq \varrho$. Therefore $w_1\mu = \lambda$ and $w_1 w\varrho = \varrho$. This implies $w_1 = w^{-1}$ and thus $\mu = w\lambda$. The last statement of the lemma follows from the fact that the multiplicity of the weight μ in V is equal to the multiplicity of the highest weight λ.

Corollary (Bott's theorem [5], [6]). *Let V be a finite-dimensional irreducible* $\mathfrak{g}$*-module. Then*

$$\dim H^i(\mathfrak{n}_-, V) = \operatorname{card} W^{(i)}.$$

Proof. We know that

$$H^i(\mathfrak{n}_-, V) = \operatorname{Ext}^i_{\mathfrak{n}_-}(\mathbf{C}, V) = \operatorname{Tor}_i^{\mathfrak{n}_-}(V^*, \mathbf{C})^* = \operatorname{Tor}_i^{\mathfrak{n}_-}(\mathbf{C}, (V^\tau)^*)^*.$$

Let us construct the resolution $\{B_i^{V_1}\}$ for the module $V_1 = (V^\tau)^*$. Then $\operatorname{Tor}_i^{\mathfrak{n}_-}(\mathbf{C}, V_1)$ will be the homologies of the complex

$$0 \longleftarrow \bar{B}_0^{V_1} \xleftarrow{\bar{d}_1} \cdots \longleftarrow \bar{B}_s^{V_1} \longleftarrow 0.$$

The algebra $\mathfrak{h}$ acts on this complex in a natural way. In view of Theorem 9.9 we have here that $\bar{B}_i^{V_1} = B_i^{V_1}/\mathfrak{n}_- B_i^{V_1}$ is a finite-dimensional space whose weights with respect to $\mathfrak{h}$ are equal to $w(\lambda+\varrho)$, $w \in W^{(i)}$, and each of these weights occurs with multiplicity one. Therefore $\dim \bar{B}_i^{V_1} = \operatorname{card} W^{(i)}$, and every $\bar{d}_i$ is the null-mapping. The corollary is proved.

§ 10. Construction of the resolution of a finite-dimensional $\mathfrak{g}$-module

The present section is devoted to the proof of Theorem 10.1, which yields a sharpening of Theorem 9.9.

Theorem 10.1. *Let V be an irreducible finite-dimensional* $\mathfrak{g}$*-module with highest weight λ. Then there exists an exact sequence of* $\mathfrak{g}$*-modules*

$$0 \longleftarrow V \xleftarrow{\varepsilon} C_0^V \xleftarrow{d_1} C_1 \longleftarrow \cdots \longleftarrow C_s^V \longleftarrow 0$$

where

$$s = \dim \mathfrak{n}_-, \quad C_k = \bigoplus_{w \in W^{(k)}} M_{w(\mu+\varrho)}.$$

First we shall present an explicit construction of the mappings d_i and ε. Let us put $\chi=\lambda+\varrho$. Then $\chi\in D$ and by Theorem 8.8′ every submodule of the module $M_{w\chi}$, $w\in W$ can be considered as a submodule of M_χ. By Theorem 8.7, any mapping $M_{w_1\chi}\to M_{w_2\chi}$ is a multiple of the canonical imbedding for $w_1<w_2$, hence it can be determined by a complex number $c_{w_1 w_2}$. Therefore, any mapping $C_i^V\to C_{i-1}^V$ can be represented by a complex matrix $(c_{w_1 w_2})$, $w_1\in W^{(i)}$ and $w_2\in W^{(i-1)}$. Thus in order to construct the mappings d_i it will suffice to define the corresponding matrices $(d^{(i)}_{w_1 w_2})$.

Definition 10.2. Let us call a quadruple (w_1, w_2, w_3, w_4) of elements of W a *square* if

$$w_1 \to w_2 \to w_4 \quad \text{and} \quad w_1 \to w_3 \to w_4$$

(see Definition 8.9).

It will be convenient to consider the finite directed graph corresponding to W, whose vertices are the members of W and in which an arc leads from w_1 to w_2 if $w_1\to w_2$.

Lemma 10.3.* *Let $w_1, w_2\in W$ and $l(w_1)-2=l(w_2)$. Then the number of elements $w'\in W$ such that $w_1\to w'\to w_2$ is equal to either zero or two.*

Lemma 10.4. *To each arrow $w_1\to w_2$ we can assign a number $s(w_1, w_2)=\pm 1$ in such a way that for every square (w_1, w_2, w_3, w_4) the product of the numbers assigned to the four arrows occurring in it is equal to -1.*

The proofs of Lemmas 10.3 and 10.4 will be given in § 11.

Now we can improve on Theorem 10.1 in the following way.

Theorem 10.1′. *With the notation of Theorem* 10.1. *we define the mapping $d_i: C_i\to C_{i-1}$ by means of the matrix $(d^{(i)}_{w_1 w_2})$, $w_1\in W^{(i)}$, $w_2\in W^{(i-1)}$, where $d^{(i)}_{w_1 w_2}=$ $=s(w_1, w_2)$ if $w_1\to w_2$ and $d^{(i)}_{w_1 w_2}=0$ otherwise. Let us denote by $\varepsilon: C_0\to V$ the natural surjection. Then the sequence*

$$0 \longleftarrow V \xleftarrow{\varepsilon} C_0^V \xleftarrow{d_1} C_1^V \longleftarrow \ldots \xleftarrow{d_s} C_s^V \longleftarrow 0 \tag{10.1}$$

is exact.

Proof. It follows immediately from Lemmas 10.3 and 10.4 that $d_i\circ d_{i+1}=0$ for $i=1, \ldots, s-1$.

We remark that $W^{(1)}=\{\sigma_\alpha, \alpha\in\Sigma\}$. Therefore, Harish-Chandra's theorem on ideals [4] implies the exactness of the sequence (10.1) at its members V and C_0.

Assume now that we have already shown the exactness of the sequence at the members $C_0, \ldots, C_{i-1}$. We shall prove that it is also exact at the member C_i, i.e. that $\operatorname{Ker} d_i=\operatorname{Im} d_{i+1}$. Let us put $K=\operatorname{Ker} d_i$. The desired equality $d_{i+1}(C_{i+1})=K$ is obviously a consequence of the following three lemmas.

* Lemma 10.3 follows easily from certain unpublished results of D.-N. Verma concerning the Möbius function on the Weyl group.

(We recall that $C_{i+1} = \bigoplus_{w \in W^{(i+1)}} M_{w\chi}$ is a free $U(\mathfrak{n}_-)$-module.)

Lemma 10.5. *Let C be a free $U(\mathfrak{n}_-)$-module with generators $f_1, \dots, f_n$ and $\zeta: C \to K$ be a homomorphism of $U(\mathfrak{n}_-)$-modules such that $\zeta(f_i)$ is a weight vector in K (with respect to $\mathfrak{h}$). Then ζ is a surjection if and only if the induced mapping $\bar{\zeta}: C/\mathfrak{n}_- C \to K/\mathfrak{n}_- K$ is surjective.*

Lemma 10.6. *The mapping*

$$\bar{d}_{i+1}: C_{i+1}/\mathfrak{n}_- C_{i+1} \to K/\mathfrak{n}_- K$$

is an injection.

Lemma 10.7.

$$\dim_{\mathbf{C}} C_{i+1}/\mathfrak{n}_- C_{i+1} = \dim_{\mathbf{C}} K/\mathfrak{n}_- K < \infty.$$

Proof of Lemma 10.5. Clearly if ζ is surjective then so is $\bar{\zeta}$. Conversely, assume that $\bar{\zeta}$ is surjective but ζ is not. Consider the weight vector f in K of weight ψ having the following properties:

a) $f \notin \operatorname{Im} \zeta$,

b) any vector f' of weight $\psi' > \psi$ belongs to $\operatorname{Im} \zeta$.

There always exists such a vector f since $K \in O$, and therefore for any weight $\psi \in \mathfrak{h}^*$ there are finitely many weights $\psi' > \psi$ such that $K^{(\psi')} \neq \{0\}$.

Let $\bar{f}$ be the image of f in $K/\mathfrak{n}_- K$. Then $\bar{f} = \sum c_i \bar{\zeta}(f_i)$. Since $\mathfrak{n}_- K$ is invariant under $\mathfrak{h}$, $\mathfrak{h}$ acts on $K/\mathfrak{n}_- K$. Therefore we can assume that $c_i \neq 0$ only for those indices i for which the weight of $\zeta(f_i)$ is equal to ψ. Moreover, $g = f - \sum c_i \zeta(f_i)$ is a weight vector lying in $\mathfrak{n}_- K$, and thus $g = \sum_{\gamma \in \Delta_+} E_{-\gamma} g_\gamma$, where the weight of g_γ is $\psi + \gamma > \psi$. According to the construction of f, $g_\gamma \in \operatorname{Im} \zeta$, and therefore $f \in \operatorname{Im} \zeta$. Lemma 10.5 is proved.

Proof of Lemma 10.6. The quotient space $C_{i+1}/\mathfrak{n}_- C_{i+1}$ is a linear space over $\mathbf{C}$ for which $\{\bar{f}_{w\chi} | w \in W^{(i+1)}\}$ forms a basis.* Since the homomorphism $\bar{d}_{i+1}$ commutes with the action of $\mathfrak{h}$ and all of the vectors $\bar{f}_{w\chi}$ have different weights it will suffice to prove that $\bar{d}_{i+1}(\bar{f}_{w\chi}) \neq 0$ for any $w \in W^{(i+1)}$.

The proof of this proposition is divided in a natural way into two steps.

Lemma 10.6a. *The irreducible modules occurring in the Jordan—Hölder decomposition of the module K are of the form $L_{w\chi}$, $l(w) > i$.*

Proof. For the proof of Lemma 10.6a we shall make use of the exact sequence

$$0 \leftarrow V \leftarrow B_0^V \leftarrow B_1^V \leftarrow \dots \leftarrow B_s^V \leftarrow 0 \tag{10.2}$$

constructed in § 9. It is clear that for all j

$$JH(B_j^V) = \bigcup_{w \in W^{(j)}} JH(M_{w\chi}) = JH(C_j^V).$$

* Let us recall, that $f_{w\chi}$ is the generator of $M_{w\chi}$ (see Def. 8.2).

Since both of the sequences (10.1) and (10.2) are exact at their members with numbers less than i, we have $JH(K)=JH(K_B)$, where K_B denotes the kernel of the mapping $B_i^V \to B_{i-1}^V$. As the sequence (10.2) is exact at its member B_i^V, K_B is equal to the image of B_{i+1}^V. Therefore

$$JH(K_B) \subset JH(B_{i+1}^V) = \bigcup_{w \in W^{(i+1)}} JH(M_{w\chi}),$$

and thus Lemma 10.6a is implied by the Corollary of Theorem 8.12.

Lemma 10.6b. *Let* $w_0 \in W$ *and the module* $M \in O$ *be given. We assume that* $l(w) \geqq l(w_0)$ *for any* $L_{w\chi}$ *occurring in* $JH(M)$. *Let* $\tau: M_{w_0\chi} \to M$ *be a homomorphism such that* $\tau(f_{w_0\chi}) \neq 0$. *Then the image of* $\tau(f_{w_0\chi})$ *in* $M/\mathfrak{n}_- M$ *is not* 0.

Proof. We shall use induction on the number of elements in $JH(M)$. Let $f \in M$ be an element of maximal weight ψ and $N \subset M$ be the submodule generated by f. Then the module N is isomorphic to a factor module of the module M_ψ. We shall distinguish two cases.

Case 1. $\tau(f_{w_0\chi}) \in N$. In this case

$$L_{w_0\chi} \subset JH(N) \subset JH(M_\psi).$$

Therefore (as it follows from Theorem 8.12) $\psi = w_1\chi$, where $w_1 \geqq w_0$. On the other hand

$$L_\psi \subset JH(N) \subset JH(M).$$

Thus, according to the condition of the lemma $w_1 = w_0$, i.e. $\psi = w_0\chi$. Since ψ is the maximal weight of M,

$$\tau(f_{w_0\chi}) \notin \mathfrak{n}_- M.$$

Case 2. $\tau(f_{w_0\chi}) \notin N$. In this case the statement of the lemma for M can be reduced to the similar statement for the module M/N. Since $JH(M/N) \subsetneqq JH(M)$, we can apply the induction hypothesis. Lemma 10.6b is proved.

To complete the proof of Lemma 10.6 it suffices to apply Lemma 10.6b to the module $M=K$.

Proof of Lemma 10.7. The module K has only a finite number of generators (as a $U(\mathfrak{n}_-)$-module). Therefore $\bar{K} = K/\mathfrak{n}_- K$ is a finite-dimensional space over **C**. Let us choose weight vectors $f_1, \ldots, f_n \in K$ whose images in $\bar{K}$ form a basis in $\bar{K}$. Let us consider the free $U(\mathfrak{n}_-)$-module C with n generators $g_1, \ldots, g_n$ and define a homomorphism of $U(\mathfrak{n}_-)$-modules $\vartheta: C \to K$ by the formula $\vartheta(g_i) = f_i$. By virtue of Lemma 10.5 ϑ is surjective.

Let us consider the exact sequence

$$0 \leftarrow V \leftarrow C_0^V \xleftarrow{d_1} \ldots \leftarrow C_i^V \xleftarrow{\vartheta} C$$

of $U(\mathfrak{n}_-)$-modules.

Since C_j^V and C are free $U(\mathfrak{n}_-)$-modules, this sequence can be augmented to a free resolution

$$0 \leftarrow V \leftarrow C_0^V \leftarrow \ldots \leftarrow C_i^V \xleftarrow{\vartheta} C \xleftarrow{\eta} D \leftarrow \ldots$$

of the $U(\mathfrak{n}_-)$-module V.

Now, consider the sequence

$$\bar{D} \xrightarrow{\bar{\eta}} \bar{C} \xrightarrow{\bar{\vartheta}} \bar{C}_i^V \xrightarrow{\bar{d}_i} \bar{C}_{i-1}^V$$

where, for any $U(\mathfrak{n}_-)$-module M, $\bar{M}$ denotes $1 \underset{U(\mathfrak{n}_-)}{\otimes} M = M/\mathfrak{n}_- M$. By definition,

$$\mathrm{Tor}_i^{\mathfrak{n}_-}(\mathbf{C}, V) = \mathrm{Ker}\,\bar{\vartheta}/\mathrm{Im}\,\bar{\eta}.$$

We shall show that $\bar{\vartheta}$ and $\bar{\eta}$ are equal to 0.

From the exact sequence

$$D \xrightarrow{\eta} C \xrightarrow{\tau} K \longrightarrow 0$$

we obtain the exact sequence

$$\bar{D} \xrightarrow{\bar{\eta}} \bar{C} \xrightarrow{\bar{\tau}} \bar{K} \longrightarrow 0.$$

But $\bar{\tau}$ is an isomorphism, hence $\bar{\eta}=0$.

Furthermore, we have the exact sequence

$$C \xrightarrow{\vartheta} C_i^V \xrightarrow{d_i} K_{i-1} \longrightarrow 0$$

and thus the sequence

$$\bar{C} \xrightarrow{\bar{\vartheta}} \bar{C}_i^V \xrightarrow{\bar{d}_i} \bar{K}_{i-1} \longrightarrow 0,$$

where $K_{i-1} = \mathrm{Ker}\, d_{i-1} = \mathrm{Im}\, d_i$. Applying Lemma 10.6 to the mapping $d_i : C_i \to K_{i-1}$ we see that $\bar{d}_i$ is an isomorphism, hence $\bar{\vartheta}=0$.

Thus we have

$$\dim \mathrm{Tor}_i^{\mathfrak{n}_-}(\mathbf{C}, V) = \dim \bar{C} = \dim (K/\mathfrak{n}_- K).$$

On the other hand, the Corollary to Theorem 9.9 shows that

$$\dim \mathrm{Tor}_i^{\mathfrak{n}_-}(\mathbf{C}, V) = \mathrm{card}\, W^{(i)} = \dim (C_i/\mathfrak{n}_- C_i).$$

Lemma 10.7 is proved.

This also completes the proof of Theorem 10.1′ and Theorem 10.1.

§ 11. Proof of the lemmas

First we shall present several results clarifying the properties of the function $l(w)$ in more detail (see [11]).

Let us put $\Phi_w = \Delta_+ \cap w(\Delta_-)$.

Lemma 11.1.

1) $l(w) = \operatorname{card} \Phi_w$.

2) If $\gamma_1, \gamma_2 \in \Phi_w$ *and* $\gamma_1 + \gamma_2 = n\gamma$ *for* $n \in \mathbf{Z}_+$, *then* $\gamma \in \Phi_w$.

Proof. Statement *1)* is proved in [11]. To verify statement *2)* it is enough to prove it separately for Δ_+ and $w\Delta_-$, and it is obvious in both cases.

Proof of Lemma 8.10. Plainly,

$$w\chi - \sigma_\gamma w\chi = \frac{2\langle w\chi, \gamma\rangle}{\langle \gamma, \gamma\rangle}\gamma.$$

Since $\chi \in \mathfrak{h}_{\mathbf{Z}}^*$ and the element χ is regular,

$$n = \frac{2\langle w\chi, \gamma\rangle}{\langle \gamma, \gamma\rangle} \in \mathbf{Z}\setminus 0.$$

We assume that $n > 0$ and prove then that $l(\sigma_\gamma w) > l(w)$. By *1)* of Lemma 11.1, it suffices to prove that $\operatorname{card} \Phi_w < \operatorname{card} \Phi_{\sigma_\gamma w}$.

a) $\gamma \notin \Phi_w$. Indeed, $\langle \chi, w^{-1}\gamma\rangle = \langle w\chi, \gamma\rangle > 0$, and therefore $w^{-1}\gamma \notin \Delta_-$.

b) $\gamma \in \Phi_{\sigma_\gamma w}$. Indeed,

$$(\sigma_\gamma w)^{-1}\gamma = w^{-1}(-\gamma) = -w^{-1}\gamma \in \Delta_-.$$

c) Let $\delta \in \Phi_w$ and $\sigma_\gamma \delta \in \Delta_+$. Then $\sigma_\gamma \delta \in \Phi_{\sigma_\gamma w}$. In fact,

$$(\sigma_\gamma w)^{-1}(\sigma_\gamma \delta) = w^{-1}\sigma_\gamma^{-1}\sigma_\gamma \delta = w^{-1}\delta \in \Delta_-.$$

d) Let $\delta \in \Phi_w$, $\sigma_\gamma \delta \in \Delta_-$. We shall show that $\delta \in \Phi_{\sigma_\gamma w}$. Assume, on the contrary that $w^{-1}\sigma_\gamma \delta \in \Delta_+$. Then $w^{-1}(-\sigma_\gamma \delta) \in \Delta_-$, and thus $-\sigma_\gamma \delta \in \Phi_w$. But $\delta + (-\sigma_\gamma \delta) = \frac{2\langle \gamma, \delta\rangle}{\langle \delta, \delta\rangle}\gamma$, while $\frac{2\langle \gamma, \delta\rangle}{\langle \delta, \delta\rangle} > 0$. By *2)* of Lemma 11.1 this implies $\gamma \in \Phi_w$, and that contradicts a).

Propositions a)—d) imply that if $n > 0$ then $l(w) < l(\sigma_\gamma w)$.

By interchanging w and $\sigma_\gamma w$ we obtain that if $n < 0$ then $l(w) > l(\sigma_\gamma w)$. Lemma 8.10 is proved.

Lemma 11.2.

1) Let $w \in W$, $\alpha \in \Sigma$. *Then* $\alpha \in \Phi_w$ *implies* $w \xrightarrow{\alpha} \sigma_\alpha w$, *and* $\alpha \notin \Phi_w$ *implies* $\sigma_\alpha w \xrightarrow{\alpha} w$.

2) If $\sigma_\alpha w \xrightarrow{\alpha} w$ *for all* $\alpha \in \Sigma$, *then* $w = 1$.

3) There exists a unique element $s \in W$ *such that* $s \to \sigma_\alpha s$ *for all* $\alpha \in \Sigma$.

This lemma follows immediately from the theorems proved in [11].

Lemma 11.3. *Let* $w_1, w_2 \in W$, $\gamma \in \Delta_+$ *and* $\alpha \in \Sigma$, $\alpha \neq \gamma$.

We put

$$\sigma_\alpha w_1 \xrightarrow{\alpha} w_1 \quad \textit{and} \quad \sigma_\alpha w_1 \xrightarrow{\gamma} w_2. \tag{11.1}$$

Then

$$w_2 \xrightarrow{\alpha} \sigma_\alpha w_2 \quad \textit{and} \quad w_1 \xrightarrow{\sigma_\alpha \gamma} \sigma_\alpha w_2. \tag{11.2}$$

Conversely, (11.2) *implies* (11.1)

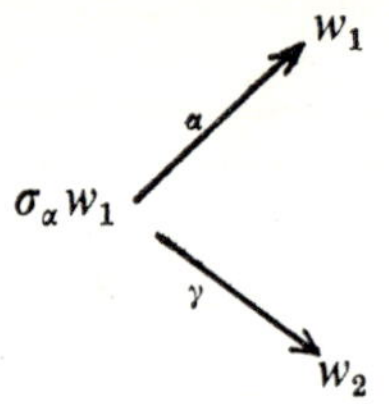

Diagram 1

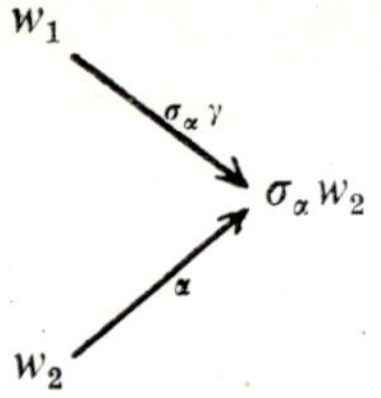

Diagram 2

Proof. We have to prove that Diagram 1 is equivalent to Diagram 2. Let $\gamma'=\sigma_\alpha\gamma$. Then $\gamma'\in\Delta_+$ and $\sigma_\alpha w_2=\sigma_\gamma w_1$. Since $l(w_1)=l(w_2)$, formula (11.2) is equivalent to $l(\sigma_\alpha w_2)=l(w_2)-1$. Let $\chi\in D$. Then, by Lemma 8.10,

$$w_2\chi-\sigma_\alpha w_1\chi = n\gamma, \quad n>0.$$

Applying σ_α to this equality we obtain

$$\sigma_\alpha w_2\chi - w_1\chi = n\gamma'.$$

Using Lemma 8.10 again we obtain that $l(\sigma_\alpha w_2)<l(w_1)$. Therefore, (11.1) implies (11.2).

We can prove similarly that (11.2) implies (11.1).

Proof of Lemma 8.11. The proof will be performed by induction on $l(w)$. If $l(\sigma_\gamma w)=l(w)-1$, then by definition $\sigma_\gamma w>w$. Since $l(w)$ and $l(\sigma_\gamma w)$ are of different parities, it remains to consider the case in which $l(\sigma_\gamma w)\leqq l(w)-3$.

Let $\alpha\in\Sigma$ be a root such that $w\to\sigma_\alpha w$. Obviously, $\alpha\neq\gamma$ and $\sigma_\alpha\sigma_\gamma w=\sigma_{\gamma'}\sigma_\alpha w$, where $\gamma'=\sigma_\alpha\gamma\in\Delta_+$.

Here we have

$$l(\sigma_{\gamma'}\sigma_\alpha w) = l(\sigma_\alpha\sigma_\gamma w) \leqq l(\sigma_\gamma w)+1 \leqq l(w)-2 < l(\sigma_\alpha w).$$

By the induction hypothesis there exists a chain

$$\sigma_\alpha w\to w_1\to w_2\to \dots \to w_2\to\sigma_\alpha\sigma_\gamma w$$

and thus also a chain

$$w_{-1}\xrightarrow{\gamma_0} w_0\xrightarrow{\gamma_1} w_1\longrightarrow\dots\longrightarrow w_k\xrightarrow{\gamma_k+1}\sigma_\alpha\sigma_\gamma w \qquad (11\ 3)$$

(where $w_{-1}=w$, $w_0=\sigma_\alpha w$ and $\gamma_0=\alpha$). There are two possibilities.

Case 1. $\sigma_\alpha\sigma_\gamma w\to\sigma_\gamma w$. Then there exists a chain

$$w=w_{-1}\to w_0\to w_1\to\dots\to w_k\to\sigma_\alpha\sigma_\gamma w\to\sigma_\gamma w,$$

hence $w<\sigma_\gamma w$.

Case 2. $\sigma_\gamma w \to \sigma_\alpha \sigma_\gamma w$. Let i be the largest number such that in the chain (11.3) $\gamma_i = \alpha$. Applying Lemma 11.3 several times we obtain that

$$\sigma_\alpha w_k \xrightarrow{\sigma_\alpha \gamma_{k+1}} \sigma_\gamma w,$$

$$\sigma_\alpha w_{k-1} \xrightarrow{\sigma_\alpha \gamma_k} \sigma_\alpha w_k, \ \dots, \ \sigma_\alpha w_i \xrightarrow{\sigma_\alpha \gamma_{i+1}} \sigma_\alpha w_{i+1}.$$

But $\sigma_\alpha w_i = w_{i-1}$. Hence we obtain a chain

$$w = w_1 \xrightarrow{\gamma_0} w_0 \xrightarrow{\gamma_1} \dots \longrightarrow w_{i-1} \xrightarrow{\sigma_\alpha \gamma_{i+1}} \sigma_\alpha w_{i+1} \longrightarrow \dots \xrightarrow{\sigma_\alpha \gamma_{k+1}} \sigma_\gamma w.$$

Therefore, $w < \sigma_\gamma w$, and Lemma 8.11 is proved.

Proof of Lemma 9.8. We want to prove that if $w \in W$, $\Phi \subset \Delta_+$ and $\varrho - w\varrho = |\Phi|$, then $\Phi = \Phi_w = \Delta_+ \cap w\Delta_-$. If $w = e$, this is obvious. We shall perform the proof by induction on $l(w)$. Let $l(w) = i > 0$. We choose an $\alpha \in \Sigma$ such that $l(\sigma_\alpha w) = i - 1$, i.e. $\alpha \in \Phi_w$. Then

$$|\sigma_\alpha \Phi| = \sigma_\alpha \varrho - \sigma_\alpha w \varrho = \varrho - \sigma_\alpha w \varrho - \alpha.$$

$\Phi \subset \Delta_+$ implies $\alpha \notin \sigma_\alpha \Phi$. Therefore, putting $\sigma_\alpha w = w'$, we have

$$\varrho - w'\varrho = \varrho - \sigma_\alpha w \varrho = |\sigma_\alpha \Phi \cup \{\alpha\}|.$$

We shall show that $\alpha \in \Phi$. Indeed, assume that this is not true. Then $\sigma_\alpha \Phi \cup \{\alpha\} \subset \Delta_+$ and by the induction hypothesis $\Phi_{w'} = \sigma_\alpha \Phi \cup \{\alpha\}$, i.e. $\alpha \in \Phi_{w'}$. According to Lemma 11.2. 1 $\sigma_\alpha w \to w$, which contradicts the choice of α.

Thus $\alpha \in \Phi$. Let us put $\Phi = \Phi - \{\alpha\}$. Then $\varrho - \sigma_\alpha w \varrho = |\sigma_\alpha \Phi'|$ and $\sigma_\alpha \Phi' \subset \Delta_+$. By the induction hypothesis $\Phi_{w'} = \sigma_\alpha \Phi'$, i.e. $\Phi = \sigma_\alpha \Phi_{w'} \cup \{\alpha\}$. It is easy to verify that $\sigma_\alpha \Phi_{w'} \cup \{\alpha\} = \Phi_w$. Lemma 9.8 is proved.

Proof of Lemma 10.3. We shall prove this by induction on $l(w_1)$. Since $l(w_1) \geqq 2$, we can choose an $\alpha \in \Sigma$ such that $w_1 \to \sigma_\alpha w_1$. There are two possibilities.

Case 1. $w_2 \to \sigma_\alpha w_2$. Let us assign to every chain

$$w_1 \xrightarrow{\gamma_1} w \xrightarrow{\gamma_2} w_2 \tag{11.4}$$

a chain

$$\sigma_\alpha w_1 \xrightarrow{\delta_1} w' \xrightarrow{\delta_2} \sigma_\alpha w_2 \tag{11.5}$$

in the following way.

a) If $\gamma_1 \neq \alpha$, then put $\delta_1 = \sigma_\alpha \gamma_1$, $\delta_2 = \sigma_\alpha \gamma_2$, $w' = \sigma_\alpha w$. It follows from Lemma 11.3 that

$$\sigma_\alpha w_1 \xrightarrow{\delta_1} w' \xrightarrow{\delta_2} \sigma_\alpha w_2.$$

b) If $\gamma_1 = \alpha$, then put $\delta_1 = \gamma_2$, $\delta_2 = \alpha$, $w' = w_2$. Using Lemma 11.3 again, it is easy to see that we have constructed a one-to-one correspondence between the chains of form (11.4) and the chains of form (11.5). As $l(\sigma_\alpha w_1) < l(w_1)$, we can apply the induction hypothesis.

Case 2. $\sigma_\alpha w_2 \to w_2$. If $w_1 \xrightarrow{\gamma_1} w \xrightarrow{\gamma_2} w_2$, then by Lemma 11.3, we have either $\gamma_1 = \alpha$ or $\gamma_2 = \alpha$. The same lemma shows us that $w_1 \to \sigma_\alpha w_2$ is equivalent to $\sigma_\alpha w_1 \to w_2$. Therefore, either there is no chain of form (11.5) or there exist exactly two chains $w_1 \to \sigma_\alpha w_1 \to w_2$ and $w_1 \to \sigma_\alpha w_2 \to w_2$. Lemma 10.3 is proved.

Proof of Lemma 10.4. For any element $w \in W$ we put $I(w) = \{w' \in W \mid w' \geqq w\}$. We shall prove by induction on $l(w)$ that $s(w_1, w_2)$ can be defined for $w_1, w_2 \in I(w)$ in such a way that the condition of the lemma be satisfied for any square (w_1, w_2, w_3, w_4) with $w_1 \in I(w)$. This will imply the statement of Lemma 10.4 since $I(s) = W$ (the element $s \in W$ is defined in Lemma 11.2. 3).

Thus assume $w \in W$. We choose an $\alpha \in \Sigma$ such that $w \to \sigma_\alpha w$. By the induction hypothesis, the function $s(w_1, w_2)$ has been defined for $w_1, w_2 \in I(\sigma_\alpha w)$.

1) Let

$$w = w_0 \xrightarrow{\gamma_1} w_1 \longrightarrow \dots \xrightarrow{\gamma_k} w_k,$$

where $\gamma_i \neq \alpha$ for all i. Then by Lemma 11.3 $\sigma_\alpha w_0 \to \sigma_\alpha w_1 \to \dots \to \sigma_\alpha w_k$ and $w_1 \to \sigma_\alpha w_i$ for all i. In particular, $\sigma_\alpha w_i \in I(\sigma_\alpha w)$ for all i.

2) Let $w' \in I(w) \setminus I(\sigma_\alpha w)$ and let

$$w = w_0 \xrightarrow{\gamma_1} w_1 \longrightarrow \dots \xrightarrow{\gamma_k} w_k = w'$$

be an arbitrary chain leading from w to w'. Then $\gamma_i \neq \alpha$ for all i, because otherwise $w' \in I(\sigma_\alpha w)$ would hold in view of *1)*. In particular, we have $w' \to \sigma_\alpha w'$ and $\sigma_\alpha w' \in I(\sigma_\alpha w)$.

3) Let us now define the function $s(w_1, w_2)$ for all arrows $w_1 \xrightarrow{\gamma} w_2$, $w_1 \in I(w)$. Here we assume that $s(w_1, w_2)$ has already been defined if $w_1 \in I(\sigma_\alpha w)$.

Let $w_1 \in I(w) \setminus I(\sigma_\alpha w)$. If $w_2 = \sigma_\alpha w_1$, then we put $s(w_1, w_2) = 1$. Let $w_2 \neq \sigma_\alpha w_1$. Then Lemma 11.3.2 implies

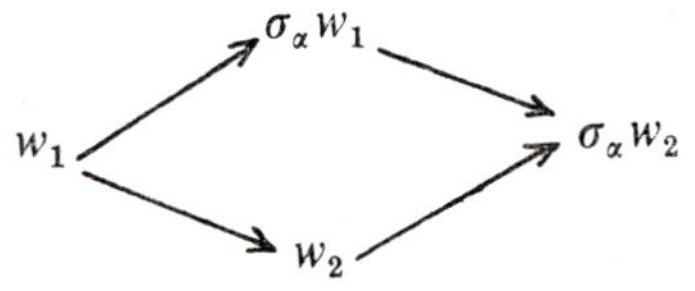

where $\sigma_\alpha w_1, \sigma_\alpha w_2 \in I(\sigma_\alpha w)$. In this case we put

$$s(w_1, w_2) = -s(w_1, \sigma_\alpha w_2) s(w_2, \sigma_\alpha w_2) s(\sigma_\alpha w_1, \sigma_\alpha w_2). \tag{11.6}$$

4) Now we prove that the function s defined in this manner satisfies the conditions of Lemma 10.4. For every square A

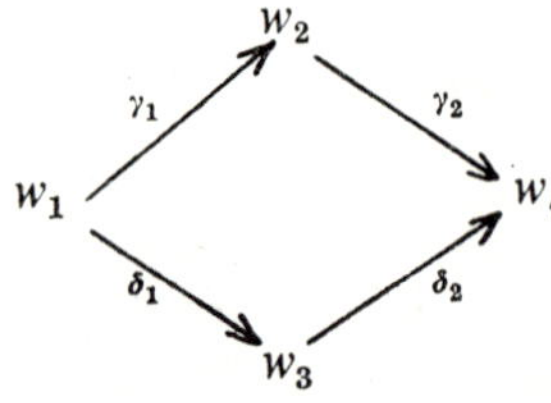

we put

$$p(A) = s(w_1, w_3)s(w_1, w_4)s(w_2, w_4)s(w_3, w_4).$$

We have to prove that $p(A) = -1$. There are the following possibilities.

a) $w_1 \in I(\sigma_\alpha w)$. In this case $p(A) = -1$ by the induction hypothesis.

b) $w_1 \notin I(\sigma_\alpha w)$ and $\gamma_1 = \delta = \alpha$. Then $p(A) = -1$ in view of formula (11.6).

c) $w_1 \notin I(\sigma_\alpha w)$ and $\gamma_1 \neq \alpha$, $\delta_1 \neq \alpha$. Since $w_2 \neq w_3$, we have $\gamma_2 \neq \delta_2$, and therefore one of the elements γ_2, δ_2 must be different from α. Then by Lemma 11.3 $w_4 \to \sigma_\alpha w_4$, and thus $\gamma_2 \neq \alpha$ and $\delta_2 \neq \alpha$. According to Lemma 11.3 we have five squares

$$A_1 = (w_1, w_2, \sigma_\alpha w_1, \sigma_\alpha w_2),$$

$$A_2 = (w_1, w_3, \sigma_\alpha w_1, \sigma_\alpha w_3),$$

$$A_3 = (w_2, w_4, \sigma_\alpha w_2, \sigma_\alpha w_4),$$

$$A_4 = (w_3, w_4, \sigma_\alpha w_3, \sigma_\alpha w_4),$$

$$A_5 = (\sigma_\alpha w_1, \sigma_\alpha w_2, \sigma_\alpha w_3, \sigma_\alpha w_4).$$

Here $p(A_i) = -1$, $i = 1, \ldots, 5$, in view of a) and b). It is easy to verify that $p(A) \cdot \prod_{i=1}^{5} p(A_i) = 1$, hence $p(A) = -1$, too.

d) $\gamma_1 = \alpha$, $\delta_2 \neq \alpha$.

By Lemma 11.3, $w_3 \to \sigma_\alpha w_3$, $w_4 \to \sigma_\alpha w_4$ and we have three squares

$$A_1 = (w_1, w_2, w_3, \sigma_\alpha w_3),$$

$$A_2 = (w_3, w_4, \sigma_\alpha w_3, \sigma_\alpha w_4),$$

$$A_3 = (w_2, w_4, \sigma_\alpha w_3, \sigma_\alpha w_4).$$

In view of a) and b) we have $p(A_i) = -1$, $i = 1, 2, 3$. Since $p(A)p(A_2) = p(A_1)p(A_3)$, we also have $p(A) = -1$.

Thus we have considered all the possibilities for the roots $\gamma_1, \gamma_2, \delta_1, \delta_2$ (up to interchanging γ_i with δ_i and w_2 with w_3). The proof of Lemma 10.4 is complete.

APPENDIX

Let $\chi, \psi \in \mathfrak{h}^*$, $\gamma_1, \ldots, \gamma_k \in \Delta_+$. We shall say that the sequence $\gamma_2, \ldots, \gamma_k$ satisfies condition (A) for the pair (χ, ψ) if

1) $\chi = \sigma_{\gamma_k} \ldots \sigma_{\gamma_1} \psi$,

2) Let $\chi_0 = \psi$ and $\chi_i = \sigma_{\gamma_i} \ldots \sigma_{\gamma_1} \psi$. Then for all i we have $\chi_{i-1} - \chi_i = n_i \gamma_i$, where

$$n_i = \frac{2\langle \chi_{i-1}, \gamma_i \rangle}{\langle \gamma_i, \gamma_i \rangle} \in \mathbf{Z}_+.$$

In the Appendix we are going to prove the following theorem.

Theorem A1. *Let* $L_\chi \in JH(M_\psi)$. *Then there exists a sequence* $\gamma_1, \ldots, \gamma_k$ *which satisfies condition* (A) *for the pair* (χ, ψ).

In the proof of Theorem A1 we shall use the following lemma.

Lemma A2. *Let* $\chi \in \mathfrak{h}^*$. *Then*

$$JH(M_\chi) = \bigcup_{\substack{\varphi \sim \chi \\ \varphi \leqq \chi}} c_\varphi L_\varphi, \quad c_\varphi \in \mathbf{Z}_+,$$

where $c_\chi = 1$.

This lemma follows easily from Propositions 8.5 and 8.6.

Proof of Theorem A1. Let $\lambda \in \mathfrak{h}^*_{\mathbf{Z}}$. Let us denote by F_λ an irreducible finite-dimensional $\mathfrak{g}$-module for which λ is an extremal weight. We recall that if $\mu \in P(F_\lambda)$, then $\|\mu\| \leqq \|\lambda\|$ and $\|\mu\| = \|\lambda\|$ implies $\mu \sim \lambda$.

Definition A3. Let $\chi \in \mathfrak{h}^*$, $\lambda \in \mathfrak{h}^*_{\mathbf{Z}}$. The pair (χ, λ) is called *admissible* if there are no $w \in W$ and $\mu \in P(F_\lambda)$ such that

a) $w\chi < \chi$,

b) $w(\chi + \mu) \geqq \chi + \lambda$,

c) $w(\chi + \mu) \sim \chi + \lambda$.

The meaning of Definition A3 is illuminated by the following proposition, whose proof we postpone until later.

Proposition A4. *Suppose* $\chi, \psi \in \mathfrak{h}^*$, $\lambda \in \mathfrak{h}^*_{\mathbf{Z}}$ *and the pair* (χ, λ) *is admissible. Furthermore, let* $L_\chi \in JH(M_\psi)$. *Then there exists a weight* $\mu \in P(F_\lambda)$ *such that* $L_{\chi+\lambda} \in JH(M_{\psi+\mu})$.

Proposition A4 includes all the information about the modules M_χ and L_χ that we need to prove Theorem A1. The subsequent arguments will only concern the geometric structure of the space $\mathfrak{h}^*$.

Lemma A5. *1) There exists a constant* $c_1 > 0$ *with the following property: if* $\varphi > \psi$, $\varphi + \vartheta_1 < \psi + \vartheta_2$ $(\varphi, \psi, \vartheta_1, \vartheta_2 \in \mathfrak{h}^*)$, *then* $c_1(\|\vartheta_1\| + \|\vartheta_2\|) > \|\varphi - \psi\|$.

2) There exists a constant c_2 *with the following property: let* $\varphi \in \mathfrak{h}^*_{\mathbf{Z}}$; *then one can find a sequence of weights* $0 = \varphi_0, \varphi_1, \ldots, \varphi_k = \varphi$, $\varphi_i \in \mathfrak{h}^*_{\mathbf{Z}}$, *such that*

$$\|\varphi_i - \varphi_{i-1}\| < c_2 \quad \text{and} \quad d(\varphi_i, [0, \varphi]) < c_2,$$

where $[0, \varphi]$ *is the segment in* $\mathfrak{h}^*$ *connecting* 0 *with* φ, *and* $d(\ ,\ )$ *denotes the distance in* $\mathfrak{h}^*_{\mathbf{R}}$.

3) Let $\varphi, \psi \in \bar{C}^+$, $w \in W$. *Then* $\langle w\varphi, \psi \rangle \leqq \langle \varphi, \psi \rangle$. *If in addition* $\varphi, \psi \in C^+$, *then the equality is satisfied for* $w = e$ *only.*

4) If C *is an arbitrary Weyl chamber,* $\varphi, \psi \in C$ *and* $\varphi \sim \psi$, *then* $\varphi = \psi$.

5) Let $\sum n_i \gamma_i = c\gamma$, *where* $\gamma_i, \gamma \in \Delta$, $n_i \in \mathbf{Z}$. *Then* $c \in \mathbf{Z}$ *as well.*

Proof. 1) It is well-known that $\langle \varrho, \alpha \rangle > 0$ for all $\alpha \in \Sigma$. Therefore there exists a $c > 0$ such that $\langle \varrho, \alpha \rangle > c\|\alpha\|$ for all $\alpha \in \Sigma$. This implies that for any $\varkappa \in K$ we have $\langle \varrho, \varkappa \rangle \geqq c\|\varkappa\|$. Moreover, $\varphi - \psi > 0$ and $\vartheta_1 - \vartheta_2 > \varphi - \psi$. Consequently,

$$\|\varrho\| (\|\vartheta_1\| + \|\vartheta_2\|) \geqq \langle \varrho, \vartheta_1 - \vartheta_2 \rangle > \langle \varrho, \varphi - \psi \rangle \geqq c \|\varphi - \psi\|.$$

Therefore we can put $c_1 = \|\varrho\| c^{-1}$.

2) It suffices to put $c_2 = 2d$, where d is the diameter of an arbitrary fundamental domain of the group $\mathfrak{h}_{\mathbf{Z}}^*$ in $\mathfrak{h}_{\mathbf{R}}^*$.

3) Let us write w in the form $w = \sigma_{\alpha_k} \dots \sigma_{\alpha_1}$ with $k = l(w)$ and put $\varphi_0 = \varphi$, $\varphi_i = w_i \varphi$, where $w_i = \sigma_{\alpha_i} \dots \sigma_{\alpha_1}$. Then $\varphi_i - \varphi_{i+1} = c_i \alpha_i$ where

$$c_i = \frac{2\langle \varphi_i, \alpha_i \rangle}{\langle \alpha_i, \alpha_i \rangle} = \frac{2\langle \varphi, w_i^{-1} \alpha_i \rangle}{\langle \alpha_i, \alpha_i \rangle}.$$

It follows from Lemma 11.2.1 that $c_i \geqq 0$. It is also clear that $c_i \neq 0$ if $\varphi \in C$. Therefore $\langle \varphi, \psi \rangle - \langle w\varphi, \psi \rangle = \sum c_i \langle \alpha_i, \psi \rangle$. This completes the proof.

4) It can be shown that $C = C^*$. Now let $\psi = w\varphi$. According to 3, $\langle \varphi, \psi \rangle \geqq \geqq \langle \psi, \psi \rangle$. But $\|\varphi\| = \|\psi\|$, hence $\varphi = \psi$.

5) We can assume that $\gamma \in \Sigma$. In this case the lemma follows from the fact that the elements of Σ form a basis in which every root has integer coordinates.

The proof of Lemma A5 is complete.

Let $\chi \in \mathfrak{h}^*$. We shall denote by $Y(\chi)$ the following proposition:

$Y(\chi)$. *For every* $\psi \in \mathfrak{h}^*$ *such that* $L_\chi \in JH(M_\psi)$ *there exists a sequence* $\gamma_1, \dots, \gamma_k \in \Delta_+$ *which satisfies condition* (A) *for the pair* (χ, ψ).

Theorem A1 says that $Y(\chi)$ is valid for all $\chi \in \mathfrak{h}^*$. We shall prove it in several steps. For any $\gamma \in \Delta_+$ we shall denote Ξ_γ the hyperplane in $\mathfrak{h}_{\mathbf{R}}^*$ orthogonal to γ.

Definition A6. Let $c_3 = 3c_1c_2$. The element $\varphi \in \mathfrak{h}^*$ is called *strongly regular* if $d(\operatorname{Re} \varphi, \Xi_\gamma) > c_3$ for all $\gamma \in \Delta_+$.

Step 1. Let $\chi, \varphi \in \mathfrak{h}^*$ be strongly regular while $\chi - \varphi \in \mathfrak{h}_{\mathbf{Z}}^*$ and $\operatorname{Re} \chi$, $\operatorname{Re} \varphi$ belong to the same Weyl chamber. Then propositions $Y(\chi)$ and $Y(\varphi)$ are equivalent.

Step 2. Let $\varphi, \chi \in \mathfrak{h}^*$ be strongly regular elements with $\chi - \varphi \in \mathfrak{h}_{\mathbf{Z}}^*$. In addition to this, let $\operatorname{Re} \chi \in C$, $\operatorname{Re} \varphi \in \sigma_\gamma C$, where C and $\sigma_\gamma C$ $(\gamma \in \Delta_+)$ are neighbouring Weyl chambers and $\langle \operatorname{Re} \chi, \gamma \rangle < 0$. Then $Y(\varphi)$ implies $Y(\chi)$.

It is easy to see that Steps 1 and 2 enable us to reduce the proof of $Y(\chi)$ for a strongly regular $\chi \in \mathfrak{h}^*$ to the case $\operatorname{Re} \chi \in C^+$. Now we are going to prove $Y(\chi)$ in this case. Let $\psi \in \mathfrak{h}^*$ and $L_\chi \in JH(M_\psi)$. It follows from Lemma A2 that $\chi \sim \psi$ and $\chi \leqq \psi$. This means that $\psi = \chi + \varkappa$, where $\varkappa \in K$. Then $\langle \varphi, \psi \rangle = \langle \chi, \chi \rangle + \langle \varkappa, \varkappa \rangle + 2\langle \chi, \varkappa \rangle$. Now

$$\langle \psi, \psi \rangle = \langle \chi, \chi \rangle, \quad \operatorname{Re} \langle \chi, \varkappa \rangle = \langle \operatorname{Re} \chi, \varkappa \rangle \geqq 0,$$

and $\langle \varkappa, \varkappa \rangle \geqq 0$. Consequently $\langle \varkappa, \varkappa \rangle = 0$, hence $\chi = \psi$.

Therefore Steps 1 and 2 prove $Y(\chi)$ for all the strongly regular weights $\chi \in \mathfrak{h}^*$. The general case is reduced to this one by means of the following Step 3.

Step 3. Let $\chi \in \mathfrak{h}^*$. Then there exists a strongly regular element $\varphi \in \mathfrak{h}^*$ such that $Y(\varphi)$ implies $Y(\chi)$.

Thus the proof of Theorem A1 reduces to the proofs of Steps 1, 2, 3 and Proposition A4.

In the proof of Steps 1, 2, 3 it will be convenient for us to make use of the following lemma.

Lemma A7. *Let C_1 and C_2 be two Weyl chambers, $\gamma, \psi \in \mathfrak{h}^*$ such that* $\operatorname{Re} \gamma \in \bar{C}_1$, $\operatorname{Re} \psi \in \bar{C}_2$, $\chi \sim \psi$ *and* $\lambda \in \mathfrak{h}^*$, $\mu \in \mathfrak{h}^*_{\mathbf{Z}}$ *be chosen so that* $\operatorname{Re} \chi + \lambda \in C_1$, $\operatorname{Re} \psi + \mu \in C_2$. *Suppose that the sequence of roots $\gamma_1, \ldots, \gamma_k \in \Delta_+$ satisfies condition* (A) *for the pair $(\chi+\lambda, \psi+\mu)$. Then it satisfies condition* (A) *for the pair (χ, ψ) as well.*

Proof. Item *2)* of condition (A) follows immediately from the fact that the sign of $\langle \chi, \gamma \rangle$ is constant in a Weyl chamber. In order to verify item *1)* we put $\chi_k = \sigma_{\gamma_k} \ldots \sigma_{\gamma_1} \psi$. Then $\chi \sim \psi \sim \chi_k$ and $\operatorname{Re} \chi$, $\operatorname{Re} \chi_k \in C_1$. By Lemma A5. 4, $\operatorname{Re} \chi = \operatorname{Re} \chi_k$. Moreover, $\chi - \chi_k \in \mathfrak{h}^*_{\mathbf{Z}}$ and consequently $\operatorname{Im} \chi = \operatorname{Im} \chi_k$, i.e. $\chi = \chi_k$.

Proof of Step 1. Lemma A5. 2 enables us to reduce the proof of Step 1 to the proof of the following proposition. Let $\chi \in \mathfrak{h}^*$, $\lambda \in \mathfrak{h}^*_{\mathbf{Z}}$ be such that $d(\operatorname{Re} \chi, \Xi) > 2c_1 c_2$, where $\Xi = \bigcup_{\gamma \in \Delta_+} \Xi_\gamma$ and $\|\lambda\| < c_2$. Then $Y(\chi+\lambda)$ implies $Y(\chi)$.

First we show that such a pair (χ, λ) is admissible. Indeed, assume that there exist $w \in W$ and $\mu \in P(F_\gamma)$ for which $w\chi < \chi$ and $w(\chi+\mu) > \chi+\lambda$. Then by Lemma A5. 1,

$$2c_2 > 2\|\lambda\| \geqq \|\lambda\| + \|\mu\| > c_1^{-1} \|\chi - w\chi\|.$$

On the other hand

$$\|\chi - w\chi\| \geqq d(\operatorname{Re} \chi, \Xi) > 2c_1 c_2.$$

Therefore, the pair (χ, λ) is indeed admissible.

Now let $L_\chi \in JH(M_\psi)$. Then according to Proposition A4, $L_{\chi+\lambda} \in JH(M_{\psi+\mu})$ for a certain $\mu \in P(F_\lambda)$. It follows from $Y(\chi+\lambda)$ that there exists a sequence of roots $\gamma_1, \ldots, \gamma_k \in \Delta_+$ which satisfies condition (A) for the pair $(\chi+\lambda, \psi+\mu)$. Since $\|\mu\| < c_2$, ψ and $\psi+\mu$ lie in the same Weyl chamber. Lemma A7 implies then that the same sequence $\gamma_1, \ldots, \gamma_k$ satisfies condition (A) for the pair (χ, ψ) as well.

Proof of Step 2. Applying Step 1, we can reduce the proof to the case when

$$3c_1 \|\varphi - \chi\| < d(\operatorname{Re} \varphi, \Xi_\delta) \tag{A.1}$$

for all

$$\delta \in \Delta_+ \setminus \gamma.$$

We put $\lambda = \varphi - \chi$ and prove that the pair (χ, λ) is admissible. Indeed, suppose we have $w \in W$ and $\mu \in P(F_\lambda)$ such that

$$w\chi < \chi, \quad w(\chi+\mu) > \chi+\lambda.$$

$w\chi < \chi$ implies $w \neq \sigma_\gamma$. Therefore,

$$\|w\chi - \chi\| > d(\operatorname{Re}\varphi, \bigcup_{\delta \in \gamma} \Xi_\delta) > 2c_1 \|\lambda\|.$$

But then, by Lemma A5. 1,

$$2\|\lambda\| \geqq \|\lambda\| + \|\mu\| \geqq c_1^{-1} \|w\chi - \chi\| > 2\|\lambda\|.$$

Thus the pair (χ, λ) is indeed admissible.

Now assume $L_\chi \in JH(M_\psi)$, $\psi \in \mathfrak{h}^*$. Then $L_{\chi+\lambda} \in JH(M_{\psi+\mu})$ for some $\mu \in P(F_\lambda)$. $Y(\chi+\lambda)$ implies the existence of a sequence $\gamma_1, \ldots, \gamma_k \in \Delta_+$ satisfying condition (A) for the pair $(\chi+\lambda, \psi+\mu)$. To complete the proof of Step 2 we have to construct a sequence of roots which satisfies condition (A) for the pair (χ, ψ).

Let $w \in W$ be such that $\psi = w\chi$. Then $w^{-1}(\psi+\mu) = \chi + w^{-1}\mu$. Since $\|w^{-1}\mu\| \leqq \|\lambda\|$, we have either $\operatorname{Re} w^{-1}(\psi+\mu) \in C$ or $\operatorname{Re} w^{-1}(\psi+\mu) \in \sigma_\gamma C$. Let us consider these two cases separately.

1) $\operatorname{Re} w^{-1}(\psi+\mu) \in C$. We have

$$w^{-1}(\psi+\mu) \sim \chi+\lambda \sim \sigma_\gamma(\chi+\lambda)$$

and

$$\operatorname{Re} \sigma_\gamma(\chi+\lambda) \in C.$$

By Lemma A5. 4,

$$\sigma_\gamma(\chi+\lambda) = w^{-1}(\psi+\mu) = \chi + w^{-1}\mu.$$

Consequently

$$\sigma_\gamma\chi - \chi = w^{-1}\mu - \sigma_\gamma\lambda.$$

Since $w^{-1}\mu$ and $\sigma_\gamma\lambda$ are weights of F_λ, we have

$$\sigma_\gamma\chi - \chi = c\gamma = \sum_{\gamma_i \in \Delta} n_i \gamma_i,$$

where $n_i \in \mathbf{Z}$. By Lemma A5. 5 $c \in \mathbf{Z}$, and in addition

$$c = -\frac{2\langle \chi, \gamma \rangle}{\langle \gamma, \gamma \rangle} > 0.$$

By Lemma A7, the sequence $\gamma_1, \ldots, \gamma_k$ satisfies condition (A) for the pair $(\sigma_\gamma\chi, \psi)$, hence the sequence $\gamma_1, \ldots, \gamma_k, \gamma$ satisfies condition (A) for the pair (χ, ψ).

2) $\operatorname{Re} w^{-1}(\psi+\mu) \in \sigma_\gamma C$. Let us put

$$\chi_i = \sigma_{\gamma_i} \ldots \sigma_{\gamma_1} \psi,$$

$$\bar{\chi}_i = \sigma_{\gamma_i} \ldots \sigma_{\gamma_1} (\psi+\mu).$$

We remark that all χ_i and $\bar{\chi}_i$ are congruent modulo $\mathfrak{h}_{\mathbf{Z}}^*$. Thus for each i we have either $\chi_i < \chi_{i-1}$ or $\chi_i > \chi_{i-1}$. If $\chi_i < \chi_{i-1}$ for all i, then the sequence $\gamma_1, \ldots, \gamma_k$ satisfies condition (A) for the pair (χ, ψ). In the opposite case we denote by i_0 the smallest index such that $\chi_{i_0} \geqq \chi_{i_0-1}$. We shall show that the sequence of roots

$$\gamma_1, \ldots, \gamma_{i_0-1}, \gamma_{i_0+1}, \ldots, \gamma_k, \gamma$$

satisfies condition (A) for the pair (χ, ψ).

First of all $\operatorname{Re}\bar{\chi}_{i_0}$ and $\operatorname{Re}\chi_{i_0-1}$ lie in the same Weyl chamber. Indeed,

$$\|\chi_{i_0-1}-\bar{\chi}_{i_0-1}\| < \|\lambda\|,$$

$$\|\chi_{i_0}-\bar{\chi}_{i_0}\| \leqq \|\lambda\|,$$

and by Lemma A2. 1,

$$\|\bar{\chi}_{i_0-1}-\bar{\chi}_{i_0}\| < 2c_1\|\lambda\|.$$

In addition to this, the real parts of the weights

$$\bar{\chi}_{i_0-1} \quad \text{and} \quad \bar{\chi}_{i_0} = \sigma_{\gamma_{i_0}}\bar{\chi}_{i_0-1}$$

as well as of the weights

$$\chi_{i_0-1} = \sigma_{\gamma_{i_0-1}}\cdots\sigma_{\gamma_1}\psi \quad \text{and} \quad \bar{\chi}_{i_0-1} = \sigma_{\gamma_{i_0-1}}\cdots\sigma_{\gamma_1}(\psi+\mu)$$

belong to different Weyl chambers. However, on account of condition (A1) the ball of radius $2c_1\|\lambda\|$ about the point $\operatorname{Re}\bar{\chi}_{i_0-1}$ intersects exactly two Wey chambers

$$\sigma_{\gamma_{i_0}}\cdots\sigma_{\gamma_k}C \quad \text{and} \quad \sigma_{\gamma_{i_0}}\cdots\sigma_{\gamma_k}\sigma_\gamma C.$$

Therefore, $\operatorname{Re}\bar{\chi}_{i_0}$ and $\operatorname{Re}\chi_{i_0-1}$ lie in the same Weyl chamber.

By Lemma A7 the sequence $\gamma_1, \ldots, \gamma_{i_0-1}$ satisfies condition (A) for the pair (χ_{i_0-1}, ψ), and the sequence $\gamma_{i_0+1}, \ldots, \gamma_k$ satisfies condition (A) for the pair $(\sigma_\gamma\chi, \chi_{i_0-1})$. In exactly the same way as in case 1) we can show that $\chi<\sigma_\gamma\chi$. Consequently the sequence of roots

$$\gamma_1, \gamma_2, \ldots, \gamma_{i_0-1}, \gamma_{i_0+1}, \ldots, \gamma_k, \gamma$$

satisfies condition (A) for the pair (χ, ψ).

Proof of Step 3. Let us first consider the case $\operatorname{Re}\chi=0$. Then the assumption $L_\chi \in JH(M_\psi)$ and Lemma A2 imply $\psi=\chi$.

Assume now that $\operatorname{Re}\chi\neq 0$. We can choose a weight $\lambda\in\mathfrak{h}_{\mathbf{Z}}^*$ in such a way that the following conditions are satisfied.

a) $\operatorname{Re}\chi$ and λ belong to the closure $\bar{C}$ of the same Weyl chamber C.

b) The weight $\chi+\lambda$ is strongly regular.

c) Let $c_4=\min\{\|\varkappa\| \mid \varkappa\in K\setminus\{0\}\}$. Then

$$\|\lambda-n\operatorname{Re}\chi\| \leqq nc_4/2c_1 \tag{A.2}$$

for some $n\in\mathbf{Z}_+$.

Lemma A8. *Let* $\nu\in P(F_\lambda)$ *such that* $\chi+\nu\sim\chi+\lambda$. *Then and* ν Re χ *lie in the same Weyl chamber and*

$$\|\nu-n\operatorname{Re}\chi\| \leqq nc_4/2c_1. \tag{A.3}$$

Proof. Let $w_1, w_2\in W$ be chosen in such a way that $w_1C=C^+$ and $w_1(\chi+\lambda)=$ $=w_2(\chi+\nu)$. Then $w_1\lambda$ is the highest weight of the representation F_λ and therefore $w_1\lambda-w_2\nu\in K$. Thus

$$\langle\varrho, w_1\lambda\rangle \geqq \langle\varrho, w_2\nu\rangle,$$

where the equality holds only if $w_2 \nu = w_1 \lambda$. On the other hand, by Lemma A2. 3,

$$\operatorname{Re}\langle \varrho, w_2 \chi \rangle \leqq \operatorname{Re}\langle \varrho, w_1 \chi \rangle.$$

Comparing these inequalities with the equality $w_1(\chi+\lambda)=w_2(\chi+\nu)$, we obtain $w_1\lambda=w_2\nu$ and $w_1\chi=w_2\chi$. Now applying the element $w_2^{-1}w_1$ to the inequality (A2) we obtain the statement of the lemma.

Now we prove that the pair (χ, λ) is admissible. Indeed, let $w \in W$ and $\mu \in P(F_\lambda)$ such that

$$w\chi < \chi, \quad w(\chi+\mu) \sim \chi+\lambda, \quad w(\chi+\mu) > \chi+\lambda.$$

Since $w\chi<\chi$, we have $\|w\chi-\chi\| \geqq c_4$. This implies

$$\|w(n+1)\chi-(n+1)\chi\| \geqq (n+1)c_4$$

and

$$w((n+1)\chi) < (n+1)\chi.$$

Applying Lemma A5. 1 we obtain that

$$c_1(\|\lambda-n\chi\|+\|\mu-n\chi\|) \geqq (n+1)c_4,$$

with contradicts (A2) and (A3). Consequently, the pair (χ, λ) is admissible.

Now let $L_\chi \in JH(M_\psi)$. Then in view of Proposition A4, $L_{\chi+\lambda} \in JH(M_{\psi+\mu})$ for some $\mu \in P(F_\lambda)$. Applying Lemma A8 again we find that ψ and $\psi+\mu$ lie in the same Weyl chamber. Let $\gamma_1, \ldots, \gamma_k \in \Delta_+$ be a sequence of roots which satisfies condition (A) for the pair $(\chi+\lambda, \psi+\mu)$. Then by Lemma A7 it also satisfies condition (A) for the pair (χ, ψ).

Proof of Proposition A4. Let $\lambda=\mu_1, \mu_2, \ldots, \mu_l$ be the weights of the module F_λ (with the corresponding multiplicities). Then, as follows from Lemma 9.10,

$$JH(M_\chi \otimes F_\lambda) = \bigcup_{i=1}^{l} JH(M_{\chi+\mu_i}).$$

Lemma A9. *Let $M_1, M_2 \in O$, while $JH(M_1) \subset JH(M_2)$ and F be a finite-dimensional* $\mathfrak{g}$*-module. Then*

$$JH(M_1 \otimes F) \subset JH(M_2 \otimes F).$$

The proof follows easily from the fact that tensor product by F is an exact functor.

Now let the pair (χ, λ) be admissible and $L_\chi \subset JH(M_\psi)$. Then

$$JH(L_\chi \otimes F_\lambda) \subset JH(M_\psi \otimes F_\lambda) = \bigcup_{i=1}^{l} JH(M_{\psi+\mu_i}).$$

Therefore, in order to show $L_{\chi+\lambda} \subset JH(M_{\psi+\mu})$, for some weight $\mu \in P(F_\lambda)$ it suffices to prove that

$$L_{\chi+\lambda} \subset JH(L_\chi \otimes F_\lambda).$$

Let M be a maximal proper submodule in M_χ such that $L_\chi=M_\chi/M$. Then

$$L_\chi \otimes F_\lambda = M_\chi \otimes F_\lambda / M \otimes F_\lambda.$$

Since $L_{\chi+\lambda} \in JH(M_\chi \otimes F_\lambda)$, it suffices to prove that $L_{\chi+\lambda} \notin JH(M \otimes F_\lambda)$.

By Lemma A2,

$$JH(M) \subset \bigcup_{\substack{\varphi\sim\chi \\ \varphi<\chi}} JH(M_\varphi),$$

consequently

$$JH(M\otimes F_\lambda) \subset \bigcup_{\substack{\varphi\sim\chi \\ \varphi<\chi}} c_\varphi JH(M_\varphi \otimes F_\lambda).$$

Thus it suffices to show that for any $\varphi\in\mathfrak{h}^*$ with $\varphi\sim\chi$, $\varphi<\chi$ and any $\mu\in P(F_\lambda)$, $L_{\chi+\lambda}\notin JH(M_{\varphi+\mu})$.

Assume, on the contrary, that $L_{\chi+\lambda}\in JH(M_{\varphi+\mu})$. Let $\varphi=w\chi$, $w\in W$. Then by lemma A2,

$$\chi+\lambda\sim\varphi+\mu = w(\varphi+w^{-1}\mu)$$

and

$$\chi+\lambda < w(\varphi+w^{-1}\mu).$$

But $w^{-1}\mu\in P(F_\lambda)$ which contradicts our assumption on admissibility.

This completes the proof of Proposition A4 and thus of Theorem A1.

Theorem A10. *Let* $\chi, \psi\in\mathfrak{h}^*$ *and* $\gamma_1, \ldots, \gamma_k\in\Delta_+$ *be a sequence of roots satisfying condition* (A) *for the pair* (χ, ψ). *Then* $\mathrm{Hom}\,(M_\chi, M_\varphi)=\mathbf{C}$.

A proof of Theorem A10 is contained in [8] and [9].

Theorems 8.8 and 8.10 follow immediately from Theorems A1 and A10.

REFERENCES

[1] Grothendieck, A., Éléments de géométrie algébrique IV (Troisième partie), *Publ. Math. I.H.E.S.* **28** (1966), 1—248.

[2] Gelfand, I. M. and Kirillov, A. A. (Гельфанд, И. М., Кириллов, А. А.), Структура тела Ли, связанного с полупростой расщепимой алгеброй Ли (The structure of the Lie division ring associated with semisimple split Lie algebras), *Функц. анализ*, **3**:1 (1969), 7—26.

[3] Bargmann, V., On a Hilbert space of analytic functions and an associated integral transform, Part I, *Comm. Pure Appl. Math.*, **14** (1961), 187—214.

[4] Harish–Chandra, On some application of the universal enveloping algebra of a semisimple Lie algebra, *Trans. Amer. Math. Soc.*, **70** (1951), 28—96.

[5] Bott, R., Homogeneous vector bundles, *Ann. of Math.*, **66** (1957), 203—248.

[6] Kostant, B. Lie algebra cohomology and the generalized Borel—Weil Theorem, *Ann. of Math.*, **74** (1961), 329—387.

[7] Verma, D.–N., Structure of certain induced representations of complex semisimple Lie algebras, *Bull. Amer. Math. Soc.*, **74** (1968), 160—166.

[8] Verma, D.–N., *Structure of certain induced representations of complex semisimple Lie algebras, Dissertation*, Yale Univ. 1966.

[9] Bernstein, I. N., Gelfand, I. M. and Gelfand, S. I. (Бернштейн И. Н., Гельфанд И. М., Гельфанд С. И.), Структура представлений, порожденных векторами старшего веса, (The structure of representations generated by a vector of highest weight), *Функц. анализ*, **5**:1 (1971) 1—9.

[10] Cartan, H. and Eilenberg, S., *Homological Algebra*, Princeton, 1956. (Картан А., Эйленберг С., *Гомологическая алгебра*, И. Л., Москва, 1960.)

[11] Bourbaki, N., *Groupes et algèbres de Lie, ch. 4—6*, Hermann, Paris 1968.

[12] Gelfand, I. M., The cohomology of infinite-dimensional Lie algebras; some questions of integral geometry, in *Actes, Congrès intern. math., 1970,* Vol. 1, pp. 95—111, Hermann, Paris 1971.

QUASI-ADMISSIBLE REPRESENTATION OF $\mathfrak{p}$-ADIC GROUPS

G. VAN DIJK

Introduction

The notion of a quasi-admissible representation of a $\mathfrak{p}$-adic group is due to Harish–Chandra. It is closely related to the notion of a quasi-simple representation of a real connected semisimple Lie group. Following some fruitful ideas of Harish–Chandra, we recently made some progress in the theory of "asymptotic expansions" as described in [2], Part IV, without assuming Conjecture III ([1(c)], § 1). Quasi-admissible representations occur very naturally there. The ideas are worked out in this note. The main applications of the theory, which we have in mind, are concerned with the asymptotic expansions of the matrix coefficients of the representations of the discrete series of G. They should belong to the $\mathfrak{p}$-adic analogue of the (Schwartz-) space $\mathscr{C}(G)$, introduced by Harish–Chandra for real semisimple Lie groups with finitely many connected components (cf. [1 (c)], § 9). We do not discuss the applications here. In § 1 we discuss admissible representations of $\mathfrak{p}$-adic groups and derive a few properties, most of which are well known. § 2 is concerned with quasi-admissible representations. The results should justify the relationship to quasi-simple representations of real groups, mentioned above. It concludes with a theorem which was inspired by ideas of Jacquet. The result turns out to be basic for the theory of the "asymptotic expansions", which is described in § 3.

§ 0. Some notations and conventions

Throughout this paper, Ω will be a $\mathfrak{p}$-adic field, i.e. a locally compact field with a non-trivial discrete valuation. Let us fix an additive Haar measure $d\omega$ on Ω. The valuation (or absolute value) on Ω is assumed to be normalized by requiring $d(\alpha\omega)=|\alpha|\,d\omega$ ($\alpha\in\Omega^*$). Let $\mathfrak{O}$ denote the ring of integers of Ω.

By G we mean a reductive $\mathfrak{p}$-adic group, i.e. the group of Ω-rational points of a connected, reductive, linear algebraic group $\mathbf{G}$ defined over Ω. Then $G\subset GL_n(\Omega)$ for a suitable $n\geqq 0$. $GL_n(\Omega)$, being an open subset of a vector space over Ω of dimension n^2 is a locally compact group. Since G is closed in $GL_n(\Omega)$, it is also locally compact. Moreover G is unimodular. We keep mainly to the notations and terminology of [2].

For a subset S of a topological space V, we shall denote by $Cl(S)$ the closure of S in V.

§ 1. Admissible representations

Admissible representations of $\mathfrak{p}$-adic groups have been introduced by Jacquet and Langlands ([3], § 2).

Let V be a complex vector space. A representation π of G on V is called *admissible* if:

(i) for every $v \in V$, the mapping $x \to \pi(x)v$ $(x \in G)$ is locally constant,

(ii) if $H \subset G$ is any *open* subgroup, then the space of vectors left fixed by $\pi(H)$ is finite-dimensional.

Let π be a representation of G on V. Denote by $C_c^\infty(G)$ the convolution algebra of the locally constant complex-valued functions on G with compact support. For any $f \in C_c^\infty(G)$ and any $v \in V$, such that the mapping $x \to \pi(x)v$ $(x \in G)$ is locally constant, we can define a linear operator $\pi(f)$ on V by

$$\pi(f)v = \int_G f(x)\pi(x)v\,\mathrm{d}x \quad (v \in V).$$

(The integral of course reduces to a finite sum.) In this way, $f \to \pi(f)$ is a representation of the algebra $C_c^\infty(G)$.

Suppose π to be admissible. Let K be any open compact subgroup of G. By $\mathscr{E}(K)$ we denote the set of equivalence classes of irreducible unitary representations of K. For $d \in \mathscr{E}(K)$, put $\alpha_d(k) = \mathrm{d}(d)\overline{\chi_d(k)}$ $(k \in K)$, where $\mathrm{d}(d)$ is the degree and χ_d the character of a representation in the class d. The definition of α_d obviously does not depend on the choice of this representation in the class d. Since K is open, we can extend α_d to a function in $C_c^\infty(G)$ by defining $\alpha_d(x) = 0$ if x is outside K. Hence $\pi(\alpha_d)$ is defined and is a projection, defined on V. Put $V(d) = \pi(\alpha_d)V$. Then $V = \oplus V(d)$ and $\dim V(d) < \infty$ for all $d \in \mathscr{E}(K)$. It is easily seen that these two properties characterize admissible representations: a continuous representation π of G on a complete locally convex Hausdorff space V is admissible if and only if it satisfies the latter conditions. Notice that the above condition (i) is equivalent to the condition: π is continuous if V is endowed with the trivial locally convex topology in which every semi-norm is continuous.

Clearly subrepresentations and quotient representations of admissible representations are admissible.

A representation π on $V \neq (0)$ with no non-trivial $\pi(G)$-stable subspace will be called *irreducible* (or: *algebraically irreducible*). Note that Schur's Lemma is valid for irreducible admissible representations: if a linear operator T on V commutes with $\pi(G)$, then it operates in every $V(d)$, hence must have eigenvectors, so that T is a scalar operator.

Let π be an admissible representation of G on V. Fix any open compact subgroup K. Consider the representation $x \to {}^t\pi(x^{-1})$ on the dual space $V^* = \prod V(d)^*$. The subspace of those $v^* \in V^*$ which are left fixed by some open subgroup of G

(or: such that $x \to {}^t\pi(x^{-1})v^*$ is locally constant) is evidently $\check{V} = \oplus V(d)^*$. Denote by $\check{\pi}(x)$ the restriction of ${}^t\pi(x^{-1})$ to $\check{V}$. Since $\check{V}(\check{d}) = V(d)^*$ for all $d \in \mathscr{E}(k)$, $\check{\pi}$ is an admissible representation, which we call the *contragredient* of π. If a subspace V_1 of V is invariant under $\pi(G)$, then the subspace $V_1^{\perp}$ of all vectors $\check{v} \in \check{V}$ which are orthogonal to V_1, i.e. $(v, \check{v}) = 0$ for all $v \in V_1$, is invariant under $\check{\pi}(G)$ and we have $(V_1^{\perp})^{\perp} = V_1$. Thus we get a one-to-one correspondence between invariant subspaces of V and $\check{V}$.

Two representations π_1 and π_2 on V_1 and V_2 respectively are said to be *equivalent* if there exists a bijective linear map $T: V_1 \to V_2$ such that $T\pi_1(x) = \pi_2(x)T$ for all $x \in G$. Let π_1, π_2 be admissible representations of G on the vector spaces V_1 and V_2 respectively and suppose there exists a non-degenerate bilinear form B on $V_1 \times V_2$ such that $B(\pi_1(x)v_1, \pi_2(x)v_2) = B(v_1, v_2)$ $(x \in G;\ v_1 \in V_1,\ v_2 \in V_2)$. Then π_2 is obviously equivalent to the contragredient of π_1, defined above.

A representation π of G on V is called *pre-unitary* if there exists on V an invariant positive-definite Hermitian form. It is then obvious that the operators $\pi(x)$ can be extended to unitary operators on the Hilbert space obtained from V by completion. We get a unitary representation of G in the classical sense. Conversely, given an open compact subgroup K of G, each *unitary* representation π of G on a Hilbert space H yields by restriction a representation π_0 on $H_0 = \oplus H(d)$, which is clearly pre-unitary and gives π after completion. The representation π_0 is admissible if $\dim H(d) < \infty$ for all $d \in \mathscr{E}(K)$. The space H_0 consists precisely of those vectors $v \in H$ such that $x \to \pi(x)v$ $(x \in G)$ is locally constant.

Let π be a pre-unitary admissible representation of G on a vector space V. Then the completion of π is topologically irreducible (no invariant *closed* subspace) if and only if π is algebraically irreducible. This is well known. A proof (even in a more general situation) will be given in § 2. Let π_1, π_2 be two admissible representations on V_1, V_2 respectively. Then we have (cf. [1 (a)], Theorem 8):

Theorem 1. *If π_1 and π_2 are pre-unitary and irreducible, then they are equivalent if and only if their completions are unitarily equivalent.*

Proof. Assume that π_1 and π_2 are equivalent. Let $T: V_1 \to V_2$ be a bijective linear map such that $T\pi_1(x) = \pi_2(x)T$ for all $x \in G$. Since π_1 and π_2 are pre-unitary and admissible, it follows, by passing to the contragredient representations, that there exists a bijective linear map $T^*: V_2 \to V_1$ such that $T^*\pi_2(x) = \pi_1(x)T^*$ $(x \in G)$ and $(Tv_1, v_2) = (v_1, T^*v_2)$ $(v_1 \in V_1, v_2 \in V_2)$. Clearly $T^*T: V_1 \to V_1$ commutes with $\pi_1(G)$, hence, since π_1 is irreducible, T^*T is a scalar, say λ (Schur's Lemma). Therefore $(T^*Tv, v) = \|Tv\|^2 = \lambda\|v\|^2$ $(v \in V_1)$. So $\lambda > 0$ and $\lambda^{-1/2}T$ can be extended to a unitary intertwining operator for the completions of π_1 and π_2. Consequently, the completions of π_1 and π_2 are unitarily equivalent. Conversely, if the completions of π_1 and π_2 are unitarily equivalent, it is obvious that π_1 and π_2 are equivalent. Hence the theorem.

Since every pre-unitary admissible representation is the direct sum of irreducible pre-unitary admissible representations, Theorem 1 remains true for arbitrary (non-necessarily irreducible) pre-unitary admissible representations.

If π is admissible, the operator $\pi(f)$ is of finite rank for all $f \in C_c^\infty(G)$ and therefore has a trace: $\mathrm{tr}\,\pi(f) = \Theta_\pi(f)$. Θ_π is called the *character* of π. Θ_π determines the class of π if π is irreducible. Indeed, let $\pi_1, \ldots, \pi_n$ be a finite set of irreducible admissible representations. Suppose no two of them are equivalent. Then their characters $\Theta_{\pi_1}, \ldots, \Theta_{\pi_n}$ are linearly independent.

(Cf. [3], Lemma 7.1; [1 (c)], Theorem 6).

§ 2. Quasi-admissible representations

1. *Definition and elementary properties*

Let K be any open compact subgroup of G. Put $H_K = C_c(G//K)$, the space of complex-valued functions f on G satisfying $f(kxk') = f(x)$ $(x \in G;\ k, k' \in K)$. Let π be any representation of G on a complex vector space V. By V_0 we shall denote the space of vectors $v \in V$ such that $x \to \pi(x)v$ $(x \in G)$ is locally constant. V_0 is $\pi(G)$-stable. A vector $v \in V_0$ is said to be H_K-finite (w.r.t. π) if $\pi(H_K)v$ is a finite-dimensional subspace of V. An element $v \in V_0$ is called *Hecke-finite* (w.r.t. π) if v is H_K-finite for every open compact subgroup K.

Theorem 2. *Let π be a (continuous) representation of G on a complete, locally convex, Hausdorff space V. Let $v \in V_0$ be Hecke-finite. Put $W_0 = \pi(C_c^\infty(G))v$. Then, given any open compact subgroup K, one has:*

(i) *W_0 is the smallest $\pi(G)$-stable subspace of V which contains v.*

(ii) $\dim W_0(d) < \infty$ *for all $d \in \mathscr{E}(K)$ and $W_0 = \oplus\, W_0(d)$.*

(iii) *Put $\mathrm{Cl}(W_0) = W$. Then $W(d) = W_0(d)$ for all $d \in \mathscr{E}(K)$. Furthermore, the map $M \to \mathrm{Cl}(M)$ is a bijection from the set of $C_c^\infty(G)$-submodules of W_0 onto the set of closed G-submodules of W. Moreover $M(d) = \mathrm{Cl}(M)(d)$ and $M = \oplus\, M(d)$.*

If $v \neq 0$, then there exists a maximal, proper, closed, G-submodule of W.

The proof is left to the reader. Compare this theorem with ([1 (a)], section 10).

The next theorem is the analogue of ([1 (c)], Lemma 77). The proof is similar to ([2], Theorem 6). Let Z be the split component of G. Fix a unitary character χ of Z. Keeping to the notations of ([2], Part I, § 1), we have:

Theorem 3. *Put $H = L^2(G, \bar{\chi})$ and denote by λ the left-regular representation of G on H. Let $f \neq 0$ be a Hecke-finite element of H_0. Denote by V the smallest closed $\lambda(G)$-stable subspace of H which contains f. Then*

$$V = \bigoplus_{1 \leq i \leq p} V_i,$$

where V_i are mutually orthogonal closed subspaces of V, which are $\lambda(G)$-stable. The restriction λ_i of λ on V_i is an irreducible χ-representation, which is square-integrable mod Z.

Corollary. Put $H=L^2(G, \bar{\chi})$. Each Hecke-finite $f\in H_0$ (w.r.t. λ), which is bi-invariant w.r.t. some open compact subgroup of G, can be written as a finite sum of matrix-coefficients of irreducible χ-representations, which are square-integrable mod Z.

Proof. Let K be an open compact subgroup of G such that f is bi-invariant with respect to K. Denote by φ_K the characteristic function of K. Put $\varphi(x)= = \int_Z \varphi_K(xz)\chi(z)dz$ $(x\in G)$. Then $\varphi\in H_0$. With the notations of Theorem 3, let E_i be the orthogonal projection of H onto V_i $(1\leq i\leq p)$.

Then: $f(x)=f^*\varphi_K(x)=(f, \lambda(x)\varphi)=\sum_{i=1}^{p}(E_i f, \lambda_i(x)E_i\varphi)$ $(x\in G)$. This proves the corollary.

Remark. 1) Notice that both $E_i f$ and $E_i\varphi$ belong to $V_{i,0}$ $(1\leq i\leq p)$.

2) By ([2], Lemma 8) every supercusp form f, which is *bi*-invariant w.r.t. some open compact subgroup of G, is Hecke-finite. Let π be a representation of G on a complex vector space V. We call π *quasi-admissible* if every $v\in V$ is contained in a $\pi(G)$-stable admissible subspace of V (i.e. in a $\pi(G)$-stable subspace W such that the restriction of π on W is admissible). By Theorem 2 (ii), this amounts to saying that $V=V_0$ and every vector $v\in V$ is Hecke-finite.

Theorem 4. *Let π be a (continuous) representation of G on a complete, locally convex, Hausdorff space V. Assume that the restriction of π on V_0 is quasi-admissible. Then π is (topologically) irreducible if and only if π_0 is (algebraically) irreducible.*

This is clear from Theorem 2 (iii).

Clearly subrepresentations and quotient representations of quasi-admissible representations are quasi-admissible.

2. *A basic result*

We conclude with a basic theorem. Choose a parabolic pair (P, A) in G with corresponding Levi decomposition $P=MN$. Let π be a representation of G on a complex vector space V. Denote by $V(N)$ the subspace of V generated by elements of the form $\pi(n)v-v$ $(n\in N, v\in V)$. The space $V(N)$ is $\pi(P)$-stable. Put $\bar{V}=V/V(N)$. The representation of P on $\bar{V}$ contains N in its kernel and yields therefore actually a representation of M, which we denote by σ:

Then we have:

Theorem 5. *If π is quasi-admissible, then σ is quasi-admissible.*

The main tool for the proof is due to H. Jacquet, who discussed similar assertions for the group $GL(n, \Omega)$ in his Montecatini talks (C.I.M.E. Summer School on the theory of group representations, Montecatini (1970)). With slight modifications, Harish–Chandra recently proved the above theorem for general reductive $\mathfrak{p}$-adic groups G. The proof can be found in a lecture of mine, given at the University of Nancy [4].

§ 3. The space $\mathscr{A}(G)$

1. *Definition and main results*

Denote by $C(G)$ the space of continuous complex-valued functions on G and let λ and ϱ be the left and right regular representation of G on $C(G)$ respectively.

Let K be any open compact subgroup of G and define H_K as before. Let $f \in C(G)$. We call f *left H_K-finite* if the space of functions of the form $h * f$ $(h \in H_K)$ is finite-dimensional. The function f is called *left Hecke-finite* if f is left H_K-finite for every open compact subgroup K.

We denote by $\mathscr{A}(G)$ the space of all $f \in C(G)$ satisfying:

(i) $f \in C(G//K)$ for some K,

(ii) f is left Hecke-finite.

Theorem 5 enables us to prove the results of ([2], Part IV) in a slightly different form, without assuming Conjecture III (1. c. § 1). The results are as follows.

Let (P, A) be any parabolic pair in G with corresponding Levi decomposition $P = MN$. Denote by $\mathfrak{a}$, $\mathfrak{n}$ the Lie algebras of A, N respectively. By a *root* α of P (or (P, A)) we mean an element α in $\mathfrak{a}^*$, the dual space of $\mathfrak{a}$, with the following property. Let $\mathfrak{n}_\alpha$ denote the set of all $X \in \mathfrak{n}$ such that $[H, X] = \alpha(H)X$ for all $H \in \mathfrak{a}$. Then $\mathfrak{n}_\alpha \neq (0)$. For any root α let ξ_α denote the corresponding character of A (or (P, A)). The set of roots of P (or (P, A)) will be denoted by $\Sigma(P)$. For $t > 0$, put

$$A^+(t) = \{a \in A\colon\ |\xi_\alpha(a)| \geqq t \quad \text{for all} \quad \alpha \in \Sigma(P)\}.$$

For $f \in C(G)$, denote by $f_{[P]}$ the restriction of f to M. Let $\varphi \in C(M)$. We write

$$\varphi \underset{P}{\approx} 0 \quad (\text{or: } \varphi \underset{(P,A)}{\approx} 0)$$

if, given any compact set ω in M, we can choose $c > 0$ such that $\varphi(ma) = 0$ for all $m \in \omega$ and all $a \in A^+(c)$.

Theorem 6. *Given $f \in \mathscr{A}(G)$ and given a parabolic pair (P, A) in G, $P \neq G$, with corresponding Levi decomposition $P = MN$, there exists exactly one element in $\mathscr{A}(M)$, called f_P, such that*

$$f_{[P]} - f_P \underset{P}{\approx} 0.$$

Let Z denote, as before, the split component of G.

Theorem 7. *If $f \in \mathscr{A}(G)$ has compact support* mod Z, *then f is a supercusp form.*

Let π be an irreducible admissible representation of G on a vector space V. The functions $x \to (\pi(x)v, \check{v})$ $(x \in G, v \in V, \check{v} \in \check{V})$ are called the matrix coefficients of π. We say that π is *supercuspidal* if some non-zero matrix coefficient f of π is a supercusp form, i.e.

(i) $\operatorname{Supp} f$ is compact mod Z.

(ii) $\int_N f(xn)\,dn = 0$ $(x \in G)$ for all parabolic subgroups $P \neq G$ (N is the unipotent radical of P and dn is some Haar measure on N).

Corollary 1. *Let π be an irreducible admissible representation of G. Then π is supercuspidal if and only if all matrix coefficients of π have compact support* mod Z.

To prove the corollary, one has to observe that the matrix coefficients of π belong to $\mathscr{A}(G)$.

Let us make the following remark, which we state in the form of a corollary.

Corollary 2. *Let π be an irreducible unitary representation of G which is square-integrable* mod Z. *Then π is supercuspidal (in the sense of* [2], *Part* I, § 3*) if and only if π has a non-zero matrix coefficient with compact support* mod Z.

For any open compact subgroup K, denote by φ_K the characteristic function of K. If f is a matrix coefficient of π, whose support is compact mod Z, the same is true for $\varphi_K * f * \varphi_K$. Since π is irreducible, one can choose K such that $f \neq 0$ implies $g = \varphi_K * f * \varphi_K \neq 0$. We may put $g(x) = (\pi(x)v, w)$ $(x \in G)$, where v and w are locally constant vectors. Now g is a matrix coefficient of the restriction π_0 to V_0 (in the notation of § 2), which is admissible by ([2], Theorem 2). Since π is unitary, we have $\check{V}_0 = \operatorname{Conj}(V_0)$, the complex conjugate space of V_0. So the remark follows from Corollary 1.

2. *Proof of Theorem* 7

Let $f \in \mathscr{A}(G)$ be such that f has compact support mod Z. Denote by ssrk G the dimension of a maximal Ω-split torus in the derived group of G. Without loss of generality we may assume that ssrk $G > 0$. Let (P, A) be any parabolic pair in G, $P \neq G$, with corresponding Levi decomposition $P = MN$. For any continuous function g on G with compact support mod Z, define

$$g^{[P]}(m) = \int_N g(mn)\,dn \quad (m \in M).$$

Put $V = C_c^\infty(G) * f$. V is stable and admissible under $\lambda(G)$. Moreover $g^{[P]} = 0$ if $g \in V(N)$. So, with the notations of § 2, $g \to g^{[P]}$ $(g \in V)$ gives rise to a natural linear map

$$\mu_P : \bar{V} \to C(M).$$

Let σ be the representation of M on $\bar{V}$, defined in § 2, section 2. Then

$$\lambda(m)\mu_P = \mu_P \sigma(m) \quad (m \in M).$$

By Theorem 5, σ is quasi-admissible.

Now choose an open compact subgroup K in G such that $f \in C(G//K)$. Put $\Phi = H_{K\cap M} * f^{[P]}$, where $H_{K\cap M} = C_c(M//K\cap M)$. Then $\dim \Phi < \infty$. Indeed, let $\bar{f}$ be the projection of f on $\bar{V}$. Then

$$\varphi = \lambda(H_{K\cap M}) f^{[P]} = \mu_P(\sigma(H_{K\cap M})\bar{f}).$$

The latter space is of finite dimension.

Let γ be the characteristic function of $K\cap M$. For $a \in A$, put $\gamma_a(m) = \gamma(am)$ ($m \in M$). Normalize the Haar measure on M such that $\operatorname{vol}(K\cap M) = 1$. Then $\gamma_{a_1 a_2} = \gamma_{a_1} * \gamma_{a_2}$ ($a_1, a_2 \in A$). We obtain a representation ϱ of A on Φ by

$$\varrho(a)\varphi = \gamma_a * \varphi \quad (\varphi \in \Phi, a \in A).$$

Actually, $\varrho(a)$ is the right translation by $a \in A$.

If $\Phi \neq (0)$, we can choose $\varphi_0 \neq 0$ in Φ and a quasi-character χ of A such that $\varrho(a)\varphi_0 = \chi(a)\varphi_0$ ($a \in A$). Thus $\operatorname{Supp} \varphi_0 = \operatorname{Supp} \varphi_0 \cdot A$. Since A/Z is *not* compact, this yields a contradiction. Therefore $\Phi = (0)$ and hence $f^{[P]} = \gamma * f^{[P]} = 0$.

Since $\mathscr{A}(G)$ is stable under left translations, the above observations hold for all left translates of f as well. Choose an open compact subgroup K' of G as in ([2], Theorem 5). We get

$$\int_N f(xn)\, dn = \int_N f(k'm'n)\, dn = 0$$

for $x \in G$, $x = k'm'n'$ ($k' \in K', m' \in M', n' \in N$). Since (P, A) was an arbitrary parabolic pair in G with $P \neq G$, this implies that f is a supercusp form and the theorem follows.

3. *Preparation for the proof of Theorem* 6

The proof of Theorem 6 needs a rather long preparation, part of which has been done in ([2], Part IV). However, for the convenience of the reader we shall give the proof more or less in detail. We assume throughout the proof that $\operatorname{ssrk} G > 0$ and that a parabolic pair (P, A) with $P \neq G$ is given. Let $P = MN$ be the corresponding Levi decomposition.

Lemma 1. (Uniqueness of f_P)
Let $\varphi \in \mathscr{A}(M)$ and suppose $\varphi \underset{P}{\approx} 0$. Then $\varphi = 0$.

Proof. Choose an open compact subgroup K in G such that $\varphi \in C(M//K\cap M)$. Put $\Phi = H_{K\cap M} * \varphi$. Then $\dim \Phi < \infty$. As before, let γ be the characteristic function of $K\cap M$ and assume $\int_M \gamma(m)\, dm = 1$. Put $\gamma_a(m) = \gamma(am)$ ($a \in A, m \in M$). Then A operates on Φ by right translations:

$$\varrho(a)\varphi = \gamma_a * \varphi \quad (a \in A, \varphi \in \Phi).$$

If $\Phi \neq (0)$, we can choose $\varphi_0 \neq 0$ in Φ and a quasi-character χ of A such that $\varrho(a)\varphi_0 = \chi(a)\varphi_0$ ($a \in A$), i.e. $\varphi_0(ma) = \chi(a)\varphi_0(m)$ for all $a \in A$, $m \in M$. Now $\varphi \underset{P}{\approx} 0$, hence $\alpha * \varphi \underset{P}{\approx} 0$ for all $\alpha \in H_{K \cap M}$. In particular $\varphi_0 \underset{P}{\approx} 0$. This implies $\varphi_0(ma)=0$ for $a \in A^+(C)$ for some $c>0$ (depending on m). Choose m such that $\varphi_0(m) \neq 0$. Then $\varphi_0(ma)=\varphi_0(m)\chi(a)\neq 0$ for all $a \in A$. Since $A^+(c)\neq\emptyset$, this gives a contradiction. Therefore $\Phi=(0)$ and hence $\varphi=0$. This completes the proof of the lemma.

Put $A^+=A^+(1)$.

Lemma 2. *Let K be any open compact subgroup of G. There exists a finite set of elements $a_1, \ldots, a_r \in A^+$ with the following property. Let S denote the semigroup generated by $a_1, \ldots, a_r$. Then given $t>0$, we can choose a finite subset B of A such that*

$$A^+(t) \subset B \cdot S \cdot AA \cap K.$$

The proof is exactly the same as the proof of ([2], Lemma 11).

Lemma 3. *Fix $f \in \mathscr{A}(G)$ and let K be any open compact subgroup of G. Put $F=H_{K\cap M} * f_{[P]}$ and let F_0 be the subspace of all $\varphi \in F$ with $\varphi \underset{P}{\approx} 0$. Then* $\dim F/F_0<\infty$.

Proof. Let $V=C^\infty(G)*f$ and denote by W the subspace of all $\varphi \in V$ such that $\varphi_{[P]} \underset{P}{\approx} 0$. Clearly W is $\lambda(M)$-stable. V is a quasi-admissible $\lambda(G)$-module, since $V \subset \mathscr{A}(G)$. Moreover $V(N) \subset W$. Indeed, let $\varphi \in V(N)$ and choose a compact set ω of M. Fix $n \in N$. Let K' be such that $\varphi \in C(G//K')$. Then we have

$$\varphi(nma)-\varphi(ma) = \varphi(man^{(ma)^{-1}})-\varphi(ma).$$

Choose $t>0$ so large that $n^{(ma)^{-1}} \in K'$ for $m \in \omega$ and $a \in A^+(t)$. Then

$$\varphi(nma)-\varphi(ma) = 0.$$

Put $V'=C^\infty(M)$ and denote by W' the $\lambda(M)$-stable subspace of all φ' such that $\varphi' \underset{P}{\approx} 0$. We have a natural map $\varphi \to \varphi_{[P]}$ of V into V', which carries W into W'.

Now consider the following sequence of $\lambda(M)$-modules:

$$V/V(N) \to V/W \to V'/W'.$$

By Theorem 5 we know that $V/V(N)$ is a quasi-admissible $\lambda(M)$-module, hence the same is true for its image in V'/W', which contains $f_{[P]}$ mod W'. This gives the lemma.

Fix $f \in \mathscr{A}(G)$ and choose an open compact subgroup K of G such that $f \in C(G//K)$. Put $F=H_{K\cap M} * f_{[P]}$ and define F_0 as before. Assume $f_{[P]} \notin F_0$. Since $f_{[P]} \in F$, we can choose a basis of the form

$$\varphi_1 = f_{[P]}, \varphi_2, \ldots, \varphi_p \quad (p = \dim F/F_0)$$

of F mod F_0. Since F and F_0 are both stable under convolution on the left by elements of $H_{K\cap M}$, we obtain a representation τ of $H_{K\cap M}$ on F mod F_0 such that

$$h * \varphi_j = \sum_j \varphi_i \tau_{ji}(h) \bmod F_0, \ \tau(h) = (\tau_{ji}(h)).$$

Put $\Phi=(\varphi_1, \ldots, \varphi_p)$.

Then, in an obvious notation,

$$h*\Phi \underset{P}{\approx} \Phi\tau(h) \quad (h\in H_{K\cap M}).$$

As before, denote by $\gamma\in H_{K\cap M}$ the characteristic function of $K\cap M$. For $a\in A$, put $\gamma_a(m)=\gamma(am)$ $(m\in M)$. Normalize the Haar measure on M such that vol $(K\cap M)=1$. Then $\gamma_{a_1a_2}=\gamma_{a_1}*\gamma_{a_2}$. Denote by χ the representation of A on F mod F_0 given by $a\to\tau(\gamma_a)$. So $\gamma_a*\Phi\underset{P}{\approx}\Phi\chi(a)$,

$$\varrho(a)\Phi \underset{P}{\approx} \Phi\chi(a) \quad (a\in A).$$

Put $\Psi_a = \varrho(a)\Phi-\Phi\chi(a) \quad (a\in A)$.

Lemma 4. *Fix a compact set* ω *in* M *and* $t>0$. *Then we can choose* $c>0$ *such that* $\Psi_a=0$ *on* $\omega A^+(c)$ *for all* $a\in A^+(t)$.

The proof is the same as the proof of ([2], Lemma 14).

4. *Proof of Theorem* 6

Fix $f\in\mathscr{A}(G)$ as above. We may assume $f_{[P]}\notin F_0$, otherwise we can choose $f_P=0$. We have to find $f_P\in\mathscr{A}(M)$ such that $f_{[P]}-f_P\underset{P}{\approx}0$. The uniqueness of f_P is guaranteed by Lemma 1. Following the method of ([2], Part IV, § 3) we find $\theta\in C(M//K\cap M)$ which is left $H_{K\cap M}$-finite and satisfies $f_{[P]}-\theta\underset{P}{\approx}0$. Let us recall this method. Let ω be a compact subset of M. By Lemma 4 we can choose $c>0$ such that $\Phi(ma)=\Phi(m)\chi(a)$ for $a\in A^+$ and $m\in A^+(c)$. Put $c_0=\max(1, c)$. Then

$$\Phi(ma) = \Phi(m)\chi(a) \quad \text{for} \quad a\in A^+(c_0),\ m\in\omega A^+(c_0),$$

or

$$\Phi(ma_1a_2) = \Phi(ma_1)\chi(a_2) \quad (m\in\omega;\ a_1, a_2\in A^+(c_0)).$$

By interchanging the role of a_1 and a_2 we get

$$\Phi(ma_1)\chi(a_1)^{-1} = \Phi(ma_2)\chi(a_2)^{-1} \quad (m\in\omega;\ a_1, a_2\in A^+(c_0)).$$

Define $\gamma_P(a)=\inf|\xi_\alpha(a)|$ $(a\in A)$, where the infimum is taken over the set Σ of all roots α of (P, A). We write $a\underset{P}{\to}\infty$ if $\gamma_P(a)\to\infty$. Define

$$\Theta(m) = \lim_{\substack{a\to\infty\\P}} \Phi(ma)\chi(a)^{-1} \quad (m\in M).$$

Θ is locally constant, hence continuous. Furthermore

$$\Theta(ma_0) = \lim_{\substack{a\to\infty\\P}} \Phi(ma_0a)\chi(a)^{-1} = \lim_{\substack{a\to\infty\\P}} \Phi(ma)\chi(a)^{-1}\chi(a_0) =$$

$$= \Theta(m)\chi(a_0) \quad \text{for} \quad m\in M,\ a_0\in A.$$

For $m \in \omega$, $a \in A^+(c_0)$ we have

$$\Theta(m) = \Phi(ma)\chi(a)^{-1}.$$

So, given ω, there exists $c_0 > 0$ such that

$$\Theta(m) = \Phi(m) \quad \text{for} \quad m \in \omega A^+(c_0).$$

Hence $\Theta \underset{P}{\approx} \Phi$. Now $\Theta = (\theta_1, \dots, \theta_p)$, so $\varphi_i \underset{P}{\approx} \theta_i$ $(1 \leqq i \leqq p)$. In particular $f_{[P]} \underset{P}{\approx} \theta_1$. Write θ for θ_1. We have

$$\Theta(kmk') = \theta(m) \quad (k, k' \in K \cap M, m \in M)$$

since Φ has this property. To prove that θ is left $H_{K \cap M}$-finite, put

$$\Theta_h = h * \Theta - \Theta\tau(h) \quad (h \in K_{K \cap M}).$$

Since $h_1 * \Phi - \Phi\tau(h) \underset{P}{\approx} 0$, we have also $\Theta_h \underset{P}{\approx} 0$. We know

$$\varrho(a)\Theta = \Theta\chi(a) \quad (a \in A).$$

Since γ_a is in the center of $H_{K \cap M}$, we get

$$\varrho(a)\Theta_h = \Theta_h\chi(a) \quad (a \in A, h \in H_{K \cap M}).$$

Without loss of generality we may assume that at least one φ_i is chosen such that $\varrho(a)\varphi_i = \chi_0(a)\varphi_i$ (mod F_0) for some quasi-character χ_0 of A. It follows that both m and ma belong to Supp Θ_h for all $a \in A$. So Supp $\Theta_h \neq \emptyset$ contradicts $\Theta_h \underset{P}{\approx} 0$. Hence $h * \Theta = \Theta\tau(h)$ for all $h \in H_{K \cap M}$. So $h * \theta = \sum_{j=1}^{p} \tau_{j1}(h)\theta_j$ for all $h \in H_{K \cap M}$. This shows that θ is left $H_{K \cap M}$-finite.

Let K' be any open compact subgroup of G such that $K' \subset K$. Put $F' = H_{K' \cap M} * f_{[P]}$ and let F_0' be defined as before. Then we have an injection $F/F_0 \to F'/F_0'$. A base of F'/F_0' can be chosen in such a way that it starts with $\varphi_1, \dots, \varphi_p$. Applying the same method again, we easily find that θ is left $H_{K' \cap M}$-finite, hence left Hecke-finite. Consequently $\theta \in \mathscr{A}(M)$. The theorem follows by choosing $f_P = \theta$.

REFERENCES

[1] Harish-Chandra, (a) Representations of a semisimple Lie group on a Banach space, *Trans. Amer. Math. Soc.*, **75** (1953), 185—243.

(b) Representations of semisimple Lie groups, III, *Trans. Amer. Math. Soc.*, **76** (1954), 234—253.

(c) Discrete series for semisimple Lie groups, II, *Acta Math.*, **116** (1966), 1—111.

[2] Harish-Chandra and van Dijk, G., *Harmonic analysis on reductive* $\mathfrak{p}$*-adic groups*, Springer Lecture Notes, Vol. **162**, 1970.

[3] Jacquet, H. and Langlands, R. P., *Automorphic forms on GL*(2), Springer Lecture Notes, Vol. **114,** 1970.

[4] van Dijk, G. et al., *Colloque sur les fonctions sphériques et la théorie des groupes, Nancy (1971),* Institut Élie-Cartan.

CONSTRUCTION OF PRIMITIVE IDEALS IN AN ENVELOPING ALGEBRA

MICHEL DUFLO

Introduction

Let $\mathfrak{g}$ be a finite dimensional Lie algebra over an algebraically closed field k of characteristic 0. In the spirit of the Kirillov's method of orbits, J. Dixmier constructed in [6] a set of primitive ideals of the enveloping algebra $U(\mathfrak{g})$ of $\mathfrak{g}$, when $\mathfrak{g}$ is solvable. In this paper we generalize this construction to an arbitrary Lie algebra. We show moreover that the set obtained is dense in the set of all primitive ideals of $U(\mathfrak{g})$ and we give an application to the study of the center $Z(\mathfrak{g})$ of $U(\mathfrak{g})$. Precisely, we prove that as an algebra, $Z(\mathfrak{g})$ is isomorphic to the algebra $I(\mathfrak{g})$ of all $\mathfrak{g}$-invariant elements of the symmetric algebra $S(\mathfrak{g})$*. These results where anounced in [11].

I. Construction of solvable polarizations

I.1. Notations. In this paper, all Lie algebras will be finite-dimensional over an algebraically closed field k of characteristic 0. Let $\mathfrak{g}$ be such a Lie algebra, and $\mathfrak{g}^*$ be the dual space. Let $g \in \mathfrak{g}^*$. We write $B_g(X, Y) = \langle g, [X, Y] \rangle$ for $X, Y \in \mathfrak{g}$, and we denote by $\mathfrak{g}(g)$ the kernel of the bilinear form B_g. More generally, if $\mathfrak{n}$ is an ideal in $\mathfrak{g}$, we consider in $\mathfrak{n}^*$ the contragradient of the adjoint representation and we denote by $\mathfrak{g}(f)$ the stabilizer of an element $f \in \mathfrak{n}^*$.

A subalgebra $\mathfrak{l}$ of $\mathfrak{g}$ is called a *polarization* at g if it is also a maximal isotropic subspace of $\mathfrak{g}$ for the form B_g, that is, if $\langle g, [\mathfrak{l}, \mathfrak{l}] \rangle = 0$ and if

$$\dim \mathfrak{l} = \tfrac{1}{2}(\dim \mathfrak{g} + \dim \mathfrak{g}(g)).$$

We shall be interested mainly in solvable polarizations, and we shall denote by $SP(g, \mathfrak{g})$ the set of solvable polarizations of $\mathfrak{g}$ at g.

Let $\mathfrak{n}$ be an ideal in $\mathfrak{g}$. Put $f = g|\mathfrak{n}$. A polarization $\mathfrak{l}$ at g is called $\mathfrak{n}$-*admissible* if $\mathfrak{l} \cap \mathfrak{n}$ is a polarization at f (see [1]).

In the sequel, the number $r = \operatorname{Inf}_{g \in \mathfrak{g}^*} \dim \mathfrak{g}(g)$ will be called the *rank* of $\mathfrak{g}$.

I.2. If $\mathfrak{g}$ is solvable, it is well known that $SP(g, \mathfrak{g})$ is never empty. In the semi-simple case, the situation is simple too.

* I am glad to acknowledge my debt to J. Dixmier who called this problem to my attention.

Lemma I.1. *Let* $\mathfrak{g}$ *be a semisimple Lie algebra. Then* $SP(g, \mathfrak{g}) \neq \emptyset$ *iff* $\dim \mathfrak{g}(g) = r$, *and in this case, the elements of* $SP(g, \mathfrak{g})$ *are Borel subalgebras.*

Proof. Put $n = \dim \mathfrak{g}$ and $r =$ rank of $\mathfrak{g}$. A solvable subalgebra of $\mathfrak{g}$ has dimension less than or equal to $\frac{1}{2}(n+r)$, and is $\frac{1}{2}(n+r)$ iff it is a Borel subalgebra. On the other hand, if $g \in \mathfrak{g}^*$, then $\dim \mathfrak{g}(g) \geqq r$; thus the dimension of $\mathfrak{l} \in SP(g, \mathfrak{g})$ is at least $\frac{1}{2}(n+r)$. This proves that if $SP(g, \mathfrak{g}) \neq \emptyset$, then $\dim \mathfrak{g}(g)$ is equal to r and that the elements of $SP(g, \mathfrak{g})$, are Borel subalgebras.

We introduce now the space E of pairs $(g, \mathfrak{h})$, where $g \in \mathfrak{g}^*$ and where $\mathfrak{h}$ is a Borel subalgebra such that $\langle g, [\mathfrak{h}, \mathfrak{h}] \rangle = 0$. Denote by G the space of the $\frac{1}{2}(n+r)$-dimensional subspaces of $\mathfrak{g}$. Then E is a (Zariski) closed subspace of $\mathfrak{g}^* \times G$, and, since G is a complete algebraic manifold, the image V of E in $\mathfrak{g}^*$ is closed. It is easy to see that V contains all $g \in \mathfrak{g}^*$ such that $\mathfrak{g}(g)$ is a Cartan subalgebra of $\mathfrak{g}$, so that in fact $V = \mathfrak{g}^*$. In particular, if $\dim \mathfrak{g}(g) = r$, then $SP(g, \mathfrak{g}) \neq \emptyset$. We shall determine $SP(g, \mathfrak{g})$ explicitly in paragraph III.5.

I.3. These two examples suggest the following result, which (for $\mathfrak{a} = 0$) was proved independently by J. Dixmier [8].

Theorem I.1. *Let* $\mathfrak{a}$ *be a solvable ideal in* $\mathfrak{g}$ *and let* $f \in \mathfrak{a}^*$. *Let* $g \in \mathfrak{g}^*$ *be such that* $g|\mathfrak{a} = f$ *and* $\dim \mathfrak{g}(g) \leqq \dim \mathfrak{g}(g')$ *for every* $g' \in \mathfrak{g}^*$ *such that* $g'|\mathfrak{a} = f$. *Then there exists* $\mathfrak{l} \in SP(g, \mathfrak{g})$.

Proof. The proof is by induction. The theorem is true if $\dim \mathfrak{g} = 1$, so we suppose $\dim \mathfrak{g} > 1$ and the theorem proved in lower dimensional cases. Let $\mathfrak{r}$ be the radical of $\mathfrak{g}$, and put $f' = g|\mathfrak{r}$. Two cases may occur.

I. Suppose that $\mathfrak{g}(f') = \mathfrak{g}$. Suppose moreover that $\ker f' \neq 0$. Then $\ker f'$ is an ideal, and we may apply the induction hypothesis to $\mathfrak{g}/\ker f'$. Going back to $\mathfrak{g}$, we obtain an element $\mathfrak{l} \in SP(g, \mathfrak{g})$.

Suppose now that $\ker f' = 0$. Then $\mathfrak{r}$ is in the center of $\mathfrak{g}$ and $\dim \mathfrak{r} \leqq 1$. Thus $\mathfrak{g}$ is reductive or semisimple. An element $\mathfrak{l} \in SP(g, \mathfrak{g})$ is obtained using Lemma I.1.

II. Suppose that $\mathfrak{g}(f') \neq \mathfrak{g}$. Put $\mathfrak{g}_1 = \mathfrak{g}(f')$ and $g_1 = g|\mathfrak{g}_1$. Then, as remarked in [1], we have:

$$\mathfrak{g}_1(g_1) = \mathfrak{g}(g) + \mathfrak{r}(f') \tag{1}$$

$$\mathfrak{r}(f') \cap \mathfrak{g}(g) = \{X \in \mathfrak{r} | \langle f', [Y, Y] \rangle = 0 \quad \text{for all} \quad Y \in \mathfrak{g}\} \tag{2}$$

Since by (2), $\mathfrak{r}(f') \cap \mathfrak{g}(g)$ depends only on f', it follows from (1) that $\dim \mathfrak{g}_1(g_1) \leqq$ $\leqq \dim \mathfrak{g}_1(g_1')$ for all $g_1' \in \mathfrak{g}_1^*$ such that $g_1'|\mathfrak{r}(f') = f'|\mathfrak{r}(f')$. In fact, let $g' \in \mathfrak{g}^*$ be such that $g'|\mathfrak{r} = f'$ and $g'|\mathfrak{g}_1 = g_1'$. In particular we get $g'|\mathfrak{a} = f$ and then, by hypothesis we have $\dim \mathfrak{g}(g) \leqq \dim \mathfrak{g}(g')$.

We now apply the induction hypothesis to get an algebra $\mathfrak{l}_1 \in SP(g_1, \mathfrak{g}_1)$. Since $\mathfrak{l}_1$ is solvable, there exists a polarization $\mathfrak{h} \in SP(f', \mathfrak{r})$ which is stable under the action of $\mathfrak{l}_1$. For instance, let

$$0 = \mathfrak{r}_0 \subset \mathfrak{r}_1 \subset \ldots \subset \mathfrak{r}_N = \mathfrak{r}$$

be a Jordan—Hölder sequence of $\mathfrak{r}$ which is stable under the action of $\mathfrak{l}_1$. Put $f_1'=f'|\mathfrak{r}_i$ for $i=1, \ldots, n$. Then the subspace $\mathfrak{h}=\sum \mathfrak{r}_i(f_i')$ is such an algebra [17].

Consider now the solvable algebra $\mathfrak{l}=\mathfrak{l}_1+\mathfrak{h}$. Then $\mathfrak{l}\in SP(g, \mathfrak{g})$ (see [1]).

I.4. This paragraph will not be used in the sequel. It is a remark on the condition imposed on g in Theorem I.1, which generalizes the Theorem of [12].

Proposition I.1. *Let V be a subspace of* $\mathfrak{g}$ *and let $f\in V^*$. Let g be an element of* $\mathfrak{g}^*$ *such that $g|V=f$ and such that* $\dim \mathfrak{g}(g) \leqq \dim \mathfrak{g}(g')$ *for all $g'\in\mathfrak{g}^*$ such that $g'|V=f$. Then* $[\mathfrak{g}(g), \mathfrak{g}(g)]\subset V$.

Proof (J. Carmona). Fix a subspace $W\subset\mathfrak{g}$ such that $\mathfrak{g}=W\oplus\mathfrak{g}(g)$. Fix $h\in\mathfrak{g}^*$ such that $h|V=0$. The set of $t\in k$ such that

$$\mathfrak{g} = W\oplus\mathfrak{g}(g+th) \tag{1}$$

is (because of the condition on the dimension of $\mathfrak{g}(g)$) a Zariski open subset of k which contains 0. Let X and Y belong to $\mathfrak{g}(g)$. Denote by $X(t)$ and $Y(t)$ the projections of X and Y on $\mathfrak{g}(g+th)$ obtained from (1). They are rational functions of t, and we have:

$$\langle g+th, [X(t), Y(t)]\rangle = 0. \tag{2}$$

Taking the derivative of (2) at 0, we get:

$$\langle h, [X, Y]\rangle+\langle g, [X, Y'(0)]\rangle+\langle g, [X'(0), Y]\rangle = 0.$$

The last two terms are 0 because X and Y belong to $\mathfrak{g}(g)$. We get: $\langle h, [X, Y]\rangle=0$. This proves the proposition, since h is arbitrary in the orthogonal complement of V in $\mathfrak{g}^*$.

The proof also works if k is merely an infinite field.

II. Construction of primitive ideals: statement of the results

II.1. We shall use induced representations of Lie algebras, with Dixmier's definition [6]. Let τ be a representation of a subalgebra $\mathfrak{h}$ of $\mathfrak{g}$ in a vector space M. The linear form $\frac{1}{2}\,\mathrm{tr}_{\mathfrak{g}/\mathfrak{h}}$ on $\mathfrak{h}$ is a one-dimensional representation, and we denote by M' the $\mathfrak{h}$-module associated with $\tau'=\tau\otimes\frac{1}{2}\,\mathrm{tr}_{\mathfrak{g}/\mathfrak{h}}$. Then the induced representation $\mathrm{Ind}(\tau, \mathfrak{g})$ of $\mathfrak{g}$ is the natural representation in the tensor product of $U(\mathfrak{g})$ and M' over $U(\mathfrak{h})$.

Suppose now that $g\in\mathfrak{g}^*$ is such that $g|\mathfrak{h}$ is a representation of $\mathfrak{h}$. Then the representation $\mathrm{Ind}(g|\mathfrak{h}, \mathfrak{g})$ is defined. Its kernel in $U(\mathfrak{g})$ is a two-sided ideal which we shall denote by $I_{g,\mathfrak{h}}$.

II.2. We shall denote by $\mathfrak{g}_0^*$ the set of all $g\in\mathfrak{g}^*$ such that $SP(g, \mathfrak{g})$ is not empty. By Theorem I.1, $\mathfrak{g}_0^*$ contains a non-empty open subset of $\mathfrak{g}^*$. The sections III and IV of this paper will be devoted to the proof of the following theorems.

Theorem II.1. *Let* $g \in \mathfrak{g}^*$. *Then:*

(i) *the ideal* $I_{g,\mathfrak{l}}$ *does not depend on* $\mathfrak{l} \in SP(g, \mathfrak{g})$.

(ii) *there exists* $\mathfrak{l} \in SP(g, \mathfrak{g})$ *such that* $\mathrm{Ind}\,(g|\mathfrak{l}, \mathfrak{g})$ *is irreducible.*

When $\mathfrak{g}$ is solvable, Theorem II.1 reduces to Theorems 7.5 and 7.9 of [6].

When $\mathfrak{l} \in SP(g, \mathfrak{g})$ we shall put $I_g = I_{g,\mathfrak{l}}$. Notice that, by (ii), I_g is a primitive ideal of $U(\mathfrak{g})$.

Theorem II.2. *Let P be a non-empty Zariski open subset of* $\mathfrak{g}^*$ *contained in* $\mathfrak{g}_0^*$. *Then*

$$\bigcap_{g \in P} I_g = 0.$$

There are other formulations of theorem II.2; namely: the set of ideals I_g with $g \in P$ is dense in the set of all primitive ideals; or: the set of irreducible representations of the form $\mathrm{Ind}\,(g|\mathfrak{l}, \mathfrak{g})$ with $g \in P$ and $\mathfrak{l} \in SP(g, \mathfrak{g})$ is complete.

We shall prove a theorem stronger than Theorem II.2 in section *V*.

III. Proof of theorems in the semisimple case

III.1. In this section $\mathfrak{g}$ is a semisimple Lie algebra with Killing form B. We choose a Cartan subalgebra $\mathfrak{t}$, and a positive system of roots Δ^+ of $\mathfrak{g}$ with respect to $\mathfrak{t}$. If $\alpha \in \mathfrak{t}^*$ is a root, we denote by $\mathfrak{g}^\alpha$ the corresponding subspace of $\mathfrak{g}$, and by H_α the corresponding element of $\mathfrak{t}$ (such that $\alpha(H_\alpha) = 2$). Put:

$$\mathfrak{n} = \sum_{\alpha \in \Delta^+} \mathfrak{g}^\alpha, \quad \mathfrak{n}^- = \sum_{-\alpha \in \Delta^+} \mathfrak{g}^\alpha$$

$$\mathfrak{b} = \mathfrak{t} + \mathfrak{n} \quad \text{and} \quad \mathfrak{b}^- = \mathfrak{t} + \mathfrak{n}^-.$$

Denote by ϱ half the sum of positive roots.

We shall identify $\mathfrak{t}^*$ and the orthogonal of $\mathfrak{n} + \mathfrak{n}^-$ in $\mathfrak{g}^*$. For instance $\varrho(H) = \frac{1}{2}\,\mathrm{tr}_{\mathfrak{g}/\mathfrak{b}}(H)$ if $H \in \mathfrak{b}$. If $g \in \mathfrak{t}^*$, the representation $\tau_g = \mathrm{Ind}\,(g|\mathfrak{b}, \mathfrak{g})$ is defined. We shall study the kernel J_g of τ_g in $U(\mathfrak{g})$.

We shall use the following result of D.–N. Verma [18].

Theorem III.1. *The representation* τ_g *is not irreducible iff there exists a positive root* α *such that* $g(H_\alpha)$ *is a strictly positive integer.*

III.2. It is well known that to each $g \in \mathfrak{t}^*$ is associated a character χ_g of the center $Z(\mathfrak{g})$ of $U(\mathfrak{g})$. In fact, if $u \in Z(\mathfrak{g})$, denote by u' the element of $U(\mathfrak{t})$ such that $u - u' \in U(\mathfrak{g})\mathfrak{n}$. Then $U(\mathfrak{t})$ is equal to the symmetric algebra $S(\mathfrak{t})$ and the scalar $\chi_g(u)$ is defined by the formula $\chi_g(u) = u'(g - \varrho)$. One knows that $\chi_g = \chi_{g'}$ iff g and g' are conjugate under the Weyl group (see [16]).

It is obvious that $\tau_g(u) = \chi_g(u)$ if u belongs to $Z(\mathfrak{g})$.

It follows that $J_g \cap Z(\mathfrak{g}) = J_{g'} \cap Z(\mathfrak{g})$ iff g and g' are conjugate under the Weyl group. Our aim is to prove that in this case we have also $J_g = J_{g'}$. We shall first need some lemmas.

III.3. Consider in the symmetric algebra $S(\mathfrak{g})$ the subspace H of harmonic polynomials on $\mathfrak{g}^*$, defined in [14]. If $x \in S(\mathfrak{g})$, there exists a unique $x_0 \in S(\mathfrak{n}^-)$ such that $x - x_0 \in S(\mathfrak{g})\mathfrak{b}$.

Lemma III.1. *Let x be a non-zero $\mathfrak{n}^-$-invariant element of H, then $x_0 \neq 0$.*

Proof. Suppose that x is a $\mathfrak{n}^-$-invariant element of H, and suppose moreover that $x \in S(\mathfrak{g})\mathfrak{b}$ (this means that x is zero on the subspace $\mathfrak{b}^\perp$ of $\mathfrak{g}^*$ orthogonal to $\mathfrak{b}$). Then we must show that $x=0$. We denote by V the algebraic manifold of principal nilpotent elements of $\mathfrak{g}^*$ (we identify $\mathfrak{g}$ and $\mathfrak{g}^*$ by means of the Killing form B). It is irreducible of dimension $n-r$ (see [14]). In the identification of $\mathfrak{g}$ to $\mathfrak{g}^*$, the space $\mathfrak{b}^\perp$ corresponds to $\mathfrak{n}$. Thus $V \cap \mathfrak{b}^\perp$ is not empty and we choose $f_0 \in V \cap \mathfrak{b}^\perp$. Denote by N^- the group corresponding to $\mathfrak{n}^-$. Then x is zero on the subset $N^-(V \cap \mathfrak{b}^\perp)$ of V. We claim that x is zero on V. By classical algebraic geometry, it is enough to prove that the mapping

$$(n, f) \rightarrow nf$$

from $N^- \times (V \cap \mathfrak{b}^\perp)$ to V is étale at the point $(1, f_0)$. Its differential at this point is the mapping

$$(X, f) \rightarrow Xf_0 + f$$

from $\mathfrak{n}^- \times \mathfrak{b}^\perp$ into $\mathfrak{g}^*$. By reason of dimension, we have just to prove that it is injective. For this, remark first that $\mathfrak{b}$ belongs to $SP(f_0, \mathfrak{g})$. Indeed $\mathfrak{b}$ has the right dimension and we have $\langle f_0, [\mathfrak{b}, \mathfrak{b}] \rangle = 0$. Suppose that the pair $(X, f) \in \mathfrak{n}^- \times \mathfrak{b}$ is such that $Xf_0 + f = 0$. For all $Y \in \mathfrak{b}$, we have:

$$0 = \langle Xf_0 + f, Y \rangle = \langle Xf_0, Y \rangle = -\langle f_0, [X, Y] \rangle,$$

from which it follows that $X \in \mathfrak{b}$. Since $X \in \mathfrak{n}^-$, we get $X=0$ and $f=0$.

Since x is zero on V and since it is harmonic, we obtain that $x=0$ by Kostant's theorem [14].

We shall denote by β the classical Poincaré—Birkhoff—Witt linear isomorphism from $S(\mathfrak{g})$ to $U(\mathfrak{g})$. Let n be a positive integer. We denote by $U_n(\mathfrak{g})$ the space of elements of $U(\mathfrak{g})$ with graduation $\leqq n$. We shall put $K = \beta(H)$. This is the space of harmonic elements of $U(\mathfrak{g})$ (see [14]). If $u \in U(\mathfrak{g})$, we denote by u_0 the element of $U(\mathfrak{n}^-)$ such that $u - u_0 \in U(\mathfrak{g})\mathfrak{b}$.

Lemma III.2. *Suppose u is a $\mathfrak{n}^-$-invariant element in K, and let n be a positive integer. If $u \notin U_n(\mathfrak{g})$, then $u_0 \notin U_n(\mathfrak{n}^-)$.*

Proof. Write $u = \beta(x)$, where x is a $\mathfrak{n}^-$-invariant element in H. Let m be the degree of x; then $m > n$. Let x_m be the homogeneous part of degree m of x; then x_m is a non zero $\mathfrak{n}^-$-invariant element in H. Hence, by Lemma III.1, $x_{m,0}$ is not zero. From this it follows that x_0 is of degree m, and that $\beta(x_0)$ does not belong to $U_{m-1}(\mathfrak{n}^-)$. Moreover, we have:

$$\beta(x)-\beta(x_0)\in U(\mathfrak{g})\mathfrak{b}+U_{m-1}(\mathfrak{g}).$$

We thus obtain: $u_0-\beta(x_0)\in U_{m-1}(\mathfrak{n}^-)$. This proves that u_0 does not belong to $U_{m-1}(\mathfrak{n}^-)$, and so, does not belong to $U_n(\mathfrak{n}^-)$.

III.4. We prove now the main result of this section.

Theorem III.2. *Let* $g\in\mathfrak{t}^*$. *Then* $J_g=U(\mathfrak{g})(Z(\mathfrak{g})\cap J_g)$.

In other words, the kernel in $U(\mathfrak{g})$ *of a representation of* $\mathfrak{g}$ *induced by a one-dimensional representation of a Borel subalgebra is generated by its intersection with* $Z(\mathfrak{g})$.

Proof. Let, as in paragraph II.1, M' be the one-dimensional $U(\mathfrak{b})$-module associated with $(g-\varrho)|\mathfrak{b}$. Choose $m\neq 0$ in M'. Denote by W the space $U(\mathfrak{g})\underset{U(\mathfrak{b})}{\otimes} M'$ of the representation τ_g. Let n be a positive integer. We write m for the element $1\otimes m$ of W, and we put $W_n=U_n(\mathfrak{g})m$. The mapping

$$u\to um$$

from $U(\mathfrak{g})$ into W induces an isomorphism of $U_n(\mathfrak{n}^-)$ onto W_n.

We choose a direct decomposition $H=\oplus H_i$ of H such that each H_i is an irreducible $\mathfrak{g}$-invariant subspace of H, of homogeneous degree n_i. Put $K_i=\beta(H_i)$. Choose in K_i a non-zero element k_i which is $\mathfrak{n}^-$-invariant. Such an element exists and is unique up to multiplication by a scalar. We claim that if $u\in U(\mathfrak{g})$ is $\mathfrak{n}^-$-invariant, it can be written in the form:

$$u=\sum k_i z_i \tag{1}$$

with elements z_i belonging to $Z(\mathfrak{g})$. One knows (see [14]) that $U(\mathfrak{g})$ is isomorphic (as a $Z(\mathfrak{g})$-module) to the tensor product $K\otimes Z(\mathfrak{g})$. In particular the elements z_i of (1) are uniquely determined. To prove (1), let (y_j) be a basis for $Z(\mathfrak{g})$. We can find elements $h_j\in K$ such that $u=\sum h_j y_j$. If $Y\in\mathfrak{n}^-$, we have:

$$[Y,u]=0=\sum [Y,h_j]y_j.$$

It follows from the result recalled above that $[Y,h_j]=0$ for all j. Then h_j is a $\mathfrak{n}^-$-invariant element in K. So it is a linear combination of elements k_i, which proves (1).

We write $J'_g=U(\mathfrak{g})(Z(\mathfrak{g})\cap J_g)$. We shall suppose that $J'_g\neq J_g$ and derive a contradiction. In fact J'_g and J_g are $\mathfrak{g}$-invariant subspaces of $U(\mathfrak{g})$. Since every finite-dimensional representation of $\mathfrak{g}$ is semisimple, J_g contains a non-zero finite-dimensional $\mathfrak{g}$-invariant subspace V such that $V\cap J'_g=0$. Then there exists an element $u\in J_g$ which does not belong to J'_g and which is $\mathfrak{n}^-$-invariant. We write $u=\sum k_i z_i$ as in (1). Put $a_i=\chi_g(z_i)$. Since u does not belong to J'_g there is at least one i such that $a_i\neq 0$. Denote by n the largest integer n_i such that $a_i\neq 0$.

Put $v = \sum_{n=n_i} a_i k_i$. Then v is a $\mathfrak{n}^-$-invariant harmonic element. Moreover, v does not belong to $U_{n-1}(\mathfrak{g})$ because $\beta^{-1}(v)$ is a non zero polynomial of homogeneous degree n. Choose a basis $b_0, b_1, \ldots, b_j, \ldots$ of $U(\mathfrak{b})$ such that $b_0 = 1$ and such that $b_j \in U(\mathfrak{b})\mathfrak{b}$ if $j \geqq 1$. Write $v = \sum v_j b_j$ where the elements v_j are in $U(\mathfrak{n}^-)$. Then, by Lemma III.2, v_0 does not belong to $U_{n-1}(\mathfrak{n}^-)$. It is clear that v_j belongs to $U_{n-1}(\mathfrak{n}^-)$ if $j \geqq 1$. The character $g - \varrho$ of $\mathfrak{b}$ defines a character of $U(\mathfrak{b})$ whose value at b_j we shall denote by c_j.

Since u belongs to J_g, we have $um = 0$. On the other hand:

$$um = \sum k_i z_i m = \sum a_i k_i m = vm \,(\mathrm{mod}\, W_{n-1}).$$

Let us compute vm:

$$vm = \sum v_j b_j m = \sum c_j v_j m = v_0 m \,(\mathrm{mod}\, W_{n-1}).$$

It then follows that $v_0 m$ belongs to W_{n-1}. But this is not possible because v_0 does not belong to $U_{n-1}(\mathfrak{n}^-)$. Theorem III.2 is proved.

Corollary III.1. (i) *Let* $g \in \mathfrak{t}^*$ *and* $g' \in \mathfrak{t}^*$. *Then* $J_g = J_{g'}$ *iff* g *and* g' *are conjugate under the Weyl group.*

(ii) *The ideal* J_g *is primitive.*

Part (ii) implies that the two-sided ideals of $U(\mathfrak{g})$ generated by the maximal ideals of $Z(\mathfrak{g})$ are primitive. This has been proved independently by J. Dixmier [7] by a different method.

Proof. (i) It is clear from Theorem III.2. (ii) It follows from (i) because, by Theorem III.1, every orbit of the Weyl group in $\mathfrak{t}^*$ contains an element g such that the representation $\mathrm{Ind}\,(g|\mathfrak{b}, \mathfrak{g})$ is irreducible.

III.5. Proof of Theorem II.1 in the semisimple case.

Let $g \in \mathfrak{g}^*$ be such that $\dim \mathfrak{g}(g) = r$. We shall first determine $SP(g, \mathfrak{g})$. Denote by Y the element of $\mathfrak{g}$ such that $B(Y, X) = \langle g, X \rangle$ for all $X \in \mathfrak{g}$. We write $Y = Y' + Y''$, where Y' is the semisimple part and Y'' the nilpotent part of Y. Let $\mathfrak{b} \in SP(g, \mathfrak{g})$. Then $\mathfrak{b}$ contains $\mathfrak{g}(g)$, which is equal to the stabilizer of Y in $\mathfrak{g}$. In particular, $\mathfrak{b}$ contains Y' and Y''. Denote by $\mathfrak{g}_1$ the stabilizer of Y' in $\mathfrak{g}$ and by $\mathfrak{m}$ the orthogonal of $\mathfrak{g}_1$ in $\mathfrak{g}$ with respect to B. Since $\mathfrak{b}$ contains Y' we have: $\mathfrak{b} = (\mathfrak{b} \cap \mathfrak{g}_1) \oplus (\mathfrak{b} \cap \mathfrak{m})$. Then $\mathfrak{b} \cap \mathfrak{g}_1$ is a Borel subalgebra of the reductive algebra $\mathfrak{g}_1$. Moreover $\mathfrak{b} \cap \mathfrak{g}_1$ contains Y'' which is a principal nilpotent element in $\mathfrak{g}_1$ (see [14]). It follows that $\mathfrak{b} \cap \mathfrak{g}_1 = \mathfrak{b}_1$, where $\mathfrak{b}_1$ is the unique Borel subalgebra of $\mathfrak{g}_1$ which contains Y''. We choose a Cartan subalgebra $\mathfrak{t}$ of $\mathfrak{g}_1$ contained in $\mathfrak{b}_1$. Then $\mathfrak{t}$ is a Cartan subalgebra of $\mathfrak{g}$. It is easy to verify that the elements of $SP(g, \mathfrak{g})$ are the Borel subalgebras of $\mathfrak{g}$ which contain $\mathfrak{b}_1$. In particular, they contain $\mathfrak{t}$.

Choose $\mathfrak{b}$ and $\mathfrak{b}'$ in $SP(g, \mathfrak{g})$. Then there exists an element w in the Weyl group of $\mathfrak{g}$ (with respect to $\mathfrak{t}$) such that $w^{-1}(\mathfrak{b}) = \mathfrak{b}'$. It is clear that $I_{g,\mathfrak{b}'} = I_{w(g),\mathfrak{b}}$. Moreover, if we put $g' = g|\mathfrak{t}$ and $w(g)' = w(g)|\mathfrak{t}$, we have: $w(g)' = w(g')$. Remark that

$I_{g,\mathfrak{b}}$ is the same as the ideal $J_{g'}$ of paragraph III.1. We have also: $I_{w(g),\mathfrak{b}}=J_{w(g)'}$. It then follows from Corollary III.1 that $I_{g,\mathfrak{b}}=I_{g,\mathfrak{b}'}$ which proves the first part of Theorem II.1.

To prove the second part, we have to find some $\mathfrak{b}\in SP(g,\mathfrak{g})$ such that Ind $(g|\mathfrak{b},\mathfrak{g})$ is irreducible. Denote by Δ the set of roots of $\mathfrak{g}$, by Δ_1 the set of roots of $\mathfrak{g}_1$, and by $\Delta_1^+\subset\Delta_1$ the set of roots contained in $\mathfrak{b}_1$. Let $\alpha\in\Delta$. Then $\alpha\in\Delta_1$ iff $\alpha(Y')=0$. We can choose an element $H\in\mathfrak{t}$ such that: 1. $\alpha(H)\in Q$ for every $\alpha\in\Delta$, 2. $\alpha(H)=$ $=\alpha(Y')$ whenever $\alpha(Y')\in Q$, 3. $\alpha(H)=0$ iff $\alpha\in\Delta_1$. Define a subset Δ^+ of Δ in the following way: the root α belongs to Δ^+ if $\alpha\in\Delta_1^+$ or if $\alpha(H)<0$. The set Δ^+ is a positive system of roots and if $\alpha\in\Delta^+$, then we have not $\alpha(Y')>0$. The Borel subalgebra $\mathfrak{b}$ corresponding to Δ^+ contains $\mathfrak{b}_1$ and thus belongs to $SP(g,\mathfrak{g})$. By Theorem III.1 the representation Ind $(g|\mathfrak{b},\mathfrak{g})$ is irreducible. This completes the proof of Theorem II.1 in the semisimple case.

III.6. Proof of Theorem II.2 in the semisimple case.

Let G be the adjoint group of $\mathfrak{g}$. Let P be a non-empty open subset of $\mathfrak{g}_0^*$. Since $I_{h(g)}=I_g$ if $g\in\mathfrak{g}_0^*$ and $h\in G$, we may suppose that P is G-invariant. Then the set $Q=P\cap\mathfrak{t}^*$ is non-empty. With the notations of paragraph III.1 we have to prove

$$\bigcap_{g\in Q} J_g = 0. \tag{1}$$

Let u be an element of the left hand side of (1). Let (h_i) be a basis for the space K of harmonic elements. Write $u=\sum h_i z_i$ with elements z_i in $Z(\mathfrak{g})$. It follows from Theorem III.2 that $\chi_g(z_i)=0$ for every i and every $g\in Q$. Since Q is Zariski dense in $\mathfrak{t}^*$, it then follows that $z_i=0$ for every i. This is well known (see [16]). It is also a particular case of Lemma III.4 below. We get $u=0$, as was to be proved.

We shall now present a more elementary proof of Theorem II.2, using neither Kostant's theorems about harmonic elements, nor finite-dimensional representations.

III.7. Second proof of Theorem II.2 in the semisimple case.

Let us prove first two lemmas. Let $x\in S(\mathfrak{g})$. Denote by x_1 the element of $S(\mathfrak{b}^-)$ such that $x-x_1\in S(\mathfrak{g})\mathfrak{n}$.

Lemma III.3. *Suppose that x is a non zero $\mathfrak{n}^-$-invariant element of $S(\mathfrak{g})$, then $x_1\neq 0$.*

Proof. We have to prove that a $\mathfrak{n}^-$-invariant polynomial x contained in $S(\mathfrak{g})\mathfrak{n}$ is 0. Denote by N^- the group corresponding to $\mathfrak{n}^-$, and by $\mathfrak{n}^\perp$ the orthogonal complement of $\mathfrak{n}$ in $\mathfrak{g}^*$. Then x is zero on $N^-(\mathfrak{n}^\perp)$. To prove that $x=0$, it remains to prove that the mapping

$$(n,f)\to nf$$

from $N^-\times\mathfrak{n}$ into $\mathfrak{g}^*$ is étale at the point $(1,f_0)$ for at least one element f_0 of $\mathfrak{n}^\perp$. As usual, we identify $\mathfrak{g}$ and $\mathfrak{g}^*$. In this identification, $\mathfrak{b}$ corresponds to $\mathfrak{n}^\perp$. Choose a regular element H in $\mathfrak{t}$. It is easily seen that the differential of the mapping

$$(n, X) \to nX$$

from $N^- \times \mathfrak{b}$ into $\mathfrak{g}$ at the point $(1, H)$ is surjective. This proves our claim.

Now, let $u \in U(\mathfrak{g})$. Denote by u_1 the element of $U(\mathfrak{b}^-)$ such that $u - u_1 \in U(\mathfrak{g})\mathfrak{n}$. The next lemma follows from Lemma III.3 exactly as Lemma III.2 follows from Lemma III.1.

Lemma III.4. *Suppose u is a $\mathfrak{n}^-$-invariant element of $U(\mathfrak{g})$ and let n be a positive integer such that u does not belong to $U_n(\mathfrak{g})$. Then u_1 does not belong to $U_n(\mathfrak{b}^-)$.*

Remark that this proves in an elementary way a lemma of D.-N. Verma ([18], Lemma 3.6)

Corollary III.2. *Let $u \in U(\mathfrak{n})$ be a $\mathfrak{n}^-$-invariant element. Then u is a scalar.*

We come now to the proof of Theorem II.2. As in paragraph III.6, we are led to prove (1). Suppose that the space $\bigcap_{g \in Q} J_g$ is not zero. Since it is a $\mathfrak{g}$-invariant subspace of $U(\mathfrak{g})$, it contains a non zero $\mathfrak{n}^-$-invariant element u. By Lemma III.4, u_1 is not zero. Let (n_i) be a basis for $U(\mathfrak{n}^-)$. We write $u_1 = \sum n_i t_i$ with elements t_i in $U(\mathfrak{t}) = S(\mathfrak{t})$. There exists i and $g \in Q$ such that $t_i(g-\varrho) \neq 0$. Put $n = \sum t_i(g-\varrho) n_i$. Then n is a non zero element in $U(\mathfrak{n}^-)$. Consider, as in the beginning of the proof of Theorem III.2, the vector m in the space of the representation Ind $(g|\mathfrak{b}, \mathfrak{g})$. Then:

$$um = u_1 m = \sum t_i(g-\varrho) n_i m = nm.$$

Since $n \neq 0$, we have $nm \neq 0$. We obtain that u does not belong to J_g. We thus get a contradiction and Theorem II.2 is proved.

IV. Proof of the theorems in the general case

IV.1. Proof of Theorem II.1.

The proof is by induction. The theorem is obvious if $\dim \mathfrak{g} = 1$. We suppose that $\dim \mathfrak{g} > 1$, and the theorem established in lower dimensional cases. We fix $g \in \mathfrak{g}_0^*$. We denote by $\mathfrak{n}$ the maximal nilpotent ideal of $\mathfrak{g}$. We put $f = g|\mathfrak{n}$ and $\mathfrak{g}_1 = \mathfrak{g}(f)$. We consider three cases.

I. Suppose that $\mathfrak{g} = \mathfrak{g}_1$ and that $\ker f = 0$. The theorem has been proved in the semisimple case (paragraph III.5). The reductive case is easily deduced from the semisimple case.

II. Suppose that $\mathfrak{g} = \mathfrak{g}_1$ and that $\ker f \neq 0$. Then $\ker f$ is an ideal in $\mathfrak{g}$ and we may apply the induction hypothesis to $\mathfrak{g}/\ker f$ and to the linear form on this algebra deduced from g. In this way, we get the theorem for $\mathfrak{g}$.

III. Suppose that $\mathfrak{g} \neq \mathfrak{g}_1$. Put $g_1 = g|\mathfrak{g}_1$. We shall show that $g_1 \in \mathfrak{g}_{1,0}^*$. We shall deduce the theorem by application of the induction hypothesis to the pair $\mathfrak{g}_1$, g_1.

Choose $\mathfrak{l} \in SP(g, \mathfrak{g})$. The algebra $\mathfrak{k} = \mathfrak{l} + \mathfrak{n}$ is solvable, and, if we put $g' = g|\mathfrak{k}$, the algebra $\mathfrak{l}$ belongs to $SP(g', \mathfrak{k})$. Conversely, choose $\mathfrak{l}' \in SP(g', \mathfrak{k})$. Then $\mathfrak{l}'$ belongs to $SP(g, \mathfrak{g})$ by reason of dimension. Moreover, since $\mathfrak{k}$ is solvable, we may apply Theorem 7.9 of [6]. Thus, the representations Ind $(g'|\mathfrak{l}, \mathfrak{k})$ and Ind $(g'|\mathfrak{l}', \mathfrak{k})$ have the same kernel in $U(\mathfrak{k})$. It follows that $I_{g,\mathfrak{l}} = I_{g,\mathfrak{l}'}$.

We know that there exist polarizations in $SP(g', \mathfrak{k})$ which are $\mathfrak{n}$-admissible (see [17], for instance). This proves that $SP(g, \mathfrak{g})$ contains $\mathfrak{n}$-admissible polarizations. Moreover, to prove Theorem II.1, it is enough to prove the following assertions:

(i) If $\mathfrak{l}$ and $\mathfrak{l}'$ are two $\mathfrak{n}$-admissible polarizations in $SP(g, \mathfrak{g})$, then $I_{g,\mathfrak{l}} = I_{g,\mathfrak{l}'}$.

(ii) There exists a $\mathfrak{n}$-admissible polarization $\mathfrak{l} \in SP(g, \mathfrak{g})$ such that the representation Ind $(g|\mathfrak{l}, \mathfrak{g})$ is irreducible.

Let $\mathfrak{l}$ be a $\mathfrak{n}$-admissible polarization in $SP(g, \mathfrak{g})$. Put $\mathfrak{h} = \mathfrak{l} \cap \mathfrak{n}$ and $\mathfrak{l}_1 = \mathfrak{l} \cap \mathfrak{g}_1$. Then $\mathfrak{l}_1$ belongs to $SP(g_1, \mathfrak{g}_1)$ and $\mathfrak{l} = \mathfrak{h} + \mathfrak{l}_1$ (see [1]). In particular, g_1 belongs to $\mathfrak{g}^*_{1,0}$. Now, by Theorems I.6.1 and I.5.3 of [10], we know that the kernel of Ind $(g|\mathfrak{l}, \mathfrak{g})$ in $U(\mathfrak{g})$ depends only on the kernel of Ind $(g_1|\mathfrak{l}_1, \mathfrak{g}_1)$ in $U(\mathfrak{g}_1)$, and that Ind $(g|\mathfrak{l}, \mathfrak{g})$ is irreducible iff Ind $(g_1|\mathfrak{l}_1, \mathfrak{g}_1)$ is also irreducible. The two assertions above follow from the induction hypothesis applied to $\mathfrak{g}_1$.

IV.2. To prove Theorem II.2, we shall need a lemma.

Let $\mathfrak{a}$ be an Abelian ideal in $\mathfrak{g}$ and let $f \in \mathfrak{a}^*$. Denote by $J_{\mathfrak{a},f}$ the left ideal of $U(\mathfrak{g})$ generated by the elements of $U(\mathfrak{a}) = S(\mathfrak{a})$ which are zero on f.

Lemma IV.1. *Let $\mathfrak{a}$ and $\mathfrak{b}$ two Abelian ideals in $\mathfrak{g}$ such that $\mathfrak{b}$ contains $\mathfrak{a}$. Let $f \in \mathfrak{a}^*$ and $u \in U(\mathfrak{g})$ be such that $u \notin J_{\mathfrak{a},f}$. Then there exists $f' \in \mathfrak{b}^*$ such that $f'|\mathfrak{a} = f$ and $u \notin J_{\mathfrak{b},f'}$.*

Proof. Let (u_i) be a basis of the right $U(\mathfrak{b})$-module $U(\mathfrak{g})$. Write $u \in U(\mathfrak{g})$ in the form $u = \sum u_i b_i$ with elements b_i in $U(\mathfrak{b})$. It is easily seen that u belongs to $J_{\mathfrak{b},f'}$ iff $b_i(f') = 0$ for all i.

Choose a subspace $\mathfrak{c}$ such that $\mathfrak{b} = \mathfrak{a} \oplus \mathfrak{c}$. Choose a basis (c_j) of $S(\mathfrak{c})$. Let $u \in U(\mathfrak{g})$ be such that $u \notin J_{\mathfrak{a},f}$. We write $u = \sum u_i c_j a_{ij}$ with elements a_{ij} in $S(\mathfrak{a})$. There exist i_0 and j_0 such that $a_{i_0 j_0}(f) \neq 0$. Put $b_i = \sum_j c_j a_{ij}$. Then $u = \sum u_i b_i$. The element $\sum_j c_j a_{i_0 j}(f)$ of $S(\mathfrak{c})$ is not zero. Choose $f'' \in \mathfrak{c}^*$ such that $\sum_j c_j(f'') a_{i_0 j}(f) \neq 0$. Denote by f' the element of $\mathfrak{b}$ such that $f'|\mathfrak{c} = f''$ and $f''|\mathfrak{a} = f$. Then $b_{i_0}(f') \neq 0$. Thus $u \notin J_{\mathfrak{b},f'}$.

IV.3. For technical reasons, we shall prove more than Theorem II.2.

Theorem IV.1. *Let $\mathfrak{g}$ be a Lie algebra. Let $\mathfrak{a}$ be an Abelian ideal in $\mathfrak{g}$. Let $u \in U(\mathfrak{g})$ and $f \in \mathfrak{a}^*$ be such that $u \notin J_{\mathfrak{a},f}$ (where $J_{\mathfrak{a},f}$ is defined as in paragraph* IV.2*). Let P be a non-empty open subset of the affine space of elements $g \in \mathfrak{g}^*$ such that $g|\mathfrak{a} = f$ and suppose that P is contained in $\mathfrak{g}^*_0$. Then there exists $g \in P$ such that $u \notin I_g$.*

Proof. The proof is by induction. The theorem is obvious if $\dim \mathfrak{g}=1$. We suppose $\dim \mathfrak{g}>1$ and the theorem proved in lower-dimensional cases. We consider five cases.

I. Suppose that $\mathfrak{g}$ is semisimple. The theorem is proved in paragraph III.7.

II. Suppose that $\mathfrak{g}$ is reductive with one-dimensional center $\mathfrak{a}$, and suppose that $\ker f=0$. Then $\mathfrak{g}=\mathfrak{s}\oplus\mathfrak{a}$, where $\mathfrak{s}$ is the semisimple algebra $[\mathfrak{g}, \mathfrak{g}]$. Choose a basis (s_i) for $U(\mathfrak{s})$ and write $u=\sum s_i a_i$ with elements a_i in $S(\mathfrak{a})$. For at least one i we have $a_i(f)\neq 0$ because $u\notin J_{\mathfrak{a},f}$. Then the element $s=\sum a_i(f)s_i$ of $U(\mathfrak{s})$ is not zero. Let $P'\subset\mathfrak{s}^*$ be the set of the restrictions to $\mathfrak{s}$ of the elements of P. Then P' is open in $\mathfrak{s}^*$ and the induction hypothesis provides us with an element $g'\in P'$ such that $s\notin I_{g'}$. Let g be the element of P such that $g|\mathfrak{s}=g'$. I claim that $u\notin I_g$. In fact, choose $\mathfrak{b}$ in $SP(g', \mathfrak{s})$. Then $\mathfrak{l}=\mathfrak{b}+\mathfrak{a}$ belongs to $SP(g, \mathfrak{g})$. The representation $\tau=\operatorname{Ind}(g|\mathfrak{l}, \mathfrak{g})$ is the tensor product of $\tau'=\operatorname{Ind}(g'|\mathfrak{b}, \mathfrak{s})$ and of the representation f of $\mathfrak{a}$. We have:

$$\tau(u) = \sum \tau'(s_i)a_i(f) = \tau'(s) \neq 0.$$

This proves the theorem in this case.

III. Suppose there exists an Abelian ideal $\mathfrak{b}$ in $\mathfrak{g}$ which contains $\mathfrak{a}$ and $f'\in\mathfrak{b}^*$ with the following properties: $f'|\mathfrak{a}=f$; the set P' of elements $g\in P$ such that $g|\mathfrak{b}=f'$ is not empty; $u\notin J_{\mathfrak{b},f'}$; and $\mathfrak{g}(f')\neq\mathfrak{g}$.

Put $\mathfrak{g}(f')=\mathfrak{g}'$. The algebra $\mathfrak{g}'$ contains $\mathfrak{b}$ and we denote by $J'_{\mathfrak{b},f'}$ the left ideal of $U(\mathfrak{g}')$ generated by the elements $b\in U(\mathfrak{b})$ such that $b(f')=0$. We choose a basis (v_i) of the right $U(\mathfrak{g}')$-module $U(\mathfrak{g})$. We write $u=\sum v_i u_i$ with elements $u_i\in U(\mathfrak{g}')$. For at least one i, say $i=i_0$, we have: $u_{i_0}\notin J'_{\mathfrak{b},f'}$, because u does not belong to $J_{\mathfrak{b},f'}$. Denote by Q the set of the restrictions to $\mathfrak{g}'$ of the elements of P'. It is a non-empty open subset in the affine space of $g'\in\mathfrak{g}'^*$ such that $g'|\mathfrak{b}=f'$. It is contained in $\mathfrak{g}_0'^*$. The induction hypothesis provides us with an element $g'\in Q$ such that $u_{i_0}\notin I_{g''}$, where $g''=g'+\frac{1}{2}\operatorname{tr}_{\mathfrak{g}/\mathfrak{g}'}$.

Choose $g\in\mathfrak{g}^*$ such that $g\in P$ and $g|\mathfrak{g}'=g'$. I claim that u does not belong to I_g. In fact, choose $\mathfrak{l}\in SP(g', \mathfrak{g}')$. Then $\mathfrak{l}$ belongs to $SP(g, \mathfrak{g})$. Put $\tau=\operatorname{Ind}(g'|\mathfrak{l}, \mathfrak{g}')$ and $\varrho=\operatorname{Ind}(g|\mathfrak{l}, \mathfrak{g})$. The representation ϱ is equivalent to $\operatorname{Ind}(\tau, \mathfrak{g})$. Let M' be the space of the representation $\tau'=\tau\otimes\frac{1}{2}\operatorname{tr}_{\mathfrak{g}/\mathfrak{g}'}$. The kernel of τ' in $U(\mathfrak{g}')$ is $I_{g''}$. There exists $m\in M'$ such that $\tau'(u_{i_0})m\neq 0$. Then, the vector

$$\varrho(u)m = \sum \varrho(v_i)\tau'(u_i)m$$

is not zero. This proves that $\varrho(u)$ is not zero. Thus u does not belong to I_g.

IV. Suppose there exists an Abelian ideal $\mathfrak{b}$ in $\mathfrak{g}$ which contains $\mathfrak{a}$ and $f'\in\mathfrak{b}^*$ with the following properties: $f'|\mathfrak{a}=f$; the set of elements $g\in P$ such that $g|\mathfrak{b}=f'$ is not empty; $u\notin J_{\mathfrak{b},f'}$; $\mathfrak{g}(f')=\mathfrak{g}$; and $\ker f'\neq 0$.

The space $\ker f'$ is an ideal in $\mathfrak{g}$. Applying the induction hypothesis to $\mathfrak{g}/\ker f'$ it is easy to prove the theorem.

V. Let $\mathfrak{n}$ be the largest nilpotent ideal of $\mathfrak{g}$ and $\mathfrak{z}$ be the center of $\mathfrak{g}$. We suppose that $\mathfrak{a}=\mathfrak{z}$, that $\dim\mathfrak{z}=1$, that $f(\mathfrak{z})\neq 0$, and that $\mathfrak{n}$ is a Heisenberg algebra with center $\mathfrak{z}\neq\mathfrak{n}$.

We choose $f'\in\mathfrak{n}^*$ such that $f'|\mathfrak{z}=f$. Let $z\in\mathfrak{z}$ be such that $f(z)=1$. The primitive ideal $I_{f'}$ of $U(\mathfrak{n})$ associated with f' is generated in $U(\mathfrak{n})$ by $z-1$. (See [5].) The left ideal $J_{\mathfrak{z},f}$ is generated in $U(\mathfrak{g})$ by $z-1$, because an element of $S(\mathfrak{z})$ which is zero on f is divisible by $z-f(z)$. It then follows that u does not belong to $U(\mathfrak{g})I_{f'}$.

We denote by $\mathfrak{s}$ the stabilizer of $\ker f'$ in $\mathfrak{g}$. We consider the semi-direct product $\mathfrak{b}=\mathfrak{s}\times\mathfrak{n}$. The natural map

$$p:\mathfrak{b}\to\mathfrak{g}$$

is surjective, and its kernel is $k(z,-z)$. Let N be the group corresponding to $\mathfrak{n}$. If $g\in\mathfrak{g}_0^*$ and if $n\in N$, we have $I_{n(g)}=I_g$. Thus, to prove the theorem, we may suppose that the set P is N-invariant. It is known that $\mathfrak{g}=\mathfrak{s}+\ker f'$. We identify $\mathfrak{s}^*$ and the orthogonal of $\ker f'$ in $\mathfrak{g}^*$. The set $P'=P\cap\mathfrak{s}^*$ is a non-empty open subset of the affine space of $g'\in\mathfrak{s}^*$ such that $g'|\mathfrak{z}=f$ (for all these assertions, see [13]). The set P' is contained in $\mathfrak{s}_0^*$. In fact, $\mathfrak{s}$ is equal to $\mathfrak{g}(f')$, and a similar statement has been checked in the case III of the proof given in paragraph IV.1.

We consider the algebra $B=U(\mathfrak{b})/U(\mathfrak{b})I_{f'}$. Denote by q the natural projection

$$q:\ U(\mathfrak{b})\to B.$$

In [10], paragraph I.3, I defined an isomorphism of algebras:

$$r:B\to U(\mathfrak{s})\otimes U(\mathfrak{n})/I_{f'}.$$

Choose an element $u'\in U(\mathfrak{b})$ such that $p(u')=u$, and put $v=r(q(u'))$. Let (n_i) be a basis for $U(\mathfrak{n})/I_{f'}$. We write $v=\sum s_i\otimes n_i$ with elements s_i in $U(\mathfrak{s})$. For at least one i, say $i=i_0$, the element s_{i_0} does not belong to $U(\mathfrak{s})(z-1)$. In fact suppose that s_i belongs to $U(\mathfrak{s})(z-1)$ for all i. Then there exist $v'\in U(\mathfrak{s})\otimes U(\mathfrak{n})/I_{f'}$ such that $v=(z-1\otimes 1)v'$. It follows from the definition of r (see [10]) that $z-1\otimes 1=r(q((z,0)-1))$. Remark that $(0,z)-1$ belongs to $I_{f'}$ (when $I_{f'}$ is identified to a subspace of $U(\mathfrak{b})$, as we have implicitly done). Hence $q((z,0)-1)= =q(z,-z)$. Thus we have: $q(u')=q(z,-z)r^{-1}(v')$. It follows that u' belongs to $U(\mathfrak{b})(z,-z)+U(\mathfrak{b})I_{f'}$. This implies that $u=p(u')$ belongs to $U(\mathfrak{g})I_{f'}$, which we know to be false.

We now apply the induction hypothesis to $\mathfrak{s}$. There exists $g'\in P'$ such that s_{i_0} does not belong to $I_{g'}$. Denote by g the element of P such that $g|\mathfrak{s}=g'$ and $g|\mathfrak{n}=f'$. I claim that u does not belong to I_g. To prove this, choose $\mathfrak{l}'\in SP(g',s)$. Then there exists $\mathfrak{h}\in SP(f',\mathfrak{n})$ which is stable under the action of $\mathfrak{l}'$, and the algebra $\mathfrak{l}=\mathfrak{l}'+\mathfrak{h}$ belongs to $SP(g,\mathfrak{g})$ (see the proof of Theorem I.1). Now, by formula (7), paragraph I.3, and by Theorem I.6.1 of [10], we can realize the representation $\varrho=\mathrm{Ind}\,(g|\mathfrak{l},\mathfrak{g})$ in the tensor product of the spaces of the representations $\sigma=\mathrm{Ind}\,(f'|\mathfrak{h},\mathfrak{n})$ and $\tau=\mathrm{Ind}\,(g'|\mathfrak{l}',\mathfrak{g})$ in such a way that

$$\varrho(u)=\sum\tau(s_i)\otimes\sigma(n_i).$$

Remark that the kernel of σ in $U(\mathfrak{n})$ is $I_{f'}$. Thus the operators $\sigma(n_i)$ are defined, and they are linearly independent. The kernel in $U(\mathfrak{s})$ of the representation τ is $I_{g'}$. Thus $\tau(s_{i_0})\neq 0$. It follows that $\varrho(u)\neq 0$; this means that u does not belong to I_g.

To complete the proof of Theorem IV.1, it remains to prove that we can always reduce the situation to one of the five cases considered above. Suppose that $\mathfrak{g}$ is not semisimple. Then $\mathfrak{g}$ contains a non zero Abelian ideal $\mathfrak{b}$, and we may suppose that $\mathfrak{b}$ contains $\mathfrak{a}$. By Lemma IV.1, there exists $f'\in\mathfrak{b}^*$ such that $f'|\mathfrak{a}=f$ and $u\notin J_{\mathfrak{b},f'}$. Obviously, the set of such f' is open in the set of linear forms on $\mathfrak{b}$ whose restriction to $\mathfrak{a}$ is f. Thus, we may suppose moreover that the set of elements $g\in P$ such that $g|\mathfrak{b}=f'$ is not empty. Suppose we are not in cases III or IV. Let $\mathfrak{n}$ be the largest nilpotent ideal of $\mathfrak{g}$ and let $\mathfrak{z}$ be the center of $\mathfrak{n}$. Consider the Abelian ideal $\mathfrak{b}=\mathfrak{z}+\mathfrak{a}$. Choose $f'\in\mathfrak{b}^*$ as above. Then $\dim\mathfrak{b}=1$ and $f(\mathfrak{b})\neq 0$. Since $\mathfrak{g}(f')=\mathfrak{g}$, it follows that $\mathfrak{b}$ is the center of $\mathfrak{g}$. We may suppose that $\mathfrak{a}=\mathfrak{z}$. If $\mathfrak{n}=\mathfrak{z}$, then $\mathfrak{g}$ is reductive with center $\mathfrak{z}$, and this is case II. Suppose now that $\mathfrak{n}\neq\mathfrak{z}$. Since we excluded case IV, $\mathfrak{n}$ does not contain characteristic Abelian ideals of dimension >1. It follows that $\mathfrak{n}$ is a Heisenberg algebra (see [3]). This is case V.

V. Application to the study of the center of the enveloping algebra

V.1. Statement of the theorem.

Let $\mathfrak{g}$ be a Lie algebra. Recall that $Z(\mathfrak{g})$ is the center of the enveloping algebra $U(\mathfrak{g})$ and that $I(\mathfrak{g})$ is the set of $\mathfrak{g}$-invariant elements of the symmetric algebra $S(\mathfrak{g})$. The Poincaré—Birkhoff—Witt mapping induces a vector space isomorphism

$$\beta: I(\mathfrak{g})\to Z(\mathfrak{g}).$$

Consider $g\in\mathfrak{g}_0^*$. We shall associate to g a character χ_g of $Z(\mathfrak{g})$. Let $u\in Z(\mathfrak{g})$. In section II, we defined the primitive ideal I_g of $U(\mathfrak{g})$. The image of u in $U(\mathfrak{g})/I_g$ lies in the center of $U(\mathfrak{g})/I_g$. This center consists only of the scalars since I_g is primitive (see [15]). Thus the image of u in $U(\mathfrak{g})/I_g$ is a scalar which we shall denote by $\chi_g(u)$.

We define now a kind of Gelfand transform of $u\in Z(\mathfrak{g})$. It should be a k-valued function on the set $\mathfrak{g}_0^*$. In fact we shall define it only on a smaller set. Let r be the rank of $\mathfrak{g}$ and denote by $\mathfrak{g}_r^*$ the set of elements $g\in\mathfrak{g}^*$ such that $\mathfrak{g}(g)$ has dimension r. It is a Zariski open subset of $\mathfrak{g}^*$ contained in $\mathfrak{g}_0^*$ (Theorem I.1). Let $u\in Z(\mathfrak{g})$. We define a k-valued function $\tilde{u}$ on $\mathfrak{g}_r^*$ by the formula

$$\tilde{u}(g) = \chi_g(u), \quad (g\in\mathfrak{g}_r^*).$$

Theorem V.1.* *Let $u\in Z(\mathfrak{g})$. Then $\tilde{u}$ is the restriction to $\mathfrak{g}_r^*$ of an element $\hat{u}$ of $I(\mathfrak{g})$.*

* In [11], I stated that $\chi_g(u)=\hat{u}(g)$ for $g\in\mathfrak{g}_0^*$. This is true in the solvable case (see [2]) and (obviously) in the semisimple case, but I do not know how to prove it in the general case. Similarly, read $\mathfrak{g}_r^*$ instead of $\mathfrak{g}_0^*$ in Theorem 2 and Lemma 3 of [11].

Of course the element $\hat{u}$ is well determined, and the mapping $u \to \hat{u}$ from $Z(\mathfrak{g})$ into $I(\mathfrak{g})$ is an algebra morphism.

When $\mathfrak{g}$ is either semisimple or solvable, the following formula is true:

$$\hat{u} = D_q^{-1}(\beta^{-1}(u))$$

where D_q is a formal sum of differential operators on $S(\mathfrak{g})$ associated with the Kirillov "function" on $\mathfrak{g}$:

$$q(X) = \left(\det \frac{\operatorname{sh} \frac{1}{2} \operatorname{ad} X}{\frac{1}{2} \operatorname{ad} X}\right)^{1/2}$$

(see [9]). I believe the formula is true in the general case, but I can only prove the easiest part of it:

Proposition V.1. *Let* $u \in Z(\mathfrak{g})$. *Then* $\hat{u}$ *and* $\beta^{-1}(u)$ *have the same term of highest degree.*

We shall prove Theorem V.1 and Proposition V.1 in the next paragraphs.

Theorem V.2. *The mapping* $u \to \hat{u}$ *defined in Theorem* V.1 *is an algebra isomorphism from* $Z(\mathfrak{g})$ *to* $I(\mathfrak{g})$.

Proof. Let n be a positive integer. Put $Z_n = Z(\mathfrak{g}) \cap U_n(\mathfrak{g})$ and $I_n = I(\mathfrak{g}) \cap S_n(\mathfrak{g})$ where $S_n(\mathfrak{g})$ is the set of polynomials of degree $\leqq n$. It is enough to prove that for all $n \in N$, the mapping $\varphi : u \to \hat{u}$ induces a linear isomorphism from Z_n to I_n. This is of course true if $n=0$. Suppose this is true for $n-1$ (with $n \geqq 1$). Then, by Proposition V.1, $\varphi(Z_n)$ is contained in I_n, and φ induces, by passage to the quotient, the same mapping as β^{-1}, from Z_n/Z_{n-1} to I_n/I_{n-1}. Thus φ induces isomorphisms from Z_{n-1} to I_{n-1} and from Z_n/Z_{n-1} to I_n/I_{n-1}. So it induces an isomorphism from Z_n to I_n.

Remark that the injectivity of the mapping $u \to \hat{u}$ is also a consequence of Theorem II.2.

The question whether $Z(\mathfrak{g})$ and $I(\mathfrak{g})$ were isomorphic was proposed to me by J. Dixmier. The semisimple case has been known for long (see [16] for instance). J. Dixmier proved it in the nilpotent case ([4]) and [5]). I proved it in the solvable case [9].

V.2. To prove Theorem V.1, we use an auxiliary algebraic space introduced by N. Conze. The proof is nearly the same as that in [2], but we have to take care of singular elements. So it is better to repeat part of [2].

Let n be the dimension and r be the rank of $\mathfrak{g}$. We denote by E the space of pairs $(g, \mathfrak{h})$ where $g \in \mathfrak{g}^*$, and $\mathfrak{h}$ is a solvable subalgebra of $\mathfrak{g}$, of dimension $\frac{1}{2}(n+r)$, such that $\langle g, [\mathfrak{h}, \mathfrak{h}] \rangle = 0$. Let G be the space of the $\frac{1}{2}(n+r)$-dimensional subspaces of $\mathfrak{g}$. Then G is a complete algebraic manifold, and E is a closed subspace of $\mathfrak{g}^* \times G$. Remark that if $g \in \mathfrak{g}_r^*$, then $(g, \mathfrak{h})$ belongs to E iff $\mathfrak{h}$ belongs to $SP(g, \mathfrak{g})$. Denote by p the projection $E \to \mathfrak{g}^*$. Then, by Theorem I.1, $p(E)$ contains $\mathfrak{g}_r^*$.

Since G is complete, $p(E)$ is closed and thus $p(E)=\mathfrak{g}^*$. It follows that there exists an irreducible component T of E such that $p(T)=\mathfrak{g}^*$.

Let $\mathfrak{a}$ be a $\frac{1}{2}(n-r)$-dimensional subset of $\mathfrak{g}$, and denote by $T_\mathfrak{a}$ the subspace of elements $(g, \mathfrak{h})\in T$ such that $\mathfrak{g}=\mathfrak{h}\oplus\mathfrak{a}$. The set $T_\mathfrak{a}$ is open in T and sets of this form cover T. Fix $\mathfrak{a}$. If $(g, \mathfrak{h})\in T_\mathfrak{a}$, the representation $\tau(g, \mathfrak{h})=\mathrm{Ind}\,(g|\mathfrak{h}, \mathfrak{g})$ can be realized in the fixed space $S(\mathfrak{a})$, and the coefficients of $\tau(g, \mathfrak{h})$ are polynomial functions on $T_\mathfrak{a}$ (see [2]). From this, it follows that if u is an element of $U(\mathfrak{g})$, the subset of the points $(g, \mathfrak{h})\in T$ where $\tau(g, \mathfrak{h})(u)$ is scalar is a closed subset of T, and that this scalar is a polynomial function on this subset.

In particular, if $u\in U(\mathfrak{g})$, the subset S of T determined by the equation $\tau(g, \mathfrak{h})(u)=0$ is closed. Then, its projection $p(S)$ is closed in $\mathfrak{g}^*$. Remark that, by Theorem II.1, $p(S)\cap\mathfrak{g}_r^*$ is exactly the set of the elements $g\in\mathfrak{g}_r^*$ such that u belongs to I_g.

Denote by Prim $U(\mathfrak{g})$ the set of primitive ideals of $U(\mathfrak{g})$, endowed with the Jacobson topology. We have just proved:

Proposition V.2. *The mapping $g\to I_g$ from $\mathfrak{g}_r^*$ to* Prim $U(\mathfrak{g})$ *is continuous.*

From Theorem II.2 and Proposition V.2 follows a stronger form of Theorem II.2:

Theorem V.3. *Let P be a dense subset of $\mathfrak{g}_r^*$. Then* $\bigcap_{g\in P} I_g=0.$

V.3. Proof of Theorem V.1.

We keep the notations of paragraph V.2. Let $u\in Z(\mathfrak{g})$. Suppose $(g, \mathfrak{h})$ belongs to the open subset $p^{-1}(\mathfrak{g}_r^*)$ of T. Then, by Theorem II.1, and by definition of $\tilde{u}$, we have $\tau(g, \mathfrak{h})(u)=\tilde{u}(g)$. Since $\tau(g, \mathfrak{h})(u)$ is scalar on the open set $p^{-1}(\mathfrak{g}_r^*)$, and since T is irreducible, it is a scalar everywhere on T. Thus, the formula $\tau(g, \mathfrak{h})(u)=\ddot{u}(g, \mathfrak{h})$ defines a polynomial function $\ddot{u}$ on T. Since $\ddot{u}(g, \mathfrak{h})=\tilde{u}(g)$ if $g=\mathfrak{g}_r^*$, it follows, as in the proof of Theorem 3 of [2] that $\tilde{u}$ is a polynomial function on $\mathfrak{g}_r^*$, that is, the restriction to $\mathfrak{g}_r^*$ of a rational function $\hat{u}$ on $\mathfrak{g}^*$. Consider now the rational function $\hat{u}\circ p$ on T. Then $\hat{u}\circ p$ and $\ddot{u}$ have the same restriction to $p^{-1}(\mathfrak{g}_r^*)$. Since T is irreducible, they are equal. Thus $\hat{u}\circ p$ is defined everywhere. It implies that $\hat{u}$ is defined everywhere, which means that $\hat{u}$ belongs to $S(\mathfrak{g})$. Of course, $\hat{u}$ is $\mathfrak{g}$-invariant.

V.4. Proof of Proposition V.1.

Let $x\in I(\mathfrak{g})$, and suppose the degree of x is d. We shall prove that the polynomial $y=\beta(x)\hat{}-x$ has degree $<d$. For this, it is enough to prove that for every element g of a Zariski dense subset of $\mathfrak{g}^*$, the polynomial

$$t\to y(tg)$$

on k has degree $<d$.

Consider the subset V of the elements of $g\in\mathfrak{g}_r^*$ for which there exists $\mathfrak{l}\in SP(g, \mathfrak{g})$ such that $g|\mathfrak{l}\neq 0$. I claim that this set is dense in $\mathfrak{g}^*$. In fact, let $g\in\mathfrak{g}_r^*$ and choose

$\mathfrak{l} \in SP(g, \mathfrak{g})$. Denote by $[\mathfrak{l}, \mathfrak{l}]^{\perp}$ the orthogonal complement of $[\mathfrak{l}, \mathfrak{l}]$ in $\mathfrak{g}^*$. If $f \in [\mathfrak{l}, \mathfrak{l}]^{\perp}$ is such that $f+g \in \mathfrak{g}_r^*$, then $\mathfrak{l} \in SP(g+f, \mathfrak{g})$, by reason of dimension. Since $\mathfrak{l}$ is solvable, we have: $\mathfrak{l} \neq [\mathfrak{l}, \mathfrak{l}]$, and we can approximate g by elements of the form $f+g$, where $f \in [\mathfrak{l}, \mathfrak{l}]^{\perp}$ is such that $f+g \in \mathfrak{g}_r^*$ and $(f+g)|\mathfrak{l} \neq 0$. Then $f+g$ belongs to V, which proves our claim.

Choose $g \in V$, then choose $\mathfrak{l} \in SP(g, \mathfrak{g})$ such that $g|\mathfrak{l} \neq 0$ and further, choose a basis $X_1, \ldots, X_n$ of $\mathfrak{g}$ such that $X_1, \ldots, X_p$ belong to ker g and $X_{p+1}, \ldots, X_n$ form a basis of $\mathfrak{l}$. If $m=(m_1, \ldots, m_n)$ belongs to $\mathfrak{N}^n$, we write $X^m = X^{m_1} \ldots X^{m_n}$, which is either an element of $S(\mathfrak{g})$ or of $U(\mathfrak{g})$. We put $|m| = m_1 + \ldots + m_n$ and write:

$$x = \sum a_m X^m$$

with elements a_m in k. Since $X_1, \ldots, X_p$ belong to ker g, we have:

$$x(tg) = \sum_{\substack{|m|=d \\ m_1=\ldots=m_p=0}} a_m X^m(g) t^d + \text{terms of lower degree in } t. \tag{1}$$

Now we compute $\beta(x)\hat{}\,(tg) = \chi_{tg}(\beta(x))$. For this, we consider the representation $\tau_t = \mathrm{Ind}\,(tg|\mathfrak{l}, \mathfrak{g})$ acting in the space $U(\mathfrak{g}) \otimes_{u(\mathfrak{l})} M'$, where, as usual, M' is the $U(\mathfrak{l})$-module associated to the representation $tg|\mathfrak{l} + \lambda$ of $\mathfrak{l}$. (We have put $\lambda = \frac{1}{2}\,\mathrm{tr}_{\mathfrak{g}/\mathfrak{l}}$.) By definition of χ_{tg}, we have $\chi_{tg}(\beta(x)) = \tau_t(\beta(x))$ when $t \neq 0$. We have:

$$\beta(x) = \sum_{|m|=d} a_m X^m \,(\mathrm{mod}\, U_{d-1}(\mathfrak{g})).$$

Fix a non-zero vector m in M'. The vectors $\tau_t(X_1^{m_1} \ldots X_p^{m_p})m$ form a basis of the space of the representation τ_t. Then $\chi_{tg}(\beta(x))$ is equal to the diagonal coefficient which corresponds to the vector $\tau(1)m = m$. Thus, we have:

$$\chi_{tg}(\beta(x)) = \sum_{\substack{|m|=d \\ m_1=\ldots=m_p=0}} a_m X^m(tg+\lambda) + \text{terms of lower degree in } t.$$

Remark that:

$$X^m(tg+\lambda) = X^m(g) t^{|m|} + \text{terms of lower degree in } t.$$

So, we have proved the formula:

$$\beta(x)\hat{}\,(tg) = \sum_{\substack{|m|=d \\ m_1=\ldots=m_p=0}} a_m X^m(g) t^d + \text{terms of lower degree in } t. \tag{2}$$

Now, comparing (1) and (2), we see that $t \to y(tg)$ has degree $<d$, which proves Proposition V.1.

REFERENCES

[1] Auslander, L. and Kostant, B., Polarizations and unitary representations of solvable Lie groups, *Inventiones Math.*, **14** (1971), 255—234.

[2] Conze, N. and Duflo, M., Sur l'algèbre enveloppante d'une algèbre de Lie résoluble, *Bull. Sci. Math.*, **94** (1970), 201—208.

[3] Dixmier, J., Sur les représentations unitaires des groupes de Lie algébriques. *Ann. Inst. Fourier (Grenoble)*, **7** (1957), 315—328.

[4] Dixmier, J., Sur l'algèbre enveloppante d'une algèbre de Lie nilpotente, *Arch. Math. (Basel)*, **10** (1959), 321—326.

[5] Dixmier, J., Représentations irréductibles des algèbres de Lie nilpotentes, *An. Acad. Brasil. Ci.*, **35** (1963), 491—519.

[6] Dixmier, J., Représentations irréductibles des algèbres de Lie résolubles, *J. Math. Pures Appl.*, **45** (1966), 1—66.

[7] Dixmier, J., Idéaux primitifs dans l'algèbre enveloppante d'une algèbre de Lie semi-simple complexe, II. *C.R. Acad. Sci. Paris*, **271** (1970), 134—136.

[8] Dixmier, J., Polarisations dans les algèbres de Lie, *Ann. Sci. École Norm. Sup.* **4** (1971), 321—335.

[9] Duflo, M., Caractères des groupes et des algèbres de Lie résolubles. *Ann. Sci. École Norm. Sup.*, **3** (1970), 23—74.

[10] Duflo, M., Sur les extensions des représentations irréductibles des groupes de Lie nilpotents, *Ann. Sci. École Norm. Sup.* **5** (1972), 71—120.

[11] Duflo, M., Représentations induites d'algèbres de Lie, *C. R. Acad. Sci. Paris*, *Sér. A*, **272** (1971), 1157—1158.

[12] Duflo, M. and Vergne, M., Une propriété de la représentation coadjointe d'une algèbre de Lie. *C. R. Acad. Sci. Paris*, *Sér. A*, **268** (1969), 583—585.

[13] Kirillov, A. A., (Кириллов А. А.) Характеры унитарных представлений групп Ли. Редукционные теоремы (Characters of unitary representations of Lie groups. Reduction theorems), *Функц. анализ*, **3**: 1(1969) 36—47.

[14] Kostant, B., Lie group representation on polynomial rings, *Amer. J. Math.*, **85** (1963), 327—404.

[15] Quillen, D., On the endomorphism ring of a simple module over an enveloping algebra, *Proc. Amer Math. Soc.* **21** (1969), 171—172.

[16] *Séminaire Sophus Lie*, Institut Henri–Poincaré, Paris (1955).

[17] Vergne, M. Construction de sous-algèbres subordonnées à un élément du dual d'une algèbre de Lie résoluble, *C. R. Acad. Sci.* **270** (1970), 704—707.

[18] Verma, D.-N., Structure of certain induced representations of complex semisimple Lie algebras, *Dissertation*, Yale University, 1966.

REPRESENTATIONS OF THE GROUP $GL(n, K)$ WHERE K IS A LOCAL FIELD

I. M. GELFAND and D. A. KAJDAN

Introduction

A. Weil and R. P. Langlands stated some remarkable conjectures concerning the non-commutative reciprocity laws. They conjectured that there is some connection between the n-dimensional representations of the Šafarevič—Weil group of a local field K and certain irreducible representations of the group $GL(n, K)$. They also supposed the existence of a global reciprocity law ([1], [2], [3]). Their conjecture becomes the classical reciprocity law for local and global fields when $n=1$.

In this paper we will state some new results about the representations of reductive groups over K, where K is a non-Archimedian local field. Some of the results concern the case of arbitrary reductive groups, some of them are related to Chevalley's groups and the other results have been obtained for $GL(n, K)$ only. A number of theorems are similar to the corresponding theorems for finite fields [4]. The results about the representations mentioned above are stated in Theorems 2, 3, 4, 5, 6. But perhaps the main result of this paper is the construction of a complex valued function $\Gamma(\pi, \tau)$, defined for "almost all" pairs π and τ, where π is a representation of the group $GL(n, K)$, τ is a representation of the group $GL(n-1, K)$. We call this function the Γ-function of the representations. For the case $n=2$ similar functions are well known [6], [3].

On the other hand, in a very interesting paper [5], Langlands constructed two functions $L(\sigma)$ and $\varepsilon(\sigma)$ where σ is a finite-dimensional representation of the Šafarevič—Weil group of the field K. The main conjecture of our paper establishes a connection between our function $\Gamma(\pi, \tau)$ and Langlands' functions $\varepsilon(\sigma)$ and $L(\sigma)$.

Let us see how the conjecture of this paper is related to Langlands' conjecture. He supposed that there exists a L-function $L(\pi)$ defined on the set of irreducible representations of an arbitrary reductive K-group G (for the representations of class 1 this function is defined in [2]). Let us consider the case $GL(n, K)$. Langlands has conjectured further that for every n-dimensional representation of the group W_K (the Šafarevič—Weil group) one can define an irreducible representation π of the group $GL(n, K)$, possibly not unique, in such a way that $L(\pi)=L(\sigma)$. In the case $n=2$ the function $L(\pi)$ is defined and for char $\mathscr{K}\neq 2$ ($\mathscr{K}$ is the residue field of K) the required map is constructed (see [3]).

Unfortunately, if $n>2$ this map is not uniquely defined by these conditions. We believe that the map constructed in this paper is well defined.

Let us denote by $\hat{W}_K$ the set of all finite-dimensional representations of the Šafarevič—Weil group and by $(\hat{W}_K)_n$ the set of its n-dimensional representations. We denote by $\Gamma(\sigma)$ the function on $\hat{W}_K$, defined by

$$\Gamma(\sigma) = \frac{L(\frac{1}{2},\sigma)}{L(\frac{1}{2},\tilde{\sigma})\varepsilon(\sigma)},$$

where $\tilde{\sigma}$ is the representation contragradient to σ, $L(\frac{1}{2},\sigma)$ and $\varepsilon(\sigma)$ are Langlands' functions. We denote by $\hat{G}L(n, K)$ the set of all irreducible representations of the group $GL(n, K)$.

Conjecture. There is an embedding $\varkappa$; $\varkappa_n:(\hat{W}_K)_n \to \hat{G}L(n, K)$ *such that*

$$\Gamma(\sigma_n \otimes \tilde{\sigma}_{n-1}) = \Gamma(\varkappa_n\sigma_n, \varkappa_{n-1}\sigma_{n-1}) \tag{I}$$

for every $\sigma_n \in (\hat{W}_K)_n$, $\sigma_{n-1} \in (\hat{W}_K)_{n-1}$. *If* char $K=0$ *or* char $K>n$, *then the map* $\varkappa_n$ *is defined uniquely by this relation.* (On the left-hand side of (I), Γ is Langlands' function mentioned above. On the right-hand side of (I), Γ is the function of pairs of the representations constructed in this paper.) The image of the map $\varkappa$ is the set of admissible irreducible nondegenerate representations of $GL(n, K)$.

If $n=2$ some simplifications are now apparent. In this case $n-1=1$ and the left-hand side of (I) is defined on the set of 2-dimensional representations of W_K. As far as the right-hand side of (I) is concerned, we have $\varkappa_{n-1}\sigma_{n-1} \in \hat{G}L(1, K) = \hat{K}^*$ and $\Gamma(\varkappa_2\sigma_2, \varkappa_1\sigma_1)$ becomes $\Gamma(\varkappa_2\sigma_2 \otimes \varkappa_1\sigma_1)$, where $\varkappa_1\sigma_1$ is regarded as a quasicharacter of the determinant. Thus in the case $n=2$ we need not consider the Γ-function of pairs of the representations.

The theorems of this paper are valid for different classes of groups. However, all these classes contain the group $GL(n, K)$. It seems to be useful to formulate the main results for $GL(n, K)$.

Let $G_n = GL(n, K)$, write Z_n for the subgroup of all upper diagonal matrices with 1 in the diagonal (i.e. a maximal unipotent subgroup of G_n); $P_n \subset G_n$ for the subgroup of matrices p of the form $p=||p_{ik}||$ where $p_{ni}=0$ if $i=1, \dots, n-1$, and $p_{nn}=1$; $M_n \subset P_n$ for the unipotent radical in P_n, i.e. matrices m of the form $m=||m_{ik}||$ where $m_{ik}=\delta_{ik}$ for $k<n$, $m_{in}=m_i$ for $i<n$, and $m_{nn}=1$; let $G_{n-1} \subset G_n$ denote the standard embedding of $GL(n-1, K)$ into $GL(n, K)$. Observe that $P_n = G_{n-1}M_n$.

Let (π, V) be an irreducible admissible representation of the group G_n in the space V (cf. Definition 2). We shall denote by $(\hat{\pi}, V)$ the representation of G_n given by the formula $\hat{\pi}(g) = \pi^{-1}(g')$, where g' is g transposed.

Theorem A (cf. Theorem 2). *There exists in* V *a nondegenerate bilinear form* $\langle\ ,\ \rangle$ *such that for every* $g \in G_n$, $\xi_1, \xi_2 \in V$ *we have* $\langle \pi(g)\xi_1, \hat{\pi}(g)\xi_2 \rangle = \langle \xi_1, \xi_2 \rangle$.

Let ψ be a nontrivial additive character on K. Put $\theta(z) = \psi(\sum_{i=1}^{n-1} z_{i,i+1})$, where $z=||z_{ik}|| \in Z_n$.

* The additive character ψ on K^+ is fixed; thus instead of $\varepsilon(\psi, \sigma)$ we write $\varepsilon(\sigma)$.

Definition. The representation (π, V) is *nondegenerate* if there exists a linear functional l on V satisfying the condition

$$l(\pi(z)\xi) = \theta(z)\, l(\xi) \qquad (*)$$

for every $z \in Z_n$ and $\xi \in V$.

Theorem B. *Every irreducible cuspidal representation of G_n is nondegenerate.*
(For the definition of a cuspidal representation see [3].)

Theorem C. *Let (π, V) be an irreducible admissible representation of G_n. Then any two linear functionals l_1 and l_2 satisfying condition $(*)$, are proportional.*

Theorem C can be formulated in another form too. Denote by $C^{\infty}_{c,\theta}$ the space of all functions $f(g)$ on G_n satisfying the following conditions: *1)* $f(z^{-1}g)=\theta(z)f(g)$, $(z\in Z_n)$ *2)* there exists an open subgroup $N_f\subset G_n$ such that $f(gn)=f(g)$ for every $g\in G_n$ and $n\in N_f$, *3)* there exists a compact set $M_f\subset G_n$ such that $\operatorname{supp} f\subset Z_n M_f$. Let us denote by $C_{\theta}^{-\infty}$ the adjoint space to $C^{\infty}_{c,\theta}$. Write $\varrho(g_0)$ for the operator $\varrho(g_0): C^{\infty}_{c,\theta}\to C^{\infty}_{c,\theta}$ given by

$$\varrho(g_0)f(g) = f(gg_0) \quad (g_0\in G_n)$$

and $\varrho'(g_0)$ for the adjoint operator in $C_{\theta}^{-\infty}$ to $\varrho(g_0)$.

Theorem D. *Let (π, V) be a nondegenerate admissible irreducible representation of G_n. Then there exists a unique $\varrho(g)$-invariant subspace $W_n(\pi)\subset C_{\theta}^{-\infty}$ (for every $g\in G_n$) such that the representation (ϱ, W) is isomorphic to (π, V).*

Now we consider the restriction of a nondegenerate representation (π, V) to $P_n\subset G_n$. Denote by $C^{\infty}_{c,\theta}(P_n)$ the space of all functions $f(p)$ $(p\in P_n)$ such that 1) $f(z^{-1}p)=\theta(z)f(p)$ $(z\in Z_n, p\in P_n)$, 2) there exists an open subgroup $N_f\subset P_n$ such that $f(pn)=f(p)$ $(p\in P_n, n\in N_f)$, 3) there exists a compact set $M_f\subset P_n$ such that $\operatorname{supp} f\subset Z_n M_f$. Denote by $C_{\theta}^{-\infty}(P_n)$ the adjoint space to $C^{\infty}_{c,\theta}(P_n)$. Write $\varrho(p_0): C^{\infty}_{c,\theta}(P_n)\to C^{\infty}_{c,\theta}(P_n)$ for the operator of translation on the right by p_0. Let (π, V) be a nondegenerate representation of G_n and l a functional on V satisfying $l(\pi(z)\xi)= =\theta(z)l(\xi)$ $(z\in Z_n, \xi\in V)$. We define the mapping $\varphi_{\pi}: V\to C_{\theta}^{-\infty}(P_n)$ in the following way: for every $\xi\in V$ we define the element $\varphi_{\pi}(\xi)\in C_{\theta}^{-\infty}(P_n)$ by the identity $\varphi_{\pi}(\xi)(p)=l(\pi(p)\xi)$. As an easy consequence of the definition, we have $\varphi_{\pi}(\pi(p_0)\xi)=\varrho'(p_0)\varphi_{\pi}(\xi)$ for every $p_0\in P_n, \xi\in V$.

Theorem E (cf. Theorem 6). *For any irreducible cuspidal representation (π, V), φ_{π} provides a P_n-isomorphism between the spaces V and $C^{\infty}_{c,\theta^{-1}}(P_n)$.*

Consequently, all cuspidal irreducible representations of G_n restricted to the subgroup P_n are isomorphic to each other.

Conjecture A. *For any nondegenerate representation (π, V) of the group G_n the mapping $\varphi_{\pi}: V\to C_{\theta}^{-\infty}(P_n)$ has no kernel.*

In proving Theorem E we use the following fact, which seems to be of independent interest.

Theorem F (cf. Proposition 2). *Any nonzero subspace of $C_{\theta}^{-\infty}(P_n)$, invariant by right translations, contains $C^{\infty}_{c,\theta^{-1}}(P_n)$.*

As easily seen, the G_{n-1}-module $C^{\infty}_{c,\theta}(P_n)$ is canonically isomorphic to $C^{\infty}_{c,\theta}(G_{n-1})$, and $C_{\theta}^{-\infty}(P_n)$ is canonically isomorphic to $C_{\theta}^{-\infty}(G_{n-1})$. Then, by virtue of Theorem D, for every irreducible admissible nondegenerate representation (τ, U) of G_{n-1}, there exists a unique subspace $W_{n-1}(\tau)\subset C_{\theta}^{-\infty}(P_n)$ such that the representation in $W_{n-1}(\tau)$ induced by right translations on elements of G_{n-1}, is isomorphic to (τ, U).

Let us turn now to the construction of Γ-functions. Let (π, V) be an irreducible cuspidal representation of G_n. Put $s_n = \sum_{i=1}^{n} (-1)^{i-1} e_{i,n-i+1}$ where e_{ik} is the matrix with 1 in the place (i, k) and zeros elsewhere. By Theorem B there exists a functional l such that $l(\pi(z)\xi) = \theta(z)l(\xi)$ for $z \in Z_n$ $\xi \in V$.

We define a functional $\hat{l}$ on V by $\hat{l}(\xi) = l(\pi(s_n)\xi)$, $(\xi \in V)$. Then for any $z \in Z_n$, $\xi \in V$ we have $\hat{l}(\hat{\pi}(z)\xi) = \theta(z)\hat{l}(\xi)$. By Theorem E the mappings φ_π and $\varphi_{\hat{\pi}}: V \to C^\infty_{c,\theta}(P_n)$ defined by the formulae $\varphi_\pi(\xi)(p) = l(\pi(p)\xi)$ and $\varphi_{\hat{\pi}}(\xi)(p) = \hat{l}(\hat{\pi}(p)\xi)$ are isomorphisms. Let $K(\pi)$ denote the operator $K(\pi) = \varphi_{\hat{\pi}} \circ \varphi_\pi^{-1}: C^\infty_{c,\theta}(P_n) \to C^\infty_{c,\theta}(P_n)$. Since by Theorem C the functional l is uniquely determined up to a multiplier, $K(\pi)$ is well defined. Since the mappings φ_π and $\varphi_{\hat{\pi}}$ are P_n-invariant we have $\varrho(g)K(\pi) = K(\pi)\varrho(g'^{-1})$ for any $g \in G_{n-1} \subset P$ where $\varrho(g)$ is right translation in $C^\infty_{c,\theta}(P_n)$. We can "improve" $K(\pi)$ so that it commutes with $\varrho(g)$. To this end we define the operator

$$A: C^\infty_{c,\theta}(P_n) \to C^\infty_{c,\theta}(P_n)$$

in the following way. Let $f(p) \in C^\infty_{c,\theta}(P_n)$. Write p in the form $p = mg_{n-1}$ $(m \in M_n, g_{n-1} \in G_{n-1} \subset P_n)$. Put $Af(p) = f(p_1)$ where $p_1 = ms_{n-1}g'^{-1}_{n-1}$. Introduce now the operator $C(\pi) = A \circ K(\pi)$; $C(\pi): C^\infty_{c,\theta}(P_n) \to C^\infty_{c,\theta}(P_n)$. Obviously $C(\pi) \circ \varrho(g) = \varrho(g) \circ C(\pi)$ for $g \in G_{n-1}$.

Let $C'(\pi)$ denote the adjoint operator acting in $C_\theta^{-\infty}(P_n)$. Let (τ, U) be an irreducible admissible nondegenerate representation of G_{n-1}. By Theorem D there exists a unique subspace $W_{n-1}(\tau) \subset \subset C_\theta^{-\infty}(G_{n-1})$ such that the representation (τ, U) is realized in $W_{n-1}(\tau)$ by right translations. Because of the uniqueness of $W_{n-1}(\tau)$ the operator $C'(\pi)$ sends $W_{n-1}(\tau)$ into itself, and therefore, since $W_{n-1}(\tau)$ is irreducible, the restriction of $C'(\pi)$ to $W_{n-1}(\tau)$ is an operator of multiplying by a number. Let us denote this number by $\Gamma(\pi, \tau)$ and call it the Γ-function of the pair of representations π and τ.

Let us now formulate the non-commutative reciprocity law. Let σ be a finite-dimensional representation of the Šafarevič—Weil group W_K of the field K. Define the function

$$\Gamma_\psi(\sigma) = \frac{L(\frac{1}{2}, \sigma)}{L(\frac{1}{2}, \hat{\sigma})\varepsilon(\psi, \sigma)}$$

where $\hat{\sigma}$ is the representation contragradient to σ. For the definition of $L(s, \delta)$ and $\varepsilon(\psi, \sigma)$, cf. [5].

Conjecture B. *There is an embedding $\varkappa_n: (\hat{W}_K)_n \to \Pi_n$ of the space $(\hat{W}_K)_n$ of all n-dimensional representations of W_K into representations of the group $GL(n, K)$ such that $\Gamma_\psi(\pi \otimes \tilde{\tau}) = \Gamma_\psi(\varkappa_n\pi, \varkappa_{n-1}\tau)$ for any $\pi \in (\hat{W}_K)_n$, $\tau \in (\hat{W}_K)_{n-1}$. The map is bijective.*

This conjecture is a more precise form of Langlands' conjecture [2] for $G = GL(n, K)$.

We are very grateful to M. I. Graev, discussions with whom were very important.

We also thank P. Deligne, E. B. Vinberg, I. N. Bernstein, G. van Dijk, and G. Harder for their useful remarks.

1. *The existence of the factor space.* Let **G** be an algebraic group over some field, which acts on the algebraic manifold **X**. We will assume that **X** is reduced.

We call a manifold **Y** with a regular map **p**:**X**→**Y** the *geometrical factor* **Y**=**X**/**G**, if the following conditions are fulfilled:

1) **p** is an epimorphism;

2) fibers of the map $\mathbf{p}$ are orbits* of $\mathbf{G}$;

3) $\mathbf{p}$ is submersive**.

Lemma 1. *Let* $\mathbf{G}$ *be an algebraic group, acting on an algebraic manifold* $\mathbf{X}$. *Then in* $\mathbf{X}$ *there exists an open nonempty* $\mathbf{G}$*-invariant nonsingular submanifold* $\mathbf{X}_0$ *for which the geometrical factor* $\mathbf{Y}=\mathbf{X}_0/\mathbf{G}$ *exists.*

The proof of this lemma, based on Theorem 5.8.1 in [8] is due to Deligne. If char $K=0$ the proof of Lemma 1 becomes simpler and follows from a theorem of Rosenlicht [12].

Remark. Lemma 1 remains true, if in the definition of the factor we add the condition:

4) $\mathfrak{O}_Y=(\mathbf{p}_*\mathfrak{O}_X)^G$.

2. *The invariant distributions on K-manifolds.* Everywhere in the sequel K is a local non-Archimedean field. If $\mathbf{X}$ is an algebraic K-manifold, then X will denote the set of its K-points. Naturally X will be considered as a topological space. We will denote by $C_c^\infty(X)$ the linear space of locally constant functions on X with compact supports. Let $C_c^\infty(X)$ have the natural topology. Then every linear subspace in $C_c^\infty(X)$ is closed.

Let $\mathbf{G}$ be an algebraic group acting on $\mathbf{X}$. For every $x\in X$ the following objects are defined:

1) $\mathbf{G}(x)\subset\mathbf{X}$, i.e. the $\mathbf{G}$-orbit of the point x;

2) $G(x)=\mathbf{G}(x)\cap X\subset X$;

3) $xG\subset X$, where xG is the set of K-points of the form xg, where $g\in G$, i.e. the G-orbit of the point x.

Definition. We will say that the pair $(\mathbf{G}, \mathbf{X})$ is *regular*, if for every $x\in X$ the map $\mathbf{G}\to\mathbf{G}(x)$ is submersive. If $(\mathbf{G}, \mathbf{X})$ is regular then $\mathbf{G}(x)\approx\mathbf{G}/\mathbf{G}_x$, where $\mathbf{G}_x$ is the stabilizer of the point x.

Let $x\in X$. Then xG is a homogeneous G-space. If there exists a nonzero invariant measure, we denote it by $|\omega_x|$. If there is no such measure, let $|\omega_x|=0$.

* Recall the definition of G-orbit. Suppose given an action of $\mathbf{G}$ on $\mathbf{X}$, i.e. a map $\varphi:\mathbf{X}\times\mathbf{G}\to\mathbf{X}$. Let us consider the submanifold in $\mathbf{X}\times\mathbf{X}$, which is the image of $\mathbf{\Delta}_X\times\mathbf{G}\hookrightarrow\mathbf{G}\times\mathbf{X}\times\mathbf{X}$ by the morphism $\mathrm{Id}\times\varphi:\mathbf{X}\times\mathbf{X}\times\mathbf{G}\to\mathbf{X}\times\mathbf{X}$; we denote it by $\mathbf{\Omega}$. The projection of $\mathbf{X}\times\mathbf{X}$ on the second component induces an $\mathbf{X}$-scheme structure on $\mathbf{\Omega}$, namely $\pi:\mathbf{\Omega}\to\mathbf{X}$. If x is a K-point of $\mathbf{X}$, then the fiber of the morphism π over x is called a $\mathbf{G}$-orbit of the point x. We shall denote it by $\mathbf{G}(x)$.

The condition *2)* means, that $\mathbf{X}\underset{\mathbf{Y}}{\times}\mathbf{X}$ and $\mathbf{\Omega}$, which are considered as submanifolds in $\mathbf{X}\times\mathbf{X}$ coincide.

Later on we shall sometimes omit the reformulation in the language of schemes.

** The map $\mathbf{p}:\mathbf{X}\to\mathbf{Y}$ is called *submersive,* if it induces an epimorphism of tangent spaces.

We shall define the space $C_0(X) \subset C_c^\infty(X)$ to consist of all finite linear combinations of functions $\varphi(x)$ of the form $\varphi(x)=f(xg)-f(x)$, where $f \in C_c^\infty(X)$, $g \in G$. As a linear subspace in $C_c^\infty(X)$, $C_0(X)$ is closed.

Let $\boldsymbol{\sigma}:\mathbf{X} \to \mathbf{X}$ be an automorphism, which satisfies the following conditions: a) the **G**-orbits remain invariant under $\boldsymbol{\sigma}$; b) $\sigma(xG)=xG$ whenever $x \in X$; c) $\sigma|\omega_x|=|\omega_x|$.

Theorem 1. *If a pair* (**G, X**) *is regular, then for every* $f(x) \in C_c^\infty(X)$, $f(\sigma x)-f(x) \in C_0(X)$.

Before we prove the theorem let us state the following proposition.

Proposition 1. *Suppose there exists a geometrical factor* $\mathbf{Y}=\mathbf{X}/\mathbf{G}$ *and the pair* (**X, G**) *is regular. Then for every* $f \in C_c^\infty(X)$ *there exists* $\mathscr{I}(x)=\int\limits_{xG} f(x)|\omega_x|$ *and* $\mathscr{I}(x) \equiv 0$ *implies* $f(x) \in C_0(X)$.

Proof. For a homogeneous **X** the space X is the union of open orbits. In this case the proof is the same as the proof of Lemma 17 in [9].

To prove the proposition in the general case let us suppose that $\mathscr{I}(x)=0$. We denote by $f_{x_0}(x)$ the restriction of $f(x)$ to $G(x_0)$. It follows from the conditions of Proposition 1 that the orbit $\mathbf{G}(x_0)$ is closed; hence $f_{x_0} \in C_c^\infty(G(x_0))$. Since $\mathbf{G}(x_0)$ is a homogeneous space and Proposition 1 is proved for this case, $\mathscr{I}(x) \equiv 0$ implies $f_{x_0} \in C_0(G(x_0))$. This means that $f_{x_0}(x)$ can be represented in the form $f_{x_0}(x) = \sum\limits_{i=1}^{N}(F_i^0(xg_{i,x_0})-F_i^0(x))$, where $F_i^0(x) \in C_c^\infty(G(x_0))$. Because $G(X_0)$ is closed, there exist functions $F'_{i,x_0}(x) \in C_c^\infty(X)$, such that $F'_{i,x_0}(x)=F^0_{i,x_0}(x)$ for $x \in G(x_0)$. Let $V'(x_0)=\{x \mid f(x) \neq \sum\limits_{i=1}^{N}(F'_{i,x_0}(xg_{i,x_0})-F'_{i,x_0}(x))\}$. It is clear that $V'(x_0)$ is an open and closed compact subset of X and $V'(x_0) \cap G(x_0)=\emptyset$. Let $V(x_0)=p^{-1}(pV'(x_0))$, where **p** is the natural projection $\mathbf{p}:\mathbf{X} \to \mathbf{Y}=\mathbf{X}/\mathbf{G}$. Since $V'(x_0)$ is a compact set, $V(x_0)$ is a closed set. Let $U(x_0)=X-V(x_0)$. The set $U(x_0)$ is open and contains $G(x_0)$. Since **p** is a projection to a geometrical factor then by the condition *3)* this map is open. Thus $V(x_0)$ is open and $U(x_0)$ is open and closed. Since $\operatorname{Supp} f(x)$ is a compact set, there exists a finite number of points $x_i \in X$ such that $\cup U(x_i) \supset \operatorname{Supp} f(x)$. Let $U_l=U(x_l)-\bigcup\limits_{j<l} U(x_j)$. Suppose further

$$F_{il}(x) = \begin{cases} F'_{i,x_l}(x), & x \in U_l; \\ 0 \quad\quad , & x \notin U_l. \end{cases}$$

Then $F_{il} \in C_c^\infty(X)$ and $f(x)=\sum\limits_{i,l}[F_{il}(xg_{i,x_l})-F_{il}(x)]$. Proposition 1 is proved.

We prove now Theorem 1. By Lemma 1 there exists a stratification $\mathbf{X}=\cup \mathbf{X}_i$, such that *1)* $\mathbf{X}_i$ is a constructible **G**-invariant K-manifold, *2)* $\overline{\mathbf{X}}_i \subset \bigcup\limits_{j \leq i} \mathbf{X}_j$, where $\overline{\mathbf{X}}_i$ is a closure of $\mathbf{X}_i$, *3)* there exists a geometrical factor $\mathbf{Y}_i=\mathbf{X}_i/\mathbf{G}$.

Let $\mathbf{Z}_i = \bigcup_{j \leq 1} \mathbf{X}_j$. According to condition *2)* $\mathbf{Z}_i$ are closed and by the conditions of the theorem they are σ-invariant. We shall prove by induction on i, that $f \in C_c^\infty(Z_i)$ implies $f^\sigma - f \in C_0(Z_i)$. Let $\bar{f}$ be the restriction of f on Z_{i-1}. By induction there exist functions $F_l(x) \in C_c^\infty(X)$ and elements $g_l \in G$ such that $f^\sigma - f = \sum_l (F_l(xg_l) - F_l(x))$, $x \in Z_{i-1}$. Further let φ be the restriction of the function $f(\sigma x) - f(x) = \sum_l (F_l(xg_l) - F_l(x))$ to X_i. Since the field is non-Archimedean, the function φ belongs to $C_c^\infty(X_i)$ and satisfies the conditions of Proposition 1. Hence there exist functions $\Phi_m \in C_c^\infty(X_i)$ and elements $g_m \in G$, such that $\varphi(x) = \sum_m (\Phi_m(xg_m) - \Phi_m(x))$ for $x \in X_i$. The support of the function Φ_m is a compact set in X_i and it does not intersect Z_{i-1}. Therefore there exist functions $\Phi'_m(x) \in C_c^\infty(X)$ such that $\Phi'_m|_{Z_{i-1}} = 0$, $\Phi'_m|_{X_i} = \Phi_m$. Finally,

$$f(\sigma x) - f(x) = \sum_l (F_l(xg_l) - F_l(x)) + \sum_m (\Phi'_m(xg_m) - \Phi'_m(x))$$

whenever $x \in Z_i$. The theorem is proved.

Let us denote by $C^{-\infty}(X)$ the dual space to $C_c^\infty(X)$. The elements of the space $C^{-\infty}(X)$ will be also called *distributions* on X.

Theorem 1 could also be stated as follows.

Theorem 1[a]. *If the pair* $(\mathbf{G}, \mathbf{X})$ *is regular, C is a G-invariant distribution on X and* $\boldsymbol{\sigma}$ *is an automorphism of* $\mathbf{X}$ *which satisfies the conditions of Theorem 1, then $C = C^\sigma$, i.e. $C(f) = C(f^\sigma)$ for every $f \in C_c^\infty(X)$.*

We will also use a generalization of Theorem 1. Let us consider an algebraic manifold $\mathbf{X}$ and an algebraic group $\mathbf{U}$ acting on $\mathbf{X}$. This action will be written on the left: $x \to u^{-1}x$. Let θ be a character of the group U. Let us denote by $C_{c,\theta}^\infty(X)$ the space of locally constant functions on X, satisfying the conditions a) $f(u^{-1}x) = \theta(u)f(x)$, b) there exists a compactum $M \subset X$ such that $\operatorname{Supp} f \subset UM$. The space dual to $C_{c,\theta}^\infty(X)$ we will denote by $C_\theta^{-\infty}(X)$. Further let the algebraic group $\mathbf{G}$ act on $\mathbf{X}$ in such a way that the actions of $\mathbf{G}$ and $\mathbf{U}$ commute. The action of the group G will be written on the right: $x \to xg$.

In $C_{c,\theta}^\infty(X)$ there is an action of the group G. Let us write $C_{c,\theta}^\infty(X)$ for the space of all finite linear combinations of elements of the form $f(xg) - f(x)$, where $f \in C_{c,\theta}^\infty(X)$.

Let us suppose that there exists a homomorphism $\boldsymbol{\theta}_0 : \mathbf{U} \to \mathbf{G}_a$ from the group $\mathbf{U}$ into the additive group $\mathbf{G}_a$ over the field K and a nontrivial fixed character ψ of $K^+ = G_a$ such that $\theta = \psi\theta_0$. We denote the kernel of $\boldsymbol{\theta}_0$ by $\mathbf{U}_0$. The group $\mathbf{U} \times \mathbf{G}$ acts on the manifold X. For every point $x \in X$ let us denote by $\mathbf{H}_x$ the stabilizer of the point x, $\mathbf{H}_x \subset \mathbf{U} \times \mathbf{G}$ and by $\overline{\mathbf{H}}_x$ its projection on $\mathbf{U}$. Let X' be the set of points $x \in X$ such that $\mathbf{H}_x \subset \mathbf{U}_0$. One can show that X' is the set of K-points of the constructible submanifold $\mathbf{X}' \subset \mathbf{X}$. Moreover $\mathbf{X}'$ is an $\mathbf{U} \times \mathbf{G}$-invariant submanifold. The map $\boldsymbol{\sigma} : \mathbf{X} \to \mathbf{X}$ is called admissible, when there exists an automorphism $\bar{\boldsymbol{\sigma}}$ of the group $\mathbf{U}$, such that

1) $\boldsymbol{\theta}_0\boldsymbol{\sigma}=\boldsymbol{\theta}_0$;

2) $\bar{\sigma}(ux)=\bar{\sigma}(u)\bar{\sigma}(x)$, i.e. the diagram

$$\begin{array}{ccc}\mathbf{U}\times\mathbf{X} & \xrightarrow{\bar{\boldsymbol{\sigma}}\otimes\boldsymbol{\sigma}} & \mathbf{U}\times\mathbf{X}\\ \downarrow & & \downarrow\\ \mathbf{X} & \xrightarrow{\boldsymbol{\sigma}} & \mathbf{X}\end{array}$$

is commutative;

3) $\boldsymbol{\sigma}\mathbf{X}'=\mathbf{X}'$.

Theorem 1′. *Let the pair* $(\mathbf{X}, \mathbf{U}\times\mathbf{G})$ *be regular,* $\boldsymbol{\sigma}:\mathbf{X}\to\mathbf{X}$ *be an admissible map,* $\sigma|_{X'}:\mathbf{X}'\to\mathbf{X}'$ *the corresponding morphism. Let us consider* $\mathbf{X}'$ *as* $\mathbf{U}_0\times\mathbf{G}$*-manifold and suppose that* $\sigma|_{X'}$ *satisfies the conditions of Theorem* 1. *Then* $f^\sigma-f\in C^\infty_{c,\theta}(X)$ *for every* $f\in C^\infty_{c,\theta}(X)$.

Another form of this theorem is as follows. If the pair $(\mathbf{X}, \mathbf{U}\times\mathbf{G})$ is regular, $C\in C_\theta^{-\infty}(X)$, C is G-invariant and $\sigma:X\to X$ satisfies the conditions of Theorem 1′, then $C^\sigma=C$.

The proof of Theorem 1′ is similar to that of Theorem 1. The only differences are

1) we must require that $\mathbf{X}_i$ (in the stratification $\mathbf{X}=\bigcup\mathbf{X}_i$) is an $\mathbf{U}\times\mathbf{G}$-invariant manifold,

2) for every i either $\mathbf{X}_i\cap\mathbf{X}'=\mathbf{X}_i$ or $\mathbf{X}_i\cap\mathbf{X}'=\emptyset$.

3. *The existence of the bilinear form.* Let $\mathbf{G}$ be an arbitrary algebraic group over a non-Archimedean local field K and π be a representation of the group G in a locally convex space V.

Definition 1. A representation π is *algebraic*, if

a) for every vector $v\in V$, Stab v is an open subgroup in G,

b) every linear subspace in V is closed.

Example. The regular representation of G in the function-space $C_c^\infty(G)$ is algebraic.

Definition 2. An algebraic representation π is *admissible,* if for every open subgroup $U\subset G$ the subspace $V^U\subset V$ of all U-invariant vectors is finite-dimensional.

Let V be the space of some representation π of the group G. Let us denote by V_0 the set of all elements of V which have open stabilizer in G. Equipped with the trivial topology, V_0 is the space of an algebraic representation π_0 of G.

Let π be an algebraic representation in V. We denote by $\tilde{V}'$ the dual space to V and by $\tilde{\pi}'$ the dual representation in the space $\tilde{V}':\tilde{\pi}'(g)=(\pi(g^{-1}))'$. The representation $\tilde{\pi}'$ is not necessarily algebraic. We will denote by V' the space $(\tilde{V}')_0$ and by π' the algebraic representation $(\tilde{\pi}')_0$. We will denote by $\tilde{V}$ the dual space to V', i.e. $\tilde{V}=(\tilde{V}')'$, and by $\tilde{\pi}$ the corresponding representation. Let us note that $V\subset\tilde{V}$, $V'\subset\tilde{V}'$. Moreover, if π is an admissible representation, then $V=(\tilde{V})_0$.

If the admissible representation π is irreducible then π' is also irreducible (see [3]).

It is easy to verify that if π is an admissible representation then for every $f \in C_c^\infty(G)$ the operator $\pi(f) = \int_G f(g)\pi(g)\,dg$ is of finite rank. Let us denote by Tr_π the distribution defined by $(\mathrm{Tr}_\pi, f) = \mathrm{Tr}\,\pi(f)$.

If π_1 and π_2 are irreducible admissible representations and $\mathrm{Tr}_{\pi_1} = \mathrm{Tr}_{\pi_2}$, then π_1 and π_2 are equivalent ([3], Lemma 7.1).

A regular pair $(\mathbf{X}, \mathbf{G})$ was defined for an action of the group $\mathbf{G}$ on the manifold $\mathbf{X}$. Now let $\mathbf{X} = \mathbf{G}$ with the adjoint action. If the pair $(\mathbf{G}, \mathbf{G})$ is regular, we call the group G *regular.*

Examples. The group $GL(n, K)$ is regular for every field K. The groups $SP(2n, K)$ and $SO(n, K)$ are regular when char $K \neq 2$; $SL(n, K)$ is regular if char K does not divide n. The exceptional groups are considered in [10].

Let $\boldsymbol{\sigma}$ be an anti-automorphism in $\mathbf{G}$, such that for every $g \in G$ the element g^σ is conjugate to g, i.e. there exists $s(g)$ such that $g^\sigma = s(g)g(s(g))^{-1}$. We suppose there exists N such that σ^{2N} is an inner automorphism. Let π be an admissible irreducible representation of the group G. We denote by $\hat{\pi}$ the representation

$$\hat{\pi}(g) = \pi^{-1}(g^\sigma).$$

Theorem 2. *Let π be an admissible irreducible representation of the regular group G. Then in the space V there exists a nondegenerate form $\langle v_1, v_2\rangle$ such that for every $g \in G$*

$$\langle \pi(g)v_1, \hat{\pi}(g)v_2\rangle = \langle v_1, v_2\rangle.$$

At first let us explain the proof in the case of a finite group G and its finite-dimensional representation. We denote by π' the representation dual to π in V', given by the formula: $\pi'(g) = (\pi(g^{-1}))'$. Since

$$\mathrm{Tr}_{\pi'}(g) = \mathrm{Tr}_\pi(g^{-1}) = \mathrm{Tr}_\pi(s(g)g^{-1}s^{-1}(g)) = \mathrm{Tr}_\pi((g^\sigma)^{-1}) = \mathrm{Tr}_{\hat{\pi}}(g)$$

the representations π' and $\hat{\pi}$ are equivalent; i.e. there exists an operator $A: V \to V'$ such, that $A^{-1}\pi'(g)A = \hat{\pi}(g)$. The required nondegenerate bilinear form is $\langle v_1, v_2\rangle = (v_1, Av_2)$, where (,) is a natural pairing of V and V'.

Now let us prove the theorem. It is sufficient to prove that π' and $\hat{\pi}$ are equivalent and this will follow if we prove that $\mathrm{Tr}_{\pi'} = \mathrm{Tr}_{\hat{\pi}}$. Let us introduce some notations. If $f(g) \in C_c^\infty(G)$ put $\hat{f}(g) = f((g^\sigma)^{-1})$ and $f^-(g) = f(g^{-1})$. For $f \in C_c^\infty(G)$ the operator $(\pi(f))'$, which is defined on $\tilde{V}'$, sends $\tilde{V}'$ into V' and therefore defines an operator in V'.

It is evident that $\hat{\pi}(f) = \pi(\hat{f})$ and $\pi'(f) = (\pi(f^-))'$. So $\mathrm{Tr}_{\hat{\pi}}(f) = \mathrm{Tr}_\pi(\hat{f})$ and $\mathrm{Tr}_{\pi'}(f) = \mathrm{Tr}_\pi(f^-)$. Thus to prove $\mathrm{Tr}_{\pi'} = \mathrm{Tr}_{\hat{\pi}}$ we must verify that

$$\mathrm{Tr}_\pi(f^-) = \mathrm{Tr}_\pi(\hat{f}) \qquad (*)$$

for every $f \in C_c^\infty(G)$. To prove this we use Theorem 1 in the case when $\mathbf{X}$ is $\mathbf{G}$

with the action $x \to g^{-1}xg$. The conditions of Theorem 1 are fulfilled. Indeed, since $\sigma x = s^{-1}(x)xs(x)$, the orbit is mapped into itself. Suppose a G-invariant measure μ exists on the orbit. Then $\mu^\sigma = \gamma\mu$, where $\gamma > 0$. Since σ^{2N} is an inner automorphism the measure μ remains invariant. This means $\gamma^{2N} = 1$ i.e. $\gamma = 1$. The distribution $\mathrm{Tr}_\pi \in C^{-\infty}(G)$ is invariant relative to the adjoint action, hence by Theorem 1^a $\mathrm{Tr}_\pi(f) = \mathrm{Tr}_\pi(f^\sigma)$ for every $f \in C_c^\infty(G)$. Since $\hat{f} = (f^-)^\sigma$, the equality (1) follows and the theorem is proved.

Remark to Theorem 2. Since the representation π and $\hat{\pi}$ are irreducible, any two bilinear π, $\hat{\pi}$-invariant forms on V are proportional. This means that in the case $\sigma^2 = 1$ it is possible to divide the representations into two classes. To do this, let $\langle v_1, v_2 \rangle$ be the invariant form constructed above. Put $\langle v_1, v_2 \rangle_1 = \langle v_2, v_1 \rangle$. Then, since $\sigma^2 = 1$, we have

$$\langle \pi(g)v_1, \hat{\pi}(g)v_2 \rangle_1 = \langle \hat{\pi}(g)v_2, \pi(g)v_1 \rangle =$$

$$= \langle \pi((g^\sigma)^{-1})v_2, \hat{\pi}((g^\sigma)^{-1})v_1 \rangle = \langle v_2, v_1 \rangle = \langle v_1, v_2 \rangle_1.$$

Thus $\langle v_2, v_1 \rangle = \varepsilon \langle v_1, v_2 \rangle$ and so $\varepsilon = \pm 1$. This means that all irreducible admissible representations can be divided into the following two classes:

a) the *even* representations with symmetric $(\pi, \hat{\pi})$-invariant form $\langle v_1, v_2 \rangle$ and

b) the *odd* representations, for which the corresponding form is skew symmetric.

Remark. For every Chevalley group with connected centre there exists the following automorphism, denoted by $' : g \to g'$.

Let H be a maximal split torus in G. Put $h' = h$, $h \in H$, $l'_\alpha(t) = l_{-\alpha}(t)$ for every root α and extend it to an anti-automorphism of the group G.

If $G = GL(n, K)$, then $'$ transforms every element g into its conjugate. We do not know for which Chevalley groups and for which fields K the element g' is conjugate to g for every $g \in G$.

4. *The simplicity of the spectrum*. Throughout this section we will suppose that $\mathbf{G}$ is a regular reductive Chevalley group with a connected reduced centre. For every such group $\mathbf{G}$ we will define an important representation, which we will call the *standard representation* of G. This representation was considered in [4] in the case of a finite field. Let $\mathbf{H}$ be a maximal split torus and $\mathbf{Z}$ be the unipotent subgroup corresponding to the positive roots (by some ordering of these roots). Let $\boldsymbol{\theta}_0 : \mathbf{Z} \to \mathbf{G}_a$ be a homomorphism defined on K. The group H acts on the set of such morphisms in the following way: for every $h \in H$ $h : \boldsymbol{\theta}_0 \to \boldsymbol{\theta}_0^h$, where $\boldsymbol{\theta}_0^h(z) = \boldsymbol{\theta}_0(hzh^{-1})$. The homomorphism $\boldsymbol{\theta}_0$ is called *nondegenerate* if $\boldsymbol{\theta}_0(e_\alpha(t)) \neq 0$ for every simple root α and $\boldsymbol{\theta}_0(e_\alpha(t)) = 0$ for every nonsimple root α. Notice that H acts transitively on the set of the nondegenerate morphisms.

Choose a nondegenerate morphism $\boldsymbol{\theta}_0$, a nontrivial character ψ on K^+, and let $\theta = \psi\theta_0$.

Consider the space $C^\infty_{c,\theta}(G)$, consisting of locally constant functions on G, which satisfy the following conditions:

1) $f(z^{-1}g)=\theta(z)f(g)$, $z\in Z$, $g\in G$;

2) $\operatorname{Supp} f\subset ZM_f$, where M_f is a compact set in G.

In the space $C^\infty_{c,\theta}(G)$ the action of G is defined by the formula $T(g_0)f(g)=f(gg_0)$. The representation T is algebraic.

The representation T is called the standard representation of the group G. Clearly T is independent of the choice of ψ and $\boldsymbol{\theta}_0$.

We shall need some results concerning Chevalley groups. As these results are true for every field, we suppose now that K is an arbitrary field.

Let us note that the definition of nondegenerate morphisms is correct for every field K and the group H acts transitively on the set of all nondegenerate morphisms.

Let us fix an $s\in G$ satisfying the following conditions:

a) $sZ's^{-1}=Z$ (for the definition of the antiautomorphism see 3 above),

b) $\theta_0(sz's^{-1})=\theta_0(z^{-1})$.

We will show that such an element s exists. Let us consider an element w of the Weyl group, which transforms positive roots into negative ones. Denote by n some element of the normalizer $N(\mathbf{H})$ of the group H which corresponds to the element w; then $nZ'n^{-1}=Z$. $\mathbf{H}$ acts transitively on the set of the nondegenerate morphisms $\mathbf{Z}\to G_a$. Thus there exists an element $h\in H$, such that $\theta_0(sz's^{-1})=$ $=\theta_0(z^{-1})$ for $s=nh$. It is easy to verify that if s_1 is another such element then $s_1=\gamma s$, where γ is an element of the centre of G.

Put $\alpha=s's^{-1}$. Then α is an element of the centre of G and $\alpha^2=1$. Indeed, the element s' also satisfies the conditions a) and b) and hence $s's^{-1}$ is an element of the centre. The antiautomorphism $'$ acts trivially on the Cartan subgroup. Since $s'=s\alpha$, so $s=\alpha s'$, i.e. $\alpha^2=1$.

Let us define the action of the group $U=Z\times Z$ on G in the following way: $(z_1z_2):g\to z_1^{-1}gz_2$.

We denote by $U_0\subset U$ the subgroup of all elements $u=(z_1z_2)\in U$ which satisfy the condition $\tilde{\theta}_0(u)=0$, where $\tilde{\theta}_0(u)\overset{\text{def}}{=}\theta_0(z_1z_2)$. Let us denote by $U_g\subset U$ the stabilizer of the element g. Let $G'=\{g\in G\,|\,U_g\subset U_0\}$. It means that $G'=\{g\in G\,|\,\forall z_1, z_2\in Z,$ $z_1^{-1}gz_2=g$ implies $\theta_0(z_1z_2)=0\}$.

Lemma 2. *For every* $g\in G'$ *there exist* $z_1, z_2\in Z$ *such that*

a) $\alpha sg's^{-1}=z_1^{-1}gz_2$,

b) $\theta_0(z_1z_2)=0$.

The elements s and α were defined above.

Proof. Lemma 2 is similar to Theorem 49 in Steinberg's Lectures [13]. This lemma follows easily from the results, which he obtained when proving Theorem 49. In the proof of this theorem Steinberg uses some results from § 10 of his lectures. Let us note that this results are valid for every field and not only for perfect ones. Denote by $h_0 \in H$ an element such that $\theta_0(h_0 z h_0^{-1}) \equiv -\theta_0(z)$. Let f be the antiautomorphism, which is given by $g^f \overset{\text{def}}{=} h_0 s g' s^{-1} h_0^{-1}$. Let $G_1 = \{g \in G \mid \forall z_1, z_2 \in Z, z_1^{-1} g z_2 = g$ implies $\theta_0(z_1 z_2^{-1}) = 0\}$. Steinberg showed, that for every $g \in G_1$ there exist $z_1, z_2 \in Z$ such that *1)* $g^f = z_1^{-1} g z_2$, *2)* $\theta_0(z_1 z_2^{-1}) = 0$. Let us show how this implies Lemma 2. Let us note that $G' = h_0 G_1$. If $g \in G'$, i.e. $g = h_0 g_1$, where $g_1 \in G_1$, then there exist z_1 and z_2 such, that $g_1^f = z_1^{-1} g z_2$, $\theta_0(z_1 z_2^{-1}) = 0$. Hence $sg's^{-1} = h_0^{-1} g^f h_0 = h_0^{-1} g_1^f h_0^f h_0 = h_0^{-1} z_1^{-1} g_1 z_2 h_0^f h_0$. But $h_0^f \cdot h_0 = \beta$, where β belongs to the centre of G. Thus $sg's^{-1} = \beta h_0^{-1} z_1^{-1} g_1 z_2 = \beta h_0^{-2} (z_1^{-1})^{h_0} h_0 g_1 z_2 = \gamma (z_1^{h_0})^{-1} g z_2$, where γ is an element of the centre of G. Since $\theta_0(z_1^{-1} z_2) = 0$ then $\theta_0(z_1^{h_0} z_2) = 0$. To prove the lemma it is sufficient to show that $\gamma = \alpha$. Let us use the proposition, which is proved above, in the case $g = s$. Since $s \in G'$ then $s(s')s^{-1} = \gamma (z_1^{h_0})^{-1} s z_2$. Because $s, s' \in N(H)$, the Bruhat decomposition implies $ss's^{-1} = \gamma s$, i.e. $s\alpha = \gamma s$. Lemma 2 is proved.

Let us denote by A the following operator in the space $C_{c,\theta}^{\infty}(G) : (Af)(g) = f(s(g')^{-1} s^{-1})$ and by A_α the operator $(A_\alpha f)(g) = f(\alpha s (g')^{-1} s^{-1})$.

Theorem 3. *Let* $C(f_1, f_2)$ *be a G-invariant bilinear form on* $C_{c,\theta}^{\infty}(G)$. *Then* $C(f_1, f_2) = C(A_\alpha f_2, A f_1)$.

Proof. Let $\mathbf{X} = \mathbf{G} \times \mathbf{G}$. Define the action of the group U on X as follows: if $u = (z_1, z_2)$, $x = (g_1, g_2)$, then $ux = (z_1 g_1, z_2 g_2)$. Define the action of G on X by $xg = (g_1 g, g_2 g)$. Let $\tilde{\theta} = \psi \tilde{\theta}_0$, that is $\tilde{\theta}(u) = \theta(z_1 z_2)$. $C_{c,\theta}^{\infty}(G)$ is a nuclear space and $C_{c,\theta}^{\infty}(G) \otimes C_{c,\theta}^{\infty}(G) = C_{c,\tilde{\theta}}^{\infty}(X)$. Therefore $C(f_1, f_2)$ can be extended to a linear functional on $C_{c,\tilde{\theta}}^{\infty}(X)$. This functional will be denoted by $C(f)$, i.e. put $C(f_1 \otimes f_2) = C(f_1, f_2)$. Let us denote by $\boldsymbol{\sigma} : \mathbf{X} \to \mathbf{X}$ the map given by the formula: $\sigma(g_1, g_2) = (\alpha s (g_2^{-1})' s^{-1}, s(g_1^{-1})' s^{-1})$. To prove the theorem it is sufficient to verify that $C(f) = C(f^\sigma)$ for every $f \in C_{c,\tilde{\theta}}^{\infty}(X)$. Now we can apply Theorem 1′. Indeed, by Lemma 2 the $U_0 \times G$-orbit on X' is mapped into itself by the action of σ; further $\sigma^4 = 1$. Therefore σ preserves the measure on every orbit. Moreover, the $\mathbf{U}_0 \times \mathbf{G}$-action on $\mathbf{X}$ is regular (see [11]). Thus by Theorem 1′ $C(f^\sigma) = C(f)$ for every G-invariant element $C \in C_{\tilde{\theta}}^{-\infty}(X)$. This proves the theorem.

Recall some notations. If the representation (π, V) is admissible, then the space dual to V we denote by $\tilde{V}'$, $\tilde{\pi}'(g) = (\pi(g^{-1})')$, where $'$ means the adjoint operator, $V' = (\tilde{V}')_0$, $\pi' = (\tilde{\pi}')_0$. Therefore $\tilde{V} = (\tilde{V}')'$ and $\tilde{\pi} = (\tilde{\pi}')'$, $V = (\tilde{V})_0$. We put $\hat{\pi}(g) = \pi(s(g')^{-1} s^{-1})$; the representation $\hat{\pi}(g)$ acts in the space V.

Let $\mathbf{G}$ be an arbitrary algebraic group and $\mathbf{Z} \subset \mathbf{G}$ be a subgroup. Let (π, V) be an admissible representation of the group $\mathbf{G}$ and θ be a (one-dimensional) character of the group Z. We denote by $\tilde{V}_\theta$ the subspace of $\tilde{V}$, consisting of all vectors $w \in \tilde{V}$ which satisfy the condition $\tilde{\pi}(z) w = \theta(z) w$ for every $z \in Z$. Similarly $\tilde{V}_\theta'$ is a subspace of $\tilde{V}'$, consisting of the vectors w' such that $\tilde{\pi}'(z) w' = \theta(z) w'$.

Let us assume now that $\mathbf{G}$ is a regular Chevalley group with a connected reduced centre and let $\mathbf{\theta}_0$ be a nondegenerate character on $\mathbf{Z}$. For such group the following theorem can be formulated.

Theorem 4. *Let* (π, V) *be an admissible and irreducible representation, then*

a) $\dim \tilde{V}_\theta \cdot \dim \tilde{V}'_\theta \leq 1$;

b) *if* $\dim \tilde{V}_\theta = \dim \tilde{V}'_\theta = 1$, *then the representations* $\hat{\pi}$ *and* π' *are equivalent.*

Remark. If the automorphism $'$ satisfies the conditions of Theorem 2, then the conditions a) and b) become $\tilde{a}$) $\dim \tilde{V}_\theta = \dim \tilde{V}'_\theta \leq 1$ and $\tilde{b}$) $\pi' \sim \hat{\pi}$ for every irreducible admissible representation (π, V) (Theorem 2). In particular the statements $\tilde{a}$) and $\tilde{b}$) are true for $GL(n, K)$. We do not know whether the equality $\dim \tilde{V}_\theta = \dim \tilde{V}'_\theta$ is always true.

Proof. Construct the following operators (Frobenius operators). Let $w \in \tilde{V}_\theta$. Define the maps $\pi_w : C^\infty_{c,\theta}(G) \to (\tilde{V})_0 = V$ and $\hat{\pi}_w : C^\infty_{c,\theta}(G) \to V$ by:

$$\pi_w f = \int_{Z \setminus G} f(g) \tilde{\pi}(g^{-1}) w \, d\bar{g},$$

$$\hat{\pi}_w f = \int_{Z \setminus G} f(g) \tilde{\hat{\pi}}(g^{-1}) w \, d\bar{g}.$$

Here $d\bar{g}$ is the invariant measure on $Z \setminus G$. $f(g)\tilde{\hat{\pi}}(g^{-1})w$ is defined on $Z \setminus G$ because $\tilde{\hat{V}}_\theta = \tilde{V}_\theta$. The corresponding operators $\pi'_{w'}$ and $\hat{\pi}'_{w'} : C^\infty_{c,\theta}(G) \to V'$ can be constructed for $w' \in \tilde{V}'_\theta$ as follows:

$$\pi'_{w'} f = \int_{Z \setminus G} f(g) \tilde{\pi}'(g^{-1}) w' \, d\bar{g},$$

$$\hat{\pi}'_{w'} f = \int_{Z \setminus G} f(g) \tilde{\hat{\pi}}'(g^{-1}) w' \, d\bar{g}.$$

Consider $\pi(\alpha)$, where α is the element defined in Lemma 2. Since the representation π is irreducible and admissible it is easy to verify that $\pi(\alpha) = \tilde{\varepsilon} \cdot \mathrm{Id}$, where $\tilde{\varepsilon} = \pm 1$.

Let us prove the theorem now. Suppose $\tilde{V}_\theta \neq \{0\}$ and $\tilde{V}'_\theta \neq \{0\}$. Choose the vectors $w \in \tilde{V}_\theta$ and $w' \in \tilde{V}'_\theta$, $w, w' \neq 0$. Consider the bilinear form $B_{w,w'}(f_1, f_2) \overset{\text{def}}{=} (\pi_w f_1, \pi'_{w'} f_2)$, where $f_1, f_2 \in C^\infty_{c,\theta}(G)$. $B_{w,w'}$ is invariant bilinear form on $C^\infty_{c,\theta}(G)$. According to Theorem 3 $B_{w,w'}(f_1, f_2) = B_{w,w'}(A_\alpha f_2, A f_1)$, i.e. $(\pi_w f_1, \pi'_{w'} f_2) = (\pi_w(A_\alpha f_2), \pi'_{w'}(A f_1))$. But $(\pi_w(A_\alpha f_2), \pi'_{w'}(A f_1)) = (\pi(\alpha)\pi_w(A f_2), \pi'_{w'}(A f_1)) = \tilde{\varepsilon}(\hat{\pi}_w f_2, \hat{\pi}'_{w'} f_1)$. Since f_1, f_2 are arbitrary in $C^\infty_{c,\theta}(G)$ we have

$$(\pi(g)\pi_w f_1, \pi'_{w'} f_2) = \tilde{\varepsilon}(\hat{\pi}_w f_2, \hat{\pi}'(g)\hat{\pi}'_{w'} f_1) \qquad (*)$$

for every $g \in G$.

This means that the representations $\pi(g)$ and $\hat{\pi}'(g)$ are equivalent (because $\pi, \pi', \hat{\pi}, \hat{\pi}'$ are irreducible). Assertion b) is proved.

Let us prove now assertion a). Because π and $\hat{\pi}'$ are equivalent, there exists an operator $S: V \to V'$, $S \neq 0$, such that $S\pi(g) = \hat{\pi}'(g) S$.

The equality $(*)$ implies that the vectors $S\pi_w f$ and $\hat{\pi}'_{w'} f$ are proportional for every $w \in \tilde{V}_\theta$, $w' \in \tilde{V}'_\theta$ and every $f \in C^\infty_{c,\theta}(G)$.

Hence for every pair $w, w' \neq 0$ there exists a number λ such that $S\pi_w = \lambda \hat{\pi}'_{w'}$. Thus the operators π_{w_1} and π_{w_2} are proportional for every $w_1, w_2 \in \tilde{V}_\theta$. Finally we obtain that if $\tilde{V}'_\theta \neq 0$, then $\dim \tilde{V}_\theta \leq 1$. Similarly $\tilde{V}_\theta \neq 0$ implies $\dim \tilde{V}'_\theta \leq 1$. Assertion a) is proved.

Corollary. Let (π, V) *be an admissible irreducible representation of the group* G *such that* $\dim \tilde{V}_\theta \neq 0$, $\dim \tilde{V}'_\theta \neq 0$. *Then the space* $C_\theta^{-\infty}(G)$ *contains a unique subspace* $B_\pi \subset C_\theta^{-\infty}(G)$ *invariant relative to the action of the group* G *by right translations and such that the action of* G *in* B_π *is isomorphic to* (π, V).

To see this, let $w' \in \tilde{V}'_\theta$ and $w' \neq 0$. Define the map $\varrho: V \to C_\theta^{-\infty}(G)$ by $(\varrho v)(f) = (v, \pi'_{w'} f)$, where $f \in C^\infty_{c,\theta}(G)$, $\pi'_{w'} f = \int\limits_{Z \setminus G} f(g)\tilde{\pi}'(g^{-1})w' d\bar{g} \in V'$. Put $B_\pi = \operatorname{Im} \varrho$. It is clear that B_π is the required subspace. To prove the uniqueness, let $B_\pi^{(1)}$ be a subspace of the space $C_\theta^{-\infty}(G)$ and $\varrho^{(1)}: V \to B_\pi^{(1)}$ be the corresponding isomorphism. Since (π, V) is an algebraic representation, $B^{(1)}$ consists of those elements $C_\theta^{-\infty}(G)$, given by locally constant functions. Consider the functional L on $B_\pi^{(1)}$ such that $L(f) = f(e)$ for $f \in B_\pi^{(1)}$. Then the formula $L^{(1)} = L\varrho^{(1)}$ defines a functional on V, $L^{(1)} \in \tilde{V}'_\theta$.

If $B_\pi^{(2)}$ is another such subspace of $C_\theta^{-\infty}(G)$ then by Theorem 4 $L^{(2)} = \lambda L^{(1)}$ i.e. $L\varrho^{(2)} = \lambda L\varrho^{(1)}$. Hence $\varrho^{(2)} = \lambda \varrho^{(1)}$ and the Corollary is proved.

The realization of the representation (π, V) in the space B_π will be called the *Bessel realization*. Let $w' \in \tilde{V}'_\theta$, $w \in \tilde{V}_\theta$, $w' w \neq 0$. Let us denote by $(\pi_w)^*: \tilde{V}' \to C_\theta^{-\infty}(G)$ the operator dual to the operator π_w introduced above and put $\mathcal{J}_\pi = (\pi_w)^*(w')$. This function $\mathcal{J}_\pi \in C_\theta^{-\infty}(G)$ will be called the *Bessel function* of the representation π.

Remark. If the admissible representation (π, V) is such that $\tilde{V}'_\theta = \{0\}$, then there is no subspace of $C_\theta^{-\infty}(G)$ isomorphic to (π, V).

5. *The representations of the group* P_n. Let $G_n = GL(n, K)$, P_n be its subgroup consisting of the matrices $\|a_{ik}\|$ such that $a_{ni} = 0$, $i = 1, \ldots, n-1$, $a_{nn} = 1$ and Z be the subgroup of those matrices $\|a_{ik}\|$ such that $a_{ik} = 0$, if $i > k$, $a_{ii} = 1$.

Let us state a simple general lemma, which we will use later on.

Let ϱ be an algebraic representation of the K-group L in the space V. Let θ be a character of the group L. We denote by $\tilde{V}'_\theta$ the subspace of the space $\tilde{V}'$, consisting of all vectors $\tilde{v}'$ such that $\tilde{\varrho}'(l)\tilde{v}' = \theta(l)\tilde{v}'$ for every $l \in L$ and by $V_\theta^\perp$ the subspace of V orthogonal to $\tilde{V}'_\theta$. Suppose that there exists a subsequence $L_i \subset L$ of open compact subgroups $L_1 \subset L_2 \subset \ldots$ such that $\bigcup\limits_i L_i = L$. Let V_θ^0 be the set of vectors $v \in V$ satisfying

$$\int\limits_{L_i} \theta(l)\varrho(l)v \, dl = 0 \qquad (*)$$

for some i. Notice that since the representation ϱ is algebraic, this integral is a finite sum.

Lemma 3. $V_\theta^\perp = V_\theta^0$.

Proof. It is clear that $V_\theta^0 \subset V_\theta^\perp$. Let us show that every $\tilde{v}' \in \tilde{V}'$ satisfying $(\tilde{v}', V_\theta^0) = 0$ is an element of $\tilde{V}_\theta'$. The vector $w = \varrho(l_0)v - \theta(l_0^{-1})v$ belongs to V_θ^0 for all $l_0 \in L$ and $v \in V$. Indeed, suppose $l_0 \in L_i$. Then

$$\int_{L_i} \theta(l)\varrho(l) w\, dl = \int_{L_i} [\theta(l)\varrho(ll_0) - \theta(ll_0^{-1})\varrho(l)] v\, dl = 0$$

Hence $(\tilde{v}', \varrho(l_0)v) = (\tilde{v}', \theta^{-1}(l_0)v)$ for every $\tilde{v}'$ such that $(\tilde{v}', V_\theta^0) = 0$ and for all $v \in V$ and $l_0 \in L$. Thus $\tilde{v}' \in \tilde{V}_\theta'$ for every $\tilde{v}'$ orthogonal to V_θ^0.

Since every linear subspace in V is closed, V_θ^0 coincides with $V_\theta^\perp$. This proves the lemma.

Theorem 5. *Let π be an algebraic representation of the group P_n in the space V. If $V \neq 0$ there exists a character θ (possibly degenerate) of the group Z_n such that $\tilde{V}_\theta' \neq 0$.*

Proof. Denote by ϱ the restriction of the representation π of the group P_n to Z_n. Since the group Z_n is the union of open compact subgroups, by Lemma 3 it is sufficient to show that there exists a character θ on Z_n satisfying $V_\theta^0 \neq V$.

Let us prove the theorem by induction on n. Denote by M_n the group of the matrices of the form: $\|a_{ik}\|$, $a_{ik} = \delta_{ik}$ $i = 1, 2, \ldots, n$, $k = 1, \ldots, n-1$, $a_{nn} = 1$; $a_{in} = m_i$, $i = 1, \ldots, n-1$. $M_n \subset Z_n \subset P_n$. It is clear that the commutative group M_n is a normal subgroup in P_n and $P_n/M_n = G_{n-1}$. Denote by ϱ_0 the restriction of π to M_n. First suppose that the representation ϱ_0 is not trivial, i.e. $\varrho_0(m) \not\equiv \mathrm{Id}$. Let $\hat{m}$ be a character of M_n. The subspace $V_{\hat{m}}^0$ is defined by the character $\hat{m}$ and the representation ϱ_0. One can prove (using Lemma 2.8.1 in [3]) that there exists an element $\hat{m}_0 \in \hat{M}$, $\hat{m}_0 \neq 0$ (that is $\hat{m}_0(m) \not\equiv 1$) such that $V_{\hat{m}_0}^0 \neq V$. Since M_n is a normal subgroup of P_n the group P_n acts on $\hat{M}_n$. This action is transitive on $\hat{M}_n - \{0\}$. Therefore $V_{\hat{m}}^0 \neq V$ for every $\hat{m} \neq 0$.

Define the character $\hat{m}_{n-1}$ of the group M_n by the formula $\hat{m}_{n-1}(m) = \psi(m_{n-1})$, where $m = (m_1, \ldots, m_{n-1})$, ψ is a nontrivial character of the group K^+. As was shown above $V_{\hat{m}_{n-1}}^0$ differs from V. We have already seen that the group P_n acts on $\hat{M}_n$. Let us embed the group P_{n-1} into the group P_n in the standard way. The action of the group P_{n-1} leaves the character $\hat{m}_{n-1}$ unchanged. Thus the restriction of the representation π to P_{n-1} preserves the subspace $V_{\hat{m}_{n-1}}^0 \subset V$ and therefore a representation $\bar{\pi}$ of the group P_{n-1} in the space $V/V_{\hat{m}_{n-1}}^0 \overset{\text{def}}{=} \bar{V}_{\hat{m}_{n-1}} \neq \{0\}$ is defined. Besides that the space $V_{\hat{m}_{n-1}}^0$ is invariant with respect to the representation ϱ_0 of the group M_n. Therefore an action of the group M_n is defined in the space $\bar{V}_{\hat{m}_{n-1}}$ and is the multiplication by $\hat{m}_{n-1}(m)$. Hence the action of the group $P_{n-1} \times M_n \supset Z_n$ is defined on $\bar{V}_{\hat{m}_{n-1}}$. By the induction hypothesis there exists a character θ_{n-1} on $Z_{n-1} \subset P_{n-1}$ such that $(\bar{V}_{\hat{m}_{n-1}})^0_{\theta_{n-1}} \neq \bar{V}_{\hat{m}_{n-1}}$. Define the character θ on Z_n as follows: $\theta(z_{n-1}) = \theta_{n-1}(z_{n-1})$ for $z_{n-1} \in Z_{n-1} \subset Z_n$ and $\theta(m) = \hat{m}_{n-1}(m)$ for $m \in M_n \subset Z_n$. Let $v \in V$ satisfy $\bar{v} \notin (\bar{V}_{\hat{m}_{n-1}})^0_{\theta_{n-1}}$ (where $\bar{v}$ is the image of v under the morphism $V \to \bar{V}_{\hat{m}_{n-1}}$); then $v \notin V^0_\theta$. This follows from the fact that the image of the integral $(*)$ is not equal to zero when $v \to \bar{v}$, because $\bar{v} \notin (\bar{V}_{\hat{m}_{n-1}})^0_{\theta_{n-1}}$.

We have investigated the case $\varrho_0(m) \not\equiv \mathrm{Id}$. If $\varrho_0(m) \equiv \mathrm{Id}$, we apply the inductive assumption to the restriction of the representation π to the group P_{n-1}. Let θ_{n-1} be a character of the group Z_{n-1} satisfying $V^0_{\theta_{n-1}} \neq V$. Denote by θ the character of Z_n, given by the formula $\theta(z_{n-1}) = \theta_{n-1}(z_{n-1})$ for $z_{n-1} \in Z_{n-1} \subset Z_n$, $\theta(m) \equiv 1$ for $m \in M_n$. Then $V^0_\theta = V^0_{\theta_{n-1}} \neq V$; theorem 5 is proved.

Corollary 1. *Every cuspidal representation of G_n is nondegenerate.*

Corollary 2. *Let $Y_n = [Z_n, Z_n]$ and* Id *be the trivial character on Y_n. Then for every algebraic representation (π, V) of the group P_n, $V^0_{\mathrm{Id}} \neq V$.*

An algebraic representation (p, V) of *the* group P_n is called *degenerate*, if for some nondegenerate character θ of the group Z (and thus for all nondegenerate θ) we have $\tilde{V}'_\theta = 0$. Similarly an algebraic representation (π, V) of the group G_n is called *degenerate*, if $\tilde{V}'_\theta = 0$ for all nondegenerate characters θ of the group Z.

Proposition 2. *If L is a P_n-invariant subspace in $C^{-\infty}_{\theta^{-1}}(P_n)$, $L \neq \{0\}$, then $L \supset C^\infty_{c,\theta}(P_n)$.*

Proof. Firstly we note that $L \neq \{0\}$ implies $L_0 \neq \{0\}$. Hence we can suppose, that the P_n-invariant subspace L consists of uniformly locally constant functions.

Let U be an open compact subgroup of P_n. Let $\varphi \in C^\infty_c(M_n)$. Denote by $\varrho(\varphi)$ the following operator in the subspace $C^\infty_\theta(P_n)$, consisting of the uniformly locally constant functions from $C^{-\infty}_\theta(P_n)$:

$$\big(\varrho(\varphi)f\big)(p) = \int_{M_n} \varphi(m) f(pm)\, dm.$$

Let us show that the function φ can be chosen in such a way that $\big(\varrho(\varphi)f\big)(p) = f(p)$, if $p \in M_n P_{n-1} U = Z_n P_{n-1} U$ and $\big(\varrho(\varphi)f\big)(p) = 0$, if $p \notin M_n P_{n-1} U$. Denote by $W \subset \hat{M}_n$ the open subset of $\hat{M}_n$, consisting of the characters $\hat{m}$ of the form $\hat{m}(m) = \hat{m}_{n-1}(m^u) \overset{\mathrm{def}}{=} \hat{m}^{u^{-1}}_{n-1}(m)$ for some $u \in U$ and by $\bar{\varphi}$ the characteristic function of the set W.

Denote by $\varphi(m)$, $m \in M$, the Fourier transform of $\bar{\varphi}$. It is easy to check that the function φ satisfies the stated requirements. Indeed,

$$\big(\varrho(\varphi)f\big)(p) = \int \varphi(m) f(pm)\, dm = \int \varphi(m) f(m^p \cdot p)\, dm =$$

$$= f(p) \cdot \int \varphi(m)\, \Theta(m^p)\, dm = f(p) \int \hat{m}_{n-1}(m^p)\, \varphi(m)\, dm =$$

$$= f(p) \cdot \bar{\varphi}(\hat{m}^{p^{-1}}_{n-1}).$$

Let us show that $\bar{\varphi}(\hat{m}_{n-1}^{p^{-1}})$ is the characteristic function of the set $M_n P_{n-1} U$. Indeed, if $\bar{\varphi}(\hat{m}_{n-1}^{p^{-1}}) \neq 1$, then there exists $u \in U$ satisfying $\hat{m}_{n-1}^{p^{-1}} = \hat{m}_{n-1}^{u}$ i.e. $\hat{m}_{n-1}^{pu} = \hat{m}_{n-1}$, and hence $pu \in M_n P_{n-1}$, so $p \in M_n P_{n-1} U$. The verification of conditions 1), 2) is trivial.

Let now L be a P_n-invariant subspace of the space $C_\theta^\infty(P_n)$. To prove Proposition 2 it is sufficient to verify that for every open compact subgroup U_n there exists an open compact subgroup U, $U \subset U_n \subset P_n$ and a function $f_u \in L$ satisfying

1) $f_u(e) = 1$;

2) $f_u(pu) = f(p)$ for $p \in P_n$, $u \in U$;

3) $\operatorname{Supp} f_u \subset Z_n U$.

Indeed it is easy to check here that every function $f \in C_{c,\theta^{-1}}^\infty(P_n)$ is a finite linear combination of translates of functions of the form f_u. We denote by $\bar{L}$ the space of functions on $P_{n-1} \subset P_n$ that are obtained by restriction of functions from L. Since $L \neq \{0\}$ and is P_n-invariant we have $\bar{L} \neq \{0\}$.

Put $U_{n-1} = U_n \cap P_{n-1}$. Since $\bar{L} \subset C_\theta^\infty(P_{n-1})$ is P_{n-1}-invariant, it contains by the induction hypothesis a function g satisfying

1) $g(e) = 1$;

2) $g(pu) = g(p)$ for every $p \in P_{n-1}$, $u \in U_{n-1}$;

3) $\operatorname{Supp} g \subset Z_{n-1} U_{n-1}$.

Let us choose a function $f \in L$ whose restriction to P_{n-1} coincides with g. Let $f'(p) = (\operatorname{mes} U_{n-1})^{-1} \cdot \int_{U_{n-1}} f(pu_{n-1})\, du_{n-1}$. We denote by V the stabilizer of f' in P_n. Since $L \subset C_\theta^\infty(P_n)$, V is an open subgroup containing U_{n-1}. Let U be an open compact subgroup, $U = V \cap U_n$; then $U \cap P_{n-1} = U_{n-1}$. Let us consider the function $\varphi_u \in C_c^\infty(M_n)$ constructed above. Then the function $\varrho(\varphi_u) f' = f_u$ satisfies conditions *1), 2), 3)* and Proposition 2 is proved.

6. It is important for us to study the restriction of irreducible admissible representations of the group G_n to P_n.

Definition. An admissible irreducible representation (π, V) of the group G_n is called *finite*, if for some (hence, for every) pair $v_1 \in V$, $v_2 \in V'$, $v_1, v_2 \neq 0$ the function

$$f_{v_1, v_2}(g) = (\pi(g) v_1, v_2)$$

is finite mod C, where C is the centre of G_n.

Theorem 6. *Let (π, V) be a finite representation of the group G_n. Then there exists a monomorphism $\varphi_\pi : V \to C_\theta^{-\infty}(P_n)$, commuting with the action of P_n and it is unique up to some factor. Moreover $\varphi_\pi V = C_{c,\theta^{-1}}^\infty(P_n)$.*

Proof. The proof of the existence of such a morphism φ_π is based on the following lemma.

Lemma 4. *Let $L_n \subset C_c^\infty(P_n)$ be a subspace which is right P_n-invariant. If $L_n \neq 0$, then the representation of P_n in L_n is nondegenerate.*

Proof of the lemma. We will prove it by induction on n. We will use the notations that were introduced when proving Theorem 5. For every fixed character $\hat{m}\in\hat{M}_n$ denote by $C_{\hat{m}}\subset C^\infty(P_n)$ the subspace of the functions $f_{\hat{m}}$ of the form:

$$f_{\hat{m}}(p)=\int_{M_n} f(m^{-1}p)\,\hat{m}(m)\,dm,\quad f\in L_n.$$

The subspace $C_{\hat{m}}$ is right P_n-invariant. Since the space L_n consists of finite functions, there exists $\hat{m}, \hat{m}(m)\not\equiv 1$, such that $C_{\hat{m}}\neq\{0\}$; let us fix this $\hat{m}\in\hat{M}_n-\{0\}$. We will show that the representation of P_n in $C_{\hat{m}}$ is nondegenerate. Since P_n acts on $\hat{M}_n-\{0\}$ transitively, then there exists $p_0\in P_n$, such that $\hat{m}^{p_0^{-1}}=\hat{m}_{n-1}$. Denote by C_1 the right-invariant subspace of functions of the form

$$f_1(p)=f_{\hat{m}}(p_0^{-1}p),\quad f_{\hat{m}}\in C_{\hat{m}}.$$

The representation by right translations in C_1 is equivalent to the representation in $C_{\hat{m}}$. By that $f_1(mp)=\hat{m}_{n-1}(m)f_1(p)$ for every $f_1\in C_1$. Hence $f_1(pm)=f_1(pmp^{-1}p)=$ $=\hat{m}^p_{n-1}(m_1)f_1(p)$. Thus, for every $f_1\not\equiv 0$, $f_1\in C_1$ there exists a $m\in M_n$ such that $f_1(pm)\not\equiv f_1(p)$. It means that the representation (ϱ_0, C_1) of the group M_n on C_1 by right translations is such that $\varrho_0(m)\neq\mathrm{Id}$. Hence, as we have shown when proving Theorem 5, $(C_1)^0_{\hat{m}^{-1}_{n-1}}\neq C_1$. Let us now describe the space $(C_1)^0_{\hat{m}^{-1}_{n-1}}$. The statement $f_1(p)\in(C_1)^0_{\hat{m}^{-1}_{n-1}}$ is equivalent to the following one: there exists a compact subgroup $K\subset M_n$ such that $\int_K \hat{m}_{n-1}(m^{-1})f_1(pm)\,dm=0$ for all $p\in P_n$. It means that

$$f_1(p)\int_K \hat{m}^p_{n-1}\hat{m}^{-1}_{n-1}(m)\,dm=0. \qquad (*)$$

But $\hat{m}^p_{n-1}=\hat{m}_{n-1}$ if and only if $p\in P_{n-1}M_n$. Hence if $(*)$ is true for some subgroup K, then $f_1(p)=0$ for $p\in P_{n-1}M_n$. Conversely let $f_1\in C_1$ and $f_1(p)=0$ for all $p\in P_{n-1}M_n$. Let $U\subset P_n$ be an open subgroup such that $f(pu)=f(p)$, $u\in U$. Denote by $\hat{M}_u$ the subset of $\hat{M}_n$, consisting of the elements of the form $\hat{m}^u_{n-1}$, $u\in U$. Let K be a compact subgroup of M_n such that $\int_K \hat{m}\hat{m}^{-1}_{n-1}(m)\,dm=0$ for all $\hat{m}\notin\hat{M}_u$.

For such a subgroup $(*)$ holds. Hence we have shown that $(C_1)^0_{\hat{m}^{-1}_{n-1}}$ consists of all functions $f_1\in C_1$ such that $f_1(p)=0$ for all $p\in P_{n-1}M_n$. Denote by L_{n-1} the space of functions on P_{n-1}, which can be obtained by restriction of the functions $f_1\in C_1$. Since $(C_1)^0_{\hat{m}^{-1}_{n-1}}\neq C_1$ it follows that $L_{n-1}\neq\{0\}$.

Since for every $f_1\in C_1$ there exists a compact neighbourhood $U_1\subset P_n$ such that $\mathrm{Supp}\,f_1\subset U_1M_n$, the subspace $L_{n-1}\subset C_c^\infty(P_{n-1})$. By induction there exists a nondegenerate character θ_{n-1} on Z_{n-1} such that $(L_{n-1})^0_{\theta_{n-1}}\neq L_{n-1}$. Let us extend it up to the nondegenerate character θ on Z_n as was done on page 110. Similarly to the proof of Theorem 5 one can show that $(C_1)^0_\theta\neq C_1$ and hence $(L_n)^0_\theta\neq L_n$. The lemma is proved.

Let the representation (π, V) satisfy the conditions of the theorem. The representation (π, V) is nondegenerate (see Corollary 1 of Theorem 5) so we have $\dim\tilde{V}'_\theta=1$. We construct the map $\varphi_\pi: V\to C_\theta^{-\infty}(P_n)$ in the following way:

$(\varphi_\pi v)(p)=(\pi(p)v, \tilde{v}'_\theta)$, where $\tilde{v}'_\theta \in \tilde{V}'_\theta$. Let us prove that φ_π is a monomorphism. Put $W=\operatorname{Ker}\varphi_\pi$. W is P_n-invariant subspace in V. Since $\dim \tilde{V}'_\theta=1$, then according to Lemma 3, V^0_θ consists of the vectors v such that $(v, \tilde{v}'_\theta)=0$. Thus $W^0_\theta=W\cap V^0_\theta=W$. It means that the representation of the group P_n in W is degenerate. Let us prove that $W=\{0\}$. If $W\neq\{0\}$ then there exist $v'\in V'$ and $w\in W$ such that $(w, v')\neq 0$. Construct the map $\Phi: W\to C^\infty(P_n)$ in the following way: $\Phi(w)(p)=(\pi(p)w, v')$. Since the representation (π, V) is finite the image of Φ lies in $C^\infty_c(P_n)$. The representation (π, W) is degenerate, hence the representation of the group P_n in $\Phi(W)$ is degenerate too. By Lemma 4 $\Phi(W)=0$. But this equality contradicts the condition $(w, v')\neq 0$. Hence φ_π is a monomorphism.

Let us now prove the uniqueness.

Let $\varphi^{(1)}_\pi: V\to C^{-\infty}_\theta(P_n)$ be another morphism, satisfying the conditions of the theorem. The representation V is admissible, hence $\varphi^{(1)}_\pi V\subset C^\infty_\theta(P_n)$. Since the map is P_n-invariant, it is sufficient to find a constant c such that for every $v\in V$ we have $(\varphi^{(1)}_\pi v)(e)=c(\varphi_\pi v)(e)$. The functional $v\to\varphi^{(1)}_\pi v(e)$ can be written in the form $\varphi^{(1)}_\pi v(e)=(l, v)$, where $l\in\tilde{V}'$. But $(l, \pi(z)v)=(\varphi^{(1)}_\pi \pi(z)v)(e)=(\varphi^{(1)}_\pi v)(z)=$ $=\theta(z)(\varphi^{(1)}_\pi v)(e)=\theta(z)(l, v)$. Then $l\in\tilde{V}'_\theta$, i.e. $l=c\tilde{v}'_\theta$. The uniqueness of φ_π is proved.

To complete the proof of Theorem 6 it is enough to check the following proposition.

Proposition 3. *If* (π, V) *is an irreducible, admissible nondegenerate finite representation of* G_n, *then* $\varphi_\pi(V)=C^\infty_{c,\theta^{-1}}(P_n)$.

Proof. Let **H** be any algebraic K-group, (τ, V_1) a nondegenerate algebraic representation of H and $(\tilde{\tau}', \tilde{V}'_1)$ the adjoint representation. Suppose that an H-invariant embedding $V_1\hookrightarrow\tilde{V}'_1$ is given such that every nonzero H-invariant subspace of $\tilde{V}'_1$ contains V_1. Let $V\subset\tilde{V}'_1$ be a H-invariant subspace, such that the representation of the group H in V is algebraic. The proof of Proposition 3 is based on the following lemma.

Lemma 5. *Assume that there exists a functional* $l(v)$ *on* V *which is invariant with respect to some compact open subgroup* (i.e. $l\in V'$) *such that*

a) *the function* $f_v(h)=l(\tau(h)v)$ *is finite for every* $v\in V$,

b) $l|_{V_1}\not\equiv 0$.

Then $V=V_1$.

We suppose that all these spaces have the trivial topology.

Proof. For simplicity of notation we shall denote by τ_1 the representations in all three spaces V, V_1 and $\tilde{V}'_1$. Consider the map $V\to C^\infty_c(H)$, given in the following way: $v\to f_v(h)=l(\tau(h))$. The kernel of this map does not contain V_1 (condition b)) and hence is equal to 0 by assumption. Thus the map $v\to f_v(h)$ is an embedding. W denotes the set of the functions $f_v(h)$, $W\subset C^\infty_c(H)$ and ϱ

the representation arising in W. Let W_1 be the subspace of W consisting of the functions $f_{v_1}(h)$, $v_1 \in V_1$, $\overline{W}$ the closure of W in $L^2(H)$ and $\overline{W}_1$ a closure of W_1. Then the unitary representation ϱ_1 realized in the space $\overline{W}_1$ is irreducible. Indeed, let us consider the functional $(w, \overline{w}_1)$ on W_1 for every $\overline{w}_1 \in \overline{W}_1$ embedding of $\overline{W}_1$ into the space dual to W_1. Since W_1 is isomorphic to V_1, by the conditions of the lemma every H-invariant subspace in $\tilde{\overline{W}}_1'$ contains W_1. Hence being a unitary representation $(\overline{W}_1, \varrho_1)$ is irreducible. Consider in $\overline{W}$ the maximal invariant subspace L, where the representation is a direct sum of representations equivalent to $(\overline{W}_1, \varrho_1)$. By Theorem 2 from [9] L is a sum of finitely many such representations. Let d denote the number of them. Consider the following map $\pi: W \to \mathrm{End}\,(\overline{W}_1)$. Namely, put

$$\pi f = \int f(h^{-1}) \varrho_1(h)\, dh,$$

where $f \in W$. According to Lemma 1 [9] and Theorem 1 [9] one can find d H-invariant maps: $\tau_i: W \to \overline{W}_1$ $(i=1, \dots, d)$ and d vectors $w_i \in \overline{W}_1$ such that $(\pi f)(\xi) = \sum_{i=1}^{d} \tau_i(f) \cdot (\xi, w_i)$ for every $\xi \in \overline{W}_1$. These maps can be chosen in such a way that $\tau_i(W_1) = 0$ for $i=2, \dots, d$ and $\tau_1(W_1) = \mathrm{Id}$. Let us notice that if $\xi \in W \subset \overline{W}_1$ then $(\pi f)(\xi) \in W_1$. But from this follows that $\tau_i(f) \in W_1$. This means that τ_1 is a projection of W in W_1. The kernel of the map τ_1 is H-invariant and does not contain W_1, hence it is equal to 0 and $W = W_1$. The lemma is proved.

Let us prove now Proposition 3. According to Proposition 2 $\varphi_\pi(V) \supset C^\infty_{c,\theta^{-1}}(P_n)$. We can use Lemma 5 in the case $H = \boldsymbol{P}_n$, $V = V$, $V_0 = \varphi_\pi^{-1}(C^\infty_{c,\theta^{-1}}(P_n))$ and $\tau = \pi$. Proposition 3 and Theorem 6 are proved. We think that the following conjecture is true.

Let (π, V) be some irreducible admissible nondegenerate representation of the group G_n. Then there exists a unique morphism $\varphi_\pi: V \to C^{-\infty}_{\theta}(P_n)$ (unique up to a factor) commuting with the action of P_n. As a consequence of Proposition 2 on p. 110 we have $C^\infty_{c,\theta^{-1}}(P_n) \subset \varphi_\pi(V)$.

Conjecture. φ_π *is a monomorphism.*

Let (π, V) be a representation such that the statement of this conjecture is true. Then we introduce the following definition.

Definition. If (π, V) is an irreducible admissible nondegenerate representation of the group G_n, then the realization of V in the space $L_\pi \subset C^{-\infty}_{\theta}(P_n)$ satisfying $\pi(p) = \varrho(p)$ for all $p \in P_n$ will be called the *Gauss model* of the representation (π, V). There ϱ acts by right translations on P_n.

Later on we will consider representations π such that this conjecture is true. As it was shown above, the conjecture is true for all finite admissible irreducible representations.

Remark. The results of this § and Theorem 2 from [9] imply that every irreducible finite representation is admissible and nondegenerate.

7. First introduce some notations. Choose a fixed nondegenerate character θ, denoted by $\theta(z)=\psi(\sum_{i=1}^{n-1} z_{i,i+1})$, $'$ is the transposition in the group $GL(n, K)=G_n$. Let $s_n \in G_n$ be an element of the form $s_n=\sum_{i=1}^{n}(-1)^{i-1}E_{i,n-i+1}$. If (π, V) is a representation of G_n, then $(\hat{\pi}, V)$ denotes the representation $\hat{\pi}(g)=\pi^{-1}(g')$.

Let π (and hence $\hat{\pi}$) be an irreducible admissible nondegenerate representation of the group G_n, for which the conjecture on p. 114 is true.

If we define $(\varphi_\pi v)(e)\stackrel{\text{def}}{=}(l, v)$ and $(\varphi_{\hat{\pi}}\pi(s_n)v)(e)\stackrel{\text{def}}{=}(\hat{l}, v)$ then $l, \hat{l}\in\tilde{V}'_\theta$. Indeed $(\tilde{\pi}'(z)l, v)=(l, \pi(z)v)=(\varphi_\pi\pi(z)v)(e)=(\varphi_\pi v)(z)=\theta(z)(\varphi_\pi v)(e)=(\theta(z)l, v)$; the fact that $\hat{l}\in\tilde{V}'_\theta$ may be checked similarly. We have $\hat{l}=cl$, since according to Theorem 4 $\dim\tilde{V}'_\theta=1$. Choose φ_π and $\varphi_{\hat{\pi}}$ such that $c=1$, hence for every $v\in V$ $(\varphi_\pi v)(e)=$ $=(\varphi_{\hat{\pi}}\pi(s_n)v)(e)$.

We have denoted by $L_\pi\subset C_\theta^\infty(P_n)$ the image of the space V under the morphism φ_π. Recall that $L_\pi\supset C^\infty_{c,\theta^{-1}}(P_n)$. Introduce the operator $K(\pi): L_\pi\to L_{\hat{\pi}}$, $K(\pi)=$ $=\varphi_{\hat{\pi}}\cdot\varphi_\pi^{-1}$ which will be of fundamental significance in the sequel. The operator $K(\pi)$ has the following properties: for every $g\in G_{n-1}\subset P_n$ we have $\varrho(g)K(\pi)=$ $=K(\pi)\varrho((g')^{-1})$. Here $\varrho(g)$ is the representation by right translations of the group P_n in the space $C_\theta^\infty(P_n)$. For every Chevalley group we have introduced the Bessel model of the representation. For $GL(n, K)$ we introduced the useful realization of representations, which was called a Gauss model.

The Gauss model of (π, V) is entirely defined by the P_n-invariant subspace $L_\pi\supset C^\infty_{c,\theta}(P_n)$ and the operator $K(\pi): L_\pi\to C_\theta^\infty(P_n)$.

Recall that $L_\pi\supset C^\infty_{c,\theta}(P_n)$, since it is P_n-invariant (see Proposition 2). The operator $K(\pi)$ has the following properties:

1) Let $L_{\hat{\pi}}\stackrel{\text{def}}{=}K(\pi)L_\pi$; $K(\pi)$ has no kernel and the space $L_{\hat{\pi}}$ is P_n-invariant.
2) $\varrho(g)K(\pi)=K(\pi)\varrho((g')^{-1})$.

Denote by $M_n\subset P_n$ the group of matrices of the form

$$\begin{pmatrix} 1 & 0 & \dots & & m_1 \\ 0 & 1 & \dots & & m_2 \\ \dots & \dots & \dots & \dots & \dots \\ 0 & 0 & \dots & 1 & m_{n-1} \\ 0 & 0 & \dots & 0 & 1 \end{pmatrix}$$

If $m\in M_n$, we put $\varrho^K(m)=K^{-1}(\pi)\varrho(m^{-1})K(\pi)$. $\varrho^K(m): L_\pi\to L_\pi$, according to property *1)*. Notice that $\varrho^K(m^1m^2)=\varrho^K(m^1)\varrho^K(m^2)$. Given $m^{(1)}, m^{(2)}\in M_n$ let $(m^{(1)}, m^{(2)})=\sum_{i=1}^{n-1} m_i^{(1)}m_i^{(2)}\in K$, $m^{(1)}\times m^{(2)}=a$, where a is a matrix of order $(n-1)$ satisfying $a_{ik}=m_i^{(1)}m_k^{(2)}$.

3) There exists a quasicharacter $\det^K$ of the group K^* such that

$$\varrho(m^{(1)})\varrho^K(m^{(2)})\det^K(1+(m^{(1)}, m^{(2)})) =$$
$$= \varrho(I+m^{(1)}\times m^{(2)})\cdot\varrho^K(m^{(2)})\varrho(m^{(1)})\cdot\varrho((1+(m^{(1)}, m^{(2)}))I)$$

if $1+m^{(1)}m^{(2)}\neq 0$, $1+m^{(1)}\times m^{(2)}\in GL(n-1, K)$.

4) There is no nontrivial subspace in L_π invariant with respect to the set $\{\varrho(p), \varrho^K(m)\}$.

Lemma 6. *Let L be a P_n-invariant subspace, $L\subset C_\theta^\infty(P_n)$ and $K:L\to C_\theta^\infty(P_n)$ an operator. If K and L satisfy conditions 1)—4), then in the space L we have an action of the nondegenerate algebraic irreducible representation (π, L) of the group G_n according to the formulas*

$$\pi(p) = \varrho(p);\quad \pi(p') = K^{-1}\varrho(p)K;\quad p\in P_n,$$

where p′ is p transposed.

The proof of Lemma 6 is identical to the proof of Theorem 2 in [4].

Let (π, V) be an irreducible admissible nondegenerate representation of the group G_n; denote $\tilde{L}_\pi = L_\pi/C_{c,\theta}^\infty(P_n)$. In this space we have a natural action of the representation ϱ_π of the group P_n.

Remark 1. As we have already proved in Theorem 6, $\tilde{L}_\pi=0$ if (π, V) is a finite representation of the group G_n.

Remark 2. Jacquet and Langlands have shown that for $GL(2, K)$, $\dim\tilde{L}_\pi\leq 2$ and the representation is finite if $\dim\tilde{L}_\pi=0$.

Denote by R_{n-1} the ring of functions $C_c^\infty(G_{n-1})$, where multiplication is defined as convolution. For any algebraic representations (τ, W) of the group G_{n-1} put $\tau(\varphi)=\int\varphi(g)\tau(g)dg$; this defines the morphism $\tau: R_{n-1}\to \mathrm{Aut}\, W$.

Let π be an admissible irreducible representation of the group G_n; ϱ_π is the representation of the group P_n in the space $\tilde{L}_\pi$. Denote by $\bar{\varrho}_\pi$ the restriction of ϱ_π to G_{n-1} and let $\mathcal{I}_\pi=\mathrm{Ker}\,\bar{\varrho}_\pi$. Denote by Π_n the set of all nondegenerate irreducible admissible representations of the group G_n.

Definition. Let Conjecture on p. 112 be true for π, $\pi\in\Pi_n$. Denote by $\Pi_{n-1}(\pi)$ the subset of Π_{n-1}, consisting of the elements τ such that $\tau(\mathcal{I}_\pi)\neq 0$. Denote by Ω_n the subset of $\Pi_n\times\Pi_{n-1}$ consisting of the pairs (π, τ), such that $\tau\in\Pi_{n-1}(\pi)$. We will now construct the function $\Gamma_\psi(\pi, \tau)$, $(\pi, \tau)\in\Omega_n$. We feel that this function of the pair of representations of G_n and G_{n-1} is of great significance in the study of representations of the group G_n and the Weil—Langlands reciprocity law.

Let us define the operator $A: C_\theta^\infty(P_n)\to C_\theta^\infty(P_n)$. Write p in the form $p=mg_{n-1}$, where $m\in M_n$, $g_{n-1}\in G_{n-1}$ and write $p_1=ms_{n-1}g^{-1}$. We define $Af=f_1$, where $f_1(p)=f(p_1)$.

Since the space $C_{\theta^{-1}}^{-\infty}(P_n)$ coincides with $C_{\theta^{-1}}^{-\infty}(G_{n-1})$ Theorem 4 implies that for every $\tau \in \Pi_{n-1}$ there exists a unique subspace $B_\tau \subset C_{\theta^{-1}}^{-\infty}(P_n)$ such that the representation τ of the group G_{n-1} is realized in B_τ.

Let us consider the operator $A\varrho(\varphi)K(\pi)$ which maps L_π into $C_{c,\theta}^{\infty}(P_n)$ for every $\varphi \in \mathscr{I}_\pi$. Let us consider $A\varrho(\varphi)K(\pi)$ as an operator from $C_{c,\theta}^{\infty}(G_{n-1})$ into the same space and denote by $\beta(\pi): C_{\theta}^{-\infty}(G_{n-1}) \to C_{\theta}^{-\infty}(G_{n-1})$ the operator conjugate to $A\varrho(\varphi)K(\pi)$. According to Theorem 4 the operator $\beta(\pi)$ sends B_τ into itself. Set $\alpha(\pi)=\beta(\pi)|_{B_\tau}$. The map $\alpha(\pi): \mathscr{I}(\pi) \to \operatorname{End} B_\tau$ satisfies the condition $\alpha(\pi)(\varphi_1)=\tau(g_0)\alpha(\pi)(\varphi)\tau(g_1)$, if $\varphi_1(g)=\varphi(g_0^{-1}gg_1)$ and $g_0, g_1 \in G_{n-1}$. Since τ is an irreducible admissible representation, the morphisms $\alpha(\pi): \mathscr{I}(\pi) \to \operatorname{End} B_\tau$ and $\bar{\tau}: \mathscr{I}(\pi) \to \operatorname{End} B_\tau$ are proportional*.

Let $\alpha(\pi)=\Gamma_\psi(\pi, \tau)\tau$. If $\tau \in \Pi_{n-1}(\pi)$, then $\tau \not\equiv 0$, hence the number $\Gamma_\psi(\pi, \tau)$ is well defined. Let us note also, that if $(\pi, \tau) \in \Omega_n$ and both π and τ can be considered as unitary representations then $|\Gamma_\psi(\pi, \tau)|=1$.

Let σ be a finite-dimensional representation of Weil group W_K. Define the function

$$\Gamma_\psi(\sigma) = \frac{L(\frac{1}{2}, \sigma)}{L(\frac{1}{2}, \tilde{\sigma})\varepsilon(\psi, \sigma)};$$

where σ is an arbitrary finite-dimensional representation of W_K, $\tilde{\sigma}$ is a contragredient representation.

The main hypothesis of this article is the following.

Conjecture 1. *There is an embedding of $(\hat{W}_K)_n$ into the space Π_n of irreducible nondegenerate admissible representations of the group $GL(n, K)$ such that*

$$\Gamma_\psi(\pi \otimes \tilde{\tau}) = \Gamma_\psi(\varkappa\pi, \varkappa\tau); \quad \varkappa: (\hat{W}_k)_n \hookrightarrow \Pi_n$$

for every $\pi \in (W_K)_n$, $\tau \in (\hat{W}_K)_{n-1}$. If char $K=0$, *the map $\varkappa$ is bijective.*

This conjecture is a more precise form of Langlands' conjecture [2] for $G=GL(n, K)$.

We are very grateful to A. Weil, the discussions with whom stimulated our interest in these problems.

* Since π is an admissible representation, this statement is the conclusion of the following simple lemma.

Lemma. Let A be a ring, $\mathscr{I}$ be a two-sided ideal of A, (π, V) be a finite-dimensional irreducible representation of A, α be the map from $\mathscr{I}$ into $\operatorname{End} V$ such that $\alpha(ajb)=\pi(a)\alpha(j)\pi(b)$ for every $a, b \in A, j \in \mathscr{I}$. Then the map α and the restriction $\bar{\pi}$ of the representation π are proportional on $\mathscr{I}$.

Proof. Since π is irreducible, the restriction $\bar{\pi}$ of the representation π to $\mathscr{I}$ is either irreducible or $\bar{\pi} \equiv 0$. It is sufficient to investigate the first case, so we will assume that $\mathscr{I}=A$. Suppose $I=\operatorname{Ker} \pi$.

It is clear that $\alpha(I)=0$. Thus it is sufficient to analyse the case $A=\operatorname{End} V$. In this case the two-sided action of $\operatorname{End} V$ is absolutely irreducible. This implies the lemma.

REFERENCES

[1] Weil, A., Über die Bestimmung Dirichletscher Reihen durch Funktionalgleichungen, *Math. Ann.,* **168** (1967), 149—156.

[2] Langlands, R. P., Problems in the theory of automorphic forms, preprint (Russian translation: *сб. Математика* **15**:2 (1971), 57—83).

[3] Jacquet, H. and Langlands, R. P., *Automorphic forms on GL*(2), Springer Lecture Notes, Vol. **114**, 1970.

[4] Gelfand, I. M. and Graev, M. I. (Гельфанд И. М., Граев М. И.), Конструкция неприводимых представлений простых алгебраических групп над конечным полем, (Construction of irreducible representations of simple algebraic groups over finite fields), *ДАН СССР*, **147** (1962), 529—532.

[5] Langlands, R. P., On Artin's *L*-functions, *Rice Univ. Studies,* **56**:2 (1970). *(Complex analysis, 1969)* 23—28.

[6] Gelfand, I. M., Graev, M. I. and Pjateckij-Šapiro, I. I. (Гельфанд И. М., Граев М. И., Пятецкий-Шапиро И. И.), *Теория представлений и автоморфные функции,* „Наука" М., 1966 (*Representation theory and automorphic functions,* W. B. Saunders, 1969).

[7] Mumford, D., *Geometric invariant theory,* (Ergebnisse der Mathematik und ihrer Grenzgebiete, Band 34) Springer, 1965.

[8] Demazure, M. and Grothendieck, A., *Schémas en groupes I. Propriétés générales des schémas en groupes.* Séminaire de Géométrie Algébrique du Bois Marie 1962/64 (SGA 3), Springer Lecture Notes, Vol. **151**, 1970.

[9] Harish-Chandra (Notes by van Dijk, G.), *Harmonic analysis on reductive* $\mathfrak{p}$*-adic groups,* Springer Lecture Notes, Vol. **162**, 1970.

[10] *Seminar on algebraic groups and related finite groups,* held at the Institute for Advanced Study, Princeton, 1968/69. By A. Borel, R. Carter, C. W. Curtis, N. Iwahori, T. A. Springer, R. Steinberg, Springer Lecture Notes, Vol. **131**, 1970.

[11] Borel, A., *Linear algebraic groups,* W. A. Benjamin, New York, 1969.

[12] Rosenlicht, M., A remark on quotient spaces, *An. Acad. Brasil,* **35** (1963), 487—489.

[13] Steinberg, R., *Lectures on Chevalley groups,* Yale University, 1967.

REPRESENTATIONS OF THE GENERAL LINEAR GROUP OVER A FINITE FIELD

S. I. GELFAND

A large part of representation theory deals with the problem of studying the representations of reductive algebraic groups over various fields [1]—[4]. These investigations have shown that many properties of these groups are independent of the field over which the group is defined — whether it is complex, real, p-adic, or finite — thus, the information gained about the representations of some of these groups is useful for all the others. In the present lecture we consider the representations of the group $GL(n, \mathscr{K})$ of non-singular matrices of order n over the finite field $\mathscr{K}$ of order q (see also [5]—[13], [15]). In studying the representations of this group we are going to use the methods of the theory of representations of semisimple groups over different fields on the one hand, and those of the theory of representations of finite groups (theory of characters) on the other.

First we introduce some notations. In the sequel $\mathscr{K}$ will denote a fixed finite field of q elements, and $\mathscr{K}_d$ — for each natural number d — its extension of degree d (which is unique up to isomorphism). Let χ be a fixed character of the additive group of the field $\mathscr{K}$. If G is a finite group, H a subgroup, and S a representation of G, then $\operatorname{Ind}_H^G (S)$ will stand for the representation of G, induced by the representation S of the subgroup H. $|X|$ will denote the number of elements of the finite set X.

At first we examine in detail the representations of the group $G_2 = GL(2, \mathscr{K})$. This group consists of all matrices $g = \begin{pmatrix} \alpha & \beta \\ \gamma & \delta \end{pmatrix}$ with elements in $\mathscr{K}$, for which $\det g = \alpha\delta - \beta\gamma \neq 0$. The number of elements of G_2 clearly equals $q(q-1)(q^2-1)$.

It is easy to construct some representations of the group G_2 [5], [6]. Introduce the linear space of complex-valued functions $f(x_1, x_2)$ on the "plane" $(x_1, x_2) \in \mathscr{K} \times \mathscr{K} - (0, 0)$. Let $\tilde{\pi}$ be a multiplicative character of the field $\mathscr{K}$ (i.e. a character of the multiplicative group $\mathscr{K}^*$ of the field $\mathscr{K}$). Let us define the action of G_2 on the space introduced above by the formula

$$[T_{\tilde{\pi}}(g) f](x) = f(xg) \tilde{\pi}(\det g),$$

where $x = (x_1, x_2)$, $xg = (\alpha x_1 + \gamma x_2, \beta x_1 + \delta x_2)$. It is easy to verify that we obtain a representation of G_2. This representation is reducible. In fact, if π is another multiplicative character of $\mathscr{K}$, then the functions f satisfying the condition

$$f(tx_1, tx_2) = \pi(t)f(x_1, x_2) \qquad (t \in \mathscr{K}^*)$$

(the "homogeneous functions of degree π") form an invariant subspace with respect to G_2. It is easier to study the representations obtained if we construct them in a somewhat different way. Namely, let P be the subgroup of G_2 consisting of matrices $\begin{pmatrix}\alpha & \beta\\ \gamma & \delta\end{pmatrix}$ with $\gamma=0$. If π_1, π_2 are two multiplicative characters of $\mathscr{K}$, we define the one-dimensional representation P by the formula

$$p = \begin{pmatrix}\alpha & \beta\\ 0 & \delta\end{pmatrix} \to \varrho_{\pi_1, \pi_2}(p) = \pi_1(\alpha)\pi_2(\delta).$$

Then the representation $T_{\pi_1, \pi_2} = \mathrm{Ind}_P^{G_2}(\varrho_{\pi_1, \pi_2})$ coincides with the restriction of the representation $T_{\pi_1 \pi_2}$ to the subspace of homogeneous functions of degree π_1. The representations T_{π_1, π_2} are irreducible for $\pi_1 \neq \pi_2$ and T_{π_1, π_2} is equivalent to $T_{\pi_1', \pi_2'}$ only if $\pi_1 = \pi_1'$, $\pi_2 = \pi_2'$ or $\pi_1 = \pi_2'$, $\pi_2 = \pi_1'$. Their dimension is obviously $q+1$. The representations $T_{\pi, \pi}$ are reducible and can be decomposed into the one-dimensional representation $T_\pi^{(1)}: g \to \pi(\det g)$ and the representation $T_\pi^{(2)}$ of dimension q, all of which are pairwise inequivalent.

The described representations do not, however, exhaust all the representations of the group G_2. Now we give the construction of the remaining ones. Their construction is more complicated, because they are not induced by any representation of a proper subgroup of the group G_2. All these representations will be act in the space V of functions $f(u)$, $u \in \mathscr{K}^*$, so that their dimension is equal to $q-1$. Fixing a multiplicative character π of the field $\mathscr{K}_2$, for which $\pi^{q+1}(\sigma) \not\equiv 1$, $\sigma \in \mathscr{K}_2^*$, define the operator $T_\pi^{(*)}(g)$, $g = \begin{pmatrix}\alpha & \beta\\ \gamma & \delta\end{pmatrix}$ in the space V by the formula

$$[T_\pi^{(*)}(g)f](u) = \sum_{v \in \mathscr{K}^*} K_\pi(g \mid u, v) f(v),$$

where

$$K_\pi(g \mid u, v) = q^{-1}\chi\left(\frac{\alpha u + \delta v}{\gamma}\right) \sum_{\substack{t \in \mathscr{K}_2 \\ N(t) = uv^{-1}\det g}} \chi\left(-\frac{1}{\gamma v}(t + t^q)\right)\pi(t),$$

if $\gamma \neq 0$ and

$$[T_\pi^{(*)}(g)f](u) = \chi(\beta\delta^{-1}u)\pi(\delta)f(\alpha\delta^{-1}u),$$

if $\gamma=0$. Here $N(t)=tt^q$, the norm of the element $t \in \mathscr{K}_2$. To verify that the above formulae define a representation, may serve as a useful exercise. The representation $T_{\pi_1}^{(*)}$ is equivalent to $T_{\pi_2}^{(*)}$ only if $\pi_1(t) = \pi_2(t)$ for all $t \in \mathscr{K}_2^*$, or $\pi_1(t) = \pi_2(t^q)$ for all $t \in \mathscr{K}_2^*$. Now we mention some further properties of representations $T_\pi^{(*)}$. First, the restriction of representations $T_\pi^{(*)}$ to the subgroup H consisting of all matrices of the form $\begin{pmatrix}\alpha & \beta\\ 0 & 1\end{pmatrix}$ is irreducible. Further, the restriction of $T_\pi^{(*)}$ to the subgroup $Z = \left\{\begin{pmatrix}1 & z\\ 0 & 1\end{pmatrix}\right\}$ does not contain the identity representation.

Finally we give formulae for the characters of the representations. We represent the classes of conjugate elements in G_2 by naming one element of each class:

$$\begin{pmatrix}\lambda & 0\\ 0 & \lambda\end{pmatrix}, \quad \lambda \in \mathscr{K}^*; \quad \begin{pmatrix}\lambda & 1\\ 0 & \lambda\end{pmatrix},$$

$$\lambda \in \mathscr{K}^*, \quad \begin{pmatrix}\lambda & 0\\ 0 & \mu\end{pmatrix} \sim \begin{pmatrix}\mu & 0\\ 0 & \lambda\end{pmatrix}, \quad \lambda \neq \mu \in \mathscr{K}^*,$$

and if ε is a fixed element in $\mathscr{K}^* \setminus (\mathscr{K}^*)^2$,

$$\begin{pmatrix}a & b\\ \varepsilon b & a\end{pmatrix} \sim \begin{pmatrix}a & -b\\ -\varepsilon b & a\end{pmatrix}, \quad a, b \in \mathscr{K}^*, \quad b \neq 0$$

The corresponding classes are denoted by $c_1(\lambda)$, $c_2(\lambda)$, $c(\lambda, \mu)$ and $c(\sigma)$, $\sigma = a + b\sqrt{\varepsilon} \in \mathscr{K}_2$. The characters of G_2 are given by the following table:

$T_\pi^{(1)}$	$T_\pi^{(2)}$	T_{π_1, π_2}	$T_\pi^{(*)}$
$\pi(\lambda^2)$	$q\pi(\lambda^2)$	$(q+1)\pi_1\pi_2(\lambda)$	$(q-1)\pi(\lambda^2)$
$\pi(\lambda^2)$	0	$\pi_1\pi_2(\lambda)$	$-\pi(\lambda^2)$
$\pi(\lambda\mu)$	$\pi(\lambda\mu)$	$\pi_1(\lambda)\pi_2(\mu)+\pi_2(\lambda)\pi_2(\mu)$	0
$\pi(\sigma\sigma^q)$	$-\pi(\sigma\sigma^q)$	0	$\pi(\sigma)+\pi(\sigma^q)$

We mention the following way of generating the characters. Choose two functions:

$$f_1(\lambda, \mu), \quad \lambda, \mu \in \mathscr{K}^*, \quad f_1(\lambda, \mu) = f_1(\mu, \lambda)$$

and

$$f_2(\sigma), \quad \sigma \in \mathscr{K}^*, \quad f_2(\sigma) = f_2(\sigma^q).$$

To each pair (f_1, f_2) we assign a function ψ on the classes of conjugate elements of G_2 by the following formulae:

$$\psi(c_1(\lambda)) = \frac{1}{2}(q+1)f_1(\lambda, \lambda) + \frac{1}{2}(-q+1)f_2(\lambda)$$

$$\psi(c_2(\lambda)) = \frac{1}{2}f_1(\lambda, \lambda) + \frac{1}{2}f_2(\lambda)$$

$$\psi(c(\lambda, \mu)) = f_1(\lambda, \mu)$$

$$\psi(c(\sigma)) = f_2(\sigma).$$

Then the characters of G_2 correspond to the following pairs (f_1, f_2):

$$(\pi(\lambda\mu), \pi(\lambda))$$

$$(\pi(\lambda\mu), -\pi(\lambda))$$

$$(\pi_1(\lambda)\pi_2(\mu) + \pi_1(\mu)\pi_2(\lambda), 0)$$

$$(0, \pi(\sigma) + \pi(\sigma^q)).$$

We can see that it is much easier to describe the characters in terms of (f_1, f_2). The mapping $(f_1, f_2) \to \psi$ is called in [13] Green's mapping (for more details see [13]).

Now we turn to the study of representations of the group $G_n = GL(n, \mathcal{K})$. We introduce some notation. A set $\varrho = (l_1, \dots, l_k)$, where the l_i are integers, $0 < l_1 \leq \dots \leq l_k$, is called a partition of the number $|\varrho| = l_1 + \dots + l_k$. An ordered set $\hat{\varrho} = (l_1, \dots, l_k)$, where the $l_i > 0$ are integers, is called an ordered partition of the number $|\varrho| = l_1 + \dots + l_k$. Let $\hat{\varrho} = (l_1, \dots, l_k)$ be an ordered partition of n. Decompose each matrix $g \in G_n$ in the following form:

$$g = \begin{pmatrix} A_{11} & \dots & A_{1k} \\ \vdots & & \vdots \\ A_{k1} & \dots & A_{kk} \end{pmatrix},$$

where A_{ik} is an $n_i \times n_k$ matrix. We introduce the following subgroups of the group G_n:

The subgroup $P_{\hat{\varrho}}$ consists of the elements $g \in G_n$ for which $A_{ij} = 0$ for $i < j$. The subgroup $Z_{\hat{\varrho}}$ consists of the elements $g \in G_n$, for which $A_{ij} = 0$ for $i < j$, $A_{ii} = E$ (the unit matrix of order n_i). The subgroup $D_{\hat{\varrho}}$ consists of the elements $g \in G_n$ for which $A_{ij} = 0$ for $i \neq j$. Clearly $D_{\hat{\varrho}} = G_{l_1} \times \dots \times G_{l_k}$ (direct product) and $P_{\hat{\varrho}} = D_{\hat{\varrho}} Z_{\hat{\varrho}}$ (semidirect product). If, in particular, $\hat{\varrho} = (1, \dots, 1)$ then the corresponding subgroups will be denoted by P_n, Z_n and D_n. These are the maximal solvable subgroup (Borel subgroup), the maximal unipotent subgroup and the maximal split torus in G_n, respectively.

Let W_n be the subgroup of G_n consisting of all permutation matrices, i.e. matrices s having exactly one entry equal to 1 in each row and column; the other entries are equal to 0. This group is isomorphic to the Weyl group W_n of the group G_n. The Bruhat decomposition states that

$$G_n = Z_n D_n W_n Z_n,$$

i.e. each element $g \in G_n$ can be written in the form $g = z_1 d s z_2$ where $z_1, z_2 \in Z_n$, $d \in D_n$, $s \in W_n$. Here d, s are uniquely determined by the element g.

Let $\hat{\varrho} = (l_1, \dots, l_k)$ be an arbitrary ordered partition, and for each i we choose a representation T_i of the group G_{n_i} in the space V_i and let T' be the representation of $P_{\hat{\varrho}}$ in the space $V_1 \otimes \dots \otimes V_k$ constructed in the following way. Let $p \in P_{\hat{\varrho}}$, $p = dz$ where $d \in D_{\hat{\varrho}}$ and $z \in Z_{\hat{\varrho}}$ and $d = \operatorname{diag}(d_1, \dots, d_k)$. Then

$$T' = T_1(d_1) \otimes \dots \otimes T_k(d_k).$$

Now put $T_1 \circ \dots \circ T_k = \operatorname{Ind}_{P_{\hat{\varrho}}}^{G_n}(T')$.

If, in particular, $\hat{\varrho} = (1, \dots, 1)$, $T_i = \pi_i(\lambda)$, then the representation $\pi_1 \circ \dots \circ \pi_n$ so constructed is called a representation of the *principal series* of the group G_n.

The advantage of the operation $\circ$ lies in the fact that it permits us to construct representations of G_n starting from representations of G_m, $m < n$. Using this operation, the description of representations of G_n can be worked out in two steps. Firstly we have to find out which representations of G_n can be obtained by the above construction and characterize the missing ones as well. Secondly, we must study the properties of the representations $T_1 \circ \ldots \circ T_k$ (irreducibility, equivalence, etc.). Here we are going to deal mainly with the first problem.

Let $\hat{\varrho}$ be an arbitrary ordered partition with $k>1$ parts. The orispheric mapping corresponding to the partition $\hat{\varrho}$ is the mapping $\theta_{\hat{\varrho}}: \mathscr{E}(G_n) \to \mathscr{E}(G_n)$, defined by

$$(\theta_{\hat{\varrho}} f)(g) = \sum_{z \in Z_{\hat{\varrho}}} f(gz)$$

($\mathscr{E}(G_n)$ is the space of functions on G_n). It is clear that if $\hat{\varrho}=(n)$, then $\theta_{\hat{\varrho}}$ is the identity mapping. Denote by ${}^\circ\mathscr{E}(G_n)$ the space of functions $f \in \mathscr{E}(G_n)$ for which

$$\theta_{\hat{\varrho}} f = 0 \quad \text{for all} \quad \hat{\varrho} \neq (n).$$

A representation T of the group G_n is called *analytic* (or *cuspidal*) if its character belongs to the space ${}^\circ\mathscr{E}(G_n)$.

We have the following theorems [12, 15].

1) Let T be an arbitrary irreducible representation of the group G_n. Then there exists a partition $\varrho=(l_1, \ldots, l_k)$ and analytic representations T_i of the groups G_{l_i} such that T is contained in $T_1 \circ \ldots \circ T_k$.

2) Let T_i be an analytic representation of G_{l_i}, $1 \leq i \leq k$ and suppose that T_i are pairwise inequivalent. Then $T=T_1 \circ \ldots \circ T_k$ is irreducible.

The above theorems show the importance of studying analytic representations that we have just alluded to.

In the following an important role will be played by the subgroup $Z_n = Z_{(1,\ldots,1)}$ and its one-dimensional representations. It is easy to see that an arbitrary one-dimensional representation θ of the group Z_n is of the form

$$z = \begin{pmatrix} 1 & a_{12} & & & * \\ & 1 & a_{23} & & \\ & & \ddots & \ddots & \\ & & 1 & & a_{n-1,n} \\ & & & \ddots & \\ 0 & & & & 1 \end{pmatrix} \to \theta(z) = \chi(\alpha_1 a_{12} + \ldots + \alpha_{n-1} a_{n-1,n})$$

where the α_i are fixed elements of $\mathscr{K}$. The representation θ is said to be *non-degenerate* if $\alpha_i \neq 0$ for all i. An equivalent definition: θ is non-degenerate if the restriction of θ to the subgroup Z_ϱ is non-trivial for any $\hat{\varrho} \neq (n)$. Now we present two theorems that will show the importance of the one-dimensional representations of Z_n [14].

Theorem 4. Every irreducible representation of the group G_n is contained in one of the representations $\mathrm{Ind}_{Z_n}^{G_n}(\theta)$ where θ is a one-dimensional representation of Z_n.

Theorem 5. *Let θ be a nondegenerate one-dimensional representation of Z_n. Then every irreducible representation of G is contained in* $\operatorname{Ind}_{Z_n}^{G_n}(\theta)$ *at most once.*

Using the Frobenius duality, Theorems 4 and 5 can be reformulated in the following way:

Theorem 4′. *Every irreducible representation T of the group G_n has in its space a vector, which is a common eigenvector of all the operators $T(z)$, $z\in Z_n$.*

Theorem 5′. *For each irreducible representation T of the group G_n and nondegenerate one-dimensional representation θ of the subgroup Z_n, the subspace of vectors x satisfying $T(z)x=\theta(z)x$ is at most one-dimensional.*

Using these theorems it is easy to establish the following Theorem 6.

Theorem 6. *The representation T of the group G_n is analytic if and only if its restriction to Z_n does not contain degenerate one-dimensional representations.*

The analytic representations of the group G_n turn out to have some nice properties (see [15]). To formulate them, we introduce first the subgroup H_n, consisting of the matrices $g=(a_{ik})$ for which $a_{nn}=1$, $a_{ni}=0$, for $1\leqq i\leqq n-1$. Let $\theta_0(z)= =\chi(a_{12}+\ldots+a_{n-1,n})$ be a nondegenerate one-dimensional representation of Z_n and $T_0=\operatorname{Ind}_{Z_n}^{H_n}(\theta_0)$.

Lemma. The representation T_0 is irreducible.

Let us mention that $\dim T_0=\varphi_n(q)=(q-1)(q^2-1)\ldots(q^{n-1}-1)$. The following theorem is basic in the study of analytic representations.

Theorem 7. *Let T be an irreducible analytic representation of the group G_n. Then the restriction of T to H_n is also irreducible.*

Actually one has a stronger assertion. Namely

Theorem 8. *The restriction of T to H_n is equivalent to the irreducible representation T_0 of the group H.*

Corollary 1. The dimension of any irreducible analytic representation of the group G_n is equal to $(q-1)(q^2-1)\ldots(q^{n-1}-1)$. Let M_n be the subgroup of H_n consisting of the elements $g\in H_n$ for which $a_{in}=0$ for $1\leqq i\leqq n-1$. Clearly, M_n is isomorphic to G_{n-1}.

Corollary 2. Let T be an irreducible analytic representation of the group G_n. Then every irreducible representation of the group $M_n\cong G_{n-1}$ is contained in the restriction of T to M_n at most once.

The further study of the analytic representations will be based on the important notion of Bessel functions introduced first in [14]. Let T be an arbitrary irreducible representation of the group G_n and θ a nondegenerate one-dimensional representation of Z_n.

The *Bessel function* of the representation T (with respect to θ) is a function $J(g)$ on the group G_n satisfying the following conditions:

1) $J(z_1 g z_2)=\theta(z_1 z_2) J(g)$,

2) the restriction of the (right) regular representation of G_n to the smallest invariant subspace $E(J)$ in $\mathscr{E}(G_n)$ containing $J \in \mathscr{E}(G_n)$ is a multiple of the representation T.

Remark. It is clear that an arbitrary function $f \in E(J)$ satisfies the condition $f(zg)=\theta(z) f(g)$. Thus, Theorem 5 implies that the restriction to $E(J)$ of the regular representation is, in fact, equivalent to T. From the definition of induced representation and the duality theorem of Frobenius it follows at once that for given θ and T the Bessel function $J(g)$ exists if and only if T is contained in $\operatorname{Ind}_{Z_n}^{G_n}(\theta)$.

From Theorem 5 it can easily be obtained that if, for given θ and T the function $J(g)$ exists, then it is determined uniquely up to a constant factor. In the sequel it will be assumed that $J(e)=1$. The Bessel function J of a representation T can easily be calculated in terms of the character of T.

Theorem 8. *Let T be an irreducible representation of the group G_n, $\psi(g)$ its character, and θ a nondegenerate one-dimensional representation of Z_n. The function*

$$J^0(g) = |Z_n|^{-1} \sum_{z \in Z_n} \theta(z^{-1}) \psi(gz) \tag{1}$$

does not vanish identically if and only if T possesses a Bessel function, and in this case the Bessel function is J^0.

Let us further remark that the function $J(g)$ has the following properties: *1)* $J(g^{-1})=\overline{J(g)}$, *2)* $J * J=(\dim T)^{-1} J$*.

The dependence of J on θ is not essential, since if J exists for some one-dimensional nondegenerate representation θ, it does exist for any other nondegenerate θ' and, in addition $J_{\theta'}(g)=J_\theta(dgd^{-1})$ for some $d \in D_n$. In what follows, we will take $\theta_0(z)$ as θ.

Consider some further properties of J. Every element $g \in G_n$ can be written, according to the Bruhat decomposition, in the form $g=z_1 d s z_2$ where $z_1, z_2 \in Z_n$ $d \in D_n$, and $s \in W_n$ is a permutation matrix. In view of property 1) of the definition of the Bessel functions, these functions are determined by their values on the elements of the form ds.

* $*$ stands for convolution on G_n,

$$(f_1 * f_2)(g) = |G_n|^{-1} \sum_{g_1 \in G_n} f_1(g_1) f_2(g_1^{-1} g).$$

Notation. Let $n=n_1+\ldots+n_k$ be a partition of the number n into positive terms, and $\lambda_1, \ldots, \lambda_k \in \mathscr{K}^*$. Denote by $g_{n_1,\ldots,n_k}(\lambda_1, \ldots, \lambda_k)$ the element in G_n of the form

$$
\begin{array}{r}
n_1 \left\{ \vphantom{\begin{matrix} a \\ a \\ a \\ a \end{matrix}} \right. \\
n_2 \left\{ \vphantom{\begin{matrix} a \\ a \\ a \\ a \end{matrix}} \right. \\
\\
n_k \left\{ \vphantom{\begin{matrix} a \\ a \\ a \\ a \end{matrix}} \right.
\end{array}
\left(\begin{array}{c|c|c|c|c}
\overbrace{\qquad 0 \qquad}^{n_k} & 0 & \cdots & \overbrace{\qquad 0 \qquad}^{n_2} & \overbrace{\begin{matrix} \lambda_1 & 0 & \ldots & 0 \\ 0 & \lambda_1 & \ldots & 0 \\ \ldots & \ldots & \ldots & \ldots \\ 0 & 0 & & \lambda_1 \end{matrix}}^{n_1} \\
\hline
0 & 0 & \cdots & \begin{matrix} \lambda_2 & 0 & \ldots & 0 \\ 0 & \lambda_2 & \ldots & 0 \\ \ldots & \ldots & \ldots & \ldots \\ 0 & 0 & \ldots & \lambda_2 \end{matrix} & 0 \\
\hline
\cdots & \cdots & \cdots & \cdots & \cdots \\
\hline
\begin{matrix} \lambda_k & 0 & \ldots & 0 \\ 0 & \lambda_k & \ldots & 0 \\ \ldots & \ldots & \ldots & \ldots \\ 0 & 0 & \ldots & \lambda_k \end{matrix} & 0 & \cdots & 0 & 0
\end{array}\right)
$$

Lemma. Let $J(ds)\neq 0$, $d\in D_n$, $s\in W_n$. Then $ds=g_{n_1,\ldots,n_k}(\lambda_1, \ldots, \lambda_k)$ for some n_i and λ_i. *In particular, if $J(h)\neq 0$, $h\in H_n$ then $h\in Z_n$.*

The above-mentioned properties of the Bessel functions have been proved for arbitrary irreducible representations. Now we consider the case of analytic representations. First we mention that analytic representations always have Bessel functions, since they are contained in a representation, induced by a one-dimensional representation of a subgroup. The following arguments will be based on a simple lemma.

Lemma. Let G be an arbitrary finite group, H a subgroup, T an irreducible representation of G and ψ its character. We assume that the restriction to H of the representation T is also irreducible. Then, for arbitrary $g_1, g_2 \in G$ we have

$$\sum_{h\in H} \psi(g_1 h)\psi(h^{-1}g_2) = |H|(\dim T)^{-1}\psi(g_1 g_2).$$

Proof. Let e_i be a basis for the space of the representation T in which $T(g)$ are unitary matrices and let $a_{ij}(g)$ be the matrix elements of the representation T in this basis. Then $\psi(g)=\sum a_{ii}(g)$, $\sum_j a_{ij}(g_1)a_{jk}(g_2)=a_{ik}(g_1g_2)$, $a_{ij}(g)=\overline{a_{ji}(g^{-1})}$. Applying these equalities we can rewrite the sum on the left-hand side in the statement of the lemma as follows:

$$\sum_{h\in H} \psi(g_1 h)\psi(h^{-1}g_2) = \sum_{i,j,k,l} a_{ij}(g_1)a_{lk}(g_2) \sum_{h\in H} a_{ji}(h)\overline{a_{lk}(h)}.$$

We know that the restriction of T to the subgroup H is irreducible, and thus, it satisfies the orthogonality relations

$$\sum_{h\in H} a_{ji}(h)\overline{a_{lk}(h)} = |H|(\dim T)^{-1}\delta_{jl}\delta_{ik}$$

where δ_{ik} is the Kronecker symbol ($\delta_{ii}=1$, $\delta_{ik}=0$ for $i\neq k$). Consequently,

$$\sum_{h\in H} \psi(g_1 h)\psi(h^{-1}g_2) = |H|(\dim T)^{-1}\psi(g_1 g_2)$$

and the lemma is proved.

From this lemma we can easily get the following

Theorem 9. *Let T be an irreducible analytic representation of the group G_n, J its Bessel function. Then for any $g_1, g_2\in G$*

$$\sum_{m\in M_n} J(g_1 m)\overline{J(g_2 m)} = |Z_{n-1}|\, J(g_1 g_2^{-1}).$$

Applying the property $J(g^{-1})=\overline{J(g)}$ of the Bessel function, the formula obtained can be rewritten as follows:

$$\sum_{m\in M_n} J(g_1 m) J(m^{-1}g_2) = |Z_{n-1}|\, J(g_1 g_2). \tag{2}$$

Having this theorem we are able to give an explicit realization of the irreducible analytic representations of the group G_n. More precisely we have the following

Theorem 10. *Let T be an irreducible analytic representation of the group G_n. Consider the space E of functions on M_n satisfying the condition*

$$f(zm) = \theta_0(z)f(m), \quad z\in Z_{n-1} = Z_n\cap M_n,$$

with the scalar product $(f_1,f_2)=|Z_{n-1}|^{-1}\sum_{m\in M_n}\overline{f_1(m)}\, f_2(m)$.

In this space we define the operator $T_1(g)$ by the formula

$$[T_1(g)f](m) = |Z_{n-1}|^{-1}\sum_{m_1\in M_n} J(mgm_1^{-1})f(m_1). \tag{3}$$

Then T_1 will be a unitary representation of G_n (in the space E), equivalent to T.

Proof. We show that formula (3) gives in fact an operator in the space E. First we show that $T_1(g)f\in E$, if $f\in E$. Indeed

$$[T_1(g)f](zm) = |Z_{n-1}|\,\theta_0(z)\sum_{m_1\in M_n} J(mgm_1^{-1})f(m_1) = \theta_0(z)[T_1(g)f](m).$$

Now we verify that the operators $T_1(g)$ give a representation on the space E, i.e. $T_1(e)=I$ (identity operator) and $T_1(g_1)T_1(g_2)=T_1(g_1 g_2)$.

By the definition of the operator $T(e)$,

$$[T_1(e)f](m) = |Z_{n-1}|^{-1} \sum_{m_1 \in M_n} J(mm_1^{-1})f(m_1).$$

Since $J(mm_1^{-1}) \neq 0$ only if $z = mm_1^{-1} \in Z_n \cap M_n$ i.e. if $m_1 = z^{-1}m$, therefore

$$[T_1(e)f](m) = |Z_{n-1}|^{-1} \sum_z J(z)f(z^{-1}m).$$

And since $J(z) = \theta_0(z)$ and $f(z^{-1}m) = \theta_0(z^{-1})f(m)$ we have

$$[T_1(e)f](m) = f(m),$$

i.e. $T_1(e) = I$.

Let us now consider the product $T_1(g_1)T_1(g_2)$. It is equal to

$$[T_1(g_1)T_1(g_2)f](m) = |Z_{n-1}|^{-2} \sum_{m_1, m_2 \in M_n} J(mg_1m_1^{-1})J(m_1g_2m_2^{-1})f(m_2).$$

According to formula (2)

$$\sum_{m_1 \in M} J(mg_1m_1^{-1})J(m_1g_2m_2^{-1}) = |Z_{n-1}|\, J(mg_1g_2m_2^{-1})$$

and thus,

$$T_1(g_1)T_1(g_2)f(m) = T_1(g_1g_2)f(m).$$

To establish that T_1 is unitary, we have to show that $(f_1, f_2) = (T_1(g)f_1, T_1(g)f_2)$ for all $g \in G_n$ and all $f_1, f_2 \in E$. We have

$$(T_1(g)f_1, T_1(g)f_2) =$$

$$= |Z_{n-1}|^{-3} \sum_{m, m_1, m_2 \in M_n} J(mgm_1^{-1})\overline{J(mgm_2^{-1})}f_1(m_1)\overline{f_2(m_2)} =$$

$$= |Z_{n-1}|^{-2} \sum_{m_1, m_2 \in M_n} f_1(m_1)\overline{f_2(m_2)}J(m_2m_1^{-1}).$$

Just as before, we have $J(m_2m_1^{-1}) \neq 0$ only for $m_2m_1^{-1} = z \in Z_{n-1}$. Thus,

$$(T_1(g)f_1, T_1(g)f_2) = |Z_{n-1}|^{-2} \sum_{\substack{z \in Z_{n-1} \\ m_2 \in M_n}} J(z)f_1(z^{-1}m_2)\overline{f_2(m_2)} = (f_1, f_2).$$

It remains only to show that the representation T_1 is equivalent to T. Let us consider the vector $f_0 \in E_1$ for which $f_0(e) = 1$ and $f_0(m) = 0$ for $m \notin Z_{n-1}$ and find the entry corresponding to f_0 in an orthonormal basis in E_1, i.e. the scalar product $(T_1(g)f_0, f_0)$. We have

$$(T_1(g)f_0, f_0) = |Z_{n-1}|^{-2} \sum_{m_1 m \in M_n} J(mgm_1^{-1})f_0(m_1)\overline{f_0(m)} =$$

$$= |Z_{n-1}|^{-2} \sum_{z, z_1 \in Z_{n-1}} J(zgz_1^{-1})\theta_0(z_1)\overline{\theta_0(z)} = J(g)$$

i.e., one of the matrix elements of representation T_1 coincides with $J(g)$. Whence it follows that T_1 contains T as a subrepresentation. On the other hand

$$\dim T_1 = |M_n| \cdot |Z_{n-1}|^{-1} = \varphi_n(g) = \dim T$$

and therefore, T_1 is equivalent to T. The proof of the theorem is complete.

We remark that the indicated realization of analytic representations of G_n coincides with the one introduced above in the case $n=2$.

Now we describe in more detail how to construct the Bessel functions of irreducible representations. For this we will need formulae for the characters of the analytic representations of G_n. We will simultaneously prove that for any n the group G_0 has analytic representations.

To describe the characters we will first need to determine the classes of conjugate elements in G_n [16, 17]. This can be done in the following way. Denote by $\mathscr{R}$ the set of all polynomials irreducible over $\mathscr{K}$ with leading coefficient 1, except the polynomial $f(t) \equiv t$. For each $f \in \mathscr{R}$ of degree p we fix a matrix $L_f \in G_p$ whose characteristic polynomial $\det(tE - L_f)$ equals f (it always exists.) For arbitrary $l \geqq 1$, $L_{f,l}$ denotes the matrix of order $l \cdot p$ of the form

$$\begin{pmatrix} L_f & 1 & 0 & \dots & 0 \\ 0 & L_f & 1 & \dots & 0 \\ \dots & \dots & \dots & \dots & \dots \\ 0 & 0 & 0 & \dots & L_f \end{pmatrix}$$

where 1 stands for the identity matrix of order p.

Let $\varrho = (l_1, \dots, l_k)$ be a partition. Denote by $L_\varrho(f)$ the matrix (of order $|\varrho| \cdot p$) $\operatorname{diag}(L_{f,l_1}, \dots, L_{f,l_k})$. Let there be given a set of different irreducible polynomials $f_1, \dots, f_N \in \mathscr{R}$ and a set of partitions $\varrho_1, \dots, \varrho_N$ such that $\sum |\varrho_i| \deg f_i = n$.

Consider the following matrix in G_n;

$$A = \operatorname{diag}\left(L_{\varrho_1}(f_1) \dots L_{\varrho_N}(f_N)\right). \tag{4}$$

Note that its characteristic polynomial equals $f_1^{|\varrho_1|} \cdots f_N^{|\varrho_N|}$.

The set of all matrices of form (4) made for all possible choices of f_i and ϱ_i satisfying the condition $\sum |\varrho_i| \deg f_i = n$ forms a complete system of representatives for the classes of conjugate elements in G_n.

The class of conjugate elements in G_n containing the element (4) will be denoted by

$$c = [f_1^{\varrho_1} \dots f_N^{\varrho_N}].$$

Recall that the element $g \in G_n$ is semisimple, if it can be brought (over some extension of $\mathscr{K}$) to diagonal form, unipotent, if all its eigenvalues are equal to 1, and regular, if the dimension of $\operatorname{Cent}_{G_n}(g)$ (as an algebraic manifold) is equal to n. These properties will not be affected if we replace g by any of its conjugates, therefore, we can speak about semisimple, unipotent and regular classes of conjugate elements. It is easily seen that a semisimple class is determined by its characteristic polynomial. We remark that the class $c = [f_1^{\varrho_1} \dots f_N^{\varrho_N}]$ is semisimple if $\varrho_i = \{1^{|\varrho_i|}\}$ for any i, unipotent if $N = 1$ and $f_1 = t - 1$, and regular if $\varrho_i = \{|\varrho_i|\}$. Every element $g \in G_n$ can uniquely be represented in the form $g = g_s g_u$ where the semisimple g_s and the unipotent g_u commute. Clearly, when g runs through a class c of conjugate elements, g_s and g_u belong to some semisimple class c_s and unipotent class c_u, respectively. For the description of the characters of G_n it is useful to introduce the following auxiliary space. Let $\varrho = (l_1, \dots, l_r)$, $|\varrho| = n$, be an arbitrary partition. Denote by E_ϱ the space of all complex valued functions

$f(x_1, \dots, x_r)$, $x_i \in \mathcal{K}_{l_i}$, such that $f(x_1, \dots, x_i, \dots, x_r) = f(x_1, \dots, x_i^q, \dots, x_r)$ for each i, and $f(x_1, \dots, x_i, \dots, x_j, \dots, x_r) = f(x_1, \dots, x_j, \dots, x_i, \dots, x_r)$ if $l_i = l_j$. It is easy to verify by direct calculation that $\sum_\varrho \dim E_\varrho = \dim K(G_n)$. This allowed Green to introduce the mapping $I: K(G_n) \to \bigoplus_\varrho E_\varrho$, Now we list some properties of this mapping.

1) If we introduce the natural scalar products in $K(G_n)$ and $\bigoplus_\varrho E_\varrho$, then I becomes a unitary mapping.

2) Let $f = \{f_\varrho\} \in \bigoplus E_\varrho$, $\varphi = If \in K(G_n)$. Then $\varphi(c)$ depends only on $f_\varrho(x_1, \dots, x_r)$.

3) Let c be a regular semisimple class, φ — its characteristic polynomial, $\varphi = \Pi \varphi_i$ a decomposition into irreducible factors, $d_i = \deg \varphi_i$, $\varrho = (d_1, \dots, d_r)$, and x_i one of the roots of φ_i. Then $(If)(c) = f(x_1, \dots, x_r)$.

The complete description of all characters is rather involved, even using the mapping I. We consider the characters of analytic representations only. To this end, let us consider the multiplicative character π of the field $\mathcal{K}_n$, satisfying the following condition:

(P) For all m dividing but not equal to n, the character $\pi^{q^m - 1}(\sigma)$ is not identically 1.

Such a multiplicative character π gives an irreducible representation T_π of the group G_n, whose character ψ_π is described as follows:

1) $\psi_\pi(c) = 0$ if the characteristic polynomial of class c is not a power of a polynomial irreducible over $\mathcal{K}$.

2) Let the class c be of the form $c = [f^\varrho]$ where f is an irreducible polynomial of degree d over $\mathcal{K}$, ϱ is a partition of m containing p parts, $md = n$, and $\sigma \in \mathcal{K}_d$ is one of the roots of f. Then

$$\psi_\pi(c) = (-1)^{p+n} \varphi_p(q^d)[\pi(\sigma) + \dots + \pi(\sigma^{q^d})], \tag{5}$$

where, as before $\varphi_p(q) = (q-1)\dots(q^{p-1} - 1)$, $\varphi_1(q) = 1$.

From the above formulae one can see that π_1 and π_2 give the same representation of the group G_n if and only if $\pi_1(\sigma) = \pi_2(\sigma^{q^r})$ for some $r \geqq 1$. Simple, but long calculation shows that the representation T_π is analytic. Formulae (3) and (5) make it possible, in principle, to find all the analytic representations of G_n. In case $n = 2$ this description coincides with the one given above. Now we give the formulae for $n = 3$.

Let $\mathcal{K}_3$ be the cubic extension of the field $\mathcal{K}$ and let $\sigma \in \mathcal{K}_3$. Set

$$S(\sigma) = \sigma + \sigma^q + \sigma^{q^2}, \quad P(\sigma) = \sigma^{q+1} + \sigma^{q^2+1} + \sigma^{q^2+q}, \quad N(\sigma) = \sigma^{q^2+q+1}.$$

It is clear that $S(\sigma)$, $P(\sigma)$, $N(\sigma) \in \mathcal{K}$ and that the three roots of the polynomial $t^3 - S(\sigma)t^2 + P(\sigma)t - N(\sigma)$ are σ, σ^q, σ^{q^2}.

Note that $S(\sigma)$ and $N(\sigma)$ are the trace and the norm of the element $\sigma \in \mathcal{K}_3$, respectively, in the extension $\mathcal{K}_3 : \mathcal{K}$. According to (5), an analytic representation of the group is given by a multiplicative character π of the field $\mathcal{K}_3$, for which $\pi^{q-1} \not\equiv 1$. This condition clearly means that the restriction of π to the set of σ's with $N(\sigma)=1$ is not identically equal to 1. The character of this representation can be determined in the following way.

Let $\sigma \in \mathscr{K}_3$, $\sigma \notin \mathscr{K}$. Denote by $c(\sigma)$ the class of conjugate elements of $[N(t-\sigma)^{\{1\}}]$.

For $\lambda \in \mathscr{K}^*$ there are three classes of conjugate elements, whose characteristic polynomial equals $(t-\lambda)^3$. These classes are

$$c_1(\lambda) = [(t-\lambda)^{\{111\}}],\ c_2(\lambda) = [(t-\lambda)^{\{12\}}],\ c_3(\lambda) = [(t-\lambda)^{\{3\}}].$$

The character ψ_π of the reducible analytic representation T_π of the group equals

$$\begin{aligned}\psi_\pi(c(\sigma)) &= \pi(\sigma)+\pi(\sigma^q)+\pi(\sigma^{q^2}),\\ \psi_\pi(c_1(\lambda)) &= (q-1)(q^2-1)\pi(\lambda),\\ \psi_\pi(c_2(\lambda)) &= (1-q)\pi(\lambda),\\ \psi_\pi(c_3(\lambda)) &= \pi(\lambda);\end{aligned}$$

$\psi(c)=0$ for the other classes c.

Now we give explicit expressions for the Bessel function $J(g)$ of the representation T_π applying formula (1).

Let z be an arbitrary element in Z_3: $z = \begin{pmatrix}1 & \zeta & \xi\\ 0 & 1 & \eta\\ 0 & 0 & 1\end{pmatrix}$

Then $\theta_0(z)=\chi(\zeta+\eta)$. Using the notations introduced above, we have:

1) If $g=g_3(\lambda)$ then $J(g)=J_3(\lambda)=\pi(\lambda)$.

2) If $g=g_{2,1}(\lambda_1,\lambda_2)$ then putting $J(g)=J_{2,1}(\lambda_1,\lambda_2)$ we have

$$J_{2,1}(\lambda_1,\lambda_2) = q^{-2} \sum_{\substack{\sigma\in\mathscr{K}_3\\ N(\sigma)=\lambda_1^2\lambda_2}} \chi(\lambda_1^{-1}\lambda_2^{-1}P(\sigma))\pi(\sigma).$$

3) If $g = g_{1,2}(\lambda_1,\lambda_2)$, then

$$J(g) = J_{1,2}(\lambda_1,\lambda_2) = q^{-2} \sum_{\substack{\sigma\in\mathscr{K}_3\\ N(\sigma)=\lambda_1\lambda_2^2}} \chi(-\lambda_2^{-1}S(\sigma))\pi(\sigma).$$

4) If $g = g_{1,1,1}(\lambda_1,\lambda_2,\lambda_3)$, then

$$J(g) = J_{1,1,1}(\lambda_1,\lambda_2,\lambda_3) = q^{-3} \sum_{\substack{\sigma\in\mathscr{K}_3\\ N(\sigma)=\lambda_1\lambda_2\lambda_3}} \pi(\sigma) \sum_{\zeta\eta=\lambda_2^{-2}\lambda_3^{-1}N(\sigma-\lambda_2)} \chi(\zeta+\eta)-$$
$$-q^{-2}\pi(\lambda_2)\delta(\lambda_1\lambda_3+\lambda_2^2),$$

where $\delta(0)=1$, $\delta(\varkappa)=0$ for $\varkappa\neq 0$. According to what we have said above, $J(ds)=0$ for the remaining ds.

By virtue of property 1 of Bessel functions, it is easy now to get the values $J(g)$ for all $g\in G_3$. We mention an interesting corollary of formula (2). Here we will suppose that $q>2$, $q\neq 3^k$.

Put $P(x)=x^3+a_1x^2+a_2x+a_3$, where $a_i\in\mathscr{K}$. For arbitrary $A, B, C\in\mathscr{K}$ we consider the set $\mathscr{P}_{A,B,C}$ of all polynomials P, for which $a_1A+a_2B+a_3C=0$. Now we determine the number of irreducible polynomials in $\mathscr{P}_{A,B,C}$. Since the characteristic of the field $\mathscr{K}$ is not equal to 3, the problem can easily be reduced to the case $B=0$. Furthermore, the case $A=B=0$ can be considered trivial. Therefore we will suppose that $A\neq 0$, $B=0$.

Let $L(x)$ $(x \in \mathscr{K})$ be the Legendre symbol $(L(0)=0, L(y^2)=1$ for $y \neq 0$, $L(x)=-1$ for the remaining x). Denote by μ_{A_1, A_2} the number of different solutions of the equation $x^3+A_1x+A_2$ lying in the field $\mathscr{K}$. Then we have the following

Theorem 11. *The number of polynomials in* $\mathscr{P}_{A,0,C}$ $(A \neq 0)$ *irreducible over* $\mathscr{K}$ *is equal to*

$$\frac{1}{3}[q^2+qL(A^2+4C^2)+1-\mu_{3A,C}].$$

REFERENCES

[1] Harish-Chandra (Notes by Mars, J.G.M.), *Automorphic forms on semisimple Lie groups,* Springer Lecture Notes, Vol. **62**, 1968.

[2] Gelfand, I. M. and Neumark, M. I. (Гельфанд И. М., Наймарк М. И.), Унитарные представления классических групп, *Труды матем. ин-та АН СССР,* **36** (1950), 1—288. (*Unitäre Darstellungen der klassischen Gruppen,* Akademie Verlag, Berlin, 1957).

[3] ——— and Pjateckij-Šapiro, I. I. (Пятецкий-Шапиро И. И.), Теория представлений и теория автоморфных функций, *Успехи мат. наук,* **14**: 2 (1959), 171—194. (Representation theory and theory of automorphic functions.)

[4] Harish-Chandra (Notes by van Dijk, G.), *Harmonic analysis on reductive* $\mathfrak{p}$*-adic groups,* Springer Lecture Notes, Vol. **162**, 1970.

[5] Gelfand, I. M. and Graev, M. I. (Гельфанд И. М., Граев М. И.), Категория представлений и задача о классификации неприводимых представлений, *ДАН СССР,* **146** (1962), 757—760.
(The category of representations and the problem of classification of irreducible representations.)

[6] Tanaka, S., Construction and classification of irreducible representations of special linear group of second order over finite field, *Osaka J. Math.,* **4** (1968), 65—84.

[7] Frobenius, G., Über Gruppencharaktere, *Sitzungsberichte der Berl. Ak.,* 1896, 985—1021.

[8] Harish-Chandra, Eisenstein series over finite fields, *preprint,* 1968.

[9] Schur, I., Untersuchungen über die Darstellungen der endlichen Gruppen durch gebrochene lineare Substitutionen, *J. für Math.,* **132** (1907), 85.

[10] Steinberg, R., The representations of $GL(3, q)$, $PGL(3, q)$, $GL(4, q)$, *Canad. J. Math.,* **3** (1951), 225—235.

[11] ———, A geometric approach to the representations of the full linear group over a Galois field, *Trans. Amer. Math. Soc.,* **74** (1951), 274—282.

[12] Green, J. A., The characters of the finite general linear groups, *Trans. Amer. Math. Soc.,* **80** (1955), 402—447.

[13] Springer, T. A., Characters of special groups, *in Seminar on algebraic groups and related finite groups,* Springer Lecture Notes, Vol. **131**, 1970.

[14] Gelfand, I. M. and Graev, M. I. (Гельфанд И. М., Граев М. И.), Конструкция неприводимых представлений простых алгебраических групп над конечным полем, *ДАН СССР,* **147** (1962), 529—532.
(The construction of irreducible representations of simple algebraic groups over a finite field.)

[15] Gelfand, S. I. (Гельфанд С. И.), Представления полной линейной группы над конечным полем, *Матем. сборник,* **83** (*125*), (1970), 15—41.
(Representations of the general linear group over a finite field.)

[16] Springer, T. A. and Steinberg, R., Conjugate classes, *in Seminar on algebraic groups and related finite groups,* Springer Lecture Notes, Vol. **131**, 1970.

[17] Dickson, L. E., *Linear groups,* New York, 1958.

DEFORMATIONS OF NON-DISCRETE UNIFORM SUBGROUPS OF SEMISIMPLE LIE GROUPS

MORIKUNI GOTO and HSIEN-CHUNG WANG

§ 1. Introduction

Let G be a topological group. A subgroup H of G is called *uniform* if there exists a compact subset C of G such that $G=CH$. Discrete uniform subgroups of a Lie group have been extensively discussed by several mathematicians. It is the aim of the present note to explain our recent results on closed non-discrete uniform subgroups of semisimple Lie groups. The main results are as follows.

Let G be a connected semisimple Lie group with a trivial center and without compact factor. Let H be a closed non-discrete uniform subgroup of G. Let us suppose that H does not contain any non-trivial normal subgroup of G, and that the identity component of H is not contained in any proper normal subgroup of G. (If, in particular, G is simple, these conditions are automatically satisfied.)

(A) *If $f: H \to G$ is a locally one-to-one continuous homomorphism, then f is one-to-one and $f(H)$ is a closed uniform subgroup of G.*

(B) *There exists a finite system of subgroups $H_1, H_2, \ldots, H_s$ of G such that any subgroup H^* of G, which is isomorphic with H, can be written as $H^* = gH_ig^{-1}$ for some g in G and i, $1 \leqq i \leqq s$.*

(C) *If G is moreover normal ($=$**R**-split) semisimple, then each injective homomorphism $f: H \to G$ can be extended to an automorphism of G.*

The outline of our results has been published in Goto—Wang [3] and the detailed proofs will appear in Goto—Wang [4].

§ 2. Deformations and rigidities of subgroups

Let P and Q be locally compact topological groups. Let $R(Q, P)$ denote the set of all continuous homomorphisms from Q into P. We introduce the compact-open topology in $R(Q, P)$. Let $R_0(Q, P)$ denote the subspace of $R(Q, P)$ composed of all topological isomorphisms. Also by Aut P and Aut Q we denote the groups of all automorphisms of P and Q, respectively. For $\alpha \in \text{Aut } P$, $\beta \in \text{Aut } Q$ and $f \in R(Q, P)$, we let

$$f \cdot (\alpha, \beta) = \alpha^{-1} \circ f \circ \beta,$$

and thus the product group Aut $P \times$ Aut Q acts on the space $R(Q, P)$ as a transformation group, leaving the subspace $R_0(Q, P)$ invariant. By the *deformation* of Q in P we mean the transformation group thus obtained. It is to be noted that the orbit space $R_0(Q, P)/\mathrm{Aut}\, Q$ can be identified with the set of all subgroups of P which are topologically isomorphic with Q.

When Q is Abelian and P is the multiplicative group of complex numbers with absolute value one, $R(Q, P)$ is the dual group of Q in the sense of Pontrjagin and is locally compact. This fact can be generalized in the following way: if P is a Lie group and if either Q is compactly generated or P is compact, then the space $R(Q, P)$ is locally compact. When Q is a compactly generated Lie group, and P is a Lie group, we can define a natural analytic set structure of finite dimension on the space $R(Q, P)$.

After this, we assume that P is a Lie group, and Q is a closed subgroup of P. For a Lie group * we adopt the notation Int * for the group of all inner automorphisms of *. By $\mathfrak{A}$ let us denote either Aut P or Int P. The deformation of Q in P is said to be *strongly rigid* with respect to $\mathfrak{A}$ if the orbit $\iota\cdot(\mathfrak{A}\times\mathrm{Aut}\, Q)$ containing the inclusion map $\iota: Q\to P$ is open in $R(Q, P)$. If $\iota\cdot(\mathfrak{A}\times\mathrm{Aut}\, Q)$ is open in $R_0(Q, P)$, the deformation is called *rigid* with respect to $\mathfrak{A}$.

Next, let us define rigidities of a more global nature. Let $s=s(Q, P, \mathfrak{A})$ denote the cardinal number of the orbit space of $R_0(Q, P)$ modulo $\mathfrak{A}\times\mathrm{Aut}\, Q$. When s is finite, Q is said to have the *finite conjugate class property* in P with respect to $\mathfrak{A}$. Here we are concerned with a connected semisimple Lie group G, and we know that the factor group $\mathrm{Aut}\, G/\mathrm{Int}\, G$ is finite. Hence if $P=G$, there is no distinction between the two finite conjugate class properties. If, in particular, $s=1$, we call Q *absolutely rigid* with respect to $\mathfrak{A}$. The absolute rigidity of course implies the (local) rigidity.

§ 3. Uniform subgroups

Let us state some general results on uniform subgroups.

(A) *Let G be a connected Lie group, K a maximal compact subgroup of G. Let H be a connected Lie subgroup of G. We let $\mathfrak{G}$, $\mathfrak{K}$ and $\mathfrak{H}$ denote the Lie algebras of G, K and H, respectively. Then H is uniform in G if and only if $G=KH$ or* $\mathfrak{G}=\mathfrak{K}+\mathfrak{H}$. This is a modification of a theorem due to D. Montgomery in [5].

(B) *Let G be a minimally almost periodic topological group and H a uniform subgroup of G. Let f be a continuous representation of G into $GL(V)$, where V is a finite-dimensional vector space over* **R** *or* **C**. *If H leaves a subspace W of V invariant and* $\det f(h)|_W=1$ *for all $h\in H$, then W is invariant under G.* This is a generalization of a theorem of A. Weil [6], and there follow the corollaries:

Corollary 1. *Let G be a connected semisimple Lie group without compact factor, and H a closed uniform subgroup of G. Then the center of H is discrete.*

Corollary 2. *Let G be a connected complex semisimple Lie subgroup of $GL(n$,* **C**), *and H a parabolic subgroup of G. Then the ring of all polynomial invariants of G coincides with that of H.*

§ 4. Pre-algebraic uniform subgroups of a semisimple linear group

Let G be a subgroup of the general linear group $GL(n, \mathbf{R})$. If there exists an algebraic subgroup A of $GL(n, \mathbf{R})$, containing G as a subgroup of finite index, then G is said to be *pre-algebraic*. A connected Lie subgroup of $GL(n, \mathbf{R})$ of a certain dimension n is called *linear*. A semisimple linear group is always pre-algebraic. For any subgroup G of $GL(n, \mathbf{R})$ we can find the smallest pre-algebraic subgroup $A(G)$ containing G. Let us call $A(G)$ the *pre-algebraic hull* of G. A subgroup G of $GL(n, \mathbf{R})$ is called $\mathbf{R}$-*triangularizable* if there is a basis of $\mathbf{R}^n$ with respect to which the elements of G are represented simultaneously by upper triangular matrices. By a result of A. Borel, in every pre-algebraic group, maximal connected $\mathbf{R}$-triangularizable subgroups are conjugate to each other with respect to inner automorphisms. It follows that a pre-algebraic subgroup L of a semisimple linear group G is uniform in G if and only if L contains a maximal $\mathbf{R}$-triangularizable subgroup of G. (Borel—Tits [2].) Using this fact we can get complete descriptions of uniform pre-algebraic subgroups of semisimple linear groups as follows.

Let G be a semisimple linear group, and let L be a uniform pre-algebraic subgroup of G. Let M be the unipotent radical (=the largest normal subgroup composed of unipotent automorphisms) of L. Then we can find a reductive subgroup F such that $L=FM$ and $F\cap M=\{e\}$ is a semi-direct product, where e denotes the identity matrix. Let A_0 denote the maximal connected $\mathbf{R}$-diagonalizable normal subgroup of F, and A a maximal connected $\mathbf{R}$-diagonalizable subgroup of G containing A_0. We denote by $\mathfrak{A}$ and $\mathfrak{A}_0$ the Lie algebras of A and A_0, respectively. $\mathfrak{A}$ is a Cartan subalgebra of the symmetric space associated to G. Then the Lie algebra $\mathfrak{M}$ of M can be written as

$$\mathfrak{M} = \sum_{\beta \in \Phi_2} \mathfrak{G}(\beta),$$

where Φ_2 is the totality of positive roots, referred to a suitable ordering, of $\mathfrak{G}$ with respect to $\mathfrak{A}$, which do not vanish on $\mathfrak{A}_0$, and $\mathfrak{G}(\beta)=\{x \in \mathfrak{G};\ [a, x]=\beta(a)x$ for $a \in \mathfrak{A}\}$. Therefore, in particular, up to inner automorphisms there are only finitely many such $\mathfrak{M}$'s.

Next, the reductive part F can be written as a product: $F=QSA_0$, where (i) S is a connected semisimple Lie subgroup without compact factor, (ii) Q is a compact subgroup, and (iii) the Lie algebra $\mathfrak{F}$ of F is decomposed into the direct sum of three ideals: $\mathfrak{F}=\mathfrak{Q}+\mathfrak{S}+\mathfrak{A}_0$. Let us call $SA_0M=J$ the *uniform radical* of L. Then the uniform radical of L is a characteristic subgroup of L and coincides with the uniform radical of the normalizer N of M. Thus M determines N and J uniquely and $N\supset L\supset J$, and L is a local semi-direct product of a compact group and J.

A parabolic subgroup of G is pre-algebraic and uniform, and conversely if G is normal semisimple, a pre-algebraic uniform subgroup is parabolic.

A Lie group P is said to be *unimodular* if a left invariant Haar measure is also right invariant. Let G be a semisimple linear group, and let L be a pre-algebraic uniform subgroup of G. If H is a closed normal subgroup of L such that the factor group L/H is unimodular, then we can prove that H contains the unipotent radical M of L. Suppose H is uniform in G. Then the above result implies that $H \supset M$ and $H=(F \cap H)M$ is a semi-direct product. The density theorem of A. Borel in [1] also follows from this for uniform lattices.

§ 5. Closed non-discrete uniform subgroups of a semisimple Lie group

Let G be a semisimple linear group. We assume that G has no compact factor and the center of G is composed of the identity e alone. We denote by Ω the totality of pre-algebraic uniform subgroups of G which do not contain any normal subgroup of G except $\{e\}$. Also we define a set Θ of closed non-discrete uniform subgroups of G by the conditions that if $H \in \Theta$ then (*a*) H does not contain any normal subgroup of G except $\{e\}$ and (*b*) the identity component of H is not contained in any proper normal subgroup of G. Then we have the following fact:

If $H \in \Theta$, then $A(H) \in \Omega$.

For a member L of Ω, we call the homomorphism π:

$$L \ni g \to \pi(g) = \mathrm{Ad}_{\mathfrak{M}}\, g \in GL(\mathfrak{M})$$

the *canonical representation* of L, where $\mathfrak{M}$ is the Lie algebra of the unipotent radical M of L. Studying the canonical representation of $A(H)=L$ in detail, we have the following result.

Let $f: H \to GL(W)$ be a linear representation of $H \in \Theta$, where W is a finite-dimensional vector space over **R**. *Then $f(M)$ consists of unipotent automorphisms of W. If f is, moreover, locally one-to-one, then f is faithful and $f(H)$ is closed in $GL(W)$.*

Let H be a member of Θ. If $f: H \to G$ is an isomorphism, then $f(H) \in \Theta$.

The remaining part of the proofs of our main results depends heavily on the following lemma.

Lemma. Let V be a finite-dimensional vector space over **R**, *and let P be a pre-algebraic subgroup of $GL(V)$. Let B be a compact subgroup of P. Let Ψ denote the set of all subgroups of P which are conjugate to B in $GL(V)$. The group* Int P *acts on Ψ in a natural manner. Then the space* $\Psi/\mathrm{Int}\, P$ *of orbits is finite.*

Using the lemma we can prove that each member of Ω has the finite conjugate class property in G.

Next, let us suppose that $H \in \Theta$ and $A(H) = L$. Then H has the finite conjugate class property in L, and more precisely

$$s(H, L, \text{Aut } L) = 1$$

$$s(H, L, \text{Int } L) \leqq i(L)$$

where $i(L)$ is a constant depending only on L.

Thus we can prove also that $H \in \Theta$ is strongly rigid and has the finite conjugate class property in G.

REFERENCES

[1] Borel, A., Density properties of certain subgroups of semisimple groups without compact components, *Ann. of Math.*, **72** (1960), 179—188.

[2] Borel, A. and Tits, J., Groupes réductifs, *Publ. Math., I.H.E.S.*, **27** (1965), 659—755.

[3] Goto, M. and Wang, H. C., Uniform subgroups of simple Lie groups,

[4] Goto, M. and Wang, H. C., Non-discrete uniform subgroups of semisimple Lie groups, *Math. Ann.*, **198** (1972), 259—286.

[5] Montgomery, D., Simply connected homogeneous spaces, *Proc. Amer. Math. Soc.*, (1950), 467—469.

[6] Weil, A., On discrete subgroups of Lie groups II, *Ann. of Math.*, **75** (1962), 578—602.

ON THE COHOMOLOGY OF $SL(2, \mathfrak{O})$

G. HARDER

Introduction

Let G_∞ be a semisimple real Lie group and let $K \subset G_\infty$ be a maximal compact subgroup. Let us denote the corresponding symmetric space by $X = K \backslash G_\infty$. Let $\Gamma \subset G_\infty$ be an arithmetically defined discrete subgroup. We want to investigate the cohomology groups $H^*(\Gamma; \mathbf{C})$. The group Γ acts discontinuously on the contractible space X and therefore we have

$$H^*(\Gamma; \mathbf{C}) = H^*(X/\Gamma; \mathbf{C}).$$

(If Γ has torsion elements we may pass to a subgroup $\Gamma' \subset \Gamma$ of finite index which has no torsion and which acts freely on X.) The cohomology groups on the right-hand side are computed by means of the de Rham complex

$$0 \to \Omega^0(X/\Gamma) \to \Omega^1(X/\Gamma) \to \ldots$$

If Γ has elements of finite order then X/Γ is not necessarily a manifold and then we define

$$\Omega^*(X/\Gamma) = \Omega^*(X)^\Gamma$$

i.e. we take the Γ-invariant differential forms.

In the case of a compact quotient one of the main tools for the investigation of these cohomology groups is the Hodge theory. This theory yields an isomorphism between the space of harmonic forms of degree ν and the cohomology group $H^\nu(X/\Gamma; \mathbf{C})$

$$\mathbf{H}^\nu(X/\Gamma; \mathbf{C}) = \{\omega \in \Omega^\nu(X/\Gamma) \mid \Delta\omega = 0\} \xrightarrow{\sim} H^\nu(X/\Gamma; \mathbf{C})$$

where $\Delta = d\delta + \delta d$ is the Laplacian operator (comp. [13] and [15]).

If moreover X is a bounded symmetric domain then we have a Kähler structure on X/Γ and from this we get a Hodge decomposition

$$H^\nu(X/\Gamma; \mathbf{C}) = \bigoplus_{p+q=\nu} H^{p,q}(X/\Gamma; \mathbf{C})$$

of the cohomology (comp. [15]). This tells us that the problem of computing the cohomology is related to the theory of automorphic forms (comp. [10] Chap. IV, Thm. 3.).

These results are not true if X/Γ is not compact. But I think that there exists a "Hodge theory" for arbitrary arithmetically defined groups Γ. Roughly speaking it should be possible to decompose the cohomology groups $H^*(X/\Gamma, \mathbf{C})$ into two parts. The first part consists of classes which can be represented by harmonic cusp forms and for this part the Hodge theory is valid. The second part consists of classes which can be represented by Eisenstein series. Here the Hodge theory fails but we have good control of this part.

In this paper we want to illustrate this general idea in the case where Γ is commensurable to $SL(2, \mathfrak{O})$ and where $\mathfrak{O}$ is the ring of integers of an algebraic number field k.

I apologize that no proofs are given here. One of the reasons for this is that I want to come back to this type of question in a more general framework. I also want to mention at this place the work of H. Garland, W. C. Hsiang [2], [3] and M. S. Raghunathan [12]. These authors obtained quite a lot of beautiful results concerning the cohomology of isotropic (i.e. X/Γ non compact) arithmetically defined groups.

§ 1

Let $\mathfrak{O}$ be the ring of integers of an algebraic number field. Let $\Gamma \subset SL(2, \mathfrak{O})$ be a congruence subgroup, and assume for simplicity that Γ has no torsion. In this section we want to describe the structure of the manifold X/Γ at "infinity".

Before doing this we have to introduce some notation. We denote the degree of our field by $n=[k:\mathbf{Q}]$. As usual r_1 is the number of real places and $2r_2$ is the number of complex places. Then $r=r_1+r_2-1$ is the rank of the group of units in $\mathfrak{O}$. Our group G_∞ is given by

$$G_\infty = SL(2, k \underset{\mathbf{Q}}{\otimes} \mathbf{R}) = \underbrace{SL(2, \mathbf{R}) \times \ldots \times SL(2, \mathbf{R})}_{r_1 \text{ factors}} \times \underbrace{SL(2, \mathbf{C}) \times \ldots \times SL(2, \mathbf{C})}_{r_2 \text{ factors}}$$

We choose our maximal compact subgroup to be

$$K = SO(2) \times \ldots \times SO(2) \times SU(2) \times \ldots \times SU(2).$$

and we have

$$K \diagdown G_\infty = H \times \ldots \times H \times Z \times \ldots \times Z = X$$

where H is an upper half-plane and Z is a three-dimensional hyperbolic space. Then

$$\dim X = 2r_1 + 3r_2 = n + r + 1.$$

It follows from reduction theory that the manifold X/Γ has the homotopy type of a compact manifold with boundary. To be more precise we know that there exists an open subset $\mathring{M} \subset X/\Gamma$ such that the closure M of $\mathring{M}$ is a compact manifold with boundary and such that we have the following properties:

(i) The inclusion

$$M \hookrightarrow X/\Gamma$$

is a homotopy equivalence.

(ii) The connected components of the boundary ∂M of M are in a natural one-to-one correspondence to the Γ-conjugacy classes of Borel subgroups of $SL(2)/k$.

This is a special case of Proposition 17.9 in [1]. We want to recall the precise description of the boundary components of ∂M given in [1] loc. cit. Let $Y \hookrightarrow \partial M$ be a connected component of the boundary and let $B \subset SL(2)/k$ be a Borel subgroup corresponding to it. Let $\alpha: B \to G_m$ be the positive root. Then α induces a homomorphism

$$\alpha_\infty: B_\infty = B(k \underset{\mathbf{Q}}{\otimes} \mathbf{R}) \to (k \underset{\mathbf{Q}}{\otimes} \mathbf{R})^* = G_m(k \underset{\mathbf{Q}}{\otimes} \mathbf{R}).$$

If we compose α_∞ with the norm homomorphism

$$\nu: (k \underset{\mathbf{Q}}{\otimes} \mathbf{R})^* \to (\mathbf{R}^+)^*$$

we get a homomorphism

$$|\alpha|: B_\infty \to (\mathbf{R}^+)^*.$$

We denote

$$B_\infty(1) = \{b \in B_\infty \,|\, |\alpha|(b) = 1\}.$$

Then we have $B_\infty(1) \cap K = B_\infty \cap K$ and $B_\infty(1) \cap \Gamma = B_\infty \cap \Gamma$ and there exists a C^∞-diffeomorphism

$$Y \xrightarrow{\sim} K \cap B_\infty \backslash B_\infty(1)/B_\infty \cap \Gamma.$$

Let us denote the unipotent radical of B by U. Then $T = B/U$ is a split torus of dimension one. Let K_T (resp. Γ_T) be the image of $K \cap B_\infty$ (resp. $B_\infty \cap \Gamma$) under the projection map $p_\infty: B_\infty \to T_\infty$. Since we assumed that Γ has no torsion it follows from Dirichlet's unit theorem that

$$K_T \backslash T_\infty(1)/\Gamma_T \xrightarrow{\sim} S^1 \times \ldots \times S^1 = (S^1)^r.$$

The projection map p_∞ induces a projection

$$Y \xrightarrow{p} K_T \backslash T_\infty(1)/\Gamma_T = (S^1)^r$$

and this map p is a locally trivial fibration with fiber

$$U_\infty/U_\infty \cap \Gamma = (S^1)^n.$$

We consider U_∞ as a vector space over $\mathbf{R}$ and we denote its dual vector space by $\check{U}_\infty$. Then it is obvious that

$$H^*(U_\infty/U_\infty \cap \Gamma; \mathbf{C}) = (\Lambda^* \check{U}_\infty) \underset{\mathbf{R}}{\otimes} \mathbf{C}.$$

The adjoint representation induces an action of Γ_T on $U_\infty/U_\infty \cap \Gamma$ and therefore

Γ_T acts also on $H^*(U_\infty/U_\infty\cap\Gamma;\mathbf{C})$. The proof of the following proposition is very easy:

Proposition 1.1. *The cohomology of Y is*

$$H(Y;\mathbf{C}) = H((S^1)^r;\mathbf{C})\otimes H(U_\infty/U_\infty\cap\Gamma;\mathbf{C})^{\Gamma_T}.$$

We want to determine the vector space of Γ_T-invariant elements in

$$H^*(U_\infty/U_\infty\cap\Gamma;\mathbf{C}) = (\Lambda^*\check{U}_\infty)\otimes\mathbf{C}.$$

Therefore we have to investigate the action of Γ_T on this space. We consider Γ_T as a subgroup of finite index in the group $\mathfrak{O}^*$ of units. Let us denote the set of imbeddings of k into $\mathbf{C}$ by

$$I_0 = \{\varphi_1, \ldots, \varphi_{r_1}, \varphi_{r_1+1}, \bar{\varphi}_{r_1+1}, \ldots, \varphi_{r_1+r_2}, \bar{\varphi}_{r_1+r_2}\}.$$

For each subset $I\subset I_0$ there exists a subspace of dimension one

$$V_I = \mathbf{C}v_I\subset\Lambda^{n-|I|}\check{U}_\infty\otimes\mathbf{C}$$

such that

$$\varepsilon V_I = \big(\prod_{i\in I}\varphi_i(\varepsilon)\big)v_I = \varphi_I(\varepsilon)\cdot V_I$$

for all $\varepsilon\in\Gamma_T\subset\mathfrak{O}^*$ and $(\Lambda^*\check{U}_\infty)\otimes_{\mathbf{R}}\mathbf{C}$ is the direct sum of the subspaces V_I.

We have to determine the subsets $I\subset I_0$ for which $\varphi_I(\varepsilon)=1$ for all $\varepsilon\in\Gamma_T$. Let us distinguish two cases:

1) $r_1\neq 0$. In this case that can only happen if $I=\emptyset$ or $I=I_0$. If on the other hand Γ satisfies suitable congruence conditions we know that $\varphi_\emptyset(\varepsilon)=\varphi_{I_0}(\varepsilon)=1$ for all $\varepsilon\in\Gamma_T$.

Therefore — if Γ satisfies some congruence conditions —

$$H^*(U_\infty/U_\infty\cap\Gamma;\mathbf{C})^{\Gamma_T} = H^0(U_\infty/U_\infty\cap\Gamma;\mathbf{C})\oplus H^n(U_\infty/U_\infty\cap\Gamma;\mathbf{C}) = \mathbf{C}\oplus\mathbf{C}\xi_U$$

where $\xi_U\in H^n(U_\infty/U_\infty\cap\Gamma;\mathbf{C})$ is a fundamental class.

2) $r_1=0$. Now it may happen that in addition

$$H^{r_2}(U_\infty/U_\infty\cap\Gamma,\mathbf{C})^{\Gamma_T}\neq(0)$$

but in all dimensions different from $0, r_2, n=2r_2$ there are no invariant classes.

This phenomenon seems to us very amusing, therefore we want to discuss shortly an example. Let $k_0/\mathbf{Q}$ be a totally real field and let k/k_0 be a totally imaginary extension. Let

$$I\subset\{\varphi_1, \bar{\varphi}_1, \ldots, \varphi_{r_2}, \bar{\varphi}_{r_2}\}$$

be a half system of imbeddings of k into $\mathbf{C}$, i.e. I contains of every pair of conjugate imbeddings precisely one. Then it is clear that for all $\varepsilon\in\Gamma_T\subset\mathfrak{O}^*$ we have $\varphi_I(\varepsilon)=1$ if Γ satisfies some congruence conditions. Therefore we have under suitable assumptions on Γ

$$((\Lambda^{r_2}\check{U}_\infty)\otimes\mathbf{C})^{\Gamma_T} = H^{r_2}(U_\infty/U_\infty\cap\Gamma;\mathbf{C})^{\Gamma_T} = \bigoplus_{I\text{ half system}} V_I.$$

§ 2

In this section we will state the main theorem of this paper and we will give a rather sketchy idea of its proof. We know that $M \hookrightarrow X/\Gamma$ is a homotopy equivalence and therefore we have

$$H^*(M; \mathbf{C}) = H^*(X/\Gamma; \mathbf{C})$$

$$H^*(M, \partial M; \mathbf{C}) = H_!^*(X/\Gamma; \mathbf{C})$$

where $H_!(\quad)$ denotes the cohomology with compact supports. From the long exact cohomology sequence we get an exact triangle

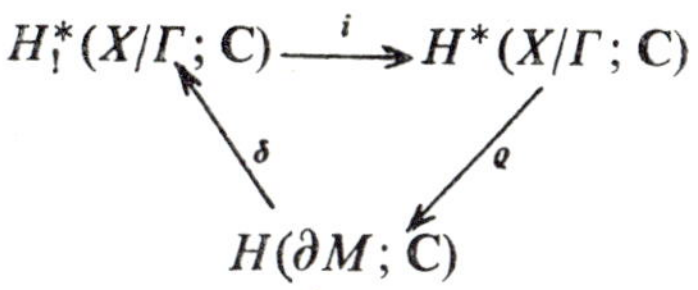

The cohomology groups $H^*(\partial M, \mathbf{C})$ are supposed to be well known, we have gathered a large amount of information about them in § 1. Our main task now is to understand the map ϱ: to be more precise we want to determine its image. Later on we will discuss the relation to the original question mentioned in the introduction.

Theorem 2.1. *The mapping*

$$\varrho_\nu : H^\nu(X/\Gamma; \mathbf{C}) \to H^\nu(\partial M; \mathbf{C})$$

is

(i) *the zero mapping for* $1 \leqq \nu \leqq r_1 + r_2 - 1 = r$,

(ii) *surjective for* $n \leqq \nu \leqq n + r - 1$.

(iii) *The mapping* ϱ_0 *is injective and it maps* $H^0(X/\Gamma; \mathbf{C}) = \mathbf{C}$ *onto the "diagonal" in* $H^0(\partial M; \mathbf{C})$.

(iv) *The mapping* ϱ_{n+r} *has a cokernel of dimension one.*

(v) *If* $r_1 = 0$ *then*

$$\dim \varrho_\nu\big(H^\nu(X/\Gamma; \mathbf{C})\big) = \frac{1}{2} \dim H^\nu(\partial M; \mathbf{C}) \textit{ for } r_2 \leqq \nu \leqq 2r_2.$$

Remarks. 1) Of course (iii) and (iv) are obvious for topological reasons.

2) For the dimensions ν not mentioned in our theorem we have $H^\nu(\partial M; \mathbf{C}) = 0$.

3) Actually our information in the case (v) is somewhat better. This will become clear in the sketch of the proof. In some cases we know precisely the image of ϱ_ν. This will play a role in an example discussed in § 3.

We want to say a few words about the proof of our theorem. Any cohomology class in $H^\nu(\partial M; \mathbf{C})$ can be represented by a harmonic form ω on ∂M. With this harmonic form we associate an Eisenstein series $E(\omega, s)$. This Eisenstein series is a differential form of degree ν on X/Γ and it depends on a complex parameter s.

Roughly speaking these Eisenstein series are constructed by the following procedure: We interpret ω as a function on a collection of Borel subgroups (we have to take one representative for each Γ-conjugacy class) with values in $\Lambda^{\nu}\check{\mathfrak{p}}$. Here $\check{\mathfrak{p}}$ is the complex dual of the space $\mathfrak{p}$ and $\mathfrak{g}_{\infty}=\mathfrak{k}\oplus\mathfrak{p}$ is the Cartan decomposition of the Lie algebra of G_{∞} with respect to K. Then $\check{\mathfrak{p}}$ is the complex cotangent space of X at the point Ke. If we look at ω from this point of view then the whole situation fits into the framework of Eisenstein series (comp. [5], Chap. II, § 2). Now we ask for the values of s for which $E(\omega, s)$ is a harmonic (resp. closed) form. To answer this question we consider the constant Fourier coefficient of $E(\omega, s)$ and we get an expression

$$E(\omega, s) \sim \omega \cdot t^{-1-s} + (M(s)\omega)t^{-1+s}$$

(comp. [5], Chap. II, § 4, 5). Here $M(s)$ is a linear transformation on $H^{\nu}(\partial M, \mathbf{C})$ and a simple calculation shows that the following happens:

(i) If $0 \leqq \deg(\omega) \leqq r$ then $E(\omega, s)$ is harmonic for $s=\pm 1$. (These classes come from the basis of the fibration of Y.)

(ii) If $n=r_1+2r_2 \leqq \deg(\omega) \leqq n+r$ then we get harmonic forms for $s=\pm 1$. (These classes are cup products of classes coming from the basis with the fundamental class of the fiber.)

(iii) If $r_1=0$ and $r_2 \leqq \deg(\omega) \leqq 2r_2-1$ then $E(\omega, s)$ is harmonic for $s=0$.

Now we want to know when these forms are closed.

(i) If $\deg(\omega) \geqq 1$ then $E(\omega, 1)$ is never closed if $\omega \neq 0$. In dimension 0 we have

$$H^0(\partial M, \mathbf{C}) = \mathbf{C}\oplus\mathbf{C}\oplus \ldots \oplus\mathbf{C} = \mathbf{C}^f$$

according to the number f of connected components. Then

$$\omega = (\omega_1, \ldots, \omega_f).$$

If $\sum_{i=1}^{f} \omega_i = 0$ then $E(\omega, 1)$ is not closed for $\omega \neq 0$. But if ω is a diagonal element $\omega=(\omega_1, \ldots, \omega_1)$ then $E(\omega, s)$ has a simple pole at $s=1$ and

$$\operatorname*{Res}_{s=1} E(\omega, s) = E'(\omega)$$

is a closed form. (It is a constant function.)

(ii) If in this case $\deg(\omega) < n+r$ then $E(\omega, 1)$ is always closed. If we denote the cohomology class represented by a closed form ξ by $[\xi]$ then we have

$$\varrho([E(\omega, 1)]) = [\omega].$$

In dimension $n+r$ we have

$$H^{n+r}(\partial M; \mathbf{C}) = \mathbf{C}\oplus\mathbf{C}\oplus \ldots \oplus\mathbf{C} = \mathbf{C}^f$$

(we have natural simultaneous orientations on ∂M). If $\omega=(\omega_1, \ldots, \omega_f)$ and $\sum_{i=1}^{f} \omega_i=0$ then $E(\omega, 1)$ is closed and we have $\varrho_{n+r}([E(\omega, 1)])=[\omega]$. But if $\sum_{i=1}^{f} \omega_i \neq 0$

then $E(\omega, 1)$ is not closed. From these considerations we get already (i)—(iv) in our theorem.

(iii) In this case the intertwining operator $M(s)$ defines a linear transformation

$$M(0): H^{\nu}(\partial M, \mathbf{C}) \to H^{\nu}(\partial M, \mathbf{C})$$

and it follows from the functional equation that $M(0)^2 = id$. Moreover it is easy to see that $M(0)$ has the eigenvalues $+1$ and -1 with the same multiplicity. Therefore we get a decomposition

$$H^{\nu}(\partial M; \mathbf{C}) = H^{\nu}_{+1} + H^{\nu}_{-1}.$$

If $[\omega] \in H_{+1}$ then $E(\omega, 0)$ is closed and $\varrho_{\nu}([E(\omega, 0)]) = 2[\omega]$. If $0 \neq [\omega] \in H_{-1}$ then $E(\omega, 0)$ is zero. This gives statement (v) in our theorem. The fact that the eigenvalues $+1$ and -1 occur with the same multiplicity can be proved without explicit knowledge of $M(0)$. But sometimes one can calculate $M(0)$ quite explicitly. (Comp. § 3.)

Actually the detailed proof is rather complicated. We need some information about the intertwining operators $M(s)$ and to get them we have to perform our calculations in the adele groups. Therefore we have to assume that Γ is a congruence subgroup.

We want to discuss some general applications of our Theorem 2.1. We have seen in the discussion of the proof of our theorem that we can find a subspace

$$H^{*}_{\text{inf}}(X/\Gamma; \mathbf{C}) \subset H^{*}(X/\Gamma; \mathbf{C})$$

such that the following is true:

(i) The classes in $H_{\text{inf}}(X/\Gamma; \mathbf{C})$ can be represented by harmonic Eisenstein series.

(ii) The mapping

$$H^{\nu}_{\text{inf}}(X/\Gamma; \mathbf{C}) \xrightarrow{\sim} \operatorname{Im} \varrho_{\nu} \subset H^{\nu}(\partial M; \mathbf{C})$$

is bijective.

Therefore we have a natural decomposition

$$H^{*}(X/\Gamma; \mathbf{C}) = \tilde{H}^{*}(X/\Gamma; \mathbf{C}) \oplus H^{*}_{\text{inf}}(X/\Gamma; \mathbf{C})$$

where

$$\tilde{H}^{*}(X/\Gamma; \mathbf{C}) = \operatorname{Im}\left(H^{*}_{!}(X/\Gamma; \mathbf{C}) \to H^{*}(X/\Gamma; \mathbf{C})\right).$$

Now we want to collect some informations about $\tilde{H}^{*}(X/\Gamma; \mathbf{C})$. A cohomology class which can be represented by a form with compact support is certainly square integrable. And a square integrable cohomology class is representable by an harmonic form (comp. [13], Thm. 24). Therefore

Corollary 2.2. *Any cohomology class in* $H^{*}(X/\Gamma; \mathbf{C})$ *can be represented by a harmonic form.*

There exists a general idea that the classes in $\tilde{H}^*(X/\Gamma; \mathbf{C})$ can be represented by harmonic cusp forms (comp. [5], p. 8). This is false and the following considerations will explain the phenomenon.

Let $X = K \diagdown G_\infty$ be any symmetric space. In the ring $\Omega^*(X)$ of differential forms we have the subring $A^*(X)$ of G_∞-invariant forms. These forms are closed and harmonic. (Comp. [9] p. 1.) If $\Gamma \subset G_\infty$ is a discrete subgroup we get a homomorphism

$$A^*(X) \to H^*(X/\Gamma; \mathbf{C}).$$

This homomorphism is injective if Γ is a uniform discrete subgroup. We shall see what happens in our case.

To describe $A^*(X)$ in our case we observe that

$$X = H \times \ldots \times H \times Z \times \ldots \times Z.$$

Let η_i be the volume form on the i'th factor and equal to one on the others and for any subset $I \subset \{1, 2, \ldots, r_1 + r_2\}$ we put

$$\eta_I = \bigwedge_{i \in I} \eta_i$$

Then $\eta_I \in A^*(X)$ and these elements form a basis of $A^*(X)$. The multiplication is given by

$$\eta_I \wedge \eta_{I'} = \begin{cases} 0 & \text{if} \quad I \cap I' \neq \emptyset \\ \pm \eta_{I \cup I'} & \text{if} \quad I \cap I' = \emptyset. \end{cases}$$

Furthermore let us denote the image of $A^*(X)$ in $H^*(X/\Gamma, \mathbf{C})$ by

$$H_A^*(X/\Gamma; \mathbf{C}) = \bigoplus_{\nu=0}^{2r_1+3r_2} H_A^\nu(X/\Gamma; \mathbf{C}).$$

Proposition 2.3.

(i) $H_A^0(X/\Gamma; \mathbf{C}) = H^0(X/\Gamma; \mathbf{C}) = H_{\mathrm{inf}}^0(X/\Gamma; \mathbf{C})$.

(ii) *If* $\nu \geqq 1$ *then*

$$H_A^\nu(X/\Gamma; \mathbf{C}) \subset \tilde{H}^\nu(X/\Gamma; \mathbf{C}).$$

(iii) *Let* $I \cap I' = \emptyset$ *and* $I \cup I' = \{1, \ldots, r_1 + r_2\}$.

Let us assume $I \neq \emptyset$. *It follows from* (ii) *that we can choose a form* ξ_I *with compact support such that* $[\xi_I] = [\eta_I]$. *Then*

$$\int_{X/\Gamma} \xi_I \wedge \eta_{I'} = \int_{X/\Gamma} \eta_I \wedge \eta_{I'}$$

and we have by definition that these integrals are equal to

$$([\xi_I]_! \cup [\omega_{I'}])(X/\Gamma)$$

where $[\xi_I]_!$ *is considered as an element of* $H_!^{|I|}(X/\Gamma; \mathbf{C})$.

(iv). *The kernel of the map*

$$A^*(X) \to H^*(X/\Gamma; \mathbf{C})$$

is equal to $A^{2r_1+3r_2}(X)$.

The proof of this proposition is easy.

Let us denote the orthogonal complement of $H_A(X/\Gamma; \mathbf{C})$ in $H^*(X/\Gamma; \mathbf{C})$ by $H^*_{\text{cusp}}(X/\Gamma; \mathbf{C})$.

Moreover we denote the space of harmonic cusp forms by

$$\mathbf{H}^\nu_{\text{cusp}}(X/\Gamma; \mathbf{C}) = \{\omega \in \Omega^\nu(X/\Gamma) | \quad \Delta\omega = 0 \quad \text{and } \omega \text{ is a cusp form}\}.$$

It is easily seen that these harmonic cusp forms are closed and we have

Theorem 2.4. *The mapping*

$$\mathbf{H}^r_{\text{cusp}}(X/\Gamma; \mathbf{C}) \to H^r_{\text{cusp}}(X/\Gamma; \mathbf{C})$$

is bijective.

This theorem together with our previous considerations shows that there is a Hodge theory for X/Γ in this special case.

To prove this theorem one observes first that the classes in $H^*_A(X/\Gamma; \mathbf{C}) + H^*_{\text{inf}}(X/\Gamma; \mathbf{C}) = H^*_{\text{Eis}}(X/\Gamma; \mathbf{C})$ are represented by Eisenstein series and then one has to show that all classes which can be represented by Eisenstein series are contained in this space.

We just want to mention that for $\nu \geqq 1$ the space

$$\tilde{H}^\nu(X/\Gamma; \mathbf{C}) = H^\nu_{\text{cusp}}(X/\Gamma; \mathbf{C}) \oplus \tilde{H}^\nu_A(X/\Gamma; \mathbf{C})$$

is precisely the space of square integrable forms, i.e. the classes in $H^\nu_{\text{inf}}(X/\Gamma; \mathbf{C})$ are not square integrable.

§ 3

In this section we want to give some special applications of our results.

Let us first consider the case of a totally real field k. In this case we have the Hodge decomposition

$$H^\nu_{\text{cusp}}(X/\Gamma; \mathbf{C}) = \bigoplus_{p+q=\nu} H^{p,q}(X/\Gamma; \mathbf{C}).$$

It follows from the work of Matsushima and Shimura [11], § 3, 4 that these cohomology groups $H^{p,q}_{\text{cusp}}(X/\Gamma; \mathbf{C})$ vanish unless we have $p+q=n$. Therefore

$$H^*_{\text{cusp}}(X/\Gamma; \mathbf{C}) = H^n_{\text{cusp}}(X/\Gamma; \mathbf{C}).$$

If we consider the Euler—Poincaré characteristic we obtain

$$\chi(X/\Gamma) = \sum_{i=0}^{2n} (-1)^i \dim H^i(X/\Gamma; \mathbf{C}) =$$

$$= (-1)^n \cdot \dim H^n_{\text{cusp}}(X/\Gamma; \mathbf{C}) + \chi_A(X/\Gamma) + \xi_{\text{inf}}(X/\Gamma) - 1$$

(we have counted $H^0(X/\Gamma; \mathbf{C})$ twice and therefore we subtract 1). We know from Theorem 2.1 and Proposition 2.4 that

$$\chi_A(X/\Gamma) = 2^n - 1$$

$$\chi_{\inf}(X/\Gamma) = 1.$$

Moreover it is known that $\chi(X/\Gamma) = [\Gamma_0 : \Gamma] \cdot \zeta_k(-1)$ where $\Gamma_0 = SL(2, \mathfrak{O})$ and ζ_k is the Dedekind ζ-function of k (comp. [4]) and we obtain the formula

$$\dim H^n_{\text{cusp}}(X/\Gamma; \mathbf{C}) = (-1)^n [\Gamma_0 : \Gamma] \zeta_k(-1) - 2^n + 1.$$

Our method does not give any information about the dimensions of the spaces $H^{p,q}(X/\Gamma, \mathbf{C})$. I think that in this case these dimensions can be computed by means of the trace formula (Labesse and Langlands [7]) and in the case $n=2$ their dimensions can be computed by the methods of Hirzebruch [5].

Finally we want to discuss the case of a totally imaginary quadratic extension of $\mathbf{Q}$ say $k = \mathbf{Q}(\sqrt{-d})$. For simplicity we assume that $d \geqq 5$ and square free. If $\Gamma \subset SL(2, \mathfrak{O})$ is a congruence subgroup it seems to be an extremely difficult problem to calculate the rank of the factor commutator group $\Gamma/[\Gamma, \Gamma]$. Of course we know

$$\operatorname{Hom}(\Gamma, \mathbf{C}) = H^1(X/\Gamma; \mathbf{C})$$

and from our general results we have the decomposition

$$H^1(X/\Gamma; \mathbf{C}) = H^1_{\text{cusp}}(X/\Gamma; \mathbf{C}) + H^1_{\inf}(X/\Gamma; \mathbf{C}).$$

It is rather difficult to get results on $H^1_{\text{cusp}}(X/\Gamma; \mathbf{C})$; what we can do is the following: Let σ be the non-trivial element in the Galois group of $k/\mathbf{Q}$ and let us assume that Γ is invariant under σ. Then σ induces an involution on X/Γ, let us denote the induced operation on the cohomology by

$$\sigma^*_\nu : H^\nu(X/\Gamma; \mathbf{C}) \to H^\nu(X/\Gamma; \mathbf{C}),$$

We consider the alternating sum of the traces

$$\operatorname{tr}(\sigma^*) = \sum_{\nu=0}^{2} (-1)^\nu \operatorname{tr}(\sigma^*_\nu) = 1 - \operatorname{tr}(\sigma^*_1) + \operatorname{tr}(\sigma^*_2).$$

The mapping σ^* respects the decomposition

$$H^*(X/\Gamma, \mathbf{C}) = H^*_{\text{cusp}}(X/\Gamma, \mathbf{C}) + H^*_{\inf}(X/\Gamma, \mathbf{C})$$

and therefore we have

$$\operatorname{tr}(\sigma^*_1) = \operatorname{tr}(\sigma^*_{1,\text{cusp}}) + \operatorname{tr}(\sigma^*_{1,\inf})$$

$$\operatorname{tr}(\sigma^*_2) = \operatorname{tr}(\sigma^*_{2,\text{cusp}}) + \operatorname{tr}(\sigma^*_{2,\inf}).$$

If $\omega \in H^1_{\text{cups}}(X/\Gamma)$ is a harmonic 1-form which is a cusp form then we may apply the $*$-operator to it and we get a harmonic 2-form $*\omega$ which is of course a cusp form (comp. [13], Chap. V). So the $*$-operator yields an isomorphism

$$*: H^1_{\text{cusp}}(X/\Gamma; \mathbf{C}) \xrightarrow{\sim} H^2_{\text{cusp}}(X/\Gamma; \mathbf{C})$$

The $*$-operator on X depends on an orientation of X and we know that σ changes the orientation on X/Γ. Therefore we have

$$* \circ \sigma^*_{1,\text{cusp}} = -\sigma^*_{2,\text{cusp}} \circ *$$

and this yields

$$\operatorname{tr}(\sigma^*_{1,\text{cusp}}) = -\operatorname{tr}(\sigma^*_{2,\text{cusp}}).$$

Our formula simplifies to

$$2\operatorname{tr}(\sigma^*_{1,\text{cusp}}) = 1 - \operatorname{tr}(\sigma^*_{1,\text{inf}}) + \operatorname{tr}(\sigma^*_{2,\text{inf}}) - \operatorname{tr}(\sigma^*).$$

The expressions on the right hand side can be computed in principle: To compute $\operatorname{tr}(\sigma^*)$ we apply the Lefschetz fixed point formula. Therefore we compute the fixed point set of the involution σ on X/Γ. This may be a difficult job, but if Γ satisfies some congruence conditions then

$$(X/\Gamma)^{\text{Gal}(k/\mathbf{Q})} = H/SL(2,\mathbf{Z}) \cap \Gamma,$$

where $H \subset X$ is the natural imbedding of the upper half-plane H into X.

The computation of $\operatorname{tr}(\sigma^*_{2,\text{inf}})$ is not so difficult in general because we understand the action of σ on ∂M.

The computation of $\operatorname{tr}(\sigma^*_{1,\text{inf}})$ is more complicated. We understand the action of σ on ∂M but we have to compute rather explicitly the operator

$$M(0): H^1(\partial M) \to H^1(\partial M)$$

and we have to understand the operation of σ^*_1 on the kernel of $M(0) - id$. We work with adele groups to carry these computations through. The final result depends on the factors $W(\chi)$ in the functional equation of the Hecke L-series

$$W(\chi) \cdot L(s,\chi) = L(1-s,\chi)$$

(comp. [8], Chap. XIV, Thm. 14, Cor. 1).

We have done the computations in the following case. Let $\{p_1 \dots p_t\} = S$ be a finite set of odd primes which are not ramified and not split in k. Let Γ be the congruence subgroup consisting of the matrices

$$\begin{pmatrix} a & b \\ c & d \end{pmatrix} \equiv I \bmod p_1 \dots p_t.$$

In this case we get

$$\operatorname{tr}(\sigma^*_{1,\text{inf}}) = -h \prod_{i=1}^{t} (p_i^2 + 1)$$

where h is the class number of k.

If $S=\emptyset$, i.e. $\Gamma=SL(2, \mathfrak{O})$ itself then this follows from a much stronger result of Serre (comp. [14], Thm. 9).

If $S\neq\emptyset$ then we have

$$(X/\Gamma)^{\{\sigma\}} = H/\Gamma\cap SL(2, \mathbf{Z})$$

and we get a rather explicit formula*

$$2\,\mathrm{tr}\,(\sigma^*_{1,\,\mathrm{cusp}}) = \frac{1}{12}\prod_{i=1}^{t} p_i(p_i^2-1) + \prod_{i=1}^{t}(p_i^2+1)h -$$

$$-\prod_{i=1}^{t}(p_i^2-1)\,2^{\delta-1}+2$$

where δ is the number of prime divisors of d. (Of course $2^{\delta-1}$ is the number of "ambiger Idealklassen", this term is $\mathrm{tr}\,(\sigma^*_{2,\,\mathrm{inf}})$.)

REFERENCES

[1] Borel, A., *Introduction aux groupes arithmétiques*, Hermann, Paris, 1969.

[2] Garland, H., The spectrum of noncompact G/Γ and the cohomology of discrete groups, *Bull. Amer. Math. Soc.*, **75** (1969), 807—811.

[3] Garland, H. and Hsiang, W. C., A square integrability criterion for the cohomology of arithmetic groups *Proc. Nat. Acad. Sci. USA,* **59** (1968), 354—360.

[4] Harder, G., A Gauss—Bonnet formula for discrete arithmetically defined groups, to appear in *Ann. Sci. École Norm. Sup.*

[5] Harish-Chandra (Notes by J. G. M. Mars), *Automorphic forms on semisimple Lie groups,* Springer Lecture Notes, Vol. **62**, 1968.

[6] Hirzebruch, F., The Hilbert modular group, *Séminaire Bourbaki,* June 1971.

[7] Labesse, J. P. and Langlands, R., Some work in preparation.

[8] Lang, S., *Algebraic number theory* Addison-Wesley, 1970.

[9] Matsushima, Y., On the Betti numbers of compact, locally symmetric Riemannian manifolds, *Osaka J. Math.*, **14** (1962), 1—20.

[10] Matsushima, Y. and Murakami, S., On vector valued harmonic forms and automorphic forms on symmetric Riemannian manifolds, *Ann. of Math.*, **78** (1963), 365—416.

[11] Matsushima, Y. and Shimura, G., On the cohomology groups attached to certain vector valued differential forms on the product of upper half-planes, *Ann. of Math.,* **78** (1963), 417—449.

[12] Raghunathan, M. S., Cohomology of arithmetic subgroups of algebraic groups, I, II, *Ann. of Math.*, **86** (1967), 409—424; **87** (1968), 279—304.

[13] de Rham, G., *Variétés différentiables,* Hermann, Paris, 1955.

[14] Serre, J.-P., Le problème des groupes de congruence pour SL_2, *Ann. of Math.*, **92** (1970), 489—527.

[15] Weil, A., *Introduction à l'étude des variétés kähleriennes,* Hermann, Paris, 1958.

* Added in proof: Unfortunately there is a mistake in our computation of the fixed point set above. If the discriminant d of our field is odd and of the form $d=q_1\ldots q_\delta$ with prime numbers q_i then we have 2^δ components of the fixed point set which are all isomorphic to $H/\Gamma\cap SL(2, \mathbf{Z})$. Therefore in this case the first term on the right-hand side of our formula has to be multiplied by 2^δ.

ON ARITHMETIC VARIETIES

D. A. KAJDAN

Introduction

Let G be an algebraic semisimple $\mathbf{Q}$-group, $G_{\mathbf{R}}$ the set of all real points of G, K a maximal compact subgroup of $G_{\mathbf{R}}$.

We assume in the sequel that $\mathrm{rank}_{\mathbf{Q}}\, G=0$ and that on the symmetric space $D=K\diagdown G$ there is an invariant complex structure. In the following, D will be considered to be equipped with this structure. Let Γ be an arithmetic subgroup of $G_{\mathbf{Q}}$ (see [1] no. 1.1), having no elements of finite order, and $X_\Gamma=D/\Gamma$. X_Γ has the natural structure of a compact smooth complex-analytic variety. Such complex manifolds $X=X_\Gamma$ will be called, in the following, *arithmetic varieties*. It is known ([2], Theorem 20) that any arithmetic variety X is projective, i.e. there is a projective algebraic $\mathbf{C}$-variety $\hat{X}^e$ such that the complex manifold $\hat{X}^e_{\mathbf{C}}$ is isomorphic to X. According to a theorem of Serre ([3], Corollary to Proposition 15), the algebraic manifold structure $\hat{X}^e$ on X is uniquely defined.*

Let $\sigma\in\mathrm{Aut}\,\mathbf{C}$ be an automorphism of the field of complex numbers. Denote by $\hat{X}^\sigma$ the algebraic $\mathbf{C}$-manifold obtained from $\hat{X}^e$ by the change of base $\sigma:\mathbf{C}\to\mathbf{C}$ (if the manifold $\hat{X}^e$ is determined in the projective space $P^N(\mathbf{C})$ by the homogeneous equations $\{\sum_j a_{j_0,\dots,j_N} X_0^{j_0}\dots X_N^{j_N}=0\}$, then the manifold $\hat{X}^\sigma$ is determined in $P^N(\mathbf{C})$ by the equations $\{\sum_j (a_{j_0,\dots,j_N})^\sigma X_0^{j_0}\dots X_N^{j_N}=0\}$). Put $X^\sigma=(\hat{X}^\sigma)_{\mathbf{C}}$. The purpose of this paper is to prove the following theorem.

The Main Theorem. For all $\sigma\in\mathrm{Aut}\,\mathbf{C}$ *the variety* X^σ *is arithmetic.*

In order to simplify the notations, it will be convenient to apply the following result.

Let X be a smooth complex manifold and X' a finite-sheeted unramified covering over X.

Lemma 0. *If* X' *is an arithmetic variety, then* X *is also an arithmetic variety.*

Proof. Since X' is an arithmetic variety, there is a semisimple group $G_{\mathbf{R}}$ and an arithmetic subgroup $\Gamma'\subset G_{\mathbf{R}}$ without elements of finite order, such that $X'=D'/\Gamma'$ where D' is the Hermitian symmetric space corresponding to the

* As shown by Shimura (see [4] Theorem 1), an arithmetic variety $\hat{X}^e$ can be defined over a number field.

group $G_{\mathbf{R}}$. Since the symmetric space D is simply connected, it coincides with the universal covering space $\tilde{X}'$ of the variety X', moreover, the image of the group $\pi_1(X')$ under the natural imbedding $j:\pi_1(X')\hookrightarrow \operatorname{Aut}(\tilde{X}')\cong \operatorname{Aut} D$ coincides with the group $\Gamma'\subset \operatorname{Aut} D$. Note that, by assumption, the universal covering $\tilde{X}$ of the variety X is isomorphic to $\tilde{X}'\cong D$, and consider the image $\Gamma=j(\pi_1(X))$ of the group $\pi_1(X)$ under the natural imbedding $j:\pi_1(X)\to \operatorname{Aut} \tilde{X}\cong$ $\cong \operatorname{Aut} D$. We have by assumption $\Gamma'\subset\Gamma$ and $[\Gamma:\Gamma']<\infty$. As Γ' is an arithmetic group, so is Γ, by definition ([1]). Lemma 0 is proved.

Corollary. In the proof of the Main Theorem we can replace the group Γ by any subgroup of finite index. In particular, in the proof of the Main Theorem it is sufficient to analyse the case when G is a simply connected group and $\Gamma\subset G_{\mathbf{Q}}$.

Notations.

1) In the sequel we will always assume that G is a simply connected simple **Q**-group and that $\Gamma\subset G_{\mathbf{Q}}$. The adjoint algebraic **Q**-group corresponding to G, will be denoted by $\tilde{G}$. Further, we denote by $\varkappa:G\to\tilde{G}$ the natural projection and by $Z=\operatorname{Ker}\varkappa_{\mathbf{Q}}$ the centre of the group $G_{\mathbf{Q}}$.

2) For any complex manifold X we denote by T_X or $T(X)$ the space of the tangent bundle to X, by p the natural projection $p:T_X\to X$, by $T_x(X)$ the fibre $p^{-1}(x)\subset T_X$ over the point $x\in X$ (i.e. the tangent space to X at the point x). By $T(X)$ we will denote also the tangent bundle over X.

For any subbundle $R\subset T(X)$ we denote by the same letter R the submanifold in T_X corresponding to R. Analogously, for any **C**-scheme $\hat{Y}$ we denote by $T(\hat{Y})$ the tangent bundle to $\hat{Y}$ and put $T_Y=T(\hat{Y})_{\mathbf{C}}$.

3) Let V be a discrete group and $U\subset V$ a subgroup such that for all $v\in V$ the intersection $v^{-1}Vv\cap U$ is a subgroup of finite index in U. Then we say that U is a *quasi-normal* subgroup in V. In this case the "topology of subgroups of finite index in U" will mean the uniform topology on V, where a basis of neighbourhoods of the identity of the group V consists of subgroups $U'\subset U$ of the group V such that $[U:U']<\infty$. The completion of V in the described topology will be denoted by $\hat{V}_U$ or simply $\hat{V}$ and called "the completion of the group V with respect to the topology of subgroups of finite index in U". We mention two obvious properties of the operation of "completion". First, for any subgroup V' in V containing U the completion $\hat{V}'_U$ is naturally isomorphic to the closure in the group $\hat{V}_U$ of the image $i(V')$ of the subgroup $V'\subset V$ under the natural morphism $i:V\to\hat{V}_U$. Secondly, for any subgroup of finite index $U'\subset U$ the completion $\hat{V}_{U'}$ is naturally isomorphic to $\hat{V}_U$.

4) Denote by $\hat{G}_{\mathbf{Q}}$ the completion of the group $G_{\mathbf{Q}}$ in the topology of subgroups of finite index in Γ. We will freely consider $G_{\mathbf{Q}}$ as a subgroup in $\hat{G}_{\mathbf{Q}}$.

5) We will write **Q**, **R**, **C**, $\mathbf{Q}_p$ for the field of rational, real, complex, p-adic numbers, respectively. If k is the field of algebraic numbers and $\mathfrak{p}$ is a prime ideal in the ring of integers $O=O_k$ of the field k, then we write $k_\mathfrak{p}$ for the local field, which is the completion of the field k with respect to the $\mathfrak{p}$-adic topology. If S is some set of prime ideals of the ring O, then we denote by O_S the subring of k, consisting of all elements of the form α/β where $\alpha, \beta \in O$ and the principal ideal (β) in O decomposes into the product of simple ideals contained in S.

6) If $\hat{X}$ is an algebraic manifold (group etc.) over ring A, then for any A-algebra B we denote by $\hat{X}_B$ the set (group etc.) of B-points of the manifold A.

The proof of the Main Theorem consists of two parts: an analytic-algebraic and a group-theoretic part. In the first part we prove the following result:

Theorem A. *There are subgroups $\Gamma^\sigma \subset G^\sigma \subset \tilde{G}_{\mathbf{R}}$ of the group $\tilde{G}_{\mathbf{R}}$ such that*

1) Γ^σ is a discrete subgroup in $\tilde{G}_{\mathbf{R}}$ without elements of finite order, and the factor space $\tilde{G}_{\mathbf{R}}/\Gamma^\sigma$ is compact;

2) the complex manifold D/Γ^σ is isomorphic to X^σ;

3) for any element $\gamma \in G^\sigma$ the intersection $\gamma^{-1}\Gamma^\sigma\gamma \cap \Gamma^\sigma$ is a subgroup of finite index in Γ^σ;

4) the natural homomorphism $\mathfrak{i}^\sigma : G^\sigma \to \hat{G}^\sigma_{\Gamma^\sigma}$ is an imbedding;

5) the completion $\hat{G}^\sigma_{\Gamma^\sigma}$ is isomorphic to a subgroup of finite index in the factor group $\hat{G}_{\mathbf{Q}}/Z$. (Here Z is the centre of the group $G_{\mathbf{Q}}$ considered as a subgroup of $\hat{G}_{\mathbf{Q}}$);

6) the group Γ^σ is dense in $G_{\mathbf{R}}$ in the Zariski topology;

7) if L is a subgroup of G^σ, containing Γ^σ, and $[L:\Gamma^\sigma]=\infty$, then the group L is dense in the connected component of the identity of the group $\tilde{G}_{\mathbf{R}}$ in the ordinary topology;

8) if $H \subset G$ is a torus and $H_{\mathbf{R}}$ is compact then there exists a $g \in \tilde{G}_{\mathbf{R}}$ such that $\varkappa(H_{\mathbf{Q}}) \cap g^{-1}G^\sigma g$ is of finite index in $\varkappa(H_{\mathbf{Q}})$;

9) for any compact Cartan subgroup $H_{\mathbf{C}} \subset G_{\mathbf{R}}$ there exists an element $g \in G_{\mathbf{R}}$ such that $g^{-1}(H_{\mathbf{R}} \cap G^\sigma)g \in G_{\mathbf{Q}}$.

In the second part of the paper we prove

Theorem B. *If Γ^σ satisfies 1. and 3.—9. then it is arithmetic.*

LIST OF CONTENTS

CHAPTER I

ANALYTIC-ALGEBRAIC PART

§ 1. Prouniversal covering manifold

Let K be an algebraically closed field, and $\hat{X}$ a connected algebraic K-variety. Consider the set $Et(\hat{X})$ of connected unramified finite-sheeted coverings $\hat{\tau}_\alpha: \hat{X}_\alpha \to \hat{X}$. Let $\ldots \hat{X}_i \xrightarrow{\eta_i} \hat{X}_{i-1} \to \ldots \to \hat{X}$ be a cofinal system in $Et(\hat{X})$. Denote by $\hat{D}_{\hat{X}}$ the proalgebraic manifold (scheme), being the projective limit of the algebraic varieties $\hat{X}_i$, by $\hat{\pi}$ the projection $\hat{\pi}: \hat{D}_{\hat{X}} \to \hat{X}$, by $\hat{\pi}_i$ the projection $\hat{\pi}_i: \hat{D}_{\hat{X}} \to \hat{X}_i$, and by $\hat{\pi}_{ji}$, $j > i$ the projection $\hat{\pi}_{ji} = \prod_{i<k\leq j} \hat{\eta}_k: \hat{X}_j \to \hat{X}_i$. In the sequel we will often write $\hat{D}$ instead of $\hat{D}_{\hat{X}}$. By definition $\hat{D}_K = \varprojlim (\hat{X}_i)_K$.

If $\hat{X} = \varprojlim \hat{X}_i$, $\hat{X}' = \varprojlim \hat{X}'_i$ are two proalgebraic varieties, then, as always, a morphism $f: \hat{X} \to \hat{X}'$ is a set of morphisms $\hat{f}_i: \hat{X}_{n_i} \to \hat{X}'_i$, $n_{i+1} > n_i$ such that for all pairs $i, j \in \mathbf{Z}^*$, $j > i$ the diagram

$$\begin{array}{ccc} \hat{X}_{n_j} & \xrightarrow{\hat{f}_j} & \hat{X}'_j \\ {\scriptstyle \hat{\pi}_{n_j, n_i}} \downarrow & & \downarrow {\scriptstyle \hat{\pi}_{j,i}} \\ \hat{X}_{n_i} & \xrightarrow[\hat{f}_i]{} & \hat{X}'_i \end{array}$$

is commutative. Taking a subsequence, we assume that the morphism $f: \hat{X} \to \hat{X}'$ is given by the set of morphisms $\hat{f}_i: \hat{X}_i \to \hat{X}'_i$.

Let $\hat{Y} = \varprojlim \hat{Y}_i$ and $\hat{Z} = \varprojlim \hat{Z}_i$ be proalgebraic varieties and $f = \varprojlim \hat{f}_i$ a morphism of $\hat{Y}$ into $\hat{Z}$. We say that $\hat{f}$ is an unramified covering, if for all sufficiently large i the morphism $\hat{f}_i: \hat{Y}_i \to \hat{Z}_i$ is an unramified covering.

Lemma 1. *Let $\hat{Y}$ and $\hat{Z}$ be proalgebraic varieties, and $\hat{f}$ an unramified covering $\hat{f}: \hat{Y} \to \hat{Z}$. Further, let $\hat{\varphi}: \hat{D}_{\hat{X}} \to \hat{Z}$ be a morphism which is also an unramified covering, $\hat{y} \in \hat{Y}_K$ and $\hat{d} \in \hat{D}_K$ points such that $\hat{f}(\hat{y}) = \hat{\varphi}(\hat{d})$. Then there is a unique unramified covering $\hat{\varphi}_1: \hat{D}_{\hat{X}} \to \hat{Y}$ such that $\hat{\varphi} = \hat{f} \circ \hat{\varphi}_1$ and $\hat{\varphi}_1(\hat{d}) = \hat{y}$.*

Proof. Let us fix a system of unramified algebraic coverings of algebraic varieties $\hat{f}_i: \hat{Y}_i \to \hat{Z}_i$ such that $f = \varprojlim \hat{f}_i$. Denote by $\hat{\pi}_i$ the natural projection $\hat{\pi}_i: \hat{D} \to \hat{X}_i$. Denote by $\hat{\pi}'_i$ and $\hat{\pi}''_i$ the natural projections $\hat{\pi}'_i: \hat{Y} \to \hat{Y}_i$, $\hat{\pi}''_i: \hat{Z} \to \hat{Z}_i$.

* More precisely, a morphism of proalgebraic schemes is a class of sets of morphisms with the natural equivalence.

Taking subsequences, we can assume that there are unramified coverings $\hat{\varphi}_i: \hat{X}_i \to \hat{Z}_i$ such that $\hat{\varphi}_i \circ \hat{\pi}_i = \hat{\pi}_i'' \circ \hat{\varphi}$. Denote by $\hat{X}_i \underset{\hat{Z}_i}{\times} \hat{Y}_i$ the fibre product of the schemes $(\hat{X}_i, \hat{\varphi}_i)$ and $(\hat{Y}_i, \hat{f}_i)$ over $\hat{Z}_i$. (Recall that the fibre product $\hat{X}_i \underset{\hat{Z}_i}{\times} \hat{Y}_i$ is the subvariety in the direct product $\hat{X}_i \times \hat{Y}_i$ consisting of all pairs $(\hat{x}, \hat{y}) \in \hat{X}_i \times \hat{Y}_i$ for which $\hat{\varphi}_i(\hat{x}) = \hat{f}_i(\hat{y})$.)

Let $\hat{p}_i'$, $\hat{p}_i''$ be the natural projections $\hat{p}_i': \hat{X}_i \underset{\hat{Z}_i}{\times} \hat{Y}_i \to \hat{X}_i$ and $\hat{p}_i'': \hat{X}_i \underset{\hat{Z}_i}{\times} \hat{Y}_i \to \hat{Y}_i$. Denote by $\hat{X}_i'$ the connected component of the point $(\hat{\pi}_i(\hat{d}), \hat{\pi}_i'(\hat{y}))$ in $\hat{X}_i \underset{\hat{Z}_i}{\times} \hat{Y}_i$ and consider $\hat{p}_i'$ as an unramified covering $\hat{p}_i': \hat{X}_i' \to \hat{X}_i$. We get from the definition of $\hat{D}_{\hat{X}}$ that there is a unique mapping $(\hat{\varphi}_1)_i: \hat{D}_{\hat{X}} \to \hat{X}_i'$ such that $\hat{\pi} = \hat{\tau}_i \circ \hat{p}_i \circ (\hat{\varphi}_1)_i$ and $(\hat{\varphi}_1)_i(\hat{d}) = (\hat{\pi}_i(\hat{d}), \hat{\pi}_i'(\hat{y}))$. Put $\hat{\varphi}_i = \hat{p}_i'' \circ (\hat{\varphi}_1)$ and $\varphi_1 = \underleftarrow{\lim}\, \hat{\varphi}_i''$. It is clear that $\hat{\varphi}_1$ is the unique morphism satisfying the conditions of the lemma. Lemma 1 is proved.

Corollary. Let $\hat{f}$ be a morphism of $\hat{D}$ into itself, such that $\hat{\pi} \circ \hat{f} = \hat{\pi}$. Then $\hat{f} \in \operatorname{Aut} \hat{D}$.

Proof. As an immediate consequence of $\hat{\pi} \circ \hat{f} = \hat{\pi}$ we get that $\hat{f}$ is an unramified covering. Let $\hat{d}$ be an arbitrary point $\hat{d} \in \hat{D}_K$. Apply Lemma 1 to the case $\hat{Y} = \hat{Z} = \hat{D}$, $\hat{f} = \hat{f}$, $\hat{\varphi} = id$, $\hat{d} = \hat{d}$, $\hat{y} = \hat{f}^{-1}(\hat{d})$. We find that there exists a morphism $\hat{f}_1: \hat{D} \to \hat{D}$ such that $\hat{f} \circ \hat{f}_1 = id$. The Corollary is proved.

Let $\hat{X}$ be again an algebraic K-variety. Put now $\hat{G}(\hat{X}) \overset{\text{def}}{=} \operatorname{Aut} \hat{D}_{\hat{X}}$ and denote by $\hat{\Gamma}(\hat{X})$ the subgroup in $\hat{G}(\hat{X})$ consisting of all automorphisms $\hat{g}: \hat{D} \to \hat{D}$ for which $\hat{\pi} \circ \hat{g} = \hat{\pi}$. We will assume that $\hat{G}(\hat{X})$ acts on $\hat{D}$ from the right. It is clear that the group $\hat{\Gamma}(\hat{X})$ is canonically (up to an inner automorphism) isomorphic to the fundamental group $\pi_1(\hat{X})$ (cf. [5] V. 7) of the abstract algebraic variety $\hat{X}$. The group $\hat{G}(\hat{X})$ is a complete locally compact topological group. (A basis of neighbourhoods of the identity in $\hat{G}(\hat{X})$ consists of the subgroups of finite index in $\hat{\Gamma}(\hat{X})$.)

From Lemma 1 it follows immediately that the group $\hat{\Gamma}(\hat{X})$ acts simply transitively on the fiber $\hat{\pi}^{-1}(\hat{x})$ over any point $\hat{x} \in X_K$. Let $\hat{x} = \hat{\pi}(\hat{d})$ and $\hat{d} \in \hat{D}$. Since $\hat{\pi}: \hat{D} \to \hat{X}$ is an unramified covering, thus for any point $\hat{d} \in \hat{D}_K$ the mapping $\hat{\pi}_*: T_{\hat{D}}(\hat{d}) \to T_{\hat{X}}(\hat{x})$ is an isomorphism. Denote by $\hat{G}_{\hat{d}}$ the subgroup in $\hat{G}(\hat{X})$, consisting of all elements $\hat{g}$ for which $\hat{d}\hat{g} = \hat{d}$. For any $\hat{g} \in \hat{G}_{\hat{d}}$ put $\hat{\Phi}(\hat{g}) = \hat{\pi}_* \hat{g} \hat{\pi}_*^{-1} \in$ $\in \operatorname{Aut} T_{\hat{X}}(\hat{x})$. It is easy to verify that the image $\Phi(\hat{G}_{\hat{d}}) \subset \operatorname{Aut} T_{\hat{X}}(\hat{x})$ depends (up to a conjugacy) only on $\hat{x} \in \hat{X}_K$. Denote it by $\hat{G}_{\hat{x}}$.

Now we turn to the construction of an algebro-geometric interpretation of the set of double cosets $\hat{\Gamma}(\hat{X}) \backslash \hat{G}(\hat{X}) / \hat{\Gamma}(\hat{X})$. Consider the set $Et_2(\hat{X})$, whose elements $(\hat{Y}, \hat{\tau}_1, \hat{\tau}_2)$ are connected proalgebraic K-varieties $\hat{Y}$ equipped with pairs of mappings $\hat{\tau}_1, \hat{\tau}_2: \hat{Y} \to \hat{X}$, which are unramified coverings. We say that two elements $(\hat{Y}', \hat{\tau}_1', \hat{\tau}_2')$, $(\hat{Y}'', \hat{\tau}_1'', \hat{\tau}_2'') \in Et_2(\hat{X})$ belong to the same class if there is an element $(\hat{Y}, \hat{\tau}_1, \hat{\tau}_2)$ in $Et_2(\hat{X})$ and unramified coverings $p': \hat{Y}' \to \hat{Y}$, $p'': \hat{Y}'' \to \hat{Y}$ such that $\hat{\tau}_i'=\hat{\tau}_i\circ\hat{p}'$, $\hat{\tau}_i''=\hat{\tau}_i\circ\hat{p}''$, $i=1, 2$. This identification is an equivalence-relation. The set of equivalence classes in $\mathrm{Et}_2(\hat{X})$ will be denoted by $\mathfrak{A}(\hat{X})$.

Lemma 2. For every $\alpha\in\mathfrak{A}(\hat{X})$ *there is a representative* $(\hat{Y}^0, \hat{\tau}_1^0, \hat{\tau}_2^0)\in\alpha\in\mathrm{Et}_2(\hat{X})$ *such that* $\hat{Y}^0$ *is an algebraic variety.*

Proof. Let $(\hat{Y}, \hat{\tau}_1, \hat{\tau}_2)\in\alpha$ and $\hat{Y}=\underline{\lim}\, \hat{Y}_i$, where $\hat{Y}_i$ is an algebraic variety. Denote, as before, by $\hat{\pi}_i'$ the projection $\hat{\pi}_i':\hat{Y}\to\hat{Y}_i$. Since $\hat{X}$ is an algebraic variety, for some i there exist morphisms $\hat{\tau}_1^0, \hat{\tau}_2^0:\hat{Y}_i\to\hat{X}$ such that $\hat{\tau}_i=\hat{\tau}_i^0\circ\hat{\pi}_i'$. By definition $(\hat{Y}_i, \hat{\tau}_1, \hat{\tau}_2)\in\alpha$. Lemma 2 is proved.

Let $\hat{g}\in G(\hat{X})$. Then $(\hat{D}_{\hat{X}}, \hat{\pi}, \hat{\pi}\circ\hat{g})\in\mathrm{Et}_2(\hat{X})$. Denote by $\hat{\psi}(g)$ the element in $\mathfrak{A}(\hat{X})$ corresponding to the class in $\mathrm{Et}_2(\hat{X})$, containing $(\hat{D}_{\hat{X}}, \hat{\pi}, \hat{\pi}\circ\hat{g})$.

Lemma 3. $\hat{\psi}$ *is a one-to-one mapping of the set* $\hat{\Gamma}(\hat{X})\diagdown\hat{G}(\hat{X})/\hat{\Gamma}(\hat{X})$ *of double cosets to* $\mathfrak{A}(\hat{X})$.

Proof. First we prove that the mapping $\hat{\psi}$ is constant on the double $\hat{\Gamma}(\hat{X})$-cosets. Let $\hat{g}\in\hat{G}(\hat{X})$ and $\hat{\gamma}_1, \hat{\gamma}_2\in\hat{\Gamma}(\hat{X})$. Since $(\hat{\pi}\times\hat{\pi}\circ\hat{\gamma}_1\hat{g}\hat{\gamma}_2)(\hat{d})=(\hat{\pi}\times\hat{\pi}\hat{g})(\hat{d}\hat{\gamma}_1)$, the elements $(\hat{D}, \hat{\pi}, \hat{\pi}\circ\hat{g})$ and $(\hat{D}, \hat{\pi}, \hat{\pi}\circ\hat{\gamma}_1\hat{g}\hat{\gamma}_2)$ in $\mathrm{Et}_2(\hat{X})$ give the same element in $\mathfrak{A}(\hat{X})$. Therefore $\hat{\psi}$ can be considered as a mapping $\hat{\psi}:\hat{\Gamma}(\hat{X})\diagdown\hat{G}(\hat{X})/\hat{\Gamma}(\hat{X})\to$ $\to\mathfrak{A}(\hat{X})$.

Now we prove the injectivity of $\hat{\psi}$. Let $\hat{g}_1$ and $\hat{g}_2$ be elements in $\hat{G}(\hat{X})$ such that $\hat{\psi}(\hat{g}_1)=\hat{\psi}(\hat{g}_2)$. This means, by definition, that there exist $(\hat{Y}, \hat{\tau}_1, \hat{\tau}_2)\in\mathrm{Et}_2(\hat{X})$ and unramified coverings $\hat{\varphi}_1, \hat{\varphi}_2:\hat{D}\to\hat{Y}$ such that $\hat{\tau}_1\circ\hat{\varphi}_i=\hat{\pi}$ and $\hat{\tau}_2\circ\hat{\varphi}_i=\hat{\pi}\circ\hat{g}_i$ for $i=1, 2$. It follows from Lemma 1 that there is a mapping $\hat{\gamma}:\hat{D}\to\hat{D}$ such that $\hat{\varphi}_2\circ\hat{\gamma}=\hat{\varphi}_1$. According to the Corollary to Lemma 1 we have $\hat{\gamma}\in\mathrm{Aut}\,\hat{D}=\hat{G}(\hat{X})$. Since $\hat{\pi}\circ\hat{\gamma}=\hat{\tau}_1\circ\hat{\varphi}_2\circ\hat{\gamma}=\hat{\tau}_1\circ\hat{\varphi}_1=\hat{\pi}$, we have $\hat{\gamma}\in\hat{\Gamma}(\hat{X})$. On the other hand $\hat{\pi}\circ\hat{g}_1=$ $=\hat{\tau}_2\circ\hat{\varphi}_1=\hat{\tau}_2\circ\hat{\varphi}_2\circ\hat{\gamma}=\hat{\pi}\circ\hat{g}_2\circ\hat{\gamma}=\hat{\pi}\circ\hat{\gamma}\circ\hat{g}_2$. Consequently, $\hat{g}_1\in\hat{\gamma}\hat{g}_2\hat{\Gamma}(\hat{X})$. The injectivity of the mapping $\hat{\psi}:\hat{\Gamma}(\hat{X})\diagdown\hat{G}(\hat{X})/\hat{\Gamma}(\hat{X})\to\mathfrak{A}(\hat{X})$ is proved. In order to complete the proof of Lemma 3 it remains to show that $\hat{\psi}$ is surjective. Let $(\hat{Y}, \hat{\tau}_1, \hat{\tau}_2)\in\mathrm{Et}_2(\hat{X})$. It follows from Lemma 1 that there are unramified coverings $\hat{\varphi}_1, \hat{\varphi}_2:\hat{D}\to\hat{Y}$ such that $\hat{\tau}_i\circ\hat{\varphi}_i=\hat{\pi}$. Applying Lemma 1 to the case $\hat{f}=\hat{\varphi}_1$ and $\hat{\varphi}=\hat{\varphi}_2$, we find that there is a morphism $\hat{g}:\hat{D}\to\hat{D}$ such that $\hat{\varphi}_2=\hat{\varphi}_1\circ\hat{g}$. Reasoning the same way as in the proof of the Corollary of Lemma 1, we find that $\hat{g}\in\mathrm{Aut}\,\hat{D}$. It is clear that $(\hat{D}, \hat{\pi}, \hat{\pi}\circ\hat{g})\in\mathrm{Et}_2(\hat{X})$ gives the same element in $\mathfrak{A}(\hat{X})$ as $(\hat{Y}, \hat{\tau}_1, \hat{\tau}_2)$. Lemma 3 is proved.

Assume now that we are given an isomorphism $\chi:K\xrightarrow{\sim}\mathbf{C}$. Then, for every algebraic K-variety $\hat{X}$, $\chi(\hat{X}_K)$ can be considered as a complex manifold. This complex manifold will be denoted by X. Analogously, for any morphism $\hat{f}:\hat{X}\to\hat{Y}$ we denote by f the corresponding morphism of the complex manifolds $f:X\to Y$. A subset Y in X is said to be an algebraic submanifold, if there is an algebraic subvariety $\hat{Y}\subset\hat{X}$ such that $Y=\chi(\hat{Y}_K)$. In this case we write $\hat{Y}=\chi^{-1}(Y)$. Let, as before, $(\hat{X}_i, \hat{\tau}_i:\hat{X}_i\to\hat{X})\in\mathrm{Et}(\hat{X})$;... $\hat{X}\xrightarrow{\eta_i}\hat{X}_{i-1}\to\ldots\to\hat{X}$ a cofinal system in $\mathrm{Et}(\hat{X})$,

$\hat{D}=\varprojlim \hat{X}_i$ and $\hat{\pi}_i$ the natural projections $\hat{\pi}_i:\hat{D}\to\hat{X}_i$. We will assume in the sequel that for any i, $\hat{\tau}_i:\hat{X}_i\to\hat{X}$ is a Galois covering.

Consider the set $\chi((\hat{D}_X)_K)$. Since $\chi(\hat{D}_K)=\varprojlim X_i$, there is on $\chi(\hat{D}_K)$ a natural structure of a complex manifold. The manifold $\hat{D}_K$, equipped with this structure, will be denoted by $\tilde{D}$. The natural projection of $\tilde{D}$ on X_i is denoted by $\tilde{\pi}_i:\tilde{D}\to X_i$. It is clear that the group $\hat{G}(\hat{X})$, acting on $\chi(\hat{D}_K)$, preserves the complex structure, i.e. it is a subgroup in $\operatorname{Aut}\tilde{D}$.

Consider the fundamental group $\pi_1(X)$ of the manifold X and denote by $\pi_1^0(X)$ the normal subgroup in $\pi_1(X)$, which is the intersection of all groups of finite index in $\pi_1(X)$.

Put $\Gamma(X)=\pi_1(X)/\pi_1^0(X)$. Denote by D_X (or simply D) the complex manifold which is the unramified covering of X, corresponding to the subgroup $\pi_1^0(X)\subset\subset\pi_1(X)$, and by π the projection $\pi:D\to X$. The group $\Gamma(X)$ acts on D_X and $D_X/\Gamma(X)\cong X$. We consider $\Gamma(X)$ as a subgroup in $\operatorname{Aut} D$. Denote by $G(X)$ the subgroup in $\operatorname{Aut} D$ consisting of all elements $g\in\operatorname{Aut} D$ for which the subgroups $g\Gamma(X)g^{-1}\cap\Gamma(X)$ and $g^{-1}\Gamma(X)g\cap\Gamma(X)$ are of finite index in $\Gamma(X)$.

Let us fix now a point $x\in X$, and two points $\tilde{d}_0\in\tilde{D}$ and $d_0\in D$ such that $\tilde{\pi}(\tilde{d}_0)=\pi(d_0)=x$. Then for any i there is a unique mapping $\varphi_i:D\to X_i$ such that $\tau_i\circ\varphi_i=\pi$ and $\varphi_i(d_0)=\tilde{\pi}_i(\tilde{d}_0)$. Put $\tilde{\varphi}=\varprojlim\varphi_i$. It is not difficult to verify that $\tilde{\varphi}$ provides an isomorphism of D with the connected component of the point $\tilde{d}_0$ in $\tilde{D}$. Further for any point $x\in X$ the set $\tilde{\varphi}(\pi^{-1}(x))$ is dense in $\tilde{\pi}^{-1}(x)$.

We will assume in the sequel that $\hat{X}$ is a complete manifold. Let Γ' be a subgroup of finite index in $\Gamma(X)$. Put $X_{\Gamma'}=D/\Gamma'$. $X_{\Gamma'}$ is a compact manifold. The imbedding $\Gamma'\hookrightarrow\Gamma(X)$ determines the unramified coverings $\tau_{\Gamma'}:X_{\Gamma'}\to X$ and $\pi_{\Gamma'}:D\to X_{\Gamma'}$; moreover the covering $\pi_{\Gamma'}:D\to X_{\Gamma'}$ is defined up to multiplication by an automorphism $\gamma:X_{\Gamma'}\to X_{\Gamma'}$ such that $\tau_{\Gamma'}\circ\gamma=\tau_{\Gamma'}$.

It follows from the existence theorem of Riemann (see [5] Exposé 12, Theorem 5.1) that there exists an algebraic unramified covering $\hat{\tau}_{\Gamma'}:\hat{X}_{\Gamma'}\to\hat{X}$ such that $X_{\Gamma'}=\chi((\hat{X}_{\Gamma'})_K)$, $\tau_{\Gamma'}=\chi((\hat{\tau}_{\Gamma'})_K)$. Let Γ'' be another subgroup of finite index in $\Gamma(X)$ and g an element of the group $G(X)$ such that $g\Gamma''g^{-1}\subset\Gamma'$. Define the mapping $\varphi_{g,\Gamma',\Gamma''}:X_{\Gamma''}\to X_{\Gamma'}$ of complex manifolds by $\varphi_{g,\Gamma',\Gamma''}=\pi_{\Gamma'}\circ g\circ(\pi_{\Gamma''})^{-1}$.

The condition $g\Gamma''g^{-1}\subset\Gamma'$ implies that the mapping $\varphi_{g,\Gamma',\Gamma''}$ is well defined. Since the manifolds $X_{\Gamma'}$, $X_{\Gamma''}$ are compact, the manifolds $\hat{X}_{\Gamma'}$, $\hat{X}_{\Gamma''}$ are complete, and it follows from the theorem of Serre (cf. [3]) that there is a unique morphism $\hat{\varphi}_{g,\Gamma',\Gamma''}:\hat{X}_{\Gamma''}\to\hat{X}_{\Gamma'}$ of algebraic manifolds such that $\varphi_{g,\Gamma',\Gamma''}=\chi(\hat{\varphi}_{g,\Gamma',\Gamma''})$.

Let Γ_i be subgroups in $\Gamma(X)$ such that $X_{\Gamma_i}=X_i$. Then Γ_i is a cofinal sequence of subgroups of finite index in $\Gamma(X)$. In fact, let Γ' be some subgroup of finite index in $\Gamma(X)$. Then $(\hat{X}_{\Gamma'},\hat{\tau}_{\Gamma'})\in\mathsf{Et}(X)$. Since $(X_i,\hat{\tau}_i)$ is a cofinal sequence in $\mathsf{Et}(X)$ by assumption, for some i there is a morphism $\hat{f}:\hat{X}_{\Gamma_i}\to\hat{X}_{\Gamma'}$ such that $\hat{\tau}_i=\hat{\tau}_{\Gamma'}\circ\hat{f}$ and $\pi_{\Gamma'}=f\circ\pi_i$. It is clear that $\Gamma_i\subset\Gamma'$.

For any $g\in G(X)$ the subgroup $\Gamma_g=\Gamma_i\cap g^{-1}\Gamma g$ is of finite index in $\Gamma(X)$.

Define the mapping $\psi: G(X) \to Et(\hat{X})$ by taking $\psi_g = (\hat{X}_{\Gamma_g}, \hat{\varphi}_{e,\Gamma(X),\Gamma_g}, \hat{\varphi}_{g,\Gamma(X),\Gamma_g})$. Consider ψ as a mapping $\psi: G(X) \to \mathfrak{A}(\hat{X})$.

Lemma 3′. *ψ is a one-to-one mapping of the set $\Gamma(X) \backslash G(X)/\Gamma(X)$ of double cosets onto $\mathfrak{A}(\hat{X})$.*

The proof of Lemma 3′ is the same as that of Lemma 3.

Let $g \in G(X)$. For any i the subgroup $\Gamma_i \cap g^{-1}\Gamma_i g$ is of finite index in $\Gamma(X)$ and consequently, as shown already, there is a $j(i)$ such that $\Gamma_{j(i)} \subset \Gamma_i \cap g^{-1}\Gamma_i g$. Consider the mappings $\hat{\varphi}_{g,\Gamma_i,\Gamma_{j(i)}}: X_{j(i)} \to \hat{X}_i$ and put $\hat{\chi}_i(g) = \hat{\varphi}_{g,\Gamma_i,\Gamma_{j(i)}} \circ \hat{\pi}_{j(i)}: \hat{D} \to \hat{X}_i$. The mappings $\hat{\chi}_i(g)$ form a projective system. Put $\hat{\chi}(g) = \varprojlim \hat{\chi}_i(g) \in \text{Aut}\, \hat{D}$.

From the definition of the mapping $\hat{\chi}(g)$ it follows immediately that for any point $d \in D$ and element $g \in G(X)$ the following equality holds:

$$\tilde{\varphi}(dg) = \tilde{\varphi}(d)\hat{\chi}(g) \qquad (*)$$

Since the image $\tilde{\varphi}(D)$ is dense in $\tilde{D}$, equality (*) implies that for any $g_1, g_2 \in G(X)$ we have $\hat{\chi}(g_1 g_2) = \hat{\chi}(g_1)\hat{\chi}(g_2)$. Therefore the mapping $\hat{\chi}$ is a homomorphism $\hat{\chi}: G(X) \to \hat{G}(\hat{X})$.

Lemma 4. *$\hat{\chi}$ is an isomorphism between $G(X)$ and the subgroup of $\hat{G}(\hat{X})$ consisting of all $\hat{g} \in \hat{G}(\hat{X})$ such that $\tilde{d}_0 \tilde{g} \in \tilde{\varphi}(D)$.*

Proof. As an immediate consequence of equality (*), $\hat{\chi}$ is a homomorphism and for any $g \in G(X)$, $\tilde{d}_0 \hat{\chi}(g) = \tilde{\varphi}(d_0 g) \in \tilde{\varphi}(D)$.

Now let $\hat{g}$ be an element of $\hat{G}(\hat{X})$ such that $\tilde{d}_0 \hat{g} \in \tilde{\varphi}(D)$. Since $\tilde{\varphi}(D)$ is the connected component of the point $\tilde{d}_0$ in $\tilde{D}$ and $\hat{G}(\hat{X}) \in \text{Aut}\, \tilde{D}$, we have $\tilde{\varphi}(D)\hat{g} = \tilde{\varphi}(D)$. Denote by $\hat{\chi}^{-1}(\hat{g})$ the automorphism D, given by the formula $\hat{\chi}^{-1}(\hat{g}) = \tilde{\varphi}^{-1} \circ \hat{g} \circ \tilde{\varphi}$. It is clear that $\hat{\chi}^{-1}(g) \in G(X)$ and $\hat{\chi}(\hat{\chi}^{-1}(g)) = \hat{g}$. Lemma 4 is proved.

Denote by G_{d_0} the subgroup of $G(X)$, consisting of all $g \in G(X)$ for which $d_0 g = d_0$. For every element $g \in G_{d_0}$ put $\Phi(g) = \pi_* \circ g_* \circ \pi_*^{-1} \in \text{Aut}\, T_X(x)$; $x = \pi(d_0)$.

Corollary 1. The homomorphism $\hat{\chi}$ induces an isomorphism $\hat{\chi}: G_{d_0} \to \hat{G}_{\chi^{-1}}(\tilde{d}_0)$. Furthermore for any $g \in G(X)$ we have the equality $\hat{\Phi}(\chi(g)) = \chi^{-1}(\Phi(g))$.*

Proof. It is clear that $\hat{\chi}(G_{d_0}) \subset \hat{G}_{\chi^{-1}}(\tilde{d}_0)$. The opposite inclusion follows immediately from Lemma 4. The last assertion of the corollary follows from equality (*). The corollary is proved.

Consider the subgroup $\Phi(G_d) \subset \text{Aut}\, T_x(X)$. It is easy to show that this subgroup does not depend on the choice of the point $d \in \pi^{-1}(x)$. Denote the subgroup by G_x.

Corollary 2. For any point $x \in X$ we have

$$G_x = \chi_* \circ \hat{G}_{\chi^{-1}(x)} \circ \chi_*^{-1}.$$

* Here χ is considered as an isomorphism $\chi: X_K \to X_{\mathbf{C}} \cong X$.

Here χ_* is the mapping of the tangent space $T_{\chi^{-1}(x)}(\hat{X})$ to $T_x(X)$, induced by the isomorphism $\chi: K \to \mathbf{C}$.

Proof. Corollary 2 follows from Corollary 1.

Consider the uniform topology on the group $G(X)$, in which a basis of neighbourhoods of the identity is the set of subgroups of finite index in $\Gamma(X)$. Denote by $\hat{G}(X)$ the completion of the group $G(X)$ in this topology. The definition of the topology on the group $\hat{G}(X)$ and the relation $\hat{\chi}(\Gamma(X)) \subset \hat{\Gamma}(\hat{X})$ imply that the homomorphism $\hat{\chi}$ extends to a homomorphism $\hat{\chi}_0: \hat{G}(X) \to \hat{G}(\hat{X})$.

Theorem 1. *The mapping $\hat{\chi}_0$ is a group isomorphism.*

Proof. Denote by $\hat{\Gamma}(X)$ the closure of the group $\Gamma(X)$ in $\hat{G}(X)$. First we show that the mapping $\hat{\chi}_0$ provides an isomorphism of $\hat{\Gamma}(X)$ and $\hat{\Gamma}(\hat{X})$. The group $\hat{\Gamma}(X)$ is isomorphic, by definition, with the completion of the group $\pi_1(X)$ with respect to the topology of subgroups of finite index. On the other hand, Lemma 1 implies immediately $\hat{\Gamma}(\hat{X}) = \pi_1(\hat{X})$.* Therefore our assertion follows from Riemann's existence theorem (cf. [5] Exposé 12). For the proof of Theorem 1 it remains to show that the mapping of the double cosets $\hat{\chi}_1: \hat{\Gamma}(X) \backslash \hat{G}(X) / \hat{\Gamma}(X) \to \hat{\Gamma}(\hat{X}) \backslash \hat{G}(\hat{X}) / \hat{\Gamma}(\hat{X})$ is a one-to-one mapping. This assertion can be obtained at once from Lemmas 3 and 3′. Theorem 1 is proved.

Corollary. a) For any subgroup $V \subset G(X)$ for which $V \cap \Gamma(X)$ is a subgroup of finite index in $\Gamma(X)$, the closure $\overline{\hat{\chi}_0(V)}$ of the image $\hat{\chi}_0(V)$ of V in $\hat{G}(\hat{X})$ is isomorphic to the completion $\bar{V}$ of the group V in the topology of subgroups of finite index in $\Gamma(X) \cap V \subset V$.

b) For any open subgroup $\hat{V}$ in $\hat{G}(\hat{X})$ the intersection $V \overset{\text{def}}{=} \hat{V} \cap \chi(G(X))$ is dense in $\hat{V}$.

In the sequel, the subgroups $V \subset G(X)$ for which the intersection $V \cap \Gamma(X)$ is of finite index in $\Gamma(X)$, will be called *admissible* subgroups.

Let $\hat{g} \in \hat{G}(\hat{X})$. We say that the proalgebraic subvariety $\hat{Y} \subset \hat{X}$ is invariant with respect to g, if the variety $\hat{\pi}^{-1}(\hat{Y}) \subset \hat{D}$ is invariant with respect to $\hat{g}$.

Lemma 5. *Let V be an admissible subgroup in $G(X)$, containing Γ, $\hat{V}$ the closure of $\hat{\chi}_0(V)$ in $\hat{G}(\hat{X})$, Z a closed analytic V-invariant submanifold in D. Put $Y = \pi(Z) \subset X$. Then Y is an algebraic submanifold in X and the algebraic subvariety $\hat{Y} = \chi^{-1}(Y) \subset \hat{X}$ is invariant with respect to the group $\hat{V}$.*

Proof. We may assume that $V \supset \Gamma(X)$. Since Z is a closed analytic $\Gamma(X)$-invariant submanifold in D, so Y is a closed analytic submanifold in X. X is a compact manifold and by Chow's lemma ([3] Proposition 14) Y is an algebraic submanifold in X. Put $\tilde{Z} = \hat{\pi}^{-1}(Y)$. We are going to show that $\tilde{Z}$ is a $\hat{V}$-invariant subvariety in $\tilde{D}$. It follows from Theorem 1 that the subgroup $\hat{\chi}_0(V)$ is dense in $\hat{G}(\hat{X})$. Therefore it is sufficient to show that the closed subvariety $\tilde{Z}$ is invariant

* $\pi_1(\hat{X})$ stands for the fundamental group of the algebraic K-manifold $\hat{X}$ ([5] Exposé V).

under the action of the group $\hat{\chi}_0(V) \subset \operatorname{Aut} \tilde{D}$. Since Z is a $\Gamma(X)$-invariant submanifold in D, we have $\tilde{\varphi}(Z) = \tilde{Z} \cap \tilde{\varphi}(D)$. Therefore $\tilde{\varphi}(Z)$ is dense in $\tilde{Z}$, and consequently, for the proof of Lemma 5 it is enough to show that the subset $\tilde{\varphi}(Z) \subset \tilde{D}$ is invariant under the action of the group $\hat{\chi}(V)$. Now this follows immediately from formula (*) and the fact that Z is a V-invariant submanifold in D. Lemma 5 is proved.

Remark. Conversely, if $\hat{Y}$ is a $\hat{V}$-invariant subvariety in $\hat{X}$, then $Z = \pi^{-1}(\hat{\chi}(\hat{Y}_K))$ is a V-invariant submanifold in D. Indeed let $\tilde{Z} = \tilde{\pi}^{-1}(\chi(Y_K))$. Then for any $g \in G(X)$

$$Z \cdot g = [\tilde{\varphi}^{-1}(\tilde{Z} \cap \tilde{\varphi}(D))] \cdot g = \tilde{\varphi}^{-1}([\tilde{Z} \cap \tilde{\varphi}(D)]\hat{\chi}(g)) = \tilde{\varphi}^{-1}(\tilde{Z} \cap \tilde{\varphi}(D)) = Z.$$

Let now $\hat{p}: T_{\hat{X}} \to \hat{X}$ be the tangent bundle over $\hat{X}$. Then the proalgebraic varieties $\hat{D}_{T\hat{X}}$ and $T_{\hat{D}\hat{X}}$ are canonically isomorphic. Therefore the group $\hat{G}(\hat{X})$ can be naturally imbedded in the group $\hat{G}(T_{\hat{X}})$. Similarly, if X is a complete complex algebraic manifold, the group $G(X)$ can be considered as a subgroup of $G(T_X)$.

Analogously let $p: T_D \to D$ be the tangent bundle over D. Then the group $G(X)$ acts on T_D. It is clear that $T_D/\Gamma \cong T_X \cong \chi(T_X)_K$.

Lemma 6. *Let V be an admissible subgroup in $G(X)$, containing $\Gamma(X)$, $\tilde{V} = \hat{\chi}_0(V)$. W is an analytic V-invariant subbundle in $T(D)$. Denote by $R = W/\Gamma \subset T(X)$ the corresponding subbundle in $T(X)$. Then R is an algebraic subbundle in T_X and the algebraic manifold $\hat{R} = \chi^{-1}(R) \subset T_{\hat{X}}$ is invariant under the action of the group $\hat{V}$.*

Proof. It is clear that R is an analytic subbundle of T_X. Since X is a compact complex-algebraic manifold, by the theorem of Serre (see [3] Theorem 3) R is an algebraic subbundle in T_X. Now the proof of Lemma 6 can be completed in the same way as Lemma 5.

Let $\sigma \in \operatorname{Aut} \mathbf{C}$. Then for every K-manifold $\hat{X}$ on the set $\chi^{\sigma}(\hat{X}_K)$ a complex manifold structure is defined, corresponding to the isomorphism $\chi^{\sigma} = \sigma \circ \chi: K \to \mathbf{C}$. We denote this complex manifold by X^{σ}. Similarly the complex manifolds $D_{X^{\sigma}}$ and $\tilde{D}_{X^{\sigma}}$ are denoted by $D^{\sigma} \xrightarrow{\pi^{\sigma}} X^{\sigma}$ and $\tilde{D}^{\sigma} \xrightarrow{\tilde{\pi}^{\sigma}} X^{\sigma}$. If $x \in X$, then put $x^{\sigma} = \chi^{\sigma} \circ \chi^{-1}(x) \in X^{\sigma}$. Similarly for any algebraic submanifold $Y \subset X$ denote by Y^{σ} the corresponding submanifold in X^{σ}.

Let V be an admissible subgroup in $G(X)$. Denote by V^{σ} the subgroup of $G(X^{\sigma})$ defined as follows. Let $\hat{\chi}_0: G(X) \to \hat{G}(\hat{X})$ be the homomorphism constructed before the formulation of Lemma 4, and $\hat{\chi}_0^{\sigma}$ the analogous homomorphism $\hat{\chi}_0^{\sigma}: G(X^{\sigma}) \to \hat{G}(\hat{X}^{\sigma})$.

Denote by $\hat{V}$ the closure in $\hat{G}(\hat{X})$ of the image $\hat{\chi}_0(V)$ of the group V, and put $V^{\sigma} = (\hat{\chi}^{\sigma})^{-1}(\hat{V} \cap \operatorname{Im} \hat{\chi}^{\sigma})$.

From the Corollary of Theorem 1 we immediately get the following assertion:

Lemma 7. *The subgroup $V^\sigma \subset G(X^\sigma)$ is admissible, and the completion of the group V^σ in the topology of subgroups of finite index in $V^\sigma \cap \Gamma(X^\sigma) \subset V^\sigma$, is isomorphic to the completion of the group V with respect to the topology of subgroups of finite index in $V \cap \Gamma(X) \subset V$.*

Remark. Lemma 7 does not imply that the groups V and V^σ are isomorphic. The Corollary of Theorem 1 and Corollaries of Lemmas 4, 5, 6 imply immediately the following Lemmas 4′, 5′ and 6′.

Lemma 4′. *For any admissible subgroup $V \subset G(X)$ and points $d \in D$, $d^\sigma \in D^\sigma$ for which $(\pi(d))^\sigma = \pi^\sigma(d^\sigma)$, there is an isomorphism $\alpha: V \cap G_d \to V^\sigma \cap G_{d^\sigma}$ of the groups $V \cap G_d$ and $V^\sigma \cap G_{d^\sigma}$ and a linear mapping of the tangent spaces $c: T_d(D) \to T_{d^\sigma}(D^\sigma)$ such that for all $g \in V \cap G_d$, $\Phi(\alpha(g)) = c[\Phi(g)]^\sigma c^{-1}$.*

Lemma 5′. *Let V be an admissible subgroup in $G(X)$. Then for any closed analytic V^σ-invariant submanifold $Z^\sigma \subset D^\sigma$ there is a closed V-invariant submanifold $Z \subset D$ such that $Z^\sigma = (\pi^\sigma)^{-1}[(\pi(Z))^\sigma]$.*

Lemma 6′. *Let V be an admissible subgroup in $G(X)$. Then for any analytic V^σ-invariant analytic subbundle $W^\sigma \subset T_{D^\sigma}$ there is an analytic $V \cap G(X)$-invariant analytic subbundle $W \subset T_D$ such that $W^\sigma = (\pi^\sigma)^*((W/\Gamma \cap V)^\sigma)$.*

In the following we will apply that the varieties X and X^σ are "algebraically indistinguishable". For convenience, we formulate explicitly some consequences of this fact as we shall apply them.

Lemma 8. a) *If W is an integrable analytic subbundle of T_X, then W is algebraic and W^σ is an integrable subbundle of T_{X^σ}.*

b) *Let $\varphi: X \to Y$ be a fibering* (i.e. a smooth proper mapping) *of compact algebraic* **C**-*varieties and ∂ a flat analytic connection in the fibering $\varphi: X \to Y$. Then ∂ is an algebraic connection and ∂^σ defines a flat connection in the fibering $\varphi^\sigma: X^\sigma \to Y^\sigma$.*

Let now ψ be a mapping $\psi: D_X \to D_Y$ such that $\varphi \circ \pi_X = \pi_Y \circ \psi$, where π_X, π_Y are the natural projections $\pi_X: D_X \to X$, $\pi_Y: D_Y \to Y$ and let ∂^* be the connection in the fibering $\psi: D_X \to D_Y$ induced by ∂. Define analogously the fibering $\psi^\sigma: D_{X^\sigma} \to D_{Y^\sigma}$ and the connection $(\partial^\sigma)^*$ in the fibering ψ^σ.

c) *Let V be an admissible subgroup of $G(X)$, which sends the fibres of ψ into fibres and which preserves the connection ∂^*. Then the subgroup V^σ sends fibres of ψ^σ into fibres and preserves the connection $(\partial^\sigma)^*$.*

Proof. We consider only the first assertion of the lemma. Parts b) and c) can be proved similarly. First we mention that the theorem of Serre ([3] Theorem 3) implies that the subbundle W is algebraic.

Now let X be an arbitrary complex manifold and $x \in X$. We denote by $\hat{O}_X$ the ring of formal power series at the point x and by $\hat{\Omega}_X = \oplus \hat{\Omega}^i_X$ the exterior algebra of the formal differential forms at the point x. Let $W \subset T(X)$ be a locally free subbundle in the tangent bundle over X and $\hat{\Omega}_X(W)$ the submodule in $\hat{\Omega}_X$ annihilating W. We say that the subbundle W is *integrable* at the point $x \in X$, if the image $d(\hat{\Omega}_X(W))$ under the exterior differentiation belongs to the ideal of the algebra $\hat{\Omega}_X$ spanned by $\hat{\Omega}_X(W)$. We say that the subbundle W is integrable, if it is integrable at every point $x \in X$. It is known (see [6] Theorem 5.2) that the bundle W is integrable if and only if it is the bundle of all tangent vectors to some foliation on X.

Now let $\hat{X}$ be an algebraic **C**-variety, $\hat{W}$ an algebraic locally free subbundle of $T(\hat{X})$ and $\hat{x} \in \hat{X}_K$. Then one can define analogously the meaning of the statement that a subbundle $\hat{W}$ is integrable at a point $\hat{x}$. Let χ be an isomorphism $\chi: K \to \mathbf{C}$, $X = \chi(\hat{X}_K)$ the corresponding complex manifold and $W = \chi(\hat{W})$ the corresponding analytic subbundle in $T(X)$. It is clear that the subbundle $\hat{W} \subset T(\hat{X})$ is integrable at the point $\hat{x} \in \hat{X}_K$ if and only if the subbundle $W \subset T(X)$ is integrable at the point $x = \chi(\hat{x}) \in X$. There is therefore a purely algebraic definition of integrability of a subbundle $W \subset T(X)$, that proves the validity of the assertion of Part a) of Lemma 8.

§ 2. Study of $G(X^\sigma)$-invariant subbundles of $T_{D(X^\sigma)}$ in case X is an arithmetic variety

Let G be a simple simply connected algebraic **Q**-group of **Q**-rank zero, K a maximal compact subgroup in $G_{\mathbf{R}}$ and $D = K \diagdown G_{\mathbf{R}}$. We assume that there is on D a $G_{\mathbf{R}}$-invariant complex structure. Such a complex structure is unique. In the following, D will be considered as a complex manifold. Denote by α the natural homomorphism $\alpha: G_{\mathbf{R}} \to \operatorname{Aut} D$.

The group $G_{\mathbf{Q}}$ is imbedded in $G_{\mathbf{R}}$ and therefore a mapping $\alpha: G_{\mathbf{Q}} \to \operatorname{Aut} D$ is defined. The kernel of the mapping α coincides with the centre Z_G of the group $G_{\mathbf{Q}}$ and consequently it is finite.

Let Γ be an arithmetic subgroup (cf. [1]) in $G_{\mathbf{Q}}$ such that its elements of finite order belong to Z_G. To avoid unnecessary notations we will not distinguish between the group Γ and its image in the factor group $G_{\mathbf{Q}}/Z_G \subset \tilde{G}_{\mathbf{Q}} \subset \operatorname{Aut} D$. The group Γ acts on D without fixed points and on the factor space $X = D/\Gamma$ there is a unique structure of a compact complex manifold. As is known (cf. [2]), there is a unique algebraic **C**-manifold $\hat{X}$ such that $\hat{X}_{\mathbf{C}} \cong X$. We will, in the sequel, identify $\hat{X}_{\mathbf{C}}$ with X. Since the intersection of all subgroups of finite index in Γ coincides with the identity e of the group Γ, the complex manifold D_X defined in § 1, coincides with D.

Let G^e be a subgroup of finite index in $\alpha(G_{\mathbf{Q}})$, containing Γ.

We now investigate the analytic linear* G^e-invariant submanifolds of T_D. First we establish the following lemma.

Lemma 9. *Let Z be a closed non-empty analytic G^e-invariant submanifold in D. Then $Z=D$.*

Proof. As is known (cf. [7]), the group $G_{\mathbf{Q}}$ is dense in $G_{\mathbf{R}}$ in the ordinary topology. Since the group G^e is a subgroup of finite index in $G_{\mathbf{Q}}$ and the group $G_{\mathbf{R}}$ is connected, the subgroup G^e is dense in $G_{\mathbf{R}}$. Consequently, Z is a $G_{\mathbf{R}}$-invariant submanifold in D. Since D is a homogeneous space and $Z\neq\emptyset$ we have thus $Z=D$. Lemma 9 is proved.

Denote by $\tilde{G}_{\mathbf{R}}$ the factor group of $G_{\mathbf{R}}$ by its centre. Let $\tilde{G}_{\mathbf{R}}=G_0\cdot\prod_{i=1}^{s}G_i$, where G_0 is a compact group, and G_i, $1\leqq i\leqq s$ are simple non-compact Lie groups. Denote by K_i a maximal compact subgroup in G_i and put $D_i=K_i\diagdown G_i$, $1\leqq i\leqq s$. Then D_i has (see [8] Chapter VIII, Proposition 5.2) a G_i-invariant complex structure and $D=\prod_{i=1}^{s}D_i$. For any point $d\in D$ denote by $D_i(d)$ the orbit of the point d under the action of the group G_i: $D_i(d)=dG_i$. It is clear that $D_i(d)$ is an analytic submanifold in D. The tangent bundle T_D decomposes into the direct sum $T_D=\bigoplus_{i=1}^{s}T_{i,D}$ where $T_{i,D}$ consists of the tangent vectors of $D_i(d)$. Each subbundle $T_{i,D}\subset T_D$ is invariant under the action of the group $G_{\mathbf{R}}$.

Lemma 10. *Let $W\subset T_D$ be an analytic G^e-invariant subbundle. Then there is a subset $J\subset[1, \ldots, s]$ such that $W=\bigoplus_{i\in J}^{s}T_{i,D}$.*

Proof. As we have shown at the beginning of the proof of Lemma 9 the group G^e is dense in $G_{\mathbf{R}}$, consequently W will be a $G_{\mathbf{R}}$-invariant subbundle in T_D. Since D is a homogeneous space, the subbundle $W\subset T_D$ is defined by the subspace $W_d\subset T_D(d)$ at an arbitrary point $d\in D$. Denote by K the stabilizer of d in $G_{\mathbf{R}}$; K is a maximal compact subgroup in $G_{\mathbf{R}}$. Since W is a $G_{\mathbf{R}}$-invariant subbundle in T_D, the space W_d is a K-invariant subspace in $T_D(d)=\bigoplus_{i=1}^{s}T_{i,D}(d)$. Therefore to prove Lemma 10 it is sufficient to establish the following assertion.

Lemma 10.1. *Any K-invariant complex subspace W_d in $\bigoplus_{i=1}^{s}T_{i,D}(d)$ is of the form $W_d=\bigoplus_{i\in J}T_{i,D}(d)$ for some subset $J\subset[1, \ldots, s]$.*

* We say that the submanifold is *linear* if for every point $d\in D$ the fibre $p_W^{-1}(d)=W\cap p^{-1}(d)$ is a linear subspace of $p^{-1}(d)$.

Proof of Lemma 10.1. Since K is a maximal compact subgroup in $G_{\mathbf{R}}$, $\alpha(K)$ is a maximal compact subgroup of the group $\prod_{i=1}^{s} G_i$, and consequently $K=\prod_{i=1}^{s} K_i$, where K_i is a maximal compact subgroup in G_i. The space T_D decomposes, by assumption, into the direct sum $T_D(d)=\bigoplus_{i=1}^{s} T_{D_i}(d)$.

Every subspace $T_{D_i}(d)\subset T_D(d)$ is K-invariant. Further the group K_j, $j\neq i$ acts on the space $T_{D_i}(d)$ trivially, and the action of the group K_i on $T_{D_i}(d)$ is isomorphic to its action on the tangent space to a K_i-invariant point in the irreducible symmetric space $K_i\diagdown G_i$. Consequently, (cf. [8] Chapter VIII, 5.1), the action of the group K_i in the space $T_{D_i}(d)$ is irreducible. This implies that any K-invariant complex subspace W in $T_D(d)$ is of the form $\bigoplus_{i\in J} T_{D_i}(d)$ for some subset $J\subset[1, \ldots, s]$. Lemma 10.1, and consequently Lemma 10 also are proved.

For any $J\subset[1, \ldots, s]$ put $T_{J,D}=\bigoplus_{i\in J} T_{i,D}$. $T_{J,D}$ is, as mentioned already, a $G_{\mathbf{R}}$-invariant subbundle in T_D. Consequently, $T_{J,D}$ is a Γ-invariant subbundle in T_D and $T_{J,D}/\Gamma$ is (see Lemma 6) an algebraic subbundle in T_X. It will be denoted by $T_{J,X}$.

Now let σ be an automorphism of the field $\mathbf{C}$, X^σ the corresponding complex manifold (see § 1). Put $D^\sigma=D_{X^\sigma}$, $\Gamma^\sigma=\Gamma(X^\sigma)$ and denote by G^σ the subgroup $(G^e)^\sigma$ in $G(X^\sigma)$. Put $T_{J,X^\sigma}=(T_{J,X})^\sigma\subset T_{X^\sigma}$ and $T_{J,D^\sigma}=(\pi^\sigma)^* T_{J,X^\sigma}\subset T_{D^\sigma}$. It is clear that T_{J,D^σ} is an analytic G^σ-invariant subbundle in T_{D^σ}. According to Lemmas 5′ and 6′, it follows from Lemmas 9 and 10 that the following assertions hold:

Lemma 9′. *Let Z^σ be a non-empty closed analytic G^σ-invariant submanifold in D^σ. Then $Z^\sigma=D^\sigma$.*

Lemma 10′. *Let $W^\sigma\subset T_{D^\sigma}$ be an analytic G^σ-invariant subbundle. Then there is a subset $J\subset[1, \ldots, s]$ such that $W^\sigma=T_{J,D^\sigma}$.*

Now apply part a) of Lemma 8 to the case $Y=X$ and $F=T_{J,X}$. We find that for any subset $J\subset[1, \ldots, s]$ the subbundles T_{J,X^σ} and T_{J,D^σ} in T_{X^σ} and T_{D^σ}, are integrable. We denote by $T_{J,X}$, T_{J,X^σ} and T_{J,D^σ} also the corresponding analytic foliations in X, X^σ and D^σ.

Lemma 11. *If $J\neq\emptyset$, $[1, \ldots, s]$ then the foliation T_{J,X^σ} has no compact leaf.*

Proof. Let Z^σ be a compact leaf of the foliation T_{J,X^σ}. Then Z^σ is a compact analytic submanifold in X^σ, and consequently, according to Chow's lemma, it is an algebraic submanifold in X^σ. Put $Z=(Z^\sigma)^{\sigma^{-1}}\subset X$. Then Z is a compact leaf of the foliation $T_{J,X}$. Consequently, $V=\pi^{-1}(X)$ is a submanifold in D, invariant under the group Γ and the groups G_i, $i\in J$. Since $J\neq\emptyset$ and the algebraic $\mathbf{Q}$-group G is simple by assumption, the strong approximation theorem (see [7]) implies that the subgroups Γ and G_i, $i\in J$ generate a dense subgroup in $G_{\mathbf{R}}$. Since V is a closed submanifold, this implies that $V=D$. This equality contradicts $J\neq[1, \ldots, s]$. Lemma 11 is proved.

Corollary. If $J\neq\emptyset$ then the foliation T_{J,D^σ} has no compact leaf.

Proof. Let V^σ be a compact leaf of the foliation T_{J,D^σ}. Put $Z^\sigma=\pi^\sigma(V^\sigma)$. Z^σ is a compact leaf in the foliation T_{J,X^σ}. According to Lemma 11, $J=[1, \ldots, s]$ and $Z^\sigma=X^\sigma$. Consequently, the leaf V^σ has to coincide with D^σ, which contradicts the compactness of V^σ. The corollary is proved.

§ 3. The existence of integrable holomorphic forms on D^σ

Now we turn to the study of the complex manifold D^σ. As our first result, we prove that on D^σ there exist holomorphic differential forms of maximal degree $\omega\neq0$, such that $\int_{D^\sigma}\omega\wedge\bar{\omega}<\infty$. To this end we will need a theorem of functional analysis. We are not going to formulate it here in its full generality, but introduce this theorem in the form convenient for our purpose.

Let Y be a complex n-dimensional manifold. Denote by $\pi_1^0(Y)$ the normal subgroup in the fundamental group $\pi_1(Y)$ of the manifold Y, which is the intersection of all subgroups of finite index in $\pi_1(Y)$, and by $D(Y)$ the complex-analytic manifold, which is the unramified covering of Y corresponding to the subgroup $\pi_1^0(Y)\subset\pi_1(Y)$. Put $\Gamma(Y)=\pi_1(Y)/\pi_1^0(Y)$. The group $\Gamma(Y)$ acts on D_Y and $Y=D_Y/\Gamma_Y$. Denote by π the projection $\pi: D_Y\to Y$.

Let Γ_l be a cofinal sequence of normal subgroups of finite index in $\Gamma(Y)$. The definition of the group $\Gamma(Y)$ implies $\bigcap_l\Gamma_l=e$. Put $Y_l=D_Y/\Gamma_l$ and denote by π_l the natural projection $\pi_l: D_Y\to Y_l$ and by τ_l the projection $\tau_l: Y_l\to Y$. Put $c_l=[\Gamma(Y)/\Gamma_l]$.

Let F be a locally free coherent analytic sheaf on Y. Then F is the sheaf of germs of the analytic sections of some analytic linear bundle $\mathbf{F}$. Consider Y as a $2n$-dimensional real manifold. Let η be a positive-definite Hermitian form on $\mathbf{F}$ with values in the $2n$-dimensional differential forms on Y. Denote by $\bar{F}$ the sheaf of germs of the measurable (locally) square integrable sections of the bundle $\mathbf{F}\to Y$. It is clear that $F\subset\bar{F}$. Denote by $H_2^0(Y, \bar{F}, \eta)$ the subspace in $H^0(Y, \bar{F})$ consisting of all sections $f\in H^0(Y, \bar{F})$ for which $\int_Y\eta(f,\bar{f})<\infty$. We introduce on $H_2^0(Y, \bar{F}, \eta)$ a scalar product $\langle\ \rangle$ by taking $\langle f_1, f_2\rangle=\int_Y\eta(f_1, f_2)$ for $f_1, f_2\in H_2^0(Y, \bar{F}, \eta)$. Thus the space $H_2^0(Y, \bar{F}, \eta)$ becomes a complete Hilbert space. Denote by $H_2^0(Y, F, \eta)$ the subspace in $H_2^0(Y, \bar{F}, \eta)$ consisting of analytic sections; $H_2^0(Y, F, \eta)=H^0(Y, F)\cap H_2^0(Y, \bar{F}, \eta)$.

Remark. In the bundle Ω_Y^n of differential forms of maximal degree on Y there is a canonical Hermitian form η given by $\eta(\omega_1, \omega_2)=\omega_1\wedge\bar{\omega}_2$, $\omega_1, \omega_2\in H^0(Y, \Omega_Y^n)$.

The bundle Ω_Y^n will be regarded as being equipped with this metric. The Hilbert space $H_2^0(Y, \Omega_Y^n)$ consists of all analytic forms $\omega \in H^0(Y, \Omega_Y^n)$ for which $\int_Y \omega \wedge \bar{\omega} < \infty$.

Lemma 12. *The subspace* $H_2^0(Y, F, \eta)$ *is closed in* $H_2^0(Y, \bar{F}, \eta)$.

Proof. Lemma 12 follows immediately from Proposition 6 of Chapter III in [9].

Remark. Recall that Y is an arbitrary complex manifold. Since together with the form η we define the Hermitian forms $\tau_l^* \eta$ and $\pi^* \eta$ on the bundles $\tau_l^* F$ and $\pi^* F$, we can apply all definitions and Lemma 12 to Y_l, D_Y etc.

Theorem 2. *Let* Y *be a compact manifold and* $h_l = \dim H^0(Y_l, \tau_l^*(F))$. *If* $\overline{\lim_{l \to \infty}} h_l / c_l > 0$ *then for any nondegenerate Hermitian form* η *on* Y *we have* $H_2^0(D_Y, \pi^* F, \pi^* \eta) \neq 0$.

Remark 1. Since Y is a compact manifold, the space $H_2^0(D_Y, \pi^* F, \pi^* \eta)$ does not depend on the choice of η.

Remark 2. The subspace $H^0(Y, F)$ is determined in the space $H^0(Y, \bar{F})$ as the solution of some elliptic system of partial differential equations. There is an analogue of Theorem 2 for any elliptic system of differential equations on a compact manifold.

Before proving Theorem 2 we present two auxiliary lemmas.

Let us fix a Lebesgue measure μ on Y. Put $\mu_l = \tau_l^*(\mu)$, $\tilde{\mu} = \pi^*(\mu)$, $\eta_l = \tau_l^*(\eta)$, $\tilde{\eta} = \pi^*(\eta)$. For any section $\gamma \in H^0(Y_l, \tau_l^*(F))$ denote by α_γ^l the non-negative function on Y_l, given by the equation $\eta_l(\gamma, \gamma) = \alpha_\gamma^l(y_l)\mu$, $y_l \in Y_l$. Define the function α^l on Y_l by

$$\alpha^l(y_l) = \max_{\gamma \in H^0(Y_l, \tau_l^*(F_l)), \langle \gamma, \gamma \rangle = 1} \alpha_\gamma^l(y_l), \quad y_l \in Y_l.$$

We call the function α^l the Plancherel measure of the subspace $H^0(Y_l, F, \tau_l^*)$ in the Hilbert space $H^0(Y_l, \bar{F}, \tau_l^*)$ corresponding to the measure μ_l.

Lemma 13. $\int_{Y_l} \alpha^l(y_l')\mu_l = h_l \stackrel{\text{def}}{=} \dim H^0(Y_l, F, \tau_l^*)$.

Proof. Let $\gamma_1, \ldots, \gamma_{h_l}$ be an orthonormal basis in $H^0(Y_l, F, \tau_l^*)$. It is easy to check that $\alpha^l(y_l) = \sum_{1 \leq i \leq h_l} \alpha_{\gamma_i}^l(y_l)$. Since we have $\int_{Y_l} \eta_l(\gamma_i, \gamma_i) = \langle \gamma_i, \gamma_i \rangle = 1$, the lemma is proved.

Remark. The factor group $\Gamma(Y)/\Gamma_l$ acts on the bundle $\tau_l^* F$ preserving the form η. Therefore the Plancherel measure of $\alpha^l(Y_l)$ is invariant under the action of the group $\Gamma(Y)/\Gamma_l$.

For sake of convenience we assume in the sequel that $\int_Y \mu = 1$.

The group $\Gamma(Y)$ acts on the space D_Y. We construct an appropriate fundamental domain M in D_Y. Consider Y as a smooth $2n$-dimensional compact real manifold. The existence of a smooth triangulation on a differentiable manifold implies that there is in Y a closed subset Z, being the union of a finite number of smooth $2n-1$-dimensional submanifolds of Y, such that $Y-Z$ is a disjoint union of contractible manifolds. Fix a subset M in D_Y such that the mapping π provides a homeomorphism of M onto $Y-Z$. Denote by $\overline{M}$ the closure of the set M in D_Y. Since Y is a compact manifold, $\overline{M}$ is a compact subset in D_Y. Further $\pi(\overline{M})=Y$.

Lemma 14. *For any relatively compact subset $K\subset D_Y$ there is a number l such that the restriction of the projection π_l to K is a homeomorphism.*

Proof. Choose a compact subset $K'\subset D_Y$ such that K belongs to the interior of K'. Since K' is a compact set, there is a finite number of different elements $\gamma_i\in\Gamma(Y)$, $1\leqq i\leqq m$ such that $K'\subset\bigcup_{i=1}^{m}\overline{M}\gamma_i$. Denote by V the interior of the set $\bigcup_{i=1}^{m}\overline{M}\gamma_i$. By construction we have $K\subset V$. Consider the elements $\gamma_{i,j}=\gamma_i\gamma_j^{-1}\in\Gamma(Y)$, $1\leqq i\neq j\leqq m$. By assumption $\bigcap_l \Gamma_l=(e)$, thus there is a number l such that $\gamma_{i,j}\notin\Gamma_l$, $1\leqq i\neq j\leqq m$. The definition of the set M implies that the restriction of π_l to V is a homeomorphism. Since $K\subset V$, the restriction of π_l to K is also a homeomorphism. Lemma 14 is proved.

From Lemma 14 we have immediately

Corollary. There is a sequence of open relatively compact sets $M_l\subset D_Y$ such that a) $M_l\subset M_{l+1}$ b) $\bigcup_l M_l=D_Y$ c) the restriction of π_l to M_l is homeomorphism of M_l to $\pi_l(M_l)\subset Y_l$.

Proof of Theorem 2. Taking a subsequence and replacing Y by Y_l we find that for all numbers l the inequality $l>c_l$ holds. Lemma 13 implies that $\int_{Y_l}\alpha^l(y_l)\mu_l>c_l=\int_{Y_l}\mu_l$. Consequently for every l there is a point $y_l^0\in Y_l$ such that $\alpha^l(y_l^0)>1$. According to the remark after Lemma 13 we can assume $y_l^0\in\pi_l(\overline{M})$. The definition of the function $\alpha^l(y_l)$ implies that there is a section $\gamma_l\in H^0(Y_l, F, \tau_l^*)$ such that $\langle\gamma_l,\gamma_l\rangle=1$ and $\alpha^l_{\gamma_l}(y_l^0)>1$. Let us fix an open relatively compact subset $M'\supset D_Y$ for which $\overline{M}\subset M'$. Proposition 5 in Chapter 3 of [9] implies that there is a positive constant b such that $\int_{\pi_l(M')}\eta(\gamma_l,\gamma_l)\mu_l>b$ for all numbers l and sections γ_l satisfying the condition $\max_{y\in\pi_l(\overline{M})}\alpha^l_{\gamma_l}(y)>1$.

Let M_l be the sequence of subsets in D_Y constructed in the Corollary to Lemma 14. Denote by $\bar{\gamma}_l$ the element of the Hilbert space $H_2^0(D_Y,\pi^*F,\hat{\eta})$ for which $\bar{\gamma}_l|_{M_l}=\pi_l^*(\gamma_l)|_{\pi_l(M_l)}$ and $\bar{\gamma}_l|_{D_Y-M_l}=0$. It is clear that $\langle\bar{\gamma}_l,\bar{\gamma}_l\rangle\leqq\langle\gamma_l,\gamma_l\rangle=1$. Therefore the sequence $\bar{\gamma}_l$ of elements of the Hilbert space $H_2^0(D_Y,\pi^*\overline{F},\hat{\eta})$ is bounded, and consequently, it has a weak limit point $\gamma \in H_2^0(D_Y, \pi^* \bar{F}, \tilde{\eta})$. Since M' is a relatively compact subset in D_Y, it follows that

$$\int_{M'} \tilde{\eta}(\gamma, \gamma) \geqq \liminf_{l\to\infty} \int_{M'} \tilde{\eta}(\bar{\gamma}_l, \bar{\gamma}_l) \geqq \liminf_{l\to\infty} \int_{\pi_l(M')} \eta_l(\gamma_l, \gamma_l) > b.$$

Consequently $\gamma \neq 0$.

Now let V be an arbitrary relatively compact subset in D_Y. Since $\bigcup_l M_l = D_Y$ there exists a natural number l_0 such that $V \subset M_l$ for $l > l_0$. Consider the element $\gamma|_V$ in $H_2^0(V, \pi^* \bar{F}, \tilde{\eta})$. By construction the element $\gamma|_V$ is a weak limit point of the sequence $\gamma_l|_V$. Since $\gamma_l|_V \in H_2^0(V, \pi^* F, \tilde{\eta})$ for $l > l_0$, Lemma 12 implies $\gamma|_V \in H_2^0(V, \pi^* F, \tilde{\eta})$. Since V is an arbitrary relatively compact set, we have $\gamma \in H_2^0(D_Y, \pi^* F, \tilde{\eta})$.

Theorem 2 is proved.

Conjecture 1. *The converse of Theorem* 2 *also holds.*
That is to say, the condition $H_2^0(D_Y, \pi^* F, \tilde{\eta}) \neq \{0\}$ implies $\overline{\lim}\, h_l/c_l > 0$.

Now we go on studying the complex manifold $D^\sigma = D_{X^\sigma}$, where X is an n-dimensional arithmetic manifold, and $\sigma \in \operatorname{Aut} \mathbf{C}$.

Theorem 3. $H_0^2(D^\sigma, \Omega_{D^\sigma}^n) \neq \{\sigma\}$.

Proof. Let X_l^σ be a cofinal sequence of finitely-sheeted unramified Galois coverings $\tau_l^\sigma : X_l^\sigma \to X^\sigma$ and let $c_l = \deg \tau_l$. Since the mappings π^σ and τ_l^σ are unramified $(\tau_l^\sigma)^*(\Omega_{X^\sigma}^n) = \Omega_{X_l^\sigma}^n$ and $(\pi^\sigma)^*(\Omega_{X^\sigma}^n) = \Omega_{D^\sigma}^n$. Therefore Theorem 2 implies that in the proof of Theorem 3 it is sufficient to establish the relation

$$\limsup_{l\to\infty} \left(\dim_{\mathbf{C}} H^0(X_l^\sigma, \Omega_{X_l^\sigma}^n)/c_l\right) > 0$$

Put $X_l = (X_l^\sigma)^{\sigma^{-1}}$ and $\tau_l = (\tau_l^\sigma)^{\sigma^{-1}} : X_l \to X$. Theorem 1 implies that X_l is a cofinal sequence of coverings of the manifold X. It is further clear that

$$\dim_{\mathbf{C}} H^0(X_l^\sigma, \Omega_{X_l^\sigma}^n) = \dim_K H^0(\hat{X}_l, \Omega_{\hat{X}_l}^0) = \dim_{\mathbf{C}} H^0(X_l, \Omega_{X_l}^n).$$

Therefore, the proof of Theorem 3 is reduced to proving the following proposition:

Proposition 1. Let $X_\Gamma = D/\Gamma$ be an arithmetic manifold, $X_l = D/\Gamma_l$ a cofinal sequence of Galois coverings, $c_l = [\Gamma/\Gamma_l]$ and $h_l^n = \dim H^0(X_l, \Omega_{X_l}^n)$. Then $\limsup_{l\to\infty} h_l^n/c_l > 0$.

Proof of Proposition 1. We introduce three auxiliary lemmas. Denote by Ω_X^i the bundle of i-dimensional holomorphic differential forms on X, by $\Omega_{X_l}^i$ the bundle of i-dimensional forms on X_l etc. Put $h_l^i = \dim H^0(X_l, \Omega_l^i)$.

Lemma 15. *There is a non-zero constant v such that for any l,* $\sum_{i=0}^{n}(-1)^i h_l^i = vc_l$.

Proof. For any l the compact complex manifold X_l is a projective arithmetic and, consequently, Kählerian manifold and according to the theorem of Hodge (see [10] § 4) $h_l^i \overset{\text{def}}{=} h^{i,0}(X_l) = h_n^{0,i}(X_l) = \dim H^i(X_l, O)$, and consequently $\sum_{i=0}^{n}(-1)^i h_l^i = \sum_{i=0}^{n}(-1)^i \dim H^i(X_l, O) = \chi(X_l)$. Let us fix now a $G_{\mathbf{R}}$-invariant Kählerian metric ϱ on D. Denote by Φ the $2n$-dimensional component of the Todd form on D, corresponding to the metric ϱ (see [10] § 16). Since ϱ is a G-invariant metric, Φ is a G-invariant $2n$-form on D, therefore Φ can be considered as a form on X_l. The Riemann—Roch theorem (see [10] 18.2) implies that $\chi(X_l) = \int_{X_l} \Phi = c_l \int_X \Phi$. Therefore $\chi(X_l) = vc_l$, where $v = \int_X \Phi$ and for the proof of Lemma 15 it is enough to show that $\int_X \Phi \neq 0$. Since Φ is a $G_{\mathbf{R}}$-invariant form, it is enough to establish the following

Lemma 15.1. $\Phi \not\equiv 0$.

Proof of Lemma 15.1. Denote by D_0 the compact Hermitian symmetric space, dual to D (see [8] Chapter V, § 2). It is the symmetric space corresponding to the compact form M of the group $G_{\mathbf{R}}$, $D_0 = M/K$, isomorphic (see [8] Chapter VIII Theorem 7.13) to the factor space of the group $G_{\mathbf{C}}$ by some parabolic subgroup $P_{\mathbf{C}} \subset G_{\mathbf{C}}$. Denote by ϱ_0 an M-invariant Kählerian metric on D_0, and by Φ_0 the $2n$-dimensional component of the Todd form on D_0 corresponding to the metric ϱ_0.

D can be considered as an open K-invariant submanifold in D_0 (see [8] Chapter VIII Theorem 7.13), further the actions of K on D and D_0 are compatible. Let $d \in D \subset D_0$ be a K-stable point and $T_D(d)$ the tangent space to D at the point d. Denote by $\mathfrak{L}$ the Lie algebra of the group $G_{\mathbf{R}}$ with the operation of commutation $[\ ,\ ]$, by $\mathfrak{M}$ the Lie algebra of the group M with the operation of the commutation $[\ ,\ ]_0$, and by $\mathfrak{K}$ the Lie algebra of the group K. Denote by $\mathfrak{P}$ the orthogonal complement to $\mathfrak{K}$ with respect to the Killing form in $\mathfrak{L}$, and by $\mathfrak{P}_0$ the complement of $\mathfrak{K}$ in $\mathfrak{M}$. It is known that $[\mathfrak{P}, \mathfrak{P}] \subset \mathfrak{K}$ and $[\mathfrak{P}_0, \mathfrak{P}_0]_0 \subset \mathfrak{K}$. Let j and j_0 be the natural isomorphisms $j: \mathfrak{P} \xrightarrow{\sim} T_D(d)$, $j_0: \mathfrak{P}_0 \to T_D(d)$.

Denote by $R_{X,Y}Z (X, Y, Z \in T_D(d))$ (see [8]) the curvature mapping (tensor) corresponding to the metric of $\mathfrak{P}$, and by $R^0_{X,Y}Z$ the curvature mapping corresponding to the metric of $\mathfrak{P}_0$. It is known (see Chapter IV, Theorem 4.2 in [8]) that $R_{X,Y}Z = j([[j^{-1}(X), j^{-1}(Y)], j^{-1}(Z)]$, $R^0_{X,Y}Z = j_0([[j_0^{-1}(X), j_0^{-1}(Y)]_0, j_0^{-1}(Z)]_0)$. Denote by $\mathfrak{P}_{\mathbf{C}}$, $\mathfrak{P}_{0\mathbf{C}}$, $\mathfrak{K}_{\mathbf{C}}$ the complexifications of the spaces $\mathfrak{P}$, $\mathfrak{P}_0$ and the Lie algebra $\mathfrak{K}$, respectively. The operation of commutation can be extended from $\mathfrak{P}$ and $\mathfrak{P}_0$ to $\mathfrak{P}_{\mathbf{C}}$ and $\mathfrak{P}_{0\mathbf{C}}$. Since the Lie algebras $\mathfrak{L}_{\mathbf{C}}$ and $\mathfrak{M}_{\mathbf{C}}$ are isomorphic, there is an isomorphism α of the linear spaces $\mathfrak{P}_{\mathbf{C}}$ and $\mathfrak{P}_{0\mathbf{C}}$, $\alpha: \mathfrak{P}_{\mathbf{C}} \to \mathfrak{P}_{0\mathbf{C}}$ such that for any $X, Y \in \mathfrak{P}_{\mathbf{C}}$ we have $[X, Y] = [\alpha(X), \alpha(Y)]_0$. This implies that $\Phi|_d = \Phi_0|_d \det(j_{\mathbf{C}} j_{0\mathbf{C}}^{-1} \alpha)$.

According to the Riemann—Roch theorem we have $\int_{D_0} \Phi_0 = \chi(D_0)$. Since in D_0 all cohomology classes can be generated by algebraic cycles*, they are all of type (k, k), and consequently $\dim H^i(D_0, O) = h^{0,i}_{D_0} = 0$ for $i \neq 0$. On the other hand $h^{0,0}_{D_0} = \dim H^0(D_0, O) = 1$ and therefore $\chi(D_0) = 1$. Consequently $\int_{D_0} \Phi_0 = 1$. Since Φ_0 is an M-homogeneous form, thus $\Phi_0|_d \neq 0$ and therefore $\Phi|_d \neq 0$. Lemma 15.1, and consequently also Lemma 15 are proved.

Now denote by Ω the $G_{\mathbf{R}}$-invariant 2-form on D corresponding to the Killing form on $\mathfrak{P}$ and by ϱ the corresponding metric.

Lemma 16. *If $i<n$ and ω is an i-dimensional form on D, such that $\int_D \omega \wedge \bar{\omega} \wedge \Omega^{n-i} < \infty$, then $\omega \equiv 0$.*

Proof. Let us represent D in the form of a bounded strongly pseudo-convex domain in $\mathbf{C}^n$ using the construction of Harish-Chandra (see [8] Chapter VIII, § 7), then the action of group K on D extends to linear unitary transformations $k \to \psi(k)$ of the space $\mathbf{C}^n$; further there is (see [8] Chapter VIII, Theorem 4.5) an imbedding of the unit circle $T = \{z = e^{i\varphi}\}$ into the centre of the group K, such that $\psi(e^{i\varphi})(z) = e^{ik\varphi} z$ for all $z \in D \subset \mathbf{C}^N$ where k is a natural number. Then the holomorphic forms on D, which are transformed by a finite-dimensional representation of the group K, are of the form $\omega = \sum_{1 \leqq j_1 \leqq \ldots \leqq j_i \leqq n} f_{j_1, \ldots, j_i} dz_{j_1} \ldots dz_{j_i}$ where $z_1, \ldots, z_n$ are the coordinates in $\mathbf{C}^n$, and the functions $f_{j_1}, \ldots, f_{j_i}$ on D can be continued to polynomials on $\mathbf{C}^n$. Now let ω be a form which satisfies the condition of Lemma 16. Then its projection to the space of holomorphic forms on D which are transformed by a fixed representation of the group K, also satisfies the condition of the lemma. Therefore we can assume that the form ω on D is of the form $\omega = \sum_{1 \leqq j_1 \leqq \ldots \leqq j_i \leqq n} f_{j_1, \ldots, j_i} dz_{j_1} \ldots dz_{j_i}$ where the functions $f_{j_1, \ldots, j_i}$ on D can be continued to polynomials on $\mathbf{C}^n$.

D is a bounded domain in $\mathbf{C}^n$, therefore the Bergman metric of D is nondegenerate. Since D is a homogeneous domain in $\mathbf{C}^n$ we can assume that $\mathfrak{P}$ coincides with the Bergman metric in D. Since the Bergman metric of a strongly pseudo-convex domain increases rapidly as we approach the boundary (see [12] Theorem 35.1), and $i<n$ by assumption, the estimates $\int_D \omega \wedge \bar{\omega} \wedge \Omega^{n-i} < \infty$ imply that the functions $f_{j_1, \ldots, j_i}$ tend to zero almost everywhere as we approach the boundary of the domain. Therefore $f_{j_1, \ldots, j_i} \equiv 0$. Lemma 16 is proved.

Lemma 17. *For $i<n$ we have $\lim_{l \to \infty} h^i_l / c_l = 0$.*

* Indeed, the algebraic manifold $G_{\mathbf{C}}/P_{\mathbf{C}}$ is the union of a finite number of orbits of the unipotent radical U_P of the group $P_{\mathbf{C}}$, each of them is biregularly equivalent to an affine space (see [11] Theorem 5.15). Since the real dimension of any complex manifold is even, the algebraic submanifolds in D_0 which are the closures of the different orbits, form a basis of the homology group of the manifold.

Proof. Denote by Ω_X the Kählerian form on X, induced by the $G_{\mathbf{R}}$-invariant form Ω on D and define the Hermitian form η^i on the fibre space corresponding to the bundle Ω_X^i by taking $\eta(\omega_1, \omega_2) = \omega_1 \wedge \bar{\omega}_2 \wedge \Omega_X^{n-i}$. Since $\pi^* \Omega_X = \Omega_D$, Lemma 16 implies that the space $H_2^0(D, \Omega_D^i, \tilde{\eta})$ equals $\{0\}$. It follows from Theorem 2 that this is possible only in the case $\lim\limits_{l \to \infty} h_l^i / c_l = 0$. Lemma 17 is proved.

Proposition 1 follows immediately from Lemmas 15 and 17. Theorem 3 is proved.

§ 4. Non-degeneracy of the Bergman metric on D^σ

Let V be a smooth complex n-dimensional manifold. Denote by $H_2^0(V)$ the linear space of holomorphic n-forms ω on V, for which $\int\limits_V \omega \wedge \bar{\omega} < \infty$, and introduce an Hermitian scalar product into the space $H_2^0(V)$ by taking $\langle \omega_1, \omega_2 \rangle = i^{n^2} \int\limits_V \omega_1 \wedge \bar{\omega}_2$ for $\omega_1, \omega_2 \in H_2^0(V)$. As shown in Proposition 6, Chapter III in [9], the space $H_2^0(V)$ is a complete Hilbert space. Let $\omega_1, \ldots, \omega_l$ be an orthonormal basis in the space $H_2^0(V)$. Put $\theta_V = i^{n^2} \sum\limits_l \omega_l \wedge \bar{\omega}_l$. As shown in Corollary 1 of Theorem 1 of the same chapter, the series $\sum \omega_l \wedge \bar{\omega}_l$ converges commutatively, and θ_V is a real-analytic non-negative $2n$-dimensional form on V, independent of the choice of the basis $\omega_1, \ldots, \omega_l, \ldots$ in the space $H_2^0(V)$. Since the form θ_V is canonically defined by the manifold structure of V, it is invariant under the action of the group Aut V on V. Denote by V' the subset in V consisting of all points $v \in V$, at which the form θ_V vanishes. For any holomorphic n-form ω on V denote by V_ω the submanifold of the zeros of ω; then V_ω is a complex submanifold of V. The definition of θ_V implies immediately that $V' = \bigcap\limits_{\omega \in H_2^0(V)} V_\omega$ and therefore V' is a closed complex submanifold in V. Let $v \in V - V'$ and $v_1, \ldots, v_n$ be a system of coordinates in a neighbourhood U of the point v. Choose a function φ on U satisfying $\theta_V = \varphi\, dv_1 \wedge \ldots \wedge dv_n \wedge d\bar{v}_1 \wedge \ldots \wedge d\bar{v}_n$ and put $\Omega_V = id'd'' \log \varphi$. Ω_V is, by definition, a closed differential form of type (1, 1) on $U - U \cap V'$, and it is easy to verify that it does not depend on the choice of the coordinate system $v_1, \ldots, v_n$ in U. Therefore Ω_V is a well-defined form of type (1, 1) on $V - V'$. Denote by ds_V^2 the Hermitian form on $V - V'$ associated to Ω_V (see [9] Chapter II.5). It is easy to check that ds_V^2 is a nonnegative form. It is called the Bergman (pseudo-)metric on V. Denote by $T'(V)$ the subset in $T(V - V')$ consisting of all vectors $\eta \in T(V - V')$ for which $ds_V^2(\eta, \eta) = 0$. For any pair of forms $\omega_1, \omega_2 \in H_2^0(V)$ denote by T_{ω_1, ω_2} the submanifold in $T(V - V')$, consisting of all vectors $\eta \in T(V)$, for which we have either $p(\eta) \in V_{\omega_2}$*, or $\frac{\partial}{\partial \eta}\left(\frac{\omega_1}{\omega_2}\right) = 0$. It is not difficult to show (see Corollary 3 of Theorem 1 of Chapter

* Here, as well as above, p denotes the natural projection $p: T(V) \to V$.

III in [9]) that $T'(V) = \bigcap_{\omega_1, \omega_2 \in H_2^0(V)} T_{\omega_1, \omega_2}$. Thus we have shown that $T'(V)$ is a closed linear complex submanifold in $T(V-V')$.

Consider now the case $V = D^\sigma$.

Theorem 4. *The Bergman pseudometric $ds^2_{D^\sigma}$ is a metric, i.e. it is defined everywhere on D^σ and is nondegenerate.*

Proof. First we show that $(D^\sigma)' = \emptyset$. Indeed, as follows from the above arguments, $(D^\sigma)'$ is a closed complex submanifold in D^σ, invariant under the group Aut D^σ. According to Lemma 9′, we have either $(D^\sigma)' = \emptyset$ or $(D^\sigma)' = D^\sigma$. Now according to Theorem 3 $H_2^0(D^\sigma) \neq 0$, thus $(D^\sigma)' \neq D^\sigma$. Consequently $(D^\sigma)' = \emptyset$.

Consider the submanifold $T'(D^\sigma) \subset T(D^\sigma)$. The above arguments imply that $T'(D^\sigma)$ is a closed linear submanifold in $T(D^\sigma)$, invariant under the action of the group Aut D^σ. Therefore, according to Lemma 10′ there is a subset $J \subset [1, \ldots, s]$ such that $T'(D^\sigma) = \bigoplus_{i \in J} T_{J, D^\sigma}$. Assume that $J \neq \emptyset$. The integrable field T_{J, D^σ} defines a foliation of D^σ. The leaf of this foliation passing through the point $d \in D^\sigma$ is denoted by $S(d)$. The relation $T'(D^\sigma) = \bigcap_{\omega_1, \omega_2 \in H_2^0(D^\sigma)} T_{\omega_1, \omega_2}$ implies that $S(d)$ is the connected component of the point d in the set

$$S'(d) = \{d' \in D^\sigma \mid \forall \omega_1, \omega_2 \in H_2^0(D^\sigma, \Omega^n)|, \omega_1/\omega_2(d) = \omega_1/\omega_2(d')\}$$

Consequently, the leaf $S(d)$ is closed. According to the Corollary of Lemma 11 it is not compact. Let us fix a Hermitian metric ϱ and a Lebesgue measure μ on X^σ, and denote by the same letters their inverse images under the projection $\pi^\sigma : D^\sigma \to X^\sigma$ i.e. we consider ϱ as a metric and μ as a measure on D^σ. Let $d \in D^\sigma$. Since the leaf $S(d)$ is closed and noncompact, there exists an infinite sequence of points $d_i \in S(d)$ such that $\varrho(d_i, d_j) > 1$ for all $i \neq j$. Choose a form $\omega \in H_2^0(D^\sigma, \Omega^n)$ for which $(\omega \wedge \bar{\omega}/\mu)(d) = 1$. This can be done since $(D^\sigma)' = \emptyset$. Consider the sequence of the numbers $a_i = (\omega \wedge \bar{\omega}/\mu)(d_i)$. Denote by $B(d_i)$ the ball $\varrho(d_0, d_i) \leq 1/2$, $d_0 \in D^\sigma$. By assumption the balls $B(d_i)$ do not intersect. Since the factor space $X^\sigma = D^\sigma/\Gamma^\sigma$ is compact, by Proposition 5 of Chapter 2 in [9] there is a constant $\gamma > 0$ such that

$$\int_{B(d_i)} \omega \wedge \bar{\omega} > \gamma |a_i|.$$

Since $\omega \in H_2^0(D^\sigma, \Omega^n)$, we have $\sum_i |a_i| < \infty$, and consequently $a_i \to 0$ for $i \to \infty$. As the factor space $X^\sigma = D^\sigma/\Gamma^\sigma$ is compact, by taking a subsequence we can assume that there are elements $\gamma_i \in \Gamma^\sigma$ such that the sequence $d_i \gamma_i$ converges to a limit $d_0 \in D^\sigma$. Since $(D^\sigma)' = \emptyset$ there is a form $\omega' \in H_2^0(D^\sigma, \Omega^n)$, $\langle \omega', \omega' \rangle = 1$ such that $(\omega' \wedge \bar{\omega}'/\mu)(d_0) = b \neq 0$. Therefore we can assume that $(\omega' \wedge \bar{\omega}'/\mu)(d_i \gamma_i) \geq c > 0$. Put $\omega'_i = \omega' \gamma_i$, then $\langle \omega'_i, \omega'_i \rangle = 1$ and $(\omega'_i \wedge \bar{\omega}'_i/\mu)(d_i) \geq c$. Since $d_i \in S(d) \subset S'(d)$

and $\omega, \omega' \in H_2^0(D^\sigma, \Omega^n)$ we have

$$\frac{(\omega_i' \wedge \overline{\omega}_i'/\mu)(d_i)}{(\omega \wedge \overline{\omega}/\mu)(d_i)} = \frac{(\omega_i' \wedge \overline{\omega}_i'/\mu)(d)}{(\omega \wedge \overline{\omega}/\mu)(d)}$$

whence we get $(\omega_i' \wedge \overline{\omega}_i'/\mu)(d) \geqq c \cdot a_i^{-1} \to \infty$. Now, this contradicts the relation $\langle \omega_i', \omega_i' \rangle = 1$. The obtained contradiction proves that $S(d) = d$. Consequently the Bergman metric $ds^2_{D^\sigma}$ of D^σ is nondegenerate. Theorem 4 is proved.

Remark. The group Aut D^σ preserves the Bergman metric, and therefore (see [14] Chapter I Theorem 8), the stabilizer of any point $d^\sigma \in D^\sigma$ in the group Aut D^σ, is compact. Consequently, the mapping φ: Aut $D^\sigma \to D^\sigma$ given by the formula $g \to d^\sigma g$, $g \in$ Aut D^σ, is proper. Therefore the orbit d^σ(Aut D^σ) of any point d^σ in D^σ is closed. Denote by $\overline{G}'^\sigma$ the closure of the group G^σ in the group Aut D^σ. $\overline{G}'^\sigma$ is a closed subgroup in Aut D^σ, consequently the orbit $d^\sigma \overline{G}'^\sigma$ is closed in D^σ for all $d^\sigma \in D^\sigma$.

§ 5. The proof of transitivity of the action of the group $\overline{G}'^\sigma$ on D^σ

To prove the homogeneity of the manifold D^σ we need some auxiliary lemmas.

Let H be an algebraic torus over **Q**. Denote by $\hat{H}$ the group of algebraic characters of H defined over **C**. We know (see [11] no. 1.2) that $\hat{H}$ is a free Abelian group, whose rank equals $\dim_{\mathbf{Q}} H$ and on which the group Gal (**C**, **Q**) acts naturally. This action can be written as $\sigma: \hat{h} \to \hat{h}^\sigma$ where $\hat{h} \in \hat{H}$, $\sigma \in$ Gal (**C**, **Q**). Further for any $h \in H_{\mathbf{Q}}$ and $\hat{h} \in \hat{H}$ we have $(\hat{h}(h))^\sigma = (\hat{h}^\sigma)(h)$.

Lemma 18. *For every* $\sigma \in$ Gal (**C**, **Q**) *there is a unique automorphism* $h \to h^\sigma$ *of the group* $H_{\mathbf{C}}$ *such that* $\hat{h}^\sigma(h^\sigma) = \hat{h}(h)$ *for all* $h \in H_{\mathbf{C}}$, $\hat{h} \in \hat{H}$. *Moreover the mapping* $h \to h^\sigma$ *is continuous.*

Proof. Let $\hat{h}_1, \ldots, \hat{h}_k$ be a basis in $\hat{H}$. Then the mapping $h \xrightarrow{\varphi} (\hat{h}_1(h), \ldots, \hat{h}_k(h))$ provides an isomorphism of the groups $H_{\mathbf{C}}$ and $\underset{(1)}{\mathbf{C}^*} \times \ldots \times \underset{(R)}{\mathbf{C}^*}$. Therefore the automorphism $h \to h^\sigma$ is uniquely determined by the condition of the lemma. Let $\varrho(\sigma)$ be the matrix of the automorphism $\hat{h} \to \hat{h}^\sigma$ of the group $\hat{H}$ in the basis $\hat{h}_1, \ldots, \hat{h}_k$: $\hat{h}^\sigma = \varrho(\sigma)\hat{h}$. Then $\varrho(\sigma)$ is an integer matrix with determinant ± 1. Therefore $(\varrho(\sigma)')^{-1} \in GL(k, \mathbf{Z})$. Let $(\varrho(\sigma)')^{-1} = (\varrho_{ij}')(\sigma)$, $1 \leqq i, j \leqq k$. For every element $h \in H_{\mathbf{C}}$ we define the element $h^\sigma \in H_{\mathbf{C}}$ by $h^\sigma = \varphi^{-1}(a_j)$, where $a_j = \prod_{1 \leqq i \leqq k} \hat{h}^{\varrho_{ij}'(\sigma)}(h)$ for all j, $1 \leqq j \leqq k$. Since the mapping $\varphi: H_{\mathbf{C}} \to (\mathbf{C}^*)^k$ is an isomorphism of topological groups, the mapping $h \to h^\sigma$ satisfies the conditions of the lemma. Lemma 18 is proved.

Remark. 1. The equality $(\hat{h}(h))^\sigma = (\hat{h}^\sigma)(h)$ for $\hat{h} \in \hat{H}$, $h \in H_{\mathbf{Q}}$ implies that $(\hat{h}(h))^\sigma = \hat{h}(h^{\sigma^{-1}})$ for $\hat{h} \in \hat{H}$, $h \in H_{\mathbf{Q}}$.

Remark 2. If $H_{\mathbf{R}}$ is a compact group then it is the unique maximal compact subgroup of the group $H_{\mathbf{C}}$. Therefore in case if $H_{\mathbf{R}}$ is a compact group, the mapping $h \to h^{\sigma}$ can be considered as an automorphism of $H_{\mathbf{R}}$.

Lemma 19. *Let $S=\{p_1, \ldots, p_k\}$ be a finite set of prime numbers, H_{∞} a maximal torus in $G_{\mathbf{R}}$, H_i a maximal torus in $G_{\mathbf{Q}_{p_i}}$, $1 \leqq i \leqq k$. Then there is a maximal torus H in G, defined over $\mathbf{Q}$, such that the subgroup $H_{\mathbf{R}} \subset G_{\mathbf{R}}$ is conjugate to H_{∞}, and the subgroups $H_{\mathbf{Q}_{p_i}} \subset G_{\mathbf{Q}_{p_i}}$ are conjugate to H_i, $1 \leqq i \leqq k$.*

Proof. Consider in $G_{\mathbf{R}} \times \prod_{i=1}^{k} G_{\mathbf{Q}_{p_i}}$ the subset V consisting of all regular elements (g_{∞}, g_i), $g \in G_{\mathbf{R}}$, $g_i \in G_{\mathbf{Q}_{p_i}}$, $1 \leqq i \leqq k$ for which g_{∞} is conjugate to an element of H_{∞} and g_i to an element of H_i. V is an open subset (see [14] Lemma 20). Therefore the strong approximation theorem implies (cf. [7]) the existence of an element $g \in G_{\mathbf{Q}} \cap V$. The centralizer H of the element $g \in G_{\mathbf{Q}}$ in G satisfies the requirements of the lemma. Lemma 19 is proved.

Theorem 5. *The group $\bar{G}'^{\sigma}$ acts transitively on D^{σ}.*

Proof. Choose a maximal torus H in G for which $H_{\mathbf{R}}$ is a compact group. This is possible according to Lemma 19. Let $d_0 \in D$ be a point which is stable under all elements $h \in H_{\mathbf{Q}}$. Put $\pi(d_0)=x \in X$ and choose a point $d_0^{\sigma} \in D^{\sigma}$ such that $\pi^{\sigma}(d_0^{\sigma})=x^{\sigma} \in X^{\sigma}$. Consider the orbit $Y^{\sigma}=d_0^{\sigma}\bar{G}'^{\sigma} \subset D^{\sigma}$ of the point d_0^{σ}. According to the remark after Theorem 4, this orbit is closed and is a C^{∞}-submanifold in D^{σ}. Denote by Z^{σ} its projection to X^{σ}. Z^{σ} is a closed, and consequently, compact C^{∞}-subset in X^{σ}. Since $[G^{\sigma}/\Gamma^{\sigma}]=\infty$ and $Z^{\sigma} \supset \pi^{\sigma}(d_0^{\sigma} G^{\sigma})$, Z^{σ} contains an infinite set of points. Now Z^{σ} is a compact manifold which contains an infinite set of points, thus $\dim Z^{\sigma}>0$, and consequently, $\dim Y^{\sigma}>0$. Denote by $T_{Y^{\sigma}}(d_0^{\sigma})$ the real tangent space to Y^{σ} at the point $d_0^{\sigma} \in Y^{\sigma}$. $T_{Y^{\sigma}}(d_0^{\sigma})$ is a real subspace in $T(d_0^{\sigma})$. The intersection $H_{\mathbf{Q}} \cap G_{d_0} \subset G^e \subset G_{\mathbf{Q}}$ is a subgroup of finite index in $H_{\mathbf{Q}}$ by assumption.

The manifold $Y^{\sigma} \subset D^{\sigma}$ is invariant under the group G^{σ}. Therefore the subspace $T_{Y^{\sigma}}(x^{\sigma}) \subset T(d_0^{\sigma})$ is invariant under all operators $\Phi_{d_0^{\sigma}}(g)$ for $g \in G^{\sigma} \cap G_{d_0^{\sigma}}$. Let α be the homomorphism $\alpha: G_{d_0} \to G_{d_0^{\sigma}}$ described in Lemma 4′. We know that $\alpha(G_{d_0} \cap G^e) \subset G_{d_0}^{\sigma} \cap G^{\sigma}$. Set $H'_{\mathbf{Q}} \overset{\text{def}}{=} H_{\mathbf{Q}} \cap G^e \subset G_{d_0}$.

Consider the representation $\Phi_{d_0^{\sigma}}(h) \overset{\text{def}}{=} \Phi_{d_0^{\sigma}}(\alpha(h))$, $h \in H_{\mathbf{Q}}$ of the group $H_{\mathbf{Q}}$ in the space $T_{d_0^{\sigma}}$. As follows from Lemma 4′, there is a $\mathbf{C}$-linear operator $c: T_{d_0} \to T_{d_0^{\sigma}}$, such that $\Phi_{d_0^{\sigma}}(h)=c(\Phi_{d_0}(h))^{\sigma} c^{-1}$ for all $h \in H_{\mathbf{Q}}$. Since Φ_{d_0} is an algebraic (over $\mathbf{C}$) representation of the group H, Remark 1 to Lemma 18 implies that the subspace $T_{Y^{\sigma}}(d_0^{\sigma}) \subset T(d_0^{\sigma})$ is invariant under the operators $c\Phi_{d_0}(h^{\sigma^{-1}})c^{-1}, h \in H_{\mathbf{Q}} \cap G_{d_0}$. Here h^{σ} is the automorphism defined in Lemma 18. Since H is a unirational manifold (cf. [15] Proposition 10), the group $H'_{\mathbf{Q}}$ is dense in the connected component $H_{\mathbf{R}}^0$ of the identity of the group $H_{\mathbf{R}}$. Therefore the subspace $T_{Y^{\sigma}}(d_0^{\sigma}) \subset T(d^{\sigma})$

is invariant under the operators $C\Phi_{d_0}(h^{\sigma-1})C^{-1}$ for all $h \in H_{\mathbf{R}}^0$. Since D admits an invariant complex structure, there is (see [8] Chapter VIII, Theorem 4.5) an element $h_0 \in H_{\mathbf{R}}^0$ such that $\Phi_{d_0}(h_0)=iE$. Since $H_{\mathbf{R}}$ is a compact group, in view of Remark 2 to Lemma 18 we have $h_0^\sigma \in H_{\mathbf{R}}^0$. Therefore the subspace $T_{Y^\sigma}(d^\sigma) \subset T(d^\sigma)$ is invariant under the operator $C\Phi_{d_0}((h_0^\sigma)^{\sigma-1})C^{-1}=iE$ and consequently, it is a complex space.

For any point $d^\sigma \in Y^\sigma$ denote by $T_{d^\sigma}(Y^\sigma)$ the tangent space to Y^σ at the point d^σ. As we have shown $T_{Y^\sigma}(d_0^\sigma)$ is a complex subspace in $T(d_0^\sigma)$. For any element $g^\sigma \in \bar{G}'^\sigma$ we have $T_{d_0^\sigma g^\sigma}(Y^\sigma)=g_*^\sigma T_{d_0^\sigma}(Y^\sigma)$ and thus $T_{d_0^\sigma g^\sigma}(Y^\sigma)$ is a complex subspace in $T_{d_0^\sigma g^\sigma}$. Therefore for any point $d^\sigma \in Y^\sigma$ the space $T_{d^\sigma}(Y^\sigma)$ is a complex subspace in $T_{d^\sigma}(D^\sigma)$. Consequently Y^σ is a closed analytic submanifold in D^σ. As we have already shown, $Y^\sigma = d_0^\sigma \bar{G}'^\sigma$ is a closed C^∞-submanifold in D^σ. Then it follows from Lemma 9′ that $d_0^\sigma \bar{G}'^\sigma = D^\sigma$. Theorem 5 is proved.

Theorem 6. *The group $\bar{G}'^\sigma$ is semisimple; the universal covering manifold X^σ is isomorphic to D^σ and is the Hermitian symmetric space, associated to the group $\bar{G}'^\sigma$.*

Proof. Since D^σ is a homogeneous complex manifold with nondegenerate Bergman metric, D^σ is isomorphic (see [16] p. 48) to a bounded domain. Since X^σ is a compact manifold, the factor space $\bar{G}'^\sigma/\Gamma^\sigma$ is compact. Consequently $\bar{G}'$ is a unimodular group. Thus D^σ is a bounded domain, homogeneous with respect to the unimodular group $\bar{G}^\sigma$. It follows from the theorem of Hano (see [17] Theorem 5), that $\bar{G}'^\sigma$ is a semisimple Lie group. According to the theorem of Borel (see [18] Theorem 5) D^σ is the Hermitian symmetric space corresponding to the group $\bar{G}^\sigma$. Since the Bergman metric on D^σ is nondegenerate, D^σ is a Hermitian symmetric space of non-compact type. Therefore D^σ is a simply connected manifold, and consequently D^σ is a universal covering space for X^σ. Theorem 6 is proved.

Corollary. $\pi_1(X) \cong \Gamma^\sigma$ and $\pi_1^0(X^\sigma)=e$.

Remark 1. Theorem 6 implies that the connected component of the identity is a subgroup of finite index in the whole group $\bar{G}'^\sigma$. Therefore, replacing the group G^e by a subgroup of finite index, we find that $\bar{G}'^\sigma$ is a connected Lie group. In the sequel we will always assume that $\bar{G}'^\sigma$ is a connected group.

Remark 2. The foliations T_{i,D^σ} are $\bar{G}'^\sigma$-invariant, and therefore we have a decomposition $D^\sigma = \prod_{i=1}^{s} D_i^\sigma$, where $D_i^\sigma = K_i^\sigma \diagdown \bar{G}_i^\sigma$ is a Hermitian symmetric space of non-compact type, $\bar{G}_i^\sigma$ is a semisimple group and K_i^σ is a maximal compact subgroup in $\bar{G}_i^\sigma$. Lemma 10′ implies immediately that D_i^σ is an indecomposable Hermitian symmetric space and consequently $\bar{G}_i^\sigma$ is a simple Lie group for every i.

§ 6. The base projective spaces

Let K be a connected compact real Lie group, H a Cartan subgroup of K. The space $Y_K \overset{\text{def}}{=} H\backslash K$ is called the *base projective space* of the group K. Recall how to introduce the invariant complex structure on Y_K. Let $\mathfrak{K}$ be the Lie algebra of the group K, $\mathfrak{K}_{\mathbf{C}}$ its complexification, and $K_{\mathbf{C}}$ a Lie group with Lie algebra $\mathfrak{K}_{\mathbf{C}}$, whose maximal compact subgroup coincides with K. Let $\mathfrak{H}\subset\mathfrak{K}$ be the Lie algebra of the torus H, $\mathfrak{H}_{\mathbf{C}}\subset\mathfrak{K}_{\mathbf{C}}$ its complexification and $H_{\mathbf{C}}\subset K_{\mathbf{C}}$ the corresponding subgroup in $K_{\mathbf{C}}$; as is known, $H_{\mathbf{C}}$ is a Cartan subgroup of $K_{\mathbf{C}}$. Let $B_{\mathbf{C}}$ be a Borel subgroup of $K_{\mathbf{C}}$, containing $H_{\mathbf{C}}$. Then from the theorem of Iwasawa (which states that $K_{\mathbf{C}}=B_{\mathbf{C}}K$, see [8] Chapter VI Theorem 5.1) it follows that $Y_K=B_{\mathbf{C}}\backslash K_{\mathbf{C}}$. Consequently, (see [11] Theorem 8.2) Y_K has a natural structure of a projective algebraic manifold. Since the different Borel subgroups in $K_{\mathbf{C}}$ are conjugate (cf. ibid.), the algebraic structure on Y_K does not depend on the choice of the Borel subgroup B. In particular, we can regard Y_K as a homogeneous complex manifold.

Remark. It is easy to show that the algebraic manifold Y_K can be defined over the field $\mathbf{Q}$.

For any element $k\in K_{\mathbf{C}}$ denote by Y^k the set of all k-stable points in Y_K. If $y\in Y^k$ and T_y is the tangent space to Y_K at the point y, then write k_y for the operator in the space T_y, induced by the automorphism $y\to y^k$ of the manifold Y_K.

For any point $y\in Y_K$ denote by B_y the stabilizer of y in $K_{\mathbf{C}}$. Denote by $p_y: B_y\to \operatorname{Aut} T_y$ the homomorphism given by the formula $p_y(k)=k_y$.

Lemma 20. *Let K be a semisimple group.*

a) *For any $k\in K_{\mathbf{C}}$ the set Y^k is not empty.*

b) *Assume that for some $y\in Y^k$ the operator k_y is compact. Then the element k decomposes into a product $k=k^s k^u$, where k^s is a compact, and k^u a unipotent element in $K_{\mathbf{C}}$, and k^s and k^u commute with each other.*

Proof. Since $Y_K=B_{\mathbf{C}}\backslash K_{\mathbf{C}}$, to prove a) it is sufficient to show that any element $k\in K_{\mathbf{C}}$ is conjugate in $K_{\mathbf{C}}$ to some element of the Borel subgroup $B_{\mathbf{C}}$. Now this fact is well known (see [11] Theorem 12). Assertion a) is proved.

Now we prove part b). By assumption we have $k\in B_y$. Represent the element k as a product $k=k^s k^u$ of a semisimple and unipotent element which commute with each other (see [18] 4.4.1). Since $p_y(k)$ is a compact, and consequently, semisimple element $p_y(k)=p_y(k^s)$. Denote by L^s the minimal algebraic $\mathbf{C}$-subgroup of B_y, containing k^s. Since k^s is a semisimple element, L^s is a semisimple group. Since the kernel of the morphism p_y is product of the centre of the group $K_{\mathbf{C}}$ and a unipotent group, the restriction of the morphism p_y to L^s is an isogeny.

Therefore the fact that $p_y(k^s)$ is a compact element, implies that k^s is a compact element. Lemma 20 is proved.

Corollary. In the notations of part b) of Lemma 20, if k is a semisimple element, then it is a compact element in $K_{\mathbf{C}}$.

Lemma 21. a) *Denote by $Z_{\mathbf{C}}$ the centre of $K_{\mathbf{C}}$. Then the group of automorphisms* Aut Y_K *of the complex manifold Y_K is a Lie group, its identity component* $\mathrm{Aut}^0\, Y_K$ *is isomorphic to $K_{\mathbf{C}}/Z_{\mathbf{C}}$, and the factor group* $\mathrm{Aut}\, Y_K/\mathrm{Aut}^0\, Y_K$ *is finite.*

b) *If K is a connected semisimple group and $K' \subset K$ a subgroup, and K acts transitively on Y_K, then $K'=K$.*

Proof. The first assertion of the lemma follows immediately from [20] Theorem III. We establish the second one. Denote by $B'_{\mathbf{C}}$ a Borel subgroup in $K'_{\mathbf{C}}$. Let $B_{\mathbf{C}}$ be a Borel subgroup in $K_{\mathbf{C}}$, containing $B'_{\mathbf{C}}$, and let $y \in Y_K$ be a point, stable under $B_{\mathbf{C}}$. Then the stabilizer $B''_{\mathbf{C}}$ of the point y in $K'_{\mathbf{C}}$ contains $B'_{\mathbf{C}}$ and lies in the group $B_{\mathbf{C}}$. Consequently, $B''_{\mathbf{C}}$ is a solvable subgroup in $K'_{\mathbf{C}}$, containing the Borel subgroup $B'_{\mathbf{C}}$. By the definition of Borel subgroups, this implies that $B''_{\mathbf{C}}=B'_{\mathbf{C}}$ and the orbit of the point $y \in Y_K$ under the action of $K'_{\mathbf{C}}$ is isomorphic to $Y_{K'}$. Since, by assumption, the group $K'_{\mathbf{C}}$ acts transitively on Y_K, we have $Y_{K'}=Y_K$. The first assertion of the lemma implies that $K'=K$. The lemma is proved.

Now we prove the following lemma:

Lemma 22. *There is an analytic imbedding $\varphi: Y_K \to P^N_{\mathbf{C}}$ of the manifold Y_K into the projective space $P^N_{\mathbf{C}}$, and a homomorphism $\varrho: \mathrm{Aut}\, Y_K \to \mathrm{Aut}\, P^N_{\mathbf{C}} \cong PGL(n+1, \mathbf{C})$ such that for any point $y \in Y_K$ and automorphism $g \in \mathrm{Aut}\, Y_K$ we have $\varphi(yg)= =\varphi(y)\varrho(g)$.*

Proof. Let $n=\dim_{\mathbf{C}} Y_K$. Denote by Ω the invertible sheaf of the germs of the n-dimensional holomorphic differential forms on Y_K. It is well known (see for example [21] Theorem 4.23) that there is a natural number l, such that the sheaf $\Omega^{\otimes -l}$ is very ample. Let $f_0, \dots, f_N$ be a basis in the space $\Gamma(Y_K, \Omega^{\otimes -l})$. Define the mapping $\varphi: Y_K \to P^N_{\mathbf{C}}$ by taking $\varphi(y)=(f_0(y), \dots, f_N(y))$. Since the sheaf $\Omega^{\otimes -l}$ is very ample, the mapping φ is well defined and is an imbedding. Every automorphism $g: Y_K \to Y_K$ canonically defines an automorphism g^* of the sheaf Ω, $g^*: \Omega \to \Omega$ and consequently an automorphism $(g^*)^{\otimes -l}$ of the sheaf $\Omega^{\otimes -l}$. Denote by $\varrho'(g)$ the automorphism of the space $\Gamma(Y_K, \Omega^{\otimes -l})$, induced by $(g^*)^{\otimes -l}$, and by $\varrho(g)$ the corresponding automorphism of the space $P^N_{\mathbf{C}}$. It is clear that the pair (φ, ϱ) satisfies the requirement of Lemma 22. The lemma is proved.

Corollary. Let X and Z be complex manifolds, and $p: Z \to X$ a locally trivial analytic fibering whose fibres are isomorphic to Y_K. If, in addition, X is an algebraic manifold, then so is Z.

Proof. Since the fibering $p:Z\to X$ is locally trivial, it is a fibering with the structure group Aut Y_K. The homomorphism ϱ:Aut $Y_K\to$Aut $P_{\mathbf{C}}^N$, constructed in Lemma 22, permits defining the associated locally trivial fibering $p':Z'\to X$ with the fibre $P_{\mathbf{C}}^N$. The mapping φ constructed in Lemma 22, permits defining the imbedding $\tilde{\varphi}:Z\to Z'$. According to Lemma 18.1 in [10], Z' is an algebraic manifold. Since $Z\cong\tilde{\varphi}(Z)$ is a closed submanifold of Z', according to Chow's lemma, Z is also an algebraic manifold. The Corollary is proved.

Lemma 23. *Let K' be another compact group and τ a homomorphism $\tau:K'\to$Aut Y_K. Then there is a point $y\in Y_K$, such that its orbit $y\tau(K')\subset Y_K$ is a complex manifold, and the stabilizer K'_y of y in K' contains a Cartan subgroup of K'.*

Proof. By Lemma 22 it is sufficient to analyse the case when K' is a connected group and τ sends K' into $K_{\mathbf{C}}\subset$Aut Y. Since any homomorphism between real semisimple algebraic groups is algebraic, τ may be extended to a homomorphism $\tau_{\mathbf{C}}:K'_{\mathbf{C}}\to K_{\mathbf{C}}$. Let $B'_{\mathbf{C}}$ be a Borel subgroup of the group $K'_{\mathbf{C}}$, $B_{\mathbf{C}}$ a Borel subgroup of the group $K_{\mathbf{C}}$ containing $\tau_{\mathbf{C}}(B'_{\mathbf{C}})$, and $y\in Y_K$ a point in $K_{\mathbf{C}}$, whose stabilizer in $K_{\mathbf{C}}$ coincides with $B_{\mathbf{C}}$. According to the theorem of Iwasawa $K'_{\mathbf{C}}=B'_{\mathbf{C}}K'$, thus the orbit yK' coincides with the orbit $yK'_{\mathbf{C}}$, which is obviously a complex submanifold of $Y_{K'}$. Lemma 23 is proved.

§ 7. Construction of the fiberings F and F' over D

As above, let $\tilde{G}_{\mathbf{R}}=G_0\times\prod_{i=1}^{s}G_i$ where G_0 is a compact group, G_i, $1\leqq i\leqq s$ are simple non-compact adjoint Lie groups. Let K be a maximal compact subgroup in $\tilde{G}_{\mathbf{R}}$. Then $K=G_0\times\prod_{i=1}^{s}K_i$, where K_i is a maximal compact subgroup in G_i. Put $G'_{\mathbf{R}}=\prod_{i=1}^{s}G_i$, $K'=\prod_{i=1}^{s}K_i$. Then $D=K\diagdown G=K'\diagdown G'$. The group $G'_{\mathbf{R}}$ is a subgroup of finite index in the group Aut D. Now let $H_{\mathbf{R}}$ be a compact Cartan subgroup of $G_{\mathbf{R}}$ lying in K. Then $H_{\mathbf{R}}=H_0\times H'$, where $H_0\subset G_0$, $H'\subset G'_{\mathbf{R}}$. Put $F=H_{\mathbf{R}}\diagdown G_{\mathbf{R}}$, $F_0=H_0\diagdown G_0\cong Y_{G_0}$, $F'=H'\diagdown G'_{\mathbf{R}}$. Then $F=F_0\times F'$, and F_0 is the base projective space, corresponding to the group G_0.

Denote by τ' the natural projection $\tau':F'\to D$, by τ the projection $\tau:F\to D$, and by τ'' the projection $\tau'':F\to F'$. Then $\tau=\tau'\tau''$. We show that on the manifold F' there is a $G'_{\mathbf{R}}$-invariant complex structure such that τ' is an analytic mapping.

Denote by D_0 the compact Hermitian symmetric space, dual to D, and consider D as an open subspace in D_0 (see [8] Chapter VIII, Theorem 7.13). Let $d_0\in D$ be a point, invariant under K'. Let $G'_{\mathbf{C}}$ be the complexification of the group $G'_{\mathbf{R}}$. The group $G'_{\mathbf{C}}$ acts transitively on D_0, the stabilizer $P'_{\mathbf{C}}$ of the point d_0 is a parabolic subgroup in G'_C and $P'_C \cap G'_R = K'$. For any Borel subgroup $B'_C \subset P'_C$ the intersection $B'_C \cap G'_R$ is a compact Cartan subgroup in K'. Since all Cartan subgroups in K' are conjugate to each other, we can choose the Borel subgroup B'_C in such a way that $B'_C \cap G'_R = H'$. Therefore the natural mapping $i: F' \cong H' \diagdown G'_R \to \to B'_C \diagdown G'_C$ of the space F' into the base projective space $B'_C \diagdown G'_C$ is an imbedding. Since $\dim_R B'_C + \dim_R G'_R = \dim_R G'_C + \dim_R H'$, thus $\dim_R F' = \dim_R B'_C \diagdown G'_C$ and consequently i is an open imbedding. Now we can consider F' as an open subset in $B'_C \diagdown G'_C$. Thus a G'_R-invariant complex structure is defined on F'. In the sequel, F' will be considered as a complex manifold. Since the natural projection $\tau'_0 : B'_C \diagdown G'_C \to P'_C \diagdown G'_C \cong D_0$ is an analytic mapping, and the projection $\tau' : F' \to D$ coincides with the restriction of τ'_0 to $F' \subset B'_C \diagdown G'_C$, thus τ' is an analytic mapping. Since the mapping τ'_0 is a locally trivial fibering with fibre $B'_C \diagdown P'_C \cong H' \diagdown K' \cong Y_{K'}$, thus τ is a locally trivial fibering with fibre $Y_{K'}$.

We introduce on the space F a complex structure determined by the decomposition of F into the direct product $F \cong F_0 \times F'$. The mapping $\tau : F \to D$ is an analytic locally trivial fibering with fibre $Y_{K'} \times F_0 \cong Y_{K' \times G_0} \cong Y_K$.

In the sequel we will use the realization of F' as a submanifold of the flag manifold over D. For any complex manifold V denote by $R(V)$ the fibering of flags over V, associated with the tangent fibering, and by $\tilde{p}$ the projection $\tilde{p} : R(V) \to V$. $R(V)$ is a complex manifold. Moreover, if V is an algebraic manifold, so is $R(V)$.

Lemma 24. There is a G'_C-covariant imbedding ψ_0 of the manifold $B'_C \diagdown G'_C$ into $R(D_0)$, such that $\tilde{p} \circ \psi_0 = \tau'_0$.

Proof. Let d be a point in $D \subset D_0$ stable under the action of the group K', P'_C the stabilizer of the point d in G'_C, and B'_C a Borel subgroup of the group G'_C, lying in P'_C. Since the group B'_C stabilizes the point $d \in D_0$, it acts on the space $\tilde{p}^{-1}(d)$, which is the manifold of flags in $T_d(D_0)$. The theorem of Lie (see [9] Chapter 3 Theorem 2.2) implies the existence of a B'_C-stable point $r \in \tilde{p}^{-1}(d)$. Since the stabilizer of a point $r \in R(D_0)$ in the group $\mathrm{Aut}\,(\tilde{p}^{-1}(d))$ is solvable, the stabilizer of the point r in the group G'_C is a solvable subgroup in G'_C containing B'_C. Now B'_C is a maximal solvable subgroup in G'_C, so we find that B'_C is the stabilizer of the point r in G'_C. Define an analytic mapping $\tilde{\psi}_0$ of the group G'_C into $R(D_0)$, by $\tilde{\psi}_0(g) = rg \in R(D_0)$, $g \in G'_C$. Since B'_C is the stabilizer of r in G'_C, the mapping $\tilde{\psi}_0$ induces an analytic G'_C-covariant imbedding ψ_0 of the manifold $B'_C \diagdown G'_C$ into $R(D_0)$. It is clear that $\tilde{p}\psi_0 = \tau_0$. Lemma 24 is proved.

We will denote by $\psi : F' \to R(D)$ the restriction of the imbedding ψ_0 to the subset $F' \subset B'_C \diagdown G'_C$.

§ 8. Construction of the fiberings F^σ and F'^σ over D^σ

The complex manifolds F and F' are, by construction, homogeneous $G_{\mathbf{R}}$-spaces. Therefore the group $\Gamma \subset G_{\mathbf{R}}$ acts on them. Because in Γ there are no elements of finite order different from e, it acts on F and F' without fixed points, and the factor spaces $Z = F/\Gamma$ and $Z' = F'/\Gamma$ have natural complex manifold structures. Since the projections $\tau: F \to D$, $\tau': F' \to D$ and $\tau'': F \to F'$ commute with the action of the group $G_{\mathbf{R}}$, they determine the analytic locally trivial fiberings $\mu: Z \to X$, $\mu': Z' \to X$ and $\mu'': Z \to Z'$, with fibres isomorphic to the base projective spaces Y_K, $Y_{K'}$ and Y_{G_0}, respectively. According to Lemma 22, Z and Z' are algebraic $\mathbf{C}$-manifolds. Therefore we can consider the algebraic $\mathbf{C}$-manifolds Z^σ and Z'^σ for any element σ of the group $\mathrm{Gal}(\mathbf{C}:\mathbf{Q})$. Denote by μ^σ, μ'^σ and μ''^σ the fiberings obtained $\mu^\sigma: Z^\sigma \to X^\sigma$, $\mu'^\sigma: Z'^\sigma \to X^\sigma$ and $\mu''^\sigma: Z^\sigma \to Z'^\sigma$. The fibres of the fiberings μ^σ, μ'^σ and μ''^σ are isomorphic to the manifolds $(Y_K)^\sigma$, $(Y_{K'})^\sigma$ and $(Y_{G_0})^\sigma$, respectively. Since all base projective spaces can be defined over the field $\mathbf{Q}$ of rational numbers, the algebraic manifold $(Y_K)^\sigma$ is isomorphic to Y_K, the manifold $(Y_{K'})^\sigma$ is isomorphic to $Y_{K'}$, and $(Y_{G_0})^\sigma$ is isomorphic to Y_{G_0}. Thus the fibres of the fiberings $\mu^\sigma: Z^\sigma \to X^\sigma$, $\mu'^\sigma: Z'^\sigma \to X^\sigma$, and $\mu''^\sigma: Z^\sigma \to Z'^\sigma$ are isomorphic to the base projective spaces Y_K, $Y_{K'}$ and Y_{G_0}, respectively. Now put $F^\sigma = D_{Z^\sigma}$, $F' = D_{Z'^\sigma}$.

Lemma 25. *The manifolds F^σ, and F'^σ are the fibre products of D^σ and Z^σ over X^σ, and D^σ and Z'^σ over X^σ, respectively:*

$$\begin{array}{ccc} Z^\sigma & \xleftarrow{\ \varrho^\sigma\ } & F^\sigma \\ {\scriptstyle \mu^\sigma}\downarrow & & \downarrow{\scriptstyle \tau^\sigma} \\ X^\sigma & \xleftarrow{\ \pi^\sigma\ } & D^\sigma \end{array} \qquad \begin{array}{ccc} Z'^\sigma & \xleftarrow{\ \varrho'^\sigma\ } & F'^\sigma \\ {\scriptstyle \mu'^\sigma}\downarrow & & \downarrow{\scriptstyle \tau'^\sigma} \\ X^\sigma & \xleftarrow{\ \pi^\sigma\ } & D^\sigma \end{array}$$

and F^σ is isomorphic to the product of F'^σ and Z^σ over Z'^σ.

$$\begin{array}{ccc} Z^\sigma & \xleftarrow{\ \varrho^\sigma\ } & F^\sigma \\ {\scriptstyle \mu'^\sigma}\downarrow & & \downarrow{\scriptstyle \mu''^\sigma} \\ Z'^\sigma & \xleftarrow{\ \varrho'^\sigma\ } & F'^\sigma \end{array}$$

Proof. We show that F^σ is a fibre product. The remaining assertions of the lemma can be obtained similarly.

Since the base projective space Y_K is a connected, simply connected manifold, from the exact homotopy sequence of the fibering $\mu^\sigma: Z^\sigma \to X^\sigma$ we see that the natural mapping $\pi_1(Z^\sigma) \to \pi_1(X^\sigma)$ is an isomorphism. Corollary 1 to Theorem 6 implies that $\pi_1^0(Z^\sigma) = e$ and consequently F^σ is a universal covering manifold of Z^σ. On the other hand, it is clear that the fibre product of D^σ and Z^σ over X^σ is an unramified covering over Z^σ. This, and the relation $\pi_0(Y_K) = \pi_1(Y_K) = e$ imply immediately that this fibre product is a connected, simply connected space, and consequently it is a simply connected covering of Z^σ. Lemma 25 is proved.

Remark. In the following, we will apply the notations τ^σ, τ'^σ, τ''^σ, ϱ^σ, ϱ'^σ introduced at the formulation of Lemma 25.

In this paragraph we show that the complex manifolds F'^σ and F^σ are homogeneous manifolds, further we show that $F'^\sigma \cong F'$, $F^\sigma \cong F$. As before, let G^e be a subgroup of finite index in the factor group of the group $G_{\mathbf{Q}}$ by its centre. The group G^e is imbedded in $\tilde{G}_{\mathbf{R}}$, and consequently, the imbeddings $j': G^e \hookrightarrow \operatorname{Aut} F'$, $j: G^e \hookrightarrow \operatorname{Aut} F$ are defined. Put $G(Z')^e = j'(G^e)$, $G(Z)^e = j(G^e)$. It is clear that $G(Z')^e \subset G(Z')$ and $G(Z)^e \subset G(Z)$ are admissible subgroups. Therefore one can define the subgroups $(G(Z')^e)^\sigma \subset G(Z'^\sigma)$ and $(G(Z)^e)^\sigma \subset G(Z^\sigma)$ (cf. § 1). Since the action of the group $G_{\mathbf{R}} \supset G^e$ on F and F' preserves the fibres of the projections $\tau: F \to D$, $\tau': F' \to D$, $\tau'': F \to F'$, according to Lemma 6′, the actions of the groups $G(Z)^\sigma$ and $G(Z')^\sigma$ on F and F' preserve the fibres of the projections

$$\tau^\sigma: F^\sigma \to D^\sigma, \quad \tau'^\sigma: F'^\sigma \to D^\sigma \quad \text{and} \quad \tau''^\sigma: F^\sigma \to F'^\sigma.$$

Therefore one can define the homomorphisms

$$\tau_*^\sigma; G(Z)^\sigma \to G(X^\sigma), \quad \tau_*'^\sigma: G(Z')^\sigma \to G(X^\sigma).$$

Lemma 26. *The homomorphisms τ_*^σ and $\tau_*'^\sigma$ provide isomorphisms of the groups $G(Z)^\sigma$ and $G(Z')^\sigma$ to the subgroup $G^\sigma \subset G(X^\sigma)$.*

Proof. The definition of the groups $G(Z)^e$ and $G(Z')^e$ implies that the homomorphisms $\tau_*: G(Z)^e \to G(X)$, $\tau_*': G(Z')^e \to G(X)$ are isomorphisms of the groups $G(Z)^e$ and $G(Z')^e$ to G^e. According to Theorem 1, this implies that the morphisms $\hat{\tau}_*: \hat{G}^e(\hat{Z}) \to \hat{G}(\hat{X})$ and $\hat{\tau}_*': \hat{G}^e(\hat{Z}') \to \hat{G}(\hat{X})$ induce isomorphisms of the groups $\hat{G}^e(\hat{Z})$ and $\hat{G}^e(\hat{Z}')$ to the subgroup $\hat{G}^e \subset \hat{G}(\hat{X})$. Applying Theorem 1 once again, we get Lemma 26.

Remark. In the sequel we will identify the groups $G(Z)^e$ and $G(Z')^e$ with G^e, and the groups $G(Z)^\sigma$ and $G(Z')^\sigma$ with G^σ.

Consider the $G_{\mathbf{R}}$-covariant analytic imbedding $\psi: F' \hookrightarrow R(D)$ described in Lemma 24. It determines an analytic imbedding $\varphi: Z' \to R(X)$. Since Z' and $R(X)$ are compact algebraic **C**-manifolds, the theorem of Serre (see [3] Proposition 15) implies that φ is an algebraic mapping. Therefore φ^σ gives an algebraic imbedding $\varphi^\sigma: Z'^\sigma \to R(X^\sigma)$. By the condition of Lemma 24, $\tilde{p} \circ \psi = \tau'$ thus $\tilde{p} \circ \varphi^\sigma = \mu'^\sigma$ (we denote with the same letter $\tilde{p}$ the projection of the flag manifold over an arbitrary manifold M, to the manifold M). Lemma 25 implies that there is an analytic proper imbedding $\psi^\sigma: F'^\sigma \to R(D^\sigma)$ such that $\tilde{p} \circ \psi^\sigma = \tau'^\sigma$ and $\pi^\sigma \circ \psi^\sigma = \varphi^\sigma \circ \varrho'^\sigma$. The actions of the groups G^σ and $\bar{G}'^\sigma$ on the manifold $R(D^\sigma)$ induce the actions of the groups G^σ and $\bar{G}'^\sigma$ on $R(D)$.

Lemma 27. a) *The submanifold* $\psi^\sigma(F'^\sigma) \subset R(D^\sigma)$ *is invariant under the action* $r \to rg$ *of the group* G^σ *on* $R(D^\sigma)$.

b) *For any* $g \in G^\sigma$ *denote by* $\bar{g}$ *the restriction of the mapping* $r \to rg$ *to* $\psi^\sigma(F'^\sigma)$. *Then*

$$(\tau'^\sigma_*)^{-1}(g) = (\psi^\sigma)^{-1}\bar{g}\psi^\sigma.$$

Proof. We analyse only part a) of Lemma 27. Part b) can be analysed similarly.

It follows from lemma 24 that the submanifold $\psi(F') \subset R(D)$ is $G'_{\mathbf{R}}$-invariant. Consequently, it is G^e-invariant. The group $G^e \subset G(X)$ can be considered as a subgroup in $G(R(X))$. Since $\pi_1(X) \cong \pi_1(R(X))$ we get that G^e is an admissible subgroup in $G(R(X))$. Therefore one can apply Lemma 5′. Assertion a) is proved.

In the sequel we will identify the manifold F'^σ with its image $\psi^\sigma(F'^\sigma)$ in $R(D^\sigma)$. According to Lemma 27 F'^σ is a G^σ-invariant submanifold in $R(D^\sigma)$. Since F'^σ is a closed submanifold in $R(D^\sigma)$, and the group G^σ is, by definition, dense in $\bar{G}'^\sigma$, F'^σ is a $\bar{G}'^\sigma$-invariant submanifold in $R(D^\sigma)$.

Theorem 7. F'^σ *is a homogeneous* $\bar{G}'^\sigma$*-manifold.*

Proof. It is clear that the restriction of the projection $\tilde{p}: R(D^\sigma) \to D^\sigma$ to F'^σ coincides with the projection $\tau'^\sigma: F'^\sigma \to D^\sigma$. As shown in the proof of Lemma 25, the fibres of the projection τ'^σ are isomorphic to $Y_{K'}$. According to Theorem 6 the manifold D^σ is a Hermitian symmetric space, associated to the group $\bar{G}'^\sigma$. Denote by K'^σ a maximal compact subgroup in $\bar{G}'^\sigma$, and by $d^\sigma \in D^\sigma$ a K'^σ-stable point. The group K'^σ acts on the fibre $(\tau'^\sigma)^{-1}(d^\sigma) \cong Y_{K'}$. According to Lemma 23 there is a point $f'^\sigma \in (\tau'^\sigma)^{-1}(d^\sigma) \subset F'^\sigma$ such that the orbit $\{f'^\sigma K'^\sigma\}$ is an analytic submanifold in the fibre $(\tau'^\sigma)^{-1}d^\sigma$, and consequently, in $R(D^\sigma)$. Denote by W^σ the orbit $\{f'^\sigma \bar{G}'^\sigma\}$ of the point f'^σ under the action of the group $\bar{G}'^\sigma$. The orbit W^σ is a closed real C^∞-submanifold in $R(D^\sigma)$.

Lemma 28. *The tangent space* $T^{\mathbf{R}}_{W^\sigma}$ *to the* C^∞*-manifold* W *at the point* f'^σ, *is a complex subspace of the tangent space* T_R *to the manifold* $R(D^\sigma)$.

Proof of Lemma 28. Denote by $\tilde{p}_*$ the natural projection of the linear space T_R onto the tangent space T_{D^σ} to the manifold D^σ at the point d^σ, and denote by T^0_R the kernel of $\tilde{p}_*$. Denote by H'^σ the stabilizer of the point f'^σ in K'^σ. The group H'^σ acts naturally on the spaces T_R and T_{D^σ} and the mapping p_* is compatible with the action of the group H'^σ. Since the group H'^σ is compact, the space T_R decomposes into the direct sum $T_R = T^0_R \oplus T^1_R$ of complex H'^σ-invariant subspaces. Moreover the mapping p_* induces an isomorphism of the complex H'^σ-spaces T^1_R and T_{D^σ}.

Consider now the real subspace $T^{\mathbf{R}}_{W^\sigma}$ in T_R. The manifold W^σ is invariant under the action of the group $H'^\sigma \subset \bar{G}'^\sigma$, thus $T^{\mathbf{R}}_{W^\sigma}$ is an H'^σ-invariant subspace in T_R. Put $T^0_W = T^{\mathbf{R}}_{W^\sigma} \cap T^0_R$, $T^1_W = T^{\mathbf{R}}_{W^\sigma} \cap T^1_R$. According to the choice of the point f'^σ, the intersection T^0_W is a complex subspace in T^0_R.

The theory of symmetric spaces implies (cf. [8] Chapter VIII, Theorem 4.5 (I)) that there is an element $k_0 \in K'^\sigma$ such that its action in the tangent space T is multiplication by $\sqrt{-1}$. Therefore k_0 acts trivially on the manifold $\varrho^{-1}(d^\sigma) \subset R(D^\sigma)$. Consequently $k_0 \in H'^\sigma$. Denote by A the operator in T_R, induced by the element k_0. By assumption, the subspaces $T_W \subset T_R^0$ and $T_R^1 \subset T_R$ are invariant under A, and $A|_{T_R^0} = id$, $A|_{T_R^1} = \sqrt{-1}\, id$. For any element $v \in T_{W^\sigma}^{\mathbf{R}}$ put $v_0 = \frac{1}{2}(v + A^2 v)$, $v_1 = \frac{1}{2}(v - A^2 v)$. The above properties of the operator A imply immediately that $v_0 \in T_{W^\sigma}^0$ and $v_1 \in T_{W^\sigma}^1$. As $v_0 + v_1 = v$, we have proved that $T_{W^\sigma}^{\mathbf{R}} = T_{W^\sigma}^0 + T_{W^\sigma}^1$. On the other hand, the group $\bar{G}'^\sigma$ acts transitively on D^σ, and so $\tilde{p}(W^\sigma) = D^\sigma$, and $\tilde{p}_*(T_{W^\sigma}^{\mathbf{R}}) = T_{D^\sigma}$. Since the restriction of $\tilde{p}_*$ to T_R^1 is an isomorphism of the spaces T_R^1 and T_{D^σ}, the equality $\tilde{p}_*(T_{W^\sigma}^{\mathbf{R}}) = T_{D^\sigma}$ implies that $T_W^1 = T_R^1$. Consequently $T_{W^\sigma} = T_{W^\sigma}^0 + T_R^1$ is a complex subspace in T_R. Lemma 28 is proved.

Since the group $\bar{G}'^\sigma$ acts on the complex manifold $R(D^\sigma)$ by analytic automorphisms, Lemma 28 implies that the tangent space to The C^∞-submanifold $W^\sigma \subset F'^\sigma$ at any point $w^\sigma \in W^\sigma$ is a complex space, and therefore W^σ is an analytic submanifold in F'^σ.

Thus we have shown that the orbit $W^\sigma = \{f'^\sigma \bar{G}'^\sigma\} \subset F'^\sigma$ is a G^σ-invariant closed analytic submanifold in F'^σ. According to Lemma 5′, there is a closed analytic G^e-invariant submanifold $W \subset F'$ such that $W^\sigma = (\varrho'^\sigma)^{-1}([\varrho'(W)]^\sigma)$. Since the group G^e is dense in $G_{\mathbf{R}}$, $W = F'$ and it follows that $W^\sigma = F'^\sigma$. Theorem 7 is proved.

Lemma 29. *The group H'^σ is a Cartan subgroup in $\bar{G}'^\sigma$.*

Proof. First we show that the group H'^σ is commutative. Denote by α the representation of the group K'^σ in the tangent space T_{D^σ} to the manifold D^σ at the point d^σ. Cartan's theorem (see [13] Chapter I, Theorem 8) implies that α is a faithful representation. Therefore it is enough to show that the group $\alpha(H'^\sigma) \subset \subset GL(n, \mathbf{C})$ is commutative. Since the group H'^σ stabilizes the point $f'^\sigma \in \tilde{p}^{-1}(d^\sigma)$, there is a Borel subgroup $B_{\mathbf{C}}$ in $GL(n, \mathbf{C})$ containing $\alpha(H'^\sigma)$. Since the group $\alpha(H'^\sigma)$ is compact, we find that it is commutative.

Now we show that H'^σ is a maximal commutative subgroup in K'^σ. The group K'^σ acts on the fibre $(\tau'^\sigma)^{-1}(d^\sigma) \cong Y_{K'}$. By assumption, (see Remark 1 to Theorem 6) the group $\bar{G}'^\sigma$ is connected. Consequently, the group K'^σ is connected. According to Lemma 21 there is a homomorphism β of the group K'^σ to $K'_{\mathbf{C}}$ such that the element $k'^\sigma \in K'^\sigma$ acts on the fibre $(\tau'^\sigma)^{-1}(d^\sigma) \cong Y_{K'}$ by the translation on the element $\beta(k'^\sigma)$.

Let H be an arbitrary commutative subgroup in K'^σ. Then $\varrho(H)$ is a commutative subgroup in the group $K'_{\mathbf{C}}$. Consequently (see [19] Theorem 10.4), there is a point $f \in (\tau'^\sigma)^{-1}(d^\sigma)$ stable under the group H. According to Theorem 7, there is an element $k_0 \in K'^\sigma$ such that $f'^\sigma k = f$. Then the subgroup kHk^{-1} stabilizes the point f'^σ, and consequently, it belongs to H'^σ, Lemma 29 is proved.

Corollary 1. The complex manifolds $Y_{K'}$ and $Y_{K'^\sigma}$ are isomorphic.

Proof. According to Lemma 29 the mapping $\bar{\beta}: k' \to f'^\sigma k'$ defines an imbedding of the complex manifold $Y_{K'^\sigma}$ into $Y_{K'}$. According to Theorem 7 the mapping $\bar{\beta}$ is an epimorphism. Corollary 1 is proved.

Corollary 2. The Lie algebras $\mathfrak{K}'$ and $\mathfrak{K}'^\sigma$ of the groups K' and K'^σ are isomorphic.

Proof. It follows from Corollary 1 and Lemma 21 that the commutator groups $[\mathfrak{K}', \mathfrak{K}']$ and $[\mathfrak{K}'^\sigma, \mathfrak{K}'^\sigma]$ are isomorphic. Now, by Remark 2 to Theorem 6, both of the symmetric spaces D and D^σ decompose into products of irreducible symmetric spaces, and the number of factors in the products coincide, thus the centres of the algebras $\mathfrak{K}'^\sigma$ and $\mathfrak{K}'$ are isomorphic. Now a reductive Lie algebra is the direct sum of its commutator and its centre. Corollary 2 is proved.

Let $G'_{\mathbf{R}} = \prod_{i=1}^{s} G_i$ be a decomposition of the group $G'_{\mathbf{R}}$ into the product of simple groups, and $\bar{G}'^\sigma = \prod_{i=1}^{s} \bar{G}_i^\sigma$ the corresponding decomposition of the group $\bar{G}'^\sigma$, introduced in Remark 2 to Theorem 6. Denote by K_i^σ a maximal compact subgroup of the group $\bar{G}_i^\sigma$. The following assertion can be proved similarly to Corollaries 1, 2:

Corollary 3. The Lie algebras of the groups K_i and K_i^σ are isomorphic.

Theorem 8. *The complex manifolds* $\mathbf{F}'^\sigma$ *and* F' *are isomorphic.*

Proof. In virtue of Lemma 29, to prove that the manifolds F'^σ and F' are isomorphic, it is sufficient to show that the groups $\bar{G}'^\sigma$ and $G'_{\mathbf{R}}$ are isomorphic. Now for this it is enough to show that the simple groups $\bar{G}_i^\sigma$ and G_i are isomorphic. Let us prove this latter statement. It follows from the definition of the group $\bar{G}_i^\sigma$ that $\dim_{\mathbf{R}} K_i \diagdown G_i = \dim_{\mathbf{R}} K_i^\sigma \diagdown \bar{G}_i^\sigma = 2 \dim_{\mathbf{C}} T_i(D)$, where K_i and K_i^σ are maximal compact subgroups of G_i and $\bar{G}_i^\sigma$. It follows from Corollary 3 of Lemma 29 that the Lie algebras of K_i and K_i^σ are isomorphic. From the Cartan classification (see [8] Chapter IX, § 4) of irreducible Hermitian symmetric spaces it follows that this is possible if and only if the groups G_i and $\bar{G}_i^\sigma$ are isomorphic. Theorem 8 is proved.

Consider now the fibering $\tau'': F \to F'$. Since $F \cong F' \times G_0$, in the fibering $\tau'': F \to F'$ there is a natural $G_{\mathbf{R}}$-invariant flat connection ∂. Applying Lemma 8 we get the flat G^σ-invariant connection ∂^σ in the fibering $\tau''^\sigma: F^\sigma \to F'^\sigma$. As we mentioned at the beginning of this §, the fibres of the mapping τ''^σ are isomorphic to the base projective space Y_{G_0}. Since the manifold F'^σ is simply connected, by giving a flat connection ∂^σ in the fibering τ''^σ we define a decomposition to direct product $F^\sigma = F'^\sigma \times Y_{G_0}$. Therefore by virtue of Theorem 8 the following assertion holds:

Theorem 8′. The complex manifold F^σ is isomorphic to F.

Denote by τ_0^σ the natural projection $\tau_0^\sigma: F^\sigma \to Y_{G_0}$. Since the connection ∂^σ is G^σ-invariant, the action of the group G^σ on F^σ defines homomorphisms of the group G^σ; $p'': G^\sigma \to G'_{\mathbf{R}} \subset \mathrm{Aut}\, F'^\sigma$, $p_0: G^\sigma \to \mathrm{Aut}\, Y_{G_0}$ such that for any element $g^\sigma \in G^\sigma$ and point $f^\sigma \in F^\sigma$ we have $\tau''^\sigma(f^\sigma g^\sigma) = \tau''^\sigma(f^\sigma)p''(g^\sigma)$ and $\tau_0^\sigma(f^\sigma g^\sigma) = \tau_0^\sigma(f^\sigma)p_0(g^\sigma)$. It follows from Lemma 21 that the group $G_{0\mathbf{C}}$ is a subgroup of finite index in the group $\mathrm{Aut}\, Y_{G_0}$. Therefore, replacing the group G^e by a subgroup of finite index, we can assume that $p_0(G^\sigma) \subset G_{0\mathbf{C}}$.

We will consider the group G^σ as a subgroup in the group $G_{0\mathbf{C}} \times G'_{\mathbf{R}} \subset \tilde{G}_{\mathbf{C}}$. The homomorphisms $p_0: G^\sigma \to G_{0\mathbf{C}}$, $p'': G^\sigma \to G'_{\mathbf{R}}$ can be obtained by restricting to G^σ the natural projections of the whole group $\tilde{G}_{\mathbf{C}}$ onto $G_{0\mathbf{C}}$ and $G'_{\mathbf{C}}$, respectively. These projections of $\tilde{G}_{\mathbf{C}}$ will be denoted by the same letters p_0 and p''.

§ 9. The proof of Theorem A

Denote by $\mathrm{Ad}: \tilde{G}_{\mathbf{C}} \to \mathrm{Aut}\, \mathfrak{G}_{\mathbf{C}}$ the adjoint representation of the group $\tilde{G}_{\mathbf{C}}$. Its restriction to the subgroups G^e and G^σ will also be denoted by Ad. Denote by U' the subset of G'_R, consisting of all compact elements. If H' is a compact Cartan subgroup in $G'_{\mathbf{R}}$, then $U' = \bigcup gH'g^{-1}$. Put $V = p''^{-1}(U') \cap G^e$, $V^\sigma = p''^{-1}(V') \cap G^\sigma$.

Let $g^\sigma \in V^\sigma$. Then, by definition the element $p''(g^\sigma)$ belongs to U', and consequently it is conjugate to some element of the compact Cartan subgroup H' in $G'_{\mathbf{R}}$. Therefore there is a point $f'^\sigma \in F'^\sigma$ such that $f'^\sigma g^\sigma = f'^\sigma$. Consequently, the action of the element g^σ on H^σ transforms the fibre $(\tau'^\sigma)^{-1}(f'^\sigma) \cong Y_{G_0}$ into itself, and acts on this fibre by translation on the element $p_0(g^\sigma) \in G_{0\mathbf{C}}$. According to Lemma 20 there is a point $f^\sigma \in F^\sigma$ such that $f^\sigma g^\sigma = f$, i.e. $g^\sigma \in G^\sigma \cap G_{f^\sigma}$ (see the definition of the group G_{f^σ} in § 1).

Denote, as before, by $\varphi^\sigma(g^\sigma)$ the action induced by the automorphism $g^\sigma \in \mathrm{Aut}\, F^\sigma$ in the tangent space T_{f^σ} to the manifold F^σ at the point f^σ. Let $f \in F$ be a point such that $(\varrho(f))^\sigma = \varrho^\sigma(f^\sigma) \in Z^\sigma$. (The mapping ϱ has been introduced in Lemma 25.) Let T_f be the tangent space to the manifold F at the point f and G_f^e the stabilizer of the point f in the group G^e. Lemma 4′ implies that there is an element $g \in G_f^e$ and an isomorphism $c: T_f \to T_{f^\sigma}$ of complex linear spaces, such that $\varphi^\sigma(g) = c(\varphi(g))^\sigma c^{-1}$. Since $F = H \diagdown G_{\mathbf{R}}$, the stabilizer of the point $f \in F$ in $G_{\mathbf{R}}$ is compact. Consequently $g \in V$ and $\varphi(g)$ is a compact element in the group of automorphisms of the linear space T_f.

Consider now the real tangent spaces $(T_f)_{\mathbf{R}}$ and $(T_{f^\sigma})_{\mathbf{R}}$ to F and F^σ at the points f and f^σ, and denote by $\psi(g)$ and $\psi(g^\sigma)$ the operators in the real spaces T_f and T_{f^σ}, obtained from $\varphi(g)$ and $\varphi^\sigma(g^\sigma)$ by restricting the scalar field from $\mathbf{C}$ to $\mathbf{R}$. Since the operators $\varphi^\sigma(g^\sigma)$ and $(\varphi(g))^\sigma$ are conjugate to each other, so are the operators $\psi^\sigma(g^\sigma)$ and $(\psi(g))^\sigma$.

Lemma 30. *The operator* Ad (g) *is conjugate to the direct sum of the operator* $\psi(g)$ *and the identity operator of dimension* $\dim_{\mathbf{R}} H$.

Proof of Lemma 30. Denote by H the stabilizer of the point $f \in H$ in $G_{\mathbf{R}}$. H is a compact Cartan subgroup in $G_{\mathbf{R}}$. Denote by $\mathfrak{G}$ the Lie algebra of the group $G_{\mathbf{R}}$, and by $\mathfrak{H}$ its subalgebra corresponding to the subgroup $H \subset G_{\mathbf{R}}$. Since the manifold F is the factor space $F \cong H \diagdown G_{\mathbf{R}}$, the space $(T_f)_{\mathbf{R}}$ is isomorphic to the space $(\mathfrak{H} \diagdown \mathfrak{G})$ and the operator $\psi(g)$ coincides with the operator in $(\mathfrak{H} \diagdown \mathfrak{G})$, induced by the operator Ad (g) in $\mathfrak{G} \subset \mathfrak{G}_{\mathbf{C}}$. Lemma 30 is proved.

Corollary 1. There is a group isomorphism $\alpha: G_f \to G_{f^\sigma}$ and an isomorphism of real linear spaces $c:(T_f)_{\mathbf{R}} \to (T_{f^\sigma})_{\mathbf{R}}$ such that for all $g \in G_f$ we have

$$\psi^\sigma(\alpha(g)) = c \circ \psi(g) \circ c^{-1}.$$

Proof. Lemma 4′ implies that there exist isomorphisms $\alpha: G_f^e \to G_{f^\sigma}^\sigma$ and $c_1:(T_f)_{\mathbf{R}} \to (T_{f^\sigma})_{\mathbf{R}}$ such that $\psi^\sigma(\alpha(g)) = c_1 \circ (\psi(g))^\sigma \circ c_1^{-1}$. Thus for the proof of Corollary 1 it is sufficient to show that the representations $g \to \psi(g)$ and $g \to (\psi(g))^\sigma$ of the group G_f^e are isomorphic. Since the group G is defined over $\mathbf{Q}$, Lemma 30 implies that both of these representations of the group G_f are fully reducible and their characters coincide. Consequently, they are equivalent. Corollary 1 is proved.

Corollary 2. The operator $\psi^\sigma(g^\sigma)$ is compact, and its characteristic polynomial has rational coefficients.

Proof. It immediately follows from Corollary 1 that the operator $\psi^\sigma(g^\sigma)$ is conjugate to the operator $\psi(g)$ for $g = \alpha^{-1}(g^\sigma) \in G_f^e \subset G_{\mathbf{Q}} \cap H$. The first assertion of the corollary follows at once from the fact that the element $g \in H'$ is compact. For the proof of the second assertion we observe that by virtue of Lemma 30 the characteristic polynomial of the operator $\psi(g)$ is equal to the characteristic polynomial of the operator Ad (g) divided by the degree of the independent variable. Since the group G is defined over $\mathbf{Q}$, the coefficients of the characteristic polynomial of the operator Ad (g) are rational. Corollary 2 is proved.

Corollary 3. Any semisimple element $g^\sigma \in V^\sigma$ is compact.

Proof. By assumption, the element $p''(g^\sigma)$ is compact, hence by virtue of Theorem 8 there is a point $f'^\sigma \in F'^\sigma$, stable under the automorphism $p''(g^\sigma) \in \text{Aut } F'^\sigma$. Consequently, the fibre $(\tau''^\sigma)^{-1} f^\sigma \cong Y_{G_0} \subset F^\sigma$ is invariant under the element $g^\sigma \in \text{Aut } F^\sigma$. It follows from Lemma 20 that there is a point $f \in (\tau''^\sigma)^{-1} f^\sigma \subset F^\sigma$ stable under the element $g^\sigma \in \text{Aut } F^\sigma$. Denote by $T_{f^\sigma}^\sigma \subset T_f^0$ the tangent space to the fibre $(\tau''^\sigma)^{-1}(f'^\sigma) \cong Y_{G_0}$ at the point f, and by $\varphi^0(g^\sigma)$ the restriction of the operator $\varphi(g^\sigma)$ to the subspace $T_{f^\sigma}^0$. Corollary 2 of Lemma 30 implies that the operator $\varphi^0(g^\sigma)$ is compact. Consider the point $f^\sigma \in F^\sigma$ as a point on the fibre $f^\sigma \in (\tau''^\sigma)^{-1}(f'^\sigma) \cong Y_{G_0}$. The operator $p_0(g^\sigma)$ acts on the fibre $(\tau''^\sigma)^{-1}(f'^\sigma)$, preserves the point f and induces in the tangent space $T^0_{f^\sigma}$ an operator $\varphi^0(g^\sigma)$. Since $\varphi^0(g^\sigma)$ is a compact operator and g^σ is a semisimple element, it follows from the Corollary to Lemma 20 that $p_0(g^\sigma)$ is a compact element in $G_{0\mathbf{C}}$. Since $p''(g^\sigma)$ is a compact element in $G'_{\mathbf{R}}$ by assumption, Corollary 3 is proved.

In the following, we will need a simple but useful assertion.

Let M_1, M_2 be topological groups, $M=M_1\times M_2$; p_1, p_2 the natural projections $p_i: M\to M_i$, $i=1, 2$ and N a closed subgroup in M. Put $N_i=N\cap M_i$.

Lemma 31. *If $p_2(N)$ is a dense subgroup in M_2, then the group N_2 is a normal subgroup in M_2.*

Proof. The fact that M_2 is a normal subgroup of M, implies immediately that N_2 is a normal subgroup of N. For any elements $m\in M$, $m_2\in M_2$ we have $mm_2m^{-1}=p_2(m)m_2(p_2(m))^{-1}$, thus N_2 is a normal subgroup of $p_2(N)$. Now N_2 is a closed subgroup in M_2, and by assumption, $p_2(N)$ is a dense subgroup in M_2, thus N_2 is a normal subgroup in M_2. Lemma 31 is proved.

Consider now the case $M_1=G_{0\mathbf{C}}$, $M_2=G'_{\mathbf{R}}$ and denote by N the closure of the subgroup G^σ in $M\cong\tilde{G}_{\mathbf{R}}=G_{0\mathbf{C}}\times G'_{\mathbf{R}}$.

Theorem 9. *The image $p_1(N)$ is a relatively compact subgroup in $G_{0\mathbf{C}}$.*

Proof. First we establish two lemmas.

Lemma 32. *The subgroup $N_2\in G'_{\mathbf{R}}$ contains a compact Cartan subgroup $H'_{\mathbf{R}}\subset G'_{\mathbf{R}}$.*

Proof of Lemma 32. Let us fix a maximal $\mathbf{Q}$-torus H in G such that $H_{\mathbf{R}}$ is a compact Cartan subgroup in $G_{\mathbf{R}}$. Let $f\in F$ be a point invariant under the subgroup $H_{\mathbf{R}}$. The group $G^e_f=G^e\cap H_{\mathbf{R}}$ is, by definition, a subgroup of finite index in the group $G_{\mathbf{Q}}\cap H_{\mathbf{R}}\cong H_{\mathbf{Q}}$. Consequently, the closure $\bar{G}^e_f$ of the group G^e_f in $H_{\mathbf{R}}$ is a subgroup of finite index in the group $H_{\mathbf{R}}$. Since the group $H_{\mathbf{R}}$ is connected, we have $\bar{G}^e_f=H_{\mathbf{R}}$.

Let us choose now a point $f^\sigma\in F^\sigma$ for which $\varrho^\sigma(f^\sigma)=(\varrho(f))^\sigma\in Z^\sigma$. Denote by L the stabilizer of f^σ in the group $G_{0\mathbf{C}}\times G'_{\mathbf{R}}\subset\text{Aut } F^\sigma$ and by $\bar{L}$ the image of the group L under the natural mapping $p: L\to\text{Aut } T_f$. It is clear that group L is isomorphic to the product $L\cong B_{0\mathbf{C}}\times H'_{\mathbf{R}}$ where $B_{0\mathbf{C}}$ is a Borel subgroup in $G_{0\mathbf{C}}$, and $H'_{\mathbf{R}}$ a compact Cartan subgroup in $G'_{\mathbf{R}}$. Put $G^\sigma_{f^\sigma}=G^\sigma\cap L$.

Denote, as above, by $(T_f)_{\mathbf{R}}$ and $(T_{f^\sigma})_{\mathbf{R}}$ the real tangent spaces to F and F^σ at the points f and f^σ, respectively. For any $g\in G^e_f$ and $g^\sigma\in G^\sigma_{f^\sigma}$ denote by $\psi(g)$ and $\psi^\sigma(g^\sigma)$ the operators in the spaces $(T_f)_{\mathbf{R}}$ and $(T_{f^\sigma})_{\mathbf{R}}$ induced by the elements g and g^σ, respectively. Corollary 1 to Lemma 30 states that there is a group isomorphism $\alpha: G^e_f\to G^\sigma_{f^\sigma}$ and an isomorphism c of real linear spaces $c:(T_f)_{\mathbf{R}}\to(T_{f^\sigma})_{\mathbf{R}}$ such that for any $g\in G^e_f$ we have the equality $\psi^\sigma(\alpha(g))=c\psi(g)c^{-1}$. Consequently, the closure of the image of the group $G^\sigma_{f^\sigma}$ in the group $\bar{L}\subset\text{Aut }(T_{f^\sigma})_{\mathbf{R}}$ is a compact subgroup in $\bar{L}$, isomorphic to $H_{\mathbf{R}}$.

Denote by $U_{0\mathbf{C}}$ the unipotent radical of the group $B_{0\mathbf{C}}$ and by v the projection $v: B_{0\mathbf{C}} \to B_{0\mathbf{C}}/U_{0\mathbf{C}} \cong H_{0\mathbf{C}}$. The fact that the closure of the image of the group $G^{\sigma}_{f^{\sigma}}$ in $\bar{L}$ is isomorphic to $H_{\mathbf{R}}$, implies that the image $vp_0(G^{\sigma}_{f^{\sigma}})$ is dense in $H_{0\mathbf{C}}$ in the Zariski topology. Since the group $G^{\sigma}_{f^{\sigma}} = \alpha(G^{e}_{f})$ is commutative, it follows that there is a Cartan subgroup $H_{0\mathbf{C}} \subset B_{0\mathbf{C}}$ such that $G^{\sigma}_{f^{\sigma}} \subset H_{0\mathbf{C}} \times H'_{\mathbf{R}}$. Now the kernel $\operatorname{Ker} p$ of the projection $p: L \to \bar{L}$ is the product of the centre of the group $G_{0\mathbf{C}}$ (it is finite) and of a unipotent group, thus the intersection $\operatorname{Ker} p \cap (H_{0\mathbf{C}} \times H'_{\mathbf{R}})$ is finite, i.e. the restriction of the projection p to $H_{0\mathbf{C}} \times H'_{\mathbf{R}} \subset L$ is an isogeny. This implies that the closure $\bar{G}^{\sigma}_{f^{\sigma}}$ of the group $G^{\sigma}_{f^{\sigma}}$ in the group $L \subset G_{0\mathbf{C}} \times G'_{\mathbf{R}}$ is also a compact subgroup, isomorphic to $H_{\mathbf{R}}$. Therefore the group $\bar{G}^{\sigma}_{\mathbf{R}}$ contains a compact Cartan subgroup $H'_{\mathbf{R}}$ of $G'_{\mathbf{R}}$. Since $\bar{G}^{\sigma}_{f^{\sigma}} \subset N$, Lemma 32 is proved.

Now we turn to the proof of Theorem 9. Since, by assumption, the group $p''(G^{\sigma})$ is dense in $\bar{G}'^{\sigma}_{\mathbf{R}} \cong G'_{\mathbf{R}}$, Lemma 31 and 32 imply that the group N_2 coincides with $G'_{\mathbf{R}}$, and consequently $N = p_1(N) \times G'_{\mathbf{R}} = N_1 \times G'_{\mathbf{R}}$.

Lemma 33. *N is a semisimple group.*

Proof of Lemma 33. Denote by R the solvable radical in N, and by Y^R the subset of Y_{G_0}, consisting of all R-stable points. It is clear that the set Y^R is invariant under N_1. It is known (cf. [19] Theorem 10.4) that Y^R is a non-empty analytic submanifold in Y_{G_0}.

Consider the analytic submanifold $Y_R \times F'^{\sigma} \subset F^{\sigma}$. By assumption, $Y_R \times F'^{\sigma}$ is a closed G^{σ}-invariant submanifold in F^{σ}. There is no closed analytic G^{e}-invariant proper submanifold in the manifold F; thus Lemma 5′ implies that $Y_R \times F'^{\sigma} = F^{\sigma}$ and consequently $Y_R = Y_{G_0}$. This is possible only in the case $R = \{e\}$. Lemma 33 is proved.

Now we pass to the proof of Theorem 9. Assume that the group N_1 is not compact. Since, according to Lemma 33, N_1 is a semisimple group, there is an open subset W in N_1 consisting of semisimple non-compact elements. Consider the subset $W \times U'$, which is open in N. Since the group N is the closure of G^{σ} in $G_{0\mathbf{C}} \times G'_{\mathbf{R}}$ the intersection $G^{\sigma} \cap \overline{W} \times U'$ is not empty. Let $g^{\sigma} \in G^{\sigma} \cap W \times U'$. Since the set $W \times U'$ consists of semisimple elements, Corollary 3 of Lemma 30 implies that g^{σ} is a compact element in $G_{0\mathbf{C}} \times G'_{\mathbf{R}}$. Now this contradicts the fact that, by assumption, in the set W there are no compact elements. This contradiction proves Theorem 9.

Remark. Since G_0 is a maximal compact subgroup in $G_{0\mathbf{C}}$, we can assume $G^{\sigma} \subset G_0 \times G'_{\mathbf{R}} = \tilde{G}_{\mathbf{R}}$.

Corollary to Theorem 9. The closure N of the group G^{σ} in $\tilde{G}_{\mathbf{R}}$ coincides with $\tilde{G}_{\mathbf{R}}$.

Proof. It is an immediate consequence of Theorem 9 that for any point $f^{\sigma} \in F^{\sigma}$ the orbit $f^{\sigma} N$ of f^{σ} under the action of the group N is closed. Using this fact one can prove, in the same way as Theorem 7, that the group N acts transitively on F^σ. Therefore the compact group $N_1 \subset G_0$ acts transitively on Y_{G_0}. Lemma 21 implies that $N_1 = G_0$. The corollary is proved.

Lemma 34. *The projection* $p_1(\Gamma^\sigma)$ *of the group* $\Gamma^\sigma \subset \tilde{G}_{\mathbf{R}}$ *on* G_0 *is dense in* G_0.

Proof. Denote by N the closure of the group $p_1(\Gamma^\sigma)$ in G_0. Assume that $N \neq G_0$. Lemmas 21 and 23 imply that in this case there is a closed N-invariant non-empty complex submanifold $U \subsetneqq Y_{G_0}$. Consider the complex submanifold $(\tau_0^\sigma)^{-1}(U^\sigma) \subset F^\sigma$. It is, by definition, Γ^σ-invariant, and invariant under the canonical flat connection in the fibering $Y_{G_0} \times F'^\sigma \xrightarrow{\tau''^\sigma} F'^\sigma$. Denote by W^σ the projection of U^σ to Z^σ. Then W^σ is a closed complex submanifold in Z^σ, invariant under the flat connection ∂^σ in the fibering $\mu''^\sigma: Z^\sigma \to Z'^\sigma$. Since Z^σ is a compact complex algebraic manifold, according to the lemma of Chow W^σ is an algebraic submanifold in Z^σ. Put $W = (W^\sigma)^{\sigma^{-1}} \subset Z$, then W will be a closed submanifold in Z, invariant under the canonical flat connection ∂ in $\mu'': Z \to Z'$. Let $z' \in Z'$ and $U = \mu''^{-1}(z') \cap \cap W \subset \mu''(z') \cong Y_{G_0}$. Since the manifold W is invariant under the connection ∂, the closed submanifold $U \subset Y_{G_0}$ will be invariant under the action of the group $p_1(\Gamma) \subset G_0$. The strong approximation theorem (cf. [7]) implies that the group $p_1(\Gamma)$ is closed in G_0. Since $U \neq \emptyset$, thus $U = Y_{G_0}$ and consequently $W = Z$, $W^\sigma = Z^\sigma$ and $U^\sigma = Y_{G_0}$. The last equality contradicts the choice of the submanifold $U^\sigma \subset Y_{G_0}$. Lemma 34 is proved.

Theorem 10. *Let* $L \subset G^\sigma$ *be a commutative subgroup, relatively compact in* $\tilde{G}_{\mathbf{R}}$. *Then there is a* **Q**-*torus* $H \subset G$, *an automorphism* $c: \mathfrak{G}_{\mathbf{R}} \to \mathfrak{G}_{\mathbf{R}}$ *and a homomorphism* $\beta: L \to H_{\mathbf{Q}}$, *such that for all* $g^\sigma \in L$ *the operators* $c^{-1} \operatorname{Ad}(g^\sigma) c$ *and* $\operatorname{Ad}(\beta(g^\sigma))$ *on* $\mathfrak{G}_{\mathbf{R}}$ *coincide. Conversely, if* $H \subset G$ *is a* **Q**-*torus such that* $H_{\mathbf{R}}$ *is a compact group, there is a commutative relatively compact subgroup* $L \subset G^\sigma$, *an automorphism* $\beta: L \to H_{\mathbf{Q}}$ *and an automorphism* $c: \mathfrak{G}_{\mathbf{R}} \to \mathfrak{G}_{\mathbf{R}}$ *such that* $\alpha(L)$ *is a subgroup of finite index in* $H_{\mathbf{Q}}$, *and for all* $g^\sigma \in L$ *the operators* $c^{-1} \operatorname{Ad}(g^\sigma) c$ *and* $\operatorname{Ad}(\beta(g^\sigma))$ *on* $\mathfrak{G}_{\mathbf{R}}$ *coincide.*

Proof. We establish only the first assertion of the theorem. The second one can be proved similarly.

Since the group $L \subset \tilde{G}_{\mathbf{R}}$ is commutative and relatively compact, there exists a point $f^\sigma \in F^\sigma$ invariant under the action of the group L. Denote by $(T_{f^\sigma})_{\mathbf{R}}$ the real tangent space to F^σ at the point f^σ, and for any $g^\sigma \in L$ denote by $\psi^\sigma(g^\sigma)$ the linear operator in the space $(T_{f^\sigma})_{\mathbf{R}}$, induced by the mapping $f^\sigma \to f^\sigma g^\sigma$, $f^\sigma \in F^\sigma$ of the manifold F^σ into itself. The following assertion can be proved the same way as Lemma 30.

Lemma 30′. *The operator* $\operatorname{Ad}(g^\sigma)$ *is conjugate to the sum of the operator* $\psi^\sigma(g^\sigma)$ *and the identity operator of dimension* $\dim_{\mathbf{R}} H_{\mathbf{R}} = \operatorname{rank}_{\mathbf{C}} G_{\mathbf{C}}$.

Put $f=(f^\sigma)^{\sigma^{-1}}$. Let $\alpha: G_f^e \to G_{f^\sigma}^\sigma$ be the homomorphism constructed in Corollary 1 of Lemma 30. $L \subset G_{f^\sigma}$ by assumption. Denote by β the restriction of the homomorphism $\alpha^{-1}: G_{f^\sigma}^\sigma \to G_f^e$ to L. The assertion of the theorem follows from Corollary 1 of Lemma 30 and from Lemma 30'. Theorem 10 is proved.

Corollary. For any element $g^\sigma \in V^\sigma$ the characteristic polynomial of the operator $\mathrm{Ad}\,(g^\sigma)$ has rational coefficients.

Proof. The corollary follows from Theorem 10 and Corollary 2 of Lemma 30.

Theorem 11. *Let M be a closed subgroup in $\tilde{G}_{\mathbf{R}}$, such that $M \supset \Gamma^\sigma$ and $[p_2(M):p_2(\Gamma^\sigma)]=\infty$. Then $M=\tilde{G}_{\mathbf{R}}$.*

Proof. Put $M_2 \overset{\mathrm{def}}{=} p_2(M) \subset G'_{\mathbf{R}}$. Since $\mathrm{Ker}\, p_2 \cong G_0$ is compact, M_2 is a closed subgroup in $G'_{\mathbf{R}}$. The factor space $K' \backslash G'_{\mathbf{R}}/p_2(\Gamma^\sigma)$ is isomorphic to the manifold X^σ, and therefore it is compact. Consequently, the factor space $G'_{\mathbf{R}}/p_2(\Gamma^\sigma)$ is also compact. Since the index $[M_2:p_2(\Gamma)]=\infty$, M_2 is not a discrete group and the connected component M^0 of the identity e in the group M_2 is not equal to e. Consider the normalizer $N(M^0)$ of the group M^0 in $G'_{\mathbf{R}}$. Since the group M^0 is connected, the normalizer of the group M_0 coincides with the normalizer of the Lie subalgebra $\mathfrak{M}_0 \subset \mathfrak{G}'$, corresponding to the subgroup M_0 in the Lie algebra $\mathfrak{G}'$ of the group $G'_{\mathbf{R}}$. Therefore the group $N(M_0)$ is algebraic. By definition, $p_2(\Gamma^\sigma)$ is contained in $N(M^0)$. The factor space $G'_{\mathbf{R}}/p_2(\Gamma^\sigma)$ is compact and the group $G'_{\mathbf{R}}$ is a connected semisimple group without compact factors, thus the group Γ^σ is dense in $G'_{\mathbf{R}}$ with respect to the Zariski topology. Therefore the inclusion relation $p_2(\Gamma^\sigma) \subset N(M^0)$ implies $N(M^0)=G'_{\mathbf{R}}$ i.e. that M^0 is a normal subgroup in $G'_{\mathbf{R}}$.

Let $G'_{\mathbf{R}}=\prod_1^s G_i$ be the decomposition of the group $G'_{\mathbf{R}}$ to the product of simple Lie groups. Since M^0 is a connected normal subgroup in $G'_{\mathbf{R}}$ and $M^0 \neq \{e\}$, there is a subset $J \subset [1, \ldots, s]$ such that $M^0=\prod_{i \in J} G_i$. Denote by p_J the natural projection of the group $G'_{\mathbf{R}}$ to $\prod_{i \notin J} G_i$. Now M_2 is a closed subgroup in $G'_{\mathbf{R}}$, and its connected identity component M^0 coincides with $\mathrm{Ker}\, p_J$, thus $p_J(M_2)$ is a discrete subgroup in $\prod_{i \notin J} G_i$. Consequently, its subgroup $p_J(p_2(\Gamma^\sigma))$ is also a discrete subgroup in $\prod_{i \notin J} G_i$. Since the factor space $G'_{\mathbf{R}}/p_2(\Gamma^\sigma)$ is compact, the discreteness of the subgroup $p_J(p_2(\Gamma^\sigma))$ implies that the factor space $\prod_{i \in J} G_i / \prod_{i \notin J} G_i \cap p_2(\Gamma^\sigma)$ is also compact. Therefore for any point $d^\sigma \in D^\sigma$ the projection $Z^\sigma \subset X^\sigma$ of the orbit $d^\sigma \prod_{i \in J} G_i \subset D^\sigma$ is a compact leaf of the foliation T_{J, X^σ}. According to Lemma 11, this is possible only in the case $J=[1, \ldots, s]$ i.e. when $M^0=G'_{\mathbf{R}}$.

Consider now the normal subgroup $N_2=M \cap G'_{\mathbf{R}} \subset M$. As the group G_0 is compact, so is the factor group M/N_2. Consequently, so is the factor group $M_2/N_2 \cong G'_{\mathbf{R}}/N_2$. By assumption, the group $G'_{\mathbf{R}}$ has no compact factors, thus

$N_2=G'_{\mathbf{R}}$. Therefore the group $N_1=G_0\cap M$ coincides with the projection $p_2(M)$ of M to G_0. Lemma 34 implies that $N_1=G_0$. Consequently, $M=\tilde{G}_{\mathbf{R}}$. The theorem is proved.

Corollary. Let $L\subset G^\sigma$ be a subgroup in G^σ, containing Γ^σ and satisfying $[L:\Gamma^\sigma]<\infty$. Then L is dense in $\tilde{G}_{\mathbf{R}}$.

Proof. The definition of the group G^σ implies that $G_0\cap G^\sigma=(e)$. Therefore $[p_2(L):p_2(\Gamma^\sigma)]=[L:\Gamma^\sigma]=\infty$. Applying Theorem 11 to the closure $\bar{L}$ of the subgroup L in $\tilde{G}_{\mathbf{R}}$ we obtain the desired result. The corollary is proved.

To finish the proof of Theorem A we need the following assertion.

Lemma 35. *Let M be an algebraic connected semisimple real group without compact factors, and $L\subset M$ its unimodular algebraic subgroup such that the factor space M/L is compact. Then $L=M$.*

Proof. First we analyse the case when L is a normal subgroup of M. Then the factor group M/L is compact by assumption. Since the group M has no compact factors, $L=M$. Now we analyse the general case. Let $\mathfrak{M}$ be the Lie algebra of the group M, $\mathfrak{L}$ its subalgebra corresponding to L, and $l\overset{\text{def}}{=}\dim\mathfrak{L}$. Denote by Ad the adjoint representation of the group M in $\mathfrak{M}$, and by Ad^l its l-th exterior power $\mathrm{Ad}^l:M\to\mathrm{Aut}\,(\overline{\wedge}^l\,\mathfrak{M})$. Put $V\overset{\text{def}}{=}\wedge^l\,\mathfrak{M}$ and denote by $v_0\in V$ some nonzero vector belonging to the line $\wedge^l\,\mathfrak{L}\subset V$. Because the group L is unimodular, the vector v_0 (and not only the line $\wedge^l\,\mathfrak{L}$) is L-stable. Since the factor space M/L is compact, the orbit $\mathrm{Ad}^l\,(M)v_0$ is compact.

Assume now that $L\neq M$. Then, as we have shown, L is not a normal subgroup in M. Consequently, $\mathfrak{L}$ is not a normal subalgebra in $\mathfrak{M}$, and the line $\Lambda^l\,\mathfrak{M}=\mathbf{R}v_0\subset V$ is not invariant under the action of the group M. Denote by W the subspace in V generated over $\mathbf{R}$ by the vectors w of the form $w=(\mathrm{Ad}^l\,m)v_0$, $m\in M$ and write ϱ for the restriction of representation Ad^l of the group M onto W. As we have shown, it follows from the assumption $L\neq M$ that the representation ϱ is not a multiple of the identity representation, i.e. $\varrho(M)\neq E$.

Denote by W' the linear space, dual to W. Since the orbit $\varrho(M)v_0=\mathrm{Ad}^l\,(M)v_0\subset W\subset V$ is compact, for any vector $w'\in W'$ the function $\langle\varrho(m)v_0, w'\rangle$ on G is bounded. As any vector $w\in W$ can be represented in the form $w=\sum_{i=1}^{N}a_i\,\varrho(m_i)v_0$, $a_i\in\mathbf{R}$, $m_i\in M$ so for any pair $w\in W$, $w'\in W'$ the function $\langle\varrho(m)w, w'\rangle$ on M is bounded. Consequently, the image $\varrho(M)\subset\mathrm{Aut}\,W$ is compact. Now recall that by assumption the group M has no compact factors. Therefore $\varrho(M)=E$. On the other hand we have already shown that in case $L\neq M$ the image $\varrho(M)\neq E$. Consequently $L=M$. Lemma 35 is proved.

Corollary. Let M be a connected semisimple algebraic real group without compact factors, and $\Delta\subset M$ a discrete subgroup such that the factor space M/Δ is compact. Then the group Δ is dense in M with respect to the Zariski topology.

Proof. Denote by L the closure of the group Δ in M in the Zariski topology. In the group L there is a discrete subgroup Δ such that the factor space $L/\Delta \subset M/\Delta$ is compact, hence L is a unimodular group. On the other hand, as $L \supset \Delta$ and the factor space M/Δ is compact by assumption, the factor space M/L is compact. Therefore Lemma 35 applies to the subgroup $L \subset M$. The corollary is proved.

Denote by $i^{\sigma}_{\mathbf{R}}$ the imbedding $i^{\sigma}_{\mathbf{R}}: G^{\sigma} \to \tilde{G}_{\mathbf{R}}$ (i.e. the group G^{σ} becomes a subgroup of $\tilde{G}_{\mathbf{R}}$ by means of $i^{\sigma}_{\mathbf{R}}$). Denote by $'\hat{G}$ the completion of the group G^{e} in the topology of subgroups of finite index in Γ. According to Lemma 7 the group $'\hat{G}$ is isomorphic to the completion of G^{σ} in the topology of subgroups of finite index in the group Γ^{σ}. Denote by $\hat{\imath}^{\sigma}$ the natural imbedding $\hat{\imath}^{\sigma}: G^{\sigma} \to {}'\hat{G}$.

Denote by $\hat{G}_{\mathbf{Q}}$ the completion of the group $G_{\mathbf{Q}}$ in the topology of the subgroups of finite index in the group $G_{\mathbf{Z}} \supset \Gamma$. Denote by Z the centre of the group $G_{\mathbf{Q}}$. By definition, the group G^{e} is a subgroup of finite index in the factor group $G_{\mathbf{Q}}/Z$. Since the centre Z of the group $G_{\mathbf{Q}}$ is finite, we have that the group $'\hat{G}$ is a subgroup of finite index in the factor group $\hat{G}_{\mathbf{Q}}/Z$. Now we show that the pair $\Gamma^{\sigma} \subset G^{\sigma} (\subset \tilde{G}_{\mathbf{R}})$ satisfies the conditions of Theorem A.

It is clear that the pair $\Gamma^{\sigma} \subset G^{\sigma}$ satisfies conditions 1, 2 and 3 of Theorem A. The validity of condition 4 follows from the Corollary of Theorem 6. The validity of condition 5 has just been established. The fact that the pair $\Gamma^{\sigma} \subset G^{\sigma}$ satisfies condition 6, follows immediately from Lemma 34 and the Corollary of Lemma 35. The validity of condition 7 follows from the Corollary of Theorem 11 and finally, the validity of conditions 8 and 9 follows from Theorem 10. Theorem A is proved.

CHAPTER II

THE GROUP-THEORETIC PART

Let G be a simple simply-connected algebraic **Q**-group and assume that the group $G_{\mathbf{R}}$ contains a compact Cartan subgroup. Let $\tilde{G}$ be the adjoint algebraic **Q**-group, $\varkappa$ the natural projection $\varkappa: G \to \tilde{G}$ and $\hat{G}_{\mathbf{Q}}$ the completion of the group $G_{\mathbf{Q}}$ in the topology of subgroups of finite index in $G_{\mathbf{Z}}$. Further, let Z denote the center of the group $G_{\mathbf{Q}}$. We consider it as a subgroup in $\hat{G}_{\mathbf{Q}}$. The aim of this chapter is to establish the validity of the following theorem.

Theorem B. Let $\Gamma^\sigma \subset G^\sigma$ be a pair of subgroups of the group $\tilde{G}_{\mathbf{R}}$ such that

1) Γ^σ is a discrete subgroup in the group $\tilde{G}_{\mathbf{R}}$ without elements of finite order, and the factor space $\tilde{G}_{\mathbf{R}}/\Gamma^\sigma$ is compact;

2) for every element $\gamma \in G^\sigma$ the intersection $\gamma^{-1}\Gamma^\sigma\gamma \cap \Gamma^\sigma$ is a subgroup of finite index in Γ^σ;

3) the natural homomorphism $\imath^\sigma: G^\sigma \to \hat{G}^\sigma_{\Gamma^\sigma}$ is an imbedding;

4) the completion $\hat{G}^\sigma_{\Gamma^\sigma}$ is isomorphic to a subgroup of finite index in the factor group $\hat{G}_{\mathbf{Q}}/Z$;

5) the group Γ^σ is dense in $\tilde{G}_{\mathbf{R}}$ in the Zariski topology;

6) if L is a subgroup in G^σ containing Γ^σ and $[L:\Gamma^\sigma]=\infty$, then the group L is dense in the connected component of the group $\tilde{G}_{\mathbf{R}}$ in the ordinary topology;

*7) for any algebraic **Q**-torus $H \subset G$ for which $H_{\mathbf{R}}$ is compact, there is an element $g \in \tilde{G}_{\mathbf{R}}$ such that the intersection $\varkappa(H_{\mathbf{Q}}) \cap g^{-1}G^\sigma g$ is a subgroup of finite index in $\varkappa(H_{\mathbf{Q}})$;**

8) for any compact Cartan subgroup $H_{\mathbf{R}} \subset \tilde{G}_{\mathbf{R}}$ there is an element $g \in \tilde{G}_{\mathbf{R}}$ such that $g^{-1}(H_{\mathbf{R}} \cap G^\sigma)g \subset \varkappa(G_{\mathbf{Q}})$.

*Then there exists a simple algebraic **Q**-group ${}^\sigma\tilde{G}$ and an isomorphism $\varphi: ({}^\sigma\tilde{G})_{\mathbf{R}} \to \tilde{G}_{\mathbf{R}}$ such that $\varphi^{-1}(G^\sigma) \subset {}^\sigma\tilde{G}_{\mathbf{Q}}$ and $\varphi^{-1}(\Gamma^\sigma)$ is an arithmetic subgroup.*

Before going into the proof of Theorem B we introduce some notation. Denote by $\hat{\varkappa}$ the natural projection $\hat{\varkappa}: \hat{G}_{\mathbf{Q}} \to \hat{G}_{\mathbf{Q}}/Z$. Let $\Gamma^\sigma \subset G^\sigma$ be a pair of subgroups of $\tilde{G}_{\mathbf{R}}$ satisfying the conditions of Theorem B. This pair will be considered fixed throughout the paper. Consider the group $\hat{G}^\sigma_{\Gamma^\sigma}$ as a subgroup in $\hat{G}_{\mathbf{Q}}/Z$ and put $\hat{G} = \hat{\varkappa}^{-1}(\hat{G}^\sigma_{\Gamma^\sigma})$. The natural imbedding of the group $\hat{\varkappa}^{-1}(\imath^\sigma(G^\sigma))$ into $\hat{G}$ will also

* We consider the group $G_{\mathbf{Q}}$ to be imbedded into $G_{\mathbf{R}}$.

be denoted by $\hat{\imath}^\sigma$. The homomorphism of the group $\hat{\kappa}^{-1}(\hat{\imath}^\sigma(G^\sigma))$ into $\tilde{G}_{\mathbf{R}}$, being the composition of the homomorphism $(\hat{\imath}^\sigma)^{-1}\circ\hat{\kappa}$ and the natural imbedding of the group G^σ into $\tilde{G}_{\mathbf{R}}$, is denoted by $i^\sigma_{\mathbf{R}}$. In the sequel we will consider only the group $\hat{\kappa}^{-1}(\hat{\imath}^\sigma(G^\sigma))$ and the group G^σ itself will not occur. Therefore we use the notation G^σ for the group $\hat{\kappa}^{-1}(\hat{\imath}^\sigma(G^\sigma))$. Similarly, the group $\hat{\kappa}^{-1}(\hat{\imath}^\sigma(\Gamma^\sigma))$ will be denoted by Γ^σ. In the sequel the group G^σ will be considered as a subgroup in $\tilde{G}_{\mathbf{R}}\times\hat{G}$ by using the imbedding $i^\sigma_{\mathbf{R}}\times\hat{\imath}^\sigma: G^\sigma\hookrightarrow\tilde{G}_{\mathbf{R}}\times\hat{G}$.

Lemma 36. *The group G^σ is a discrete subgroup of $G_{\mathbf{R}}\times\hat{G}$ and the factor space $G_{\mathbf{R}}\times\hat{G}/G^\sigma$ is compact.*

Proof. Denote by $\hat{\Gamma}$ the closure of the subgroup $\hat{\imath}^\sigma(\Gamma^\sigma)$ in the group $\hat{G}$. It follows from Lemma 7 that the group $\hat{\Gamma}$ is an open compact subgroup in $\hat{G}$, which is, according to Lemma 7, isomorphic to the completion of the group Γ in the topology given by its subgroups of finite index. Since, by definition, the image $\hat{\imath}^\sigma(G^\sigma)$ is a dense subgroup in $\hat{G}$ and the subgroup $\hat{\Gamma}\subset\hat{G}$ is open, for the proof of Lemma 35 it is sufficient to show that the subgroup $G^\sigma\cap\tilde{G}_{\mathbf{R}}\times\hat{\Gamma}$, which is isomorphic to Γ^σ, is a discrete subgroup of $\tilde{G}_{\mathbf{R}}\times\hat{\Gamma}$, and that the factor space $\tilde{G}_{\mathbf{R}}\times\hat{\Gamma}/G^\sigma\cap\tilde{G}_{\mathbf{R}}\cap\hat{\Gamma}$ is compact. Since the group $\hat{\Gamma}$ is compact, it is enough to show that the group $i_{\mathbf{R}}(\Gamma^\sigma)$ is a discrete subgroup in $G_{\mathbf{R}}$ and the factor space $\tilde{G}_{\mathbf{R}}/i_{\mathbf{R}}(\Gamma^\sigma)$ is compact. Now this follows immediately from condition 1 of Theorem B. Lemma 36 is proved.

The only result we need from the first chapter is Lemma 19. Therefore in Chapter II we may use the same letters G_0, G_i as in Chapter I for entirely different objects without confusion.

§ 10. Proof of rationality of the subgroup $G^\sigma\subset\tilde{G}_{\mathbf{R}}$

The aim of this paragraph is to prove the following theorem:

Theorem 12. *There exists a* **Q**-*form* ${}^\sigma\tilde{G}$ *of the group* $\tilde{G}$ *and an algebraic isomorphism* $\varphi:({}^\sigma\tilde{G})_{\mathbf{R}}\xrightarrow{\sim}(\tilde{G})_{\mathbf{R}}$ *such that* $i^\sigma_{\mathbf{R}}(G^\sigma)\subset\varphi({}^\sigma\tilde{G}_{\mathbf{Q}})$.

Proof. Denote by G^0 the connected component of the identity in the group $\tilde{G}_{\mathbf{R}}$ in the classical topology, and by $R\subset G^0$ the subset consisting of all regular* compact elements.

Since, by assumption, the group $\tilde{G}_{\mathbf{R}}$ contains a compact Cartan subgroup, R is a non-empty subset in $\tilde{G}_{\mathbf{R}}$, open in the classical topology. Condition 8 of Theorem B implies that for any element $\gamma\in G^\sigma\cap R$ the trace $\mathrm{Sp}(\mathrm{Ad}(\gamma))$** is rational. On the other hand, it follows from condition 6 of Theorem B that the

* An element $g\in\tilde{G}_{\mathbf{R}}$ is called *regular* if its centralizer $Z_{\tilde{G}_{\mathbf{R}}}(g)$ in $\tilde{G}_{\mathbf{R}}$ is a Cartan subgroup.

** Here, as before, Ad stands for the adjoint representation $\mathrm{Ad}:\tilde{G}_{\mathbf{R}}\to\mathrm{Aut}\,\mathfrak{G}_{\mathbf{R}}$ of the group $G_{\mathbf{R}}$.

group G^σ is dense in $G^0 \subset \tilde{G}_{\mathbf{R}}$. Therefore, to prove Theorem 12 it is enough to establish the following theorem:

Theorem 12′. *Let* $\Lambda \subset \tilde{G}_{\mathbf{R}}$ *be a dense subgroup in* $G^0 \subset \tilde{G}_{\mathbf{R}}$, *and assume that for any element* $\lambda \in \Lambda \cap R$ *the trace* $\mathrm{Sp}\,(\mathrm{Ad}\,(\lambda))$ *is rational. Then there exists a* **Q**-*form* $\tilde{G}_\Lambda$ *of the group* $\tilde{G}$ *and an algebraic isomorphism* $\varphi : (\tilde{G}_\Lambda)_{\mathbf{R}} \to (\tilde{G})_{\mathbf{R}}$ *such that* $\Lambda \subset \varphi(\tilde{G}_\Lambda)_{\mathbf{Q}}$.

Proof. We reduce Theorem 12′ to the following proposition:

Proposition 2. Let the group $\Lambda \subset G^0 \subset \tilde{G}_{\mathbf{R}}$ satisfy the conditions of Theorem 12′. Then for any $\lambda \in \Lambda$ the trace $\mathrm{Sp}\,(\mathrm{Ad}\,(\lambda))$ is rational.

First we show how to get Theorem 12′ from Proposition 2. Put $\Lambda^0 = \Lambda \cap G^0$. Since the group Λ^0 is dense in the identity component G_0 of $\tilde{G}_{\mathbf{R}}$ (in the classical topology), and the group $\tilde{G}$ is connected (in the Zariski topology), the group Λ^0 is dense in $\tilde{G}_{\mathbf{C}}$ in the Zariski topology.

For any element $\tau \in \mathrm{Gal}\,(\mathbf{C}/\mathbf{Q})$ denote by the same letter τ the induced (not necessarily continuous) homomorphism $\tau : \tilde{G}_{\mathbf{C}} \to \tilde{G}_{\mathbf{C}}$. Put $\mathrm{Ad}^\tau = \mathrm{Ad} \circ \tau : \tilde{G}_{\mathbf{C}} \to \mathrm{Aut}\, \mathfrak{G}_{\mathbf{C}}$. We denote the restrictions of the representations Ad and Ad^τ to the group $\Lambda^0 \subset \tilde{G}_{\mathbf{C}}$ also by Ad and Ad^τ. Since the group Λ^0 is dense in the group $\tilde{G}_{\mathbf{C}}$ and the adjoint representation of the group $\tilde{G}_{\mathbf{C}}$ is fully reducible ([8] Chapter 3, § 7, Exercise B.4.), thus the representations Ad and Ad^τ of the group Λ^0 are fully reducible. On the other hand, it follows from Proposition 2 that for any $\lambda \in \Lambda^0$ we have the equality $\mathrm{Sp}\,\mathrm{Ad}\,(\lambda) = \mathrm{Sp}\,\mathrm{Ad}^\tau\,(\lambda)$. Therefore Ad and Ad^τ are isomorphic representations of the group Λ^0 i.e. there is an element $f^\tau \in \mathrm{Aut}\,\mathfrak{G}_{\mathbf{C}}$ such that

$$\mathrm{Ad}^\tau(\lambda) = (f^\tau)^{-1}\,\mathrm{Ad}\,(\lambda) f^\tau \quad \text{for all } \lambda \in \Lambda^0. \qquad (*)$$

Since the group $\tilde{G}_{\mathbf{C}}$ has no centre, it can be identified with the image $\mathrm{Ad}\,(\tilde{G}_{\mathbf{C}}) \subset \mathrm{Aut}\,\mathfrak{G}_{\mathbf{C}}$. Equality $(*)$ implies that $(f^\tau)^{-1}\Lambda^0 f^\tau \subset \tilde{G}_{\mathbf{C}}$. Since the group Λ^0 is dense in $\tilde{G}_{\mathbf{C}}$ in the Zariski topology, we have $(f^\tau)^{-1}\tilde{G}_{\mathbf{C}} f^\tau \subset \tilde{G}_{\mathbf{C}}$ and consequently the mapping $g \to (f^\tau)^{-1} g f^\tau$ induces an algebraic automorphism of the group $\tilde{G}_{\mathbf{C}}$, which we denote also by f^τ. It is easy to show that the mapping $\tau \to f^\tau$ is well defined and gives a one-dimensional cocycle of the group $\mathrm{Gal}\,(\mathbf{C} : \mathbf{Q})$ with coefficients in the group $\mathrm{Aut}\,\tilde{G}_{\mathbf{C}}$. Denote by $\tilde{G}_\Lambda$ the **Q**-form of the group $\tilde{G}$ obtained by twisting with respect to the cocycle $\tau \to f^\tau$ ([22] Chapter 3, § 1).

It follows from condition $(*)$ that there is an isomorphism $\varphi_{\mathbf{C}} : (\tilde{G}_\Lambda)_{\mathbf{C}} \xrightarrow{\sim} (\tilde{G})_{\mathbf{C}}$ such that $\Lambda^0 \subset \varphi_{\mathbf{C}}(\tilde{G}_\Lambda)_{\mathbf{Q}}$. Since the group Λ^0 is, by definition, a normal subgroup in Λ, thus for any $\lambda \in \Lambda$ we have $\lambda^{-1}\Lambda^0\lambda = \Lambda^0$. Since the group Λ^0 is dense in $\tilde{G}_{\mathbf{C}}$ in the Zariski topology, the group $\varphi_{\mathbf{C}}^{-1}(\Lambda^0)$ is dense in $(\tilde{G}_\Lambda)_{\mathbf{Q}}$ in the Zariski topology. Therefore the equality $\lambda^{-1}\Lambda\lambda = \Lambda^0$ implies that $\varphi_{\mathbf{C}}^{-1}(\lambda^{-1})(\tilde{G}_\Lambda)_{\mathbf{Q}}\varphi_{\mathbf{C}}^{-1}(\lambda) = (\tilde{G}_\Lambda)_{\mathbf{Q}}$. Since $\tilde{G}_\Lambda$ is an adjoint group, we have $\varphi_{\mathbf{C}}^{-1}(\lambda) \in (G_\Lambda)_{\mathbf{Q}}$ for any $\lambda \in \Lambda$. I.e. $\Lambda \subset \varphi_{\mathbf{C}}(G_\Lambda)_{\mathbf{Q}}$.

For the deduction of Theorem 12′ from Proposition 2 it remains to verify the inclusion $\varphi_{\mathbf{C}}(\tilde{G}_\Lambda)_{\mathbf{R}} \subset \tilde{G}_{\mathbf{R}}$. The inclusion $\Lambda \subset \varphi_{\mathbf{C}}(G_\Lambda)_{\mathbf{Q}}$ implies that the closed subgroup $\varphi_{\mathbf{C}}(\tilde{G}_{\Lambda})_{\mathbf{R}} \subset \tilde{G}_{\mathbf{C}}$ contains Λ, and consequently also the identity component of the group $\tilde{G}_{\mathbf{R}}$. Therefore the restriction of $\varphi_{\mathbf{C}}$ to the subgroup $(\tilde{G}_{\Lambda})_{\mathbf{R}} \subset (\tilde{G}_{\Lambda})_{\mathbf{C}}$ induces an algebraic isomorphism $\varphi: (\tilde{G}_{\Lambda})_{\mathbf{R}} \cong \tilde{G}_{\Lambda}$. It is clear that the isomorphism φ satisfies the requirement of Theorem 12′. Thus we have shown that Theorem 12′ is a consequence of Proposition 2.

Now we pass to the proof of Proposition 2. The following two simple lemmas will prove useful.

Let T be an algebraic **C**-torus (cf. [19] § 8) and $\chi_1, \ldots, \chi_s$ a set of pairwise distinct characters $\chi_i: T \to GL_1$. We consider them as functions on T. Let $k_1, \ldots, k_s$ be a set of pairwise different integers. Consider the $l \times l$ matrix $A = (a_{ij})$, $a_{ij} = \chi_i^{k_j}$, $1 \leqq i, j \leqq s$.

Lemma 37. $\det A \neq 0$.

Proof. Consider the homomorphisms $\chi_{ij} = \chi_i \chi_j^{-1}: T \to GL_1$ and put $T_{ij} = \operatorname{Ker} \chi_{ij}$. By assumption, $\bigcup\limits_{i,j} T_{ij}$ is a proper algebraic submanifold of T, closed in the Zariski topology. Therefore there is a homomorphism $\alpha: GL_1 \to T$ of algebraic groups, such that $\operatorname{Im} \alpha \not\subset \bigcup\limits_{i,j} T_{ij}$ (cf. [19] 8.5 and 8.6). Consequently the characters $\chi_i \circ \alpha: GL_1 \to GL_1$ are pairwise distinct. Put $\chi_i' = \chi_i \circ \alpha$ and consider the matrix $A' = (a_{ij}')$, $a_{ij}' = \chi_i'^{k_j}$. It is clear that for the proof of Lemma 37 it is sufficient to show that $\det A' \neq 0$.

It is well known that every character $\chi_i': GL_1 \to GL_1$ is of the form $\chi_i'(z) = z^{t_i}$, $t_i \in \mathbf{Z}$. One can assume (possibly changing the sign of $\det A'$) that $t_1 < \ldots < t_s$ and $k_1 < \ldots < k_s$. Put $N = \sum\limits_1^s k_i t_i$. Then it is clear that the function $\det A'$ is of the form $\det A' = Z^N + A_0$, where A_0 is of the form $\sum\limits_{s_i < N} c_i z^{s_i}$. Consequently $\det A' \neq 0$. Lemma 37 is proved.

Let $l = \dim \tilde{G}_{\mathbf{C}} - \operatorname{rank} \tilde{G}_{\mathbf{C}} = \{$the number of roots in $\mathfrak{G}_{\mathbf{C}}\}$.

Denote by $\tilde{G}_{\mathbf{C}}'$ the subset of $\tilde{G}_{\mathbf{C}}$, consisting of all regular elements. For any $g \in \tilde{G}_{\mathbf{C}}'$ the operator $\operatorname{Ad} g$ in $\mathfrak{G}_{\mathbf{C}}$ has l eigenvalues $\mu_1(g), \ldots, \mu_l(g)$ different from 1, defined uniquely up to permutations.

Let $k_0 < k_1 < \ldots < k_l$ be a set of integers. For every element $g \in \tilde{G}_{\mathbf{C}}'$ write $A(k_0, \ldots, k_l)(g) = (a_{ij})$ for the matrix of order $l+1$ where $a_{0j} = 1$, $a_{ij} = \mu_i(g)^{k_j}$, $0 \leqq i, j \leqq l$. Denote by $\tilde{G}_{\mathbf{C}}'(k_0, \ldots, k_l)$ the subset of $\tilde{G}_{\mathbf{C}}'$ consisting of all elements $g \in \tilde{G}_{\mathbf{C}}'$ for which $\det A(k_0, \ldots, k_l)(g) = 0$. It is clear that $\tilde{G}_{\mathbf{C}}'(k_0, \ldots, k_l)$ is a closed algebraic manifold in $\tilde{G}_{\mathbf{C}}'$ for any set of integers $k_0 < \ldots < k_l$. Lemma 27 implies that $\tilde{G}_{\mathbf{C}}'(k_0, \ldots, k_l)$ is a proper submanifold in $\tilde{G}_{\mathbf{C}}'$. Since the group $\tilde{G}_{\mathbf{C}}$ is irreducible, we have the following assertion.

Corollary to Lemma 37. For any set of of integers $k_0 < \ldots < k_l$ we have $\dim \tilde{G}_{\mathbf{C}}'(k_0, \ldots, k_l) < \dim \tilde{G}_{\mathbf{C}}$.

Denote by R_0 the subset of R, consisting of all elements $r \in R$ such that for all integers j, $1 \leqq j \leqq l$ the element r^j is regular.

Let $\Phi(x_1, \dots, x_l)$ be a symmetric polynomial of l variables with rational coefficients.

Lemma 38. *For any element $r_0 \in R_0 \cap \Lambda$ the number $\Phi(\mu_1(r_0), \dots, \mu_l(r_0))$ is rational.*

Proof. Since $r_0 \in R_0$ thus for all j, $1 \leq j \leq l$ the element r_0^j belongs to R. Consequently, by the condition of Proposition 2, Sp Ad $(r_0^j) \in \mathbf{Q}$. This implies that all elementary symmetric functions of the numbers $\{\mu_1(r_0), \dots, \mu_l(r_0)\}$, are rational. Since the function Φ is a polynomial of the elementary symmetric functions with rational coefficients, $\Phi(\mu_1(r_0), \dots, \mu_l(r_0)) \in \mathbf{Q}$. Lemma 38 is proved.

Denote by R^n the subset of $\tilde{G}_{\mathbf{R}}$, consisting of all elements $g \in \tilde{G}_{\mathbf{R}}$ of the form $g = r_1 \dots r_n$, $r_i \in R_0$, $1 \leq i \leq n$. Put $\Lambda^n = \Lambda \cap R^n$. The proof of Proposition 2 will be carried out by induction. We write, for brevity, Sp g instead of Sp Ad (g) for $g \in \tilde{G}_{\mathbf{R}}$.

Proposition 2.n. Assume that for every $\lambda \in \Lambda^{n-1}$ the trace Sp λ is rational. Then for all $\lambda \in \Lambda^n$ the trace Sp λ is rational.

Proof of Proposition 2.n. Proposition 2.1 holds by assumption. We assume that $n > 1$. Let us fix $\lambda \in \Lambda^n$. We have, by definition, $\lambda = r_1 r_{n-1}$, where $r_1 \in R_0$, $r_{n-1} \in R^{n-1}$. As we have already shown, the set R_0 is open in $\tilde{G}_{\mathbf{R}}$ (in the ordinary topology). Therefore, so is the set $\mathbf{R}^{n-1}$. Consequently there is an open neighbourhood $U \subset R_0 \subset \tilde{G}_{\mathbf{R}}$ of the element r_1 such that $u^{-1}\lambda \in R^{n-1}$ for all $u \in U$.

Since the element r_1 is compact by the definition of the set R, there exists a sequence of natural numbers $0 < i_1 < i_2 < \dots < i_t < \dots$, such that $r_i^{i_t} \to r_1$ as $t \to \infty$. Write, as before, l for the number of roots in the Lie algebra $\mathfrak{G}_{\mathbf{C}}$. The limit relation $r_1^{i_t} \to r_1$ $(t \to \infty)$ implies that there are natural numbers $1 < k_1 < \dots < k_l$ such that $r_1^{i_j} \in U$, $1 \leq j \leq l$. Let us fix now a neighbourhood $U_1 \subset U$ of the element r_1 such that $u_1^{i_j} \subset U$ for all $u_1 \in U_1$, $1 \leq j \leq l$. Since according to the Corollary to Lemma 37 dim $\tilde{G}'_{\mathbf{C}}(1, k_1, \dots, k_l) <$ dim $\tilde{G}_{\mathbf{C}}$ and, by assumption, the group Λ is dense in the identity component G^0 of of the group $\tilde{G}_{\mathbf{R}}$ in the classical topology, there is an element λ_1 in $U_1 \cap \Lambda$ such that $\lambda_1^{-1} \notin \tilde{G}'_{\mathbf{C}}(1, k_1, \dots, k_l)$. Put $k_0 = 1$ and $\mu_0 = 1$.

Let $m =$ dim $\tilde{G}_{\mathbf{C}}$. Choose a basis $f_1, \dots, f_m$ in $\mathfrak{G}_{\mathbf{C}}$ for which Ad $\lambda_1^{-1}(f_i) = \mu_i(\lambda_1^{-1})$, $1 \leq i \leq l$, Ad $\lambda_1^{-1}(f_i) = f_i$, $l < i \leq m$. Write the matrix of Ad λ in the basis $f_1, \dots, f_m$, Ad $\lambda = (a_{\alpha\beta})$, $1 \leq \alpha$, $\beta \leq m$. Put $a_0 = \sum_{\alpha = l+1}^{m} a_{\alpha\alpha}$, $a_i = a_{ii}$, $1 \leq i \leq l$. Then Sp $\lambda = \sum_{i=0}^{l} a_i$. Since $\lambda_1 \in U_1$ we have $\lambda\lambda_1^{-k_j} \in \Lambda^{n-1}$, $0 \leq j \leq l$. Consequently Sp $\lambda\lambda_1^{-k_j}$ is a rational number (by the condition of Proposition 2.n). On the other hand, Sp $\lambda\lambda_1^{-k_j} = \sum_{i=0}^{l} a_i \mu_i^{k_j}$, where $\mu_i \overset{\text{def}}{=} \mu_i(\lambda_1^{-1})$. According to the choice of $\lambda_1 \in \tilde{G}_{\mathbf{R}}$ the determinant $|(\mu_i^{k_j})| \neq 0$, $0 \leq i, j \leq l$, thus every number a_i can be expressed in the form $\dfrac{P_i(\mu_1, \dots, \mu_l)}{|(\mu_i^{k_j})|}$, where the polynomial P_i has rational coefficients. Consequently, the number Sp $\lambda = \sum_{i=0}^{l} a_i$ can be represented in the form

$\mathrm{Sp}\,\lambda = \dfrac{Q_1(\mu_1, \dots, \mu_l)}{Q_2(\mu_1, \dots, \mu_l)}$, where Q_1 Q_2 are symmetric polynomials with rational coefficients. Therefore, by Lemma 38 $\mathrm{Sp}\,\lambda$ is a rational number. Proposition 2.n is proved.

Now we complete the proof of Proposition 2.

From the definition of the set R_0 it follows immediately that the inclusion $r_0 \in R_0$ implies $r_0^{-1} \in R_0$. Therefore the union $\bigcup_n R^n$ is a subgroup in $\tilde{G}_{\mathbf{R}}$. Since the set R_0 is open, the subgroup $\bigcup_n R^n \subset \tilde{G}_{\mathbf{R}}$ is open, and consequently $G^0 \subset \bigcup_n R^n$.

Thus Proposition 2 follows from the validity of Propositions 2.n for all n, $1 \leqq n \leqq \infty$ and consequently it is proved. Theorem 12 is proved.

Applying Theorem 12 we can reformulate conditions 7 and 8 of Theorem **B**. Denote by ${}^\sigma i$ the natural imbedding ${}^\sigma i: G^\sigma \to {}^\sigma\tilde{G}_{\mathbf{Q}}$.

Condition 7′. Let $H \subset G$ a maximal torus such that the group $H_{\mathbf{R}}$ is compact. Then there is a homomorphism of algebraic **Q**-groups $\varphi: H \to {}^\sigma G$ with a finite kernel, such that the intersection $\varphi(H_{\mathbf{Q}}) \cap {}^\sigma i(G^\sigma)$ is of finite index in the image $\varphi(H_{\mathbf{Q}})$ and the representation $\mathrm{Ad} \circ \varphi$ of the group H is equivalent to the representation Ad.

Condition 8′. Let $\gamma \in G^\sigma$ be an element such that the image ${}^\sigma i_{\mathbf{R}}(\gamma)$ is a regular compact element of the group $\tilde{G}_{\mathbf{R}}$. Denote by H^γ the (unique) maximal torus in ${}^\sigma G$ for which ${}^\sigma i(\gamma) \subset (H^\gamma)_{\mathbf{Q}}$. Then there are a maximal torus $H \subset G$ and an isogeny $\varphi: H \to H'$ such that the representation $\mathrm{Ad} \circ \varphi$ of the torus H is equivalent to the representation H.

We have proved in Theorem 12 that $i_{\mathbf{R}}(G^\sigma) \subset {}^\sigma\tilde{G}_{\mathbf{Q}}$. Now we are going to prove that by replacing the group Γ^σ by a subgroup of finite index, we can obtain the inclusion $i_{\mathbf{R}}(\Gamma^\sigma) \subset {}^\sigma\tilde{G}_{\mathbf{Z}}$.

§ 11. Study of the representations of the group $\hat{G}_{\mathbf{Q}}$

Denote by A the adele ring (see [23] no. 2) over **Q**, and by $\hat{\mathbf{Q}} \subset A$ the subring of finite adeles. As is known (see [23] no. 3), the group $G_{\hat{\mathbf{Q}}}$ is a locally compact group, isomorphic to the restricted direct product ([23] no. 3) of all the algebraic p-adic groups $G_{\mathbf{Q}_p}$ where p are prime numbers. Denote by π_p' the natural projection $\hat{\pi}_p': G_{\mathbf{Q}} \to G_{\mathbf{Q}_p}$. The diagonal imbedding $\mathbf{Q} \hookrightarrow \hat{\mathbf{Q}}$ defines a homomorphism $j: G_{\mathbf{Q}} \to G_{\hat{\mathbf{Q}}}$. Since the image $j(\Gamma)$ of the group Γ in $G_{\hat{\mathbf{Q}}}$ belongs to a profinite subgroup $G_{\hat{\mathbf{Z}}} \subset G_{\hat{\mathbf{Q}}}$, the homomorphism j extends to a continuous homomorphism $\hat{j}: \hat{G}_{\mathbf{Q}} \to G_{\hat{\mathbf{Q}}}$.

Lemma 39. *The homomorphism $\hat{j}$ is an epimorphism, and its kernel* $\mathrm{Ker}\,\hat{j}$ *is a compact normal subgroup in $\hat{G}$.*

Proof. First we establish the second assertion of the lemma. Consider the subgroup $\hat{j}^{-1}(G_{\hat{Z}}) \subset \hat{G}_{Q}$. The definition of the group $\hat{G}_Q$ implies that the group $\hat{j}^{-1}(G_{\hat{Z}})$ is isomorphic to the completion of the group G_Z in the topology of subgroups of finite index. Therefore the subgroup $\hat{j}^{-1}(G_{\hat{Z}}) \subset G_{\hat{Q}}$ is compact, and so is the kernel $\operatorname{Ker} \hat{j} = \hat{j}^{-1}(0) \subset \hat{j}^{-1}(G_{\hat{Z}})$. Therefore the mapping $\hat{j}$ is proper and consequently the image $\hat{j}(\hat{G}_Q)$ is a closed subgroup in $G_{\hat{Q}}$.

Now we prove the first assertion of the lemma. By virtue of the strong approximation theorem (cf. [7]) the image $j(G_Q)$ is dense in $G_{\hat{Q}}$, and consequently the homomorphism $\hat{j}: \hat{G}_Q \to G_{\hat{Q}}$ is an epimorphism. Lemma 39 is proved.

Denote by $\hat{\pi}: \hat{G} \to G_{\hat{Q}}$ the restriction of the homomorphism $\hat{j}$ to $\hat{G}$ and put $I = \operatorname{Ker} \hat{\pi} \subset \hat{G}$. Lemma 39 implies that the image $\hat{\pi}(\hat{G})$ is a subgroup of finite index $G'_{\hat{Q}}$ in $G_{\hat{Q}}$ and that I is a compact normal subgroup in $\hat{G}$. The homomorphism $\pi'_p \circ \pi: \hat{G} \to G_{Q_p}$ will be denoted by π_p.

In the following we will apply a useful assertion:

Lemma 40. *Let R be a semisimple simply connected algebraic p-adic group without compact factors. Then a) any open normal subgroup in R coincides with R, and b) every subgroup of finite index in R coincides with R.*

Proof. Denote by R^u the normal subgroup in R generated by the unipotent elements. Theorem B in [7] implies that $R^u = R$.

Now we prove part a). Let R^0 be an open normal subgroup in R. Any unipotent element $u \in R$ can be conjugated in such a way that it falls into an arbitrary neighbourhood of the identity of the group, thus $u \in R^0$. Consequently $R^0 \supset R^u$. Assertion a) is proved. Now let R' be a subgroup of finite index in R. Then there is a normal subgroup of finite index $R^0 \subset R$ belonging to R'. Since any subgroup of finite index in R is open, according to part a), $R^0 = R$. Consequently $R' = R$. Lemma 40 is proved.

Let G and H be two groups, $H_1, H_2 \subset H$ subgroups in H such that $[H_1, H_2] = e$. Let ϱ_1, ϱ_2 be homomorphisms of the group G into H_1, H_2. The homomorphism $\varrho: G \to H$, given by the formula $\varrho(g) = \varrho_1(g)\varrho_2(g)$, $g \in G$, will be called their *tensor product* $\varrho = \varrho_1 \otimes \varrho_2$.

Let M be an algebraic p-adic group.

Definition. The homomorphism $\varrho: \hat{G} \to M$ is *algebraic* if it is of the form $\varrho = \varrho_p \circ \pi_p$ where ϱ_p is an algebraic homomorphism $\varrho_p: G_p \to M$. (Similarly, a homomorphism $\varrho: G_{\hat{Q}} \to M$ is *algebraic* if it is of the form $\varrho = \varrho_p \circ \pi'_p$ where ϱ_p is an algebraic homomorphism.)

A homomorphism $\varrho: \hat{G} \to M_p$ is *compact*, if the image $\varrho(\hat{G})$ is compact, and finite if the image $\varrho(\hat{G})$ is finite.

Now we turn to study the set of homomorphisms ϱ of the group $\hat{G}$ into M. Consider first the case where ϱ is a homomorphism of $G_{\hat{Q}}$ into M, i.e. $\varrho(I) = e$.

Lemma 41. *Any homomorphism $\varrho: G'_{\hat{Q}} \to M$ can be represented in the form of a tensor product $\varrho = \varrho_1 \otimes \varrho_2$, where ϱ_1 is an algebraic, and ϱ_2 is a finite homomorphism.*

Proof. It is shown in the proof of Theorem 16.2 in [24] that there is an algebraic homomorphism $\varrho_1: G_Q \to M$ and a subgroup of finite index G_0 in $G'_{\hat{Q}}$, such that $\varrho(g) = \varrho_1(g)$ for $g \in G_0$. We can assume that G_0 is a normal subgroup in $G_{\hat{Q}}$. Let $L = \varrho_1(G_{\hat{Q}})$, $L_0 = \varrho_1(G_0) \subset M$. The algebraic group G is connected, and G_0 is a subgroup of finite index in $G_{\hat{Q}}$. Therefore the group G_0 is dense in $G_{\hat{Q}}$ in the Zariski topology. Since the homomorphism ϱ_1 is algebraic, the centralizer $Z(L)$ of the group L in M coincides with the centralizer $Z(L_0) \subset M$. Since ϱ and ϱ_1 are homomorphisms, and G_0 is a normal subgroup in $G_{\hat{Q}}$, we have

$$\varrho_1(g)\varrho_1(g_0)\varrho_1(g^{-1}) = \varrho_1(gg_0g^{-1}) = \varrho(gg_0g^{-1}) =$$

$$= \varrho(g)\varrho(g_0)\varrho(g^{-1}) = \varrho(g)\varrho_1(g_0)\varrho(g^{-1})$$

for any $g \in G_{\hat{Q}}$, $g_0 \in G_0$. From this it follows that for any $g \in G_{\hat{Q}}$ the element $\varrho_2(g) \overset{\text{def}}{=} \varrho_1(g)\varrho(g)$ belongs to $Z(L_0) = Z(L)$. Therefore the mapping $g \to \varrho_2(g)$ is a (finite) homomorphism of the group G_0 into M. Further $\varrho = \varrho_1 \otimes \varrho_2$. Lemma 41 is proved.

Decompose now the group G_{Q_p} into a product $G_{Q_p} = G_0^p \times G'^p$, where G_0^p is a compact p-adic group, and G'^p a semisimple algebraic group decomposing into $G'^p = \prod_{i=1}^{s_p} G_i^p$, where G_i^p are simply-connected simple algebraic noncompact p-adic groups. Put $I_p = \pi_p^{-1}(G_0^p)$. The factor group $\hat{G}/I_p$ is, by definition, a subgroup of finite index in the simply-connected semisimple algebraic group G'^p, having no compact factors. Then, by virtue of Lemma 40, the factor group $\hat{G}/I_p$ is isomorphic to G'^p.

Now let $\varrho: \hat{G} \to M$ be an arbitrary continuous homomorphism.

Lemma 42. *$\varrho(\hat{G})$ is a closed subgroup in M, and $\varrho(I_p)$ is a compact set in M.*

Proof. Denote by M' the closure of $\varrho(\hat{G})$ in M. M' is an analytic p-adic Lie group. Since I is a compact normal subgroup in $\hat{G}$, thus $\varrho(I)$ is a compact normal subgroup in M'. Let $\overline{M} = M'/\varrho(I)$ and r' be the natural projection $r': M' \to \overline{M}$. Put $\bar{\varrho} = r' \circ \varrho$. The mapping $\bar{\varrho}: \hat{G} \to \overline{M}$ is trivial on I and we will consider it as a homomorphism $\bar{\varrho}: G'_{\hat{Q}} \to \overline{M}$.

Denote by $\mathfrak{M}$ the Lie algebra of the group $\overline{M}$, and consider the homomorphism $\mathrm{Ad} \circ \bar{\varrho}: \hat{G} \to GL(\mathfrak{M})$. It follows from Lemma 41 that the homomorphism $\mathrm{Ad} \circ \bar{\varrho}$ is tensor product of an algebraic and a finite homomorphism. Consequently the image $\mathrm{Ad} \circ \bar{\varrho}(\hat{G})$ of the group $\hat{G}$ in $GL(\mathfrak{M})$ is closed, and the image $\mathrm{Ad} \circ \bar{\varrho}(I_p)$ is compact. Denote by $L \subset \overline{M}$ the kernel of the homomorphism Ad in $\overline{M}$ and by $\mathfrak{L} \subset \mathfrak{M}$ the Lie subalgebra corresponding to L in $\mathfrak{M}$.

Lemma 42.1. *The group L is finite.*

Proof. Let us fix an open compact subgroup $L_0 \subset L$ on which the mapping $\lg: L_0 \to \mathfrak{L}$ is defined. Since the mapping lg is one-to-one on its domain of definition, the group L_0 belongs to the centre of $\overline{M}$. Denote by r'' the natural projection $r'': \overline{M} \to \overline{M}/L_0$. Since L_0 is a compact subgroup containing a neighbourhood of the identity of L, the image $r''(L)$ is a discrete subgroup in $r''(\overline{M})$. Denote by $G^0_{\hat{Q}} \subset G_{\hat{Q}}$ the kernel of the homomorphism $\mathrm{Ad} \circ \bar{\varrho}$. Since the homomorphism $\mathrm{Ad} \circ \bar{\varrho}$ is the tensor product of an algebraic and a finite homomorphism, the group $G^0_{\hat{Q}}$ is a subgroup of finite index in the restricted direct product of some algebraic simply connected semisimple l-adic groups, which coincide for $l \neq p$ with the group G_{Q_l}. It is clear that the image $\bar{\varrho}(G^0_{\hat{Q}})$ of the group $G^0_{\hat{Q}}$ belongs to L, and so the image $r''\bar{\varrho}(G^0_{\hat{Q}})$ is a discrete group. Since r'' and $\bar{\varrho}$ are continuous homomorphisms, the kernel $\mathrm{Ker}\, r''\bar{\varrho}$ is an open normal subgroup in $G^0_{\hat{Q}}$. Since the group $\hat{G}^0_Q$ is a subgroup of finite index in the restricted product of simple simply connected algebraic groups, and only a finite number of them are compact, it follows easily from Lemma 40 that every open normal subgroup in the group $G^0_{\hat{Q}}$ is of finite index. Therefore the kernel $\mathrm{Ker}\, r''\bar{\varrho}$ is a subgroup of finite index in $G^0_{\hat{Q}}$. Consequently the image $r''\bar{\varrho}(\hat{G}^0)$ is finite.

By assumption $\bar{\varrho}(\hat{G}'_Q)$ is dense in $\overline{M}$. On the other hand according to Lemma 41 the homomorphism $\mathrm{Ad} \circ \bar{\varrho}: G'_Q \to GL(\mathfrak{M})$ is the product of an algebraic and a finite homomorphism. Therefore the image of the group $\hat{G}^0_Q$ in $GL(\mathfrak{M})$ is closed and dense in the image $\mathrm{Ad}\,(\overline{M})$ of $\overline{M}$ in $GL(\mathfrak{M})$. Consequently, it coincides with the image $\mathrm{Ad}\,(\overline{M})$ of the group $\overline{M}$. Since the image $\bar{\varrho}(G'_{\hat{Q}})$ is dense in $\overline{M}$, the subgroup $\bar{\varrho}(G^0_{\hat{Q}})$ is dense in L. Consequently the group $r''\bar{\varrho}(\hat{G}^0)$ is dense in the group L/L_0, and the finiteness of the image $r''\bar{\varrho}(\hat{G}^0)$ implies that the group L/L_0 is finite.

Consider the inverse image $G^1_{\hat{Q}} \overset{\text{def}}{=} \varrho^{-1}(L_0) \subset G^0_{\hat{Q}}$. Then the finiteness of the index $[L:L_0]$ implies that the group $\hat{G}^1_Q$ is a subgroup of finite index in $\hat{G}^0_Q$. Therefore this group is a subgroup of finite index in the restricted direct product of algebraic simply connected semisimple l-adic groups, coinciding for $l \neq p$ with the group $G_{\hat{Q}_l}$. Now this implies, in view of Lemma 38 and Theorem B from [7] that the factor group $G^1_{\hat{Q}}/[G^1_{\hat{Q}}, G^1_{\hat{Q}}]$ of the group $G^1_{\hat{Q}}$ by its commutator subgroup, is finite. As we have shown earlier, the group L_0 belongs to the centre of the group $\overline{M}$ and consequently it is commutative. We have shown, on the other hand, that the image $\bar{\varrho}(G^1_{\hat{Q}})$ is dense in L_0. Therefore the finiteness of the group $G^1_{\hat{Q}}/[G^1_{\hat{Q}}, G^1_{\hat{Q}}]$ implies that the group L_0 is finite. Since the factor group L/L_0 is finite, so is the group L. Lemma 42.1 is proved.

Now we complete the proof of Lemma 42. The group L is finite, the image $\bar{\varrho}(\hat{G})$ is dense in $\overline{M}$ and the image $\mathrm{Ad} \circ \bar{\varrho}(\hat{G})$ coincides with the subgroup $\mathrm{Ad}\,(\overline{M}) \subset \mathrm{Aut}\, \mathfrak{M}$, thus the image $\bar{\varrho}(\hat{G})$ is equal to $\overline{M}$. The compactness of $\mathrm{Ad} \circ \bar{\varrho}(I_p)$ follows from that of $\bar{\varrho}(I_p)$.

The group $\varrho(\hat{G}) \subset M'$ contains $\varrho(I)$ and, as shown already, the image $r'(\varrho(\hat{G})) = r'(M') = \overline{M}$. Therefore $\varrho(\hat{G}) = M'$, i.e. $\varrho(\hat{G})$ is closed in M. Now we prove the second assertion of Lemma 42. Since I is a compact normal subgroup in $\hat{G}$, so $\varrho(I)$ is a compact normal subgroup in the group M', and the compactness of $\varrho(I_p)=(r')^{-1}(I_p))$ follows from that of the group $\bar{\varrho}(I_p)$. Lemma 42 is proved.

The homomorphisms $i_{\mathbf{R}}^{\sigma}: G^{\sigma}\to G_{\mathbf{R}}$ and $\varrho\circ\hat{\imath}^{\sigma}: G^{\sigma}\to M'$ define an imbedding $i_{\mathbf{R}}^{\sigma}\times(\varrho\circ\hat{\imath}^{\sigma}): G^{\sigma}\to G_{\mathbf{R}}\times M'$.

Corollary. The image $\left(i_{\mathbf{R}}^{\sigma}\times(\varrho\circ\hat{\imath}^{\sigma})\right)(G^{\sigma})$ is dense in the group $G_{\mathbf{R}}\times M'$.

Proof. Denote by U the image $\varrho(\hat{\Gamma})\subset M'$. Lemma 42 implies that U is an open compact subgroup in M'. Denote by G_U^{σ} the inverse image

$$(\hat{\imath}^{\sigma})^{-1}\circ\varrho^{-1}\left(U\cap\varrho\circ\hat{\imath}^{\sigma}(G^{\sigma})\right)\subset G^{\sigma}.$$

By definition we have $G_U^{\sigma}\subset\Gamma^{\sigma}$, and $[G_U^{\sigma}:\Gamma^{\sigma}]=[\varrho^{-1}(U):\hat{\Gamma}]$.

Remark now that $[\varrho^{-1}(U):\hat{\Gamma}]\geqq\left[\pi\left(\varrho^{-1}(U)\right):\pi(\hat{\Gamma})\right]\geqq\left[\pi\left(\varrho^{-1}(U)\right):G_{\hat{\mathbf{Z}}}\right]\geqq[G_{\mathbf{Q}_l}:G_{\mathbf{Z}_l}]$ for any prime $l\neq p$. Since the groups $G_{\mathbf{Q}_l}$ are non-compact for almost all prime numbers l, thus $[\varrho^{-1}(U):\hat{\Gamma}]=\infty$.

Denote by $\bar{G}_U^{\sigma}$ the closure of the group $\left(i_{\mathbf{R}}^{\sigma}\times(\varrho\circ i^{\sigma})\right)(G_U^{\sigma})$ in the group $G_{\mathbf{R}}\times U$. Condition 6 of Theorem B implies that the image of the group $\bar{G}_U^{\sigma}$ under the natural projection $p_1: G_{\mathbf{R}}\times U\to G_{\mathbf{R}}$ is dense in $G_{\mathbf{R}}$. Since the group U is compact, the projection p_1 is a proper mapping, and consequently the image $p_1(\bar{G}_U^{\sigma})$ is closed. Therefore the image $p_1(\bar{G}_U^{\sigma})$ coincides with the whole group $G_{\mathbf{R}}$. The group U is an image of the group $\hat{\Gamma}$, and consequently it is profinite. On the other hand, for any natural number n the subgroup of the group $G_{\mathbf{R}}$ generated by the elements of the form g^n, $g\in G_{\mathbf{R}}$, coincides with $G_{\mathbf{R}}$. Therefore the fact that the image $p_1(\bar{G}_U^{\sigma})$ is equal to $G_{\mathbf{R}}$, implies that $\bar{G}_U^{\sigma}\supset G_{\mathbf{R}}$. Since, by definition, the image of the group $\bar{G}_U^{\sigma}$ under the projection $p_2: G_{\mathbf{R}}\times U\to U$ is dense in U, the group $\bar{G}_U^{\sigma}$ coincides with the whole group $G_{\mathbf{R}}\times U$. Now we remark that by the definition of the group M', the image $\varrho\circ\hat{\imath}^{\sigma}(G^{\sigma})$ is dense in M', therefore the fact that the image $\left(i_{\mathbf{R}}^{\sigma}\times(\varrho\circ\hat{\imath}^{\sigma})\right)(G_U^{\sigma})$ of the group G_U^{σ} is dense in $G_{\mathbf{R}}\times\bar{U}$, implies that so is the image $\left(i_{\mathbf{R}}^{\sigma}\times(\varrho\circ\hat{\imath}^{\sigma})\right)(G^{\sigma})$ in $G_{\mathbf{R}}\times M'$. The corollary is proved.

Theorem 13. *Any continuous homomorphism $\varrho:\hat{G}\to M$ of the group $\hat{G}$ into an algebraic p-adic group M can be represented in the form of a tensor product $\varrho=\varrho_1\otimes\varrho_2$ where ϱ_1 is an algebraic and ϱ_2 a compact homomorphism.*

Remark. We can assume, without loss of generality, that the image $\varrho(\hat{G})$ is dense in M in the Zariski topology.

Proof. Let ϱ be a non-compact homomorphism. Denote by M' the image $\varrho(\hat{G})$ of the group $\hat{G}$ in M and by N the image $\varrho(I_p)\subset M'$. According to Lemma 42 the subgroup $M'\subset M$ is closed, and the subgroup N is compact. Since I_p is a normal subgroup in $\hat{G}$, thus N is a normal subgroup in M'. Denote by r the natural homomorphism $r: M'\to M'/N$.

Since N is a compact p-adic group, there exists a sequence of normal subgroups of finite index $N \supset N_1 \supset \ldots \supset N_l$ such that $\bigcap N_l = e$ and each subgroup N_l is invariant under all automorphisms of the group N.*

Denote by π_l the natural projection $\pi_l : N \to N/N_l$, by ψ the homomorphism $\psi : M' \to \operatorname{Aut} N$, given by the formula $\psi(m')n = m'^{-1}nm'$, $m' \in M'$, $n \in N$ and by ψ_l the homomorphism $\psi_l : M' \to \operatorname{Aut} N/N_l$ induced by the homomorphism ψ. Put $M_0 = \operatorname{Ker} \psi$, $M_l = \operatorname{Ker} \psi_l$.

Lemma 43. *The natural homomorphism $i : M_0/M_0 \cap N \to M'/N$ is an isomorphism.*

Proof. It is clear that i is an imbedding. We verify that it is an epimorphism. Denote by i_l the imbedding $i_l : M_l/M_l \cap N \to M'/N$. Since N is compact, $\operatorname{Im} i = \bigcap_l \operatorname{Im} i_l$ and it is sufficient to show that the homomorphism i_l is surjective for any l. Since M_l is a subgroup of finite index in M', thus $\operatorname{Im} i_l$ is a subgroup of finite index in the group M'/N, and consequently the inverse image $(\varrho'')^{-1}(\operatorname{Im} i_p)$ is a subgroup of finite index in the group $\hat{G}/I_p$. The group $\hat{G}/I_p$ is, by definition, a subgroup of finite index in the group $\prod_{i=1}^{S_p} G_i^p$. By virtue of Lemma 40 the group $(\varrho'')^{-1}(\operatorname{Im} i_l)$ coincides with the group $\prod_{i=1}^{S_p} G_i^p$, and consequently, with $\hat{G}/I_p$. Therefore $\operatorname{Im} i_l = M'/N$. Lemma 43 is proved.

Denote by M_0' and $\tilde{N}$ the closure of the subgroups M_0 and N in M with respect to the Zariski topology. Since N is, by assumption, a normal subgroup of the group $Z(M_0)$ which is contained in the centralizer $Z(\tilde{M}_0)$ of the group M_0, $\tilde{N}$ is a normal subgroup of M contained in the centralizer $Z(\tilde{M}_0)$ of the subgroup $\tilde{M}_0$ in M (see [19] Chapter 1, no. 2.1). Denote by A the algebraic group being the factor group of M by $\tilde{N}$ (see [19] Chapter 1, Theorem 6.8), and by $\tilde{r}$ the natural projection $\tilde{r} : M \to A$. Since, by assumption, (see the Remark to Theorem 13) the group M' is dense in M in the Zariski topology, Lemma 43 implies that the image $\tilde{r}(M_0)$ is a dense subgroup in A with respect to the Zariski topology. Consequently, the restriction $\tilde{r}_0$ of the morphism $\tilde{r}$ to $\tilde{M}_0$ is an epimorphism of algebraic groups.

Denote by i the natural mapping $M'/N \to M/\tilde{N}$ and by ϱ' the homomorphism of the group $\hat{G}$ into A, given by the formula $\varrho' = i \circ r \circ \varrho = \tilde{r} \circ \varrho$. Since, by definition, we have $\varrho'(I_p) = e$ we can consider ϱ' as a continuous homomorphism of the algebraic semisimple simply connected p-adic group G'^p, having no compact factors, into the group A.

Lemma 44. *Let G_1 be an algebraic semisimple simply connected p-adic group without compact factors, G_2 some algebraic p-adic group and $v : G_1 \to G_2$ a continuous homomorphism of the group G_1 into G_2. Then v is an algebraic homomorphism.*

* One can set, for example, $N_l = (N)^{l!}$.

Proof. Denote by $\mathfrak{G}_i$, $i=1, 2$ the Lie algebra of the group G_i, and by $\bar{v}$ the homomorphism $\bar{v}\colon \mathfrak{G}_1 \to \mathfrak{G}_2$ of Lie algebras, induced by the homomorphism $v\colon G_1 \to G_2$ (see [25] Part II, Chapter V, § 1, Lemma 1). Then for any $g \in G_1$ and $\mathfrak{g}_1 \in \mathfrak{G}_1$ we have $\bar{v}(\mathrm{Ad}\, g(\mathfrak{g}_1)) = \mathrm{Ad}\, v(g)(\bar{v}(\mathfrak{g}_1))$*. Denote by $\mathfrak{G}_2' \subset \mathfrak{G}_2$ the image $\bar{v}(\mathfrak{G}_1)$ of the Lie algebra $\mathfrak{G}_1$ in $\mathfrak{G}_2$. Since the commutator of the Lie algebra $\mathfrak{G}_1$ coincides with itself, the commutator of the Lie algebra $\mathfrak{G}_2'$ coincides with $\mathfrak{G}_2'$. Therefore (see [19] Chapter II, Corollary 7.9) the Lie algebra $\mathfrak{G}_2' \subset \mathfrak{G}_2$ is algebraic, i.e. there is a connected algebraic subgroup $G_2' \subset G_2$ with Lie algebra $\mathfrak{G}_2'$.

Denote by Z the algebraic subgroup in G_2 consisting of all elements $g \in G_2$ for which $\mathrm{Ad}\, g(\mathfrak{g}_2') = \mathfrak{g}_2'$ for all $\mathfrak{g}_2' \in \mathfrak{G}_2'$. Since the group G_2' is connected, the group Z coincides with the centralizer $Z(G_2')$ of the group G_2' in G_2.

Consider the group $G_1 \times G_2'$ and denote its projections to G_1 and G_2' by r_1 and r_2'. Analogously denote by $\bar{r}_1$, $\bar{r}_2$ the projections of the Lie algebras $\mathfrak{G}_1 \times \mathfrak{G}_2'$ to $\mathfrak{G}_1$ and $\mathfrak{G}_2'$. Consider the subalgebra $\mathfrak{G}_0 \subset \mathfrak{G}_1 \times \mathfrak{G}_2'$ consisting of all elements of the form $(\mathfrak{g}_1, \bar{v}(\mathfrak{g}_1)) \in \mathfrak{G}_1 \times \mathfrak{G}_2$, $\mathfrak{g}_1 \in \mathfrak{G}_1$. Since the Lie algebra $\mathfrak{G}_0$ coincides with its commutator, there is a connected algebraic subgroup $G_0 \subset G_1 \times G_2$ such that its Lie algebra coincides with $\mathfrak{G}_0$. Denote by r_1' the restriction of the projection r_1 to G_0. Since the mapping r_1' induces an isomorphism $\bar{r}_1'\colon \mathfrak{G}_0 \to \mathfrak{G}_1$ of Lie algebras, the group G_0 is connected and the group G_1 is connected and simply connected, the mapping r_1' is an isomorphism. Put $v' = r_2' \circ (r_1')^{-1}\colon G_1 \to G_2$. It is clear that the algebraic homomorphism v' induces a morphism $\bar{v}$ of Lie algebras.

Consider now the mapping $v''\colon G_1 \to G_2$ given by the formula $v''(g) = v^{-1}(g) v'(g)$. Let $\mathfrak{g}_1 \in \mathfrak{G}_1$, $g \in G_1$. Then

$$\mathrm{Ad}\, v''(g)(v(\mathfrak{g}_1)) = \mathrm{Ad}\,(v(g^{-1})) \circ \mathrm{Ad}\,(v'(g))(\bar{v}(\mathfrak{g}_1)) =$$

$$= \mathrm{Ad}\,(v(g^{-1}))\bar{v}(\mathrm{Ad}\, g(\mathfrak{g}_1)) = v(\mathfrak{g}_1),$$

and consequently $v''(g) \in Z = Z(G_2')$ for all $g \in G_1$. Since, by definition, $v'(g) \in G_2'$, the mapping $v''\colon G_1 \to G_2$ is a continuous homomorphism of groups and $v = v' \otimes v''$. By construction, the mapping of Lie algebras $\bar{v}''\colon \mathfrak{G}_1 \to \mathfrak{G}_2$ corresponding to the homomorphism v'', is the zero mapping. Therefore (see [25] Part II, Chapter V, § 2, Corollary 3 of Theorem 1) the kernel $\mathrm{Ker}\, v''$ is an open normal subgroup of the group G_1. According to Lemma 40 $\mathrm{Ker}\, v'' = G_1$, and consequently $v = v'$. Lemma 44 is proved.

Denote by $\bar{r}_0$ the mapping of Lie algebras $\bar{r}_0\colon \mathfrak{M}_0 \to \mathfrak{A}$ induced by the homomorphism $\bar{r}_0\colon M_0 \to A$. Lemma 43 implies that $\bar{r}_0$ induces an isomorphism of the factor algebra $\mathfrak{M}_0/\mathfrak{N}_0$ with $\mathfrak{A}$. Since $\mathfrak{A}$ is a semisimple Lie algebra, and by definition, the Lie algebra $\mathfrak{N}_0$ belongs to the centre of $\mathfrak{M}_0$, thus, according to the theorem of Levi and Malcev (see [25] Part I, Chapter VI, § 4, Theorem 1), the Lie algebra $\mathfrak{M}_0$ is isomorphic to the direct sum $\mathfrak{M}_0 = \mathfrak{A}' \oplus \mathfrak{N}_0$ where $\mathfrak{A}' = [\mathfrak{M}_0, \mathfrak{M}_0]$

* As always, we denote by Ad the adjoint action of the group on its Lie algebra.

is the commutator of the Lie algebra $\mathfrak{M}_0$. Consequently the mapping $\bar{r}_0$ induces an isomorphism $\bar{r}_1$ of Lie algebras $\bar{r}_1\colon \mathfrak{A}' \to \mathfrak{A}$. Denote by $\mathfrak{G}'$ the Lie algebra of the group $\prod_{i=1}^{S_p} G_i^p$ and by $\bar{\varrho}'$ the homomorphism of Lie algebras $\bar{\varrho}'\colon \mathfrak{G}' \to \mathfrak{A}$ induced by the homomorphism of groups $\varrho'\colon \prod_{i=1}^{S_p} G_i^p \to A$. Define a morphism $\bar{\varrho}_0$ of the Lie algebra $\mathfrak{G}'$ into $\mathfrak{A}'$ by taking $\bar{\varrho}_0\colon (\bar{r}_1)^{-1} \circ \bar{\varrho}'\colon \mathfrak{G}' \to \mathfrak{A}'$. Since the group $\prod_{i=1}^{S_p} G_i^p$ is simply connected, there is (see the proof of Lemma 44) an algebraic homomorphism of groups $\varrho_0\colon G'^p \to A' \subset \tilde{M}_0$ inducing a morphism $\bar{\varrho}_0$ of Lie algebras. We have shown earlier that the morphism ϱ' is algebraic. Since, by definition, the algebraic morphisms ϱ' and $\tilde{p} \circ \varrho_0$ of the connected group G'^p into A define the same morphism of the Lie algebra $\mathfrak{G}'$ into $\mathfrak{A}$, they coincide.

Denote by ϱ_1 the homomorphism $\varrho_1\colon \hat{G} \to M$ given by the formula $\varrho_1 = \varrho_0 \circ \pi_p$. By definition, ϱ_1 is an algebraic homomorphism. Define the mapping $\varrho_2\colon \hat{G} \to M$ by taking $\varrho_2(g) = \varrho(g)\varrho_1(g)^{-1}$, $g \in \hat{G}$. Since the morphisms ϱ' and $\tilde{r} \circ \varrho_0'$ coincide, so do the morphisms $\tilde{r} \circ \varrho$ and $\tilde{r} \circ \varrho_1\colon \hat{G} \to A$. Consequently $\varrho_2(g) \in \operatorname{Ker} \tilde{p} = \tilde{N} \subset \subset Z(M_0)$ for all $g \in \hat{G}$. Since, by assumption $\varrho_1(g) \in \tilde{M}_0$ for all $g \in \hat{G}$, the mapping $\varrho_2\colon \hat{G} \to \tilde{M}$ is a homomorphism of groups and $\varrho = \varrho_1 \otimes \varrho_2$.

If the homomorphism $\varrho_2\colon \hat{G} \to Z(\tilde{M}_0)$ is compact, then Theorem 13 is proved. If it is not compact, then we can apply to it the same arguments as to ϱ. The group $\tilde{M}_0$ is not commutative (since the image $M_0/M_0 \cap N \cong M/N$ of the subgroup $M_0' \subset \tilde{M}$ is the image $\varrho'(G'^p)$ of the semisimple group G'^p which is not trivial by assumption). Thus the group $Z(\tilde{M}_0)$ does not coincide with M. Therefore our process of decomposing the morphism ϱ into a tensor product stops. Theorem 13 is proved.

Corollary. Let M be a simple adjoint noncompact algebraic p-adic group and $\varrho\colon \hat{G} \to M$ a continuous non-compact homomorphism such that the image $\varrho(\hat{G})$ is dense in M in the Zariski topology. Then ϱ is an algebraic homomorphism.

Proof. Let us represent the homomorphism ϱ in the form $\varrho = \varrho_1 \otimes \varrho_2$, where ϱ_1 is an algebraic and ϱ_2 a compact homomorphism. Denote by M_1 the algebraic subgroup $\varrho_1(\hat{G}) \subset M$. M_1 is a normal subgroup in $\varrho(\hat{G})$, and consequently, since $\varrho(\hat{G})$ is dense in M with respect to the Zariski topology, it is a normal subgroup also in M. As ϱ is, by assumption, a non-compact homomorphism, we have $M_1 \neq \{e\}$, and since the algebraic group M is simple, $M_1 = M$. Consequently, the image $\varrho_2(\hat{G})$ belongs to the centre of the group M. Since M is an adjoint group, $\varrho_2(\hat{G}) = e$ and $\varrho = \varrho_1$. The corollary is proved.

Let Λ be an arbitrary finite set of prime numbers, containing p. Denote by Z_Λ the subring in $\mathbf{Q}$, consisting of all rational numbers, whose denominators split into products of prime numbers contained in Λ. Denote by $\hat{Z}_\Lambda$ the closure of the subring $Z_\Lambda \subset \mathbf{Q}$ in $\hat{\mathbf{Q}}$. The group $G_{\hat{Z}_\Lambda} \subset G_{\hat{\mathbf{Q}}}$ consist of all elements $\hat{g} \in G_{\hat{\mathbf{Q}}}$ such that for any prime number $l \notin \Lambda$ the image $\pi_l'(\hat{g})$ belongs to the standard maximal compact subgroup $G_{Z_l} \subset G_{Q_l}$. Put $\hat{G}_A = \pi^{-1}(G_{\hat{Z}_A}) \subset \hat{G}$ and $G_A^\sigma = (\hat{\imath}^\sigma)^{-1}(\hat{G}_A \cap \hat{\imath}^\sigma(G^\sigma))$. By Lemma 36 the group G_A^σ can be considered as a discrete subgroup $G_A^\sigma \subset G_{\mathbf{R}} \times \prod_{p \in A} G_{\mathbf{Q}_p}$, further the factor space $G_A^\sigma \backslash G_{\mathbf{R}} \times \prod_{p \in A} G_{\mathbf{Q}_p}$ is compact.

Therefore (see [26] Theorem 2) the group G_A^σ has a finite system of generators.

Consider now the homomorphisms of the group $\hat{G}_A$ into an algebraic p-adic group. We can define, just as before, the notion of algebraic and compact homomorphisms.

Theorem 13′. *Any continuous homomorphism $\varrho: \hat{G}_A \to M$ can be represented in the form of a tensor product $\varrho = \varrho_1 \otimes \varrho_2$ where ϱ_1 is an algebraic and ϱ_2 a compact homomorphism.*

Theorem 13′ can be proved in the same way as Theorem 13.

§ 12. Some results on representations of the group G^σ

First we formulate a general result on the homomorphisms of discrete into algebraic groups.

Let K be a local (i.e. a non-discrete locally compact) field of zero characteristic, $\overline{\mathbf{Q}}$ the field of algebraic numbers.

Let R_K be a semisimple K-group. It follows from Corollary 18.8 in [19] that there is a subfield $k_0 \subset K$ which is a finite extension of the field $\mathbf{Q}$ and an algebraic k_0-group R, such that the group $R \underset{k_0}{\otimes} K$ is isomorphic to the group R_K. In the sequel these two groups will be identified.

Denote by $\mathfrak{R}$ the Lie algebra of the group R, and by Ad the adjoint representation $\mathrm{Ad}: R \to GL(\mathfrak{R})$.

Let Δ be a group with a finite number of generators $\delta_1, \ldots, \delta_n \in \Delta$ and $A_j(\delta) = \left\{ \prod_{i=1}^{N_j} \delta_{l_{i,j}}^{n_{i,j}} = e \right\}$ a complete system of relations. Denote by $S(\Delta, R)$ (or simply S) the algebraic k_0-subscheme in the direct product $R^n \overset{\text{def}}{=} \underset{(1)}{R} \times \ldots \times \underset{(n)}{R}$ given by the system of equations $A_j(r_1, \ldots, r_n) = \left\{ \prod_{i=1}^{N_j} r_{k_{i,j}}^{n_{i,j}} = e \right\}$. It is clear that the set S_K of K-points of the scheme S is naturally isomorphic to the set $\mathrm{Hom}(\Delta, R_K)$. The topology in the field K induces in S_K a structure of a locally compact topological space.

Now we give the action of the group R on R^n by the formula

$$(r_1, \ldots, r_n) r \overset{\text{def}}{=} (r^{-1} r_1 r, \ldots, r^{-1} r_n r).$$

This action preserves the ideal S and so it defines the action of the group R on S. For every point $s \in S_K$ denote by $\Omega_s \subset S$ the algebraic constructive subset of S, which is the orbit of the point $s \in S_K$ under the action of the group R, and by Ω_s^0 the subset of S_K consisting of the points $s' \in S$ of the form $s' = sr$, $r \in R_K$.

For every point $s \in S_K$ denote by the same letter the corresponding homomorphism $s: \Delta \to R_K$ and by R_s the Δ-module which is isomorphic to R_K as a group, and on which Δ acts by the homomorphism $\mathrm{Ad} \circ s: \Delta \to GL(R_K)$.

Lemma 45. *Let $s \in S_K$ be a point such that the orbit Ω_s^0 is an open set in S_K. Then there is a finite extension* k *of the field* k_0 *such that $\Omega_s^0 \cap S_{\mathsf{k}} \neq \emptyset$.*

Proof. Denote by $K_0 = K \cap \overline{\mathbf{Q}}$ the subfield in K consisting of all elements algebraic over k_0, or which is the same over $\mathbf{Q}$. Let X be an arbitrary algebraic k_0-manifold. The topology on X_K induced by the topology of the field K, will be called the *classical* topology on X_K.

Lemma 45.1. *The subset X_{K_0} is dense in X_K in the classical topology.*

Proof of Lemma 45.1. Let $x \in X_K$ and $v \subset X_K$ be an open neighbourhood of the point x in the classical topology. Denote by $X' \subset X$ the minimal closed algebraic k_0-submanifold for which $v \subset X'_K$*. Replacing the manifold X by X', we can assume that the neighbourhood v is contained in no proper k_0-submanifold in X, further it can be assumed that k_0 is an irreducible manifold. Let $d = \dim X$. According to Noether's normalization theorem (see [19] Chapter AG, 3.7) there exist a closed k_0-submanifold $X_0 \subset X$, a k_0-morphism $\varphi: X - X_0 \to A^d$ of the manifold $X - X_0$ to the affine d-dimensional space A^d, and an open k_0-subset $Z \subset A^d$ such that for any point $z \in Z_K$ the fibre $\varphi^{-1}(z)$ is not empty and consists of a finite number of points. Denote by $X_1 \subset X - X_0$ the k_0-subset consisting of the points $x \in X_K$ which satisfy at least one of the following three conditions: a) x is a non-singular point in X_0, b) $\varphi(x) \notin Z_K$, or c) the differential of the mapping φ is degenerate at the point x. It is not difficult to show that X_1 is a closed proper k_0-submanifold in $X - X_0$. Put $Y = X - (X_0 \cup X_1)$. By assumption, the set $v \cap Y_K$ is a non-empty open subset in Y_K. It follows from conditions a)—c) that the image $\varphi(v \cap Y_K)$ is an open subset in A_K^d. Since the field K_0 is dense in K in the classical topology, there exists a point $z_0 \in \varphi(v \cap Y_K) \cap A_{K_0}^d$. Let $x_0 \in Y_K$ be a point such that $z_0 = \varphi(x_0)$. By definition the fibre $\varphi^{-1}(z_0)$ consists of a finite number of points, and therefore $x_0 \in Y_{\bar{\mathsf{k}}_0}$ Consequently, $x_0 \in Y_{K_0} \subset X_{K_0}$. Since the point $x \in X_K$ and the neighbourhood v have been chosen arbitrarily, we find that the set X_{K_0} is dense in X_K in the classical topology. Lemma 45.1 is proved.

Lemma 45 follows immediately from Lemma 45.1.

Consider now a simple non-compact component F_p of the group $G_{\mathbf{Q}_p}$. As mentioned already, there is a finite extension $\mathsf{k}_0 \subset \mathbf{Q}_p$ of the field $\mathbf{Q}$ of rational numbers and a simple algebraic group F over k_0, such that the group F_p is isomorphic to the group $F_{\mathbf{Q}_p}$. In the sequel the groups F_p and $F_{\mathbf{Q}_p}$ will be identified.

* The existence of such a manifold X' follows immediately from Hilbert's theorem on the finiteness of the bases in any ideal of the ring of polynomials $\mathsf{k}_0[X_1, \ldots, X_n]$.

Let $\tilde{F}$ be the adjoint group corresponding to F and β the natural projection $\beta\colon G_{\mathbf{Q}_p} \to \tilde{F}_{\mathbf{Q}_p}$ defined over the field k_0. Put $\hat{s} = \beta \circ i_p$ and $s = \hat{s} \circ \hat{\imath}^\sigma\colon G^\sigma \to \tilde{F}_{\mathbf{Q}_p}$.

Lemma 46. *There is a finite extension* $\mathsf{k} \subset \mathbf{Q}_p$ *of the field* k_0, *and an element* $f \in \tilde{F}_{\mathbf{Q}}$ *such that* $s(G^\sigma) \subset f^{-1}\tilde{F}_{\mathsf{k}} f$.

Proof. Denote by F^0 the closure of the image $s(G^\sigma)$ in the group $\tilde{F}_{\mathbf{Q}_p}$. Theorem 5 in Chapter III of [22] implies (see also [7]) that F^0 is a subgroup of finite index in $\tilde{F}_{\mathbf{Q}_p}$. Since the group G^σ is the union of its subgroups G^σ_A, (see the end of § 11), there is a finite set A prime of numbers such that the image $s(G^\sigma_A)$ is dense in F_0.

Lemma 46.1. *There is a finite extension* $\mathsf{k} \subset \mathbf{Q}_p$ *of the field* k_0 *such that* $s(G^\sigma_A) \subset \tilde{F}_{\mathsf{k}}$.

Proof. We establish the validity of the conditions of Lemma 45 in the case when $\Delta = G^\sigma_A$ and $R = \tilde{F}$. Remark that $s(\Gamma^\sigma) \subset \tilde{F}_{\mathbf{Z}_p}$ and the image $s(G^\sigma_A)$ is, by assumption, dense in F_0. Let $s'\colon G^\sigma_A \to \tilde{F}_{\mathbf{Q}_p}$ be a homomorphism so near to s that a) $s'(\Gamma^\sigma) \subset \tilde{F}_{\mathbf{Z}_p}$ and b) the image $s'(G^\sigma_A)$ is dense in F_0. Condition a) implies that the image $s'(\Gamma^\sigma)$ belongs to a profinite group. Therefore there is a continuous homomorphism $\hat{s}'\colon \hat{G}_A \to \tilde{F}_{\mathbf{Q}_p}$ such that $s' = \hat{s}' \circ \hat{\imath}^\sigma$. Condition b) implies that the image $\hat{s}'(\hat{G}_A)$ coincides with F_0. According to Theorem 13′ the homomorphism $\hat{s}'$ can be represented in the form of a tensor product $\hat{s}' = s_1' \otimes s_2'$, where s_1' is an algebraic, and s_2' a compact homomorphism of the group $\hat{G}$ to $\tilde{F}_{\mathbf{Q}_p}$. Then the image $s_2'(\hat{G}_A)$ will be a compact normal subgroup in F_0. Since the group F_0 has no non-trivial compact normal subgroups, $\hat{s}' = s_1'$, and consequently the homomorphism $\hat{s}'\colon \hat{G}_A \to \tilde{F}_{\mathbf{Q}_p}$ is algebraic. Since there are only a finitely many inconjugate algebraic homomorphisms $\varphi\colon G_{\mathbf{Q}_p} \to \tilde{F}_{\mathbf{Q}_p}$ of the semisimple group $G_{\mathbf{Q}_p}$ into $\tilde{F}_{\mathbf{Q}_p}$, the condition that the homomorphism s' is sufficiently near to s, implies that s and s' are conjugate. Thus the condition of Lemma 45 is fulfilled. Lemma 46.1 is proved.

Now we complete the proof of Lemma 46. Let $\gamma \in G^\sigma$. The definition of the group G^σ_A implies that the intersection $G^\sigma_{\gamma, A} = \gamma^{-1} G^\sigma_A \gamma \cap G^\sigma_A$ is of finite index in the group G^σ_A. Therefore the closure of the image $s(G^\sigma_{\gamma, A})$ in $\tilde{F}_{\mathbf{Q}_p}$ is of finite index in $\tilde{F}_{\mathbf{Q}_p}$, and consequently $s(G^\sigma_{\gamma, A})$ is dense in $\tilde{F}_{\mathbf{Q}_p}$ in the Zariski topology. Since $s(G^\sigma_{\gamma, A}) \subset \tilde{F}_{\mathsf{k}}$ and $s(\gamma)^{-1}(s(G^\sigma_{\gamma, A}))s(\gamma) \subset \tilde{F}_{\mathsf{k}}$, the density of the group $s(G^\sigma_{\gamma, A})$ in $\tilde{F}_{\mathbf{Q}_p}$ with respect to the Zariski topology implies that the automorphism $g \to s(\gamma)^{-1} g s(\gamma)$ of the group $\tilde{F}_{\mathbf{Q}_p}$ is defined over k. Since $\tilde{F}$ is an adjoint group, this implies that $s(\gamma) \in \tilde{F}_{\mathsf{k}}$. Lemma 46 is proved.

Lemma 47. Ker $s = (e)$.

Proof. Put $V = \operatorname{Ker} s \subset G^\sigma$. First we establish the following auxiliary statement:

Lemma 47.1. *If* $V \cap \Gamma^\sigma = (e)$, *then* $V = (e)$.

Proof of Lemma 47.1. Let $g \in V$ and $\gamma \in g^{-1}\Gamma^\sigma g \cap \Gamma^\sigma \subset G^\sigma$. Then $g\gamma g^{-1}\gamma^{-1} \in \Gamma^\sigma$ and $s(g\gamma g^{-1}\gamma^{-1}) = e$. According to the condition of Lemma 47.1, this implies that

$g\gamma g^{-1}\gamma^{-1}=e$. Since this equality holds for all $\gamma\in g^{-1}\Gamma^\sigma g\cap\Gamma^\sigma$, thus g belongs to the centralizer of the group $g^{-1}\Gamma^\sigma g\cap\Gamma^\sigma$. Since the group $g^{-1}\Gamma^\sigma g\cap\Gamma^\sigma$ is of finite index in Γ^σ, according to the remark to condition 5 of Theorem B, the group $i^\sigma_{\mathbf{R}}(g^{-1}\Gamma^\sigma g\cap\Gamma^\sigma)$ is dense in the group $\tilde{G}_{\mathbf{R}}$ in the Zariski topology. Therefore $i^\sigma_{\mathbf{R}}(g)=e$, i.e. $g=e$. Lemma 47.1 is proved.

Now we consider the group $F_{\mathbf{Q}_p}\subset G_{\mathbf{Q}_p}$ as a subgroup of the group $G_{\hat{\mathbf{Q}}}$. Put $U=\pi(\hat{\Gamma})\subset G_{\hat{\mathbf{Q}}}$ and denote by Γ^σ_F the subgroup in G^σ given by the formula $\Gamma^\sigma_F=(\pi\circ\hat{\imath}^\sigma)^{-1}(UF_{\mathbf{Q}_p}\cap(\pi\circ\hat{\imath}^\sigma)(G^\sigma))$. It is clear that $\Gamma^\sigma\subset\Gamma^\sigma_F$. Since U, and consequently $UF_{\mathbf{Q}_p}$ also, is an open subgroup in $G_{\hat{\mathbf{Q}}}$, in view of the strong approximation theorem (cf. [7]) we get $[\Gamma^\sigma_F:\Gamma^\sigma]=\infty$. Therefore condition 6 of Theorem B implies that $i^\sigma_{\mathbf{R}}(\Gamma^\sigma_F)$ is a dense subgroup in $\tilde{G}_{\mathbf{R}}$. Put $W=V\cap\Gamma^\sigma_F$. It is clear that W is a normal subgroup in Γ^σ_F. Since, by definition, $(\beta\circ\pi_p\circ\hat{\imath}^\sigma)(W)=(e)$, we have $(\pi\circ\hat{\imath}^\sigma)(W)\subset UF_{\mathbf{Q}_p}\cap\operatorname{Ker}(\beta\circ\pi'_p)\subset U$ and consequently $W\subset\Gamma$. Consider the group $i^\sigma_{\mathbf{R}}(W)\subset\tilde{G}_{\mathbf{R}}$. On one hand, the group $i^\sigma_{\mathbf{R}}(W)$ belongs to the group $i^\sigma_{\mathbf{R}}(\Gamma)$, and consequently it is discrete. On the other hand, it is a normal subgroup of the everywhere dense subgroup $i^\sigma_{\mathbf{R}}(\Gamma^\sigma_N)\subset\tilde{G}_{\mathbf{R}}$. This implies that the group $i^\sigma_{\mathbf{R}}(W)$ centralizes the identity component of the group $\tilde{G}_{\mathbf{R}}$. Since $\tilde{G}$ is a connected adjoint algebraic group, thus $i^\sigma_{\mathbf{R}}(W)=(e)$. Consequently $W=(e)$. Now it follows from Lemma 47.1 that $V=(e)$. Lemma 47 is proved.

Now let k be a finite extension of the field $\mathbf{Q}$, M an algebraic k-group and ϱ a homomorphism $\varrho: G^\sigma\to M_k$. Since the subgroup $\Gamma^\sigma\subset G^\sigma$ has a finite number of generators, there exists a finite set Λ of simple ideals of the field such that $\varrho(\Gamma^\sigma)\subset M_{O_\Lambda}$. (Here, as always, O_Λ denotes the subring of the field k, consisting of all elements of the form α/β, where α, β are integers of the field k, and the principal ideal (β), generated by β in the ring O of integers of the field k, decomposes into a product of prime ideals contained in Λ.) For any prime ideal $\mathfrak{p}$ of the field k we write $k_{\mathfrak{p}}$ for the completion of the field k in the $\mathfrak{p}$-adic topology, $O_{\mathfrak{p}}$ for the ring of integers in $k_{\mathfrak{p}}$, $i_{\mathfrak{p}}$ for the natural imbedding $i_{\mathfrak{p}}: M_k\hookrightarrow M_{k_{\mathfrak{p}}}$ and $\varrho_{\mathfrak{p}}$ for the homomorphism $\varrho_{\mathfrak{p}}=i_{\mathfrak{p}}\circ\varrho: G^\sigma\to M_{k_{\mathfrak{p}}}$. If $\mathfrak{p}\notin\Lambda$, then, by assumption, $\varrho_{\mathfrak{p}}(\Gamma^\sigma)\subset M_{O_{\mathfrak{p}}}$. Since $M_{O_{\mathfrak{p}}}$ is a profinite group, for any $\mathfrak{p}\notin\Lambda$ the homomorphism $\varrho_{\mathfrak{p}}$ can be extended to a continuous homomorphism $\hat{\varrho}_{\mathfrak{p}}: \hat{G}\to M_{\mathfrak{p}}$.

Definition. A homomorphism $\varrho: G^\sigma\to M_k$ is called *degenerate*, if there is a finite set Λ' of prime ideals in k, $\Lambda'\supset\Lambda$, such that for all prime ideals $\mathfrak{p}\notin\Lambda'$ the homomorphism $\hat{\varrho}_{\mathfrak{p}}$ is compact.

Theorem 14. *If ϱ is a degenerate homomorphism, then* Ker $\varrho\neq e$.

Proof. Denote by $M^{k/\mathbf{Q}}$ the algebraic $\mathbf{Q}$-group obtained from the group M by the operation $R_{k/\mathbf{Q}}$ of restriction of the field. Since the groups M_k and $(M^{k/\mathbf{Q}})_{\mathbf{Q}}$ are canonically isomorphic, it is enough to analyse the case $k=\mathbf{Q}$. Since any algebraic group can be imbedded into the group of unimodular matrices $GL(n)$ for sufficiently large n (see [19] Proposition 1.10), we can assume that $M=GL(n)$.

For any abstract group: R we denote by $cd_{\mathbf{Q}}(R)$ the largest natural number i for which there is an $(R, \mathbf{Q})$-module B such that $H^i(R, B) \neq 0$. If such an i does not exist, we set $cd_{\mathbf{Q}}(R) = \infty$. Denote by G_ϱ^σ the image $\varrho(G^\sigma) \subset M_{\mathbf{Q}}$. Theorem 14 is an immediate consequence of the following two assertions:

Lemma 48. $cd_{\mathbf{Q}}(G^\sigma) = \infty$.

Lemma 49. *If ϱ is a degenerate homomorphism then* $cd_{\mathbf{Q}}(G_\varrho^\sigma) < \infty$.

Proof of Lemma 48. Let H be a maximal torus in G such that H_R is a compact Cartan subgroup in G_R. According to condition 7 of Theorem B there is a subgroup $L \subset G^\sigma$ and an imbedding $\alpha: L \to H_{\mathbf{Q}}$ such that $\alpha(L)$ is a subgroup of finite index in $H_{\mathbf{Q}}$. Since H is a unirational manifold, thus $H_{\mathbf{Q}}$ contains a free Abelian subgroup C with an infinite number of generators. Since $\alpha(L)$ is a subgroup of finite index in $H_{\mathbf{Q}}$, we can assume that $C \subset \alpha(L)$. Consider the subgroup $\alpha^{-1}(C) \subset G^\sigma$. Since α is an imbedding, thus $\alpha^{-1}(C) \cong C$, $cd(\alpha^{-1}(C)) = \infty$. Therefore (see [27] Chapter 4, Proposition 4.1) we have $cd_{\mathbf{Q}} G^\sigma = \infty$. Lemma 48 is proved.

Proof of Lemma 49. Consider $M_{\mathbf{Q}}$ as a discrete subgroup in M_A. If ϱ is a nondegenerate homomorphism, then for every $p \in \Lambda$ there is a compact subgroup K_p in $M_{\mathbf{Q}_p}$ such that $G_\varrho^\sigma \subset M_{\mathbf{R}} \times \prod_{p \in \Lambda} M_{\mathbf{Q}_p} \times \prod_{p \notin \Lambda} K_p$. Choose for some $p \notin \Lambda$ a subgroup of finite index $K_p' \subset K_p$ containing no elements of finite order except the identity. Then the subgroup $'G_\varrho^\sigma \overset{\text{def}}{=} G_\varrho^\sigma \cap (i_p)^{-1}(K_p)$ is a subgroup of finite index in G_ϱ^σ without elements of finite order.

Denote by φ the natural projection

$$\varphi: M_{\mathbf{R}} \times \prod_{p \in \Lambda} M_{\mathbf{Q}_p} \times \prod_{p \in \Lambda} K_p \to M_{\mathbf{R}} \times \prod_{p \in \Lambda} M_{\mathbf{Q}_p}.$$

Since $G_\varrho^\sigma \in M_{\mathbf{Q}}$, we have $\operatorname{Ker} \varphi \cap G_\varrho^\sigma = (e)$. Since $\prod_{p \notin \Lambda} K_p$ is a compact group, G_ϱ^σ can be considered as a discrete subgroup in $M_{\mathbf{R}} \times \prod_{p \in \Lambda} M_{\mathbf{Q}_p}$. It follows from theorem 1 in [26] that $cd_{\mathbf{Q}}{'G_\varrho^\sigma} < \infty$. Since $'G_\varrho^\sigma$ is a subgroup of finite index in G_ϱ^σ, Proposition 6.3 in Chapter 4 in [27] implies that $cd_{\mathbf{Q}} G_\varrho^\sigma \leq cd_{\mathbf{Q}}{'G_\varrho^\sigma}$. Therefore $cd_{\mathbf{Q}} G_\varrho^\sigma < \infty$. Lemma 47, and consequently Theorem 14 is proved.

Corollary. Let M be an absolutely simple adjoint $\mathbf{Q}$-group and $\varrho: G_{\mathbf{Q}} \to M_{\mathbf{Q}}$ a monomorphism of groups such that the image $\varrho(G_{\mathbf{Q}})$ is dense in M in the Zariski topology. Then the homomorphism ϱ is algebraic.

Proof. It follows from Theorem 14 that ϱ cannot be a degenerate homomorphism. Therefore there is a prime number p such that the homomorphism ϱ can be represented in the form $\varrho = \hat{\varrho}_p \circ \hat{\imath}$, where $\hat{\imath}$ is the natural imbedding $\hat{\imath}: G_{\mathbf{Q}} \to \hat{G}$, and $\hat{\varrho}_p$ a continuous homomorphism $\hat{\varrho}_p: \hat{G} \to M_{\mathbf{Q}_p}$ and the image $\hat{\varrho}_p(\hat{G}) \subset M_{\mathbf{Q}_p}$ is not compact. The conditions of the corollary imply that $M_{\mathbf{Q}_p}$ is a simple group, and the image is dense in $\hat{\varrho}_p(\hat{G})$ in the Zariski topology.

According to the Corollary of Theorem 13, the homomorphism $\hat{\varrho}_p$ is algebraic, then so is the homomorphism ϱ. The corollary is proved.

§ 13. Completion of the proof of Theorem B

Consider the subgroup $\Gamma^\sigma \hookrightarrow {}^\sigma G_{\mathbf{Q}}$. (See Theorem 12.) Since the group Γ^σ has a finite number of generators there is a finite set Λ of prime numbers such that $\Gamma^\sigma \subset {}^\sigma G_{\mathbf{Z}_\Lambda}$. Enlarging the set Λ one can achieve (cf. [28], no. 6) that for any $p \in \Lambda$ the group $G_{\mathbf{Q}_p}$ is quasi-decomposable, and consequently it has no compact factors.

Let us fix a prime number $p \notin \Lambda$. Since $p \notin \Lambda$ the imbedding ${}^\sigma i_p \colon G^\sigma \to {}^\sigma G_{\mathbf{Q}_p}$ can be continued to a homomorphism ${}^\sigma\pi_p \colon \hat{G} \to {}^\sigma\tilde{G}_{\mathbf{Q}_p}$.

Theorem 15. *For any prime number $p \notin \Lambda$ there is an isogeny $\psi \colon G_{\mathbf{Q}_p} \to {}^\sigma\tilde{G}_{\mathbf{Q}_p}$ such that ${}^\sigma\pi_p = \psi \circ \pi_p$.*

Proof. Since the group ${}^\sigma\tilde{G}_{\mathbf{Q}_p}$ is a form of the group $\tilde{G}_{\mathbf{Q}_p}$, they have the same dimension d. Set $r = \operatorname{rank}_{\mathbf{Q}_p} G_{\mathbf{Q}_p}$. We denote by M the image ${}^\sigma\pi_p(\hat{G}) \subset {}^\sigma\tilde{G}_{\mathbf{Q}_p}$.

According to Lemma 19, there is a maximal torus $H \subset G$ such that $H_{\mathbf{R}}$ is a compact Cartan subgroup and $\operatorname{rank}_{\mathbf{Q}_p} H_{\mathbf{Q}_p} = r$. According to Condition 7′ of Theorem B there are a subgroup of finite index $H_0 \subset H_{\mathbf{Q}}$ and a homomorphism $\beta \colon H_0 \to i_{\mathbf{R}}(G^\sigma)$ such that for all $h \in H_0$ the operators $\operatorname{Ad}(h)$ and $\operatorname{Ad}(\hat{\varrho}(h))$ acting in the space $\mathfrak{G}_{\mathbf{R}}$, are conjugate.

Since $\operatorname{rank}_{\mathbf{Q}_p} H_{\mathbf{Q}_p} = r$, there is a subgroup $D \subset H_{\mathbf{Q}} \subset H_{\mathbf{Q}_p}$ isomorphic to the free Abelian group with r generators such that D, as a subgroup in $G_{\mathbf{Q}_p}$, is discrete. Since H_0 is a subgroup of finite index in $H_{\mathbf{Q}}$, we can assume that $D \subset H_0$. Consider the image $\beta(D)$ as a subgroup in $M \subset {}^\sigma G_{\mathbf{Q}_p}$. The discreteness of the subgroup $D \subset H_{\mathbf{Q}_p}$ and the conjugacy of the operators $\operatorname{Ad}(d) \in GL(\mathfrak{G}_{\mathbf{Q}_p})$ and $\operatorname{Ad}(\beta(d)) \in GL({}^\sigma\mathfrak{G}_{\mathbf{Q}_p})$* imply that $\beta(D)$ is a discrete subgroup in M isomorphic to the free Abelian group with r generators.

According to Theorem 13, the homomorphism ${}^\sigma\hat{\pi}_p$ is the tensor product of an algebraic homomorphism ${}^\sigma\pi' \colon \hat{G} \to {}^\sigma\tilde{G}_{\mathbf{Q}_p}$ and a compact homomorphism ${}^\sigma\pi'' \colon \hat{G} \to {}^\sigma\tilde{G}_{\mathbf{Q}_p}$. Put $M' = {}^\sigma\pi'(\hat{G})$ and $M'' = {}^\sigma\pi''(\hat{G})$. Since ${}^\sigma\pi'$ is an algebraic homomorphism, M' is an algebraic subgroup in ${}^\sigma G_{\mathbf{Q}_p}$ and there is an algebraic epimorphism $\psi \colon G_{\mathbf{Q}_p} \to M'$ such that ${}^\sigma\pi' = \psi \circ \pi_p$. Therefore the group M' is semi-simple. Since ${}^\sigma\pi''$ is a compact homomorphism, M'' is a compact subgroup of the group ${}^\sigma\tilde{G}_{\mathbf{Q}_p}$. The intersection $M' \cap M''$ being a central normal subgroup in M', is finite. Consider now the subset $M'M''$ in ${}^\sigma G_{\mathbf{Q}_p}$ consisting of all elements

* ${}^\sigma\mathfrak{G}$ stands for the Lie algebra of the group ${}^\sigma G$.

g of the form $g=m'm''$, $m'\in M'$, $m''\in M''$. By the definition of tensor product, $[M', M'']=(e)$ thus $M'M''$ is a subgroup in ${}^{\sigma}G_{\mathbf{Q}_p}$. By assumption, we have $M\subset M'M''$. Put $\overline{M}'\stackrel{\text{def}}{=}M'/M'\cap M''$ and denote by v the natural projection $v: M'M''\to M'M''/M''\cong\overline{M}'$. Since M'' is a compact group, v is a proper mapping and the subgroup $D'\stackrel{\text{def}}{=}v\circ\beta(D)$ is a discrete subgroup in the semisimple group $\overline{M}'$. Consequently, in the semisimple p-adic group $\overline{M}'$ there exists a discrete subgroup D', isomorphic to the free Abelian group with r generators. This implies immediately that $\operatorname{rank}_{\mathbf{Q}_p} M'\geqq r$. On the other hand the mapping $\psi: G_{\mathbf{Q}_p}\to M'$ is an epimorphism by assumption, the group $G_{\mathbf{Q}_p}$ has no compact factors and $\operatorname{rank}_{\mathbf{Q}_p} G_{\mathbf{Q}_p}=r$. Therefore the morphism $\psi: G_{\mathbf{Q}_p}\to M'$ is an isogeny, and $\dim_{\mathbf{Q}_p} M'=\dim_{\mathbf{Q}_p} G_{\mathbf{Q}_p}=d=\dim_{\mathbf{Q}_p} {}^{\sigma}G_{\mathbf{Q}_p}$. Consequently, M' is a subgroup of finite index in ${}^{\sigma}\tilde{G}_{\mathbf{Q}_p}$ and the group M'' belongs to the centre of the group ${}^{\sigma}\tilde{G}_{\mathbf{Q}_p}$. Since ${}^{\sigma}\tilde{G}$ is an adjoint group, $M''=(e)$. Theorem 15 is proved.

Corollary. Denote by $\tilde{G}$ the adjoint algebraic group corresponding to the simply connected group G, and by t the natural projection $t: G\to\tilde{G}$. Then for every prime number $p\notin A$ there is an algebraic isomorphism of $\mathbf{Q}_p$-groups $\psi: {}^{\sigma}\tilde{G}_{\mathbf{Q}_p}\to\tilde{G}_{\mathbf{Q}_p}$ such that for all $\gamma\in G^{\sigma}$ we have $\psi\circ{}^{\sigma}i_p(\gamma)=t\circ\pi_p\circ\hat{\imath}^{\sigma}(\gamma)$.

Now let p be an arbitrary prime number, $\tilde{F}_{\mathbf{Q}_p}$ a simple non-compact component in the group $\tilde{G}_{\mathbf{Q}_p}$, β the natural projection $\beta: G_{\mathbf{Q}_p}\to\tilde{F}_{\mathbf{Q}_p}$. Put $s=\beta\circ\pi_p\circ\hat{\imath}^{\sigma}: G^{\sigma}\to\tilde{F}_{\mathbf{Q}_p}$.

Theorem 16. *There exists a homomorphism of algebraic groups* $\psi: {}^{\sigma}\tilde{G}_{\mathbf{Q}_p}\to\tilde{F}_{\mathbf{Q}_p}$ *such that for all* $\gamma\in G^{\sigma}$ *we have* $\psi\circ{}^{\sigma}i_p(\gamma)=s(\gamma)$.

Proof. Since $\tilde{F}_{\mathbf{Q}_p}$ is a simple $\mathbf{Q}_p$-group, there exists (cf. [27] Chapter X, no. 1.3) a finite extension $k_{\mathfrak{p}}$ of the field $\mathbf{Q}_p$ and an absolutely simple $k_{\mathfrak{p}}$-group E such that $\tilde{F}=R_{k_{\mathfrak{p}}/\mathbf{Q}_p}$. The group $\tilde{F}_{\mathbf{Q}_p}$ is canonically isomorphic to the group $E_{k_{\mathfrak{p}}}$, and we will identify these groups. Since the group E is simple, it can be defined over some finite extension $k\subset k_{\mathfrak{p}}$ of the field $\mathbf{Q}$. Lemma 45 implies the existence of a finite extension $k'\subset k_{\mathfrak{p}}$ of the field k and an element $e_0\in E_{k_{\mathfrak{p}}}$ such that $e_0^{-1}s(G^{\sigma})e_0\subset E_k$. Replacing the field k by k' and the imbedding $\gamma\to s(\gamma)\subset E_{k_{\mathfrak{p}}'}$, $\gamma\in G^{\sigma}$ by $\gamma\to e_0^{-1}s(\gamma)e_0$, we can assume that $s(G^{\sigma})\subset E_k$. According to Lemma 46 the mapping $s: G^{\sigma}\to E_k$ is a monomorphism. Therefore, according to Theorem 14, the homomorphism $s: G^{\sigma}\to E_k$ is nondegenerate, and consequently there is an infinite set of prime ideals $\mathfrak{g}$ of the field k such that the induced homomorphism $s_{\mathfrak{g}}: G^{\sigma}\to E_{k_{\mathfrak{q}}}$ extends to a continuous homomorphism $\hat{s}_{\mathfrak{g}}: \hat{G}\to E_{k_{\mathfrak{g}}}$ which is not compact.

Let us fix a prime ideal $\mathfrak{g}$ of the field k having this property and such that the group $G_{\mathbf{Q}_{\mathfrak{q}}}$ has no compact factors, where $\mathfrak{q}=N_{k/\mathbf{Q}}\mathfrak{g}$.

Since, by definition, the closure of the image $\pi_p\circ\hat{\imath}^{\sigma}(G^{\sigma})\subset G_{\mathbf{Q}_p}$ in the classical topology is a subgroup of finite index in $G_{\mathbf{Q}_p}$, the image $\pi_p\circ\hat{\imath}^{\sigma}(G^{\sigma})$ is dense in $G_{\mathbf{Q}_p}$ in the Zariski topology. Therefore the image $s(G^{\sigma})\subset E_k$ is dense in E_k in the Zariski topology, and consequently the image $\hat{s}_{\mathfrak{g}}(\hat{G})$ is dense in $E_{k_{\mathfrak{g}}}$ in the Zariski topology. Since, by the condition imposed on $\mathfrak{g}$, the homomorphism $\hat{s}_{\mathfrak{g}}$ is not compact, according to the Corollary to Theorem 13, it is algebraic, i.e. there exists an algebraic homomorphism $s^{\mathfrak{g}}\colon G_{\mathbf{Q}_{\mathfrak{q}}}\to E_{\mathbf{Q}_{\mathfrak{q}}}$ such that $\hat{s}_{\mathfrak{g}}=s_{\mathfrak{g}}\circ\pi_{\mathfrak{g}}$. Hence, by Theorem 15 there exists an algebraic homomorphism $\varphi\colon {}^{\sigma}\tilde{G}_{\mathbf{Q}_{\mathfrak{q}}}\to E_{k_{\mathfrak{g}}}$ such that for all $\gamma\in G^{\sigma}$ we have $\varphi\circ i_{\mathfrak{g}}(\gamma)=s_{\mathfrak{g}}(\gamma)$. Since $G^{\sigma}\subset{}^{\sigma}G_{\mathbf{Q}}$ the image of ${}^{\sigma}i_{\mathfrak{g}}(G^{\sigma})\subset{}^{\sigma}\hat{G}_{\mathbf{Q}}$ is dense in ${}^{\sigma}\tilde{G}_{\mathbf{Q}}$ in the Zariski topology and $s_{\mathfrak{g}}(G^{\sigma})\subset E_k\subset E_{k_{\mathfrak{g}}}$ thus the algebraic homomorphism φ is defined over the field k. Since the field k is contained in the field $k_{\mathfrak{g}}$, the homomorphism φ defines an algebraic homomorphism $\psi\colon{}^{\sigma}\tilde{G}_{\mathbf{Q}_p}\to E_{k_p}$. This construction implies that for any $\gamma\in G^{\sigma}$ we have the equality $\psi\circ{}^{\sigma}i_p(\gamma)=s(\gamma)$. Theorem 16 is proved.

Decompose the group $G_{\mathbf{Q}_p}$ to the product $G_{\mathbf{Q}_p}=G'\times G^{\circ}$, where G' is a semisimple p-adic group without compact factors, and G° is a compact group. Denote by β' the projection of the group $G_{\mathbf{Q}_p}$ to G'. Since the group G' decomposes into the direct product of simple non-compact p-adic groups, as a consequence of Theorem 16 there is an algebraic homomorphism $\psi\colon{}^{\sigma}\tilde{G}_{\mathbf{Q}_p}\to G'$ such that for every $\gamma\in G^{\sigma}$ we have the equality

$$\psi\circ{}^{\sigma}i_p(\gamma)=\beta'\circ\pi_p\circ\hat{\imath}^{\sigma}(\gamma). \qquad (*)$$

Since the closure of the group $\pi_p\circ\hat{\imath}^{\sigma}(G^{\sigma})$ in the group $G_{\mathbf{Q}_p}$ is a subgroup of finite index in the group $G_{\mathbf{Q}_p}$, the mapping ψ is an epimorphism. Denote by ${}^{\sigma}G^{\circ}$ the kernel of the homomorphism $\psi\colon{}^{\sigma}\tilde{G}_{\mathbf{Q}_p}\to G'$.

Since the algebraic group ${}^{\sigma}\tilde{G}_{\mathbf{Q}_p}$ is semisimple, it decomposes into a direct product ${}^{\sigma}\tilde{G}_{\mathbf{Q}_p}={}^{\sigma}G'\times{}^{\sigma}G^{\circ}$, further the homomorphism ψ provides an isomorphism $\psi\colon{}^{\sigma}G'\to G'$. Let ${}^{\sigma}\pi'$, ${}^{\sigma}\pi^{\circ}$ be projections of the group ${}^{\sigma}\tilde{G}_{\mathbf{Q}_p}$ on the groups ${}^{\sigma}G'$ and ${}^{\sigma}G^{\circ}$, respectively.

Lemma 50. *The image $({}^{\sigma}\pi'\circ{}^{\sigma}i_p)(\Gamma^{\sigma})$ is relatively compact in the group ${}^{\sigma}G'$ and the closure $\bar{\mathrm{G}}$ of the image $({}^{\sigma}\pi'\circ{}^{\sigma}i_p)(G^{\sigma})$ in ${}^{\sigma}G'$ is a subgroup of finite index in ${}^{\sigma}G'$.*

Proof. Since the mapping ψ is an isomorphism of the groups ${}^{\sigma}G'$ and G', it is sufficient to show that the image $(\psi\circ{}^{\sigma}\pi'\circ{}^{\sigma}i_p)(\Gamma^{\sigma})$ is relatively compact, and the image $(\psi\circ{}^{\sigma}\pi'\circ{}^{\sigma}i_p)(G^{\sigma})$ is dense in the group G'. But equality $(*)$ implies that these images equal $(\beta'\circ\pi_p\circ\hat{\imath}^{\sigma})(\Gamma^{\sigma})$ and $(\beta'\circ\pi_p\circ\hat{\imath}^{\sigma})(G^{\sigma})$, respectively. Consequently the image $(\psi\circ{}^{\sigma}\pi'\circ{}^{\sigma}i_p)(\Gamma^{\sigma})$ belongs to the compact group $(\beta'\circ\pi_p)(\hat{\Gamma})$ and the closure of the image $(\psi\circ{}^{\sigma}\pi'\circ{}^{\sigma}i_p)(G^{\sigma})$ is a subgroup of finite index in G'. Lemma 40 implies that the image $(\psi\circ{}^{\sigma}\pi_p\circ{}^{\sigma}i_p)(G^{\sigma})$ is dense in G'. Lemma 50 is proved.

Put $r=\operatorname{rank}_{\mathbf{Q}_p}G_{\mathbf{Q}_p}$, ${}^{\sigma}r=\operatorname{rank}_{\mathbf{Q}_p}{}^{\sigma}G_{\mathbf{Q}_p}$.

Theorem 17. ${}^{\sigma}r\leqq r$.

Proof. Denote by M the closure of the image $({}^\sigma\pi^\circ \circ {}^\sigma i_p)(G^\sigma)$ of the group G^σ in ${}^\sigma G^\circ$, by N the closure of the image

$$(\hat{\imath}_{\mathbf{R}} \times {}^\sigma i_p)(G^\sigma) \subset \tilde{G}_{\mathbf{R}} \times \bar{G} \times M \subset \tilde{G}_{\mathbf{R}} \times {}^\sigma G' \times M \subset \tilde{G}_{\mathbf{R}} \times {}^\sigma G_{\mathbf{Q}_p}$$

of G^σ in the group $\tilde{G}_{\mathbf{R}} \times {}^\sigma G_{\mathbf{Q}_p}$. Put $N' = N \cap \tilde{G}_{\mathbf{R}} \times {}^\sigma G'$.

Lemma 51. *There is a maximal torus T in the group $\tilde{G}_{\mathbf{R}} \times {}^\sigma G_{\mathbf{Q}_p}$ such that the intersection $T \cap N$ is of finite index in T.*

Proof. According to Lemma 19 there exists a maximal torus H in the group G such that the group $H_{\mathbf{R}}$ is compact. Let $\varphi\colon H \to {}^\sigma\tilde{G}$ be a homomorphism satisfying condition 7′ of Theorem B. Put $T = \varphi(H_{\mathbf{R}} \times H_{\mathbf{Q}_p}) \subset G_{\mathbf{R}} \times {}^\sigma G_{\mathbf{Q}_p}$. Since H is a unirational algebraic manifold (cf. [19] Theorem 18.2), the closure of the image of the group $H_{\mathbf{Q}}$ under the natural imbedding $H_{\mathbf{Q}} \hookrightarrow H_{\mathbf{R}} \times H_{\mathbf{Q}_p}$, is a subgroup of finite index in the group $H_{\mathbf{R}} \times H_{\mathbf{Q}_p}$. Therefore the closure of the image $(\hat{\imath}_{\mathbf{R}} \times {}^\sigma i_p)(\varphi(H_{\mathbf{Q}}))$ of the group $\varphi(H_{\mathbf{Q}})$ in the group $\tilde{G}_{\mathbf{R}} \times {}^\sigma \tilde{G}_{\mathbf{Q}_p}$ contains a subgroup of finite index in the maximal torus T. Since the intersection $\varphi(H_{\mathbf{Q}}) \cap G^\sigma$ is of finite index in the group $\varphi(H_{\mathbf{Q}})$, the intersection $\varphi(H_{\mathbf{Q}}) \cap G^\sigma$ is of finite index in T. Lemma 51 is proved.

Corollary. Put $T' = T \cap \tilde{G}_{\mathbf{R}} \times G'$, $T^\circ = T \cap {}^\sigma G^\circ$. Then the intersection $T' \cap N'$ is of finite index in T', and the intersection $T^\circ \cap M$ is of finite index in T°.

Lemma 52. *The group N is a subgroup of finite index in the group $\tilde{G}_{\mathbf{R}} \times {}^\sigma G' \times M$.*

Proof. Since, according to Lemma 50, the image ${}^\sigma\pi' \circ {}^\sigma i_p(\Gamma^\sigma) \subset {}^\sigma G'$ is relatively compact, and consequently it is contained in a profinite subgroup $K \subset {}^\sigma G'$, there exists a continuous homomorphism $\hat{\varrho}\colon \hat{G} \to {}^\sigma G'$ such that $\hat{\varrho} \circ \hat{\imath}^\sigma = {}^\sigma\pi' \circ {}^\sigma i_p$. Since, by definition, the image ${}^\sigma\pi' \circ {}^\sigma i_p(G^\sigma)$ is dense in $\bar{G}$, according to Lemma 42, the image $(i^\sigma_{\mathbf{R}} \times {}^\sigma\pi' \circ {}^\sigma i_p)(G^\sigma)$ is dense in the group $\tilde{G}_{\mathbf{R}} \times \bar{G}$. Consequently, N' is a normal subgroup in the group $\tilde{G}_{\mathbf{R}} \times \bar{G}$. According to Lemma 50, $\bar{G}$ is a subgroup of finite index in the group ${}^\sigma G'$, and according to the Corollary of Lemma 51 N' contains a subgroup of finite index in the maximal torus $T_0 \subset \tilde{G}_{\mathbf{R}} \times G'$, hence N' is a subgroup of finite index in the group $G_{\mathbf{R}} \times {}^\sigma G'$. Since, by definition, the image ${}^\sigma\pi^\circ(N)$ is dense in M, N is a subgroup of finite index in the group $\tilde{G}_{\mathbf{R}} \times {}^\sigma G' \times M$. Lemma 52 is proved.

Lemma 53. *The group M is an open subgroup in the group ${}^\sigma G^\circ$.*

Proof. Denote by ${}^\sigma\mathfrak{G}^\circ$ the Lie algebra of the group ${}^\sigma G^\circ$ and by $\mathfrak{M}$ its subalgebra corresponding to the group M (cf. [25] Part II, Chapter V, § 1). Since the group ${}^\sigma\hat{\imath}_{\mathbf{R}}(G^\sigma)$ is dense (cf. Condition 5 of Theorem B) in ${}^\sigma G_{\mathbf{R}}$ in the Zariski topology, the group ${}^\sigma i_p(G^\sigma)$ is dense in ${}^\sigma G_{\mathbf{Q}_{p'}}$, and consequently the group $M \supset {}^\sigma\pi \circ (G^\sigma)$ is dense in ${}^\sigma G^\circ$ in the Zariski topology. Since the normalizer of the Lie subalgebra $\mathfrak{M} \subset {}^\sigma\mathfrak{G}^\circ$ in the group ${}^\sigma G^\circ$ is an algebraic subgroup containing

M, it coincides with the whole group ${}^\sigma G^\circ$. Consequently, the Lie subalgebra $\mathfrak{M}$ is normal subalgebra in the Lie algebra ${}^\sigma\mathfrak{G}^\circ$. On the other hand, Lemma 51 implies that the group M contains the subgroup $M\cap T^\circ$ of finite index in the maximal torus T° of the group ${}^\sigma G''$. Therefore the Lie algebra $\mathfrak{M}$ contains the Cartan subalgebra $\mathfrak{T}\subset{}^\sigma\mathfrak{G}''$ corresponding to the torus T°. Consequently $\mathfrak{M}={}^\sigma\mathfrak{G}^\circ$ and M is an open subgroup of the group ${}^\sigma G^\circ$. Lemma 53 is proved.

Now we complete the proof of Theorem 17. Let T_p be a torus in ${}^\sigma G_{\mathbf{Q}_p}$ such that $\operatorname{rank}_{\mathbf{Q}_p} T_p={}^\sigma r$. Since, according to Lemma 53, the group ${}^\sigma G'\times M$ contains a neighbourhood of the identity of the group ${}^\sigma G_{\mathbf{Q}_p}$, there exists a regular element $t\in T_p\cap{}^\sigma G'\times M$. Lemma 20 in [14] implies the existence of an open neighbourhood U of the element t in the group ${}^\sigma G_{\mathbf{Q}_p}$ such that any element $u\in U$ is regular and conjugate inside the group ${}^\sigma G_{\mathbf{Q}_p}$ to some element of the torus T_p. Write, as before, R for the open subset of the group $\tilde{G}_{\mathbf{R}}$ consisting of all regular compact elements. In consequence of Lemma 53 the intersection $(i^\sigma_{\mathbf{R}}\times{}^\sigma i_p)(G^\sigma)\cap R\times U$ is not empty. Let $\gamma\in G^\sigma$ be an element for which $i^\sigma_{\mathbf{R}}(\gamma)\in R$, $\sigma_{i_p}(\gamma)\in U$. Since $\sigma_i(\gamma)\in{}^\sigma G_{\mathbf{Q}}$ is a regular element, there is a unique maximal torus $H^\gamma\subset{}^\sigma G$ defined over $\mathbf{Q}$, for which $\gamma\in H^\gamma$. The inclusion ${}^\sigma i_p(\gamma)\in U$ implies that the torus $(H^\gamma)_{\mathbf{Q}_p}$ is isomorphic to T_p. Consequently $\operatorname{rank}_{\mathbf{Q}_p} H^\gamma={}^\sigma r$. On the other hand, condition 8′ of Theorem B (see § 10) implies the existence of a $\mathbf{Q}$-torus $H\subset G$ isogeneous to H^γ. Therefore $\operatorname{rank}_{\mathbf{Q}_p} H=\operatorname{rank}_{\mathbf{Q}_p} H'={}^\sigma r$, and consequently ${}^\sigma r=\operatorname{rank}_{\mathbf{Q}_p} H\leqq\operatorname{rank}_{\mathbf{Q}_p} G=r$. Theorem 17 is proved.

Corollary. One has the equality ${}^\sigma r=r$, and the group ${}^\sigma G^\circ$ is compact.

Proof. Since the group G° is compact, we have $\operatorname{rank}_{\mathbf{Q}_p} G'=\operatorname{rank}_{\mathbf{Q}_p}\tilde{G}_{\mathbf{Q}}=r$. On the other hand, the group ${}^\sigma G'$ is isomorphic to the group G'. Therefore $r^\sigma=\operatorname{rank}_{\mathbf{Q}_p}{}^\sigma G=\operatorname{rank}_{\mathbf{Q}_p}{}^\sigma G'+\operatorname{rank}_{\mathbf{Q}_p}{}^\sigma G''=\operatorname{rank}_{\mathbf{Q}_p} G'+\operatorname{rank}_{\mathbf{Q}_p}{}^\sigma G''=r+\operatorname{rank}_{\mathbf{Q}_p}{}^\sigma G''$. It follows from Theorem 17 that ${}^\sigma r=r$ and $\operatorname{rank}_{\mathbf{Q}_p}{}^\sigma G''=0$. Now Godement's theorem (cf. [11] Corollary 9.4) implies that the group ${}^\sigma G^\circ_{\mathbf{Q}_p}$ is compact. The corollary is proved.

Now we complete the proof of Theorem B. Denote by $\bar{i}$ the natural mapping $\bar{i}:{}^\sigma G_{\mathbf{Q}}\to{}^\sigma G_{\hat{\mathbf{Q}}}$, by K the closure of the image $\bar{i}({}^\sigma G_{\mathbf{Z}})$ of the arithmetic subgroup ${}^\sigma G_{\mathbf{Z}}$ in the group ${}^\sigma G_{\hat{\mathbf{Q}}}$. The subgroup $K\subset G_{\hat{\mathbf{Q}}}$ is known to have the following properties: a) it is compact, b) the inverse image $(\bar{i})^{-1}(K\cap\bar{i}({}^\sigma G_{\hat{\mathbf{Q}}}))$ is an arithmetic subgroup, c) for any compact subgroup $K'\subset{}^\sigma G_{\hat{\mathbf{Q}}}$ the intersection $K\cap K'$ is a subgroup of finite index in K'. Consider now the closure K^σ of the image $\bar{i}(\Gamma^\sigma)\subset\subset i({}^\sigma G_{\mathbf{Q}})$ in the group ${}^\sigma G_{\hat{\mathbf{Q}}}$. Theorem 17 implies that the group K^σ is compact. Consider the subgroup $\Gamma_0\overset{\text{def}}{=}(\bar{i})^{-1}(K\cap K^\sigma\cap\bar{i}({}^\sigma G_{\mathbf{Q}}))\subset{}^\sigma G_{\mathbf{Q}}$. Property c) of the subgroup K implies that the intersection $\Gamma_0^\sigma\overset{\text{def}}{=}\Gamma_0\cap\Gamma^\sigma\subset{}^\sigma G_{\mathbf{Q}}$ is a subgroup of finite index in the group Γ^σ. Therefore the factor space ${}^\sigma G_{\mathbf{R}}/\Gamma_0^\sigma$ is compact. On the other hand, $\Gamma_0^\sigma\subset\Gamma_0\subset G_{\mathbf{Z}}$ where $G_{\mathbf{Z}}$ is, by property b) of the group K, an arithmetic group. Since the group $G_{\mathbf{Z}}$ is discrete in ${}^\sigma G_{\mathbf{R}}$, thus it follows from the compactness of the factor space ${}^{\sigma}G_{\mathbf{R}}/\Gamma_0^{\sigma}$ that the group Γ_0^{σ} is a subgroup of finite index in $G_{\mathbf{Z}}$. Consequently, Γ_0^{σ} is an arithmetic group. As we observed, the group Γ_0^{σ} is a subgroup of finite index in the group Γ^{σ}, thus Γ^{σ} is an arithmetic group. Theorem B is proved.

REFERENCES

[1] Borel, A., Reduction theory for arithmetic groups, in *Algebraic groups and discontinuous subgroups,* Providence, 1966. (Russian translation in *Арифметические группы и автоморфные функции*, М. „Мир" 1969.)

[2] Šafarevič, I. R. (Шафаревич И. Р.), *Основы алгебраической геометрии,* М. „Наука" 1972.

[3] Serre, J.-P., Géométrie algébrique et géométrie analytique, *Ann. Inst. Fourier,* **6** (1956), 1—42.

[4] Shimura, G., Algebraic varieties without deformation and Chow variety, *J. Math. Soc. Japan,* **20** (1968), 336—341.

[5] *Revêtements étalés et groupe fondamental (SGA* 1*)*, Dirigé par A. Grothendieck, Springer Lectures Notes, Vol. **224**, 1971.

[6] Sternberg, S., *Lectures on differential geometry,* Prentice Hall, Englewood Cliffs 1964. (Стернберг С., *Лекции по дифференциальной геометри*, М. „Мир" 1970.)

[7] Platonov, V. P. (Платонов В. П.), Проблема сильной аппроксимации и гипотеза Кнезера-Титса для алгебраических групп, *Изв. АН СССР,* **33** (1969), 1211—1220; Дополнение к работе "Проблема сильной аппроксимации и гипотеза Кнезера-Титса для алгебраических групп", *Изв. АН СССР*, **34** (1970), 775—777.
(The problem of strong approximation and the Kneser-Tits conjecture for algebraic groups.)

[8] Helgason, S., *Differential geometry and symmetric spaces,* Academic Press 1962. (Хелгасон С., *Дифференциальная геометрия и симметрические пространства*, М. „Мир" 1964.)

[9] Weil, A., *Introduction à l'étude des variétés kähleriennes,* Hermann, Paris 1958. (Вейль А., *Введение в теорию кэлеровых многообразий*, М. Изд. иностр. лит. 1961.)

[10] Chern, S., *Complex manifolds,* Univ. of Chicago, 1955/56. (Чжень Шен-шень, *Комплексные многообразия*, М. Изд. иностр. лит. 1961.)

[11] Borel, A. and Tits, J., Groupes réductifs, *Publ. Math. I.H.E.S.,* **27** (1965). (*Математика, сб. перев.*, 1967, **11**:1, 43—111, **11**:23—31.)

[12] Hörmander, L., L^2-estimates and existence theorems for the $\bar{\partial}$ operator, *Acta Math.,* **113** (1965), 89—152. (*Математика, сб. перев.*, **10**:2(1966), 59—116.)

[13] Bochner, S. and Martin, W. T., *Several complex variables,* Princeton, 1948. (Бохнер С., Мартин У., *Функции многих комплексных перетенных*, М. Изд. иностр. лит. 1951.)

[14] Harish-Chandra and van Dijk, G., *Harmonic analysis on reductive p-adic groups,* Springer Lecture Notes, Vol. **162**, 1970.

[15] Rosenlicht, M., Some rationality questions on algebraic groups, *Annali di Mat.* (IV), **43** (1957), 25—50.

[16] Pjateckij-Šapiro, I. I. (Pyatetskii-Shapiro, I.), *Automorphic functions and the geometry of classical domains*, Mathematics and its applications Vol. 8, Gordon and Breach, Science Publishers, 1969.

[17] Hano, J., On Kaehlerian homogeneous spaces of unimodular Lie groups, *Amer. J. Math.,* **79** (1957), 885—900.

[18] Borel, A., Kählerian coset spaces of semisimple Lie groups, *Proc. Nat. Acad. Sci. USA,* **40** (1954), 1147—1151.

[19] Borel, A. (Notes by Bass, H.), *Linear algebraic groups,* Benjamin, 1969. (Борель А., *Линейные алгебраические группы*, М. „Мир" 1972.)

[20] Wang, H.-Ch., Homogeneous complex structure, *Amer. J. of Math.*, **76** (1954), 1—32.

[21] Griffiths, Ph. A. and Schmid, W., Locally homogeneous complex manifolds, *Acta Math.*, **123** (1969), 253—302. (*Усп. мат. наук*, **26**:5(1971).)

[22] Serre, J.-P., *Cohomologie Galoisienne*, Springer Lecture Notes, Vol. **5**, 1965.

[23] Tamagawa, T., Adeles, *in Algebraic groups and discontinuous subgroups,* Providence, 1966. (*Арифметические группы и автоморфные функции*, M. „Мир" 1969.)

[24] Bass, H., Milnor, J. and Serre, J.-P., Solution of the congruence subgroup problem for SL_n $(n \geqq 3)$ and Sp_{2n} $(n \geqq 2)$, *Publ. Math. I.H.E.S.*, **33** (1967). (*Математика Сб. перев.*, **14**:6 (1970), 64—128.)

[25] Serre, J.-P., *Lie algebras and Lie groups*, New-York—Amsterdam, Benjamin, 1966. (Серр Ж.-П, *Алгебры Ли и группы Ли*, M. „Мир" 1969.)

[26] Serre, J.-P., Cohomologie des groupes discrets, *C. R. Acad. Sci. Paris,* **268**, 268—271.

[27] *Algebraic number theory* (Edited by Cassels, J. W. S. and Fröhlich, A.), Proceedings of an instructional conference organized by the London Mathematical Society, Academic Press, 1967. (*Алгебраическая теория чисел*, „Мир" 1969.)

[28] Langlands, R. P., Problems in the theory of automorphic forms, *preprint*.

[29] Kajdan, D. A., Arithmetic varieties and their fields of quasi-definition, *in Actes, Congrès intern. math., 1970*. Vol. **2**, pp. 321—325.

THE METHOD OF ORBITS IN REPRESENTATION THEORY

A. A. KIRILLOV

Let G be a Lie group, $\mathfrak{g}$ its Lie algebra, $\mathfrak{g}^*$ the dual space of $\mathfrak{g}$. The group G acts on $\mathfrak{g}^*$ (the so-called coadjoint representation) and we denote by $O(G)$ the family of G-orbits in $\mathfrak{g}^*$. The method of orbits is based on the following "experimental fact": the set $\hat{G}$ of equivalence classes of irreducible unitary representations of G is strongly connected with the set $O(G)$ of G-orbits in $\mathfrak{g}^*$. All classical problems of representation theory (classification and construction of representations, and calculation of their generalized and infinitesimal characters, decomposition of a representation restricted to or induced from a subgroup, decomposition of tensor product) find intuitive and convenient interpretation in terms of orbits.

The method of orbits was first proposed by the author for nilpotent and completely solvable groups in [1]; the possibility of generalizing this method to other groups was also mentioned there. Further progress has been achieved in the works of B. Kostant [2], [3], B. Kostant and L. Auslander [4], P. Bernat [5], L. Pukánszky [6], [7], M. Duflo [8], [9], [10], M. Vergne [11], [12] and the author [13], [14], [15], [16], [17].

A complete construction of representation theory of Lie groups based on the method of orbits has not yet been accomplished and the present note is a survey of results obtained in this direction.

§ 1. Orbits of G in $\mathfrak{g}^*$

Recall the definition of coadjoint representation. The Lie group G is a smooth manifold and the Lie algebra $\mathfrak{g}$ can be realized as the space of all right-invariant vector fields (first order differential operators, annihilating the constants) on G. The space $\mathfrak{g}^*$ can thus be identified with the space of all right-invariant differential 1-forms on G. The group G acts on $\mathfrak{g}$ and $\mathfrak{g}^*$ by left translations.

Example. Let $G=GL(n, K)$ be the group of all non-singular matrices with elements in the field K ($K=\mathbf{R}$ or $\mathbf{C}$). Every right-invariant vector field on G is of the form $v_A(X)=AX$, where A is some fixed matrix. The left translation by Y sends the field v_A into the field $v_{YAY^{-1}}$.

Every right-invariant 1-form on G is of the form $\omega_B(X) = \mathrm{Re}\,\mathrm{tr}\,(BdX \cdot X^{-1})$. The value of the form $\omega_B(X)$ at the vector $v_A(X)$ does not depend on X and equals tr BA. The left translation by Y translates the form ω_B into the form $\omega_{YBY^{-1}}$. Thus, the orbits of the group G in $\mathfrak{g}^*$ are, in the given example, classes of similar matrices.

It turns out that by far not every manifold can be the orbit of a group G in $\mathfrak{g}^*$.

Theorem 1. *On every G-orbit* Ω *in* $\mathfrak{g}^*$ *there exists a closed non-degenerate G-invariant 2-form* B_Ω.

Corollary. Every orbit has even dimension.

Thus, every orbit can be considered as the phase space of some Hamiltonian mechanical system.

As mentioned first by B. Kostant in [2], the process, described below, of construction of a representation of the group G by a G-orbit in $\mathfrak{g}^*$, is analogous to the operation of quantization of a mechanical system (see details in [3] and [17]).

§ 2. Construction of a representation by an orbit

The majority of irreducible unitary representations of Lie groups can be obtained as induced representations in the sense of Mackey or in a more general sense (holomorphically induced representations, representations in L^2-cohomology).

Recall this construction. Let H be a closed subgroup in G and U a unitary representation of H on the Hilbert space V.

Denote by X the homogeneous manifold of all right H-cosets in G. In every class $x \in X$ we choose a representative $s(x) \in G$, so that every element $g \in G$ can be uniquely written in the form

$$g = h \cdot s(x), \quad h \in H, \quad x \in X. \tag{1}$$

Such a mapping $s: X \to G$ will be called a *section*. In general, continuous sections may not exist. Nevertheless, one can always find a measurable section s, which will also be smooth everywhere outside some submanifold of lower dimension in X.

The right-invariant Haar measure $d_r g$ on G in "coordinates" h and x, defined by formula (1), can be written in the form

$$d_r g = \frac{\Delta_G(h)}{\Delta_H(h)} d_r h \, d\mu_s(x), \tag{2}$$

where $d_r h$ is the right-invariant Haar measure on H, and the function Δ_G on the group G is defined by the formula $\Delta_G(g) = \dfrac{d_r(gg_1)}{d_r g_1}$. The measure μ_s on X

depends on the choice of the section s. It is quasi-invariant under the group G, and one has the equality

$$\frac{d\mu_s(xg)}{d\mu_s(x)} = \frac{\Delta_H(h)}{\Delta_G(h)}, \tag{3}$$

where the element $h=h(x, g)$ is defined by the relation

$$s(x)g = h \cdot s(xg). \tag{4}$$

If a G-invariant measure μ exists on X, then μ_s is independent of s and proportional to μ.

Define the representation T of the group G in the space $L^2(X, \mu_s, V)$ of all square integrable V-valued functions on X by the formula

$$[T(g)f](x) = A(x, g)f(x, g), \tag{5}$$

where

$$A(x, g) = \left[\frac{\Delta_H(h)}{\Delta_G(h)}\right]^{1/2} U(h),$$

and $h=h(x, g)$ is given by (4).

An easy calculation, based on (3), shows that T is a unitary representation. It is said to be induced by the representation U of the subgroup H, and is denoted by the symbol Ind (G, H, U).

One can show that every unitary representation of G on $L^2(X, \mu, V)$, given by the formula (5) is equivalent to a representation which is induced from the subgroup H. This follows from the following criterion of Mackey:

A representation T of G on the Hilbert space L is equivalent to a representation induced from a subgroup H if and only if there exists a projection measure P on $X=G/H$ such that

$$T(g)^{-1}P(E)T(g) = P(Eg), \quad E \subset X, \quad g \in G.$$

Another realization of this representation can be obtained as follows. Consider the space $L(G, H, U)$ of all measurable V-valued functions on G, satisfying the condition

$$F(hg) = \left[\frac{\Delta_H(h)}{\Delta_G(h)}\right]^{1/2} U(h)F(g). \tag{6}$$

It can be verified that the quantity

$$\|F\|^2 = \int_X \|F(s(x))\|_V^2 \, d\mu_s(x) \tag{7}$$

does not depend on the choice of the section $s: X \to G$ and defines in $L(G, H, U)$ a Hilbert space structure. This Hilbert space is denoted by $L^2(G, H, U)$.

The group G acts on $L^2(G, H, U)$ by right translations and the unitary representation obtained is equivalent to Ind (G, H, U).

We now remark that if the group $\bar{H}$ is connected, then condition (6) for smooth functions F (constituting a dense subset in $L^2(G, H, U)$) can be rewritten in the form of a system of differential equations

$$L(X)F = dU_0(X)F, \quad X \in \mathfrak{h}, \tag{8}$$

where $\mathfrak{h}$ is the Lie algebra of the subgroup H, $L(X)$ is the right-invariant vector field on G, corresponding to the element $X \in \mathfrak{h}$, and dU_0 is the Lie algebra representation of $\mathfrak{h}$ corresponding to the group representation $U_0(h) = \left[\frac{\Delta_H(h)}{\Delta_G(h)}\right]^{1/2} U(h)$ of H.

The advantage of writing (8) is that it permits a "complex continuation". Namely if $\mathfrak{g}_C$ is the complexification of the Lie algebra $\mathfrak{g}$ and U_0 is the representation of some complex subalgebra $\mathfrak{h} \subset \mathfrak{g}_C$ in the space V, then system (8) remains meaningful (since the mapping $X \rightarrow L(X)$ extends to $\mathfrak{g}_C$ by linearity). If $\mathfrak{h} = \bar{\mathfrak{h}}$, then the subalgebra $\mathfrak{h} \subset \mathfrak{g}_C$ is the complexification of the real subalgebra $\mathfrak{h}_0 = \mathfrak{h} \cap \mathfrak{g}$ and system (8) is equivalent to the same system for $\mathfrak{h}_0$ and condition (6) for the connected subgroup H_0 corresponding to the subalgebra $\mathfrak{h}_0 \subset \mathfrak{g}$.

In some sense the opposite case occurs if $\mathfrak{h} + \bar{\mathfrak{h}} = \mathfrak{g}_C$. One can show in this case that the space of solutions of system (8) can be identified with the space of all holomorphic V-valued functions on the complex manifold G/K, where K is the subgroup in G, corresponding to the subalgebra $\mathfrak{k} = \mathfrak{h} \cap \mathfrak{g}$.

The representation of the group G, obtained in the space of solutions of such a system, is said to be holomorphically induced and denoted by Ind $(G, \mathfrak{h}, U_0)$. In the general case, when $\mathfrak{h} + \bar{\mathfrak{h}}$ is a subalgebra lying strictly between $\mathfrak{h}$ and $\mathfrak{g}_C$, the space of solutions of system (8) can be realized as a space of partially holomorphic functions on some manifold of mixed type. Namely, the manifold considered is a fibre space with a real base and a complex fibre and the structure group of the fibre bundle is the group of the holomorphic transformations of the fibre.

Furthermore, a generalization of the representation construction is possible using the concepts of Lie algebra cohomology or cohomology with coefficients in sheaves.

The difference between ordinary and L^2-cohomology is that in the latter case we consider only cocycles which are square integrable with respect to some scalar product and we consider cocycles as cohomological if their difference is equal to a square integrable coboundary. This generalization is set forth in more detail in [18] and [19].

Now we show the scheme of the construction of representation of G by a G-orbit $\Omega \subset \mathfrak{g}^*$. Let F be a point of Ω, G_F the stabilizer of this point in G, and $\mathfrak{g}_F$ the corresponding subalgebra in $\mathfrak{g}$. We call the subalgebra $\mathfrak{h} \subset \mathfrak{g}$ *subordinate* to the functional F, if F vanishes on the commutator $[\mathfrak{h}, \mathfrak{h}]$. In this case the mapping $X \rightarrow i\langle F, X \rangle$ is a one-dimensional representation of the algebra $\mathfrak{h}$.

Let H_0 be the connected subgroup of G, corresponding to the subalgebra $\mathfrak{h}$. We assume that the subgroup H generated by H_0 and G_F is closed in G, and that the local one-dimensional representation $\tilde{U}$ given by the formula

$$\tilde{U}(\exp X) = e^{2\pi i\langle F, X\rangle}$$

extends to a one-dimensional unitary representation U of the group H.

If the representation $T = \text{Ind}(G, H, U)$ is irreducible, then we call it the representation corresponding to the orbit Ω. It is seen that this definition contains certain arbitrariness (the choice of the functional $F \in \Omega$ and the subalgebra $\mathfrak{h}$ subordinate to F, the extension of $\tilde{U}$ to U), and requires the validity of additional conditions (closedness of H, existence of the extension U, irreducibility of T).

Therefore, the construction described does not assign a representation of G to every orbit Ω, on one hand, and to some orbits it assigns a whole set of representations on the other hand.

As it turns out, that part of arbitrariness and additional conditions which depends on the extension of U_0 to U, can be completely described in terms of the topology of the orbit Ω.

Theorem 2. *The representation U_0 of the group H_0 extends to a representation U of the group H if and only if the canonical form B_Ω on Ω has integer-valued integrals over all two-dimensional cycles in Ω. All possible extensions are parametrized by the characters of the group $\pi_1(\Omega)$ (the fundamental group of Ω).*

In particular, if the orbit Ω has Betti number zero in dimensions 1 and 2, then the representation U_0 has a unique extension U.

For the irreducibility of the representation T there is a simple necessary condition, which is also sufficient in many (although not in all) cases. Namely that the codimension of $\mathfrak{h}$ in $\mathfrak{g}$ should be equal to the half of the dimension of the orbit Ω: $\dim \mathfrak{g} - \dim \mathfrak{h} = \frac{1}{2} \dim \Omega$.

A stronger condition, necessary and sufficient for the irreducibility of T in case of exponential groups (see below), was proposed by L. Pukánszky. It can be written as $F + \mathfrak{h}^{\perp} \subset \Omega$ and means that all functionals F_1, coinciding with F on $\mathfrak{h}$, belong to Ω. This condition is natural, as it will become clear in the next section.

The described construction and its modifications (based on the above-mentioned generalizations of the concept of induced representation) permit us to construct all irreducible unitary representations for many classes of groups.

Thus, for exponential groups all irreducible unitary representations turn out to be monomial, (i.e. induced from a one-dimensional representation of some subgroup) and our construction establishes in this case a one-to-one correspondence between orbits and representations. (Recall that Lie group G is called *exponential* if the canonical mapping $\exp: \mathfrak{g} \to G$ is a homeomorphism; all such groups are solvable and can be characterized by the following condition: the operators of the adjoint representation of the Lie algebra have no purely imaginary eigenvalues, different from zero.)

A theorem is announced in [4] for solvable Lie groups of type 1, stating that every irreducible unitary representation is holomorphically induced by a one-dimensional representation of a suitable subgroup, and the set $\hat{G}$ is canonically parametrized by the characters of the fundamental groups of G-orbits in $\mathfrak{g}^*$. Here the condition of integrality is lacking, since for solvable groups of type 1 the form B_Ω is always cohomologous to zero.

For compact simply connected Lie groups our construction establishes a one-to-one correspondence between $\hat{G}$ and the family of all integer orbits (i.e. orbits satisfying the condition of theorem 2) of maximal dimension in $\mathfrak{g}^*$.

The progress made in recent years in the theory of representations of semi-simple Lie groups justifies the hope that the correspondence between orbits and representations takes places also for non-compact semisimple groups (see the survey [19]).

In the general case there is little known yet. There are examples showing that subalgebras of the required dimension, subordinate to F, may not exist. In the simplest of these examples the group G consists of all linear automorphisms of $\mathbf{R}^{2n}$ preserving a non-singular antisymmetric form and leaving fixed some non-zero vector $x \in \mathbf{R}^{2n}$. This group is a direct product of a simple subgroup isomorphic to $\mathrm{Sp}(2n-2, \mathbf{R})$ and a nilpotent normal subgroup H_n, called the *Heisenberg group*.

Let us consider now the orbit Ω in $\mathfrak{g}^*$ passing through the functional F which vanishes on $\mathfrak{sp}(2n-2, \mathbf{R})$ and is different from zero on the centre of the Lie algebra $\mathfrak{h}_n$ of the group H_n. Let T be the representation of G corresponding to the orbit Ω (the correspondence in question can be made by the consideration of either the generalized or the infinitesimal character of the representation — see below, sections 4 and 5), remains irreducible if we restrict it to the nilpotent radical N of the group G, and its restriction to the semisimple subgroup $\mathrm{Sp}(2n-2, \mathbf{R})$ coincides with the spin representation of this group.

T cannot be obtained by inducement (either in a generalized sense) from any one-dimensional representation of any subgroup, because no subalgebra of the required dimension is subordinate to F in $\mathfrak{g}_{\mathbf{C}}$.

As this example shows, we may find a reasonable correspondence between orbits and representations even though the representations are not monomial.

We also remark that sometimes we recognize connections between orbits and representations for classes of groups which are broader than Lie groups (algebraic groups over finite or discontinuous locally compact fields, infinite-dimensional groups (see e.g. [21], [22])).

§ 3. Functorial properties of the correspondence between orbits and representations

The following problems play an important role in representation theory and its applications.

1. *Let H be a subgroup of G and T an irreducible representation of G. The restriction of T to H is, generally, reducible. Find its decomposition into irreducible components.*

2. *Let H be a subgroup of G and U an irreducible representation of H. The induced representation $T = \operatorname{Ind}(G, H, U)$ is, in general, reducible. Find its decomposition into irreducible components.*

A special case of the first problem is the problem of decomposition of the tensor product of two irreducible representations, and a special case of the second one is the problem of decomposition of representations realized in terms of functions, vector or tensor fields on homogeneous manifolds.

In cases when there is a sufficiently good correspondence between orbits and representations, it is natural to expect an intuitive solution of the problems mentioned in terms of orbits; this is indeed so. We formulate the result for the case of nilpotent Lie groups, where it is particularly simple.

Let $\mathfrak{h}$ be the subalgebra of $\mathfrak{g}$ corresponding to the subgroup $H \subset G$. Denote by p the natural projection of $\mathfrak{g}^*$ onto $\mathfrak{h}^*$, associating to the functional F on $\mathfrak{g}$ its restriction to $\mathfrak{h}$.

Theorem 3. *If the irreducible representation T of the group G corresponds to a G-orbit $\Omega \subset \mathfrak{g}^*$, then its restriction to H decomposes into a continuous sum of irreducible representations of H corresponding to the H-orbits in $\mathfrak{h}^*$ lying in $p(\Omega)$.*

If the irreducible representation U of the subgroup H corresponds to an H-orbit $\omega \subset \mathfrak{h}^*$, then the representation $T = \operatorname{Ind}(G, H, U)$ decomposes into a continuous sum of irreducible representations of G corresponding to the G-orbits in $\mathfrak{g}^*$ lying in the G-saturation of $p^{-1}(\Omega)$.

We leave to the reader the formulation of the corollaries of Theorem 3 relating to the above-mentioned particular cases of problems 1 and 2.

We also mention that for one-dimensional U the condition of irreducibility of the representation $T = \operatorname{Ind}(G, H, U)$, appearing in Theorem 3, coincides with that of Pukánszky.

It is a very interesting problem to extend this theorem to larger classes of groups (in the first place, to exponential groups)*. Preliminary results, obtained in this direction, show that the situation essentially depends on the topological properties of the fibres of the mapping p.

* This problem is now solved for exponential groups (see [23]).

§ 4. Generalized characters

Infinite-dimensional unitary representations do not have characters in the ordinary sense, since one cannot assign to a unitary operator in an infinite dimensional space a reasonable trace. It can happen, however, that for some class of "fundamental" or "test" functions φ on the group G, the operators

$$T(\varphi) = \int_G \varphi(g)\, dg$$

have a trace. This trace is a linear functional on the space of basic functions and, thus, defines a generalized function on the group G. It is interesting that in some cases (e.g. for irreducible unitary representations of semi-simple Lie groups) the generalized characters are of the form

$$\operatorname{tr} T(\varphi) = \int_G \chi_T(g)\varphi(g)\, d_x g,$$

where χ_T is some locally integrable function on the group G which is continuous (and also smooth) on some open subset G_0 of full measure in G. If it is so, then the number $\chi_T(g)$ is naturally considered as the trace of the operator $T(g)$ for $g \in G_0$.

It is known (see [20]) that if the group G is of type 1, then every irreducible unitary representation T of this group has a generalized character. This means that the space $\mathcal{D}_T(G)$ of all functions φ for which the operator $T(\varphi)$ is of trace class is sufficiently large in the sense that the operators $T(\varphi)$, $\varphi \in \mathcal{D}_T(G)$ form a dense subset of the set of all bounded operators in the strong topology.

(Actually this condition is equivalent to the existence of at least one function for which the operator $T(\varphi)$ is of trace class and is non-zero.)

The space $\mathcal{D}_T(G)$ of test functions depends on the kind of the group G and the representation T. For semisimple and nilpotent groups all spaces $\mathcal{D}_T(G)$ contain the subspace $C_0^\infty(G)$ of infinitely differentiable functions with compact support. For solvable groups, as will be seen below, $\mathcal{D}_T(G)$ does not necessarily have this property.

We mention that a unitary representation T having generalized character, is uniquely determined by this character up to equivalence.

Now we pass to the description of the connections between characters and orbits. To every G-orbit $\Omega \subset \mathfrak{g}^*$ one can assign a generalized function I_Ω on G as follows.

Let V be an open subset in G, covered by the canonical coordinate system. In other words, if $W \subset \mathfrak{g}$ is the domain of regularity of the exponential mapping $\exp: \mathfrak{g} \to G$, then $V \subset \exp U$. For a solvable, compact and complex semisimple group G we can choose V in such a way that $G \setminus V$ is of measure zero.

On the orbit Ω there is a canonical nondegenerate 2-form B_Ω. Put

$$\beta = \frac{1}{k!} B_\Omega \wedge \ldots \wedge B_\Omega,$$

where $k = \frac{1}{2} \dim \Omega$. We get a nondegenerate differential form of degree $2k = \dim \Omega$, giving a volume element on Ω. Now we define the generalized function I_Ω as the linear functional on $\mathscr{D}(V)$, given by the formula

$$\langle I_\Omega, \varphi \rangle = \int_\Omega \Big(\int_{\mathfrak{g}} \varphi(\exp X) e^{i\langle F, X\rangle} dX \Big) \beta(F).$$

It turns out that in many cases the generalized function I_Ω is essentially equal to the generalized character χ_T of some irreducible unitary representation $\boldsymbol{T}$. Namely, we have

$$\chi_T = I_\Omega \cdot p_\Omega^{-1}, \tag{Φ}$$

where $p_\Omega(g)$ is some analytic function on V, invariant under inner automorphisms of G and satisfying the condition $p_\Omega(e) = 1$.

Formula (Φ) is proved for all exponential Lie groups, and for p_Ω an explicit formula is given [10]. For orbits of maximal dimension, p_Ω does not depend on Ω and is given by the formula

$$p_\Omega(\exp X) = \det\big(s\hbar\,(\operatorname{ad} X)\big),$$

where $s\hbar$ denotes the function $x \to \dfrac{\operatorname{sh}(x/2)}{x/2}$.

The same formula with the same factor p_Ω holds for representations of compact groups [15], the principal series of representations of real semisimple groups [9] and the discrete series of representations of the group $SL(2, \mathbf{R})$ [17].

There are many reasons (see [15] and [16]) for believing that formula (Φ) holds for a Lie group G if only it holds for a maximal semisimple subgroup S of G. On the other hand, formula (Φ) does not hold for degenerate series of representations of the group $SL(3, \mathbf{R})$. It is a very interesting problem to obtain a full description of the "domain of validity" of formula (Φ).

Another interesting direction is to find the connections between this formula and the Feynman integrals and the quasi-classical approach in quantum mechanics.

In the cases, when formula (Φ) holds for all representations corresponding to orbits of maximal dimension, and the factor p_Ω is independent of Ω, we obtain a simple and intuitive interpretation of the so-called Plancherel formula for the group G.

This formula gives a decomposition of the δ-function, concentrated at the identity of the group G, into characters of irreducible representations. Under the above assumptions the Plancherel formula turns out to be the decomposition of the Euclidean measure on $\mathfrak{g}^*$ into the canonical measures on the orbits of general position. This permits one to give a simple universal expression for the Plancherel measure. For details see [13].

We also remark that in the cases formula (Φ) holds it helps to clarify the nature of the generalized function χ_T. In particular, if the orbit Ω is closed in $\mathfrak{g}^*$, then I_Ω, and consequently also χ_T are defined on the whole space $C_0^\infty(V)$. If Ω is not closed, then the integral defining I_Ω can diverge. In this case we have to impose further equality-type conditions on the functions in $C_0^\infty(V)$ in defining the space of test functions.

The group G of affine orientation-preserving transformations of the real line can serve as a good illustration. This group is exponential. It has two infinite dimensional irreducible representations T^+ and T^-, and a series of one-dimensional representations.

The set $\mathfrak{g}^*$ which is the two-dimensional real plane splits under the action of G to two open half-planes Ω^+ and Ω^- and a family of fixed points, constituting a line. It can be verified that a function belongs to $\mathcal{D}_{T^\pm}(G)$ if and only if it is annihilated by all the one-dimensional representations. This is a corollary of the following simple fact of the analysis: a function f, smooth on $\mathbf{R}^2$ and rapidly decreasing at infinity, is integrable on the half plane $y>0$ according to the measure $y^{-1}dx\,dy$ if and only if $f(x, 0)\equiv 0$. Very interesting description of the C^*-algebra of this group is recently obtained by Do Ngok Zyep [24] using the Brown—Duglas—Fillmor's Ext-functor.

§ 5. Infinitesimal characters

Let $U(\mathfrak{g})$ be the enveloping algebra of the Lie algebra $\mathfrak{g}$. As is known, $U(\mathfrak{g})$ can be realized as the algebra of all right-invariant differential operators on the Lie group G.

If T is a unitary representation of the group G on the Hilbert space V, then on some dense subspace V^∞ (the so-called *Gårding space*) a representation T^∞ of the algebra $U(\mathfrak{g})$ is defined. If $\xi\in V$ and $\varphi\in\mathcal{D}(G)$, then $T(\varphi)\xi\in V^\infty$ and we have the equality

$$T^\infty(X)T(\varphi)\xi = T(X\varphi)\xi, \quad X\in U(\mathfrak{g}).$$

Let $Z(\mathfrak{g})$ be the center of the algebra $U(\mathfrak{g})$. The elements of $Z(\mathfrak{g})$ are called the *Laplace operators* on G. For any irreducible unitary representation T of the group G the operators $T^\infty(X)$, $X\in Z(\mathfrak{g})$ are scalar. Thus, any irreducible unitary representation T of the G defines a homomorphism λ_T of the algebra $Z(\mathfrak{g})$ to $\mathbf{C}$, which is called the infinitesimal character of T. Recall the well-known result of I. M. Gelfand about the structure of $Z(\mathfrak{g})$. Let σ be a linear mapping of the space $P(\mathfrak{g}^*)$ of all polynomial functions on $\mathfrak{g}^*$ into $U(\mathfrak{g})$, uniquely defined by the following condition: if $X\in\mathfrak{g}$, then $\sigma(X^k)=X^k$. Here X^k on the left-hand side of the equality is understood as the polynomial function, taking at the point $F \in \mathfrak{g}^*$ the value $\langle F, X \rangle^k$, and X^k on the right-hand side denotes the k-th power of $X \in \mathfrak{g} \subset U(\mathfrak{g})$.

The theorem of Gelfand. The mapping σ *is an isomorphism of* $P(\mathfrak{g}^*)$ *onto* $U(\mathfrak{g})$*; the inverse image of* $Z(\mathfrak{g})$ *under this isomorphism is the set* $I(\mathfrak{g}^*)$ *of all G-invariant polynomials on* $\mathfrak{g}^*$.

It turns out that the infinitesimal characters λ_T of irreducible unitary representations T can be expressed very easily in terms of the corresponding orbits. Namely, the following equality holds for any G-invariant polynomial P on $\mathfrak{g}^*$:

$$\lambda_T(\sigma(P)) = P(\Omega_T),$$

where $P(\Omega_T)$ denotes the value of this polynomial at any point of the orbit Ω_T corresponding to the representation T.

This fact is proved for exponential groups in [8], for compact groups in [15]. Its validity in the general case has not yet been established. We remark that this general result would imply the validity of the conjecture, formulated long ago, saying that σ is an isomorphism between $I(\mathfrak{g}^*)$ and $Z(\mathfrak{g})$, not only as spaces, but as algebras, too. (See also, in this regard, the report of M. Duflo in the present volume.)

REFERENCES

[1] Kirillov, A. A. (Кириллов А. А.), Унитарные представления нильпотентных грипп Ли, *Успехи матем. наук*, **17**:4 (1962), 57—110. (Unitary representations of nilpotent Lie groups.)

[2] Kostant, B., Orbits, symplectic structures and representation theory, in *Proc. of US—Japan Seminar in Differential Geometry (Kyoto, 1965)*, Tokyo, 1966.

[3] ———, Quantization and unitary representation. Part I. Prequantization, in *Lectures in Modern Analysis and Application* III.

[4] ———, and Auslander, L., Quantization and representation of solvable Lie groups, *Bull. Amer. Math. Soc.*, **73** (1963), 692—695.

[5] Bernat, P., Sur les représentations unitaires des groupes de Lie résolubles, *Ann. Sci. École Norm. Sup.*, **82** (1965), 37—99.

[6] Pukánszky, L., On the unitary representations of exponential groups, *J. Functional Analysis*, **2** (1968), 73—113.

[7] ———, Characters of algebraic solvable groups, *J. Functional Analysis*, **3** (1969), 435-491.

[8] Duflo, M., Caractères des groupes et des algèbres de Lie résolubles, *Ann. Sci. École Norm. Sup.* **3** (1970), 23—74.

[9] ———, (М. Дюфло), Представления основной серии полупростой группы Ли, *Функц. анализ*, **4**:2 (1970), 28—42. (Representations of the principal series of a semisimple Lie group.)

[10] ———, Représentations induites d'algèbres de Lie, *C. R. Acad. Sci. Paris*, **272** (1971), 1157—1158.

[11] Vergne, M., Construction de sous-algèbres subordonnées à un élément du dual d'une algèbre de Lie résoluble, *C. R. Acad. Sci. Paris*, **270** (1970), 173—175, 704—707.

[12] ———, La structure de Poisson sur l'algèbre symétrique d'une algèbre de Lie nilpotente, *C. R. Acad. Sci. Paris,* **269** (1970), 950—952.

[13] Kirillov, A. A. (Кириллов А. А.), О мере Планшереля для нильпотентных групп Ли, *Функц. анализ,* **1**:4 (1967), 84—85.
(On the Plancherel measure for nilpotent Lie groups.)

[14] ———, Метод орбит в теории унитарных представлений групп Ли, *Функц. анализ,* **2**:1 (1968), 98—100.
(The method of orbits in the theory of unitary representations of Lie groups.)

[15] ———, Характеры унитарных представлений групп Ли, *Функц. анализ,* **2**:2 (1968), 40—55.
(Characters of unitary representations of Lie groups.)

[16] ———, Характеры унитарных представлений групп Ли. Редукцонные теоремы. *Функц. анализ,* **3**:1 (1969), 36—47.
(Characters of unitary representations of Lie groups. Reduction theorems.)

[17] ———, *Лекции по теории представлений групп Ли*, IV *(Представления групп и механика)*, Изд-во МГУ, М. 1971.
(Lectures on the representation theory of Lie groups, IV (Group representations and mechanics).)

[18] ———, Конструкция унитарных неприводимых представлений групп Ли, *Вестник МГУ*, сер. матем., 1970, № 2, 41—51.
(The construction of unitary irreducible representations of Lie groups.)

[19] Schiffman, G., Un analogue du théoreme de Borel—Weil—Bott dans le cas non-compact, *Séminaire Bourbaki*, Exp. 398, 1971.

[20] Dixmier, J., *Les C*-algèbres et leurs représentations*, Gauthier-Villars, Paris, 1969.

[21] Kirillov, A. A., (Кириллов А. А.), Представления бесконечномерной унитарной группы, *ДАН СССР*, **212** (1973), 288—290.
(Representations of an infinite-dimensional unitary group.)

[22] ———, Предстабления некоторых бесконечномерных групп Ли, *Вестник МГУ*, сер. матем., 1974, №1, 75—83.
(Representations of certain infinite-dimensional Lie groups.)

[23] Busjackaja, I. K. (Бусяцкая И. К.), Некоторые свойства представлений экспоненциальных групп Ли, *Успехи матем. наук*, **28**:4 (1973), 211.
(Certain properties of representations of exponential Lie groups.)

[24] Do Ngok Zyep (До Нгок Зьеп), О структуре групповой C*-алгебры группы аффинных преобразований прямой, *Функц. анализ.* **9**:1 (1975).
(On the structure of the group-C*-algebra of the group of affine transformations of a line.)

ON THE EXISTENCE AND IRREDUCIBILITY OF CERTAIN SERIES OF REPRESENTATIONS*

BERTRAM KOSTANT

Introduction

From the way the subject grew up, and for other reasons, one initially (and often fruitfully) tends to study the theory of (say) Banach space representations π of a semisimple Lie group G from the point of view of analysis.

However the fact that the restriction $\pi|K$ decomposes so nicely, where K is the maximal compact subgroup of G (assume G has finite center and π is irreducible) and the fact that the space X of K-finite vectors is in a natural way a U-module, where U is the enveloping algebra of the Lie algebra $\mathfrak{g}$ of G, (Theorem of Harish–Chandra) opens the door to an algebraic approach to such representations. This is particularly true when one has fine structure theorems for U.

In this paper we are mainly concerned with U-modules X with the property that (1) dim $X^{\mathfrak{k}}=1$ and (2) $UX^{\mathfrak{k}}=X$. Here $\mathfrak{k}$ is the Lie algebra of K and the superscript denotes the space of invariants. For simplicity, call such modules *cyclic* U-modules. Any irreducible U-module Z such that $Z^{\mathfrak{k}}\neq 0$ is cyclic in this sense. The application to G is then clearly to representations which admit a K-fixed vector, so-called class 1-representations. This, of course, is directly related to Gelfand's theory of spherical functions.

Among the cyclic U-modules to which the theory developed here applies are those which arise from induced representations by a character (not necessarily unitary) on a maximal parabolic subgroup $B\subseteqq U$. That is, if $B=MAN$ is the usual decomposition of B then any $\lambda\in\mathfrak{a}'$ (the complex dual to the Lie algebra $\mathfrak{a}_{\mathbf{R}}$ of A) defines a character on A and hence on B, $B\to b^{\lambda}$ by defining $b^{\lambda}=1$ if $b\in MN$. For any $\lambda\in\mathfrak{a}'$ one then defines a U-module (induced) X_{λ} by letting X_{λ} be the space of all K-finite analytic functions f on G such that $f(ab)=b^{\lambda'}f(a)$. Here λ' is the contragredient to λ (see § 1.7.2). It is easy to see that $X_{\lambda}^{\mathfrak{k}}$ is a one-dimensional space spanned by an element 1_{λ}. The cyclic U-module referred to above is the U-module $X_{\lambda}^{e}=U1_{\lambda}$ generated by 1_{λ}.

Now let Γ denote the set of equivalence classes of irreducible K-modules (actually in the paper we do not use K but rather an extension K_{θ} of K). For any $\gamma\in\Gamma$ the multiplicity of γ in X_{λ} is independent of λ and is written $\boldsymbol{l}(\gamma)$. How-

* This research was supported in part by NSF Research Grant GP 28969.

ever in X_λ^e the multiplicity of γ is a much more subtle thing. It may, however, be given by the rank of a certain $\boldsymbol{l}(\gamma)\times\boldsymbol{l}(\gamma)$ matrix $P^\gamma(\lambda)$. Thus if $p^\gamma(\lambda)$ is the determinant of $P^\gamma(\lambda)$ one has that $p^\gamma(\lambda)\neq 0$ for all $\gamma\in\Gamma$ if and only if $X_\lambda=X_\lambda^e$. One of the key points of the paper is that for any $\gamma\in\Gamma$ we can explicitly determine the function $\lambda\to p^\gamma(\lambda)$ on $\mathfrak{a}'$. In fact it is a polynomial whose prime factors are of the first degree so that the zeros of $p^\gamma(\lambda)$ are certain hyperplanes in $\mathfrak{a}'$ explicitly given.

But the knowledge of $p^\gamma(\lambda)$ does more than determine whether or not $X_\lambda=X_\lambda^e$. In fact one determines whether or not X_λ is irreducible. The reason for this is as follows: From the general representation theory of U one may associate, in a natural way, to every $\lambda\in\mathfrak{a}'$ an irreducible U-module Z_λ such that $Z_\lambda^{\mathfrak{k}}\neq 0$; and in fact one obtains in this way (up to equivalence) any irreducible U-module Z such that $Z^{\mathfrak{k}}\neq 0$. As in the case of X_λ^e the multiplicity of $\gamma\in\Gamma$ in Z_λ varies with λ in a non-trivial way. (One notes that for certain λ, Z_λ may be finite-dimensional). But in fact the multiplicity of γ in Z_λ is exactly the rank of the $\boldsymbol{l}(\gamma)\times\boldsymbol{l}(\gamma)$ matrix $R^\gamma(\lambda)=P^\gamma(-\bar{\lambda}-2\varrho)^*P^\gamma(\lambda)$. Here ϱ is one half the positive roots and * denotes the Hermitian adjoint matrix. One shows easily then that X_λ is irreducible if and only if $R^\gamma(\lambda)$ is non-singular for all $\gamma\in\Gamma$ (in which case X_λ is equivalent to Z_λ). Thus taking determinants one has that X_λ is irreducible if and only if $r^\gamma(\lambda)=$ $=p^\gamma(-\lambda-2\varrho)p^\gamma(\lambda)\neq 0$ for all $\gamma\in\Gamma$. If S denotes the subset of all $\lambda\in\mathfrak{a}'$ such that X_λ is irreducible and S^c its complement in $\mathfrak{a}'$ then the knowledge of $p^\gamma(\lambda)$ completely determines S^c (and hence S) and shows it to be an explicit countable union of certain hyperplanes in $\mathfrak{a}'$. (Helgason in [6b] has observed that the set S^c can be described very nicely in terms of the so-called c-function. As noted in [6b] however, trying to give an analytic proof of the irreducibility of X_λ for all $\lambda\in S$ using the c-function apparently runs into some integration difficulties.) In a subsequent paper we will show how the c-function can be expressed as a sum over $\gamma\in\Gamma$ of the numbers $\dim\gamma\cdot\mathrm{tr}\,P_0^\gamma(\lambda)/P^\gamma(-\lambda-2\varrho)$. Here P_0^γ is a slight modification of P^γ. In the rank 1 case the summands reduce to $\dim\gamma\cdot p^\gamma(\lambda)/p(-\gamma-2\varrho)$. For $G=SO(n,1)$ this formula for the c-function reduces to a theorem of Kummer giving the value of the hypergeometric function $F(a,b,c;z)$ at $z=-1$ for $c=1+a-b$. Now by examining S^c one makes the observation that the hyperplanes occurring in S^c make a regular pattern *but only* outside a naturally defined convex region in $\mathfrak{a}'$ which we call the critical strip CS (generalizing terminology for $Sl(2,\mathbf{R})$). Moreover CS is symmetric about the point $-\varrho$ and $CS\subseteq S$ so that X_λ is irreducible for all $\lambda\in CS$. But now the fact that $-\varrho+iv\ CS$ for all $v\in\mathfrak{a}'_{\mathbf{R}}$ (the real dual) then establishes the irreducibility of the unitary spherical principal series for all real semisimple groups.

The notion of the unitary (not necessarily spherical) principal series was pioneered by Gelfand and Neumark in [3], where irreducibility for the complex classical groups was proved. For all complex semisimple groups this was established by Želobenko, Wallach and in the spherical case by Varadarajan, Rao and Parthasarthy [14]. For real groups irreducibility was proved by Bruhat for "almost all" elements of the unitary principal series. However the most complete results in the unitary case were obtained by Knapp and Stein [8a]. See also Knapp [8c].

In this paper we deal only with the class or spherical representations, However we are not restricted to the unitary case, (i.e., λ need not be equal to $-\varrho+i\nu$). Also the method in this paper enables one to settle similar irreducibility questions when MAN is replaced by an arbitrary parabolic. It becomes a question of considering certain minors in the P^{γ} matrices. This will be done in a subsequent paper.

The general notion of complementary series in the complex case was investigated by Kunze [11]. The method here leads to the notion of spherical complementary series and the proof of irreducibility of such representations in the general case. For other results on the complementary series (non-spherical but rank 1) see [8a].

The situation with the complementary series is dealt with in this paper along the following lines: One shows that for any $\lambda\in\mathfrak{a}'$ the irreducible U-module Z_λ has a U-invariant pre-Hilbert inner product if and only if the matrices $R^{\gamma}(\lambda)$ are positive semi-definite for all $\gamma\in\Gamma$. Let $P\subseteqq\mathfrak{a}'$ be the set of all λ with this property. Using Nelson's theorem the problem of determining all irreducible unitary representations of G with a K-fixed vector is then nothing more than the problem of determining the set P. Let H denote the larger set of all $\lambda\in\mathfrak{a}'$ such that all the matrices $R^{\gamma}(\lambda)$ are just Hermitian. The problem of determining H is trivial. One has that $\lambda\in H$ if and only if μ is a Weyl group conjugate to $-\bar{\mu}$ where $\mu=\lambda+\varrho$. That is, if and only if there is a Weyl group element σ such that $\sigma\mu_1=-\mu_1$ and $\sigma\mu_2=\mu_2$ where μ_1 and μ_2 are respectively the real and imaginary parts of μ. Thus $-\varrho+i\mathfrak{a}'_{\mathbf{R}}\subseteqq H$ and, for example, if -1 is in the Weyl group, $\mathfrak{a}'_{\mathbf{R}}\subseteqq H$. Now the set H is connected. In fact if $\lambda\in H$ then the line segment L_λ joining λ to $-\varrho$ lies in H. But now the point is as follows: We have already established that $R^{\gamma}(\lambda)$ is non-singular for any $\lambda\in CS$ and CS is convex. But $R^{\gamma}(-\varrho)$ is clearly positive-definite. Thus if $\lambda\in H\cap CS$ the eigenvalues of the non-singular Hermitian matrices R^{γ} along the line segment L_λ cannot change sign. Thus $\lambda\in P\cap CS$. The argument extends to the closure and hence we have determined $P\cap\overline{CS}$ to be just the set $H\cap\overline{CS}$. We have defined the complementary series of classes of class one to be the unitary representations corresponding to all the real $\lambda\in P\cap\overline{CS}$. That is all real $\lambda\in\overline{CS}$ such that $\lambda+\varrho$ and $-(\lambda+\varrho)$ are Weyl group conjugate.

We have not determined P outside of $\overline{CS}$ except in the rank one case. The rank one case is easy to deal with for the following reason. We have established in the paper that if G has rank one then $\boldsymbol{l}(\gamma)=1$ for all $\gamma\in\Gamma$. (For another proof of this fact see [12]). Since $\boldsymbol{l}(\gamma)=1$ the $\boldsymbol{l}(\gamma)\times\boldsymbol{l}(\gamma)$ matrices $R^{\gamma}(\lambda)$ reduce to their determinants $r^{\gamma}(\lambda)$. Hence it is only a question of knowing which λ have the property that $r^{\gamma}(\lambda) \geqq 0$ for all $\gamma \in \Gamma$. Since we have explicitly computed $r^{\gamma}(\lambda)$ this is easy and the answer is that except for possibly 0 and -2ϱ one has $P \subseteqq \overline{CS}$. The theorem can then be stated that if rank $G=1$ then except for possibly the trivial representation the principal and the complementary series of class one exhaust all irreducible unitary representations of G with a K-fixed vector. Note that in this case, since -1 is in the Weyl group, the complementary series is given simply by all real $\lambda \in \overline{CS}$. For other results on the complementary series, see [3], [11], [8a] and [8b].

The results of this paper were announced in [9b]. See [9b] for a further summary of these results.

In the overall algebraic approach this paper owes much to the paper [14] of Varadarajan, Rao and Parthasarathy. In [14] the authors deal with the case where G is complex. The techniques here and in [14] both involve reduction to the rank 1 case. When G is complex this just means $SL(2, \mathbf{C})$. Here it means an arbitrary G of rank 1 and most of the paper is devoted to developing the theory for such a group. The principal point of the difficulty (and from our point of view the principal source of interest) arises from the fact that twice a root can be a root. That is $\mathfrak{g}^{2\alpha} \neq 0$. This induces a 2-parameter (i, j)-structure on the representation of K involved and this causes ramifications throughout the paper. In fact the most painful part of the paper and the key result, was obtaining (see § 2.8) the 2-parameter product formula for $p_{v(i,j)}$.

In [14] one has a formula of the form $R^{\gamma}(\lambda) = Q^{\gamma}(\lambda) P^{\gamma}(\lambda)$ with $Q^{\gamma}(\lambda)$ unknown. Here we have found $Q^{\gamma}(\lambda) = P^{\gamma}(-\bar{\lambda} - 2\varrho)^{*}$. Another novelty here is the connection between positive semi-definiteness of $R^{\gamma}(\lambda)$ and the unitarity of the corresponding representation of G leading, as mentioned above, to the complementary series.

The major reference in this paper is the Kostant—Rallis paper [10]. The paper rests on [10] in the same sense that [14] rests on [9a].

The notation in the paper is somewhat different from that used in this introduction. The real group whose representations we are concerned with only makes its major appearance at the very end. The letter G throughout denotes the complexified adjoint group. Also the notation in the paper is different from [9b]. In [9b] X_{λ} is defined so that ϱ is the center of the critical strip CS. Here X_{λ} is defined so that $-\varrho$ plays this role.

Finally I have to thank R. Parthasarathy for writing up my lectures at the Tata Institute. This paper is essentially those notes. I would also like to thank Ann Kostant for typing and other assistances in the preparation of this manuscript. Also I have to thank J. Lepowsky for going over the manuscript and making a number of improvements and simplifications.

TABLE OF CONTENTS

CHAPTER I

MODULES AND MULTIPLICITIES

§ 1. $U(\mathfrak{g})$-Birkhoff—Witt basis

We recall first a few basic facts about universal enveloping algebras of Lie algebras. If $\mathfrak{g}$ is a Lie algebra, then its universal enveloping algebra $U(\mathfrak{g}) = \frac{T(\mathfrak{g})}{I}$, where $T(\mathfrak{g})$ is the tensor algebra over $\mathfrak{g}$, and I is the ideal generated by all elements of the form $x\otimes y - y\otimes x - [x, y]$ for all $x, y \in \mathfrak{g}$. One knows that the composite map $\mathfrak{g}\to T(\mathfrak{g})\to U(\mathfrak{g})$ is an injection. Thus $\mathfrak{g}\subseteq U(\mathfrak{g})$. Furthermore, $U(\mathfrak{g})$ has the property that given any linear map $\varrho:\mathfrak{g}\to A$ (A an associative algebra) such that $\varrho([x, y]) = \varrho(x)\varrho(y) - \varrho(y)\varrho(x)$, then there exists a unique extension, also denoted by ϱ, of ϱ to all of $U(\mathfrak{g})$ such that $\varrho(xy) = \varrho(x)\varrho(y)$ for all $x, y \in U(\mathfrak{g})$. In particular, every $U(\mathfrak{g})$-module can be considered as a $\mathfrak{g}$-module and conversely.

Regarding the structure of $U(\mathfrak{g})$ as a vector space, one has the Birkhoff—Witt theorem, which says that if $\{x_1, x_2, \ldots, x_n\}$ is any basis of $\mathfrak{g}$, then the set of all elements

$$\{x_1^{i_1} x_2^{i_2} \cdots x_n^{i_n}\}$$

in $U(\mathfrak{g})$ forms a basis of $U(\mathfrak{g})$.

If $\mathfrak{h}$ is a subalgebra of $\mathfrak{g}$, then $U(\mathfrak{h})\subseteq U(\mathfrak{g})$.

If $\mathfrak{g}=\mathfrak{a}+\mathfrak{b}$, where $\mathfrak{a}$ and $\mathfrak{b}$ are subalgebras of $\mathfrak{g}$ and $\mathfrak{a}\cap\mathfrak{b}=0$, then $U(\mathfrak{g})$ as a vector space is isomorphic to $U(\mathfrak{a})\otimes U(\mathfrak{b})$. This can be seen as follows. Choose a basis $x_1, x_2, \ldots, x_i, x_{i+1}, \ldots, x_n$ of $\mathfrak{g}$ where $x_1, x_2, \ldots, x_i \in \mathfrak{a}$ and $x_{i+1}, \ldots, x_n \in \mathfrak{b}$. We know that $\{x_1^{\alpha_1} x_2^{\alpha_2} \cdots x_i^{\alpha_i}\}$ forms a basis of $U(\mathfrak{a})$ and similarly $\{x_{i+1}^{\alpha_{i+1}} \cdots x_n^{\alpha_n}\}$ forms a basis of $U(\mathfrak{b})$. Thus $\{x_1^{\alpha_1} \cdots x_i^{\alpha_i} \otimes x_{i+1}^{\alpha_{i+1}} \cdots x_n^{\alpha_n}\}$ forms a basis for the tensor product $U(\mathfrak{a})\otimes U(\mathfrak{b})$. Under the natural map $U(\mathfrak{a})\otimes U(\mathfrak{b})\to U(\mathfrak{g})$ the above basis maps bijectively onto the basis $\{x_1^{\alpha_1} x_2^{\alpha_2} \cdots x_n^{\alpha_n}\}$ of $U(\mathfrak{g})$. Thus the map $U(\mathfrak{a})\otimes U(\mathfrak{b})\to U(\mathfrak{g})$ is an isomorphism. The same argument applies to prove more generally that if $\mathfrak{g}=\mathfrak{g}_1+\mathfrak{g}_2+\ldots+\mathfrak{g}_l$, where each $\mathfrak{a}_j$ is a subalgebra and the sum is a direct sum as vector spaces, then $U(\mathfrak{g}) = U(\mathfrak{g}_1)\otimes U(\mathfrak{g}_2)\otimes\cdots\otimes U(\mathfrak{g}_l)$.

§ 2. The p_u map and representations

We use much of the notation of [10]. Thus (and henceforth) $\mathfrak{g}_{\mathbf{R}}=\mathfrak{k}_{\mathbf{R}}+\mathfrak{p}_{\mathbf{R}}$ is a Cartan decomposition of a real reductive Lie algebra $\mathfrak{g}_{\mathbf{R}}$ and $\mathfrak{g}=\mathfrak{k}+\mathfrak{p}$ is the corresponding complexification, θ, equal to 1 on $\mathfrak{k}$ and -1 on $\mathfrak{p}$, is the corresponding involution. Also, $\mathfrak{a}_{\mathbf{R}}$ is a maximal Abelian subspace of $\mathfrak{p}_{\mathbf{R}}$, so that the complexification $\mathfrak{a}$ is a Cartan subspace of $\mathfrak{p}$. Let $\mathfrak{g}=\mathfrak{a}+\mathfrak{m}+\sum\mathfrak{g}^{\varphi}$ be the root-space decomposition of $\mathfrak{g}$ with respect to $\mathfrak{a}$, where $\mathfrak{m}$ denotes the centralizer of $\mathfrak{a}$ in $\mathfrak{k}$. With respect to some ordering in Δ, $(=\Delta(\mathfrak{a},\mathfrak{g}))$, let Δ_+ denote the set of positive roots in Δ and let $\mathfrak{n}$ denote the subalgebra $\sum_{\varphi\in\Delta_+}\mathfrak{g}^{\varphi}$. It is easily seen that $\mathfrak{n}$ is a nilpotent subalgebra of $\mathfrak{g}$.

Lemma 1.2.1. One has a decomposition (the Iwasawa decomposition) $\mathfrak{g}=\mathfrak{k}+\mathfrak{a}+\mathfrak{n}$ (direct sum).

Proof. Let $x+y+z=0$, where $x\in\mathfrak{k}$, $y\in\mathfrak{a}$ and $z\in\mathfrak{n}$. Then one has $\theta x+\theta y+\theta z=0$, i.e. $x-y+\theta z=0$. Hence $2y+z-\theta z=0$. But since $\theta|\mathfrak{a}=-$identity, and $\theta(\mathfrak{n})\subseteq\sum_{\varphi\in-\Delta_+}\mathfrak{g}^{\varphi}$, one has $2y=0$, $z=0$ (and $\theta z=0$); thus one has also $x=0$. Thus $\mathfrak{k}+\mathfrak{a}+\mathfrak{n}$ is a direct sum. Now, given any element $x\in\mathfrak{g}$ one can write $x=z_1+z_2+\sum_{\varphi<0}y_{\varphi}+\sum_{\varphi>0}y_{\varphi}$, where $z_1\in\mathfrak{a}$, $z_2\in\mathfrak{m}$ and $y_{\varphi}\in\mathfrak{g}^{\varphi}(\varphi\in\Delta)$. Hence

$$x=\Big(z_2+\sum_{\varphi<0}(y_{\varphi}+\theta y_{\varphi})\Big)+z_1+\Big(-\sum_{\varphi<0}\theta y_{\varphi}+\sum_{\varphi>0}y_{\varphi}\Big)$$

and hence $x\in\mathfrak{k}+\mathfrak{a}+\mathfrak{n}$. Q.E.D.

We denote by U the universal enveloping algebra of $\mathfrak{g}$ and by $U(\mathfrak{k})$, $U(\mathfrak{a})$ and $U(\mathfrak{n})$ the universal enveloping algebras of $\mathfrak{k}$, $\mathfrak{a}$ and $\mathfrak{n}$, considered as subalgebras of U. Then one has $U=U(\mathfrak{k})\otimes U(\mathfrak{a})\otimes U(\mathfrak{n})$. Since $\mathfrak{a}$ is Abelian, $U(\mathfrak{a})=S(\mathfrak{a})$, the symmetric algebra on $\mathfrak{a}$, which again can be considered as the algebra of polynomial functions on $\mathfrak{a}'$, the dual of $\mathfrak{a}$. Thus for any $p\in U(\mathfrak{a})$ and $\lambda\in\mathfrak{a}'$, one has a uniquely defined complex number $p(\lambda)$.

If V is a U-module and $0\neq v\in V$, we say that v is a weight vector for the weight $\lambda\in\mathfrak{a}'$ if $x\cdot v=\langle\lambda,x\rangle v$ for all $x\in\mathfrak{a}$. In this case it is easy to see that for any $p\in U(\mathfrak{a})$, $p\cdot v=p(\lambda)v$.

Proposition 1.2.2. To each $u\in U$ there exists a unique element p_u of $U(\mathfrak{a})$ such that $u-p_u\in\mathfrak{k}U+U\mathfrak{n}$.

Proof. Since $U=U(\mathfrak{k})\otimes U(\mathfrak{a})\otimes U(\mathfrak{n})$, we have $U=(\mathfrak{k}U+U\mathfrak{n})\oplus U(\mathfrak{a})$. By taking p_u to be the component of u in $U(\mathfrak{a})$ with respect to the above decomposition, the proposition follows. Q.E.D.

The map $u\to p_u$ is a linear map.

Now if V is a finite-dimensional irreducible U-module there exists a weight

$\lambda \in \mathfrak{a}'$ such that $\lambda+\varphi$ is not a weight for any $\varphi \in \Delta_+$. Let $V^\lambda \subseteqq V$ denote the corresponding weight space. Also, let $V^\mathfrak{n}$ be the subspace annihilated by $\mathfrak{n}$.

Proposition 1.2.3. One has $V^\lambda = V^\mathfrak{n}$ and $V^\mathfrak{n}$ is an irreducible $\mathfrak{a}+\mathfrak{m}$-module.

Proof. It is clear that $V^\mathfrak{n}$ is stable under $\mathfrak{a}+\mathfrak{m}$. But now $V^\mathfrak{n}$ is irreducible under $\mathfrak{a}+\mathfrak{m}$ since any highest weight vector for $\mathfrak{a}+\mathfrak{m}$ in $V^\mathfrak{n}$ is a highest weight vector $\mathfrak{g}$ in V. But there is exactly one such vector, up to a scalar, so that $V^\mathfrak{n}$ is an irreducible $\mathfrak{a}+\mathfrak{m}$-module. Since $\mathfrak{a}$ is central in $\mathfrak{a}+\mathfrak{m}$, one has $V^\mathfrak{n} \subseteqq V^\mu$ for some weight $\mu \in \mathfrak{a}'$. But clearly $V^\lambda \subseteqq V^\mathfrak{n}$. Thus $\lambda=\mu$ and $V^\mathfrak{n}=V^\lambda$. Q.E.D.

Assume that $V^\mathfrak{k} \neq 0$. Then $(V')^\mathfrak{k} \neq 0$ where V' denotes the complex dual vector space of V, with the contragredient action of $\mathfrak{g}$, that is, the action of $\mathfrak{g}$ defined by

$$\langle v, x\cdot\alpha\rangle = \langle -x\cdot v, \alpha\rangle, \quad \text{for} \quad v\in V, \alpha\in V' \quad \text{and} \quad x\in\mathfrak{g}.$$

Now suppose ξ' is a nonzero element in $(V')^\mathfrak{k}$ and η a nonzero element in $V^\mathfrak{n}$.

Lemma 1.2.4. One has $\langle \eta, \xi'\rangle \neq 0$.

Proof. Suppose $\langle \eta, \xi'\rangle = 0$. Then

$$0 = \langle \eta, U(\mathfrak{k})\cdot\xi'\rangle = \langle U(\mathfrak{k})\cdot\eta, \xi'\rangle = \langle U(\mathfrak{k})U(\mathfrak{a})U(\mathfrak{n})\eta, \xi'\rangle = \langle U\cdot\eta, \xi'\rangle = \langle V, \xi'\rangle,$$

a contradiction. Q.E.D.

Corollary 1.2.5. Both $V^\mathfrak{n}$ and $(V')^\mathfrak{k}$ are one-dimensional, and hence so are $V^\mathfrak{k}$ and $(V')^\mathfrak{n}$ (by interchanging V and V').

We now assume that ξ' is chosen such that $\langle \eta, \xi'\rangle = 1$.

Proposition 1.2.6. One has $\langle u\cdot\eta, \xi'\rangle = p_u(\lambda)$ for any $u\in U$.

Proof. Recalling the definition of the map $u \to p_u$, we have

$$\langle u\cdot\eta, \xi'\rangle = \langle p_u\cdot\eta, \xi'\rangle = p_u(\lambda)\langle\eta, \xi'\rangle = p_u(\lambda). \qquad \text{Q.E.D.}$$

Now choose $\xi \in V^\mathfrak{k}$ so that $\langle \xi, \xi'\rangle = 1$.

Lemma 1.2.7. If $u\in U^\mathfrak{k}$ (=set of all elements in U which commute with all elements in $\mathfrak{k}$) then $u\cdot\xi = p_u(\lambda)\xi$.

Proof. Let $\eta=\eta_1+\alpha\xi$, where η_1 lies in the sum V_1 of the primary subspaces of V corresponding to the non-trivial irreducible representations of $\mathfrak{k}$, and $\alpha\in\mathbf{C}$. Now $\langle V_1, \xi'\rangle = 0$ and $u\cdot V_1 \subseteqq V_1$, since $u\in U^\mathfrak{k}$. Hence $1=\langle\eta, \xi'\rangle = \alpha\langle\xi, \xi'\rangle = \alpha$, so that $\eta=\eta_1+\xi$; furthermore, $\langle u\cdot\xi, \xi'\rangle = \langle u\cdot\eta, \xi'\rangle = \langle p_u(\lambda)\xi, \xi'\rangle$ by Proposition 1.2.6. Thus $u\cdot\xi = p_u(\lambda)\xi$ since $V^\mathfrak{k}$ and $(V')^\mathfrak{k}$ are one-dimensional non-singularly paired spaces. Q.E.D.

Lemma 1.2.8. For $u\in U^\mathfrak{k}$ and $v\in U$, one has $p_{uv}=p_u p_v$.

Proof. One has $v \equiv p_v \pmod{\mathfrak{k}U + U\mathfrak{n}}$. Since $u \in U^{\mathfrak{k}}$, one has $u\mathfrak{k}U = \mathfrak{k}uU$, so that

$$uv \equiv up_v \pmod{\mathfrak{k}U + U\mathfrak{n}},$$

showing that $p_{uv} = p_{(up_v)}$.

Now if $u = xy + wz + p_u$, where $x \in \mathfrak{k}$, $y, w \in U$ and $z \in \mathfrak{n}$, one has

$$up_v = x(yp_v) + w(zp_v) + p_u p_v.$$

Thus the lemma will follow, by linearity, if we show that $zp_v \in U\mathfrak{n}$. But this follows from the fact that $\mathfrak{a}$ normalizes $\mathfrak{n}$. Q.E.D.

§ 3. The irreducible module Z_λ

Let $\varrho \in \mathfrak{a}'$ be the element defined by $\varrho(x) = \frac{1}{2}\operatorname{tr}(\operatorname{ad} x|\mathfrak{n})$ for all $x \in \mathfrak{a}$. (One can easily check that ϱ is explicitly given by $\varrho = \frac{1}{2}\sum_{\varphi \in \Delta_+} (\dim \mathfrak{g}^\varphi)\varphi$.) Let $T: \mathfrak{a}' \to \mathfrak{a}'$ be defined by $T(\lambda) = \lambda + \varrho$. Let W be the Weyl group of $(\mathfrak{a}, \mathfrak{g})$ operating on $\mathfrak{a}'$. We define W_T by

$$W_T = \{T^{-1}\sigma T \mid \sigma \in W\}.$$

Then clearly W_T is a group of transformations of $\mathfrak{a}'$; it is called the *translated* Weyl group. The transformation T defines an algebra automorphism, also denoted by T, of the algebra $S(\mathfrak{a})(=U(\mathfrak{a}))$ of polynomial functions on $\mathfrak{a}'$, defined by

$$(Tp)(\lambda) = p(T^{-1}\lambda) = p(\lambda - \varrho).$$

Similarly, we may regard W and W_T as acting on $U(\mathfrak{a})$. We now have the following:

Theorem 1.3.1 (Harish–Chandra). *$U\mathfrak{k} \cap U^{\mathfrak{k}} = \mathfrak{k}U \cap U^{\mathfrak{k}}$ is a two-sided ideal in $U^{\mathfrak{k}}$. If $(U(\mathfrak{a}))^{W_T}$ is the set of W_T-invariants in $U(\mathfrak{a})$, then one has an exact sequence*

$$0 \to I \xrightarrow{i} U^{\mathfrak{k}} \xrightarrow{p} (U(\mathfrak{a}))^{W_T} \to 0,$$

where $I = U\mathfrak{k} \cap U^{\mathfrak{k}}$, i is the inclusion map and p is the map $u \to p_u$. (For a proof of this theorem, see Chapter X, Lemma 6.13 [6a].)

We now study certain irreducible representations of $\mathfrak{g}$ (or, which is the same, irreducible representations of $U = U(\mathfrak{g})$) which are not necessarily finite-dimensional.

Lemma 1.3.2. Let Z be an irreducible U-module. Suppose $Z^{\mathfrak{k}} \neq 0$. Then $Z^{\mathfrak{k}}$ is one-dimensional.

Proof. We first note that since Z is algebraically irreducible, the commuting ring $\mathscr{E}$ to the image of U in $\operatorname{End} Z$ consists of the scalars. Indeed $\mathscr{E}$ is a division ring over $\mathbf{C}$ and $\dim U$ countable $\Rightarrow \dim Z$ countable $\Rightarrow \dim \mathscr{E}$ countable, since the map $\mathscr{E} \to Z$, $u \to u \cdot z$, for some $0 \neq z \in Z$, is injective.

On the other hand, every field F over $\mathbf{C}$ is either $\mathbf{C}$ itself or else has uncountable dimension over $\mathbf{C}$. (Indeed, if $X \in F$ is transcendental over $\mathbf{C}$, then $\left\{\frac{1}{X-\lambda} \,\middle|\, \lambda \in \mathbf{C}\right\}$ is an uncountable linearly independent set in F over $\mathbf{C}$.) Thus if $x \in \mathscr{E}$, then $\mathbf{C}(x) = \mathbf{C}$, so that $x \in \mathbf{C}$. (This argument is due to Dixmier.)

Now $U^{\mathfrak{k}}$ leaves $Z^{\mathfrak{k}}$ stable. Let V be an arbitrary finite-dimensional subspace of $Z^{\mathfrak{k}}$, and let $\tau \in \operatorname{End} V$. By the last paragraph and the Jacobson density theorem, we have that there exists $a \in U$ such that $a \cdot v = \tau v$ for all $v \in V$. But for all $y \in \mathfrak{k}$, one has $((\operatorname{ad} y)^n a) \cdot v = 0$ for all $n = 1, 2, \ldots$, since v and $a \cdot v \in Z^{\mathfrak{k}}$; and so $(k \cdot a) \cdot v = \tau v$ for all $k \in \exp \operatorname{ad} \mathfrak{k}$. Integrating over the compact group $\exp \operatorname{ad} \mathfrak{k}_{\mathbf{R}}$, we get $x \cdot v = \tau v$ for all $v \in V$, where $x \in U^{\mathfrak{k}}$. In particular, if $\dim Z^{\mathfrak{k}} > 1$ then the image of $U^{\mathfrak{k}}$ in $\operatorname{End} Z^{\mathfrak{k}}$ is non-commutative.

(It follows easily at this point that $Z^{\mathfrak{k}}$ is irreducible under $U^{\mathfrak{k}}$.)

The ideal $I = U\mathfrak{k} \cap U^{\mathfrak{k}}$ acts trivially on $Z^{\mathfrak{k}}$, so that $Z^{\mathfrak{k}}$ can be considered as a module over $U^{\mathfrak{k}}/I$, which is commutative. Hence $\dim Z^{\mathfrak{k}} = 1$ by the above.

Q.E.D.

Remark 1.3.3. The above proof can be generalized to give analogous information about the primary subspaces of Z associated with the non-trivial finite dimensional irreducible representations of $\mathfrak{k}$.

Corollary 1.3.4. There exists $\lambda \in \mathfrak{a}'$ such that one has $u \cdot \xi = p_u(\lambda)\xi$ for all $u \in U^{\mathfrak{k}}$, $\xi \in Z^{\mathfrak{k}}$.

Proof. Let $f: U^{\mathfrak{k}} \to \mathbf{C}$ be the ring homomorphism defined by $u \cdot \xi = f(u)\xi$ for $u \in U^{\mathfrak{k}}$. Then f vanishes on I and hence induces a ring homomorphism $f: U^{\mathfrak{k}}/I \to \mathbf{C}$. Now consider the exact sequence (Theorem 3.3.1) $0 \to I \to U^{\mathfrak{k}} \to (U(\mathfrak{a}))^{W_T} \to 0$, where the map $U^{\mathfrak{k}} \to (U(\mathfrak{a}))^{W_T}$ is given by $u \to p_u$. The homomorphism $f: U^{\mathfrak{k}}/I \to \mathbf{C}$ can be considered as a homomorphism of $(U(\mathfrak{a}))^{W_T}$ into $\mathbf{C}$ (by defining for $u \in U^{\mathfrak{k}}$, $f(p_u) = f(u)$). But since W_T operates as a reductive algebraic group every ring homomorphism of $(U(\mathfrak{a}))^{W_T}$ into $\mathbf{C}$ is obtained by evaluating on a unique W_T-orbit in $\mathfrak{a}'$.

Thus there exists $\lambda \in \mathfrak{a}'$ (unique up to conjugation by W_T) such that $u \cdot \xi = p_u(\lambda)\xi$ for all $u \in U^{\mathfrak{k}}$, $\xi \in Z^{\mathfrak{k}}$. Q.E.D.

Thus, with every irreducible U-module Z such that $Z^{\mathfrak{k}} \neq 0$, we have associated an orbit of W_T in $\mathfrak{a}'$. The following theorem shows that this sets up a one-one correspondence between the set of all equivalence classes of such irreducible U-modules and all orbits of W_T in $\mathfrak{a}'$.

Theorem 1.3.5. *Let $\lambda \in \mathfrak{a}'$ be arbitrary. Then there exists uniquely (up to equivalence) an irreducible U-module Z_λ such that* (i) $Z_\lambda^{\mathfrak{k}} \neq 0$ *and* (ii) *for any $u \in U^{\mathfrak{k}}$, the action of u on $Z_\lambda^{\mathfrak{k}}$ is scalar multiplication by $p_u(\lambda)$. Thus we have a one-one correspondence between the set of equivalence classes of irreducible U-modules Z with $Z^{\mathfrak{k}} \neq 0$ and the set of W_T-orbits in $\mathfrak{a}'$.*

Proof. Define $L_\lambda = \{u \in U \mid p_{(vu)_0}(\lambda) = 0 \text{ for all } v \in U\}$, where we have denoted by w_0 the component in $U^{\mathfrak{k}}$ of an element $w \in U$, with respect to the decomposition $U = U^{\mathfrak{k}} \oplus [\mathfrak{k}, U]$. Clearly L_λ is a proper left ideal in the ring U. We assert that the U-module $Z_\lambda = U/L_\lambda$ has the desired properties. For this we first show that

(i) $\mathfrak{k} \subseteqq L_\lambda$;
(ii) for all $u \in U^{\mathfrak{k}}$, $u - p_u(\lambda) \in L_\lambda$;
(iii) L_λ is a maximal left ideal in U.

Supposing (i), (ii) and (iii) are proved, it follows from (iii) that $Z_\lambda = U/L_\lambda$ is an irreducible U-module. Then (i) implies that $1 + L_\lambda \in Z_\lambda^{\mathfrak{k}}$. Now Lemma 1.3.2 implies that $Z_\lambda^{\mathfrak{k}}$ is one-dimensional, so that it is generated by $1 + L_\lambda$. Then (ii) implies that $u(1 + L_\lambda) = p_u(\lambda)(1 + L_\lambda)$. Hence, up to isomorphism, $Z_\lambda = U/L_\lambda$ has the desired properties.

It remains to prove (i), (ii) and (iii).

One has $\mathfrak{k} \subseteqq L_\lambda$ if and only if for all $x \in \mathfrak{k}$ and $u \in U$, $p_{(ux)_0}(\lambda) = 0$. But $(ux)_0 \in U\mathfrak{k} \cap U^{\mathfrak{k}}$ since $U\mathfrak{k}$ is stable under the map $v \to v_0$ and according to Theorem 1.3.1, $U\mathfrak{k} \cap U^{\mathfrak{k}}$ is in the kernel of the map $u \to p_u$. This proves (i).

To show (ii) we have to show that for all $u \in U^{\mathfrak{k}}$ and $v \in U$, $p_{(v(u - p_u(\lambda)))_0}(\lambda) = 0$. But this is clear since $(vu)_0 = v_0 u$, and $p_{v_0 u} = p_{v_0} p_u$ by Lemma 1.2.8. Thus (ii) is proved.

To show (iii), suppose that M is a left ideal of U such that $L_\lambda \subset M \subset U$. Let $u \in M - L_\lambda$. The fact that $u \notin L_\lambda$ implies that there exists $v \in U$ such that $p_{(vu)_0}(\lambda) \neq 0$. We assert that $(vu)_0 \in M$. For this, it is enough to show that M is stable under the adjoint action of $\mathfrak{k}$. But this is obvious, for on the one hand, M being a left ideal, $\mathfrak{k}M \subseteqq M$, and on the other hand, $M\mathfrak{k} \subseteqq U\mathfrak{k} \subseteqq L_\lambda \subseteqq M$.

Of course, one has $(vu)_0 \in U^{\mathfrak{k}}$. Hence by (ii) $(vu)_0 - p_{(vu)_0}(\lambda) \in L_\lambda \subseteqq M$. Thus one has

$$0 \neq p_{(vu)_0}(\lambda) = (vu)_0 - ((vu)_0 - p_{(vu)_0}(\lambda)) \in M.$$

But this implies $M = U$. This contradiction proves that L_λ is a maximal left ideal in U.

Now, we prove the uniqueness of Z_λ. For this, suppose that Z is an irreducible U-module such that $Z^{\mathfrak{k}} \neq 0$ and such that for any $u \in U^{\mathfrak{k}}$, the action of u on $Z^{\mathfrak{k}}$ is multiplication by the scalar $p_u(\lambda)$. Choose a non-zero element $\xi \in Z^{\mathfrak{k}}$. Since Z is irreducible, if M is the annihilator of ξ then M is a maximal left ideal of U, and one has $Z \cong U/M$. Thus to show that $Z \cong Z_\lambda$, it is enough to show that $M \subseteqq L_\lambda$. This is proved exactly as (iii) is proved above, except that in place of the relation $L_\lambda \subseteqq M$, we now have directly that $U\mathfrak{k} \subseteqq M$ and $u - p_u(\lambda) \in M$ for all $u \in U^{\mathfrak{k}}$.

Q.E.D.

Remark 1.3.6. The last part of the above proof has shown that if M is any proper left ideal in U containing $U\mathfrak{k}$ and $u - p_u(\lambda)$ for all $u \in U^{\mathfrak{k}}$, then $M \subseteqq L_\lambda$.

§ 4. The decomposition $U = U\mathfrak{k} + H^* J^*$

In this section we obtain certain decompositions of the universal enveloping algebra U of $\mathfrak{g}$ which arise from the decomposition $\mathfrak{g}=\mathfrak{k}+\mathfrak{p}$. Let $S(\mathfrak{g})$ denote the symmetric algebra of $\mathfrak{g}$.

There exists a unique linear map $\beta: S(\mathfrak{g}) \to U$ such that $\beta(x^k)=x^k$ for $x \in \mathfrak{g}$. If $x_1, x_2, \ldots, x_n \in \mathfrak{g}$, then $\beta(x_1 x_2 \cdots x_n)=\frac{1}{n!}\sum_{\sigma} x_{\sigma(1)} x_{\sigma(2)} \cdots x_{\sigma(n)}$, where the summation is over all permutations σ of $\{1, 2, \ldots, n\}$. Note that β is a linear isomorphism which commutes with the actions of $G(=\mathrm{Int}\,\mathfrak{g})$ on $S(\mathfrak{g})$ and on U. We recall from [10] that K_θ is the subgroup of elements of G which commute with θ, and that K is the identity component of K_θ.

We now define S^*, J^* and $H^* \subseteq U$ by setting $S^*=\beta(S)$, $J^*=\beta(J)$ and $H^*=\beta(H)$. Here $S=S(\mathfrak{p})$ is the symmetric algebra of $\mathfrak{p}$, $J=S^K$, and H is the space of harmonics in S (see [10], Chapter III). Note that S^* is spanned by $\{x^k \mid x \in \mathfrak{p}, k \in \mathbf{Z}_+\}$, and that H^* is spanned by $\{x^k \mid x \in \mathfrak{n}\}$, where $\mathfrak{n}$ is the variety of nilpotents in $\mathfrak{p}$ (see [10], Chapter III). Moreover, we have that $J=S^{K_\theta}$ (see [10], Proposition 10).

Lemma 1.4.1. One has (i) $U=U\mathfrak{k} \oplus S^*$ and (ii) $U=\mathfrak{k}U \oplus S^*$.

Proof. Let $x_1, \ldots, x_d$ be a basis of $\mathfrak{p}$ and $x_{d+1}, \ldots, x_m$ a basis of $\mathfrak{k}$. Then by the Birkhoff—Witt theorem, we have $U=U\mathfrak{k} \oplus S^{**}$, where S^{**} is spanned by $\{x_1^{j_1} \cdots x_d^{j_d} \mid j_1, \ldots, j_d \in \mathbf{Z}_+\}$. Now U has a natural filtration $U_{(i)}$ $(i \in \mathbf{Z}_+)$, where $U_{(i)}$ is the subspace of U spanned by $\{y_1 \cdots y_j \mid y_1, \ldots, y_j \in \mathfrak{g}$ and $j \leqq i\}$. One has in fact $U_{(i)}=(U\mathfrak{k})_{(i)} \oplus (S^{**})_{(i)}$, where $(U\mathfrak{k})_{(i)}=U\mathfrak{k} \cap U_{(i)}$ and $(S^{**})_{(i)}=S^{**} \cap U_{(i)}$. We now prove by induction that $U_{(i)}=(U\mathfrak{k})_{(i)} \oplus S^*_{(i)}$, where $S^*_{(i)}=S^* \cap U_{(i)}$. To do this, assume that $i>0$ and that for $j<i$, $U_{(j)}=(U\mathfrak{k})_{(j)} \oplus S^*_{(j)}$. (Note that $U_{(0)}=\mathbf{C} \cdot 1$, $(U\mathfrak{k})_0=0$ and $S^*_{(0)}=\mathbf{C} \cdot 1$, so that $U_{(0)}=(U\mathfrak{k})_0 \oplus S^*_{(0)}$.) Now we prove that $U_{(i)}=(U\mathfrak{k})_{(i)}+S^*_{(i)}$. Since $U_{(i)}=(U\mathfrak{k})_{(i)}+S^{**}_{(i)}$ it is enough to prove that $S^{**}_{(i)} \subseteq (U\mathfrak{k})_{(i)}+S^*_{(i)}$. For this, clearly it is enough to prove that if $y_1, y_2, \ldots, y_i \in \mathfrak{p}$, then $y_1 y_2 \cdots y_i \in (U\mathfrak{k})_{(i)}+S^*_{(i)}$. We denote the multiplication in $S(\mathfrak{g})$ by "$\circ$". We have

$$y_1 \circ y_2 \circ \cdots \circ y_i \in S \quad \text{and} \quad \beta(y_1 \circ y_2 \circ \cdots \circ y_i) = \frac{1}{i!}\sum_{\sigma} y_{\sigma(1)} y_{\sigma(2)} \cdots y_{\sigma(i)},$$

where σ runs through all permutations of $1, 2, \ldots, i$. Thus $\beta(y_1 \circ y_2 \circ \cdots \circ y_i) \in S^*_{(i)}$. Now consider the element $y_1 y_2 \cdots y_i - \beta(y_1 \circ y_2 \circ \cdots \circ y_i) \in U_{(i)}$. We have

$$y_1 y_2 \cdots y_i - \beta(y_1 \circ y_2 \circ \cdots \circ y_i) = \frac{1}{i!}\sum_{\sigma} (y_1 y_2 \cdots y_i - y_{\sigma(1)} y_{\sigma(2)} \cdots y_{\sigma(i)}).$$

Using the fact that if $u \in U_{(k)}$ and $v \in U_{(l)}$ then $uv - vu \in U_{(k+l-1)}$, one can easily show that for each σ, $y_1 y_2 \cdots y_i - y_{\sigma(1)} y_{\sigma(2)} \cdots y_{\sigma(i)} \in U_{(i-1)}$. Thus

$$y_1 y_2 \cdots y_i - \beta(y_1 \circ y_2 \circ \cdots \circ y_i) \in U_{(i-1)}.$$

Hence

$$y_1 y_2 \cdots y_i \in S^*_{(i)} + U_{(i-1)} = S^*_{(i)} + (U\mathfrak{k})_{(i-1)} + S^*_{(i-1)} \subseteqq (U\mathfrak{k})_{(i)} + S^*_{(i)}.$$

Thus, indeed, $U_{(i)} = (U\mathfrak{k})_{(i)} + S^*_{(i)}$. To show that $(U\mathfrak{k})_{(i)} \cap S^*_{(i)} = 0$, it is enough to show that $\dim S^*_{(i)} = \dim S^{**}_{(i)}$, since $(U\mathfrak{k})_{(i)} + S^*_{(i)} = U_{(i)} = (U\mathfrak{k})_{(i)} \oplus S^{**}_{(i)}$. But if $S_{(i)} \subseteqq S$ is defined by $S_{(i)} = \sum_{j=0}^{i} S_j$, where S_j are the graded components, then $\dim S_{(i)} = \dim S^*_{(i)}$ and $\dim S_{(i)} = \dim S^{**}_{(i)}$, the former because β is a filtration-preserving isomorphism and the latter in view of the Birkhoff—Witt theorem. This completes the proof of (i). Similarly, one can prove (ii). Q.E.D.

Lemma 1.4.2. One has

(i) $U = U\mathfrak{k} \oplus H^* J^*$;
(ii) $U = U\mathfrak{k} \oplus J^* H^*$;
(iii) $U = \mathfrak{k}U \oplus H^* J^*$;
(iv) $U = \mathfrak{k}U \oplus J^* H^*$.

Proof. We prove only (i), the other three cases are proved similarly. Let $\alpha: H \otimes J \to H^* J^*$ be the map given by $\alpha(u \otimes v) = \beta(u)\beta(v)$. The space $S \otimes S$ is naturally filtered inducing a filtration on $H \otimes J$. (In fact $S \otimes S$ and $H \otimes J$ are naturally graded.) We first prove inductively that $U_{(i)} = (U\mathfrak{k})_{(i)} + \alpha((H \otimes J)_i)$. This is obviously true if $i = 0$ and hence we may assume $i > 0$ and the statement is true for $j < i$. Let $D_{(i)} = (U\mathfrak{k})_{(i)} + \alpha((H \otimes J)_i)$. By Lemma 1.4.1 it is enough to prove that $S^*_{(i)} \subseteqq D_{(i)}$. But for this it is enough to show that if $u \in H_{(k)}$ and $v \in J_{(i-k)}$ then $\beta(u \circ v) \in D_{(i)}$ (since $S = H \circ J$). But now clearly $\beta(u)\beta(v) \in D_{(i)}$. On the other hand the standard argument of Lemma 1.4.1 shows that $\beta(u \circ v) - \beta(u)\beta(v) \in U_{(i-1)}$. But by induction $U_{(i-1)} = D_{(i-1)} = D_{(i)}$. Thus $\beta(u \circ v) \in D_{(i)}$ or $D_{(i)} = U_{(i)}$. Next we assert that the sum is direct. That is $U_{(i)} = (U\mathfrak{k})_{(i)} \oplus \alpha((H \otimes J)_i)$. But using the fact that $S = H \otimes J$, this is clear since then $\dim \alpha((H \otimes J)_i) = \dim (H \otimes J)_i = \dim S_{(i)} = \dim S^*_{(i)}$. Now since $\alpha(H \otimes J) = H^* J^*$ it follows that $U = U\mathfrak{k} \oplus H^* J^*$.

Remark. 1.4.3. If $(H^* J^*)_{(i)} = H^* J^* \cap U_{(i)}$ note that $(H^* J^*)_{(i)} = \alpha((H \otimes J)_i)$ and hence one has $U_{(i)} = (U\mathfrak{k})_{(i)} \oplus (H^* J^*)_{(i)}$. This follows from the proof of Lemma 1.4.2. together with the fact $\alpha((H \otimes J)_i) \subseteqq (H^* J^*)_{(i)}$.

Also from the fact that $\dim (H^* J^*)_{(i)} = \dim S_{(i)}$ and the fact that $S = J \otimes H$, it follows that the map $H^* \otimes J^* \to H^* J^*$ defined by $u \otimes v \to uv$ is an isomorphism. Similarly, the map $J^* \otimes H^* \to J^* H^*$ defined by $u \otimes v \to uv$ is an isomorphism.

Remark 1.4.4. Since β is a G-isomorphism of $S(\mathfrak{g})$ onto U, it follows that $J^* \subseteqq U^{\mathfrak{k}}$. Thus, by Lemma 1.4.2, we have $U = U\mathfrak{k} + H^* U^{\mathfrak{k}} = U\mathfrak{k} + U^{\mathfrak{k}} H^* = \mathfrak{k}U + H^* U^{\mathfrak{k}} = \mathfrak{k}U + U^{\mathfrak{k}} H^*$. Moreover, all these equalities remain true when $U^{\mathfrak{k}}$ is replaced by U^{K_0}, since $J = S^{K_0}$.

§ 5. The matrix $R^{\gamma}(\lambda)$ and K multiplicities in Z_λ

Definition 1.5.1. A module X for U is called *admissible* if

(i) $\dim X^{\mathfrak{k}}=1$

and

(ii) X is completely reducible as a $\mathfrak{k}$-module.

Definition 1.5.2. In this paper a *cyclic* module for U is an admissible module X for U such that $U\cdot X^{\mathfrak{k}}=X$.

We note that (ii) follows from (i) and the condition that $U\cdot X^{\mathfrak{k}}=X$. Indeed, U is a completely reducible $\mathfrak{k}$-module, and the map from U to X given by $u\to u\cdot\xi$ (ξ a non-zero element of $X^{\mathfrak{k}}$) is a $\mathfrak{k}$-module map.

Given any $\lambda\in\mathfrak{a}'$ the unique irreducible U-module Z_λ of Theorem 1.3.5 is a cyclic module for U.

We denote by $\mathscr{A}$ the set of equivalence classes of all admissible modules for U and by $\mathscr{C}$ the set of equivalence classes of all cyclic modules for U.

Now suppose X is a module for U such that $X^{\mathfrak{k}}$ is one-dimensional. Let $\xi\in X^{\mathfrak{k}}$, $\xi\neq 0$. Then the proof of Corollary 1.3.4 also proves that there exists $\lambda\in\mathfrak{a}'$ such that for any $u\in U^{\mathfrak{k}}$, $u\cdot\xi=p_u(\lambda)\xi$.

Now, given any $\lambda\in\mathfrak{a}'$, we denote by $\mathscr{A}(\lambda)$ the set of equivalence classes of all admissible modules X for U such that the action of any element $u\in U^{\mathfrak{k}}$ on $X^{\mathfrak{k}}$ is multiplication by the scalar $p_u(\lambda)$. We denote by $\mathscr{C}(\lambda)$ the set of all classes of modules in $\mathscr{A}(\lambda)$ which are cyclic. Then one has $\{Z_\lambda\}\in\mathscr{C}(\lambda)$, where $\{Z_\lambda\}$ denotes the class of Z_λ.

Recall (see the proof of Theorem 1.3.5) that we have $Z_\lambda=U/L_\lambda$, where $L_\lambda=\{u\in U\,|\,p_{(vu)_0}(\lambda)=0$ for all $v\in U\}$. We have $\mathfrak{k}\subseteq L_\lambda$ and for all $u\in U^{\mathfrak{k}}$, $u-p_u(\lambda)\in L_\lambda$. We now define $L_\lambda^{\min}$ to be the ideal of U generated by the set $\mathfrak{k}\cup\{u-p_u(\lambda)\,|\,u\in U^{\mathfrak{k}}\}$. Observe that $L_\lambda^{\min}\subseteq L_\lambda$. Moreover, note that for any $Y\in\mathscr{C}(\lambda)$, if $0\neq\xi\in Y^{\mathfrak{k}}$ and if $L\subseteq U$ is the annihilator of ξ in U, then $L_\lambda^{\min}\subseteq L\subseteq L_\lambda$ ($L\subseteq L_\lambda$ by Remark 1.3.6). We put $Y_\lambda=U/L_\lambda^{\min}$.

Note that K_θ acts on U by restriction of the action of G on U.

Lemma 1.5.3. The ideals $L_\lambda^{\min}$ and L_λ are stable under the action of K_θ, so that the U-modules $Y_\lambda=U/L_\lambda^{\min}$ and $Z_\lambda=U/L_\lambda$ have a canonical K_θ-module structure.

Proof. First we prove that $L_\lambda^{\min}$ is stable under K_θ. The sets $\mathfrak{k}$ and $\{u-p_u(\lambda)\,|\,u\in U^{\mathfrak{k}}\}$ generate $L_\lambda^{\min}$, and so it is enough to prove that K_θ leaves these sets stable. Clearly $\mathfrak{k}$ is stable under K_θ. Now for any element $u\in U^{\mathfrak{k}}$ and any $a\in K_\theta$, one has $a\cdot(u-p_u(\lambda))=a\cdot u-p_u(\lambda)$ and $a\cdot u\in U^{\mathfrak{k}}$. We assert that for any $u\in U^{\mathfrak{k}}$ and $a\in K_\theta$, $p_u(\lambda)=p_{a\cdot u}(\lambda)$. To see this, consider the decomposition (Lemma 1.4.2) $U=\mathfrak{k}U\oplus H^*J^*$. Observe that $\mathfrak{k}U$ and H^*J^* are stable under the adjoint action of $\mathfrak{k}$. Hence one has $U^{\mathfrak{k}}=\mathfrak{k}U\cap U^{\mathfrak{k}}\oplus H^*J^*\cap U^{\mathfrak{k}}$. Since $H^*J^*\cong$

$\cong H^* \otimes J^*$ and since the space $(H^*)^{\mathfrak{k}} = \mathbf{C} \cdot 1$, it follows that $H^* J^* \cap U^{\mathfrak{k}} = J^*$. Thus one has $U^{\mathfrak{k}} = \mathfrak{k} U \cap U^{\mathfrak{k}} \oplus J^*$. Hence for $u \in U^{\mathfrak{k}}$, writing $u = u_1 + u_2$, where $u_1 \in \mathfrak{k} U \cap U^{\mathfrak{k}}$ and $u_2 \in J^*$, one has $p_u = p_{u_1} + p_{u_2}$, since obviously $p_{u_1} = 0$. Now if $a \in K_\theta$, then a leaves $U^{\mathfrak{k}}$ and $\mathfrak{k} U \cap U^{\mathfrak{k}}$ stable. Moreover, since the map $\beta \colon S(\mathfrak{g}) \to U$ is a G-isomorphism, $J^* = \beta(J)$ is the set of invariants for the action of K_θ on $S^* = \beta(S)$. Thus one has $a \cdot u = a \cdot u_1 + a \cdot u_2 = a \cdot u_1 + u_2$, and so $p_{a \cdot u} = p_{u_2} = p_u$, proving the assertion. Thus K_θ leaves $L_\lambda^{\min}$ stable.

To prove that K_θ leaves L_λ stable, it is enough to show that $L_\lambda^{\min}$ is contained in a unique maximal left ideal of U, for then, for any $a \in K_\theta$, $L_\lambda^{\min} \subseteqq a \cdot L_\lambda$ and $a \cdot L_\lambda$ is maximal. But L_λ is the unique maximal left ideal containing $L_\lambda^{\min}$ by Remark 1.3.6. Q.E.D.

We note that $U^{\mathfrak{k}} = \mathfrak{k} U \cap U^{\mathfrak{k}} \oplus J^*$, which has been established in the above proof.

Proposition 1.5.4. For $\lambda \in \mathfrak{a}'$, let $X \in \mathscr{C}(\lambda)$, and let $\xi \in X^{\mathfrak{k}}$, $\xi \neq 0$. The map $f \colon H^* \to X$ given by $u \to u \cdot \xi$ is a $\mathfrak{k}$-linear map between the $\mathfrak{k}$-modules H^* and X. Moreover, f is a surjection.

Proof. For $u \in H^*$ and $x \in \mathfrak{k}$, one has

$$f([x, u]) = [x, u] \cdot \xi = (xu - ux) \cdot \xi = (xu) \cdot \xi - (ux) \cdot \xi = x \cdot (u \cdot \xi) - u \cdot (x \cdot \xi) =$$

$$= x \cdot (u \cdot \xi) = x \cdot f(u),$$

showing that f is a $\mathfrak{k}$-linear map.

Now $X = U \cdot \xi = (U\mathfrak{k} + H^* U^{\mathfrak{k}}) \cdot \xi$ by Remark 1.4.4. But $U\mathfrak{k} \cdot \xi = 0$, and every element of $U^{\mathfrak{k}}$ acts on ξ by a scalar. Thus $X = H^* \cdot \xi$, showing that f is surjective. Q.E.D.

For $\lambda \in \mathfrak{a}'$, consider the unique (up to equivalence) irreducible U-module $Z_\lambda \in \mathscr{C}(\lambda)$. By Lemma 1.5.3, Z_λ is a K_θ-module. Also, H^* is a K_θ-module, and it is clear that the map $f \colon H^* \to Z_\lambda$ of Proposition 1.5.4 is a K_θ-linear map. Thus we have the following corollary. (We recall from [10] that M_θ is the centralizer of $\mathfrak{a}$ in K_θ, and that Γ is the set of equivalence classes of irreducible holomorphic finite-dimensional K_θ-modules admitting an M_θ-fixed vector, or equivalently, the set of equivalence classes of irreducible K_θ-modules occurring in H^*. We recall further that for all $\gamma \in \Gamma$, $l(\gamma) = \dim V_\gamma^{M_\theta} =$ the multiplicity of γ in H^*, where V^γ denotes a fixed module in the class γ (see [10]).)

Corollary 1.5.5. Let $0 \neq \xi \in Z_\lambda^{\mathfrak{k}}$. Then the map $f \colon H^* \to Z_\lambda$, defined by setting $f(u) = u \cdot \xi$, is a K_θ-linear surjection, so that Z_λ is completely reducible as a K_θ-module. Hence $Z_\lambda = \coprod_{\gamma \in \Gamma} (Z_\lambda)^\gamma$, where $(Z_\lambda)^\gamma$ denotes the primary component of Z_λ of type γ. Furthermore, $m(\gamma) \leqq l(\gamma)$, where $m(\gamma)$ is the multiplicity of γ in Z_λ.

Remark. Unlike $l(\gamma)$ note that $m(\gamma)$ depends, of course, on λ and should be written $m_\lambda(\gamma)$. However λ will generally be understood so we simply write $m(\gamma)$.

Our next result concerns another special case of Proposition 1.5.4.

Lemma 1.5.6. The class of the U-module $Y_\lambda = U/L_\lambda^{\min}$ is a member of $\mathscr{C}(\lambda)$. The map $f: H^* \to Y_\lambda$ of Proposition 1.5.4 is a bijection. Moreover, $\{Y_\lambda\}$ is the unique element of $\mathscr{C}(\lambda)$ with this property.

Proof. By definition, $L_\lambda^{\min}$ is the ideal of U generated by $\mathfrak{k} \cup \{u - p_u(\lambda) \mid u \in U^{\mathfrak{k}}\}$. Note that $C_\lambda = \{u - p_u(\lambda) \mid u \in U^{\mathfrak{k}}\}$ is a two-sided ideal of $U^{\mathfrak{k}}$, and that one has

$$L_\lambda^{\min} = UC_\lambda + U\mathfrak{k}.$$

But $U = U\mathfrak{k} \oplus H^* J^*$ (Lemma 1.4.2), and so $L_\lambda^{\min} = (U\mathfrak{k} \oplus H^* J^*) C_\lambda + U\mathfrak{k}$. Note that $\mathfrak{k} C_\lambda = C_\lambda \mathfrak{k}$. Moreover, since $1 \in J^* \subseteq U^{\mathfrak{k}}$, $J^* C_\lambda = C_\lambda$. Thus one has $L_\lambda^{\min} = H^* C_\lambda + U\mathfrak{k}$.

Now $U^{\mathfrak{k}} = \mathfrak{k}U \cap U^{\mathfrak{k}} \oplus J^*$, and $\mathfrak{k}U \cap U^{\mathfrak{k}} \subseteq C_\lambda$. Hence if we define $J_\lambda^* = J^* \cap C_\lambda$, then $C_\lambda = \mathfrak{k}U \cap U^{\mathfrak{k}} \oplus J_\lambda^* = U\mathfrak{k} \cap U^{\mathfrak{k}} \oplus J_\lambda^*$, so that $L_\lambda^{\min} = H^* J_\lambda^* + U\mathfrak{k}$.

But $J_\lambda^* = \{u - p_u(\lambda) \mid u \in J^*\}$, so that $J^* = \mathbf{C} \cdot 1 \oplus J_\lambda^*$. Moreover, since $H^* J^* \cong$ $\cong H^* \otimes J^*$, one has

$$U = U\mathfrak{k} \oplus H^* J^* = U\mathfrak{k} \oplus H^*(\mathbf{C} \cdot 1 \oplus J_\lambda^*) = H^* \oplus H^* J_\lambda^* \oplus U\mathfrak{k} = H^* \oplus L_\lambda^{\min}.$$

It should be observed that this direct sum is not a U-module direct sum, but is a $\mathfrak{k}$-module direct sum. Thus $Y_\lambda = U/L_\lambda^{\min}$, considered as a $\mathfrak{k}$-module, is isomorphic to H^*. In particular, $(Y_\lambda)^{\mathfrak{k}}$ is one-dimensional, and Y_λ as a $\mathfrak{k}$-module is completely reducible. Moreover, the coset $1 + L_\lambda^{\min} \in (Y_\lambda)^{\mathfrak{k}}$, so that $U \cdot (Y_\lambda)^{\mathfrak{k}} = Y_\lambda$. This shows that $\{Y_\lambda\} \in \mathscr{C}(\lambda)$.

To prove that the map $f: H^* \to Y_\lambda$ defined by $f(u) = u \cdot \xi$, where $0 \neq \xi \in (Y_\lambda)^{\mathfrak{k}}$, is a bijection, we can assume without loss that $\xi = 1 + L_\lambda^{\min}$. But now the fact that $f: H^* \to Y_\lambda$ is a bijection follows from the decomposition $U = H^* \oplus L_\lambda^{\min}$.

Now suppose $\{Y_\lambda'\} \in \mathscr{C}(\lambda)$ and that $f': H^* \to Y_\lambda'$ is a bijection, where $f'(u) = u \cdot \xi$, for some non-zero $\xi \in (Y_\lambda')^{\mathfrak{k}}$. Let L be the kernel of f, so that $Y_\lambda' \cong U/L$. But we know that $L_\lambda^{\min} \subseteq L$. Thus the natural map $U/L_\lambda^{\min} \to U/L$ gives a map $\eta: Y_\lambda \to Y_\lambda'$ which is clearly U-linear. It is then clear that the following diagram is commutative:

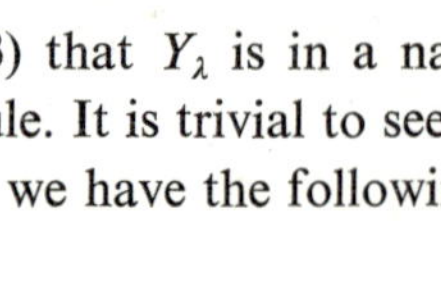

where $f: H^* \to Y_\lambda$ is the map defined by $f(u) = u \cdot (1 + L_\lambda^{\min})$. Since f and f' are both bijective, it follows that $\eta: Y_\lambda \to Y_\lambda'$ is a U-isomorphism. Q.E.D.

We have seen (Lemma 1.5.3) that Y_λ is in a natural way a K_θ-module. Also H^*, as we know, is a K_θ-module. It is trivial to see that the bijection f of Lemma 1.5.6 is in fact K_θ-linear. Thus we have the following:

Corollary 1.5.7. If $\gamma \in \Gamma$, then the multiplicity of γ in Y_λ is $\mathfrak{l}(\gamma)$, and $Y_\lambda = \coprod_{\gamma \in \Gamma} (Y_\lambda)^\gamma$.

We recall that $m(\gamma)$=multiplicity of γ in Z_λ. The following is an immediate consequence of Lemma 1.5.6 and Corollary 1.5.5:

Corollary 1.5.8. One has $\mathfrak{l}(\gamma)=m(\gamma)$ for all $\gamma \in \Gamma$ if and only if there is only one element in $\mathscr{C}(\lambda)$.

Let $\gamma \in \Gamma$. Henceforth, we denote by V_γ the representation space of a fixed member of γ. Let V'_γ denote the dual space of V_γ and $\gamma' \in \Gamma$ the corresponding class, which is contragredient to γ. Put $E_\gamma = \mathrm{Hom}_{K_\theta}(V_\gamma, H^*)$ and $E'_\gamma = \mathrm{Hom}_{K_\theta}(V'_\gamma, H^*)$. Note that $\dim E_\gamma$=multiplicity of γ in $H^* = \mathfrak{l}(\gamma) = \mathfrak{l}(\gamma')$=multiplicity of γ' in $H^* = \dim E'_\gamma$.

Let $\{v_i\}$ $(i=1, 2, \ldots, \dim V_\gamma)$ be a basis of V_γ and let $\{v'_i\}$ be the dual basis of V'_γ. For any $\sigma \in E_\gamma$ and $\nu \in E'_\gamma$, we define $z(\nu, \sigma) = \sum_{i=1}^{\dim V_\gamma} \nu(v'_i)\sigma(v_i)$, where the product $\nu(v'_i)\sigma(v_i)$ is, of course, the product in U. It can be easily checked that $z(\nu, \sigma)$ is independent of the basis of V_γ, and that $z(\nu, \sigma)$ is invariant under the action of K_θ, so that $z(\nu, \sigma) \in U^{\mathfrak{k}}$.

Now let $\sigma_1, \ldots, \sigma_{\mathfrak{l}(\gamma)}$ be any basis of E_γ and $\nu_1, \ldots, \nu_{\mathfrak{l}(\gamma)}$ any basis of E'_γ. Let $(z(i,j))$ denote the matrix with entries in $U^{\mathfrak{k}}$ given by $z(i,j) = z(\nu_i, \sigma_j)$. Now let $R^\gamma = (p_{z(i,j)})$, so that R^γ is an $\mathfrak{l}(\gamma) \times \mathfrak{l}(\gamma)$ matrix with entries in $U(\mathfrak{a})$. Observe that by changing the basis $\sigma_1, \ldots, \sigma_{\mathfrak{l}(\gamma)}$ of E_γ and $\nu_1, \ldots, \nu_{\mathfrak{l}(\gamma)}$ of E'_γ, R^γ becomes $AR^\gamma B$, where A and B are non-singular matrices with constant entries.

For $\lambda \in \mathfrak{a}'$, we denote by $R^\gamma(\lambda)$ the matrix $(p_{z(i,j)}(\lambda))$.

Remark 1.5.9. For any $\sigma \in E_\gamma$ and $\nu \in E_\gamma$, $z(\nu, \sigma) \in U(\mathfrak{g}_0)$, where $\mathfrak{g}_0 = [\mathfrak{g}, \mathfrak{g}]$. (This is because $H^* \subseteqq U(\mathfrak{g}_0)$, H^* being spanned by powers of nilpotent elements.) Thus $p_{z(\nu,\sigma)} \in U(\mathfrak{a}_0)$, where $\mathfrak{a}_0 = \mathfrak{a} \cap [\mathfrak{g}, \mathfrak{g}]$. Thus if $\mathfrak{a}_0^\perp = \{\lambda \in \mathfrak{a}' \mid \lambda | \mathfrak{a}_0 = 0\}$, then for any $\lambda \in \mathfrak{a}'$ and $\lambda_0 \in \mathfrak{a}_0^\perp$, $p_{z(\nu,\sigma)}(\lambda+\lambda_0) = p_{z(\nu,\sigma)}(\lambda)$; in particular, $R^\gamma(\lambda) = R^\gamma(\lambda+\lambda_0)$.

One has that U is completely reducible under the action of K_θ. Let U^{K_θ} denote the set of K_θ-invariants in U, and let $V \subseteqq U$ denote the sum of the primary subspaces of U corresponding to the non-trivial irreducible representations of K_θ. For all $u \in U$, let u_* denote the U^{K_θ}-component of u with respect to the decomposition $U = U^{K_\theta} \oplus V$.

Lemma 1.5.10. One has that $U^{K_\theta} = \mathfrak{k}U \cap U^{K_\theta} \oplus J^*$, and that for all $u \in U$, $u_0 - u_* \in \mathfrak{k}U \cap U^{\mathfrak{k}}$.

Proof. Since $U^{\mathfrak{k}} = \mathfrak{k}U \cap U^{\mathfrak{k}} \oplus J^*$, and since both components are K_θ-stable, one has $U^{K_\theta} = \mathfrak{k}U \cap U^{K_\theta} \oplus J^* \cap U^{K_\theta} = \mathfrak{k}U \cap U^{K_\theta} \oplus J^*$. The second assertion follows from the fact that $V \cap U^{\mathfrak{k}} \subseteqq \mathfrak{k}U \cap U^{\mathfrak{k}}$, which in turn follows from the decomposition $U^{\mathfrak{k}} = \mathfrak{k}U \cap U^{\mathfrak{k}} \oplus J^*$ and the fact that $J^* = (S^*)^{K_\theta}$. Q.E.D.

Corollary 1.5.11. For all $u \in U$, one has that $p_{u_0} = p_{u_*}$.

Lemma 1.5.12. One has that $m(\gamma)$=rank $R^\gamma(\lambda)$ for all $\gamma \in \Gamma$, $\lambda \in \mathfrak{a}'$.

Proof. Let $E_\gamma^0 = \{\sigma \in E_\gamma | \sigma(V_\gamma) Z_\lambda^{\mathfrak{k}} = 0\}$. That is, if I = kernel of the map $f: H^* \to Z_\lambda$, $(f(u) = u \cdot \xi, 0 \neq \xi \in (Z_\lambda)^{\mathfrak{k}})$, then $E_\gamma^0 = \{\sigma \in E_\gamma | \sigma(V_\gamma) \subseteq I\}$. One has that the multiplicity of γ in Z_λ = multiplicity of γ in H^* − multiplicity of γ in I. That is, $m(\gamma) = l(\gamma) - \dim E_\gamma^0$.

Now, suppose $\sigma \in E_\gamma$. We assert that $\sigma \in E_\gamma^0$ if and only if for every $\varrho \in E'_\gamma$, $p_{z(\varrho, \sigma)}(\lambda) = 0$.

Indeed, $\sigma \in E_\gamma^0$ if and only if $\sigma(V_\gamma) \subseteq I$, i.e., if and only if $\sigma(V_\gamma) \subseteq L_\lambda$, i.e., if and only if for every $u \in \sigma(V_\gamma)$ and every $v \in U$, one has $p_{(vu)_*}(\lambda) = 0$ (by Corollary 1.5.11). But $U = \mathfrak{k}U + U^{K_\theta} H^*$, by Remark 1.4.4. For any element $w \in \mathfrak{k}U$ one has $wu \in \mathfrak{k}U$ and $(wu)_* \in \mathfrak{k}U \cap U^{K_\theta}$, so that $p_{(wu)_*}(\lambda) = 0$. Thus $\sigma \in E_\gamma^0$ if and only if for all $u \in \sigma(V_\gamma)$ and $v \in U^{K_\theta} H^*$, one has $p_{(vu)_*}(\lambda) = 0$. But if $v_1 \in U^{K_\theta}$ and $v_2 \in H^*$, one has $(v_1 v_2 u)_* = v_1 (v_2 u)_*$. Hence $p_{(v_1 v_2 u)_*}(\lambda) = p_{v_1}(\lambda) p_{(v_2 u)_*}(\lambda)$ (see Lemma 1.2.8). Thus $\sigma \in E_\gamma^0$ if and only if for every $u \in \sigma(V_\gamma)$ and $v \in H^*$, one has $p_{(vu)_*}(\lambda) = 0$. Now consider the map $\alpha: H^* \otimes \sigma(V_\gamma) \to U$ defined by $\alpha(v \otimes u) = vu$. Clearly, α is K_θ-linear. We have $H^* = \coprod_{\mu \in \Gamma} H_\mu^*$, where H_μ^* denotes the primary component of H^* of type μ. If μ is not isomorphic to γ', then by Schur's Lemma, $H_\mu^* \otimes \sigma(V_\gamma)$ does not contain any K_θ-invariant ($\neq 0$). Hence

$$\big(\alpha(H^* \otimes \sigma(V_\gamma))\big)_* = \big(\alpha(H^*(\gamma') \otimes \sigma(V_\gamma))\big)_*.$$

Thus we have that $\sigma \in E_\gamma^0$ if and only if for every $u \in \sigma(V_\gamma)$ and every $v \in H^*(\gamma')$, $p_{(vu)_*}(\lambda) = 0$, i.e., if and only if for every $\varrho \in E'_\gamma$, one has $p_{(vu)_*}(\lambda) = 0$ for every $u \in \sigma(V_\gamma)$ and every $v \in \varrho(V'_\gamma)$.

The K_θ-linear maps σ and ϱ give rise to a K_θ-linear map $\chi: V'_\gamma \otimes V_\gamma \to U^{K_\theta}$ given by $\chi(y \otimes x) = (\varrho(y)\sigma(x))_*$. But by Schur's lemma, $\chi(V'_\gamma \otimes V_\gamma)$ is at most one-dimensional. Thus for any $u \in \sigma(V_\gamma)$ and $v \in \varrho(V'_\gamma)$, one has $p_{(vu)_*} = c p_{z(\varrho, \sigma)}$ for a scalar c. Thus we have that $\sigma \in E_\gamma^0$ if and only if $p_{z(\varrho, \sigma)}(\lambda) = 0$ for every $\varrho \in E'_\gamma$, which was our assertion.

Now let $k = \dim E_\gamma^0$. Choose a basis $\sigma_1, \ldots, \sigma_k, \sigma_{k+1}, \ldots, \sigma_{l(\gamma)}$ of E_γ such that $\sigma_1, \ldots, \sigma_k$ is a basis of E_γ^0, and a basis $\varrho_1, \ldots, \varrho_{l(\gamma)}$ of E'_γ. We can assume without loss that the matrix $R^\gamma(\lambda)$ obtained from these bases is the same as the matrix $R^\gamma(\lambda)$ of the lemma. Now since $\sigma_1, \ldots, \sigma_k \in E_\gamma^0$, one has that if $z_{(i,j)} = z_{(\varrho_i, \sigma_j)}$, then $p_{z(i,j)}(\lambda) = 0$ for all i and all $j \leq k$. Thus rank $R^\gamma(\lambda) \leq l(\gamma) - k$. It will follow that rank $R^\gamma(\lambda) = l(\gamma) - k$ if we prove that the vectors

$$C_j = \big(p_{z(1,j)}(\lambda), p_{z(2,j)}(\lambda), \ldots, p_{z(l(\gamma),j)}(\lambda)\big)_{j=k+1,\ldots,l(\gamma)}$$

are linearly independent. Accordingly, suppose $\sum_{j=k+1}^{l(\gamma)} t_j C_j = 0$, where the t_j are scalars. Then one has $\sum_{j=k+1}^{l(\gamma)} t_j p_{z(i,j)}(\lambda) = 0$ for all $i = 1, \ldots, l(\gamma)$. But the map $E'_\gamma \times E_\gamma \to U(\mathfrak{a})$ sending (ϱ, σ) into $p_{z(\varrho, \sigma)}$ is bilinear, so that $p_{z(\varrho_i, \sigma_0)}(\lambda) = 0$ for all $i = 1, \ldots, l(\gamma)$, where $\sigma_0 = \sum_{j=k+1}^{l(\gamma)} t_j \sigma_j$. But since ϱ_i, $i = 1, \ldots, l(\gamma)$, form a basis of E'_γ, we have that $p_{z(\varrho,\sigma_0)}(\lambda)=0$ for all $\varrho\in E'_\gamma$, which implies that $\sigma_0\in E^0_\gamma$. But this is impossible, unless $\sigma_0=0$, since σ_0 is a linear combination of $\sigma_{k+1},\ldots,\sigma_{l(\gamma)}$. Thus, we indeed have that rank $R^\gamma(\lambda)=l(\gamma)-k=\dim E_\gamma-\dim E^0_\gamma=m(\gamma)$.

Q.E.D.

Corollary 1.5.13. For all $\gamma\in\Gamma$, $\lambda\in\mathfrak{a}'$, one has that $l(\gamma)=m(\gamma)$ if and only if $r^\gamma(\lambda)\neq 0$, where $r^\gamma(\lambda)=\det R^\gamma(\lambda)$.

§ 6. The Hermitian structure in Z_λ

The transpose operation t in U, denoted $u\to u^t$ for $u\in U$, is defined as the unique linear map satisfying the following conditions:

(i) For any scalar $\alpha\in U$, $\alpha^t=\alpha$,

(ii) for $x\in\mathfrak{g}$, $x^t=-x$, and

(iii) for $u, v\in U$, $(uv)^t=v^t u^t$.

Now suppose that V is a $\mathfrak{g}$-module, possibly infinite-dimensional. Let V' be the complex dual of V, considered as a $\mathfrak{g}$-module by defining $x\cdot v'$ for $x\in\mathfrak{g}$, $v'\in V'$, by requiring $\langle v, x\cdot v'\rangle=\langle -x\cdot v, v'\rangle$ for any $v\in V$. The $\mathfrak{g}$-module structures on V and V' extend uniquely to U-module structures on V and V'. Then for $u\in U$, $x\in V$ and $v'\in V'$, one obviously has

$$\langle u\cdot x, v'\rangle=\langle x, u^t\cdot v'\rangle.$$

Remark 1.6.1. Recall (see § 4) the linear map $\beta: S(\mathfrak{g})\to U$, uniquely characterized by the property that for $x\in\mathfrak{g}$, $\beta(x^k)=(\beta(x))^k$ for all integers $k\geqq 0$. In fact, β is explicitly defined by $\beta(x_1\ldots x_k)=\frac{1}{k!}\sum_\sigma x_{\sigma(1)}\cdots x_{\sigma(k)}$, the summation being taken over all permutations σ of $\{1, 2, \ldots, k\}$ (the elements $x_1, \ldots, x_k$ are arbitrary elements of $\mathfrak{g}$). Clearly, one has $(\beta(x_1\cdots x_k))^t=(-1)^k\beta(x_1\cdots x_k)$. In particular, for any element u in $S_i^*\subseteqq S^*$ (recall that $S^*=\beta(S)$, where $S=S(\mathfrak{p})\subseteqq S(\mathfrak{g})$), where S_i^* is the subspace of S^* spanned by $\{\beta(x_1\ldots x_i)|x_1,\ldots,x_i\in\mathfrak{p}\}$, one has $u^t=(-1)^i u$. Note that if θ denotes the unique algebra automorphism of U extending the Lie algebra isomorphism $\theta:\mathfrak{g}\to\mathfrak{g}$, then one also has $\theta(u)=(-1)^i u$ for $u\in S_i^*$. Thus $\theta|S^*=t|S^*$.

Remark 1.6.2. If $p\in U(\mathfrak{a})$, then from the fact that $U(\mathfrak{a})\subseteqq S^*$ and the fact that $t|S^*=\theta|S^*$, it follows immediately that for any $\lambda\in\mathfrak{a}'$, $p^t(\lambda)=p(-\lambda)$.

It is well known that the Weyl group W of $(\mathfrak{a},\mathfrak{g})$ acts simply transitively on the set of all Weyl chambers in $\mathfrak{a}_{\mathbf{R}}\cap[\mathfrak{g},\mathfrak{g}]$. We denote by $\varkappa$ the unique element of the Weyl group which maps the fixed positive Weyl chamber D into its negative. For any $\lambda\in\mathfrak{a}'$, we denote by λ' the element $-\varkappa\cdot\lambda$.

Lemma 1.6.3. For any $u\in U^t$ and any $\lambda\in\mathfrak{a}'$, one has $p_u(\lambda)=p_{u^t}(\lambda')$.

Proof. One has $u^t = p_{u^t} \pmod{\mathfrak{k}U + U\mathfrak{n}}$.

Note that since $\theta|\mathfrak{a} = -$ identity, if $\mathfrak{n}^* = \sum_{\varphi \in \Delta_+} \mathfrak{g}^{-\varphi}$, then $\theta(\mathfrak{n}) = \mathfrak{n}^*$ and $\varkappa \cdot \mathfrak{n}^* = \mathfrak{n}$. Thus $\theta(u^t) \equiv \theta(p_{u^t}) \pmod{\mathfrak{k}U + U\mathfrak{n}^*}$. Now applying $\varkappa$, one has $\varkappa \cdot \theta(u^t) \equiv \varkappa \cdot \theta(p_{u^t}) \pmod{\mathfrak{k}U + U\mathfrak{n}}$. Hence, $p_{\varkappa \cdot \theta(u^t)} = \varkappa \cdot \theta(p_{u^t})$. Clearly, $(U^{\mathfrak{k}})^t = U^{\mathfrak{k}}$, and since $\varkappa$ and θ leave $\mathfrak{k}$ stable, $\varkappa \cdot \theta(U^{\mathfrak{k}}) \subseteq U^{\mathfrak{k}}$. One has $U^{\mathfrak{k}} = \mathfrak{k}U \cap U^{\mathfrak{k}} \oplus J^*$. Writing $u = u_1 + u_2$, where $u_1 \in \mathfrak{k}U \cap U^{\mathfrak{k}}$ and $u_2 \in J^*$, one has $p_u = p_{u_2}$. Also, $u^t = u_1^t + u_2^t$ and $u_1^t \in U\mathfrak{k} \cap U^{\mathfrak{k}} = \mathfrak{k}U \cap U^{\mathfrak{k}}$; moreover, since $u_2 \in J^* \subseteq S^*$, one has $u_2^t = \theta(u_2)$. Thus $\varkappa \cdot \theta(u^t) = \varkappa \cdot \theta(u_1^t) + \varkappa \cdot \theta(\theta(u_2)) = \varkappa \cdot \theta(u_1^t) + \varkappa \cdot u_2 = \varkappa \cdot \theta(u_1^t) + u_2$, and $\varkappa \cdot \theta(u_1^t) \in \mathfrak{k}U \cap U^{\mathfrak{k}}$. Hence $p_{\varkappa \cdot \theta(u^t)} = p_{u_2} = p_u$. Thus, from the equation $p_{\varkappa \cdot \theta(u^t)} = \varkappa \cdot \theta(p_{u^t})$, one has $p_u = \varkappa \cdot \theta(p_{u^t})$. Hence for $\lambda \in \mathfrak{a}'$, $p_u(\lambda) = \varkappa \cdot \theta(p_{u^t})(\lambda) = \theta(p_{u^t})(\varkappa \cdot \lambda) = p_{u^t}(-\varkappa \cdot \lambda) = p_{u^t}(\lambda')$. Q.E.D.

Since $\mathfrak{g} = \mathfrak{g}_{\mathbf{R}} + i\mathfrak{g}_{\mathbf{R}}$, one has $U \cong U_{\mathbf{R}} \otimes \mathbf{C}$, where $U_{\mathbf{R}}$ is the universal enveloping algebra of $\mathfrak{g}_{\mathbf{R}}$. The isomorphism is canonical and is obtained as follows: The composite of the inclusion maps $\mathfrak{g}_{\mathbf{R}} \to \mathfrak{g} \to U$ is a Lie algebra homomorphism of $\mathfrak{g}_{\mathbf{R}}$ into U, where U is considered as a Lie algebra over $\mathbf{R}$. Hence there exists a homomorphism $U_{\mathbf{R}} \to U$, and hence a homomorphism $U_{\mathbf{R}} \otimes \mathbf{C} \to U$, which is easily seen to be an isomorphism. By means of this isomorphism, we identify $U_{\mathbf{R}}$ with its image in U, so that one has $U = U_{\mathbf{R}} + iU_{\mathbf{R}}$.

We now define the operation of conjugation in U, denoted $u \to \bar{u}$. If $u \in U$ and if $u = u_1 + iu_2$ where $u_1, u_2 \in U_{\mathbf{R}}$, then we define $\bar{u} = u_1 - iu_2$. If $u, v \in U$, one clearly has $\overline{uv} = \bar{u}\bar{v}$ and $\overline{u+v} = \bar{u} + \bar{v}$. If $\alpha \in \mathbf{C}$ and $u \in U$, then $\overline{\alpha u} = \bar{\alpha}\bar{u}$.

Let $\mathfrak{a}'_{\mathbf{R}}$ denote the real subspace of $\mathfrak{a}'$ defined as follows:

$$\mathfrak{a}'_{\mathbf{R}} = \{\alpha \in \mathfrak{a}' \,|\, \alpha(X) \in \mathbf{R} \text{ for every } x \in \mathfrak{a}_{\mathbf{R}}\}.$$

Then one has $\mathfrak{a}' = \mathfrak{a}'_{\mathbf{R}} + i\mathfrak{a}'_{\mathbf{R}}$. Let $\lambda \in \mathfrak{a}'$, and let $\lambda = \lambda_1 + i\lambda_2$, where $\lambda_1, \lambda_2 \in \mathfrak{a}'_{\mathbf{R}}$. Then we define $\bar{\lambda} = \lambda_1 - i\lambda_2$.

We now define $u^s = (\bar{u})^t$ for all $u \in U$. Then for all $u, v \in U$ and $\alpha \in \mathbf{C}$, one has $(u+v)^s = u^s + v^s$, $(uv)^s = v^s u^s$ and $(\alpha u)^s = \bar{\alpha} u^s$.

One easily sees that for any $u \in U$, $(\overline{u^t}) = (\bar{u})^t$. Recall (see § 3) that $\varrho \in \mathfrak{a}'$ denotes the linear form on $\mathfrak{a}$ defined by $\varrho(x) = \frac{1}{2}$ (trace ad $x|\mathfrak{n}$) for every $x \in \mathfrak{a}$.

Lemma 1.6.4. Let $\lambda \in \mathfrak{a}'$. Writing $\mu = \lambda + \varrho$, one has $p_u(\lambda) = \overline{p_{u^s}(\lambda)}$ for all $u \in U^{\mathfrak{k}}$ if and only if μ and $-\bar{\mu}$ are conjugate under an element of the Weyl group $W = W(\mathfrak{a}, \mathfrak{g})$.

Proof. For all $u \in U$, one has $p_{\bar{u}} = \overline{p_u}$, since $\mathfrak{k}$ and $\mathfrak{n}$ are stable under conjugation. Also, for all $p \in U(\mathfrak{a})$ and $\nu \in \mathfrak{a}'$, one has $\overline{p(\nu)} = \bar{p}(\bar{\nu})$. Hence $p_u(\lambda) = p_{u^t}(\lambda') = \overline{p_{u^s}}(\lambda') = \overline{p_{u^s}(\bar{\lambda}')}$ for all $u \in U^{\mathfrak{k}}$, by Lemma 1.6.3. Thus $p_u(\lambda) = \overline{p_{u^s}(\lambda)}$ for all $u \in U^{\mathfrak{k}}$ if and only if $p_u(\lambda) = p_u(\bar{\lambda}')$ for all $u \in U^{\mathfrak{k}}$, i.e. if and only if $\lambda \underset{W_T}{\sim} \bar{\lambda}'$ (by Theorem 1.3.1), i.e., if and only if $\lambda + \varrho \underset{W}{\sim} -\overline{\varkappa \cdot \lambda} + \varrho = \varkappa \cdot (-\bar{\lambda} - \varrho) = \varkappa \cdot (-(\overline{\lambda + \varrho}))$. Q.E.D.

Let V be a vector space over $\mathbf{C}$ of arbitrary dimension (finite or infinite). By a *Hermitian form* on V we mean a map from $V \times V$ into $\mathbf{C}$, denoted by $\langle x, y \rangle$, satisfying the following conditions:

(i) $\langle \alpha x, y \rangle = \alpha \langle x, y \rangle$ for all $x, y \in V, \alpha \in \mathbf{C}$

(ii) $\langle x_1 + x_2, y \rangle = \langle x_1, y \rangle + \langle x_2, y \rangle$ for all $x_1, x_2, y \in V$

and

(iii) $\langle x, y \rangle = \overline{\langle y, x \rangle}$ for all $x, y \in V$.

Given a Hermitian form $\langle\ ,\ \rangle$ on V, we speak of the *Hermitian structure* on V with respect to $\langle\ ,\ \rangle$.

Now suppose that Z is a U-module and suppose in addition that Z has a Hermitian structure given by a Hermitian form $\langle\ ,\ \rangle$ on Z. We say that the Hermitian structure on Z is U-invariant if for every $x, y \in Z$ and $u \in U$, $\langle u \cdot x, y \rangle = \langle x, u^s \cdot y \rangle$.

We shall now concern ourselves with the problem of finding necessary and sufficient conditions on $\lambda \in \mathfrak{a}'$ so that the unique irreducible U-module Z_λ (see Theorem 1.3.5) admits a non-trivial U-invariant Hermitian structure.

Theorem 1.6.5. *For any* $\lambda \in \mathfrak{a}'$, *writing* $\mu = \lambda + \varrho$, *there exists a non-trivial U-invariant Hermitian structure on* Z_λ *if and only if* μ *is conjugate to* $-\bar{\mu}$ *under the Weyl group* W. *If such a form exists, it is nonsingular and is unique up to multiplication by a non-zero real scalar; moreover, it can be chosen so that* $\langle \xi_\lambda, \xi_\lambda \rangle = 1$, *where* ξ_λ *is a given non-zero element in* $(Z_\lambda)^{\mathfrak{k}}$, *and it is explicitly given by* $\langle u \cdot \xi_\lambda, v \cdot \xi_\lambda \rangle = p_{(v^s u)_0}(\lambda)$.

Proof. Suppose there exists a non-trivial U-invariant Hermitian form on Z_λ. Let $W = \{x \in Z_\lambda | \langle x, y \rangle = 0 \text{ for every } y \in Z_\lambda\}$. Clearly, W is stable under U. Hence either $W = 0$ or $W = Z_\lambda$. Since the Hermitian structure is non-trivial we conclude that $W = 0$. Thus the Hermitian structure is non-singular.

Now, Z_λ is completely reducible as a $\mathfrak{k}$-module, and one has $(Z_\lambda)^{\mathfrak{k}} \oplus V = Z_\lambda$, where V denotes the sum of the primary $\mathfrak{k}$-submodules of Z_λ of non-trivial type. Note that $\mathfrak{k} \cdot V = V$, where $\mathfrak{k} \cdot V$ is the subspace of Z_λ spanned by the set $\{x \cdot v | x \in \mathfrak{k}$ and $v \in V\}$. One has

$$\langle (Z_\lambda)^{\mathfrak{k}}, V \rangle = \langle (Z_\lambda)^{\mathfrak{k}}, \mathfrak{k} \cdot V \rangle = \langle \mathfrak{k}^s \cdot (Z_\lambda)^{\mathfrak{k}}, V \rangle = 0.$$

Hence the Hermitian form is nonsingular on $(Z_\lambda)^{\mathfrak{k}}$. Thus for $0 \neq \xi_\lambda \in (Z_\lambda)^{\mathfrak{k}}$, $\langle \xi_\lambda, \xi_\lambda \rangle \neq 0$; since $\langle \xi_\lambda, \xi_\lambda \rangle$ is real, we can assume without loss that $\langle \xi_\lambda, \xi_\lambda \rangle = 1$, by altering the Hermitian form by a real scalar, if necessary. Since for $u \in U^{\mathfrak{k}}$ one has $u \cdot \xi_\lambda = p_u(\lambda) \xi_\lambda$, one obtains

$$p_u(\lambda) = \langle u \cdot \xi_\lambda, \xi_\lambda \rangle = \langle \xi_\lambda, u^s \cdot \xi_\lambda \rangle = \langle \xi_\lambda, p_{u^s}(\lambda) \xi_\lambda \rangle = \overline{p_{u^s}(\lambda)}.$$

Thus by Lemma 1.6.4 we have $\mu \underset{W}{\sim} -\bar{\mu}$.

Now, if $\langle\ ,\ \rangle$ denotes any U-invariant Hermitian form on Z_λ such that $\langle\xi_\lambda, \xi_\lambda\rangle=1$, one has that for any $u, v\in U$, $\langle u\cdot\xi_\lambda, v\cdot\xi_\lambda\rangle=\langle v^s u\cdot\xi_\lambda, \xi_\lambda\rangle=\langle(v^s u)_0\cdot\xi_\lambda, \xi_\lambda\rangle+$ $+\langle(v^s u-(v^s u)_0)\cdot\xi_\lambda, \xi_\lambda\rangle$. But $v^s u-(v^s u)_0\in[\mathfrak{k}, U]$, and $[\mathfrak{k}, U]\cdot\xi_\lambda\subseteq V$. But, as has already been proved, $\langle V, (Z_\lambda)^{\mathfrak{k}}\rangle=0$. Thus

$$\langle u\cdot\xi_\lambda, v\cdot\xi_\lambda\rangle = \langle(v^s u)_0\cdot\xi_\lambda, \xi_\lambda\rangle = \langle p_{(v^s u)_0}(\lambda)\xi_\lambda, \xi_\lambda\rangle = p_{(v^s u)_0}(\lambda),$$

proving the uniqueness.

Conversely, assume that μ and $-\bar{\mu}$ are W-conjugate. Then by Lemma 1.6.4 we have $p_u(\lambda)=\overline{p_{u^s}(\lambda)}$ for any $u\in U^{\mathfrak{k}}$. Now, we define a map $Z_\lambda\times Z_\lambda\to\mathbf{C}$, denoted by $\langle x, y\rangle$, by the condition that if $x=u\cdot\xi_\lambda$ and $y=v\cdot\xi_\lambda$, where $u, v\in U$ and $\xi_\lambda\in(Z_\lambda)^{\mathfrak{k}}$ $(\xi_\lambda\neq 0)$, then $\langle x, y\rangle=p_{(v^s u)_0}(\lambda)$. Note that $\langle\ ,\ \rangle$ is well defined. We assert that $\langle\ ,\ \rangle$ is a U-invariant Hermitian form on Z_λ. Clearly, one has the relations $\langle\alpha x, y\rangle=\alpha\langle x, y\rangle$, $\langle x_1+x_2, y\rangle=\langle x_1, y\rangle+\langle x_2, y\rangle$. To check that $\langle y, x\rangle=\overline{\langle x, y\rangle}$, assume $x=u\cdot\xi_\lambda$ and $y=v\cdot\xi_\lambda$. Then $\langle y, x\rangle=p_{(u^s v)_0}(\lambda)$, and

$$\overline{\langle x, y\rangle} = \overline{p_{(v^s u)_0}(\lambda)} = \overline{p_{((u^s v)^s)_0}(\lambda)} = \overline{p_{((u^s v)_0)^s}(\lambda)} = p_{(u^s v)_0}(\lambda),$$

since μ and $-\bar{\mu}$ are W-conjugate (Lemma 1.6.4). Thus, indeed, $\langle y, x\rangle=\overline{\langle x, y\rangle}$.

Finally, to check that $\langle\ ,\ \rangle$ is U-invariant, let $x=u\cdot\xi_\lambda$ and $y=v\cdot\xi_\lambda$. Then for any $w\in U$, one has $\langle w\cdot x, y\rangle=\langle wu\cdot\xi_\lambda, v\cdot\xi_\lambda\rangle=p_{(v^s wu)_0}(\lambda)$, and also $\langle x, w^s\cdot y\rangle=$ $=\langle u\cdot\xi_\lambda, w^s v\cdot\xi_\lambda\rangle=p_{(v^s wu)_0}(\lambda)$, proving that $\langle w\cdot x, y\rangle=\langle x, w^s\cdot y\rangle$. Q.E.D.

Recall that for any $\lambda\in\mathfrak{a}'$, Z_λ can be considered in a natural way as a K_θ-module. We now assume that Z_λ carries a U-invariant Hermitian structure.

We define $K_{\theta,\mathbf{R}}=\{a\in K_\theta\,|\,a \text{ leaves } \mathfrak{g}_{\mathbf{R}} \text{ stable}\}$.

We assert that $K_{\theta,\mathbf{R}}$ leaves the Hermitian structure of Z_λ invariant, i.e., that for every $a\in K_{\theta,\mathbf{R}}$ and every $x, y\in Z_\lambda$, $\langle a\cdot x, a\cdot y\rangle=\langle x, y\rangle$. This can be proved as follows: For any element a of $K_{\theta,\mathbf{R}}$, consider $\langle a\cdot(u\cdot\xi), a\cdot(v\cdot\xi)\rangle$, where $\xi\in Z_\lambda^{\mathfrak{k}}(\xi\neq 0)$ and $u, v\in U$. One has

$$\langle a\cdot(u\cdot\xi), a\cdot(v\cdot\xi)\rangle = \langle(a\cdot u)\cdot\xi, (a\cdot v)\cdot\xi\rangle = p_{((a\cdot v)^s(a\cdot u))_0}(\lambda).$$

But since $a\in K_{\theta,\mathbf{R}}$ $a\cdot U_{\mathbf{R}}=U_{\mathbf{R}}$; hence $(a\cdot v)^s=a\cdot v^s$, so that $\big((a\cdot v)^s(a\cdot u)\big)_0=$ $=\big(a\cdot(v^s u)\big)_0=a\cdot(v^s u)_0$. But for any $w\in U^{\mathfrak{k}}$, $a\in K_{\theta,\mathbf{R}}$, one has $p_{a\cdot w}=p_w$ (see the proof of Lemma 1.5.3). Thus one has

$$\langle a\cdot(u\cdot\xi), a\cdot(v\cdot\xi)\rangle = p_{(v^s u)_0}(\lambda)$$

$$=\langle u\cdot\xi, v\cdot\xi\rangle,$$

which proves the assertion.

Recall that Z_λ is completely reducible as a K_θ-module and that $Z_\lambda=\coprod_{\gamma\in\Gamma}(Z_\lambda)^\gamma$, where $(Z_\lambda)^\gamma$ denotes the primary component of Z_λ of type $\gamma \in \Gamma$, and Γ as usual denotes the set of equivalence classes of irreducible, finite-dimensional holomorphic K_θ-representations admitting an M_θ-fixed vector. We now assert that if $\gamma_1, \gamma_2 \in \Gamma$ and $\gamma_1 \neq \gamma_2$, then $(Z_\lambda)^{\gamma_1}$ is orthogonal to $(Z_\lambda)^{\gamma_2}$ with respect to the U-invariant Hermitian form on Z_λ. This can be seen as follows: Suppose $(Z_\lambda)^{\gamma_1}$ is not orthogonal to $(Z_\lambda)^{\gamma_2}$. Then choose a suitable copy V_{γ_1} of γ_1 in $(Z_\lambda)^{\gamma_1}$ and a copy V_{γ_2} of γ_2 in $(Z_\lambda)^{\gamma_2}$ such that if B denotes the restriction to $V_{\gamma_1} \times V_{\gamma_2}$ of the Hermitian form on Z_λ, then $B \neq 0$. We will now arrive at the contradiction $\gamma_1 = \gamma_2$. Since γ_1 and γ_2 are holomorphic K_θ-representations, it is enough to show that $V_{\gamma_1} \cong V_{\gamma_2}$ as $K_{\theta,\mathbf{R}}$-modules. For this, let $\langle\, , \rangle$ be a $K_{\theta,\mathbf{R}}$-invariant nonsingular Hermitian form on V_{γ_2}, and define a map $\tau: V_{\gamma_1} \to V_{\gamma_2}$ by setting, for $x \in V_{\gamma_1}$, $\langle \tau(x), v\rangle = B(x, v)$ for every $v \in V_{\gamma_2}$. Then it is easily checked that τ gives a $K_{\theta,\mathbf{R}}$-isomorphism of V_{γ_1} onto V_{γ_2}. This proves that $(Z_\lambda)^{\gamma_1}$ is orthogonal to $(Z_\lambda)^{\gamma_2}$.

Let $\gamma \in \Gamma$. Since the Hermitian structure on Z_λ is nonsingular and since $Z_\lambda^\gamma \perp Z_\lambda^{\gamma_0}$ if $\gamma \neq \gamma_0$, where $\gamma_0 \in \Gamma$, we conclude that the Hermitian structure on Z_λ restricts to a non-singular Hermitian structure on Z_λ^γ.

Since $K_{\theta,\mathbf{R}}$ is compact, the action of $K_{\theta,\mathbf{R}}$ on V_γ leaves invariant a positive definite Hermitian form on V_γ, which is unique up to a positive real scalar by Schur's lemma.

For a fixed $v \in V_\gamma$, the mapping $u \to \langle u, v\rangle$ $(u \in V_\gamma)$ is complex linear and hence induces a mapping $v \to v^s$ of V_γ into V_γ'. This mapping is injective and has the property that $(\alpha v)^s = \bar{\alpha} v^s$ for all $\alpha \in \mathbf{C}$, $v \in V_\gamma$. We also denote by $v \to v^s$ the inverse mapping of V_γ' into V_γ.

Using the map $v \to v^s$ of V_γ' into V_γ, we define a conjugate linear isomorphism $\sigma \to \sigma^s$ of $E_\gamma (= \mathrm{Hom}_{K_\theta}(V_\gamma, H^*))$ onto $E_\gamma' (= \mathrm{Hom}_{K_\theta}(V_\gamma', H^*))$, as follows: We know that in U, H^* is generated as a $\mathbf{C}$-vector space by the set $\{e^n | e \in \mathfrak{p}, n \geqq 0, e \text{ nilpotent}\}$. From this it follows that if $u \in H^*$, then $u^s \in H^*$; indeed, $(e^n)^s = (e^s)^n$, and $e^s \in \mathfrak{p}$ and is nilpotent. Now if $\sigma \in E_\gamma$, we define $\sigma^s \in E_\gamma'$ by $\sigma^s(v) = (\sigma(v^s))^s$ for any $v \in V_\gamma'$.

We denote by $\mathfrak{a}^*$ the set $\{\lambda \in \mathfrak{a}' | \mu \underset{W}{\sim} -\bar{\mu}\}$, where $\mu = \lambda + \varrho$ and W is the Weyl group of $(\mathfrak{a}, \mathfrak{g})$. Then $\mathfrak{a}^*$ is just the set of all $\lambda \in \mathfrak{a}'$ such that Z_λ admits a U-invariant Hermitian structure. Let $\mathfrak{a}_+^* = \{\lambda \in \mathfrak{a}^* | Z_\lambda$ admits a U-invariant positive-definite Hermitian structure$\}$.

Recall (see § 5) the matrix $R^\gamma(\lambda) = (p_{z(i,j)}(\lambda))$, where $z(i,j) = \sum_k v_i(v_k')\sigma_j(v_k)$. Here $\sigma_1, \ldots, \sigma_{l(\gamma)}$ is a basis of E_γ, $v_1, \ldots, v_{l(\gamma)}$ is a basis of E_γ', $v_1, \ldots, v_n$ is a basis of V_γ and $v_1', \ldots, v_n'$ is the dual basis of V_γ'.

Lemma 1.6.6. For any $\gamma \in \Gamma$, let $\sigma_1, \sigma_2, \ldots, \sigma_{l(\gamma)}$ be any basis of E_γ, so that $\sigma_1^s, \sigma_2^s, \ldots, \sigma_{l(\gamma)}^s$ is a basis of E_γ'. With respect to these bases, the matrices $R^\gamma(\lambda)$ are Hermitian for each $\lambda \in \mathfrak{a}^*$; moreover, if $\lambda \in \mathfrak{a}^*$, then $\lambda \in \mathfrak{a}_+^*$ if and only if $R^\gamma(\lambda)$ is positive-semidefinite for each $\gamma \in \Gamma$.

Proof. Let $v_1, \ldots, v_n$ be an orthonormal basis of V_γ. If $v'_1, \ldots, v'_n$ is the dual basis of V'_γ, then it is easy to see that $v'_i = v^s_i$. For all $\lambda \in \mathfrak{a}^*$, one has

$$P_{z(i,j)}(\lambda) = P_{z(\sigma^s_i, \sigma_j)}(\lambda) = p_{\sum\limits_l (\sigma^s_i(v^s_l)\sigma_j(v_l)}(\lambda) = p_{\sum\limits_l (\sigma_i(v_l))^s \sigma_j(v_l)}(\lambda) =$$

$$= p_{\sum\limits_l ((\sigma_j(v_l))^s (\sigma_i(v_l))^s}(\lambda) = \overline{p_{\sum\limits_l (\sigma_j(v_l))^s \sigma_i(v_l)}(\lambda)} =$$

(see Lemma 1.6.4)

$$= \overline{p_{z(j,i)}(\lambda)},$$

showing that $R^\gamma(\lambda)$ is Hermitian.

Now, since $R^\gamma(\lambda)$ is Hermitian, we can choose a unitary matrix $L = (L_{ij})$ (i.e., a matrix L such that $L^{-1} = {}^t\bar{L}$) such that $L^{-1} R^\gamma(\lambda)\ L$ is a real diagonal matrix, say D. Now let $\sigma'_i = \sum\limits_k L_{ki}\sigma_k$. Then $(\sigma'_i)^s = \sum\limits_k \overline{L_{ki}}\sigma^s_k$. Now

$$D_{ij} = \sum_{k,l} \overline{L_{ki}} p_{z(\sigma^s_k, \sigma_l)}(\lambda) L_{lj} = p_{\sum\limits_{k,l} \overline{L_{ki}} L_{lj} z(\sigma^s_k, \sigma_l)}(\lambda) =$$

$$= p_{z(\sum\limits_k \overline{L_{ki}}\sigma^s_k, \sum\limits_l L_{lj}\sigma_l)}(\lambda) = p_{z((\sigma'_i)^s, \sigma'_j)}(\lambda).$$

Thus, with respect to the basis $\sigma'_1, \ldots, \sigma'_{l(\gamma)}$ of E_γ and the basis $(\sigma'_1)^s, \ldots, (\sigma'_{l(\gamma)})^s$ of E'_γ, the matrix $R^\gamma(\lambda)$ is real diagonal. We may assume without loss that $\sigma_i = \sigma'_i$, so that $R^\gamma(\lambda) = D$.

We know that for any $\sigma \in E_\gamma$, $v \in E'_\gamma, v \in V_\gamma$ and $v' \in V'_\gamma$, $p_{(v(v')\sigma(v))_*}$ is a scalar multiple of $p_{z(v,\sigma)}$. (See the discussion immediately preceding Lemma 1.5.10 for the definition of u_* $(u \in U)$.)

Now assume that $0 \neq \xi_\lambda \in (Z_\lambda)^{K_\theta}$ and suppose $\langle \xi_\lambda, \xi_\lambda \rangle = 1$. Then for $u, v \in V_\gamma$, one has

$$\langle \sigma_i(u) \cdot \xi_\lambda, \sigma_j(v) \cdot \xi_\lambda \rangle = \langle (\sigma_j(v))^s \sigma_i(u) \cdot \xi_\lambda, \xi_\lambda \rangle.$$

But

$$(\sigma_j(v))^s \sigma_i(u) \cdot \xi_\lambda \equiv ((\sigma_j(v))^s \sigma_i(u))_* \cdot \xi_\lambda \ (\text{mod} \coprod_\gamma Z^\gamma_\lambda, \gamma \text{ non-trivial}).$$

But

$$(Z_\lambda)^K \perp (\coprod_\gamma Z^\gamma_\lambda, \gamma \text{ non-trivial}).$$

Thus

$$\langle \sigma_i(u) \cdot \xi_\lambda, \sigma_j(v) \cdot \xi_\lambda \rangle = \langle ((\sigma_j(v))^s \sigma_i(u))_* \cdot \xi_\lambda, \xi_\lambda \rangle = p_{((\sigma_j(v))^s \sigma_i(u))_*}(\lambda) = \alpha p_{z(j,i)}(\lambda),$$

for some scalar α. Thus

$$\langle \sigma_i(u) \cdot \xi_\lambda, \sigma_j(v) \cdot \xi_\lambda \rangle = 0 \quad \text{if} \quad i \neq j, \text{ i. e.,}$$

$$(\sigma_i(V_\gamma) \cdot \xi_\lambda) \perp (\sigma_j(V_\gamma) \cdot \xi_\lambda).$$

Now let $i = j$, and assume that $\sigma_i(V_\gamma) \cdot \xi_\lambda \neq 0$. Then writing $V_{\gamma,i} = \sigma_i(V_\gamma) \cdot \xi_\lambda$, the restriction to $V_{\gamma,i}$ of the Hermitian structure on Z_λ is a $K_{\theta,\mathbf{R}}$-invariant non-singular Hermitian structure on $V_{\gamma,i}$. But there exists a $K_{\theta,\mathbf{R}}$-invariant positive definite Hermitian form B on V_γ and hence on $V_{\gamma,i}$, and, by Schur's lemma, any

$K_{\theta,\mathbf{R}}$-invariant Hermitian form on $V_{\gamma,i}$ is of the form cB, where c is a real scalar. Thus, for any $u \in V_\gamma$, $\langle \sigma_i(u)\cdot\xi_\lambda, \sigma_i(u)\cdot\xi_\lambda\rangle = cB(\sigma_i(u)\cdot\xi_\lambda, \sigma_i(u)\cdot\xi_\lambda)$, where c is a non-zero real scalar independent of u.

In particular, either $\langle \sigma_i(u)\cdot\xi, \sigma_i(u)\cdot\xi\rangle > 0$ for all non-zero $u \in V_\gamma$, or

$$\langle \sigma_i(u)\cdot\xi, \sigma_i(u)\cdot\xi\rangle < 0$$

for all non-zero $u \in V_\gamma$. Since $p_{z(i,i)}(\lambda) = \sum_k p_{((\sigma_i(u_k))^s\sigma_i(u_k))_0}(\lambda)$ (where $\{u_k\}$ is an orthonormal basis of V_γ) $= \sum_k \langle \sigma_i(u_k)\cdot\xi_\lambda, \sigma_i(u_k)\cdot\xi_\lambda\rangle$, it follows that for any non-zero $u \in V_\gamma$, $\langle \sigma_i(u)\cdot\xi_\lambda, \sigma_i(u)\cdot\xi_\lambda\rangle$ and $p_{z(i,i)}(\lambda)$ are either both >0 or both <0.

We can assume without loss that $p_{z(i,i)}(\lambda) \neq 0$ for $i = 1, 2, \ldots, \boldsymbol{l}$ and $=0$ for $i = \boldsymbol{l}+1, \ldots, \boldsymbol{l}(\gamma)$. Then we know that $\boldsymbol{l} = m(\gamma)$ (Lemma 1.5.12), and that $\sigma_i(V_\gamma)\cdot\xi_\lambda \neq 0$ for $i = 1, \ldots, \boldsymbol{l}$ and $=0$ for $i = \boldsymbol{l}+1, \ldots, \boldsymbol{l}(\gamma)$. Also, we assume that for $i = 1, \ldots, k$, $p_{z(i,i)}(\lambda) > 0$ and that for $i = k+1, \ldots, m(\gamma)$, $p_{z(i,i)}(\lambda) < 0$.

Now, one has $(Z_\lambda)^\gamma = V_{\gamma,1} \oplus \cdots \oplus V_{\gamma,m(\gamma)}$. It is immediate now that the restriction to $V_{\gamma,i}$ of the U-invariant Hermitian form on Z_λ (with $\langle \xi_\lambda, \xi_\lambda\rangle = 1$) is positive-definite for $i = 1, \ldots, k$ and negative-definite for $i = k+1, \ldots, m(\gamma)$. Moreover, if $i \neq j$, $V_{\gamma,i}$ is orthogonal to $V_{\gamma,j}$.

Thus, $R^\gamma(\lambda)$ is positive-semidefinite if and only if the restriction to $(Z_\lambda)^\gamma$ of the U-invariant form on Z_λ (such that $\langle \xi_\lambda, \xi_\lambda\rangle = 1$) is positive definite. Hence $R^\gamma(\lambda)$ is positive-semidefinite for all γ if and only if the U-invariant Hermitian form on Z_λ (such that $\langle \xi_\lambda, \xi_\lambda\rangle = 1$) is positive-definite, i.e., if and only if $\lambda \in \mathfrak{a}^*_+$

Q.E.D.

§ 7. The matrix $P^\gamma(\lambda)$ and induced module X_λ

For every subset $S \subseteq G$, we define $S_{\mathbf{R}} = \{a \in S \mid a\cdot\mathfrak{g}_{\mathbf{R}} \subseteq \mathfrak{g}_{\mathbf{R}}\}$. One has $G_{\mathbf{R}} = G^0_{\mathbf{R}} F$. where $G^0_{\mathbf{R}}$ denotes the connected component of the identity in $G_{\mathbf{R}}$ and F the set of elements of order 2 in A; recall that A is the analytic subgroup of G with Lie algebra $\mathrm{ad}_{\mathfrak{g}}\,\mathfrak{a}$ (see [10]). One has $A_{\mathbf{R}} = A^0_{\mathbf{R}} F$, where $A^0_{\mathbf{R}}$ denotes the connected component of the identity in $A_{\mathbf{R}}$. Also, one has $K_{\theta,\mathbf{R}} = K_{\mathbf{R}} F$ (this follows from the fact that $K_\theta = KF$ (see [10])). We observe that $K_{\mathbf{R}} = \exp \mathrm{ad}_{\mathfrak{g}}\,\mathfrak{k}_{\mathbf{R}}$. Indeed, on the one hand, $K_{\mathbf{R}}$ is clearly compact, and on the other hand $\exp \mathrm{ad}_{\mathfrak{g}}\,\mathfrak{k}_{\mathbf{R}}$ is maximal compact, while $\exp \mathrm{ad}_{\mathfrak{g}}\,\mathfrak{k}_{\mathbf{R}} \subseteq K_{\mathbf{R}}$. Let N be the analytic subgroup of G with Lie algebra $\mathfrak{n} = \sum_{\varphi \in \Delta_+} \mathfrak{g}^\varphi$. Then $N_{\mathbf{R}}$ is connected, and the map $\tau\colon K_{\theta,\mathbf{R}} \times A^0_{\mathbf{R}} \times N_{\mathbf{R}} \to G_{\mathbf{R}}$ defined by $(k, a, n) \to kan$ is an analytic diffeomorphism onto $G_{\mathbf{R}}$. Also, the map $\tau^0\colon K_{\mathbf{R}} \times A^0_{\mathbf{R}} \times N_{\mathbf{R}} \to G^0_{\mathbf{R}}$ which is the restriction of τ is an analytic diffeomorphism onto $G^0_{\mathbf{R}}$. Thus we can identify the analytic manifolds $G_{\mathbf{R}}$ and $G^0_{\mathbf{R}}$ with $K_{\theta,\mathbf{R}} \times A^0_{\mathbf{R}} \times N_{\mathbf{R}}$ and $K_{\mathbf{R}} \times A^0_{\mathbf{R}} \times N_{\mathbf{R}}$, respectively.

Let $\mathfrak{m}$ be the centralizer of $\mathfrak{a}$ in $\mathfrak{k}$, as in § 2, and M the analytic subgroup of G with Lie algebra $\mathfrak{m}$. Then $M_\theta = MF$ (where M_θ is the centralizer of $\mathfrak{a}$ in K_θ) and $M_{\theta,\mathbf{R}} = M_{\mathbf{R}} F$. Moreover, $M_{\mathbf{R}}$ is connected.

Now let $B=MAN$. Since M centralizes $\mathfrak{a}$ and hence normalizes $\mathfrak{n}$, it is clear that B is a subgroup of G. Furthermore, $B_{\mathbf{R}}=M_{\theta,\mathbf{R}}A^0_{\mathbf{R}}N_{\mathbf{R}}$. Since $M_{\theta,\mathbf{R}}$ centralizes $\mathfrak{a}$, $\mathfrak{n}$ is stable under $M_{\theta,\mathbf{R}}$, so that $N_{\mathbf{R}}$ is stable under $M_{\theta,\mathbf{R}}$. Hence $A^0_{\mathbf{R}}N_{\mathbf{R}}$ is a normal subgroup of $B_{\mathbf{R}}$, and $N_{\mathbf{R}}$ is a normal subgroup of $B_{\mathbf{R}}$ and of $A^0_{\mathbf{R}}N_{\mathbf{R}}$. Clearly, $B_{\mathbf{R}}$ is a closed subgroup of $G_{\mathbf{R}}$ and hence is a Lie subgroup. Moreover, $M_{\theta,\mathbf{R}}A^0_{\mathbf{R}}$ is a direct product of groups, and $B_{\mathbf{R}}/N_{\mathbf{R}}\cong M_{\theta,\mathbf{R}}A^0_{\mathbf{R}}$.

Let $\mathfrak{a}'_0=\{\lambda\in\mathfrak{a}'\,|\,\lambda|\mathfrak{a}\cap\mathfrak{z}=0\}$, where $\mathfrak{z}$=center of $\mathfrak{g}$. For any $\lambda\in\mathfrak{a}'_0$, let $\chi_\lambda\colon A^0_{\mathbf{R}}\to\mathbf{C}$ be the character defined by $\chi_\lambda(\exp \operatorname{ad} x)=e^{\langle\lambda,x\rangle}$, where $x\in\mathfrak{a}_{\mathbf{R}}$. Extend χ_λ to a character, also denoted by χ_λ, of $M_{\theta,\mathbf{R}}A^0_{\mathbf{R}}$ by defining $\chi_\lambda(ma)=\chi_\lambda(a)$. Composing χ_λ with the canonical map $B_{\mathbf{R}}\to B_{\mathbf{R}}/N_{\mathbf{R}}\cong M_{\theta,\mathbf{R}}A^0_{\mathbf{R}}$, we get a character, also denoted by χ_λ, on $B_{\mathbf{R}}$. For $b\in B_{\mathbf{R}}$, we write b^λ for $\chi_\lambda(b)$.

Now, let X be the space of all complex-valued analytic functions f on $G_{\mathbf{R}}$ such that $f(ab)=f(a)$ for $a\in G_{\mathbf{R}}$ and $b\in B_{\mathbf{R}}$ and such that f is $K_{\theta,\mathbf{R}}$-finite in the following sense: For any $a\in K_{\theta,\mathbf{R}}$, if we define $a\cdot f$ by $a\cdot f(g)=f(a^{-1}g)$ for $g\in G_{\mathbf{R}}$, then $\{a\cdot f\,|\,a\in K_{\theta,\mathbf{R}}\}$ spans a finite-dimensional vector subspace of the space of all complex-valued analytic functions on $G_{\mathbf{R}}$.

X can be considered as the space of all $K_{\theta,\mathbf{R}}$-finite complex-valued analytic functions on the homogeneous space $L=G_{\mathbf{R}}/B_{\mathbf{R}}=K_{\theta,\mathbf{R}}/M_{\theta,\mathbf{R}}$ (since $K_{\theta,\mathbf{R}}$ operates on L, "$K_{\theta,\mathbf{R}}$-finite" makes sense). By the Frobenius reciprocity theorem, given any irreducible representation γ of $K_{\theta,\mathbf{R}}$ in a finite-dimensional complex vector space V_γ, the multiplicity of γ in X is the dimension of $(V_\gamma)^{M_{\theta,R}}$ (complex dimension). The representation γ of $K_{\theta,\mathbf{R}}$ in V_γ can be extended uniquely to a holomorphic representation of K_θ in V_γ, and X becomes a K_θ-module. Thus one has the following:

Lemma 1.7.1. The K_θ-module X has the decomposition $X=\coprod_{\gamma\in\Gamma}X^\gamma$, where X^γ denotes the primary component of X of type γ; moreover, the multiplicity of γ in X is $l(\gamma)(=\dim(V_\gamma)^{M_\theta})$.

Definition 1.7.2. For $\lambda\in\mathfrak{a}'_0$, we define X_λ to be the space of all complex-valued analytic functions on $G_{\mathbf{R}}$ which are $K_{\theta,\mathbf{R}}$-finite and which satisfy $f(ab)=b^{\lambda'}f(a)$, for $a\in G_{\mathbf{R}}$ and $b\in B_{\mathbf{R}}$, where $\lambda'=-\varkappa\cdot\lambda$ (see § 6 for the definition of $\varkappa$).

Let $1_\lambda\in X_\lambda$ be defined by $1_\lambda(kan)=a^{\lambda'}$ for k, a, n in $K_{\theta,\mathbf{R}}$, $A^0_{\mathbf{R}}$ and $N_{\mathbf{R}}$, respectively. Now if $f\in X$, it is easily checked that $f1_\lambda\in X_\lambda$. Also, if $h\in X_\lambda$, then $h=(hl_\lambda^{-1})1_\lambda$, where 1_λ^{-1} is defined by $1_\lambda^{-1}(g)=(1_\lambda(g))^{-1}$ for $g\in G_{\mathbf{R}}$; clearly, $h1_\lambda^{-1}\in X$. Thus the mapping $f\to f1_\lambda$ is a bijection of X onto X_λ. This is clearly a $K_{\theta,\mathbf{R}}$-linear map. Thus one has a decomposition $X_\lambda=\coprod(X_\lambda)^\gamma$, where $(X_\lambda)^\gamma$ denotes the primary component of X_λ of type γ. Here γ denotes an equivalence class of finite-dimensional irreducible representations of $K_{\theta,\mathbf{R}}$. Also V_γ denotes a fixed $K_{\theta,\mathbf{R}}$-module in the class of γ. As remarked already, γ extends uniquely to a holomorphic representation of K_θ in V_γ, and hence X_λ becomes a K_θ-module. Thus, we have the following:

Lemma 1.7.3. The K_θ-module X_λ has the decomposition $X_\lambda = \coprod_{\gamma\in\Gamma} (X_\lambda)^\gamma$, where $(X_\lambda)^\gamma$ is the primary component of X_λ of type γ; moreover, the multiplicity of γ in X_λ is $\boldsymbol{l}(\gamma)$.

Now X_λ can be considered as a U-module in the following way: For $f\in X_\lambda$ and $x\in\mathfrak{g}_{\mathbf{R}}$, we define $(x\cdot f)(g)=\frac{\mathrm{d}}{\mathrm{d}t}\left(f((\exp -tx)g)\right)\big|_{t=0}$, for $g\in G_{\mathbf{R}}$. Then it is easily checked that for $g\in G_{\mathbf{R}}$ and $b\in B_{\mathbf{R}}$, $(x\cdot f)(gb)=b^{\lambda'}(x\cdot f)(g)$. Moreover, if $a\in K_{\theta,\mathbf{R}}$, writing as usual $a\cdot x$ for $\mathrm{Ad}\,a(x)$, one has

$$(a\cdot(x\cdot f))(g) = (x\cdot f)(a^{-1}g) =$$

$$= \frac{\mathrm{d}}{\mathrm{d}t} f((\exp - tx)a^{-1}g)\big|_{t=0} = \frac{\mathrm{d}}{\mathrm{d}t} f(a^{-1}a(\exp - tx)a^{-1}g)\big|_{t=0} =$$

$$= \frac{\mathrm{d}}{\mathrm{d}t} f(a^{-1}(\exp - ta\cdot x)g)\big|_{t=0} = \frac{\mathrm{d}}{\mathrm{d}t}\{(a\cdot f)((\exp - ta\cdot x)g)\}\big|_{t=0} =$$

$$= ((a\cdot x)\cdot(a\cdot f))(g),$$

showing that $a\cdot(x\cdot f)=(a\cdot x)\cdot(a\cdot f)$. As a varies over $K_{\theta,\mathbf{R}}$, $a\cdot x$ lies in the finite dimensional space $\mathfrak{g}$ and $a\cdot f$ also lies in a finite dimensional space; from this it follows that $x\cdot f$ is $K_{\theta,\mathbf{R}}$-finite.

Thus $\mathfrak{g}_{\mathbf{R}}$ acts on X_λ and hence $\mathfrak{g}$ acts on X_λ, so that U acts on X_λ.

Lemma 1.7.4. The U-module X_λ is an admissible module, i.e., $\{X_\lambda\}\in\mathscr{A}$. (We will prove below that actually $\{X_\lambda\}\in\mathscr{A}(\lambda)$.)

Proof. We must prove:

(i) $\dim (X_\lambda)^{\mathfrak{k}}=1$ and
(ii) X_λ is completely reducible as a $\mathfrak{k}$-module.

Now (ii) is obvious.

To prove (i), it is enough to prove that $(X_\lambda)^{\mathfrak{k}_{\mathbf{R}}}$ is one-dimensional. Note that $1_\lambda\in X_\lambda^{\mathfrak{k}}$. Now let $f\in X_\lambda$ and suppose that for every $x\in\mathfrak{k}_{\mathbf{R}}$, $x\cdot f=0$. Since f is analytic, we then conclude that $f|K_{\mathbf{R}}$ is constant, say c. One has $K_{\theta,\mathbf{R}}=K_{\mathbf{R}}F$ and $F\subseteq M_{\theta,\mathbf{R}}$. Thus for $a\in F$ and $k\in K_{\mathbf{R}}$, one has $f(ka)=a^{\lambda'}f(k)=1\cdot c=c$. Thus $f|K_{\theta,\mathbf{R}}=c$. Now $G_{\mathbf{R}}=K_{\theta,\mathbf{R}}A_{\mathbf{R}}^0N_{\mathbf{R}}$, and for $k\in K_{\theta,\mathbf{R}}$, $a\in A_{\mathbf{R}}^0$ and $n\in N_{\mathbf{R}}$, one has $f(kan)=f(k)(an)^{\lambda'}= =c(an)^{\lambda'}=(c\cdot 1_\lambda)(kan)$. Thus $f=c\cdot 1_\lambda$; proving (i). Q.E.D.

Lemma 1.7.5. One has $\{X_\lambda\}\in\mathscr{A}(\lambda)$.

Before proving Lemma 1.7.5, we first prove the following auxiliary lemma:

Lemma 1.7.6. Let X be an admissible module for U. Let X' be the dual of X, considered as a module for U with respect to the contragredient action of $\mathfrak{g}$.

Now let $\lambda \in \mathfrak{a}'$. Then $\{X\} \in \mathscr{A}(\lambda)$ if there exists $\eta \in X'$ such that

(i) $\mathfrak{n} \cdot \eta = 0$,

(ii) for every $x \in \mathfrak{a}$, $x \cdot \eta = \langle \lambda', x \rangle \eta$, where as usual $\lambda' = -\varkappa \cdot \lambda$

and

(iii) $\langle \eta, X^{\mathfrak{k}} \rangle \neq 0$.

Proof. For any $u \in U$, one has $u^t \equiv p_{u^t}$ (mod $\mathfrak{k}U + U\mathfrak{n}$). Hence, taking the transpose of both sides, $u \equiv p^t_{u^t}$ (mod $U\mathfrak{k} + \mathfrak{n}U$). Thus, choosing $\xi \in X^{\mathfrak{k}}$ such that $\langle \eta, \xi \rangle \neq 0$ (this is possible in view of (iii)), one has $\langle \eta, u \cdot \xi \rangle = \langle \eta, p^t_{u^t} \cdot \xi \rangle = \langle p_{u^t} \cdot \eta, \xi \rangle = p_{u^t}(\lambda') \langle \eta, \xi \rangle$, in view of (ii) and the fact that $p_{u^t} \in U(\mathfrak{a})$. On the other hand, since $\{X\} \in \mathscr{A}$, there exists $\lambda_0 \in \mathfrak{a}'$ such that $\{X\} \in \mathscr{A}(\lambda_0)$. Then for $u \in U^{\mathfrak{k}}$, one has from above,

$$\begin{aligned} p_{u^t}(\lambda')\langle \eta, \xi \rangle &= \langle \eta, u \cdot \xi \rangle \\ &= \langle \eta, p_u(\lambda_0) \xi \rangle \\ &= p_u(\lambda_0) \langle \eta, \xi \rangle. \end{aligned}$$

Hence, for any $u \in U^{\mathfrak{k}}$, $p_{u^t}(\lambda') = p_u(\lambda_0)$. But for any $u \in U^{\mathfrak{k}}$, one has $p_{u^t}(\lambda') = p_u(\lambda)$ (see Lemma 1.6.3). Thus for any $u \in U^{\mathfrak{k}}$, one has $p_u(\lambda) = p_u(\lambda_0)$, and so $\{X\} \in \mathscr{A}(\lambda)$. Q.E.D.

Proof of Lemma 1.7.5. Consider X'_λ, the dual of X_λ, as a U-module in the usual way (i.e., with respect to the contragredient action of $\mathfrak{g}$ on X'_λ). Let $\delta \in X'_\lambda$ be defined by $\langle \delta, f \rangle = f(e)$, where $f \in X_\lambda$ and e is the identity element of $G_{\mathbf{R}}$.

Now, for $x \in \mathfrak{a}_{\mathbf{R}} + \mathfrak{n}_{\mathbf{R}}$, one has

$$\langle x \cdot \delta, f \rangle = \langle \delta, -x \cdot f \rangle = (-x \cdot f)(e) =$$

$$= \frac{\mathrm{d}}{\mathrm{d}t} (f(\exp tx)) \Big|_{t=0} = \frac{\mathrm{d}}{\mathrm{d}t} (f(e)(\exp tx)^{\lambda'}) \Big|_{t=0},$$

since $\exp tx \in B_{\mathbf{R}}$. Now if $x \in \mathfrak{a}_{\mathbf{R}}$, then $(\exp tx)^{\lambda'} = e^{\langle \lambda', tx \rangle} = e^{t\langle \lambda', x \rangle}$. Hence

$$\langle x \cdot \delta, f \rangle = \frac{\mathrm{d}}{\mathrm{d}t} (f(e) e^{t\langle \lambda', x \rangle}) \Big|_{t=0} = f(e) \langle \lambda', x \rangle = \langle \langle \lambda', x \rangle \delta, f \rangle.$$

Thus $x \cdot \delta = \langle \lambda', x \rangle \delta$. Similarly, if $x \in \mathfrak{n}_{\mathbf{R}}$, then $x \cdot \delta = 0$. This shows that $\mathfrak{n} \cdot \delta = 0$ and that for any $x \in \mathfrak{a}$, $x \cdot \delta = \langle \lambda', x \rangle \delta$. Also, $\langle \delta, (X_\lambda)^{\mathfrak{k}} \rangle \neq 0$ since $1_\lambda \in (X_\lambda)^{\mathfrak{k}}$ and $\langle \delta, 1_\lambda \rangle = 1$. Now the result follows immediately from Lemma 1.7.6. Q.E.D.

Now let X^e_λ be the U-submodule of X_λ generated by 1_λ. Then X^e_λ is a cyclic module for U, and in fact $\{X^e_\lambda\} \in \mathscr{C}(\lambda)$. Furthermore, X^e_λ is a K_θ-module, since for $a \in K_\theta$ and $u \in U$, one has $a \cdot (u \cdot 1_\lambda) = (a \cdot u) \cdot 1_\lambda$, where $a \cdot u = \mathrm{Ad}\, a(u)$.

Corollary 1.7.7. One has that X_λ is irreducible if and only if for every $\gamma \in \Gamma$, $\boldsymbol{l}(\gamma) = m(\gamma)$ (see § 5), or equivalently (by Corollary 1.5.13), $r^\gamma(\lambda) \neq 0$, where $r^\gamma(\lambda) = \det R^\gamma(\lambda)$.

Proof. We know that the multiplicity of γ in X_λ is $\boldsymbol{l}(\gamma)$. Now suppose that X_λ is irreducible. Then $\{X_\lambda\}=\{Z_\lambda\}$, so that $m(\gamma)=\boldsymbol{l}(\gamma)$. Conversely, suppose $m(\gamma)=\boldsymbol{l}(\gamma)$ for all $\gamma\in\Gamma$. Then we know that $\mathscr{C}(\lambda)$ consists of only one element, namely, $\{Z_\lambda\}$. Thus $\{X_\lambda^e\}=\{Z_\lambda\}$. Hence the multiplicity of γ in $X_\lambda^e=m(\gamma)=$ $=\boldsymbol{l}(\gamma)=$the multiplicity of γ in X_λ. Thus $X_\lambda^e=X_\lambda$, and so X_λ is irreducible.

Q.E.D.

Let Y be any cyclic module for U, i.e., $\{Y\}\in\mathscr{C}\subseteq\mathscr{A}$.

Definition 1.7.8. We say that Y is *effective* if (i) there exists a non-zero element $\eta\in Y'$ such that $\mathfrak{n}\cdot\eta=0$, (ii) $\mathfrak{a}$ leaves stable the one-dimensional subspace generated by η and (iii) $U\cdot\eta$ is non-singularly paired to Y under the duality between Y and Y', i.e., if $\psi\in Y$ is such that $\langle u\cdot\eta,\psi\rangle=0$ for every $u\in U$, then $\psi=0$.

Note that there exists $\mu\in\mathfrak{a}'$ such that for every $x\in\mathfrak{a}$, $x\cdot\eta=\langle\mu,x\rangle\eta$. We may write $\mu=\lambda'$, for a certain $\lambda\in\mathfrak{a}'$. (See § 1.6.)

Since Y is cyclic, $Y=U/L$, where L is the annihilator of $Y^{\mathfrak{k}}$.

For all $u\in U$, we now define $p^u\in U(\mathfrak{a})$ by $p^u(\lambda)=p_{u^t}(\lambda')$, for all $\lambda\in\mathfrak{a}'$.

Lemma 1.7.9. With the above notation, assume Y is effective. Then

(i) $\langle\eta, Y^{\mathfrak{k}}\rangle\neq0$ (so that $\{Y\}\in\mathscr{C}(\lambda)$ by Lemma 1.7.6).
(ii) L is stable under K_θ, so that Y is a K_θ-module.
(iii) $u\in L$ if and only if for all $a\in K_\theta$, $p^{a\cdot u}(\lambda)=0$.

Proof. If $\langle\eta, Y^{\mathfrak{k}}\rangle=0$, then $0=\langle U(\mathfrak{a})U(\mathfrak{n})\cdot\eta, U(\mathfrak{k})\cdot Y^{\mathfrak{k}}\rangle$. Transposing $U(\mathfrak{k})$ one has $0=\langle U\cdot\eta, Y^{\mathfrak{k}}\rangle$, a contradiction. This proves (i). Before proving (ii) and (iii), we first prove the following lemma:

Lemma 1.7.10.

1) For any $a\in G$ and $u\in U$, $(a\cdot u)^t=a\cdot u^t$.
2) For any $a\in M_\theta$ and $u\in U$, $p_{a\cdot u}=p_u$.
3) For any $a\in M_\theta$ and $u\in U$, $p^{a\cdot u}=p^u$.

Proof. To prove *1)*, consider the map $u\to u^\tau$ of U into U defined by $u^\tau=a^{-1}\cdot(a\cdot u)^t$. For $x\in\mathfrak{g}$, one has $x^\tau=a^{-1}\cdot(a\cdot x)^t=a^{-1}\cdot(-a\cdot x)=-x$. For $u,v\in U$, one clearly has $(uv)^\tau=v^\tau u^\tau$. Now from the definition of the transpose map t, it follows that $u^\tau=u^t$ for every $u\in U$. Hence $a^{-1}\cdot(a\cdot u)^t=u^t$, i.e., $(a\cdot u)^t=$ $=a\cdot u^t$.

To prove *2)*, note that a acts as the identity on $\mathfrak{a}$ and hence leaves $\mathfrak{n}$ stable. Also, a leaves $\mathfrak{k}$ stable. Now, one has $u\equiv p_u$ (mod $\mathfrak{k}U+U\mathfrak{n}$). Applying a to both sides, we get $a\cdot u=a\cdot p_u$ (mod $\mathfrak{k}U+U\mathfrak{n}$). Hence $p_{a\cdot u}=a\cdot p_u$. But since a acts as the identity on $\mathfrak{a}$, a acts as the identity on $U(\mathfrak{a})$ also. Hence $a\cdot p_u=p_u$. Thus one has $p_{a\cdot u}=p_u$.

Finally, $p^{a\cdot u}(\lambda)=p_{(a\cdot u)^t}(\lambda')=p_{a\cdot u^t}(\lambda')=p_{u^t}(\lambda')=p^u(\lambda)$, proving *3)*. Q.E.D.

Now we return to the proof of Lemma 1.7.9. In view of Lemma 1.7.10, and the fact that $K_\theta = KF$ and that $F \subseteqq M_\theta$, assertion (iii) of Lemma 1.7.9, is the same as the following assertion:

(iii′) $u \in L$ if and only if for every $a \in K$, $p^{a \cdot u}(\lambda) = 0$.

Note that under the adjoint action of G on U, U is a G-module. This induces a $\mathfrak{g}$-module structure on U, and the action is given by $x \cdot u = [x, u] = xu - ux$ for $x \in \mathfrak{g}$ and $u \in U$. Since U is the universal enveloping algebra of $\mathfrak{g}$, the $\mathfrak{g}$-module structure on U extends to a U-module structure on U. Now, assertion (iii′) is equivalent to saying that $u \in L$ if and only if $p^v(\lambda) = 0$ for every v in the K-module generated by u. But the K-module generated by u is the same as the $U(\mathfrak{k})$-module generated by u. Thus assertion (iii′) is equivalent to the assertion that $u \in L$ if and only if for every $v \in U(\mathfrak{k}) \cdot u$, $p^v(\lambda) = 0$.

Now we prove the last assertion. For any $v \in U$ one has $v \equiv p^t_{vt}$ (mod $U\mathfrak{k} + \mathfrak{n}U$). Choosing $0 \neq \xi \in Y^{\mathfrak{k}}$, one has $\langle \eta, v \cdot \xi \rangle = \langle \eta, p^t_{vt} \cdot \xi \rangle = \langle p_{vt} \cdot \eta, \xi \rangle = p_{vt}(\lambda') \langle \eta, \xi \rangle = p^v(\lambda) \langle \eta, \xi \rangle$. Thus $p^{U(\mathfrak{k}) \cdot u}(\lambda) = 0$ if and only if $\langle \eta, (U(\mathfrak{k}) \cdot u) \cdot \xi \rangle = 0$. Since $\xi \in Y^{\mathfrak{k}}$, $(U(\mathfrak{k}) \cdot u) \cdot \xi = U(\mathfrak{k}) u \cdot \xi$ (where $U(\mathfrak{k})u$ denotes multiplication in U). Thus $p^{U(\mathfrak{k}) \cdot u}(\lambda) = 0$ if and only if $\langle \eta, U(\mathfrak{k}) u \cdot \xi \rangle = 0$, i.e., if and only if $\langle U(\mathfrak{k}) \cdot \eta, u \cdot \xi \rangle = 0$, i.e., if and only if $\langle U \cdot \eta, u \cdot \xi \rangle = 0$, since $U \cdot \eta = U(\mathfrak{k}) U(\mathfrak{a}) U(\mathfrak{n}) \cdot \eta = U(\mathfrak{k}) \cdot \eta$. Now, noting that $U \cdot \eta$ is non-singularly paired to Y, one has that $\langle U \cdot \eta, u \cdot \xi \rangle = 0$ if and only if $u \cdot \xi = 0$. Thus we finally obtain $p^{U(\mathfrak{k}) \cdot u}(\lambda) = 0$ if and only if $u \cdot \xi = 0$, i.e., if and only if $u \in L$. This proves our last assertion, hence (iii′), and hence (iii).

Now (ii) is an immediate consequence of (iii). Q.E.D.

We assert that X^e_λ is an effective U-module. It is trivial to check that δ, the Dirac measure at the identity of $G_{\mathbf{R}}$, serves as η. For example, let us check that $U \cdot \delta$ is non-singularly paired to X^e_λ. Suppose $f \in X^e_\lambda$ is such that $\langle U \cdot \delta, f \rangle = 0$. This means that all the derivatives of f at e (=identity of $G_{\mathbf{R}}$) vanish. Since f is analytic, this means that f is zero on $G^0_{\mathbf{R}}$. Now $G_{\mathbf{R}} = G^0_{\mathbf{R}} F$ and $F \subseteqq M_{\theta, \mathbf{R}} \subseteqq B_{\mathbf{R}}$. Since for $a \in G_{\mathbf{R}}$ and $b \in B_{\mathbf{R}}$, $f(ab) = f(a) b^{\lambda'}$, it follows that f vanishes identically on $G_{\mathbf{R}}$.

Corollary 1.7.11. For all $u \in U$, one has $u \cdot 1_\lambda = 0$ if and only if $p^{K_\theta \cdot u}(\lambda) = 0$.

Now, for any $u \in U$, let Q_u be the K_θ-module generated by u in U. Let $(Q_u)^{M_\theta}$ be the subspace of Q_u fixed by M_θ and R_u the subspace of Q_u generated by $\{a \cdot q - q \mid q \in Q_u \text{ and } a \in M_\theta\}$. Then $Q_u = (Q_u)^{M_\theta} \oplus R_u$ (see Lemma 1.5.1). Since one has $p^{a \cdot v} = p^v$, for any $a \in M_\theta$, it follows that $p^{R_u}(\lambda) = 0$. Thus from Corollary 1.7.11, it follows that $u \cdot 1_\lambda = 0$ if and only if $p^{(Q_u)^{M_\theta}}(\lambda) = 0$. This implies the following:

Let $v_1, \ldots, v_{l(\gamma)}$ be a basis of $(V_\gamma)^{M_\theta}$. Then for any $\sigma \in E_\gamma = \mathrm{Hom}_{K_\theta}(V_\gamma, H^*)$, one has that $\sigma(V_\gamma) \cdot 1_\lambda = 0$ if and only if $p^{\sigma(v_i)}(\lambda) = 0$ for $i = 1, \ldots, l(\gamma)$.

This fact implies the following theorem, whose proof is analogous to the last part of the proof of Lemma 1.5.12.

Theorem 1.7.12. *Let* $\sigma_1, \ldots, \sigma_{l(\gamma)}$ *be the basis of* E_γ *and let* $v_1, \ldots, v_{l(\gamma)}$ *be a basis of* $(V_\gamma)^{M_\theta}$. *Let* P^γ *be the matrix with coefficients in* $U(\mathfrak{a})$ *defined by* $P^\gamma=(P^\gamma_{ij})$, *where* $P^\gamma_{ij}=p^{\sigma_j(v_i)}$. *Then rank* $P^\gamma(\lambda)=n_\lambda(\gamma)$, *where* $n_\lambda(\gamma)$ *denotes the multiplicity of* γ *in* X^e_λ.

Remark. Unlike the matrix $R^\gamma(\lambda)$, which depends only on the W_T-orbit of λ, the matrix $P^\gamma(\lambda)$ may be different for different λ in the same W_T-orbit. This is essentially because X^e_λ depends on λ, and not just on the W_T-orbit of λ.

§ 8. The formula $R^\gamma(\lambda)=P^\gamma(-\bar{\lambda}-2\varrho)^* P(\lambda)$

In this section, we will prove that $R^\gamma(\lambda)=(P^\gamma(-\bar{\lambda}-2\varrho))^* P^\gamma(\lambda)$ (for any matrix T, T^* denotes the conjugate transpose of T). Our proof uses a certain bilinear pairing on $X_\lambda \times X_{-\lambda-2\varrho}$ which is $\mathfrak{g}$-invariant, i.e., if $y\in\mathfrak{g}$, $f\in X_\lambda$ and $g\in X_{-\lambda-2\varrho}$, then $\langle y\cdot f, g\rangle+\langle f, y\cdot g\rangle=0$.

The homogeneous space $L=G_{\mathbf{R}}/B_{\mathbf{R}}$ is an analytic manifold. Let T_0 denote the tangent space at $o\in L$, where o denotes the coset $B_{\mathbf{R}}$. With respect to the left action of $G_{\mathbf{R}}$ on $G_{\mathbf{R}}/B_{\mathbf{R}}$, o is invariant under $B_{\mathbf{R}}$. Thus $B_{\mathbf{R}}$ acts on T_0 and hence on T_0', the (real) dual of T_0. Hence $B_{\mathbf{R}}$ acts on $\Lambda^d(T_0')$, the dth exterior power of T_0', where d=dimension of T_0.

Since $\Lambda^d(T_0')$ is a one-dimensional real vector space, the action of $B_{\mathbf{R}}$ on $\Lambda^d(T_0')$ is given by a real character χ on $B_{\mathbf{R}}$, i.e., there exists a homomorphism $\chi\colon B_{\mathbf{R}}\to\mathbf{R}^*$ ($\mathbf{R}^*$ is the multiplicative group of non-zero real numbers) such that for $b\in B_{\mathbf{R}}$ and $u\in\Lambda^d(T_0')$, $b\cdot u=\chi(b)u$.

Let E be the homogeneous complex line bundle on $G_{\mathbf{R}}/B_{\mathbf{R}}$ induced by the character $|\chi|\colon B_{\mathbf{R}}\to\mathbf{R}^*$ defined by $|\chi|(b)=|\chi(b)|$. Now E has a natural real analytic structure, and there is a natural left action of $G_{\mathbf{R}}$ as automorphisms of the bundle E which lifts the action of $G_{\mathbf{R}}$ on $G_{\mathbf{R}}/B_{\mathbf{R}}$. Hence $G_{\mathbf{R}}$ acts on the space $\tilde{X}$ of analytic sections of the bundle E. In fact if da denotes the unique $K_{\theta,\mathbf{R}}$-invariant measure on $G_{\mathbf{R}}/B_{\mathbf{R}}$ such that $\int da=1$ then we may regard $\tilde{X}$ as the space of all measures μ on $G_{\mathbf{R}}/B_{\mathbf{R}}$ of the form $\mu=\varphi\, da$ where φ is a complex-valued real analytic function on $G_{\mathbf{R}}/B_{\mathbf{R}}$. Furthermore $\mu\to\int\mu$ is a linear functional on $\tilde{X}$ and the action of $G_{\mathbf{R}}$ on $\tilde{X}$ is the natural action of $G_{\mathbf{R}}$ on measures so that one has $\int a\cdot\mu=\int\mu$ for all $a\in G_{\mathbf{R}}$.

Lemma 1.8.3. For any $b\in B_{\mathbf{R}}$ one has $|\chi|(b)=b^{2\varrho}$.

Proof. For any $b\in B_{\mathbf{R}}$ one has $|\chi|(b)=|(\det \mathrm{Ad}\, b|\mathfrak{g}_{\mathbf{R}})^{-1}(\det \mathrm{Ad}\, b|\mathfrak{b}_{\mathbf{R}})|$ where $\mathfrak{b}_{\mathbf{R}}$ is the Lie algebra of $B_{\mathbf{R}}$. This is clear since $T_0\cong\mathfrak{g}_{\mathbf{R}}/\mathfrak{b}_{\mathbf{R}}$. But $\det \mathrm{Ad}\, b\,|\mathfrak{g}_{\mathbf{R}}=1$ so one has $|\chi|(b)=|\det \mathrm{Ad}\, b|\mathfrak{b}_{\mathbf{R}}$. Now $B_{\mathbf{R}}=M_{\theta,\mathbf{R}}A^0_{\mathbf{R}}N_{\mathbf{R}}$. But if $b\in M_{\theta,\mathbf{R}}N_{\mathbf{R}}$ one has $|\chi|(b)=1$ since $M_{\theta,\mathbf{R}}$ is compact and $N_{\mathbf{R}}$ is unipotent (recall we are taking absolute values). If $b\in A_{\mathbf{R}}$ obviously $|\chi|(b)=b^{2\varrho}$ since $\mathfrak{b}_{\mathbf{R}}=\mathfrak{m}_{\mathbf{R}}+\mathfrak{a}_{\mathbf{R}}+\mathfrak{n}_{\mathbf{R}}$. This proves the lemma. Q.E.D.

Now let $\tilde{X}_{-2\varrho}$ denote the space of all complex-valued real analytic functions f on $G_{\mathbf{R}}$ such that $f(ab)=b^{-2\varrho}f(a)$ for all $a\in G_{\mathbf{R}}, b\in B_{\mathbf{R}}$. It follows easily then from Lemma 1.8.3 that if $u\in\Lambda^d(T_0')$ denotes the value of *da* (regarded as a section of E) at 0 and $f\in\tilde{X}_{-2\varrho}$ then a section $\mu_f\in\tilde{X}$ of E is well defined by putting $\mu_f(gB_{\mathbf{R}})=f(g)g\cdot u$. Furthermore the map $\tilde{X}_{-2\varrho}\to\tilde{X}$ given by $f\to\mu_f$ is a $G_{\mathbf{R}}$-isomorphism and $\int\mu_f=\int_{K_{\mathbf{R}}}f\,dh$ where dh is normalized Haar measure on $K_{\mathbf{R}}$. Thus for any $a\in G_{\mathbf{R}}$ and $f\in\tilde{X}_{-2\varrho}$ one has $\int_{K_{\mathbf{R}}}f\,dh=\int_{K_{\mathbf{R}}}a\cdot f\,dh$. By differentiation, then, one has $\int_{K_{\mathbf{R}}}x\cdot f\,dh=0$ for all $x\in\mathfrak{g}$, $f\in\tilde{X}_{-2\varrho}$. But now the space of all $f\in\tilde{X}_{-2\varrho}$ which are $K_{\theta,\mathbf{R}}$-finite is exactly $X_{-2\varrho}$ since clearly $(-2\varrho)'=-2\varrho$. Also $X_{-2\varrho}$ is U-stable and $\int_{K_{\mathbf{R}}}x\cdot f\,dh=0$ for all $x\in\mathfrak{g}$, $f\in X_{-2\varrho}$.

Definition 1.8.4. For $f\in X_\lambda$ and $g\in X_{-\lambda-2\varrho}$, we define $\langle f,g\rangle=\int_{K_{\theta,\mathbf{R}}}fg\,dh$, where dh denotes Haar measure on $K_{\theta,\mathbf{R}}$ of total volume 1, as in Remark 1.8.1.

Note that $fg\in X_{-2\varrho}$. Then, from above, for $x\in\mathfrak{g}$, $\int_{K_{\theta,\mathbf{R}}}x\cdot(fg)\,dh=0$. Thus

$$\langle x\cdot f,g\rangle+\langle f,x\cdot g\rangle=\int_{K_{\theta,\mathbf{R}}}((x\cdot f)g+f(x\cdot g))\,dh=\int_{K_{\theta,\mathbf{R}}}x\cdot(fg)\,dh=0,$$

and so $\langle x\cdot f,g\rangle=\langle f,-x\cdot g\rangle$. Thus for $u\in U$, $\langle u\cdot f,g\rangle=\langle f,u^t\cdot g\rangle$.

Remark. 1.8.5. Observe that if $f\in X_\lambda$, then $\bar{f}\in X_{\bar{\lambda}}$. Now when $\lambda=-\varrho$, we have $-\lambda-2\varrho=-\varrho$, and so we get a positive-definite Hermitian form B on $X_{-\varrho}$ by setting $B(f,g)=\langle f,\bar{g}\rangle$, where $\langle\ ,\ \rangle$ is the bilinear form of Definition 1.8.4. That is, $B(f,g)=\int_{K_{\theta,\mathbf{R}}}f\bar{g}\,dh$. From the fact that $\langle u\cdot f,g\rangle=\langle f,u^t\cdot g\rangle$ for $u\in U$, it follows immediately that B is U-invariant, i.e., that $B(u\cdot f,g)=B(f,u^s\cdot g)$.

Now let $\gamma\in\Gamma$. Recall that $E_\gamma=\mathrm{Hom}_{K_\theta}(V_\gamma,H^*)$ and that $E_\gamma'=\mathrm{Hom}_{K_\theta}(V_\gamma',H^*)$. Now fix $\varepsilon\in E_\gamma$, $\varepsilon'\in E_\gamma'$, $v\in V_\gamma$ and $v'\in V_\gamma'$. Consider $\langle\varepsilon(v)\cdot 1_\lambda,\varepsilon'(v')\cdot 1_{-\lambda-2\varrho}\rangle$, where $1_\lambda\in X_\lambda$ is the function satisfying $1_\lambda(kb)=b^{\lambda'}$ for $k\in K_{\theta,\mathbf{R}}$ and $b\in B_{\mathbf{R}}$. One has

$$\langle\varepsilon(v)\cdot 1_\lambda,\varepsilon'(v')\cdot 1_{-\lambda-2\varrho}\rangle=\langle(\varepsilon'(v'))^t\varepsilon(v)\cdot 1_\lambda,1_{-\lambda-2\varrho}\rangle=$$

$$=\langle((\varepsilon'(v'))^t\varepsilon(v))_0\cdot 1_\lambda,1_{-\lambda-2\varrho}\rangle$$

(using the fact that $\mathfrak{k}$ annihilates 1_λ and $1_{-\lambda-2\varrho}$)

$$=p_{((\varepsilon'(v'))^t\varepsilon(v))_0}(\lambda).$$

We assert that

$$p_{((\varepsilon'(v'))^t\varepsilon(v))_0}(\lambda)=\frac{\langle v',v\rangle}{\dim V_\gamma}p_{z((\varepsilon')^t,\varepsilon)}(\lambda),$$

where $(\varepsilon')^t\colon V_\gamma'\to H^*$ is defined by $(\varepsilon')^t(v)=(\varepsilon'(v))^t$.

This can be seen as follows: Let π denote the projection map $U \to S^*$ with respect to the decomposition $U = \mathfrak{k}U \oplus S^*$. Note that π is a K_θ-homomorphism. Moreover, observe that if $u \in U^{\mathfrak{t}}$, then $\pi(u) \in J^*$, so that $\pi(u)$ is K_θ-invariant. Now consider the two K_θ-linear maps B_1 and B_2 from $V'_\gamma \otimes V_\gamma$ into S^* defined by $B_1(v' \otimes v) = \langle v', v\rangle \pi\big(z((\varepsilon')^t, \varepsilon)\big)$ and $B_2(v' \otimes v) = \pi\big([(\varepsilon'(v'))^t \varepsilon(v)]_0\big)$. Clearly, both $B_1(V'_\gamma \otimes V_\gamma)$ and $B_2(V'_\gamma \otimes V_\gamma)$ are contained in the space of K_θ-invariants in $\pi\big((\varepsilon')^t(V'_\gamma)\varepsilon(V_\gamma)\big)$. In view of Schur's lemma, we can conclude that there exists a scalar c such that

$$\pi\big([(\varepsilon'(v'))^t \varepsilon'(v)]_0\big) = c\langle v', v\rangle \pi\big(z((\varepsilon')^t, \varepsilon)\big). \qquad (*)$$

Now let $v_1, \ldots, v_n$ be a basis of V_γ and $v'_1, \ldots, v'_n$ the dual basis of V'_γ. Then one has by definition

$$z\big((\varepsilon')^t, \varepsilon\big) = \sum_i \big(\varepsilon'(v'_i)\big)^t \varepsilon(v_i) = \sum_i \big((\varepsilon'(v'_i))^t \varepsilon(v_i)\big)_0,$$

since $z\big((\varepsilon')^t, \varepsilon\big)$ is in $U^{\mathfrak{t}}$. Now applying π to both sides and using $(*)$ one easily sees that $c = \dfrac{1}{\dim V_\gamma}$. Thus we get

$$\pi\big([(\varepsilon'(v'))^t \varepsilon(v)]_0\big) = \frac{\langle v', v\rangle}{\dim V_\gamma} \pi\big(z((\varepsilon')^t, \varepsilon)\big).$$

Our assertion now follows by taking p of both sides and observing that for $u \in U$, $p_u = p_{\pi(u)}$.

Thus we have

$$\frac{\langle v', v\rangle}{\dim V_\gamma} p_{z((\varepsilon')^t, \varepsilon)}(\lambda) =$$

$$= p_{((\varepsilon'(v'))^t \varepsilon(v))_0}(\lambda) = \langle \varepsilon(v) \cdot 1_\lambda, \varepsilon'(v') \cdot 1_{-\lambda-2\varrho}\rangle.$$

Now writing $f = \varepsilon(v) \cdot 1_\lambda$ and $g = \varepsilon'(v') \cdot 1_{-\lambda-2\varrho}$, and da for normalized Haar measure on $K_{\theta,\mathbf{R}}$ one has

$$\langle \varepsilon(v) \cdot 1_\lambda, \varepsilon'(v') \cdot 1_{-\lambda-2\varrho}\rangle = \int_{K_{\theta,\mathbf{R}}} f(a) g(a)\, da.$$

Let $\delta \in X'_\lambda$ be defined by $\langle \delta, f\rangle = f(e)$, where X'_λ is the dual of X_λ. Now

$$f(a) = (a^{-1} \cdot f)(e) = \langle \delta, a^{-1} \cdot f\rangle = \langle \delta, a^{-1} \cdot (\varepsilon(v) \cdot 1_\lambda)\rangle =$$

$$= \langle \delta, \varepsilon(a^{-1} \cdot v) \cdot 1_\lambda\rangle = \langle (\varepsilon(a^{-1} \cdot v))^t \cdot \delta, 1_\lambda\rangle = p^{\varepsilon(a^{-1} \cdot v)}(\lambda).$$

Similarly, $g(a) = p^{\varepsilon'(a^{-1} \cdot v)}(-\lambda - 2\varrho)$. Thus we have

$$\langle \varepsilon(v) \cdot 1_\lambda, \varepsilon'(v') \cdot 1_{-\lambda-2\varrho}\rangle = \int_{K_{\theta,\mathbf{R}}} f(a) g(a)\, da =$$

$$= \int_{K_{\theta,\mathbf{R}}} p^{\varepsilon(a^{-1} \cdot v)}(\lambda) = p^{\varepsilon'(a^{-1} \cdot v')}(-\lambda - 2\varrho)\, da.$$

We now define $T_\lambda: E_\gamma \to V'_\gamma$ and $T'_\lambda: E'_\gamma \to V_\gamma$ as follows: for $\varepsilon \in E_\gamma$, $T_\lambda(\varepsilon) \in V'_\gamma$ is the unique element satisfying $\langle T_\lambda(\varepsilon), v\rangle = p^{\varepsilon(v)}(\lambda)$ for any $v \in V$; similarly, for $\varepsilon' \in E'_\gamma$, $T'_\lambda(\varepsilon')$ is the unique element in V_γ satisfying $\langle v', T'_\lambda(\varepsilon')\rangle = p^{\varepsilon'(v')}(\lambda)$ for any $v' \in V'_\gamma$.

Hence we have

$$\langle \varepsilon(v) \cdot 1_\lambda, \varepsilon'(v') \cdot 1_{-\lambda-2\varrho}\rangle = \int_{K_{\theta,\mathbf{R}}} p^{\varepsilon(a^{-1}\cdot v)}(\lambda)\, p^{\varepsilon'(a^{-1}\cdot v')}(-\lambda-2\varrho)\, da =$$

$$= \int_{K_{\theta,\mathbf{R}}} \langle T_\lambda(\varepsilon), a^{-1}\cdot v\rangle \langle a^{-1}\cdot v', T'_{-\lambda-2\varrho}(\varepsilon')\rangle\, da.$$

But one knows that if $u, v \in V_\gamma$ and $u', v' \in V'_\gamma$, then

$$\int_{K_{\theta,\mathbf{R}}} \langle u', a^{-1}\cdot v\rangle \langle a^{-1}\cdot v', u\rangle\, da = \frac{\langle u', u\rangle \langle v', v\rangle}{\dim V_\gamma}$$

(see [16], p. 73).

Thus we have

$$\langle \varepsilon(v)\cdot 1_\lambda, \varepsilon'(v')\cdot 1_{-\lambda-2\varrho}\rangle = \frac{1}{\dim V_\gamma} \langle T_\lambda(\varepsilon), T_{-\lambda-2\varrho}(\varepsilon')\rangle \langle v', v\rangle.$$

But we have already seen that

$$\langle \varepsilon(v)\cdot 1_\lambda, \varepsilon'(v')\cdot 1_{-\lambda-2\varrho}\rangle = \frac{\langle v', v\rangle}{\dim V_\gamma} p_{z((\varepsilon')^t, \varepsilon)}(\lambda).$$

Thus we have

$$p_{z((\varepsilon')^t, \varepsilon)}(\lambda) = \langle T_\lambda(\varepsilon), T'_{-\lambda-2\varrho}(\varepsilon')\rangle.$$

We assert that $T'_\lambda(\varepsilon') \in (V_\gamma)^{M_\theta}$ and $T_\lambda(\varepsilon) \in (V'_\gamma)^{M_\theta}$ for any $\lambda \in \mathfrak{a}'$. Indeed, if $a \in M_\theta$, then we have, for any $v' \in V'_\gamma$

$$\langle v', a\cdot T'_\lambda(\varepsilon')\rangle = \langle a^{-1}\cdot v', T'_\lambda(\varepsilon')\rangle = p^{\varepsilon'(a^{-1}\cdot v')}(\lambda) = p^{a^{-1}\cdot\varepsilon'(v')}(\lambda) = p^{\varepsilon'(v')}(\lambda)$$

(by Lemma 1.7.10) $= \langle v', T'_\lambda(\varepsilon')\rangle$, showing that $a\cdot T'_\lambda(\varepsilon') = T'_\lambda(\varepsilon')$. The other assertion is proved similarly.

Fix a $K_{\theta,\mathbf{R}}$-invariant positive-definite Hermitian form on V_γ, and let $v_1, \dots, v_{l(\gamma)}$ be an orthonormal basis of $(V_\gamma)^{M_\theta}$. Then $v_1^s, \dots, v_{l(\gamma)}^s$ is a basis of $(V'_\gamma)^{M_\theta}$ (see § 6 for the definition of the map $v \to v^s$). Identifying $(V'_\gamma)^{M_\theta}$ with the dual of $(V_\gamma)^{M_\theta}$ in the obvious way, $v_1^s, \dots, v_{l(\gamma)}^s$ is just the dual basis to $v_1, \dots, v_{l(\gamma)}$. Clearly,

$$T_\lambda(\varepsilon) = \sum_i \langle T_\lambda(\varepsilon), v_i\rangle v_i^s \quad \text{and} \quad T'_{-\lambda-2\varrho}(\varepsilon') = \sum_i \langle v_i^s, T'_{-\lambda-2\varrho}(\varepsilon')\rangle v_i,$$

so that we have

$$\langle T_\lambda(\varepsilon), T_{-\lambda-2\varrho}(\varepsilon')\rangle = \sum_i \langle T_\lambda(\varepsilon), v_i\rangle \langle v_i^s, T'_{-\lambda-2\varrho}(\varepsilon')\rangle.$$

Thus we have

$$p_{z((\varepsilon')^t, \varepsilon)}(\lambda) = \langle T_\lambda(\varepsilon), T'_{-\lambda-2\varrho}(\varepsilon')\rangle = \sum_i \langle T_\lambda(\varepsilon), v_i\rangle \langle v_i^s, T'_{-\lambda-2\varrho}(\varepsilon')\rangle.$$

Now let $\varepsilon_1, \ldots, \varepsilon_{l(\gamma)}$ be a basis of E_γ. Then $\varepsilon_1^s, \ldots, \varepsilon_{l(\gamma)}^s$ is a basis of E'_γ. We use these bases for computing the matrices $R^\gamma(\lambda)$ and $P^\gamma(\lambda)$.

Substituting $\varepsilon=\varepsilon_j$ and $\varepsilon'=(\varepsilon_k^s)^t$ in the above equation, we get

$$p_{z(\varepsilon_k^s, \varepsilon_j)}(\lambda) = \sum_i \langle T_\lambda(\varepsilon_j), v_i\rangle \langle v_i^s, T'_{-\lambda-2\varrho}((\varepsilon_k^s)^t)\rangle.$$

That is,

$$\begin{aligned}(R^\gamma(\lambda))_{kj} &= \sum_i p^{\varepsilon_j(v_i)}(\lambda)\, p^{(\varepsilon_k^s)^t(v_i^s)}(-\lambda-2\varrho) = \\ &= \sum_i p^{\varepsilon_j(v_i)}(\lambda)\, \overline{p^{\varepsilon_k(v_i)}}(-\lambda-2\varrho) = \sum_i p^{\varepsilon_j(v_i)}(\lambda)\, \overline{p^{\varepsilon_k(v_i)}(-\bar\lambda-2\varrho)} = \\ &= \sum_i (P^\gamma(\lambda))_{ij} \overline{(P^\gamma(-\bar\lambda-2\varrho))_{ik}} = \sum_i (P^\gamma(\lambda))_{ij} ((P^\gamma(-\bar\lambda-2\varrho))^*)_{ki} = \\ &= ((P^\gamma(-\bar\lambda-2\varrho))^* P^\gamma(\lambda))_{kj}.\end{aligned}$$

This proves that

$$R^\gamma(\lambda) = (P^\gamma(-\bar\lambda-2\varrho))^* P^\gamma(\lambda)$$

for all $\lambda \in \mathfrak{a}'_0$, with respect to the above choices of bases. We assert that this formula remains true for all $\lambda \in \mathfrak{a}'$. Indeed, let $\lambda \in \mathfrak{a}'_0$ and $\lambda_0 \in \mathfrak{a}_0^\perp = \{\lambda \in \mathfrak{a}' | \lambda|\mathfrak{a}_0 = 0\}$. Then $R^\gamma(\lambda+\lambda_0) = R^\gamma(\lambda)$ by Remark 1.5.9, and similarly, $P^\gamma(\lambda+\lambda_0) = P^\gamma(\lambda)$ and $P^\gamma(-(\overline{\lambda+\lambda_0})-2\varrho) = P^\gamma(-\bar\lambda-2\varrho)$. This proves the assertion. Hence we have proved:

Theorem 1.8.6. *With respect to a choice of any basis σ_i of E_γ and the corresponding basis σ_i^s of E'_γ as in Lemma* 1.6.6 *and any orthonormal basis in $(V_\gamma)^{M_\theta}$, one has*

$$R^\gamma(\lambda) = (P^\gamma(-\bar\lambda-2\varrho))^* P^\gamma(\lambda)$$

for all $\lambda \in \mathfrak{a}'$.

CHAPTER II

THE DETERMINANT OF $P^{\gamma}(\lambda)$ AND APPLICATIONS TO REPRESENTATION THEORY

§ 1. Two Transitivity and the decomposition of T under K^{z}

Let $\mathfrak{d}_0$ be a finite-dimensional real vector space and $q_0': \mathfrak{d}_0 \otimes \mathfrak{d}_0 \to \mathbf{R}$ a positive-definite symmetric bilinear form on $\mathfrak{d}_0$. Let $\mathfrak{d}$ denote the complexification of $\mathfrak{d}_0$ and let $q': \mathfrak{d} \otimes \mathfrak{d} \to \mathbf{C}$ denote the unique extension of $q_0': \mathfrak{d}_0 \otimes \mathfrak{d}_0 \to \mathbf{R}$. Let $O(\mathfrak{d})$ denote the group of automorphisms of $\mathfrak{d}$ leaving q' invariant and $O(\mathfrak{d}, \mathbf{R})$ the subgroup of $O(\mathfrak{d})$ which leaves $\mathfrak{d}_0$ stable. Let $SO(\mathfrak{d})$ and $SO(\mathfrak{d}, \mathbf{R})$ denote the subgroups of $O(\mathfrak{d})$ and $O(\mathfrak{d}, \mathbf{R})$ respectively, consisting of the automorphisms of determinant 1. Then $SO(\mathfrak{d})$ and $SO(\mathfrak{d}, \mathbf{R})$ are connected, and $SO(\mathfrak{d})$ is a complex reductive Lie group of which $SO(\mathfrak{d}, \mathbf{R})$ is a maximal compact subgroup. Let q be a non-zero element of $\mathfrak{d} \otimes \mathfrak{d}$ which is invariant under the action of $O(\mathfrak{d})$. This exists and is unique up to a scalar multiple, in view of Schur's lemma. Let $\mathfrak{d}'$ be the dual of $\mathfrak{d}$, and let T and T' denote the symmetric algebras on $\mathfrak{d}$ and $\mathfrak{d}'$, respectively. We recall the standard algebra homomorphism $\partial: T \to \mathrm{End}\,(T')$ by which elements of T can be considered as differential operators on T; for all $x \in \mathfrak{d}$, ∂_x is the unique derivation of T' such that $\partial_x y = \langle x, y \rangle$ for all $y \in \mathfrak{d}'$.

Definition 2.1.1. We define H' to be the space of all $f \in T'$ which satisfy the "Laplace equation" $\partial_q f = 0$. Elements of H' are called *harmonics.*

Remark 2.1.2. We will now interpret H' as the harmonics in the sense of [10], when the dimension n of $\mathfrak{d}_0$ is greater than 1. For all $n > 1$, let $SO(n, 1)$ be the subgroup of $SL(n+1, \mathbf{R})$ which leaves invariant the quadratic form $-x_0^2 + x_1^2 + \dots + x_n^2$ on $\mathbf{R}^{n+1}$. The Lie algebra $\mathfrak{g}_{\mathbf{R}}$ of $SO(n, 1)$ is semisimple. If $\mathfrak{g}_{\mathbf{R}} = \mathfrak{k}_{\mathbf{R}} + \mathfrak{p}_{\mathbf{R}}$ is a Cartan decomposition of $\mathfrak{g}_{\mathbf{R}}$, then it is well known that $\mathfrak{k}_{\mathbf{R}}$ is isomorphic to $\mathfrak{so}(n, \mathbf{R})$, the Lie algebra of the rotation group $SO(n, \mathbf{R})$, and $\mathfrak{p}_{\mathbf{R}}$ is isomorphic to $\mathbf{R}^n$ in such a way that the adjoint action of $\mathfrak{k}_{\mathbf{R}}$ on $\mathfrak{p}_{\mathbf{R}}$ corresponds to the action of $\mathfrak{so}(n, \mathbf{R})$ on $\mathbf{R}^n$. Noting that $J_+ = Jq'$, where q' corresponds to q it now follows that the definition of a harmonic polynomial in [10] corresponds to a harmonic polynomial in the sense of Definition 2.1.1.

Now let $n \geqq 1$, and let J' denote the ring of invariants for the action of $O(\mathfrak{d})$ on T'. We note that if $n \geqq 2$, then J' is also the ring of invariants for the action of $SO(\mathfrak{d})$ on T'. Then one knows, and also one has from [10], that $T' = J' \otimes H'$ if $n > 1$ (this formula clearly also holds if $n = 1$). Let $\beta: \mathfrak{d} \to \mathfrak{d}'$ be the isomorphism induced by q', and let β also denote the isomorphism $T \to T'$ extending the isomorphism $\beta: \mathfrak{d} \to \mathfrak{d}'$. Define $H = \beta^{-1}(H')$. Let J be the ring of invariants in T for the action of $O(\mathfrak{d})$ on T. Then we also have $T = J \otimes H$. We call H the space of *harmonics* in T.

Remark 2.1.3. Let $n > 1$. Since $J'_+ = J'q'$, one knows, also it follows from [10], that H is the space spanned by $\{x^k | x \in \mathfrak{n}, k \in \mathbf{Z}_+\}$, where $\mathfrak{n}$ is the variety of nilpotents with respect to the Cartan decomposition considered in Remark 2.1.2. In this case, $\mathfrak{n} = \{x \in \mathfrak{d} | q'(x \otimes x) = 0\}$.

Definition 2.1.4. A subgroup R of $SO(\mathfrak{d}, \mathbf{R})$ is said to be *two-transitive* on the unit sphere in $\mathfrak{d}_0$ if $R = SO(\mathfrak{d}, \mathbf{R})$ when $n \leqq 2$ and when $n > 2$, 1) R is transitive on the unit sphere in $\mathfrak{d}_0$ and 2) if Q_f is the isotropy subgroup of R at any point f in the unit sphere in $\mathfrak{d}_0$, then Q_f is transitive on the unit sphere in $f^\perp$, where $f^\perp$ denotes the orthocomplement of f in $\mathfrak{d}_0$.

Remark. In view of condition 1), we can replace 2) by the condition that Q_f is transitive on the unit sphere in $f^\perp$ for some point $f \in \mathfrak{d}_0$.

Lemma 2.1.5. Let $R \subseteqq SO(\mathfrak{d})$ be a subgroup such that $R_{\mathbf{R}}$ is two-transitive on the unit sphere in $\mathfrak{d}_0$, where $R_{\mathbf{R}} = R \cap SO(\mathfrak{d}, \mathbf{R})$. Let T be the symmetric algebra on $\mathfrak{d}$ and $W \subseteqq T$ an irreducible $SO(\mathfrak{d})$-submodule. Then W is an irreducible R-module.

Proof. We can assume $\dim \mathfrak{d} > 2$, for otherwise $R_{\mathbf{R}} = SO(\mathfrak{d}, \mathbf{R})$. Let $f \in \mathfrak{d}_0$ and let $\mathfrak{e}_0$ be the orthocomplement of $\mathbf{R}f$ in $\mathfrak{d}_0$. Let $\mathfrak{e} = \mathfrak{e}_0 + i\mathfrak{e}_0$. We can regard $SO(\mathfrak{e}, \mathbf{R})$ as the isotropy subgroup of $SO(\mathfrak{d}, \mathbf{R})$ at f.

Since $T = J \otimes H$, we can assume without loss that $W \subseteqq H$. It is a classical fact that $W^{SO(\mathfrak{e}, \mathbf{R})}$, the space of elements of W fixed by $SO(\mathfrak{e}, \mathbf{R})$, is one dimensional (a special case of the K-fixed vector theorem). Now let $Q_{\mathbf{R}}$ denote the isotropy subgroup of $R_{\mathbf{R}}$ at f. Since $R_{\mathbf{R}}$ is two-transitive on the unit sphere in $\mathfrak{d}_0$, $Q_{\mathbf{R}}$ is transitive on the unit sphere in $\mathfrak{e}_0$. But this implies that for any point $y \in \mathfrak{d}_0$, the two orbits $SO(\mathfrak{e}, \mathbf{R}) \cdot y$ and $Q_{\mathbf{R}} \cdot y$ coincide. Now, identifying T with the polynomial algebra on $\mathfrak{d}_0$, $W^{SO(\mathfrak{e}, \mathbf{R})} = \{u \in W | u$ is constant on $SO(\mathfrak{e}, \mathbf{R}) \cdot y$ for any $y \in \mathfrak{d}_0\} = \{u \in W | u$ is constant on $Q_{\mathbf{R}} \cdot y$ for any $y \in \mathfrak{d}_0\} = W^{Q_{\mathbf{R}}}$. Thus $W^{Q_{\mathbf{R}}}$ is one-dimensional.

Since $R_{\mathbf{R}}$ is transitive on the unit sphere in $\mathfrak{d}_0$, we have $SO(\mathfrak{d}, \mathbf{R}) = R_{\mathbf{R}} SO(\mathfrak{e}, \mathbf{R})$. Now let u be a non-zero element of $W^{SO(\mathfrak{e}, \mathbf{R})}$. Then $SO(\mathfrak{d}, \mathbf{R}) \cdot u = R_{\mathbf{R}} \cdot u$. But since W is an irreducible $SO(\mathfrak{d})$-module, $SO(\mathfrak{d}, \mathbf{R}) \cdot u$ spans W. Thus $R_{\mathbf{R}} \cdot u$ spans W.

Let $\bar{R}_{\mathbf{R}}$ denote the closure of $R_{\mathbf{R}}$ in $SO(\mathfrak{d}, \mathbf{R})$. Then $\bar{R}_{\mathbf{R}}$ is compact, since $SO(\mathfrak{d}, \mathbf{R})$ is compact, and thus W is completely reducible as an $\bar{R}_{\mathbf{R}}$-module, hence also as an $R_{\mathbf{R}}$-module.

Now suppose W is not irreducible as an $R_{\mathbf{R}}$-module. Then there exists a decomposition $W=W_1\oplus W_2$, where W_1 and W_2 are non-zero $R_{\mathbf{R}}$-submodules of W. Let $u=u_1+u_2$ be the decomposition of u with respect to $W=W_1\oplus W_2$. Observe that u_1 and u_2 are both different from zero since $R_{\mathbf{R}}\cdot u$ spans W. Since u is $Q_{\mathbf{R}}$-invariant, it follows that u_1 and u_2 are also $Q_{\mathbf{R}}$-invariant. But this is a contradiction since $W^{Q_{\mathbf{R}}}$ is one-dimensional. Q.E.D.

From now on, let $\mathfrak{g}_{\mathbf{R}}$ be a reductive Lie algebra and $\mathfrak{g}_{\mathbf{R}}=\mathfrak{k}_{\mathbf{R}}+\mathfrak{p}_{\mathbf{R}}$ a Cartan decomposition. Let $\mathfrak{g}=\mathfrak{k}+\mathfrak{p}$ denote the complexification. Let $\mathfrak{a}_{\mathbf{R}}\subseteq\mathfrak{p}_{\mathbf{R}}$ be a maximal Abelian subalgebra, so that $\mathfrak{a}=\mathfrak{a}_{\mathbf{R}}+i\mathfrak{a}_{\mathbf{R}}$ is a Cartan subspace. We assume throughout § 2.1 and 2.2 that $\mathfrak{a}_{\mathbf{R}}\cap[\mathfrak{g}_{\mathbf{R}},\mathfrak{g}_{\mathbf{R}}]$ is *one-dimensional*, that is that, $\mathfrak{g}_{\mathbf{R}}$ is of *rank one*. Then one knows that there exists $\alpha\in\mathfrak{a}'$ (α taking real values on $\mathfrak{a}_{\mathbf{R}}$) such that $\mathfrak{g}=\sum_{j=-2}^{2}\mathfrak{g}^{j\alpha}$; here $\mathfrak{g}^{2\alpha}$ and $\mathfrak{g}^{-2\alpha}$ can be zero and $\mathfrak{g}^{0\alpha}=\mathfrak{a}+\mathfrak{m}$, where $\mathfrak{m}$, as usual, is the centralizer of $\mathfrak{a}$ in $\mathfrak{k}$. Let $w\in\mathfrak{a}_{\mathbf{R}}\cap[\mathfrak{g}_{\mathbf{R}},\mathfrak{g}_{\mathbf{R}}]$ be the unique element such that $\alpha(w)=1$. Thus the eigenvalues of ad w in $\mathfrak{g}$ are 0, ± 1 and possibly ± 2.

Let B be the Killing form of $\mathfrak{g}$, and for $j\neq 0$, let B_j be the bilinear form on $\mathfrak{g}^{j\alpha}$ defined by $B_j(u,v)=B(u,\theta v)$. (Note that $\theta\mathfrak{g}^{j\alpha}=\mathfrak{g}^{-j\alpha}$.) Clearly B_j is symmetric.

Remark. One has that B_j is negative-definite on $\mathfrak{g}_{\mathbf{R}}^{j\alpha}$. Indeed, B is negative-definite on $\mathfrak{k}_{\mathbf{R}}\cap[\mathfrak{g}_{\mathbf{R}},\mathfrak{g}_{\mathbf{R}}]$ and positive-definite on $\mathfrak{p}_{\mathbf{R}}\cap[\mathfrak{g}_{\mathbf{R}},\mathfrak{g}_{\mathbf{R}}]$. Hence the bilinear form $u, v\to B(u,\theta v)$ is negative-definite on all of $[\mathfrak{g}_{\mathbf{R}},\mathfrak{g}_{\mathbf{R}}]$.

Proposition 2.1.6. If $e\in\mathfrak{g}^{j\alpha}$, $j=1,2$, then

$$[e,\theta e]=\frac{jB_j(e,e)}{\langle w,w\rangle}w.$$

(We write $\langle u,v\rangle$ for $B(u,v)$.)

Proof. We first note that $[e,\theta e]\in\mathfrak{a}\cap[\mathfrak{g},\mathfrak{g}]$. It suffices to show $u=[e,\theta e]\in\mathfrak{a}$. But this is clear since $\theta u=-u$ so that $u\in\mathfrak{p}$ and u commutes with $\mathfrak{a}$ so that $u\in\mathfrak{a}$. But now $[e,\theta e]=\lambda w$, where λ is a complex number. But one has

$$\lambda\langle w,w\rangle=\langle[e,\theta e],w\rangle=\langle e,[\theta e,w]\rangle=j\langle e,\theta e\rangle=jB_j(e,e).$$

Hence

$$\lambda=\frac{jB_j(e,e)}{\langle w,w\rangle}.$$

Q.E.D.

Recall that M is the analytic subgroup of G (the adjoint group of $\mathfrak{g}$) corresponding to the Lie algebra $\mathrm{ad}_{\mathfrak{g}}\,\mathfrak{m}$. Since M centralizes $\mathfrak{a}$, M maps each $\mathfrak{g}^{j\alpha}$ onto itself. Obviously M leaves the bilinear form B_j invariant and thus $M_{\mathbf{R}}$ acts on the unit sphere in $\mathfrak{g}_{\mathbf{R}}^{j\alpha}$ (relative to $-B_j$).

Theorem 2.1.7. *One has that* $M_{\mathbf{R}}$ *is transitive on the unit sphere in* $\mathfrak{g}_{\mathbf{R}}^{j\alpha}$, *if* $\dim\mathfrak{g}^{j\alpha}>1$.

Proof. Let e_j be a non-zero element in $\mathfrak{g}_{\mathbf{R}}^{j\alpha}$. Then since B_j is negative-definite on $\mathfrak{g}_{\mathbf{R}}^{j\alpha}$ it follows from Proposition 2.1.6 that e_j, θe_j and w span a *TDS*, say $\mathfrak{u}$. From the representation theory of a *TDS*, $[\mathfrak{g}_{\mathbf{R}}^0, e_j]=\mathfrak{g}_{\mathbf{R}}^{j\alpha}$.

Now $\mathfrak{g}_{\mathbf{R}}^0=\mathfrak{a}_{\mathbf{R}}+\mathfrak{m}_{\mathbf{R}}$. But $[\mathfrak{a}_{\mathbf{R}}, e_j]=\mathbf{R}e_j$. On the other hand, for any $x\in\mathfrak{m}_{\mathbf{R}}$, one has $B_j([x, e_j], e_j)=B([x, e_j], \theta e_j)=B(x, [e_j, \theta e_j])=0$, since $x\in\mathfrak{k}$ and $[e_j, \theta e_j]\in\mathfrak{p}$. Thus $[\mathfrak{m}_{\mathbf{R}}, e_j]\subseteq e_j^{\perp}$ (relative to B_j). But $[\mathfrak{a}_{\mathbf{R}}+\mathfrak{m}_{\mathbf{R}}, e_j]=[\mathfrak{g}_{\mathbf{R}}^0, e_j]=\mathfrak{g}_{\mathbf{R}}^{j\alpha}$ so that $[\mathfrak{m}_{\mathbf{R}}, e_j]=e_j^{\perp}$. From this it is immediate that for any e_j on the unit sphere in $\mathfrak{g}_{\mathbf{R}}^{j\alpha}$, $M_{\mathbf{R}}\cdot e_j$ contains an open neighborhood of e_j in the unit sphere in $\mathfrak{g}_{\mathbf{R}}^{j\alpha}$. But this unit sphere is connected since $\dim\mathfrak{g}_{\mathbf{R}}^{j\alpha}>1$. Thus we can conclude that $M_{\mathbf{R}}\cdot e_j$ is the unit sphere in $\mathfrak{g}_{\mathbf{R}}^{j\alpha}$. Q.E.D.

Remark. The previous result is quite general and does not depend on rank 1.

We now once and for all fix the following notation.

Let $e_1\in\mathfrak{g}_{\mathbf{R}}^{\alpha}$ be fixed so that $B_1(e_1, e_1)=-\dfrac{\langle w, w\rangle}{2}$, and let $e_{-1}=-\theta(e_1)$. Note that $e_{-1}\in\mathfrak{g}_{\mathbf{R}}^{-\alpha}$. Let $\mathfrak{u}$ be the *TDS* spanned by $\{w, e_1, e_{-1}\}$.

We have $\langle e_1, e_{-1}\rangle=\dfrac{\langle w, w\rangle}{2}$ and by Proposition 2.1.6, $[e_1, e_{-1}]=\dfrac{w}{2}$. Thus $(2w, 2e_1, 2e_{-1})$ is an S-triple. Put $z=i(e_1-e_{-1})$ and $w'=(e_1+e_{-1})$. Note that $z\in i\mathfrak{k}_{\mathbf{R}}$ and $w'\in\mathfrak{p}_{\mathbf{R}}$. A straightforward computation shows that $[iw', w]=-z$ and $[iw', z]=w$. Hence $\exp\operatorname{ad}\left(\dfrac{\pi}{2}iw'\right)(w)=-z$, $\exp\operatorname{ad}\left(\dfrac{\pi}{2}iw'\right)(z)=w$ and of course $\exp\operatorname{ad}\left(\dfrac{\pi}{2}iw'\right)(w')=w'$. Let U denote the analytic subgroup of G with Lie algebra $\operatorname{ad}_{\mathfrak{g}}\mathfrak{u}$. Then $\exp\operatorname{ad}\left(\dfrac{\pi}{2}iw'\right)\in U$. Let $c=\exp\operatorname{ad}\left(\dfrac{\pi}{2}iw'\right)$, so that $cz=w$, $cw=-z$ and $cw'=w'$ and let $b=c^2$, so that $bw=-w$, $bz=-z$ and $bw'=w'$.

Let $\mathfrak{g}_i\subseteq\mathfrak{g}$ be the sum of all the $2i+1$-dimensional $\mathfrak{u}$-irreducible submodules of $\mathfrak{g}$. Note that since $\mathfrak{u}$ is spanned by the S-triple $(2w, 2e, 2e_{-1})$ and since the eigenvalues of $2w$ in $\mathfrak{g}$ are 0, ± 2 and (possibly) ± 4, it follows from the representation theory of a *TDS* that $\mathfrak{g}=\mathfrak{g}_0+\mathfrak{g}_1+\mathfrak{g}_2$. In particular, all the irreducible $\mathfrak{u}$-submodules of $\mathfrak{g}$ are odd-dimensional, which implies that $U\cong PSL(2, \mathbf{C})$ (i.e., the center of U is trivial). Also $\mathfrak{u}$ is a principal *TDS*. See [10].

Let $\mathfrak{h}\subseteq\mathfrak{g}$ be the unique linear complement of $\mathfrak{a}+\mathfrak{u}$ in $\mathfrak{g}$ such that $\mathfrak{h}\perp(\mathfrak{a}+\mathfrak{u})$ (with respect to the Killing form). It is clear that $\mathfrak{h}=(\mathfrak{h}\cap\mathfrak{k})+(\mathfrak{h}\cap\mathfrak{p})$ and that $\mathfrak{h}$ is a $\mathfrak{u}$-submodule of $\mathfrak{g}$. Put $\mathfrak{h}_i=\mathfrak{h}\cap\mathfrak{g}_i$ $(i=0, 1, 2)$ and $h_i^j=\mathfrak{h}_i\cap\mathfrak{g}^{j\alpha}$ $(i=0, 1, 2,$ $j=0, \pm 1, \pm 2)$. Clearly $\mathfrak{g}^0\cap\mathfrak{h}=\mathfrak{m}$, and $\mathfrak{h}$ is the $\mathfrak{u}$-submodule generated by $\mathfrak{m}$. We write $\mathfrak{m}_0$, $\mathfrak{m}_1$ and $\mathfrak{m}_2$ for $\mathfrak{h}_0^0$, $\mathfrak{h}_1^0$ and $\mathfrak{h}_2^0$, respectively.

Now put $\tau=\theta b$. Observe that $\theta b=b\theta$. For this, first note that $\theta b\theta=b^{-1}$, from the definition of b and c. But $b|\mathfrak{u}=b^{-1}|\mathfrak{u}$, and since $b, b^{-1}\in U\cong PSL(2, \mathbf{C})$, it follows that $b=b^{-1}$. One thus has $\tau^2=1$.

Proposition 2.1.8. One has $\tau=(-1)^{i+j}$ on $\mathfrak{h}_i^j$.

Proof. Assume $u \in \mathfrak{h}_i^j$ and $j \geqq 0$. (The proof for the case $j \leqq 0$ is similar.) Then $u = (\operatorname{ad} e_1)^j v$, $v \in \mathfrak{m}_i$. Now $\tau u = (\operatorname{ad} \tau e_1)^j \tau v$. But substituting $z = i(e_1 - e_{-1})$ and $w' = e_1 + e_{-1}$ in the equations $bz = -z$ and $bw' = w'$, we see that $be_1 = e_{-1}$ and $be_{-1} = e_1$, so that $\tau e_1 = -e_1$ and $\tau e_{-1} = -e_{-1}$. Thus $\tau u = (-1)^j (\operatorname{ad} e_1)^j \tau v = (-1)^j (\operatorname{ad} e_1)^j bv$, since $v \in \mathfrak{k}$ and $\theta = 1$ on $\mathfrak{k}$.

Now observe that $\mathfrak{t} = \mathbf{C}w$ is a Cartan subalgebra of $\mathfrak{u} \cong \mathfrak{sl}(2, \mathbf{C})$. Since $b|\mathbf{C}w = -1$, b defines the non-trivial element of the Weyl group of $(\mathfrak{u}, \mathfrak{t})$. But now, consider the pair $(\mathfrak{sl}(2, \mathbf{C}), \mathfrak{t}_0)$ where $\mathfrak{t}_0$ is the Cartan subalgebra $\mathbf{C}\begin{pmatrix} 1 & 0 \\ 0 & -1 \end{pmatrix}$. One knows that the $(n+1)$-dimensional irreducible representation of $SL(2, \mathbf{C})$ is its representation on the space T^n of homogeneous polynomials of degree n in $\mathbf{C}[X, Y]$. From this one can easily check that if β is any element of $SL(2, \mathbf{C})$, e.g., $\begin{pmatrix} 0 & 1 \\ -1 & 0 \end{pmatrix}$, representing the non-trivial element of the Weyl group of $(\mathfrak{sl}(2, \mathbf{C}), \mathfrak{t}_0)$, then β acts as $(-1)^n$ on the zero eigenspace of $\mathfrak{t}_0$ in T^{2n}, namely $\mathbf{C}(X^n Y^n)$. Moreover, observe that this implies the corresponding statement for $PSL(2, \mathbf{C})$.

Now through suitable isomorphisms it follows immediately that the action of b in the zero eigenspace of $\mathfrak{t}$ in a $(2i+1)$-dimensional irreducible representation of U is scalar multiplication by $(-1)^i$. In particular, $bv = (-1)^i v$, so that

$$\tau u = (-1)^j (\operatorname{ad} e_1)^j ((-1)^i v) = (-1)^{i+j} u. \qquad \text{Q.E.D.}$$

Let $\mathfrak{m}^\tau$ be the $+1$ eigenspace of τ in $\mathfrak{m}$. That is, $\mathfrak{m}^\tau = \{x \in \mathfrak{m} | \tau x = x\}$. Then the following corollary is an immediate consequence of Proposition 2.1.8.

Corollary 2.1.9. One has $\mathfrak{m}^\tau = \mathfrak{m}_0 + \mathfrak{m}_2$.

Let $\mathfrak{d}^{-\tau}$ be the (-1)-eigenspace of τ in $\mathfrak{g}^\alpha$ ($\mathfrak{g}^{j\alpha}$ is stable under τ since $\tau w = w$). We assert that $\mathfrak{d}^{-\tau} = \mathfrak{h}_2^1 + \mathbf{C}e_1$. This is clear in view of the fact that $\mathfrak{g}^\alpha = \mathfrak{h}_1^1 \oplus \mathfrak{h}_2^1 \oplus \mathbf{C}e_1$, the fact that $\tau e_1 = -e_1$ (which we observed in the proof of Proposition 2.1.8) and Proposition 2.1.8. Note that $[\mathfrak{m}^\tau, \mathfrak{d}^{-\tau}] \subseteqq \mathfrak{d}^{-\tau}$. Also, $\mathfrak{g}_{\mathbf{R}}$ is τ-stable, so that $\mathfrak{d}^{-\tau}$ is defined over $\mathbf{R}$.

Let M^τ be the analytic subgroup of G with Lie algebra $\operatorname{ad} \mathfrak{m}^\tau$.

Proposition 2.1.10. One has that $M_{\mathbf{R}}^\tau$ is transitive on the unit sphere of $(\mathfrak{d}^{-\tau})_{\mathbf{R}}$ (relative to $-B_1|\mathfrak{d}_{\mathbf{R}}^{-\tau}$) if $\dim \mathfrak{d}^{-\tau} > 1$.

Proof. Consider the map $\mu: M_{\mathbf{R}}^\tau \to S$, ($S$ = unit sphere in $\mathfrak{d}_{\mathbf{R}}^{-\tau}$) given by $a \to a \cdot (\lambda e_1)$, where λ is a suitable scalar such that $\lambda e_1 \in S$. We can identify the tangent spaces at the points $1 \in M_{\mathbf{R}}^\tau$ and $\lambda e_1 \in S$ with $\mathfrak{m}_{\mathbf{R}}^\tau$ and $(\mathfrak{h}_2^1)_{\mathbf{R}}$, respectively, (note that $\mathfrak{h}_2^1 \perp \mathbf{C}e_1$) in a natural way. Then the differential of μ at $1 \in M_{\mathbf{R}}^\tau$, $\mathfrak{d}\mu_{(1)}: \mathfrak{m}_{\mathbf{R}}^\tau \to (\mathfrak{h}_2^1)_{\mathbf{R}}$, is just the map $x \to [x, \lambda e_1]$. Thus to prove the proposition it is enough to show that $[\mathfrak{m}_{\mathbf{R}}^\tau, e_1] = (\mathfrak{h}_2^1)_{\mathbf{R}}$.

But since $\mathfrak{m}^\tau = \mathfrak{m}_{\mathbf{R}}^\tau + i\mathfrak{m}_{\mathbf{R}}^\tau$, it is enough to show that $[\mathfrak{m}^\tau, e_1] = \mathfrak{h}_2^1$. But $\mathfrak{m}^\tau = \mathfrak{m}_0 + \mathfrak{m}_2$ and by representation theory of a *TDS*, $[\mathfrak{m}_2, e_1] = \mathfrak{h}_2^1$. Q.E.D.

We will now prove that $M_{\mathbf{R}}^{\tau}$ is actually two-transitive on the unit sphere in $\mathfrak{d}_{\mathbf{R}}^{-\tau}$.

Theorem 2.1.11. *If the dimension of* $\mathfrak{d}^{-\tau}$ *is greater than* 2, *then* $M_{\mathbf{R}}^{\tau}$ *is two-transitive on the unit sphere in* $\mathfrak{d}_{\mathbf{R}}^{-\tau}$.

We prove several lemmas before proving the theorem. First observe that $\mathfrak{m}_0=\{x\in\mathfrak{m}^{\tau}|[x,e_1]=0\}$. Let M_0 be the analytic subgroup of G with Lie algebra ad $\mathfrak{m}_0$. The isotropy subgroup of $M_{\mathbf{R}}^{\tau}$ at e_1 contains $M_{0,\mathbf{R}}$ and hence, to prove the theorem, it is enough to show that $M_{0,\mathbf{R}}$ is transitive on the unit sphere in $(\mathfrak{h}_2^1)_{\mathbf{R}}$. But this is equivalent to showing that $M_{0,\mathbf{R}}$ is transitive on the unit sphere in $(\mathfrak{h}_2^2)_{\mathbf{R}}$, or the unit sphere in $(\mathfrak{m}_2)_{\mathbf{R}}$.

Definition. If $x\in\mathfrak{g}$, we define $x^*=[e_1,x]$. We write x^{**} instead of $(x^*)^*$.

We recall now a certain fact from the representation theory of a *TDS*. Let (x,e,f) be an S-triple spanning a *TDS* $\mathfrak{v}$. Let π be an irreducible finite-dimensional representation of $\mathfrak{v}$ on the space V. Let $v\in V$ be an eigenvector for x. Now suppose that q is the minimal j such that $\pi(e)^{j+1}(v)=0$ and p the minimal i such that $\pi(f)^{i+1}(v)=0$. Then $\pi(f)\pi(e)(v)=q(p+1)v$.

Noting that $(2w, 2e_1, 2e_{-1})$ is an S-triple spanning $\mathfrak{u}$, we get the following lemma.

Lemma 2.1.12. Let $x\in\mathfrak{m}_2$. Then $[e_{-1},x^{**}]=x^*$ and $\frac{2}{3}[e_{-1},x^*]=x$.

Lemma 2.1.13. Let $x,y,\in\mathfrak{m}_2$. Then

$$[x,y^{**}]=[x^{**},y]=2/3[y^*,x^*].$$

Proof.

$$[x,y^{**}]=2/3[[e_{-1},x^*],y^{**}]=2/3\{[[e_{-1},y^{**}],x^*]+[e_{-1},[x^*,y^{**}]]\}.$$

But since 3α is not a root, $[x^*,y^{**}]=0$, showing that

$$[x,y^{**}]=2/3[[e_{-1},y^{**}],x^*]=2/3[y^*,x^*],$$

by Lemma 2.1.12.

By the anti-symmetry in $[y^*,x^*]$ one also obtains the symmetry $[x,y^{**}]=[x^{**},y]$.

Q.E.D.

Lemma 2.1.14. Let $x,y\in\mathfrak{m}_2$. Then

$$[x,y]^{**}=-[x,y^{**}]=-[x^{**},y]=2/3[x^*,y^*].$$

Proof.

$$[x,y]^*=[e_1,[x,y]]=[x^*,y]+[x,y^*].$$

Hence

$$[x,y]^{**}=[e_1,[x,y]^*]=[x^{**},y]+2[x^*,y^*]+[x,y^{**}]=$$

$$=-2/3[x^*,y^*]+2[x^*,y^*]-2/3[x^*,y^*]$$

(by Lemma 2.1.13),

$$=2/3[x^*,y^*]=-[x^{**},y]=-[x,y^{**}]$$

(by Lemma 2.1.13). Q.E.D.

Definition 2.1.15. For $x \in \mathfrak{m}^\tau$, we define x_i ($i=0, 2$) to be the component of x in $\mathfrak{m}_i$ with respect to the decomposition $\mathfrak{m}^\tau = \mathfrak{m}_0 + \mathfrak{m}_2$.

Lemma 2.1.16. Let $x, y \in \mathfrak{m}_2$. Then

$$[[x, y]_0, y^{**}] = -2[[x, y]_2, y^{**}].$$

Proof.

$$[[x, y]_2, y^{**}] = [([x, y]_2)^{**}, y] = [[x, y]^{**}, y] = [-[x, y^{**}], y]$$

(by Lemma 2.1.14)

$$= -\{[[x, y], y^{**}] - [x, [y^{**}, y]]\} = -[[x, y], y^{**}]$$

(since $[y^{**}, y] = 0$ by Lemma 2.1.13)

$$= -[[x, y]_0, y^{**}] - [[x, y]_2, y^{**}].$$ Q.E.D.

For $x \in \mathfrak{g}$, we define $x_* = [e_{-1}, x]$. If $x \in \mathfrak{m}$ note then that since $\theta(e_{-1}) = -e_1$ one has $x_* = -\theta(x^*)$ and $x_{**} = \theta(x^{**})$.

Now let $y \in \mathfrak{m}_2$ be such that $B_2(y^{**}, y^{**}) = -\dfrac{\langle w, w\rangle}{2}$. Then

$$\langle y^{**}, y_{**}\rangle = -\frac{\langle w, w\rangle}{2}$$

and hence by Proposition 2.1.6 $[y^{**}, y_{**}] = -w$. Thus $(w, y^{**}, -y_{**})$ forms an S-triple. Using the representation theory of a *TDS*, we have the following:

Lemma 2.1.17. Assume $y \in \mathfrak{m}_2$ and $B_2(y^{**}, y^{**}) = -\dfrac{\langle w, w\rangle}{2}$. Then for any $x \in \mathfrak{m}_2$, $[[x^*, y_{**}], y^{**}] = -x^*$.

Lemma 2.1.18. Let $x, y \in \mathfrak{m}_2$ be such that $B_2(x^{**}, y^{**}) = 0$. Then $[x^*, y_{**}] = -3/2[x, y]_*$.

Proof. Consider the element $[x_*, y^{**}]$. Now $x_* = [e_{-1}, x]$. But then

$$\langle [x_*, y^{**}], e_{-1}\rangle = 0$$

since $\langle y^{**}, x_{**}\rangle = B_2(x^{**}, y^{**}) = 0$. But on the other hand $[x_*, y^{**}] \in \mathfrak{g}^\alpha$ and $\tau([x_*, y^{**}]) = -[x_*, y^{**}]$ since $\tau x_* = -x_*$ and $\tau y^{**} = y^{**}$ by Proposition 2.1.8. Hence we conclude that $[x_*, y^{**}] \in \mathfrak{h}_2^1$. Now

$$[x_*, y^{**}]^* = [e_1, [x_*, y^{**}]] = [[e_1, x_*], y^{**}]$$

since $[e_1, y^{**}] = 0$. But from the representation theory of a *TDS*, $[e_1, x_*] = \dfrac{3}{2}x$, and by Lemma 2.1.14, $[x, y^{**}] = -[x, y]^{**}$. Thus we have

$$[x_*, y^{**}]^* = -3/2[x, y]^{**} = (-3/2[x, y]^*)^*.$$

But the mapping $x \to x^*$ is injective on $\mathfrak{h}_2^1$. Hence $[x_*, y^{**}] = -3/2[x, y]^*$. The lemma now follows by applying $-\theta$ to both sides of this equation. Q.E.D.

The following lemma is the main tool in the proof of Theorem 2.1.11.

Lemma 2.1.19. Let $x, y \in \mathfrak{m}_2$. Suppose $B_2(x^{**}, y^{**})=0$ and $B_2(y^{**}, y^{**})= = -\frac{\langle w, w\rangle}{2}$. Then

$$[[x, y]_0, y^{**}] = -8/9\, x^{**}.$$

Proof. By Lemma 2.1.16, we have

$$[[x, y]_0, y^{**}] = -2[[x, y]_2, y^{**}].$$

But $[x, y]_2=2/3[e_1, [x, y]_*]$, using Lemma 2.1.12. Thus

$$\begin{aligned}[[x, y]_0, y^{**}] &= -2(2/3)\,[[e_1,[x, y]_*], y^{**}] = \\ &= -4/3[e_1, [[x, y]_*, y^{**}]] = -4/3[[x, y]_*, y^{**}]^* = \\ &= -4/3[-2/3[x^*, y_{**}], y^{**}]^*\end{aligned}$$

(by Lemma 2.1.18)

$$= -8/9\, x^{**}$$

(by Lemma 2.1.17). Q.E.D.

Now for proving Theorem 2.1.11, let $u, u' \in \mathfrak{g}_{\mathbf{R}}^{2\alpha}$ be of the same length and linearly independent. Choose x and y in $(\mathfrak{m}_2)_{\mathbf{R}}$ such that

$$B_2(x^{**}, y^{**}) = 0,\ B_2(x^{**}, x^{**}) = B_2(y^{**}, y^{**}) = -\frac{\langle w, w\rangle}{2},$$

and x^{**} and y^{**} span the same plane as u and u'. Then by Lemma 2.1.19, we have

$$\operatorname{ad}(9/8[x, y]_0)(x^{**}) = y^{**}$$

and

$$\operatorname{ad}(9/8[x, y]_0)(y^{**}) = -x^{**}.$$

Writing $z=9/8[x, y]_0$, it follows that

$$\exp \operatorname{ad}(tz)(x^{**}) = (\cos t)x^{**}+(\sin t)y^{**}$$

and

$$\exp \operatorname{ad}(tz)(y^{**}) = -(\sin t)x^{**}+(\cos t)y^{**}.$$

This implies that the one-parameter group $\{\exp(t \operatorname{ad} z)\}_{-\infty<t<+\infty}$ acts transitively on any circle with center at the origin in the plane defined by x^{**} and y^{**}.

But $z \in \mathfrak{m}_{0,\mathbf{R}}$, and hence $\exp(t \operatorname{ad} z) \in M_{0,\mathbf{R}}$. Thus $u' \in M_{0,\mathbf{R}} \cdot u$.

This completes the proof of Theorem 2.1.11. Q.E.D.

Remark 2.1.20. If dimension of $\mathfrak{d}^{-\tau}=1$ or 2, then the mapping $M_{\mathbf{R}}^{\tau} \to SO(\mathfrak{d}^{-\tau}, \mathbf{R})$ is a surjection. This follows from Proposition 2.1.10.

Proposition 2.1.21. One has $c^{-1}\theta c=\tau$, so that if $\mathfrak{g}^1$ and $\mathfrak{g}^{-1}$ denote respectively the $+1$ eigenspace and the -1 eigenspace of τ in $\mathfrak{g}$, then c^{-1} interchanges $\mathfrak{g}^1$ and $\mathfrak{k}$ and also $\mathfrak{g}^{-1}$ and $\mathfrak{p}$.

Proof. One has $\theta c \theta = c^{-1}$, from the definition of c. Thus $c^{-1}\theta c = \theta(\theta c^{-1}\theta)c = \theta c^2 = \theta b = \tau$. Q.E.D.

Now $\tau w = \theta b w = w$. Also, clearly $\tau|(\mathfrak{p}\cap\mathfrak{z}) = \theta b|(\mathfrak{p}\cap\mathfrak{z}) = -$identity, where $\mathfrak{z}$=center of $\mathfrak{g}$. Using Proposition 2.1.8 it is evident that $\mathfrak{g}^{-1} = \mathfrak{d}^{-\tau} \oplus (\mathfrak{m}_1 + \mathfrak{z}\cap\mathfrak{p}) \oplus \oplus \theta\mathfrak{d}^{-\tau}$, where $\mathfrak{g}^{-1}$ is the -1 eigenspace of τ in $\mathfrak{g}$. We now define $\mathfrak{d} = c^{-1}\mathfrak{d}^{-\tau}$.

Proposition 2.1.22. The only eigenvalues of z in $\mathfrak{p}$ are 0 and ± 1. Precisely, $\mathfrak{p} = \mathfrak{d} \oplus c^{-1}(\mathfrak{m}_1 + \mathfrak{z}\cap\mathfrak{p}) \oplus \tau\mathfrak{d}$ and z acts as $+1$ in $\mathfrak{d}$, as 0 in $c^{-1}(\mathfrak{m}_1 + \mathfrak{z}\cap\mathfrak{p})$ and as -1 in $\tau\mathfrak{d}$.

Proof. In fact, $\mathfrak{p} = c^{-1}(\mathfrak{g}^{-1}) = c^{-1}(\mathfrak{d}^{-\tau} + (\mathfrak{m}_1 + \mathfrak{z}\cap\mathfrak{p}) + \theta\mathfrak{d}^{-\tau})$, and $z = c^{-1}w$. But w acts as $+1$ in $\mathfrak{d}^{-\tau}$, as 0 in $\mathfrak{m}_1 + \mathfrak{z}\cap\mathfrak{p}$ and as -1 in $\theta\mathfrak{d}^{-\tau}$. Hence z acts as $+1$ in $\mathfrak{d}$, as 0 in $c^{-1}(\mathfrak{m}_1 + \mathfrak{z}\cap\mathfrak{p})$ and as -1 in $c^{-1}\theta\mathfrak{d}^{-\tau} = c^{-1}\theta c(c^{-1}\mathfrak{d}^{-\tau}) = \tau\mathfrak{d}$. Q.E.D.

Notice that if $f \in \mathfrak{d}$, then f is nilpotent, since $[z, f] = f$. Thus $\mathfrak{d} \subseteq \mathcal{N}$ (=the set of nilpotent elements in $\mathfrak{p}$). In particular, $\mathfrak{d}\cap\mathfrak{p}_{\mathbf{R}} = \{0\}$, so that $\mathfrak{d}$ is not defined over $\mathbf{R}$.

We now fix some more notation. Put $\mathfrak{d}_0 = c^{-1}(\mathfrak{d}_{\mathbf{R}}^{-\tau})$, so that $\mathfrak{d} = \mathfrak{d}_0 + i\mathfrak{d}_0$. Let B_z be the bilinear form on $\mathfrak{d}$ defined by $B_z(u, v) = B_1(cu, cv)$. Then one has $B_z(u, v) = B_1(cu, cv) = B(cu, \theta cv) = B(u, c^{-1}\theta cv) = B(u, \tau v)$. Obviously, B_z is negative-definite on $\mathfrak{d}_0$.

Put $e = c^{-1}e_1$. Since e_1 is nilpotent and since $e_1 \in \mathfrak{u}$, a principal normal *TDS*, we know that $\dim \mathfrak{g}^{e_1} = \dim \mathfrak{a} + \dim \mathfrak{m}$ (see [7]). Hence also $\dim \mathfrak{g}^{e} = \dim \mathfrak{a} + \dim \mathfrak{m}$ so that e is principal nilpotent in $\mathfrak{p}$. Indeed note that $e_1 \in \mathfrak{d}_{\mathbf{R}}^{-\tau}$, so that $e \in \mathfrak{d}_0 \subseteq \mathfrak{p}$.

Note that $\mathfrak{m}^{\tau} + \mathbf{C}\cdot w$ is the $+1$ eigenspace of τ in $\mathfrak{g}^{w} = \mathfrak{a} + \mathfrak{m}$. Hence $c^{-1}(\mathfrak{m}^{\tau}) + \mathbf{C}\cdot z$ is the $+1$ eigenspace of θ in $\mathfrak{g}^{z}$. But observe that the $+1$ eigenspace of θ in $\mathfrak{g}^{z}$ is just the centralizer of z in $\mathfrak{k}$. Thus $c^{-1}(\mathfrak{m}^{\tau}) + \mathbf{C}\cdot z = \mathfrak{k}^{z}$.

We define $\mathfrak{r}^{z} = c^{-1}(\mathfrak{m}^{\tau})$. Since $\mathfrak{m}^{\tau}$ is orthogonal to $\mathbf{C}\cdot w$ under the Killing form of $\mathfrak{g}$, it follows that $\mathfrak{r}^{z}$ is orthogonal to $\mathbf{C}\cdot z$ under the Killing form of $\mathfrak{g}$.

Let R^{z} denote the analytic subgroup of G with Lie algebra ad $\mathfrak{r}^{z}$.

Theorem 2.1.23. *One has that* $R^{z}_{\mathbf{R}}$ *is two-transitive on the unit sphere in* $\mathfrak{d}_0$ *if* $\dim \mathfrak{d}_0 > 2$. *If* $\dim \mathfrak{d}_0 = 2$ *or* 1, $R^{z}_{\mathbf{R}} \to SO(\mathfrak{d}, \mathbf{R})$ *is a surjection.*

Proof. This follows from Theorem 2.1.11 and Remark 2.1.20 on applying c^{-1} and observing that $c^{-1}(\mathfrak{m}^{\tau}_{\mathbf{R}}) = \mathfrak{r}^{z}_{\mathbf{R}}$. The fact that $c^{-1}(\mathfrak{m}^{\tau}_{\mathbf{R}}) = \mathfrak{r}^{z}_{\mathbf{R}}$ can be proved as follows: $c = \exp \operatorname{ad}\left(i\frac{\pi w'}{2}\right)$ and $w' = e_1 + e_{-1} \in \mathfrak{p}_{\mathbf{R}}$. But $i\frac{\pi w'}{2} \in i\mathfrak{p}_{\mathbf{R}} \subseteq \mathfrak{k}_{\mathbf{R}} + i\mathfrak{p}_{\mathbf{R}}$, which is a (compact) real form of $\mathfrak{g}$, so that c leaves $\mathfrak{k}_{\mathbf{R}} + i\mathfrak{p}_{\mathbf{R}}$ stable. Note that $\mathfrak{m}^{\tau}_{\mathbf{R}} = \mathfrak{m}^{\tau}\cap(\mathfrak{k}_{\mathbf{R}} + i\mathfrak{p}_{\mathbf{R}})$ and $\mathfrak{r}^{z}_{\mathbf{R}} = \mathfrak{r}^{z}\cap(\mathfrak{k}_{\mathbf{R}} + i\mathfrak{p}_{\mathbf{R}})$. Thus $c^{-1}(\mathfrak{m}^{\tau}_{\mathbf{R}}) = c^{-1}(\mathfrak{m}^{\tau})\cap(\mathfrak{k}_{\mathbf{R}} + i\mathfrak{p}_{\mathbf{R}}) = \mathfrak{r}^{z}\cap(\mathfrak{k}_{\mathbf{R}} + i\mathfrak{p}_{\mathbf{R}}) = \mathfrak{r}^{z}_{\mathbf{R}}$. Q.E.D.

The following theorem is classical.

Theorem 2.1.24. *Let $\mathfrak{d}_0$ be a real vector space with a positive definite bilinear form B. Let B also denote the unique extension to $\mathfrak{d}=\mathfrak{d}_0\otimes\mathbf{C}$ of the bilinear form B on $\mathfrak{d}_0$. Let $SO(\mathfrak{d})$ denote the rotation group of $\mathfrak{d}$ relative to B, and let $O(\mathfrak{d})$ denote the full orthogonal group. Let T be the symmetric algebra on $\mathfrak{d}$ and let L be the space of harmonics in T. Let $L_j=L\cap T_j$.*

If $\dim\mathfrak{d}>2$, *then for every j, L_j is irreducible as an $SO(\mathfrak{d})$-module.*

If $\dim\mathfrak{d}=2$, *then for each $j>0$, L_j has two irreducible $SO(\mathfrak{d})$-components, denoted L_j^+ and L_j^-.*

If $\dim\mathfrak{d}=1$, *then $L_j=0$ if $j>1$, $L_0=\mathbf{C}$ and $L_1=\mathfrak{d}$.*

Moreover, the components described above are mutually non-isomorphic $SO(\mathfrak{d})$-modules if $\dim\mathfrak{d}>1$, *and mutually non-isomorphic $O(\mathfrak{d})$-modules if* $\dim\mathfrak{d}=1$.

(For a proof of this theorem, see [17].)

Remark 2.1.25. Consider the case of Theorem 2.1.24 in which $\dim\mathfrak{d}=2$. Let e_1, e_2 be an orthonormal basis of $\mathfrak{d}_0$ and let $\mathscr{M}=\mathbf{C}(e_1+ie_2)\cup\mathbf{C}(e_1-ie_2)$. One can easily prove that $\mathscr{M}$ is precisely the set $\{x\in\mathfrak{d}\,|\,B(x,x)=0\}$. By Remark 2.1.3, the space L of harmonics in T is $\mathbf{C}\cdot 1+\sum_{j>0}\mathbf{C}(e_1+ie_2)^j+\sum_{j>0}\mathbf{C}(e_1-ie_2)^j$. We assert that the spaces L_j^+ and L_j^- of Theorem 2.1.24 are just the spaces $\mathbf{C}(e_1+ie_2)^j$ and $\mathbf{C}(e_1-ie_2)^j$. For this, it is enough to prove that $SO(\mathfrak{d})(e_1+ie_2)\subseteqq\mathbf{C}(e_1+ie_2)$ and $SO(\mathfrak{d})(e_1-ie_2)\subseteqq\mathbf{C}(e_1-ie_2)$. This is done by direct verification.

Now let J be the ring of invariants for the action of $O(\mathfrak{d})$ on T, and let q be the unique (up to scalar) non-zero element of $T_2\cap J$. Since $T=J\otimes L$ and since $J=\sum_{i>0}\mathbf{C}q^i$, we now have the following theorem:

Theorem 2.1.26. *Under the action of $SO(\mathfrak{d})$ on $\mathfrak{d}$, one has the following decomposition of T into irreducible $SO(\mathfrak{d})$-modules*:

If $\dim\mathfrak{d}>2$, *then* $T=\sum_{i,j}q^iL_j$.

If $\dim\mathfrak{d}=2$, *then* $T=\sum_{i\geqq 0,\,j>0}q^iL_j^++\sum_{i\geqq 0,\,j>0}q^iL_j^-+\sum_{i\geqq 0}\mathbf{C}q^i$.

If $\dim\mathfrak{d}=1$, *then* $T=\sum_{i\geqq 0,\,j=0,1}q^iL_j$.

Observe that Theorem 2.1.23 and Lemma 2.1.5 imply that Theorem 2.1.26 remains valid if we replace $SO(\mathfrak{d})$ by R^z. Note that K^z operates on $\mathfrak{d}$, since $\mathfrak{d}$ is the $+1$ eigenspace of z in $\mathfrak{p}$. Hence K^z also operates on T, so that $\mathfrak{k}^z$ operates on T. For $x\in\mathfrak{k}^z$, we denote by ad x, or sometimes by x, the action of x on T.

Remark 2.1.27. The element $q\in T_2$, which is an invariant for R^z, is not an invariant for $\mathfrak{k}^z$. In fact, since ad $z=1$ on $\mathfrak{d}=T_1$, one has ad $z=j$ on T_j. In particular, ad $z(q)=2q\neq 0$.

Since $\mathfrak{k}^z=\mathfrak{r}^z+\mathbf{C}z$, $\mathfrak{k}^z$ leaves each q^iL_j stable, we now have the following theorem:

Theorem 2.1.28. *Let* $s = \dim \mathfrak{g}^{2\alpha}$. *One has a unique decomposition of T into irreducible* K^z*-modules and it is given by*

$$T = \sum_{i \geqq 0,\, j \geqq 0} q^i L_j \quad \text{if} \quad s > 1,$$

$$T = \sum_{i \geqq 0,\, j \geqq 0} q^i L_j^+ + \sum_{i \geqq 0,\, j > 0} q^i L_j^- + \sum_{i \geqq 0} \mathbf{C} q^i$$

if $s = 1$, *and*

$$T = \sum_{i \geqq 0,\, j = 0, 1} q^i L_j \quad \text{if} \quad s = 0.$$

In each case, any two factors in the decomposition are non-isomorphic. In particular, any irreducible K^z*-module in T occurs with multiplicity* 1.

Proof. First note that $\dim \mathfrak{d} = s+1$, since $\dim \mathfrak{d} = \dim \mathfrak{d}^{-\tau} = \dim \mathfrak{h}_2^1 + \dim \mathbf{C}e_1 = \dim \mathfrak{g}^{2\alpha} + 1$. If $s \leqq 1$ then R^z operates like $SO(\mathfrak{d})$ so the result follows from Remark 2.1.27. Hence we may assume $s > 1$. Now suppose $q^i L_j$ and $q^{i'} L_{j'}$ are isomorphic as K^z-modules. One has $q^i L_j \subseteqq T_{2i+j}$ and $q^{i'} L_{j'} \subseteqq T_{2i'+j'}$. Hence, by Remark 2.1.27, $2i+j = 2i'+j'$. On the other hand, $q^i L_j$ and $q^{i'} L_{j'}$ are also isomorphic as R^z-modules. Hence $j = j'$ since $\dim L_j \neq \dim L_{j'}$ for $j \neq j'$. Thus $i = i'$.
Q.E.D.

We define Γ^z to be the set of all irreducible representations of K^z occurring in T. For $\gamma^z \in \Gamma^z$, we denote by T_{γ^z} the unique corresponding irreducible K^z-submodule of T.

Remark 2.1.29. Assume that $\dim \mathfrak{d}^{-\tau} > 2$. Recall that for the action of $M_{\mathbf{R}}^{\tau}$ on $\mathfrak{d}_{\mathbf{R}}^{-\tau}$, $M_{0,\mathbf{R}}$ is transitive on the unit sphere in $(\mathfrak{h}_2^1)_{\mathbf{R}}$, ($(\mathfrak{h}_2^1)_{\mathbf{R}}$ is the orthocomplement of e_1 in $\mathfrak{d}_{\mathbf{R}}^{-\tau}$ relative to B_1). Note that $c^{-1} = 1$ on $\mathfrak{m}_0$, since $\mathfrak{m}_0$ centralizes $\mathfrak{u}$ and $c^{-1} \in U$. Thus for the action of $R_{\mathbf{R}}^z$ on $\mathfrak{d}$, $M_{0,\mathbf{R}}$ is transitive on the unit sphere in $\mathfrak{b}_0$, where $\mathfrak{b}_0$ is the orthocomplement of $\mathbf{R}e$ in $\mathfrak{d}_0$. From a classical fact it follows that if $W \subseteqq T$ is an irreducible $SO(\mathfrak{d})$-module and if $H = (SO(\mathfrak{d}))^e$ is the isotropy group of $SO(\mathfrak{d})$ at e, then W^H, the subspace of all elements of W fixed by H, is one-dimensional. (See the proof of Lemma 2.1.5.) Then from the double transitivity of $R_{\mathbf{R}}^z$ on the unit sphere in $\mathfrak{d}_0$, it follows (see, again the proof of Lemma 2.1.5) that $W^{M_{0,\mathbf{R}}}$ is one-dimensional. Hence for every $\gamma^z \in \Gamma^z$, $(T_{\gamma^z})^{M_0} = 1$. (This fact also holds if $\dim \mathfrak{d}^{-\tau} = 1$ or 2 since W itself is one-dimensional and M_0 operates trivially in $\mathfrak{d}$).

§ 2. The (i, j) decomposition of H under K_θ in the rank 1 case

We will now set up a one-one correspondence between the sets Γ and Γ^z.

Lemma 2.2.1. Any non-zero element in $\mathfrak{d}_0$ is principal nilpotent.

Proof. This follows from the fact that $R_{\mathbf{R}}^z$ is transitive on the unit sphere in $\mathfrak{d}_0$ when $\dim \mathfrak{d}_0 > 1$ and $e \in \mathfrak{d}_0$ is principal nilpotent. Q.E.D.

Lemma 2.2.2. For the action of z on S, the symmetric algebra on $\mathfrak{p}$, the highest eigenvalue in S_k is k and the corresponding eigenspace is T_k.

Proof. By Proposition 2.1.22, we have $\mathfrak{p}=\mathfrak{d}+c^{-1}(\mathfrak{m}_1+\mathfrak{p}\cap\mathfrak{z})+\tau\mathfrak{d}$, and $\mathfrak{d}$ is the $+1$ eigenspace of $\operatorname{ad} z$ in $\mathfrak{p}$, $c^{-1}(\mathfrak{m}_1+\mathfrak{p}\cap\mathfrak{z})$ the 0 eigenspace of $\operatorname{ad} z$ in $\mathfrak{p}$ and $\tau\mathfrak{d}$ the -1 eigenspace of $\operatorname{ad} z$ in $\mathfrak{p}$. The result follows immediately from this. Q.E.D.

Note that all the eigenvalues of z in S are integers.

Definition 2.2.3. If V is any K_θ-module (or K-module) contained in S, we define V_z to be the eigenspace of z in V corresponding to the highest eigenvalue of z in V.

Since T is generated by powers x^k where $x\in\mathfrak{d}_0\subseteqq\mathscr{N}$ it follows that $T\subseteqq H$. The following result leads to a characterization of the subspace T in H.

Theorem 2.2.4. *Let $V\subseteqq H_j(H_j=H\cap S_j)$ be a non-zero K_θ-module. Then the highest eigenvalue of z in V is j, so that $V\cap T_j\neq 0$ and $V_z=V\cap T_j$.*

Proof. Let n be the highest eigenvalue of z in V. Since $V\subseteqq S_j$ and j is the highest eigenvalue of z in S_j, it follows that $n\leqq j$. Suppose $n<j$, i.e., suppose $V\cap T_j=0$. Now choose a K_θ-submodule $W\subseteqq H_j$ so that $H_j=V\oplus W$. Since e is a nilpotent element, $e^j\in H_j$. Write $e^j=v+w$, $v\in V$ and $w\in W$. Since $e\in\mathfrak{d}$, $z(e^j)=je^j$. Hence $z(v)=jv$ and $z(w)=jw$. Since the highest eigenvalue of z in V is less than j we conclude $v=0$ so that $e^j\in W$. Hence for any $a\in K_\theta$, $a\cdot e^j=(a\cdot e)^j\in W$. But as a varies over K_θ, $a\cdot e$ varies over the set of all principal nilpotent elements, (see [10]) and we know that $\{f^j|f \text{ principal nilpotent}\}$ spans H_j. Thus $W=H_j$. But this contradicts the fact that V is non-zero. Q.E.D.

Remark. Note that if one knows that the set of all principal nilpotent elements is a single K-orbit, then the theorem is valid if V is just a K-module. We assert that this is the case when $\dim \mathfrak{p}\cap[\mathfrak{g},\mathfrak{g}]>2$. (Remember that all our present considerations are in the rank 1 case.) Indeed, $\mathscr{N}$, the set of all nilpotent elements in $\mathfrak{p}$, is given by $\mathscr{N}=\{y\in\mathfrak{p}|f(y)=0 \text{ for every } f\in J'_+S'\}$. Let $q'\in(S')_2$ be the K_θ-invariant quadratic form which defines the restriction to $\mathfrak{p}$ of the Killing form of $\mathfrak{g}$. Then it is easy to see that (in the rank 1 case) $\mathscr{N}=\{y\in\mathfrak{p}\cap[\mathfrak{g},\mathfrak{g}]|q'(y)=0\}$. Choosing a basis $x_1,\ldots,x_m$ of $\mathfrak{p}_{\mathbf{R}}\cap[\mathfrak{g},\mathfrak{g}]$ such that $B(x_i,x_j)=\delta_{ij}$, we get $\mathscr{N}=\{\sum c_ix_i|\sum c_i^2=0\}$, which shows that $\mathscr{N}$ is an irreducible algebraic set if $m\geqq 3$ (since the polynomial $\sum X_i^2$ is irreducible in that case). To complete the proof of the assertion, note that, in general, if $\mathscr{N}_i$ is any irreducible component of $\mathscr{N}$, then the set of principal nilpotent elements in $\mathscr{N}_i$ forms a single K-orbit (see Theorem 3, § 1.4. [10]).

Corollary 2.2.5 *(to Theorem* 2.2.4*)*. If $V\subseteqq H$ is any irreducible K_θ-module, then V is necessarily homogeneous. More precisely, if j is the highest eigenvalue of z in V, then $V\subseteqq H_j$.

Proof. Suppose p_i is a non-zero projection map of V into H_i with respect to the decomposition $H=\sum_i H_i$. Then $V \cong p_i(V)$, and hence j is the highest eigenvalue of z in $p_i(V)$. But, by Theorem 2.2.4, i is the highest eigenvalue of z in $p_i(V)$. Hence $i=j$. Q.E.D.

Lemma 2.2.6. Let $V \subseteq S$ be any irreducible K-module. Then V_z is an irreducible K^z-module.

Proof. Clearly, V_z is stable under K^z. To show that V_z is irreducible under $\mathfrak{k}^z$, first note that all the eigenvalues of z in $\mathfrak{k}$ are real.

Let $\mathfrak{k}_+$ (resp., $\mathfrak{k}_-$) denote the sum of the eigensubspaces of $\mathfrak{k}$ with respect to the positive (resp., negative) eigenvalues of z in $\mathfrak{k}$. Then $\mathfrak{k}=\mathfrak{k}_- \oplus \mathfrak{k}^z \oplus \mathfrak{k}_+$ is a direct sum of subalgebras, so that $U(\mathfrak{k})=U(\mathfrak{k}_-)\otimes U(\mathfrak{k}^z)\otimes U(\mathfrak{k}_+)$. Now let $v \in V_z$, $v \neq 0$. Then $V_z \subset U(\mathfrak{k})\cdot v$, since V is irreducible under $\mathfrak{k}$. Hence $V_z \subset U(\mathfrak{k}_-)U(\mathfrak{k}^z)\cdot v$, so that $V_z = U(\mathfrak{k}^z)\cdot v$. Q.E.D.

Lemma 2.2.7. Let $s=\dim \mathfrak{g}^{2\alpha}$. If $s>0$, then s is odd.

Proof. It suffices to show that $\mathfrak{d}_{\mathbf{R}}^{-\tau}$ is even-dimensional. Now $[e_1, \mathfrak{d}_{\mathbf{R}}^{-\tau}]=$ $=[e_1, (\mathfrak{h}_2^1)_{\mathbf{R}}]=(\mathfrak{h}_2^2)_{\mathbf{R}}=\mathfrak{g}_{\mathbf{R}}^{2\alpha}$. But since $M_{\mathbf{R}}^{\tau}$ is transitive on the unit sphere of $\mathfrak{d}_{\mathbf{R}}^{-\tau}$, we have that $[x, \mathfrak{d}_{\mathbf{R}}^{-\tau}]=\mathfrak{g}_{\mathbf{R}}^{2\alpha}$, for every non-zero $x \in \mathfrak{d}_{\mathbf{R}}^{-\tau}$. Thus if $0 \neq u \in \mathfrak{g}_{\mathbf{R}}^{2\alpha}$, then the bilinear form $B_u(x, y)=B_2([x, y], u)$ is non-singular and alternating on $\mathfrak{d}_{\mathbf{R}}^{-\tau}$. Hence $\mathfrak{d}_{\mathbf{R}}^{-\tau}$ is even-dimensional. Q.E.D.

Lemma 2.2.8. Let $V \subseteq S$ be any irreducible K_θ-module. Then V_z is an irreducible K^z-module.

Proof. We can assume without loss that $V \subseteq H_j$. By Lemma 2.2.4, $V_z \subseteq T_j$. Now T_j is one-dimensional if $\mathfrak{d}$ is one-dimensional. Thus we can assume $\dim \mathfrak{d}>1$. Then we assert that $K_\theta = K$, so that the Lemma is an immediate consequence of Lemma 2.2.6.

The assertion can be proved as follows: let $\mathfrak{h}_1$ be a Cartan subalgebra of $\mathfrak{m}$ defined over $\mathbf{R}$, so that $\mathfrak{h}_1+\mathfrak{a}$ is a Cartan subalgebra of $\mathfrak{g}$ defined over $\mathbf{R}$. Let P be the set of all roots φ relative to $\mathfrak{h}_1+\mathfrak{a}$ such that $\mathfrak{g}^\varphi \subseteq \mathfrak{g}^{2\alpha}$. Since $\mathfrak{h}_1+\mathfrak{a}$ leaves $\mathfrak{g}^{2\alpha}$ stable, clearly $\mathfrak{g}^{2\alpha}=\sum_{\varphi \in P} \mathfrak{g}^\varphi$. Note that if $\varphi \in P$, then $-\varphi^\theta \in P$, where φ^θ is defined by $\varphi^\theta(x)=\varphi(\theta x)$, for $x \in \mathfrak{h}_1+\mathfrak{a}$. Since $\mathfrak{g}^{2\alpha}$ is odd-dimensional (by Lemma 2.2.7), we then conclude that there exists a non-zero element $e_2 \in \mathfrak{g}_{\mathbf{R}}^{2\alpha}$ such that $[\mathfrak{h}_1, e_2]=0$ since there exists φ such that $\theta\varphi=-\varphi$. Note that $e_2+\theta e_2 \in \mathfrak{k}_{\mathbf{R}}$ and hence $e_2+\theta e_2$ is semisimple. Let $\mathfrak{h}_2=\mathfrak{h}_1+\mathbf{C}(e_2+\theta e_2)$. Then $\mathfrak{h}=\mathfrak{h}_2+(\text{center } \mathfrak{g})\cap \mathfrak{a}$ is a Cartan subalgebra of $\mathfrak{g}$. Note that $\mathfrak{h}_2 \subseteq \mathfrak{k}$.

We claim that if σ is the unique nontrivial element of F (the group of elements of order 2 in $\exp \operatorname{ad} \mathfrak{a}$), then $\sigma=1$ on $\mathfrak{h}$. Indeed, since $\sigma \in \exp \operatorname{ad} \mathfrak{a}$, σ acts as a scalar c_j on $\mathfrak{g}^{j\alpha}$. But clearly $c_2=c_1^2$. Moreover, since $\sigma^2=1$, $c_1=\pm 1$ and hence $c_2=1$. But since σ commutes with θ, one also has $c_{-1}=c_1$ and $c_{-2}=c_2$. Thus

$\sigma(e_2)=e_2$ and $\sigma(\theta e_2)=\theta e_2$. Clearly, σ=identity on $\mathfrak{h}_1$ and on (center $\mathfrak{g}$)$\cap\mathfrak{a}$. This proves the claim.

Thus, σ is in the centralizer of $\mathfrak{h}$ in G. But one knows that the centralizer of $\mathfrak{h}$ in G is exp ad $\mathfrak{h}$=exp ad $\mathfrak{h}_2$. But since $\mathfrak{h}_2\subseteqq\mathfrak{k}$, we conclude that $\sigma\in K$. Q.E.D.

We recall that Γ is the set of equivalence classes of finite-dimensional irreducible K_θ-modules V such that $V^{M_\theta}\neq\{0\}$.

Theorem 2.2.9. *One has* $\boldsymbol{l}(\gamma)=1$ *for all* $\gamma\in\Gamma$, *i.e., every irreducible representation of* K_θ *in the harmonics occurs with multiplicity one. Moreover, for each* $\gamma\in\Gamma$, *there exists a unique* $\gamma^z\in\Gamma^z$ *such that* $H_\gamma\cap T=T_{\gamma^z}=H_{\gamma,z}$, *the eigenspace of* z *in* H_γ *corresponding to the highest eigenvalue of* z *in* H_γ. *The correspondence* $\gamma\rightarrow\gamma^z$ *is a bijection between* Γ *and* Γ^z.

Proof. For $\gamma\in\Gamma$, let $V_\gamma\subseteqq H$ denote an irreducible K_θ-module of type γ. By Lemma 2.2.8, $V_{\gamma,z}$ is an irreducible K^z-module. By Corollary 2.2.5, $V_\gamma\subseteqq H_j$, where j is the highest eigenvalue of z in V_γ, and moreover $V_{\gamma,z}=V_\gamma\cap T$. Thus $V_\gamma\cap T=T_{\gamma^z}$, for a unique $\gamma^z\in\Gamma^z$.

Now let $\gamma_1,\gamma_2\in\Gamma$, and let V_{γ_1} and V_{γ_2} be irreducible K_θ-submodules of H of types γ_1 and γ_2 respectively. Assume $V_{\gamma_1}\neq V_{\gamma_2}$. Then clearly $V_{\gamma_1,z}\neq V_{\gamma_2,z}$. Thus $V_{\gamma_1,z}$ and $V_{\gamma_2,z}$ are two *distinct* irreducible K^z-modules contained in T. By Theorem 2.1.28, $V_{\gamma_1,z}$ and $V_{\gamma_2,z}$ are non-isomorphic. Hence $\gamma_1\neq\gamma_2$, and so $\boldsymbol{l}(\gamma)=1$.

Now for any $\gamma^z\in\Gamma^z$, let $u\in T_{\gamma^z}$. By Theorem 2.1.28, $T_{\gamma^z}\subseteqq T_l$ for some $\boldsymbol{l}$. Note that since $\mathfrak{d}$ consists of nilpotent elements, T consists of harmonics. Thus $u\in T_{\gamma^z}\subseteqq H_l$. Let $H_l=H_{\gamma_1}+\cdots+H_{\gamma_j}$ be the decomposition of H_l into irreducible K_θ-modules, and write $u=u_1+\cdots+u_j$ with respect to this decomposition. Since $z\cdot u=lu$, one has also $z\cdot u_i=lu_i$ $(i=1,\ldots,j)$. But by Theorem 2.2.4, for each H_{γ_i}, the highest eigenvalue of z in H_{γ_i} is $\boldsymbol{l}$. Hence $u_i\in H_{\gamma_i,z}(=H_{\gamma_i}\cap T_l=T_{\gamma_i^z})$, so that $u\in H_{\gamma_1,z}+\cdots+H_{\gamma_j,z}$, and so $T_{\gamma^z}\subseteqq T_{\gamma_1^z}+\cdots+T_{\gamma_j^z}$. Thus in view of Theorem 2.1.28, $T_{\gamma^z}=T_{\gamma_i^z}=H_{\gamma_i,z}$ for some i.

This proves that the correspondence $H_\gamma\rightarrow H_{\gamma,z}=H_\gamma\cap T=T_{\gamma^z}$ is a bijection. Q.E.D.

We will now make some observations which will be useful later.

Remark. 2.2.10. For any $\gamma\in\Gamma$, there are two (not necessarily distinct) distinguished one-dimensional spaces in H_γ: (i) $(H_\gamma)^{M_\theta}$ is one dimensional since $\boldsymbol{l}(\gamma)=1$ and (ii) $(H_{\gamma,z})^{M_0}$ is one-dimensional, in view of Remark 2.1.29.

Recall (see § III 3, [10]) that if $\gamma\in\Gamma$ and V_γ is an irreducible K_θ-module of type $\gamma\in\Gamma$, then to any element $x\in\mathfrak{p}$ there is associated a subspace $V_\gamma(x)$ of V_γ which is $\boldsymbol{l}(\gamma)$-dimensional if x is regular (Theorem 20, [10]).

Lemma 2.2.11. One has (i) $(H_\gamma)^{M_\theta}=H_\gamma(w)$ and (ii) $(H_{\gamma,z})^{M_0}=H_\gamma(e)$.

Proof. (i) follows from Theorem 20, [10]. To prove (ii), first recall (Theorem 20, [10]) that if K_θ^e is the centralizer of e in K_θ, then $H_\gamma(e)\subseteqq(H_\gamma)^{K_\theta^e}$. Since M_0 centralizes e and $M_0\subseteqq K_\theta$ one has $H_\gamma(e)\subseteqq(H_\gamma)^{M_0}$.

Since $[z, e]=e$ we have by Theorem 21, [10] that the action of z in $H_\gamma(e)$ is diagonalizable and the eigenvalues of z in $H_\gamma(e)$ are precisely the degrees $d_i(\gamma)$ in which γ occurs in H. But by Theorem 2.2.9, $\boldsymbol{l}(\gamma)=1$. Thus, assuming $H_\gamma \subseteq H_j$, we conclude that $z=j$ on $H_\gamma(e)$. Then Theorem 2.2.4 implies that $H_\gamma(e)\subseteq H_{\gamma, z}$.

Hence we conclude that $H_\gamma(e)\subseteq (H_{\gamma,z})^{M_0}$. But by Theorem 20, [10], $H_\gamma(e)$ is one-dimensional. Since, by Remark 2.1.29, $(H_{\gamma,z})^{M_0}$ is also one-dimensional, (ii) follows.

Q.E.D.

Lemma 2.2.12. For $t \in \mathbf{R}$, let $a(t)=\exp(t \operatorname{ad} z)$. Then, in the Grassmannian of one-dimensional subspaces of H_γ,

$$\lim_{t\to+\infty} a(t)\cdot H_\gamma(w) = H_\gamma(e).$$

Proof. One has $w=cz=-i(ce_1-ce_{-1})=-i(e-ce_{-1})$, so that $iw=(e-ce_{-1})$. Also one has $[tz, e]=te$ and $[tz, ce_{-1}]=-tce_{-1}$. Thus $\exp(t \operatorname{ad} z)(iw)=\exp(t)e-$ $-\exp(-t)ce_{-1}$, and so $\exp(t \operatorname{ad} z)\big(i(\exp(-t))w\big)=e-\exp(-2t)ce_{-1}$. Thus

$$\lim_{t\to+\infty} \exp(t \operatorname{ad} z)\big(i(\exp(-t))w\big) = e.$$

Hence, in the Grassmannian of one-dimensional subspaces of H_γ,

$$\lim_{t\to+\infty}\big\{H_\gamma\big(\exp(t \operatorname{ad} z)\big(i(\exp(-t))w\big)\big)\big\} = H_\gamma(e).$$

That is,

$$\lim_{t\to+\infty}\big\{\exp(t \operatorname{ad} z)\cdot H_\gamma\big(i(\exp(-t))w\big)\big\} = H_\gamma(e).$$

However, $i(\exp(-t))w$ is a regular semisimple element, so that by Theorem 20, [10], $H_\gamma\big(i\exp(-t)w\big)=(H_\gamma)^{M_\theta}=H_\gamma(w)$.

Thus, indeed,

$$\lim_{t\to+\infty} a(t)\cdot H_\gamma(w) = H_\gamma(e). \qquad \text{Q.E.D.}$$

Remark 2.2.13. Let $(H_\gamma^{M_\theta})^\perp$ denote the unique M_θ-submodule of H_γ which is complementary to $H_\gamma^{M_\theta}$. We assert that $H_\gamma(e)\nsubseteq(H_\gamma^{M_\theta})^\perp$, so that $p_u\neq 0$ for a non-zero element $u\in H_\gamma^*(e)$ if $p_v\neq 0$ for a non-zero element $v\in(H_\gamma^*)^{M_\theta}$. (Of course, they only differ by a scalar multiple, in view of Lemma 1.7.10.)

This can be proved as follows:

Note that relative to a $K_{\theta,\mathbf{R}}$-invariant positive-definite Hermitian form on H_γ, $(H_\gamma^{M_\theta})^\perp$ is the orthocomplement of $H_\gamma^{M_\theta}$ in H_γ. Since $z\in i\mathfrak{k}_{\mathbf{R}}$, z operates on H_γ as a Hermitian operator, so that if $H_\gamma(e)\perp H_\gamma(w)$, then $a(t)\cdot H_\gamma(w)\perp a(-t)\cdot H_\gamma(e)$. But since $[z, e]=e$, one has by Theorem 21, [10] that $z\cdot H_\gamma(e)\subseteq H_\gamma(e)$, which implies that $a(-t)\cdot H_\gamma(e)=H_\gamma(e)$. Hence $a(t)\cdot H_\gamma(w)\perp H_\gamma(e)$, which implies $H_\gamma(e)\perp H_\gamma(e)$ by Lemma 2.2.12. But this is impossible, and so $H_\gamma(e)\nsubseteq(H_\gamma^{M_\theta})^\perp$.

Remark 2.2.14. Let $\beta: S(\mathfrak{g})\to U$ be the linear isomorphism such that $\beta(x^k)=x^k$ for all $x\in\mathfrak{g}$. For all $\gamma\in\Gamma$, let $v(\gamma)$ be a non-zero element of $(H_\gamma^*)^{M_\theta}$, and for all $\gamma^z\in\Gamma^z$, let $w(\gamma^z)$ be a non-zero element of the one-dimensional space $\beta\big((T_{\gamma^z})^{M_0}\big)=$

$=(\beta(T_{\gamma^z}))^{M_0}$ (see Remark 2.1.29). Under the bijection $\gamma\to\gamma^z$ between Γ and Γ^z (see Theorem 2.2.9), we have that $p_{v(\gamma)}$ and $p_{w(\gamma^z)}$ differ by a non-zero scalar multiple, in view of the above remark and Lemma 2.2.11. Hence we have that $\{\mathbf{C}p_{v(\gamma)}|\gamma\in\Gamma\}=\{\mathbf{C}p_{w(\gamma^z)}|\gamma^z\in\Gamma^z\}$. The decomposition of T into irreducible K^z-modules is given in Theorem 2.1.28. If $s>1$, then $(T_{\gamma^z})^{M_0}=(T_{\gamma^z})^H$, where $H=(SO(\mathfrak{d}))^e$, by Remark 2.1.29, so that $w(\gamma^z)$ ranges through the elements of the form $\beta(q^i\boldsymbol{l}_j)$ $(i,j\geqq 0)$, where $\boldsymbol{l}_j$ is a non-zero element of $(L_j)^H$. If $s=0$, then $(T_{\gamma^z})^{M_0}=T_{\gamma^z}$ is one-dimensional, and $w(\gamma^z)$ ranges through the elements of the form $\beta(q^i\boldsymbol{l}_j)$ $(i\geqq 0, j=0, 1)$, where $\boldsymbol{l}_j$ is a non-zero element of L_j. Finally, if $s=1$, then $(T_{\gamma^z})^{M_0}=T_{\gamma^z}$ is one-dimensional, and $w(\gamma^z)$ ranges through the elements of the form $\beta(q^i\boldsymbol{l}_j^+)$ $(i\geqq 0, j>0)$, $\beta(q^i\boldsymbol{l}_j^-)$ $(i\geqq 0, j>0)$ and $\beta(q^i)$, where $\boldsymbol{l}_j^+$ is a non-zero element of L_j^+ and $\boldsymbol{l}_j^-$ is a non-zero element of L_j^-.

§ 3. The formula for $n_\varphi(\gamma)$ in terms of the $d(\delta_i)$

In this section, $\mathfrak{g}$ will denote a complex reductive Lie algebra of arbitrary rank and the rest of the notation will be as usual. We have $\mathfrak{g}=\mathfrak{g}^0+\sum_{\varphi\in\Delta}\mathfrak{g}^\varphi$, where $\mathfrak{g}^0=\mathfrak{a}+\mathfrak{m}$. Let $\Delta^1=\{\alpha\in\Delta|\frac{1}{2}\alpha\notin\Delta\}$.

Definition 2.3.1. For every $\varphi\in\Delta^1$ we define $\mathfrak{g}_\varphi=\mathfrak{g}^0+\sum_{j=-2}^{+2}\mathfrak{g}^{j\varphi}$.

(Note that 2φ may or may not be a root. If 2φ is not a root, of course we have $\mathfrak{g}^{2\varphi}=0$.)

Remark 2.3.2. Let $y\in(\ker\varphi)\cap\mathfrak{a}_{\mathbf{R}}$ be chosen so that $\psi(y)\neq 0$ for every root $\psi\neq j\varphi$, $(j=\pm 1, \pm 2)$. Then clearly $\mathfrak{g}_\varphi=\mathfrak{g}^y$, where $\mathfrak{g}^y$ denotes the centralizer of y in $\mathfrak{g}$. Thus $\mathfrak{g}_\varphi$ is reductive and clearly $\mathfrak{g}_\varphi=\mathfrak{g}_{\varphi,\mathbf{R}}+i\mathfrak{g}_{\varphi,\mathbf{R}}$, where $\mathfrak{g}_{\varphi,\mathbf{R}}=\mathfrak{g}_\varphi\cap\mathfrak{g}_{\mathbf{R}}$. Moreover, $\mathfrak{g}_\varphi$ is stable under θ, so that $\mathfrak{g}_\varphi=\mathfrak{k}_\varphi+\mathfrak{p}_\varphi$, where $\mathfrak{k}_\varphi=\mathfrak{k}\cap\mathfrak{g}_\varphi$ and $\mathfrak{p}_\varphi=\mathfrak{p}\cap\mathfrak{g}_\varphi$. Now, $\mathfrak{a}\subseteqq\mathfrak{p}_\varphi$ and $\mathfrak{a}$ is a Cartan subspace for $\mathfrak{g}_\varphi=\mathfrak{k}_\varphi+\mathfrak{p}_\varphi$. Since the root space decomposition of $\mathfrak{g}_\varphi$ with respect to $\mathfrak{a}$ is given by $\mathfrak{g}_\varphi=(\mathfrak{a}+\mathfrak{m})+\sum_{j=-2}^{+2}\mathfrak{g}^{j\varphi}$, it follows that $\mathfrak{g}_\varphi$ is of rank 1, i.e., that $[\mathfrak{g}_\varphi,\mathfrak{g}_\varphi]\cap\mathfrak{a}$ is one-dimensional.

Let $G_\varphi\subseteqq G$ be the analytic subgroup with Lie algebra $\operatorname{ad}\mathfrak{g}_\varphi$ and let $K_{\varphi,\theta}=\{a\in G_\varphi|a\theta=\theta a\}$. Let $M_{\varphi,\theta}$ be the centralizer of $\mathfrak{a}$ in $K_{\varphi,\theta}$.

Lemma 2.3.3. One has $M_{\varphi,\theta}=M_\theta$.

Proof. Clearly, $M\subseteqq M_{\varphi,\theta}\subseteqq M_\theta$. Since $M_\theta=MF$ it is enough to show that $F\subseteqq M_{\varphi,\theta}$. By definition, $K_{\varphi,\theta}=K_\theta\cap G_\varphi$ and so $M_{\varphi,\theta}=M_\theta\cap G_\varphi$. Hence it is enough to show that $F\subseteqq G_\varphi$. But this is obvious since $F\subseteqq\exp\operatorname{ad}\mathfrak{a}\subseteqq G_\varphi$. Q.E.D.

Now, for $\varphi\in\Delta^1$, let G^φ denote the adjoint group of $\mathfrak{g}_\varphi$. Let π denote the homomorphism of G_φ into G^φ given by $a\to a|\mathfrak{g}_\varphi$ for $a\in G_\varphi$. Clearly, one has an exact sequence

$$e \to \text{centre}\,(G_\varphi) \to G_\varphi \xrightarrow{\pi} G^\varphi \to e.$$

Let K_θ^φ, M_θ^φ, K^φ, M^φ and F^φ play the roles of K_θ, M_θ, K, M and F for the case $\mathfrak{g}_\varphi = \mathfrak{k}_\varphi + \mathfrak{p}_\varphi$.

Lemma 2.3.4. One has $\pi(K_{\varphi,\theta}) = K_\theta^\varphi$ and $\pi(M_\theta) = M_\theta^\varphi$.

Proof. Note that for $a \in G_\varphi$, $\pi(\theta a \theta) = \theta \pi(a) \theta$. (Observe that the θ on the right hand side of the above relation denotes the map θ for the case $\mathfrak{g}_\varphi = \mathfrak{k}_\varphi + \mathfrak{p}_\varphi$, and this map equals $\theta | \mathfrak{g}_\varphi$.) Hence $\pi(K_{\varphi,\theta}) \subseteqq K_\theta^\varphi$. Now, clearly, $\pi(M_{\varphi,\theta}) \subseteqq M_\theta^\varphi$. Moreover, obviously, $\pi(K_{\varphi,\theta}) \supseteqq K^\varphi$ and $\pi(M_{\varphi,\theta}) \supseteqq M^\varphi$. Since one has $K_\theta^\varphi = F^\varphi K^\varphi$ and $M_\theta^\varphi = F^\varphi M^\varphi$, it is now enough to prove that $\pi(F) \supseteqq F^\varphi$. Since $\mathfrak{g}^\varphi$ is of rank 1, F^φ has a unique nontrivial element. But choosing $\sigma \in F$ such that $\sigma = -1$ on $\mathfrak{g}^\varphi$, it is obvious that $\pi(\sigma)$ is the unique nontrivial element of F^φ. (That such a σ exists follows, for example, by choosing a simple root system $\Pi = \{\alpha_1, \ldots, \alpha_{r_0}\}$ of $\Delta(\mathfrak{a}, \mathfrak{g})$ such that $\varphi \in \Pi$ and then noting that the map $\sigma \to (a_1^\sigma, \ldots, a_{r_0}^\sigma)$, where a_i^σ is the scalar defining the action of σ on $\mathfrak{g}^{\alpha_i}$, is an isomorphism of A onto $\mathbf{C}^* \times \mathbf{C}^* \times \cdots \times \mathbf{C}^*$ (r_0 times), and F maps onto the 2^{r_0} elements of order 2 in $(\mathbf{C}^*)^{r_0}$.) Q.E.D.

Let Γ_φ denote the set of equivalence classes of irreducible $K_{\varphi,\theta}$-modules admitting an M_θ-fixed vector.

Proposition 2.3.5. If L denotes the kernel of the surjection $\pi: K_{\varphi,\theta} \to K_\theta^\varphi$, then $L \subseteqq M_\theta$.

Proof. Clearly $L = K_{\varphi,\theta} \cap \text{center}\,(G_\varphi)$. But, obviously $K_{\varphi,\theta} \cap \text{centre}\,(G_\varphi) \subseteqq M_{\varphi,\theta} = M_\theta$. Q.E.D.

Remark 2.3.6. For $\delta \in \Gamma_\varphi$, let V_δ denote the representation space of δ. Since $L \subseteqq \text{center}\,(K_{\varphi,\theta})$, L acts as scalars on V_δ by Schur's lemma. But since $L \subseteqq M_\theta$, there exists an L-fixed vector in V_δ, and so L acts trivially on V_δ. Hence Γ_φ can be identified with the usual Γ for the case $\mathfrak{g}_\varphi = \mathfrak{k}_\varphi + \mathfrak{p}_\varphi$.

Now fix $\gamma \in \Gamma$ and $\varphi \in \Delta^1$. Let V_γ be the representation space of γ and let W be the $K_{\varphi,\theta}$-submodule of V_γ generated by $(V_\gamma)^{M_\theta}$.

Theorem 2.3.7. One has that W decomposes into a direct sum of $l(\gamma)$ irreducible $K_{\varphi,\theta}$-modules. If $W = \sum_{i=1}^{l(\gamma)} W_i$ is any such decomposition, then $W_i \cap (V_\gamma)^{M_\theta}$ is one-dimensional, so that if $v_i \in W_i \cap (V_\gamma)^{M_\theta}$ is non-zero ($i = 1, \ldots, l(\gamma)$), then $\{v_i\}$ is a basis of $(V_\gamma)^{M_\theta}$.

Proof. Let $W = \sum_{i=1}^{k} W_i$ be a direct sum decomposition of W into irreducible $K_{\varphi,\theta}$-modules. We can assume that for all i such that $1 \leqq i \leqq m \leqq k$, W_i has an M_θ-fixed vector, and for $m < i \leqq k$, W_i has no M_θ-fixed vector. Then clearly

$(V_\gamma)^{M_\theta}=\sum_{i=1}^{m}(W_i)^{M_\theta}$, which implies by the definition of W that $W=\sum_{i=1}^{m} W_i$. This shows that each W_i $(i=1, \ldots, k)$ has an M_θ-fixed vector.

Now by Remark 2.3.6 and Theorem 2.2.9 (applied to the case $\mathfrak{g}_\varphi=\mathfrak{k}_\varphi+\mathfrak{p}_\varphi$), we see that $(W_i)^{M_\theta}$ is one-dimensional for all i. Since $(V_\gamma)^{M_\theta}$ is $l(\gamma)$-dimensional and $(V_\gamma)^{M_\theta}=\sum_{i=1}^{k}(W_i)^{M_\theta}$, we conclude that $k=l(\gamma)$. Q.E.D.

Now, for each $\varphi\in\Delta^1$, choose $w_\varphi\in\mathfrak{a}\cap[\mathfrak{g}_\varphi, \mathfrak{g}_\varphi]$ such that $\langle\varphi, w_\varphi\rangle=1$.

Recall the subalgebra $\mathfrak{g}^*\subseteq\mathfrak{g}$ (see [7], § II. 2). Observe that since $\mathfrak{a}$ is a Cartan subalgebra of $\mathfrak{g}^*$, $\mathfrak{g}^*\cap\mathfrak{g}_{\mathbf{R}}^{\varphi}$ is one-dimensional for every $\varphi\in\Delta$. Now for $\varphi>0$ choose $e_\varphi\in\mathfrak{g}^*\cap\mathfrak{g}_{\mathbf{R}}^{\varphi}$ such that $\langle e_\varphi, \theta e_\varphi\rangle=-\frac{\langle w_\varphi, w_\varphi\rangle}{2}$ (relative to the Killing form of $\mathfrak{g}$) and define $e_{-\varphi}=-\theta e_\varphi$. It is easy to prove from the above condition that $(2w_\varphi, 2e_\varphi, 2e_{-\varphi})$ is a S-triple, and from this fact in turn that relative to the Killing form of $\mathfrak{g}_\varphi$, $\langle e_\varphi, \theta e_\varphi\rangle=-\frac{\langle w_\varphi, w_\varphi\rangle}{2}$. Thus w_φ, e_φ and $e_{-\varphi}$ play the role of the elements w, e_1, and e_{-1} of § 2.1 for the rank 1 case $\mathfrak{g}_\varphi=\mathfrak{k}_\varphi+\mathfrak{p}_\varphi$. Let $\mathfrak{u}_\varphi$ be the *TDS* spanned by w_φ, e_φ and $e_{-\varphi}$. Now define $z_\varphi=i(e_\varphi-e_{-\varphi})$ and $w'_\varphi=e_\varphi+e_{-\varphi}$. Let $c_\varphi=\exp \operatorname{ad}_{\mathfrak{g}}\left(i\frac{\pi}{2}w'_\varphi\right)$, so that one has $c_\varphi^{-1}w_\varphi=z_\varphi$. Now let $e^\varphi=c_\varphi^{-1}e_\varphi$.

We now fix $\gamma\in\Gamma$ and $\varphi\in\Delta^1$. With the notation of Theorem 2.3.7, each W_i corresponds to some $\delta_i\in\Gamma_\varphi$. By Corollary 2.2.5 (applied to the case $\mathfrak{g}_\varphi=\mathfrak{k}_\varphi+\mathfrak{p}_\varphi$), the highest eigenvalue of z_φ in W_i is $d(\delta_i)$. (Recall that if $\gamma\in\Gamma$ and if $\sigma_1, \ldots, \sigma_{l(\gamma)}$ is a basis of $\operatorname{Hom}_{K_\theta}(V_\gamma, H')$ such that $\sigma_i(V_\gamma)\subseteq H'_{d_i(\gamma)}$, then by definition, $d(\gamma)=\sum_{i=1}^{l(\gamma)} d_i(\gamma)$. In the rank one case, $d(\gamma)$ is just the degree in which γ occurs in the harmonics.)

For each i $(=1, \ldots, l(\gamma))$ we have two (not necessarily distinct) distinguished one-dimensional subspaces, $W_i(w_\varphi)$ and $W_i(e^\varphi)$, contained in W_i. (See Remark 2.2.10 and Lemma 2.2.11). In each $W_i(e^\varphi)$, choose a non-zero element u_i and let X_φ be the $l(\gamma)$-dimensional subspace of V_γ generated by $\{u_i\}$. Note that $\sum_{i=1}^{l(\gamma)} W_i(w_\varphi)=(V_\gamma)^{M_\theta}$. As an immediate consequence of Lemma 2.2.12, we have the following lemma:

Lemma 2.3.8. In the Grassmannian of $l(\gamma)$-dimensional subspaces of V_γ, we have

$$X_\varphi = \lim_{t\to\infty} a_\varphi(t)\cdot(V_\gamma)^{M_\theta},$$

where

$$a_\varphi(t) = \exp(t \operatorname{ad} z_\varphi).$$

Now let $\mathfrak{a}_\varphi$ be the kernel of φ, and choose $y\in\mathfrak{a}_{\varphi,\mathbf{R}}$ such that $\psi(y)\neq 0$ for any root $\psi\in\Delta$ different from $\pm\varphi$ and $\pm 2\varphi$. Note that $\mathfrak{g}^y=\mathfrak{g}_\varphi$, where $\mathfrak{g}^y$ denotes the centralizer of y in $\mathfrak{g}$. Put $e=e^{\varphi}+y$. Since e^{φ} is principal nilpotent in $\mathfrak{g}_{\varphi}$, it follows from Theorem 7, [10] that e is a regular element, i.e., that $e\in\mathscr{R}$.

We now assume furthermore that $y\in\mathfrak{a}_{\varphi,\mathbf{R}}$ is chosen so that $y+w_{\varphi}$ is regular, in addition to the condition that $\psi(y)\neq 0$ for all $\psi\in\Delta$ different from $\pm\varphi$ and $\pm 2\varphi$. (That such a choice is possible is seen from the fact that the set $\{y+tw_{\varphi}|t\in\mathbf{R}-(0),\ y\in\mathfrak{a}_{\varphi,\mathbf{R}}$ and $\psi(y)\neq 0$ for any $\psi\in\Delta$ different from $\pm\varphi$ and $\pm 2\varphi\}$ is an open set in $\mathfrak{a}_{\mathbf{R}}$ and hence meets $\mathscr{R}\cap\mathfrak{a}_{\mathbf{R}}$, which is a nonempty Zariski open set in $\mathfrak{a}_{\mathbf{R}}$.)

Proposition 2.3.9. One has $X_{\varphi}=V_{\gamma}(e)$.

Proof. As in the proof of Lemma 2.2.12, we have $e^{\varphi}=\lim\limits_{t\to\infty} a_{\varphi}(t)\left(\dfrac{iw_{\varphi}}{\exp t}\right)$.

Let $y_t=\dfrac{iw_{\varphi}}{\exp t}+y$. Since $z_{\varphi}\in\mathfrak{g}_{\varphi}=\mathfrak{g}^{y}$, $a_{\varphi}(t)(y)=y$ for every t. Hence

$$\lim_{t\to\infty} a_{\varphi}(t)\left(\frac{iw_{\varphi}}{\exp t}+y\right)=\lim_{t\to\infty} a_{\varphi}(t)(y_t)=e^{\varphi}+y=e.$$

Since $y+w_{\varphi}$ is regular, $y_t=y+\dfrac{iw_{\varphi}}{\exp t}$ is regular except for finitely many t. (To see this, consider the nonempty Zariski open subset $(y+\mathbf{C}w_{\varphi})\cap\mathscr{R}$ of $y+\mathbf{C}w_{\varphi}$.) In particular, $V_{\gamma}(y_t)$ is $l(\gamma)$-dimensional for all but finitely many t. Since $\lim\limits_{t\to\infty} a_{\varphi}(t)(y_t)=e$, we have $\lim\limits_{t\to\infty} V_{\gamma}(a_{\varphi}(t)(y_t))=V_{\gamma}(e)$, i.e., $\lim\limits_{t\to\infty} a_{\varphi}(t)V_{\gamma}(y_t)=$ $=V_{\gamma}(e)$.

But since all but finitely many y_t are regular semisimple elements, $V_{\gamma}(y_t)=$ $=(V_{\gamma})^{K_{\theta}^{y_t}}=(V_{\gamma})^{M_{\theta}}$, by [10], Theorem 20 and Lemma 20. Thus $\lim\limits_{t\to\infty} a_{\varphi}(t)\cdot(V_{\gamma})^{M_{\theta}}=$ $=V_{\gamma}(e)$. But by Lemma 2.3.8, $\lim\limits_{t\to\infty} a_{\varphi}(t)(V_{\gamma})^{M_{\theta}}=X_{\varphi}$, so that $V_{\gamma}(e)=X_{\varphi}$.

Q.E.D.

Recall that $F_{\gamma}=\operatorname{Hom}_{K_{\theta}}(V'_{\gamma}, H')$ and that the map $\beta_y\colon F_{\gamma}\to V_{\gamma}(y)$ is an isomorphism if $y\in\mathfrak{p}$ is a regular element (see [10], § III. 3). Now $\beta_e(F_{\gamma})=X_{\varphi}$. Choose a basis $\sigma_1,\ldots,\sigma_{l(\gamma)}$ of F_{γ} such that $\beta_e(\sigma_j)=u_j$, and choose a basis $v_1,\ldots,v_{l(\gamma)}$ of $(V_{\gamma})^{M_{\theta}}$ such that $v_i\in W_i\cap(V_{\gamma})^{M_{\theta}}$ (see Theorem 2.3.7). Since $(V_{\gamma})^{M_{\theta}}$ is non-singularly paired to $(V'_{\gamma})^{M_{\theta}}$, we can choose a basis $g_1,\ldots,g_{l(\gamma)}$ of $(V'_{\gamma})^{M_{\theta}}$ dual to the basis $v_1,\ldots,v_{l(\gamma)}$.

Let W' be the $K_{\varphi,\theta}$-submodule of V'_{γ} generated by $(V'_{\gamma})^{M_{\theta}}$.

Lemma 2.3.10. One has that W is non-singularly paired to W'.

Proof. Let W_0 be a $K_{\varphi,\theta}$-submodule of V_{γ} which is a complement of W in V_{γ}. Similarly, let W'_0 be a $K_{\varphi,\theta}$-complement of W' in V'_{γ}. Since $(V_{\gamma})^{M_{\theta}}\subseteq W$ and $(V'_{\gamma})^{M_{\theta}}\subseteq W'$, W_0 and W'_0 have no M_{θ}-fixed vectors. On the other hand, any $K_{\varphi,\theta}$-irreducible component of W (resp., W') has an M_{θ}-fixed vector. Thus, using Schur's lemma, we conclude that $W\perp W'_0$, and $W_0\perp W'$, so that indeed W and W' are non-singularly paired. Q.E.D.

Now let W_i' be the $K_{\varphi,\theta}$-submodule generated by $g_i \in (V_\gamma')^{M_\theta}$.

Lemma 2.3.11. One has $W' = \sum_{i=1}^{l(\gamma)} W_i'$, a direct sum decomposition into irreducible $K_{\varphi,\theta}$-submodules. Also, $W_i \perp W_j'$ if $i \neq j$, and W_i and W_i' are non-singularly paired.

Proof. For each i, let $X_i \subseteq W'$ be the orthocomplement of $\sum_{j \neq i} W_j$. Then X_i is an irreducible $K_{\varphi,\theta}$-submodule which is non-singularly paired with W_i, and $W' = \sum X_i$ is a direct sum. Moreover, $(X_i)^{M_\theta}$ is one-dimensional, and so it must be spanned by g_i. Hence $X_i = W_i'$. Q.E.D.

Remark 2.3.12. Now, by Remark 2.2.13, we know that $W_i(e^\varphi) \nsubseteq (W_i^{M_\theta})^\perp$. Hence from the above lemma, $\langle g_i, u_i \rangle \neq 0$. (Recall that $u_1, \ldots, u_{l(\gamma)}$ is a basis of X_φ with $u_i \in W_i(e^\varphi)$.) Also, from Lemma 2.3.11, $\langle g_j, u_i \rangle = 0$ if $i \neq j$. But $u_i = \beta_e(\sigma_i)$. Thus, using the definition of the map β_e (see [7], § III.3), we see that

$$0 \neq \langle g_i, u_i \rangle = \langle g_i, \beta_e(\sigma_i) \rangle = \sigma_i(g_i)(e)$$

and

$$0 = \langle g_j, u_i \rangle = \langle g_j, \beta_e(\sigma_i) \rangle = \sigma_i(g_j)(e),$$

if $i \neq j$.

Let $D_\gamma = \det\{\sigma_i(g_j)\}$. It is clear that the polynomial D_γ depends only on γ, up to non-zero scalar multiple, and that the degree of D_γ is $d(\gamma) = \sum_{i=1}^{l(\gamma)} d_i(\gamma)$.

We assert that $D_\gamma(x) \neq 0$ for every regular element $x \in \mathfrak{a}$, so that in particular, $D_\gamma \neq 0$. Indeed, for every such x, we have $\sigma_i(g_j)(x) = \langle g_j, \beta_x(\sigma_i) \rangle$. But $\beta_x(\sigma_i)$ $(i = 1, \ldots, l(\gamma))$ is a basis of $(V_\gamma)^{M_\theta}$ (see [10], Theorem 20). Since $(V_\gamma)^{M_\theta}$ and $(V_\gamma')^{M_\theta}$ are non-singularly paired under the duality between V_γ and V_γ' we have that $\det\{\langle g_j, \beta_x(\sigma_i) \rangle\} \neq 0$, proving the assertion.

Proposition 2.3.12a. There exist non-negative integers $n_\varphi(\gamma)$, for $\varphi \in \Delta_+^1$, such that $D_\gamma | \mathfrak{a} = c \prod_{\varphi \in \Delta_+^1} \varphi^{n_\varphi(\gamma)}$, where c is a non-zero scalar. In particular, $d(\gamma) = \sum_{\varphi \in \Delta_+^1} n_\varphi(\gamma)$.

Proof. An element $x \in \mathfrak{a}$ is regular if and only if $\varphi(x) \neq 0$ for every $\varphi \in \Delta_+^1$, and so $\mathfrak{a} - \mathscr{R} \cap \mathfrak{a}$ is precisely the set of zeros of $\prod_{\varphi \in \Delta_+^1} \varphi$. But from the above assertion, $D_\gamma(x) \neq 0$ for all $x \in \mathfrak{a} \cap \mathscr{R}$. Thus, by Hilbert's Nullstellensatz, $D_\gamma | \mathfrak{a}$ divides $(\prod_{\varphi \in \Delta_+^1} \varphi)^m$ for some m. Since the $\varphi \in \Delta_+^1$ are distinct prime polynomials, the result follows. Q.E.D.

We want to compute the integers $n_\varphi(\gamma)$.

First recall the definition of y. See the paragraph following Lemma 2.3.8.

Lemma 2.3.13. Let $\varphi \in \Delta_+^1$. Consider the polynomial $D_\gamma(tw_\varphi + y)$. Suppose $D_\gamma(tw_\varphi + y) = \sum a_n t^n$ and suppose a_m is the first nonvanishing coefficient. Then $m = n_\varphi(\gamma)$.

Proof. One has

$$D_\gamma(tw_\varphi+y) = c\prod_{\psi\in\Delta_+^1}(\psi(tw_\varphi+y))^{n_\psi(\gamma)} = c(\varphi(tw_\varphi+y))^{n_\varphi(\gamma)}\prod_{\psi\in\Delta_+^1-(\varphi)}(\psi(tw_\varphi+y))^{n_\psi(\gamma)}.$$

But $\varphi(w_\varphi)=1$ and $\varphi(y)=0$, whereas for $\psi\in\Delta_+^1-(\varphi)$, $\psi(y)\neq 0$.

This shows that the first nonvanishing coefficient in $D_\gamma(tw_\varphi+y)$ is that of $t^{n_\varphi(\gamma)}$. Q.E.D.

Theorem 2.3.14. *Let* $\gamma\in\Gamma$ *and* $\varphi\in\Delta_+^1$. *Let* $\delta_1,\dots,\delta_{l(\gamma)}\in\Gamma_\varphi$ *correspond to* $W_1,\dots,W_{l(\gamma)}$ *in the decomposition* $W=W_1+\cdots+W_{l(\gamma)}$. (See Theorem 2.3.7.) *Then*

$$n_\varphi(\gamma) = d(\delta_1)+\cdots+d(\delta_{l(\gamma)}).$$

Proof. Let $(S^\varphi)'$, $(J^\varphi)'$ and $(H^\varphi)'$ play the roles of S', J' and H', respectively, for the case $\mathfrak{g}_\varphi=\mathfrak{k}_\varphi+\mathfrak{p}_\varphi$.

If $f\in S'$, so that f is a polynomial function on $\mathfrak{p}$, then $f|\mathfrak{p}_\varphi$ is a polynomial function on $\mathfrak{p}_\varphi$, so that $f|\mathfrak{p}_\varphi\in(S^\varphi)'$. Let $\pi: S'\to(S^\varphi)'$ denote the restriction map. Observe that π is a $K_{\varphi,\theta}$-map.

Let $T_{-y}:(S^\varphi)'\to(S^\varphi)'$ denote the translation map defined by $T_{-y}(f)(x)=f(x+y)$. Since $\mathfrak{g}_\varphi=\mathfrak{g}^y$, $K_{\varphi,\theta}$ fixes y, and hence T_{-y} is a $K_{\varphi,\theta}$-map. Thus $T_{-y}\circ\pi: S'\to(S^\varphi)'$ is a $K_{\varphi,\theta}$-surjection. We have $\sigma_i(g_j)(tw_\varphi+y)=(T_{-y}\circ\pi)(\sigma_i g_j)(tw_\varphi)$.

Since W_i and W_i' are non-singularly paired and since W_i corresponds to $\delta_i\in\Gamma_\varphi$, W_i' corresponds to $\delta_i'\in\Gamma_\varphi$. Now choose a $K_{\varphi,\theta}$-isomorphism $\varrho_i: W_i'\to(H^\varphi)'_{\delta_i'}$. (Recall (Theorem 2.2.9) that every $\delta\in\Gamma_\varphi$ occurs in the harmonics $(H^\varphi)'$ with multiplicity one.) Let $g_i'=\varrho_i(g_i)$. Note that $(H^\varphi)'_{\delta_i'}$ consists of homogeneous harmonic polynomials of degree $d(\delta_i)$ (observe that $d(\delta_i)=d(\delta_i')$).

Assertion. There exist $p_{ij}\in(J^\varphi)'$ such that $(T_{-y}\circ\pi)(\sigma_i(g_j))=p_{ij}g_j'$.

This follows from the fact that $(S^\varphi)'=(J^\varphi)'\otimes(H^\varphi)'$ and the fact that $T_{-y}\circ\pi$ is a $K_{\varphi,\theta}$-homomorphism.

Let λ_{ij} be the constant term in the polynomial p_{ij}, i.e., $\lambda_{ij}=p_{ij}(0)=p_{ij}(e')$, for any nilpotent element $e'\in p_\varphi$ (see [10]). Since e^φ is nilpotent, we then have

$$(T_{-y}\circ\pi)(\sigma_i(g_j))(e^\varphi) = p_{ij}(e^\varphi)g_j'(e^\varphi) = \lambda_{ij}g_j'(e^\varphi). \tag{*}$$

We assert

(i) $g_j'(e^\varphi)\neq 0$, $g_j'(w_\varphi)\neq 0$.

(ii) $\lambda_{ij}=0$ if $i\neq j$ and $\lambda_{ii}\neq 0$.

Putting $i=j$ in (*), we get $\lambda_{ii}g_i'(e^\varphi)=\sigma_i(g_i)(e^\varphi+y)\neq 0$, by Remark 2.3.12, noting that $e=e^\varphi+y$ by definition. This proves that $g_i'(e^\varphi)\neq 0$ and $\lambda_{ii}\neq 0$. Since, by Remark 2.3.12, $\lambda_{ij}g_j'(e^\varphi)=\sigma_i(g_j)(e^\varphi+y)=0$ if $i\neq j$, it now follows that $\lambda_{ij}=0$.

To prove that $g_j'(w_\varphi)\neq 0$, first observe that if β_{w_φ}: $\mathrm{Hom}_{K_{\varphi,\theta}}(W_j',(H^\varphi)')\to W_j$ is the usual map for the case $\mathfrak{g}_\varphi = \mathfrak{k}_\varphi + \mathfrak{p}_\varphi$, then

$$\begin{aligned} g'_j(w_\varphi) &= \varrho_j(g_j)(w_\varphi) \\ &= \langle \beta_{w_\varphi}(\varrho_j), g_j \rangle \text{ (by the definition of } \beta_{w_\varphi}) \\ &\neq 0, \end{aligned}$$

in view of Theorem 20, [10] and the fact that $W_j^{M_\theta}$ and $(W'_j)^{M_\theta}$ are nonsingularly paired.

The following assertion leads now to the proof of Theorem 2.3.14:

Assertion. If $\sigma_i(g_j)(tw_\varphi + y) = \sum_n b_n t^n$, then the minimal m such that $b_m \neq 0$ is at least $d(\delta_j)$. It is equal to $d(\delta_j)$ if and only if $i = j$.

Proof. One has

$$\begin{aligned} \sigma_i(g_j)(tw_\varphi + y) &= (T_{-y} \circ \pi)(\sigma_i(g_j))(tw_\varphi) = p_{ij}(tw_\varphi) g'_j(tw_\varphi) = \\ &= p_{ij}(tw_\varphi) g'_j(w_\varphi) t^{d(\delta_j)}, \end{aligned}$$

since g'_j is homogeneous of degree $d(\delta_j)$. This proves the first part of the assertion.

The second part now follows from the fact that $g'_j(w_\varphi) \neq 0$ and the fact that the constant term λ_{ij} of $p_{ij}(tw_\varphi)$ is non-zero if and only if $i = j$.

Thus each term in the jth column of the matrix $\sigma_i(g_j)(tw_\varphi + y)$ is a polynomial in t divisible by $t^{d(\delta_j)+1}$ except for the diagonal term. This, however, is divisible by $t^{d(\delta_j)}$. Thus the first n such that $a_n \neq 0$, where $D_\gamma(tw_\varphi + y) = \sum a_j t^j$, equals $\sum_{j=1}^{l(\lambda)} d(\delta_j)$. This together with Lemma 2.3.13 proves Theorem 2.3.14. Q.E.D.

We now vary $\varphi \in \Delta_+^1$ and write $\delta_1^\varphi, \ldots, \delta_{l(\gamma)}^\varphi$ for the $K_{\varphi,\theta}$-modules occurring in the irreducible decomposition of the $K_{\varphi,\theta}$-module generated by $(V_\gamma)^{M_\theta}$. Recall that the degree of D_γ (and also of $D_\gamma|\mathfrak{a}$) is $d(\gamma)$. Since $d(\gamma) = \sum_{\varphi \in \Delta_+^1} n_\varphi(\gamma)$, we get the following corollary to Theorem 2.3.14:

Corollary 2.3.15. One has $d(\gamma) = \sum_{\varphi \in \Delta_+^1} \left(\sum_{i=1}^{l(\gamma)} d(\delta_i^\varphi) \right)$.

Also, since the δ's occurring in the decomposition of W' are the duals of the δ's occurring in the decomposition of W, and since $d(\delta) = d(\delta')$, we get the following corollary:

Corollary 2.3.16. One has $n_\varphi(\gamma) = n_\varphi(\gamma')$.

Let $\{U_{(k)}\}$ be the filtration in $U = U(\mathfrak{g})$, where $U_{(k)}$ is the subspace of U spanned by $\{x_1 x_2 \ldots x_l | l \leq k \text{ and } x_i \in \mathfrak{g}\}$. Fix $\gamma \in \Gamma$ and let $d_1(\gamma), \ldots, d_{l(\gamma)}(\gamma)$ be the degrees in which γ occurs in the harmonics. Let $v_1, \ldots, v_{l(\gamma)}$ be a basis of $(V_\gamma)^{M_\theta}$, and let $g'_1, \ldots, g'_{l(\gamma)}$ be the basis of $(V'_\gamma)^{M_\theta}$ which is dual to the basis $v_1, \ldots, v_{l(\gamma)}$. Let $\sigma'_1, \ldots, \sigma'_{l(\gamma)}$ be a basis of $\mathrm{Hom}_{K_\theta}(V'_\gamma, H')$ such that $\sigma'_i(g'_j) \in H'_{d_i(\gamma)}$.

Now let $\pi: S \to S'$ be the isomorphism extending the isomorphism $\pi: \mathfrak{p} \to \mathfrak{p}'$ which is induced by a non-singular symmetric K_θ-invariant bilinear form $B_\mathfrak{p}$ on $\mathfrak{p}$ which extends the Killing form on $[\mathfrak{g}, \mathfrak{g}] \cap \mathfrak{p}$, and which is positive-definite on $\mathfrak{p}_\mathbf{R}$; then it is also clear that $[\mathfrak{g}, \mathfrak{g}] \cap \mathfrak{p} \perp (\text{centre } \mathfrak{g}) \cap \mathfrak{p}$. Recall the linear isomorphism $\beta: S(\mathfrak{g}) \to U$ having the property $\beta(1)=1$ and $\beta(x^k)=x^k$ for $x \in \mathfrak{g}$. Now choose a basis $\varepsilon_1, \ldots, \varepsilon_{l(\gamma)}$ of $\mathrm{Hom}_{K_\theta}(V_\gamma, H^*)$ so that $\varepsilon_k(v_m) = \beta\pi^{-1}(\sigma_k'(g_m'))$.

Recall the map $u \to p_u$, defined with respect to the decomposition $U = \mathfrak{k}U + U(\mathfrak{a}) + U\mathfrak{n}$.

Definition 2.3.17. We define $p_\gamma = \det\{p_{\varepsilon_k(v_m)}\} \in U(\mathfrak{a})$.

Theorem 2.3.18. *One has* $p_\gamma \equiv c \prod_{\varphi \in \Delta_+^1} w_\varphi^{n_\varphi(\gamma)} \pmod{U_{(d(\gamma)-1)}}$, *where* c *is a non-zero scalar. (In particular,* $p_\gamma \neq 0$.)

Proof. First consider $\det\{\beta^{-1}(\varepsilon_k(v_m))\}$. We have $\beta^{-1}(\varepsilon_k(v_m)) = \pi^{-1}(\sigma_k'(g_m'))$, so that

$$\det\{\beta^{-1}(\varepsilon_k(v_m))\} = \pi^{-1}(\det\{\sigma_k'(g_m')\}). \tag{*}$$

Let $\mathfrak{b} \subseteq \mathfrak{p}$ be the orthogonal complement (in $\mathfrak{p}$) of $\mathfrak{a}$, with respect to $B_\mathfrak{p}$. Let $\mathfrak{b}' = \pi(\mathfrak{b})$. Let I and I' be the ideals generated by $\mathfrak{b}$ in S and by $\mathfrak{b}'$ in S', respectively. Also, let S and S' be identified with $S(\mathfrak{a}) \otimes S(\mathfrak{b})$ and $S'(\mathfrak{a}) \otimes S'(\mathfrak{b})$, respectively, so that $S(\mathfrak{a}) \subseteq S$ and $S'(\mathfrak{a}) \subseteq S'$. Note that one has $S = S(\mathfrak{a}) \oplus I$ and $S' = S'(\mathfrak{a}) \oplus I'$. It follows from Proposition 2.3.12a that

$$\det\{\sigma_k'(g_m')\} \equiv c \prod_{\varphi \in \Delta_+^1} \varphi^{n_\varphi(\gamma')} \pmod{I'} \equiv c \prod_{\varphi \in \Delta_+^1} \varphi^{n_\varphi(\gamma)} \pmod{I'}$$

where c is a non-zero scalar. Hence

$$\pi^{-1}(\det\{\sigma_k'(g_m')\}) = c \prod_{\varphi \in \Delta_+^1} (\pi^{-1}\varphi)^{n_\varphi(\gamma)} \pmod{I}.$$

By definition of π, $\pi^{-1}\varphi = \dfrac{w_\varphi}{\langle w_\varphi, w_\varphi \rangle}$. Thus

$$\pi^{-1}(\det\{\sigma_k'(g_m')\}) \equiv c \prod_{\varphi \in \Delta_+^1} w_\varphi^{n_\varphi(\gamma)} \pmod{I},$$

where c is some non-zero constant. Thus, from (*),

$$\det\{\beta^{-1}(\varepsilon_k(v_m))\} \equiv c \prod_{\varphi \in \Delta_+^1} w_\varphi^{n_\varphi(\gamma)} \pmod{I}. \tag{**}$$

Now let E be the ideal in $S(\mathfrak{g})$ generated by $\mathfrak{k} + \mathfrak{b}$, and let $S(\mathfrak{g})$ be identified with $S(\mathfrak{k}) \otimes S$, so that $S \subseteq S(\mathfrak{g})$. Then (**) implies

$$\det\{\beta^{-1}(\varepsilon_k(v_m))\} \equiv c \prod_{\varphi \in \Delta_+^1} w_\varphi^{n_\varphi(\gamma)} \pmod{E}. \tag{***}$$

Clearly, $\mathfrak{k} + \mathfrak{b}$ is orthogonal to $\mathfrak{a}$ under the Killing form, and we have $\mathfrak{g} = (\mathfrak{k} \oplus \mathfrak{b}) \oplus \mathfrak{a}$. But since $\mathfrak{g} = \mathfrak{k} \oplus \mathfrak{a} \oplus \mathfrak{n}$ and $\mathfrak{k} + \mathfrak{n}$ is orthogonal to $\mathfrak{a}$, we conclude that $\mathfrak{k} + \mathfrak{b} = \mathfrak{k} + \mathfrak{n}$. Thus E is the ideal generated by $\mathfrak{k} + \mathfrak{n}$.

For $u \in U_{(j)}$, let $[u]_j \in S_j(\mathfrak{g})$ be the unique element defined by requiring $\beta([u]_j) - u \in U_{(j-1)}$.

Observe that if $u \in U_{(j)}$, then $p_u \in U_{(j)}$. We assert that $[u]_j - [p_u]_j \in E$. To prove this, let $v = u - p_u$, so that $v \in \mathfrak{k}U + U\mathfrak{n}$. Write $v = \sum x_i u_i + \sum v_i y_i$, where $x_i \in \mathfrak{k}$, $y_i \in \mathfrak{n}$ and u_i, $v_i \in U_{(j-1)}$. We denote the product in $S(\mathfrak{g})$ by $\circ$. Writing $u_i = \beta(u_i')$ and $v_i = \beta(v_i')$, where

$$u_i', v_i' \in S_{(j-1)}(\mathfrak{g}) = \sum_{i=1}^{j-1} S_j(\mathfrak{g})$$

we have

$$v = \Sigma_i\{\beta(x_i)\beta(u_i') + \beta(v_i')\beta(y_i) - \beta(x_i \circ u_i') - \beta(v_i' \circ y_i)\} + \\ + \Sigma_i\{\beta(x_i \circ u_i') + \beta(v_i' \circ y_i)\}.$$

But $\beta(x_i)\beta(u_i') - \beta(x_i \circ u_i') \in U_{(j-1)}$, and similarly, $\beta(v_i')\beta(y_i) - \beta(v_i' \circ y_i) \in U_{(j-1)}$, so that $[v]_j = \sum_i [\beta(x_i \circ u_i')]_j + \sum_i [\beta(v_i' \circ y_i)]_j$. But writing $u_i' = t_0 + t_1 + \cdots + t_{j-1}$ and $v_i' = s_0 + s_1 + \cdots + s_{j-1}$ with t_k, $s_k \in S_k(\mathfrak{g})$ it is clear that $[\beta(x_i \circ u_i')]_j = x_i \circ t_{j-1}$ and $[\beta(v_i' \circ y_i)]_j = s_{j-1} \circ y_i$. This shows that $[v]_j \in E$, proving the assertion.

Now, observe that since $\sigma_k'(g_m') \in S'_{d_k(\gamma)}$, one has

$$[\varepsilon_k(v_m)]_{d_k(\gamma)} = [\beta\pi^{-1}(\sigma_k'(g_m'))]_{d_k(\gamma)} = \pi^{-1}(\sigma_k'(g_m')) = \beta^{-1}(\varepsilon_k(v_m)).$$

Thus, from (***),

$$\det\{[\varepsilon_k(v_m)]_{d_k(\gamma)}\} \equiv c \prod_{\varphi \in \Delta_+^1} w_\varphi^{n_\varphi(\gamma)} \,(\mathrm{mod}\, E).$$

Hence, from the assertion above,

$$\det\{[p_{\varepsilon_k(v_m)}]_{d_k(\gamma)}\} \equiv c \prod_{\varphi \in \Delta_+^1} w_\varphi^{n_\varphi(\gamma)} \,(\mathrm{mod}\, E).$$

But $p_{\varepsilon_k(v_m)} \in U(\mathfrak{a})$. Under β, the algebras $S(\mathfrak{a})$ and $U(\mathfrak{a})$ can be identified, and it is easy to see that if $u \in U(\mathfrak{a})$ is a polynomial of degree j, then $[u]_j$ is just the leading term of u. Hence

$$\det\{[p_{\varepsilon_k(v_m)}]_{d_k(\gamma)}\} = [\det\{p_{\varepsilon_k(v_m)}\}]_{d(\gamma)},$$

and so

$$[\det\{p_{\varepsilon_k(v_m)}\}]_{d(\gamma)} \equiv c \prod_{\varphi \in \Delta_+^1} w_\varphi^{n_\varphi(\gamma)} \,(\mathrm{mod}\, E).$$

But since $[\det\{p_{\varepsilon_k(v_m)}\}]_{d(\gamma)} \in S(\mathfrak{a})$, one actually has

$$[\det\{p_{\varepsilon_k(v_m)}\}]_{d(\gamma)} = c \prod_{\varphi \in \Delta_+^1} w_\varphi^{n_\varphi(\gamma)}.$$

Hence, applying β to both sides,

$$\det\{p_{\varepsilon_k(v_m)}\} \equiv c \prod_{\varphi \in \Delta_+^1} w_\varphi^{n_\varphi(\gamma)} \,(\mathrm{mod}\, U_{(d(\gamma)-1)}). \qquad \text{Q.E.D.}$$

§ 4. The product formula $p_\gamma = \Pi p_{(\varphi)}$

In this section, we will obtain a formula for p_γ in terms of its analogues in the rank one case. Let $\Sigma \subseteq \Delta_+^1$ be a system of simple roots. Fix $\alpha \in \Sigma$ and let $\Delta_+^{(\alpha)} = \Delta_+^1 - \{\alpha\}$, so that $\Delta_+^1 = \{\alpha\} \cup \Delta_+^{(\alpha)}$. Put $\mathfrak{n}_\alpha = \mathfrak{g}^\alpha + \mathfrak{g}^{2\alpha}$ and $\mathfrak{n}^\alpha = \sum_{\psi \in \Delta_+^{(\alpha)}} (\mathfrak{g}^\psi + \mathfrak{g}^{2\psi})$. Note that $\Delta_+^{(\alpha)}$ consists of all $\psi \in \Delta_+^1$ such that if $\psi = \sum_{\beta \in \Sigma} m_\beta \beta$, then for some $\beta \neq \alpha$, $m_\beta \neq 0$. From this, it is easy to conclude that $[\mathfrak{g}_\alpha, \mathfrak{n}^\alpha] \subseteq \mathfrak{n}^\alpha$.

Let $U(\mathfrak{g}_\alpha)$ denote the universal enveloping algebra of $\mathfrak{g}_\alpha$, considered as a subalgebra of U. Let S_α^*, J_α^* and H_α^* play the roles of S^*, J^* and H^* for $\mathfrak{g}_\alpha = \mathfrak{k}_\alpha + \mathfrak{p}_\alpha$. Let $U(\mathfrak{n}^\alpha)$ be the universal enveloping algebra of $\mathfrak{n}^\alpha$, considered as a subalgebra of U.

Proposition 2.4.1. The map $J_\alpha^* \otimes H_\alpha^* \otimes U(\mathfrak{n}^\alpha) \to U$ given by $u \otimes v \otimes w \to uvw$ is an injection. Identifying $J_\alpha^* \otimes H_\alpha^* \otimes U(\mathfrak{n}^\alpha)$ with its image in U, we have

$$U = \mathfrak{k}U \oplus (J_\alpha^* \otimes H_\alpha^* \otimes U(\mathfrak{n}^\alpha)),$$

thus giving a decomposition

$$U = \mathfrak{k}U \oplus J_\alpha^* H_\alpha^* \oplus J_\alpha^* H_\alpha^* U(\mathfrak{n}^\alpha)\mathfrak{n}^\alpha.$$

Proof. Let $\xi: U \to U/\mathfrak{k}U$ be the natural map. To prove the proposition, it is enough to prove that the map $J_\alpha^* \otimes H_\alpha^* \otimes U(\mathfrak{n}^\alpha) \to U/\mathfrak{k}U$ given by $u \otimes v \otimes w \to \xi(uvw)$ is a bijection.

Let $\mathfrak{s} = \mathfrak{a} + \mathfrak{n}$ and $\mathfrak{s}_\alpha = \mathfrak{a} + \mathfrak{n}_\alpha$, so that $\mathfrak{s} = \mathfrak{s}_\alpha + \mathfrak{n}^\alpha$. Let $U(\mathfrak{s})$ and $U(\mathfrak{s}_\alpha)$ be the universal enveloping algebras of $\mathfrak{s}$ and $\mathfrak{s}_\alpha$, considered as subalgebras of U. We have $U(\mathfrak{s}) = U(\mathfrak{s}_\alpha) \otimes U(\mathfrak{n}^\alpha)$. Since $U = \mathfrak{k}U \oplus U(\mathfrak{s})$, ξ maps $U(\mathfrak{s})$ isomorphically onto $U/\mathfrak{k}U$. Considering $U/\mathfrak{k}U$ as a right U-module, one has for $u, v \in U$, $\xi(uv) = \xi(u) \cdot v$. Thus we have an isomorphism $\xi_1: U(\mathfrak{s}_\alpha) \otimes U(\mathfrak{n}^\alpha) \to U/\mathfrak{k}U$, defined by $\xi_1(u \otimes v) = \xi(u) \cdot v = \xi(uv)$.

But since we have $U(\mathfrak{g}_\alpha) = \mathfrak{k}_\alpha U(\mathfrak{g}_\alpha) \oplus U(\mathfrak{s}_\alpha)$ and also $U(\mathfrak{g}_\alpha) = \mathfrak{k}_\alpha U(\mathfrak{g}_\alpha) \oplus J_\alpha^* H_\alpha^*$ (see Lemma 1.4.2), there is a bijection $\pi: J_\alpha^* H_\alpha^* \to U(\mathfrak{s}_\alpha)$ such that $u - \pi(u) \in \mathfrak{k}_\alpha U(\mathfrak{g}_\alpha)$ for $u \in J_\alpha^* H_\alpha^*$. Thus we have an isomorphism $\xi_2: J_\alpha^* H_\alpha^* \otimes U(\mathfrak{n}^\alpha) \to U/\mathfrak{k}U$, defined by $\xi_2(u \otimes v) = \xi_1(\pi u \otimes v)$. But $\xi_1(\pi u \otimes v) = \xi(\pi u) \cdot v = \xi(u) \cdot v$, and hence $\xi_2(u \otimes v) = \xi(u) \cdot v = \xi(uv)$. But we know that the map $J_\alpha^* \otimes H_\alpha^* \to J_\alpha^* H_\alpha^*$ given by $u \otimes v \to uv$ is an isomorphism.

This proves that the map $J_\alpha^* \otimes H_\alpha^* \otimes U(\mathfrak{n}^\alpha) \to U/\mathfrak{k}U$ given by $u \otimes v \otimes w \to \xi(uvw)$ is a bijection, and hence the proposition is proved. Q.E.D.

Remark 2.4.2. For $u \in U$, if u' denotes the component of u in $J_\alpha^* H_\alpha^*$ with respect to the decomposition $U = \mathfrak{k}U \oplus J_\alpha^* H_\alpha^* \oplus J_\alpha^* H_\alpha^* U(\mathfrak{n}^\alpha)\mathfrak{n}^\alpha$, then $p_u = p_{u'}$. (One has only to observe that $\mathfrak{n}^\alpha \subseteq \mathfrak{n}$.)

Now fix $\gamma \in \Gamma$. Let $v_1, \ldots, v_{l(\gamma)}$ be a basis of $\operatorname{Hom}_{K_\theta}(V_\gamma, H)$, and let $H_{\gamma,i} = v_i(V_\gamma)$, so that $H_\gamma = \sum_i H_{\gamma,i}$. We assume that $H_{\gamma,i}$ consists of homogeneous harmonics of degree $d_i(\gamma)$. Now choose a basis $\sigma_1, \ldots, \sigma_{l(\gamma)}$ of $\operatorname{Hom}_{K_\theta}(V'_\gamma, H')$ such that under the pairing between H and H', $\sigma_i(V'_\gamma)$ is orthogonal to $v_j(V_\gamma)$ if $i \neq j$ and $\sigma_i(V'_\gamma)$ is non-singularly paired to $v_i(V_\gamma)$. Furthermore, changing σ_i by a scalar multiple if necessary, we may assume that $\langle \sigma_i(g_k), v_j(v_l)\rangle = 0$ if $i \neq j$ and $\langle \sigma_i(g_k), v_i(v_l)\rangle = 1$ if $k = l$ and 0 if $k \neq l$, where $\{v_i\}$ is a basis of $(V_\gamma)^{M_\theta}$ as in Theorem 2.3.7 (but replacing $\varphi \in \Delta^1$ by $\alpha \in \Sigma$) and g_i is the dual basis of $(V'_\gamma)^{M_\theta}$.

Let $\varepsilon_1, \ldots, \varepsilon_{l(\gamma)}$ be the basis of $\operatorname{Hom}_{K_\theta}(V_\gamma, H^*)$ defined by $\varepsilon_i = \beta \circ v_i$, where β is the usual map $S(\mathfrak{g}) \to U$, and let $H^*_{\gamma,i} = \varepsilon_i(V_\gamma)$.

Let R_j be the $K_{\alpha,\theta}$-module generated by $\varepsilon_i(v_j) \in H^*_{\gamma,i}$. Clearly, R_j is isomorphic to the $K_{\alpha,\theta}$-module W_j generated by v_j. Let p_{δ_j} be the unique (up to scalar) element in $U(\mathfrak{a})$ corresponding to δ_j, where $\delta_j \in \Gamma_\alpha$ is the class of the $K_{\alpha,\theta}$-module W_j, i.e., if H^*_{α,δ_j} is the unique irreducible component of type δ_j in H^*_α, then $p_{\delta_j} = p_u$ for some non-zero $u \in (H^*_{\alpha,\delta_j})^{M_\theta}$. Let $\varkappa_\alpha$ be the element of the Weyl group W of $(\mathfrak{a}, \mathfrak{g})$ which is the reflection with respect to the hyperplane $\ker \alpha \subseteq \mathfrak{a}$, and let $\tilde{\varkappa}_\alpha$ be the element $T_{-\varrho}\varkappa_\alpha T_\varrho$ in the translated Weyl group W_T.

Proposition 2.4.3. One has that $p_{\varepsilon_i(v_j)} = p_{\delta_j} r_{ij}$, where $r_{ij} \in U(\mathfrak{a})$ is invariant under $\tilde{\varkappa}_\alpha$.

Proof. Let $\tau: U \to J^*_\alpha H^*_\alpha$ be the projection map with respect to the decomposition $U = \mathfrak{k}U \oplus J^*_\alpha H^*_\alpha \oplus J^*_\alpha H^*_\alpha U(\mathfrak{n}^\alpha)\mathfrak{n}^\alpha$. Observe that (i) for $u \in U$, $p_u = p_{\tau u}$ and (ii) for $u \in J^*_\alpha H^*_\alpha \subseteq U(\mathfrak{g}^\alpha) \subseteq U$, the operation of taking p is the same whether it is with respect to $\mathfrak{g} = \mathfrak{k} + \mathfrak{p}$ or with respect to $\mathfrak{g}_\alpha = \mathfrak{k}_\alpha + \mathfrak{p}_\alpha$. (For this, one has only to note that $\mathfrak{k}_\alpha \subseteq \mathfrak{k}$ and $\mathfrak{n}_\alpha \subseteq \mathfrak{n}$.) Moreover, note that $K_{\alpha,\theta}$ leaves $\mathfrak{k}U$, $J^*_\alpha H^*_\alpha$ and $J^*_\alpha H^*_\alpha U(\mathfrak{n}^\alpha)\mathfrak{n}^\alpha$ stable, the last being a consequence of the fact that $[\mathfrak{g}_\alpha, \mathfrak{n}^\alpha] \subseteq \mathfrak{n}^\alpha$. Thus τ is a $K_{\alpha,\theta}$-map, and so $\tau(R_j) \subseteq J^*_\alpha H^*_{\alpha,\delta_j}$. In particular, $\tau(\varepsilon_i(v_j)) = q_{ij} v'_j$, where $q_{ij} \in J^*_\alpha$ and $v'_j = \varrho(v_j)$, ϱ being a $K_{\alpha,\theta}$-isomorphism of W_j onto H^*_{α,δ_j}. Hence $p_{\varepsilon_i(v_j)} = p_{\tau(\varepsilon_i(v_j))} = p_{q_{ij}v'_j} = p_{q_{ij}} p_{v'_j}$ (see Lemma 1.2.8). We have $p_{v'_j} = p_{\delta_j}$, up to scalar.

Let ϱ_α play the role of ϱ for $\mathfrak{g}_\alpha = \mathfrak{k}_\alpha + \mathfrak{p}_\alpha$, and let $T_{\varrho_\alpha}: \mathfrak{a}' \to \mathfrak{a}'$ be the translation map $\varphi \to \varphi + \varrho_\alpha$. Note that the unique nontrivial element of $W(\mathfrak{a}, \mathfrak{g}_\alpha)$ is $\varkappa_\alpha$. By Harish–Chandra's theorem (1.3.1), $p_{q_{ij}} \in (U(\mathfrak{a}))^{T_{-\varrho_\alpha}\varkappa_\alpha T_{\varrho_\alpha}}$. But using the fact that $\varkappa_\alpha$ permutes the set of roots $\Delta_+ - \{\alpha, 2\alpha\}$, one easily sees that $T_{-\varrho_\alpha}\varkappa_\alpha T_{\varrho_\alpha} = T_{-\varrho}\varkappa_\alpha T_\varrho$, i.e., that $T_{-\varrho_\alpha}\varkappa_\alpha T_{\varrho_\alpha} = \tilde{\varkappa}_\alpha$. Setting $r_{ij} = p_{q_{ij}}$, we have the proposition. Q.E.D.

For $f \in \mathfrak{a}'$, let $T_f: U(\mathfrak{a}) \to U(\mathfrak{a})$ be the map defined by $(T_f p)(g) = p(g - f)$, for $p \in U(\mathfrak{a})$ and $g \in \mathfrak{a}'$. Clearly, for $w \in \mathfrak{a}$, $T_f w = w - \langle f, w\rangle$. For $\varphi \in \Delta^1_+$, let $\varrho_\varphi = \frac{1}{2}\{(\dim \mathfrak{g}^\varphi)\varphi + (\dim \mathfrak{g}^{2\varphi})2\varphi\}$. Note that ϱ_φ plays the role of ϱ for the case $\mathfrak{g}_\varphi = \mathfrak{k}_\varphi + \mathfrak{p}_\varphi$.

For $\varphi \in \Delta^1_+$, define $p_\varphi = p_{\delta^\varphi_1} \cdots p_{\delta^\varphi_{l(\gamma)}}$, where $\delta^\varphi_1, \ldots, \delta^\varphi_{l(\gamma)} \in \Gamma_\varphi$ correspond to the $K_{\varphi,\theta}$-irreducible components occurring in the decomposition of the $K_{\varphi,\theta}$-module generated by $(V_\gamma)^{M_\theta}$. Now define $p_{(\varphi)} = T_{\varrho_\varphi - \varrho} p_\varphi$.

Remark 2.4.4. Note that each $p_{\delta_i^\varphi}$ is a polynomial in w_φ. In particular, p_φ and hence $p_{(\varphi)}$ are polynomials in w_φ.

Notice that for $\alpha \in \sum$, $T_{\varrho_\alpha - \varrho}(w_\alpha) = w_\alpha$, since $(\varrho_\alpha - \varrho)(w_\alpha) = \varkappa_\alpha(\varrho_\alpha - \varrho)(w_\alpha) = (\varrho_\alpha - \varrho)(-w_\alpha) = 0$. Thus, in view of the above remark, $p_{(\alpha)} = p_\alpha$.

In view of Proposition 2.4.3, $p_{\delta_1^\alpha} \cdots p_{\delta_{l(\gamma)}^\alpha}$ divides $p_\gamma = \det\{p_{\varepsilon_i(v_j)}\}$, i.e., p_α divides p_γ. The following proposition generalizes this.

Proposition 2.4.5. For any $\varphi \in \Delta_+^1$, $p_{(\varphi)}$ divides p_γ.

Proof. For $\varphi \in \Delta_+^1$, define $O(\varphi) = \sum m_i$, if $\varphi = \sum_{\alpha_i \in \Sigma} m_i \alpha_i$. Note that if $O(\varphi) = 1$, then $\varphi \in \Sigma$ and so the proposition is true. Now we prove the proposition by induction on $O(\varphi)$. Thus, assume that $O(\varphi) > 1$ and that the proposition is true for all $\psi \in \Delta_+^1$ such that $O(\psi) < O(\varphi)$. Using the fact that for some $\alpha \in \Sigma$, $\langle \varphi, w_\alpha \rangle$ is positive, one can find a root $\psi \in \Delta_+^1$ such that $O(\psi) < O(\varphi)$ and $\varphi = \varkappa_\alpha \cdot \psi$ for some $\alpha \in \Sigma$. Note that $\psi \neq \alpha$. But we know that $p_\gamma = r_\alpha p_\alpha$, where r_α is invariant under the action of $\tilde{\varkappa}_\alpha$ on $U(\mathfrak{a})$; also, by induction hypothesis, $p_{(\psi)}$ divides $r_\alpha p_\alpha$. Since p_α is a polynomial in w_α, while $p_{(\psi)}$ is a polynomial in w_ψ, and since $w_\alpha \neq w_\psi$, p_α and $p_{(\psi)}$ are mutually prime. Thus $p_{(\psi)}$ divides r_α, and so $\tilde{\varkappa}_\alpha \cdot p_{(\psi)}$ divides r_α.

We assert that $\tilde{\varkappa}_\alpha \cdot p_{(\psi)} = p_{(\varphi)}$ (up to scalar). This can be seen as follows: Since $\varkappa_\alpha \cdot \psi = \varphi$, one has $\varkappa_\alpha \cdot \mathfrak{g}_\psi = \mathfrak{g}_\varphi$, $\varkappa_\alpha \cdot \mathfrak{k}_\psi = \mathfrak{k}_\varphi$ and $\varkappa_\alpha \cdot \mathfrak{p}_\psi = \mathfrak{p}_\varphi$. Moreover, $\varkappa_\alpha \cdot \mathfrak{a} = \mathfrak{a}$ and $\varkappa_\alpha \cdot \mathfrak{n}_\psi = \mathfrak{n}_\varphi$. Thus, for $u \in U(\mathfrak{g}_\psi)$, $\varkappa_\alpha \cdot p_u = p_{\varkappa_\alpha \cdot u}$. Also, $\varkappa_\alpha K_{\psi,\theta} \varkappa_\alpha^{-1} = K_{\varphi,\theta}$ and $\varkappa_\alpha M_\theta \varkappa_\alpha^{-1} = M_\theta$. Thus in the K_θ-module V_γ, $\varkappa_\alpha \cdot (V_\gamma)^{M_\theta} = (V_\gamma)^{M_\theta}$, and $\varkappa_\alpha$ maps the $K_{\psi,\theta}$-module generated by $(V_\gamma)^{M_\theta}$ isomorphically onto the $K_{\varphi,\theta}$-module generated by $(V_\gamma)^{M_\theta}$. Furthermore, it is clear that if $\sum W_{i,\psi}$ is a $K_{\psi,\theta}$-irreducible decomposition of the $K_{\psi,\theta}$-module generated by $(V_\gamma)^{M_\theta}$ and if $W_{i,\varphi} = \varkappa_\alpha \cdot W_{i,\psi}$, then $\sum W_{i,\varphi}$ is a $K_{\varphi,\theta}$-irreducible decomposition of the $K_{\varphi,\theta}$-module generated by $(V_\gamma)^{M_\theta}$. It is now easy to see that $\varkappa_\alpha \cdot p_\psi = p_\varphi$ (up to scalar).

But

$$\tilde{\varkappa}_\alpha \cdot p_{(\psi)} = T_{-\varrho} \varkappa_\alpha T_\varrho T_{\varrho_\psi - \varrho} p_\psi = T_{-\varrho} \varkappa_\alpha T_{\varrho_\psi} p_\psi = T_{-\varrho} \varkappa_\alpha T_{\varrho_\psi} \varkappa_\alpha^{-1} \varkappa_\alpha \cdot p_\psi = \\ = T_{-\varrho} T_{\varkappa_\alpha \cdot \varrho_\psi} \varkappa_\alpha \cdot p_\psi = T_{\varrho_\varphi - \varrho} p_\varphi = p_{(\varphi)}.$$

This completes the proof of the assertion.

Thus $p_{(\varphi)}$ divides p_γ. Q.E.D.

Theorem 2.4.6. *There exists a non-zero scalar c such that*

$$p_\gamma = c \prod_{\varphi \in \Delta_+^1} p_{(\varphi)}.$$

Proof. If $\varphi, \psi \in \Delta_+^1$ and $\varphi \neq \psi$, then $p_{(\varphi)}$ and $p_{(\psi)}$ are mutually prime since they are polynomials in w_φ and w_ψ, respectively. Thus $\prod_{\varphi \in \Delta_+^1} p_{(\varphi)}$ divides p_γ. Now, to show that $p_\gamma = c \prod_{\varphi \in \Delta_+^1} p_{(\varphi)}$, it is enough to show that the leading terms of p_γ and $\prod_{\varphi \in \Delta_+^1} p_{(\varphi)}$ have the same degree. By Theorem 2.3.18, the leading term of p_γ

is of degree $\sum n_\varphi(\gamma)=d(\gamma)$. On the other hand, applying Theorem 2.3.18, to the case $\mathfrak{g}_\varphi=\mathfrak{k}_\varphi+\mathfrak{p}_\varphi$ and replacing γ by δ_i^φ, one sees that the leading term of $p_{\delta_i}^\varphi$ is of degree $d(\delta_i^\varphi)$. Hence the leading term of $p_{(\varphi)}$, which is the same as the leading term of p_φ, is of degree $\sum d(\delta_i^\varphi)=n_\varphi(\gamma)$ (Theorem 2.3.14). Thus $\prod_{\varphi\in\Delta_+^1} p_{(\varphi)}$ is also of degree $\sum n_\varphi(\gamma)=d(\gamma)$. Q.E.D.

Remark 2.4.7. Let $\varphi\in\Delta_+^1$, and suppose δ^φ is any $K_{\varphi,\theta}$-irreducible representation with an M_θ-fixed vector. Then for some $\gamma\in\Gamma$, δ^φ occurs in the $K_{\varphi,\theta}$-module generated by $(V_\gamma)^{M_\theta}$.

To see this, let $W_{\delta^\varphi}\subseteqq H_\varphi$ be the unique $K_{\varphi,\theta}$-irreducible module of type δ^φ. Since H_φ (resp., H) is spanned by powers of nilpotent elements in $\mathfrak{p}_\varphi$ (resp., $\mathfrak{p}$), and since any nilpotent element in $\mathfrak{p}_\varphi$ is a nilpotent element of $\mathfrak{p}$, we have $H_\varphi\subseteqq H$. Let $v\in(W_{\delta^\varphi})^{M_\theta}$ be a non-zero element and choose some irreducible K_θ-module $V_\gamma\subseteqq H$ such that v has a non-zero component, say v_0, in V_γ. Now observe that $v_0\in(V_\gamma)^{M_\theta}$ and that the $K_{\varphi,\theta}$-module generated by v_0 is irreducible of type δ^φ.

As a consequence, we have the following corollary to Theorem 2.4.6:

Corollary 2.4.8. For $\lambda\in\mathfrak{a}'$, $p_\gamma(\lambda)\neq 0$ for every $\gamma\in\Gamma$ if and only if $(T_{\varrho_\varphi}p_{\delta^\varphi})(\mu)\neq 0$ for every $\varphi\in\Delta_+^1$ and every $\delta^\varphi\in\Gamma_\varphi$, where $\mu=\lambda+\varrho$.

Proof. Remark 2.4.7 and Theorem 2.4.6 imply that $p_\gamma(\lambda)\neq 0$ for every γ if and only if $(T_{\varrho_\varphi-\varrho}p_{\delta^\varphi})(\lambda)\neq 0$ for every $\varphi\in\Delta_+^1$ and every $\delta^\varphi\in\Gamma_\varphi$. But

$$(T_{\varrho_\varphi-\varrho}p_{\delta^\varphi})(\lambda)=(T_{\varrho_\varphi}p_{\delta^\varphi})(\lambda+\varrho)=(T_{\varrho_\varphi}p_{\delta^\varphi})(\mu).$$

Q.E.D.

§ 5. The expression for p_u using the Hopf structure

Theorem 2.4.6 of the last section shows that for any $\gamma\in\Gamma$, the polynomial p_γ can be expressed in terms of similar polynomials in the rank 1 case. We will now compute explicitly the polynomials p_γ in the rank 1 case.

Let $\delta: U\to U\otimes U$ be the algebra homomorphism with the property that for any $x\in\mathfrak{g}$, $\delta(x)=x\otimes 1+1\otimes x$. Recall that if $U(\mathfrak{g}\oplus\mathfrak{g})$ is the universal enveloping algebra of $\mathfrak{g}\oplus\mathfrak{g}$ and $\varrho: U(\mathfrak{g}\oplus\mathfrak{g})\to U\otimes U$ is the homomorphism extending the Lie algebra homomorphism of $\mathfrak{g}\oplus\mathfrak{g}$ into $U\otimes U$ defined by $(x,y)\to x\otimes 1+1\otimes y$, then ϱ is an isomorphism. Composing ϱ with the map $U\to U(\mathfrak{g}\oplus\mathfrak{g})$ which is induced by the inclusion map $x\to(x,x)$ of $\mathfrak{g}$ into $\mathfrak{g}\oplus\mathfrak{g}$, we get the map δ. Let $x_1,\ldots,x_k\in\mathfrak{g}$. If S is a subset of $(1,2,\ldots,k)$ let $X_S=x_{i_1}x_{i_2}\cdots x_{i_m}$, where $i_1,\ldots,i_m$ are the elements of S written in increasing order. (If S is empty, $X_S=1$.) Then it can be easily checked that

$$\delta(x_1\cdots x_k)=\sum_{S\subseteqq(1,\ldots,k)} X_S\otimes X_{S^c}, \qquad (*)$$

where S^c denotes the complement of S in $(1,\ldots,k)$. In particular, note that for

$x \in \mathfrak{g}$, one has

$$\delta(x^k) = \sum_{r=0}^{k} \binom{k}{r} x^r \otimes x^{k-r}, \text{ where } \binom{k}{r} = \frac{k!}{r!(k-r)!}.$$

Remark. 2.5.1. Consider $U(\mathfrak{a})$ as the algebra of polynomial functions on $\mathfrak{a}'$. Let $p \in U(\mathfrak{a})$ and let $\delta(p) = \sum p_i' \otimes p_i''$, where $p_i', p_i'' \in U(\mathfrak{a})$. Then for any λ, $\nu \in \mathfrak{a}'$, one has

$$p(\lambda + \nu) = \sum p_i'(\lambda) p_i''(\nu).$$

To prove this, let $\alpha_1 : U(\mathfrak{a}) \otimes U(\mathfrak{a}) \to \mathbf{C}$ be the algebra homomorphism defined by $\alpha_1(f \otimes g) = f(\lambda) g(\nu)$. Then $\alpha_1 \circ \delta : U(\mathfrak{a}) \to C$ is an algebra homomorphism, and for $x \in \mathfrak{a}$ we have $(\alpha_1 \circ \delta)(x) = \alpha_1(x \otimes 1 + 1 \otimes x) = x(\lambda) + x(\nu) = x(\lambda + \nu)$. Thus $\alpha_1 \circ \delta$ coincides on $\mathfrak{a}$ with the algebra homomorphism $\alpha_2 : U(\mathfrak{a}) \to C$ defined by $p \to p(\lambda + \nu)$. Hence one has $\alpha_2 = \alpha_1 \circ \delta$.

Remark 2.5.2. Let $U_{(0)} \subset U_{(1)} \subseteqq \cdots \subseteqq U_{(k)} \subseteqq \cdots$ be the filtration in U, and define a filtration $(U \otimes U)_{(0)} \subseteqq (U \otimes U)_{(1)} \subseteqq \cdots \subseteqq (U \otimes U)_{(l)} \subseteqq \cdots$ in $U \otimes U$ by $(U \otimes U)_{(l)} = \sum_i U_{(i)} \otimes U_{(l-i)}$. (Note that this is just the filtration of the enveloping algebra $U(\mathfrak{g} \oplus \mathfrak{g})$ when identified with $U \otimes U$.) It is immediate from $(*)$ that $\delta(U_{(k)}) \subseteqq (U \otimes U)_{(k)}$.

Lemma 2.5.3. Let $u \in U$ and let $\delta(u) = \sum_i u_i' \otimes u_i''$.

Then

$$\delta(p_u) = \sum_i p_{u_i'} \otimes p_{u_i''}.$$

Proof. Let $U_0 = \mathfrak{k}U + U\mathfrak{n}$, so that $U = U_0 \oplus U(\mathfrak{a})$. Then one has $U \otimes U = (U_0 \otimes U_0 + U(\mathfrak{a}) \otimes U_0 + U_0 \otimes U(\mathfrak{a})) \oplus U(\mathfrak{a}) \otimes U(\mathfrak{a})$. Let Γ denote the projection map $U \otimes U \to U(\mathfrak{a}) \otimes U(\mathfrak{a})$. Then one clearly has $\Gamma(u \otimes v) = p_u \otimes p_v$ for $u, v \in U$.

Using $(*)$, it is obvious that $\delta(U_0) \subseteqq U_0 \otimes U + U \otimes U_0$. Hence for $u \in U$, $\Gamma(\delta(u - p_u)) = 0$, i.e., $\Gamma(\delta(u)) = \Gamma(\delta(p_u))$. But observe that $\delta(p_u) \in U(\mathfrak{a}) \otimes U(\mathfrak{a})$, so that $\Gamma(\delta(p_u)) = \delta(p_u)$. Thus,

$$\delta(p_u) = \Gamma(\delta(u)) = \Gamma(\sum u_i' \otimes u_i'') = \sum p_{u_i'} \otimes p_{u_i''}. \qquad \text{Q.E.D.}$$

From Remark 2.5.1 we now have the following:

Corollary 2.5.4. If $u \in U$ and λ, $\nu \in \mathfrak{a}'$, and if

$$\delta(u) = \sum u_i' \otimes u_i'', \text{ then } p_u(\lambda + \nu) = \sum p_{u_i'}(\lambda) p_{u_i''}(\nu).$$

Now consider the algebra homomorphism δ_k: $U \to \underbrace{U \otimes U \otimes \cdots \otimes U}_{k \text{ times}}$ defined by setting $\delta_k(x) = x \otimes 1 \otimes \cdots \otimes 1 + 1 \otimes x \otimes 1 \otimes \cdots \otimes 1 + \cdots + 1 \otimes 1 \otimes \cdots \otimes 1 \otimes x$, for $x \in \mathfrak{g}$. (Identifying $\underbrace{U \otimes U \otimes \cdots \otimes U}_{k \text{ times}}$ with the universal enveloping algebra of $\underbrace{\mathfrak{g} \oplus \mathfrak{g} \oplus \cdots \oplus \mathfrak{g}}_{k \text{ times}}$, δ_k is the homomorphism induced by $x \to (x, x, \ldots, x)$.) Then it is easily proved that if $\mathfrak{b}$ is any subalgebra of $\mathfrak{g}$, and if $U(\mathfrak{b})$ is the universal enveloping algebra of $\mathfrak{b}$ then $\delta_k(U(\mathfrak{b}))\subseteq U(\mathfrak{b})\otimes\cdots\otimes U(\mathfrak{b})$. In particular, $\delta_k(U(\mathfrak{a}))\subseteq U(\mathfrak{a})\otimes \otimes U(\mathfrak{a})\otimes\cdots\otimes U(\mathfrak{a})$. Let $U_+=\mathfrak{g}U$, so that $U=U_+\otimes \mathbf{C}\cdot 1$, and $U\otimes U\otimes\cdots\otimes U= =\{\mathbf{C}\cdot 1\otimes U\otimes\cdots\otimes U+U\otimes\mathbf{C}\cdot 1\otimes\cdots\otimes U+\cdots+U\otimes U\otimes\cdots\otimes U\otimes \mathbf{C}\cdot 1\}\oplus U_+\otimes \otimes U_+\otimes\cdots\otimes U_+$. Now let $\delta_{k,+}: U\to U_+\otimes U_+\otimes\cdots\otimes U_+$ be the map defined by $\delta_{k,+}(u)$=component of $\delta_k(u)$ in $U_+\otimes U_+\otimes\cdots\otimes U_+$ with respect to the above decomposition.

Let $v\in\mathfrak{a}'$, and define a linear map $\boldsymbol{l}_{v,k}:\underbrace{U\otimes U\otimes\cdots\otimes U}_{k \text{ times}}\to \mathbf{C}$ by

$$\boldsymbol{l}_{v,k}(u_1\otimes\cdots\otimes u_k)=\prod_i p_{u_i}(v). \text{ Let } S_{v,k}=\boldsymbol{l}_{v,k}\circ\delta_{k,+}.$$

Proposition 2.5.5. Let $u\in U_{(n)}$ and let $v\in\mathfrak{a}'$. Then for all scalars t,

$$p_u(tv)=p_u(0)+\sum_{m=1}^{n} S_{v,m}(u)\binom{t}{m},$$

where

$$\binom{t}{m}=\frac{t(t-1)\dots(t-m+1)}{m!}.$$

Proof. Arguing as in Remark 2.5.1, one proves easily that for $p\in U(\mathfrak{a})$ and $v\in\mathfrak{a}'$, one has $p(kv)=\boldsymbol{l}_{v,k}(\delta_k(p))$. The extension of Lemma 2.5.3, namely, that if $u\in U$ and $\delta_k(u)=\sum_i u_i^{(1)}\otimes u_i^{(2)}\otimes\cdots\otimes u_i^{(k)}$, where $u_i^{(1)},\dots,u_i^{(k)}\in U$, then $\delta_k(p_u)= =\sum_i p_{u_i^1}\otimes p_{u_i^2}\otimes\cdots\otimes p_{u_i^k}$, is also proved along the same lines as the proof of Lemma 2.5.3. Combining one gets for $u\in U$,

$$p_u(kv)=\boldsymbol{l}_{v,k}(\delta_k(u)). \tag{I}$$

Let S be a subset of $(1,2,\dots,k)$. For $i\in(1,2,\dots,k)$, let $U_+(i)$ (resp. $U(i)$) denote U_+ (resp. U) if $i\in S$ and $\mathbf{C}\cdot 1$ (resp. $\mathbf{C}\cdot 1$) if $i\notin S$. Now let $U_S=U(1)\otimes \otimes U(2)\otimes\cdots\otimes U(k)$ and let $U_{S,+}=U_+(1)\otimes U_+(2)\otimes\cdots\otimes U_+(k)$. Note that if S consists of m elements then U_S is isomorphic to $\underbrace{U\otimes U\otimes\cdots\otimes U}_{m \text{ times}}$ in the obvious way.

One has $\underbrace{U\otimes U\otimes\cdots\otimes U}_{k \text{ times}}=\underbrace{(U_+\oplus\mathbf{C}\cdot 1)\otimes\cdots\otimes(U_+\oplus\mathbf{C}\cdot 1)}_{k \text{ times}}=\sum_{S\subseteq(1,\dots,k)} U_{S,+}$ (direct sum). Note that $U_S=\sum_{T\subseteq S} U_{T,+}$ (direct sum). Now, for $v\in\underbrace{U\otimes U\otimes\cdots\otimes U}_{k \text{ times}}$, let v_S denote the component in U_S and let $v_{S,+}$ denote the component in $U_{S,+}$. We have, clearly, $\boldsymbol{l}_{v,k}(v)=\sum_{S\subseteq(1,\dots,k)}\boldsymbol{l}_{v,k}(v_{S,+})$.

Now suppose S consists of m elements $(0<m\leqq k)$. Let U_S be identified with $\underbrace{U\otimes U\otimes\cdots\otimes U}_{m \text{ times}}$ in the obvious way and let $\eta: U_S\to\underbrace{U\otimes U\otimes\cdots\otimes U}_{m \text{ times}}$ be the precise isomorphism making this identification. (For example, when $S=(1,2,\dots,m)$, $U_S=\underbrace{U\otimes U\otimes\cdots\otimes U}_{m \text{ times}}\otimes\underbrace{\mathbf{C}\cdot 1\otimes\cdots\otimes\mathbf{C}\cdot 1}_{k-m \text{ times}}$, and the map η is given by $u_1\otimes u_2\otimes\cdots$

$\cdots \otimes u_m \otimes 1 \otimes \cdots \otimes 1 \to u_1 \otimes u_2 \otimes \cdots \otimes u_m$.) With this identification, it is obvious from the definition that for $v_S \in U_S$,

$$\boldsymbol{l}_{v,m}(\eta(v_S)) = \boldsymbol{l}_{v,k}(v_S). \tag{II}$$

Now let $u \in U$ and let $v = \delta_k(u)$. We assert that

$$\eta(v_S) = \delta_m(u). \tag{III}$$

We prove this when $S = (1, 2, \ldots, m)$, and the method is the same for any S. The map $\eta: U_S \to \underbrace{U \otimes U \otimes \cdots \otimes U}_{m \text{ times}}$ is defined by $u_1 \otimes u_2 \otimes \cdots \otimes u_m \otimes 1 \otimes \cdots \otimes 1 \to$ $\to u_1 \otimes u_2 \otimes \cdots \otimes u_m$. Let $x \in \mathfrak{g}$. If $v = \delta_k(x)$, then $v = x \otimes 1 \otimes \cdots \otimes 1 + 1 \otimes x \otimes \cdots \otimes 1 +$ $+ \cdots + 1 \otimes 1 \otimes \cdots \otimes x \in \underbrace{U \otimes U \otimes \cdots \otimes U}_{k \text{ times}}$. Then v_S is the sum of the first m terms of the above equality, and $\eta(v_S) = x \otimes 1 \otimes \cdots \otimes 1 + 1 \otimes x \otimes \cdots \otimes 1 + \cdots + 1 \otimes 1 \otimes \cdots \otimes x \in$ $\in \underbrace{U \otimes U \otimes \cdots \otimes U}_{m \text{ times}}$, which by definition of δ_m equals $\delta_m(x)$. This proves the assertion since both sides of III define homomorphisms of U.

Now let $u \in U$ and let $v = \delta_k(u)$. Then by (I),

$$p_u(kv) = \boldsymbol{l}_{v,k}(\delta_k(u)) = \boldsymbol{l}_{v,k}(v) = \sum_{S \subseteqq (1, \ldots, k)} \boldsymbol{l}_{v,k}(v_{S,+}). \tag{$*$}$$

Now, for a fixed m such that $0 < m \leqq k$, there are $\binom{k}{m}$ subsets $S \subseteqq (1, 2, \ldots, k)$ such that S consists of m elements. Each such S contributes to the right-hand side of $(*)$ a term $\boldsymbol{l}_{v,k}(v_{S,+})$, which by (II) equals $\boldsymbol{l}_{v,m}(\eta(v_{S,+}))$. But by (III), $\delta_m(u) = \eta(v_S)$, and from this one easily sees that $\delta_{m,+}(u) = \eta(v_{S,+})$. Thus, the contribution to the right-hand side of $(*)$ from each subset S of m elements is $\boldsymbol{l}_{v,m}(\delta_{m,+}(u)) = S_{v,m}(u)$. Note that when S is the empty set the contribution is the constant term $p_u(0)$ of u. Then from $(*)$ one sees that

$$p_u(kv) = p_u(0) + \sum_{0 < m \leqq k} S_{v,m}(u) \binom{k}{m}. \tag{$**$}$$

Now since $u \in U_{(n)}$, one has that for any k,

$$\delta_k(u) \in \sum U_{(i_1)} \otimes U_{(i_2)} \otimes \cdots \otimes U_{(i_k)},$$

the summation being taken over all k-tuples $\{i_1, \ldots, i_k\}$ of integers $\geqq 0$ such that $i_1 + i_2 + \cdots + i_k = n$. In particular, if $k > n$ and $S \subseteqq (1, 2, \ldots, k)$ is a subset containing more than n elements, then $v_{S,+} = 0$, where $v = \delta_k(u)$. Hence $S_{v,m}(u) = 0$ for $m > n$. Thus for all $k > n$, from $(**)$,

$$p_u(kv) = p_u(0) + \sum_{0 < m \leqq n} S_{v,m}(u) \binom{k}{m}.$$

Writing t for k, the two expressions above are polynomials of degree n in t. Since they coincide for all integers $k>n$, we conclude that

$$p_u(tv) = p_u(0) + \sum_{0<m\leqq n} S_{v,m}(u)\binom{t}{m} \text{ for all } t. \qquad \text{Q.E.D.}$$

Remark 2.5.6. Now let us specialize Proposition 2.5.5 to the rank one case. Assume $u \in U([\mathfrak{g},\mathfrak{g}]) \cap U_{(n)}$. Then $p_u \in U(\mathfrak{a} \cap [\mathfrak{g},\mathfrak{g}])$. But $U(\mathfrak{a} \cap [\mathfrak{g},\mathfrak{g}])$ is isomorphic to the polynomial ring $\mathbf{C}[w]$ where $0 \neq w \in \mathfrak{a} \cap [\mathfrak{g},\mathfrak{g}]$. Thus if $v \in \mathfrak{a}'$ where $\langle v, w\rangle \neq 0$, any element p in $U(\mathfrak{a} \cap [\mathfrak{g},\mathfrak{g}])$ is determined by the polynomial function $t \to p(tv)$. However the polynomial function $t \to \binom{t}{m}$ is achieved by the binomial element $\left(\begin{matrix}\frac{w}{\langle v, w\rangle} \\ m\end{matrix}\right)$ in $U(\mathfrak{a} \cap [\mathfrak{g},\mathfrak{g}])$. Thus from Proposition 2.5.5 we have the formula

$$p_u = p_u(0) + \sum_{0<m\leqq n} S_{v,m}(u)\left(\begin{matrix}\frac{w}{\langle v, w\rangle} \\ m\end{matrix}\right).$$

§ 6. Parametrizing finite dimensional representations of class one

In this section we investigate all finite-dimensional irreducible G-modules with a K_θ-fixed vector, where $G = \mathrm{Ad}\,\mathfrak{g}$, $\mathfrak{g} = \mathfrak{k} + \mathfrak{p}$, $K = \exp \mathrm{ad}\,\mathfrak{k}$ and $\mathfrak{a} \subseteqq \mathfrak{p}$ are as usual. Let $\mathfrak{m}$ be, as usual, the centralizer of $\mathfrak{a}$ in $\mathfrak{k}$, and let $\mathfrak{h}_m \subseteqq \mathfrak{m}$ be a Cartan subalgebra of $\mathfrak{m}$, so that $\mathfrak{h} = \mathfrak{a} + \mathfrak{h}_m$ is a Cartan subalgebra of $\mathfrak{g}$. Let $\Delta(\mathfrak{h},\mathfrak{g})$ be the set of roots of $\mathfrak{g}$ with respect to $\mathfrak{h}$. Let $\bar{\Delta} = \{\psi \in \Delta(\mathfrak{h},\mathfrak{g}) \mid \psi|\mathfrak{a} \neq 0\}$. Then for $\psi \in \bar{\Delta}$, $\psi|\mathfrak{a} \in \Delta(\mathfrak{a},\mathfrak{g}) = \Delta$ and the map $\bar{\Delta} \to \Delta$ defined by $\psi \to \psi|\mathfrak{a}$ is a surjection. We may choose the system of positive roots $\Delta_+(\mathfrak{h},\mathfrak{g})$ in such a way that for $\psi \in \Delta_+(\mathfrak{h},\mathfrak{g}) \cap \bar{\Delta}$, we have $\psi|\mathfrak{a} \in \Delta_+$. Now for any $\psi \in \Delta_+(\mathfrak{h},\mathfrak{g})$, let $h_\psi \in [\mathfrak{g}^\psi, \mathfrak{g}^{-\psi}] \subseteqq \mathfrak{h}$ be the unique element defined by $\psi(h_\psi) = 2$. Let I be the subgroup of $\mathfrak{h}'$ ($=$dual of $\mathfrak{h}$) generated by $\Delta(\mathfrak{h},\mathfrak{g})$, and let $I_+ = \{v \in I \mid \langle v, h_\psi\rangle \in \mathbf{Z}_+ \text{ for all } \psi \in \Delta_+(\mathfrak{h},\mathfrak{g})\}$. From the Cartan—Weyl theory of representations of $\mathfrak{g}$, one knows that given any $\mu \in I_+$, there exists one and only one irreducible finite-dimensional (complex) representation of G with dominant weight μ. For every $\mu \in I_+$, let W_μ be the corresponding irreducible finite-dimensional G-module and let $\eta_\mu \in W$ be the highest weight vector, i.e., η_μ is the unique (up to scalar) non-zero element of W_μ such that for $y \in \mathfrak{h}$, $y \cdot \eta_\mu = \mu(y)\eta_\mu$. Let $I'_+ = \{\mu \in I_+ \mid (W_\mu)^{K_\theta} \neq 0\}$.

Theorem 2.6.1 *(Helgason). Let $\mu \in I_+$. Let W_μ be the finite-dimensional irreducible G-module with highest weight μ, and let $\eta_\mu \in W_\mu$ be the highest weight vector. Then $\mu \in I'_+$ if and only if η_μ is M_θ-fixed.*

Proof. Assume M_θ fixes η_μ. To prove that $W_\mu^{K_\theta} \neq \{0\}$, it suffices to show that the K_θ-fixed vector $\int_{K_{\theta,\mathbf{R}}} a \cdot \eta_\mu \, da$ is not zero. Hence it is enough to show that

$\langle a\cdot\eta_\mu, \eta'\rangle \geqq 0$ for all $a\in K_{\theta,\mathbf{R}}$, where η' is a lowest weight vector in the dual module W'_μ of W_μ such that $\langle\eta_\mu, \eta'\rangle = 1$.

From the Gelfand—Bruhat decomposition and using the notation of § 1.7 it follows that $C=(N_-)_{\mathbf{R}} M_{\theta,\mathbf{R}} A^0_{\mathbf{R}} N_{\mathbf{R}} \subseteqq G_{\mathbf{R}}$ is a dense open set in $G_{\mathbf{R}}$ where $N_- = \theta N\theta$. Hence it follows from the Iwasawa decomposition (by projecting on $G_{\mathbf{R}}/A^0_{\mathbf{R}}N_{\mathbf{R}}$) that $C\cap K_{\theta,\mathbf{R}}$ is a dense open set in $K_{\theta,\mathbf{R}}$. Now suppose $a\in C\cap K_{\theta,\mathbf{R}}$. Let $a=n_1 m(\exp \operatorname{ad} y)n_2$ where $n_1\in(N_-)_{\mathbf{R}}$, $m\in M_{\theta,\mathbf{R}}$, $y\in\mathfrak{a}_{\mathbf{R}}$ and $n_2\in N_{\mathbf{R}}$. Then

$$\langle a\cdot\eta_\mu, \eta'\rangle = \langle n_1 m(\exp \operatorname{ad} y)n_2\cdot\eta_\mu, \eta'\rangle = \langle m(\exp \operatorname{ad} y)n_2\cdot\eta_\mu, n_1^{-1}\cdot\eta'\rangle =$$

$$= \langle m(\exp \operatorname{ad} y)\cdot\eta_\mu, \eta'\rangle = e^{\langle y,\mu\rangle}\langle m\cdot\eta_\mu, \eta'\rangle = e^{\langle y,\mu\rangle}\langle\eta_\mu, \eta'\rangle = e^{\langle y,\mu\rangle} > 0$$

since $\langle y, \mu\rangle$ is real. This shows that $W_\mu^{K_\theta}\neq 0$.

Conversely, assume that $W_\mu^{K_\theta}\neq 0$. One knows that the $M_\theta A$-module generated by η_μ is irreducible. (Note that $M_\theta A = MA$ is connected.) But A acts as scalars on this module, since A is in the center of $M_\theta A$. Hence $M_\theta\cdot\eta_\mu$ spans an irreducible M_θ-module.

Let $E: W_\mu \to W_\mu^{K_\theta}$ be the projection with respect to the decomposition of W_μ into K_θ-primary subspaces. Then $E\eta_\mu\neq 0$, since η_μ is a cyclic vector for W_μ regarded as a K_θ-module. Hence $M_\theta\cdot\eta_\mu$ spans a non-zero irreducible M_θ-module of the same type as that spanned by $M_\theta\cdot E\eta_\mu$. But $E\eta_\mu$ is fixed by M_θ, and hence so is η_μ. Q.E.D.

Now, let $J\subseteqq\mathfrak{a}'$ be the subgroup generated by Δ. Then, clearly, for any $\mu\in I$, $\mu|\mathfrak{a}\in J$. For $\varphi\in\Delta^1_+$ let $w_\varphi\in\mathfrak{a}$, $e_\varphi\in\mathfrak{g}^\varphi$ and $e_{-\varphi}\in\mathfrak{g}^{-\varphi}$ be chosen as usual, see the definition of $\mathfrak{g}^*$ in [10], so that $(2w_\varphi, 2e_\varphi, 2e_{-\varphi})$ forms an S-triple. Now let $J_+ = \{\lambda\in J | \langle\lambda, w_\varphi\rangle \geqq 0 \text{ for all } \varphi\in\Delta^1_+\}$.

Remark. We assert that if $\mu\in I_+$, then $\mu|\mathfrak{a}\in J_+$. Indeed, let $\eta_\mu\in W_\mu$ be as usual, so that $\mathfrak{n}'\cdot\eta_\mu = 0$, where $\mathfrak{n}' = \sum_{\psi\in\Delta_+(\mathfrak{h},\mathfrak{g})} \mathfrak{g}^\psi$. Note that $\mathfrak{n}\subseteqq\mathfrak{n}'$, so that $\mathfrak{n}\cdot\eta_\mu=0$, and so $e_\varphi\cdot\eta_\mu = 0$ for all $\varphi\in\Delta^1_+$. Then from the representation theory of a *TDS*, $(2w_\varphi)\cdot\eta_\mu = m\eta_\mu$, where m is a non-negative integer. Thus $\langle\mu, w_\varphi\rangle\geqq 0$, which shows that $\mu|\mathfrak{a}\in J_+$.

Theorem 2.6.2. *Let* $\lambda\in I_+$. *Then* $\lambda\in I'_+$ *if and only if* (i) $\lambda|\mathfrak{h}_m = 0$ *and* (ii) $\lambda|\mathfrak{a}\in 2J_+$.

Proof. Let $(\alpha_1, \alpha_2, \ldots, \alpha_{r_0})$ be the system of simple roots in Δ_+ and let $(x_1, x_2, \ldots, x_{r_0})$ be the dual basis of $\mathfrak{a}\cap[\mathfrak{g},\mathfrak{g}]$. Let $A_i = \exp \operatorname{ad}(\mathbf{C}x_i)\subseteqq A$, so that one has $A = A_1\times\cdots\times A_{r_0}$. (Recall that the map $A\to\mathbf{C}^*\times\cdots\times\mathbf{C}^*$ defined by $a\to(a^{\alpha_1}, \ldots, a^{\alpha_{r_0}})$ is a bijection, where $a^{\alpha_i} = e^{\langle\alpha_i, x\rangle}$ if $a = \exp \operatorname{ad} x$ $(x\in\mathfrak{a})$.) Recall that F is generated by $\{\tau_j\}$, where $\tau_j = \exp \operatorname{ad}(i\pi x_j)$.

Now suppose $\lambda\in I'_+$, so that by Theorem 2.6.1, M_θ fixes η_λ. This implies that 1) F fixes η_λ and 2) for every $y\in\mathfrak{h}_m$, $\langle\lambda, y\rangle = 0$. Then 1) implies that τ_j fixes η_λ for every j, i.e., $\exp \operatorname{ad}(i\pi x_j)\cdot\eta_\lambda = \eta_\lambda$ for every j; hence, $\langle\lambda, x_j\rangle\in 2\mathbf{Z}$ for every j, i. e., $\lambda|\mathfrak{a}\in 2J$. Thus, by the remark preceding the theorem, $\lambda|\mathfrak{a}\in 2J\cap J_+ = 2J_+$.

Conversely, assume that $\lambda \in I_+$ and that (i) $\lambda | \mathfrak{h}_m = 0$ and (ii) $\lambda | \mathfrak{a} \in 2J_+$.

Let $\Delta_0 = \{\psi \in \Delta(\mathfrak{h}, \mathfrak{g}) | \psi | \mathfrak{a} = 0\}$. Clearly, $\mathfrak{m} = \sum_{\psi \in \Delta_0} \mathfrak{g}^\psi$, and hence the restriction map $\psi \to \psi | \mathfrak{h}_m$ maps Δ_0 bijectively on $\Delta(\mathfrak{h}_m, \mathfrak{m})$. Then choosing $\Delta_+(\mathfrak{h}_m, \mathfrak{m}) = \{\psi | \mathfrak{h}_m | \psi \in \Delta_0 \cap \Delta_+(\mathfrak{h}, \mathfrak{g})\}$, it is clear that $\Delta_+(\mathfrak{h}_m, \mathfrak{m})$ is a system of positive roots of $\Delta(\mathfrak{h}_m, \mathfrak{m})$.

Let V be the MA-module generated by η_λ. Then V is an irreducible MA-module (by Remark 1.2.3); in fact, V is irreducible even as an M-module since A acts as scalars on V. We assert that $\lambda | \mathfrak{h}_m$ is the highest weight of $\mathfrak{h}_m$ in V. Since $\mathfrak{m} = \sum_{\psi \in \Delta_0} \mathfrak{g}^\psi$, V is the span of the set

$$\{e_{-\psi_1} e_{-\psi_2} \cdots e_{-\psi_j}(\eta_\lambda) | \psi_i \in \Delta_0 \cap \Delta_+(\mathfrak{h}, \mathfrak{g}), e_{-\psi_i} \in \mathfrak{g}^{-\psi_i}\}.$$

Note that if $\mu = \lambda - (\psi_1 + \cdots + \psi_j)$, then $e_{-\psi_1} \cdots e_{-\psi_j}(\eta_\lambda) = 0$ if μ is not a weight of $\mathfrak{h}$ in W_λ and $e_{-\psi_1} \cdots e_{-\psi_j}(\eta_\lambda) \in (W_\lambda)^\mu$ if μ is a weight. This implies that the weights of $\mathfrak{h}_m$ in V are of the form $(\lambda - (\psi_1 + \cdots + \psi_j)) | \mathfrak{h}_m$, where $\psi_i \in \Delta_0 \cap \Delta_+(\mathfrak{h}, \mathfrak{g})$, i.e., of the form $(\lambda | \mathfrak{h}_m) - (\psi_1 + \psi_2 + \cdots + \psi_j)$, where $\psi_i \in \Delta_+(\mathfrak{h}_m, \mathfrak{m})$. Thus, relative to $\Delta_+(\mathfrak{h}_m, \mathfrak{m})$, $\lambda | \mathfrak{h}_m$ is the highest weight of $\mathfrak{h}_m$ in V.

Hence, the assumption (i), namely that $\lambda | \mathfrak{h}_m = 0$, implies that V is the trivial one-dimensional M-module, i.e., that η_λ is fixed by M. Now, by assumption (ii), namely that $\lambda | \mathfrak{a} \in 2J_+$, it follows that $\tau_j^\lambda (= e^{\langle \lambda, i\pi x_j \rangle}) = 1$ for every j, and hence that $\sigma^\lambda = 1$ for $\sigma \in F$, since the τ_j generate F. Since for any $a \in A$, $a \cdot \eta_\lambda = a^\lambda \eta_\lambda$, this implies that F fixes η_λ. Thus M_θ fixes η_λ, since $M_\theta = MF$. The theorem now follows from Theorem 2.6.1. Q.E.D.

Remark 2.6.3. Note that Theorem 2.6.2 implies in particular that the map $\lambda \to \lambda | \mathfrak{a}$ of I'_+ into $2J_+$ is an injection. We assert that it is actually a bijection. To see this, let $H_m = \exp \operatorname{ad} \mathfrak{h}_m$. Since $\mathfrak{h}_m \subseteqq \mathfrak{k}$, one has that for $a \in H_m$, $a^\theta = a$, while for $a \in A$, $a^\theta = a^{-1}$. Thus, $H_m \cap A \subseteqq F$. Now if $\mu \in 2J_+$, then, as seen in the proof of the above theorem, $a^\mu = 1$ for $a \in F$. Thus the character $a \to a^\mu$ of A agrees with the trivial character on $H_m \cap A$, and hence extends to a character of H which is trivial on H_m. But the character group of H can be identified with I, where $\lambda \in I$ corresponds to the character $a \to a^\lambda$ ($a^\lambda = e^{\langle \lambda, x \rangle}$ if $a = \exp \operatorname{ad} x$, for $x \in \mathfrak{h}$). Thus there exists $\lambda \in I$ such that (i) $\lambda | \mathfrak{h}_m = 0$ and (ii) $\lambda | \mathfrak{a} = \mu$. We assert that $\lambda \in I'_+$. In view of Theorem 2.6.2, it is enough to show that $\lambda \in I_+$.

Let $\psi \in \Delta_+(\mathfrak{h}, \mathfrak{g})$. By the definition of I_+, we have only to show that it $h_\psi \in [\mathfrak{g}^\psi, \mathfrak{g}^{-\psi}] \subseteqq \mathfrak{h}$ is the unique element such that $\langle \psi, h_\psi \rangle = 2$, then $\langle \lambda, h_\psi \rangle \geqq 0$. Buf if $h'_\psi \in \mathfrak{h} \cap [\mathfrak{g}, \mathfrak{g}]$ is chosen so that for $h \in \mathfrak{h}$, $\langle h, h'_\psi \rangle = \langle \psi, h \rangle$, then one knows that $\langle h'_\psi, h'_\psi \rangle$ is a positive real number that $h_\psi = \dfrac{2h'_\psi}{\langle h'_\psi, h'_\psi \rangle}$. Hence it is enough to show that $\langle \lambda, h'_\psi \rangle \geqq 0$.

Note that if ψ^θ is defined by setting $\langle \psi^\theta, h \rangle = \langle \psi, \theta h \rangle$ for $h \in \mathfrak{h}$, then $\psi^\theta \in \Delta(\mathfrak{h}, \mathfrak{g})$. Now, $\langle \lambda, h'_\psi \rangle = \langle \lambda^\theta, \theta h'_\psi \rangle$. But since $\lambda | \mathfrak{h}_m = 0$, $\lambda^\theta = -\lambda$. Also, one has $\theta h'_\psi = h'_{\psi^\theta}$.

Thus $\langle\lambda, h'_\psi\rangle = \langle -\lambda, h'_{\psi^\theta}\rangle$, and so

$$\langle\lambda, h'_\psi\rangle = \left\langle \lambda, \frac{1}{2}(h'_\psi - h'_{\psi^\theta})\right\rangle. \qquad (*)$$

Note that $\theta\left(\frac{h'_\psi - h'_{\psi^\theta}}{2}\right) = -\left(\frac{h'_\psi - h'_{\psi^\theta}}{2}\right)$. Hence $\frac{1}{2}(h'_\psi - h'_{\psi^\theta}) \in \mathfrak{a} \cap [\mathfrak{g}, \mathfrak{g}]$. If $\psi|\mathfrak{a}=0$, then $h'_\psi - h'_{\psi^\theta}=0$; hence we can assume, in view of $(*)$, that $\psi|\mathfrak{a}\neq 0$. Then $\psi|\mathfrak{a}\in\Delta_+$. We have for any $x\in\mathfrak{a}$, $\left\langle x, \frac{1}{2}(h'_\psi - h'_{\psi^\theta})\right\rangle = \frac{1}{2}\langle\psi-\psi^\theta, x\rangle = \psi(x)$. Thus if $\varphi=\psi|\mathfrak{a}$, then w_φ is a positive real multiple of $\frac{h'_\psi - h'_{\psi^\theta}}{2}$. But since $\varphi=\psi|\mathfrak{a}\in\Delta_+$ and since $\lambda|\mathfrak{a}\in 2J_+$, we have $\langle\lambda, w_\varphi\rangle \geqq 0$. Hence also $\left\langle\lambda, \frac{h'_\psi - h'_{\psi^\theta}}{2}\right\rangle \geqq 0$, and so from $(*)$ we have $\langle\lambda, h'_\psi\rangle \geqq 0$.

We will generally identify I'_+ with $2J_+$ and hence speak about representations of G with highest weight $\lambda\in 2J_+$.

§ 7. The explicit formula for $p_{u(k,m)}$

We now begin to apply Proposition 2.5.5 to compute explicitly the polynomials $p_\gamma\in U(\mathfrak{a})$ in the rank one case.

Lemma 2.7.1. Let $u\in U$, $\lambda\in 2J_+$. Assume $z, w\in\mathfrak{g}$ have the following properties:

1) $w\in D\subseteqq \mathfrak{a}_{\mathbf{R}}\cap[\mathfrak{g}, \mathfrak{g}]$, where D is the positive Weyl chamber defined by Δ_+,

2) $z\in\mathfrak{k}$ is conjugate to $-w$,

3) $[z, u]=cu$ for some scalar c.

Then c and $\langle\lambda, w\rangle$ are real and if $c>\langle\lambda, w\rangle$, one has $p_u(\lambda)=0$.

Proof. All the eigenvalues of ad w on $\mathfrak{g}$ and hence on U are real. Since $-w$ and z are conjugate, the same is true of z, so that c is real.

Let W_λ be the irreducible G-module with a K_θ-fixed vector such that if $\eta_\lambda\in W_\lambda$ is the highest weight vector then for $x\in\mathfrak{a}$, $x\cdot\eta_\lambda=\langle\lambda, x\rangle\eta_\lambda$. (Such a module exists in view of Remark 2.6.3.) Let $\xi'_\lambda\in(W'_\lambda)^{K_\theta}$ be chosen so that for $u\in U$, $\langle u\cdot\eta_\lambda, \xi'_\lambda\rangle = p_u(\lambda)$. (See Proposition 1.2.6.)

Since $[z, u]=cu$, we have $cu^t=[z,u]^t=[z,u^t]$ (because $z^t=-z$) and hence $c(u^t\cdot\xi'_\lambda)=[z,u^t]\cdot\xi'_\lambda = z\cdot(u^t\cdot\xi'_\lambda)$, since $z\cdot\xi'_\lambda=0$. This shows that c is an eigenvalue of z in W'_λ if $u^t\cdot\xi'_\lambda\neq 0$. Since z is conjugate to $-w$, c is an eigenvalue of $-w$. But $\langle\lambda, w\rangle$ is the highest eigenvalue of w in W_λ, and hence $\langle\lambda, w\rangle$ is the highest eigenvalue of $-w$ in W'_λ.

Since by assumption $c>\langle\lambda, w\rangle$, c is not an eigenvalue of $-w$. Thus we conclude that $u^t\cdot\xi'_\lambda=0$. Hence $p_u(\lambda)=0$, since $p_u(\lambda)=\langle u\cdot\eta_\lambda, \xi'_\lambda\rangle = \langle\eta_\lambda, u^t\cdot\xi'_\lambda\rangle$.

Q.E.D.

From now on, we assume $\dim \mathfrak{a} \cap [\mathfrak{g}, \mathfrak{g}] = 1$. We recall our earlier notation in the rank 1 case: $\mathfrak{g} = \mathfrak{a} + \mathfrak{m} + \sum_{j=\pm 1, \pm 2} \mathfrak{g}^{j\alpha}$ ($\mathfrak{g}^{\pm 2\alpha}$ may be zero); $w \in \mathfrak{a} \cap [\mathfrak{g}, \mathfrak{g}]$ is chosen such that $\alpha(w) = 1$, the bilinear form B_j on $\mathfrak{g}^{j\alpha}$, defined by $B_j(x, y) = \langle x, \theta y \rangle$ is negative-definite on $\mathfrak{g}_{\mathbf{R}}^{j\alpha}$; $e_1 \in \mathfrak{g}_{\mathbf{R}}^{\alpha}$ is chosen such that $B_1(e_1, e_1) = -\frac{\langle w, w \rangle}{2}$; $e_{-1} = -\theta(e_1) \in \mathfrak{g}_{\mathbf{R}}^{-\alpha}$; one has $\langle e_1, e_{-1} \rangle = \frac{\langle w, w \rangle}{2}$ and $[e_1, e_{-1}] = \frac{w}{2}$; $\mathfrak{v}$ is the *TDS* spanned by w, e_1 and e_{-1}; $z = i(e_1 - e_{-1}) \in \mathfrak{k}$; $w' = (e_1 + e_{-1}) \in \mathfrak{p}$; $c = \exp \operatorname{ad} \left(\frac{\pi}{2} i w' \right)$; one has $cw = -z$ and $cz = w$; $b = c^2$.

Under the adjoint action $\mathfrak{v}$ in $\mathfrak{g}$, $\mathfrak{g} = \mathfrak{g}_0 + \mathfrak{g}_1 + \mathfrak{g}_2$, where $\mathfrak{g}_i$ is the primary component corresponding to the $(2i+1)$-dimensional $\mathfrak{v}$-irreducible submodules. Furthermore $\mathfrak{h} \subseteq \mathfrak{g}$ is the unique linear complement of $\mathfrak{a} + \mathfrak{v}$ in $\mathfrak{g}$ which is orthogonal to $\mathfrak{a} + \mathfrak{v}$ (with respect to a non-singular Aut $\mathfrak{g}$-invariant bilinear form). Clearly $\mathfrak{h} = (\mathfrak{h} \cap \mathfrak{k}) + (\mathfrak{h} \cap \mathfrak{p})$. We have $\mathfrak{h}_i^j = \mathfrak{g}^{j\alpha} \cap \mathfrak{h}_i$, where $\mathfrak{h}_i = \mathfrak{h} \cap \mathfrak{g}_i$. We have defined $\mathfrak{d}^{-\tau}$ to be the -1 eigenspace of τ in $\mathfrak{g}^{\alpha}$, so that $\mathfrak{d}^{-\tau} = \mathfrak{h}_2^1 \oplus \mathbf{C} e_1$, and we have defined $\mathfrak{d} = c^{-1} \mathfrak{d}^{-\tau}$ and $\mathfrak{d}_0 = c^{-1}(\mathfrak{d}_{\mathbf{R}}^{-\tau})$. Then $\mathfrak{d}$ is the $+1$ eigenspace of z in $\mathfrak{p}$. We have $\mathfrak{b} = c^{-1} \mathfrak{h}_2^1 =$ the orthocomplement of $\mathbf{C} e$ in $\mathfrak{d}$, where $e = c^{-1} e_1 \in \mathfrak{d}_0$. Finally, the bilinear form B_z on $\mathfrak{d}$ defined by $B_z(u, v) = B_1(cu, cv)$ is negative-definite on $\mathfrak{d}_0$, and one has $B_z(u, v) = B(u, \tau v)$.

We now define $\mathfrak{d}_z = i \mathfrak{d}_0$, so that B_z is positive-definite on $\mathfrak{d}_z$. Let $g = -ie$ and let $\mathfrak{b}_z = \mathfrak{b} \cap \mathfrak{d}_z$. Note that $\dim_{\mathbf{R}} \mathfrak{b}_z = \dim \mathfrak{h}_2^1 = s$, where $s = \dim \mathfrak{g}^{2\alpha}$. Choose an orthogonal basis $h_1, \ldots, h_s$ of $\mathfrak{b}_z$ such that $B_z(h_i, h_i) = \frac{\langle w, w \rangle}{2}$. Note that g, $h_1, \ldots, h_s$ is an orthogonal basis of $\mathfrak{d}_z$.

Let $\mathfrak{c}$ be the subspace of $\mathfrak{k}$ spanned by $[\mathfrak{d}, \mathfrak{d}]$. Since 3 and 4 are not eigenvalues of z in $\mathfrak{g}$, $[\mathfrak{d}, [\mathfrak{d}, \mathfrak{d}]] = 0 = [[\mathfrak{d}, \mathfrak{d}], [\mathfrak{d}, \mathfrak{d}]]$. Thus, if we put $\mathfrak{e} = \mathfrak{c} + \mathfrak{d}$, then $\mathfrak{e}$ is a Lie subalgebra of $\mathfrak{g}$, and if Q is the (associative) subalgebra of U generated by $\mathfrak{d}$, then Q is the universal enveloping algebra of $\mathfrak{e}$. Note that $\mathfrak{c}$ is contained in the centre of $\mathfrak{e}$. Further, if $Q_j = \{u \in Q \mid [z, u] = ju\}$, then Q_j is the span of $\{u_1 u_2 \ldots u_j \mid u_i \in \mathfrak{d}\}$, and $Q = \sum_j Q_j$ (direct sum). Observe that $Q_1 = \mathfrak{d}$. Clearly, $Q_i Q_j \subseteq Q_{i+j}$.

Lemma 2.7.2. One has $p_u(2\alpha) = 0$ if $u \in Q_j$ and $j > 2$.

Proof. This is an immediate consequence of Lemma 2.7.1, since $2\alpha \in 2J_+$ and $\langle 2\alpha, w \rangle = 2$. Q.E.D.

Lemma 2.7.3. Let $x \in [\mathfrak{g}, \mathfrak{g}]$. Then $p_x = \frac{\langle x, w \rangle}{\langle w, w \rangle} w$.

Proof. From the decomposition $\mathfrak{g} = \mathfrak{k} + \mathfrak{a} + \mathfrak{n}$, one has the decomposition

$$[\mathfrak{g}, \mathfrak{g}] = \mathfrak{k} \cap [\mathfrak{g}, \mathfrak{g}] + Cw + \mathfrak{n}.$$

Note that $\mathfrak{k}\cap[\mathfrak{g}, \mathfrak{g}]+\mathfrak{n}$ is orthogonal to w (under the Killing form). Writing $x=x_1+cw$, where $x_1 \in \mathfrak{k}\cap[\mathfrak{g}, \mathfrak{g}]+\mathfrak{n}$ and c is a scalar, we have $p_x=cw$. But $\langle x, w\rangle=$ $=\langle x_1, w\rangle+c\langle w, w\rangle=c\langle w, w\rangle$, so that $c=\dfrac{\langle x, w\rangle}{\langle w, w\rangle}$. Thus

$$p_x = \frac{\langle x, w\rangle}{\langle w, w\rangle} w. \qquad \text{Q.E.D.}$$

Lemma 2.7.4. One has $B_z(g, g) = \dfrac{\langle w, w\rangle}{2}$ and $\langle g, w\rangle = \dfrac{\langle w, w\rangle}{2}$.

Proof. Consider $\langle e_1, z\rangle$. One has

$$\langle e_1, z\rangle = \langle e_1, i(e_1 - e_{-1})\rangle = -i\langle e_1, e_{-1}\rangle = -i\frac{\langle w, w\rangle}{2}.$$

Since $c^{-1}e_1=e$, $c^{-1}z=-w$ and since $g=-ie$, we have

$$\langle g, w\rangle = \langle -ic^{-1}e_1, -c^{-1}z\rangle = i\langle e_1, z\rangle = \frac{\langle w, w\rangle}{2}.$$

For the other part of the lemma, one has

$$B_z(g, g) = -B_z(e, e) = -B_1(ce, ce) = -B_1(e_1, e_1) = \frac{\langle w, w\rangle}{2}. \qquad \text{Q.E.D.}$$

Lemma 2.7.5. One has $p_g = \dfrac{w}{2}$ and $p_{h_i} = 0$.

Proof. We have by Lemma 2.7.3, $p_{h_i} = \dfrac{\langle h_i, w\rangle}{\langle w, w\rangle} w = 0$. But $\langle h_1, w\rangle = 0$ since $\langle \mathfrak{b}, \mathfrak{v}\rangle = \langle c\mathfrak{b}, c\mathfrak{v}\rangle = \langle \mathfrak{h}_2^1, \mathfrak{v}\rangle = 0$. Also,

$$p_g = \frac{\langle g, w\rangle}{\langle w, w\rangle} w = \frac{w}{2}, \text{ since } \langle g, w\rangle = \frac{\langle w, w\rangle}{2}. \qquad \text{Q.E.D.}$$

We have the decomposition of $\mathfrak{g}=\mathfrak{a}+\mathfrak{m}+\mathfrak{n}+\mathfrak{n}_-$, where $\mathfrak{n}_- = \sum_{\alpha\in\Delta_+} \mathfrak{g}^{-\alpha}$. Note that $\mathfrak{a}$ is orthogonal to $\mathfrak{m}+\mathfrak{n}+\mathfrak{n}_-$ and $\mathfrak{m}$ is orthogonal to $\mathfrak{a}+\mathfrak{n}+\mathfrak{n}_-$. We have $\theta(\mathfrak{n})=\mathfrak{n}_-$ and hence $\mathfrak{n}+\mathfrak{n}_-$ is stable under θ. Thus

$$\mathfrak{g} = \mathfrak{a}+\mathfrak{m}+(\mathfrak{n}+\mathfrak{n}_-)\cap\mathfrak{k}+(\mathfrak{n}+\mathfrak{n}_-)\cap\mathfrak{p}.$$

Furthermore, $(\mathfrak{n}+\mathfrak{n}_-)\cap\mathfrak{p}$ is a linear complement of $\mathfrak{a}$ in $\mathfrak{p}$ and is orthogonal to $\mathfrak{a}$, and $(\mathfrak{n}+\mathfrak{n}_-)\cap\mathfrak{p}=\{y-\theta y \mid y\in\mathfrak{n}\}$.

Lemma 2.7.6. Let $y\in\mathfrak{n}$ and let $v=y-\theta y$. Then for any $x\in\mathfrak{g}$,

$$p_{vx} = \frac{2\langle x, [w, y]\rangle}{\langle w, w\rangle} w.$$

Proof. One has $v=2y-(y+\theta y)$ and $y+\theta y \in \mathfrak{k}$. Thus $p_{vx}=p_{2yx}-p_{(y+\theta y)x}=2p_{yx}$, since $(y+\theta y)x \in \mathfrak{k}U$. Also, $yx=[y, x]+xy$ and $xy \in U\mathfrak{n}$. Thus $p_{yx}=p_{[y,x]}$. Hence

$$p_{vx} = 2p_{[y,x]} = 2\frac{\langle w, [y,x]\rangle}{\langle w, w\rangle} w \text{ (by Lemma 2.7.3)}$$

$$= 2\frac{\langle x, [w,y]\rangle}{\langle w, w\rangle} w,$$

by the invariance of the Killing form. Q.E.D.

Recall that $\mathfrak{b}_z=\mathfrak{b}\cap\mathfrak{d}_z$, where $\mathfrak{b}$ is the orthocomplement of $\mathfrak{e}$ in $\mathfrak{d}$ and $\mathfrak{d}_z=i\mathfrak{d}_0=ic^{-1}(\mathfrak{d}_{\mathbf{R}}^{-\tau})$. Since $h_1, \ldots, h_s$ is an orthogonal basis of $\mathfrak{b}_z$ (relative to B_z), writing $q_j=ch_j$, it is clear that $q_1, \ldots, q_s$ is an orthogonal basis of $i(\mathfrak{h}_2^1)_{\mathbf{R}}$ (relative to B_1).

Let V_i be the irreducible 5-dimensional $\mathfrak{v}$-submodule of $\mathfrak{g}$ containing q_i, so that $\mathfrak{h}_2=\sum_{i=1}^{s} V_i$ (direct sum). As noted previously V_i is stable under θ since $V_i\cap\mathfrak{m}\neq 0$.

Lemma 2.7.7. The Killing form is non-singular on each V_i. If $i\neq j$, V_i is orthogonal to V_j.

Proof. Suppose $i\neq j$. To show the orthogonality, it is enough to prove that there exists a non-zero vector in V_i which is orthogonal to V_j. We assert that q_i is orthogonal to V_j. Since $\mathfrak{g}^\alpha$ is orthogonal to $\mathfrak{a}+\mathfrak{m}$ and also to $\mathfrak{g}^{k\alpha}$ for $k\neq -1$, it is enough to show that $\langle q_i, V_j\cap\mathfrak{g}^{-\alpha}\rangle=0$. But $V_j\cap\mathfrak{g}^{-\alpha}$ is the one-dimensional space spanned by θq_j, and $\langle q_i, \theta q_j\rangle=B_1(q_i, q_j)=0$.

To prove that the Killing form is non-degenerate on V_i, observe that $\langle q_i, \theta q_i\rangle=B_1(q_i, q_i)\neq 0$. The desired result is then a consequence of the irreducibility of ad $\mathfrak{v}$ on V_i. Q.E.D.

Each V_j is a U-module, where (here) $U=\exp \operatorname{ad} \mathfrak{v}\subseteq G$. Since $h_i=c^{-1}q_i$ and since $c\in\exp \operatorname{ad} \mathfrak{v}$, we have $h_i\in V_i$. Thus we can write h_i uniquely as

$$h_i = z_i+(x_i+x_i')+(y_i+y_i'),$$

where $z_i\in\mathfrak{m}\cap V_i$, $x_i\in\mathfrak{g}^\alpha\cap V_i$, $x_i'\in\mathfrak{g}^{-\alpha}\cap V_i$, $y_i\in\mathfrak{g}^{2\alpha}\cap v_i$ and $y_i'\in\mathfrak{g}^{-2\alpha}\cap V_i$. Then, using the fact that $\theta h_i=-h_i$, one easily concludes that $z_i=0$, $x_i'=-\theta x_i$ and $y_i'=-\theta y_i$.

Thus, each h_i can be uniquely written as $h_i=(x_i-\theta x_i)+(y_i-\theta y_i)$, where $x_i\in V_i\cap\mathfrak{g}^\alpha$ and $y_i\in V_i\cap\mathfrak{g}^{2\alpha}$.

Lemma 2.7.8. One has $y_j=ix_j^*$ (where $x_j^*=[e_1, x_j]$) and

$$\langle x_j, \theta x_j\rangle = -\langle y_j, \theta y_j\rangle = \frac{\langle w, w\rangle}{8}.$$

Proof. Since $h_j \in \mathfrak{d}$, $[z, h_j]=h_j$. Thus we have

$$h_j = [i(e_1 - e_{-1}), (x_j - \theta x_j) + (y_j - \theta y_j)]. \qquad (*)$$

Comparing the decomposition of the right-hand side of $(*)$ with respect to the decomposition $\mathfrak{g}=\mathfrak{a}+\mathfrak{m}+\mathfrak{g}^{\alpha}+\mathfrak{g}^{-\alpha}+\mathfrak{g}^{2\alpha}+\mathfrak{g}^{-2\alpha}$, with the decomposition $h_j=x_j-\theta x_j+y_j-\theta y_j$, we see that $y_j=i[e_1, x_j]=ix_j^*$.
Now,

$$\langle y_j, \theta y_j\rangle = -\langle [e_1, x_j], \theta[e_1, x_j]\rangle = \langle [e_1, x_j], [e_{-1}, \theta x_j]\rangle = $$
$$= -\langle [e_{-1}, [e_1, x_j]], \theta x_j\rangle$$

(by the invariance of the Killing form)

$$= -\langle x_j, \theta x_j\rangle,$$

since by Lemma 2.1.12, $[e_{-1}, [e_1, x_j]]=x_j$.

Finally, $\dfrac{\langle w, w\rangle}{2} = B_z(h_j, h_j) = \langle h_j, \tau h_j\rangle$ (where $\tau = b\theta$). But $h_j = x_j - \theta x_j + y_j - \theta y_j$, and by Lemma 2.1.8,

$$\tau(x_j - \theta x_j) = -(x_j - \theta x_j) \text{ and } \tau(y_j - \theta y_j) = y_j - \theta y_j.$$

Thus

$$\frac{\langle w, w\rangle}{2} = \langle h_j, \tau h_j\rangle = \langle x_j - \theta x_j + y_j - \theta y_j, -(x_j - \theta x_j) + (y_j - \theta y_j)\rangle =$$

$$= \langle x_j, \theta x_j\rangle + \langle -\theta x_j, -x_j\rangle + \langle y_j, -\theta y_j\rangle + \langle -\theta y_j, y_j\rangle$$

(using the fact that $\langle \mathfrak{g}^{k\alpha}, \mathfrak{g}^{i\alpha}\rangle = 0$ if $k \neq -i$)

$$= 2\langle x_j, \theta x_j\rangle - 2\langle y_j, \theta y_j\rangle =$$

$$= 4\langle x_j, \theta x_j\rangle, \text{ since } \langle y_j, \theta y_j\rangle = -\langle x_j, \theta x_j\rangle.$$

Thus

$$\langle x_j, \theta x_j\rangle = \frac{\langle w, w\rangle}{8}. \qquad \text{Q.E.D.}$$

Lemma 2.7.9. One has $p_{h_i h_j}=0$ if $i \neq j$ and $p_{h_i^2}=\dfrac{w}{4}$.

Proof. As usual, write $h_i=(x_i-\theta x_i)+(y_i-\theta y_i)$. Put $y=x_i+y_i$, so that $h_i=y-\theta y$. Now, by Lemma 2.7.6,

$$p_{h_i h_j} = 2\frac{\langle h_j, [w, y]\rangle}{\langle w, w\rangle} w.$$

But $[w, y] = [w, x_i+y_i] = x_i + 2y_i \in V_i$, and by Lemma 2.7.7, $\langle h_j, x_i+2y_i\rangle = 0$ if $i \neq j$. Hence $p_{h_i h_j} = 0$ if $i \neq j$.

Also,

$$p_{h_i^2} = 2\frac{\langle h_i, [w, y]\rangle}{\langle w, w\rangle} w = \frac{2\langle h_i, x_i+2y_i\rangle}{\langle w, w\rangle} w.$$

But one has

$$\langle h_i, x_i+2y_i\rangle = \langle x_i+y_i-\theta x_i-\theta y_i, x_i+2y_i\rangle = \langle -\theta x_i, x_i\rangle + \langle -\theta y_i, 2y_i\rangle =$$

$$= \frac{-\langle w, w\rangle}{8} + 2\frac{\langle w, w\rangle}{8} = \frac{\langle w, w\rangle}{8}.$$

Thus

$$p_{h_i^2} = \frac{2}{\langle w, w\rangle}\frac{\langle w, w\rangle}{8} w = \frac{w}{4}. \qquad \text{Q.E.D.}$$

Lemma 2.7.10. One has $p_{h_i g} = 0 = p_{g h_i}$ and

$$p_{g^2} = \frac{w(w-1)}{4}.$$

Proof. We have $g = -ie = -ic^{-1}e_1$. But $c^{-1}z = -w$ and $c^{-1}w' = w'$, i.e., $c^{-1}(i(e_1-e_{-1})) = -w$ and $c^{-1}(e_1+e_{-1}) = w'$. Hence $c^{-1}e_1 = \frac{iw+w'}{2}$, and so $g = \frac{w-iw'}{2} = \frac{w-i(e_1+e_{-1})}{2}$. Note that $g \in \mathfrak{u}$, Since $\langle \mathfrak{h}, \mathfrak{u}\rangle = 0$, it is now immediate that $p_{h_i g} = 0$, using Lemma 2.7.6.

But since $[g, h_i] \in \mathfrak{k}$, $p_{[g, h_i]} = 0$ and hence $p_{gh_i} = p_{h_i g}$.

Now

$$g^2 = \left(\frac{w-i(e_1+e_{-1})}{2}\right) g = \frac{w}{2} g - \frac{i}{2}(e_1-\theta e_1) g. \qquad (*)$$

By Lemma 2.7.6,

$$p_{(e_1-\theta e_1)g} = 2\frac{\langle g, [w, e_1]\rangle}{\langle w, w\rangle} w = \frac{2\langle g, e_1\rangle}{\langle w, w\rangle} w.$$

Since $g = \frac{w-i(e_1+e_{-1})}{2}$ and since $\langle w, e_1\rangle = 0 = \langle e_1, e_1\rangle$, we have $\langle g, e_1\rangle = = -\frac{i}{2}\langle e_{-1}, e_1\rangle = -\frac{i}{2}\frac{\langle w, w\rangle}{2}$. Thus $p_{(e_1-\theta e_1)g} = -\frac{iw}{2}$, and so $p_{-\frac{i}{2}(e_1-\theta e_1)g} = -\frac{w}{4}$.

On the other hand, $\frac{w}{2} g = \frac{w}{2}\left(\frac{w-i(e_1+e_{-1})}{2}\right) = \frac{w^2}{4} - \frac{i}{4} w(e_1+e_{-1})$. Hence $p_{\frac{w}{2}g} = p_{\frac{w^2}{4}} - \frac{i}{4} p_{w(e_1+e_{-1})}$. Since $\frac{w^2}{4} \in U(\mathfrak{a})$, $\frac{p_{w^2}}{4} = \frac{w^2}{4}$. Also, $p_{w(e_1+e_{-1})} = p_{(e_1+e_{-1})w}$, since $[e_1+e_{-1}, w] \in \mathfrak{k}$. By Lemma 2.7.6, $p_{(e_1+e_{-1})w} = \frac{2\langle w, [w, e_1]\rangle}{\langle w, w\rangle} w = 0$. Hence $p_{\frac{w}{2}g} = \frac{w^2}{4}$. Thus, from $(*)$,

$$p_{g^2} = \frac{w^2}{4} - \frac{w}{4} = \frac{w(w-1)}{4}. \qquad \text{Q.E.D.}$$

Recall that $T=S(\mathfrak{d})\subseteq S(\mathfrak{p})$. We denote the product of two elements u and v in T by $u\circ v$ and we denote by $u^{(n)}$ the product (in T) of u with itself n times.

Since $\mathfrak{d}=\mathbf{C}g\oplus\mathfrak{b}$, $T=S(\mathbf{C}g)\otimes S(\mathfrak{b})$. Let $O(\mathfrak{d})$ (resp., $O(\mathfrak{b})$) denote the subgroup of $GL(\mathfrak{d})$ (resp., $GL(\mathfrak{b})$) leaving invariant the bilinear form B_z (resp., $B_z|\mathfrak{b}$). Note that $O(\mathfrak{b})$ can be regarded as the isotropy subgroup of $O(\mathfrak{d})$ at g. We have $T^{O(\mathfrak{b})}=S(\mathbf{C}g)\otimes S(\mathfrak{b})^{O(\mathfrak{b})}$. Consider the quadratic element $h=h_1^{(2)}+\ldots+h_s^{(2)}$. Then $\{h^{(m)}|m\in Z_+\}$ forms a basis of $S(\mathfrak{b})^{O(\mathfrak{b})}$. Thus $\{g^{(k)}\circ h^{(m)}|m, k\in Z_+\}$ forms a basis of $T^{O(\mathfrak{b})}$. Put $u(k, m)=\beta(g^{(k)}\circ h^{(m)})$, where $\beta: S(\mathfrak{g})\to U$ is the usual map. Let $T^*=\beta(T)$. Observe that $\{u(k, m)|k, m\in Z_+\}$ forms a basis of $(T^*)^{O(\mathfrak{b})}$ under the action of $O(\mathfrak{d})$ on T^* defined by requiring $\beta: T\to T^*$ to be an $O(\mathfrak{d})$-isomorphism. We want to compute $p_{u(k,m)}$.

Theorem 2.7.11. *For all* $s>0$, *one has*

$$p_{u(k,m)} = v[w(w-2)\cdots(w-2m+2)][(w-2m)(w-2m-1)\cdots(w-2m-k+1)],$$

where

$$v = \frac{s(s+2)\cdots(s+2m-2)}{2^{2m+k}}.$$

The proof of this theorem is divided into a number of lemmas.

Recall that $Q\subseteq U$ is the enveloping algebra of $\mathfrak{e}=\mathfrak{c}+\mathfrak{d}$, where $\mathfrak{c}$ is the subspace of $\mathfrak{k}$ spanned by $[\mathfrak{d}, \mathfrak{d}]$. As remarked already, $\mathfrak{c}$ is central in $\mathfrak{e}$. Since $\mathfrak{e}$ is a Lie subalgebra, β maps $S(\mathfrak{e})$ onto Q. By the Birkhoff—Witt theorem, $Q=\mathfrak{c}Q\oplus T^*$. Let $\varphi: Q\to T^*$ be the projection map with respect to the above decomposition. Clearly we have $[\mathfrak{e}, \mathfrak{e}]=\mathfrak{c}$. Hence $\mathfrak{e}/\mathfrak{c}$ is Abelian. Identifying $\mathfrak{e}/\mathfrak{c}$ with the vector space $\mathfrak{d}$ by the isomorphism $\mathfrak{d}\to\mathfrak{e}\to\mathfrak{e}/\mathfrak{c}$, we can regard $S(\mathfrak{d})=T$ as the universal enveloping algebra of $\mathfrak{e}/\mathfrak{c}$. Thus the map $\mathfrak{e}\to\mathfrak{e}/\mathfrak{c}$ induces a homomorphism $\pi: Q\to T$, and we have an exact sequence $0\to\mathfrak{c}Q\to Q\xrightarrow{\pi}T\to 0$.

Lemma 2.7.12. One has $\beta\circ\pi=\varphi$, so that for $u\in Q$, $p_u=p_{\beta\circ\pi(u)}$.

Proof. The kernel of both φ and π is $\mathfrak{c}Q$. Hence it is enough to show that $\beta\circ\pi|T^*=$identity. But T^* is spanned by $\{x^k|x\in\mathfrak{d}\}$, and on this set clearly $\beta\circ\pi$ is the identity. Q.E.D.

Lemma 2.7.13. One has $p_{u(k,m)} = \lambda p_{g^k h_1^{2m}}$, where λ is the scalar $\dfrac{2^m m!}{(2m)!}s(s+2)\cdots\cdots(s+2(m-1))$.

Proof. Observe that when $s=1$ one has $\lambda=1$. We have, clearly,

$$p_{g^k h_1^{2m}} = p_{\varphi(g^k h_1^{2m})}.$$

But

$$\varphi(g^k h_1^{2m}) = \beta\circ\pi(g^k h_1^{2m}) = \beta(g^{(k)}\circ h_1^{(2m)}) = u(k, m).$$

Thus, the lemma is proved when $s=1$. We now assume $s>1$.

Recall that $M_{0,\mathbf{R}}$ is transitive on the unit sphere in $\mathfrak{b}_0(=\mathfrak{b}\cap\mathfrak{d}_0)$ (see Remark 2.1.29). Thus the invariants in $S(\mathfrak{b})$ for the actions of $M_{0,\mathbf{R}}$ and $O(\mathfrak{b})$ on $S(\mathfrak{b})$ are the same. Now observe that there exists a scalar λ such that

$$\lambda \int_{M_{0,\mathbf{R}}} a\cdot h_1^{(2m)}\, da = h^{(m)}. \tag{$*$}$$

Now regard $S(\mathfrak{b})$ as the polynomial ring on $\mathfrak{b}_0'$, and let $x_1, \ldots, x_s \in \mathfrak{b}_0'$ be the dual basis to the h_i. Let

$$\partial = \frac{\partial^2}{\partial x_1^2} + \cdots + \frac{\partial^2}{\partial x_s^2}.$$

We have

$$\frac{\partial^2}{\partial x_l^2}(h^{(m)}) = \frac{\partial}{\partial x_l}(mh^{(m-1)}2h_l) = 2m\{(m-1)h^{(m-2)}2h_l^{(2)} + h^{(m-1)}\}.$$

Hence,

$$\partial h^{(m)} = 2m\sum_l \{(m-1)h^{(m-2)}2h_l^{(2)} + h^{(m-1)}\} =$$

$$= 2m(2(m-1)h^{(m-1)} + sh^{(m-1)}) = 2m(s+2(m-1))h^{(m-1)}.$$

Hence, by induction,

$$\partial^m h^{(m)} = 2^m m!(s+2(m-1))(s+2(m-2))\ldots(s+2)s.$$

Note that ∂ commutes with the action of $M_{0,\mathbf{R}}$ on $S(\mathfrak{b})$. We have

$$\partial(\lambda \int_{M_{0,\mathbf{R}}} a\cdot h_1^{(2m)}\, da) = \lambda \int_{M_{0,\mathbf{R}}} \partial(a\cdot h_1^{(2m)})\, da = \lambda \int_{M_{0,\mathbf{R}}} a\cdot \partial h_1^{(2m)}\, da =$$

$$= \lambda \int_{M_{0,\mathbf{R}}} a\cdot 2m(2m-1)h_1^{(2m-2)}\, da = 2m(2m-1)\lambda \int_{M_{0,\mathbf{R}}} a\cdot h_1^{(2m-2)}\, da.$$

Hence, by induction

$$\partial^m(\lambda \int_{M_{0,\mathbf{R}}} a\cdot h_1^{(2m)}\, da) = (2m)!(\lambda \int_{M_{0,\mathbf{R}}} 1\, da) = (2m)!\,\lambda.$$

Hence, from $(*)$,

$$\lambda(2m)! = 2^m m!\, s(s+2)\cdots(s+2(m-1)).$$

Thus

$$\lambda = \frac{2^m m!}{(2m)!} s(s+2)\cdots(s+2(m-1)).$$

Let $v \in Q$ be defined by $v = \lambda \int_{M_{0,\mathbf{R}}} a\cdot g^k h_1^{2m}\, da$. We have,

$$p_v = \lambda \int_{M_{0,\mathbf{R}}} p_{a\cdot g^k h_1^{2m}}\, da = \lambda \int_{M_{0,\mathbf{R}}} p_{g^k h_1^{2m}}\, da$$

(by Lemma 1.7.10)

$$= \lambda p_{g^k h_1^{2m}}. \tag{I}$$

Note that since $M_{0,\mathbf{R}}$ leaves $\mathfrak{d}$ stable, the map $\mathfrak{e}\to\mathfrak{e}/\mathfrak{c}$ is an $M_{0,\mathbf{R}}$-map, and hence the map $\pi: Q\to T$ is also an $M_{0,\mathbf{R}}$-map. Thus

$$\pi(v) = \lambda \int_{M_{0,\mathbf{R}}} a\cdot\pi(g^k h_1^{2m})\,da = \lambda \int_{M_{0,\mathbf{R}}} a\cdot(g^{(k)}\circ h_1^{(2m)})\,da,$$

since $\pi(g^k h_1^{2m})=g^{(k)}\circ h_1^{(2m)}$. But for $a\in M_{0,\mathbf{R}}$, $a\cdot g=g$, so that

$$\lambda \int_{M_{0,\mathbf{R}}} a\cdot(g^{(k)}\circ h_1^{(2m)})\,da = g^{(k)}\circ\lambda \int_{M_{0,\mathbf{R}}} a\cdot h_1^{(2m)}\,da = g^{(k)}\circ h^{(m)}.$$

Thus $\pi(v)=g^{(k)}\circ h^{(m)}$. Hence $\beta(\pi(v))=u(k,m)$, and so by Lemma 2.7.12, $u(k,m)==\varphi(v)$ and $p_{u(k,m)}=p_v$. Thus by (I), $p_{u(k,m)}=\lambda p_{g^k h_1^{2m}}$. Q.E.D.

Now to compute $p_{g^k h_1^{2m}}$, we start by computing p_{g^k}.

Lemma 2.7.14. One has $p_{g^j}(\alpha)=0$ for all $j>1$.

Proof. When $j=2$, we know that $p_{g^2}=\dfrac{w(w-1)}{4}$, by Lemma 2.7.10, which implies that $p_{g^2}(\alpha)=0$. Assume now that $j>2$ and that $p_{g^k}(\alpha)=0$ for $2\leqq k<j$. By Lemma 2.7.2, we know that $p_{g^j}(2\alpha)=0$. By Corollary 2.5.4, we have that if $\delta(g^j)=\sum_i u_i'\otimes u_i''$, then

$$p_{g^j}(2\alpha) = \sum_i p_{u_i'}(\alpha)\,p_{u_i''}(\alpha).$$

Since

$$\delta(g^j) = \sum_{i=0}^{j}\binom{j}{i} g^i\otimes g^{j-i},$$

we have

$$p_{g^j}(2\alpha) = \sum_{i=0}^{j}\binom{j}{i} p_{g^i}(\alpha)\,p_{g^{j-i}}(\alpha) = 2p_{g^j}(\alpha),$$

using the induction hypothesis.

This shows that $p_{g^j}(\alpha)=0$. Q.E.D.

Lemma 2.7.15. For $k\geqq 1$,

$$p_{g^k} = \frac{w(w-1)\cdots(w-k+1)}{2^k}.$$

Proof. By Remark 2.5.6, we have

$$p_{g^k} = \sum_{j=0}^{k} S_{\alpha,j}(g^k)\begin{pmatrix}\dfrac{w}{\langle\alpha,w\rangle}\\ j\end{pmatrix} = \sum_{j=0}^{k} S_{\alpha,j}(g^k)\binom{w}{j}. \qquad (*)$$

We have $\delta_j(g^k)=\sum c_\mu g_\mu$, where the summation runs through all j-tuples $\mu=(\mu_1,\dots,\mu_j)$ of nonnegative integers such that $\mu_1+\dots+\mu_j=k$, $g_\mu=g^{\mu_1}\otimes\dots\otimes g^{\mu_j}$ and c_μ are scalars. Hence $\delta_{j,+}(g^k)=\sum c_\mu g_\mu$, where μ now runs through all

j-tuples $(\mu_1, \ldots, \mu_j)$ of positive integers such that $\mu_1+\cdots+\mu_j=k$. Then

$$S_{\alpha,j}(g^k) = (l_{\alpha,j}\circ\delta_{j,+})(g^k) = \sum_{\mu} c_\mu p_{g^{\mu_1}}(\alpha)\ldots p_{g^{\mu_j}}(\alpha).$$

Using Lemma 2.7.14, it is now clear that if $j<k$, then since in each term $\mu_i>1$ for some i, $S_{\alpha,j}(g^k)=0$. Also, when $j=k$, we have $\delta_{k,+}(g^k)=k!\ \underbrace{g\otimes g\otimes\cdots\otimes g}_{k\text{ times}}$, so that $S_{\alpha,k}(g^k)=k!(p_g(\alpha))^k=k!\frac{1}{2^k}$, since by Lemma 2.7.5, $p_g=\frac{w}{2}$ and $\frac{w}{2}(\alpha)=$ $=1/2$. Thus, from $(*)$, we have

$$p_{g^k} = \frac{k!}{2^k}\,\frac{w(w-1)\cdots(w-k+1)}{k!} = \frac{w(w-1)\cdots(w-k+1)}{2^k}. \quad \text{Q.E.D.}$$

By Remark 2.5.6, we have

$$p_{g^k h_1^{2m}} = \sum_{n=0}^{k+2m} S_{2\alpha,n}(g^k h_1^{2m})\binom{w/2}{n}.$$

Lemma 2.7.16. If $n<m$, $S_{2\alpha,n}(g^k h_1^{2m}) = 0$. If $n\geqq m$,

$$S_{2\alpha,n}(g^k h_1^{2m}) = \binom{n}{m} S_{2\alpha,n-m}(g^k)\, S_{2\alpha,m}(h_1^{2m}).$$

Proof. We have

$$\delta_n(g^k h_1^{2m}) =$$

$$= (g\otimes 1\otimes\cdots\otimes 1+1\otimes g\otimes\cdots\otimes 1+\cdots+1\otimes 1\otimes\cdots\otimes g)^k(h_1\otimes 1\otimes\cdots\otimes 1+$$
$$+1\otimes h_1\otimes\cdots\otimes 1+\cdots+1\otimes 1\otimes\cdots\otimes h_1)^{2m} = \sum_{i\in I} u_{1i}\otimes u_{2i}\otimes\ldots\otimes u_{ni},$$

where $u_{ji}=g^{t_{ji}}h_1^{s_{ji}}$, where t_{ji} and s_{ji} being non-negative integers such that $\sum_{j=1}^{n} t_{ji}=k$ and $\sum_{j=1}^{n} s_{ji}=2m$, and where I is a suitably chosen index set. For $i\in I$, let $c_i=p_{u_{1i}}(2\alpha)p_{u_{2i}}(2\alpha)\cdots p_{u_{ni}}(2\alpha)$. Let I_1 be the subset of all $i\in I$ such that the elements u_{ji} $(j=1, 2, \ldots, n)$ are one of the three elements g, g^2 and h_1^2. Then we assert that

$$S_{2\alpha,n}(g^k h_1^{2m}) = \sum_{i\in I_1} c_i. \qquad \text{(I)}$$

Indeed, we have $\delta_{n,+}(g^k h_1^{2m})=\sum_{i\in I'} u_{1i}\otimes u_{2i}\otimes\cdots\otimes u_{ni}$, where I' is the subset of all $i\in I$ such that $t_{ji}+s_{ji}>0$ for each j $(j=1, 2, \ldots, n)$. By Lemma 2.7.2, if $t_{ji}+s_{ji}>2$, then $p_{u_{ji}}(2\alpha)=0$, and for $i\in I'$, by Lemma 2.7.5 and 2.7.10, $p_{u_{ji}}(2\alpha)=0$ unless $u_{ji}=g$, g^2 or h_1^2. This proves our assertion. In particular, note that if $n<m$, then the condition $\sum_{j=1}^{n} s_{ji}=2m$ implies that for any i, there exists a j such that $s_{ji}>2$. Thus $c_i=0$ for all i, and hence $S_{2\alpha,n}(g^k h_1^{2m})=0$.

Now, assume $n\geqq m$. Let S be a subset of $(1, 2, \ldots, n)$. As in the proof of Proposition 2.5.5, let U_S (resp. $U_{S,+}$) $\subseteqq \underbrace{U\otimes U\otimes\cdots\otimes U}_{n\text{ times}}$ be defined by $U_S=U(1)\otimes U(2)\otimes\cdots\otimes U(n)$, (resp. $U_{S,+}=U_+(1)\otimes U_+(2)\otimes\cdots\otimes U_+(n)$), where, for $l\in(1, 2, \ldots, n)$,

$U(l)=U$ (resp. $U_+(l)=U_+$) if $l\in S$ and $U(l)=\mathbf{C}\cdot 1$ (resp. $U_+(l)=\mathbf{C}\cdot 1$) if $l\notin S$. We have $\underbrace{U\otimes U\otimes\cdots\otimes U}_{n \text{ times}}=\sum_{S\subseteq(1,\ldots,n)} U_{S,+}$ and $U_S=\sum_{T\subseteq S} U_{T,+}$. For an element $x\in U\otimes U\otimes\cdots\otimes U$, let $x_{S,+}$ denote the component of x in $U_{S,+}$ and let $x_S(=\sum_{T\subseteq S} x_{T,+})$ be the component in U_S. Now suppose S consists of m elements and let U_S be identified with $\underbrace{U\otimes U\otimes\cdots\otimes U}_{m \text{ times}}$ through the obvious isomorphism η_S. (For example, when $S=(1, 2, \ldots, m)$, we have $U_S=\underbrace{U\otimes U\otimes\cdots\otimes U}_{m \text{ times}}\otimes\mathbf{C}\cdot 1\otimes\cdots\otimes\mathbf{C}\cdot 1$, and the map η_S is given by

$$u_1\otimes u_2\otimes\cdots\otimes u_m\otimes 1\otimes\cdots\otimes 1\to u_1\otimes u_2\otimes\cdots\otimes u_m.)$$

We have for $v\in U_S$ and $\nu\in\mathfrak{a}'$, $\boldsymbol{l}_{\nu,n}(v)=\boldsymbol{l}_{\nu,m}(\eta_S(v))$ and for $v\in\underbrace{U\otimes U\otimes\cdots\otimes U}_{n \text{ times}}$, $\boldsymbol{l}_{\nu,n}(v)=$ $=\sum_S \boldsymbol{l}_{\nu,n}(v_{S,+})$. Let S' denote the complement of S in $(1, 2, \ldots, n)$, and let $i_S\colon U_S\to\underbrace{U\otimes U\otimes\cdots\otimes U}_{n \text{ times}}$ be the inclusion map. Then clearly $i_S\otimes i'_S\colon U_S\otimes U'_S\to$ $\to\underbrace{U\otimes U\otimes\cdots\otimes U}_{n \text{ times}}$ is an isomorphism, and we identify $U_S\otimes U'_S$ with $\underbrace{U\otimes U\otimes\cdots\otimes U}_{n \text{ times}}$ by this isomorphism. Now suppose S consists of m elements. Then we have clearly $\boldsymbol{l}_{\nu,m}(x)\boldsymbol{l}_{\nu,n-m}(x')=\boldsymbol{l}_{\nu,n}(x\otimes x')$.

Let $v=\delta_n(g^k)$ and $w=\delta_n(h_1^{2m})$, so that $\delta_n(g^k h_1^{2m})=vw$. Let S be a subset of $(1, 2, \ldots, n)$ consisting of m elements, and let S' be the complement of S in $(1, 2, \ldots, n)$, so that $(i_S\otimes i_{S'})(w_{S,+}\otimes v_{S',+})\in U_+\otimes\cdots\otimes U_+$. Using assertion (I) it is easily checked that

$$\delta_{n,+}(g^k h_1^{2m}) = \sum (i_S\otimes i_{S'})(w_{S,+}\otimes v_{S,m})+y, \tag{$*$}$$

where S runs through all subsets of $(1, 2, \ldots, n)$ consisting of m elements, S' is the complement of S in $(1, 2, \ldots, n)$, and y is such that $\boldsymbol{l}_{2\alpha,n}(y)=0$. Moreover, as remarked in the proof of Proposition 2.5.5, $\eta_S(w_{S,+})=\delta_{m,+}(h_1^{2m})$ and similarly, $\eta_{S'}(v_{S',+})=\delta_{n-m,+}(g^k)$. In particular, we have

$$\begin{aligned}&\boldsymbol{l}_{2\alpha,n}\big(i_S\otimes i_{S'}(w_{S,+}\otimes v_{S',+})\big) =\\ &= \boldsymbol{l}_{2\alpha,m}\big(\delta_{m,+}(h_1^{2m})\big)\,\boldsymbol{l}_{2\alpha,n-m}\big(\delta_{n-m,+}(g^k)\big) = S_{2\alpha,m}(h_1^{2m})\,S_{2\alpha,n-m}(g^k).\end{aligned}$$

But from ($*$),

$$S_{2\alpha,n}(g^k h_1^{2m}) = \boldsymbol{l}_{2\alpha,n}\big(\delta_{n,+}(g^k h_1^{2m})\big) = \sum \boldsymbol{l}_{2\alpha,n}\big(i_S\otimes i_{S'}(w_{S,+}\otimes v_{S',+})\big).$$

On the right-hand side the summation is taken over all subsets of $S\subseteq(1, 2, \ldots, n)$ consisting of m elements, and each such S, which are $\binom{n}{m}$ in number, contributes the value $S_{2\alpha,m}(h_1^{2m})S_{2\alpha,n-m}(g^k)$. Thus we have

$$S_{2\alpha,n}(g^k h_1^{2m}) = \binom{n}{m} S_{2\alpha,n-m}(g^k)\,S_{2\alpha,m}(h_1^{2m}).$$

Q.E.D.

Lemma 2.7.17. We have

$$p_{g^k h_1^{2m}} = \frac{(2m)!}{2^{2m}}\binom{w/2}{m}\sum_{j=0}^{k} S_{2\alpha,j}(g^k)\binom{w/2-m}{j}.$$

Proof. Write $u=g^k h_1^{2m}$. We have, by Remark 2.5.6,

$$p_{g^k h_1^{2m}} = \sum_{n=0}^{k+2m} S_{2\alpha,n}(u)\binom{w/2}{n} = \sum_{n=m}^{k+2m}\binom{n}{m} S_{2\alpha,n-m}(g^k)\,S_{2\alpha,m}(h_1^{2m})\binom{w/2}{n},$$

by Lemma 2.7.16. Now

$$\delta_m(h_1^{2m}) = \sum \frac{(2m)!}{i_1!\,i_2!\dots i_m!} h_1^{i_1}\otimes h_1^{i_2}\otimes\cdots\otimes h_1^{i_m},$$

where the summation runs over all m-tuples $(i_1, i_2, \dots, i_m)$ of nonnegative integers such that $i_1+i_2+\cdots+i_m=2m$. Hence

$$S_{2\alpha,m}(h_1^{2m}) = \sum \frac{(2m!)}{i_1!\,i_2!\dots i_m!} p_{h_1^{i_1}}(2\alpha)\cdots p_{h_1^{i_m}}(2\alpha),$$

where the summation runs over all m-tuples $(i_1, i_2, \dots, i_m)$ of positive integers such that $i_1+i_2+\cdots+i_m=2m$. For any such m-tuple $(i_1, i_2, \dots, i_m)$, if $(i_1, i_2, \dots, i_m)\neq$ $\neq(2, 2, \dots, 2)$, then by Lemmas 2.7.2 and 2.7.5, $p_{h_1^{i_1}}(2\alpha)\cdots p_{h_1^{i_m}}(2\alpha)=0$. Moreover, by Lemma 2.7.9, $p_{h_1^2}=\frac{w}{4}$, so that $p_{h_1^2}(2\alpha)=\langle w/4, 2\alpha\rangle=1/2$. Thus

$$S_{2\alpha,m}(h_1^{2m}) = \frac{(2m)!}{2^m}\left(\frac{1}{2}\right)^m = \frac{(2m)!}{2^{2m}},$$

and so

$$p_{g^k h_1^{2m}} = \sum_{n=m}^{k+2m}\binom{n}{m} S_{2\alpha,n-m}(g^k)\frac{(2m)!}{2^{2m}}\binom{w/2}{n} =$$

$$= \sum_{n=m}^{k+2m}\binom{n}{m} S_{2\alpha,n-m}(g^k)\frac{(2m)!}{2^{2m}}\binom{w/2}{m}\binom{w/2-m}{n-m}\binom{n}{m}^{-1} =$$

$$= \frac{(2m)!}{2^{2m}}\binom{w/2}{m}\sum_{n=m}^{k+2m} S_{2\alpha,n-m}(g^k)\binom{w/2-m}{n-m} = \frac{(2m)!}{2^{2m}}\binom{w/2}{m}\sum_{j=0}^{k+m} S_{2\alpha,j}(g^k)\binom{w/2-m}{j}.$$

But clearly, if $j>k$, $\delta_{j,+}(g^k)=0$, and hence $S_{2\alpha,j}(g^k)=0$. Thus we have

$$p_{g^k h_1^{2m}} = \frac{(2m)!}{2^{2m}}\binom{w/2}{m}\sum_{j=0}^{k} S_{2\alpha,j}(g^k)\binom{w/2-m}{j}. \qquad \text{Q.E.D.}$$

Finally, in $U(\mathfrak{a}\cap[\mathfrak{g},\mathfrak{g}])$, we have two expressions for p_{g^k}: first,

$$p_{g^k} = \frac{w(w-1)\cdots(w-k+1)}{2^k}$$

by Lemma 2.7.15, and second, $p_{g^k}=\sum_{j=0}^{k} S_{2\alpha,j}(g^k)\binom{w/2}{j}$ by Remark 2.5.6. But $U(\mathfrak{a}\cap[\mathfrak{g},\mathfrak{g}])=\mathbf{C}[w]$, the polynomial ring in one indeterminate. Thus we have the formal identity

$$\frac{w(w-1)\cdots(w-k+1)}{2^k} \equiv \sum_{j=0}^{k} S_{2\alpha,j}(g^k)\binom{w/2}{j}.$$

Substituting $w-2m$ for w in this identity we have

$$\frac{(w-2m)(w-2m-1)\cdots(w-2m-k+1)}{2^k} = \sum_{j=0}^{k} S_{2\alpha,j}(g^k)\binom{\frac{w}{2}-m}{j}.$$

Thus, the assertion of Lemma 2.7.17 can be rewritten as

$$p_{g^k h_1^{2m}} = \frac{(2m)!}{2^{2m}}\binom{w/2}{m}\frac{(w-2m)(w-2m-1)\cdots(w-2m-k+1)}{2^k}$$

i.e., as

$$p_{g^k h_1^{2m}} = \frac{(2m)!}{2^{2m}}\,\frac{w(w-2)\cdots(w-2m+2)}{2^m m!}\,\frac{(w-2m)(w-2m-1)\cdots(w-2m-k+1)}{2^k}.$$

Now substituting for $p_{g^k h_1^{2m}}$ in Lemma 2.7.13, we get

$$p_{u(k,m)} = v[w(w-2)\cdots(w-2m+2)]\,[(w-2m)(w-2m-1)\cdots(w-2m-k+1)],$$

where

$$v = \frac{s(s+2)\cdots(s+2m-2)}{2^{2m+k}}.$$

This completes the proof of Theorem 2.7.11.

§ 8. The key result; the explicit formula for $p_{v(i,j)}$

We now consider a different basis of $(T^*)^{O(\mathfrak{b})}$. Let $q=g^{(2)}+h_1^{(2)}+\cdots+h_s^{(2)}$, so that $T=\sum q^{(i)}\circ L_j$, where L_j is the space of usual harmonics in T homogeneous of degree j. We have, therefore, $T^{O(\mathfrak{b})}=\sum q^{(i)}\circ(L_j)^{O(\mathfrak{b})}$.

Lemma 2.8.1. Assume $s>0$. Let $[j/2]$ be the greatest integer which is less than or equal to $j/2$. Let

$$\boldsymbol{l}_j = \sum_{m=0}^{[j/2]} a_m g^{(j-2m)}\circ h^{(m)},$$

where $a_0 = 1$ and for

$$1 \leq m \leq \left[\frac{j}{2}\right],\quad a_m = \frac{(-1)^m}{2^m m!}\,\frac{j(j-1)\cdots(j-2m+1)}{s(s+2)\ldots(s+2m-2)}.$$

Then $\boldsymbol{l}_j$ spans $(L_j)^{O(\mathfrak{b})}$.

Proof. Let

$$l'_j = \sum_{m=0}^{[j/2]} a'_m g^{(j-2m)} \circ h^{(m)},$$

where a'_m are arbitrary scalars. Since $g^{(j-2m)} \circ h^{(m)} \in (T_j)^{O(\mathfrak{b})}$, clearly $l'_j \in (T_j)^{O(\mathfrak{b})}$. Now, $l'_j \in (L_j)^{O(\mathfrak{b})}$ if and only if $\partial l'_j = 0$, where

$$\partial = \frac{\partial^2}{\partial g^2} + \frac{\partial^2}{\partial h_1^2} + \cdots + \frac{\partial^2}{\partial h_s^2}.$$

It is easy to check that

$$\frac{\partial^2}{\partial g^2}(l'_j) = \sum_{m=1}^{[j/2]} a'_{m-1}(j-2m+2)(j-2m+1)\, g^{(j-2m)} \circ h^{(m-1)}$$

and that

$$\frac{\partial^2}{\partial h_i^2}(l'_j) = \sum_{m=1}^{[j/2]} a'_m g^{(j-2m)} \circ \{2mh^{(m-1)} + 4m(m-1)h^{(m-2)} h_i^2\}.$$

The latter implies that

$$\frac{\partial^2}{\partial h_1^2} + \cdots + \frac{\partial^2}{\partial h_s^2}(l'_j) = \sum_{m=1}^{[j/2]} 2ma'_m(s+2(m-1))\, g^{(j-2m)} \circ h^{(m-1)}.$$

Thus

$$\partial l'_j = \sum_{m=1}^{[j/2]} \left(a'_{m-1}(j-2m+2)(j-2m+1) + a'_m 2m(s+2(m-1))\right) g^{(j-2m)} \circ h^{(m-1)}.$$

Hence $l'_j \in (L_j)^{O(\mathfrak{b})}$ if and only if

$$a'_m = -a'_{m-1} \frac{(j-2m+2)(j-2m+1)}{2m(s+2(m-1))},$$

i.e., if and only if

$$a'_m = a'_0 \frac{(-1)^m}{2^m m!} \frac{j(j-1)\dots(j-2m+1)}{s(s+2)\cdots(s+2m-2)}.$$

This proves the lemma since $L_j^{O(\mathfrak{b})}$ is one-dimensional. Q.E.D.

Since $T^{O(\mathfrak{b})} = \sum q^{(i)} \circ (L_j)^{O(\mathfrak{b})}$, it follows from Lemma 2.8.1 that when $s>0$, $\{q^{(i)} \circ l_j\}$ is a basis of $T^{O(\mathfrak{b})}$. When $s=0$, note that $\mathfrak{b}=0$ and $L_j=0$ if $j>1$. We define $l_0=1$ and $l_1=g$, so that $\{q^{(i)} \circ l_j\}$ is still a basis of $T^{O(\mathfrak{b})}$. We define $v(i,j) = \beta(q^{(i)} \circ l_j)$, so that $\{v(i,j)\}$ is a basis of $(T^*)^{O(\mathfrak{b})}$. Note that $v(0,j) = \beta(l_j)$.

Remark 2.8.2. When $s=0$, $p_{v(i,j)} = \dfrac{w(w-1)\cdots(w-d+1)}{2^d}$, where $d=2i+j$. This follows from Lemma 2.7.15, on observing that when $s=0$, $v(i,j)=g^d$. (Note that when $s=0$, $j=0$ or 1.)

We now begin to compute $p_{v(i,j)}$ when $s>0$. It follows from Lemma 2.8.1 that

$$p_{v(0,j)} = \sum_{m=0}^{[j/2]} a_m p_{u(j-2m,m)}.$$

Proposition 2.8.3. One has $p_{v(0,j)}=\dfrac{w(w-2)\cdots(w-2j+2)}{2^j}$.

Proof. Let $f=g-ih_1$. Note that $B_z(f,f)=0$, since $B_z(f,f)=B_z(g,g)-2iB_z(g,h_1)-B_z(h_1,h_1)$, and since $B_z(g,g)=B_z(h_1,h_1)=\dfrac{\langle w,w\rangle}{2}$ and $B_z(g,h_1)=0$. Thus $f^{(j)}\in L_j$, by Remark 2.1.3.

We have $f^2=g^2-i(gh_1+h_1g)-h_1^2$. By Lemma 2.7.10, $p_{gh_1}=p_{hg_1}=0$, so that $p_{f^2}=p_{g^2}-p_{h_1^2}=\dfrac{w(w-1)}{4}-\dfrac{w}{4}=\dfrac{w(w-2)}{4}$. Hence $p_{f^2}(2\alpha)=0$, and so $p_{f^i}(2\alpha)=0$ for all $i\geqq 2$, by Lemma 2.7.2. Now, by Remark 2.5.6, $p_{f^k}=\sum_{j=0}^{k} S_{2\alpha,j}(f^k)\binom{w/2}{j}$.
We have $\delta_{k,+}(f^k)=k!\,\underbrace{f\otimes f\otimes\cdots\otimes f}_{k\text{ times}}$, and for $j<k$, $\delta_{j,+}(f^k)=\sum\dfrac{k!}{i_1!\,i_2!\ldots i_j!}$ $f^{i_1}\otimes f^{i_2}\otimes\cdots\otimes f^{i_j}$, where the summation runs over all j-tuples $(i_1,\ldots,i_j)$ of positive integers such that $i_1+\cdots+i_j=k$. It follows from the fact that $p_{f^i}(2\alpha)=0$ for $i\geqq 2$ that $S_{2\alpha,j}(f^k)=0$ if $j<k$. Thus $p_{f^k}=S_{2\alpha,k}(f^k)\binom{w/2}{k}=k!\,(p_f(2\alpha))^k\binom{w/2}{k}$. But $p_f=p_{g-ih_1}=p_g-ip_{h_1}=w/2$, by Lemma 2.7.5, so that $p_f(2\alpha)=1$. Thus $p_{f^k}=k!\binom{w/2}{k}=\dfrac{w(w-2)\cdots(w-2k+2)}{2^k}$.

We assert that $p_{f^j}=p_{\beta(l_j)}$. To prove this, let $l'_j=\int_{M_{0,\mathbf{R}}} a\cdot f^{(j)}\,da$ if $s>1$, and let $l'_j=\dfrac{f^{(j)}+(\bar f)^{(j)}}{2}$ if $s=1$, where $\bar f=g+ih_1$. When $s>1$, $M_{0,\mathbf{R}}$ is transitive on the unit sphere in $\mathfrak{b}_0$ and hence $S(\mathfrak{b})^{O(\mathfrak{b})}=S(\mathfrak{b})^{M_{0,\mathbf{R}}}$, so that $T^{O(\mathfrak{b})}=S(\mathbf{C}g)\otimes S(\mathfrak{b})^{M_{0,\mathbf{R}}}=T^{M_{0,\mathbf{R}}}$. When $s=1$, $O(\mathfrak{b})$ consists of the identity element and the element of $GL(\mathfrak{d})$ defined by $g\to g$ and $h_1\to -h_1$. Thus it follows that in all cases $l'_j\in(L_j)^{O(\mathfrak{b})}$. Thus, if we show that the component of l'_j in $\mathbf{C}g^{(j)}$ with respect to the decomposition $T_j=\mathbf{C}g^{(j)}\oplus(T\circ\mathfrak{b})_j$ is $g^{(j)}$, ($T\circ\mathfrak{b}$ is the ideal in T generated by $\mathfrak{b}$), then it follows in view of Lemma 2.8.1 that $l'_j=l_j$. When $s=1$, $l'_j=\dfrac{(g-ih_1)^{(j)}+(g+ih_1)^{(j)}}{2}$, which has component $g^{(j)}$ in $Cg^{(j)}$. When $s>1$, note that $M_{0,\mathbf{R}}$ acts trivially on $\mathbf{C}g^{(j)}$ and leaves $(T\circ g)_j$ stable. Also, the component of $(g-ih_1)^{(j)}$ in $\mathbf{C}g^{(j)}$ is $g^{(j)}$. Thus it follows that $\int_{M_{0,\mathbf{R}}} a\cdot(g-ih_1)^{(j)}\,da$ has component $g^{(j)}$ in $\mathbf{C}g^{(j)}$.

Hence $l_j=\int_{M_{0,\mathbf{R}}} a\cdot f^{(j)}\,da$ if $s>1$ and $l_j=\dfrac{f^{(j)}+(\bar f)^{(j)}}{2}$ when $s=1$. Note that the proof that $p_{f^j}=\dfrac{w(w-2)\cdots(w-2j+2)}{2^j}$ goes through without change to prove also that $p_{\bar f^j}=\dfrac{w(w-2)\cdots(w-2j+2)}{2^j}$. Thus, when $s=1$, upon applying β one has $2\beta(\mathbf{l}_j)=f^j+\bar{f}^j$ and hence we have $p_{\beta(\mathbf{l}_j)}=p_{f^j}$. When $s>1$ the relation $\mathbf{l}_j=\int\limits_{M_{0,\mathbf{R}}} a\cdot f^{(j)}\,da$ implies (on applying β) that in T^* we have $\beta(\mathbf{l}_j)=\int\limits_{M_{0,\mathbf{R}}} a\cdot f^j\,da$. By Lemma 1.7.10, we have $p_{a\cdot f^j}=p_{f^j}$ for all $a\in M_{0,\mathbf{R}}$, and hence $p_{\beta(\mathbf{l}_j)}= = \int\limits_{M_{0,\mathbf{R}}} p_{a\cdot f^j}\,da=\int\limits_{M_{0,\mathbf{R}}} P_{f^j}\,da=p_{f^j}$.

This completes the proof of the assertion and hence the proposition. Q.E.D.

Lemma 2.8.4. One has the identity

$$\frac{w(w-2)\cdots(w-2j+2)}{2^j}=$$

$$=\sum_{m=0}^{[j/2]} b_m(\{w(w-2)\cdots(w-2m+2)\}\{(w-2m)(w-2m-1)\cdots(w-j+1)\}),$$

where

$$b_m=(-1)^m\,\frac{j(j-1)\cdots(j-2m+1)}{2^{j+m}m!}.$$

(When $m=0$, $b_0=\frac{1}{2^j}$, and the corresponding term in the sum is

$$\frac{w(w-1)\cdots(w-j+1)}{2^j}.)$$

Proof. By Lemma 2.8.1, we have

$$p_{\beta(\mathbf{l}_j)}=\sum_{m=0}^{[j/2)} a_m p_{u(j-2m,m)},$$

where $a_0=1$, and for $m>1$,

$$a_m=\frac{(-1)^m}{2^m m!}\,\frac{j(j-1)\cdots(j-2m+1)}{s(s+2)\cdots(s+2m-2)}.$$

The result now follows from substituting the expressions for $p_{\beta(\mathbf{l}_j)}$ and $p_{u(j-2m,m)}$ from Proposition 2.8.3 and Theorem 2.7.11, respectively. Q.E.D.

Definition 2.8.5. Let a, b be non-negative integers such that $b\geqq 2a$. We define $C_a(X)$, $D_{a,b}(X)$ and $F_{a,b}(X)$ to be formal polynomials in X by setting $C_a(X)=X(X-2)\cdots(X-2a+2)$, $D_{a,b}(X)=(X-2a)(X-2a-1)\cdots(X-b+1)$ and $F_{a,b}(X)=C_a(X)D_{a,b}(X)$. It is to be understood that when $a=0$, $C_a=1$ and when $2a=b$, $D_{a,b}(X)=1$. Thus for example we can rewrite Theorem 2.7.11 as $p_{u(k,m)}=vF_{m,2m+k}(w)$, where $v=\frac{s(s+2)\cdots(s+2m-2)}{2^{2m+k}}$. Also, Lemma 2.8.4 can be written as

$$\frac{C_j(w)}{2^j}=\sum_{m=0}^{[j/2]} b_m F_{m,j}(w),$$

where

$$b_m=(-1)^m\,\frac{j(j-1)\cdots(j-2m+1)}{2^{j+m}m!}.$$

Now consider $v(i, 0)$. We have $v(i, 0)=\beta(q^{(i)})=\beta((g^{(2)}+h)^{(i)})$. But $(g^{(2)}+h)^{(i)}= = \sum_{m=0}^{i}\binom{i}{m} g^{(2i-2m)} \circ h^{(m)}$, so that $v(i, 0)= \sum_{m=0}^{i}\binom{i}{m} u(2i-2m, m)$. Hence, from Theorem 2.7.11,

$$p_{v(i,0)} = \frac{1}{2^{2i}} \sum_{m=0}^{i} \binom{i}{m} s(s+2)\cdots(s+2m-2) F_{m,2i}(w). \qquad (*)$$

Lemma 2.8.6. One has

$$p_{v(i,0)} = \frac{1}{2^{2i}} [w(w-2)\cdots(w-2i-2)] [(w-1+s)(w-3+s)\cdots(w-2i+1+s)].$$

Proof. We have $F_{m,2i}(w)=[w(w-2)\cdots(w-2m+2)][(w-2m)(w-2m-1)\cdots \cdots(w-2i+1)]=EO_m$ where $E=w(w-2)\cdots(w-2m+2)(w-2m)(w-2m-2)\cdots \cdots(w-2i+2)$ and $O_m=(w-2m-1)(w-2m-3)\cdots(w-2i+1)$. (Note that E is independent of m.) Thus, from $(*)$, we have

$$p_{v(i,0)} = \frac{E}{2^{2i}} \sum_{m=0}^{i} \binom{i}{m} s(s+2)\cdots(s+2m-2) O_m = EQ,$$

where

$$Q = \frac{1}{2^{2i}} \sum_{m=0}^{i} \binom{i}{m} s(s+2)\cdots(s+2m-2) O_m.$$

We can write

$$Q = \frac{i!}{(-2)^i} \sum_{m=0}^{i} \binom{s/-2}{m} \begin{pmatrix} \dfrac{w-2i+1}{-2} \\ i-m \end{pmatrix}.$$

One can easily prove the well-known identity

$$\binom{X+Y}{i} = \sum_{j=0}^{i} \binom{X}{j}\binom{Y}{i-j}.$$

Thus we have

$$Q = \frac{i!}{(-2)^i} \begin{pmatrix} \dfrac{w-2i+1+s}{-2} \\ i \end{pmatrix} =$$

$$= \frac{i!}{(-2)^i} (w-2i+1+s)(w-2i+1+s+2)\cdots(w-2i+1+s+2i-2)\frac{1}{(-2)^i i!} =$$

$$= \frac{1}{2^{2i}} (w-1+s)(w-3+s)\cdots(w-2i+1+s).$$

Substituting for E and Q in $p_{v(i,0)}=EQ$, we get the desired result. Q.E.D.

Remark 2.8.7. Note that the above proof establishes the following formal identity in w and s:

$$\sum_{m=0}^{i} \binom{i}{m} s(s+2)\cdots(s+2m-2)\, F_{m,2i}(w) =$$

$$= [w(w-2)\cdots(w-2i+2)]\,[(w-1+s)(w-3+s)\cdots(w-2i+1+s)]$$

for every i.

The main result we need is

Theorem 2.8.8. *For all* $s \geqq 0$ *one has*

$$p_{v(i,j)} = \frac{1}{2^d}\left[w(w-2)\cdots(w-2(i+j)+2)\right][(w-1+s)(w-3+s)\cdots(w-2i+1+s)],$$

where $d=2i+j$; *recall when* $s=0$, *then* j *takes only the values* $j=0$ *or* 1.

Proof. The case $s=0$ follows from Remark 2.8.2. Now assume $s>0$. We have

$$q^{(i)} = (g^{(2)}+h)^{(i)} = \sum_{k=0}^{i} \binom{i}{k} g^{(2i-2k)} \circ h^{(k)}$$

and

$$l_j = \sum_{n=0}^{[j/2]} a_n g^{(j-2n)} \circ h^{(n)}$$

by Lemma 2.8.1, where

$$a_n = \frac{(-1)^n}{2^n n!}\,\frac{j(j-1)\cdots(j-2n+1)}{s(s+2)\cdots(s+2n-2)}.$$

Thus

$$q^{(i)} \circ l_j = \sum_{k=0}^{i}\sum_{n=0}^{[j/2]} a_n \binom{i}{k} g^{(2i+j-2k-2n)} \circ h^{(k+n)} = \sum_{n=0}^{[j/2]}\sum_{m=n}^{i+m} a_n \binom{i}{m-n} g^{(d-2m)} \circ h^{(m)}.$$

Hence

$$p_{v(i,j)} = \sum_{n=0}^{[j/2]}\sum_{m=n}^{i+n} a_n \binom{i}{m-n} p_{u(d-2m,\,m)}.$$

By Theorem 2.7.11, $p_{u(d-2m,\,m)} = v_m F_{m,d}(w)$, where

$$v_m = \frac{s(s+2)\cdots(s+2m-2)}{2^d}.$$

We have

$$a_n v_m = \frac{(-1)^n}{2^{n+d} n!} j(j-1)\cdots(j-2n+1)\,\lambda_{n,m},$$

where $\lambda_{n,m} = (s+2n)(s+2n+2)\cdots(s+2m-2)$. Substituting, we have

$$p_{v(i,j)} = \sum_{n=0}^{[j/2]} \frac{(-1)^n}{2^{n+d} n!} j(j-1)\cdots(j-2n+1) \sum_{m=n}^{i+n} \lambda_{n,m} \binom{i}{m-n} F_{m,d}(w). \tag{I}$$

In order to complete the proof of Theorem 2.8.8, we observe the following:

Lemma 2.8.9. One has

1) $C_m(w) = C_n(w)\, C_{m-n}(w-2n)$.

2) $D_{m,d}(w) = D_{m,2(i+n)}(w)\, D_{i+n,d}(w)$.

3) $D_{m,2(i+n)}(w) = D_{m-n,2i}(w-2n)$.

Proof. All are immediate from Definition 2.8.5. Q.E.D.

Now *2)* and *3)* of Lemma 2.8.9 together give the following:

4) $D_{m,d}(w) = D_{m-n,2i}(w-2n)\, D_{i+n,d}(w)$.

Since, by definition, $F_{m,d}(w) = C_m(w) D_{m,d}(w)$, we have from *1)* and *4)* the following:

$$F_{m,d}(w) = C_n(w)\, F_{m-n,2i}(w-2n)\, D_{i+n,d}(w).$$

Thus we have from (I)

$$p_{v(i,j)} = \sum_{n=0}^{[j/2]} \frac{(-1)^n}{2^{n+d} n!} j(j-1)\cdots(j-2n+1)\, C_n(w)\, D_{i+n,d}(w) \cdot$$

$$\cdot \sum_{m=n}^{i+n} \lambda_{n,m} \binom{i}{m-n} F_{m-n,2i}(w-2n). \tag{II}$$

Writing $m-n=k$, $w_1 = w-2n$ and $s_1 = s+2n$, we have $\lambda_{n,m} = s_1(s_1+2)\cdots(s_1+2k-2)$ and

$$\sum_{m=n}^{i+n} \lambda_{n,m} \binom{i}{m-n} F_{m-n,2i}(w-2n) = \sum_{k=0}^{i} \binom{i}{k} s_1(s_1+2)\cdots(s_1+2k-2) F_{k,2i}(w_1) =$$

$$= [w_1(w_1-2)\cdots(w_1-2i+2)]\,[(w_1-1+s_1)(w_1-3+s_1)\cdots(w_1-2i+1+s_1)],$$

by Remark 2.8.7. Note that $w_1+s_1 = w+s$. Write $B = (w-1+s)(w-3+s)\cdots(w-2i+1+s)$. We now have

$$C_n(w)\, D_{i+n,d}(w) \sum_{m=n}^{i+m} \lambda_{n,m} \binom{i}{m-n} F_{m-n,2i}(w-2n) = C_n(w) D_{i+n,d}(w) C_i(w-2n) B. \tag{III}$$

But $C_n(w) C_i(w-2n) = C_{i+n}(w)$ (by Lemma 2.8.9) $= C_i(w) C_n(w-2i)$, and

$$D_{i+n,d}(w) = D_{n,d-2i}(w-2i).$$

Thus, from (III),

$$C_n(w)\, D_{i+n,d}(w) \sum_{m=n}^{i+n} \lambda_{n,m} \binom{i}{m-n} F_{m-n,2i}(w-2n) =$$

$$= C_i(w) C_n(w-2i)\, D_{n,d-2i}(w-2i)\, B = C_i(w)\, F_{n,j}(w-2i)\, B.$$

Substituting in (II), we obtain

$$p_{v(i,j)} = C_i(w)\,B \sum_{n=0}^{[j/2]} \frac{(-1)^n}{2^{n+d}n!} j(j-1)\cdots(j-2n+1)\,F_{n,j}(w-2i) =$$

$$= C_i(w)\,B\,\frac{(w-2i)(w-2i-2)\cdots(w-2i-2j+2)}{2^d},$$

using Lemma 2.8.4. Substituting for $C_i(w)$ and B, we obtain

$$p_{v(i,j)} = \frac{1}{2^d}\,[w(w-2)\cdots(w-2(i+j)+2)][(w-1+s)(w-3+s)\cdots(w-2i+1+s)].$$

This completes the proof of Theorem 2.8.8. Q.E.D.

§ 9. Application; the critical strip and the criterion for irreducibility and Weyl chambers

We return to the case of arbitrary rank. Let $\gamma\in\Gamma$, and let $v_1, \ldots, v_{l(\gamma)}$ be a basis of $(V_\gamma)^{M_\theta}$. Let $\varepsilon_1, \ldots, \varepsilon_{l(\gamma)}$ be a basis of $E_\gamma = \mathrm{Hom}_{K_\theta}(V_\gamma, H^*)$. We recall that for all $u\in U$, we have by definition that $p^u = \varkappa\theta p_{u^t}$ (see § 1.7.6 and § 1.7.8). We recall further that P^γ is the matrix with coefficients in $U(\mathfrak{a})$ defined by $P^\gamma = (P^\gamma_{ij})$, where $P^\gamma_{ji} = p^{\varepsilon_i(v_j)}$. We define $p^\gamma = \det P^\gamma$, and we recall that $p_\gamma = \det\{p_{\varepsilon_i(v_j)}\}$.

Now choose $\bar\varkappa\in K$ which represents the Weyl group element $\varkappa$. One has that $\bar\varkappa\theta$ preserves the decomposition $U = (\mathfrak{k}U + U\mathfrak{n})\oplus U(\mathfrak{a})$, and so $\varkappa\theta p_{u^t} = p_{\bar\varkappa\theta u^t}$ for all $u\in U$. If $u\in S^*$, then by Remark 1.6.1 we have $\theta u^t = u$, so that $p_{\bar\varkappa\theta u^t} = p_{\bar\varkappa u}$. Thus $p^u = p_{\bar\varkappa u}$, and so $p^{\varepsilon_i(v_j)} = p_{\bar\varkappa\varepsilon_i(v_j)} = p_{\varepsilon_i(\bar\varkappa\cdot v_j)}$. Clearly, $\bar\varkappa M_\theta\bar\varkappa^{-1} = M_\theta$ and hence $\bar\varkappa\cdot V_\gamma^{M_\theta} = V_\gamma^{M_\theta}$. Thus $\det\{p_{\varepsilon_i(\bar\varkappa\cdot v_j)}\} = \det\{p_{\varepsilon_i(v_j)}\}\det\{\bar\varkappa\,|\,(V_\gamma)^{M_\theta}\}$, i.e.,

$$p^\gamma = p_\gamma \det(\bar\varkappa\,|\,(V_\gamma)^{M_\theta}).$$

Hence p^γ is a non-zero scalar multiple of p_γ.

Now let $\lambda\in\mathfrak{a}'$(=dual of $\mathfrak{a}$), and consider $r^\gamma(\lambda) = \det R^\gamma(\lambda)$. By Theorem 1.8.6, $R^\gamma(\lambda) = (P^\gamma(-\bar\lambda-2\varrho))^* P^\gamma(\lambda)$, with respect to suitable bases. Hence

$$r^\gamma(\lambda) = \overline{p^\gamma(-\bar\lambda-2\varrho)}\,p^\gamma(\lambda) = \bar p^\gamma(-\lambda-2\varrho)\,p^\gamma(\lambda) = \bar p_\gamma(-\lambda-2\varrho)\,p_\gamma(\lambda)$$

(up to a non-zero scalar). But from the expression of the polynomials p_γ in the rank 1 case given by Theorem 2.8.8 (see Remark 2.9.2 below) and from the formula for p_γ (in the general case) given in Theorem 2.4.6, it follows that $p_\gamma = \bar p_\gamma$. Thus we have:

Lemma 2.9.1. Up to a scalar, $r^\gamma(\lambda) = p_\gamma(-\lambda-2\varrho)\,p_\gamma(\lambda)$ for all $\lambda\in\mathfrak{a}'$ and $\gamma\in\Gamma$.

Remark 2.9.2. Our assertion that in the rank 1 case the polynomials p_γ are the polynomials of Theorem 2.8.8 follows from Remark 2.2.14, when $s>1$ or when $s=0$. We now prove this assertion when $s=1$. First note that $L_j^+ = \mathbf{C}f^{(j)}$, where $f = g - ih_1$, and $L_j^- = \mathbf{C}\bar f^{(j)}$, where $\bar f = g + ih_1$ (see Remark 2.1.25). Let σ

be the unique nontrivial element of $O(\mathfrak{b})$, and observe that $\sigma(f)=\bar{f}$. Hence $f^{(j)}+\bar{f}^{(j)}\in(L_j)^{O(\mathfrak{b})}$, and so $f^{(j)}+\bar{f}^{(j)}$ is a non-zero multiple of $\boldsymbol{l}_j$. In view of Remark 2.2.14, it is now sufficient to show that $p_{\beta(q^{(i)}\circ f^{(j)})}=p_{\beta(q^{(i)}\circ \bar{f}^{(j)})}$. To prove this, first note that $q^{(i)}\circ f^{(j)}=(g^{(2)}+h_1^{(2)})^{(i)}\circ(g-ih_1)^{(j)}$ and $q^{(i)}\circ\bar{f}^{(j)}=$ $=(g^{(2)}+h_1^{(2)})^{(i)}\circ(g+ih_1)^{(j)}$. Since they differ only for odd powers of h_1 it is enough to prove that $p_{\beta(g^{(k)}\circ h_1^{(m)})}=0$ if m is odd, or equivalently, by Lemma 2.7.12, that $p_{g^k h_1^m}=0$ if m is odd. But this follows from Remark 2.5.6 and the fact that each coefficient $S_{2\alpha,j}(g^k h_1^m)$ vanishes. (For this, note that $\delta_{j,+}(g^k h_1^m)$ is a sum of terms of the form $u_1\otimes u_2\otimes\cdots\otimes u_j$, where at least one u_i is of the form $g^s h_1^t$ with t odd. (See §2.5.) Now the fact that $S_{2\alpha,j}(g^k h_1^m)=0$ follows from the fact that $p_{g^s h_1^t}(2\alpha)=0$ if t is odd, which itself is a consequence of Lemma 2.7.2, Lemma 2.7.5 and Lemma 2.7.10.)

Recall that for every $\lambda\in\mathfrak{a}_0'$, where $\mathfrak{a}_0'=\{\lambda\in\mathfrak{a}'\,|\,\lambda|(\text{center }\mathfrak{g})\cap\mathfrak{a}=0\}$, we have a U-module X_λ (see Definition 1.7.2). We will now determine those $\lambda\in\mathfrak{a}_0'$ such that X_λ is U-irreducible. Let $\mathfrak{a}_I'=\{\lambda\in\mathfrak{a}_0'\,|\,X_\lambda \text{ is } U\text{-irreducible}\}$. By Corollary 1.7.7, if $\lambda\in\mathfrak{a}_0'$, then $\lambda\in\mathfrak{a}_I'$ if and only if for every $\gamma\in\Gamma$, $r^\gamma(\lambda)\neq0$; thus, by Lemma 2.9.1, $\lambda\in\mathfrak{a}_I'$ if and only if for every $\gamma\in\Gamma$, $p_\gamma(\lambda)\neq0$ and $p_\gamma(-\lambda-2\varrho)\neq0$.

Corollary 2.4.8 implies the following:

Proposition 2.9.3. Let $\lambda\in\mathfrak{a}'$. Then $r^\gamma(\lambda)\neq0$ for every $\gamma\in\Gamma$ if and only if $(T_{\varrho_\varphi}p_{\delta^\varphi})(\mu)\neq0$ and $(T_{\varrho_\varphi}p_{\delta^\varphi})(-\mu)\neq0$ for every $\varphi\in\Delta_+^1$ and every $\delta^\varphi\in\Gamma_\varphi$, where $\mu=\lambda+\varrho$. In particular, if $\lambda\in\mathfrak{a}_0'$, then $\lambda\in\mathfrak{a}_I'$ if and only if these conditions hold.

Remark 2.9.4. Recall that Γ_φ plays the role of Γ for the case $\mathfrak{g}_\varphi=\mathfrak{k}_\varphi+\mathfrak{p}_\varphi$ (see Remark 2.3.6). Thus, for a fixed $\varphi\in\Delta_+^1$, the polynomials p_{δ^φ} $(\delta^\varphi\in\Gamma_\varphi)$ are the polynomials of the form (see Remark 2.9.2)

$$[w_\varphi(w_\varphi-2)\cdots(w_\varphi-2(i+j)+2)]\,[(w_\varphi-1+s_\varphi)(w_\varphi-3+s_\varphi)\cdots(w_\varphi-2i+1+s_\varphi)],$$

where $s_\varphi=\dim\mathfrak{g}^{2\varphi}$, and where the pair (i,j) runs through the set $\{(i,j)\,|\,i,j\in\mathbf{Z},$ $i\geqq0,\ j\geqq0\}$ if $s_\varphi>0$ (i.e., if $2\varphi\in\Delta$) and the set $\{(i,j)\,|\,i\in\mathbf{Z},\ i\geqq0$ and $j=0$ or $1\}$ if $s_\varphi=0$ (i.e., if $2\varphi\notin\Delta$). We have

$$T_{\varrho\varphi}(w_\varphi)=w_\varphi-\langle\varrho_\varphi,w_\varphi\rangle=w_\varphi-\left(s_\varphi+\frac{\dim\mathfrak{g}^\varphi}{2}\right)=w_\varphi-k_\varphi,$$

writing $k_\varphi=s_\varphi+\dfrac{\dim\mathfrak{g}^\varphi}{2}$. Thus, as δ^φ runs through Γ_φ, the polynomials $T_{\varrho_\varphi}p_{\delta^\varphi}$ run through all the polynomials of the form

$$[(w_\varphi-k_\varphi)(w_\varphi-k_\varphi-2)\cdots(w_\varphi-k_\varphi-2(i+j)+2)]\cdot$$
$$\cdot[(w_\varphi-k_\varphi-1+s_\varphi)(w_\varphi-k_\varphi-3+s_\varphi)\cdots(w_\varphi-k_\varphi-2i+1+s_\varphi)],$$

the pair (i,j) running through the sets indicated above.

Definition 2.9.5. For every $\varphi \in \Delta^1_+$, let

$$I_\varphi = \left\{ t \in \mathbf{R} \,\middle|\, |t| < \frac{\dim \mathfrak{g}^\varphi}{2} \right\} \quad \text{if} \quad 2\varphi \notin \Delta$$

and

$$I_\varphi = \left\{ t \in \mathbf{R} \,\middle|\, |t| < \frac{\dim \mathfrak{g}^\varphi}{2} + 1 \right\} \quad \text{if} \quad 2\varphi \in \Delta.$$

Definition 2.9.6. Let $CS = \{\lambda \in \mathfrak{a}' \,|\, \text{real part of } \langle \lambda + \varrho, w_\varphi \rangle \in I_\varphi \text{ for all } \varphi \in \Delta^1_+\}$. ($CS$ stands for "critical strip".)

Remark 2.9.7. Let $\mathfrak{a}_0^\perp = \{\lambda \in \mathfrak{a}' \,|\, \lambda \,|\, \mathfrak{a} \cap [\mathfrak{g}, \mathfrak{g}] = 0\}$. Then from the fact that for $\varphi \in \Delta^1_+$, $w_\varphi \in \mathfrak{a} \cap [\mathfrak{g}, \mathfrak{g}]$, it follows that if $\lambda \in CS$, then $\lambda + \mathfrak{a}_0^\perp \subseteqq CS$.

The necessary and sufficient condition for irreducibility of X_λ (or rather reducibility) is now given in

Theorem 2.9.8. *Let $\lambda \in \mathfrak{a}'$. Then $r^\gamma(\lambda) = 0$ for some $\gamma \in \Gamma$ if and only if there exists $\varphi \in \Delta^1_+$ such that*

1) $\langle \lambda + \varrho, w_\varphi \rangle \notin I_\varphi$

and

2) $\langle \lambda + \varrho, w_\varphi \rangle - l_\varphi \in n_\varphi \mathbf{Z}$, *where* $l_\varphi = \dfrac{\dim \mathfrak{g}^\varphi}{2}$

if $2\varphi \notin \Delta$ *and* $l_\varphi = \dfrac{\dim \mathfrak{g}^\varphi}{2} + 1$ *if* $2\varphi \in \Delta$, *and* $n_\varphi = 1$ *or* 2 *according as* $2\varphi \notin \Delta$ *or* $2\varphi \in \Delta$. *In particular, if* $\lambda \in \mathfrak{a}_0'$, *then* $\lambda \notin \mathfrak{a}_I'$ *if and only if the above condition is satisfied.*

Proof. We first prove the following:

Lemma 2.9.9. Let $\lambda \in \mathfrak{a}'$. Then there exists $\gamma \in \Gamma$ such that $p_\gamma(\lambda) = 0$ if and only if there exists $\varphi \in \Delta^1_+$ such that

$$\langle \lambda + \varrho, w_\varphi \rangle - l_\varphi \in n_\varphi \mathbf{Z}_+$$

(where $\mathbf{Z}_+ = \{i \in \mathbf{Z} \,|\, i \geqq 0\}$).

Proof. By Corollary 2.4.8, there exists $\gamma \in \Gamma$ such that $p_\gamma(\lambda) = 0$ if and only if there exists $\varphi \in \Delta^1_+$ and $\delta^\varphi \in \Gamma_\varphi$ such that $(T_{\varrho_\varphi} p_{\delta^\varphi})(\lambda + \varrho) = 0$. By Remark 2.9.4, this happens if and only if there exists $\varphi \in \Delta^1_+$ and a pair of integers (i, j) such that

(i) $2\varphi \in \Delta$ and $i, j \geqq 0$ or
(ii) $2\varphi \notin \Delta$ and $i \geqq 0$ and $j = 0$ or 1

with the property that

$$[(a - k_\varphi)(a - k_\varphi - 2) \cdots (a - k_\varphi - 2(i+j) + 2)] \cdot$$
$$\cdot [(a - k_\varphi - 1 + s_\varphi)(a - k_\varphi - 3 + s_\varphi) \cdots (a - k_\varphi - 2i + 1 + s_\varphi)] = 0,$$

where $a = \langle \lambda + \varrho, w_\varphi \rangle$.

We assert that the last condition is equivalent to the condition that either

(i) there exists $\varphi \in \Delta^1_+$ such that $2\varphi \in \Delta$ with the property that $a-k_\varphi-1+$ $+s_\varphi \in 2\mathbf{Z}_+$, or

(ii) there exists $\varphi \in \Delta^1_+$ such that $2\varphi \notin \Delta$ with the property that $a-k_\varphi \in \mathbf{Z}_+$.

This follows easily on observing that if $2\varphi \in \Delta$, then $s_\varphi (=\dim \mathfrak{g}^{2\varphi})$ is odd (see Lemma 2.2.7).

The lemma is now immediate, for when $2\varphi \in \Delta$, $l_\varphi = k_\varphi + 1 - s_\varphi$, and when $2\varphi \notin \Delta$, $l_\varphi = k_\varphi$. Q.E.D.

Now, to prove Theorem 2.9.8, first note that for any $\lambda \in \mathfrak{a}'$ and any $\gamma \in \Gamma$, $r^\gamma(\lambda)=0$ if and only if either $p_\gamma(\lambda)=0$ or $p_\gamma(-\lambda-2\varrho)=0$, by Lemma 2.9.1. Thus Lemma 2.9.9 implies that there exists $\gamma \in \Gamma$ such that $r^\gamma(\lambda)=0$ if and only if there exists $\varphi \in \Delta^1_+$ such that either

(i) $\langle \lambda+\varrho, w_\varphi \rangle - l_\varphi \in n_\varphi \mathbf{Z}_+$

or

(ii) $\langle -\lambda-\varrho, w_\varphi \rangle - l_\varphi \in n_\varphi \mathbf{Z}_+$.

We assert that for any $\varphi \in \Delta^1_+$, the condition that (i) or (ii) above holds is equivalent to the conditions of the theorem, namely that

1) $\langle \lambda+\varrho, w_\varphi \rangle \notin I_\varphi$

and

2) $\langle \lambda+\varrho, w_\varphi \rangle - l_\varphi \in n_\varphi \mathbf{Z}$.

To see this, first note that I_φ is the open interval in $\mathbf{R}$ of length $2l_\varphi$ with centre 0. Now assume either (i) or (ii) holds. If (i) holds, then *1)* and *2)* are immediate. If (ii) holds, *1)* follows trivially, and to prove *2)*, it is enough to prove that $2l_\varphi \in n_\varphi \mathbf{Z}$, since we can write $\langle \lambda+\varrho, w_\varphi \rangle - l_\varphi$ as $\langle \lambda+\varrho, w_\varphi \rangle + l_\varphi - 2l_\varphi$. If $s_\varphi=0$, then $l_\varphi = \dfrac{\dim \mathfrak{g}^\varphi}{2}$ and $n_\varphi=1$, so that indeed $2l_\varphi \in n_\varphi \mathbf{Z}$. If $s_\varphi \neq 0$, then $l_\varphi = \dfrac{\dim \mathfrak{g}^\varphi}{2}+1$ and $n_\varphi=2$, and so it is enough to show that $\dim \mathfrak{g}^\varphi$ is even. But as observed in the proof of Lemma 2.2.8, if $\mathfrak{g}_{\varphi,0}=[\mathfrak{g}_\varphi, \mathfrak{g}_\varphi]$, $\mathfrak{k}_{\varphi,0}=\mathfrak{g}_{\varphi,0} \cap \mathfrak{k}_\varphi$ and $\mathfrak{p}_{\varphi,0}=\mathfrak{g}_{\varphi,0} \cap \mathfrak{p}_\varphi$, then there exists a Cartan subalgebra $\mathfrak{h}_\varphi$ of $\mathfrak{g}_{\varphi,0}$ such that $\mathfrak{h}_\varphi \subseteq \mathfrak{k}_{\varphi,0}$. Clearly, if $\psi \in \Delta(\mathfrak{h}_\varphi, \mathfrak{g}_{\varphi,0})$, then either $\mathfrak{g}^{\pm\psi}_{\varphi,0} \subseteq \mathfrak{k}_{\varphi,0}$ or $\mathfrak{g}^{\pm\psi}_{\varphi,0} \subseteq \mathfrak{p}_{\varphi,0}$. Thus $\mathfrak{p}_{\varphi,0}$ is even-dimensional. Since we have $\mathfrak{g}_{\varphi,0}=\mathfrak{k}_{\varphi,0} \oplus \mathfrak{p}_{\varphi,0}$ and also $\mathfrak{g}_{\varphi,0}=\mathfrak{k}_{\varphi,0} \oplus \mathbf{C}w_\varphi \oplus \mathfrak{g}^\varphi \oplus \mathfrak{g}^{2\varphi}$, it follows that $\mathbf{C}w_\varphi \oplus \mathfrak{g}^\varphi \oplus \mathfrak{g}^{2\varphi}$ is even-dimensional. Since, by Lemma 2.2.7, $\mathfrak{g}^{2\varphi}$ is odd-dimensional, it now follows that $\mathfrak{g}^\varphi$ is even-dimensional.

Finally, we have to prove that if *1)* and *2)* hold, then either (i) or (ii) holds. Assuming *1)* and *2)*, it is enough to prove that if $\langle \lambda+\varrho, w_\varphi \rangle - l_\varphi \notin n_\varphi \mathbf{Z}_+$, then (ii) holds. Since $\langle \lambda+\varrho, w_\varphi \rangle \notin I_\varphi$ and since $\langle \lambda+\varrho, w_\varphi \rangle - l_\varphi \notin n_\varphi \mathbf{Z}_+$, it follows that in fact $\langle \lambda+\varrho, w_\varphi \rangle \leqq -l_\varphi$ (using, of course, the assumption that $\langle \lambda+\varrho, w_\varphi \rangle - l_\varphi \in n_\varphi \mathbf{Z}$);

hence, $\langle \lambda+\varrho, w_\varphi \rangle + l_\varphi \leqq 0$. But since, as remarked already, $2l_\varphi \in n_\varphi \mathbf{Z}$, and since $\langle \lambda+\varrho, w_\varphi \rangle - l_\varphi \in n_\varphi \mathbf{Z}$, it follows that $\langle \lambda+\varrho, w_\varphi \rangle + l_\varphi \in n_\varphi \mathbf{Z}$. Thus in fact

$$\langle -\lambda-\varrho, w_\varphi \rangle - l_\varphi \in n_\varphi \mathbf{Z}_+ .$$

Q.E.D.

Remark. Helgason recognized that the condition given in Theorem 2.9.8 may be expressed very nicely in terms of the c-function. See [6a].

From Definition 2.9.6, the following is an immediate consequence of Theorem 2.9.8:

Corollary 2.9.10. For $\lambda \in CS$, $r^\gamma(\lambda) \neq 0$ for any $\gamma \in \Gamma$. In particular, for $\lambda \in CS \cap \mathfrak{a}_0'$, X_λ is irreducible.

§ 10. Irreducibility if $\lambda+\varrho \in C$ and cyclicity if $\lambda+\varrho \in -C$

We assume throughout this section that $\mathfrak{g}_{\mathbf{R}}$ is semisimple.

We define the complex Weyl chamber C by $C = \{\lambda \in \mathfrak{a}' \mid \text{real part of } \lambda \in D\}$, where $D \subseteqq \mathfrak{a}_{\mathbf{R}}$ is the positive Weyl chamber associated with $\Delta_+(\mathfrak{a}, \mathfrak{g})$. Thus one has

$$C = \{\lambda \in \mathfrak{a}' \mid \text{real part of } \langle \lambda, w_\varphi \rangle \geqq 0 \quad \text{for all } \varphi \in \Delta_+^1\}.$$

Lemma 2.10.1. One has $p^\gamma(\lambda) \neq 0$ for any $\gamma \in \Gamma$ if $\mu \in -C$, where $\mu = \lambda+\varrho$.

Proof. By Lemma 2.9.9, $p^\gamma(\lambda)=0$ for some $\gamma \in \Gamma$ if and only if for some $\varphi \in \Delta_+^1$, $\langle \mu, w_\varphi \rangle - l_\varphi \in n_\varphi \mathbf{Z}_+$. But since $\mu \in -C$, the real part of $\langle \mu, w_\varphi \rangle \leqq 0$ for all $\varphi \in \Delta_+^1$, and the lemma follows. Q.E.D.

Remark 2.10.2. Recall that for any $\lambda \in \mathfrak{a}'$, we have two distinguished cyclic U-modules $Z_\lambda = U/L_\lambda$ and $Y_\lambda = U/L_\lambda^{\min}$. In fact for any $X = U/L$ in $\mathscr{C}(\lambda)$, one has $L_\lambda^{\min} \subseteqq L \subseteqq L_\lambda$. (See Chapter I, § 5.) If L is stable under K_θ and if for any $\gamma \in \Gamma$, the multiplicity of γ in X is $\boldsymbol{l}(\gamma)$ (resp., if the multiplicity of γ in X is rank $R^\gamma(\lambda)$) then by Corollary 1.5.7 (resp., Lemma 1.5.12), we have $X = Y_\lambda$ (resp., $X = Z_\lambda$). Now let $\lambda \in \mathfrak{a}'$ and assume $\lambda+\varrho \in -C$. Then by Lemma 2.10.1, rank $P^\gamma(\lambda) = \boldsymbol{l}(\gamma)$, and by Theorem 1.7.12, the multiplicity of γ in X_λ^e is $\boldsymbol{l}(\gamma)$. Thus $X_\lambda^e = Y_\lambda$. (Note that the fact the multiplicity of $\gamma \in \Gamma$ in X_λ^e is $\boldsymbol{l}(\gamma)$ implies, in view of Lemma 1.7.3, that $X_\lambda = X_\lambda^e$.) Now suppose $\lambda+\varrho \in C$; then, by the definition of C, $\bar{\lambda}+\varrho \in C$, so that $(-\bar{\lambda}-2\varrho)+\varrho = -\bar{\lambda}-\varrho \in -C$. Hence, by Lemma 2.10.1, $P^\gamma(-\bar{\lambda}-2\varrho)$ is non-singular for every $\gamma \in \Gamma$. Since by Theorem 1.8.6 we have $R^\gamma(\lambda) = (P^\gamma(-\bar{\lambda}-2\varrho))^* P^\gamma(\lambda)$, it follows that rank $R^\gamma(\lambda) = \text{rank } P^\gamma(\lambda)$. Thus $X_\lambda^e = Z_\lambda$.

Since every element of $\mathfrak{a}'$ is W_T-conjugate to an element λ such that $\lambda+\varrho \in C$ and to an element ν such that $\nu+\varrho \in -C$ (recall that W_T is the translated Weyl group), we now have the following theorem, using Theorem 1.3.5 and Remark 2.10.2:

Theorem 2.10.3. *For every $\lambda \in \mathfrak{a}'$ such that $\lambda+\varrho \in C$, $X_\lambda^e \subseteq X_\lambda$ is an irreducible U-module, and one obtains, up to equivalence, all irreducible U-modules Z such that $Z^{\mathfrak{k}} \neq 0$ in this way. Moreover, for every $\lambda \in \mathfrak{a}'$ such that $\lambda+\varrho \in -C$, one has $X_\lambda^e = X_\lambda = Y_\lambda$, so that the induced module X_λ is U-cyclic. Finally, if X is any U-module cyclically generated by $X^{\mathfrak{k}}$, where* $\dim X^{\mathfrak{k}} = 1$, *then X is equivalent to a quotient of X_λ for some $\lambda \in \mathfrak{a}'$ such that $\lambda+\varrho \in -C$.*

§ 11. The main results in the principal and complementary series

We assume throughout this section that $\mathfrak{g}_{\mathbf{R}}$ is semisimple, and that G_0 is an arbitrary connected Lie group with Lie algebra $\mathfrak{g}_{\mathbf{R}}$. Henceforth, the notation $\mu \underset{W}{\sim} \nu$ denotes that μ and $\nu \in \mathfrak{a}'$ are W-conjugate.

Recall the irreducible modules Z_λ associated with each $\lambda \in \mathfrak{a}'$ (see Theorem 1.3.5). Recall also (Theorem 1.6.5) that Z_λ admits a U-invariant Hermitian structure if and only if $\lambda \in \mathfrak{a}^* = \{\lambda \in \mathfrak{a}' | \mu \underset{W}{\sim} -\bar{\mu}$, where $\mu = \lambda+\varrho\}$. Finally, recall that $\mathfrak{a}_+^* = \{\lambda \in \mathfrak{a}^* | Z_\lambda$ admits a U-invariant positive-definite Hermitian structure$\}$.

Remark 2.11.1. Let $\lambda \in \mathfrak{a}_+^*$, and let $\mathscr{H}^\lambda$ denote the Hilbert space completion of Z_λ. Now we may use Nelson's theorem (see [13]) to obtain an irreducible unitary representation π^λ of G_0 on $\mathscr{H}^\lambda$. Indeed, the element $u \in U$ obtained by taking the Casimir operator of $\mathfrak{g}_{\mathbf{R}}$ minus twice the Casimir operator of $\mathfrak{k}_{\mathbf{R}}$ is of the form $u = \sum_{i=1}^{\dim \mathfrak{g}_{\mathbf{R}}} x_i^2$ ($x_i \in \mathfrak{g}_{\mathbf{R}}$, $x_i \neq 0$). But u is a scalar on the γ-primary subspace of Z_λ for all $\gamma \in \Gamma$, and hence u as an operator on Z_λ is essentially self-adjoint. Thus we obtain a unitary representation of the simply connected group G_S corresponding to $\mathfrak{g}_{\mathbf{R}}$, and this representation is irreducible, since Z_λ is $\mathfrak{g}_{\mathbf{R}}$-irreducible (see Harish–Chandra [5]). But since the center of G_S lies in the subgroup corresponding to $\mathfrak{k}_{\mathbf{R}}$, and since $Z_\lambda^{\mathfrak{k}_{\mathbf{R}}} \neq 0$, the center of G_S operates trivially, defining the irreducible unitary representation π^λ of G_0.

Let $K_0 \subseteq G_0$ be the subgroup corresponding to $\mathfrak{k}_{\mathbf{R}}$.

Remark 2.11.2. If $\lambda \in \mathfrak{a}_+^*$, then Z_λ corresponds to the irreducible unitary representation π^λ of G_0. To clarify the word "corresponds", we recall that for any unitary representation of G_0 on a Hilbert space $\mathscr{H}$, there is induced a U-module structure on the space $\mathscr{H}^\infty$ of C^∞ elements in $\mathscr{H}$. We will say that a U-module X corresponds to the given unitary representation of G_0 if X is U-equivalent to a U-submodule Y of $\mathscr{H}^\infty$, where Y is dense in $\mathscr{H}$. Clearly, Z_λ corresponds to π^λ. Now it follows easily from [5] that if Z_λ corresponds to the unitary representation π of G_0 on $\mathscr{H}$, then

1) π is irreducible,
2) all the elements of Y are well behaved and
3) Y is the set of all K_0-finite vectors in $\mathscr{H}$.

Furthermore, if Z_λ also corresponds to π_1, then π and π_1 are unitarily equivalent.

Proposition 2.11.3. For any $\lambda \in \mathfrak{a}_+^*$, π^λ is an irreducible unitary representation of G_0. Moreover, π^ν is equivalent to π^λ ($\nu \in \mathfrak{a}_+^*$) if and only if λ and ν are W_T-equivalent. Finally, for any $\lambda \in \mathfrak{a}_+^*$, one has $(\mathscr{H}^\lambda)^{K_0} \neq 0$, and one thus obtains up to equivalence all irreducible unitary representations of G_0 on a Hilbert space $\mathscr{H}$ such that $\mathscr{H}^{K_0} \neq 0$.

Proof. In view of the above remarks, we need only show that every irreducible unitary representation π of G_0 on a Hilbert space $\mathscr{H}$ such that $\mathscr{H}^{K_0} \neq 0$ is of the form π^λ. But by [5], the space of K_0-finite vectors is an irreducible U-module with a $\mathfrak{k}_{\mathbf{R}}$-fixed vector, and hence is equivalent to Z_λ for some $\lambda \in \mathfrak{a}'$. Then $\lambda \in \mathfrak{a}_+^*$, since the restriction of the inner product on $\mathscr{H}$ to the space of K_0-finite vectors is a $\mathfrak{g}_{\mathbf{R}}$-invariant positive-definite Hermitian form. Thus π is equivalent to π^λ.

Q.E.D.

Theorem 2.11.4. *If $\lambda \in \overline{CS}$ (=the closure of the critical strip CS), then Z_λ corresponds to a (necessarily irreducible) unitary representation of G_0 if and only if $\lambda \in \mathfrak{a}^*$. In particular, for $\lambda \in CS$ such that $\lambda+\varrho \underset{W}{\sim} -(\overline{\lambda+\varrho})$, the induced representation X_λ is U-irreducible, and corresponds to a unitary irreducible representation of G_0.*

Proof. We first assert that the set $\mathfrak{a}_u' = \{\lambda \in \overline{CS} | \lambda+\varrho \underset{W}{\sim} -(\overline{\lambda+\varrho})\}$ is connected. Indeed, if $\lambda \in \overline{CS}$ and $\lambda+\varrho \underset{W}{\sim} -(\overline{\lambda+\varrho})$, then for $0 \leqq t \leqq 1$, the element $\lambda_t = t\lambda + (1-t)(-\varrho)$ satisfies the condition $\lambda_t+\varrho \underset{W}{\sim} -(\overline{\lambda_t+\varrho})$; but since CS is convex and since $-\varrho \in CS$, we have $\lambda_t \in CS$ for $0 \leqq t < 1$. This proves our claim.

Now, for any $\gamma \in \Gamma$, we fix a basis of $\mathrm{Hom}_{K_\theta}(V_\gamma, H^*)$ and a basis of $\mathrm{Hom}_{K_\theta}(V_\gamma', H^*)$ as in Lemma 1.6.6, so that for all $\lambda \in \mathfrak{a}_u'$, the matrices $R^\gamma(\lambda)$ are Hermitian. Moreover, since for $\lambda \in CS$, $r^\gamma(\lambda) \neq 0$, the matrices $R^\gamma(\lambda)$ are nonsingular. *We now assert that for any $\lambda \in \mathfrak{a}_u'$, $R^\gamma(\lambda)$ is a positive-semidefinite Hermitian matrix.* This will follow if we show that for $\lambda \in \mathfrak{a}_u'$, the curve $t \to R^\gamma(\lambda_t)$ $(0 \leqq t < 1)$, where $\lambda_t = t\lambda + (1-t)(-\varrho)$, consists of positive-definite matrices. One can easily prove that the set of positive-definite Hermitian matrices in $GL(n, \mathbf{C})$ is a connected component of the set of all nonsingular Hermitian matrices. Thus, our assertion will be proved if we show that the matrix $R^\gamma(-\varrho)$ is a positive-definite Hermitian matrix. But by Remark 1.8.5, $X_{-\varrho}$ admits a U-invariant positive-definite Hermitian form, and since $-\varrho \in CS \cap \mathfrak{a}_0'$, it follows from Corollary 2.9.10 that $X_{-\varrho} = Z_{-\varrho}$. Thus it follows from Lemma 1.6.6 that $R^\gamma(-\varrho)$ is a positive-definite Hermitian matrix. This proves our assertion. More directly $R^\gamma(-\varrho)$ is non-singular by Corollary 2.9.10 and hence is positive-definite by the relation $R^\gamma(-\varrho) = P^\gamma(-\varrho)^* P^\gamma(-\varrho)$.

The theorem is now an immediate consequence of Lemma 1.6.6. Q.E.D.

Among all the elements $\lambda \in \overline{CS} \cap \mathfrak{a}^*$, there are two extreme types. Writing $\lambda = \lambda_1 + i\lambda_2$, where $\lambda_1, \lambda_2 \in \mathfrak{a}'_{\mathbf{R}} = \{\varphi \in \mathfrak{a}' \mid \bar{\varphi} = \varphi\}$, then $\lambda \in \mathfrak{a}^*$ if and only if $\lambda_1 + \varrho + i\lambda_2 \underset{W}{\sim} -(\lambda_1 + \varrho) + i\lambda_2$. The extreme cases are:

a) $\lambda + \varrho$ is pure imaginary, that is $\lambda = -\varrho + i\lambda_2$. Any such $\lambda \in \mathfrak{a}'$ is automatically an element of $CS \cap \mathfrak{a}^*_+$. The representations π^λ obtained in this way range precisely through the principal series of class one of $G_{\mathbf{R}}$.

b) $\lambda + \varrho$ is real, that is, $\lambda = \lambda_1$. In this case, $\lambda \in \mathfrak{a}^*$ if and only if $\lambda + \varrho \underset{W}{\sim} -(\lambda + \varrho)$, and $\lambda \in \overline{CS}$ if and only if $\langle \lambda + \varrho, w_\varphi \rangle \in \bar{I}_\varphi$ for all $\varphi \in \Delta^1_+$ (see Definition 2.9.5).

Definition 2.11.5. Let $CI = \{\lambda \in \mathfrak{a}'_{\mathbf{R}} \mid \langle \lambda + \varrho, w_\varphi \rangle \in I_\varphi$ for all $\varphi \in \Delta^1_+\}$. (CI stands for "critical interval".)

Thus the elements λ described in (b) consist of the elements $\lambda \in \overline{CI}$ such that $\lambda + \varrho \underset{W}{\sim} -(\lambda + \varrho)$. Generalizing a classical notion for $G_0 = SL(2, \mathbf{R})$, we make the following definition:

Definition 2.11.6. *The complementary series of class one* is the family of representations π^λ as λ ranges through the elements of CI such that $\lambda + \varrho \underset{W}{\sim} -(\lambda + \varrho)$.

Remark 2.11.7. Note that if $-1 \in W$, then every element of CI defines a member of the complementary series. In particular, this holds if $\mathfrak{g}_{\mathbf{R}}$ has rank 1.

For the class one case we now have the following theorem for the principal and complementary series.

Theorem 2.11.8. *Let $\lambda \in \mathfrak{a}'$. If $\lambda = -\varrho + i\lambda_2$, where $\lambda_2 \in \mathfrak{a}'_{\mathbf{R}}$, then the irreducible U-module Z_λ $(= X_\lambda)$ corresponds to a unitary irreducible representation of G_0 (necessarily π^λ). That is, every member of the principal series of class one of G_0 is irreducible. If $\lambda \in \overline{CI}$, then the irreducible U-module Z_λ corresponds to a unitary irreducible representation of G_0 (necessarily π^λ) if and only if $\mu \underset{W}{\sim} -\mu$, where $\mu = \lambda + \varrho$. These representations constitute the complementary series of class one of G_0. Moreover, if in addition $\lambda \in CI$, then π^λ is in the complementary series of class* 1 *and the induced module $X_\lambda (= Z_\lambda)$ is irreducible and corresponds to π^λ.*

Remark. It is interesting that the result above (for the principal series) apparently does not generalize to the degenerate case (see [4]) and also to the non-class one case (see [8b]).

In the rank 1 case we can determine all irreducible unitary representations of class one.

Theorem 2.11.9. *Suppose that* $\mathfrak{g}_{\mathbf{R}}$ *has rank* 1. *Let $\lambda \in \mathfrak{a}'$, and suppose that Z_λ admits a U-invariant positive definite Hermitian structure (i.e., $\lambda \in \mathfrak{a}^*_+$). Then either* 1) *$\lambda \in \overline{CS}$ or* 2) *$\lambda = 0$ or -2ϱ. In case* 1), *either $\lambda \in -\varrho + i\mathfrak{a}'_{\mathbf{R}}$ or $\lambda \in \overline{CI}$, and in case* 2) *$Z_\lambda$ is the trivial one-dimensional representation of U. In particular, the irreducible unitary representations of G_0 with a non-zero K-fixed vector consist of the principal series of class one, the complementary series of class* *one, the unique (see Remark* 2.11.10) *representation* π^{λ_0} *where* $\lambda_0 \in \overline{CI} - CI$, *and the trivial representation (which may agree with* π^{λ_0}). *Furthermore, the complementary series is defined for all* $\lambda \in \overline{CI}$.

Proof. A priori, λ satisfies the Hermitian condition $\mu \underset{W}{\sim} -\bar{\mu}$, where $\mu = \lambda + \varrho$. Writing $\lambda = \lambda_1 + i\lambda_2$, where $\lambda_1, \lambda_2 \in \mathfrak{a}'_{\mathbf{R}}$ and noting that the Weyl group W consists of $\pm$ identity (of $\mathfrak{a}'$), it follows that either $\lambda \in \mathfrak{a}'_{\mathbf{R}}$ or $\lambda \in -\varrho + i\mathfrak{a}'_{\mathbf{R}}$. Thus, to prove the theorem, it is sufficient to show that for all $\lambda \in \mathfrak{a}'_{\mathbf{R}}$ such that $\lambda \notin \overline{CI}$, we have $\lambda \notin \mathfrak{a}^*_+$ unless $\lambda = 0, -2\varrho$.

For a given $\lambda \in \mathfrak{a}'_{\mathbf{R}}$, to show that Z_λ does not admit a positive-definite structure, it is enough to show that for some $\gamma \in \Gamma$, $r^\gamma(\lambda) < 0$, where $r^\gamma(\lambda)$ is the unique entry of the matrix $R^\gamma(\lambda)$ of Lemma 1.6.6. We have $r^\gamma(\lambda) = \overline{p^\gamma}(-\lambda - 2\varrho)p^\gamma(\lambda)$, where p^γ is the unique entry of the matrix p^γ of Theorem 1.8.6. Now as remarked already (see § 9), we have for a suitable pair of integers i and j and a non-zero scalar $c \in \mathbf{C}$,

$$p^\gamma = c\big[w(w-2)\cdots(w-2(i+j)+2)\big]\cdot[(w-1+s)(w-3+s)\cdots(w-2i+1+s)]$$

and

$$\overline{p^\gamma} = \bar{c}\big[w(w-2)\cdots(w-2(i+j)+2)\big]\cdot[(w-1+s)(w-3+s)\cdots(w-2i+1+s)].$$

(Here as γ runs through Γ, the pair (i,j) runs through $\{(i,j)\,|\, i,j \in \mathbf{Z}_+\}$ if $s(=\dim \mathfrak{g}^{2\alpha}) > 0$ and the set $\{(i,j)\,|\, i,j \in \mathbf{Z}_+,\ j=0 \text{ or } 1\}$ if $s=0$.)

First consider the case $s=0$. In this case $CI = \left\{\varphi \in \mathfrak{a}'_{\mathbf{R}} \,\middle|\, |\langle \varphi + \varrho, w\rangle| < \dfrac{\dim \mathfrak{g}^\alpha}{2}\right\}$.

Thus it is enough to show that if $|\langle \lambda + \varrho, w\rangle| > \dfrac{\dim \mathfrak{g}^\alpha}{2}$, then for some $\gamma \in \Gamma$, $r^\gamma(\lambda) < 0$. Choosing γ such that $i(\gamma) = 0$ and $j(\gamma) = 1$, we have

$$r^\gamma(\lambda) = c\bar{c}\langle w, \lambda\rangle\langle w, -\lambda - 2\varrho\rangle.$$

Thus, if $|\langle \lambda + \varrho, w\rangle| > \dfrac{\dim \mathfrak{g}^\alpha}{2}$, then $r^\gamma(\lambda) < 0$. $\left(\text{Note that } \langle w, \varrho\rangle = \dfrac{\dim \mathfrak{g}^\alpha}{2}\right)$.

When $s > 0$, $CI = \left\{\varphi \in \mathfrak{a}'_{\mathbf{R}} \,\middle|\, |\langle \varphi + \varrho, w\rangle| < \dfrac{\dim \mathfrak{g}^\alpha}{2} + 1\right\}$. Note that we have $\langle \varrho, w\rangle = \dfrac{\dim \mathfrak{g}^\alpha}{2} + s$. Thus $CI = \{\varphi \in \mathfrak{a}'_{\mathbf{R}} \,|\, (-\dim \mathfrak{g}^\alpha - s - 1) < \langle \varphi, w\rangle < (-s+1)\}$. Note that $\langle -2\varrho, w\rangle \leqq (-\dim \mathfrak{g}^\alpha - s - 1) < (-s+1) \leqq 0$. If $\lambda \notin \overline{CI}$ and if $\lambda \neq -2\varrho, 0$, then we have two cases:

(i) $\langle \lambda, w\rangle < \langle -2\varrho, w\rangle$ or $\langle \lambda, w\rangle > 0$,

or

(ii) $\langle -2\varrho, w\rangle < \langle \lambda, w\rangle < (-\dim \mathfrak{g}^\alpha - s - 1)$ or $(-s+1) < \langle \lambda, w\rangle < 0$.

In case (i), choosing γ such that $i(\gamma)=0$ and $j(\gamma)=1$, we have $r^{\gamma}(\lambda)= =c\bar{c}\langle\lambda, w\rangle\langle-\lambda-2\varrho, w\rangle$, so that $r^{\gamma}(\lambda)<0$.

In case (ii), choosing γ such that $i(\gamma)=1$ and $j(\gamma)=0$, we have

$$r^{\gamma}(\lambda) = c\bar{c}x(x-1+s)(-x-\dim \mathfrak{g}^{\alpha}-2s)(-x-\dim \mathfrak{g}^{\alpha}-s-1),$$

where $x=\langle\lambda, w\rangle$, and it is easy to see that $r^{\gamma}(\lambda)<0$. Q.E.D.

Remark 2.11.10. Note that for $\mathfrak{g}_{\mathbf{R}}$ of rank 1, Theorem 2.11.9 points out a single irreducible unitary representation, distinguished in the following sense: The complement of CI in $\overline{CI}$ consists of two points which are conjugate under $\tilde{W}$, and hence this orbit defines a single irreducible unitary representation π^{λ} of G_0. In case $s=0$ or 1 (the orthogonal and unitary cases), this representation is just the trivial representation. However, if $s=3$ or 7 (the symplectic and F_4 cases), this representation is not trivial, and it is distinguished among all non-trivial irreducible unitary representations π^{λ} of G_0 with a K_0-fixed vector in that it is the only one whose subspace Z_{λ} of K_0-finite vectors is not U-equivalent to the induced module X_{λ}, i.e., the only one such that X_{λ} is not irreducible.

One notes also that in the latter case and only in this case (i.e., $s=3$ or 7), the identity representation is not in the „closure" of the complementary series, and hence is in fact isolated from all other unitary irreducible representations. This fact should perhaps be related with results of D. A. Kajdan [7].

REFERENCES

[1] Bruhat, F., Sur les représentations induites des groupes de Lie, *Bull. Soc. Math. France* **84** (1956), 97—205.

[2] Gelfand, I. M. and Graev, M. I. (Гельфанд И. М., Граев, М. И.), Унитарные представления вещественной унимодулярной группы (основные невырожденные серии), *Изв. АН СССР*, **17** (1953) 189—248. (English translation: Unitary representations of the real unimodular group (principal nondegenerate series), *Amer. Math. Soc. Transl.*, (2) **2** (1956), 147—205.)

[3] ——— and Neumark, M. A. (Наймарк М. А.), Унитарные представления классических групп, *Труды матем. ин-та АН СССР*, **36** (1950), 1—288. (*Unitäre Darstellungen der klassischen Gruppen*, Akademie Verlag, Berlin, 1957.)

[4] Gross, K., The dual of a parabolic subgroup and a degenerate principal series of Sp $(n, \mathbf{C})$ *Amer. J. Math.*, **93** (1971), 398—428.

[5] Harish-Chandra, Representations of a semisimple Lie group on a Banach Space, I, *Trans. Amer. Math. Soc.*, **75** (1953), 185—243.

[6a] Helgason, S., *Differential geometry and symmetric spaces*, Academic Press Inc., New York—London, 1962.

[6b] ———, A duality for symmetric spaces with applications to group representations, *Advances in Mathematics*, **5** (1970), 1—154.

[7] Kajdan, D. A. (Каждан Д. А.), О связи дуального пространства группы со строением ее замкнутых подгрупп, *Функц. анализ*, **1**:1 (1967), 71—74.
(On the relation of the dual space of a group to the structure of its closed subgroups.)

[8a] Knapp, A. W. and Stein, E. M., The existence of the complementary series, *in Problems in Analysis*, Princeton, 1970.

[8b] ———, Intertwining operators for semisimple groups, *Annals of Math.*, **93** (1971), 489—577.

[8c] ———, Determination of intertwining operators, *Proc. of Symposia in Pure Mathematics*, AMS, **26** (1973).

[9a] Kostant, B., Lie group representations on a polynomial ring, *Amer. J. Math.*, **85** (1963), 327—404.

[9b] ———, On the existence and irreducibility of certain series of representations, *Bull. Amer. Math. Soc.*, **75** (1969), 627—642.

[10] ——— and Rallis, S., Orbits and Lie group representations associated to symmetric spaces, *Amer. J. Math.*, **93** (1971), 753—809.

[11] Kunze, R., Analytic continuation of intertwining operators, *preprint* (1968).

[12] Lepowsky, J., Multiplicity formulas for certain semisimple Lie groups, *Bull. Amer. Math. Soc.*, **77** (1971), 600—605.

[13] Nelson, E., Analytic vectors, *Ann. of Math.*, (2) **70** (1969), 572—615.

[14] Parthasarathy, K. R., Ranga Rao, R. and Varadarajan, V. S., Representations of complex semisimple Lie groups and Lie algebras, *Ann. of Math.*, (2) **85** (1967), 383—429.

[15] Wallach, N., Cyclic vectors and irreducibility for non-unitary principal series representations, *Trans. Amer. Math. Soc.*, **158** (1971), 107—113.

[16] Weil, A., *L'intégration dans les groupes topologiques et ses applications*, Paris, 1938.

[17] Weyl, H., *The classical groups*, Princeton, 1946.

[18] Želobenko, D. P. (Желобенко Д. П.), Анализ неприводимости в классе элементарных представлений полупростой комплексной группы Ли, *Изв. АН СССР*, **32** (1968), 108—133.
(The analysis of irreducibility in the class of elementary representations of a semisimple complex Lie group.)

L-INDISTINGUISHABLE REPRESENTATIONS AND TRACE FORMULA FOR $SL(2)$

JEAN-PIERRE LABESSE

Introduction

The results announced in this paper rest to a large extent on research done by R. P. Langlands and the author in Bonn in May and June 1971. We shall only sketch the proofs, and state some conjectures.

Let F be a global field and A be the ring of adeles of F. Put $G=GL(2)$ and $S=SL(2)$. Let us denote by L_0^2 the space of cusp forms on S_A/S_F and by ϱ the natural representation of S_A on L_0^2. If g is an element of G_A one can define a representation ϱ^g of S_A on L_0^2 by

$$\varrho^g(s) = \varrho(gsg^{-1}) \quad s \in S_A.$$

The problem to be solved here answers a question asked by R. P. Langlands: "Is the representation ϱ^g equivalent to ϱ, and if not how do they differ?" We shall show that in general they are not equivalent and give some explicit expression of their "difference". The result and the method to prove it were guessed by Langlands. It is a direct application of Selberg's Trace Formula for $SL(2)$, and even merely a way of writing it. The complete knowledge of this Trace Formula leads to some conjectures, namely the description of the spectrum of ϱ in terms of the spectrum of the natural representation of G_A in the space of cusp forms on G_A/G_F studied by Jacquet and Langlands in [4].

§ 1. Arithmetical equivalence

The group S_A is an invariant subgroup in G_A, so that if g is an element of G_A and U any unitary representation of S_A, we can define a unitary representation U^g of S_A by

$$U^g(s) = U(gsg^{-1}), \quad s \in S_A.$$

The two representations U and U^g will be called arithmetically equivalent. If U is irreducible then U and U^g will also sometimes be called L-indistinguishable; in fact from some arithmetical points of view they seem indistinguishable despite the fact they are not in general equivalent in the ordinary sense. One must remark that if U occurs in the restriction V to S_A of some representation $\tilde{V}$ of G_A, then U^g will occur in V with the same multiplicity for any $g \in G_A$.

There is an obvious local analogue of this global arithmetical equivalence. For instance a member of the discrete series of representations of $SL(2, \mathbf{R})$ and its complex conjugate are arithmetically equivalent in the local sense.

It is well known that the representation ϱ of S_A in L_0^2, splits into a direct sum of irreducible representations with finite multiplicities. Let us denote by $\hat{S}_A$ the set of equivalence classes of irreducible unitary representations of S_A, and by $m(U)$ the multiplicity of $U \in \hat{S}_A$ in ϱ, so that

$$\varrho = \sum_{U \in \hat{S}_A} m(U)U.$$

In the same way, given any $g \in G_A$ we have

$$\varrho^g = \sum_{U \in \hat{S}_A} m(U)U^g = \sum_{U \in \hat{S}_A} m(U^{g^{-1}})U.$$

We see that the comparison of ϱ and ϱ^g amounts to comparing $m(U)$ and $m(U^{g^{-1}})$. Our problem can be restated in the following way: do U and $U^{g^{-1}}$ occur with the same multiplicity (may be zero) in ϱ? We are thus led to study the continuous functions

$$g \to m(U^g)$$

from G_A to the set $\mathbf{N}$ of natural numbers. If this function is constant (resp. non-constant) the representation U will be called stable (resp. unstable).

Denoting by Z the centre of G, we have

Lemma 1. *If g belongs to $S_A Z_A G_F$, then ϱ^g is equivalent to ϱ.*

If g lies in S_A or in Z_A the statement is obvious. If g lies in G_F the conjugation by g defines an automorphism of the space S_A/S_F, and so determines an intertwining operator between ϱ and ϱ^g, hence the lemma.

As a corollary we see that the function $g \to m(U^g)$ is constant on the invariant subgroup $S_A Z_A G_F$ of G_A. If we denote by I the group of ideles of F, by I^2 the group of square ideles and by F^* the multiplicative group of F, then $G_A/S_A Z_A G_F$ is isomorphic via the determinant mapping to the compact abelian group $I/I^2 F^*$. Now the function $g \to m(U^g)$ induces a discrete valued continuous function on a compact group and hence assumes only finitely many values.

Now let ε be an element of $\mathscr{E}$ the group of characters of $I/I^2 F^*$. We can define

$$m_\varepsilon(U) = \int_{G_A/S_A Z_A G_F} \varepsilon(\det g) m(U^g)\, d\dot{g},$$

where $d\dot{g}$ is the Haar measure with total mass one.

We have the following

Lemma 2. (i) *A representation $U \in \hat{S}_A$ is stable if and only if $m_\varepsilon(U) = 0$ for any non-trivial $\varepsilon \in \mathscr{E}$.*

(ii) *A representation* $U \in \hat{S}_A$ *is unstable if and only if* $m_\varepsilon(U) \neq 0$ *for some non-trivial* $\varepsilon \in \mathscr{E}$.

(iii) $m(U) = \sum_{\varepsilon \in \mathscr{E}} m_\varepsilon(U)$, *the series being only a sum over a finite subgroup of* $\mathscr{E}$ *depending on* U.

The study of the numbers $m_\varepsilon(U)$ will now be made using Selberg's trace formula.

§ 2. The trace formula

We shall denote by $\mathscr{D}(S_A)$ the space of Schwartz—Bruhat functions with compact support on S_A. It follows from [1] that for any $f \in \mathscr{D}(S_A)$, the operator $\varrho(f)$ is of trace class and its trace is given by Selberg's Trace Formula. Given $f \in \mathscr{D}(S_A)$ and $g \in G_A$ we define a function $f^g \in \mathscr{D}(S_A)$ by

$$f^g(s) = f(gsg^{-1});$$

we have $\operatorname{tr} \varrho^g(f) = \operatorname{tr} \varrho(f^{g^{-1}})$; moreover the function

$$g \to \operatorname{tr} \varrho(f^g)$$

is continuous and constant on the cosets modulo $S_A Z_A G_F$, so that one can define

$$I_\varepsilon(f) = \int_{G_A/S_A Z_A G_F} \varepsilon(\det g) \operatorname{tr} \varrho(f^g)\, d\dot{g}.$$

This integral is merely a finite sum, and is zero except for finitely many values of $\varepsilon \in \mathscr{E}$. On the other hand we have

$$\operatorname{tr} \varrho(f) = \sum_{U \in \hat{S}_A} m(U) \operatorname{tr} U(f),$$

the series being absolutely convergent. It is then clear that

$$I_\varepsilon(f) = \sum_{U \in \hat{S}_A} m_\varepsilon(U) \operatorname{tr} U(f)$$

and that

$$\operatorname{tr} \varrho(f) = \sum_{\varepsilon \in \mathscr{E}} I_\varepsilon(f).$$

The values of the factors $m_\varepsilon(U)$ will be deduced from the explicit knowledge of the integrals $I_\varepsilon(f)$ for sufficiently many functions f, in application of the principle that a representation is determined by its trace. Before we explain the computation of the integrals $I_\varepsilon(f)$, let us state the result.

§ 3. The formula for $I_\varepsilon(f)$

Throughout this paragraph we fix a nontrivial character $\varepsilon \in \mathscr{E}$. Global class field theory associates to ε a separable quadratic extension L of F. We shall denote by E the kernel of the norm mapping $N_{L/F}$. The algebraic group E is an F-anisotropic torus of dimension one; we shall denote by $\Theta(\varepsilon)$ the dual group of the compact group E_A/E_F. If ψ is a nontrivial character of A/F, then using the Weil representation one can define for any character θ of E_A a representation $U(\psi, \theta)$ of S_A (cf. Shalika and Tanaka [5]). The equivalence class of $U(\psi, \theta)^g$ depends only on $d = \det(g)$ and we shall denote it by $U(\psi, \theta, d)$; moreover two such equivalence classes for d and d' in I coincide if and only if $d'd^{-1} \in N_{L/F}(L_A^*)$. We recall that $N_{L/F}(L_A^*)$ is an open subgroup of I. Let us now define

$$U_\theta^+ = \sum_{\substack{d \in I/N_{L/F}(L_A^*) \\ \varepsilon(d)=1}} U(\psi, \theta, d)$$

and

$$U_\theta^- = \sum_{\substack{d \in I/N_{L/F}(L_A^*) \\ \varepsilon(d)=-1}} U(\psi, \theta, d).$$

The representations U_θ^+ and U_θ^- are independent of the choice of ψ; moreover their intertwining number is zero. We can now state the

Theorem: *Let ε be a non-trivial character of I/I^2F^*, and let $f \in \mathscr{D}(S_A)$, then:*

(i) *The operators $U_\theta^+(f)$ and $U_\theta^-(f)$ are of trace class.*

(ii) *The function $\theta \to [\operatorname{tr} U_\theta^+(f) - \operatorname{tr} U_\theta^-(f)]$ is a Schwartz—Bruhat function on the dual $\hat{E}_A$ of E_A.*

(iii) $$I_\varepsilon(f) = \frac{1}{4} \sum_{\substack{\theta \in \Theta(\varepsilon) \\ \theta \neq 1}} [\operatorname{tr} U_\theta^+(f) - \operatorname{tr} U_\theta^-(f)].$$

The proof of assertion (iii) will be outlined in § 4; let us only observe now that assertion (ii) implies the absolute convergence of the series. To prove (i) and (ii) we may assume that f is a decomposable function $f = \bigotimes_v f_v$ where for almost all v (the places of F) $f_v = \chi_v$ the characteristic function of the standard maximal compact subgroup K_v of $S_v = SL(2, F_v)$. Now if v is a finite place of F and if U_v is an irreducible unitary representation of S_v, then $U_v(\chi_v) = 0$ unless U_v contains the trivial representation of K_v. The representation $U(\psi, \theta, d)$ is a tensor product of representations $U(\psi_v, \theta_v, d_v)$ and if v is finite then $U(\psi_v, \theta_v, d_v)(\chi_v) = 0$ if θ_v is ramified (i.e. if θ_v is nontrivial on the maximal compact subgroup of E_v). If v is finite and if θ_v is unramified then $\operatorname{tr} U(\psi_v, \theta_v, d_v)(\chi_v)$ is independent of θ_v and can only assume the values $+1$ or 0 if the local Haar measure is such that $\operatorname{vol}(K_v) = 1$, depending on the choice of ψ_v, d_v and the structure of L_v. In any case $\theta_v \to \operatorname{tr} U(\psi_v, \theta_v, d_v)(f_v)$ is a Schwartz—Bruhat function on $\hat{E}_v$. We deduce from these facts that $\theta \to \operatorname{tr} U(\psi, \theta, d)(f)$ is a Schwartz—Bruhat function on $\hat{E}_A$.

But one can show that outside a finite subgroup of $I/N_{L/F}(L_A^*)$, independent of θ then $U(\psi, \theta, d)(f)=0$, and hence the assertions (i) and (ii) are proved.

Corollary: Let $U\in\hat{S}_A$ then $m_\varepsilon(U)=0$ or $|m_\varepsilon(U)|\geqq 1/4$.

Using a slightly generalized form of the Lemma 16.1.1, p. 195 in [4] one see that if $m_\varepsilon(U)\neq 0$ then U is equivalent to an irreducible part of some $U(\psi, \theta, d)$, and the corollary is proved. More precise information can be obtained from the knowledge of the decomposition of the representations $U(\psi, \theta, d)$ and of the intertwining operators between $U(\psi, \theta, d)$ and $U(\psi, \theta', d')$, so that one obtains a complete classification of the unstable representations and the exact value of $m_\varepsilon(U)$. Let us simply say here that the local representations $U(\psi_v, \theta_v, d_v)$ are irreducible if $\theta_v=1$ or if $\theta_v^2\neq 1$ and that the intertwining number between $U(\psi_v, \theta_v, d_v)$ and $U(\psi_v, \theta_v', d_v')$ is zero unless $d_v=d_v'$ and $\theta_v=\theta_v'$ or $\theta_v^{-1}=\theta_v'$ in which case they are equivalent. If $\theta_v\neq 1$ and $\theta_v^2=1$ then $U(\psi_v, \theta_v, d_v)$ splits into two inequivalent irreducible parts. One must remark that if $U\in\hat{S}_A$ and if $m_\varepsilon(U)\neq 0$ for some $\varepsilon\in\mathscr{E}$ then there exists at least one $g\in G_A$ such that $m(U^g)\neq 0$ and we thus get a new proof (but in a less explicit way) of the result of Shalika and Tanaka [5] according to which certain representations of S_A attached to global characters of a nonsplit F-torus occur in ϱ. (See also [4] p. 396.)

§ 4. Computation of $I_\varepsilon(f)$

We shall denote by r the natural representation of S_A on the space L^2 of square integrable functions on S_A/S_F; the representation ϱ is the restriction of r to the invariant subspace L_0^2. The orthogonal complement of L_0^2 in L^2 is the direct sum of the one dimensional space of constant functions, and of a space L_1^2 on which r induces a representation equivalent to a continuous integral of representations of the global principal series (i.e. representations associated to unitary Hecke characters of a F-split torus). This spectral decomposition is constructed by the use of Eisenstein series (cf. [3], see also [1], [2] and [4]) in an explicit way, and one finds that if δ denotes the unit representation of S_A, then

$$\operatorname{tr}\varrho(f)+\delta(f) = \int_{S_A/S_F} [K_f(s, s)-H_f(s, s)]\, ds$$

where

$$K_f(x, y) = \sum_{\gamma\in S_F} f(x\gamma y^{-1})$$

is the kernel of $r(f)$ in L^2 and where $H_f(x, y)$ is the kernel of the operator $r(f)$ restricted to L_1^2. Let us denote by S_e the set of elliptic elements in S_F (i.e. the set of elements whose eigenvalues do not lie in F) and by S_p its complementary set, the set of parabolic elements. The above integral can be split into two convergent parts. In fact it is easy to show that the integral

$$J(f) = \int_{S_A/S_F} \{\sum_{\gamma\in S_e} f(s\gamma s^{-1})\}\, ds$$

is absolutely convergent. The other part is

$$C(f) = \int_{S_A/S_F} \{ \sum_{\gamma \in S_p} f(s\gamma s^{-1}) - H_f(s, s)\} \, ds$$

and is called the complementary term (cf. [1]). We now introduce the integrals

$$J_\varepsilon(f) = \int_{G_A/Z_A E_F} \varepsilon(\det g) \sum_{\gamma \in S_e} f(gs\gamma g^{-1}) \, dg.$$

The invariant measures are chosen in such a way that

$$J(f) = \sum_{\varepsilon \in \mathscr{E}} J_\varepsilon(f).$$

Now fix some non-trivial $\varepsilon \in \mathscr{E}$, and taking over the notations of § 3 choose some embedding of L^* in G_F; it is not difficult to prove that

$$J_\varepsilon(f) = \frac{1}{4} \operatorname{vol}(E_A/E_F) \sum_{G_A/S_A L_A^*} \varepsilon(\det g) \int_{S_A/E_A} \sum_{\substack{\gamma \in E_F \\ \gamma \neq \pm 1}} f(gs\gamma s^{-1} g^{-1}) \, ds.$$

We now introduce local factors:

$$F_{f_v}^\varepsilon(e_v) = \sum_{g \in G_v/S_v L_v^*} \varepsilon_v(\det g) \int_{S_v/E_v} a_v(e_v) f_v(gse_v s^{-1} g^{-1}) \, ds$$

where

$$a_v(e_v) = b_v(e_v) |N_{L/F}(e_v - e_v^{-1})|_v^{1/2}$$

and

$$b_v \begin{pmatrix} x & y \\ z & t \end{pmatrix} = \varepsilon_v(y).$$

The function $e_v \to F_{f_v}(e_v)$ is *a priori* defined only when $e_v \neq \pm 1$ but it can be continued in a Schwartz—Bruhat function on E_v. This can be checked directly or by computing its Fourier transform, namely

$$\hat{F}_{f_v}^\varepsilon(\theta_v) = \lambda(L_v/F_v, \psi_v) \sum_{d \in F_v^*/N_{L/F}(L_v^\times)} \varepsilon_v(d) \operatorname{tr} U(\psi_v, \theta_v, d)(f_v)$$

(cf. [2]) where $\lambda(L_v/F_v, \psi_v)$ is a factor of absolute value one defined in [4] p. 6. Using the assertions (i) and (ii) of the theorem in § 3, we have then proved that

$$\hat{F}_f^\varepsilon(\theta) = \prod_v \hat{F}_f^\varepsilon(\theta_v) = [\operatorname{tr} U_\theta^+(f) - \operatorname{tr} U_\theta^-(f)]$$

is a Schwartz—Bruhat function. (We recall that the factors $\lambda(L_v/F_v, \psi_v)$ have a product equal to 1.) Now applying Poisson's summation formula we obtain

$$J_\varepsilon(f) = \frac{1}{4} \sum_{\theta \in \Theta(\varepsilon)} [\operatorname{tr} U_\theta^+(f) - \operatorname{tr} U_\theta^-(f)] - \left(F_f^\varepsilon(1) + F_f^\varepsilon(-1)\right) \frac{\operatorname{vol}(E_A/E_F)}{4}$$

if the base field is of characteristic zero or $p \neq 2$; if the characteristic is 2 one should read $F_f^\varepsilon(1)$ instead of $\left(F_f^\varepsilon(1) + F_f^\varepsilon(-1)\right)$.

Let $K = \prod_v K_v$ be the standard maximal compact subgroup of S_A. If $s = k \begin{pmatrix} t & 0 \\ 0 & t^{-1} \end{pmatrix} \begin{pmatrix} 1 & u \\ 0 & 1 \end{pmatrix}$ is an Iwasawa decomposition of $s \in S_A$, a Haar measure on

S_A is given by $ds=|t|^2 dk\, d^*t\, du$. The measures dk and d^*t are left arbitrary but du will be the Tamagawa measure on A (i.e. vol $(A/F)=1$). If we now choose the measure on E_A in such a way that vol $(E_A/E_F)=2$ we find

$$F_f^{\varepsilon}(\pm 1) = \prod_v F_{f_v}^{\varepsilon}(\pm 1) = \lim_{\sigma=1} \int_I f^K \begin{pmatrix} \pm 1 & t \\ 0 & \pm 1 \end{pmatrix} \varepsilon(t) |t|^{\sigma} d^*t$$

where

$$f^K(s) = \int_K f(ksk^{-1})\, dk.$$

We now have to study the complementary term $C(f)$. One can exhibit expressions $C_\varepsilon(f)$ such that

$$C(f) = \sum_{\varepsilon \in \mathscr{E}} C_\varepsilon(f)$$

with

$$C_\varepsilon(f^g) = \varepsilon(\det g) C_\varepsilon(f) \quad \text{for any} \quad g \in G_A.$$

Let (φ_i) be a sequence of Schwartz—Bruhat functions on S_A converging to the constant function 1 in L^2 defined like in [1] p. 235 and put

$$C'(f, \varphi_i) = \int_{S_A/S_F} \varphi_i(s) \sum_{\gamma \in S_p} f(s\gamma s^{-1})\, ds.$$

The evaluation of such integrals leads to the computation of the residue at $z=1$ of the analytic continuation of the functions

$$Z(f, z; a) = \int_{u \in A} \int_{t \in I/F^*} \sum_{\eta \in F^\times} f^K \begin{pmatrix} a & (a-a^{-1})u + t^2\eta \\ 0 & a^{-1} \end{pmatrix} \frac{|t|^{z+1}}{R(u)^{z+1}} d^*t\, du$$

if $\mathrm{Re}\,(z)>1$ where $R(u)$ is defined in [1] p. 244. One sees immediately that

$$Z(f, z; a) = \sum_{\varepsilon \in \mathscr{E}} \tilde{Z}(f, z, \varepsilon; a)$$

where

$$\tilde{Z}(f, z, \varepsilon; a) = \frac{1}{2} \int_{u \in A} \int_{t \in I} f^K \begin{pmatrix} a & (a-a^{-1})u + t \\ 0 & a^{-1} \end{pmatrix} \frac{\varepsilon(t) |t|^{(z+1)/2}}{R(u)^{z+1}} d^*t\, du$$

if $\mathrm{Re}\,(z)>1$. These integrals are introduced in [1] p. 243 with similar notation and are studied in detail when $\varepsilon=1$, the trivial character. When $\varepsilon \neq 1$ one can show that if $a \neq a^{-1}$ then $\tilde{Z}(f, z, \varepsilon; a)$ is holomorphic at $z=1$ and hence gives no contribution. If $\varepsilon \neq 1$ and $a=a^{-1}$ then $Z(f, z, \varepsilon; a)$ has a simple pole at $z=1$ and gives a contribution to $C'(f, \varphi_i)$ which with our choice of the Haar measures is

$$\lim_{z=1} \frac{1}{2} \int_{t \in I} f^K \begin{pmatrix} a & t \\ 0 & a \end{pmatrix} |t|^{(z+1)/2} \varepsilon(t)\, d^*t$$

and is in $C_\varepsilon(f)$. (This contribution is independent of φ_i because we assume that $\int_{S_A/S_F} \varphi_i(s)\, ds = \mathrm{vol}\,(S_A/S_F)$ for all i.)

The computation of the remaining part

$$C''(f, \varphi_i) = -\int_{S_A/S_F} \varphi_i(s) H_f(s, s)\, ds$$

can be done exactly as in [1]. In the result appears, among other terms, the expression $\left(-\frac{1}{4}\operatorname{tr}\hat{f}(\varepsilon)M(\varepsilon)\right)$ where $\hat{f}(\varepsilon)$ is the convolution operator by f in the space of the representation of the global principal series for S_A relative to the character ε; and $M(\varepsilon)$ is an intertwining operator of that representation. One can show that this term is in $C_\varepsilon(f)$ and that if $\varepsilon \neq 1$ then

$$C_\varepsilon(f) = \frac{1}{2}\lim_{\sigma=1} \int_I \sum_{\substack{a^2=1\\ a\in F}} f^K\begin{pmatrix} a & t\\ 0 & a\end{pmatrix} \varepsilon(t)\,|t|^\sigma\, d^*t - \frac{1}{4}\operatorname{tr}\left(\hat{f}(\varepsilon)\, M(\varepsilon)\right).$$

Assertion (iii) of the Theorem in § 3 then is the consequence of the relation

$$\operatorname{tr}\hat{f}(\varepsilon)\, M(\varepsilon) = \operatorname{tr} U^+_{\theta_0}(f) - \operatorname{tr} U^-_{\theta_0}(f),$$

where θ_0 is the unit element in $\Theta(\varepsilon)$.

§ 5. Some conjectures

We do not give here the expression for the "stable" term of the trace formula. Let us simply say it is very close to the complete trace formula for G_A. I am led to formulate the following conjectures:

(i) A representation $U \in \hat{S}_A$ occurs in ϱ^g for some $g \in G_A$ if and only if U occurs in the restriction to S_A of some representation $\tilde{U}$ of G_A occurring in the space of cusp forms for G_A relative to some character of the centre.

(ii) The multiplicity one theorem is true for the representation of $S_A G_F$ on $L^2_0(S_A G_F/G_F) = L^2_0(S_A/S_F)$.

(iii) If $U \in \hat{S}_A$ is unstable $m(U)=1$ if $m_\varepsilon(U) \geqq 0$ for all $\varepsilon \in \mathscr{E}$ and $m(U)=0$ otherwise.

REFERENCES

[1] Duflo, M. and Labesse, J. P., Sur la Formule des Traces de Selberg, *Ann. Sci. École Norm Sup.* 4 série, t. **4**, fasc. *2*, (1971), 193—284.

[2] Gelfand, I. M., Graev, M. I. and Pjateckij-Šapiro, I. I. (Гельфанд И. М., Граев М. И. Пятецкий-Шапиро И. И.), *Теория представлений и автоморфные функции*, „Наука" M., 1966. (*Representation theory and automorphic function*, W. B. Saunders, 1969.)

[3] Godement, R., Analyse spectrale des fonctions modulaires, *Séminaire Bourbaki* 1964/65, № **278** (rédaction de mars 1966).

[4] Jacquet, H. and Langlands, R. P., *Automorphic forms on GL*(2), Springer Lecture Notes, Vol. **114,** 1970.

[5] Shalika, J. A. and Tanaka, S., On an explicit construction of a certain class of automorphic forms, *Amer. J. Math.*, **91**(1969), 1049—1076.

ON THE ANALOGY BETWEEN SEMISIMPLE LIE GROUPS AND CERTAIN RELATED SEMI-DIRECT PRODUCT GROUPS[1]

GEORGE W. MACKEY[2]

§ 1. Introduction

The semisimple Lie group $G=SL(2, \mathbf{C})$ has $K=SU(2)$ as maximal compact subgroup and the corresponding homogeneous space G/K is three-dimensional and has an essentially unique G-invariant Riemannian structure. G is then the simply-connected covering group of the connected component of the group of all isometries of a non-compact three-dimensional Riemannian space. In this sense it is an analogue of the simply-connected covering group E of the group of isometries of ordinary three-dimensional Euclidean space.

Algebraically G and E have very different structures. While G modulo its two element centre is a non-compact simple group, E is a semi-direct product of a three-dimensional normal vector subgroup and the compact group $SU(2)$. This difference in structure is reflected in a difference in the ease with which the irreducible unitary representations of the two groups may be determined. Those of E may be obtained systematically by applying a general theorem while those of G can only be determined by special ad hoc methods which do not generalize. Indeed for most semisimple Lie groups the problem of finding *all* equivalence classes of irreducible unitary representations is still open.

On the other hand, the physical meaning of the unitary representations of E together with the fact that G/K is a possible model for physical space suggests that there might be a much closer relationship between the unitary representation theories of G and E than the above considerations would lead one to expect. The irreducible unitary representations of E fall into two classes; those which are finite-dimensional and trivial on the translation subgroup and those which

[1] This paper is a considerable expansion of the talk actually given by the author at the Budapest summer school seminar in August 1971. Only the material in sections one and two was presented in detail in the seminar. That in sections four, seven, nine, and eleven was briefly and vaguely alluded to and was only partially worked out in the author's mind at the time. The ideas and results in the other sections were found later.

[2] Some of the work described in this paper was done in the fall and winter of the academic year 1970—71 during the tenure of a fellowship from the John Simon Guggenheim Memorial Foundation. During the fall, the author benefited from the hospitality of the I.H.E.S. in Bures-sur-Yvette and during the winter from that of the Tata Institute of Fundamental Research in Bombay.

are infinite dimensional and are non-trivial on the translation subgroup. The former are just the result of lifting the irreducible unitary representations $D_j\left(j=0, \frac{1}{2}, 1, \frac{3}{2}\ldots\right)$ of $SU(2)$ up to E. The latter are obtained as follows. Let T be the subgroup of E generated by the translations and the (commutative) subgroup of $SU(2)$ which leaves the positive z axis fixed. Then for every $r>0$ and every $n=0, \pm 1/2, \pm 1, \ldots, x, y, z, \theta \rightarrow e^{irz} e^{2in\theta}=\chi_{r,n}(x, y, z, \theta)$ defines a one-dimensional unitary representation of T such that the induced representation $U^{\chi_{r,n}}$ is irreducible. $U^{\chi_{r_1,n_1}}$ and $U^{\chi_{r_2,n_2}}$ are equivalent if and only if $r_1=r_2$ and $n_1=n_2$, and every irreducible unitary representation of E which does not reduce to the identity on the translation subgroup is equivalent to some $U^{\chi_{r,n}}$.

The parameters r, n which label the infinite-dimensional irreducible unitary representations of E have a physical interpretation which may be explained as follows. The Hilbert space of states for a quantum mechanical particle of spin j is the space of a unitary representation of E because of the invariance of physical laws under the isometries of space. This representation turns out[3] to be the representation of E induced by the representation D_j of the subgroup $SU(2)$ and application of the general theory of induced representations shows that it has the following structure

$$U^{D_j} = \int_0^\infty [U^{\chi_{r,-j}} \oplus U^{\chi_{r,-j+1}} \oplus \ldots U^{\chi_{r,j}}]\, dr.$$

In particular if $r_1<r_2$ we have a well defined subrepresentation

$$\int_{r_1}^{r_2} [U^{\chi_{r,-j}} \ldots U^{\chi_{r,j}}]\, dr \quad \text{of} \quad U^{D_j}$$

and we can ask for the physical significance of having the state vector of thn particle lie in the corresponding subspace. In turns out that the state vectors is this subspace are precisely those for which the total momentum of the particle i- certain to be between r_1 and r_2. Thus we may think of the "infinitesimal subrepree sentation"

$$U^{\chi_{r,-j}} \oplus U^{\chi_{r,-j+1}} \ldots U^{\chi_{r,j}}$$

as defining the states of momentum r. When $j=0$ this "infinitesimal subrepresentation" is irreducible. Otherwise we may ask for the physical significance of having the state vector be in the subspace $U^{\chi_{r,-j+k}}$ for some $k=0, 1, 2, \ldots, 2j$. One finds that the states in $U^{\chi_{r,-j+k}}$ are those for which the component of the spin angular momentum along the direction of motion is equal to $-j+k$. This component of the spin angular momentum is called the *helicity* of the particle. Thus we may think of the infinite-dimensional irreducible unitary representations of E as

[3] Further details about the connection with physics will be found in another article by the author, see p. 7.

parametrized by the possible momenta and helicities of a quantum mechanical particle.

Suppose now that we replace the usual Euclidean model for physical space by G/K where $G=SL(2, \mathbf{C})$ and $K=SU(2)$ and work out the quantum mechanics of a single free particle. One will then have a canonical unitary representation of $G=SL(2, \mathbf{C})$ in the Hilbert space of states and for a particle of spin j this can be shown to be the representation U^{D_j} of G induced by the $2j+1$ irreducible unitary representations D_j of $SU(2)$. Since curving space while leaving it infinite does not produce easily detectable physical effects, one expects that U^{D_j} will decompose in this case also into irreducibles corresponding to particles with definite helicity and total momentum. Now $\sum_{j=0}^{\infty} jD_j$ is the regular representation of K so that $\sum_{j=0}^{\infty} jU^{D_j}$ is the regular representation of G. Thus one expects not only to find that $SL(2, \mathbf{C})$ has a family of irreducible unitary representations analogous to the $U^{\chi_{r,n}}$ of E, but also that these are sufficient to decompose the regular representation of $SL(2, \mathbf{C})$. In other words the physical interpretation suggests that there ought to exist a "natural" one-to-one correspondence between almost all of the irreducible unitary representations of E and almost all the irreducible unitary representations of $SL(2, \mathbf{C})$ — in spite of the rather different algebraic structures of these groups.

This is in fact the case. The well-known principal series of unitary irreducible representations of $SL(2, \mathbf{C})$ of Gelfand and Neumark [6] may be defined as follows. Let T be the subgroup of $SL(2, \mathbf{C})$ consisting of all matrices of the form $\begin{pmatrix} \lambda & 0 \\ a & 1/\lambda \end{pmatrix}$ with $\lambda \neq 0$ and let χ be any character of the multiplicative group C^* of all non-zero complex numbers. Then $\begin{pmatrix} \lambda & 0 \\ a & 1/\lambda \end{pmatrix} \to \chi(\lambda)$ defines a character χ' of T. The induced representations $U^{\chi'}$ of $SL(2, \mathbf{C})$ are all irreducible and constitute the principal series. As shown by Gelfand and Neumark these representations suffice to decompose the regular representation and so constitute "almost all" of the unitary irreducible representations of $SL(2, \mathbf{C})$. Now the characters of $\mathbf{C}^*$ are just the functions $\varrho e^{i\theta} \to e^{i\lambda \log \varrho} e^{in\theta}$ where λ is a real number and n is an integer. However, U^{χ_1} and U^{χ_2} are equivalent if and only if $\chi_1 = \overline{\chi_2}$. Thus λ, n and $-\lambda$, $-n$ define the same χ and we get every χ just once by letting λ range over all non-negative real numbers while n ranges over all the integers. If we ignore those with $\lambda=0$ (which are of Plancherel measure zero in any case) we have exactly the same parametrization in the two cases.

Actually the correspondence between the irreducible representations of the two groups goes beyond a mere coincidence of parametrizations. The relevant representations of each group are induced by one-dimensional representations of four-dimensional subgroups and these four-dimensional subgroups, while different in structure, are geometric analogues of one another. In fact we can use the general theory of semi-direct products and the above described analogy to motivate the construction of the principal series.

Recall what the general semi-direct product theory tells us about finding the most general irreducible unitary representation of a semi-direct product $N \stackrel{s}{\circ} H$ of separable locally compact groups N and H where N is commutative and H is compact. Choose a character $\chi \in \hat{N}$, and an irreducible unitary representation L of the subgroup H_χ of H consisting of all h with $[\chi]h = \chi$ where $\chi \to [\chi]h$ is the automorphism of $\hat{N}$ defined by the automorphism $n \to hnh^{-1}$ of N. Form the representation $\chi L: nh \to \chi(n) L_h$ of $N \stackrel{s}{\circ} H_\chi$. Then the induced representation $U^{\chi L}$ is irreducible and every irreducible unitary representation of $N \stackrel{s}{\circ} H$ is equivalent to some $U^{\chi L}$. If $[\chi_1]h = \chi_2$ for some h then every $U^{\chi_1 L}$ is equivalent to some $U^{\chi_2 L^1}$ but if χ is confined to be in a subset A of $\hat{N}$ which meets each H orbit just once then $U^{\chi_1 L^1} \cong U^{\chi_2 L^2}$ if and only if $\chi_1 = \chi_2$ and $L^1 \cong L^2$.

Now $SL(2, \mathbf{C})$ not only has an obvious analogue of the rotation subgroup $(SU(2))$, it also has an analogue of the translation subgroup. Let $\tilde{N}$ denote the group of all matrices of the form $\begin{pmatrix} \lambda & 0 \\ a & 1/\lambda \end{pmatrix}$ where $\lambda > 0$ and a is arbitrary. Then every element in $SL(2, \mathbf{C})$ is uniquely a product of an element in $\tilde{N}$ and an element in $SU(2)$. Now $\tilde{N}$ is neither normal nor commutative. However, it is solvable, three-dimensional and simply connected and because of the factorization just described there exists, for each s_1, s_2 in $S = SU(2)/(SL(2, \mathbf{C}))$, a unique $n \in \tilde{N}$ with $s_1 n = s_2$. Let us accept $\tilde{N}$ as the nearest analogue to a translation group that exists for our curved Riemannian space and try to copy the construction of $U^{\chi L}$ above. The most general character is $\begin{pmatrix} \lambda & 0 \\ a & 1/\lambda \end{pmatrix} \to e^{i\sigma \log \lambda}$. In defining an analogue of H_χ we must remember that $\tilde{N}$ is not normal and so define H_χ as the group of all h in $SU(2)$ which *normalize* $\tilde{N}$ and leave our character fixed. This is of course just the group of all $\begin{pmatrix} e^{i\theta} & 0 \\ 0 & e^{-i\theta} \end{pmatrix}$ in $SU(2)$. The most general L is then of the form $\begin{pmatrix} e^{i\theta} & 0 \\ 0 & e^{-i\theta} \end{pmatrix} \to e^{in\theta}$ and we obtain a character χL of $\tilde{N} H_\chi$ by setting $\chi L\left[\begin{pmatrix} \lambda & 0 \\ a & 1/\lambda \end{pmatrix} \begin{pmatrix} e^{i\theta} & 0 \\ 0 & e^{-i\theta} \end{pmatrix}\right] = e^{i\sigma \log \lambda} e^{in\theta}$. Inducing these we get analogues of the irreducible unitary representations of E provided by the semi-direct product theory. But $\tilde{N} H_\chi$ is just the subgroup T of all $\begin{pmatrix} \lambda & 0 \\ a & 1/\lambda \end{pmatrix}$ with a arbitrary and $\lambda \neq 0$ and our induced representations are just the members of the principal series.

Now let G be an arbitrary connected semisimple Lie group with a finite centre and let K be a maximal compact subgroup. The homogeneous space G/K will be a C_∞ manifold S and we may choose an origin s_0 in S such that K is the subgroup leaving s_0 fixed. Let V denote the tangent space to S at s_0 and for each $k \in K$ let A_k denote the linear transformation of V into itself defined by the action of k on S. Then $k \to A_k$ is a homomorphism of K into the group of automorphisms of the additive group of V and one may form the corresponding semi-direct product $V \overset{s}{\circ} K$. There will exist K invariant positive-definite bilinear forms in V and the choice of **one** of these makes S into a Riemannian space and V into a Euclidean space which approximates S locally. G will be a group of isometries of S and $V \overset{s}{\circ} K$ a "naturally corresponding" group of isometries of V. In view of the facts outlined above about the special case in which $G = SL(2, \mathbf{C})$ it is natural to wonder to what extent one can find a similar correspondence between "most" of the irreducible unitary representations of G and those of the semi-direct product $V \overset{s}{\circ} K$.

In this paper, we shall show that such a correspondence does exist in a number of other special cases and present evidence allowing one to hope for a general theorem along these lines.[4] As might be expected, things are much simpler when the group G has a complex analytic structure and it is only in this case that one can hope to have a really close analogue of the results for $SL(2, \mathbf{C})$. We devote sections two and three to the discussion of this case. The eight sections following these are concerned with the much more difficult and interesting case in which G is not complex and so has several non-conjugate Cartan subgroups. In addition to discussing $G = SL(n, \mathbf{R})$ in some detail, we prove some general theorems about conjugate classes in semi-direct products, and show that the well-known duality between Cartan subgroups and equivalence classes of irreducible unitary representations for semisimple Lie groups has a very precise analogue for certain semi-direct products. These dualities enable us to reduce the problem of setting up a correspondence between the equivalence classes of irreducible unitary representations to the more elementary one of setting up a correspondence between the conjugate classes. We also introduce a notion of "truncated inducing" as a more general alternative to holomorphic inducing and show how it can be used to construct irreducible unitary representations of real groups which are not in the principal series. The discrete series, when it exists, presents special problems, which can be overcome in part by bringing in certain projective representations of the associated semi-direct product.

[4] The semi-direct product $V \overset{s}{\circ} K$ is a special case of what Segal [16] and Wigner and Inönü [9] call a "contraction" of the Lie group G. Such a contraction can be regarded as a limit of Lie groups isomorphic to G and one can attempt to exploit this fact to set up correspondences between the representations of G and those of its contractions. There are a number of papers in the physical literature which make this attempt and exhibit such correspondences (usually in infinitesimal form) in special cases of interest in physics. We mention especially Mukunda [13] and [14] and Hermann [8]. Further references will be found in [13] which among other things contains a treatment of the principal series for $SL(3, \mathbf{C})$ whose results overlap those we obtain by specializing the results of section two. The precise relationship between the rather fragmentary infinitesimal results in this literature and the program which we outline is not clear at the present time.

§ 2. The special case in which $G=SL(n, \mathbf{C})$

Let $G=SL(n, \mathbf{C})$ be the group of all $n\times n$ complex matrices with determinant one and let S denote the set of all matrices in G which are self-adjoint and positive definite. Then S becomes a G-space if we set $T(A)=TAT^*$. Clearly S is a transitive G-space and the subgroup K leaving the identity I fixed is the group of all $T\in G$ with $TT^*=I$, that is the group $SU(n)$ of all unitary matrices in G. Thus we may identify $G/K=SL(n, \mathbf{C})/(SU(n))$ with S and it is routine to verify that we may identify the tangent space V to S at I with the vector space of all self-adjoint $n\times n$ matrices of trace zero in such a manner that for all $u\in K$, $a\in V$, $u(a)=uau^{-1}$.

Let us apply the general semi-direct product theory to find the irreducible unitary representations of $V \overset{s}{\circ} K$. We note first that setting $l_b(a)=\text{Trace}\,(ab^*)$ sets up a one-to-one real linear mapping of V on V^* so that we may identify $\hat{V}$ with V by defining $\chi_b(a)=e^{2\pi i\,\text{Trace}(ab^*)}$ for all $a, b\in V$. Moreover,

$$\chi_b(uau^{-1}) = e^{2\pi i\,\text{Trace}\,(uau^{-1}b^*)} = e^{2\pi i\,\text{Trace}\,(au^{-1}b^*u)} = e^{2\pi i\,\text{Trace}\,(a(ubu^{-1})^*)}.$$

Thus the action of K on $\hat{V}$ is defined by $(\chi_b)u=\chi_{u^{-1}bu}$ and the orbits of K in $\hat{V}$ are just the orbits of the self-adjoint matrices under conjugation by unitary matrices. By elementary linear algebra every orbit contains a diagonal matrix so that we need only look at those χ_b for which b is diagonal. Moreover, all members of $\hat{V}$ except those on a subset of lower dimension (and hence of Haar measure zero) lie on the same K-orbit with a diagonal matrix whose entries are all distinct.

Thus if we content ourselves for the moment with enough irreducible unitary representations of $V \overset{s}{\circ} K$ to decompose the regular representation it will suffice to look at those χ_b for which b is a diagonal matrix with all unequal entries. For such a χ_b, H_{χ_b} is just the group of all diagonal matrices in $SU(n)$. Moreover, $\chi_{b'}$ and χ_b lie in the same K-orbit if and only if the elements in b are a permutation of the elements of b'. Thus we get a family of inequivalent irreducible unitary representations of $V \overset{s}{\circ} K$ parametrized by the unordered $n-1$ tuples of distinct real numbers and the ordered $n-1$ tuples of integers. These are "almost all" irreducible unitary representations of $V \overset{s}{\circ} K$ in the sense that they suffice to decompose the regular representation.

Now recall the definition of the principal series of $SL(n, \mathbf{C})$ defined by Gelfand and Neumark [7]. Let T be the subgroup of $SL(n, \mathbf{C})$ consisting of all matrices with zero entries above the main diagonal and let D be the subgroup of all diagonal matrices. If $d(\|a_{ij}\|)$ is the diagonal matrix whose jth diagonal element is a_{jj} then d defines a homomorphism of T on D and if χ is any one-dimensional character $\chi\circ d$ defines a one-dimensional character χ' of T. The induced representations $U^{\chi'}$ constitute the principal series. As shown by Gelfand and Neumark these representations are all irreducible and $U^{\chi'_1}$ and $U^{\chi'_2}$ are equivalent if and only if χ_1 and χ_2 are transforms of one another under the normalizer of D. This normalizer is a semi-direct product of D and the group W of all permutation matrices. Now D is isomorphic to the direct product of $n-1$ replicas of the multiplicative group of the non-zero complex numbers so the possible χ's are parametrized by the $n-1$ tuples of characters of this group, i.e. by $n-1$ tuples of real numbers and $n-1$ tuples of integers. Because of equivalence under W distinct members of the principal series are parametrized by unordered $n-1$ triples of real numbers and ordered $n-1$ triples of integers. If we ignore the null set of those for which these $n-1$ real numbers are not distinct, we have a family of unitary irreducible representations sufficient to decompose the regular representation of $SL(n, \mathbf{C})$ and parametrized in just the same way as our family which sufficed to decompose the regular representation of the semi-direct product $V \overset{s}{\circ} K$.

As in the case $n=2$ our correspondence goes further than a mere coincidence of parametrization. Again we may introduce the subgroup $\tilde{N}$ of all matrices which are positive on the main diagonal and zero below it and observe that it is an analogue of the subgroup V of all translations in the sense that for each s_1 and s_2 in $S=SL(n, \mathbf{C})/SU(n)$ there is one and only one member of $\tilde{N}$ which takes s_1 into s_2. Again we can "copy" the construction of the irreducible unitary representations of $V \overset{s}{\circ} K$ by choosing characters of $\tilde{N}$ extending to $\tilde{N}K_\chi$ and inducing. Here K_χ is the subgroup of the normalizer of N in K which leaves χ fixed (in fact the whole normalizer). In this way we can motivate the construction of the principal series and we find that all of our observations about $SL(2, \mathbf{C})$ hold equally well for $SL(n, \mathbf{C})$ for any n.

§ 3. The principal series in the general case

Now let G be an arbitrary connected semisimple Lie group with a finite center and no compact components. Let K be a maximal compact subgroup. The Iwasawa structure theorem tells us that G admits a closed nilpotent subgroup N and a closed vector subgroup A such that A normalizes N and every element x of G may be written uniquely in the form kan where $n \in N$, $a \in A$, $k \in K$. The product $NA=\tilde{N}$ then acts as an analogue of the translations in the sense that given any two elements s_1 and s_2 in G/K there exists a unique element x of $\tilde{N}$ such that $s_1 x=s_2$. Let V denote the Lie algebra of $\tilde{N}$. It is naturally isomorphic to the tangent space to G/K at the origin so that the semi-direct product G' of V and K defined by the natural action of K on V is the semi-direct product whose representations we wish to compare with those of G. Let V_A denote the subalgebra of V corresponding to the subgroup A of $\tilde{N}$ and let V_N be the subalgebra corresponding to N. Regarding V as a vector group, let $V_N^\perp$ denote the subgroup of the character group $\hat{V}$ of V consisting of characters which are trivial on V_N. Then the natural isomorphism of A with V_A defines an isomorphism of $\hat{A}$ with $V_N^\perp \subseteq \hat{V}$. For each

$\chi \in V_N^{\perp}$, let K'_χ denote the subgroup of K consisting of all elements which leave χ fixed and let K_χ denote the subgroup of K consisting of all elements which normalize NA and leave fixed the character χ^0 of NA defined by the image of χ in $\hat{A}$. If M is the centralizer of A in K then $M \subseteqq K_\chi \cap K'_\chi$ for all χ. Let us say that χ is *regular* if $K_\chi = K'_\chi = M$.

Now according to the general theory of semi-direct products we may associate an irreducible unitary representation of G' with each pair consisting of a $\chi \in V_N^{\perp}$ and an irreducible unitary representation L of K'_χ. It is the unitary representation $U^{\chi L}$ of G' induced by the representation $vh \to \chi(v) L_h$ of $V \circledS K'_\chi$. For those χ which are regular we may assign to each $U^{\chi L}$ an analogously defined unitary representation of G. Indeed we have $K'_\chi = K_\chi = M$ and M normalizes NA so we may form the representation $nam \to \chi^0(a) L_m$ of NAM and induce to G where χ^0 as above is the image of χ in $\hat{A}$. These representations are all members of the principal series of unitary representations of G and it has been shown by Bruhat [3] that they are irreducible for almost all χ. Thus we get a natural one-to-one correspondence between certain induced unitary representations of G' and certain members of the principal series of G. When $G = SL(n, \mathbf{C})$ this correspondence clearly coincides with that described above. It seems plausible that in the general case it will share the following properties of the $SL(n, \mathbf{C})$ correspondence.

1) For almost all χ and all L, $U^{\chi L}$ is equivalent to $U^{\chi_1 L}$ if and only if $U^{\chi^0 L}$ is equivalent to $U^{\chi_1^0 L}$.

2) χ is regular for almost all χ.

3) Every character in $\hat{V}$ is in the same K orbit with some $\chi \in V_N^{\perp}$.

If so it defines a natural one-to-one correspondence between almost all equivalence classes of irreducible unitary representations of G' and almost all equivalence classes of irreducible unitary representations of G lying in the principal series. In particular, for those semisimple Lie groups G almost all of whose equivalence classes of irreducible unitary representations lie in the principal series, we will have a natural one-to-one correspondence between almost all equivalence classes of irreducible unitary representations of G and almost all equivalence classes of irreducible unitary representations of the associated semi-direct product G'. These of course include the semisimple Lie groups with a complex structure.

It is easy to see that our correspondence has properties *1)* , *2)* and *3)* when $G = SL(n, \mathbf{R})$ as well as when $G = SL(n, \mathbf{C})$ and it is probably not difficult to check other cases as well. At the present writing we have not done this nor have we made a serious attempt to prove the suggested general theorem. We plan to return to these matters in the near future.

§ 4. The special case in which $G = SL(n, \mathbf{R})$

As remarked above, we cannot expect the correspondence described in section 3 to encompass almost all equivalence classes of irreducible unitary representations of G except when G is a complex analytic group. Nevertheless, when G is not complex one can still hope to understand almost all equivalence classes of irreducible unitary representations of G in terms of those of the associated semi-direct product G'. We shall give details for $SL(n, \mathbf{R})$ and then speculate about the general case.

When $G = SL(n, \mathbf{R})$ we may take K to be the subgroup $SO(n)$ of orthogonal matrices in $SL(n, \mathbf{R})$, N to be the subgroup of all matrices which are one on the diagonal and zero above and A to be the subgroup of all diagonal matrices which have positive diagonal elements. Then M will be the subgroup of all matrices which are ± 1 on the diagonal and zero elsewhere so that AM is simply the group of all diagonal matrices in $SL(n, \mathbf{R})$. It is a Cartan subgroup of G and the members of the principal series for G are obtained by lifting its characters to NAM and inducing to G. Like most non-complex groups, however, G has Cartan subgroups not conjugate to AM and for each conjugacy class y there exists a family of equivalence classes of irreducible unitary representations parametrized by its characters (more precisely by finite equivalence classes of these characters). Indeed, if t is the largest integer such that $2t \leqq n$ there are just t mutually non conjugate Cartan subgroups which are not conjugate to AM. The l-th is the product $D_l K_l$ where D_l is the group of all diagonal matrices in $SL(n, \mathbf{R})$ with $a_{11} = a_{22}, a_{33} = a_{44}, \ldots, a_{2l-1, 2l-1} = a_{2l, 2l}$ and K_l is the group of all matrices of this form

$$\begin{matrix} \cos\theta_1 & \sin\theta_1 & & & & & & & & \\ -\sin\theta_1 & \cos\theta_1 & & & & & & & & \\ & & \cos\theta_2 & \sin\theta_2 & & & & & & \\ & & -\sin\theta_2 & \cos\theta_2 & & & & & & \\ & & & & \ddots & & & & & \\ & & & & & \cos\theta_l & \sin\theta_l & & & \\ & & & & & -\sin\theta_l & \cos\theta_l & & & \\ & & & & & & & \pm 1 & & \\ & & & & & & & & \ddots & \\ & & & & & & & & & \pm 1 \end{matrix}$$

where all non specified elements are zero and there are an even number of minus ones. When $n = 3$, $t = 1$ and every Cartan subgroup is conjugate either to the group of all matrices of the form $\begin{pmatrix} \lambda_1 & 0 & 0 \\ 0 & \lambda_2 & 0 \\ 0 & 0 & \lambda_3 \end{pmatrix}$ with $\lambda_1 \lambda_2 \lambda_3 = 1$ or to the group $D_1 K_1$ of all matrices of the form $\begin{pmatrix} \lambda\cos t & \lambda\sin t & 0 \\ -\lambda\sin t & \lambda\cos t & 0 \\ 0 & 0 & 1/\lambda^2 \end{pmatrix}$ where $\lambda > 0$.

In addition to the principal series for $SL(3, \mathbf{R})$ one has a series of irreducible unitary representations of $SL(3, \mathbf{R})$ which may be described as follows. Consider the group G_0 of all matrices of the form $\begin{pmatrix} a & b & 0 \\ c & d & 0 \\ 0 & 0 & \lambda \end{pmatrix}$ where $\lambda = \frac{1}{ad-bc}$ and $ad-bc \neq 0$. Note that G_0 is isomorphic to $GL(2, \mathbf{R})$ and normalizes the group N_0 of all matrices of the form $\begin{pmatrix} 1 & 0 & 0 \\ 0 & 1 & 0 \\ x & y & 1 \end{pmatrix}$. Note also that if $SL^{\pm}(2, \mathbf{R})$ denotes the group of all 2×2 real matrices with determinant ± 1 then $GL(2, \mathbf{R})$ is isomorphic to $SL^{\pm}(2, \mathbf{R}) \times D_1$. Let L^k, $k = \pm\frac{1}{2}, \pm 1, \pm\frac{3}{2} \ldots$ denote the members of the so-called discrete series of irreducible unitary representations of $SL(2, \mathbf{R})$ and let $\tilde{L}^k$ denote the unitary representation of $SL^{\pm}(2, \mathbf{R})$ induced by L^k. Then $\tilde{L}^k$ and $\tilde{L}^{-k}$ are equivalent but $\tilde{L}^k$ and $\tilde{L}^{k^1}$ are inequivalent whenever $|k| \neq |k'|$. If χ is an arbitrary character of D_1 and $k = \frac{1}{2}, 1, \frac{3}{2}, \ldots$ then $\tilde{L}^k \times \chi$ defines an irreducible unitary representation of G_0 which may be lifted to $N_0 G_0$ and then induced to $G = SL(3, \mathbf{R})$. It is not difficult to show that these induced unitary representations of G are all irreducible and, as shown by Gelfand and Graev in [5] almost every irreducible unitary representation of G is equivalent either to a member of the principal series or to one of the representations just defined. Note that (apart from possible equivalences) these new representations are parametrized by the pairs $2k, \chi$ where $2k$ is a non-negative integer and $\chi \in \hat{D}_1$; that is, by characters of the Cartan subgroup $D_1 K_1$.

There is an obvious generalization of this construction in which G_0 is the direct product of l replicas of $GL(2, \mathbf{R})$ and a commutative group and which allows us to associate a family of irreducible unitary representations to each Cartan subgroup $D_l K_l$ of $SL(n, \mathbf{R})$. Moreover, it follows from the results of [5] that almost every irreducible unitary representation of $G = SL(n, \mathbf{R})$ is equivalent to a member of one of these families.

Let us now compare these facts about $G = SL(n, \mathbf{R})$ with the irreducible unitary representation theory of the associated semi-direct product G'. This semi-direct product is quite analogous to that associated with $SL(n, \mathbf{C})$. One simply replaces the vector space of all $n \times n$ complex self adjoint matrices of trace zero by the vector space V of all $n \times n$ real symmetric matrices of trace zero, and $SU(n)$ by $SO(n)$. We have $G' = V \mathbin{\mathring{\circ}} SO(n)$ where just as before $u(a) = uau^{-1}$ for each $a \in V$, $u \in K = SU(n)$. Also, as before, we may identify $\hat{V}$ with V by defining $\chi_b(a) = e^{2\pi i \, \mathrm{Trace}\,(ab^*)}$ for all $a, b \in V$ and we find that $(\chi_b)(u) = \chi_{ubu^{-1}}$. Thus the orbits of K in $\hat{V}$ are just the orbits of real symmetric matrices under conjugation by orthogonal matrices of determinant one. By elementary linear algebra, every orbit contains a diagonal matrix so we need only look at those χ_b for which b is diagonal. Now for almost all $\chi_b \in \hat{V}$ there is a χ_b in the same orbit with b diagonal and with all diagonal elements different. For such χ, K_χ is the subgroup of all diagonal matrices with only plus and minus one's on the diagonal and, as indicated in chapter three, the corresponding equivalence classes of irreducible unitary representations of G' are one-to-one correspondences with almost all members of the principal series for G. When b is diagonal but the diagonal elements are not all different, K_{χ_b} will be a larger subgroup of K and we may classify the irreducible unitary representations of G' into families according to the conjugacy class of K_{χ_b}. Clearly there will be one such class for each partition of n into smaller summands. If $n=n_1+n_2+\ldots+n_k$ $(n_1\geqq n_2\geqq n_3\geqq\ldots\geqq n_k)$ is a typical such partition, the corresponding K_{χ_b} will be isomorphic to a subgroup of index 2 of $O(n_1)\times O(n_2)\times\ldots\times O(n_k)$. The simplest cases (after $n_1=n_1=\ldots=n_k=1$) are those in which $n_1=n_2=\ldots=n_t=2$, $n_{t+1}=\ldots=n_k=1$ where $t=1, 2, \ldots$ and $2t\leqq n$. The corresponding K_{χ_b} is a subgroup of index 2 of $O(2)^t\times O(1)^{n-2t}$ and hence is a finite extension of a group isomorphic to the subgroup K_t of $SL(n, \mathbf{R})$ described above. The corresponding family of irreducible unitary representations of G' (apart from grouping into finite equivalence classes) will be parametrized by pairs consisting of a character χ_b of V and a character of K_t where b is a diagonal matrix $\|b_{ij}\|$ for which $b_{11}=b_{22}$, $b_{33}=b_{44}$, $\ldots$, $b_{2t-1,2t-1}=b_{2t,2t}$ and there are no other coincidences among the b_{ij}; that is, by almost all characters of a group isomorphic to the subgroup D_tK_t of G. This approximate coincidence of parameter spaces suggests the existence of a one-to-one correspondence between almost all equivalence classes of irreducible unitary representations of G in the series associated with the Cartan subgroup D_tK_t of G and "most" of the equivalence classes of irreducible unitary representations in the family defined by the subgroup K_χ whose partition of n has t two's and $n-2t$ ones. This correspondence turns out to exist[5] and to be natural in the sense that corresponding characters of D_tK_t and its isomorphic image in G' define corresponding irreducible representations of G and G'.

While the correspondence just described does provide a natural one-to-one correspondence between almost all equivalence classes of irreducible unitary representations of G and of G' it is less satisfactory than that described in section two in several respects. Above all it is a mere coincidence of parametrization with no evident relationship between the constructions of corresponding representations. In particular, it is difficult to see how the semi-direct product theory could be used to suggest the construction of the irreducible representations of G which are not in the principal series. Also we have given no fundamental reason for allowing the diagonal matrix b in χ_b to have twofold degeneracies but not higher ones and for excluding certain irreducible representations of K_{χ_b}. Fortunately these deficiencies can all be removed. We give details in the sections to follow.

[5] For $n\geqq 3$. As explained in section eleven, the case $n=2$ is special.

§ 5. The correspondence between the irreducible unitary representations of $SL(3, \mathbf{R})$ and $V \overset{s}{\circ} SO(3)$, a closer look

Let us consider the semi-direct product $V \overset{s}{\circ} SO(3)$ associated with $SL(3, \mathbf{R})$. The only admissible partition other than that defining the principal series is $3=2+1$. The most general unitary representation of $G'=V \overset{s}{\circ} SO(3)$ in the corresponding family is obtained by choosing a diagonal matrix

$$b = \begin{pmatrix} b_{11} & 0 & 0 \\ 0 & b_{11} & 0 \\ 0 & 0 & -2b_{11} \end{pmatrix}$$

with $b_{11} \neq 0$, an irreducible unitary representation L of the group K_{χ_b} and inducing the unitary irreducible representation $\chi_b L$ from $V \overset{s}{\circ} K_{\chi_b}$ to G'. Now K_{χ_b} is a semi-direct product of the compact commutative group K_0 of all matrices of the form $\begin{pmatrix} \cos\theta & \sin\theta & 0 \\ -\sin\theta & \cos\theta & 0 \\ 0 & 0 & 1 \end{pmatrix}$ and the two element group generated by $\begin{pmatrix} 1 & 0 & 0 \\ 0 & -1 & 0 \\ 0 & 0 & -1 \end{pmatrix}$. Applying the semi-direct product theory we find that K_{χ_b} has two one-dimensional irreducible representations which are trivial on K_0 and that all other irreducible unitary representations of K_{χ_b} are those induced by the characters of K_0 other than the identity. Now if φ_n is the character $\begin{pmatrix} \cos\theta & \sin\theta & 0 \\ -\sin\theta & \cos\theta & 0 \\ 0 & 0 & 1 \end{pmatrix} \to e^{in\theta}$ and $L=U^{\varphi_n}$ then $\chi_b L$ is just the representation of $V \overset{s}{\circ} K_{\chi_b}$, induced by the character $\chi_b \varphi_n$ of $V \overset{s}{\circ} K_0$ so $U^{\chi_b L}=U^{\chi_b \varphi_n}$. Let V_0 be the subgroup of V consisting of all matrices of the form $\begin{pmatrix} \lambda & 0 & 0 \\ 0 & \lambda & 0 \\ 0 & 0 & -2\lambda \end{pmatrix}$. Then χ_b is uniquely determined by its restriction χ_b^0 to V_0.

Moreover, V_0 and K_0 commute. Thus when $L=U^{\varphi_n}$ for $n \neq 0$ the irreducible unitary representation $U^{\chi_0 L}$ of G' may also be defined as follows. Take the character $\chi_b \times \varphi_n$ of the commutative group $V_0 K_0$. Notice that $V=V_0^{\perp} \oplus V_0$ where $V_0^{\perp}$ is normalized by K_0 so that $V \overset{s}{\circ} K_0$ has $V_0 K_0$ as a quotient group. Lift $\chi_b \times \varphi_n$ to $V \overset{s}{\circ} K_0$ and induce up to G'.

Having formulated the construction of $U^{\chi_b L}$ in this manner we can attempt to do something analogous in $G=SL(3, \mathbf{R})$ starting with the corresponding character χ of the isomorphic Cartan subgroup $D_1 K_1$ of G. The analogue of V is the group $\tilde{N}$ of all matrices in $SL(3, \mathbf{R})$ which are zero above, and positive on, the main diagonal and this has a unique maximal subgroup which is normalized by K_1 and hence by $D_1 K_1$, namely the intersection of all of its transforms by elements of K_1. The group generated by this subgroup and $D_1 K_1$ is a semi-direct product of the group N_0 of all matrices of the form $\begin{pmatrix} 1 & 0 & 0 \\ 0 & 1 & 0 \\ x & y & 1 \end{pmatrix}$ and $D_1 K_1$. Hence we can lift χ from D_1K_1 to this larger group and induce to G. The procedure is quite close to that used in defining the representation $U^{\chi_b L}$ of G' starting with $\chi_b \times \varphi_n$ but differs in that we are inducing from a four-dimensional subgroup instead of a six-dimensional one. As might be expected, we find accordingly that we do *not* get an irreducible representation.

Let us see what we do get and compare it with what we had hoped for. It will be useful to induce first to the semi-direct product of N_0 and the subgroup G_0 of all matrices of the form $\begin{pmatrix} a & b & 0 \\ c & d & 0 \\ 0 & 0 & \lambda \end{pmatrix}$ where $\lambda = 1/(ad-bc)$ and $ad-bc \neq 0$. Since G_0 normalizes N_0, it follows that we get the same result by inducing χ from $D_1K_1 \subseteq G_0$ to G_0 and then lifting this induced representation to N_0G_0. Let $\chi = \chi_1 \times \Psi_k$ where $\Psi_k \begin{pmatrix} \cos\theta & \sin\theta & 0 \\ -\sin\theta & \cos\theta & 0 \\ 0 & 0 & 1 \end{pmatrix} = e^{ik\theta}$. Thus χ induced up to $G^0 \cong D_1 \times SL^{\pm}(2, \mathbf{R})$ is just $\chi_1 \times$(representation of $SL^{\pm}(2, \mathbf{R})$ induced by Ψ_k). But by the known theory of $SL(2, \mathbf{R})$ and the discussion of the relationship between $SL(2, \mathbf{R})$ and $SL^{\pm}(2, \mathbf{R})$ at the beginning of section four we see that if k is odd, the representation of $SL^{\pm}(2, \mathbf{R})$ induced by Ψ_k is the direct sum of $\tilde{L}^{1/2} \oplus \tilde{L}^{3/2} \oplus \ldots \oplus \tilde{L}^{|k|/2}$ and a direct integral of those members of the principal series which reduce on the centre to the non-trivial character. Similarly, if k is even and not zero, the representation of $SL^{\pm}(2, \mathbf{R})$ induced by Ψ_k is the direct sum of $\tilde{L}^1 \oplus \tilde{L}^2 \oplus \ldots \oplus \tilde{L}^{|k|/2}$ and a direct integral of those members of the principal series which reduce to the identity on the centre. Moreover, one verifies easily that if M is a member of the principal series for $SL^{\pm}(2, \mathbf{R})$ then $\chi_1 \times M$ lifted to N_0G_0 and induced to G is a member of the principal series for G. Hence $\chi_1 \times \Psi_k$ lifted to $N_0D_1K_1$ and induced to G is the direct sum of a direct integral of members of the principal series and of the representations $U^{\chi_1 \times \tilde{L}^j}$ for $j = |k|/2, |k|/2-1, \ldots, 1$ or $1/2$. But $U^{\chi_1 \times \tilde{L}^{|k|/2}}$ is just the irreducible unitary representation of G we had hoped to get by the induction in question. In other words, proceeding by analogy with the construction of the corresponding irreducible representations of G' has led us to a reducible unitary representation W which contains the irreducible unitary representation of G we were seeking as a discrete direct summand.

We now make the following key observation. For $k = 2, 3, \ldots$ the restriction of $U^{\chi_1 \times \tilde{L}^{|k|/2}}$ to $SO(3) \subseteq SL(3, \mathbf{R})$ is quasi-equivalent[6] to the corresponding restriction of the corresponding irreducible unitary representation of $G' = V \mathbin{\mathring{s}} SO(3)$. Moreover, no other irreducible constituent of W (discrete or continuous) has a restriction to $SO(3)$ which is quasi-equivalent to this one. Indeed (using the

[6] That is, has the same irreducible constituents with possibly different multiplicities. In more general contexts one defines two unitary representations to be quasi-equivalent if each proper subrepresentation of either has a non-trivial intertwining operator with the other.

restriction theorem of [11]) we compute easily that for any $j=\pm 1, \pm 2, \ldots$

$$U^{\chi_1\times\tilde{L}^{|j|/2}} \upharpoonright SO(3) \simeq 2D^{|j|} \oplus 2D^{|j|+1} \oplus 4D^{|j|+2} \oplus 4D^{|j|+3} \oplus 6D^{|j|+4} \oplus 6D^{|j|+5} \oplus \ldots$$

and that

$$U^{\chi_b\times\Phi_n} \upharpoonright SO(3) \simeq D^{|n|} \oplus D^{|n|+1} \oplus \ldots.$$

Moreover, the restriction to $SO(3)$ of any representation of the principal series for $G=SL(3, \mathbf{R})$ is the representation induced by a character of the four element subgroup $Z_2\times Z_2$ generated by $\begin{pmatrix}-1 & 0 & 0\\ 0 & -1 & 0\\ 0 & 0 & 1\end{pmatrix}$ and $\begin{pmatrix}1 & 0 & 0\\ 0 & -1 & 0\\ 0 & 0 & -1\end{pmatrix}$ and is quasi-equivalent to $D^0\oplus D^2\oplus D^3$ or to $D^1\oplus D^2\oplus D^3$ according to whether this character is trivial or not. It follows that $U^{\chi_1\times\tilde{L}^{|k|/2}}$ for $k=2, 3, \ldots$ may be defined as the unique irreducible constituent of the induced representation W corresponding to $U^{\chi_b\times\varphi_h}$ whose restriction to $SO(3)$ does not contain D^{k-1}. On the other hand, this same definition applied to $U^{\chi_b\times\varphi_k}$ yields $U^{\chi_b\times\varphi_k}$ all over again. In other words, we can modify the inducing construction in such a manner that (barring the mysterious case $k=1$) we can define corresponding irreducible unitary representations of G' and G in a uniform way.

In the next section we shall show that the indicated modification of the inducing process is equivalent to one with a much less *ad hoc* appearance.

§ 6. Truncated induced representations

Let G be a separable locally compact group and let K be a closed subgroup of G. Let L be an arbitrary unitary representation of K. For each unitary representation W of G form the restriction $W\upharpoonright K$ of W to K and let H_L' denote the largest subspace of $H(W)$ on which $W\upharpoonright K$ is disjoint from L. Let $H_L=\bigcap_{x\in G} W_x(H_L')$. Then H_L is a closed invariant subspace of $H(W)$. Let W/L denote the subrepresentation of W defined by H_L. We shall call it the *truncation* of W by L. In the special case in which K is compact and L is finite-dimensional we may describe H_L in concrete terms as follows. The vector φ is in H_L if and only if

$$\int_K W_{kx}(\varphi)\,\bar{\chi}(k)\,dk = 0,$$

where χ is the character of L. Moreover, if W is a direct integral of irreducibles W^λ, L is finite-dimensional and K is compact, it is easy to see that W/L is the direct integral of those W^λ for which $W^\lambda\upharpoonright K$ is disjoint from χ.

Now suppose that K is compact, L is finite-dimensional with character χ and W is the representation U^M of G induced by a unitary representation M of a closed subgroup G_0. Then the space of $W=U^M$ is the set of all Borel functions f from G to $H(M)$ which satisfy the identity

$$f(kx) = M_k f(x) \quad \text{for all } k\in K \text{ and all } x\in G$$

and in addition a certain square integrability condition. We identify functions equal almost everywhere and define $U_x^M(f)$ to be the function $y \to \frac{f(xy)\sqrt{\varrho(xy)}}{\sqrt{\varrho(x)}}$, where ϱ is a positive real-valued function on G such that $x \to \varrho(xy)/\varrho(x)$ is a function on G/G_0 which defines the Radon—Nikodym derivative of a quasi-invariant measure on G/G_0 with respect to translation by y. One verifies easily that the function f lies in the subspace $H(U^M)_L$ if and only if $\int_K f(xky)\sqrt{\varrho(xky)}\,\bar{\chi}(k)\,dk = 0$ for almost all pairs $x, y \in G \times G$. Let us say that a vector or a scalar-valued Borel function g on G is *disjoint from* χ if $\int_K g(xky)\,\bar{\chi}(k)\,dk$ exists and is zero for almost all x and y in G. For any χ the functions disjoint from χ form a two-sided invariant subspace of the space of all Borel functions having values in any given vector space. Suppose finally that in the definition of induced representations as outlined above, we put one extra restriction on f; namely that $f\sqrt{\varrho}$ must be disjoint from χ. We get then a subrepresentation of the induced representation U^M and it is clear that this subrepresentation is just U^M/L, the truncation of U^M by L.

When formulated in terms of restricting the allowable functions the notion of truncated inducing is similar in spirit to "holomorphic inducing". In holomorphic inducing, one restricts the functions in the space of U^M to those which are holomorphic with respect to a suitable complex structure on suitable submanifolds of G/G_0. Holomorphic inducing has been much used in constructing irreducible unitary representations of semisimple Lie groups outside of the principal series and also in constructing irreducible unitary representations of solvable Lie groups [4], [1]. The notion has been studied systematically to some extent in [2], and [4].

The notion of truncated inducing is presented here for the first time and has not yet been studied systematically. There is reason to believe that it can be shown to include holomorphic inducing as a special case. In any event, it can be applied to construct irreducible unitary representations of many groups for which holomorphic inducing does not work; for example, compact Lie groups and the symmetric groups.

§ 7. Conjugate classes in semi-direct products

In setting up the correspondence between equivalence classes of irreducible representations of $G = SL(n, \mathbf{R})$ and $G' = V \overset{s}{\circ} SO(n)$ we used the division of almost all of the equivalence classes for G into a finite number of families, one for each conjugacy class of Cartan subgroups. We then found that the equivalence classes of irreducible unitary representations of G' fell naturally into families parametrized by conjugacy classes of subgroups of $SO(n)$ and that a certain (apparently rather arbitrarily chosen) subset of these families corresponded one-to-one to the families for G. In order to justify this choice on a priori grounds and in other ways to prepare ourselves to seek generalizations of the results of sections four and five, we must study the conjugate classes in semi-direct products.

Let $G=N\overset{s}{\circ}H$ be a semi-direct product of the separable locally compact groups N and H where N is commutative. A theory describing the irreducible unitary representations of G in terms of the characters of N and the irreducible unitary representations of certain closed subgroups of H has been known for over twenty years and is described briefly in section one for the special case in which H is compact. It seems to have escaped notice, however, that there is a dual theory describing the conjugate classes. Since it is not difficult, we give a complete account here. Let Φ be the canonical map $n, h\to h=\Phi(n, h)$ of G on H. Then $\Phi(x_1)$ and $\Phi(x_2)$ are conjugate in H whenever x_1 and x_2 are conjugate in G and hence Φ defines a mapping $\tilde{\Phi}$ of the conjugate classes of G onto those of H. The basic question to be answered then is this: Given an element h_0 of H, find all conjugate classes in G having the conjugate class of h_0 as its map by $\tilde{\Phi}$. Such a conjugate class always contains an element of the form nh_0 where $n\in N$ and our only problem is to discover when n_1h_0 and n_2h_0 are conjugate.

Of course n_1h_0 and n_2h_0 are conjugate if and only if there exist $n\in N$ and $h\in H$ such that $(nh)(n_1h_0)=(n_2h_0)(nh)$ or equivalently $nh(n_1)hh_0=n_2h_0(n)h_0h$. But this is so if and only if $hh_0=h_0h$ and $h_0(n)n^{-1}=h(n_1)n_2^{-1}$. It follows at once that n_1h_0 and n_2h_0 are conjugate if and only if there exists h in the centralizer of h_0 in H such that $h(n_1)=n_2$ modulo the subgroup of all elements of N of the form $h_0(n)n^{-1}$. In other words, we note that $Z_H(h_0)$ the centralizer of h_0 in H leaves invariant the range R_{h_0} of the endomorphism $n\to h_0(n)n^{-1}$ of N and hence that $Z_H(h_0)$ acts as a group of automorphisms of the quotient group N/R_{h_0}. The conjugate classes in question then correspond one-to-one to the orbits of this automorphism group. We have proved then

Theorem A. *Let J be a cross-section for the conjugate classes of H. For each $h_0\in J$ let Ψ_{h_0} denote the canonical map of N on N/R_{h_0} where R_{h_0} is the subgroup of all $h_0(n)n^{-1}$. Let $Z_H(h_0)$ denote the centralizer of h_0 in H. Then for each $h\in Z_H(h_0)$, $h(\Psi_{h_0}(n))$ depends only on $h(n)$ so we may define $h(\Psi_{h_0}(n))=h(\Psi_{h_0}(n))$. Let J'_{h_0} be a cross-section for the $Z_H(h_0)$ orbits in N/R_{h_0}. Let J''_{h_0} be a subset of N which has just one element in common with each $\Psi_{h_0}^{-1}(j)$ for $j\in J'_{h_0}$. Then the set of all nh with $n\in J''_{h_0}$ $h\in J'$ is a cross-section for the conjugacy classes of G.*

We are interested here mainly in the important special case in which H is compact and N is a finite-dimensional real vector space. In that case N has a positive-definite inner product which is invariant under $n\to h(n)$ for all n and we can conclude that N is a direct sum of the vector subgroups R_{h_0} and N_{h_0} where h_0 is the centralizer of h_0 in N. More generally, let us define the action of H on N to be *semisimple with respect to the identity* if for each $h_0\in H$, N is the direct product of the centralizer of h_0 in N and the range of the endomorphism $n \to h_0(n)n^{-1}$. Whenever this condition holds, the orbits of $Z_H(h_0)$ in N/R_{h_0} are canonically identifiable with the orbits of $Z_H(h_0)$ in N_{h_0}. In particular, every element in G is conjugate to an element of the form nh where h is in H, n is in N and n and h commute. Moreover, we can prove the following close analogue of the theorem determining the irreducible unitary representations.

Theorem B. *Let $G=N \overset{s}{\circ} H$ where N and H are separable locally compact groups, N is commutative and the action of H on N is semisimple with respect to the identity. Let J be a subset of N which meets each H-orbit on N just once. For each $j \in J$ let H_j be the centralizer of j in H and let C_j be a set which meets each conjugate class in H_j just once. Then each conjugate class of G contains one and only one element of the form of jh where $j \in J$, $h \in C_j$.*

Proof. By the remark preceding the statement of the theorem, every conjugate class contains an element of the form nh where $n \in N$ and h is in H_n the centralizer of n. Let us set $hnh^{-1}=h(n)$ for $n \in N$, $h \in H$. Since $h_1 nh_1^{-1}=h_1(n)h_1 hh_1^{-1}=h_1(n)h_2$ where $h_2=h_1 hh_1^{-1}$ we see that we may always choose $n \in J$. Suppose now that $n_1 h_1$ is conjugate to $n_2 h_2$ where h_i is in H_{n_i} for $i=1, 2$. We shall show that n_1 and n_2 are in the same orbit so that if $n_2=j_2 \in J$ then $j_i=j_2$. Indeed conjugacy implies that n and h exist in N and H respectively so that $(nh)(n_1 h_1)=(n_2 h_2)(nh)$ and this means that $nh(n_1)hh_1=n_2 h_2(n)h_2 h$ and hence that $nh(n_1)=n_2 h_2(n)$ and $hh_1=h_2 h$. Now we may write $n=n'n''$ where $n' \in N_{h_2}$ $n'' \in R_{h_2}$ and conclude that $h_2(n)=h_2(n')h_2(n'')=n'h_2(n'')$ where $h_2(n'') \in R_{h_2}$. Substituting this above we find that $n'n''n(n_1)=n_2 n'h_2(n'')$ so that $h(n_1)n_2^{-1}=n_2(n'')n''^{-1}$. But since $hh_1=h_2 h$ we conclude that $h_2 h(n_1)=hh_1(n_1)=h(n_1)$ so that $h(n_1) \in N_{h_2}$. Thus $h(n_1)n_2^{-1}$ and $h_2(n'')n''^{-1}$ are in complementary subspaces and hence both are the identity. Thus $h(n_1)=n_2$ and n_1 and n_2 are in the same orbit as was to be proved.

To conclude we have only to show that if h_1 and h_2 are members of H_j for $j \in J$ then jh_1 and jh_2 are conjugate in G if and only if h_1 and h_2 are conjugate in H_j. But jh_1 and jh_2 are conjugate if and only if there exist $n \in N$ and $h \in H$ with

$$nhjh_1 = jh_2 nh \quad \text{or equivalently} \quad nh(j) = jh_2(n) \quad \text{and} \quad hh_1 = h_2 h.$$

If $hh_1 h^{-1}=h_2$ for some $h \in H_j$ then the second condition holds and the first reduces to $n=h_2(n)$. We have only to take $n=e$ to see that jh_1 and jh_2 are conjugate. Conversely, if they are conjugate then we have $nh(j)=j_2 h_2(n)$ and $hh_1=h_2 h$ and we may write $n=n'n''$ where $n' \in Nh_2$, $n'' \in R_{h_2}$. Our first equation becomes $n''h(j)=jh_2(n'')$. Moreover, from $hh_1=h_2 h$ we conclude that $h_2 h(j)=hh(j)= =h(j)$ so that $h(j) \in N_{h_2}$. It follows that $h(j)j^{-1}$ and $h_2(n'')n''^{-1}$ are both the identity. Thus $h \in H_j$ so h_1 and h_2 are conjugate in H_j and the proof is complete.

Just as the representation theorem tells us that we may obtain the equivalence classes of irreducible unitary representations of G by finding the H-orbits in $\hat{N}$ and the equivalence classes of irreducible unitary representations of the subgroups H_χ of H, so the present theorem tells us that we may find the conjugate classes of G by finding the H-orbits in N and the conjugate classes in the subgroups H_n of H.

§ 8. The duality between conjugate classes and equivalence classes of irreducible unitary representations for certain semi-direct products

Let $G = N \overset{s}{\circ} K$ where N is a finite-dimensional real vector group and K is a compact group. By the theory of the last section, we know that every conjugate class in G contains an element of the form nk where $n \in N$, $k \in K$ and $k(n) = n$. Moreover, if K_n is the centralizer of n in K then the conjugacy class of K_n is uniquely determined by that of nk. Thus the conjugate classes in G fall into families parametrized by certain conjugacy classes of closed subgroups of K. Given a subgroup H of K which occurs as a centralizer of an element of N let N_H denote the centralizer of H in N. Then N_H and H are the centralizers of one another and the elements of N_H whose centralizers are subgroups other than H lie in proper subspaces of N_H. Moreover, in the cases of greatest interest to us, there will be only a finite number of such proper subspaces. Let N_H^0 denote the subset of N_H consisting of all n whose centralizer is exactly H. Then every element in $N_H^0 \times H$ will have a conjugate class belonging to the family parametrized by the conjugacy class of H. Let $\tilde{H}$ denote the normalizer of H in K. One verifies without difficulty that two elements in N_H^0 are on the same K-orbit if and only if they are in the same $\tilde{H}/H$-orbit and thus that the conjugate classes in the family parametrized by the conjugate class of H correspond one-to-one with the pairs consisting of an $\tilde{H}/H$-orbit in N_H^0 and a conjugate class in H.

Now every equivalence class of irreducible unitary representations of G contains an element of the form $U^{\chi L}$ where χ is a character of N and L is an irreducible unitary representation of the subgroup K_χ of K consisting of all k with $[\chi]k = \chi$. Moreover, the conjugacy class of K_χ is uniquely determined by the equivalence class of $U^{\chi L}$. Thus the equivalence classes of irreducible unitary representations of G also fall into families parametrized by certain conjugacy classes of closed subgroups of K. In addition, the closed subgroups of K which occur are precisely the same as those which occur in dealing with the conjugate classes in G. To see this, let B be any K-invariant inner product in N. Then B defines an isomorphism of N with $\hat{N}$ which is equivariant with respect to the K-actions. Hence each K_χ is equal to some K_n and vice versa. It follows that we have a natural one-to-one correspondence between the families of conjugate classes and the families of equivalence classes of irreducible unitary representations.

Let H be the centralizer of a member of N and let us consider the family of equivalence classes of irreducible unitary representations of G which it parametrizes. Each such equivalence class will contain a member of the form $U^{\chi L}$ where χ is a member of the character group of $\hat{N}$ such that $K_\chi = H$ and L is an irreducible unitary representation of H. Let $(\hat{N})_H$ denote the group of all $\chi \in \hat{N}$ with $[\chi]h = \chi$ for all $h \in H$. Then $(\hat{N})_H$ is canonically identifiable with the dual $\hat{N}_H$ of N_H and all members χ of $(\hat{N})_H$ occur in the subset $(\hat{N})_H^0$ for which $K_\chi = H$ except those lying in certain proper subspaces. Just as above, χ_1 and $\chi_2 \in (\hat{N})_H^0$ lie in the same K-orbit if an only if they lie in the same $\tilde{H}/H$-orbit. Thus the equivalence classes of irreducible unitary representations in the family parametrized by the conjugacy class of H are in one-to-one correspondence with the pairs consisting of an $\tilde{H}/H$-orbit in $\hat{N}_H^0$ and an equivalence class of irreducible unitary representations of H. Note that this parametrization differs from that for the corresponding family of conjugate classes only in that N_H is replaced by $\hat{N}_H$ and in that conjugate classes in H are replaced by equivalence classes of irreducible unitary representations of H.

Now whenever H is a compact connected semisimple Lie group, it is well known that H admits a maximal closed commutative subgroup T_H which meets every conjugate class and is unique to within conjugacy. Let W be the quotient of the normalizer of T_H by its centralizer. It is also known that W is finite, that two elements of T_H are conjugate on H if and only if they lie in the same W-orbit and that the equivalence classes of irreducible unitary representations of H are in natural one-to-one correspondence with the orbits of W in the dual $\hat{T}_H$ of T_H. Suppose that the H parametrizing a family of conjugate classes in G admits such a commutative subgroup of T_H. Then the conjugate classes in the family are parametrized by the $H'/H \times W$-orbits in the (usually dense) subset $N_H^0 \times T_H$ of G. Moreover, the equivalence classes of irreducible unitary representations in the corresponding family are parametrized by the $H'/H \times W$-orbits in the (usually dense) subset $\hat{N}_H^0 \times \hat{T}_H$ of the dual group $\hat{N}_H \times \hat{T}_H$ of $N_H \times T_H$.

If our semi-direct product is such that every H which occurs admits such a T_H, then, speaking crudely, we have shown the following: The conjugate classes of G are parametrized by the orbits of certain automorphism groups acting on certain closed commutative subgroups and the equivalence classes of irreducible unitary representations are parametrized by the orbits of the dual actions on the duals of these closed subgroups.

Of course to draw such a conclusion about G it is not necessary that H have the rather special properties described above. It will suffice if each H has a finite number of closed commutative subgroups $T_H^1, T_H^2, \ldots, T_H^k$ which collectively play the role of T_H. Then we may replace the subgroups $N_H \times T_H$ by $N_H \times T_H^j$ with H and j varying separately. In other words, a duality principle of the sort indicated in the last paragraph holds for G whenever it holds for each H which arises.

There is evidence suggesting that a similar duality holds for a rather broad class of groups. We hope to study this question at a later date.

§ 9. The correspondence between the conjugate classes in $SL(n, \mathbf{R})$ and those in the associated semi-direct product

Recall that the semi-direct product associated with $G=SL(n, \mathbf{R})$ is $G'=V \overset{s}{\circ} SO(n)$ where V is the vector space of all $n \times n$ real matrices with trace zero and that every orbit of $SO(n)$ in V contains a diagonal matrix. Thus every centralizer in $SO(n)$ of an element of V is conjugate to the centralizer of a diagonal element. Applying the theory of section eight we find that we have just as many families of conjugate classes in G' as there are partitions of n. If $n=n_1+n_2+\ldots+n_\nu$ where $n_1 \geqq n_2 \geqq \ldots \geqq n_\nu$ is one such partition then the corresponding conjugacy class of subgroups of $SO(n)$ contains the centralizer in $SO(n)$ of a diagonal matrix $((d_{ij}))$ with $d_{11}=d_{22}=\ldots=d_{n_1,n_1}$, $d_{n_1+1,n_1+1}=d_{n_1+2,n_1+2}=\ldots$ $\ldots=d_{n_2,n_2}$, $d_{n_2+1,n_2+1}=d_{n_2+2,n_2+2}=\ldots=d_{n_3,n_3}$, etc., and no other equalities among the diagonal elements. This centralizer H is clearly isomorphic to a subgroup of index two in $O(n_1) \times O(n_2) \times \ldots \times O(n_\nu)$ and is easily seen to admit a duality principle of the sort described at the end of section eight. Let t denote the number of n_j which are odd. We may choose T_H^1 to be isomorphic to the direct product of $\frac{n-t}{2}$ replicas of a one-dimensional torus and $t-1$ replicas of the two element group Z_2. All other T_H^j will then be of lower dimension. Furthermore the centralizer V_H of H in V is the subspace of V consisting of all diagonal matrices with $d_{11}=$ $=d_{22}=\ldots=d_{n_1,n_1}$, $d_{n_1+1,n_1+1}=d_{n_1+2,n_1+2}=\ldots=d_{n_2,n_2}$, etc. and trace zero and is accordingly an $(r-1)$-dimensional vector group. Thus the commutative subgroup $V_H \times T_H^1$ has dimension $\frac{n+2r-t}{2}-1$. Since $r-t$ is the number of even n_j it follows that $2(r-t)+t=2r-t \leqq n$ and equals n if and only if every n_j is one or two. In that case $t=2r-n$ and the dimension of $V_H \times T_H$ is $n-1$. Thus for each $r=1, 2, \ldots, n$ we have a $V_H \times T_H$ of dimension $n-1$ and every other $V_H \times T_H$ is of lower dimension.

Now consider $G=SL(n, \mathbf{R})$. Then the commutative component A of the Iwasawa decomposition may be taken to be the group of all diagonal matrices with positive elements and determinant one and the compact component to be $SO(n)$.

For each partition $n=n_1+\ldots+n_r$ of n let A_H denote the centralizer in A of the corresponding subgroup H of $SO(n)$ and let T_H be the commutative subgroup of H defined above. Then $A_H T_H$ is isomorphic to $V_H \times T_H$ and in fact canonically since the exponential map applied to V is an isomorphism of V_H onto A_H. Moreover the $A_H T_H$ of dimension $n-1$ are all Cartan subgroups of G and every Cartan subgroup of G is conjugate to one of them.

Taking into account the action of the automorphism groups H'/H and W and their counterparts in G we now verify easily that (modulo a lower-dimensional exceptional subset) we have a canonical one-to-one map of the conjugate classes of G' onto those of G. Moreover this map respects the division into families parametrized by the Cartan subgroups and their isomorphic replicas in G'.

We may now understand the correspondence between equivalence classes of irreducible unitary representations for G and G' described in section four as a consequence of a much more elementary and straightforward correspondence between conjugate classes in the two groups combined with the fact that there exists a duality between irreducible representations and conjugate classes in the two cases. In particular we understand the restriction to partitions with $n_i=1$ or 2: it corresponds to ignoring the exceptional conjugate classes lying in families associated with non-maximal commutative subgroups which do not correspond to Cartan subgroups of G. On the other hand one can hope to extend our correspondence to include at least part of the degenerate series for G by bringing in the exceptional conjugate classes.

§ 10. Some speculations about the general case

As in section three let G once again denote an arbitrary connected semisimple Lie group with a finite centre and no compact components. Let K, N, A, $\tilde{N}$ etc. be defined as in that section so that $G=NAK$ is an Iwasawa decomposition, $G'=V \overset{s}{\circ} K$ is the associated semi-direct product and $V=V_N \oplus V_A$ where V_A and A are canonically isomorphic via the exponential mapping. We wish to set up a natural one-to-one correspondence between almost all equivalence classes of irreducible unitary representations of G and "most" equivalence classes of irreducible unitary representations of G'. Our experience with $G=SL(n, \mathbf{R})$ suggests the path to be followed. We must apply the theory of section seven and eight to G' verifying that the subgroups H which occur as centralizers of elements of V satisfy the appropriate hypotheses so that the compact commutative subgroups T_H exist. Hopefully we will be able to prove that the subgroups V_H may always be taken to be inside V_A. If so, each V_H will have a canonical image and we can hope that this image will commute with H considered as a subgroup of $K \subseteq G$. If so, the commutative subgroups $V_H \times T_H$ will have canonical isomorphic images in G and we can attempt to show that those of maximal dimension are Cartan subgroups of G and include to within conjugacy all Cartan subgroups of G. If all of this goes through, the duality principles for G and G' tell us that almost all equivalence classes of irreducible unitary representation are parametrized by finite subsets of the duals of the same finite set of locally compact commutative groups. If finally we can establish that the finite equivalence relations between characters are concordant in the two cases, we will have the desired one-to-one correspondence.

Let G be a group for which the hopes expressed in the preceding paragraph are realized. The problem remains of defining corresponding representations in a parallel fashion as was done for $SL(3, \mathbf{R})$ in section five and one can hope to generalize the considerations of section five by proceeding as follows. Let H be a closed subgroup of K which normalizes an element of V and suppose that $V_H \times T_H$ is of maximal dimension. The most general equivalence class of irreducible unitary representations in the family parameterized by the conjugacy class of H is that of the induced representation $U^{\chi L}$, where χ is a character of V whose isotropy group in K is H, and L is an irreducible unitary representation of H. Using the canonical maps of V_H and H with $A \subseteqq G$ and $K \subseteqq G$ respectively we obtain a closed subgroup $A_H H$ of AK and a unitary representation $\chi^1 L^1$ of $A_H H$. Now let $N_H = \bigcap_{h \in H} hNh^{-1}$. This will be normalized by H and presumably also by A_H so that we may form the semi-direct product $N_H \overset{s}{\circ} A_H H$ and lift $\chi' L'$ to it. If the situation for $SL(3, \mathbf{R})$ is a reliable guide, the unitary representation $U^{(\chi' L')^0}$ of G induced by $\chi' L'$ lifted to $N_H \overset{s}{\circ} A_H H$ will contain just one irreducible constituent whose restriction to K is quasi-equivalent to the restriction of $U^{\chi L}$ to K; that is, to the representation of K induced by the representation L of H. Moreover it will contain it discretely and with multiplicity one. If so, we can define both $U^{\chi L}$ and $U^{(\chi' L')^0}$ in a uniform way as truncated induced representations and so achieve our aim.

§ 11. The discrete series

There is one case in which it is easy to see that the considerations of section five do not generalize without modification. That is the case in which $V_H = 0$, $H = K$ so that $V_H \times T_H$ is compact and the corresponding family of unitary irreducible representations of G is the so-called discrete series. When $G = SL(n, \mathbf{R})$ this can only happen when $n = 2$ since otherwise T_H is not of maximal dimension. In any event, the relevant family of unitary irreducible representations of G' is just the family of finite-dimensional representations obtained by lifting the irreducible unitary representations of K. Clearly the restrictions of these to K cannot be quasi-equivalent to the restriction to K of any irreducible unitary representation of G. On the other hand, it follows from the work of Harish-Chandra that every member of the discrete series appears as a discrete subrepresentation of an induced representation of the form U^L where L is an irreducible unitary representation of K. This suggests that we might still define the members of the discrete series by truncated inducing, but that guidance as to how to truncate must be obtained from the representation theory of the semi-direct product G' in another way.

Now our whole investigation was suggested by the physical significance of the irreducible unitary representations of the isometry group of space. Let us return to this source of inspiration and recall that in applications to physics projective unitary representations are just as important as ordinary representations. Indeed in the passage from Galilean relativity to the special relativity of

Einstein there is a natural correspondence between projective representations of the Galilean group and ordinary representations of the inhomogeneous Lorentz group. Now whenever G is semisimple, the equivalence classes of projective multipliers σ for G correspond one-to-one in a natural way to those for K and hence to those equivalence classes of projective multipliers for G' which are obtained by lifting those for K to G'. On the other hand, there are many cases in which the vector space V in $G' = V \overset{s}{\circ} K$ admits a non-trivial antisymmetric bilinear form B which is K-invariant. Given such a B let $\sigma_B(n_1 k_1, n_2 k_2) = e^{iB(n_1 n_2)}$. Then σ_B is readily verified to be a non-trivial projective multiplier for G' which is also non-trivial on V and hence does not correspond to any non-trivial projective multiplier for G. To what on G do the projective unitary representations of G' with multiplier σ_B correspond? Since σ_B is trivial on K, it seems likely that they correspond to ordinary representations. Could they perhaps correspond to the discrete series?

Consider the case in which B is non-degenerate. Such a B exists whenever G/K admits an invariant complex structure. In this case, the irreducible unitary σ_B-representations of G' are especially easy to analyze since by the generalized Stone—von Neumann uniqueness theorem [10] there is to within equivalence a unique irreducible σ_B-representation of V. By the theory developed in [12], we may obtain the irreducible unitary σ_B-representations of $G' = V \overset{s}{\circ} K$ as follows. Let W be an irreducible σ_B-representation of V. Because of the uniqueness just alluded to, the σ_B-representation $v \to W_{k(v)}$ will be equivalent to W for all $k \in K$. Hence there exists a unitary operator M_k unique up to a multiplicative constant such that $W_{k(v)} = M_k W_v M_k^{-1}$ for all $v \in V$, $k \in K$. If K is simply-connected the M_k may be chosen so that $k \to M_k$ is a representation of K and in any case it will be a σ-representation for some projective multiplier σ. Let π be an arbitrary irreducible projective representation of K with multiplier $1/\sigma$. Then $v, k \to W_v M_k \otimes \pi_k$ will be an irreducible unitary σ_B-representation of G'. Moreover, the mapping $\pi \to WM \otimes \pi$ sets up a one-to-one correspondence between the equivalence classes of irreducible $1/\sigma$-unitary representations of K and the equivalence classes of irreducible σ_B-representations of K. For simplicity let us confine our discussion to the case in which σ is a trivial multiplier (which of course is always so when K is simply-connected). Then the equivalence classes of irreducible unitary σ_B-representations of G' are parametrized by the equivalence classes of irreducible unitary representations of K. However, unlike the family of ordinary representations obtained by simply lifting the irreducible unitary representations of K these σ_B-representations are infinite-dimensional. Moreover, their restrictions to K are infinite-dimensional ordinary representations which could conceivably be quasi-equivalent to corresponding restrictions of corresponding members of the discrete series for G. If so, and if the unitary representations $M \otimes \pi$ of K all belong to different quasi-equivalent classes we may hope to realize the members of the discrete series for G as truncated induced representations where the truncating to be done is determined by looking at the quasi-equivalence classes of the representations $M \otimes \pi$ of K. The results of Okamoto and Narasimhan in [15] encourage one to believe in the feasibility of this program. Further encouragement is provided by looking at the case in which $G = SL(2, \mathbf{R})$. This case is atypical in that K is commutative and neither semisimple nor simply connected. Nevertheless, the results are suggestive. We have $G' = V \overset{s}{\circ} SO(2)$ and verify without difficulty that B exists and is unique up to a multiplicative constant. The irreducible unitary representations of $SO(2)$ are all one-dimensional and are described by their characters $\theta \to e^{in\theta} = \chi_n(t)$ where $n, \pm 1, \pm 2, \ldots$. The unitary representation M turns out to be $\chi_0 \oplus \chi_2 \oplus \chi_4 \oplus \ldots$ or $\chi_0 \oplus \chi_{-2} \oplus \chi_{-4} \oplus \ldots$ and we get one case from the other by replacing B by $-B$. Thus $M \otimes \chi_n = \chi_n \oplus \chi_{n+2} \oplus \ldots$ or $\chi_n \oplus \chi_{n-2} \oplus \chi_{n-4} \ldots$. If we use the B for which M is $\chi_0 \oplus \chi_2 \oplus \chi_4$ for $n = 1, 2, \ldots$ and $-B$ for $n = -1, -2, \ldots$, we find that the corresponding $M \otimes \chi_n$'s coincide exactly with the restrictions to $SO(2)$ of the members of the discrete series for $SL(2, \mathbf{R})$.

§ 12. Concluding remarks

We have not yet ventured to formulate a precise conjecture along the lines suggested by the speculations in the preceding sections because we have not studied enough examples to be reasonably sure that we have not overlooked some important phenomena that a really general result must take into account. However, we feel sure that some such result exists and that a routine if somewhat lengthy investigation will tell us what it is. We also feel that a further study of the apparently rather close relationship between the representation theory of a semisimple Lie group and that of its associated semi-direct product will throw valuable light on the much more difficult semisimple case. Many questions suggest themselves; for example how are characters of corresponding representations related? The reader will easily think of others.

In dealing with the p-adic analysis of the semisimple Lie groups, one can use notions from the theory of algebraic groups to define an action of a maximal compact subgroup on a finite-dimensional p-adic vector space which is the analogue of the action of K on the tangent space to the origin of G/K in the semisimple case. It is at least conceivable that the study of such semi-direct products could be helpful in the further development of the p-adic theory.

REFERENCES

[1] Auslander, L. and Kostant, B., Quantization and representations of solvable Lie groups, *Bull. Amer. Math. Soc.,* **73** (1967), 692—695.

[2] Blattner, R. J., On induced representations II. Infinitesimal induction, *Amer. J. Math.,* **83** (1961), 499—512.

[3] Bruhat, F., Sur les représentations induites des groupes de Lie, *Bull. Soc. Math. France,* **84** (1956), 97—205.

[4] Dixmier, J., Représentations induites holomorphes des groupes résolubles algébriques, *Bull. Soc. Math. France,* **94** (1966), 181—206.

[5] Gelfand, I. M. and Graev, M. I. (Гельфанд И. М., Граев М. И.), Унитарные представления вещественной унимодулярной группы (основные невырожденные серии), *Изв. АН СССР,* **17** (1953), 189—248.
(Unitary representations of real unimodular groups (principal nondegenerate series).)

[6] Gelfand, I. M. and Neumark, M. A. (Гельфанд И. М. Наймарк М. А.), Унитарные представления группы Лоренца, *Изв. АН СССР,* **55** (1947), 567—570.
(Unitary representations of the Lorentz group.)

[7] ———, ———, Унитарные представления классических групп, *Труды Матем. ин-та АН СССР,* **36** (1950). (*Unitäre Darstellungen der klassischen Gruppen,* Akademie Verlag, Berlin 1957.)

[8] Hermann, R., The Gell-Mann formula for representations of semisimple groups, *Comm. Math. Phys.,* **2** (1966), 155—164.

[9] Inönü, E. and Wigner, E. P., On the contraction of groups and their representations, *Proc. Nat. Acad. Sci. U.S.A.,* **39** (1953), 510—524.

[10] Mackey, G. W., On a theorem of Stone and von Neumann, *Duke Math. J.,* **16** (1949), 313—326.

[11] ———, Induced representations of locally compact groups I, *Ann. of Math.,* **55** (1952), 101—139.

[12] ———, Unitary representations of group extension I, *Acta Math.,* **99** (1958), 265—311.

[13] Mukunda, N., Expansion of Lie groups and representations of $SL(3, \mathbf{C})$, *J. Math. Phys.,* **10** (1969), 897—911.

[14] ———, Unitary representations of the group $SL(3, \mathbf{C})$ in an $SU(3)$ basis, *J. Math. Phys.,* **11** (1970), 1759—1771.

[15] Narasimhan, M. S. and Okamoto, K., An analogue of the Borel—Weil—Bott theorem for Hermitian symmetric pairs of non compact type, *Ann. of Math.,* **91** (1970), 486—511.

[16] Segal, I. E., A class of operator algebras which are determined by group, *Duke Math. J.,* **18** (1951), 221—265.

ON THE ACTION OF UNIPOTENT GROUPS IN THE SPACE OF LATTICES

G. A. MARGULIS

Let T^t be a one-parameter group consisting of unipotent linear transformations of the Euclidean space R^n. The group T^t induces a group of continuous transformations in the space A_n of all lattices of full dimension on R^n (in the natural topology of A_n). We remark that A_n can be identified with $GL(n, R)/GL(n, Z)$.

Theorem 1. *For any lattice $Z^n \subset R^n$ of full dimension there exists a $\delta(Z^n) > 0$ such that the set of all t satisfying the condition that the intersection of the $\delta(Z^n)$-neighbourhood of* 0 *in R^n and $T^t Z^n$ consists of* 0 *alone, contains arbitrarily large positive numbers.*

By virtue of Mahler's compactness criterion, Theorem 1 can be reformulated in the following way:

Theorem 1′. *The trajectories of the semigroup T^t ($t \geqq 0$) in A_n do not tend to infinity in A_n; i.e., for any $x \in A_n$ there exists a compactum $K(x) \subset A_n$ such that $T^t x \in K(x)$ holds for arbitrarily large positive values of t.*

Theorem 1 was formulated first by Pjateckij–Šapiro in [1]. The formulation of Theorem 1 coincides with the formulation of Theorem 2 in [2]*.

§ 1. Some lemmas on polynomials

We shall denote by M_n the space of positive polynomials of one variable of degree not exceeding n.

Lemma 1. *For any $k > 1$ and any n there exist positive $\varepsilon_1(k, n)$ and $\varepsilon_2(k, n)$ such that any polynomial $P(x) \in M_n$ satisfying $P(1) = 1$ and $P(x) \leqq 1$ for $0 \leqq x \leqq 1$, we have $\varepsilon_1(k, n) < P(x) < \varepsilon_2(k, n)$ on at least one of the intervals* $[k, k^2]$, $[k^3, k^4]$, … …, $[k^{2n+1}, k^{2n+2}]$ (this interval may depend on P).

Lemma 2. *For any $k > 1$ and any n there exists a positive number $\bar{\varepsilon}(k, n)$ such that for any $P(x) \in M_n$ satisfying $P(0) = 1$ and $P(1) < \bar{\varepsilon}(k, n)$, we have $P(x) = \bar{\varepsilon}(k, n)$ for some $x \in [1, k]$.*

* The formulation of Theorem 2′ in [2] is not precise.
It should be changed to that of Theorem 1′ of the present paper.

Lemmas 1 and 2 are easy consequences of the compactness of the set of all non-negative polynomials of degree not exceeding n and uniformly bounded on some interval, and of Bezout's theorem. Applying the Lagrange interpolation formula one easily gets estimates for $\varepsilon_1(k, n)$, $\varepsilon_2(k, n)$ and $\bar{\varepsilon}(k, n)$.

Further on we shall need Lemmas 1′ and 2′. They follow immediately from Lemmas 1 and 2 by affine substitution.

Lemma 1′. For any $k>1$ and any n there exist positive numbers $\varepsilon_1(k, n)$ and $\varepsilon_2(k, n)$ such that whenever $P(x)\in M_n$ and t_1, t_2, c satisfy the conditions *1)* $t_1<t_2$, *2)* $c>0$, *3)* $P(t_2)=c$, *4)* $P(x)\leqq c$ if $t_1\leqq x\leqq t_2$, then the values of $P(x)$ are greater than $c\varepsilon_1(k, n)$ and smaller than $c\varepsilon_2(k, n)$, on at least one of the intervals $[t_1+k(t_2-t_1), t_1+k^2(t_2-t_1)], \ldots, [t_1+k^{2n+1}(t_2-t_1), t_1+k^{2n+2}(t_2-t_1)]$.

Lemma 2′. For any $k>1$ and any n there exists a positive number, $\bar{\varepsilon}(k, n)$ such that for any $P(x)\in M_n$ and any t_1, t_2, c satisfying the conditions *1)* $t_1<t_2$, *2)* $c>0$, *3)* $P(t_1)=c$, *4)* $P(t_2)<c\bar{\varepsilon}(k, n)$, at a suitable point $x\in[t_2, t_1+k(t_2-t_1)]$ we have $P(x)=c\bar{\varepsilon}(k, n)$.

Definition 1. The *length* $d(B)$ of a partially ordered set B is the maximal number of elements in linearly ordered subsets of B.

Definition 2. An *n-equipment* of a set B means a mapping $f: B\to M_n$.

The image of $b\in B$ under the mapping f will be written as $P_b(x)$.

Let us now formulate the Main Lemma. Let n and m be natural numbers, $d>0$, $t_0>0$, $k>1$, and $h(x)$ a positive function on the real line. Write $A(n, m, d, t_0, k, h(x))$ for the set of all n-equipped partially ordered sets B satisfying the following conditions:

A1. $d(B)\leqq m$.

A2. If L is any (possibly empty) linearly ordered subset of B, $c>0$, $x_0\in[t_0, kt_0]$, and $P_l(x_0)<c$ for $l\in L$, then there exists a maximal linearly ordered subset $\hat{L}$, $L\subset\hat{L}\subset B$, such that $P_l(x_0)<h(c)$ for $l\in\hat{L}$.

A3. For any $b\in B$ there exists an $x(b)$, $0\leqq x(b)\leqq t_0$ such that $P_b(x(b))\geqq d$.

A4. For any x and a the set of all $b\in B$ satisfying $P_b(x)<a$ is finite.

Main Lemma. For any natural n and m, any $k>1$, any $d>0$ and any positive function $h(x)$ there exist positive constants $F_1(n, m, d, k, h(x))$ and $F_2(n, m, d, k, h(x))$ such that to any $t_0>0$ and $B\in A(n, m, d, t_0, k, h(x))$ there exists an $\tilde{x}\in[t_0, kt_0]$ and a maximal linearly ordered set $\bar{L}\subset B$ such that

$$F_1(n, m, d, k, h(x)) < P_l(\tilde{x}) < F_2(n, m, d, k, h(x))$$

whenever $l\in\bar{L}$.

Proof. We shall prove the lemma by induction on m. For $m=1$ and any n, d, k, $h(x)$ the lemma follows easily from Lemma 2′. Suppose that the hypothesis of the lemma is proved for $m=s$ (and any n, d, k, $h(x)$). We prove it for $m=s+1$.

Let $B \in A(n, s+1, d, t_0, k, h(x))$. Put $k_1 = (1+k)/2$. Consider the set H of all $b \in B$ satisfying $P_b(t_0) < d\bar{\varepsilon}(k_1, n)$. Suppose first that H is not empty. By condition A4, H is a finite set. By virtue of Lemma 2′ for any $b \in H$, the set $X(b)$ of all $x \in [t_0, k_1 t_0]$ such that $P_b(x) = d\bar{\varepsilon}(k_1, n)$ is non-empty. For $b \in H$ put $x(b) = \inf_{x \in X(b)} x$. Put $z = \sup_{b \in H} x(b)$. Since H is finite there exists a $\bar{b}$ such that $x(\bar{b}) = z$.

Let $\tilde{k} > 1$ be a number such that

$$t_0 + \tilde{k}^{2n+2}(k_1 t_0 - t_0) = k t_0, \quad \text{i.e.,} \quad \tilde{k} = \sqrt[2n+2]{\frac{k-1}{k_1-1}}.$$

By Lemma 1′ on one of the intervals

$$[t_0 + \tilde{k}(z-t_0), t_0 + \tilde{k}^2(z-t_0)], [t_0 + \tilde{k}^3(z-t_0), t_0 + \tilde{k}^4(z-t_0)], \dots,$$
$$[t_0 + \tilde{k}^{2n+1}(z-t_0), t_0 + \tilde{k}^{2n+2}(z-t_0)]$$

the values of the polynomial $P_{\bar{b}}(x)$ are greater than $\varepsilon_1(\tilde{k}, n) d\bar{\varepsilon}(k_1, n)$ and smaller than $\varepsilon_2(\tilde{k}, n) d\bar{\varepsilon}(k_1, n)$. Let $[t_1, t_2]$ be this interval, where $t_1 = t_0 + \tilde{k}^{2n+1}(z-t_0)$ and $t_2 = t_0 + \tilde{k}^{2n+2}(z-t_0)$. Observe that $[t_1, t_2] \subset [t_0, kt_0]$ and $t_2 - t_0 = \tilde{k}(t_1 - t_0)$.

Write $B(\bar{b}) \subset B$ for the set of all elements which are comparable with $\bar{b}$ (i.e. either greater or smaller than $\bar{b}$) and different from $\bar{b}$. Let $f: B(\bar{b}) \to M_n$ be the mapping, associating $\tilde{P}_b(x) = P_b(x+t_0)$ to $b \in B(\bar{b})$. We shall make some observations which will lead to the conclusion that the set $B(\bar{b})$ equipped with f is an element of $A(n, s, d\bar{\varepsilon}(k_1, n), t_1 - t_0, \tilde{k}, \tilde{h}(x))$, where

$$\tilde{h}(x) = h\left[\max\left(x, \varepsilon_2(\tilde{k}, n)\, d\bar{\varepsilon}(k_1, n)\right)\right].$$

Remark 1. The length of $B(\bar{b})$ is strictly less than the length of B.

Remark. 2. Let L be a linearly ordered subset of $B(\bar{b})$, $c > 0$, $x_0 \in [t_1 - t_0, \tilde{k}(t_1 - t_0)]$, and $\tilde{P}_l(x_0) < c$ for $l \in L$. Put $\bar{L} = L \cup \bar{b}$. The set $\bar{L}$ is a linearly ordered subset of B. Since $t_2 - t_0 = \tilde{k}(t_1 - t_0)$ it follows that $\tilde{x}_0 = x_0 + t_0 \in [t_1, t_2] \subset [t_0, kt_0]$. Therefore $P_{\bar{b}}(\tilde{x}_0) < \varepsilon_2(k, n) d\bar{\varepsilon}(k_1, n)$. It is clear, on the other hand, that $l \in L$ implies $P_l(\tilde{x}_0) < c$. Consequently, $\tilde{P}_l(x_0) < \max\left(c, \varepsilon_2(\tilde{k}, n) d\bar{\varepsilon}(k_1, n)\right)$ for $l \in \bar{L}$. Since B satisfies A2 and $\tilde{x}_0 \in [t_0, kt_0]$, there exists a maximal linearly ordered subset $\bar{\bar{L}}$, $\bar{L} \subset \bar{\bar{L}} \subset B$ such that

$$P_l(\tilde{x}_0) < h\left(\max\left(c, \varepsilon_2(\tilde{k}, n)\, d\bar{\varepsilon}(k_1, n)\right)\right) \quad \text{for} \quad l \in \bar{\bar{L}}.$$

Put $\hat{L} = \bar{\bar{L}} \setminus \bar{b}$. Then $L \subset \hat{L}$, and $\hat{L}$ is easily seen to be a maximal linearly ordered subset of $B(\bar{b})$. One gets easily, on the other hand, that

$$\tilde{P}_l(x_0) = P_l(\tilde{x}_0) < h\left(\max\left(c, \varepsilon_2(\tilde{k}, n)\, d\bar{\varepsilon}(k_1, n)\right)\right) = \tilde{h}(c) \quad \text{for} \quad l \in \hat{L}.$$

Remark 3. Since $x(\bar{b}) = \sup_{b \in H} x(b)$, it follows from the definition of H that for any $b \in B(b)$ there exists an $x(b)$, $0 \leqq x(b) \leqq t_1 - t_0$ such that $\tilde{P}_b(x(b)) \geqq d\bar{\varepsilon}(k_1, n)$.

Now $B(\bar{b}) \in A(n, s, d\bar{\varepsilon}(k_1, n), t_1 - t_0, \tilde{k}, \tilde{h}(x))$ and the induction hypothesis implies the existence of an $\tilde{x} \in [t_1, t_2]$ and a maximal linearly ordered $L_1 \subset B(\bar{b})$ such that for every $l \in L_1$

$$F_1(n, s, d\bar{\varepsilon}(k_1, n), \tilde{k}, \tilde{h}(x)) < P_l(x) < F_2(n, s, d\bar{\varepsilon}(k_1, n), \tilde{k}, \tilde{h}(x)).$$

Putting $\bar{L} = L_1 \cup \bar{b}$, and

$$F_1(n, s+1, d, k, h(x)) = \min\left[\varepsilon_1(\tilde{k}, n) d\bar{\varepsilon}(k_1, n), F_1(n, s, d\bar{\varepsilon}(k_1, n), \tilde{k}, \tilde{h}(x))\right],$$

$$F_2(n, s+1, d, k, h(x)) = \max\left[\varepsilon_2(\tilde{k}, n) d\bar{\varepsilon}(k_1, n), F_2(n, s, d\bar{\varepsilon}(k_1, n), \tilde{k}, \tilde{h}(x))\right]$$

we get the statement of the lemma for $m=s+1$, and for B with the further assumption that the set H of all $b \in B$ satisfying $P_b(t_0) < d\bar{\varepsilon}(k_1, n)$ is not empty.

Let us now consider the case when the set H is empty. Put $\tilde{x}=t_0$ and observe that condition A2 on B implies that for the empty set L, the point $\tilde{x}=t_0$ and $c=1$ (or any $c>0$) there exists a maximal linearly ordered subset $\tilde{L} \subset B$ such that $\frac{d}{2} \bar{\varepsilon}(k_1, n) < P_l(\tilde{x}) < h(1)$ for every $l \in \tilde{L}$ (we may take $h(c)$ for any $c>0$ rather than $h(1)$). Thus, we get the lemma also for empty H. The estimates we have for $F_1(n, m, d, k, h(x))$ and $F_2(n, m, d, k, h(x))$ are different from those in the case of non-empty H, but that is not important.

§ 2. Proof of Theorem 1

Let S be a lattice in R^n. Write S_R for the minimal linear subspace of the space R^n containing S. We call S a full sublattice of Z^n if S is a sublattice of Z^n satisfying $S_R \cap Z^n = S$. The set of all full sublattices of non-zero dimension of Z^n will be denoted by $S(Z^n)$.

We recall the following well-known theorem on lattices, due to Minkowski (cf. [3]).

For every n there exists a $c(n)$ such that any lattice Z^n lying in the Euclidean space R^n, contains a non-zero element z satisfying $\|z\| < c(n) d(Z^n)^{1/n}$ where $d(Z^n)$ is the determinant of the lattice.

Remark. The theorem of Minkowski is usually given in a more general and complete form, which implies immediately that $c(n) = 2\sqrt[n]{s(n)}$, where $s(n)$ denotes the volume of the unit ball in R^n.

The following lemma can be obtained easily by induction from the theorem of Minkowski.

Lemma 3. *For every n there exists a $\hat{c}(n)$ such that any lattice Z^k lying in the Euclidean space R^k, where $k \leqq n$, contains a sequence of full sublattices S_i $(1 \leqq i \leqq k)$ with the following properties:*

1. S_i *is of dimension i,*
2. $S_k = Z^k$,
3. $S_{i+1} \supset S_i$,
4. *the determinant $d(S_i)$ of S_i is less than $\hat{c}(n) d(Z^k)^{i/k}$.*

Let T^t be, as above, a one-parameter unipotent group of linear transformations of R^n.

Lemma 4. *If S is a lattice in R^n, then $d^2(T^t S)$ as a function of t is a polynomial of degree not exceeding* $2n^2$.

Proof. Let $e_1, \ldots, e_k$ be a basis in $S (k \leqq n)$. Then

$$d^2(T^t S) = |a_{ij}(t)| \qquad (1 \leqq i \leqq k, 1 \leqq j \leqq k)$$

where $a_{ij}(t) = (T^t e_i, T^t e_j)$. Since T^t is a one-parameter unipotent group, $a_{ij}(t)$ is a polynomial of degree not exceeding $2n$ and consequently $d^2(T^t S)$ is a polynomial of degree not exceeding $2n^2$.

Let Z^n be a lattice of dimension n in the space R^n. We introduce in the set $S(Z^n)$ a partial ordering, by writing $S_1 > S_2$ if $S_1 \supset S_2$. The length of the set $S(Z^n)$ is then obviously equal to n. For $s \in S(Z^n)$ put $P_s(x) = d^2(T^x s)$. By virtue of Lemma 4 the polynomials $P_s(x)$ define a $2n^2$-equipment of the set $S(Z^n)$. We shall prove now that with the equipment defined in this way $S(Z^n) \in \in A(2n^2, n, d, t_0, k, h(x))$, with arbitrary $t_0 > 0$ and $k > 1$; here $d = \inf_{s \in S(Z^n)} d^2(s)$ and $\tilde{h}(x) = \hat{c}(n)^2 \max(x, d^2(Z^n))$.

The conditions A1, A3 and A4 are obviously satisfied. Let us verify A2. Let $S_1 < S_2 < \ldots < S_p$ be a strictly increasing sequence in $S(Z^n)$ and let $d^2(S_i) < c$ for any $i (1 \leqq i \leqq p, c > 0)$.

Applying Lemma 3 to each factor lattice S_{i+1}/S_i (S_{i+1}/S_i is considered as a lattice in $(S_{i+1})_R/(S_i)_R$) we find that there exists a sequence $S_i \subset S_i^1 \subset \ldots \subset S_i^{h_i} = S_{i+1}$ $(h_i = \dim S_{i+1} - \dim S_i)$ such that

1) $S_i^l \subset S(Z^n)$,

2) $\dim S_i^{l+1} - \dim S_i^l = 1$,

3) $d(S_i^l) < \hat{c}(n)\, d(S_{i+1}/S_i)^{\frac{l}{h_i}}\, d(S_i) = \hat{c}(n)\, d(S_{i+1})^{\frac{l}{h_i}}\, d(S_i)^{\frac{h_i - l}{h_l}} <$

$$< \hat{c}(n) \sqrt{c}^{\frac{l}{h_i}} \sqrt{c}^{\frac{h_i - l}{h_i}} = \hat{c}(n) \sqrt{c}.$$

We construct analogously a sequence S_0^l for the lattice S_1 and a sequence S_l^p for the lattice Z^n/S^p (and show in the same way that $d(S_0^l) < \hat{c}(n)\sqrt{c}$ and $d(S_p^l) < < \hat{c}(n) \max(\sqrt{c}, d(Z^n))$. We take the union of all these sequences S_i^l $(0 \leqq i \leqq p)$. The unified sequence $\bar{S}_k$ is easily seen to be a maximal linearly ordered subset of $S(Z^n)$, further $d^2(\bar{S}_k) < \hat{c}(n)^2 \max(c, d^2(Z^n)) = h(c)$. Thus we have proved that $S(Z^n)$ satisfies A2.

Let us prove now Theorem 1. Let t_0 be an arbitrary large number. As a consequence of the Main Lemma one can find constants $F_1(Z^n)$ and $F_2(Z^n)$ such that there exist a $\bar{t}$, $t_0 < \bar{t} < 2t_0$ and a sequence of full sublattices of the lattice $Z^n = S_n \supset S_{n-1} \supset \ldots \supset S_1$ such that S_i is of dimension i and $F_1(Z^n) < d^2(T^{\bar{t}} S_i) < < F_2(Z^n)$ (put $F_1(Z^n) = F_1(2n^2, n, \tilde{d}, 2, \tilde{h}(x))$ and $F_2(Z^n) = F_2(2n^2, n, \tilde{d}, 2, \tilde{h}(x))$).

Let $z \in Z^n$ and $z \neq 0$. Then either $z \in S_1$, or for some i we have $z \in S_{i+1}$ and $z \notin S$. If $z \in S_1$, we have $d(T^i S_{i+1}) \leqq \|T^i z\| d(T^i S_i)$. Consequently

$$\|T^i z\| \geqq \frac{d(T^i S_{i+1})}{d(T^i S_i)} \geqq \sqrt{\frac{F_1(Z^n)}{F_2(Z^n)}}.$$

when $z \notin S_1$. On the other hand, for $z \in S_1$ we get easily

$$\|T^i z\| \geqq d(T^i S_1) \geqq \sqrt{F_1(Z^n)}$$

Because of this inequality, if we take

$$\varepsilon = \min\left(\sqrt{F_1(Z^n)}, \sqrt{\frac{F_1(Z^n)}{F_2(Z^n)}}\right),$$

the lattice $T^i Z^n$ intersects the ε-neighbourhood of 0 in R^n in 0 only. Theorem 1 is proved.

REFERENCES

[1] Pjateckij-Šapiro, I. I. (Пятецкий-Шапиро И. И.), Дискретные подгруппы полупростых групп, *Труды Моск. матем. об-ва*, **18** (1968), 3—18.
(Discrete subgroups of semisimple goups.)

[2] Margulis, G. A. (Маргулис Г. А.), К проблеме арифметичности дискретных групп. *ДАН СССР,* **187** (1969), 518—520.
(On the problem of arithmeticity of discrete groups.)

[3] Cassels, J. W. S., *An introduction to the geometry of numbers,* Springer-Verlag, 1959, (Дж. Касселс, *Введение в геометрию чисел,* Москва, изд-во „Мир”, 1965.)

NON-UNIFORM LATTICES IN SEMISIMPLE ALGEBRAIC GROUPS

G. A. MARGULIS

The aim of the present paper is to give the proof of some theorems announced by the author [40], and also to present a proof of an analogue of Mostow's theorem [44] on the extension of isomorphisms of uniform subgroups, in the case of non-uniform lattices. The precise formulations will be given in § 2. We note here, however, that the main result of this paper is Theorem 17.43 (see also § 2) which is a slight strengthening of the basic theorem in [40].

Now we briefly sketch the contents of the paper.

§ 1 contains, without proofs, some well-known results of the theory of algebraic groups and the theory of discrete subgroups. The appropriate references can be found in section IX of § 1.

In § 2 we formulate the main results of the paper.

In § 3 assertions will be formulated (Lemmas 3.3 and 3.4) about the action of unipotent groups in the space of lattices. These assertions will be essential in the following (§ 11). Their proof is given in [42].

In § 4 we prove some simple corollaries to the theorems of Auslander and Zassenhaus ([9] and [13]) on the intersection of discrete subgroups of a Lie group with solvable radical, and also to the theorem of Mostow ([15] and [18]) on discrete subgroups of solvable Lie groups. Furthermore, we prove in § 4 the rather simple Lemma 4.8 on the intersection of lattices (definition see in § 1) in semisimple algebraic groups with direct summands of these groups. The author has recently learned that an analogous statement was proved by Prasad and Raghunathan in [39].

§ 5 presents, with proofs, some well-known results on discrete subgroups of simply-connected nilpotent Lie groups and unipotent algebraic groups (see [10], [24], [28], [31] and [43]).

Let G be a Lie group and Γ a discrete subgroup. In § 6 we will study the following two classes of closed subgroups:

1) The class of subgroups having property P, i.e. subgroups H of the group G such that for an arbitrary compact $K \subset G$ there are finitely many elements $\gamma_1, \ldots, \gamma_n \in \Gamma$ such that $(K \cdot H) \cap \Gamma = \bigcup_{i=1}^{n} \gamma_i (H \cap \Gamma)$.

2) The class of integer subgroups i.e. subgroups for which the quotient space $H/H\cap\Gamma$ is compact.

In § 6 we will prove some simple assertions about these classes of subgroups. In particular, it will be proved that H has property P if and only if the natural mapping $H/H\cap\Gamma\to G/\Gamma$ is proper.

From § 7 on — except § 8 — it will always be supposed, without saying it each time explicitly, that G is a semisimple algebraically connected real algebraic group such that 1) the centre of the complexification of G is trivial, 2) G has no compact factors, while Γ is a non-uniform lattice in G (definition in § 1), and the pair (G, Γ) is irreducible (definition in § 2).

Let u be a unipotent subgroup of the group Γ. Denote by $\overline{U}$ the Zariski closure of the subgroup u in the group G, by $N(\overline{U})$ and $C(\overline{U})$ the normalizer and centralizer of $\overline{U}$ in G, and by $\overline{S}N(\overline{U})$ the subgroup of the group $N(\overline{U})$, consisting of the elements which have determinant ± 1 at the natural action of $N(\overline{U})$ on the Lie algebra of the group $\overline{U}$. In § 7 we study various connections between the subgroups $\overline{S}N(\overline{U})$ and $C(\overline{U})$ and the subgroup Γ. The basic result in § 7 is Lemma 7.6. We note that almost all assertions of §§ 6 and 7 have earlier been proved by the author in [41] but the proofs given in this paper are more detailed.

In § 8 we define the notion of quasi-algebraic **R**-group and arithmetic subgroup of a quasi-algebraic **R**-group. All statements proved in § 8 are, in fact, well known but the author was unable to find the necessary references.

In § 9 we will prove some auxiliary statements mainly concerning the behaviour of the trajectories of unipotent subgroups in the quotient spaces of an algebraic group with respect to an algebraic subgroup.

§ 10 is devoted to the proof of some auxiliary algebraic statements. The basic results in § 10 are contained in Lemma 10.3 (a modification of Lemma 1.6 in the paper [6] of Garland and Raghunathan) which will be applied in § 11, and Lemmas 10.6, 10.7, 10.24 and 10.29 which, in turn, will be applied in § 13. In § 10 some well-known results dealing mainly with the parabolic and orispheric (definition in § 1) subgroups of algebraic groups are presented without proofs. In reading § 10 the reader should keep in mind the remarks in section 11 of § 18.

§ 11 is devoted mainly to the study of maximal unipotent subgroups of the group Γ. The basic theorem here is Theorem 11.13, which says that for an arbitrary maximal unipotent subgroup U of the group Γ the subgroup $\overline{U}$ ($\overline{U}$ is the Zariski closure of U in G) is an orispheric subgroup of the group G. In the proof of Theorem 11.13 we will make an essential use of Lemmas 3.3 and 3.4. Theorem 11.13 will prove to be essential in connection with the facts that *1)* the normalizer of an arbitrary orispheric subgroup is a parabolic subgroup and *2)* the intersection of any two parabolic subgroups contains "sufficiently many" elements.

For a unipotent subgroup U of the group Γ, $\tilde{A}(U)$ denotes the Zariski closure of the subgroup $\bar{S}N(\bar{U})\cap\Gamma$ in group G, and $(\tilde{\tilde{A}}U)$ the algebraically connected component of the unit element of the group $\tilde{A}(U)$.

In § 12 we prove assertions which state, roughly speaking, that for a rather wide class of unipotent subgroups U_1 and U_2, the subgroup $\tilde{\tilde{A}}(U_1)\cap\tilde{\tilde{A}}(U_2)\cap\Gamma$ contains sufficiently many elements. Unfortunately, this class does not contain the maximal unipotent subgroups U with $\dim \tilde{A}(U)=\dim \bar{U}$.

Therefore, in § 13 such maximal unipotent subgroups will be studied. § 13 is one of the most technically involved. The basic result here is Theorem 13.20 which will be deduced from Lemma 13.13 and 13.19. We also remark that from § 13 on it will be assumed that the real rank of the group G is greater than 1.

In § 14 we give a construction of representation theory which seems to be of independent interest. This construction will be used in proving Theorem 14.15, which, in turn, will be applied in the proofs of the basic results of this paper. § 14 is one of the basic sections in this paper.

In § 15 we will study the representations of discrete subgroups, which, roughly speaking, can be decomposed into a sufficiently "non-trivial" semi-direct product.

In § 16 it will be proved (Theorem 16.17) that the group G can be defined over the field of rational numbers in such a way that the subgroup Γ be contained in the group $G_{\mathbf{Q}}$, of the **Q**-rational points of the group G. Theorem 16.17 is deduced from Theorem 16.11, which, in turn, follows from Theorems 16.4 and 14.15 and the assertions of §§ 12 and 15. Theorem 16.4 is itself a rather simple corollary to Theorems 13.20 and 11.13 and the statement of § 12. Theorem 16.17 coincides with part *1)* of the basic theorem (Theorem 17.43) and is substantially used in the proof of the remaining parts of the basic theorem.

In § 17 we prove statements concerning, essentially, the intersections of parabolic subgroups of the group G with the group Γ. From these statements and Theorem 16.17 we obtain the Basic Theorem.

§ 18 is an appendix to §§ 10 and 13.

In § 19 we prove (see § 2 and Theorems 19.5 and 19.6) the rigidity of the group Γ and the analogue of the theorem of Mostow ([44]) on the extension of isomorphisms of uniform subgroups. In the proofs these theorems we substantially use Theorem 14.15, the Basic Theorem and also the assertions of § 15.

The author wishes to express his thanks to I. N. Bernstein, E. B. Vinberg, D. A. Kajdan, and I. I. Pjateckij–Šapiro for their valuable remarks and for their interest in this work. The author is particularly indebted to B. Ju. Weissfeiler whose help was inestimable at the different stages of writing this work.

LIST OF CONTENTS

§ 0. Notation and terminology

I. We will denote by (g_1, g_2) the commutator $g_1 g_2 g_1^{-1} g_2^{-1}$ of two elements of the group and, for subgroups A and B, by (A, B) the subgroup generated by the commutators of the form (a, b), where $a \in A$ and $b \in B$. $[\mathfrak{g}_1, \mathfrak{g}_2]$ denotes the commutator of the elements of a Lie algebra and for subalgebras $\mathfrak{A}$ and $\mathfrak{B}$ of a Lie algebra, $[\mathfrak{A}, \mathfrak{B}]$ denotes the subalgebra generated by the commutators of the form $[\mathfrak{a}, \mathfrak{b}]$, where $\mathfrak{a} \in \mathfrak{A}$ and $\mathfrak{b} \in \mathfrak{B}$.

For subsets A and B of a group we will denote by $A \cdot B$ the set of all elements of the form $a \cdot b$ where $a \in A$, $b \in B$, and by A^{-1} the subset of all elements of the form a^{-1}, where $a \in A$. For an arbitrary group G and subgroup $H \subset G$ the natural mapping $G \to G/H$ $(G \to H\backslash G)$ will mean the mapping under which $g \in G$ corresponds to the right (left) coset $gH (Hg)$. If H and Γ are subgroups of the group G, the natural mapping $H/H \cap \Gamma \to G/\Gamma$ means the mapping π' given by the formula

$$\pi' = \pi \circ \tilde{\pi}^{-1}, \tag{1}$$

where π and $\tilde{\pi}$ are the natural mappings $G \longrightarrow G/\Gamma$ and $H \longrightarrow H/H \cap \Gamma$ (π' is well defined, since for arbitrary $h_1, h_2 \in H$ with $\tilde{\pi}(h_1) = \tilde{\pi}(h_2)$ we have $\pi(h_1) = \pi(h_2)$).

For an arbitrary subgroup H of a group G, $N(H)$ denotes the normalizer of H in G and $C(H)$ the centralizer of H in G.

If G_1 and G_2 are groups, $H_1 \subset G_1$ and $H_2 \subset G_2$ subgroups and f is a homomorphism $G_1 \xrightarrow{f} G_2$ such that $f(H_1) \subset H_2$, then the natural mapping of G_1/H_1 to G_2/H_2 means the mapping defined by the formula

$$\pi' = \pi_2 \circ f \circ \pi_1^{-1} \tag{2}$$

where π_1 and π_2 are the natural mappings $G_1 \xrightarrow{\pi_1} G_1/H_1$ and $G_2 \xrightarrow{\pi_2} G_2/H_2$ (the correctness of the definition of π' follows from the relation $f(H_1) \subset H_2$).

All the group-theoretical and set-theoretical notions (centre, centralizer, normalizer, image, etc.) occurring in this paper are always meant in the sense of abstract group theory and set theory, irrespective of the additional structure defined on the group or set.

Let G be a group, and H a subgroup of G. Then, as a rule, the normalizer of the subgroup H in the group G will be denoted by $N_G(H)$ or simply by $N(H)$, and the centralizer of H in G by $C_G(H)$ or simply by $C(H)$.

For an arbitrary element g of the group G and arbitrary subset $A \subset G$, gAg^{-1} always denotes the union $\bigcup_{a \in A} gag^{-1}$.

Let G be a group and A a subset of G. In the paper we shall use expressions like "the minimal subgroup (normal subgroup) of the group G containing A". This will mean that among all the subgroups (normal subgroups) containing A we take the unique minimal one.

II. For an arbitrary complex (real) linear space V we denote by $GL(V)$ the group of all non-singular linear transformations of the space.

To avoid misunderstanding, we emphasize that in the present paper an algebraic mapping of one algebraic manifold to another means a mapping which can be realized by rational functions, while a regular algebraic mapping is a mapping realizable by everywhere defined rational functions.

The field of real numbers will always be denoted by **R**, the field of complex numbers by **C**, and the field of rational numbers by **Q**.

§ 1. Some well-known results

I. Let G be a locally compact topological group. The subgroup Γ of G is called *discrete* if there is a neighbourhood of the identity whose intersection with Γ consists only of the identity. If Γ is a discrete subgroup in G then G/Γ denotes the quotient-space of G by the subgroup Γ with the usual quotient topology. Let μ be a right-invariant Haar measure on the group G. Then μ induces a measure on G/Γ in a natural way which will also be denoted by μ. We call the discrete subgroup Γ a *lattice* if the volume of the quotient space G/Γ with respect to the measure μ is finite. (This definition does not depend on the choice of μ since two arbitrary right-invariant Haar measures on G are proportional.) A discrete subgroup is called *uniform* if the factor space is compact. It is clear that every uniform discrete subgroup is a lattice.

If there is any lattice in G, then, as it was shown by Siegel [11], G is a unimodular group, i.e. there exists a two-sided invariant measure on G. Note that the converse is not true, i.e. an arbitrary unimodular group may contain no lattices.

Let P be a closed normal subgroup and Γ a discrete subgroup of the group G. Denote by φ the natural homomorphism $G \to G/P$. We list here, without proofs, some simple and well-known propositions.

Lemma 1.1.1. If Γ is a lattice in G and $\varphi(\Gamma)$ is discrete in G/P, then $P \cap \Gamma$ is a lattice in P, and $\varphi(\Gamma)$ is a lattice in G/P.

Lemma 1.1.2. If $P \cap \Gamma$ is a lattice in P and $\varphi(\Gamma)$ is a lattice in G/P, then Γ is a lattice in G.

Lemma 1.1.3. If $P \cap \Gamma$ is a uniform subgroup in P then $\varphi(\Gamma)$ is discrete.

Lemma 1.1.4. If Γ is a uniform subgroup in G and $\varphi(\Gamma)$ is discrete, then $P \cap \Gamma$ is a uniform subgroup in P.

Lemma 1.1.5. If $P \cap \Gamma$ is a uniform subgroup in P and $\varphi(\Gamma)$ is a uniform subgroup in G/P then Γ is a uniform subgroup in G.

Lemma 1.1.6. The quotient group of a connected, simply connected Lie group with respect to a connected closed normal subgroup is a connected, simply connected Lie group.

II. A complex (real) linear algebraic group is a subgroup of the group $GL(n, \mathbf{C})(GL(n, \mathbf{R}))$ (the group of all complex (real) matrices of order n with nonzero determinant) characterized by the condition that some given polynomials of the matrix coefficients vanish. Complex and real linear algebraic groups will often be called algebraic **C**-groups and algebraic **R**-groups, respectively. Two linear algebraic groups (complex or real) are said to be *isomorphic* if there exists an isomorphism between them (in the sense of abstract group theory) which is, at the same time, an algebraic mapping regular in both directions.

For an algebraic **C**-group (**R**-group) G and an arbitrary subring k of the field **C** (field **R**) G_k denotes the set of points of the group G defined over k (i.e. the elements of G with matrix coefficients belonging to k) whose determinant is a unit of the ring k (i.e. an invertible element of k).

If the polynomials defining the linear algebraic group can be chosen so that all their coefficients belong to some field K, we say that this group is *defined over K*. If the algebraic **R**-group G is defined over the field **Q** of rational numbers, then the subgroup $G_{\mathbf{Z}} \subset G$ where **Z** is the ring of integers, is called an *arithmetic subgroup* of the group G. Let G_1 and G_2 be two isomorphic algebraic **R**-groups, and f an algebraic isomorphism from G_1 onto G_2, and $G_{\mathbf{Z}}$ an arithmetic subgroup in G_1. Then the subgroup $f(G_{\mathbf{Z}})$ is also called *arithmetic*. Two subgroups are said to be *commensurable* if their intersection has finite index in both of them. A subgroup commensurable with an arithmetic one is also called arithmetic. Note that an arithmetic subgroup, as one can easily see, is always discrete.

A **Q**-form of an algebraic **R**-group G is a pair $F=(\tilde{G},f)$, where $\tilde{G}$ is an algebraic **R**-group defined over **Q** and f an isomorphism $\tilde{G}\to G$. We denote $f(\tilde{G}_{\mathbf{Q}})$ by $F_{\mathbf{Q}}$ and $f(\tilde{G}_{\mathbf{Z}})$ by $F_{\mathbf{Z}}$. For a given **Q**-form of the algebraic group G and an algebraic subgroup H (i.e. a subgroup which is at the same time an algebraic submanifold) we may say that H is defined over **Q**, if $f^{-1}(H)$ can be determined in $\tilde{G}$ by polynomials with rational coefficients.

For an arbitrary subset $M\subset G$, we denote by $A(M)$ the Zariski closure of the subset M (i.e. $A(M)$ is the minimal algebraic submanifold of the group G containing M). It is easy to see that if M is a subgroup then so is $A(M)$.

Let $H\subset GL(n, \mathbf{R})$ be a real algebraic group. Then we denote by $H_{\mathbf{C}}$ the complexification of the group H i.e., the Zariski closure of H in $GL(n, \mathbf{C})$ (here $GL(n, \mathbf{C})$ is considered as a complex algebraic manifold). It is clear that $H_{\mathbf{C}}$ is a complex algebraic group.

One can see easily that an arbitrary algebraic **R**-group (**C**-group) is a Lie group at the same time. Thus, on every algebraic **R**-group (**C**-group) there are two topologies: the Zariski topology and the topology of the Lie group. We say that the subset M of an algebraic **R**-group (**C**-group) is algebraically connected (closed, open, etc.) if it is connected (closed, open, etc.) in the Zariski topology. If M is connected (closed, open etc.) in the topology of the Lie group, then we will say that M is topologically connected (closed, open etc.). It is easy to see that the algebraic connectedness of the group G is equivalent to the fact that it does not contain any proper algebraic subgroup (i.e. $\neq G$) of finite index in G. It is well known [27] that the topological connectedness of an algebraic **R**-group G implies the fact that G is algebraically connected as well, while the algebraic connectedness does not imply the topological connectedness (it can be shown only [27] that the topologically connected component of the identity of an algebraic **R**-group G is a subgroup of finite index in G). If, however, G is an algebraic **C**-group, then its topological connectedness is equivalent to its algebraic connectedness. Therefore we will often (but not always) speak about complex algebraic groups without specifying whether algebraic or topologic connectedness is involved.

The following propositions are well known:

Lemma 1.2.1. The adjoint representation Ad of an algebraic **R**-group (**C**-group) G is algebraic (i.e. Ad is a regular algebraic mapping of G into $GL(\mathfrak{G})$. Here $\mathfrak{G}$ is the Lie algebra of the group G and $GL(\mathfrak{G})$ is the algebraic **R**-group (**C**-group) of all non-degenerate linear transformations of the real (complex) vector space $\mathfrak{G}$).

Lemma 1.2.2. In an **R**-group G the algebraically connected component of the identity contains the topologically connected component of the identity.

Lemma 1.1.3. The Zariski closure of a nilpotent subgroup of an algebraic **R**-group (**C**-group) is a nilpotent algebraic **R**-group (**C**-group).

Lemma 1.2.4. The Zariski closure of a solvable subgroup of an algebraic **R**-group (**C**-group) is a solvable algebraic **R**-group (**C**-group).

Lemma 1.2.5. The algebraic closure of a normal subgroup of an algebraic **R**-group (**C**-group) is an algebraic normal subgroup.

Lemma 1.2.6. Let f be a regular algebraic mapping of an algebraic **R**-group (**C**-group) into another one. If the restriction of f to a Zariski-dense subgroup (i.e. a subgroup dense in the Zariski topology) is a homomorphism, then f is a homomorphism itself.

Lemma 1.2.7. Let G be an algebraic **R**-group (**C**-group), G^0 the algebraically connected component of the unity in the group G, and $H \subset G$ an algebraic subgroup. Then

1. if H has a finite index in G, then $H \supset G^0$,
2. G^0 has a finite index in G.

Lemma 1.2.8. In an algebraic **R**-group (**C**-group) the normalizer of any algebraic subgroup is an algebraic subgroup.

Lemma 1.2.9. In an algebraic **R**-group (**C**-group) the centralizer of any subset is an algebraic subgroup.

Lemma 1.2.10. Let G be an algebraic **R**-group, and $H \subset G$ its algebraic subgroup. Then $\dim H = \dim G$ implies that H is a subgroup of finite index in G.

Lemma 1.2.11. If an algebraically connected **R**-group (**C**-group) is defined over some field $K \subset \mathbf{R} (K \subset \mathbf{C})$, then G_K is Zariski dense in $G_{\mathbf{R}} (G_{\mathbf{C}})$.

Lemma 1.2.12. An algebraically connected component of an algebraic **R**-group (**C**-group) defined over the field K is itself defined over K.

Lemma 1.2.13. Let G be an algebraic **R**-group (**C**-group) defined over the field $K \subset \mathbf{R} (K \subset \mathbf{C})$, and H an algebraic subgroup in G defined over the same field K. Then the normalizer and centralizer of H in G are also defined over K.

Lemma 1.2.14. Let G be an algebraic **R**-group (**C**-group). Then the algebraic closure of any topologically connected subgroup of the group is algebraically connected.

Lemma 1.2.15. The centre of an arbitrary real algebraic group is contained in the centre of its complexification.

Lemma 1.2.16. In algebraic **R**-group (**C**-group) the normalizer of any topologically connected Lie subgroup is an algebraic subgroup.

Lemma 1.2.17. Let G_1 and G_2 be complex algebraic groups and $f: G_1 \to G_2$ a one-to-one algebraic homomorphism. Then f is an isomorphism in the sense of algebraic groups (for real algebraic groups the analogous statement is not true).

Lemma 1.2.18. Let G be an algebraic **R**-group (**C**-group); and $A \subset G$, $B \subset G$ algebraic subgroups, $A_1 \subset A$ and $B_1 \subset B$ subsets. If A_1 is Zariski dense in A and B_1 is Zariski dense in B, then the subgroup generated by A_1 and B_1 is Zariski dense in the minimal algebraic subgroup containing $A \cup B$.

Lemma 1.2.19. Let G be an algebraic **R**-group (**C**-group), $A \subset G$ an algebraic subgroup of G, $\Lambda \subset G$ a Zariski dense subgroup. If $\Lambda \cap A$ is Zariski dense in A, then the normal subgroup of the group Λ generated by $\Lambda \cap A$ is Zariski dense in the minimal algebraic normal subgroup of the group G containing A.

III. Let $G \subset GL(n, \mathbf{R})(GL(n, \mathbf{C}))$ be an algebraic **R**-group (**C**-group). The elements of the group will be considered at the same time as elements of the group $GL(n, \mathbf{R})(GL(n, \mathbf{C}))$.

An element u of the group G is called *unipotent* if all its eigenvalues are equal to one. Throughout the paper, for a unipotent element $u \in G$, $\ln u$ will denote the nilpotent element of the Lie algebra $\mathfrak{G}$ of the group G for which $\exp(\ln u) = u$ (the existence and uniqueness of such an element can be proved easily). A subgroup is called *unipotent*, if all its elements are unipotent. For an arbitrary subgroup $H \subset G$ we denote by H^u the set of all unipotent elements in H. The unipotent radical of a group is its maximal unipotent normal subgroup (note that the unipotent radical is uniquely determined). We will say that an algebraic unipotent subgroup of the group G is *orispheric*, if it coincides with the unipotent radical of its normalizer.

The following lemmas are well known.

Lemma 1.3.1. Every unipotent algebraic **R**-group (**C**-group) is a topologically and consequently also an algebraically connected, simply connected, nilpotent Lie group, and every unipotent subgroup of an algebraic group is nilpotent.

Lemma 1.3.2. The unipotent radical of an algebraic **R**-group (**C**-group) is a topologically, and consequently also an algebraically connected algebraic normal subgroup in this group.

Lemma 1.3.3. The set of all unipotent elements of an algebraic **R**-group (**C**-group) is a topologically connected algebraic submanifold, and consequently, the algebraic closure of any unipotent subgroup is unipotent.

Lemma 1.3.4. An algebraic homomorphism of an algebraic **R**-group (**C**-group) into another, carries over any unipotent element into a unipotent one.

Lemma 1.3.5. Let G be an algebraic **R**-group (**C**-group) defined over some field $K \subset \mathbf{R}(K \subset \mathbf{C})$. Then the unipotent radical of G is defined over K.

Lemma 1.3.6. Let G be an algebraic **R**-group and U the unipotent radical of the group G. Then the unipotent radical of the complexification $G_{\mathbf{C}}$ of the group G coincides with the complexification $U_{\mathbf{C}}$ of the subgroup U.

Lemma 1.3.7. Let G be an algebraic **R**-group (**C**-group) and $H \subset G$ an algebraic subgroup. If $G^u = H^u$, then the unipotent radical of the group G coincides with the unipotent radical of the subgroup H.

Lemma 1.3.8. Let G_1 and G_2 be algebraic **R**-groups (**C**-groups), f an algebraic homomorphism of G_1 into G_2 whose kernel is unipotent. Then the inverse image $f^{-1}(u)$ of any unipotent element $u \in G_2$ consists of unipotent elements only.

Lemma 1.3.9. Let $U \subset GL(n, \mathbf{R})(GL(n, \mathbf{C}))$ be an arbitrary unipotent group. Then in $\mathbf{R}^n$ ($\mathbf{C}^n$) there exists a non-zero vector x such that $Ax = x$ for any $A \in U$.

IV. An algebraic **R**-group (**C**-group) G is called *semisimple* (resp. *reductive*) if G has no non-trivial algebraically connected solvable (resp. unipotent) algebraic normal subgroup. An algebraic **R**-group (**C**-group) G is called *simple* (resp. *almost simple*) if G contains no proper algebraic normal subgroups except the group $\{e\}$ (resp. the finite subgroups).

The algebraic **R**-group (**C**-group) G is the *direct product* of the algebraic normal subgroups $G_1, \ldots, G_m$ if the mapping of multiplication $G_1 \times \ldots \times G_m \to G$ is an isomorphism between algebraic groups; G is the *almost direct product* of the groups G_i, if this mapping is an isogeny (i.e. an epimorphism with finite kernel). G is the *semi-direct product* of the algebraic subgroup H and the algebraic normal subgroup N, if the mapping given by multiplication induces an isomorphism between the algebraic manifolds $H \times N$ and G.

The following lemmas are well known.

Lemma 1.4.1. An algebraic **R**-group (**C**-group) is semisimple if and only if it is the almost direct product of non-commutative almost simple algebraic **R**-groups (**C**-groups).

Lemma 1.4.2. If G is an algebraically connected semisimple algebraic **R**-group whose complexification has a trivial centre, then *1)* G decomposes into a direct product of algebraically connected simple algebraic **R**-groups $G = \bigoplus_{i=1}^{m} G_i$ *2)* any algebraic normal subgroup of the group G coincides with the product of certain G_i *3)* any normal subgroup (not necessarily algebraic) of the group G contains, as a subgroup of finite index, the product of certain groups G_i^0, where G_i^0 denotes the topologically connected component of the identity in the group G_i.

Lemma 1.4.3. The centre of any semisimple algebraic **R**-group (**C**-group) is finite.

Lemma 1.4.4. Any algebraic representation of an arbitrary reductive algebraic **R**-group (**C**-group) is completely reducible.

Lemma 1.4.5. The Lie algebra $\mathfrak{G}$ of an arbitrary semisimple algebraic **R**-group (**C**-group) G is semisimple (i.e. $\mathfrak{G}$ contains no non-trivial solvable ideal).

Lemma 1.4.6. Any topologically closed subgroup of finite index (and, in particular, the topologically connected component of the identity) of an arbitrary semisimple (resp. simple) algebraic **R**-group (**C**-group), is a semisimple (resp. simple) Lie group.

V. An algebraically connected commutative algebraic subgroup T of an algebraic **R**-group (**C**-group) $G \subset GL(n, \mathbf{R})(GL(n, \mathbf{C}))$ is called a *torus* if it consists of semisimple elements (i.e. matrices which can be brought to diagonal form over the field of complex numbers). The torus T is said to be *K-split* or *K-decomposable* if it is defined over K and can be brought to diagonal form over K.

The following lemmas are well known.

Lemma 1.5.1. Every algebraically connected nilpotent algebraic **R**-group (**C**-group) is uniquely decomposable into a direct product (in the sense of section IV) $N = T \times U$, where T is a torus and U is a unipotent algebraic subgroup.

Lemma 1.5.2. Any semisimple element of an algebraically connected algebraic **C**-group belongs to at least one torus of the group G.

Lemma 1.5.3. An algebraic homomorphism of an algebraic **R**-group (**C**-group) into another, carries semisimple elements always into semisimple ones.

Lemma 1.5.4. Every element g of an arbitrary algebraic **R**-group (**C**-group) can be written as a product $g = t \cdot u$ where *1)* $t \in G$, $u \in G$ *2)* t and u commute, *3)* t is semisimple and u is unipotent.

Lemma 1.5.5. Any algebraically connected solvable algebraic **R**-group (**C**-group) can be decomposed into a semidirect product (in the sense of section IV) of a torus T and a unipotent algebraic subgroup U.

Lemma 1.5.6. Let G be an algebraic **R**-group (**C**-group), T an arbitrary torus in G, $N(T)$ the normalizer of T in G, and $C(T)$ the centralizer of T in G. Then the quotient group $N(T)/C(T)$ is finite.

VI. Let H be a Lie-group, $\varrho_{\mathfrak{H}}$ an arbitrary Euclidean metric in the Lie algebra $\mathfrak{H}$ of the group H, and ϱ_H the right-invariant metric in H corresponding to $\varrho_{\mathfrak{H}}$. Then, as is well known the following assertions hold.

Lemma 1.6.1. There is an $\varepsilon > 0$ for which the ε-neighbourhood (in the metric ϱ_H) V of the identity in the group H has the following properties: a) there is a canonical system of coordinates (see [12]) on V, b) for any $h_1, h_2 \in V$ we have

$$\varrho_H(e, h_1h_2h_1^{-1}h_2^{-1}) < \frac{1}{2}\min\left(\varrho_H(e, h_1), \varrho_H(e, h_2)\right)$$

Lemma 1.6.2. If H is a semisimple algebraic **R**-group, then for any unipotent subgroup $U \subset H$ and any positive c_1 and c_2 there is an element h of the group H for which $u \in U$ and $\varrho_H(e, u) < c_2$ imply $\varrho_H(e, huh^{-1}) < c_1$.

VII. Let us take on the group G a structure α of an algebraic **R**-group (algebraic **C**-group) and a structure β of a Lie group. We will say that α and β are *compatible* if the Lie group structure induced by α on G coincides with the structure β.

The following lemma is well known.

Lemma 1.7.1. On the quotient group G/N of an algebraic **C**-group by an algebraic normal subgroup N, one can define a structure of an algebraic **C**-group such that *1)* this structure is compatible with the natural Lie group structure of G/N, *2)* the natural mapping $\varphi: G \to G/N$ is a regular algebraic mapping.

VIII. The algebraic subgroup P of a connected complex algebraic group G is called *parabolic* if P contains some Borel subgroup of the group G (i.e. a maximal connected solvable subgroup in G). The algebraic subgroup P of an algebraically connected real algebraic group G is called parabolic if the complexification $P_{\mathbf{C}}$ of the group P is a parabolic subgroup in the complexification $G_{\mathbf{C}}$ of the group G.

The following propositions are well known.

Lemma 1.8.1. Let G be an algebraically connected complex (real) algebraic group and $P \subset G$ an arbitrary parabolic subgroup. Then G/P is a complex (real) algebraic manifold which is compact both in algebraic and topological sense.

Lemma 1.8.2. The normalizer of any maximal unipotent subgroup in an arbitrary complex (real) algebraically connected algebraic group G is a parabolic subgroup in G.

Lemma 1.8.3. Let G be an algebraically connected algebraic **C**-group (**R**-groups, and $P_1 \subset G$ and $P_2 \subset G$ two arbitrary parabolic subgroups. Then $P_1 \cap P_2$ contain) the centralizer of some maximal (maximal **R**-split) torus of the group G.

IX. In this section we make some remarks concerning the propositions stated in the previous sections of § 1.

Remark 1. Lemmas 1.2.1—1.2.10, 1.2.12—1.2.19, 1.3.1—1.3.3, 1.3.5 and 1.3.6 are elementary in character. Therefore, we only mention that either they are proved in [1], [3], and [27], or follow immediately from the propositions proved in [1], [3] and [27].

Remark 2. Concerning Lemma 1.2.11 see [29] and also [3] and [5].

Remark 3. For Lemmas 1.3.4 and 1.5.3 see [1] and [3].

Remark 4. Lemma 1.3.7 follows from Lemma 10.8 of § 10 while for Lemma 10.8 see Remark 18.9.

Remark 5. As for Lemmas 1.5.1, 1.5.2, 1.5.4, 1.5.5 and 1.5.6, see [1] or [3].

Remark 6. Lemma 1.3.8 is an immediate consequence of Lemmas 1.3.4, 1.5.3 and 1.5.4.

Remark 7. For Lemmas 1.4.1—1.4.6 see [1], [3] or [5].

Remark 8. As for Lemma 1.6.1, see [12].

Remark 9. Lemma 1.6.2 easily follows from Lemma 10.4 (concerning Lemma 10.4 see Remark 18.9), from the fact that every unipotent subgroup of the group H is contained in a maximal unipotent subgroup, and from Lemmas 10.16 and 10.17 (concerning Lemmas 10.16 and 10.17 see Remark 18.8).

Remark 10. For Lemma 1.7.1 see [27] or [3].

Remark 11. For Lemma 1.8.1 see [3].

Remark 12. Lemma 1.8.2 is an easy consequence of Lemmas 10.18 and 10.16, while Lemma 1.8.3 follows from Lemmas 10.8 and 10.17 (concerning Lemmas 10.8, 10.16 and 10.17 see Remarks 18.9 and 18.8).

§ 2. Formulation of the main results

The pair (G, Γ) consisting of a Lie group G and a discrete subgroup Γ will be called *irreducible* if for any decomposition $G=G_1\times G_2$ of G into a direct product of be called the closed subgroups G_1 and G_2 with dim $G_1>0$ and dim $G_2>0$, the subgroup $(G_1\cap\Gamma)\times(G_2\cap\Gamma)$ is not commensurable with Γ.

Suppose G is a semisimple algebraically connected real algebraic group such that *1)* the centre of the complexification $G_{\mathbf{C}}$ of G is trivial, *2)* G has no compact factors, *3)* the real rank of G (i.e. the dimension of a maximal **R**-split torus of G) is greater than 1, and Γ is a discrete subgroup for which *1)* the quotient space G/Γ has finite volume but is not compact (i.e. Γ is a non-uniform lattice in G), *2)* the pair (G, Γ) is irreducible. Under these assumptions, we formulate the main result of the present paper:

Main Theorem (*Theorem* 17.43). *There is a* **Q**-*form F of the group G for which*
1) $\Gamma\subset F_{\mathbf{Q}}$,
2) the group G is simple over **Q** (with respect to the **Q**-form F),
3) for any non-trivial (i.e. different from G) *parabolic subgroup P of G defined*

over **Q** (with respect to the **Q**-form F) *the subgroups* $P \cap F_Z$ *and* $P \cap \Gamma$ *are commensurable,*

4) the **Q**-*rank of the group* G* (with respect to the **Q**-form F) *is greater than zero.*

Theorem 19.5. Γ *is a rigid subgroup of the group* G (for the definition of a rigid subgroup see the beginning of § 19.).

Theorem 19.6 *Let* G' *be an algebraically connected semisimple algebraic group, such that 1)* G' *has no compact factors, 2) the centre of the complexification of* G' *is trivial, and* Γ' *is a lattice in* G'. *Then every isomorphism* φ (in the sense of abstract group theory) *of the group* Γ *onto the group* Γ' *can be extended to an algebraic isomorphism of the groups* G *and* G'.

We make the following remarks:

Remark 1. It is not difficult to show that in Theorem 19.5 the condition requiring the centre of the complexification of the group G to be trivial can be omitted.

Remark 2. In [6] Garland and Ragunathan proved that if G is a semisimple real algebraic group for which *1)* the real rank of G is equal to 1, *2)* $G \neq SL(2, \mathbf{R})$, *3)* G contains no compact factors, and if Γ is a lattice in G, then Γ is a rigid subgroup of the group G. This result and Theorem 19.5 (see note 1) stated above prove Selberg's conjecture stating the rigidity of non-uniform lattices (see [34]) provided that G and Γ satisfy the following conditions:

1) G is a semisimple real algebraic group different from $SL(2, \mathbf{R})$, *2)* G contains no compact factors, *3)* Γ is an irreducible non-uniform lattice in G (i.e. the pair (G, Γ) is irreducible). On the other hand it is not difficult to show that the case when (G, Γ) is reducible, can be reduced to the case of irreducible pairs. Thus, we obtain that Selberg's conjecture on the rigidity of non-uniform lattices is proved for all G and Γ, for which *1)* G is a real semisimple algebraic group different from $SL(2, \mathbf{R})$, *2)* G has no compact factors, *3)* Γ is a non-uniform lattice in G, *4)* for any algebraic homomorphism

$$f: G \to SL(2, \mathbf{R})/\pm 1$$

the subgroup $f(\Gamma)$ is not discrete.

Remark 3. Theorem 19.6 is an analogue of the theorem of Mostow [44] on the extension of isomorphisms of the fundamental groups of compact locally symmetric spaces. Let us note that Mostow's method does not immediately apply to the case of non-uniform lattices.

Remark 4. The main theorem is interesting in connection with the problem of the arithmeticity of discrete subgroups.

* i.e. the dimension of a maximal **Q**-split torus in G.

Remark 5. A theorem analogous to Theorem 19.5 was proved earlier by Raghunathan (see [31] and [32]) in the case where Γ is an arithmetic subgroup. We also remark that rigidity of uniform subgroups of semisimple algebraic groups has been investigated in [34] and [35] of A. Weil and A. Selberg, now considered as classical works.

Remark 6. The assumption in the Main Theorem that the real rank of G is different from 1, is essential, as examples constructed by E. B. Vinberg and V. S. Makarov show (even if not for all groups of real rank 1) (see [37]).

Remark 7. In [24] I. I. Pjateckij–Šapiro obtained results for the case where Γ is a discrete subgroup of Hecke type, which are closely related to the main theorem of the present work.

Remark 8. Let G^1 and G^2 be two connected semisimple complex algebraic groups with no centre, defined over $\mathbf{Q}$ and simple over $\mathbf{Q}$, and let S_1 and S_2 be two non-empty sets of prime numbers. Let $G^1_{S_1}(G^2_{S_2})$ denote the subgroup of all points in $G^1_{\mathbf{Q}}(G^2_{\mathbf{Q}})$ which are integer points for every $p \notin S_1 (p \notin S_2)$. Then, although this is not carried out here, one can prove the following statements in the same way as Theorems 19.5 and 19.6:

1) $G^1_{S_1}(G^2_{S_2})$ is a rigid subgroup of $G^1(G^2)$

2) any isomorphism (in the sense of abstract groups) φ of $G^1_{S_1}$ onto $G^2_{S_2}$ extends to an algebraic isomorphism of G^1 onto G^2.

Remark 9. Applying the method of Raghunathan [31] one can easily prove the statements of § 8 with exception of the case when the groups G^1 and G^2 are of $\mathbf{Q}$-rank 1. The methods of Raghunathan in [32], on the other hand, cannot be applied to the investigation of $G^1_{S_1}$ and $G^2_{S_2}$ because of their analytic character (though for groups G^1 and G^2 having real ranks greater than 1, statement *1)* in Remark 8 can be derived from the results of [3]).

§ 3. The action of unipotent groups in the space of lattices

Let L be a finite-dimensional real Euclidean space. A quasi-lattice* is a discrete subgroup in L. We denote the space of all quasi-lattices in L by $W(L)$ or simply by W, if it is clear which space is in question. For any quasi-lattice $\Delta \in W$, $\Delta_{\mathbf{R}}$ denotes the linear subspace of L spanned by Δ, and $d(\Delta)$ the volume of the quotient space $\Delta_{\mathbf{R}}/\Delta$ (with respect to the natural measure induced by the measure on $\Delta_{\mathbf{R}}$ which, in turn, is induced by the Euclidean metric on L). The term *determinant of the quasi-lattice* Δ will stand for $d(\Delta)$. The *dimension* of the quasi-lattice Δ is the dimension of the space $\Delta_{\mathbf{R}}$. We denote the dimension

* In [42], instead of the term "quasi-lattice" the term "lattice" is used. We prefer "quasi-lattice" to avoid confusion with lattices in topological groups (see § 1).

of the quasi-lattice Δ by dim Δ. We say that Δ is of "full dimension" if dim $\Delta =$ $=$dim L. Let us introduce in W the following topology: a subset $U \subset W$ is open if together with a quasi-lattice $\Delta \in U$ it contains, for some $\varepsilon > 0$, the set of all quasi-lattices of the form $T\Delta$, where T is a linear transformation of L for which $\|T-E\| < \varepsilon$ (E is the identity transformation). It is easy to see that the restriction of the topology introduced to the space of quasi-lattices of full dimension (i.e. lattices in the terminology of [33]) coincides with the ordinary topology in the space of lattices (see [33]). A basis of a quasi-lattice Δ is a set of elements $\delta_1, \dots, \delta_k \in \Delta$ ($k=$dim Δ) such that any element δ of the quasi-lattice Δ can be uniquely represented in the form $\delta = m_1\delta_1 + \dots + m_k\delta_k$, where $m_1, \dots, m_k$ are integers. It is easy to see that in any quasi-lattice there is a basis.

The following well-known criterion of compactness is due to Mahler: a set $M \subset W$ is relatively compact if and only if there are positive constants D and ε such that *1)* the ε-neighbourhood of the zero in the space L does not contain any non-zero element from any quasi-lattice belonging to M, *2)* $d(\Delta) < D$ if $\Delta \in M$.

Remark. Strictly speaking, Mahler's criterion of compactness is usually proved in the case of quasi-lattices of full dimension. It is easy to see, however, that the proof of this criterion (see [33]) can be carried over without difficulty to the case of quasi-lattices of any dimension. The following theorem of Minkowski is well known (see [33]):

For any n there is a constant $c(n)$ such that in any lattice Δ (i.e. quasi-lattice of full dimension) lying in a Euclidean space H of dimension not exceeding n, there is a basis $e_1, \dots, e_k$ ($k=$dim H), such that for any l

$$\prod_{i=1}^{l} \|e_i\| < c(n)\, d(\Delta)^{l/k}.$$

Since any quasi-lattice $\Delta \in W(L)$ is of full dimension in the space $\Delta_{\mathbf{R}}$, the theorem of Minkowski just formulated implies immediately the following

Lemma 3.1. For any n there is a constant $c(n)$ such that in any quasi-lattice Δ lying in a Euclidean space H of dimension not exceeding n, there is a basis $e_1, \dots, e_k$ ($k=$dim Δ) such that for any l ($1 \leq l \leq k$)

$$\prod_{i=1}^{l} \|e_i\| < c(n)\, d(\Delta)^{l/k}.$$

From Mahler's criterion of compactness and Lemma 3.1 one easily gets

Lemma 3.2. For any relatively compact set $M \subset W$ there are positive constants d_1 and d_2 such that in any quasi-lattice in M there is a basis consisting of elements, each of norm less than d_1 and greater than d_2.

Let U^t be a one-parameter group of unipotent linear transformations of the space L. In [42] the following two equivalent assertions are proved:

Lemma 3.3. If $\Delta \in W(L)$ and $\dim \Delta = \dim L$, then $U^t(\Delta)$ does not tend to infinity in $W(L)$ as $t \to +\infty$ (i.e. there is a compact $K \subset W(L)$, such that the set of all t for which $U^t(\Delta) \in K$, contains arbitrarily large positive numbers).

Lemma 3.4. For any quasi-lattice $\Delta \in W(L)$ of full dimension there is a $\delta(\Delta) > 0$, such that the set of all t, for which the intersection of $U^t(\Delta)$ with the $\delta(\Delta)$-neighbourhood of zero of the space L consists of the zero only, contains arbitrarily large positive numbers.

§ 4. Discrete subgroups in direct and semi-direct products

Lemma 4.1. Let G be a connected Lie group, $Q \subset G$ a connected, simply connected closed normal subgroup, $\varphi: G \to G/Q$ the natural mapping. Then for any discrete subgroup $\Gamma \subset G$ the connected component of the identity in the closure of the group $\varphi(\Gamma)$ is solvable.

In case Q is commutative, this lemma has been proved by Zassenhaus [9]. In the stated form the lemma is due to Auslander [13] and [14].

Let now G be an algebraic **R**-group, $\Gamma \subset G$ a Zariski dense lattice, G_0 the topologically connected component of the identity in the group G, R the solvable radical of the group G (i.e. the maximal algebraically connected solvable algebraic normal divisor), U the unipotent radical of the group G, finally f and h the natural mappings $f: G \to G/R$ and $h: G \to G/U$. It is well known (see [20]) that G/U is a reductive algebraic **R**-group. Therefore it is easy to see that the following lemma is true.

Lemma 4.2. If G is algebraically connected, then $h(R)$ is contained in the centre of the group G/U.

For a subgroup H of an arbitrary Lie group we denote by $I(H)$ the connected component of the identity in the closure of the subgroup H. It follows from Lemmas 1.3.1 and 1.3.2 that U is a connected, simply connected nilpotent Lie group. Therefore, according to Lemma 4.1 the group $I(h(G_0 \cap \Gamma))$ is solvable. On the other hand, since G is an algebraic **R**-group, G_0 is a subgroup of finite index in G. Thus, it is easy to see that

$$I(h(\Gamma)) = I(h(G_0 \cap \Gamma)). \tag{1}$$

Equality (1) follows from the facts that

1) as G/G_0 is finite, $h(\Gamma)/h(G_0 \cap \Gamma)$ is also finite;

2) as $h(\Gamma)/h(G_0 \cap \Gamma)$ is finite, $I(h(G_0 \cap \Gamma))$ is a subgroup of finite index in $I(h(\Gamma))$;

3) since $I(h(G_0 \cap \Gamma))$ is a closed subgroup of finite index in $I(h(\Gamma))$, the connectedness of the group $I(h(\Gamma))$ implies (1). On the basis of (1) and the solvability of the group $I(h(G_0 \cap \Gamma))$ we have

Lemma 4.3. The group $I(h(\Gamma))$ is solvable.

Lemma 4.4. The group $f(\Gamma)$ is a discrete subgroup of the group G/R.

Proof. In view of Lemma 1.2.7 we can suppose without loss of generality that G is algebraically connected. Denote by R_0 the topologically connected component of the identity in the group R, and by f_0 the natural mapping $f_0: G \to G/R_0$. Since R is a real algebraic group, R/R_0 is finite. Therefore, as one can easily see, the discreteness of the subgroup $f(\Gamma) \subset G/R$ is equivalent to that of the subgroup $f_0(\Gamma) \subset G/R_0$. Suppose now that the statement of the lemma is false, i.e. that $f(\Gamma)$ is not discrete. This implies, as we have just remarked, that the group $f_0(\Gamma)$ is not discrete. Since a closed subgroup of a Lie group is a Lie subgroup, we have

$$\dim I(f_0(\Gamma)) > 0. \tag{2}$$

Denote the Lie algebra of the group $I(\Gamma \cdot R_0)$ by $\mathfrak{I}$. It is easy to see that $I(\Gamma \cdot R_0)$ will be normalized by the subgroup Γ. Therefore for any $\gamma \in \Gamma$ we have

$$\operatorname{Ad} \gamma \mathfrak{I} = \mathfrak{I}. \tag{3}$$

On the other hand, Γ is Zariski dense in G by assumption. Therefore, in view of Lemma 1.2.1, (3) implies that for any $g \in G$

$$\operatorname{Ad} g \mathfrak{I} = \mathfrak{I}. \tag{4}$$

From (4) and the topological connectedness of the group $I(\Gamma \cdot R_0)$ it follows that $I(\Gamma \cdot R_0)$ is a normal subgroup in G. Let us denote by $\hat{I}(\Gamma \cdot R_0)$ the algebraic closure of the subgroup $I(\Gamma \cdot R_0)$. Observe that

1) since $I(\Gamma \cdot R_0)$ is a normal subgroup in G, according to Lemma 1.2.5, $\hat{I}(\Gamma \cdot R_0)$ is a normal subgroup of G;

2) since $I(\Gamma \cdot R_0)$ is topologically connected, in view of Lemma 1.2.2 $\hat{I}(\Gamma \cdot R_0)$ is algebraically connected;

3) since R is algebraically connected and R/R_0 is finite, R coincides with the algebraic closure of the subgroup R_0 and, consequently, $\hat{I}(\Gamma \cdot R_0) \supset R$;

4) in view of (2) we have $\dim I(\Gamma \cdot R_0) > \dim R_0$ and since $\dim R = \dim R_0$ and $\hat{I}(\Gamma \cdot R_0) \supset I(\Gamma \cdot R_0)$ we have $\dim \hat{I}(\Gamma \cdot R_0) > \dim R$.

Thus, $\hat{I}(\Gamma \cdot R_0)$ is an algebraically closed normal subgroup in G containing R as a proper subgroup. Now, since R is the solvable radical, $\hat{I}(\Gamma \cdot R_0)$ is not solvable and consequently, according to Lemma 1.2.4, the group $I(\Gamma \cdot R_0)$ is not solvable.

Recall that for an arbitrary subgroup H its commutator is denoted by (H, H). As we have supposed that G is algebraically closed, we have by Lemma 4.2.

$$h((\Gamma \cdot R_0, \Gamma \cdot R_0)) = (h(\Gamma \cdot R_0), h(\Gamma \cdot R_0)) = (h(\Gamma) \cdot h(R_0), h(\Gamma) \cdot h(R_0)) = \\ = (h(\Gamma), h(\Gamma)). \tag{5}$$

(5) immediately implies

$$U \cdot (\Gamma \cdot R_0, \Gamma \cdot R_0) = U \cdot (\Gamma, \Gamma). \tag{6}$$

From (6) and the topological connectedness of U (U is topologically connected by Lemmas 1.3.1 and 1.3.2) it follows that

$$I(U\cdot(\Gamma,\Gamma)) = I(U\cdot(\Gamma\cdot R_0, \Gamma\cdot R_0)) \supset I(U)\cdot I((\Gamma\cdot R_0, \Gamma\cdot R_0)) \supset$$
$$\supset U\cdot(I(\Gamma\cdot R_0), I(\Gamma\cdot R_0)) \tag{7}$$

(the latter inclusion holds because for any subgroup H we have $I((H, H)) \supset \supset (I(H), I(H))$). As we have already shown, $I(\Gamma\cdot R_0)$ is not solvable and consequently $(I(\Gamma\cdot R_0), I(\Gamma\cdot R_0))$ is not solvable either, thus (7) implies that the group $I(U\cdot(\Gamma,\Gamma))$ is not solvable. Therefore the group $I(U, \Gamma)$ is not a solvable group (since $(\Gamma, \Gamma)\subset\Gamma$). On the other hand, Lemma 4.3 and the nilpotence of the group U (U is nilpotent according to Lemma 1.3.1) imply at once that $I(U\cdot\Gamma)$ is solvable. This is a contradiction. The lemma is proved.

As an immediate consequence of Lemmas 4.4 and 1.1.1, we have

Lemma 4.5. $R_0\cap\Gamma$ is a lattice in R_0.

Now we formulate some of Mostow's results (see [15] and [18]) in the form of the following lemma:

Lemma 4.6. Let Q be a connected solvable Lie group. Then

1) any lattice in the group Q is a uniform discrete subgroup;

2) if Λ is a uniform discrete subgroup of the group Q then $N/N\cap\Lambda$ is compact, where N is the maximal connected closed nilpotent normal subgroup of the group Q (which — as is not difficult to show — is uniquely defined).

Denote by N the maximal topologically connected closed nilpotent normal divisor of the group R_0. Since N is defined uniquely for R_0 and R_0 is a normal subgroup in G, N is a normal subgroup of G. We remark further that $N\supset U$, since U is connected and nilpotent (according to Lemmas 1.3.1 and 1.3.2). Thus, according to Lemmas 4.5 and 4.6 we have the following

Lemma 4.7. There exists in G a topologically connected nilpotent normal subgroup N for which

1) $N/N\cap\Gamma$ is compact;

2) N contains the unipotent radical U of the group G.

Lemma 4.8. Let G be an algebraically connected semisimple real algebraic group. Suppose that G has no compact factors and the centre of the complexification of the group G is trivial, and let Γ be a lattice in G. Then, if the pair (G, Γ) is irreducible, then the intersection of Γ with any non-trivial algebraic normal subgroup of the group G consists of the identity alone.

Proof. According to Lemma 1.4.2, G can be written as the following direct product: $G=\bigoplus_{i\in I} S_i$, (in the sense of IV of § 1), where I is a finite set and for any i S_i is a simple algebraically connected algebraic **R**-group. We consider the groups S_i naturally imbedded into G. For any $i\in I$ denote by α_i the natural projection $\alpha_i: G \to S_i$. Since $\bigoplus_{i \in I} S_i$ is a direct product in the sense of IV of § 1, for any i S_i is a algebraic subgroup of the group G, and α_i is a regular algebraic homomorphism.

Suppose now that the assertion of the lemma is not true. Let $N \subset G$ be a non-trivial algebraic normal subgroup, $\gamma_0 \in N \cap \Gamma$ and $\gamma_0 \neq e$. In view of Lemma 1.4.2 we have $N = \bigoplus_{i \in I_N} S_i$ where $I_n \subset I$. Denote by $I(\gamma_0)$ the set of all $i \in I$ for which $\alpha_i(\gamma_0) \neq e$, and put

$$G_{\gamma_0} = \bigotimes_{i \in I(\gamma_0)} S_i, \quad G^*_{\gamma_0} = \bigotimes_{i \in I - I(\gamma_0)} S_i.$$

Clearly $G = G_{\gamma_0} \times G^*_{\gamma_0}$. Let α be the natural projection $\alpha: G \to G_{\gamma_0}$. Since $\bigoplus_{i \in I} S_i$ is a direct product in the sense of IV of § 1, G_{γ_0} is an algebraic subgroup of the group G, and α is an algebraic homomorphism.

Since Γ is a lattice in G, and G is an algebraically connected semisimple algebraic **R**-group having no compact factors, Γ is Zariski dense in G (see [2]). Now denote by $\alpha(\Gamma)^0$ the topologically connected component of the identity in the topological closure of the subgroup $\alpha(\Gamma)$. Since *1)* it is easy to see that $\alpha(\Gamma)^0$ is normalized by the subgroup $\alpha(\Gamma)$, *2)* $\alpha(\Gamma)$ is Zariski dense in G_{γ_0} (because Γ is algebraically dense in G, and α is an algebraic homomorphism), according to Lemma 1.2.16 and the topological connectedness of $\alpha(\Gamma)^0$, $\alpha(\Gamma)^0$ is a normal subgroup in G_{γ_0}. Therefore, according to Lemma 1.4.2, $\alpha(\Gamma)^0$ contains $\bigoplus_{i \in I_0} S_i^0$ as a subgroup of finite index, where $I_0 \subset I$, and S_i^0 is the topologically connected component of the identity in the group S_i. Since $\alpha(\Gamma)^0$ is topologically connected, we have $\alpha(\Gamma) = \bigoplus_{i \in I_0} S_i^0$.

Assume now that $\alpha(\Gamma)$ is not discrete. Since any closed subgroup of a Lie group is a Lie subgroup, we have $\alpha(\Gamma)^0 \neq e$, and consequently I_0 is non-empty. Let $i_0 \in I_0$. Since $\alpha(\Gamma) \subset G_{\gamma_0}$, we have $\alpha(\Gamma)^0 \subset G_{\gamma_0}$, and consequently $I_0 \subset I(\gamma_0)$. Therefore $i_0 \in I(\gamma_0)$, whence, according to the definition of the subgroup G_{γ_0}, we get

$$\alpha_{i_0}(\gamma_0) \neq e. \tag{8}$$

Since the centre of the group G is trivial (according to Lemma 1.2.15 and the fact that the centre of the complexification of G is trivial), so is the centre of the group S_{i_0}. Now S_{i_0} is algebraically connected (since G is algebraically connected). Therefore, since $S_{i_0}^0$ is a subgroup of finite index in S_{i_0}, the algebraic closure of the subgroup $S_{i_0}^0$ in S_{i_0} coincides with S_{i_0}. Since the centre of the group S_{i_0} is trivial, this implies that the centralizer of $S_{i_0}^0$ in S_{i_0} is trivial. Thus (8) implies

$$(\alpha_{i_0}(\gamma_0), S_{i0}^0) \neq e, \tag{9}$$

where $(\alpha_{i_0}(\gamma_0), S_{i_0}^0)$ denotes the commutator of the element $\alpha_{i_0}(\gamma_0)$ with $S_{i_0}^0$ (i.e. the subgroup of the group G generated by all commutators of the form (γ_0, s), where $s\in S_i^0$). On the other hand, it is clear that $\alpha(\Gamma)^0\supset S_{i_0}^0$ (since $\alpha(\Gamma)^0=\bigoplus_{i\in I_0} S_i^0$ and $i_0\in I_0$). Now since $\alpha_{i_0}(S_{i_0}^0)=S_{i_0}^0$, according to (9) we have

$$(\alpha_{i_0}(\gamma_0), \alpha_{i_0}(\alpha(\Gamma)^0))\supset(\alpha_{i_0}(\gamma_0), \alpha_{i_0}(S_{i_0}^0)) = (\alpha_{i_0}(\gamma_0), S_{i_0}^0) \neq e. \tag{10}$$

α_{i_0} is a homomorphism, thus (10) implies

$$(\gamma_0, \alpha(\Gamma)^0) \neq e. \tag{11}$$

Now it is easy to see that

$$(\gamma_0, \alpha(\Gamma)^0)\supset\overline{(\gamma_0, \alpha(\Gamma))} \tag{12}$$

where $\overline{(\gamma_0, \alpha(\Gamma))}$ is the topological closure of the commutator $(\gamma_0, \alpha(\Gamma))$. Further, since $\gamma_0\in G_{\gamma_0}$ and $G=G_{\gamma_0}\times G_{\gamma_0}^*$ we have

$$(\gamma_0, \Gamma) = (\gamma_0, \alpha(\Gamma)). \tag{13}$$

We note that, in view of the topological connectedness of $\alpha(\Gamma)^0$, the commutator $(\gamma_0, \alpha(\Gamma)^0)$ is easily seen to be topologically connected. Therefore it follows from (11), (12) and (13) that the topological closure $\overline{(\gamma_0, \Gamma)}$ of the commutator (γ_0, Γ) contains a non-trivial connected subgroup of the group S_{i_0}. But this contradicts the discreteness of the subgroup Γ. Thus, the assumption of the non-discreteness of $\alpha(\Gamma)$ leads to a contradiction. Consequently, $\alpha(\Gamma)$ is discrete. From this and the fact that Γ is a lattice in G it follows, according to Lemma 1.1.1, that $G_{\gamma_0}^*\cap\Gamma$ is a lattice in $G_{\gamma_0}^*$. Therefore, since $G_{\gamma_0}^*$ has no compact factors (because G has no compact factors) and it is algebraically connected (since G is algebraically connected) and semisimple (because $G_{\gamma_0}^*=\bigoplus_{i\in I-I(\gamma_0)} S_i$), the subgroup $G_{\gamma_0}^*\cap\Gamma$ is Zariski dense in $G_{\gamma_0}^*$ (see [2]). This implies the existence of a $\gamma_1\in G_{\gamma_0}^*\cap\Gamma$, such that for any $i\in I-I(\gamma_0)$ we have $\alpha_i(\gamma_1)\neq e$. Similarly to the proof of the fact that $G_{\gamma_0}^*\cap\Gamma$ is a lattice in $G_{\gamma_0}^*$, we can prove that $G_{\gamma_0}\cap\Gamma$ is a lattice in G_{γ_0}. Thus we have shown that $G_{\gamma_0}\cap\Gamma$ is a lattice in G_{γ_0} and $G_{\gamma_0}^*\cap\Gamma$ is a lattice in $G_{\gamma_0}^*$. Therefore, since $G=G_{\gamma_0}\times G_{\gamma}^*$, in view of Lemma 1.1.2 we have that $(G_{\gamma_0}\cap\Gamma)\times(G_{\gamma_0}^*\cap\Gamma)$ is a lattice in G and, consequently, $(G_{\gamma_0}\cap\Gamma)\times(G_{\gamma_0}^*\cap\Gamma)$ is a subgroup of finite index in Γ. Now in order to get a contradiction to the irreducibility of the pair (G, Γ) it remains to note that

1) the subgroups G_{γ_0} and $G_{\gamma_0}^*$ are algebraic (since $G_{\gamma_0}=\bigoplus_{i\in I(\gamma_0)} S_i$ and $G_{\gamma_0}^*=\bigoplus_{i\in I-I(\gamma_0)} S_i$);

2) since $\gamma_0\in N$ we have $I(\gamma_0)\subset I_N$ whence $I-I(\gamma_0)\neq\emptyset$ and, consequently, by the algebraic connectedness of all S_i $(i\in I)$ we have $\dim G_{\gamma_0}^*>0$;

3) since $\gamma_0\neq e$, we have $I(\gamma_0)\neq\emptyset$, and consequently, by the algebraic connectedness of all S_i we have $\dim G_{\gamma_0}>0$. The lemma is proved.

§ 5. Unipotent discrete groups and their finite-dimensional representations

Let Z be a connected, simply connected n-dimensional nilpotent Lie group, and $\mathfrak{Z}$ its Lie algebra. We formulate some of A. I. Malcev's and C. Chevalley's results ([10] and [28]) in the form of the following lemma:

Lemma 5.1. *1)* The exponential mapping $\mathfrak{z}\to\exp\mathfrak{z}$ of $\mathfrak{Z}$ into Z is a homeomorphism;

2) if Γ is a discrete uniform subgroup in Z then there exist n one-parameter subgroups $x_i(t)$ such that

a) for every $z\in Z$, $z=x_1(t_1)\cdot\ldots\cdot x_n(t_n)$, and this representation is unique (the dot on the right-hand side denotes group-multiplication),

b) if $Z_i=\{x_i(t_i)\cdot\ldots\cdot x_n(t_n)\}$, $t_i,\ldots,t_n$ are arbitrary real numbers, then Z_i is a closed connected normal subgroup in Z and Z_{i-1}/Z_i belongs to the centre of the group Z/Z_i,

c) $\Gamma=\{x_1(m_1)\cdot\ldots\cdot x_n(m_n)\}$ where m_i $(i=1,\ldots,n)$ are integers.

3) The commutator (Z, Z) of the group Z is a connected, simply connected closed normal subgroup of Z.

Since the exponential mapping of $\mathfrak{Z}$ into Z is a homeomorphism, the inverse mapping of Z into $\mathfrak{Z}$ denoted by ln, is well defined and is also a homeomorphism.

A *subring* of the Lie algebra $\mathfrak{Z}$ is a subset which, together with $\mathfrak{z}$, contains $-\mathfrak{z}$, and together with $\mathfrak{z}_1$ and $\mathfrak{z}_2$ contains $\mathfrak{z}_1+\mathfrak{z}_2$ and $[\mathfrak{z}_1,\mathfrak{z}_2]$.

For a nilpotent group Γ, $l(\Gamma)$ denotes its *class*, i.e. the length of the lower central series (or which is the same — since Γ is nilpotent — the length of the upper central series).

Lemma 5.2. There is an integer b such that if $\Gamma\subset Z$ is a subgroup, $\overline{\ln\Gamma}$ is the subring in $\mathfrak{Z}$ generated by $\ln\Gamma$, then $b\,\overline{\ln\Gamma}\subset\ln\Gamma$, where $b\,\overline{\ln\Gamma}=\bigcup_{\alpha\in\overline{\ln\Gamma}} b\alpha$ (further b only depends on $l(Z)$, not on Z).

Proof. Put $(z_1,\ldots,z_{k-1},z_k)=((z_1,\ldots,z_{k-1}),z_k)$ and

$$[\mathfrak{z}_1,\ldots,\mathfrak{z}_{k-1},\mathfrak{z}_k]=[[\mathfrak{z}_1,\ldots,\mathfrak{z}_{k-1}],\mathfrak{z}_k]$$

where $z_1,\ldots,z_k\in Z$, $\mathfrak{z}_1,\ldots,\mathfrak{z}_k\in\mathfrak{Z}$, and call k the *weight* of the commutators $(z_1,\ldots,z_{k-1},z_k)$ and $[\mathfrak{z}_1,\ldots,\mathfrak{z}_{k-1},\mathfrak{z}_k]$. The formula of Campbell and Zassenhaus implies immediately that for any $z_1,z_2\in Z$

$$\ln(z_1,z_2)=[\ln z_1,\ln z_2]+A \tag{1}$$

where A is a rational linear combination of commutators, each of weight greater than 2, of $\ln z_1$ and $\ln z_2$ (A is a finite sum since $\mathfrak{Z}$ is nilpotent) and the coefficients in this linear combination have denominators not exceeding some constant depending on $l(Z)$ only. By induction on l, we get for any $z_1,\ldots,z_{l-1},z_l\in Z$

$$\ln(z_1,\ldots,z_{l-1},z_l)=[\ln z_1,\ldots,\ln z_{l-1},\ln z_l]+A, \tag{2}$$

where A is a rational linear combination of commutators, each of weight greater than l, of $\ln z_1, \ldots, \ln z_{l-1}, \ln z_l$, and the coefficients have denominators less then a constant s depending on $l(Z)$ only. (Such an s exists since $(z_1, \ldots, z_{l-1}, z_l)=e$ whenever $l>l(Z)$.)

Rewrite (2) in the form

$$[\ln z_1, \ldots, \ln z_{l-1}, \ln z_l] = \ln (z_1, \ldots, z_{l-1}, z_l) - A. \tag{3}$$

From (3) and the Campbell—Hausdorff formula by induction on p we find that for any $z_1, \ldots, z_m$

$$\ln (z_1, \ldots, z_m) - (\ln z_1 + \ldots + \ln z_m) = A_p + B_p \tag{4}$$

holds, where A_p is a rational linear combination of the logarithms of the commutators of $z_1, \ldots, z_m$ of weight not greater than p, and B_p is a rational linear combination of the commutators of $\ln z_1, \ldots, \ln z_m$ whose weight is greater than p, furthermore the denominators of the coefficients in A_p and B_p are less than some constant r_p depending only on $l(Z)$ and p (and not depending on $z_1, \ldots, z_m$). Since Z is a nilpotent group, for some p_0 we have $B_{p_0}=0$. Therefore

$$\ln z_1 + \ldots + \ln z_m = \ln (z_1 \ldots z_m) + A, \tag{5}$$

where A is a rational linear combination of the logarithms of the commutators of $z_1, \ldots, z_m$, whose weight is greater than 1, and the denominators of the coefficients in A are less than $r=r_{p_0}$.

For a subset $L \subset \mathfrak{Z}$, denote $\overbrace{L+\ldots+L}^{t}$ by L^t and $\overbrace{[L, \ldots, L]}^{t}$ by ${}^tL$. It follows from (5) that for any subgroup $\Gamma \subset Z$

$$R \bigcup_{1 \leqq t < \infty} (\ln \Gamma)^t \subset \ln \Gamma, \tag{6}$$

where $R=r!$. We obtain in a similar way that

$$\tilde{R} \bigcup_{1 \leqq t < \infty} {}^t(\ln \Gamma) \subset \ln \Gamma, \tag{7}$$

where $\tilde{R}$ is some integer depending only on $l(Z)$ (it is sufficient to replace (4) and (5) by analogous formulae for $[\ln z_1, \ldots, \ln z_m]$ and note that for $t>l(Z)$ we have ${}^t\mathfrak{Z}=0$). Since

$$\overline{\ln \Gamma} = \bigcup_{1 \leqq t_1 < \infty} \Big(\bigcup_{1 \leqq t_2 < \infty} {}^{t_2}(\ln \Gamma)\Big)^{t_1},$$

from (6) and (7) we conclude

$$R\tilde{R}\, \overline{\ln \Gamma} \subset \overline{\ln \Gamma}. \tag{8}$$

Now, in order to complete the proof of the lemma, it remains to put $b=R\tilde{R}$ and remark that R and $\tilde{R}$ depend only on $l(Z)$.

Denote by $\mathfrak{Z}(\Gamma_{\mathbf{R}}) \subset \mathfrak{Z}$ the linear subspace consisting of all real linear combinations of the elements in the set $\ln \Gamma$. It follows from Lemma 5.2 that $\mathfrak{Z}(\Gamma_{\mathbf{R}})$ is a subalgebra of the algebra $\mathfrak{Z}$. Therefore, on applying the Campbell—Hausdorff formula and using the nilpotence of the group Z we find that $\exp(\mathfrak{Z}(\Gamma_{\mathbf{R}}))$ is a subgroup which will be denoted by $\Gamma_{\mathbf{R}}$. Since exp is a homeomorphism — according to Lemma 5.1 — $\Gamma_{\mathbf{R}}$ is a connected, simply connected closed subgroup of the group Z.

Lemma 5.3. The quotient space $\Gamma_{\mathbf{R}}/\Gamma$ is compact.

Proof. We prove the lemma by induction on $l(\Gamma)$. Let $l(\Gamma)=1$, i.e. the group Γ commutative. We show that in this case $\Gamma_{\mathbf{R}}$ is commutative. In fact let $\gamma_1, \gamma_2 \in \Gamma$. Let us formulate an equality, analogous to (5), for the commutator of $[\ln \gamma_1, \ln \gamma_2]$ (i.e. express $[\ln \gamma_1, \ln \gamma_2]$ by $\ln(\gamma_1, \gamma_2)$ and a linear combination of the logarithms of the commutators of z_1 and z_2). Then we see that since γ_1 and γ_2 commute, any commutator of γ_1 and γ_2 equals e, and consequently $[\ln \gamma_1, \ln \gamma_2]=0$. From this we have that the Lie algebra $\mathfrak{Z}(\Gamma_{\mathbf{R}})$ is commutative, and consequently the group $\Gamma_{\mathbf{R}}$ is commutative. On the other hand, as it has already been remarked, $\Gamma_{\mathbf{R}}$ is connected and simply connected. Thus, $\Gamma_{\mathbf{R}}$ is a connected, simply connected commutative Lie group. Therefore, according to Lemma 5.1, one can assume that $\Gamma_{\mathbf{R}}$ is a finite-dimensional real linear space and Γ is a quasi-lattice in this space (for the definition of quasi-lattices see § 3). From this, we get easily that $\Gamma_{\mathbf{R}}/\Gamma$ is compact (for $\ln \Gamma$ is generated by $\mathfrak{Z}(\Gamma_{\mathbf{R}})$ and consequently, Γ is a quasi-lattice of full dimension in the linear space $\Gamma_{\mathbf{R}}$).

Now assume that the lemma has already been proved for all Γ for which $l(\Gamma)<m$. Consider the case $l(\Gamma)=m$. Write Γ' for the commutator of the group Γ, $\Gamma'_{\mathbf{R}}$ the subgroup defined similarly to $\Gamma_{\mathbf{R}}$, and $\mathfrak{Z}(\Gamma'_{\mathbf{R}})$ for the Lie algebra of the group $\Gamma'_{\mathbf{R}}$. Since for any $\gamma_1, \gamma_2 \in \Gamma$ we have $\operatorname{Ad}\gamma_1(\ln \gamma_2)=\ln(\gamma_1\gamma_2\gamma_1^{-1})$, it follows that for any $\gamma \in \Gamma$ we have

$$\operatorname{Ad}\gamma(\mathfrak{Z}(\Gamma'_{\mathbf{R}})) = \mathfrak{Z}(\Gamma'_{\mathbf{R}}). \tag{9}$$

Now we note that

1) for any $z \in Z$ we have $\ln(\operatorname{Ad} z)=\operatorname{ad}(\ln z)$, where $\ln(\operatorname{Ad} z)$ denotes a nilpotent linear transformation (i.e. a transformation whose eigenvalues are all 0) such that $\exp(\ln(\operatorname{Ad} z))=\operatorname{Ad} z$ (such a transformation exists and is unique because of the unipotence of the transformation $\operatorname{Ad} z$);

2) $\ln(\operatorname{Ad} z)$ is a polynomial of the linear transformation $\operatorname{Ad} z$. Therefore, it follows from (9) that for any $\gamma \in \Gamma$

$$(\operatorname{ad}(\ln \gamma))(\mathfrak{Z}(\Gamma'_{\mathbf{R}})) = \mathfrak{Z}(\Gamma'_{\mathbf{R}}), \tag{10}$$

and since $\ln \Gamma$ is generated by $\mathfrak{Z}(\Gamma_{\mathbf{R}})$ as a linear space, $\mathfrak{Z}(\Gamma'_{\mathbf{R}})$ is an ideal in $\mathfrak{Z}(\Gamma_{\mathbf{R}})$. Consequently, by the connectedness of the groups $\Gamma_{\mathbf{R}}$ and $\Gamma'_{\mathbf{R}}$, $\Gamma'_{\mathbf{R}}$ is a normal subgroup in $\Gamma_{\mathbf{R}}$. On the other hand, since $\exp: \mathfrak{Z} \to Z$ is a homeomorphism (according to Lemma 5.1), $\Gamma'_{\mathbf{R}}$ is a closed subgroup in the group $\Gamma_{\mathbf{R}}$. Therefore $\Gamma_{\mathbf{R}}/\Gamma'_{\mathbf{R}}$ is a Lie group.

Denote $\Gamma_{\mathbf{R}}/\Gamma'_{\mathbf{R}}$ by Z_0. Let f be the natural mapping $f:\Gamma_{\mathbf{R}}\to Z_0$, and φ the natural mapping $\varphi:\mathfrak{Z}(\Gamma_{\mathbf{R}})\to\mathfrak{Z}(Z_0)$, where $\mathfrak{Z}(Z_0)$ is the Lie algebra of the group Z_0. We remark that

1) since $\Gamma_{\mathbf{R}}$ and $\Gamma'_{\mathbf{R}}$ are connected and $\Gamma_{\mathbf{R}}$ is simply connected, according to Lemma 1.1.6, Z_0 is connected and simply connected.

2) since $\Gamma_{\mathbf{R}}$ is nilpotent, Z_0 is nilpotent.

Thus Z_0 is a connected, simply connected nilpotent Lie group. Therefore, the mapping ln, the inverse mapping of the exponential mapping $\exp:\mathfrak{Z}(Z_0)\to Z_0$, is well defined. Thus, as is easily seen, for any $z\in\Gamma_{\mathbf{R}}$

$$\varphi(\ln z) = \ln f(z). \tag{11}$$

Since $\Gamma_{\mathbf{R}}$ is generated by $\ln\Gamma$, (11) implies that $\mathfrak{Z}(Z_0)$ is generated by $\ln f(\Gamma)$. On the other hand, since $\Gamma'\subset\Gamma'_{\mathbf{R}}$ and Γ/Γ' is commutative, $f(\Gamma)$ is also commutative. Using the same argument as in proving that the commutativity of Γ implies the commutativity of $\Gamma_{\mathbf{R}}$, we obtain that Z_0 is commutative. In the same way we find that

1) $\mathfrak{Z}(Z_0)$ is generated by $\ln f(\Gamma)$;

2) Z_0 is a commutative connected, simply connected Lie group;

3) since $l(\Gamma')<l(\Gamma)$, by the hypothesis of the induction we have that $\Gamma'_{\mathbf{R}}/\Gamma'$ is compact, and consequently, $f(\Gamma)$ is discrete according to Lemma 1.1.3. Therefore, since the assertion of the lemma has already been proved for commutative groups, $Z_0/f(\Gamma)$ is compact. On the other hand, as remarked already, $\Gamma'_{\mathbf{R}}/\Gamma'$ is also compact. Now the compactness of $Z_0/f(\Gamma)$ and $\Gamma'_{\mathbf{R}}/\Gamma'$ implies, according to Lemma 1.1.5, that $\Gamma_{\mathbf{R}}/\Gamma$ is compact. The lemma is proved.

Let now G be an algebraic **R**-group and $U\subset G$ a discrete unipotent subgroup. Denote by $\bar{U}$ the Zariski closure of U in G (i.e. $\bar{U}=A(U)$). Since, by Lemma 1.3.3, the set of unipotent elements of the group G is an algebraic submanifold, the group $\bar{U}$ is unipotent. Therefore, according to Lemma 1.3.1 $\bar{U}$ is a connected, simply connected nilpotent Lie group.

Write $\mathfrak{U}$ for the Lie algebra of the group U. By an easy consequence of the unipotence of the algebraic **R**-group $\bar{U}$ the logarithmic mapping $\ln:\bar{U}\to\mathfrak{U}$ is an algebraic mapping regular in both directions. Therefore, the set $\ln U$ is Zariski-dense in $\mathfrak{U}$, and consequently $\mathfrak{U}$ is generated by $\ln U$ as a linear space. From this and Lemma 5.3 we get

Lemma 5.4. The quotient space $\bar{U}/U$ is compact.

Lemma 5.5. Let N be a unipotent algebraic **R**-group, $Q\subset N$ an algebraic subgroup, and f the natural mapping $f:N\to N/Q$. If $Q\neq N$, then there exists a one-parameter subgroup $n(t)\subset N$ such that $f(n(t))$ tends to infinity in N/Q as $t\to\infty$.

Proof. Denote by N' the commutator of the group N, by $\tilde{N}$ the quotient group N/N', and by h the natural mapping $h:N\to\tilde{N}$. According to Lemma 1.3.1 N is a connected, simply connected nilpotent Lie group. But then it follows from Lemma 5.1 that N' is a connected, simply connected closed normal subgroup in N. Therefore, according to Lemma 1.1.6 $\tilde{N}$ is a connected, simply connected commutative Lie group.

Denote by $\mathfrak{N}$, $\tilde{\mathfrak{N}}$ and $\mathfrak{Q}$ the Lie algebras of the groups N, $\tilde{N}$ and Q, by h and φ the natural mappings $N \to \tilde{N}$ and $\mathfrak{N} \to \tilde{\mathfrak{N}}$. Clearly

$$h \circ \exp_N = \exp_{\tilde{N}} \circ \varphi, \tag{12}$$

where $\exp_N$ and $\exp_{\tilde{N}}$ denote the exponential mappings $\exp_N : \mathfrak{N} \to N$ and $\exp_N : \tilde{\mathfrak{N}} \to \tilde{N}$. As has already been remarked, N and $\tilde{N}$ are connected, simply connected nilpotent Lie groups, thus, according to Lemma 5.1 $\exp_N$ and $\exp_{\tilde{N}}$ are homeomorphisms. From this and formula (12) it follows that

$$h(Q) = \exp_{\tilde{N}}\big(\varphi(\ln_N Q)\big), \tag{13}$$

where $\ln_N$ denotes the mapping inverse to $\exp_N$. On the other hand, since $\exp_N$ is a homeomorphism, $\exp_N \mathfrak{Q}$ is a closed submanifold in N of dimension $\dim \mathfrak{Q} = \dim Q$. Therefore, in view of the connectedness of the manifold Q (Q is connected according to Lemma 1.3.1), we have

$$\exp_N \mathfrak{Q} = Q. \tag{14}$$

From (13) and (14) we obtain the equality

$$h(Q) = \exp_{\tilde{N}}\big(\varphi(\mathfrak{Q})\big). \tag{15}$$

Since $\exp_{\tilde{N}}$ is a homeomorphism, (15) implies that $h(Q)$ is a connected closed subgroup of the group $\tilde{N}$.

Since $\tilde{N}$ is a connected, simply connected commutative Lie group, in view of Lemma 5.1 one can assume that $\tilde{N}$ is a finite-dimensional real linear space. Thus, $h(Q)$ is a connected closed subgroup in the linear space $\tilde{N}$. Therefore, as easily seen, $h(Q)$ is a linear subspace of the space $\tilde{N}$ (this can also be seen directly from (15)).

Since $Q \neq N$, by assumption, and the group N is nilpotent, we have

$$h(Q) \neq \tilde{N} \tag{16}$$

($h(Q) = \tilde{N}$ implies in fact $\tilde{N} = N' \cdot Q$, and consequently, by the Corollary 10.3.3 of the book [8], we have $Q = N$). Thus Q is in $\tilde{N}$ a linear subspace of positive codimension. This guarantees the existence of a one-parameter subgroup $\tilde{n}(t) \subset \tilde{N}$ such that

$$k\big(\tilde{n}(t)\big) \text{ tends to infinity in } \tilde{N}/h(Q) \text{ as } t \to +\infty, \tag{17}$$

where k denotes the natural mapping $k : \tilde{N} \to \tilde{N}/h(Q)$. Let $n(t) \subset N$ be a one-parameter subgroup for which $h\big(n(t)\big) = \tilde{n}(t)$ (the existence of $n(t)$ is obvious, since the mapping h is an epimorphism). Denote by m the natural mapping $m : N/Q \to \tilde{N}/h(Q)$. Clearly the mapping m is continuous. Therefore (17) implies that $n(t)$ is the desired subgroup (since $k(\tilde{n}(t))=m(f(n(t))))$. The lemma is proved.

Lemma 5.6. Let N be an algebraic unipotent subgroup of the group G, such that $N \supset U$ and N/U is compact (G and U are the same as before Lemma 5.4). Then $N=\overline{U}$, where $\overline{U}$ is the Zariski closure of U in G.

Proof. Since N/U is compact and $\overline{U} \supset U$, $N/\overline{U}$ is also compact. Now we observe that $\overline{U}$ is an algebraic subgroup of the unipotent algebraic **R**-group N. Therefore, according to Lemma 5.5, the compactness of $N/\overline{U}$ implies $N=\overline{U}$. The lemma is proved.

Let G be, as above, an algebraic **R**-group, $\mathfrak{G}$ its Lie algebra, $\varrho_{\mathfrak{G}}$ some Euclidean metric on $\mathfrak{G}$, and ϱ_G the corresponding right-invariant metric on G. Denote by X the set of all discrete unipotent subgroups of the group G. Since in an algebraic **R**-group any two maximal unipotent groups are conjugate to each other (see [5] and Lemma 10.20 of the present paper), there exists an $N<\infty$ such that for any subgroup $U \in X$ we have $l(U)<N$ where $l(U)$ denotes the class of the nilpotent group U (U is nilpotent according to Lemma 1.3.1). Therefore, according to Lemma 5.2, there exists an integer b such that if $U \in X$, then $b\,\overline{\ln U} \subset \ln U$ (for $u \in U$, $\ln u$ stands for the unique element of the Lie algebra of the algebraic closure $\overline{U}$ of the subgroup U, for which $\exp(\ln u)=u$). Denote $b\,\overline{\ln U}$ by Δ_U. Clearly we have *1)* $\Delta_U+\Delta_U=\Delta_U$ *2)* since U is discrete and $\Delta_U \subset \overline{\ln U}$, Δ_U is also discrete. Therefore, Δ_U is a quasi-lattice in $\mathfrak{G}$. Let f be the mapping $f: X \to W$ associating to $U \in X$ the lattice Δ_U (W denotes the space of all quasi-lattices in $\mathfrak{G}$). We introduce a topology on the space W, in the way described in § 3.

Lemma 5.7. For any subset $L \subset X$ for which $f(L)$ is relatively compact in W, there are constants $\varepsilon>0$ and $c<\infty$ such that if $U \in L$, then *1)* $\varrho_G(e, u)<\varepsilon$ if $u \in U$ and $u \neq e$, *2)* there exist $u_1, \ldots, u_k \in U$ such that a) $\varrho_G(e, u_i)<c$ for any i $(1 \leqq i \leqq k)$, b) any algebraic subgroup of the group G containing all $u_1, \ldots, u_k$ also contains $\overline{U}$ ($\overline{U}$ is the Zariski closure of U in G).

Proof. Since G is an algebraic **R**-group, one can consider the elements of the group G and the Lie algebra $\mathfrak{G}$ as matrices. Note that for any $u \in U$ ($U \in X$), $\ln u$ is, by definition, an element of the Lie algebra $\mathfrak{U}$ of the group $\overline{U}$ such that $\exp(\ln u)=u$. Therefore, in view of the unipotence of the group $\overline{U}$ (that follows from Lemma 1.3.3), for any $u \in U$ $\ln u$ is a nilpotent matrix. From this it follows that $\ln u$ is a polynomial in u. Therefore $\varrho_{\mathfrak{G}}(0, \ln u)$ is small if only $\varrho_G(e, u)$ is small. This implies *1)* in view of Mahler's criterion for compactness. Since $f(L)$ is relatively compact in W, by Lemma 3.2 there exists a constant d such that for any subgroup $U \in L$ there are elements $u_1, \ldots, u_k$ having the following properties: *1)* $\bigcup_{1 \leqq i \leqq k} \ln u_i$ generates $\mathfrak{U}$ as a real linear space, *2)* for any i $(1 \leqq i \leqq k)$ the norm of the element $\ln u_i$ (i.e. $\varrho_{\mathfrak{G}}(0, \ln u_i)$) is less than d. Let now H be the subgroup of the group G generated by $\bigcup_{1 \leqq i \leqq k} u_i$, $\overline{H}$ its algebraic closure, and $\mathfrak{H}$

the Lie algebra of the group $\overline{H}$. Then, as a consequence of Lemmas 1.3.1 and 1.3.3, we have that $\overline{H}$ is a connected, simply connected nilpotent Lie group (and also unipotent). Therefore, according to Lemma 5.1, for any u_i there exists an $\mathfrak{h}_i \in \mathfrak{H}$ such that $\exp \mathfrak{h}_i = u_i$. Note that $\mathfrak{h}_i$,as well as $\ln u_i$, are nilpotent matrices (since $\ln u_i \in \mathfrak{U}$, $\mathfrak{h}_i \in H$, and U and $\overline{H}$ are unipotent). Therefore, $\ln u_i$ and $\mathfrak{h}_i$ coincide. Consequently, $\mathfrak{H} \supset \bigcup_{1 \leq i \leq k} \ln u_i$ and so $\mathfrak{H} \supset \mathfrak{U}$, whence in view of the connectedness of the group we obtain that $H \subset \overline{U}$ ($\overline{U}$ is connected according to Lemmas 1.3.1 and 1.3.3.). Now it remains to notice that if $\varrho_{\mathfrak{G}}(0, \ln u) < d$, then $\varrho_G(e, u) < c$, where c depends only on d, and that any algebraic subgroup of the group G containing $\bigcup_{1 \leq i \leq k} u_i$ also contains H.

Now we prove a lemma on finite-dimensional representations of discrete unipotent subgroups of a group G. A finite-dimensional representation of a group Q in an algebraic group is called *unipotent* if the image of Q under this representation is unipotent.

Lemma 5.8. A unipotent representation T of an arbitrary discrete unipotent subgroup $U \subset G$ in a real linear space can be extended in a unique way, to an algebraic unipotent representation of the group $\overline{U}$ ($\overline{U}$ is the Zariski closure of U in G).

Proof. According to Lemma 5.4 $\overline{U}/U$ is compact. Thus, in view of Lemmas 1.3.1 and 1.3.3, there exist one-parameter subgroups $x_i(t) \subset \overline{U}$ ($1 \leq i \leq q$, $q = \dim \overline{U}$) satisfying conditions a), b) and c) of *2)* of Lemma 5.1. Denote by h the mapping $h: \mathbf{R}^q \to \mathfrak{U}$ ($\mathbf{R}^q$ is the q-dimensional real linear space) which maps $(t_1, \ldots, t_q) \in \mathbf{R}^q$ to $\ln (x_1(t_1) \cdot \ldots \cdot x_q(t_q)) \in \mathfrak{U}$. Since G is an algebraic $\mathbf{R}$-group, we can consider the elements of the group $\overline{U}$ and the Lie algebra $\mathfrak{U}$ as matrices. Then, according to the unipotence of the group $\overline{U}$, the exponential and logarithmic mappings $\exp: \mathfrak{U} \to U$ and $\ln: \overline{U} \to \mathfrak{U}$ are polynomials. Thus we obtain that *1)* $x_i(t_i)$ is a polynomial of t_i; *2)* $\ln (x_1(t_1) \ldots x_q(t_q))$ is a polynomial of $x_1(t_1) \ldots x_q(t_q)$. On the other hand, it is clear that the product is a polynomial of the factors $x_1(t_1), \ldots, x_q(t_q)$. These imply that the mapping h is a polynomial. Now, according to Lemma 5.1 the mapping is one-to-one. Since the inverse mapping of a one-to-one polynomial mapping of a real linear space to another is a polynomial itself, h^{-1} is polynomial.

For any $t = (t_1, \ldots, t_q) \in \mathbf{R}^q$ put

$$P(t) = \exp [t_1 \ln T(x_1(1))] \ldots \exp [t_q \ln T(x_q(1))]^*. \tag{18}$$

Denote by L the linear space of all linear transformations of the space of the representation T. Then it is clear that $P(t)$ represents a polynomial mapping of R^q to L (since $\ln T(x_i(1))$ is nilpotent for any i). Now we mention that since T

* As above, for a unipotent linear transformation A, $\ln A$ denotes the unique nilpotent transformation for which $\exp (\ln A) = A$.

is a representation and it is easy to see that $h^{-1} \circ \ln$ is the inverse of the mapping $\mathbf{R}^q \to \bar{U}$ which maps $(t_1, \ldots, t_q) \in \mathbf{R}^q$ into the product $x_1(t_1) \ldots x_q(t_q)$, thus, according to condition c) of *2)* of Lemma 5.1, for any $u \in U$ we have

$$T(u) = P(h^{-1}(\ln u)). \tag{19}$$

For any $\bar{u} \in \bar{U}$, put $\tilde{T}(\bar{u}) = P(h^{-1}(\ln \bar{u}))$. Since the mappings P, h^{-1} and ln are regular algebraic mappings, the mapping $\tilde{T}$ is also regular and algebraic. Now, in view of (19) the restriction of $\tilde{T}$ to U coincides with T. Therefore, since U is Zariski dense in $\bar{U}$ and T is a representation, according to Lemma 1.2.6 $\tilde{T}$ is also a representation. The uniqueness and unipotence of the algebraic extension of T are consequences of the fact that U is Zariski dense in $\bar{U}$, and of Lemma 1.3.3.

Lemma 5.8 can be reformulated as follows:

Lemma 5.9. A unipotent representation T of an arbitrary discrete unipotent subgroup $U \subset G$ in a real algebraic group H can uniquely be extended to an algebraic unipotent representation of the real algebraic group $\bar{U}$.

Lemma 5.10. Let W be a real unipotent algebraic group, $W_{\mathbf{C}}$ the complexification of W. Then any continuous (in the sense of Lie group theory) unipotent representation T of the group W in a complex algebraic group Q can uniquely be extended to a unipotent algebraic representation $T_{\mathbf{C}}$ of the complex algebraic group $W_{\mathbf{C}}$ in the complex algebraic group Q.

Proof. Denote by $\mathfrak{W}$, $\mathfrak{W}_{\mathbf{C}}$ and $\mathfrak{Q}$ the Lie algebras of the groups W, $W_{\mathbf{C}}$ and Q. Since T is continuous in the sense of Lie group theory, T induces a representation $\tilde{T}$ of the real Lie algebra $\mathfrak{W}$ in the complex Lie algebra $\mathfrak{Q}$. Now $\tilde{T}$ can be extended to a representation $\tilde{T}_{\mathbf{C}}$ of the complex Lie algebra $\mathfrak{W}_{\mathbf{C}}$ in the complex Lie algebra $\mathfrak{Q}$. Since W is unipotent, $W_{\mathbf{C}}$ is a complex unipotent algebraic group. Therefore the logarithmic mapping $\ln\colon W_{\mathbf{C}} \to \mathfrak{W}_{\mathbf{C}}$ is well defined. For any $w \in W_{\mathbf{C}}$, set

$$T_{\mathbf{C}}(w) = \exp(\tilde{T}_{\mathbf{C}}(\ln w)). \tag{20}$$

Now we observe that *1)* since T is a continuous unipotent representation, the Lie algebra $\tilde{T}(\mathfrak{W})$, and consequently the Lie algebra $\tilde{T}_{\mathbf{C}}(\mathfrak{W})$ too, are nilpotent *2)* in complex algebraic unipotent groups and complex Lie algebras the mappings ln and exp are regular and algebraic. Therefore, one can see from (20) that $T_{\mathbf{C}}$ is a regular algebraic mapping of $W_{\mathbf{C}}$ to Q. On the other hand, since for any $w \in W$ we have $\ln w \in \mathfrak{W}$, the restriction of $T_{\mathbf{C}}$ to W coincides with T. Therefore, it follows from Lemma 1.2.6 that $T_{\mathbf{C}}$ is a representation. The unipotence of $T_{\mathbf{C}}$ is a consequence of the unipotence of T and Lemma 1.3.3. The proof is complete.

From Lemmas 5.9 and 5.10 we get

Lemma 5.11. Let G be a real algebraic group, $U \subset G$ a discrete unipotent group, $\overline{U}$ the algebraic closure of U in G, $\overline{U}_{\mathbf{C}}$ the complexification of $\overline{U}$ (it is easy to see that this latter coincides with the algebraic closure of U in the complexification of the group G). Then any unipotent representation T of the group U into the complex algebraic group Q can be uniquely extended to an algebraic representation $T_{\mathbf{C}}$ of the complex algebraic group $\overline{U}_{\mathbf{C}}$ in the complex algebraic group Q.

Remark. Let us point out that $T_{\mathbf{C}}$ is an algebraic representation of $\overline{U}_{\mathbf{C}}$, the latter being considered as a complex algebraic group.

§ 6. Property P and integer subgroups

Let G be a Lie group, Γ a discrete subgroup, and π the natural mapping $\pi: G \to G/\Gamma$.

Definition 6.1. We say that the closed subgroup $H \subset G$ has property P if for any compact $K \subset G$ there exist finitely many elements $\gamma_1, \ldots, \gamma_n \in \Gamma$ such that

$$(K \cdot H) \cap \Gamma = \bigcup_{i=1}^{n} \gamma_i \cdot (H \cap \Gamma),$$

(i.e. the set $(K \cdot H) \cap \Gamma$ can be represented as the union of finitely many left cosets of Γ with respect to the subgroup $H \cap \Gamma$).

Remark 6.2. In [41] p. 607 it is noted that H has property P if

$$(K \cdot H) \cap \Gamma = \bigcup_{i=1}^{n} g_i \cdot (H \cap \Gamma),$$

where $g_i \in G$. Now it is easy to see that all g_i are elements of the group Γ. Therefore the definition given above and definition 2 of [41] of property P are equivalent.

Remark 6.3. It is clear that for any subgroup $H \subset G$ and any compact K the set $(K \cdot H) \cap \Gamma$ is the union of some right cosets with respect to the subgroup $H \cap \Gamma$. Therefore, if $(K \cdot H) \cap \Gamma \subset \bigcup_{i=1}^{n} \gamma_i \cdot (H \cap \Gamma)$ for any compact K, then H has property P.

Lemma 6.4. The following two conditions are equivalent:

a) $H \subset G$ has property P,

b) the natural mapping $\pi': H/H \cap \Gamma \to G/\Gamma$ is proper.

Proof. I. a)$\Rightarrow$b). Let $K \subset G/\Gamma$ be an arbitrary compact set, and $\overline{K}$ a compact set in G for which $\pi(\overline{K}) = K$ (the existence of $\overline{K}$ is clear). Denote by $\tilde{\pi}$ the natural mapping $\tilde{\pi}: H \to H/H \cap \Gamma$. Formula (1) of § 0 implies that

$$\pi'^{-1}(K) = \tilde{\pi}(\pi^{-1}(K) \cap H) = \tilde{\pi}((\bar{K} \cdot \Gamma) \cap H). \tag{1}$$

Since H has property P, there are elements $\gamma_1, \dots, \gamma_n \in \Gamma$ such that

$$(\bar{K}^{-1} \cdot H) \cap \Gamma = \bigcup_{i=1}^{n} \gamma_i \cdot (H \cap \Gamma). \tag{2}$$

Let $h \in (\bar{K} \cdot \Gamma) \cap H$. Then $h = k \cdot \gamma$, where $k \in \bar{K}$, and $\gamma \in \Gamma$, and consequently $k^{-1} \cdot h = \gamma$. Therefore, applying (2) we obtain

$$k^{-1} \cdot h \in (\bar{K}^{-1} \cdot H) \cap \Gamma = \bigcup_{i=1}^{n} \gamma_i \cdot (H \cap \Gamma). \tag{3}$$

From (3) it follows immediately that

$$h \in \bar{K} \cdot \Big(\bigcup_{i=1}^{n} \gamma_i \cdot (H \cap \Gamma)\Big) = \Big(\bigcup_{i=1}^{n} \bar{K} \cdot \gamma_i\Big) \cdot (H \cap \Gamma). \tag{4}$$

Thus

$$(\bar{K} \cdot \Gamma) \cap H \subset \Big(\bigcup_{i=1}^{n} \bar{K} \cdot \gamma_i\Big) \cdot (H \cap \Gamma). \tag{5}$$

As a consequence of (1) and (5) we have

$$\pi'^{-1}(K) \subset \tilde{\pi}\Big(\Big(\bigcup_{i=1}^{n} \bar{K} \cdot \gamma_i\Big) \cdot (H \cap \Gamma)\Big) = \tilde{\pi}\Big(\bigcup_{i=1}^{n} \bar{K} \cdot \gamma_i\Big). \tag{6}$$

Therefore, since $\bigcup_{i=1}^{n} \bar{K} \cdot \gamma_i$ is compact, $\pi'^{-1}(K)$ is compact, too.

II. b)⇒a). Let $\bar{K} \subset G$ be an arbitrary compact set. In view of (1) we have

$$\pi'^{-1}(\pi(\bar{K}^{-1})) = \tilde{\pi}((\bar{K}^{-1} \cdot \Gamma) \cap H). \tag{7}$$

On the other hand, since the mapping π' is proper $\pi'^{-1}(\pi(\bar{K}^{-1}))$ is compact. Therefore it follows from (7) that there exists a compact set $\tilde{K} \subset H$ such that

$$(\bar{K}^{-1} \cdot \Gamma) \cap H \subset \tilde{K} \cdot (H \cap \Gamma). \tag{8}$$

Since the subgroup Γ is discrete, the set $(\tilde{K}^{-1} \cdot \bar{K}^{-1}) \cap \Gamma$ is finite. Denote the elements of this set by $\gamma_1, \dots, \gamma_n$.

Let $\gamma \in (\bar{K} \cdot H) \cap \Gamma$. Then $\gamma = k \cdot h$ where $k \in \bar{K}$, and $h \in H$, and consequently $k^{-1} \cdot \gamma \in (\bar{K}^{-1} \cdot \Gamma) \cap H$. Therefore, in view of (8), we have

$$k^{-1} \cdot \gamma = \tilde{k} \cdot \tilde{\gamma}, \tag{9}$$

where $\tilde{k} \in \tilde{K}$, and $\tilde{\gamma} \in H \cap \Gamma$. From (9) we obtain that $\tilde{k}^{-1} \cdot k^{-1} = \tilde{\gamma} \cdot \gamma^{-1}$, whence $\tilde{\gamma} \cdot \gamma^{-1} \in (\tilde{K}^{-1} \cdot K^{-1}) \cap \Gamma$. Therefore, we have

$$\tilde{\gamma}\gamma^{-1} = \gamma_i \tag{10}$$

for some i $(1 \leq i \leq n)$. From (10) we get $\gamma = \gamma_i^{-1} \cdot \tilde{\gamma}$, and since $\tilde{\gamma} \in H \cap \Gamma$, we have

$$\gamma \in \gamma_i^{-1} \cdot (H \cap \Gamma). \tag{11}$$

Since γ is an arbitrary element in $(\bar{K}\cdot H)\cap\Gamma$ it follows from (11) that

$$(\bar{K}\cdot H)\cap\Gamma\subset\bigcap_{i=1}^{n}\gamma_i^{-1}\cdot(H\cap\Gamma). \tag{12}$$

Since $\bar{K}$ was arbitrary, (12) and remark 6.3 imply that H has property P.

Lemma 6.5. If H_1 and H_2 have property P, then $H_1\cap H_2$ also has property P.

Proof. Let $K\subset G$ be an arbitrary compact set. Since H_1 and H_2 have property P, we have $(K\cdot H_1)\cap\Gamma=\bigcup_{k=1}^{n}\gamma_{1k}\cdot(H_1\cap\Gamma)$ and $(K\cdot H_2)\cap\Gamma=\bigcup_{l=1}^{m}\gamma_{2l}(H_2\cap\Gamma)$, where $\gamma_{1k}\in\Gamma$ and $\gamma_{2l}\in\Gamma$ for any k and l. Therefore

$$(K\cdot(H_1\cap H_2))\cap\Gamma\subset((K\cdot H_1)\cap\Gamma)\cap((K\cdot H_2)\cap\Gamma)=$$

$$=\bigcup_{k=1}^{n}\bigcup_{l=1}^{m}[(\gamma_{1k}\cdot(H_1\cap\Gamma))\cap(\gamma_{kl}\cdot(H_2\cap\Gamma))]=\bigcup_{k=1}^{n}\bigcup_{l=1}^{m}\gamma_{kl}\cdot(H_1\cap H_2\cap\Gamma)$$

(the last equality holds because

$$(\gamma_{1k}\cdot(H_1\cap\Gamma))\cap(\gamma_{2l}\cdot(H_2\cap\Gamma))=\gamma_{kl}(H_1\cap H_2\cap\Gamma)$$

for any $\gamma_{kl}\in(\gamma_{1k}\cdot(H_1\cap\Gamma))\cap(\gamma_{2l}\cdot(H_2\cap\Gamma))$, which, in view of Remark 6.3, proves the lemma.

Definition 6.6. The closed subgroup $H\subset G$ is called an *integer* subgroup, if $H/H\cap\Gamma$ is compact.

Lemma 6.7. Any integer subgroup has property P.

Proof. Any continuous mapping of a compact set is proper, so for any integer subgroup H the natural mapping $H/H\cap\Gamma\to G/\Gamma$ is proper. From this and Lemma 6.4 the assertion of our lemma follows.

Lemma 6.8. If H_1 has property P, and H_2 is integer, then $H_{12}=H_1\cap H_2$ is integer.

Proof. Denote by π_2' and π_{12}' the natural mappings $\pi_2'\colon H_2/H_2\cap\Gamma\to G/\Gamma$ and $\pi_{12}'\colon H_{12}/H_{12}\cap\Gamma\to G/\Gamma$ and by $\tilde{\pi}_2$ and $\tilde{\pi}_{12}$ the natural mappings $\tilde{\pi}_2\colon H_2\to H_2/H_2\cap\Gamma$ and $\tilde{\pi}_{12}\colon H_{12}\to H_{12}/H_{12}\cap\Gamma$. Formula (1) of § 1 implies that

$$\pi_{12}'(H_{12}/H_{12}\cap\Gamma)=\pi(H_{12})\subset\pi(H_2). \tag{13}$$

On the other hand, since H_1 has property P by assumption, and so has H_2 in view of Lemma 6.7, thus, according to Lemma 6.5 the subgroup H_{12} has property P. Now, in virtue of Lemma 6.4, π_{12}' is a proper mapping. This and formula (13) imply the assertion of the lemma, since H_2 is integer.

Lemma 6.9. Let $H_1\subset G$ have property P. If H_2 is a closed subgroup of G, such that $H_2\subset H_1$ and $H_2\cap\Gamma=H_1\cap\Gamma$, then H_2 has property P.

Proof. Let $K \subset G$ be a compactum. Then

$$(K \cdot H_2) \cap \Gamma \subset (K \cdot H_1) \cap \Gamma = \bigcup_{i=1}^{n} \gamma_i \cdot (H_1 \cap \Gamma) = \bigcup_{i=1}^{n} \gamma_i \cdot (H_2 \cap \Gamma).$$

From this and Remark 6.3 we get the lemma.

Lemma 6.10. Let $H_1 \subset G$ have property P, $H_2 \subset H_1$ be an integer subgroup, H_2 a normal subgroup in H_1, and ϱ the natural homomorphism $\varrho: H_1 \to H_1/H_2$. If for $M \subset H_1$ *1)* M has property P, *2)* the restriction of ϱ to M is an isomorphism (in Lie group sense), then $\varrho(M \cap \Gamma)$ is a subgroup of finite index in $\varrho(H_1 \cap \Gamma)$.

Proof. Since the restriction of ϱ to M is an isomorphism, we can identify M with H_1/H_2, and we can also assume that ϱ is a homomorphism of H_1 onto M. Since the subgroup H_2 is integer, we get $H_2 = K \cdot (H_2 \cap \Gamma)$, where $K \subset H_2$ is compact. Therefore

$$H_2 = H_2^{-1} = (H_2 \cap \Gamma)^{-1} \cdot K^{-1} = (H_2 \cap \Gamma) \cdot K^{-1}. \tag{14}$$

Let $\gamma \in H_1 \cap \Gamma$. Then $\gamma = h_\gamma \cdot \varrho(\gamma)$, where $h_\gamma \in H_2$. Now, in view of (14), $h_\gamma = \hat{\gamma} \cdot k_\gamma$, where $\hat{\gamma} \in H_2 \cap \Gamma$ and $k_\gamma \in K^{-1}$. Thus $\gamma = \hat{\gamma} \cdot k_\gamma \cdot \varrho(\gamma)$, and consequently $\hat{\gamma}^{-1} \cdot \gamma = k_\gamma \cdot \varrho(\gamma)$. Therefore

$$k_\gamma \cdot \varrho(\gamma) \in (K \cdot M) \cap \Gamma. \tag{15}$$

Now M has property P. Therefore

$$(K \cdot M) \cap \Gamma = \bigcup_{i=1}^{n} \gamma_i \cdot (M \cap \Gamma). \tag{16}$$

From (15) and (16) it follows that for some i we have $k_\gamma \cdot \varrho(\gamma) = \gamma_i \cdot m_\gamma$, where $m_\gamma \in M \cap \Gamma$, and consequently $\varrho(\gamma) \cdot m_\gamma^{-1} = k_\gamma^{-1} \cdot \gamma_i$. Therefore

$$\varrho(\gamma) \cdot m_\gamma^{-1} \in \big(\varrho(H_1 \cap \Gamma)\big) \cap \Big(\bigcup_{i=1}^{n} K \cdot \gamma_i\Big). \tag{17}$$

Now, according to Lemma 1.3 $\varrho(H_1 \cap \Gamma)$ is discrete, and it is clear that $\bigcup_{i=1}^{n} K \cdot \gamma_i$ is compact. Thus the set $\big(\varrho(H_1 \cap \Gamma)\big) \cap \big(\bigcup_{i=1}^{n} K \cdot \gamma_i\big)$ is finite. Therefore, as can be seen from (17), for every $\gamma \in H_1 \cap \Gamma$ there exist elements $m_\gamma \in M \cap \Gamma$ such that $\varrho(\gamma) \cdot m_\gamma^{-1}$ belongs to the finite set $\big(\varrho(H_1 \cap \Gamma)\big) \cap \big(\bigcup_{i=1}^{n} K \cdot \gamma_i\big)$, which proves the lemma.

Lemma 6.11. Let H_1, H_2, ϱ and M satisfy the conditions of Lemma 6.10. Then the subgroup generated by $M \cap \Gamma$ and $H_2 \cap \Gamma$ is of finite index in $H_1 \cap \Gamma$.

Proof. In view of Lemma 6.10, for some finite set $B \subset H_1 \cap \Gamma$ we have $\varrho(H_1 \cap \Gamma) = \varrho(B) \cdot \varrho(M \cap \Gamma)$. Therefore, for any $\gamma \in H_1 \cap \Gamma$, there is a b_γ such that

$$\varrho(\gamma) \cdot \varrho(b_\gamma) \in \varrho(M \cap \Gamma) \quad \text{or} \quad \varrho(\gamma \cdot b_\gamma) \in \varrho(M \cap \Gamma). \tag{18}$$

It is easy to see on the other hand, that if $\tilde{\gamma}\in H_1\cap\Gamma$ and $\varrho(\tilde{\gamma})=\varrho(\tilde{m})$, where $\tilde{m}\in M\cap\Gamma$, then $\tilde{\gamma}\cdot\tilde{m}^{-1}\in H_2\cap\Gamma$, and consequently $\tilde{\gamma}=\tilde{m}\cdot\tilde{h}$, where $\tilde{h}\in H_2\cap\Gamma$. Therefore, (18) implies that

$$\gamma\cdot b_\gamma\in(M\cap\Gamma)\cdot(H_2\cap\Gamma). \tag{19}$$

Now from (19) we have at once that

$$\Gamma\cdot B\subset(M\cap\Gamma)\cdot(H_2\cap\Gamma) \tag{20}$$

which proves the lemma.

Lemma 6.12. Let $H\subset G$ be an integer subgroup. If $\tilde{H}$ is a closed subgroup of G such that $\tilde{H}\subset H$ and $\tilde{H}\cap\Gamma=H\cap\Gamma$, then $\tilde{H}$ is an integer subgroup.

Proof. By Lemmas 6.7 and 6.9 $\tilde{H}$ has property P. Therefore, since $\tilde{H}=H\cap\tilde{H}$, in view of Lemma 6.8 $\tilde{H}$ is an integer subgroup.

Lemma 6.13. Let $H\subset G$ have property P. Let π and $\tilde{\pi}$ be the natural mapping $\pi\colon G\to G/\Gamma$ and $\tilde{\pi}\colon H\to H/H\cap\Gamma$, and $B\subset G$ a subset. If $\pi(B)$ is relatively compact in G/Γ, then $\tilde{\pi}(B\cap H)$ is relatively compact in $H/H\cap\Gamma$.

Proof. Let π' be the natural mapping $\pi'\colon H/H\cap\Gamma\to G/\Gamma$. In view of formula (1) of § 0, we have

$$\pi'(\tilde{\pi}(B\cap H))=\pi(\tilde{\pi}^{-1}(\tilde{\pi}(B\cap H)))=\pi((B\cap H)\cdot(H\cap\Gamma))=$$
$$=\pi(B\cap H)\subset\pi(B). \tag{21}$$

Since H has property P, π' is a proper mapping, according to Lemma 6.4. Therefore it follows from (21) and the relative compactness of $\pi(B)$ that $\tilde{\pi}(B\cap H)$ is relatively compact. The lemma is proved.

§ 7. The groups $\bar{S}N(\bar{U})$ and $C(\bar{U})$

From now on — except § 8 — we always assume that G is a semisimple algebraically connected real algebraic group such that *1)* the centre of the complexification G_{C} of G is trivial, (and consequently, in view of Lemma 1.2.15, so is the centre of G), *2)* G has no compact factors; further $\Gamma\subset G$ is a discrete subgroup of G such that *1)* the factor space is of finite volume but it is non-compact (i.e. Γ is in G a non-uniform lattice), *2)* the pair (G, Γ) is irreducible. The natural mapping $G\to G/\Gamma$ will always be denoted by π. We will assume further that on the Lie algebra $\mathfrak{G}$ of the group G a Euclidean metric $\varrho_{\mathfrak{g}}$ is introduced, and on the group G a right-invariant metric ϱ_G corresponding to $\varrho_{\mathfrak{g}}$ is given.

For any discrete unipotent subgroup U of the group G we denote by $\bar{U}$ the Zariski closure of U in G, by $\mathfrak{U}$ the Lie algebra of the group $\bar{U}$, and by $\Delta_U\subset\mathfrak{G}$ the quasi-lattice which corresponds to U as described in § 5 (preceding Lemma 5.7). From the definition of the group U and the quasi-lattice Δ_U it follows that for any algebraic automorphism T of G we have $T(\bar{U})=\overline{T(U)}$ and $\Delta_{T(U)}=$ $=\tilde{T}(\Delta_U)$, where $\tilde{T}$ is the automorphism of the Lie algebra $\mathfrak{G}$ corresponding to T. Therefore

$$\overline{gUg^{-1}} = g\bar{U}g^{-1} \tag{1}$$

and

$$\Delta_{gUg^{-1}} = (\mathrm{Ad}\, g)\Delta_U \tag{2}$$

for any $g \in G$.

A unipotent subgroup U of the group Γ is called *regular*, if $U=\bar{U}\cap\Gamma$. The set of all regular unipotent subgroups of the group Γ is denoted by $B(\Gamma)$. It follows from (1) that

$$\gamma U\gamma^{-1}\in B(\Gamma), \quad \text{if} \quad U\in B(\Gamma) \quad \text{and} \quad \gamma\in\Gamma. \tag{3}$$

As mentioned in § 5 (before Lemma 5.4), $\ln U$ generates $\mathfrak{U}$ as a linear space. Therefore, since $\Delta_U \supset b \ln U$ (for the definition of the number b see § 5 before Lemma 5.7), the following lemma holds:

Lemma 7.1. Δ_U generates $\mathfrak{U}$ as linear space, that is to say, Δ_U is a quasi-lattice of full dimension in $\mathfrak{U}$.

Lemma 7.2. For any positive constant a, the set B_a of all $U\in B(\Gamma)$ for which the determinant of the quasi-lattice Δ_U is less than a is finite (the determinant of the quasi-lattice Δ_U is taken with respect to the metric $\varrho_{\mathfrak{g}}$).

Proof. Denote by $\tilde{B}_a$ the set of all lattices of the form Δ_U where $U\in B_a$. From the discreteness of the group Γ, in view of Mahler's criterion for compactness, the compactness of the set $\tilde{B}_a$ follows. Therefore, according to Lemma 5.7, there exists a constant $c>0$ such that in every subgroup $U\in B_a$ there is a finite subset $H(U)$ having the following properties: a) $H(U)\subset H_c$, where H_c is the set of all $\gamma\in\Gamma$ for which $\varrho_G(e, \gamma)<c$, b) every algebraic subgroup of the group G containing $H(U)$ also contains $\bar{U}$.

If $H(U_1)=H(U_2)$, then property b) implies $\bar{U}_1=\bar{U}_2$ and, since the subgroups U_1 and U_2 are regular, we have $U_1=\bar{U}_1\cap\Gamma=\bar{U}_2\cap\Gamma=U_2$. Therefore, the cardinality of the set B_a does not exceed the cardinality of the set of all subsets of H_c. Now H_c is finite, by the discreteness of Γ. Consequently B_a is finite, and the lemma is proved.

Let $N(\bar{U})$ denote the normalizer of $\bar{U}$ in G. The group $N(\bar{U})$ is acting naturally on the Lie algebra $\mathfrak{U}$ of the group $\bar{U}$ (the adjoint representation of $N(\bar{U})$ restricted to $\mathfrak{U}$). The subgroup of the group $N(\bar{U})$ consisting of the elements having determinant ± 1 under this action will be denoted by $\bar{S}N(\bar{U})$.

Lemma 7.3. For an arbitrary subgroup $U\in B(\Gamma)$, the subgroup $\bar{S}N(\bar{U})$ has property P.

Proof. Let $K\subset G$ be a compactum. Consider the set F of all quasi-lattices f that can be written in the form $f=(\mathrm{Ad}\, m(f))\Delta_U$, where $m(f)\in(K\cdot\bar{S}N(\bar{U}))\cap\Gamma$.

Since K is compact, and for any f, we have $m(f)\in K\cdot \bar{S}N(\bar{U})$, there exists a constant $c(K)$ depending only on K such that the determinant $d(f)$ of any quasi-lattice f is less than $c(K)$. Therefore, in case of any quasi-lattice $f\in\Gamma$, (2), (3) and the relations $m(f)\in\Gamma$ and $U\in B(\Gamma)$ imply the inclusion

$$m(f)Um^{-1}(f)\in B_{c(K)}. \tag{4}$$

From (2), (4) and Lemma 7.2 the finiteness of the set F follows. Denote the elements of this set by $f_1, \dots, f_n$. Then for any f_i $(1\leqq i\leqq n)$ there exists an $m_i\in(K\cdot\bar{S}N(\bar{U}))\cap\Gamma$ such that $(\mathrm{Ad}\, m_i)\Delta_U=f_i$. Now let m be an arbitrary element of the set $(K\cdot\bar{S}N(\bar{U}))\cap\Gamma$. Then for some i we have $(\mathrm{Ad}\, m)\Delta_U=(\mathrm{Ad}\, m_i)\Delta_U$, whence

$$(\mathrm{Ad}\,(m_i^{-1}m))\Delta_U = \Delta_U. \tag{5}$$

Now, according to Lemma 7.1, Δ_U is a quasi-lattice of full dimension in $\mathfrak{U}$. Therefore (5) implies

$$(\mathrm{Ad}\,(m_i^{-1}m))\mathfrak{U} = \mathfrak{U} \tag{6}$$

and

$$\det(\mathrm{Ad}\,(m_i^{-1}m)_{\mathfrak{U}}) = \pm 1, \tag{7}$$

where $\mathrm{Ad}\,(m_i^{-1}m)_{\mathfrak{U}}$ is the restriction of $\mathrm{Ad}\,(m_i^{-1}m)$ to $\mathfrak{U}$.

As it has already been remarked in § 5 (before Lemma 5.4), $\bar{U}$ is topologically connected. Therefore, (6) implies

$$m_i^{-1}m\in N(\bar{U}). \tag{8}$$

From (7) and (8) we get

$$m_i^{-1}m\in \bar{S}N(\bar{U}). \tag{9}$$

On the other hand, $m_i\in\Gamma$ and $m\in\Gamma$, and so

$$m_i^{-1}m\in\Gamma. \tag{10}$$

(9) and (10) imply

$$m\in m_i\cdot(\bar{S}N(\bar{U})\cap\Gamma). \tag{11}$$

Since m is an arbitrary element of the set $(K\cdot\bar{S}N(\bar{U}))\cap\Gamma$, (11) implies the inclusion

$$(K\cdot\bar{S}N(\bar{U}))\cap\Gamma\subset\bigcup_{i=1}^{n} m_i\cdot(\bar{S}N(\bar{U})\cap\Gamma). \tag{12}$$

From (12) and Remark 6.3 the statement of the lemma follows.

Lemma 7.4. If the solvable subgroup R of the group G is generated by unipotent elements, then it is unipotent.

Proof. It is easy to see that it can be assumed, without restriction of generality, that R is generated by finitely many unipotent elements. Let $u_1, \dots, u_k$ denote these elements, U_i the subgroup of R generated by the element u_i $(1\leqq i\leqq k)$, $\bar{U}_i$ the algebraic closure of U_i in G and $\bar{R}$ the algebraic closure of R in G. It follows from Lemmas 1.3.3 and 1.3.1 that, for any i, the subgroup $\overline{U}_i$ is algebraically connected. On the other hand, since $U_i \subset R$ for any i, we have

$$\overline{U}_i \subset \overline{R}. \tag{13}$$

Denote by R^0 the algebraically connected component of the identity of the group R. From (13) and the algebraic connectedness of the subgroup $\overline{U}_i$ it follows that for any i, $u_i \in R^0$.

Thus, since N is generated by the elements u_i, we have

$$R \subset R^0. \tag{14}$$

From (14) and the fact that R^0 is a Zariski closed subset of R (R^0 is algebraically closed by definition), it follows that

$$\overline{R} = R^0. \tag{15}$$

(15) and Lemma 1.2.4 imply that $\overline{R}$ is an algebraically connected solvable algebraic **R**-group. Therefore, according to Lemma 1.5.5 $\overline{R}$ is a semi-direct product of a torus T and an algebraic unipotent subgroup V. Since the elements are unipotent, for any i we have $u_i \in V$. From this and the fact that R is generated by the elements u_i, one gets

$$R \subset V. \tag{16}$$

From (16) it automatically follows that R is a unipotent subgroup. The lemma is proved.

In [25], in fact, the following lemma is proved:

Lemma 7.5. One can choose in Γ a non-empty finite set H having the following properties:

1) any element of H is different from e and is unipotent,

2) for any neighbourhood $V \subset G$ of the identity there exists a compactum $K(V) \subset G/\Gamma$ such that whenever $\pi(g) \notin K(V)$, the intersection $(g\Gamma g^{-1}) \cap V$ contains a unipotent element $\gamma_g(V)$ (different from the identity) for which $g^{-1}\gamma_g(V)g = \gamma h \gamma^{-1}$ holds for some $h \in H$ and $\gamma \in \Gamma$.

Remark. In [41] the formulation of Lemma 5.6 contains a misprint: $\pi(g) \in K(V)$ should read $\pi(g) \notin K(V)$.

Lemma 7.6. Let $U \in B(\Gamma)$ be a subgroup for which $(C(\overline{U}) \cap \Gamma)^u \subset U$ ($C(\overline{U})$ is the centralizer of $\overline{U}$ in G, and $(C(\overline{U}) \cap \Gamma)^u$ is the set of unipotent elements in $C(\overline{U}) \cap \Gamma$). If $N \subset N(\overline{U})$ is a subset such that $(\mathrm{Ad}\, N)\Delta_U$ is relatively compact in the space of quasi-lattices ($(\mathrm{Ad}\, N)\Delta_U = \bigcup_{n \in N} (\mathrm{Ad}\, n)\Delta_U$) then $\pi(N)$ is relatively compact in G/Γ.

Proof. By virtue of Lemma 1.6.1, there exists an ε_0-neighbourhood $\tilde{V}$ of the identity $e \in G$ such that for any $g_1, g_2 \in \tilde{V}$ the following inequality holds:

$$\varrho_G(e, g_1 g_2 g_1^{-1} g_2^{-1}) < \frac{1}{2} \min\left(\varrho_G(e, g_1), \varrho_G(e, g_2)\right). \tag{17}$$

Since $N \subset N(\overline{U})$, for any $n \in N$ we have

$$nUn^{-1} \subset \overline{U}. \tag{18}$$

Now this set $\bigcup_{n \in N} \Delta_{nUn^{-1}}$ is compact by formula (2) and the condition of the Lemma. Therefore, according to Lemma 5.7 there exists a $c>0$ such that for any $n \in N$ there is a finite subset $U(n) \subset nUn^{-1}$ having the following properties:

1) if $u \in U(n)$, then $\varrho_G(e, u) < c$,

2) any algebraic unipotent subgroup of G containing $U(n)$ also contains $\overline{nUn^{-1}}$. Hence, in view of the Lemma 1.6.2 and formula (1), there is an element $g_0 \in G$ such that if $u \in U(n)$ then

$$\varrho_G(e, g_0 u g_0^{-1}) < \varepsilon_0 \quad \text{and consequently} \quad g_0 U(n) g_0^{-1} \subset \tilde{V}, \tag{19}$$

3) any algebraic subgroup of G containing $g_0 U(n) g_0^{-1}$ also contains $g_0 nUn^{-1} g_0^{-1}$. On the other hand, since the set $(\mathrm{Ad}\, N)\Delta_U$ is compact, so is the set $(\mathrm{Ad}\,(g_0 N))\Delta_U = \mathrm{Ad}\, g_0((\mathrm{Ad}\, N)\Delta_U)$. Therefore, in view of formula (2) the set $\bigcup_{n \in N} \Delta_{g_0 nUn^{-1} g_0^{-1}}$ is also compact. This and Lemma 5.7 imply the existence of an $\varepsilon > 0$ ($\varepsilon < \varepsilon_0$) such that if $n \in N$ and $u \in nUn^{-1}$ and $u \neq e$, then

$$\varrho_G(e, g_0 u g_0^{-1}) > \varepsilon. \tag{20}$$

Now if $\pi(N)$ is not relatively compact in G/Γ, neither is $\pi(g_0 N)$. Therefore for any compact $K \subset G$ there is a $n(K) \in N$ such that $\pi(g_0 \cdot n(K)) \notin K$. Denote $n(K(\hat{V}))$ by $\hat{n}$, where $\hat{V} \subset G$ is an ε-neighbourhood of the identity (for the definition of the compactum $K(\hat{V})$ see Lemma 7.5). Then, according to Lemma 7.5, the intersection $g_0 \hat{n} \Gamma \hat{n}^{-1} g_0^{-1} \cap \hat{V}$ contains a unipotent element $\gamma_{g_0 \hat{n}}(\hat{V})$ (different from the identity), which we denote by $\hat{\gamma}$.

Since $\hat{\gamma} \in \hat{V}$ and $\hat{V} \subset \tilde{V}$, (because $\varepsilon < \varepsilon_0$), we have

$$\hat{\gamma} \subset \hat{V}. \tag{21}$$

For any two subsets $A, B \subset G$, we denote by $\{A, B\}$ the set of all commutators of the form $aba^{-1}b^{-1}$ where $a \in A$ and $b \in B$. Put $U_{-1} = g_0 U(\hat{n}) g_0^{-1}$, $U_0 = \hat{\gamma}$ and $U_{i+1} = \{U_i, U_{i-1}\}$ for $i \geqq 0$. Denote $\sup_{u \in U_i} \varrho_G(e, u)$ by d_i. Since we have *1)* $\varrho_G(e, \hat{\gamma}) < \varepsilon$ by the choice of $\hat{\gamma}$, *2)* $\hat{\gamma} \in \hat{V}$ according to (21), *3)* $g_0 U(\hat{n}) g_0^{-1} \subset \tilde{V}$ according to (19), applying (17) we can get the following inequality by induction on i:

$$d_i < \frac{\varepsilon}{2^i} \tag{22}$$

for any $i \geqq 0$. On the other hand, both $\hat{\gamma}$ and U_{-1} belong to the discrete subgroup $g_0 \hat{n} \Gamma \hat{n} g_0^{-1}$. Therefore, for any i

$$U_i \subset g_0 \hat{n} \Gamma \hat{n}^{-1} g_0^{-1}. \tag{23}$$

Now, (22), (23) and the discreteness of Γ imply the existence of an integer k with

$$U_k \neq e \quad \text{and} \quad U_{k+1} = e, \tag{24}$$

further $k > 0$, since $\hat{\gamma} \neq e$. Let us choose in U_k some element $\hat{u}$ (different from the identity). Then, according to (24) and (22) we have

$$\hat{u} \in C(U_{-1}) \quad \text{and} \quad \varrho_G(e, \hat{u}) < \varepsilon, \tag{25}$$

where $C(U_{-1})$ denotes the centralizer of the set U_{-1}.

We claim that the element $\hat{u}$ is unipotent. Since $\hat{u}$ belongs to the subgroup L generated by the union $\hat{\gamma} \cup U_{-1}$, it is enough to show that L is unipotent. The same way as inequality (22) has been proved shows that any commutator M of elements from $\hat{\gamma} \cup U_{-1}$ satisfies the following inequality:

$$\varrho_G(e, M) < \frac{\varepsilon_0}{2^{\omega(M)}}, \tag{26}$$

where $\omega(M)$ is the weight of the commutator M (for the definition of the weight of a commutator see the beginning of the proof of Lemma 5.2).

On the other hand $\hat{\gamma} \cup U_{-1} \subset g_0 \hat{n} \Gamma \hat{n}^{-1} g_0^{-1}$, and since Γ is discrete, it follows from (26) that there exists an integer $\hat{k}$ such that

$$M = e, \quad \text{if only} \quad \omega(M) > \hat{k}. \tag{27}$$

Denote by $Z_i(L)$, $(1 \leqq i \leqq k)$ the subgroup of L generated by all possible commutators of elements in $\hat{\gamma} \cup U_{-1}$ of weight greater than $\hat{k} - i$. From (27) it follows that $Z_1(L)$ commutes with $\hat{\gamma} \cup U_{-1}$, and consequently, belongs to the centre of L, i.e.

$$\big(L, Z_1(L)\big) = e. \tag{28}$$

It is clear that

$$(\gamma, M) \subset Z_1(L), \quad \text{if only} \quad \gamma \in \hat{\gamma} \cap U_{-1} \quad \text{and} \quad \omega(M) > k-2. \tag{29}$$

Denote by f the natural mapping $f: L \to L/Z_1(L)$. From (29) it follows at once that

$$\big(f(\gamma), f(M)\big) = e, \quad \text{if only} \quad \gamma \in \tilde{\gamma} \cup U_{-1} \quad \text{and} \quad \omega(M) > k-2. \tag{30}$$

This and the definition of the subgroups L and $Z_2(L)$ imply that $\big(f(L), f(Z_2(L)\big) = e$, and consequently $\big(L, Z_2(L)\big) \subset Z_1(L)$. With the same argument we find for all $i > 1$

$$\big(L, Z_i(L)\big) \subset Z_{i-1}(L). \tag{31}$$

On the other hand it is clear that $Z_{\hat{k}}(L) = L$. Thus (28) and (31) imply that L is nilpotent. (Here we have actually repeated the proof of Zassenhaus' lemma which states that a discrete subgroup generated by elements belonging to a sufficiently small neighbourhood of the identity, is always nilpotent.) Now L is generated by unipotent elements. Therefore, according to Lemma 7.4, L is unipotent, and consequently, so is the element $\hat{u}$, proving the assertion claimed.

From (25) it follows at once that

$$g_0 U(\hat{n}) g_0^{-1} = U_{-1} \subset C(\hat{u}), \tag{32}$$

where $C(\hat{u})$ is the centralizer of $\hat{u}$ in G. On the other hand, in view of Lemma 1.29, $C(\hat{u})$ is an algebraic subgroup. Thus, from (32) and the definition of the set $U(\hat{n})$ it follows that

$$g_0 \hat{n} \overline{U} \hat{n}^{-1} g_0^{-1} \subset C(\hat{u}), \tag{33}$$

and consequently

$$\hat{u} \in C(g_0 \hat{n} \overline{U} \hat{n}^{-1} g_0^{-1}). \tag{34}$$

(By passing from (33) to (34) one has to apply formula (1).) But $\hat{\gamma} \cup U_{-1} \subset \subset g_0 \hat{n} \Gamma \hat{n}^{-1} g_0^{-1}$, and so

$$\hat{u} \in L \subset g_0 \hat{n} \Gamma \hat{n}^{-1} g_0^{-1}. \tag{35}$$

Since $\hat{u} \neq e$, from (35), (25) and (20) it follows that

$$\hat{u} \notin g_0 \hat{n} U \hat{n}^{-1} g_0^{-1}, \tag{36}$$

and consequently

$$\hat{n}^{-1} g_0^{-1} \hat{u} g_0 \hat{n} \notin U. \tag{37}$$

Now (34) and (35) imply

$$\hat{n}^{-1} g_0^{-1} \hat{u} g_0 \hat{n} \notin C(\overline{U}) \cap \Gamma. \tag{38}$$

As U is a regular subgroup, (37) implies that

$$\hat{n}^{-1} g_0^{-1} \hat{u} g_0 \hat{n} \notin \overline{U}. \tag{39}$$

Since $\hat{u}$ is unipotent, it follows from (38) and (39) that

$$(C(\overline{U}) \cap \Gamma)^u \not\subset \overline{U}. \tag{40}$$

Thus the assumption that $\pi(N)$ is not compact has led to formula (40) which contradicts the condition of the lemma. The lemma is proved.

Lemma 7.7. If U is a maximal unipotent subgroup of the group G, then $U \in B(\Gamma)$ and $(C(\overline{U}) \cap \Gamma)^u \subset U$.

Proof. As we have already remarked several times, $\overline{U}$ is a unipotent subgroup. Thus the maximality of U implies at once that $\overline{U} \cap \Gamma = U$, i.e. $U \in B(\Gamma)$. Let us now assume that U is maximal and $(C(\overline{U}) \cap \Gamma)^u$ contains an element u_0 outside U. Denote by U_0 the subgroup generated by u_0 and U. Since u_0 commutes with U, and U is nilpotent, according to Lemma 1.3.1 we have that U_0 is nilpotent. Therefore, since u_0 and U are unipotent, Lemma 7.4 implies that U_0 is also unipotent. But this would contradict the maximality of U (as $u_0 \in U_0$ and $u_0 \notin U$). The lemma is proved.

From Lemmas 7.6 and 7.7 we have immediately the following:

Lemma 7.8. The assertion of Lemma 7.6 holds if U is a maximal unipotent subgroup of Γ.

If $U \in B(\Gamma)$ and $C(\bar{U}) \subset \bar{U}$, then $(C(\bar{U}) \cap \Gamma)^u \subset \bar{U} \cap \Gamma = U$. Therefore we have:

Lemma 7.9. The assertion of Lemma 7.6 holds if only for the subgroup $U \in B(\Gamma)$ we have $C(\bar{U}) \subset \bar{U}$.

Lemma 7.10. For any subgroup $U \in B(\Gamma)$ the subgroup $C(\bar{U})$ has property P.

Proof. It easily follows from Lemmas 1.3.3, 1.3.1, 5.4 and from 2c of Lemma 5.1 that there exists in U a finite generating system. Denote these generators by $u_1, \ldots, u_l$. Let $K \subset G$ be a compactum. Then it is easy to see that the set E of distinct l-tuples $[mu_1m^{-1}, \ldots, mu_lm^{-1}]$, where $m \in (K \cdot C(\bar{U})) \cap \Gamma$, is compact as well as discrete (in the product topology of $G \times \ldots \times G$), and therefore it is finite.

On the other hand, if for the pair m_1, m_2 the l-tuples $[m_1 u_1 m_1^{-1}, \ldots, m_1 u_l m_1^{-1}]$ and $[m_2 u_1 m_2^{-1}, \ldots, m_2 u_l m_2^{-1}]$ coincide, then

$$m_1 m_2^{-1} \in C(\bar{U}) \cap \Gamma \tag{41}$$

(since *1)* $u_1, \ldots, u_l$ are generators of U, *2)* U is Zariski dense in $\bar{U}$ and *3)* $m_1, m_2 \in \Gamma$). From (41) and the finiteness of the set E it follows at once that

$$(K \cdot C(\bar{U})) \cap \Gamma = \bigcup_{i=1}^{l} \gamma_i (C(\bar{U}) \cap \Gamma), \tag{42}$$

where $\gamma_i \in \Gamma$. From (42) and Remark 6.3 the statement of the lemma follows.

Lemma 7.11. Let $U \in B(\Gamma)$ be a subgroup satisfying $(C(\bar{U}) \cap \Gamma)^u \subset U$. Then $C(\bar{U})$ is an integer subgroup (i.e. $C(\bar{U})/C(\bar{U}) \cap \Gamma$ is compact).

Proof. From the inclusion $C(\bar{U}) \subset N(\bar{U})$ and Lemma 7.6 we get that $\pi(C(\bar{U}))$ is relatively compact in G/Γ. On the other hand, according to Lemma 7.10, $C(\bar{U})$ has property P. It is easy to see that $C(\bar{U})/C(\bar{U}) \cap \Gamma$ is the inverse image of $\pi(C(\bar{U}))$ under the natural mapping $C(\bar{U})/C(\bar{U}) \cap \Gamma \to G/\Gamma$. Consequently, the relative compactness of $\pi(C(\bar{U}))$ implies that $C(\bar{U})/C(\bar{U}) \cap \Gamma$ is compact. The lemma is proved.

From Lemmas 7.7 and 7.11 we get

Lemma 7.12. The assertion of Lemma 7.11 holds if U is a maximal unipotent subgroup of Γ.

Lemma 7.13. Let U be an arbitrary unipotent subgroup of Γ. Then we have $N(\bar{U}) \cap \Gamma = \bar{S}N(\bar{U}) \cap \Gamma$.

Proof. Put $U' = \overline{U} \cap \Gamma$. It is easy to see that $\overline{U'} = \overline{U}$ (because $\overline{U} \subset U' \subset U$). Therefore $U' \in B(\Gamma)$. Thus it follows from (1) that for any $\gamma \in N(\overline{U}) \cap \Gamma$ we have

$$\gamma U' \gamma^{-1} = \gamma(\overline{U'} \cap \Gamma)\gamma^{-1} = \gamma \overline{U'} \gamma^{-1} \cap \gamma \Gamma \gamma^{-1} =$$
$$= \overline{\gamma U' \gamma^{-1}} \cap \Gamma = \overline{U'} \cap \Gamma = \overline{U} \cap \Gamma = U'. \tag{43}$$

From (43) and (2) we conclude that for any $\gamma \in N(\overline{U}) \cap \Gamma$ the quasi-lattice $\Delta_{U'}$ is invariant under the mapping Ad γ. Now according to Lemma 7.1 $\Delta_{U'}$ is a quasi-lattice of full dimension in $\mathfrak{U}$. Consequently for any $\gamma \in N(\overline{U}) \cap \Gamma$ the restriction of Ad γ to $\mathfrak{U}$ has determinant ± 1. The lemma is proved.

§ 8. Quasialgebraic R-groups

A quasialgebraic **R**-group is a topologically closed subgroup of the group $GL(n, \mathbf{R})$ which is a subgroup of finite index in its algebraic closure. For a quasialgebraic **R**-group $G \subset GL(n, \mathbf{R})$, $[G]$ denotes the Zariski closure of G in $GL(n, \mathbf{R})$. Since a closed subgroup of a Lie group is a Lie group, any quasialgebraic **R**-group is a Lie group. It follows from this that for any quasialgebraic **R**-group a) $\dim G = \dim [G]$, b) G contains the topologically connected component of the identity of the group $[G]$. The Zariski topology on the quasialgebraic **R**-group G means the topology induced on G by the Zariski topology of the algebraic **R**-group $[G]$. A regular algebraic (or simply algebraic) mapping of a quasialgebraic **R**-group $G_1 \subset GL(n_1, \mathbf{R})$ into a quasialgebraic **R**-group $G_2 \subset GL(n_2, \mathbf{R})$ (or into an algebraic manifold M) is a mapping which extends to a regular algebraic mapping of $[G_1]$ into $[G_2]$ (of $[G_1]$ into M).

An algebraic subgroup of a quasialgebraic **R**-group G is the intersection of G with an algebraic subgroup of the group $[G]$. A quasialgebraic subgroup of a quasialgebraic **R**-group G is a subgroup H, which a) is topologically closed, b) is of finite index in its algebraic closure (i.e. H has to be a quasialgebraic **R**-group itself). The subgroup Γ of a quasialgebraic **R**-group is called *arithmetic*, if Γ is an arithmetic subgroup in $[G]$. We will say that a quasialgebraic **R**-group G is defined over some field K, if the real algebraic group $[G]$ is defined over K.

The maximal unipotent normal subgroup in G is called the *unipotent radical* of the quasialgebraic **R**-group G. Since the unipotent radical of an arbitrary algebraic **R**-group is topologically connected (according to Lemma 1.3.2), the unipotent radical of an arbitrary quasialgebraic **R**-group G coincides with the unipotent radical of the group $[G]$.

The following two lemmas are well known (see [27]):

Lemma 8.1. Let G_1 and G_2 be algebraic **C**-groups, and f an algebraic homomorphism $f: G_1 \to G_2$. Then

1) the image $f(G_1)$ of the group G_1 under the mapping f is an algebraic subgroup of the group G_2,

2) if G_1 and G_2 are defined over some field K, and f is defined over the same field K (i.e. it is given by polynomials with coefficients in K), then a) the group $f(G_1)$ is defined over K, b) the kernel of the mapping f is defined over K (i.e. it can be given in G_1 by polynomial equations with coefficients in K),

3) $\dim f(G_1) = \dim G_1 - \dim (\operatorname{Ker} f)$.

Lemma 8.2. If the algebraic **C**-group G is defined over **R**, then $\dim_{\mathbf{R}} G_{\mathbf{R}} = \dim_{\mathbf{C}} G$, where $G_{\mathbf{R}}$ denotes the subgroup of real points of the group G, $\dim_{\mathbf{R}} G_{\mathbf{R}}$ the real dimension of $G_{\mathbf{R}}$, and $\dim_{\mathbf{C}} G$ the complex dimension of G.

Lemma 8.3. The image of a quasialgebraic **R**-group G_1 under an algebraic homomorphism f of the group G_1 into a quasialgebraic **R**-group G_2 is a quasialgebraic subgroup of the group G_2.

Proof. Denote by $[G_1]_{\mathbf{C}}$ and $[G_2]_{\mathbf{C}}$ the complexifications of the groups $[G_1]$ and $[G_2]$, and by $\hat{f}$ and $\hat{f}_{\mathbf{C}}$ the regular algebraic mappings $\hat{f}: [G_1] \to [G_2]$ and $\hat{f}_{\mathbf{C}}: [G_1]_{\mathbf{C}} \to [G_2]_{\mathbf{C}}$ which are the extensions of the mapping f. From Lemma 1.2.6 we get immediately that $\hat{f}$ and $\hat{f}_{\mathbf{C}}$ are homomorphisms.

Denote by H, $\hat{H}$ and $\hat{H}_{\mathbf{C}}$ the kernels of the mappings f, $\hat{f}$ and $\hat{f}_{\mathbf{C}}$. We clearly have

$$H = \hat{H}_{\mathbf{C}} \cap G_1 \tag{1}$$

and

$$\hat{H} = \hat{H}_{\mathbf{C}} \cap [G_1]. \tag{2}$$

Since G_1 is a topologically closed subgroup of finite index in $[G_1]$, in view of (1) and (2) H is a topologically closed subgroup of finite index in $\hat{H}$. Therefore

$$\dim_{\mathbf{R}} H = \dim_{\mathbf{R}} \hat{H}. \tag{3}$$

Since $\hat{f}_{\mathbf{C}}$ is the algebraic extension of $\hat{f}$, which is a homomorphism of an algebraic **R**-group into another algebraic **R**-group, it is clear that $\hat{f}_{\mathbf{C}}$ is defined over **R**. Therefore, in view of Lemma 8.1, $\hat{H}_{\mathbf{C}}$ is defined over **R**. On the other hand, $\hat{H}$ is the subgroup of real points of the group $\hat{H}_{\mathbf{C}}$ (in view of (2) and the fact that $[G_1]$ is the set of all real points of the group $[G_1]_{\mathbf{C}}$ according to the definition of complexification). Therefore, in view of Lemma 8.2 we have

$$\dim_{\mathbf{R}} \hat{H} = \dim_{\mathbf{C}} \hat{H}_{\mathbf{C}}. \tag{4}$$

Since $[G_1]_{\mathbf{C}}$ is the complexification of the algebraic **R**-group $[G_1]$ we get that $[G_1]_{\mathbf{C}}$ is defined over **R**. Therefore, in view of Lemma 8.2 we have

$$\dim_{\mathbf{R}} [G_1] = \dim_{\mathbf{C}} [G_1]_{\mathbf{C}}. \tag{5}$$

Further, as we have already mentioned,

$$\dim_{\mathbf{R}} G_1 = \dim_{\mathbf{R}} [G_1]. \tag{6}$$

Since the homomorphism $\hat{f}_{\mathbf{C}}$ is defined over **R**, according to Lemma 8.1 $\hat{f}_{\mathbf{C}}([G_1])_{\mathbf{C}}$ is an algebraic **C**-group defined over **R**. Denote $\hat{f}_{\mathbf{C}}([G_1]_{\mathbf{C}})$ by $\tilde{G}$, and the subgroup of all real points of the group $\tilde{G}$ by $\tilde{G}_{\mathbf{R}}$. Since $\tilde{G}$ is defined over $\mathbf{R}$, we have in view of Lemma 8.2

$$\dim_{\mathbf{R}} \tilde{G}_{\mathbf{R}} = \dim_{\mathbf{C}} \tilde{G}. \tag{7}$$

Now, in view of Lemma 8.1

$$\dim_{\mathbf{C}} \tilde{G} = \dim_{\mathbf{C}} [G_1]_{\mathbf{C}} - \dim_{\mathbf{C}} \hat{H}_{\mathbf{C}}. \tag{8}$$

On the other hand it is clear that there exists a neighbourhood U of the identity in the group G_1, such that $f(U)$ is a submanifold (not closed) in G_2, and for the dimension $\dim_{\mathbf{R}} f(U)$ we have

$$\dim_{\mathbf{R}} f(U) = \dim_{\mathbf{R}} G_1 - \dim_{\mathbf{R}} H. \tag{9}$$

From (3), (4), (5), (6), (7), (8) and (9) it follows

$$\dim_{\mathbf{R}} f(U) = \dim_{\mathbf{R}} G_1 - \dim_{\mathbf{R}} H = \dim_{\mathbf{C}} [G_1]_{\mathbf{C}} - \dim_{\mathbf{C}} \hat{H}_{\mathbf{C}} =$$
$$= \dim_{\mathbf{C}} \tilde{G} = \dim_{\mathbf{R}} \tilde{G}_{\mathbf{R}}. \tag{10}$$

Since *1)* $f(U)$ is a submanifold in G_2, *2)* it is easily seen that $f(G_1)$ belongs to $\tilde{G}_{\mathbf{R}}$. From (10) it follows that $f(U)$ contains some neighbourhood of the identity in $\tilde{G}_{\mathbf{R}}$. Now it follows from this that $f(G_1)$ contains the topologically connected component of the identity of the group $\tilde{G}_{\mathbf{R}}$, whence we have, in turn, that $f(G_1)$ is a quasialgebraic subgroup of the group G_2. The lemma is proved.

Definition 8.4. Let X be an algebraic submanifold of the space $\mathbf{R}^n(\mathbf{C}^n)$, and $e_1, \ldots, e_n$ a basis in $\mathbf{R}^n(\mathbf{C}^n)$. We say that in the basis $e_1, \ldots, e_n$, X is defined over $\mathbf{Q}$ if X consists of all common zeros of some polynomials of $(x_1, \ldots, x_n)$ (these are the coordinates in $e_1, \ldots, e_n$) with coefficients in $\mathbf{Q}$ where $\mathbf{Q}$ is the field of rational numbers.

Lemma 8.5. Let X be an algebraic submanifold in $\mathbf{R}^n(\mathbf{C}^n)$, and $e_1, \ldots, e_n$ a basis in $\mathbf{R}^n(\mathbf{C}^n)$. If the set of rational points of X (in the basis $e_1, \ldots, e_n$) is Zariski dense in X, then X is defined over $\mathbf{Q}$ in the basis $e_1, \ldots, e_n$.

Proof. Denote by $X_{\mathbf{Q}}$ the set of all points of the manifold X which are rational in the basis $e_1, \ldots, e_n$, by I the ideal in $\mathbf{R}(x_1, \ldots, x_n)$ $(\mathbf{C}(x_1, \ldots, x_n))$ corresponding to X and, for any natural number k, by $I_k \subset I$ the set of polynomials of the ideal I of degree not exceeding k. Fix some natural number k_0. If $P \in I_{k_0}$, then the coefficients of P satisfy a system of linear equations with rational coefficients and finitely many unknowns, which we denote by S (the system S can be obtained by substituting all points of the set $X_{\mathbf{Q}}$ into P). On the other hand, since $X_{\mathbf{Q}}$ is Zariski dense in X, any polynomial of degree not exceeding k_0, whose coefficients satisfy the system S, belongs to I_{k_0}.

Since the system S contains only finitely many unknowns, there is a finite subsystem $S_0 \subset S$ equivalent to S. Since the coefficients of the finite system S_0 are rational, there is a basis E in the linear space of its solutions which consists of rational points. Now it follows that the polynomials corresponding to the elements of the basis E form a basis in I_{k_0}. Thus, we have proved that for any k_0 the linear space I_{k_0} is generated by polynomials with rational coefficients. From this and the relation $I = \bigcup_{0 \leq k_0 \leq \infty} I_{k_0}$ the assertion of the lemma follows.

Denote by End $\mathbf{R}^n$(End $\mathbf{C}^n$) the linear space of all linear transformations of the space $\mathbf{R}^n(\mathbf{C}^n)$. Let $e_1, \ldots, e_n$ be a basis in $\mathbf{R}^n(\mathbf{C}^n)$. For any i and j, $(1 \leq i \leq n, 1 \leq j \leq n)$, denote by $E_{ij} \in$ End $\mathbf{R}^n$(End $\mathbf{C}^n$) the linear transformation mapping e_i into e_j and e_k into zero for $k \neq i$. It is clear that the E_{ij} form a basis in the linear space End $\mathbf{R}^n$(End $\mathbf{C}^n$). This basis will be called "the basis in End $\mathbf{R}^n$(End $\mathbf{C}^n$) corresponding to the basis $e_1, \ldots, e_n$".

Definition 8.6. Let $H \subset GL(n, \mathbf{R})(GL(n, \mathbf{C}))$ be a quasialgebraic $\mathbf{R}$-group (algebraic $\mathbf{C}$-group), and $e_1, \ldots, e_n$ a basis in $\mathbf{R}^n(\mathbf{C}^n)$. We say that H is defined over $\mathbf{Q}$ in the basis $e_1, \ldots, e_n$ if the algebraic closure $[H]$ of the group H considered as an algebraic submanifold of the space End $\mathbf{R}^n$(End $\mathbf{C}^n$) is defined over $\mathbf{Q}$ (in the sense of Definition 8.4) in the basis of the space End $\mathbf{R}^n$(End $\mathbf{C}^n$) corresponding to the basis $e_1, \ldots, e_n$.

The element h of a quasialgebraic $\mathbf{R}$-group (algebraic $\mathbf{C}$-group)

$$H \subset GL(n, \mathbf{R})(\text{resp } GL(n, \mathbf{C}))$$

is called rational (integer) in the basis $e_1, \ldots, e_n$, if the matrix corresponding to the transformation h has rational (integer) coefficients.

Lemma 8.7. Let $H \subset GL(n, \mathbf{R})(GL(n, \mathbf{C}))$ be a quasialgebraic $\mathbf{R}$-group (algebraic $\mathbf{C}$-group) and $e_1, \ldots, e_n$ a basis in $\mathbf{R}^n(\mathbf{C}^n)$. If the set of all rational elements (in the basis $e_1, \ldots, e_n$) of the group H is Zariski dense in H, then H is defined over $\mathbf{Q}$ in the basis $e_1, \ldots, e_n$.

Proof. In view of Lemma 8.5 the group $[H]$ considered as an algebraic submanifold in the space $\mathbf{R}^n(\mathbf{C}^n)$ is defined over $\mathbf{Q}$ in the basis of the space End $\mathbf{R}^n$(End $\mathbf{C}^n$) corresponding to the basis $e_1, \ldots, e_n$. Now this is equivalent to the assertion of the lemma.

From Lemma 1.2.7 we immediately have the following:

Lemma 8.8. Let G be an algebraic $\mathbf{C}$-group ($\mathbf{R}$-group), G^0 the algebraically connected component of the identity in the group G, and $\Gamma \subset G$ a Zariski dense subgroup. If Γ_0 is any subgroup of finite index in Γ, then the algebraic closure $A(\Gamma_0)$ of the subgroup contains G^0.

By a rational character of an algebraic $\mathbf{C}$-group $G \subset GL(n, \mathbf{C})$, we mean an algebraic homomorphism $f: G \to \mathbf{C}^*$, where $\mathbf{C}^*$ is the group of non-zero complex numbers with respect to multiplication. Let G be an algebraic $\mathbf{C}$-group defined over some field K ($K \subset \mathbf{C}$). We say that the rational character f of G is defined over K, if f is realized by polynomials with coefficients in K. For an algebraic $\mathbf{C}$-group G defined over $\mathbf{Q}$, $\mathbf{Q}(G)$ denotes the group of its rational characters defined over $\mathbf{Q}$.

Let $H \subset GL(n, \mathbf{R})$ be a quasialgebraic $\mathbf{R}$-group and $e_1, \ldots, e_n$ a basis in $\mathbf{R}^n$. Denote by $H_{\mathbf{C}}$ the complexification of the group $[H]$, by $H_{\mathbf{C}}^0$ the algebraically connected component of the identity in the group $H_{\mathbf{C}}$, and by $H_{\mathbf{Z}}$ the subgroup of integer elements (in the basis $e_1, \ldots, e_n$) of the group H.

Lemma 8.9. If $H_{\mathbf{Z}}$ is Zariski dense in H, then $H_{\mathbf{Z}}$ is a lattice in H (i.e. $v(H/H_{\mathbf{Z}}) < \infty$).

Proof. Since $H_{\mathbf{Z}}$ is Zariski dense in H, in view of Lemma 8.7 H is defined over $\mathbf{Q}$ in the basis $e_1, \ldots, e_n$. Therefore, it is easy to see that $H_{\mathbf{C}}$ is also defined over $\mathbf{Q}$ in the basis $e_1, \ldots, e_n$. First we show that $\mathbf{Q}(H_{\mathbf{C}}^0) = 1$. Indeed let f be an arbitrary rational character of the group $H_{\mathbf{C}}^0$ defined over $\mathbf{Q}$. Since f is defined over $\mathbf{Q}$, it is clear that $f(H_{\mathbf{Z}} \cap H_{\mathbf{C}}^0)$ consists only of rational numbers whose denominators do not exceed some constant. Since $f(H_{\mathbf{Z}} \cap H_{\mathbf{C}}^0)$ forms a group with respect to multiplication, we have the inclusion

$$f(H_{\mathbf{Z}} \cap H_{\mathbf{C}}^0) \subset \{\pm 1\}. \tag{11}$$

Since *1)* $H_{\mathbf{Z}}$ is Zariski dense in H, *2)* H is Zariski dense in $H_{\mathbf{C}}$ (this follows from the definition of $H_{\mathbf{C}}$), *3)* $H_{\mathbf{C}}^0$ is of finite index in $H_{\mathbf{C}}$ (in view of Lemma 1.2.7), Lemma 8.8 implies that $H_{\mathbf{Z}} \cap H_{\mathbf{C}}^0$ is Zariski dense in $H_{\mathbf{C}}^0$. Therefore, since the mapping f is algebraic, $f(H_{\mathbf{Z}} \cap H_{\mathbf{C}}^0)$ is Zariski dense in $f(H_{\mathbf{C}}^0)$. This, together with (11), implies

$$f(H_{\mathbf{C}}^0) \subset \{\pm 1\}. \tag{12}$$

Now $H_{\mathbf{C}}^0$ is algebraically connected. Therefore $f(H_{\mathbf{C}}^0)$ is also algebraically connected. Thus, (12) implies

$$f(H_{\mathbf{C}}^0) = 1 \tag{13}$$

and, consequently

$$\mathbf{Q}(H_{\mathbf{C}}^0) = 1. \tag{14}$$

Since $[H]$ coincides with the subgroup of all real points of the group $H_{\mathbf{C}}$ (here $[H]$ stands for the algebraic closure of H in $GL(n, \mathbf{R})$), applying the theorems of Borel—Harish-Chandra and Mostow—Godement (see [4] Theorem 9.4) we obtain from (14) that $[H]_{\mathbf{Z}}$ is a lattice in $[H]$, where $[H]_{\mathbf{Z}}$ is the subgroup of integer elements (in the basis $e_1, \ldots, e_n$) of the group $[H]$. Since H is of finite index in $[H]$ (according to the definition of quasialgebraic $\mathbf{R}$-groups), the above statement implies the assertion of the lemma.

Lemma 8.10. If $H_{\mathbf{Z}}$ is Zariski dense and all unipotent elements of the subgroup $H_{\mathbf{Z}}$ belong to the unipotent radical of the group H, then the quotient space $H/H_{\mathbf{Z}}$ is compact.

Proof. Lemma 8.10 can be proved exactly the same way as Lemma 8.9. The only change is the application of Theorem 11.8 in [4], rather than that of Theorem 9.4 in [4].

§ 9. Some auxiliary lemmas of topological character

Lemma 9.1. Let H be a separable Lie group (in the topological sense), $N\subset H$ a closed normal subgroup, $P\subset H$ a closed subgroup, and φ the natural homomorphism $\varphi: H\to H/N$. If $\varphi(P)$ is a closed subgroup of the Lie group H/N, then the natural mapping $\varphi': N/N\cap P\to H/P$ is proper.

Proof. First we observe that any closed subgroup of a Lie group is a Lie subgroup, thus N, P and $N\cap P$ are Lie subgroups of the group H, and $\varphi(P)$ is a Lie subgroup of the group H/N. Therefore, for some (in fact arbitrary) relatively compact neighbourhood $W_0\subset P$ of the identity we have the following equality:

$$\dim\varphi(W_0) = \dim P - \dim(N\cap P). \tag{1}$$

Since H is separable, so is P, and there exists a denumerable set of numbers $h_1, \ldots, h_n, \ldots$ for which

$$P = \bigcup_{i=1}^{\infty}(h_i\cdot W_0). \tag{2}$$

Now *1)* since $\varphi(P)$ is a Lie subgroup of the group H/N, $\varphi(P)$ is a submanifold of the manifold H/N, *2)* since W_0 is relatively compact in P, it is easy to see that $\varphi(W_0)$ is a (not closed) submanifold of the manifold H/N.

Therefore, since *1)* a manifold cannot be represented as the union of a denumerable set of submanifolds of smaller dimension, *2)* for any i we have $\dim\varphi(h_i\cdot W_0)=\dim W_0$ (because $\varphi(h_i\cdot W_0)=\varphi(h_i)\cdot\varphi(W_0)$), (1) and (2) imply

$$\dim\varphi(P) = \dim\varphi(W_0) = \dim(P)-\dim(N\cap P). \tag{3}$$

Consequently, $\varphi(W_0)$ is a submanifold of full dimension in $\varphi(P)$. Whence it automatically follows that $\varphi(W_0)$ is an open subset in $\varphi(P)$.

Let $K\subset H$ be an arbitrary compact set. Since $\varphi(W_0)$ is open, there exist finitely many $p_1, \ldots, p_m\in P$, such that

$$\varphi(K)\cap\varphi(P)\subset\bigcup_{i=1}^{m}(\varphi(p_i)\cdot\varphi(W_0)) = \varphi\left(\bigcup_{i=1}^{m}(p_i\cdot W_0)\right). \tag{4}$$

Denote by $\tilde{K}$ the closure of the set $\bigcup_{i=1}^{m}(p_i\cdot W_0)$. Since W_0 is relatively compact in P, $\tilde{K}$ is compact.

For any two subsets $A, B\subset H$ we clearly have

$$(A\cdot N)\cap(B\cdot N) = \varphi^{-1}(\varphi(A)\cap\varphi(B)). \tag{5}$$

(4) and (5) imply

$$\begin{aligned}(K^{-1}\cdot P)\cap N&\subset K^{-1}\cdot(P\cap(K\cdot N))\subset K^{-1}\cdot(P\cap((P\cdot N)\cap(K\cdot N))) =\\ &= K^{-1}\cdot(P\cap\varphi^{-1}(\varphi(P)\cap\varphi(K)))\subset K^{-1}\cdot(P\cap\varphi^{-1}(\varphi(\tilde{K}))) =\\ &= K^{-1}\cdot(P\cap(\tilde{K}\cdot N))\subset K^{-1}\cdot[\tilde{K}\cdot((\tilde{K}^{-1}\cdot P)\cap N)] =\\ &= K^{-1}\cdot\tilde{K}\cdot((\tilde{K}^{-1}\cdot P)\cap N) = K^{-1}\cdot\tilde{K}\cdot(P\cap N).\end{aligned} \tag{6}$$

(The last equality holds because $\tilde{K}\subset P$.)

Let now $K' \subset H/P$ be an arbitrary compact set. Then, denoting by φ_1 and φ_2 the natural mappings $\varphi_1: H \to H/P$ and $\varphi_2: N \to N/N \cap P$, it is clear that there exists a compact set $K \subset H$ such that $\varphi_1(K^{-1}) = K'$. We have in view of (6) and formula (1) § 0 the following

$$\varphi'^{-1}(K') = \varphi_2(N \cap \varphi_1^{-1}(K')) = \varphi_2(N \cap (K^{-1} \cdot P)) \subset$$
$$\subset \varphi_2(K^{-1} \cdot \tilde{K} \cdot (P \cap N)) = \varphi_2(K^{-1} \cdot \tilde{K}). \tag{7}$$

(The last equality holds because $P \cap N$ is the kernel of the mapping φ_2.) Now since K^{-1} and $\tilde{K}$ are compact, so is $\varphi'^{-1}(K')$. The lemma is proved.

Lemma 9.2. Let H be an algebraic **R**-group, $N \subset H$ an algebraic normal subgroup, and $P \subset H$ an algebraic subgroup. Then the natural mapping $\varphi: N/N \cap P \to H/P$ is proper.

Proof. Let $H_{\mathbf{C}}$ and $N_{\mathbf{C}}$ be the complexifications of the groups H and N, and $\varphi_{\mathbf{C}}$ the natural homomorphism $H_{\mathbf{C}} \to H_{\mathbf{C}}/N_{\mathbf{C}}$. According to Lemma 1.7.1 we can introduce a structure α of an algebraic **C**-group on $H_{\mathbf{C}}/N_{\mathbf{C}}$ such that *1)* this structure is compatible with the natural Lie group structure β on $H_{\mathbf{C}}/N_{\mathbf{C}}$, *2)* $\varphi_{\mathbf{C}}$ is a regular algebraic mapping of algebraic **C**-groups. This structure α induces in a natural way a structure $\alpha_{\mathbf{R}}$ of an algebraic **R**-group on $H_{\mathbf{C}}/N_{\mathbf{C}}$.

Let us equip $H_{\mathbf{C}}$ with the natural real algebraic group structure (using the complex algebraic group structure of $H_{\mathbf{C}}$).

Since α is compatible with β, so is $\alpha_{\mathbf{R}}$.

Introduce now on $H_{\mathbf{C}}$ the natural real algebraic group structure δ (by the complex algebraic group structure given on $H_{\mathbf{C}}$). Since $\varphi_{\mathbf{C}}$ is a regular algebraic mapping of $H_{\mathbf{C}}$ to $H_{\mathbf{C}}/N_{\mathbf{C}}$ in the sense of algebraic **C**-groups, it turns out to be a regular algebraic mapping in the sense of algebraic **R**-groups (on $H_{\mathbf{C}}$ the structure δ is given and on $H_{\mathbf{C}}/N_{\mathbf{C}}$ the structure $\alpha_{\mathbf{R}}$).

It is easy to see that H is an algebraic subgroup of $H_{\mathbf{C}}$ with respect to the structure δ. It is obvious, on the other hand, that δ restricted to H, coincides with the algebraic **R**-group structure given on H. Then both P and N are algebraic subgroups of H with respect to δ. This fact and Lemma 8.3 imply — since $\varphi_{\mathbf{C}}$ is an algebraic mapping — that $\varphi_{\mathbf{C}}(P)$ is a quasi-algebraic subgroup of $H_{\mathbf{C}}$, further, since $\alpha_{\mathbf{R}}$ is compatible with β, we get that the subgroup $\varphi_{\mathbf{C}}(P)$ is closed in $H_{\mathbf{C}}$ with respect to the structure β.

$H_{\mathbf{C}}$ is a complex algebraic group and α is compatible with β, therefore $H_{\mathbf{C}}$ is a separable Lie group in the structure β. Since $\varphi_{\mathbf{C}}(P)$ is β-closed in $H_{\mathbf{C}}$, this implies in view of Lemma 9.1, that the mapping

$$\varphi'_{\mathbf{C}}: N/N \cap P \to H_{\mathbf{C}}/P$$

is proper (P and N are topologically closed in $H_{\mathbf{C}}$ since they are algebraic with respect to δ). This statement is equivalent to the following: to every compact set

$K \subset H_{\mathbf{C}}$ there exists a compact set $\hat{K} \subset N$ such that

$$N \cap (K \cdot P) \subset \hat{K} \cdot (N \cap P), \tag{8}$$

in particular, for every compact set $K \subset H$. Now this means that φ' is proper. The lemma is proved.

Lemma 9.3. Let H be an algebraic **R**-group, V an algebraic unipotent normal subgroup of the group H, $P \subset H$ an algebraic subgroup, and φ the natural mapping $\varphi: H \to H/P$. If for any one-parameter subgroup $v(t) \subset V$, $\varphi(v(t))$ does not tend to infinity in H/P as $t \to +\infty$, then P contains V.

Proof. Suppose, on the contrary, that $P \not\supset V$. Then

$$P \cap V \neq P. \tag{9}$$

By assumption V is an algebraic normal subgroup of the group H. Since P is also an algebraic subgroup of the group H, $P \cap V$ is an algebraic subgroup of the group V. Therefore, it follows from (9) and Lemma 5.5 that there exists a one-parameter subgroup $v(t) \subset V$ such that $f(v(t))$ tends to infinity in $V/P \cap V$ as $t \to +\infty$, where f denotes the natural mapping $f: V \to V/P \cap V$. Then, Lemma 9.2 implies that $\varphi(v(t))$ tends to infinity in H/P as $t \to +\infty$, which is a contradiction. The lemma is proved.

Lemma 9.4. Let H be a Lie group, $Q \subset H$ a subgroup, and φ the natural mapping $\varphi: H \to H/Q$. If Q is finite then the mapping φ is proper.

Proof. Let $K \subset H/Q$ be an arbitrary compact set, and $\tilde{K} \subset H$ a compact set for which $\varphi(\tilde{K}) \subset K$ (the existence of $\tilde{K}$ is clear). It is easy to see that $\varphi^{-1}(K) \subset \tilde{K} \cdot Q$ and since $\tilde{K}$ is compact and Q is a finite subgroup, $\varphi^{-1}(K)$ is also compact. The lemma is proved.

Lemma 9.5. Let H be an algebraic **R**-group, and $u(t) \subset H$ a non-trivial one-parameter unipotent subgroup. Then $u(t)$ tends to infinity in H as $t \to \infty$.

Proof. Since H is an algebraic **R**-group, we can assume that for some n, H is an algebraic subgroup of the group $GL(n, \mathbf{R})$ and $u(t)$ is a one-parameter unipotent subgroup in $GL(n, \mathbf{R})$. Then it is easy to see that $u(t)$ tends to infinity in $GL(n, \mathbf{R})$ as $t \to \infty$. Now, since the topology on H is induced by the topology on $GL(n, \mathbf{R})$, the assertion of the lemma follows.

The following lemma is well known [17].

Lemma 9.6. Let H be an algebraic **R**-group (**C**-group), $V \subset H$ the unipotent radical of the group H, and ψ the natural mapping $\psi: H \to H/V$. Then one can introduce a structure of an algebraic **R**-group (**C**-group) on H/V such that *1)* this structure is compatible with the natural Lie group structure of H/V, *2)* ψ is a regular algebraic mapping. Moreover there exists in H a reductive algebraic subgroup F such that *1)* the restriction of ψ to F is an isomorphism of algebraic **R**-groups, *2)* H is a semi-direct product of F and V (in the sense of IV, § 1).

Remark 9.7. An algebraic subgroup F of the group H having properties *1)* and *2)* (which are equivalent, as is easy to show) is called a *Levi subgroup* of the group H. It is well known that *1)* any two Levi subgroups of the group H are conjugate to each other by some element of V, *2)* any semisimple element of the group H belongs to some Levi subgroup.

Lemma 9.8. Let H, V, P and φ be the same as in Lemma 9.3. If *1)* H contains unipotent elements not lying in V, *2)* for any one-parameter unipotent subgroup $u(t)\subset H$, $\varphi(u(t))$ does not tend to infinity in H/P as $t\to+\infty$, than $\dim P>\dim V$.

Proof. Let $u_0\in H$ be a unipotent element not lying in V. Denote by U_0 the subgroup generated by the element u_0, and by $\overline{U}_0$ the Zariski closure of U_0 in H. In view of Lemma 1.3.3, $\overline{U}_0$ is a unipotent algebraic subgroup of the group H. Therefore, as a consequence of Lemma 1.3.1 and *1)* of Lemma 5.1, there exists a one-parameter unipotent subgroup $u(t)\subset\overline{U}_0$, for which $u(1)=u_0$. Note that $u_0\notin\overline{U}$ and $u_0\in u(t)$ imply

$$u(t)\not\subset\overline{U}. \tag{10}$$

Denote by ψ the natural mapping $\psi: H\to H/V$. According to Lemma 9.6, ψ is an algebraic mapping. Therefore, (10) and Lemma 1.3.4 imply that $\psi(u(t))$ is a non-trivial one-parameter unipotent subgroup in H/V. Thus, applying Lemma 9.5 we get

$$\psi(u(t)) \quad \text{tends to infinity in} \quad H/V \quad \text{as} \quad t\to\infty. \tag{11}$$

Now assume $\dim P=\dim V$. Then, in view of Lemma 1.2.10, $\overline{U}$ is a subgroup of finite index in $\tilde{A}(U)$, and consequently, $\psi(P)$ is a finite subgroup. Therefore, according to Lemma 9.4, (11) implies that

$$\varphi(u(t)) \quad \text{tends to infinity in} \quad H/P \quad \text{as} \quad t\to\infty. \tag{12}$$

Now, since $u(t)$ is unipotent, (12) contradicts the condition of the lemma. Consequently $\dim P>\dim V$. The lemma is proved.

Lemma 9.9. Let H be a Lie group, Γ_1 and Γ_2 discrete subgroups of the group H, and ψ the natural mapping $\psi\colon H\to H/\Gamma_2$. If $\Gamma_1\supset\Gamma_2$ and $\psi(\Gamma_1)$ is relatively compact in Γ_2, then the set Γ_1/Γ_2 is finite.

Proof. Since $\psi(\Gamma_1)$ is relatively compact in R/Γ_2, we have for some compact $\tilde{K}\subset R/\Gamma_2$ the inclusion

$$\psi(\Gamma_1)\subset\tilde{K}. \tag{13}$$

Let $K\subset R$ be a compact set for which $\psi(K)=\tilde{K}$ (the existence of K is clear). Then, (13) implies

$$\Gamma_1\subset K\cdot\Gamma_2. \tag{14}$$

Denote by F the set $K\cap\Gamma_1$. Since the subgroup Γ_1 is discrete and K is compact, the set F is finite. Let now γ be an arbitrary element of the subgroup Γ_1. By (14) $\gamma=k\cdot\tilde{\gamma}$, where $k\in K$ and $\tilde{\gamma}\in\Gamma_2$. Then *1)* $\gamma\tilde{\gamma}^{-1}\in K$, *2)* since $\Gamma_2\subset\Gamma_1$ we have $\gamma\tilde{\gamma}^{-1}\in\Gamma_1$. Therefore

$$\gamma\tilde{\gamma}^{-1}\in F \quad\text{or}\quad \gamma\in F\cdot\Gamma_2. \tag{15}$$

Since $\gamma\in\Gamma_1$ was arbitrary, (15) and the finiteness of the set F imply the assertion of the lemma. The lemma is proved.

§ 10. Some auxiliary statements of algebraic character

Lemma 10.1. Let H be a complex Lie group, $\mathfrak{H}$ its Lie algebra and $h\in H$ an arbitrary element. Consider the root decomposition of $\mathfrak{H}$ with respect to the element h:

$$\mathfrak{H}=\mathfrak{H}_1\oplus(\bigoplus_{\lambda\neq 1}\mathfrak{H}_\lambda) \tag{1}$$

($\mathfrak{H}_\lambda$ consists of all $\mathfrak{h}\in\mathfrak{H}$, for which $(\operatorname{Ad} h-\lambda E)^N\mathfrak{h}=0$ for some natural number N) and denote by $\hat{\mathfrak{H}}\subset\mathfrak{H}$ the subalgebra generated by $\bigcup_{\lambda\neq 1}\mathfrak{H}_\lambda$. Then $\hat{\mathfrak{H}}$ is an ideal in $\mathfrak{H}$.

Proof. In view of (1) it is sufficient to show that for any λ we have $[\mathfrak{H}_\lambda,\hat{\mathfrak{H}}]\subset\hat{\mathfrak{H}}$. If $\lambda\neq 1$, then $[\mathfrak{H}_\lambda,\hat{\mathfrak{H}}]\subset\hat{\mathfrak{H}}$, since $\mathfrak{H}_\lambda\subset\hat{\mathfrak{H}}$, and $\hat{\mathfrak{H}}$ is a subalgebra. And if $\lambda=1$, then $[\mathfrak{H}_1,\hat{\mathfrak{H}}]\subset\hat{\mathfrak{H}}$, since *1)* $[\mathfrak{H}_1,\mathfrak{H}_\lambda]\subset\mathfrak{H}_\lambda$ (because $[\mathfrak{H}_{\lambda_1},\mathfrak{H}_{\lambda_2}]\subset\mathfrak{H}_{\lambda_1\lambda_2}$, *2)* $\bigcup_{\lambda\neq 1}\mathfrak{H}_\lambda$ generates $\hat{\mathfrak{H}}$. The proof is complete.

The following lemma is well known.

Lemma 10.2. Let H be an algebraically connected reductive complex (real) algebraic group. Then the centralizer of any reductive algebraic subgroup of the group H, is a reductive algebraic subgroup.

Lemma 10.3. Let H be an algebraically connected complex (real) semisimple algebraic group, $V\subset H$ a non-trivial unipotent algebraic subgroup. Denote by W the unipotent radical of the centralizer $C(V)$ of the subgroup V in the group H, and by $C(W)$ the centralizer of W in H. Let $t\in H$ be a semisimple element in the intersection $C(V)\cap C(W)$. If V is contained in no non-trivial algebraic normal subgroup of the group H, then t belongs to the centre of the group H.

Proof. First we prove the lemma in case H is a complex group. By assumption $t\in C(V)$. Then, since t is semisimple, in view of Remark 9.7 there exists in $C(V)$ a Levi subgroup N containing t. Denote by $\mathfrak{N}$ the Lie algebra of the group N. Since $t\in N$, we have

$$(\operatorname{Ad} t)\mathfrak{N}\subset\mathfrak{N}. \tag{2}$$

Consider the root decomposition of the Lie algebra $\mathfrak{H}$ in the group H with respect to the element t (which exists since H is a complex group)

$$\mathfrak{H} = \mathfrak{H}_1 \oplus (\bigoplus_{\lambda \neq 1} \mathfrak{H}_\lambda), \tag{3}$$

and for any λ denote by $\mathfrak{C}_\lambda$ the intersection of $\mathfrak{H}_\lambda$ with the Lie algebra $\mathfrak{C}(V)$ of the group $C(V)$. Since $t \in C(V)$, we have the decomposition

$$\mathfrak{C}(V) = \mathfrak{C}_1 \oplus (\bigoplus_{\lambda \neq 1} \mathfrak{C}_\lambda). \tag{4}$$

Since $t \in C(W)$, we have the inclusion

$$\mathfrak{W} \subset \mathfrak{H}_1, \tag{5}$$

where $\mathfrak{W}$ denotes the Lie algebra of the Lie group W. Since N is a Levi subgroup in $C(V)$, and W is the unipotent radical of $C(V)$, we have

$$\mathfrak{C}(V) = \mathfrak{N} \oplus \mathfrak{W}. \tag{6}$$

As an easy consequence of (2), (4), (5) and (6)

$$\mathfrak{C}_\lambda \subset \mathfrak{N} \quad \text{for any} \quad \lambda \neq 1. \tag{7}$$

Let $C(N)$ denote the centralizer of N in H. Since $t \in N$, we have $C(N) \subset C(t)$, where $C(t)$ is the centralizer of t in H. Therefore $\operatorname{Ad} C(N)$ commutes with $\operatorname{Ad} t$. From this for any λ we get

$$(\operatorname{Ad} C(N))\mathfrak{H}_\lambda = \mathfrak{H}_\lambda. \tag{8}$$

It follows from (7) that for any $c \in C(N)$, any $\lambda \neq 1$ and any $\mathfrak{h} \in \mathfrak{C}_\lambda$ we have $(\operatorname{Ad} c)\mathfrak{h} = \mathfrak{h}$. Therefore, we have for any $\lambda \neq 1$

$$(\operatorname{Ad} C(N))\mathfrak{C}_\lambda = \mathfrak{C}_\lambda. \tag{9}$$

Since N is a Levi subgroup, by definition, it is a reductive algebraic subgroup. Therefore, in view of Lemma 10.2 $C(N)$ is also a reductive algebraic subgroup. On the other hand, in view of Lemma 1.2 the adjoint representation Ad of the group H is algebraic. Since $C(N)$ is reductive and Ad is algebraic, according to Lemma 1.4.4 the restriction of Ad to $C(N)$ is completely reducible. Therefore, it follows from (8) and (9) that there exists, for any $\lambda \neq 1$, a subspace $\hat{\mathfrak{C}}_\lambda \subset \mathfrak{H}_\lambda$ such that

$$\mathfrak{H}_\lambda = \hat{\mathfrak{C}}_\lambda \otimes \mathfrak{C}_\lambda \quad \text{and} \quad (\operatorname{Ad} C(N))\hat{\mathfrak{C}}_\lambda = \hat{\mathfrak{C}}_\lambda. \tag{10}$$

Since $N \subset C(V)$, we have $V \subset C(N)$. Therefore (10) implies

$$(\operatorname{Ad} V)\hat{\mathfrak{C}}_\lambda = \hat{\mathfrak{C}}_\lambda \quad \text{for any} \quad \lambda \neq 1. \tag{11}$$

Now V is a unipotent group. Thus, in view of Lemmas 1.2.1 and 1.3.4 $\operatorname{Ad} V$ is also unipotent. Assume now $\mathfrak{C}_{\lambda_0} \neq \mathfrak{H}_{\lambda_0}$ for some $\lambda_0 \neq 1$. Then $\hat{\mathfrak{H}}_{\lambda_0} \neq 0$. Since the group $\operatorname{Ad} V$ is unipotent, in view of (11) and Lemma 1.3.9 there exists a non-zero linear subspace $\mathfrak{C}'_{\lambda_0} \subset \hat{\mathfrak{C}}_{\lambda_0}$ such that $(\operatorname{Ad} v)\mathfrak{c} \subset \mathfrak{c}$ for any $v \in V$ and $\mathfrak{c} \in \mathfrak{C}'_{\lambda_0}$.

Then, it is easy to see that $\mathfrak{C}'_\lambda \subset \hat{\mathfrak{C}}_{\lambda_0}(V)$, which is impossible since *1)* $\mathfrak{C}'_{\lambda_0} \subset \hat{\mathfrak{C}}_{\lambda_0}$, *2)* $\hat{\mathfrak{C}}_{\lambda_0} \cap \mathfrak{C}_{\lambda_0} = 0$ (in view of (10)), *3)* $\mathfrak{C}'_{\lambda_0} \neq 0$, *4)* $\mathfrak{C}_{\lambda_0} = \mathfrak{H}_{\lambda_0} \cap \mathfrak{C}(V)$ (by the definition of $\mathfrak{C}_\lambda$). Thus the assumption that $\mathfrak{C}_{\lambda_0} \neq \mathfrak{H}_{\lambda_0}$ holds for some $\lambda_0 \neq 1$ led to a contradiction. Consequently

$$\mathfrak{C}_\lambda = \mathfrak{H}_\lambda \quad \text{for every} \quad \lambda \neq 1. \tag{12}$$

Assume now that t does not belong to the centre of the group H. Then, for some $\lambda \neq 1$, $\mathfrak{H}_\lambda \neq 0$ and consequently

$$\hat{\mathfrak{H}} \neq 0, \tag{13}$$

where $\hat{\mathfrak{H}} \subset \mathfrak{H}$ is the subalgebra generated by $\bigcup\limits_{\lambda \neq 1} \mathfrak{H}_\lambda$. On the other hand, we have from (4) and (12) that

$$\mathfrak{C}(V) \supset \hat{\mathfrak{H}}. \tag{14}$$

In view of Lemma 10.1 $\hat{\mathfrak{H}}$ is an ideal in $\mathfrak{H}$. Therefore, (13) and (14) imply that $\mathfrak{C}(V)$ contains a nontrivial ideal $\hat{\mathfrak{H}}$. Note that since $\mathfrak{C}(V) \supset \hat{\mathfrak{H}}$, we have

$$V \subset C(\hat{\mathfrak{H}}), \tag{15}$$

where $C(\hat{\mathfrak{H}})$ is the centralizer of $\hat{\mathfrak{H}}$ in H (i.e. the set of all $h \in H$ for which $(\mathrm{Ad}\, h)\mathfrak{c} = \mathfrak{c}$ for any $\mathfrak{c} \in \hat{\mathfrak{H}}$).

Since H is algebraically connected and $\hat{\mathfrak{H}}$ is an ideal in $\mathfrak{H}$, $(\mathrm{Ad}\, H)\hat{\mathfrak{H}} = \hat{\mathfrak{H}}$. Now it is easy to see that in this case $C(\hat{\mathfrak{H}})$ is an algebraic normal subgroup in H ($C(\hat{\mathfrak{H}})$ is algebraic, since the adjoint representation of the group H is algebraic). On the other hand, since H is semisimple, (13) implies that

$$C(\hat{\mathfrak{H}}) \neq H. \tag{16}$$

Since *1)* $C(\hat{\mathfrak{H}})$ is an algebraic normal subgroup in H, *2)* $V \neq e$ (by assumption), (15) and (16) imply that V is contained in the nontrivial algebraic normal subgroup $C(\hat{\mathfrak{H}})$ of the group H, which contradicts the condition of the lemma. Thus, the assumption t does not belong to the centre of the group H, has led to a contradiction. Consequently the lemma is proved in the complex case.

Consider now the case when H is a real group. Let $H_{\mathbf{C}}$, $V_{\mathbf{C}}$ and $W_{\mathbf{C}}$ be the complexifications of the groups H, V and W, $C(V_{\mathbf{C}})$ the centralizer of $V_{\mathbf{C}}$ in $H_{\mathbf{C}}$, and $\tilde{W}$ the unipotent radical of the group $C(V_{\mathbf{C}})$. Since $V_{\mathbf{C}}$ is the complexification of V, $V_{\mathbf{C}}$ is defined over $\mathbf{R}$. Therefore, in view of Lemma 1.2.13, $C(V_{\mathbf{C}})$ is defined over $\mathbf{R}$. From this, according to Lemma 1.3.5 it follows, that $\tilde{W}$ is defined over $\mathbf{R}$. Therefore, the subgroup $\tilde{W}_{\mathbf{R}} \subset \tilde{W}$ of real points of the group $\tilde{W}$ is Zariski dense in $\tilde{W}$ in view of Lemma 1.2.12, since $\tilde{W}$ is connected (in view of Lemma 1.3.2). On the other hand *1)* since $V_{\mathbf{C}}$ is the complexification of V, we have $C(V_{\mathbf{C}}) = C_{\mathbf{C}}(V)$ where $C_{\mathbf{C}}(V)$ is the centralizer of V in $H_{\mathbf{C}}$, *2)* it is clear that $C_{\mathbf{C}}(V) \cap H = C(V)$, *3)* it is also clear that $\tilde{W}_{\mathbf{R}} = \tilde{W} \cap H$. From this we get that $\tilde{W}_{\mathbf{R}}$ is a unipotent normal subgroup in $C(V)$. Therefore

$$\tilde{W}_{\mathbf{R}} \subset W \tag{17}$$

(and further $\tilde{W}_{\mathbf{R}} = W$). Since $\tilde{W}_{\mathbf{R}}$ is algebraically dense in $\tilde{W}$, from (17) it follows that $W \cap \tilde{W}$ is algebraically dense in $\tilde{W}$. Therefore

$$\text{if} \quad t \in C(W), \quad \text{then} \quad t \in C(\tilde{W}), \tag{18}$$

where $C(\tilde{W})$ is the centralizer of $\tilde{W}$ in $H_{\mathbf{C}}$. On the other hand, $V_{\mathbf{C}}$ belongs to no algebraic normal subgroup Q of the group $H_{\mathbf{C}}$, since in the opposite case the intersection $Q \cap H$ would be a nontrivial algebraic normal subgroup in H ($Q \cap H$ is nontrivial because *1)* $Q \supset V_{\mathbf{C}} \supset V \neq e$ *2)* $H \cap Q \neq H$, since $Q \neq H_{\mathbf{C}}$ and H is Zariski dense in $H_{\mathbf{C}}$). This fact and (18) imply, since the assertion of the lemma has already been proved in the complex case, that the assertion of the lemma holds for H. The proof of the lemma is complete.

Remark. Lemma 10.3 can also be deduced easily from Lemma 1.6 of the work [6] of Garland and Raghunathan.

In the sequel we will make use of, though not always refer to, the following well-known

Lemma 10.4. Let H be a semisimple real algebraic group, V an arbitrary maximal unipotent subgroup in H, $T \subset H$ an arbitrary maximal **R**-split torus of the group H, normalizing V, Σ the root system of H corresponding to T. For any $\alpha \in \Sigma$, $\mathfrak{U}_\alpha$ denotes the root subspace corresponding to the root α, and $U_\alpha \subset H$ the set $\exp(\mathfrak{U}_\alpha)$. Then, in Σ there is a subset Σ^+ such that *1)* $\alpha \in \Sigma^+$ if and only if $U_\alpha \subset V$, *2)* for any $\alpha \in \Sigma$ we have either $\alpha \in \Sigma^+$ or $-\alpha \in \Sigma^+$, *3)* if $\alpha_1 \in \Sigma^+$ and $\alpha_2 \in \Sigma^+$, then $\alpha_1 + \alpha_2 \in \Sigma^+$.

The set Σ^+ described in Lemma 10.4 is called the set of positive roots corresponding to V.

Lemma 10.5. We preserve the notation of Lemma 10.4. The root $\tilde{\alpha} \in \Sigma^+$ is called a highest root if $\tilde{\alpha} + \beta$ is not a root for any $\beta \in \Sigma^+$. Denote by $Z(V)$ the centre of the group V and by $\hat{\Sigma}$ the set of all highest roots in Σ^+. Then *1)* $Z(V)$ coincides with the subgroup generated by $\bigcup_{\tilde{\alpha} \in \hat{\Sigma}} U_{\tilde{\alpha}}$, *2)* for any $\tilde{\alpha} \in \hat{\Sigma}$, $U_{\tilde{\alpha}}$ is a subgroup, *3)* if H is simple then the highest root is unique.

Remark. Statements *2)* and *3)* of Lemma 10.5 are well known, *1)* will be proved in § 18.

Lemma 10.6. Let H be a simple real algebraic group, V an arbitrary maximal unipotent subgroup of the group H, $Z(V)$ the centre of the group V, $C(Z(V))$ the centralizer of $Z(V)$ in H and finally T a maximal **R**-split torus of the group H. If all unipotent elements of the subgroup $C(Z(V))$ belong to V, then either *1)* $\dim T = 1$, i.e., the real rank of the group is equal to 1, or *2)* $\dim T = 2$ and the root system of the group H is of the type A_2 with respect to the torus T.

First we mention that we have not assumed the group H to be absolutely simple.*

Since *1)* for any maximal unipotent subgroup V of the group H there is a maximal $\mathbf{R}$-split torus normalizing V, *2)* any two maximal $\mathbf{R}$-split tori of the group H, are conjugate to each other by some element of H, thus we can assume without loss of generality that T normalizes V. Let Σ be the system of roots of the group H with respect to the torus T, $\Sigma^+ \subset \Sigma$ the set of positive roots corresponding to V. Write $\tilde{\alpha} \in \Sigma^+$ for the highest root in Σ^+ ($\tilde{\alpha}$ is unique because H is simple).

In view of Lemma 10.5 the centre $Z(V)$ of the maximal unipotent subgroup V coincides with the root subgroup $U_{\tilde{\alpha}}$ corresponding to $\tilde{\alpha}$ ($U_{\tilde{\alpha}}$ is a subgroup since $\tilde{\alpha}$ is a highest root). Therefore, since $(C(Z(V)))^u \subset V$ by the condition of the lemma, it is easily seen that for any root $\alpha \in \Sigma^- = \Sigma - \Sigma^+$ for which $-\alpha \neq \tilde{\alpha}$, $\alpha + \tilde{\alpha}$ is a root. Now, since the length of the root $\tilde{\alpha}$ is not less than the length of any other root $\alpha \in \Sigma$ (because $\tilde{\alpha}$ is a highest root), it follows that $\tilde{\alpha}$ is not orthogonal to any root $\alpha \in \Sigma$. Now we show that in this case for any $\alpha \in \Sigma^+$ not proportional to $\tilde{\alpha}$, the angle $\widehat{(\tilde{\alpha}, \alpha)}$ between the roots $\tilde{\alpha}$ and α (in the usual sense) is equal to $\pi/3$. Indeed, $\tilde{\alpha}$ is a highest root, and so the angle of $\tilde{\alpha}$ with any $\alpha \in \Sigma^+$ is $\leqq \pi/2$. Therefore, if there is a root $\alpha_0 \in \Sigma^+$ not proportional to $\tilde{\alpha}$, such that $\widehat{(\tilde{\alpha}, \alpha_0)} \neq \pi/3$ then we have three possibilities:

1) $\widehat{(\tilde{\alpha}, \alpha_0)} = \dfrac{\pi}{2}$

2) $\widehat{(\tilde{\alpha}, \alpha_0)} = \dfrac{\pi}{4}, \quad \|\tilde{\alpha}\| = \sqrt{2}\, \|\alpha_0\|$

3) $\widehat{(\tilde{\alpha}, \alpha_0)} = \dfrac{\pi}{6}, \quad \|\tilde{\alpha}\| = \sqrt{3}\, \|\alpha_0\|,$

where $\|\alpha\|$ denotes the length of the root α. Now we show that in any of these cases there is a root $\alpha' \in \Sigma^+$ orthogonal to $\tilde{\alpha}$. Indeed, in case *1)* $\alpha' = \alpha_0$, in case *2)* $\alpha' = \tilde{\alpha} - 2\alpha_0$, in case *3)* $\alpha' = \tilde{\alpha} - 2\alpha_0$ (that α' is a root in all the three cases follows from the properties of a root system). Thus we have shown that in our case the angle $\widehat{(\tilde{\alpha}, \alpha)}$ between $\tilde{\alpha}$ and any $\alpha \in \Sigma^+$ not proportional to α, is equal to $\pi/3$. Now it is easy to see then that all roots $\alpha \in \Sigma$ are of the same length, in fact of the length of $\tilde{\alpha}$.

Now we show that in our case the following situation cannot occur: $\alpha_1 \in \Sigma^+$, $\alpha_2 \in \Sigma^+$, $\alpha_1 + \alpha_2$ is a root, $\alpha_1 \neq k\alpha_2$, $\alpha_1 \neq k\tilde{\alpha}$, $\alpha_2 \neq k\tilde{\alpha}$, $\alpha_1 + \alpha_2 \neq \tilde{\alpha}$. Let us assume that the contrary holds. Then, as shown already, all roots are of the same

* A real algebraic group is said to be *absolutely simple* if its complexification is simple.

length, thus $\alpha_1+\alpha_2\neq\tilde{\alpha}$ implies that the root $\alpha_1+\alpha_2$ is not proportional to $\tilde{\alpha}$. Therefore

$$(\widehat{\tilde{\alpha}, \alpha_1+\alpha_2}) = \frac{\pi}{3}. \tag{19}$$

Similarly, since α_1 and α_2 are not proportional to $\tilde{\alpha}$, we have

$$(\widehat{\tilde{\alpha}, \alpha_1}) = (\widehat{\tilde{\alpha}, \alpha_2}) = \frac{\pi}{3}. \tag{20}$$

Now an easy elementary geometric argument shows that (19) and (20) cannot hold simultaneously if α_1 and α_2 are not proportional. We now have a contradiction, i.e., we have shown that if the sum of any two of the roots $\alpha_1, \alpha_2 \in \Sigma^+$ not proportional to $\tilde{\alpha}$ and each other is a root, then $\alpha_1+\alpha_2=\tilde{\alpha}$.

Assume now that in the conditions of the lemma we have dim $T\neq 1$. In this case, since H is simple, in Σ^+ there are two different simple roots α_1 and α_2 such that $\alpha_1+\alpha_2$ is a root. It follows from the above that $\tilde{\alpha}=\alpha_1+\alpha_2$, i.e. the highest root is the sum of two simple roots. Consequently, in this case the system Σ is of type A_2. The lemma is proved.

Lemma 10.7. Let H be an algebraically connected semisimple real algebraic group, such that the centre of its complexification is trivial*, and let V_1 and V_2 be maximal unipotent subgroups of the group H. Denote by $Z(V_1)$ the centre of the group V_1, by $C(Z(V_1))$ the centralizer of $Z(V_1)$ in H, by $A\subset V_1$ the minimal algebraic subgroup containing $(V_1\cap V_2)\cup Z(V_1)$. Then, if *1)* all unipotent elements of the group $C(Z(V_1))$ belong to V_1, *2)* $V_1\cap V_2\neq e$, *3)* $V_1\neq V_2$, *4)* $V_1\cap V_2$ does not belong to any nontrivial algebraic normal subgroup of H, then we have $A\neq V_1$ and A is an orispheric unipotent subgroup of the group H.

Proof. First we prove the lemma in case H is simple. Since $(C(Z(V_1)))^u\subset V_1$ (by assumption), in view of Lemma 10.6 the following two cases are possible:

a) the real rank of the group H is equal to 1,

b) the root system of H corresponding to an arbitrary maximal **R**-split torus, is of type A_2.

It is easy to see that in any simple real algebraic group of real rank 1 any two maximal unipotent subgroups with nontrivial intersection coincide. (This can be deduced easily from Lemma 10.4 and Lemmas 10.17 and 10.16). Therefore, since we have $V_1\cap V_2\neq e$ and $V_1\neq V_2$, case a) is impossible. Now we prove the lemma in case b). In view of Lemmas 1.8.2 and 1.8.3 in H there is a maximal **R**-split torus normalizing both V_1 and V_2. Let Σ be the system of roots of the group H with respect to T, Σ_1^+ the set of positive roots corresponding to the subgroup V_1, and Σ_2^+ the set of positive roots corresponding to the subgroup V_2.

* The condition that the centre of the complexification is trivial, is not essential.

Since we are considering case b), Σ is of type A_2. Therefore in Σ_1^+ there are three roots: α_1, α_2 and $\alpha_1+\alpha_2$.

Since $V_1\cap V_2\neq e$ and $V_1\neq V_2$, we have $\Sigma_1^+\neq\Sigma_1^+\cap\Sigma_2^+\neq\emptyset$. Therefore, it is easy to see that we have the following four possibilities:

1) $\Sigma_1^+\cap\Sigma_2^+ = \{\alpha_1\}$,

2) $\Sigma_1^+\cap\Sigma_2^+ = \{\alpha_2\}$,

3) $\Sigma_1^+\cap\Sigma_2^+ = \{\alpha_1, \alpha_1+\alpha_2\}$,

4) $\Sigma_1^+\cap\Sigma_2^+ = \{\alpha_2, \alpha_1+\alpha_2\}$,

Before analysing these cases, we mention that in view of Lemma 10.5 $Z(V_1)=$ $=U_{\alpha_1+\alpha_2}$. In case *1)* $V_1\cap V_2=U_{\alpha_1}$. Therefore, it is easy to see that in this case A coincides with the subgroup $\tilde{B}$ generated by $U_{\alpha_1}\cup U_{\alpha_1+\alpha_2}$. Now $\tilde{B}$ is easily seen to be orispheric. In case *3)* A also equals $\tilde{B}$. Cases 2 and 4 can be analysed similarly to cases *1)* and *3)*. Thus the lemma is proved for simple groups.

Consider the general case. Since H is an algebraically connected semisimple real algebraic group with no centre in the complexification, according to Lemma 1.4.2 H decomposes into the direct product (in the sense of § 1) of simple algebraic **R**-groups: $H=\bigoplus_{i=1}^{n} H_i$. Since V_1 and V_2 are maximal unipotent subgroups in H, it is easy to see that $V_1=\bigoplus_{i=1}^{n} V_1^i$ and $V_2=\bigoplus_{i=1}^{n} V_2^i$, where $V_1^i=V_1\cap H_i$ and $V_2^i=V_2\cap H_i$. Denote by A^i the minimal algebraic subgroup in H_i containing $(V_1^i\cap V_2^i)\cup$ $\cup(Z(V_1^i)\cap H_i)$. It is easy to see that

$$A = \bigoplus_{i=1}^{n} A^i \tag{21}$$

Since $V_1\cap V_2$ contains no nontrivial algebraic normal subgroup of the group H, for any i $(1\leqq i\leqq n)$ we have

$$V_1^i\cap V_2^i \neq e \tag{22}$$

Since $V_1\neq V_2$, we have $V_1^{i_0}\neq V_2^{i_0}$ for some i_0 and since we have the lemma for simple groups, we obtain $A^{i_0}\neq V_1^{i_0}$. This and (21) imply

$$A \neq V_1. \tag{23}$$

Let us divide the interval of natural numbers from 1 to n into two sets B_1 and B_2 in the following way: $i\in B_1$ if $V_1^i=V_2^i$ and $i\in B_2$ if $V_1^i\neq V_2^i$. If $i\in B_1$, then A^i is orispheric since in this case A^i coincides with V_1^i. Now if $i\in B_2$, then A^i is orispheric in view of (22) and the fact that the lemma has already been proved for simple groups. Now this and (21) imply that A is orispheric. Therefore, in view of (23) the lemma is proved for all H.

The following lemma is well known.

Lemma 10.8. Let F be an arbitrary real (complex) algebraic group, and W the unipotent radical of the group F. Then *1)* W belongs to all maximal unipotent subgroups of the group F, *2)* in F there are two maximal unipotent subgroups V_1 and V_2 for which $V_1 \cap V_2 = W$.

In the sequel of this §, H will denote an algebraically connected real semisimple algebraic group. Before going into further discussion we mention that the results quoted below have been proved by their authors not only for real groups.

B. Ju. Weissfeiler (see [7]) proved the following

Lemma 10.9. An algebraic unipotent subgroup V of a group H is orispheric if and only if V is the unipotent radical of some parabolic subgroup of the group H.

Two parabolic subgroups of the group H are called *opposite* (or opposing each other) if their intersection is a Levi subgroup in both of them. For any parabolic subgroup P we denote by $R(P)$ the solvable radical (i.e. the maximal algebraically connected solvable algebraic normal subgroup) of the group P, and by $R_u(P)$ the unipotent radical of the group P. Now we formulate some results in connection with H which can be found basically in [3] and [5] of A. Borel and J. Tits in the form of the following lemmas:

Lemma 10.10. For any parabolic subgroup P of the group H there always exists a parabolic subgroup P' opposing to P.

Lemma 10.11. Any parabolic subgroup of the group H coincides with its normalizer.

Lemma 10.12. For any minimal parabolic subgroup P of the group H there exists a minimal parabolic subgroup P' opposing P. Furthermore, any parabolic subgroup opposing a minimal parabolic subgroup, is minimal.

Lemma 10.13. If P_1 and P_2 are opposite parabolic subgroups, then the set of all $h \in H$ such that P_1 and hP_2h^{-1} are opposite, is Zariski dense and open in H.

Lemma 10.14. Two parabolic subgroups P_1 and P_2 of the group H are opposite if and only if

$$R_u(P_1) \cap P_2 = R_u(P_2) \cap P_1 = e.$$

Lemma 10.16. The intersection of any two parabolic subgroups of the group H contains the centralizer of some maximal $\mathbf{R}$-split torus of the group H.

Lemma 10.17. The normalizer of any maximal unipotent subgroup of the group H is a minimal parabolic subgroup of the group H, and conversely, the unipotent radical of any minimal parabolic subgroup of the group H is a maximal unipotent subgroup of H.

Lemma 10.18. Let P be an arbitrary minimal parabolic subgroup of the group H. Then the centralizer of any maximal $\mathbf{R}$-split torus of H contained in P, is a Levi subgroup in P, and conversely, any Levi subgroup in P is the centralizer of some maximal $\mathbf{R}$-split torus contained in P.

Lemma 10.19. Any two minimal parabolic subgroups of the group H are conjugate by an element of H.

Lemma 10.20. Any two maximal unipotent subgroups of an arbitrary real algebraic group F are conjugate by an element of F.

Lemma 10.21. Any parabolic subgroup of H coincides with the normalizer of its unipotent radical.

Lemma 10.22. Let P be an arbitrary parabolic subgroup in H, P_1 and P_2 parabolic subgroups in H opposing P. Then *1)* P_1 and P_2 are conjugate by a unique element of $R_u(P)$, *2)* if $P \cap P_1 = P \cap P_2$, then $P_1 = P_2$.

Lemma 10.23. Let F be an arbitrary real algebraic group, and $N \subset F$ an algebraic normal subgroup. If N contains some maximal unipotent subgroup of the group F, then all unipotent elements of the group F are contained in N.

Now we prove the following

Lemma 10.24. Let $F \subset H$ be an algebraic subgroup containing some maximal unipotent subgroup of the group H. Denote by W the unipotent radical of the group F, and by $N(W)$ the normalizer of W in H. Then *1)* $N(W)$ is a parabolic subgroup in H, *2)* W coincides with the unipotent radical of the group $N(W)$, *3)* all unipotent elements of the group $N(W)$ are contained in F.

Proof. In view of Lemma 10.8, W can be represented as the intersection $W = V_1 \cap V_2$ of two maximal unipotent subgroups of the group F. Since F contains a maximal unipotent subgroup of the group F, in view of Lemma 10.20 V_1 and V_2 are maximal unipotent subgroups of the group H. Therefore, in view of Lemma 10.17 $N(V_1)$ and $N(V_2)$ are parabolic subgroups in H, where $N(V_1)$ and $N(V_2)$ denote the normalizers of the subgroups V_1 and V_2 in H, and consequently, in view of Lemma 10.16 the intersection $N(V_1) \cap N(V_2)$ contains the centralizer $C(T)$ of some maximal **R**-split torus T of H. Denote by Q the minimal algebraic subgroup of the group H containing the union $C(T) \cup V_1 \cup V_2$. Since *1)* V_1 and V_2 normalize W (because $V_1 \subset F$ and $V_2 \subset F$ and W is a unipotent radical in F), *2)* $C(T)$ normalizes W (because $C(T)$ normalizes both V_1 and V_2, and $W = V_1 \cap V_2$), we have

$$Q \subset N(W). \tag{24}$$

Since the algebraic subgroup Q contains $C(T) \cup V_1$, in view of Lemmas 10.17 and 10.18 Q contains a minimal parabolic subgroup $N(V_1)$. Therefore, Q itself is a parabolic subgroup in H. Thus, by (24), W is an algebraic unipotent normal subgroup in the parabolic subgroup Q. On the other hand, W is the intersection of the two maximal unipotent subgroups V_1 and V_2 of the group Q (V_1 and V_2 are maximal in Q since they are maximal in H), and consequently, in view of

Lemma 10.8 $W \supset R_u(Q)$. The last two sentences imply

$$W = R_u(Q). \tag{25}$$

As a consequence of (24), (25) and Lemma 10.21, we have

$$Q = N(W). \tag{26}$$

From (24), (25) and (26) parts *1)* and *2)* of our lemma follow.

Now we prove part *3)*. Denote by D the minimal algebraic subgroup of the group H containing $V_1 \cup V_2$. It is clear that

$$D \subset F. \tag{27}$$

Since $C(T)$ normalizes both V_1 and V_2, $C(T)$ normalizes D, and consequently Q normalizes D. Therefore, in view of (26), $N(W)$ normalizes D. Now, since D contains the maximal unipotent subgroup V_1 of the group $N(W)$ (V_1 is maximal in $N(W)$ since V_1 is maximal in H), in view of Lemma 10.23 we obtain that D contains all unipotent elements of the group H. This and (27) imply part *3)*. The lemma is proved.

Let T be an arbitrary maximal **R**-split torus of H. Now we set forth the well-known description of all parabolic subgroups of H normalized by the torus T. In view of Lemma 10.9 this will produce at the same time a description of all orispheric subgroups of H normalized by the torus T.

Denote by Σ the root system of the group H with respect to the torus T. Let us define an ordering ψ on Σ, and denote by $\Sigma^+ \subset \Sigma$ the set of positive roots with respect to ψ, by Σ^- the set of negative roots with respect to ψ (i.e. $\Sigma^- = \Sigma = \Sigma^+$), and by Σ_0 the set of simple roots in Σ^+. Just as above, for any $\alpha \in \Sigma$, $\mathfrak{U}_\alpha$ denotes the root subspace corresponding to the root α, and $U_\alpha \subset H$ the set $\exp(\mathfrak{U}_\alpha)$ (note that U_α is not necessarily a subgroup for all $\alpha \in \Sigma$). For any subset D of the set Σ_0 write *1)* Σ_D for the subset of the set Σ^+ consisting of roots α of the form $\alpha = \sum_{i=1}^{k} n_i d_i$, where $d_i \in D$ and $n_i \geqq 0$ ($1 \leqq i \leqq k$, k is the number of elements in D), *2)* $D^- \subset \Sigma^-$ for the set of roots of form $-\alpha$, where $\alpha \in D$, *3)* $\tilde{\Sigma}_D \subset \Sigma^+$ for the set $\Sigma^+ - \Sigma_{\Sigma_0 - D}$, *4)* $\hat{\Sigma}_D \subset \Sigma$ for the set $\Sigma^+ \cup \Sigma_{D^-}$. Further we denote by V_A, for any subset $A \subset \Sigma$, the minimal algebraic subgroup of H containing $\bigcup_{\alpha \in A} U_\alpha$, and by W_A the minimal algebraic subgroup of H containing $C(T) \cup (\bigcup_{\alpha \in A} U_\alpha)$, where $C(T)$ is the centralizer of the torus T in H. Put now

$$P_D = W_{\hat{\Sigma}_D} \tag{28}$$

$$H_D = V_{\tilde{\Sigma}_D} \tag{29}$$

$$S_D = V_{\Sigma_D}. \tag{30}$$

We note that, by replacing ψ by its opposite and D by D^-, we can define P_{D^-}, H_{D^-} and S_{D^-} analogously.

Lemma 10.25. For any $D \subset \Sigma_0$, the subgroup P_D is parabolic, and any parabolic subgroup of H normalized by the torus T, coincides with P_D, if the ordering ψ and the set $D \subset \Sigma_0$ are appropriately chosen.

The parabolic subgroups P_D are called *standard parabolic subgroups.* It is well known (see [5]) that any parabolic subgroup is conjugate to one and only one standard parabolic subgroup.

It is easy to see that the following lemma holds:

Lemma 10.26. For any $D \subset \Sigma_0$, H_D coincides with the unipotent radical of the group P_D.

From Lemmas 10.25, 10.26 and 10.9 we get

Lemma 10.27. For any $D \subset \Sigma_0$ the subgroup H_D is orispheric, and any orispheric subgroup of the group H normalized by the torus T coincides with H_D if the ordering ψ and the set D are appropriately chosen.

Denote V_{Σ^+} and V_{Σ^-} by V^+ and V^-. It is easy to see that the following lemma holds:

Lemma 10.28. For any $D \subset \Sigma_0$ *1)* $N(H_D) \cap V^- = S_{D^-}$, where $N(H_D)$ is the normalizer of H_D in H, *2)* $N(H_D) = P_D$.

In § 18 the following lemma will be proved.

Lemma 10.29. The minimal algebraic normal subgroup of the group $V^+(V^-)$ containing $S_D(S_{D^-})$, coincides with $H_D(H_{D^-})$.

Lemma 10.30. Let V_1 and V_2 be maximal unipotent subgroups of the group H, and F an algebraic subgroup of H, containing V_1. Denote by $N \subset V_2$ the minimal algebraic normal subgroup of the group V_2, containing the intersection $F \cap V_2$. Then, if *1)* $V_1 \cap V_2 = e$, *2)* not all unipotent elements of the group H are contained in F, *3)* not all unipotent elements of the group F are contained in V_1, then a) $N \neq e$, b) $N \neq V_2$, c) N is an orispheric subgroup.

Proof. Since V_2 is unipotent, in view of Lemma 10.24 we can assume, without loss of generality, that F is a parabolic subgroup. Since $F \supset V_1$, and V_1 is a maximal unipotent subgroup in H, from Lemmas 10.16, 10.17 and 10.18 it follows easily that

$$F \supset N(V_1) \tag{31}$$

where $N(V_1)$ is the normalizer of V_1 in H. In view of Lemmas 10.16 and 10.17 there exists a maximal **R**-split torus T normalizing both V_1 and V_2. By virtue of (31) $T \subset F$, and consequently T normalizes F. This, together with Lemmas 10.25, 10.27, 10.28 and 10.29 implies the assertion of the lemma.

We keep the notations introduced before Lemma 10.25. Then it is easy to establish the following

Lemma 10.31. The parabolic subgroups P_{D_1} and P_{D_2} are opposite if and only if $D_2 = D_1^-$.

Analogously to the definition of $\bar{S}N(\bar{U})$ in § 7, we define $\bar{S}N(V)$ for any algebraic unipotent subgroup V of the group H. Namely, $\bar{S}N(V)$ is the set of all elements of the group $N(V)$ ($N(V)$ is the normalizer of V in H) of determinant ± 1 under the natural action of $N(V)$ on the Lie algebra $\mathfrak{B}$ of the group V.

It is not difficult to prove the following

Lemma 10.32. For any D we have

$$\bar{S}N(H_D) \cap N(H_{D^-}) = N(H_D) \cap \bar{S}N(H_{D^-}) = \bar{S}N(H_D) \cap \bar{S}N(H_{D^-}).$$

Lemma 10.33. Let V_1 and V_2 be two orispheric subgroups of the group H such that $N(V_1)$ and $N(V_2)$ are opposite parabolic subgroups. Then

$$\bar{S}N(V_1) \cap N(V_2) = N(V_1) \cap \bar{S}N(V_2) = \bar{S}N(V_1) \cap \bar{S}N(V_2).$$

Proof. Since $N(V_1)$ and $N(V_2)$ are parabolic subgroups, according to Lemma 10.16 $N(V_1) \cap N(V_2)$ contains some maximal **R**-split torus T of H. After this it remains to apply Lemmas 10.25, 10.31, 10.28 and 10.33.

§ 11. Orisphericality of maximal unipotent subgroups and similar questions

In this paragraph the notations and the terminology of § 7 apply. Recall that G and Γ satisfy the conditions formulated at the beginning of § 7.

For any subset $H \subset G$, $\hat{H}$ stands for $A(H \cap \Gamma)$ (as above, $A(M)$ denotes the Zariski closure of the set M). For any unipotent subgroup U of Γ, $\tilde{A}(U)$ denotes $A(\bar{S}N(\bar{U}) \cap \Gamma)$, $\tilde{\tilde{A}}(U)$ the algebraically connected component of the identity in the group $\tilde{A}(U)$, and T_U the restriction of the adjoint representation of the group $\bar{S}N(\bar{U})$ to $\mathfrak{U}$. The adjoint representation of the group G is algebraic according to Lemma 1.2.1, thus for any algebraic subgroup $Q \subset \bar{S}N(\bar{U})$, in view of Lemma 8.3, $T_U(Q)$ is a quasialgebraic subgroup of the group $GL(\mathfrak{U})$ ($GL(\mathfrak{U})$ is the algebraic **R**-group of all nondegenerate linear transformations of $\mathfrak{U}$).

Recall that $B(\Gamma)$ denotes the set of all regular unipotent subgroups of the group Γ (i.e. the unipotent subgroups U of the group Γ for which $U = \bar{U} \cap \Gamma$).

Lemma 11.1. Let the unipotent subgroup $U \in B(\Gamma)$ satisfy $(C(\bar{U}) \cap \Gamma)^u \subset U$. If H is an algebraic subgroup of G such that *1)* $H \subset \bar{S}N(\bar{U})$ *2)* H has property P, then a) $T_U(\hat{H} \cap \Gamma)$ is an arithmetic subgroup in $T_U(\hat{H})$ (in the sense of § 8) b) $v(T_U(\hat{H})/T_U(\hat{H} \cap \Gamma)) < \infty$ c) the algebraic closure of the subgroup $T_U(\hat{H} \cap \Gamma)$ in $GL(\mathfrak{U})$ contains the group $T_U(\hat{H})$. If, in addition, $T_U(\hat{H} \cap \Gamma)$ does not contain unipotent elements not belonging to the unipotent radical of the group $T_U(\hat{H})$, then d) the quotient space $T_U(\hat{H})/T_U(\hat{H} \cap \Gamma)$ is compact.

Before proving Lemma 11.1 we mention, without proof, the following obvious

Lemma 11.2. Let X_1 and X_2 be real (complex) algebraic manifolds, B a subset in X_1, and f a regular algebraic mapping $f: X_1 \to X_2$. If B is Zariski dense in some algebraic submanifold M of X_1, then $f(B)$ is Zariski dense in $f(M)$ (i.e. $f(B)$ is dense in $f(M)$ in the topology which is induced on $f(M)$ by the Zariski topology of X_2).

Proof of Lemma 11.1. Since $U \in B(\Gamma)$, in view of (1) and (3) of § 7, for any $\gamma \in \bar{S}N(\bar{U}) \cap \Gamma$ we have $\gamma U \gamma^{-1} = \overline{\gamma U \gamma^{-1}} \cap \Gamma = \gamma \bar{U} \gamma^{-1} \cap \Gamma = \bar{U} \cap \Gamma = U$. Therefore, according to formula (2) of § 7

$$(\operatorname{Ad} \gamma) \Delta_U = \Delta_U \tag{1}$$

for any $\gamma \in \bar{S}N(\bar{U}) \cap \Gamma$. Introduce a basis $e_1, \ldots, e_n$ in the quasi-lattice Δ_U. Now (1) shows that the elements of the subgroup $T_U(\bar{S}N(\bar{U}) \cap \Gamma)$ can be written in the basis $e_1, \ldots, e_n$ as matrices with integer elements and determinant ± 1. Consequently

$$T_U(\hat{H} \cap \Gamma) \subset A_Z, \tag{2}$$

where A_Z denotes the subgroup of $T_U(\hat{H})$ consisting of the elements which can be written in the basis $e_1, \ldots, e_n$ as matrices with integer elements and determinant ± 1.

Since H is an algebraic subgroup, we have

$$\hat{H} = A(H \cap \Gamma) \subset H. \tag{3}$$

As an immediate consequence of (3),

$$\hat{H} \cap \Gamma = H \cap \Gamma. \tag{4}$$

Now H has property P. Therefore, according to Lemma 6.9, (3), (4) and the fact that $\hat{H}$ is topologically closed (since it is algebraic) imply that $\hat{H}$ has property P.

Since for any $g \in T_U^{-1}(A_Z)$ we have $(\operatorname{Ad} g)\Delta_U = \Delta_U$, it follows from Lemma 7.6 that the set $\pi(T_U^{-1}(A_Z))$ is relatively compact in G/Γ (as in § 7, π is the natural mapping $\pi: G \to G/\Gamma$). Therefore, since $\hat{H}$ has property P, in view of Lemma 6.13 the set $\hat{\pi}(\hat{H} \cap T_U^{-1}(A_Z))$ is relatively compact in $\hat{H}/\hat{H} \cap \Gamma$, where $\hat{\pi}$ denotes the natural mapping $\hat{\pi}: \hat{H} \to \hat{H}/\hat{H} \cap \Gamma$. Denote by π_T the natural mapping $\pi_T: \hat{H}/\hat{H} \cap \Gamma \to T_U(\hat{H})/T_U(\hat{H} \cap \Gamma)$, and by π_A the natural mapping

$$\pi_A: T_U(\hat{H}) \to T_U(\hat{H})/T_U(\hat{H} \cap \Gamma).$$

Note that in view of formula (2) of § 0, we have $\pi_T = \pi_A \circ T_U \circ \hat{\pi}^{-1}$. Therefore, taking into account (2) we get

$$\begin{aligned} \pi_T[\hat{\pi}[T_U^{-1}(A_Z) \cap \hat{H}]] &= \pi_A[T_U[\hat{\pi}^{-1}[\hat{\pi}[T_U^{-1}(A_Z) \cap \hat{H}]]]] = \\ &= \pi_A[T_U[(T_U^{-1}(A_Z) \cap \hat{H}) \cdot (\hat{H} \cap \Gamma)]] = \pi_A[(T_U(T_U^{-1}(A_Z) \cap \hat{H})) \cdot (T_U(\hat{H} \cap \Gamma))] \subset \\ &\subset \pi_A[(T_U(T_U^{-1}(A_Z)) \cap T_U(\hat{H})) \cdot A_Z] = \pi_A[(A_Z \cap T_U(\hat{H})) \cdot A_Z] \subset \\ &\subset \pi_A(A_Z \cap T_U(\hat{H})) \cdot \pi_A(A_Z) = \pi_A(A_Z). \end{aligned} \tag{5}$$

Since $\hat{\pi}[T_U^{-1}(A_Z)\cap\hat{H}]$ is relatively compact in $\hat{H}/\hat{H}\cap\Gamma$, it follows from (5) that $\pi_A(A_Z)$ is relatively compact in $T_U(\hat{H})/T_U(\hat{H}\cap\Gamma)$. From this, relation (2) and Lemma 9.9 we get that the set $A_Z/T_U(\hat{H}\cap\Gamma)$ is finite.

In view of Lemma 11.2 and the definition of $\hat{H}$, $T_U(\hat{H}\cap\Gamma)$ is Zariski dense in $T_U(\hat{H})$. Therefore, in view of (2) A_Z is Zariski dense in $T_U(\hat{H})$. Part c) is proved. From this and Lemma 8.7 we conclude that $T_U(\hat{H})$ is defined over $\mathbf{Q}$ (in the sense of Definition 8.6) in the basis $e_1, \ldots, e_n$. Therefore, in view of the finiteness of $A_Z/T_U(\hat{H}\cap\Gamma)$, $T_U(\hat{H}\cap\Gamma)$ is an arithmetic subgroup in $T_U(\hat{H})$. Thus, part a) is proved. Since A_Z is algebraically dense in $T_U(\hat{H})$, Lemma 8.9 and the finiteness of $A_Z/T_U(\hat{H}\cap\Gamma)$ imply part b).

Now we prove part d). If $T_U(\hat{H}\cap\Gamma)$ does not contain unipotent elements different from the identity, then, in view of the finiteness of $A_Z/T_U(\hat{H}\cap\Gamma)$, neither does A_Z (since no non-zero power of any unipotent element different from the identity, equals the identity). Therefore, since A_Z is Zariski dense in $T_U(\hat{H})$, $T_U(\hat{H})/A_Z$ is compact by Lemma 8.10. From this and the finiteness of $A_Z/T_U(\hat{H}\cap\Gamma)$ we obtain that $T_U(\hat{H})/T_U(\hat{H}\cap\Gamma)$ is compact. The proof of the lemma is complete.

Lemma 11.3. Let H_1 and H_2 be separable (in the topological sense) Lie groups, and f a continuous homomorphism of H_1 to H_2. Then, if f is an isomorphism in the abstract group-theoretical sense, then f^{-1} is continuous, and consequently it is isomorphism in the sense of the theory of Lie groups.

Proof. It is easy to see that for any sufficiently small neighbourhood $W\subset H_1$ of the identity, $f(W)$ is a submanifold (not closed) of the Lie group H_2, and further, since $\operatorname{Ker} f=e$, we have

$$\dim f(W) = \dim W = \dim H_1. \tag{6}$$

Since H_1 is separable, H_1 is the union of countably many translate of the set $W: H_1 = \bigcup_{i=1}^{\infty} h_i \cdot W$. It is clear that since $f(W)$ is a submanifold, for any $i(1 \leqq i < \infty)$, $f(h_i \cdot W) = f(h_i)\cdot f(W)$ is also a submanifold, further we have in view of (6)

$$\dim f(h_i\cdot W) = \dim f(W) = \dim H_1. \tag{7}$$

Therefore, since

$$f(H_1) = f\left(\bigcap_{i=1}^{\infty} h_i\cdot W\right) = \bigcap_{i=1}^{\infty} f(h_i\cdot W)$$

and $H_2=f(H_1)$ (since f is an isomorphism), H_2 is the union of countably many submanifolds, each of dimension $\dim H_1$. Therefore, since a manifold cannot be represented as a union of countably many submanifolds of smaller dimension, we have $\dim f(W) = \dim H_2$, and consequently, since $f(W)$ is a submanifold, $f(W)$ is an open subset in H_2. Since f is one-to-one and W can be chosen arbitrarily small, it follows that f^{-1} is continuous at the identity. Now, since f^{-1} is a homomorphism, we get that f^{-1} is continuous everywhere. The lemma is proved.

Lemma 11.4. Let $U \in B(\Gamma)$ satisfy $(C(\bar{U}) \cap \Gamma)^u \subset U$. Then, for any algebraic subgroup H of G satisfying $H \subset \bar{S}N(\bar{U})$ and having property P, we have $v(\hat{H}/\hat{H} \cap \Gamma) < \infty$.

Proof. As shown in the proof of Lemma 11.1, $\hat{H}$ has property P (see (3) and (4)). Therefore, since $C(\bar{U})$ is an integer subgroup (according to Lemma 7.11), in view of Lemma 6.8 $\hat{H} \cap C(\bar{U})$ is also an integer subgroup.

As we have mentioned at the beginning of the §, for any algebraic subgroup $Q \subset \bar{S}N(\bar{U})$, $T_U(Q)$ is a quasialgebraic subgroup of the group $GL(\mathfrak{U})$. Therefore, since every quasialgebraic subgroup is a Lie subgroup, $T_U(\hat{H})$ is a Lie group. Since $\bar{U}$ is topologically connected (in view of Lemmas 1.3.3 and 1.3.1), it is easy to see that the kernel of representation T_U coincides with $C(\bar{U})$. Therefore, T_U induces the homomorphism $\varphi: \hat{H}/\hat{H} \cap C(\bar{U}) \to T_U(\hat{H})$. Now we observe that *1)* since $C(\bar{U}) = \mathrm{Ker}\, T_U$, φ is an isomorphism in the abstract group-theoretical sense, *2)* since the representation T_U is continuous, φ is a continuous mapping of the Lie group $\hat{H}/\hat{H} \cap C(\bar{U})$ into the Lie group $T_U(\hat{H})$. Therefore, in view of Lemma 11.3 φ is an isomorphism of Lie groups. Consequently, we can identify $T_U(\hat{H})$ and $\hat{H}/\hat{H} \cap C(\bar{U})$ as Lie groups, and assume that $T_U(\hat{H})$ is the factor group (in the sense of Lie group theory) of the Lie group $\hat{H}$ by the closed subgroup $\hat{H} \cap C(\bar{U})$.

According to b) of Lemma 11.1, $T_U(\hat{H} \cap \Gamma)$ is a lattice in $T_U(\hat{H})$. Now, as shown already, $\hat{H} \cap C(\bar{U})$ is an integer subgroup, i.e. $\hat{H} \cap C(\bar{U})/\hat{H} \cap C(\bar{U}) \cap \Gamma$ is compact. Therefore in view of Lemma 1.1.2, $\hat{H} \cap \Gamma$ is a lattice in $\hat{H}$ (we can apply Lemma 1.1.2, since we have shown that $T_U(\hat{H})$ is the quotient group $\hat{H}/\hat{H} \cap C(\bar{U})$ in the sense of Lie groups). The lemma is proved.

Lemma 11.5. Let $U \in B(\Gamma)$ satisfy $(C(\bar{U}) \cap \Gamma)^u \subset U$. If H is an algebraic subgroup such that *1)* $H \subset \bar{S}N(\bar{U})$, *2)* H has property P, then $\hat{H}$ contains the unipotent radical $R_u(H)$ of the group H.

Proof. Assume, on the contrary, that $\hat{H} \not\supset R_u(H)$. Then, in view of Lemma 9.3 there exists a one-parameter subgroup $u(t) \subset R_u(H)$ such that $\varphi(u(t))$ tends to infinity in $H/\hat{H}$ as $t \to +\infty$, where φ denotes the natural mapping $\varphi: H \to H/\hat{H}$. Therefore, since $H \cap \Gamma \subset \hat{H}$ (because $\hat{H} = A(H \cap \Gamma)$), thus $\tilde{\varphi}(u(t))$ tends to infinity in $H/H \cap \Gamma$, where $\tilde{\varphi}$ is the natural mapping $\tilde{\varphi}: H \to H/H \cap \Gamma$. Since H has property P, it follows that $\pi(u(t))$ tends to infinity in G/Γ as $t \to \infty$. Thus, on the strength of Lemma 7.6, we have for $t \to +\infty$

$$(\mathrm{Ad}\, u(t))\Delta_U \tag{8}$$

tends to infinity in $R(\mathfrak{U})$, where $R(\mathfrak{U})$ is the subspace of all quasi-lattices in the Lie algebra $\mathfrak{U}$ of the group $\bar{U}$. Now *1)* since $u(t)$ is a unipotent group (because $u(t) \subset R_u(H)$), in view of Lemmas 1.2.1 and 1.3.4, $\mathrm{Ad}\, u(t)$ is a one-parameter group of unipotent linear transformations, *2)* since $H \subset N(\bar{U})$, $\mathfrak{U}$ is invariant under $\mathrm{Ad}\,u(t)$ for all t; *3)* according to Lemma 7.1 Δ_U is in $\mathfrak{U}$ a quasi-lattice of full dimension, i.e. $\dim \Delta_U = \dim \mathfrak{U}$. Therefore (8) contradicts Lemma 3.3. Thus, starting from $\hat{H} \not\supset R_u(H)$ we found a contradiction. The lemma is proved.

Lemma 11.6. For any unipotent subgroup $U \subset \Gamma$, $\bar{S}N(\overline{U})$ and $C(\overline{U})$ are algebraic subgroups in G.

Proof. That $C(\overline{U})$ is algebraic, follows from Lemma 1.2.9. Since $\overline{U}$ is algebraic, Lemma 1.2.8 implies that the subgroup $N(\overline{U})$ is also algebraic. Now since $N(\overline{U})$ and the adjoint representation Ad of G are both algebraic (Ad is algebraic according to Lemma 1.2.1), one has that $\bar{S}N(\overline{U})$ is also algebraic. The lemma is proved.

From Lemmas 11.1, 11.4, 11.5, 11.6 and 7.3 we get the following

Lemma 11.7. Let $U \in B(\Gamma)$ satisfy $(C(\overline{U}) \cap \Gamma)^u \subset U$. Then *1)* $T_U(\tilde{A}(U) \cap \Gamma)$ is an arithmetic subgroup in $T_U(\tilde{A}(U))$, *2)* $v(\tilde{A}(U)/\tilde{A}(U) \cap \Gamma) < \infty$, *3)* $\tilde{A}(U)$ contains the unipotent radical of the group $\bar{S}N(\overline{U})$ (for the definition of $\tilde{A}(\overline{U})$ see the beginning of this §).

In view of Lemma 7.7 the following lemma holds:

Lemma 11.8. The assertions of Lemmas 11.1, 11.4, 11.5 and 11.7 hold for any maximal unipotent subgroup U of the group Γ.

If $U \in B(\Gamma)$, ~~and~~ $C(\overline{U}) \subset \overline{U}$, then

$$(C(\overline{U}) \cap \Gamma)^u \subset C(\overline{U}) \cap \Gamma \subset \overline{U} \cap \Gamma = U$$

(the last equality holds because $U \in B(\Gamma)$). Therefore we have the following

Lemma 11.9. The assertions of Lemmas 11.1, 11.4, 11.5 and 11.7 are valid, if $U \in B(\Gamma)$ satisfies $C(\overline{U}) \subset \overline{U}$.

Lemma 11.10. For any unipotent subgroup $U \subset \Gamma$, the unipotent radical V of $\bar{S}N(\overline{U})$ contains the unipotent radical V_0 of the group $C(\overline{U})$.

Proof. Since *1)* $C(\overline{U})$ is an algebraic normal subgroup in $\bar{S}N(\overline{U})$ ($C(\overline{U})$ is algebraic in view of Lemma 11.6), *2)* the unipotent radical V_0 of the group $C(\overline{U})$ is invariant under all algebraic automorphisms of the group $C(\overline{U})$ (since the unipotent radical of an algebraic group is uniquely defined), thus V_0 is a unipotent normal subgroup of $\bar{S}N(\overline{U})$. Therefore V contains V_0 (because V is the unipotent radical of the group $\bar{S}N(\overline{U})$).

As an immediate consequence of Lemma 5.5, we have

Lemma 11.11. Let W be a unipotent algebraic **R**-group, $H \subset W$ an algebraic subgroup. If the quotient space W/H is compact, then $H = W$.

Lemma 11.12. For any unipotent subgroup $U \subset \Gamma$ we have *1)* $(N(\overline{U}))^u = (\bar{S}N(\overline{U}))^u$ (as above, H^u denotes the set of all unipotent elements in H), *2)* the unipotent radical of $N(\overline{U})$ coincides with the unipotent radical of $\bar{S}N(\overline{U})$.

Proof. Lemma 1.2.1 implies that the natural action φ of the group $N(\bar{U})$ on the Lie algebra $\mathfrak{U}$ of the group $\bar{U}$, is an algebraic representation of the group $N(\bar{U})$. Therefore, in view of Lemma 1.3.4, $\varphi[(N(\bar{U}))^u]\subset[\varphi(N(\bar{U}))]^u$, whence, since the determinant of any unipotent matrix is equal to 1, statement *1)* follows. Now, from *1)* and Lemma 1.3.7 we have *2)*. The lemma is proved.

Theorem 11.13. For any maximal unipotent subgroup U of the group Γ, its algebraic closure $\bar{U}$ is an orispheric unipotent subgroup of the group G.

Proof. Denote by V the unipotent radical of the group $\bar{S}N(\bar{U})$, and by $\tilde{V}$ the unipotent radical of the group $\tilde{A}(U)$. Now, *1)* $\tilde{A}(U)\subset\bar{S}N(\bar{U})$ (because $\bar{S}N(\bar{U})$ is an algebraic subgroup according to Lemma 11.6), *2)* in view of Lemma 11.7 $\tilde{A}(U)\supset V$, thus V is a unipotent normal subgroup in $\tilde{A}(U)$, and consequently

$$V\subset\tilde{V}. \tag{9}$$

According to Lemma 11.7 $\tilde{A}(U)\cap\Gamma$ is a lattice in $\tilde{A}(U)$. It is clear on the other hand that $\tilde{A}(U)\cap\Gamma$ is Zariski dense in $\tilde{A}(U)$ (since $\tilde{A}(U)=A(\bar{S}N(\bar{U})\cap\Gamma)$. Thus, $\tilde{A}(U)\cap\Gamma$ is a Zariski dense lattice in $\tilde{A}(U)$. Therefore, in view of Lemma 4.7, there exists in $\tilde{A}(U)$ a topologically connected nilpotent normal subgroup N such that $N/N\cap\Gamma$ is compact and

$$N\supset\tilde{V}. \tag{10}$$

Denote by $\bar{N}$ the Zariski closure of N in G. Since N is a nilpotent normal subgroup in $\tilde{A}(U)$, in view of Lemmas 1.2.3 and 1.2.5 $\bar{N}$ is an algebraic nilpotent normal subgroup in $\tilde{A}(U)$. On the other hand, since N is topologically connected, Lemma 1.2.14 implies that $\bar{N}$ is algebraically connected. Thus, $\bar{N}$ is an algebraically connected nilpotent algebraic group. Then, according to Lemma 1.5.1, $\bar{N}=$ $=U_N\times T_N$, where U_N is an algebraic unipotent subgroup, and T_N is a torus. If $t\in T_N$, then *1)* since t belongs to a nilpotent algebraic normal subgroup of the group $\tilde{A}(U)$, the restriction of Ad t to the Lie algebra $\mathfrak{U}$ of the group $\tilde{A}(U)$ is a unipotent linear transformation, *2)* since t belongs to a torus, t is semisimple, and consequently, in view of Lemmas 1.2.1 and 1.5.3, the restriction of Ad t to $\mathfrak{U}$ is a semisimple linear transformation (i.e. it can be brought to diagonal form over the field of complex numbers). *1)* and *2)* show that the restriction of Ad t to $\mathfrak{U}$ is the identity linear transformation. Therefore t commutes with all elements of the topologically (and also algebraically) connected component of the group $\tilde{A}(U)$, and in particular with $\tilde{V}$ and $\bar{U}$ ($\tilde{V}$ and $\bar{U}$ are topologically connected in view of Lemmas 1.3.1, 1.3.2 and 1.3.3, and $\bar{U}\subset\tilde{A}(U)$, since $\bar{U}\subset\bar{S}N(\bar{U})$ in view of Lemma 11.12, and consequently

$$\bar{U} = A(U)\subset A(\bar{U}\cap\Gamma)\subset A(\bar{S}N(\bar{U})\cap\Gamma) = \tilde{A}(U)).$$

Therefore, since t was arbitrary, we have

$$(T_N, \tilde{V}) = e \quad\text{and}\quad (T_N, \bar{U}) = e, \tag{11}$$

where $(T_N, \tilde{V})$ and $(T_N, \overline{U})$ denote the commutator of T_N with $\tilde{V}$ and of T_N with $\overline{U}$ (since (9) shows that $\overline{U}$ belongs to $\tilde{V}$, the right-hand part in (11) can be deduced from the left-hand part). From (11), (9) and Lemma 11.12 one has that

$$(T_N, R_u(N(\overline{U}))) = e, \tag{12}$$

where $R_u(N(\overline{U}))$ is the unipotent radical of $N(\overline{U})$.

In view of Lemma 7.5 and the maximality of U, $U \neq e$. Therefore, since G and Γ satisfy, by assumption, the conditions formulated at the beginning of § 7, in view of Lemma 4.8, U is contained in no non-trivial algebraic normal subgroup of G, and since $\overline{U} \supset U$, the same can be said about $\overline{U}$. Therefore, since T_N consists of semisimple elements, (11) and (12) and Lemma 10.3 imply that T_N belongs to the centre of the group G. Now the centre of the group G is trivial (see the beginning of § 7). Therefore $T_N = e$, and since $\overline{N} = U_N \times T_N$, $\overline{N}$ is unipotent. Thus, $\overline{N}$ is an algebraic unipotent normal subgroup in $\tilde{A}(U)$, and consequently

$$N \subset \overline{N} \subset \tilde{V}. \tag{13}$$

From (10) and (13) we have

$$N = \tilde{V}. \tag{14}$$

As a consequence of (14) and the compactness of the quotient space $N/N \cap \Gamma$, the quotient space $\tilde{V}/\tilde{V} \cap \Gamma$ is compact.

Assume now that $\overline{U}$ is not an orispheric subgroup of G. Then, in view of Lemma 11.12 and the definition of orispheric subgroups we have

$$\overline{U} \subset V, \quad \text{but} \quad \overline{U} \neq V. \tag{15}$$

(15) and (9) imply

$$\overline{U} \subset \tilde{V} \tag{16}$$

but

$$\overline{U} \neq \tilde{V}. \tag{17}$$

Since U is a maximal unipotent subgroup of Γ and $\tilde{V}$ is unipotent, (16) yields

$$\tilde{V} \cap \Gamma = \overline{U} \cap \Gamma. \tag{18}$$

From (18) and the compactness of the quotient space $\tilde{V}/\tilde{V} \cap \Gamma$ (which has been shown earlier) we get that the quotient space $\tilde{V}/\overline{U}$ is compact ($\overline{U} \subset \tilde{V}$ in view of (16)). Thus, in view of Lemma 11.11

$$\tilde{V} = \overline{U}. \tag{19}$$

Now (19) contradicts (17). Thus, the assumption that $\overline{U}$ is not orispheric, led to a contradiction. The theorem is proved.

Lemma 11.14. Let $U \in B(\Gamma)$ satisfy $C(\overline{U}) \subset \overline{U}$. If H is an algebraic subgroup of G, such that *1)* $H \subset SN(\overline{U})$, *2)* H has property P, *3)* all unipotent elements of $\hat{H} \cap \Gamma$ belong to U, then $\hat{H}$ is an integer subgroup (i.e. $\hat{H}/\hat{H} \cap \Gamma$ is compact).

Proof. As shown already in the proof of Lemma 11.4, in view of the topological connectedness of $\bar{U}$, $C(\bar{U})$ coincides with the kernel of the representation T_U. Therefore, since we have, by assumption, *1)* $C(\bar{U})\subset\bar{U}$, *2)* $(\hat{H}H\cap\Gamma)^u\subset U$, on applying Lemma 1.3.8 and the fact that the representation T_U is algebraic (in view of Lemma 1.2.1) we obtain that

$$(T_u(\hat{H}\cap\Gamma))^u\subset T_u((\hat{H}\cap\Gamma)^u)\subset T_u(\hat{H}\cap U). \tag{20}$$

Since $\hat{H}$ normalizes U (because $\hat{H}\subset\bar{S}N(\bar{U})$) $\hat{H}\cap U$ belongs to the unipotent radical of the group H. Therefore $T_U(\hat{H}\cap U)$ belongs to the unipotent radical of the group $T_U(\hat{H})$ (in view of Lemma 1.3.4, since the representation T_U is algebraic). Thus, (20) implies that $T_U(\hat{H}\cap\Gamma)$ does not contain any unipotent elements not lying in $T_U(\hat{H})$. Therefore, according to Lemma 11.9 and d) of Lemma 11.1, the quotient space $T_U(\hat{H})/T_U(\hat{H}\cap\Gamma)$ is compact. From this point on the proof of the lemma is the same as that of Lemma 11.4 (the only change is the application of Lemma 1.1.5 rather than 1.1.2).

From Lemmas 11.14 and 7.3 we have

Lemma 11.15. Let $U\in B(\Gamma)$ satisfy $C(\bar{U})\subset\bar{U}$. If $\bar{S}N(\bar{U})\cap\Gamma\subset U$, then the subgroup $\tilde{A}(U)$ is integer.

Lemma 11.16. Let V be an orispheric unipotent subgroup of an arbitrary semisimple complex (real) algebraic group H with trivial centre. If V is contained in no non-trivial algebraic normal subgroup of H, then $C(V)\subset V$, where $C(V)$ is the centralizer of V in H.

Proof. It is easy to see that the unipotent radical W of the subgroup $C(V)$ belongs to the unipotent radical of the normalizer $N(V)$ of the subgroup V in G (this can be proved the same way as Lemma 11.10). Therefore, since the subgroup V is orispheric, we have $W\subset V$, and consequently $C(V)\subset C(W)$. This and Lemma 10.3 imply (since the centre of H is trivial) that $C(V)$ contains no semisimple elements different from the identity. Now, in view of Lemma 1.2.9, $C(V)$ is an algebraic subgroup of G. Therefore, since $C(V)$ contains no semisimple elements except the identity, in view of Lemma 1.5.4 $C(V)$ is unipotent. Consequently, $C(V)=W\subset V$. The lemma is proved.

Remark. It is not difficult to prove Lemma 11.16 directly (without applying Lemma 10.3).

In the course of the proof of Theorem 11.13 (see the paragraph after formula (12)) we have actually proved the following

Lemma 11.17. For any unipotent subgroup $U\neq e$ of the group Γ, $\bar{U}$ is contained in no non-trivial algebraic normal subgroup of G. In particular (in view of Lemma 7.5), this is true for any maximal unipotent subgroup U of Γ.

From Theorem 11.13 and Lemmas 11.16 and 11.17 we get

Lemma 11.18. For any maximal unipotent subgroup U of the group Γ we have the inclusion $C(\overline{U})\subset\overline{U}$.

Lemma 11.19. Let U be a maximal unipotent subgroup of the group Γ. If H is an algebraic subgroup such that *1)* $H\subset\bar{S}N(\overline{U})$ *2)* H has property P, then $\hat{H}$ is an integer subgroup (i.e. $\hat{H}/\hat{H}\cap\Gamma$ is compact).

Proof. Assume, on the contrary, that $\hat{H}$ is not integer. Then, in view of the maximality of U, Lemmas 11.18, 11.14 and 7.7 ensure that $\hat{H}\cap\Gamma$ contains unipotent elements not lying in U. Consequently, there is an $u_0\in(\hat{H}\cap\Gamma)^u$ such that $u_0\notin U$. Denote by U_0 the subgroup generated by the union $u_0\cup U$. Since $u_0\in\Gamma$, $U\subset\Gamma$, and $u_0\notin U$, we have

$$\Gamma\supset U_0\supset U \quad\text{and}\quad U_0\neq U. \tag{21}$$

Denote by $\tilde{U}_0$ the subgroup generated by the union $u_0\cup\overline{U}$. Since $U\subset\overline{U}$,

$$U_0\subset\tilde{U}_0. \tag{22}$$

Since $u_0\in\hat{H}$ and $\hat{H}\subset\bar{S}N(\overline{U})$, thus u_0 normalizes $\overline{U}$. Therefore, $\overline{U}$ is a normal subgroup in $\tilde{U}_0$. Denote by f the natural homomorphism $f:\tilde{U}_0\to\tilde{U}_0/\overline{U}$. Since $\tilde{U}_0$ is generated by $u_0\cup\overline{U}$, $\tilde{U}_0/\overline{U}$ is generated by the single element $f(u_0)$, and consequently, the group $\tilde{U}_0/\overline{U}$ is commutative. Now $\overline{U}$ is nilpotent, as mentioned several times. Therefore the commutativity of $\tilde{U}_0/\overline{U}$ implies that $\tilde{U}_0$ is solvable. Now $\tilde{U}_0$ is generated by unipotent elements. Therefore, since $\tilde{U}_0$ is solvable, Lemma 7.4 implies that $\tilde{U}_0$ is unipotent. From this and (22) we get that U_0 is also unipotent. Now in view of (21), this contradicts the fact that U is a maximal unipotent subgroup of Γ. The lemma is proved.

From Lemmas 11.19 and 7.3 we get

Lemma 11.20. For any maximal unipotent subgroup U of the group Γ, $\tilde{A}(U)$ is an integer subgroup.

Lemma 11.21. If U is a maximal unipotent subgroup of the group Γ such that $N(\overline{U})$ contains unipotent elements not lying in $\overline{U}$, then $\dim\tilde{A}(U)>\dim\overline{U}$.

Proof. Let $u(t)$ be an arbitrary one-parameter unipotent subgroup in $N(\overline{U})$. Then, in view of Lemma 11.12

$$u(t)\subset\bar{S}N(\overline{U}). \tag{23}$$

By Lemmas 3.3, 1.2.1 and 1.3.4, $(\mathrm{Ad}\,u(t))\Delta_U$ does not tend to infinity with $t\to+\infty$ in the space of quasi-lattices. Thus, in view of (23) and Lemmas 7.6 and 7.3, $\pi_S(u(t))$ does not tend to infinity in $\bar{S}N(\overline{U})/\bar{S}N(\overline{U})\cap\Gamma$, where π_S denotes the natural mapping $\pi_S:\bar{S}N(\overline{U})\to\bar{S}N(\overline{U})/\bar{S}N(\overline{U})\cap\Gamma$. Therefore, since $\tilde{A}(U)\supset\supset\bar{S}N(\overline{U})\cap\Gamma$, $\pi_A(u(t))$ does not tend to infinity in $\bar{S}N(\overline{U})/\tilde{A}(U)$ with $t\to+\infty$, where π_A is the natural mapping $\pi_A:\bar{S}N(\overline{U})\to\bar{S}N(\overline{U})/\tilde{A}(U)$. From this we get immediately, using $N(\overline{U})\subset\bar{S}N(\overline{U})$, that $\tilde{\pi}_A(u(t))$ does not tend to infinity in

$N(\overline{U})/\tilde{A}(U)$ as $t \to +\infty$, where $\tilde{\pi}_A$ is the natural mapping $\tilde{\pi}_A : N(\overline{U}) \to N(\overline{U})/\tilde{A}(U)$. Thus *1)* $N(\overline{U})$ contains unipotent elements not lying in $\overline{U}$, *2)* $\overline{U}$ is the unipotent radical of $N(\overline{U})$ (in view of Theorem 11.13 and the fact that U is maximal), *3)* for any one-parameter unipotent group $u(t) \subset N(\overline{U})$, $\tilde{\pi}_A(u(t))$ tends to infinity in $N(\overline{U})/\tilde{A}(U)$ as $t \to +\infty$, *4)* $N(\overline{U})$ and $\tilde{A}(U)$ are easily seen to be algebraic subgroups. From the above and Lemma 9.8 the assertion of the lemma follows (with the notations of Lemma 9.8 $H = N(\overline{U})$, $V = \overline{U}$, $P = \tilde{A}(U)$, $\varphi = \pi_A$).

§ 12. Intersections of the subgroups $\tilde{\tilde{A}}(U)$

We are going to use the notations of §§ 7 and 11. Recall that G and Γ satisfy the conditions formulated at the beginning of § 7.

The following lemma is well known.

Lemma 12.1. Let H be an arbitrary semisimple real (complex) algebraic group, and $V \subset H$ an arbitrary maximal unipotent subgroup. Then the set of all $h \in H$ for which $hVh^{-1} \cap V = e$ is Zariski dense and open in H.

Remark. Lemma 12.1 follows from Lemmas 1.2.7, 10.12, 10.13, 10.17, 10.19 and 10.20.

Lemma 12.2. Let U be an arbitrary unipotent subgroup of the group Γ, and H an integer subgroup of G (for the definition of integer subgroups see § 6). Then $H \cap N(\overline{U}) = H \cap \overline{S}N(\overline{U})$.

Proof. For any $n \in N(\overline{U})$, write $k(n)$ for the determinant of the restriction of Ad n to the Lie algebra $\mathfrak{U}$ of the group $\overline{U}$. It follows from the definition of $\overline{S}N(\overline{U})$ that $k(n) = \pm 1$ if and only if $n \in \overline{S}N(\overline{U})$.

Assume now that $H \cap N(\overline{U}) \neq H \cap \overline{S}N(\overline{U})$. Then in $H \cap N(\overline{U})$ there is an element n_0 for which $k(n_0) \neq \pm 1$. Now, since $H \cap N(\overline{U})$ is a subgroup, we have

$$\inf_{n \in H \cap N(\overline{U})} |k(n)| = 0. \tag{1}$$

Now H is an integer subgroup. Therefore $H = K \cdot (H \cap \Gamma)$, where $K \subset H$ is compactum. Let h be an arbitrary element of the subgroup H. Then $h = k(h) \cdot \gamma(h)$, where $k(h) \in K$, and $\gamma(h) \in \Gamma$. Therefore

$$(\mathrm{Ad}\, h) \Delta_U = (\mathrm{Ad}\, k(h)) (\mathrm{Ad}\, \gamma(h)) \Delta_U. \tag{2}$$

Recall that for a quasi-lattice Δ, $d(\Delta)$ denotes its determinant. Since $\gamma(h)$ is an element of the discrete subgroup Γ, and $k(h)$ is an element of the compactum K, it is easy to conclude from (2) that there exists a positive constant $c(K, \Gamma)$ depending only on K and Γ such that for any $h \in H$

$$d((\mathrm{Ad}\, h) \Delta_U) > c(K, \Gamma)\, d(\Delta_U). \tag{3}$$

Now, according to Lemma 7.1, Δ_U is a quasi-lattice of full dimension in $\mathfrak{U}$, thus (3) contradicts (1). Thus, the assumption $H\cap N(\overline{U})\neq H\cap \overline{S}N(\overline{U})$ led to a contradiction. The lemma is proved.

Recall (see § 7) that $B(\Gamma)$ denotes the set of all regular unipotent subgroups of the group Γ (i.e. unipotent subgroups U for which $\overline{U}\cap\Gamma=U$).

Lemma 12.3. Let $U\in B(\Gamma)$ and H an integer subgroup. Then $H\cap N(\overline{U})$ is an integer subgroup.

Proof. According to Lemma 7.3, $\overline{S}N(\overline{U})$ has property P. Therefore, in view of Lemma 6.8 $H\cap\overline{S}N(\overline{U})$ is an integer subgroup. From this and lemma 12.2 the assertion of the lemma follows.

Recall (see § 11) that $\tilde{A}(U)$ stands for $A(\overline{S}N(\overline{U})\cap\Gamma)$, and $\tilde{\tilde{A}}(U)$ for the algebraically connected component of the identity in the group $\tilde{A}(\overline{U})$.

Lemma 12.4. Let U_1 and U_2 be arbitrary maximal unipotent subgroups of the group Γ. Then the intersection $\tilde{\tilde{A}}(U_1)\cap\tilde{\tilde{A}}(U_2)$ is an integer subgroup.

Proof. According to Lemma 11.20, $\tilde{A}(U_1)$ and $\tilde{A}(U_2)$ are integer subgroups. Now $\tilde{\tilde{A}}(U_1)$ and $\tilde{\tilde{A}}(U_2)$ are algebraically connected components of the identity, and consequently, by virtue of Lemma 1.2.7, $\tilde{\tilde{A}}(U_1)$ is of finite index in $\tilde{A}(U_1)$ and $\tilde{\tilde{A}}(U_2)$ is of finite index in $\tilde{A}(U_2)$. These imply that $\tilde{\tilde{A}}(U_1)$ and $\tilde{\tilde{A}}(U_2)$ are integer subgroups. Now, from this and Lemmas 6.7 and 6.8 the assertion of our lemma follows.

Lemma 12.5. Let $U_1\in B(\Gamma)$ and $U_2\in B(\Gamma)$ be subgroups such that *1)* $\overline{U}_1$ and $\overline{U}_2$ are orispheric unipotent subgroups of G, *2)* the normalizers $N(\overline{U}_1)$ and $N(\overline{U}_2)$ are opposite parabolic subgroups of G. Then $\tilde{\tilde{A}}(U_1)$ is the semi-direct product (in the sense of § 1) of $\tilde{\tilde{A}}(U_1)\cap\tilde{\tilde{A}}(U_2)$ and $\overline{U}_1$.

Proof. We mention first that, since *1)* $\overline{U}_1\subset\tilde{A}(U_1)$ (because $\overline{U}_1=A(U_1\cap\Gamma)$ and $\tilde{A}(U_1)=A(\overline{S}N(\overline{U}_1)\cap\Gamma)$ *2)* $\overline{U}_1$ is algebraically connected (by Lemmas 1.3.3 and 1.3.1), we have

$$\tilde{\tilde{A}}(U_1)\supset\overline{U}_1. \tag{4}$$

From condition *2)* and Lemma 10.33 we have

$$\overline{S}N(\overline{U}_1)\cap N(\overline{U}_2)=N(\overline{U}_1)\cap\overline{S}N(U_1)=\overline{S}N(\overline{U}_1)\cap\overline{S}N(\overline{U}_2). \tag{5}$$

In view of Lemma 7.3 $\overline{S}N(\overline{U}_1)$ and $\overline{S}N(\overline{U}_2)$ have property P. Therefore, according to Lemma 6.5, $\overline{S}N(\overline{U}_1)\cap\overline{S}N(\overline{U}_2)$ has property P. This and (5) immediately imply that

$$\overline{S}N(U_1)\cap N(U_2)\quad\text{has property}\quad P. \tag{6}$$

If an algebraic group F is the semi-direct product (in the sense of § 1) of an algebraic subgroup H and an algebraic normal subgroup N, then we will write

$F=H\tilde{\times}N$. Since $N(\bar{U}_1)$ and $N(\bar{U}_2)$ are opposite parabolic subgroups, $N(\bar{U}_1)\cap\cap N(\bar{U}_2)$ is a Levi subgroup in $N(\bar{U}_1)$. Therefore, since *1)* $\bar{U}_1$ coincides with the unipotent radical of the group $N(\bar{U}_1)$ (for $\bar{U}_1$ is orispheric), *2)* $\bar{S}N(\bar{U}_1)\supset\bar{U}_1$, we have

$$\bar{S}N(\bar{U}_1) = (\bar{S}N(\bar{U}_1)\cap N(\bar{U}_2))\tilde{\times}\bar{U}_1. \tag{7}$$

Note that *1)* $\bar{U}_1$ is an integer subgroup (in view of Lemma 5.4), *2)* $U_1=\bar{U}_1\cap\Gamma$ (since $U_1\in B(\Gamma)$). Therefore, it follows from (6), (7) and Lemma 6.11 that the subgroup S_Γ generated by $\bar{S}N(U_1)\cap N(\bar{U}_2)\cap\Gamma$ and U_1 is of finite index in $\bar{S}N(\bar{U}_1)\cap\Gamma$.

Since $\bar{S}N(\bar{U}_1)$ normalizes $\bar{U}_1$, and $U_1=\bar{U}_1\cap\Gamma$ (because $U_1\in B(\Gamma)$), (7) leads to

$$S_\Gamma = (\bar{S}N(\bar{U}_1)\cap N(U_2)\cap\Gamma)\cdot U_1 \tag{8}$$

(the dot stands for group multiplication). Denote by $A(S_\Gamma)$ the algebraic closure of the subgroup S_Γ and by A_1 the algebraic closure of the subgroup $\bar{S}N(\bar{U}_1)\cap\cap\bar{S}N(\bar{U}_2)\cap\Gamma$. Then from (7) and (8) it easily follows that

$$A(S_\Gamma) = A_1\tilde{\times}\bar{U}_1. \tag{9}$$

From the definition of subgroups $\tilde{A}(U_1)$, $\tilde{A}(U_2)$ and A_1 we have that

$$\tilde{A}(U_1)\cap\tilde{A}(U_2)\supset A_1. \tag{10}$$

Since $\tilde{A}(U_1)\subset\bar{S}N(\bar{U}_1)$, and $\tilde{A}(U_2)\subset N(\bar{U}_2)$, in view of (7) we can define the algebraic subgroup D by the formula

$$D = (\tilde{A}(U_1)\cap\tilde{A}(U_2))\tilde{\times}\bar{U}_1. \tag{11}$$

Since S_Γ is a subgroup of finite index in $\bar{S}N(\bar{U}_1)\cap\Gamma$, we get that $A(S_\Gamma)$ is a subgroup of finite index in $A(\bar{S}N(\bar{U}_1)\cap\Gamma)=\tilde{A}(U_1)$. From this, (9), (10) and (11) we get that D is an algebraic subgroup of finite index in $\tilde{A}(U_1)$.

Denote by A^0 the algebraically connected component of the identity in the group $\tilde{A}(U_1)\cap\tilde{A}(U_2)$, and by D^0 the algebraically connected component of the identity in the group D. Since $\bar{U}_1$ is algebraically connected (in view of Lemmas 1.3.3 and 1.3.1), (11) implies

$$D^0 = A^0\tilde{\times}\bar{U}_1. \tag{12}$$

Since D is an algebraic subgroup of finite index in $\tilde{A}(U_1)$, and D^0 is the algebraically connected component of the identity in D, on account of Lemma 1.2.7, D^0 is an algebraically connected algebraic subgroup of finite index in $\tilde{A}(U_1)$, and consequently,

$$D^0 = \tilde{\tilde{A}}(U_1). \tag{13}$$

From (12) and (13) we have

$$\tilde{\tilde{A}}(U_1) = A^0\tilde{\times}U_1. \tag{14}$$

Since $\tilde{\tilde{A}}(U_1)\subset \bar{S}N(U_1)$, and $\tilde{\tilde{A}}(U_2)\subset N(\bar{U}_2)$, in view of (7) we can define the algebraic subgroup $\tilde{D}$ by the formula

$$\tilde{D} = (\tilde{\tilde{A}}(U_1)\cap \tilde{\tilde{A}}(U_2)) \times \bar{U}_1. \tag{15}$$

From the definition of subgroups A^0, $\tilde{\tilde{A}}(U_1)$ and $\tilde{\tilde{A}}(U_2)$ we have

$$A^0 \subset \tilde{\tilde{A}}(U_1)\cap \tilde{\tilde{A}}(U_2). \tag{16}$$

From (14), (15) and (16) we get

$$\tilde{\tilde{A}}(U_1)\subset \tilde{D}. \tag{17}$$

On the other hand, it can be seen from (4) and (15) that

$$\tilde{D}\subset \tilde{\tilde{A}}(U_1). \tag{18}$$

From (17), (18) and (15) the assertion of the lemma follows.

Lemma 12.6. Let U_1 and U_2 be maximal unipotent subgroups of the group Γ. If $\bar{U}_1\cap \bar{U}_2=e$, then $\bar{U}_1\cap N(\bar{U}_2)=\bar{U}_2\cap N(\bar{U}_1)=e$.

Proof. In view of Lemma 5.4 $\bar{U}_2$ is an integer subgroup. Therefore, according to Lemma 12.3 $\bar{U}_2\cap N(\bar{U}_1)$ is an integer subgroup.

Since *1)* by Lemma 1.3.3 $\bar{U}_2$ is an algebraic unipotent subgroup, *2)* in view of Lemma 1.2.8 $N(\bar{U}_1)$ is an algebraic subgroup, we get that $\bar{U}_2\cap N(\bar{U}_1)$ is an algebraic unipotent subgroup. Therefore, according to Lemma 1.3.1 $\bar{U}_2\cap N(\bar{U}_1)$ is a connected, simply connected nilpotent Lie group, and consequently, in view of Lemma 5.1 $\bar{U}_2\cap N(\bar{U}_1)$ is homeomorphic to a Euclidean space.

Assume now that

$$\bar{U}_2\cap N(\bar{U}_1) \neq e. \tag{19}$$

From (19) and the two previous paragraphs it follows that $\bar{U}_2\cap N(\bar{U}_1)\cap\Gamma\neq e$. Let $u_0\neq e$ be an element in $\bar{U}_2\cap N(\bar{U}_1)\cap\Gamma$, and U_0 the subgroup generated by $u_0\cup\bar{U}_1$. Since $u_0\in\bar{U}_2\cap\Gamma$, $u_0\neq e$, and $\bar{U}_1\cap\bar{U}_2=e$, we have $u_0\notin\bar{U}_1$, and consequently

$$u_0\in U_0\cap\Gamma, \quad \text{but} \quad u_0\notin U_1. \tag{20}$$

Since $u_0\in N(\bar{U}_1)$, u_0 normalizes $\bar{U}_1$, we get that $\bar{U}_1$ is a normal subgroup in U_0. It is easy to see, that $U_0/\bar{U}_1$ is generated by a single element, and consequently, it is commutative. Therefore, since $\bar{U}_1$ is nilpotent (by virtue of Lemmas 1.3.3 and 1.3.1), U_0 is solvable. On the other hand, *1)* u_0 is unipotent (because $u_0\in\bar{U}_2$), *2)* $\bar{U}_1$ is unipotent, and consequently, U_0 is generated by unipotent elements. These last two statements and Lemma 7.4 imply that U_0 is unipotent. Therefore, $U_0\cap\Gamma$ is also unipotent. From this, (20) and the relation $U_0\cap\Gamma\supset\bar{U}_1\cap\Gamma\supset U_1$ we conclude that U_1 is not a maximal unipotent subgroup in Γ. Thus (19) has led to contradiction. The lemma is proved.

Lemma 12.7. Let U_1 and U_2 be maximal unipotent subgroups of the group Γ. If $\overline{U}_1 \cap \overline{U}_2 = e$ then $\tilde{\tilde{A}}(U_1)$ is the semidirect product (in the sense of § 1) of $\tilde{\tilde{A}}(U_1) \cap \cap \tilde{\tilde{A}}(U_2)$ and $\overline{U}_1$ (and of course a similar statement holds true for $\tilde{\tilde{A}}(U_2)$.)

Proof. In view of Lemma 12.6 we have

$$\overline{U}_2 \cap N(\overline{U}_1) = \overline{U}_1 \cap N(\overline{U}_2) = e. \tag{22}$$

According to Theorem 11.13 $\overline{U}_1$ and $\overline{U}_2$ are orispheric unipotent subgroups of the group G. Therefore, by Lemma 10.9 $N(\overline{U}_1)$ and $N(\overline{U}_2)$ are parabolic subgroups in G, further

$$\overline{U}_1 = R_u(N(\overline{U}_1)) \tag{23}$$

and

$$\overline{U}_2 = R_u(N(\overline{U}_2)), \tag{24}$$

where $R_u(N(\overline{U}_1))$ and $R_u(N(\overline{U}_2))$ denote the unipotent radicals of the groups $N(\overline{U}_1)$ and $N(\overline{U}_2)$ ((23) and (24) are true not because of Lemma 10.9 but because $\overline{U}_1$ and $\overline{U}_2$ are orispheric). As a consequence of (22), (23), (24) and Lemma 10.14, we have that the parabolic subgroups $N(\overline{U}_1)$ and $N(\overline{U}_2)$ are opposite. From the above said we get that U_1 and U_2 satisfy the conditions of Lemma 12.5 (it should only be noted that $U_1 \in B(\Gamma)$ and $U_2 \in B(\Gamma)$, in view of the maximality of U_1 and U_2 and Lemma 7.7), whence the assertion of our lemma follows.

§ 13. Ultramaximal unipotent subgroups

In this section we will use the notations of sections 7 and 11. Recall that G and Γ satisfy the conditions formulated at the beginning of § 7, and that π is the natural projection $\pi : G \to G/\Gamma$. Further, we will always assume (not only in this section) that the real rank of the group G (i.e., the dimension of a maximal **R**-split torus of G), is greater than or equal to two.

The following lemma is well known (see Corollary 10.3.1 in [8]).

Lemma 13.1. Any proper subgroup of an arbitrary nilpotent group is a proper subgroup of its normalizer.

Lemma 13.2. Let V be a unipotent subgroup in G, and $N(V)$ the normalizer of V in G. If all unipotent elements of the group $N(V)$ belong to V, then V is a maximal unipotent subgroup in G.

Proof. If V is not maximal, then V is a proper subgroup of some unipotent subgroup $\tilde{V} \subset G$. Now $\tilde{V}$ is nilpotent, according to Lemma 1.3.1. Therefore, in view of Lemma 13.1 V is a proper subgroup in $\tilde{N}(V)$, where $\tilde{N}(V)$ is the normalizer of V in $\tilde{V}$. On the other hand, since $(N(V))^U \subset V$, and $\tilde{V}$ is unipotent, we have $\tilde{N}(V) \subset V$ which is a contradiction. The lemma is proved.

Definition 13.3. A unipotent subgroup U of the group Γ is called *ultramaximal*, if:

1) U is a maximal unipotent subgroup in Γ;

2) $\dim \tilde{A}(U)=\dim \bar{U}$ (recall that $\tilde{A}(U)=A(\bar{S}N(\bar{U})\cap\Gamma)$).

Lemmas 11.21 and 13.2 imply the following

Lemma 13.4. For any ultramaximal unipotent subgroup U of the group Γ, $\bar{U}$ is a maximal unipotent subgroup of the group G.

Recall that on the Lie algebra $\mathfrak{G}$ of the group G a Euclidean metric $\varrho_{\mathfrak{G}}$ is introduced, and on the group G the right invariant metric ϱ_G corresponding to $\varrho_{\mathfrak{G}}$.

Lemma 13.5. Let V be an arbitrary maximal unipotent subgroup of G, $\mathfrak{B}$ the Lie algebra of V, and T an arbitrary maximal **R**-split torus of G normalizing V. Then there exists a one-parameter subgroup $t(s)$ of the torus T such that

$$\lim_{s\to+\infty}\ \inf_{\mathfrak{v}\in\mathfrak{B}} \frac{\varrho_{\mathfrak{G}}(0, (\mathrm{Ad}\,(t(s)))\mathfrak{v})}{\varrho_{\mathfrak{G}}(0, \mathfrak{v})}=\infty.$$

Proof. Lemma 13.5 is, in fact, well known. Therefore, we mention only that we can take $\exp(s\mathfrak{t})$ as $t(s)$, where $\mathfrak{t}$ is an arbitrary element of the Lie algebra of the torus T contained in the open Weyl chamber, which corresponds to the maximal unipotent subgroup V.

Recall the following

Lemma 13.6 (Zassenhaus [9]). There exists in G a sufficiently small neighbourhood W of the unity such that any discrete subgroup of the group G generated by elements contained in the closure of W is nilpotent.

Lemma 13.7. If all maximal unipotent subgroups of the group Γ are ultramaximal, then in Γ there exist two maximal unipotent subgroups U_1 and U_2, such that $\bar{U}_1\cap\bar{U}_2\neq e$ and $\bar{U}_1\neq\bar{U}_2$ (recall that the real rank of G is greater than or equal to two).

Proof. Let U_1 be an arbitrary maximal unipotent subgroup of Γ. Since all maximal unipotent subgroups of Γ are ultramaximal, in view of Lemma 13.4 $\bar{U}_1$ is a maximal unipotent subgroup of G. Therefore, in view of Lemmas 10.17 and 10.16, there is in G a maximal **R**-split torus T normalizing $\bar{U}_1$. Denote by $\tilde{T}$ the algebraically connected component of the identity in the intersection of T and $\bar{S}N(\bar{U}_1)$. The real rank of G is greater than 1, and $\dim N(\bar{U}_1)-\dim \bar{S}N(\bar{U}_1)\leqq 1$ (because $\bar{S}N(\bar{U}_1)$ is the kernel of an algebraic homomorphism of $N(\bar{U}_1)$ into the multiplicative group of real numbers), thus, by Lemma 1.2.7

$$\dim \tilde{T}\geqq 1. \tag{1}$$

Since T is an **R**-split torus in G, and $\bar{S}N(\bar{U})$ is an algebraic subgroup in G

(in view of Lemma 11.6), $\tilde{T}$ is an **R**-split torus in G. From this and (1) we get that the torus $\tilde{T}$ is not compact in the topological sense.

Since U_1 is ultramaximal (for, by assumption, all maximal unipotent subgroups of Γ are ultramaximal), we have

$$\dim \tilde{A}(U_1) = \dim \overline{U}_1. \tag{2}$$

Now $\tilde{A}(U_1)$ and $\overline{U}_1$ are algebraic subgroups (by their definition). Therefore, it follows from (2) and Lemma 1.2.10 that $\tilde{A}(U_1)/\overline{U}_1$ is finite. Now, since $\tilde{A}(U)= =A(\bar{S}N(\overline{U}_1)\cap\Gamma)$, and $\overline{U}_1=A(U_1)$, $\bar{S}N(\overline{U}_1)\cap\Gamma/U_1$ is also finite. Therefore, the natural mapping π': $\bar{S}N(\overline{U}_1)/U_1\to\bar{S}N(\overline{U}_1)/\bar{S}N(\overline{U}_1)\cap\Gamma$ is proper. Denote by π_1 and $\bar{\pi}_1$ the natural mappings $\pi_1:\bar{S}N(\overline{U}_1)\to\bar{S}N(\overline{U}_1)/U_1$ and $\bar{\pi}_1:\bar{S}N(\overline{U}_1)\to\bar{S}N(\overline{U}_1)/\overline{U}_1$.

Since $\overline{U}_1$ is a maximal unipotent subgroup of G, and T normalizes $\overline{U}_1$, in view of Lemmas 10.17 and 10.18 the centralizer $C(T)$ of the torus T is a Levi subgroup in $N(\overline{U}_1)$. This implies, since $\tilde{T}$ is not compact, that $\bar{\pi}_1(\tilde{T})$ is not compact in $\bar{S}N(\overline{U}_1)/\overline{U}_1$ (we say that the subset A of the topological space X is compact in X, if it is a relatively compact subset in X). Now since $\overline{U}_1\supset U_1$, we have that $\pi_1(\tilde{T})$ is not compact in $\bar{S}N(\overline{U}_1)/U_1$. Therefore, since the mapping π' is proper, $\pi'(T)$ is not compact in $\bar{S}N(\overline{U}_1)/\bar{S}N(\overline{U}_1)\cap\Gamma$. From this and Lemma 7.3 we get that $\pi(\tilde{T})$ is not compact in G/Γ.

Denote by $\mathfrak{U}_1$ the Lie algebra of the algebraic unipotent group $\overline{U}_1$, and by ln the logarithmic mapping $\ln:\overline{U}_1\to\mathfrak{U}_1$. Let b be the integer constant from § 5 occurring in the definition of the quasi-lattice Δ_U (see the section preceding Lemma 5.7). From the definition of b and the quasi-lattice Δ_U we have

$$b\ln U_1\subset\Delta_{U_1}\subset\ln U_1. \tag{3}$$

Let $W\subset G$ be the neighbourhood of the identity occurring in Lemma 13.6, and $\tilde{W}\subset W$ a neighbourhood of the identity, such that for any $w\in\tilde{W}$, $w^b\in W$ and let $K(\tilde{W})\subset G/\Gamma$ be the compactum of Lemma 7.5. Denote by $\mathfrak{W}\subset\mathfrak{U}_1$ the neighbourhood of zero in $\mathfrak{U}_1$, defined by the equality $\mathfrak{W}=\ln(W\cap\overline{U}_1)$. Recall that for a quasi-lattice Δ, $d(\Delta)$ stands for the determinant of Δ (see § 3).

Since $\pi(\tilde{T})$ is not compact in G/Γ, and $\tilde{T}\subset N(\overline{U}_1)$, thus, by virtue of Lemma 7.8 $(\mathrm{Ad}\,\tilde{T})\Delta_{U_1}$ is non-compact in the space of quasi-lattices. On the other hand, since $\tilde{T}\subset\bar{S}N(\overline{U}_1)$, and Δ_{U_1} is a quasi-lattice of full dimension in $\mathfrak{U}_1$ (in view of Lemma 7.1), thus

$$d((\mathrm{Ad}\,t)\Delta_{U_1}) = d(\Delta_{U_1}) \quad \text{for any} \quad t\in\tilde{T}. \tag{4}$$

For any quasi-lattice $\Delta\subset\mathfrak{G}$ write $\|\Delta\|$ for the infimum $\inf\limits_{\delta\in\Delta-0}\varrho_{\mathfrak{G}}(0,\delta)$. Now $(\mathrm{Ad}\,\tilde{T})\Delta_{U_1}$ is not compact in the space of quasi-lattices, thus, in view of (4) and Mahler's criterion of compactness (see § 3) we obtain that

$$\inf_{t\in\tilde{T}}\|(\mathrm{Ad}\,t)\Delta_{U_1}\| = 0. \tag{5}$$

In consequence of Lemma 13.5 there is in T a one-parameter subgroup $t(s)$, such that

$$\lim_{s\to+\infty} \inf_{\mathfrak{u}\in\mathfrak{U}_1} \frac{\varrho_{\mathfrak{G}}(0, \operatorname{Ad}(t(s))\mathfrak{u})}{\varrho_{\mathfrak{G}}(0, \mathfrak{u})} = \infty. \tag{6}$$

It follows from (6) that for any quasi-lattice $\Delta\subset\mathfrak{U}_1$ we have

$$\lim_{s\to+\infty} \frac{\|(\operatorname{Ad}(t(s)))\Delta\|}{\|\Delta\|} = \infty. \tag{7}$$

Since any subneighbourhood of a neighbourhood obeying the conclusion of Lemma 13.6 obeys this conclusion itself, we can assume, without loss of generality, that for some $\tilde{\varepsilon}>0$ $\mathfrak{W}$ is an $\tilde{\varepsilon}$-neighbourhood of zero in $\mathfrak{U}_1$. From (6) and (7) we obtain that for any $\tilde{t}\in T$ there exists a number $s(t)$ such that

$$\|\operatorname{Ad}(t(s(\tilde{t})))((\operatorname{Ad}\tilde{t})\Delta_{U_1})\| = \tilde{\varepsilon}. \tag{8}$$

Denoting the product $\tilde{t}\cdot t(s(\tilde{t}))$ by $\varphi(\tilde{t})\in T$ for any $\tilde{t}\in T$, we conclude from (8) that

$$\|(\operatorname{Ad}(\varphi(\tilde{t})))\Delta_{U_1}\| = \tilde{\varepsilon} \quad \text{for any} \quad \tilde{t}\in T. \tag{9}$$

(5) implies the existence of a sequence of elements $\tilde{t}_1, \ldots, \tilde{t}_n$ of the torus T for which

$$\lim_{n\to\infty} s(\tilde{t}_n) = \infty. \tag{10}$$

Since Δ_{U_1} is a quasi-lattice of full dimension in $\mathfrak{U}_1$ (in view of Lemma 7.1), it follows from (10), (6) and (4) that

$$\lim_{n\to\infty} d((\operatorname{Ad}(\varphi(\tilde{t}_n)))\Delta_{U_1}) = \infty. \tag{11}$$

For any element g of the group G, the distance $\varrho_G(e, g)$ is called the *norm* of g. Before completing the proof of Lemma 13.7 we formulate the following Lemmas 13.8 and 13.9.

Lemma 13.8. For any $c>0$ and any compactum $K\subset G/\Gamma$ there is a constant $a(c, K)$, such that if *1)* $g\in G$, *2)* $\pi(g)\in K$, *3)* U is a unipotent subgroup of the group $g\Gamma g^{-1}$ generated by elements of norm not exceeding c, then there is in $g\Gamma g^{-1}$ a maximal unipotent subgroup $\hat{U}$ containing U and generated by elements of norm not exceeding $a(c, K)$.

Lemma 13.9. For any $a>0$ there is a constant $k(a)$, such that if a discrete unipotent subgroup U of the group G is generated by elements of norm not exceeding a, then the determinant $d(\Delta_U)$ of the quasi-lattice Δ_U is less than $k(a)$.

Now we return to the proof of Lemma 13.7. Put $\varepsilon_0 = \sup_{w\in W} \varrho_G(e, w)$ (recall that W is defined in Lemma 13.6). For any n denote $\varphi(\tilde{t}_n)$ by φ_n. Then, in view of Lemma 13.9, from (11) it follows that for all n greater than some N, the subgroup $\varphi_n U_1 \varphi_n^{-1}$ cannot be generated by elements of norm not exceeding

$a(\varepsilon_0, K(\tilde{W}))$, where $a(\varepsilon_0, K(\tilde{W}))$ is defined in Lemma 13.8 (recall that $\tilde{W} \subset W$ is a neighbourhood of the identity such that $w^b \in W$ for any $w \in \tilde{W}$, where b is the same as in § 5, and $K(\tilde{W}) \subset G/\Gamma$ is the compactum mentioned in Lemma 7.5). Let $\tilde{n}$ be an arbitrary natural number greater than N. We analyse two cases:

$$1)\ \pi(\varphi_{\tilde{n}}) \subset K(\tilde{W}), \qquad 2)\ \pi(\varphi_{\tilde{n}}) \not\subset K(\tilde{W}).$$

Case 1. Since $\varphi_{\tilde{n}}$ stands for $\varphi(\tilde{t}_{\tilde{n}})$, in view of (9) we have that $\|(\mathrm{Ad}\,\varphi_{\tilde{n}})\Delta_{U_1}\| = \tilde{\varepsilon}$. Therefore, since *1)* $\tilde{\varepsilon}$ is the radius of the neighbourhood $\mathfrak{W} \subset \mathfrak{U}_1$, *2)* $\mathfrak{W} = \ln(W \cap \overline{U}_1)$, in view of (3) and the definition of the symbol $\|\Delta\|$, we see that the subgroup $\varphi_{\tilde{n}} U_1 \varphi_{\tilde{n}}^{-1}$ contains an element u_0 on the boundary of the neighbourhood W. Now $\varrho_G(e, u_0) < \varepsilon_0$ (as above, $\varepsilon_0 = \sup_{w \in W} \varrho_G(e, w)$). Therefore, since in the analysed case we have $\pi(\varphi_{\tilde{n}}) \subset K(\tilde{W})$, according to Lemma 13.8 u_0 is contained in a maximal unipotent subgroup U' of the group $\varphi_{\tilde{n}} \Gamma \varphi_{\tilde{n}}^{-1}$ generated by elements of norm not exceeding $a(\varepsilon_0, K(\tilde{W}))$. On the other hand, since $\tilde{n} > N$, $\varphi_{\tilde{n}} U_1 \varphi_{\tilde{n}}^{-1}$ cannot be generated by elements of norm not exceeding ε_0. The last two sentences imply that

$$U' \neq \varphi_{\tilde{n}} U_1 \varphi_{\tilde{n}}^{-1}. \tag{12}$$

On the other hand, U', as well as $\varphi_{\tilde{n}} U_1 \varphi_{\tilde{n}}^{-1}$ contains u_0. Therefore

$$U' \cap (\varphi_{\tilde{n}} U_1 \varphi_{\tilde{n}}^{-1}) \neq e. \tag{13}$$

Put $U_2 = \varphi_{\tilde{n}}^{-1} U' \varphi_{\tilde{n}}$. Since U' is a maximal unipotent subgroup of the group $\varphi_{\tilde{n}} \Gamma \varphi_{\tilde{n}}^{-1}$, U_2 is a maximal unipotent subgroup of the group Γ. Now it follows from (12) and (13) that

$$U_1 \neq U_2 \quad \text{and} \quad U_1 \cap U_2 \neq e. \tag{14}$$

Thus, in Γ we have found two maximal unipotent subgroups U_1 and U_2 satisfying (14). As mentioned several times, for any unipotent subgroup U the subgroup $\overline{U}$ is unipotent. Therefore, it follows from (14) and the maximality of U_1 and U_2 that

$$\overline{U}_1 \neq \overline{U}_2 \quad \text{and} \quad \overline{U}_1 \cap \overline{U}_2 \neq e. \tag{15}$$

Thus, the lemma is proved in Case 1.

Case 2. As in Case 1, we observe that $\|(\mathrm{Ad}\,\varphi_{\tilde{n}})\Delta_{U_1}\| = \tilde{\varepsilon}$. Since *1)* $\tilde{\varepsilon}$ is the radius of the neighbourhood $\mathfrak{W} \subset \mathfrak{U}_1$, *2)* $\mathfrak{W} = \ln(W \cap \overline{U}_1)$ *3)* $\tilde{W}^b \subset W$, in view of (3) and the definition of the symbol $\|\Delta\|$, the subgroup $\varphi_{\tilde{n}} U_1 \varphi_{\tilde{n}}^{-1}$ contains an element u_0 lying on the boundary of the neighbourhood W, but does not contain elements from $\tilde{W}$ (except the identity). On the other hand, in the analysed case we have $\pi(\varphi_{\tilde{n}}) \not\subset K(\tilde{W})$. Therefore, according to Lemma 7.5, the intersection $(\varphi_{\tilde{n}} \Gamma \varphi_{\tilde{n}}^{-1}) \cap \tilde{W}$ contains a non-identity unipotent element $\tilde{u}$. Since $\tilde{u} \in \tilde{W}$, $\tilde{u} \neq e$, and $(\varphi_{\tilde{n}} U_1 \varphi_{\tilde{n}}^{-1}) \cap \tilde{W}$ consists of the identity alone,

$$\tilde{u} \notin \varphi_{\tilde{n}} U_1 \varphi_{\tilde{n}}^{-1}. \tag{16}$$

Denote by $\hat{U}$ the subgroup generated by the union $\tilde{u} \cup u_0$. Then, it follows from (16) that

$$\hat{U} \not\subset \varphi_{\tilde{n}} U_1 \varphi_{\tilde{n}}^{-1}. \tag{17}$$

On the other hand, $\hat{U}$, as well as $\varphi_{\tilde{n}} U_1 \varphi_{\tilde{n}}^{-1}$ contains u_0. Therefore,

$$\hat{U} \cap (\varphi_{\tilde{n}} U_1 \varphi_{\tilde{n}}^{-1}) \neq e. \tag{18}$$

Since *1)* $\tilde{u}$, as well as u_0 is contained in the closure of the neighbourhood W ($\tilde{u} \in W$, since $\tilde{u} \in \tilde{W}$ and $\tilde{W} \subset W$) and in the discrete subgroup $\varphi_{\tilde{n}} \Gamma \varphi_{\tilde{n}}^{-1}$, *2)* W is the neighbourhood mentioned in the Lemma 13.6, by virtue of Lemma 13.6 $\hat{U}$ is a nilpotent subgroup. From this, the unipotence of $\tilde{u}$ and u_0, and Lemma 7.4 we get that $\hat{U}$ is unipotent. Let U' be some maximal unipotent subgroup of the group $\varphi_{\tilde{n}} U_1 \varphi_{\tilde{n}}^{-1}$ containing $\hat{U}$. Put $U_2 = \varphi_{\tilde{n}}^{-1} U' \varphi_{\tilde{n}}$. It follows from (17) and (18) that

$$U_1 \neq U_2 \quad \text{and} \quad \overline{U}_1 \cap \overline{U}_2 \neq e. \tag{19}$$

After this, the maximality of U_2 and the validity of (15) can be established in the same way as in Cases 1. Thus, the lemma is proved in Case 2. The proof of Lemma 13.7 is complete.

Remark. In the analysis of Case 1 and 2 we have used, without stating explicitly, the relation $\varphi_{\tilde{n}} U_1 \varphi_{\tilde{n}}^{-1} \subset U_1$ (since $\varphi_{\tilde{n}} \in T$ and T normalizes $\overline{U}_1$).

Now we pass to the proof of Lemmas 13.8 and 13.9.

Proof of Lemma 13.8. Let $\tilde{K} \subset G$ be a compact set for which $\pi(\tilde{K}) = K$ (the existence of $\tilde{K}$ is clear). If $g \in G$ and $\pi(g) \in K$, then $g = k(g) \cdot \gamma(g)$, where $k(g) \in \tilde{K}$ and $\gamma(g) \in \Gamma$. Let U be an arbitrary unipotent subgroup of the group $g \Gamma g^{-1}$, and let $\pi(g) \in K$. Then $k^{-1}(g) U k(g) = \gamma(g)(g^{-1} \Gamma g) \gamma^{-1}(g) \subset \Gamma$. Thus, U is conjugate to some unipotent subgroup of Γ by an element of the compactum $\tilde{K}$. After this, the lemma can easily be deduced from the following remarks: *1)* The number of all unipotent subgroups of the discrete group Γ, generated by elements of norm not exceeding some given number, is finite, *2)* any unipotent subgroup of Γ is contained in some maximal unipotent subgroup of Γ, *3)* in view of Lemmas 5.4 and 5.1 any unipotent subgroup (and consequently the maximal ones, too) of the discrete group Γ can be generated by a finite number of elements.

Proof of Lemma 13.9. Let U be a discrete unipotent subgroup of the group G generated by elements $u_1, \dots, u_k$ of norm not exceeding a, and let ln be the logarithmic mapping of $\overline{U}$ into the Lie algebra $\mathfrak{U}$ of the group $\overline{U}$. It is easy to see that there exists a constant $k_1(a)$ such that

$$\varrho_{\mathfrak{G}}(0, \ln u_i) < k_1(a) \tag{20}$$

for any i $(1 \leqq i \leqq k)$. Since $u_1, \dots, u_k$ generate a unipotent subgroup U, it follows easily from the Campbell—Zassenhaus formula that $\ln u$ belongs to the real linear subspace spanned by all $\ln u_1, \ldots, \ln u_k$. Therefore, in view of (3) and Lemma 7.1, $\ln u_1, \ldots, \ln u_k$ form a basis in $\mathfrak{U}$. This and (3) imply that the determinant of the quasi-lattice Δ_U is not greater than b^k-times the determinant of the quasi-lattice spanned on $\ln u_1, \ldots, \ln u_k$. From this fact and (20) the assertion of the lemma follows easily.

Lemma 13.10. Let U be an arbitrary maximal unipotent subgroup of the group Γ, and $Z(\overline{U})$ the centre of the group $\overline{U}$. Then *1)* $Z(\overline{U}) \cap \Gamma$ is Zariski dense in $Z(\overline{U})$, *2)* the factor space $Z(\overline{U})/Z(\overline{U}) \cap \Gamma$ is compact.

Proof. In view of Lemma 11.18

$$Z(\overline{U}) = C(\overline{U}). \tag{21}$$

From (21) and Lemma 7.12, part *2)* of the lemma follows. Now part *2)*, Lemma 5.6 and the fact that $Z(\overline{U})$ is an algebraic subgroup, imply part *1)*. The lemma is proved.

Remark. It is not difficult to show that the assertion of the lemma holds for all unipotent subgroups of the group Γ, and more generally, an analogous assertion holds for all uniform discrete subgroups of an arbitrary algebraic unipotent group.

Definition 13.11. Two orispheric subgroups of a group G are called *opposite*, if their normalizers are opposite parabolic subgroups.

Recall that for any subset F of an arbitrary algebraic group, F^u stands for the set of all unipotent elements in F.

Definition 13.12. Let H be a semisimple real (complex) algebraic group, V a maximal unipotent subgroup in H, $Z(V)$ the centre of V, and $C(Z(V))$ the centralizer of $Z(V)$ in H. The group H is *rich* if $(C(Z(V)))^u \not\subset V$ and it is *poor* if $(C(Z(V)))^u \subset V$.

Remark. Since any two maximal unipotent subgroups in an arbitrary algebraic **R**-group (**C**-group) are conjugate (in view of Lemma 10.20), a group H will be rich or poor independently of the choice of V.

Lemma 13.13. Let U_1 and U_2 be two ultramaximal unipotent subgroups of the group Γ. If *1)* the group G is rich, *2)* $\overline{U}_1 \cap \overline{U}_2 = e$, then there are two non-unit unipotent subgroups U_1' and U_2' of the group Γ such that a) $\overline{U}_1' \subset \overline{U}_1$ b) $U_2' \subset \overline{U}_2$, c) $\overline{U}_1'$ and $\overline{U}_2'$ are not maximal unipotent subgroups of G, d) $\overline{U}_1'$ and $\overline{U}_2'$ are opposite orispheric subgroups of the group G.

Proof. We observe, first of all, that in view of Lemma 13.4 $\overline{U}_1$ and $\overline{U}_2$ are maximal unipotent subgroups of G. Therefore, since two maximal unipotent subgroups in G are conjugate to each other (in view of Lemma 10.20), and the group G is rich, we get

$$(C(Z(\overline{U}_1)))^u \not\subset \overline{U}_1 \quad \text{and} \quad (C(Z(\overline{U}_2)))^u \not\subset \overline{U}_2, \tag{21}$$

where $Z(\bar{U}_1)$ and $Z(\bar{U}_2)$ are the centres of the groups $\bar{U}_1$ and $\bar{U}_2$, and $C(Z(\bar{U}_1))$ and $C(Z(\bar{U}_2))$ are the centralizers of the subgroups $Z(\bar{U}_1)$ and $Z(\bar{U}_2)$ in the group G.

Since $\bar{U}_2$ is unipotent, $\bar{U}_2$ is nilpotent by Lemma 1.3.1. Therefore $Z(\bar{U}_2)\neq e$. On the other hand $Z(\bar{U}_2)$ is unipotent. These last two sentences and the semi-simplicity of G imply that

$$(C(Z(\bar{U}_2)))^u \neq G^u. \tag{22}$$

As we have already mentioned, $\bar{U}_1$ and $\bar{U}_2$ are maximal unipotent subgroups of G. Therefore, since $(C(Z(\bar{U}_2)))^u \not\subset \bar{U}_2$ in view of (21), thus on account of (22) and Lemma 10.30 the minimal algebraic normal subgroup N_1 of the group $\bar{U}_1$ containing the intersection $C(Z(\bar{U}_2))\cap\bar{U}_1$ has the following properties: *1)* $N_1\neq e$ *2)* $N_1\neq\bar{U}_1$, *3)* N_1 is an orispheric subgroup.

Denote $\Gamma\cap Z(\bar{U}_2)$ by U_3. Since U_2 is a maximal unipotent subgroup of Γ, as a consequence of Lemma 13.10, we have $\bar{U}_3=Z(\bar{U}_2)$. Therefore, according to Lemma 7.10 the subgroup $C(Z(\bar{U}_2))$ has property P. On the other hand, in view of Lemma 5.4 the subgroup $\bar{U}_1$ is integer (for the definition of integer subgroups see § 6). These last two sentences and Lemma 6.8 imply that $\bar{U}_1\cap C(Z(\bar{U}_2))$ is an integer subgroup. Now this and Lemma 5.6 imply that $\bar{U}_1\cap C(Z(\bar{U}_2))\cap\Gamma$ is Zariski dense in $\bar{U}_1\cap C(Z(\bar{U}_2))$. On the other hand, by the definition of $\bar{U}_1$, U_1 is Zariski dense in $\bar{U}_1$. These two assertions and Lemma 1.2.19 imply that the normal subgroup of the group $\bar{U}_1\cap\Gamma=U_1$ generated by $\bar{U}_1\cap C(Z(\bar{U}_2))\cap\Gamma$ ($\bar{U}_1\cap\Gamma=U_1$, since U_1 is an ultramaximal, and consequently maximal unipotent subgroup of Γ), is Zariski dense in the minimal algebraic normal subgroup of the group $\bar{U}_1$ containing $C(Z(\bar{U}_1))\cap\bar{U}_2$, i.e. in N_1. From this we get that $N_1\cap\Gamma$ is Zariski dense in N_1. Put $U_1'=N_1\cap\Gamma$ and note that *1)* $\bar{U}_1'=N_1$ since $N_1\cap\Gamma$ is Zariski dense in the algebraic subgroup N_1, *2)* $\bar{U}_1'=N_1$ is orispheric, *3)* $\bar{U}_1'=N_1\neq e$, *4)* $\bar{U}_1'=N_1\neq\bar{U}_1$ and consequently $\bar{U}_1'$ is not a maximal unipotent subgroup in G.

Taking now the subgroup U_2 instead of U_1, define the subgroups N_2 and U_2' analogously to N_1 and U_1', and note that N_2 and U_2' have similar properties as N_1 and U_1' i.e. *1)* $\bar{U}_2'=N_2$, *2)* $\bar{U}_2'=N_2$ is orispheric, *3)* $\bar{U}_2'=N_2\neq e$, *4)* $\bar{U}_2'=N_2\neq\bar{U}_2$ (we apply here (21)). Therefore, in order to complete the proof of the lemma it remains to show that the orispheric subgroups $\bar{U}_1'$ and $\bar{U}_2'$ are opposite. To this end, we formulate the following well-known

Lemma 13.14. Let H be a real algebraic group, and T a maximal **R**-split torus of the group H. Then there exists an algebraic anti-automorphism I_T of the group H having the following properties:

1) $I_T^2=E$, i.e. the anti-automorphism I_T is an involution,

2) for any $t\in T$, $I_T(t)=t$, i.e. all elements of the torus T remain fixed under the action of I_T.

Remark. Since the author is not able to give any reference, the proof of Lemma 13.14 will be presented in § 18.

Now we continue the proof of Lemma 13.13. Since $\overline{U}_1$ and $\overline{U}_2$ are maximal unipotent subgroups of G, in view of Lemmas 10.17 and 10.16 there is a maximal **R**-split torus T of the group normalizing both $\overline{U}_1$ and $\overline{U}_2$. Let Σ be the root system of the group G with respect to the torus T. For any $\alpha \in \Sigma$ we denote by $\mathfrak{G}_\alpha$ the root subspace corresponding to the root α, and by G_α the set $\exp(\mathfrak{G}_\alpha)$ (note that G_α is not necessarily a subgroup).

According to Lemma 13.14 there exists an algebraic anti-automorphism I_T of the group G leaving the elements of the torus T fixed. Then, for any $\alpha \in \Sigma$ we can show that

$$I_T(G_\alpha) = G_{-\alpha} \tag{23}$$

(equality (23) is immediate from the definition of the sets G_α and the fact that I_T is an algebraic anti-automorphism leaving all elements of the torus T fixed). Since *1)* $\overline{U}_1$ and $\overline{U}_2$ are maximal unipotent subgroups of G normalized by the maximal **R**-split torus T, *2)* $\overline{U}_1 \cap \overline{U}_2 = e$, thus it is easily seen from (23) that $I_T(\overline{U}_1) = \overline{U}_2$ (here we have applied the classification of all orispheric subgroups normalized by some maximal **R**-split torus, introduced in § 10). Then it is easy to see from the definition of subgroups N_1 and N_2 that the following equality holds: $I_T(N_1) = N_2$, whence, in turn, it follows that $I_T(N(N_1)) = N(N_2)$, where $N(N_1)$ and $N(N_2)$ denote the normalizers of the subgroups N_1 and N_2 in the group G.

Since the subgroups $\overline{U}_1$ and $\overline{U}_2$ are normalized by the torus T, their centres $Z(\overline{U}_1)$ and $Z(\overline{U}_2)$ are also normalized by T. Therefore, as an easy consequence of the definition of the subgroups N_1, N_2, $N(N_1)$ and $N(N_2)$, we get that all of these subgroups are normalized by T. On the other hand, as shown before, the subgroups N_1 and N_2 are orispheric, thus, in view of Lemma 10.9 their normalizers $N(N_1)$ and $N(N_2)$ are parabolic subgroups. Thus, from the last two sentences, the above mentioned classification, the equality $I_T(N(N_1)) = N(N_2)$ and equality (23) it follows easily that $N(N_1)$ and $N(N_2)$ are opposite parabolic subgroups. Consequently, $N(N_1)$ and $N(N_2)$ are opposite orispheric subgroups. Now in order to complete the proof of the lemma, it remains only to recall that $\overline{U}'_1 = N_1$ and $\overline{U}'_2 = N_2$.

Remark. In the sequel we show that the assertion of Lemma 13.13 will hold true in case the group G is poor (assuming, of course, that the real rank of G is greater than or equal to two).

Lemma 13.15. Assume that *1)* the group G is poor, *2)* all maximal unipotent subgroups of the group Γ are ultramaximal. Then there is in Γ a unipotent subgroup $U \neq e$, such that *1)* $\overline{U}$ is an orispheric subgroup of G, *2)* $\overline{U}$ is not a maximal unipotent subgroup of G (recall that the real rank of G is greater than 1).

Proof. According to Lemma 13.7 there are in Γ two maximal unipotent subgroups U_1 and U_2 such that

$$\overline{U}_1 \cap \overline{U}_2 \neq e \quad \text{and} \quad \overline{U}_1 \neq \overline{U}_2. \tag{24}$$

By the assumption of the lemma all maximal unipotent subgroups of Γ are ultramaximal, thus, in view of Lemma 13.4 $\overline{U}_1$ and $\overline{U}_2$ are maximal unipotent subgroups of G. On the other hand, by virtue of Lemma 5.4 $\overline{U}_1$ and $\overline{U}_2$ are integer subgroups, whence applying Lemmas 6.7 and 6.8 we obtain that $\overline{U}_1 \cap \overline{U}_2$ is an integer subgroup. Since $\overline{U}_1 \cap \overline{U}_2$ is not compact (in view of (24) and Lemmas 1.3.3 and 1.3.1) we have therefore

$$\overline{U}_1 \cap \overline{U}_2 \cap \Gamma \neq e. \tag{25}$$

Since G and Γ satisfy the conditions formulated at the beginning of § 7, applying Lemma 4.8 we get from (25) that $\overline{U}_1 \cap \overline{U}_2$ is contained in no nontrivial algebraic normal subgroup of the group G.

Since G is poor, the above-mentioned fact and (24) imply, in view of Lemma 10.7, that the minimal algebraic subgroup A of the group G containing $(\overline{U}_1 \cap \overline{U}_2) \cup \cup Z(\overline{U}_1)$, is orispheric and not equal to $\overline{U}_1$ ($Z(\overline{U}_1$ is the centre of the group $\overline{U}_1$).

As mentioned already, $\overline{U}_1 \cap \overline{U}_2$ is an integer subgroup. Therefore, in view of Lemma 5.6 $\overline{U}_1 \cap \overline{U}_2 \cap \Gamma$ is Zariski dense in $\overline{U}_1 \cap \overline{U}_2$. On the other hand, from the maximality of $\overline{U}_1$ and Lemma 13.10 it follows that $Z(\overline{U}) \cap \Gamma$ is Zariski dense in $Z(\overline{U}_1)$. The last two statements and Lemma 1.2.18 imply that the subgroup generated by $(\overline{U}_1 \cap \overline{U}_2 \cap \Gamma) \cup (Z(\overline{U}_1) \cap \Gamma)$ is Zariski dense in A, and consequently $A \cap \Gamma$ is Zariski dense in A. Put $U = A \cap \Gamma$ and remark that *1)* $\overline{U} = A$, since $A \cap \Gamma$ is Zariski dense in the algebraic subgroup A, *2)* $\overline{U} = A$ is orispheric, *3)* $\overline{U}$ is not a maximal unipotent subgroup of the group G (since $\overline{U} = A$, $A \subset \overline{U}_1$ and $A \neq \overline{U}_1$), *4)* $\overline{U} = A \neq e$ (since *1)* $A \supset \overline{U}_1 \cap \overline{U}_2$, *2)* in view of (24) $\overline{U}_1 \cap \overline{U}_2 \neq e$). Thus, U is the desired subgroup. The lemma is proved.

Lemma 13.16. Let H be an algebraically connected semisimple real algebraic group, V_1 an arbitrary maximal unipotent subgroup of H, V an orispheric subgroup of H, contained in V_1. Denote by $N(V_1)$ and $N(V)$ the normalizers of V_1 and V in H. Then $N(V_1) \subset N(V)$.

Proof. In view of Lemmas 10.9 and 10.17 the normalizers $N(V)$ and $N(V_1)$ are parabolic subgroups in H. Therefore, according to Lemma 10.16 $N(V_1) \cap N(V)$ contains the centralizer $C(T)$ of some maximal **R**-split torus T. Thus, applying the remark before Lemma 10.25 of § 10, in view of Lemma 10.27 we have that

$$V = H_D, \quad V_1 = H_{D_1}, \tag{26}$$

and

$$D_1 \supset D. \tag{27}$$

Since V_1 is unipotent, (27) implies that $D_1 \cap D^- = \emptyset$ (D^- is the set of all α, for which $-\alpha \in D$). From this and Lemma 10.28 by virtue of (26) and (27) it easily follows that

$$V_1 \subset N(V). \tag{28}$$

On the other hand,

$$C(T) \subset N(V). \tag{29}$$

Since V_1 is a maximal unipotent subgroup in H, we have, in view of Lemma 10.18, that $C(T)$ is a Levi subgroup in $N(V_1)$. On the other hand, since V_1 is a maximal unipotent subgroup in G, V_1 is the unipotent radical in $N(V_1)$. These last two assertions and inclusions (28) and (29) imply the statement of the lemma.

Before formulating the next lemma, we recall that for any algebraic unipotent subgroup V of an algebraic group H, $Z(V)$ denotes the centre of V, and that the definition of the symbol $\bar{S}N(V)$ has been already given at the end of § 10.

Lemma 13.17. Let H be a poor, algebraically connected, semisimple, real algebraic group, whose complexification has a trivial centre, furthermore, let V_1 and V_2 be maximal unipotent subgroups of H, and V a non-unit orispheric subgroup of H contained in V_1. Denote by W_1 the intersection $\bar{S}N(V) \cap V_2$, by $\tilde{V}$ the minimal algebraic subgroup containing $W_1 \cup Z(V_2)$, by W_2 the intersection $\bar{S}N(\tilde{V}) \cap V_1$, and by W the minimal algebraic subgroup containing $W_1 \cup W_2$. If *1)* $V_1 \cap V_2 = e$, *2)* V is contained in no nontrivial algebraic normal subgroup of the group H, *3)* $W_1 \neq e$ and W_1 is contained in no nontrivial algebraic normal subgroup of H, *4)* $V \neq V_1$, then a) W is an orispheric subgroup, b) the set A of all $h \in H$ for which the orispheric subgroups hWh^{-1} are opposing to V, is not empty and is a Zariski-open subset in H.

Proof. We prove the lemma first in case H is simple. Then, since H is poor, in view of Lemma 10.6 the following two cases exhaust all possibilities:

a) the real rank of H is equal to one,

b) the root system of H with respect to an arbitrary maximal **R**-split torus is of type A_2.

Now we analyse these two cases.

Case a). Since the real rank of H is equal to 1, the well-known classification of parabolic subgroups given in § 10 guarantees that any nontrivial parabolic subgroup of H is a minimal parabolic subgroup. Therefore, in view of Lemmas 10.9 and 10.17 any non-unit orispheric subgroup of H is a maximal unipotent subgroup in H. From this and the condition of the lemma conclude that V is a maximal unipotent subgroup in H. Now, since $V \subset V_1$, we have $V = V_1$ that contradicts part *4)* of the lemma. Thus, in case a) the situation described in the condition of the lemma cannot occur.

Case b). From Lemmas 10.17 and 10.16 it follows that the intersection of normalizers $N(V_1)$ and $N(V_2)$ contains some maximal **R**-split torus T of the group H. Denote by Σ the root system of H with respect to T, and by $\Sigma^+ \subset \Sigma$

the set of positive roots corresponding to V_1 (see Lemma 10.4). In case b) Σ is of type A_2 and consequently Σ^+ consists of three roots: α_1, α_2 and $\alpha_1+\alpha_2$.

Now V is orispheric and belongs to the maximal unipotent subgroup V_1, and T normalizes V_1, thus, according to Lemma 13.16 T normalizes V. Therefore, since *1)* $V\neq e$, *2)* $V\neq V_1$, *3)* V is orispheric, by virtue of Lemma 10.27 two cases are possible

1) $V=H_D$ where $D=\{\alpha_1, \alpha_1+\alpha_2\}$

2) $V=H_D$ where $D=\{\alpha_2, \alpha_1+\alpha_2\}$

(here and in the following we apply the notations of § 10).

Consider case *1)* (case *2)* is considered analogously). Since $V_1\cap V_2=e$, and V_1 and V_2 are maximal unipotent subgroups,

$$V_2 = H_{\Sigma-\Sigma^+}. \tag{29}$$

Applying Lemma 10.28 we easily deduce from (29) that

$$W_1 = \bar{S}N(V)\cap V_2 = U_{-\alpha_2}, \tag{30}$$

where $U_{-\alpha_2}=\exp(\mathfrak{U}_{-\alpha_2})$, and $\mathfrak{U}_{-\alpha_2}$ is the root subspace corresponding to the root $-\alpha_2$. As an easy consequence of (29) we have

$$Z(V_2) = U_{-\alpha_1-\alpha_2}. \tag{31}$$

From (30) and (31) we get

$$\tilde{V} = V_D, \quad \text{where} \quad D = \{-\alpha_2, -\alpha_1-\alpha_2\}. \tag{32}$$

(We keep the notation introduced before Lemma 10.25 of § 10.)

Analogously to (30), we get

$$W_2 = U_{\alpha_1}. \tag{33}$$

From (30) and (33) it follows that

$$W = V_D, \quad \text{where} \quad D = \{\alpha_1, -\alpha_2\}. \tag{34}$$

Since $V=H_D$, where $D=\{\alpha_1, \alpha_1+\alpha_2\}$, it is easy to see that

$$V = V_D, \quad \text{where} \quad D = \{\alpha_1, \alpha_1+\alpha_2\}. \tag{35}$$

Let ${}_{\mathbf{R}}W(T, H)=N(T)/C(T)$ denote the relative Weyl group of the group H. Since Σ is of type A_2, it is easy to see from the consideration of system A_2 that there exists a $w\in{}_{\mathbf{R}}W(T, H)$ such that the transformation $\tilde{w}: \Sigma\to\Sigma$ corresponding to w transforms α_1 to $-\alpha_1-\alpha_2$ and $-\alpha_2$ to $-\alpha_1$. This implies the existence of an $h_0\in H$ such that

$$h_0Wh_0^{-1} = W_D, \quad \text{where} \quad D = \{-\alpha_1, -\alpha_1-\alpha_2\}. \tag{36}$$

From (35), (36) and Lemma 10.28 it follows easily that $h_0Wh_0^{-1}$ and V are opposite orispheric subgroups. Consequently, the set A is nonempty. Thus, we get from Lemma 10.13 (and Definition 13.11) that A is Zariski open in H. Thus the lemma is proved in case b). Now, since case a) has been examined earlier, we have proved the lemma in case H is simple.

Consider now the general case. Since H is an algebraically connected semisimple real algebraic group, whose complexification has no centre, according to Lemma 1.4.2 H decomposes into a direct product (in sense of § 1) of simple algebraic $\mathbf{R}$-groups: $H=\bigoplus_{i=1}^{n} H_i$. For any i $(1\leqq i\leqq n)$, we denote the intersections $V_1\cap H_i$, $V_2\cap H_i$ and $V\cap H_i$ by V_1^i, V_2^i, and V^i. We remark now that *1)* since V_1 and V_2 are maximal unipotent subgroups in H, it is easy to verify the following expansions: $V_1=\bigoplus_{i=1}^{n} V_1^i$ and $V_2=\bigoplus_{i=1}^{n} V_2^i$, further for any i $(1\leqq i\leqq n)$ V_1^i and V_2^i are maximal unipotent subgroups in H_i, *2)* since V is orispheric, it is easy to see that for any i $(1\leqq i\leqq n)$ V^i is orispheric, *3)* since V is contained in no nontrivial algebraic normal divisor of the group H, for any i $(1\leqq i\leqq n)$ we have $V^i\neq e$, *4)* since $V_1\cap V_2=e$, for any i $(1\leqq i\leqq n)$ we have $V_1^i\cap V_2^i=e$, *5)* since the group H is poor, it is easy to see that for any i $(1\leqq i\leqq n)$ the group H_i is poor, *6)* since $V_1=\bigoplus_{i=1}^{n} V_1^i$, it is easy to see that for any i $(1\leqq i\leqq n)$ $\bar{S}N(V_1)\cap H_i=\bar{S}N(V_1^i)\cap H_i$, and consequently, since $V_2\cap H_i=V_2^i$, we have

$$\bar{S}N(V_1)\cap V_2\cap H_i = \bar{S}N(V_1^i)\cap V_2^i.$$

From *6)* and the fact that $W_1=\bar{S}N(V)\cap V_2$ is non-identity and is contained in no nontrivial algebraic normal subgroup of H, it follows that for any i $(1\leqq i\leqq n)$ $\bar{S}N(V^i)\cap V_2^i\neq e$. The above implies that for any i $(1\leqq i\leqq n)$, H_i, V_1^i, V_2^i and V^i satisfy the conditions of the lemma. Since the lemma has already been proved in the case of simple groups, this in turn, implies that the assertion of the lemma holds for H, too. The proof of the lemma is complete.

Lemma 13.18. Let H be an algebraically connected semisimple real algebraic group, V an orispheric, but not maximal unipotent, subgroup of H, $\tilde{V}$ a maximal unipotent subgroup of H. Then $\bar{S}N(V)\cap\tilde{V}\neq e$.

Proof. Assume that

$$\bar{S}N(V)\cap\tilde{V} = e. \tag{37}$$

Since $\tilde{V}$ is unipotent, it is easy to see (cf. Lemma 11.12) that $\tilde{V}\cap N(V)=\tilde{V}\cap\bar{S}N(V)$ ($N(V)$ is the normalizer of V in H). Therefore, it follows from (37) that

$$N(V)\cap\tilde{V} = e. \tag{38}$$

Since $\tilde{V}$ is a maximal unipotent subgroup of H, all unipotent elements of the normalizer $N(\tilde{V})$ belong to $\tilde{V}$. Therefore, since $V\cap\tilde{V}=e$ (in view of (38)) and V is unipotent,

$$N(\tilde{V})\cap V = e. \tag{39}$$

Denote by $R_u(N(V))$ and $R_u(N(\tilde{V}))$ the unipotent radicals of the groups $N(V)$ and $N(\tilde{V})$. Since V is orispheric, we have

$$V = R_u(N(V)). \tag{40}$$

Since $\tilde{V}$ is a maximal unipotent subgroup of H, it is easy to see that

$$\tilde{V}=R_u(N(\tilde{V})). \tag{41}$$

On the other hand, since V is orispheric and $\tilde{V}$ is a maximal unipotent subgroup of H, in view of Lemmas 10.9 and 10.17 $N(V)$ and $N(\tilde{V})$ are parabolic subgroups in H. From this, (38), (39), (40), (41) and Lemma 10.14 it follows that $N(V)$ and $N(\tilde{V})$ are opposite parabolic subgroups. Therefore, since $\tilde{V}$ is a maximal unipotent subgroup in H, from Lemmas 10.12 and 10.17 and equality (41) we conclude that V is a maximal unipotent subgroup in H. But this contradicts the condition of the lemma. The lemma is proved.

We recall again that G and Γ satisfy the conditions formulated at the beginning of this section and § 7.

Lemma 13.19. Assume that *1)* the group G is poor, *2)* all maximal unipotent subgroups of Γ are ultramaximal. Then there are in Γ two non-unit unipotent subgroups U_1' and U_2' such that *1)* $\overline{U}_1'$ and $\overline{U}_2'$ are not maximal unipotent subgroups in G, *2)* $\overline{U}_1'$ and $\overline{U}_2'$ are opposite orispheric subgroups in G.

Proof. Since the conditions of the lemma coincide with those of Lemma 13.15, there exists in Γ a non-unit unipotent subgroup U such that $\overline{U}$ is an orispheric but not maximal unipotent subgroup of the group G. It is clear further, that we can assume $U\in B(\Gamma)$, i.e. that $U=\overline{U}\cap\Gamma$ (for, in the opposite case, we could replace U by $\overline{U}\cap\Gamma$).

Let U_1 be an arbitrary maximal unipotent subgroup of Γ containing U. Since all maximal unipotent subgroups in Γ are ultramaximal (by assumption), on account of Lemma 13.4 $\overline{U}_1$ is a maximal unipotent subgroup of the group G.

Since G and Γ satisfy the conditions formulated at the beginning of § 7, according to the theorem of A. Borel on the algebraic density of discrete subgroups applied several times (see [2]) we have that Γ is Zariski dense in G. Therefore, since G is semisimple, and $\overline{U}_1$ is unipotent, in view of Lemma 12.1 there exists a $\gamma_0\in\Gamma$ such that $(\gamma_0\overline{U}_1\gamma_0^{-1})\cap\overline{U}_1=e$. Denote $\gamma_0 U_1\gamma_0^{-1}$ by U_2 and remark that $\overline{U}_2=\gamma_0\overline{U}_1\gamma_0^{-1}$ (see (1) of § 7).

Setting $H=G$, $V_1=\overline{U}_1$, $V_2=\overline{U}_2$ and $V=\overline{U}$, we show that H, V_1, V_2 and V satisfy the conditions of Lemma 13.17. To this end, we make the following remarks:

Remark 1. Since $V_1=\overline{U}_1$, $V_2=\overline{U}_2=\gamma_0\overline{U}_1\gamma_0^{-1}$ and $\overline{U}_1\cap(\gamma_0\overline{U}_1\gamma_0^{-1})=e$, thus $V_1\cap V_2=e$.

Remark 2. Since $U\neq e$, further G and Γ satisfy the conditions formulated at the beginning of § 7, in view of Lemma 4.8 U is not contained in any nontrivial algebraic normal subgroup of the group G, and, therefore, neither is $V\subset\overline{U}$.

Remark 3. Since $V=\overline{U}$ is an orispheric subgroup but not a maximal unipotent subgroup of the group $H=G$, and $V_2=\overline{U}_2=\gamma_0\overline{U}_1\gamma_0^{-1}$ is a maximal unipotent subgroup of the group $H=G$ (because $\overline{U}_1$ is a maximal unipotent subgroup in G), we have, in view of Lemma 13.18,

$$W_1 = \bar{S}N(V)\cap V_2 \neq e. \tag{42}$$

Remark 4. Since $U\in B(\Gamma)$, by Lemma 7.3 $\bar{S}N(\overline{U})$ has property P. On the other hand, in view of Lemma 5.4 the subgroup $\overline{U}_2$ is integer. The last two sentences and Lemma 6.8 imply that the subgroup $W_1=\bar{S}N(V)\cap V_2=\bar{S}N(\overline{U})\cap\overline{U}_2$ is integer. It is clear, on the other hand, that W_1 is an algebraic unipotent group. The last two statements, (42) and Lemma 5.6 imply that

$$W_1\cap\Gamma \neq e. \tag{43}$$

Since G and Γ satisfy the conditions formulated at the beginning of § 7, in view of Lemma 4.8 it follows from (43) that W_1 is contained in no nontrivial algebraic normal subgroup of the group $H=G$.

Remark 5. Since $V=\overline{U}$ is an orispheric, but not maximal unipotent, subgroup of the group $H=G$ and $V_1=\overline{U}_1$ is a maximal unipotent subgroup of the group $H=G$, we have $V\neq V_1$.

Remark 6. Since G is poor, and satisfies the conditions formulated at the beginning of § 7, and $H=G$, H is a poor algebraically connected semisimple real algebraic group whose complexification has a trivial centre.

It follows from Remarks 1—6 that H, V_1, V_2 and V satisfy the conditions of Lemma 13.17.

As mentioned in Remark 4, the algebraic unipotent group W_1 is an integer subgroup. Therefore, in view of Lemma 5.6 $W_1\cap\Gamma$ is Zariski dense in W_1. On the other hand, since $V_2=\overline{U}_2$ is a maximal unipotent subgroup in $H=G$, by virtue of Lemma 13.10 $Z(V_2)\cap\Gamma$ is Zariski dense in $Z(V_2)$ ($Z(V_2)$ is the centre of the group V_2). The last two statements and Lemma 1.2.18 imply that the subgroup generated by $W_1\cap\Gamma$ and $Z(V_2)\cap\Gamma$ is Zariski dense in the minimal algebraic subgroup $\tilde{V}$ containing $W_1\cup Z(V_2)$. Therefore $\tilde{V}\cap\Gamma$ is Zariski dense in $\tilde{V}$.

Denote $\tilde{V}\cap\Gamma$ by U_3 and remark that $\overline{U}_3=\tilde{V}$ (since $\tilde{V}\cap\Gamma$ is Zariski dense in $\tilde{V}$), and that $U_3\in B(\Gamma)$ (since $U_3=\tilde{V}\cap\Gamma=\overline{U}_3\cap\Gamma$). Therefore, according to Lemma 7.3 $\bar{S}N(\tilde{V})=\bar{S}N(\overline{U}_3)$ has property P. Repeating the arguments used for W_1, it is easy to show that $W_2\cap\Gamma$ is Zariski dense in W_2, where $W_2=\bar{S}N(\tilde{V})\cap V_1$. On the other hand, as mentioned already, $W_1\cap\Gamma$ is Zariski dense in W_1. The last two assertions and Lemma 1.2.18 imply that $W\cap\Gamma$ is Zariski dense in W, where W is the minimal algebraic subgroup containing $W_1\cup W_2$.

Since H, V_1, V_2 and V satisfy the conditions of Lemma 13.17, we have a) W is an orispheric subgroup, b) the set A of all $h\in H$ for which the orispheric subgroups hWh^{-1} and V are opposite, is non-empty and is a Zariski open subset in H. Since Γ is Zariski dense in H and A is non-empty and Zariski open in H, there is a $\gamma_1\in\Gamma$ such that $\gamma_1W\gamma_1^{-1}$ and V are opposite. Put now $U_1'=V\cap\Gamma$ and $U_2'=(\gamma_1W\gamma_1^{-1})\cap\Gamma$ (since $U\in B(\Gamma)$, $U_1'=U$). We show that U_1' and U_2' are the desired subgroups. We mention to this end that a) $V=\overline{U}'_1$ since $U\subset\Gamma$, $V=\overline{U}$, and $U'_1=V\cap\Gamma$, b) since $W\cap\Gamma$ is Zariski dense in W, we have $\overline{U}_2=\gamma_1 W\gamma_1^{-1}$, c) $\overline{U}'_1=V$ is an orispheric, but not maximal unipotent, subgroup in G, d) since $\overline{U}'_2=\gamma_1 W\gamma_1^{-1}$ is an orispheric subgroup opposing to V, in view of Lemmas 10.17 and 10.12 it is not a maximal unipotent subgroup of the group G. These remarks show that U'_1 and U'_2 are the desired subgroups. The lemma is proved.

Before passing to Theorem 13.20 we recall that the definition of $B(\Gamma)$ is given in § 7 and that the real rank of the group G is greater than or equal to two.

Theorem 13.20 Assume that all maximal unipotent subgroups of the group Γ are ultramaximal. Then there are in Γ non-unit $U^1\in B(\Gamma)$ and $U^2\in B(\Gamma)$ such that *1)* $\overline{U}^1$ and $\overline{U}^2$ are not maximal unipotent subgroups of the group G, *2)* $\overline{U}^1$ and $\overline{U}^2$ are opposite orispheric subgroups of the group G.

Proof. We distinguish two cases:

a) the group G is rich,

b) the group G is poor.

Case a). Let U_1 be an arbitrary maximal unipotent subgroup of the group Γ. We mentioned several times that Γ is Zariski dense in G (in view of the conditions of § 7 and Borel's theorem on the algebraic density of discrete subgroups). From this, the semisimplicity of G and Lemma 12.1 we obtain that there exists $\gamma\in\Gamma$ such that $\gamma\overline{U}_1\gamma^{-1}\cap U_1=e$. Put $U_2=\gamma U_1\gamma^{-1}$. Since all maximal unipotent subgroups of Γ are ultramaximal, and U_1 is maximal, U_1 and U_2 satisfy the conditions of Lemma 13.13 (recall that in the analysed case the group G is rich). Therefore in case a) the assertion of the theorem follows from Lemma 13.13 and the following

Remark 1. We can assume that in Lemma 13.13 the subgroups U'_1 and U'_2 are regular (i.e. $U'_1\in B(\Gamma)$ and $U'_2\in B(\Gamma)$), because in the opposite case we could replace U'_1 and U'_2 by $\overline{U}'_1\cap\Gamma$ and $\overline{U}'_2\cap\Gamma$ (we note that since U_1 and U_2 are maximal, $\overline{U}'_1\cap\Gamma\subset U_1$ and $\overline{U}'_2\cap\Gamma\subset U_2$).

Case b). In this case the assertion of the theorem follows from Lemma 13.19 and a remark analogous to Remark 1.

§ 14. A construction of representation theory and its applications

Let I be an arbitrary finite-dimensional complex linear space. For any natural number k we denote by $G_k(I)$ the set of all k-dimensional linear subspaces of I. It is well known that one can introduce on $G_k(I)$ a complex algebraic manifold structure in a canonical way. For the sake of definiteness we describe one possible way of obtaining that structure.

Consider an arbitrary basis in I and denote by A the set of basis vectors. For any $D \subset A$, denote by I_D the linear subspace of I, generated by the basis vectors belonging to D, and for any D of cardinality k write $G_k(I, D)$ for the subset of $G_k(I)$ consisting of all k-dimensional subspaces whose intersection with I_{A-D} consists of the zero alone. It is easy to see that $G_k(I) = \bigcup G_k(I, D)$, the union is taken for all $D \subset A$ of cardinality k. Now for any D of cardinality k, write f_D for the natural mapping $f_D: G_k(I, D) \to \mathrm{End}\,(I_D, I_{A-D})$ ($\mathrm{End}\,(I_D, I_{A-D})$ is the linear space of all linear mappings of I_D to I_{A-D}) associating to every $s \in G_k(I, D)$ the linear transformation $f_D(s) \in \mathrm{End}\,(I_D, I_{A-D})$ with the graph s, where we identify I and $I_D \times I_{A-D}$). It is easy to see that for any D of cardinality k, f_D is a one-to-one mapping, and that for any D_1 and D_2 of cardinality k, the mapping

$$f_{D_2} f_{D_1}^{-1}: f_{D_1}(G_k(I, D_1) \cap G_k(I, D_2)) \to f_{D_2}(G_k(I, D_1) \cap G_k(I, D_2))$$

is an algebraic mapping regular in both directions. Therefore, the system of pairs $(G_k(I, D), f_D)$ gives a structure of complex algebraic manifold on $G_k(I)$. It is easy to show that the structure does not depend on the choice of the basis.

Now denote by $G(I)$ the complex algebraic manifold which is the disjoint union of all $G_k(I)$ $(0 \leqq k \leqq \dim I)$. It is easy to see that the following assertions are valid:

Lemma 14.1. For any $s \in G(I)$ there is a Zariski open subset $V_s \subset G(I)$ containing s, and a set $(f_1, \dots, f_p)$ of regular algebraic mappings of V_s into I, such that for any $v \in V_s$, $f_i(v) \in v$ for all i $(1 \leqq i \leqq p)$, and $\bigcup_{1 \leqq i \leqq p} f_i(v)$ generates v as a linear space.

Lemma 14.2. For any linear subspace $F \subset I$ and natural number l, denote by $G(F, l)$ the set of all $s \in G(I)$, for which the dimension of the intersection $s \cap F$ is equal to l. Then for any F and l *1)* $\bigcup_{\tilde{l} > l} G(F, \tilde{l})$ is an algebraic submanifold in $G(I)$, *2)* the mapping $G(F, l) \to G(F)$ associating to $s \in G(F, l)$ the intersection $s \cap F$ is a regular algebraic mapping.

Now let H be an algebraically connected complex algebraic group, $\Lambda \subset H$ a Zariski dense subgroup in H, T an algebraic representation of group H in a finite-dimensional complex linear space L, $N \subset L$ and $M \subset L$ linear subspaces. Suppose that we are given a non-empty Zariski open subset $V \subset H$ and a regular algebraic mapping $\varphi: V \to G(N)$, such that

1) if $h \in V$, then $\varphi(h) \subset ((T_h M) \cap N)$,

2) N is generated by the union $\bigcup_{h \in V} \varphi(h)$ as a linear space.

Lemma 14.3 There exists a finite set $\lambda_1, \dots, \lambda_n \in \Lambda$ such that for any $h \in H$ $\bigcup_{h^{-1}\lambda_i \in V} \varphi(h^{-1}\lambda_i)$ generates N as a linear space ($\bigcup_{h^{-1}\lambda_i \in V}$ denotes a union taken over all λ_i $(1 \leqq i \leqq n)$ for which $h^{-1}\lambda_i \in V$).

Proof. For any group or set B and natural number n, denote by B^n the direct product $\overbrace{B\times\ldots\times B}^{n}$. Consider for any natural number k the set $\tilde{U}_k \subset V^k$ of all k-tuples $(v_1, \ldots, v_k) \in V^k$, for which $\bigcup_{1\leq j\leq k} \varphi(v_j)$ generates N as a linear space. Since *1)* it is easy to see that the subset in $G(N)^k$ consisting of all k-tuples $(s_1, \ldots, s_k) \in G(N)^k$ for which $\bigcup_{1\leq j\leq k} s_j$ generates N as a linear space, is Zariski open in $G(N)^k$, *2)* φ is a regular algebraic mapping of V into $G(N)$, the set $\tilde{U}_k$ is Zariski open in V^k.

Denote by J_k the interval of natural numbers from 1 to k, and for any $j \in J_k$ by π_j the natural projection of the direct product $H^k = H\times\ldots\times H$ onto the j-th factor. For any subset $A \subset J_k$ and any subset $B \subset H$, $B^A \subset H^k$ will denote the subset of all $m \in H^k$, for which *1)* $\pi_j(m) = e$, if $j \notin A$ *2)* $\pi_j(m) \in B$, if $j \in A$ (in particular, H^A is the product of all factors in the product $H^k = H\times\ldots\times H$ whose index belongs to A). For any $A \subset J_k$, denote by $\tilde{U}_A$ the subset of all $v \in V^A$ for which $\bigcup_{j\in A} \varphi(\pi_j(v))$ generates N as a linear space.

Write now $U_k \subset H^k$ for the set of all k-tuples $(h_1, \ldots, h_k) \subset H^k$ for which $\bigcup_{h_j \in V} \varphi(h_j)$ generates N as a linear space ($\bigcup_{h_j\in V}$ stands for a union taken for all h_j for which $h_j \in V$). It is easy to see that the following equality holds:

$$U_k = \bigcup_{A\in 2^{J_k}} (\tilde{U}_A \times H^{J_k - A}) \tag{1}$$

where 2^{J_k} denotes the set of all subsets of J_k.

As we have already shown, $\tilde{U}_j$ is Zariski open in V^k for any j, therefore for any $A \subset J_k$, $\tilde{U}_A$ is Zariski open in V^A. Now since V^A is Zariski open in H^A (for V is Zariski open in H), $\tilde{U}_A$ is also Zariski open in H^A for any $A \subset J_k$. From this and equality (1) it follows that for any k, U_k is Zariski open in H^k.

Since N is finite-dimensional, it follows immediately from the condition of the lemma that U_k is non-empty for some k. Let k_0 be such that $U_{k_0} \neq \emptyset$. Since H is algebraically connected, so is H^{k_0} and U_{k_0} is open in the Zariski topology, we have

$$\dim H^{k_0} - \dim (H^{k_0} - U_{k_0}) \geqq 1. \tag{2}$$

On the other hand, it is easy to see that for any integers k and l we have the inclusion

$$(H^{kl} - U_{kl}) \subset (H^k - U_k)^l, \tag{3}$$

where $(H^k - U_k)^l$ is considered as a subset in $(H^k)^l = H^{kl}$. Now (2) and (3) imply that there exists an l such that

$$\dim H^{k_0 l} - \dim (H^{k_0 l} - U_{k_0 l}) > \dim H. \tag{4}$$

Let ψ be the mapping of the direct product $\psi: H\times H^{k_0 l} \to H^{k_0 l}$ defined in the following way: if $h \in H$ and $(h_1, \ldots, h_{k_0 l}) \in H^{k_0 l}$, then $\psi(h, h_1, \ldots, h_{k_0 l}) =$

$=(h^{-1}h_1, \ldots, h^{-1}h_{k_0 l})$. Since H is an algebraic group, ψ is a regular algebraic mapping.

It is well known that the image $f(M)$ of any algebraic submanifold M of a complex algebraic manifold X_1 under a regular algebraic mapping f of X_1 into a complex algebraic manifold X_2 has the following properties: *1)* the topological dimension of the set $f(M)$ does not exceed the dimension of M, *2)* the dimension of the algebraic closure of the set $f(M)$ coincides with the dimension of the set $f(M)$. Therefore, since *1)* ψ is a regular algebraic mapping, *2)* $U_{k_0 l}$ is Zariski open in $H^{k_0 l}$, $H^{k_0 l}-U_{k_0 l}$ is an algebraic submanifold in $H^{k_0 l}$, and consequently $H\times(H^{k_0 l}-U_{k_0 l})$ is an algebraic submanifold in $H\times H^{k_0 l}$, we have

$$\dim \hat{U}_{k_0 l} \leqq \dim (H^{k_0 l}-U_{k_0 l})+\dim H, \tag{5}$$

where $\hat{U}_{k_0 l}$ denotes the algebraic closure of the subset $\psi(H\times(H^{k_0 l}-U_{k_0 l}))$ in $H^{k_0 l}$. From (4) and (5) $\dim \hat{U}_{k_0 l}<\dim H^{k_0 l}$ follows. Therefore

$$H^{k_0 l}-\hat{U}_{k_0 l} \neq \emptyset. \tag{6}$$

Remark now that *1)* $H^{k_0 l}-\hat{U}_{k_0 l}$ is Zariski open in $H^{k_0 l}$ (since $\hat{U}_{k_0 l}$ is algebraically closed), *2)* since Λ is Zariski dense in H, $\Lambda^{k_0 l}$ is Zariski dense in $H^{k_0 l}$. Therefore, in consequence of (6) there exists a $\lambda=(\lambda_1, \ldots, \lambda_{k_0 l})\in\Lambda^{k_0 l}$ such that $\lambda\notin\hat{U}_{k_0 l}$, and consequently

$$\lambda\notin\psi(H\times(H^{k_0 l}-U_{k_0 l})). \tag{7}$$

Now (7) is equivalent to the fact that for any $h\in H$

$$(h^{-1}\lambda_1, \ldots, h^{-1}\lambda_{k_0 l})\notin(H^{k_0 l}-U_{k_0 l}), \tag{8}$$

and consequently for any $h\in H$

$$(h^{-1}\lambda_1, \ldots, h^{-1}\lambda_{k_0 l})\in U_{k_0 l}. \tag{9}$$

From (9) and the definition of the set $U_{k_0 l}$ it follows that $(\lambda_1, \ldots, \lambda_{k_0 l})$ is the desired set. The lemma is proved.

If $h_1, h_2\in H$ and $h_1^{-1}h_2\in V$, then put $\varphi(h_1, h_2)=T_{h_1}\varphi(h_1^{-1}h_2)$ and remark that

$$\varphi(h_1, h_2) = T_{h_1}\varphi(h_1^{-1}h_2)\subset T_{h_1}(N\cap T_{h_1^{-1}h_2}M) = T_{h_1}N\cap T_{h_2}M, \tag{10}$$

and that for any $h\in H$

$$T_h\varphi(h_1, h_2) = T_h(T_{h_1}\varphi(h_1^{-1}h_2)) = T_{hh_1}\varphi(h_1^{-1}h_2) = \varphi(hh_1, hh_2). \tag{11}$$

Denote by $\tilde{M}$ the linear subspace of M generated by the union $\bigcup_{\lambda^{-1}\in\Lambda\cap V}\varphi(\lambda, e)$ ($\varphi(\lambda, e)$ belongs to M in view of (10)). (11) implies that if $\lambda_1, \lambda_2\in\Lambda$ and $\lambda_1^{-1}\lambda_2\in V$ then

$$\varphi(\lambda_1, \lambda_2) = T_{\lambda_2}\varphi(\lambda_2^{-1}\lambda_1, e)\subset T_{\lambda_2}\tilde{M}. \tag{12}$$

Combining (10) and (12) we get that if $\lambda_1, \lambda_2\in\Lambda$ and $\lambda_1^{-1}\lambda_2\in V$, then

$$\varphi(\lambda_1, \lambda_2)\subset T_{\lambda_1}N\cap T_{\lambda_2}\tilde{M}. \tag{13}$$

Now we define the space L_Λ. If $x\in L$, $\lambda\in\Lambda$ and $x\in T_\lambda N$, then $(x,\lambda)_N$ denotes the element of the direct sum $L_N=\bigoplus_{\lambda\in\Lambda} T_\lambda N$ which corresponds to x under the natural embedding of $T_\lambda N$ into L_N, and if $x\in T_\lambda\tilde{M}$ then $(x,\lambda)_{\tilde{M}}$ denotes the analogously defined element of $L_{\tilde{M}}=\bigoplus_{\lambda\in\Lambda} T_\lambda\tilde{M}$. Further, we will assume that L_N and $L_{\tilde{M}}$ are naturally embedded into the direct sum $L_N\oplus L_{\tilde{M}}$. Let now L_1 be the factor space of the space $L_N\oplus L_{\tilde{M}}$ with respect to the subspace generated by all elements of the form $(x,\lambda_1)_N-(x,\lambda_2)_{\tilde{M}}$, where $\lambda_1^{-1}\lambda_2\in V$ and $x\in\varphi(\lambda_1,\lambda_2)$.

Remark 14.4. Let d and $\hat{d}$ be the natural mappings $d{:}L_N\oplus L_{\tilde{M}}\to L_\Lambda$ and $\hat{d}{:}L_N\oplus L_{\tilde{M}}\to L$ (d is a natural mapping of a linear space onto its factor space, and $\hat{d}$ is a linear mapping defined by the formulae $\hat{d}((x,\lambda)_N)=x$ and $\hat{d}((y,\lambda)_{\tilde{M}})=y$, where $\lambda\in\Lambda$, $x\in T_\lambda N$, and $y\in T_\lambda\tilde{M}$). From the definition of the space L_Λ and the mappings d and $\hat{d}$ we see that the kernel of the mapping $\hat{d}$ contains the kernel of the mapping d. Therefore there exists precisely one mapping $d'{:}\,L_\Lambda\to L$ satisfying the relation $d'\circ d=\hat{d}$. We do not know whether d' is always a monomorphism.

Lemma 14.5. The space L_Λ is finite-dimensional.

Proof. Let $\lambda_1,\ldots,\lambda_n\in\Lambda$ be a set satisfying the conclusion of Lemma 14.3. Then for any $h\in H$, the union $\bigcup_{h^{-1}\lambda_i\in V}\varphi(h^{-1}\lambda_i)$ generates N. On the other hand, in view of (11) we have the following equality:

$$\bigcup_{h^{-1}\lambda_i\in V}\varphi(h,\lambda_i)=T_h\bigcup_{h^{-1}\lambda_i\in V}\varphi(h^{-1}\lambda_i). \tag{14}$$

Therefore, for any $h\in H$, $\bigcup_{h^{-1}\lambda_i\in\Lambda}\varphi(h,\lambda_i)$ generates T_hN.

As in Remark 14.4, let d be the natural mapping $d{:}L_N\oplus L_{\tilde{M}}\to L_\Lambda$. Since for any $\lambda\in\Lambda$, $T_\lambda N$ is generated by the union $\bigcup_{\lambda^{-1}\lambda_i\in V}\varphi(\lambda,\lambda_i)$, and $\varphi(\lambda,\lambda_i)$ belongs to $T_{\lambda_i}\tilde{M}$ (in view of (12)), it follows from the definition of the space L_Λ and the mapping d that $\bigcup_{1\le i\le n} d(T_{\lambda_i}\tilde{M})$ generates $d(\bigoplus_{\lambda\in\Lambda}T_\lambda N)$. Therefore the space $d(L_N)$ is finite-dimensional. We remark now that *1)* since $\tilde{M}$ is, by definition, generated by the union $\bigcup_{\lambda^{-1}\in\Lambda\cap V}\varphi(\lambda,e)$, in view of (11) for any $\lambda\in\Lambda$, $T_\lambda\tilde{M}$ is generated by the union $\bigcup_{\tilde{\lambda}^{-1}\lambda\in\Lambda\cap V}\varphi(\tilde{\lambda},\lambda)$, *2)* according to (10), for any $\tilde{\lambda},\lambda\in\Lambda$ $(\tilde{\lambda}^{-1}\lambda\in V)$, $\varphi(\tilde{\lambda},\lambda)\subset T_{\tilde{\lambda}}N$. Therefore, we have the inclusion

$$d(\bigoplus_{\lambda\in\Lambda}T_\lambda\tilde{M})\subset d(\bigoplus_{\lambda\in\Lambda}T_\lambda N). \tag{15}$$

Since $d(L_N)$ is finite-dimensional, it follows from (15) that $d(L_{\tilde{M}})$ is finite-dimensional, and consequently $d(L_N\oplus L_{\tilde{M}})$ is also finite-dimensional. Now in order to complete the proof of the lemma it remains only to note that d is an epimorphism. The lemma is proved.

Now we define the representation P of the group Λ in the space $L_N \oplus L_{\tilde{M}} = (\bigoplus_{\lambda \in \Lambda} T_\lambda N) \oplus (\bigoplus_{\lambda \in \Lambda} T_\lambda \tilde{M})$. It is clear that in order to determine P it is sufficient to show how P acts on $(x, \lambda)_N$ and $(y, \lambda)_{\tilde{M}}$, when λ runs through Λ, x through the space $T_\lambda N$, and y through the space $T_\lambda \tilde{M}$. Put now for any $\tilde{\lambda} \in \Lambda$

$$P_{\tilde{\lambda}}(x, \lambda)_N = (T_{\tilde{\lambda}}(x), \tilde{\lambda}\lambda)_N \tag{16}$$

and

$$P_{\tilde{\lambda}}(y, \lambda)_{\tilde{M}} = (T_{\tilde{\lambda}}(y), \tilde{\lambda}\lambda)_{\tilde{M}}. \tag{17}$$

As an immediate consequence of formulae (11), (16) and (17), we have that the subspace of $L_N \oplus L_{\tilde{M}}$ generated by the elements of the form $(x, \lambda_1)_N - (x, \lambda_2)_{\tilde{M}}$ ($\lambda_1, \lambda_2 \in \Lambda$ and $x \in \varphi(\lambda_1, \lambda_2)$) is invariant by P_λ, ie., P_λ maps the kernel of the mapping d into itself (for the definition of d see Remark 14.4). Therefore, since d is an epimorphism, P induces on L_Λ a representation of the group Λ which will be denoted by $\tilde{P}$. Now, before studying the representation $\tilde{P}$ we give Definitions 14.6 and 14.7 and Lemmas 14.8, 14.9, 14.10.

Definition 14.6. Let X be a real (complex) algebraic manifold (not necessarily affine), $B \subset X$ a subset, and f a mapping of B into a real (complex) algebraic manifold M. Let f be called a *regular* algebraic mapping if it can be extended to an algebraic mapping of $A(B)$ into M, which is regular on B, where $A(B)$ is the algebraic closure of B in X.

Definition 14.7. Let Q be a real (complex) algebraic group, B a subgroup of Q, S a representation of B in a finite-dimensional real (complex) linear space W. The representation S is called *algebraic*, if S is a regular algebraic mapping (in the sense of Definition 14.6) of the set B to End W (End W is the linear space of all linear transformations of the space W).

Let Q, B, S and W be the same as in Definition 14.7. For any $w \in W$, denote by f_w the mapping of B into W defined by the formula $f_w(b) = S_b(w)$ (S_b is the operator corresponding to b in the representation S).

Lemma 14.8. If there is in W a subset D such that *1)* D generates W as a linear space, *2)* for any $d \in D$ the mapping f_d is a regular algebraic mapping, then the representation S is algebraic.

Proof. Since D generates W as a linear space, there is in W a basis $d_1, \ldots, d_n$ consisting of elements belonging to D. Now condition *2)* implies that in the basis $d_1, \ldots, d_n$ the matrix-coefficients of representation S are regular rational functions on B (i.e. they can be continued on the algebraic closure of B in X to rational functions which are regular on B). From this we get easily that representation S is algebraic.

The following lemma can be proved easily:

Lemma 14.9. Let I be a finite-dimensional complex linear space, Y a complex algebraic manifold, and $(\tilde{k}_1, \ldots, \tilde{k}_n)$ a set of regular algebraic mappings of Y

into I such that for any $y \in Y$, $\bigcup_{1 \leqq r \leqq n} \tilde{k}_r(y)$ generates I as a linear space. Then it is possible to choose in Y a non-empty Zariski open subset W and a subset $(k_1, \ldots, k_m)$ of $(\tilde{k}_1, \ldots, \tilde{k}_n)$ $(m = \dim I)$ such that for any $w \in W$ the vectors $k_1(w), \ldots, k_m(w)$ form a basis in the space I.

Lemma 14.10. Let I be a finite-dimensional complex linear space, X a complex algebraic manifold, and $(d_1, \ldots, d_q)$ a set of regular algebraic mappings of X into $G(I)$ such that for any $x \in X$, $\bigcup_{1 \leqq l \leqq q} d_l(x)$ generates I as a linear space. Then there is in X a non-empty Zariski open subset W and a set $(k_1, \ldots, k_m)$ $(m = \dim I)$ of regular algebraic mappings of W into I, such that *1)* for any j one can select an $l(j)$ such that for any $w \in W$ $k_j(w) \in d_{l(j)}(w)$, *2)* for any $w \in W$ the vectors $k_1(w), \ldots, k_j(w), \ldots, k_m(w)$ form a basis in the space I.

Proof. Fix some $x_0 \in X$. It follows from Lemma 14.1 that for any l $(1 \leqq l \leqq q)$ there is a Zariski open subset $W_l \subset X$ containing x_0 and a set $(k_{1,l}, \ldots, k_{p(l),l})$ of regular algebraic mappings of W_l into I such that *1)* for any $w \in W_l$, $k_{i,l}(w) \in$ $\in d_l(w)$ for all i $(1 \leqq i \leqq p(l))$, *2)* $\bigcup_{1 \leqq i \leqq p(l)} k_{i,l}(w)$ generates $d_l(w)$ as a linear space (for W_l it is sufficient to take the inverse image of $V_{d_l(x_0)}$ under the mapping d_l, and for the mappings $k_{i,l}$, the composition of mapping d_l and the mappings f_i of Lemma 14.1). Let $\tilde{W}$ be the intersection $\bigcup_{1 \leqq l \leqq q} W_l$, and $(\tilde{k}_1, \ldots, \tilde{k}_n)$ $(n = \sum_{1 \leqq l \leqq q} p(l))$ the set of regular algebraic mappings of $\tilde{W}$ into I consisting of the restriction to $\tilde{W}$ of all mappings $k_{i,l}$ $(1 \leqq l \leqq q,\ 1 \leqq i \leqq p(l))$. Since for any $x \in X$, $\bigcup_{1 \leqq l \leqq q} d_l(x)$ generates I as a linear space, it follows from the definition of the set $(\tilde{k}_1, \ldots, \tilde{k}_n)$ that for any $w \in \tilde{W}$, $\bigcup_{1 \leqq r \leqq n} \tilde{k}_r(w)$ generates I as a linear space. On the other hand, since for any r, $\tilde{k}_r$ is the restriction to $\tilde{W}$ of one of the mappings $k_{i,l}$, for any r there is an $l(r)$ such that for any $w \in W$

$$\tilde{k}_r(w) \in d_{l(r)}(w). \tag{18}$$

Note further that since $W = \bigcap_{1 \leqq l \leqq q} W_l$, and all W_l are Zariski open and contain x_0, W is Zariski open and non-empty. The last three statements and Lemma 14.9 imply the assertion of the lemma.

We return now to the space L_Λ and the representation $\tilde{P}$.

Lemma 14.11. The representation $\tilde{P}$ of the group Λ is algebraic (in the sense of Definition 14.7).

Proof. In the proof of this lemma we will apply the notations introduced between Lemma 14.2 and Definition 14.6.

Let f be the mapping $f: N \times \Lambda \to L_\Lambda$ assigning to the pair $(z, \lambda) \in N \times \Lambda$ the point $d((T_\lambda(z), \lambda)_N) \in L_\Lambda$. Assume that, considering $N \times \Lambda$ as a subspace of the complex algebraic manifold $N \times H$, we have proved that f is a regular algebraic mapping in the sense of Definition 14.6. In view of inclusion (15), $d(\bigoplus_{\lambda\in\Lambda} T_\lambda N)$ generates L_Λ as a linear space (since $L_\Lambda = d(L_N \oplus L_{\tilde{M}})$, $L_N = \bigoplus_{\lambda\in\Lambda} T_\lambda N$, and $L_{\tilde{M}} = \bigoplus_{\lambda\in\Lambda} T_\lambda \tilde{M}$). Now, since f is regular and algebraic, in view of Lemma 14.8 it follows that representation $\tilde{P}$ is algebraic (we also recall that for any $z\in N$ and $\lambda_1, \lambda_2 \in \Lambda$,

$$\tilde{P}_{\lambda_1}(f(z, \lambda_2)) = \tilde{P}_{\lambda_1}(d((T_{\lambda_2}(z), \lambda_2)_N)) = d(P_{\lambda_1}(T_{\lambda_2}(z), \lambda_2(_N) =$$
$$= d(T_{\lambda_1\lambda_2}(z), (\lambda_1\lambda_2)_N) = f(z, \lambda_1\lambda_2)).$$

Now, we will show that f is a regular algebraic mapping. Let $\lambda_1, \dots, \lambda_n$ be a set satisfying the conclusion of Lemma 14.3. Denote by V' the set of all $h\in H$, for which $h^{-1}\lambda_i \in V$ for any i $(1\leqq i\leqq n)$. Since V is a Zariski open subset of H, and it is easy to see that V' coincides with the intersection $\bigcup_{1\leqq i\leqq n} \lambda_i V^{-1}$, so V' is in H a Zariski open subset. Define for any i $(1\leqq i\leqq n)$ the mapping φ_i of the set V' into $G(N)$ by the following formula:

$$\varphi_i(h) = \varphi(h^{-1}\lambda_i) \quad \text{where} \quad h\in V'. \tag{19}$$

Since φ is a regular algebraic mapping, and H is an algebraic group, the mapping φ_i is regular algebraic for any i $(1\leqq i\leqq n)$. Therefore, since for any $h\in V'$ the union $\bigcup_{1\leqq i\leqq n} \varphi_i(h) = \bigcup_{1\leqq i\leqq n} \varphi(h^{-1}\lambda_i)$ generates N as a linear space according to Lemma 14.3, in view of Lemma 14.10 there is in V' a non-empty Zariski open subset W and a set $(k_1, \dots, k_m)$ $(m=\dim N)$ of regular algebraic mappings of W into N such that *1)* there is a function $i(j)$ such that for any j $(1\leqq j\leqq m)$ $k_{i(j)}(w)\in\varphi_{i(j)}(w)$ for all $w\in W$, *2)* for any $w\in W$ the vectors $k_1(w), \dots, k_m(w)$ form a basis in the space N (we can apply Lemma 14.10, since V' is a Zariski open subset in X and every Zariski open subset of a complex algebraic manifold is a complex algebraic manifold itself). Consider for any $w\in W$ the expansion of an arbitrary $z\in N$ in the basis $k_1(w), \dots, k_m(w)$ and denote by $c_j(z, w)$ $(1\leqq j\leqq m)$, the coefficients in this expansion. Since $k_1, \dots, k_m$ are algebraic mappings, regular on W, it is easy to see that for any j $(1\leqq j\leqq m)$, $c_j(z, w)$ is a regular rational function on $N\times W$.

Now, *1)* $\varphi(h_1, h_2) = T_{h_1}\varphi(h_1^{-1}h_2)$ if only $h_1^{-1}h_2\in V$, *2)* for any $w\in W$ and any i $(1\leqq i\leqq n)$, $w^{-1}\lambda_i\in V$ (since $W\subset V'$ and $V' = \bigcap_{1\leqq i\leqq n} \lambda_i V^{-1}$), *3)* for any j, $k_j(w)\in\varphi_{i(j)}(w)$ for all $w\in W$, thus, in view of (12) and (19) for any j and any $w\in W$ we have

$$T_w(k_j(w))\in T_w(\varphi(w^{-1}\lambda_{i(j)})) = \varphi(w, \lambda_{i(j)})\subset T_{\lambda_{i(j)}}\tilde{M}. \tag{20}$$

Denote by α_i $(1\leqq i\leqq n)$, the natural imbedding of $T_{\lambda_i}\tilde{M}$ into $\bigoplus_{1\leqq i\leqq n} T_{\lambda_i}\tilde{M}$. Then, in view of (20) one can define the mapping $\theta'\colon N\times W\to \bigoplus_{1\leqq i\leqq n} T_{\lambda_i}\tilde{M}$ by the following formula:

$$\theta(z, w) = \sum_{1\leqq j\leqq m} c_j(z, w)\alpha_{i(j)}(T_w(k_j(w))), \tag{21}$$

where $z\in N$, and $w\in W$. Since the functions $c_j(z, w)$ and the mappings $k_j(w)$ are regular algebraic, T is an algebraic representation, and all $\alpha_{i(j)}$ are linear mappings, thus it follows from (21) that θ is a regular algebraic mapping.

We will assume that $\bigoplus_{1\leqq i\leqq n} T_{\lambda_i}\tilde{M}$ is naturally imbedded into $L_{\tilde{M}}$, and consequently also into $L_N\oplus L_{\tilde{M}}$. Then, in consequence of the definition of the mappings $\alpha_{i(j)}$ and the symbol $(x, \lambda)_N$, for any j $(1\leqq j\leqq n)$ and any $\lambda\in W\cap\Lambda$ we have

$$(T_\lambda(k_j(\lambda)), \lambda_{i(j)})_{\tilde{M}} = \alpha_{i(j)}(T_\lambda(k_j(\lambda))). \tag{22}$$

According to (20), for any $\lambda\in W\cap\Lambda$ and any j $(1\leqq j\leqq m)$, we have $T_\lambda(k_j(w))\in$ $\in\varphi(\lambda, \lambda_{i(j)})$. Therefore it follows from the definition of the space L_Λ and the mapping d, and from formulae (21) and (22), that for any $\lambda\in W\cap\Lambda$ and any j $(1\leqq j\leqq m)$

$$d\big((T_\lambda(k_j(\lambda)), \lambda)_N\big) = d\big((T_\lambda(k_j(\lambda), \lambda_{i(j)})_{\tilde{M}}\big) =$$

$$= d\big(\alpha_{i(j)}(T_\lambda(k_j(\lambda)))\big) = d\big(\theta(k_j(\lambda), \lambda)\big) \tag{23}$$

holds. Since *1)* the mapping d is linear, *2)* the mapping θ is linear in z, *3)* the transformation T_λ is linear, in view of formula (21) we obtain from (23) that for any $z\in N$ and any $\lambda\in W\cap\Lambda$

$$d\big((T_\lambda z, \lambda)_N\big) = d\big(\theta(z, \lambda)\big). \tag{24}$$

Since, by definition, $f(z, \lambda)=d\big((T_\lambda z, \lambda)_N\big)$ for any $z\in Z$ and $\lambda\in\Lambda$, it follows from (24) and the fact that θ is a regular algebraic mapping on $N\times W$ that $f(z, \lambda)$ is a regular algebraic mapping on $N\times(W\cap\Lambda)$. Now, as we have shown already at the beginning of the proof, $\tilde{P}_{\lambda_1}\big(f(z, \lambda_2)\big)=f(z, \lambda_1\lambda_2)$ for any $z\in N$ and $\lambda_1, \lambda_2\in\Lambda$. Therefore, since $\tilde{P}_{\lambda_1}$ is a linear transformation for any $\lambda_1\in\Lambda$ and f is a regular algebraic mapping on $N\times(W\cap\Lambda)$, we have that for any $\lambda_1\in\Lambda$, f is a regular algebraic mapping on $N\times(\lambda_1 W\cap\Lambda)$. Now, since W is a non-empty Zariski open subset in H, and Λ is Zariski dense in H, we conclude that f is a regular algebraic mapping on all of $N\times\Lambda$. The proof of the lemma is complete.

It is easy to see that an algebraic representation of any subgroup of a complex algebraic group extends to an algebraic representation of its algebraic closure, therefore Lemma 14.11 and the Zariski density of Λ in H imply the following

Lemma 14.12. The representation $\tilde{P}$ extends to an algebraic representation P' of the group H.

Lemma 14.13. Let $\hat{T}$ be a finite-dimensional representation of the subgroup Λ in a complex linear space $\hat{L}$. Suppose that for any $\lambda\in\Lambda$ one can define mappings $\tilde{k}_\lambda$ and $\tilde{\tilde{k}}_\lambda; \tilde{k}_\lambda: T_\lambda N\to\hat{L}$, $\tilde{\tilde{k}}_\lambda: T_\lambda\tilde{M}\to\hat{L}$, such that the following conditions hold: *1)* For any $\lambda\in\Lambda$, $\tilde{k}_\lambda$ and $\tilde{\tilde{k}}_\lambda$ are linear, *2)* for any $\lambda_1, \lambda_2\in\Lambda$ and $z\in N$, $\tilde{\tilde{k}}_{\lambda_1\lambda_2}(T_{\lambda_1\lambda_2}(z))=\hat{T}_{\lambda_1}(\tilde{\tilde{k}}_{\lambda_2}(T_{\lambda_2}(z)))$, *3)* for any $\lambda_1, \lambda_2\in\Lambda$ and $y\in\tilde{M}$, $\tilde{\tilde{k}}_{\lambda_1\lambda_2}(T_{\lambda_1\lambda_2}(y))=$

$=\hat{T}_{\lambda_1}(\tilde{\tilde{k}}_{\lambda_2}(T_{\lambda_2}(y)))$, *4)* if $\lambda_1, \lambda_2 \in \Lambda$, $\lambda_1^{-1}\lambda_2 \in V$ and $x \in \varphi(\lambda_1, \lambda_2)$ then $\tilde{k}_{\lambda_1}(x) = \tilde{\tilde{k}}_{\lambda_2}(x)$, *5)* $\bigcup_{\lambda \in \Lambda} \tilde{k}_\lambda(T_\lambda N)$ generates $\hat{L}$ as a linear space. Then the representation $\hat{T}$ can be extended to an algebraic representation T' of the group H.

Proof. It follows from conditions *1)* and *4)* that the families of mappings $\tilde{k}_\lambda$ and $\tilde{\tilde{k}}_\lambda$ induce a linear mapping $\hat{k}: L_\Lambda \to \hat{L}$ ($\hat{k}$ is uniquely defined by the conditions that for any $\lambda \in \Lambda$ and $x \in T_\lambda N$ ($x \in T_\lambda \tilde{M}$) we have

$$\hat{k}(d((x, \lambda)_N)) = \tilde{k}_\lambda(x) \quad (\hat{k}(d((x, \lambda)_{\tilde{M}})) = \tilde{\tilde{k}}_\lambda(x)).$$

From conditions *2)* and *3)* and the definition of the representation $\tilde{P}$ we get that for any $\lambda \in \Lambda$, $\hat{k} \circ \tilde{P}_\lambda = \hat{T}_\lambda \circ \hat{k}$, and from condition *5)* — that $\hat{k}$ is an epimorphism. Therefore, $\hat{T}$ is quotient representation of the representation $\tilde{P}$ by the kernel of mapping $\hat{k}$ (i.e. a representation which induces $\tilde{P}$ on the quotient space $L_\Lambda/\mathrm{Ker}\,\hat{k}$). According to Lemma 14.12, $\tilde{P}$ can be extended to an algebraic representation P' of the group H. Since $\mathrm{Ker}\,\hat{k}$ is invariant under $\tilde{P}_\lambda$ for all $\lambda \in \Lambda$, and Λ is Zariski dense in H, furthermore the representation P' is algebraic, we have that $P'_h(\mathrm{Ker}\,\hat{k}) = \mathrm{Ker}\,\hat{k}$ for any $h \in H$. Therefore P' induces on $\hat{L} = L_\Lambda/\mathrm{Ker}\,\hat{k}$ an algebraic representation T' of the group H, which is easily seen to be the extension of $\hat{T}$. The lemma is proved.

For any two linear subspaces $\overline{N}, \overline{M} \subset L$, $V(\overline{N}, \overline{M}) \subset H$ denotes the set of all $h_0 \in H$, for which $\dim((T_{h_0}\overline{M}) \cap \overline{N}) = \min_{h \in H}(\dim((T_h\overline{M}) \cap \overline{N}))$. Since the representation T is algebraic, it follows from Lemma 14.2 that for any $\overline{N}$, $\overline{M} \subset L$ *1)* $V(\overline{N}, \overline{M})$ is a Zariski open subset in H, *2)* the mapping $\psi: V(\overline{N}, \overline{M}) \to G(\overline{N})$ assigning to $h \in V(\overline{N}, \overline{M})$ the intersection $(T_h\overline{M}) \cap \overline{N}$, is a regular algebraic mapping. Therefore from Lemma 14.13 and the remark which will be made after the formulation of Lemma 14.14 we can easily get the following:

Lemma 14.14. Let $\overline{V}$ be a non-empty Zariski open subset in H, contained in $V(\overline{N}, \overline{M})$. Assume that $\bigcup_{h \in \overline{V}}((T_h\overline{M}) \cap \overline{N})$ generates $\overline{N}$ as a linear space, and denote by $\tilde{\tilde{M}}$ the linear subspace of $\overline{M}$ generated by the union $\bigcup_{\lambda^{-1} \in \Lambda \cap \overline{V}}((T_\lambda\overline{N}) \cap \overline{M})$. Let $\hat{T}$ be a finite-dimensional representation of the subgroup Λ in a complex linear space $\hat{L}$. Assume that for any $\lambda \in \Lambda$ one can define mappings $\tilde{k}_\lambda$ and $\tilde{\tilde{k}}_\lambda$; $\tilde{k}_\lambda: T_\lambda\overline{N} \to \hat{L}$ and $\tilde{\tilde{k}}: T_\lambda\tilde{\tilde{M}} \to \hat{L}$, in such a way that they satisfy the following conditions: *1)* for any $\lambda \in \Lambda$, $\tilde{k}_\lambda$ and $\tilde{\tilde{k}}_\lambda$ are linear, *2)* for any $\lambda_1, \lambda_2 \in \Lambda$ and $z \in \overline{N}$, $\tilde{k}_{\lambda_1\lambda_2}(T_{\lambda_1\lambda_2}(z)) = \hat{T}_{\lambda_1}(\tilde{k}_{\lambda_2}(T_{\lambda_2}(z)))$, *3)* for any $\lambda_1, \lambda_2 \in \Lambda$ and $y \in \tilde{\tilde{M}}$, $\tilde{\tilde{k}}_{\lambda_1\lambda_2}(T_{\lambda_1\lambda_2}(y)) = \hat{T}_{\lambda_1}(\tilde{\tilde{k}}_{\lambda_2}(T_{\lambda_2}(y)))$ *4)* if $\lambda_1, \lambda_2 \in \Lambda$, $\lambda_1^{-1}\lambda \in \overline{V}$ and $x \in T_{\lambda_1}\overline{N} \cap T_{\lambda_2}\overline{M}$ then $\tilde{k}_{\lambda_1}(x) = \tilde{\tilde{k}}_{\lambda_2}(x)$, *5)* $\bigcup_{\lambda \in \Lambda} \tilde{k}_\lambda(T_\lambda N)$ generates $\hat{L}$ as a linear space. Then the representation $\hat{T}$ can be extended to an algebraic representation T' of the group H.

Remark. Recall that $\tilde{M}$ is defined by N, M, V and φ as a linear subspace of the space M, generated by the union $\bigcup_{\lambda^{-1}\in\Lambda\cap V} \varphi(\lambda, e) = \bigcup_{\lambda^{-1}\in\Lambda\cap V} T_\lambda\varphi(\lambda^{-1})$. Then it is easy to see that the $\tilde{M}$ constructed to $\bar{N}$, $\bar{M}$, $\bar{V}$ and $\bar{\psi}$ coincides with $\tilde{\bar{M}}$ of Lemma 14.14.

Let H be, as above, a complex algebraic group and $\Lambda\subset H$ a Zariski dense subgroup. Let R and S be two algebraically connected algebraic subgroups of H such that $R\cap\Lambda$ is Zariski dense in R, and $S\cap\Lambda$ is Zariski dense in S. Denote by $V(R, S)$ the set of all $h_0\in H$ for which $\dim((h_0 Sh_0^{-1})\cap R) = \min_{h\in H} \dim((hSh^{-1})\cap R)$. For any $\lambda\in\Lambda\cap V(R, S)$ denote by $(R, S)_\lambda$ the connected component of the identity in the intersection $(\lambda S\lambda^{-1})\cap R$, and for any $\lambda_1, \lambda_2\in\Lambda$ for which $\lambda_1^{-1}\lambda_2\in V(R, S)$, denote by $(R, S)_{\lambda_1, \lambda_2}$ the connected component of the identity in the intersection $\lambda_1 R\lambda_1^{-1}\cap\lambda_2 S\lambda_2^{-1}$ (recall that for complex algebraic groups, algebraic and topological connectedness coincide).

Theorem 14.15. *Let $\tilde{T}$ be a representation of the subgroup Λ in a connected complex algebraic group D with trivial center, and $\bar{V}$ a nonempty Zariski open subset in H contained in $V(R, S)$. Assume that the following conditions are satisfied: 1) For any $\lambda\in\Lambda\cap\bar{V}$ the intersection $\Lambda\cap(R, S)_\lambda$ is Zariski dense in $(R, S)_\lambda$, 2) there exists in $R\cap\Lambda$ a subgroup Λ_R of finite index such that the restriction of $\tilde{T}$ to Λ_R can be extended to an algebraic representation T^R of the group R, 3) in $S\cap\Lambda$ there is a subgroup Λ_S of finite index, such that the restriction of $\tilde{T}$ to Λ_S can be extended to an algebraic representation T^S of the group S, 4) $\tilde{T}(\Lambda)$ is Zariski dense in D, 5) the minimal algebraic normal subgroup of the group D containing the union $\bigcup_{\lambda\in\Lambda\cap\bar{V}} (T^R(R, S)_\lambda)$, coincides with D. Then the representation $\tilde{T}$ can be extended to an algebraic representation $\tilde{T}^H$ of the group H.*

Proof. The restrictions of $\tilde{T}$ to Λ_R and Λ_S extend to algebraic representations T^R and T^S of the subgroups R and S, thus for any $\lambda\in\Lambda$ the restriction of $\tilde{T}$ to $\lambda\Lambda_R\lambda^{-1}$ and $\lambda\Lambda_S\lambda^{-1}$ can be extended to algebraic representations $T^{R,\lambda}$ and $T^{S,\lambda}$ of the subgroups $\lambda R\lambda^{-1}$ and $\lambda S\lambda^{-1}$. Now, *1)* $R\cap\Lambda$ and $S\cap\Lambda$ are Zariski dense in R and S, respectively, *2)* R and S are algebraically connected, *3)* Λ_R and Λ_S are of finite index in $R\cap\Lambda$ and $S\cap\Lambda$, respectively, thus, in view of Lemma 8.8 Λ_R and Λ_S are Zariski dense in R and S, respectively. This implies that for any $\lambda\in\Lambda$ the subgroups $\lambda\Lambda_R\lambda^{-1}$ and $\lambda\Lambda_S\lambda^{-1}$ are Zariski dense in $\lambda R\lambda^{-1}$ and $\lambda S\lambda^{-1}$, respectively. Therefore, for any $\lambda\in\Lambda$ the representations $T^{R,\lambda}$ and $T^{S,\lambda}$ are uniquely defined for any $\lambda\in\Lambda$ (since two algebraic representations of an algebraic group coinciding on a Zariski dense subgroup, coincide on the whole group). It follows from the uniqueness of $T^{R,\lambda}$ and $T^{S,\lambda}$ that for any $\lambda_1, \lambda_2\in\Lambda$

$$T^{R,\lambda_1\lambda_2}(\lambda_1\lambda_2 h\lambda_2^{-1}\lambda_1^{-1}) = \tilde{T}_{\lambda_1}\circ(T^{R,\lambda_2}(\lambda_2 h\lambda_2^{-1}))\circ\tilde{T}_{\lambda_1}^{-1} \tag{25}$$

for any $h\in R$, and

$$T^{S,\lambda_1\lambda_2}(\lambda_1\lambda_2 h\lambda_2^{-1}\lambda_1^{-1}) = \tilde{T}_{\lambda_1}\circ(T^{S,\lambda_2}(\lambda_2 h\lambda_2^{-1}))\circ\tilde{T}_{\lambda_1}^{-1} \tag{26}$$

for any $h \in S$ (it should only be mentioned that for $h \in R \cap \Lambda (S \cap \Lambda)$ the validity of (25) ((26)) follows from the definition of the representations $T^{R,\lambda}(T^{S,\lambda})$).

It is easy to see that for any $\lambda_1, \lambda_2 \in \Lambda$, for which $\lambda_1^{-1}\lambda_2 \in \overline{V}$, we have

$$(R, S)_{\lambda_1, \lambda_2} = \lambda_1 (R, S)_{\lambda_1^{-1}\lambda_2} \lambda_1^{-1}. \tag{27}$$

By the condition of the theorem, for any $\lambda \in \Lambda \cap \overline{V}$ the intersection $\Lambda \cap (R, S)_\lambda$ is Zariski dense in $(R, S)_\lambda$. From this and (27) we obtain that for any $\lambda_1, \lambda_2 \in \Lambda$ for which $\lambda_1^{-1}\lambda_2 \in \overline{V}$, the intersection $\Lambda \cap (R, S)_{\lambda_1, \lambda_2}$ is Zariski dense in $(R, S)_{\lambda_1, \lambda_2}$. Remark now that *1)* since Λ_R and Λ_S are of finite index in $R \cap \Lambda$ and $S \cap \Lambda$, respectively, and $(R, S)_{\lambda_1, \lambda_2} \subset \lambda_1 R \lambda_1^{-1} \cap \lambda_2 S \lambda_2^{-1}$, for any $\lambda_1, \lambda_2 \in \Lambda$ for which $\lambda_1^{-1}\lambda_2 \in \overline{V}$, the intersection $(\lambda_1 \Lambda_R \lambda_1^{-1} \cap \lambda_2 \Lambda_S \lambda_2^{-1}) \cap (R, S)_{\lambda_1, \lambda_2}$ is of finite index in $\Lambda \cap (R, S)_{\lambda_1, \lambda_2}$, *2)* for any $\lambda_1, \lambda_2 \in \Lambda$ for which $\lambda_1^{-1}\lambda_2 \in \overline{V}$, the subgroup $(R, S)_{\lambda_1, \lambda_2}$ is connected (by definition). These and Lemma 8.8 imply that for any $\lambda_1, \lambda_2 \in \Lambda$ for which $\lambda_1^{-1}\lambda_2 \in \overline{V}$, the intersection $(\lambda_1 \Lambda_R \lambda_1^{-1} \cap \lambda_2 \Lambda_S \lambda_2^{-1}) \cap (R, S)_{\lambda_1, \lambda_2}$ is Zariski dense in $(R, S)_{\lambda_1, \lambda_2}$. Since the representations $T^{R,\lambda}$ and $T^{S,\lambda}$ are algebraic, this implies that for any $\lambda_1, \lambda_2 \in \Lambda$ for which $\lambda_1^{-1}\lambda_2 \in \overline{V}$, and any $h \in (R, S)_{\lambda_1, \lambda_2}$, we have

$$T^{R, \lambda_1}(h) = T^{S, \lambda_2}(h). \tag{28}$$

(It is sufficient to mention that for $h \in (\lambda_1 \Lambda_R \lambda_1^{-1} \cap \lambda_2 \Lambda_S \lambda_2^{-1}) \cap (R, S)_{\lambda_1, \lambda_2}$ (28) follows from the definition of $T^{R,\lambda}$ and $T^{S,\lambda}$).

Denote by $\mathfrak{H}$, $\mathfrak{R}$, $\mathfrak{S}$, $\mathfrak{D}$ the Lie algebras of the groups H, R, S and D. Since the representations $T^{R,\lambda}$ and $T^{S,\lambda}$ are algebraic, for any $\lambda \in \Lambda$, $T^{R,\lambda}$ and $T^{S,\lambda}$ induce the homomorphisms $\tilde{k}^0_\lambda : (\mathrm{Ad}\,\lambda)\mathfrak{R} \to \mathfrak{D}$ and $\tilde{\tilde{k}}^0_\lambda : (\mathrm{Ad}\,\lambda)\mathfrak{S} \to \mathfrak{D}$ of the Lie algebras $(\mathrm{Ad}\,\lambda)\mathfrak{R}$ and $(\mathrm{Ad}\,\lambda)\mathfrak{S}$ into the Lie algebra $\mathfrak{D}$. Then (25), (26) and (28) imply

$$\tilde{k}^0_{\lambda_1\lambda_2}((\mathrm{Ad}\,(\lambda_1\lambda_2))z) = (\mathrm{Ad}\,(\tilde{T}_{\lambda_1}))\,(\tilde{k}^0_{\lambda_2}((\mathrm{Ad}\,\lambda_2)z)) \tag{29}$$

for any $\lambda_1, \lambda_2 \in \Lambda$ and $z \in \mathfrak{R}$, and

$$\tilde{\tilde{k}}^0_{\lambda_1\lambda_2}((\mathrm{Ad}\,(\lambda_1\lambda_2))y) = (\mathrm{Ad}\,(\tilde{T}_{\lambda_1}))\,(\tilde{\tilde{k}}^0_{\lambda_2}((\mathrm{Ad}\,\lambda_2)y)) \tag{30}$$

for any $\lambda_1, \lambda_2 \in \Lambda$ and $y \in \mathfrak{S}$, and

$$\tilde{k}^0_{\lambda_1}(x) = \tilde{\tilde{k}}^0_{\lambda_2}(x) \tag{31}$$

for any $\lambda_1, \lambda_2 \in \Lambda$ for which $\lambda_1^{-1}\lambda_2 \in \overline{V}$, and any $x \in ((\mathrm{Ad}\,\lambda_1)\mathfrak{R}) \cap ((\mathrm{Ad}\,\lambda_2)\mathfrak{S})$ (here we applied the obvious fact that the Lie algebra of the subgroup $(R, S)_{\lambda_1, \lambda_2}$ coincides with $((\mathrm{Ad}\,\lambda_1)\mathfrak{R}) \cap ((\mathrm{Ad}\,\lambda_2)\mathfrak{S})$).

Write $\overline{N} \subset \mathfrak{R}$ for the linear subspace generated by $\bigcup_{\lambda \in \overline{V} \cap \Lambda} (((\mathrm{Ad}\,\lambda)\mathfrak{S}) \cap \mathfrak{R})$, T for the adjoint representation of the group H, $\hat{T}_0$ for the representation of the group Λ in the space $\mathfrak{D}$, defined by the formula $\hat{T}_0 = \mathrm{Ad}_D \circ \tilde{T}$ where Ad D denotes the adjoint representation of D, $\hat{L}$ for the linear subspace generated by $\bigcup_{\lambda \in \Lambda} \tilde{k}_\lambda((\mathrm{Ad}\,\lambda)N)$; put $\overline{M} = \mathfrak{S}$, and write $\tilde{\tilde{M}}$ for the linear subspace of the space

$\overline{M}$ generated by the union $\bigcup_{\lambda^{-1}\in\Lambda\cap V}((T_\lambda\overline{N})\cap\overline{M})$, and $\tilde{k}_\lambda$ and $\tilde{\tilde{k}}_\lambda$ for the restrictions of $\tilde{k}^0_\lambda$ and $\tilde{\tilde{k}}^0_\lambda$ to $T_\lambda\overline{N}$ and $T_\lambda\tilde{\tilde{M}}$, respectively. It follows from (29) that $\hat{L}$ is invariant under $\hat{T}_0(\lambda)$ for any $\lambda\in\Lambda$. Denote by $\hat{T}$ the restriction of $\hat{T}_0$ to $\hat{L}$. Our notations are brought into accord with the notations of Lemma 14.14. We make the following remarks:

1) It is clear that for any $\lambda\in\Lambda$, $\tilde{k}_\lambda$ and $\tilde{\tilde{k}}_\lambda$ are linear.

2) By the definition of $\hat{L}$, the union $\bigcup_{\lambda\in\Lambda}\tilde{k}_\lambda((\operatorname{Ad}\lambda)\overline{N})$ generates $\hat{L}$ as a linear space.

3) Since $\dim((hSh^{-1})\cap R)=\dim((\operatorname{Ad}h)\mathfrak{S})\cap\mathfrak{R})$ for any $h\in H$, we get that $V(R, S)$ coincides with $V(\overline{N}, \overline{M})$ defined before Lemma 14.14.

4) $\tilde{k}_\lambda$ and $\tilde{\tilde{k}}_\lambda$ are, by definition, restrictions of $\tilde{k}^0_\lambda$ and $\tilde{\tilde{k}}^0_\lambda$.

5) For any $\tilde{\lambda}\in\overline{V}\cap\Lambda$ we have

$$(T_\lambda\overline{M})\cap\overline{N}=((\operatorname{Ad}\tilde{\lambda})\overline{M})\cap\overline{N}=((\operatorname{Ad}\tilde{\lambda})\mathfrak{S})\cap\overline{N}\supset$$
$$\supset((\operatorname{Ad}\tilde{\lambda})\mathfrak{S})\cap\Big(\bigcup_{\lambda\in\overline{V}\cap\Lambda}((\operatorname{Ad}\lambda)\mathfrak{S})\cap\mathfrak{R})\Big)\supset((\operatorname{Ad}\tilde{\lambda})\mathfrak{S})\cap\mathfrak{R} \tag{32}$$

(the last inclusion holds in view of the relation $\tilde{\lambda}\in\overline{V}\cap\Lambda$). In consequence of (32) and the definition of $\overline{N}$, $\bigcup_{\lambda\in\overline{V}\cap\Lambda}(T_\lambda\overline{M})\cap\overline{N}$ generates $\overline{N}$. The above remarks, formulae (29), (30) and (31) and Lemma 14.14 imply that the representation $\hat{T}$ of the subgroup Λ can be extended to an algebraic representation $\hat{T}^H$ of the group H.

Put $k=\tilde{k}_e$ (e is the identity of the group Λ), i.e. k is a homomorphism of the Lie algebra $\mathfrak{R}$ into the Lie algebra $\mathfrak{D}$ inducing T^R. It is clear that for any $\lambda\in\Lambda$ the Lie algebra of the subgroup $(R, S)_\lambda$ coincides with the intersection $(((\operatorname{Ad}\lambda)\mathfrak{S})\cap\mathfrak{R})$. Therefore, since *1)* by the condition of the theorem, the minimal algebraic normal subgroup of the group D containing $\bigcup_{\lambda\in\Lambda\cap V}(T^R(R, S)_\lambda)$, coincides with D, *2)* for any $\lambda\in\overline{V}\cap\Lambda$ the subgroup $(R, S)_\lambda$, and consequently also the subgroup $T^R(R, S)_\lambda$, is connected, *3)* any linear subspace in the Lie algebra $\mathfrak{A}$ of an arbitrary complex algebraic group A (and also real) invariant under $\operatorname{Ad}a$ for any $a\in A$, is an ideal in $\mathfrak{A}$, we get that the minimal linear subspace $\mathfrak{D}^0$ containing $\bigcup_{\lambda\in\Lambda\cap V}\tilde{k}((\operatorname{Ad}\lambda)\mathfrak{S})\cap\mathfrak{R}$ and invariant under $\operatorname{Ad}d$ for any $d\in D$ coincides with $\mathfrak{D}$ (for, in the opposite case $\mathfrak{D}^0$ would be a Lie algebra of the nontrivial algebraic normal subgroup D^0 of D, and D^0 would contain $\bigcup_{\lambda\in\Lambda\cap V}T^R(R, S)_\lambda$ in view of the connectedness of all $(R, S)_\lambda$). Now since *1)* $\hat{L}$ is invariant under $\operatorname{Ad}(\tilde{T}_\lambda)$ for all $\lambda\in\Lambda$ (because $\hat{L}$ is invariant under $\hat{T}_0(\lambda)$ for any $\lambda\in\Lambda$), *2)* $\tilde{T}(\Lambda)$ is Zariski-dense in D (by the condition of the theorem), *3)* $\hat{L}\supset\overline{N}\supset\bigcup_{\lambda\in\overline{V}\cap\Lambda}(((\operatorname{Ad}\lambda)\mathfrak{S})\cap\mathfrak{R})$ (by the definition of $\hat{L}$ and $\overline{N}$), *4)* the adjoint representation Ad_D of the group D is algebraic (in view of Lemma 1.2.1), it follows then that

$$\hat{L} = D. \tag{33}$$

Now we can complete the proof of the theorem. It is easy to see that Lemmas 1.2.1, 1.2.17, and 8.1 imply the following simple

Lemma 14.16. Let G_1 and G_2 be complex algebraic groups, $\Delta \subset G_1$ a Zariski dense subgroup, f a representation of Δ in G_1. Assume that G_2 is connected and the centre of G_2 is trivial. Denote by Ad_{G_2} the adjoint representation of the group G_2. If $\mathrm{Ad}_{G_2} \circ f$ extends to an algebraic representation of the group G_1, then f can also be extended to an algebraic representation of the group G_1.

As we have already shown, $\hat{T}$ extends to an algebraic representation $\hat{T}^H$ of the group H. Now D is connected and the centre of D is trivial (by assumption). Thus, formula (33) and Lemma 14.16 imply the assertion of the theorem.

§ 15. Extension of representations of a class of subgroups

Recall that the commutator (A, B) of two subgroups A and B is the subgroup generated by all commutators of the form $aba^{-1}b^{-1}$, where $a \in A$ and $b \in B$. If A and B are subgroups of an algebraic group H, then the algebraic commutator of the subgroups A and B will mean the algebraic closure in H of the commutator (A, B). The algebraic commutator of the subgroups A and B will be denoted by $(\overline{A, B})$. The following lemma can easily be established.

Lemma 15.1. Let A and B be algebraic subgroups in a complex (real) algebraic group $\tilde{A} \subset A$ and $\tilde{B} \subset B$ subgroups algebraically dense in A and B, respectively. Then the commutator $(\tilde{A}, \tilde{B})$ of $\tilde{A}$ and $\tilde{B}$ is Zariski dense in the algebraic commutator $(\overline{A, B})$ of A and B.

Although we will not apply it, we still mention that for any complex (real) algebraic group H and any unipotent algebraic normal subgroup V of the group H, the algebraic commutator $(\overline{H, V})$ coincides with (H, V).

Lemma 15.2. Let F be a connected complex algebraic group, $V \subset F$ a unipotent algebraic normal subgroup, $\Omega \subset F$ a subgroup algebraically dense in F, such that $\Omega \cap V$ is Zariski dense in V, and T a representation of Ω in a complex algebraic group H. Then there is in Ω a subgroup S such that *1)* $S \subset (\Omega, \Omega \cap V)$, *2)* S is algebraically dense in $(\overline{F, V})$, *3)* for any $s \in S$ the element $T(s)$ of the group H is unipotent, *4)* S is a normal subgroup in Ω.

Proof. First we make two remarks: *1)* since V is unipotent, V is nilpotent according to Lemma 1.3.1, *2)* since V is a normal subgroup in F, $\Omega \cap V$ is a normal subgroup in Ω. Therefore, $T(\Omega \cap V)$ is a nilpotent normal subgroup in $T(\Omega)$. Denote by N the algebraic closure of $T(\Omega \cap N)$ in H, and by N^0 the algebraically connected component of the identity in the group N. Now *1)* since

$T(\Omega \cap V)$ is nilpotent, thus, according to Lemma 1.2.3, N and consequently N^0 also are nilpotent, *2)* since $T(\Omega)$ normalizes $T(\Omega \cap V)$, $T(\Omega)$ normalizes N and consequently also N^0. Since N^0 is an algebraically connected nilpotent algebraic **C**-group, N^0 decomposes into the direct product $N^0 = R \times W$, where R is a torus, and W a unipotent algebraic subgroup. Therefore, since *1)* $T(\Omega)$ normalizes N^0, *2)* any subgroup normalizing N^0 also normalizes R (for R is the unique maximal torus in N^0), $T(\Omega)$ normalizes R, i.e. $T(\Omega) \subset N(R)$, where $N(R)$ is the normalizer of R in H. In view of Lemma 1.5.6 the group $N(R)/C(R)$ is finite $\big(C(R)$ is the centralizer of R in $H\big)$. Now since $T(\Omega) \subset N(R)$, it follows that $\tilde{\Omega} = T(\Omega) \cap \cap C(R)$ is of finite index in $T(\Omega)$, and that $\tilde{\Omega}$ is a normal subgroup in $T(\Omega)$.

Put $\Omega_1 = T^{-1}(\tilde{\Omega})$ and $\Omega_2 = \Omega \cap V \cap T^{-1}(N_0)$. Since $\tilde{\Omega}$ is of finite index in $T(\Omega)$, the subgroup Ω_1 is of finite index in Ω and since $T(\Omega \cap V) \subset N$, and N^0 is of finite index in N (in view of Lemma 1.2.7), we get that Ω_2 is a subgroup of finite index in $\Omega \cap V$. Therefore, since *1)* F and V are algebraically connected (V is algebraically connected in view of Lemma 1.3.1) *2)* Ω is Zariski dense in F, and $\Omega \cap V$ in V (by the condition of the lemma), in view of Lemma 8.8 Ω_1 is Zariski dense in F, and Ω_2 in V. From this and Lemma 15.1 we obtain that the commutator $S = (\Omega_1, \Omega_2)$ is Zariski dense in $\overline{(F, V)}$.

Since *1)* $T(\Omega)$ normalizes N^0, *2)* $T(\Omega_1) = \tilde{\Omega} \subset T(\Omega)$, *3)* W is the unipotent radical in N^0 (because $N^0 = R \times W$, and R is a torus), $T(\Omega_1)$ normalizes W. Therefore, we have

$$\big(T(\Omega_1), W\big) \subset W. \tag{1}$$

Now $T(\Omega_1) = \tilde{\Omega} = T(\Omega) \cap C(R)$ belongs to $C(R)$. Therefore

$$\big(T(\Omega_1), R\big) = e. \tag{2}$$

From (1), (2) and the relation $N^0 = R \times W$ we get

$$\big(T(\Omega_1), N^0\big) \subset W. \tag{3}$$

Since $\Omega_2 = \Omega \cap V \cap T^{-1}(N^0)$, thus $T(\Omega_2) \subset N^0$. From this and (3) we have

$$\big(T(\Omega_1), T(\Omega_2)\big) \subset W. \tag{4}$$

From (4) it is immediate that

$$T(S) = T\big((\Omega_1, \Omega_2)\big) \subset W, \tag{5}$$

and consequently, for any $s \in S$, $T(s)$ is unipotent.

Thus, in order to complete the proof of the lemma, it remains to show part *4)*. To do this we mention that *1)* since $\tilde{\Omega}$ is a normal subgroup in $T(\Omega)$, $\Omega_1 = T^{-1}(\tilde{\Omega})$ is a normal subgroup in Ω, *2)* since V is a normal subgroup in F, and $T(\Omega)$ normalizes N^0, Ω normalizes $\Omega_2 = \Omega \cap V \cap T^{-1}(N^0)$, i.e. Ω_2 is a normal subgroup in Ω. Therefore $S = (\Omega_1, \Omega_2)$ is a normal subgroup in Ω. The proof of the lemma is complete.

Lemma 15.3. Let F be an algebraically connected real algebraic group, $F_{\mathbf{C}}$ the complexification of F, $V \subset F$ a unipotent algebraic normal subgroup, $L \subset F$ an algebraic subgroup such that F is the semidirect product of L and V (in the sense of IV of § 1), Ω a discrete subgroup of F, and T a representation of Ω in a complex algebraic group H. Denote by $\tilde{V}$ the algebraic closure of $T(\Omega \cap V)$ in H, by $\tilde{L}$ the algebraic closure of $T(\Omega \cap L)$ in H, by $\tilde{V}^0$ the algebraically connected component of the identity of the group $\tilde{V}$, and by $C(\tilde{V}^0)$ the centralizer of $\tilde{V}^0$ in H. If *1)* the volume of the quotient space $L/L \cap \Omega$ is finite, *2)* Ω is Zariski dense in F, $\Omega \cap V$ in V, and $\Omega \cap L$ in L, *3)* there is in $\Omega \cap V$ a subgroup Φ of finite index such that $T(\Phi)$ is unipotent, *4)* $\tilde{L} \cap C(\tilde{V}^0) = e$, then there is a unique algebraic representation φ of the complex algebraic group $F_{\mathbf{C}}$ in the complex algebraic group H such that the restriction of φ to a suitable subgroup Ω_φ of finite index of the group Ω, coincides with the restriction of T to Ω_φ (we emphasize that Ω_φ is not given in advance, but it is constructed from φ).

Proof. Since *1)* V is algebraically connected in view of its unipotence and Lemma 1.3.1, *2)* $\Omega \cap V$ is Zariski dense in V, *3)* Φ is a subgroup of finite index in $\Omega \cap V$, according to Lemma 8.8 Φ is Zariski dense in V. On the other hand, $T(\Phi)$ is unipotent by assumption. Thus, in view of Lemma 5.11, the restriction of T to Φ can be extended to an algebraic representation φ_V of the complex algebraic group $V_{\mathbf{C}}$ in H, where $V_{\mathbf{C}}$ denotes the complexification of V. Denote by $\tilde{\varphi}_V$ the homomorphism corresponding to φ_V of the complex Lie algebra $\mathfrak{V}_{\mathbf{C}}$ into the complex Lie algebra $\mathfrak{H}$, where $\mathfrak{V}_{\mathbf{C}}$ and $\mathfrak{H}$ denote the Lie algebras of the groups $V_{\mathbf{C}}$ and H.

Let ω be an arbitrary element of Ω. Since V is a normal subgroup in F, $\Omega \cap V$ is a normal subgroup in Ω. Therefore, since Φ is a subgroup of finite index in $\Omega \cap V$, $\omega^{-1}\Phi\omega$ is a subgroup of a finite index in $\Omega \cap V$. Since the intersection of two subgroups of finite index is also a subgroup of finite index, $\Phi \cap \omega^{-1}\Phi\omega$ is a subgroup of finite index in $\Omega \cap V$. Therefore, since V is algebraically connected, as mentioned several times, and $\Omega \cap V$ is Zariski dense in V, in view of Lemma 8.8 $\Phi \cap \omega^{-1}\Phi\omega$ is Zariski dense in V, and consequently also in $V_{\mathbf{C}}$.

Now we show that for any $v \in V_{\mathbf{C}}$ and any $\omega \in \Omega$ we have

$$\varphi_V(\omega v \omega^{-1}) = T_\omega \circ \varphi_V(v) \circ T_\omega^{-1}. \tag{6}$$

Indeed, if $v \in \Phi \cap \omega^{-1}\Phi\omega$, or in other words $v \in \Phi$ and $\omega v \omega^{-1} \in \Phi$, then (6) holds in view of the fact that φ_V is an extension of the restriction of T to Φ. Now $\Phi \cap \omega^{-1}\Phi\omega$ is Zariski dense in $V_{\mathbf{C}}$. On the other hand, since φ_V is an algebraic homomorphism, it is easy to see that the homomorphisms φ_1 and φ_2 of $V_{\mathbf{C}}$ into H, defined by the formulae $\varphi_1(v) = \varphi_V(\omega v \omega^{-1})$ and $\varphi_2(v) = T_\omega \circ \varphi_V(v) \circ T_\omega^{-1}$, $(v \in V_{\mathbf{C}})$, are algebraic. Thus φ_1 and φ_2 are algebraic homomorphisms coinciding on the Zariski dense subgroup $\Phi \cap \omega^{-1}\Phi\omega$. Therefore φ_1 and φ_2 coincide everywhere on V_C, i.e. we have proved (6). As an immediate consequence of (6), we have

$$\tilde{\varphi}_V((\mathrm{Ad}\,\omega)\mathfrak{v}) = (\mathrm{Ad}\,(T_\omega))(\tilde{\varphi}_V(\mathfrak{v})) \tag{7}$$

for any $\mathfrak{v} \in \mathfrak{V}_C$ and any $\omega \in \Omega$.

Now we show that

$$\varphi_V(V_C) = \tilde{V}^0. \tag{8}$$

To this purpose we mention first that since V_C is a complex algebraic group and φ_V is an algebraic homomorphism, by virtue of Lemma 8.1, $\varphi_V(V_C)$ is an algebraic subgroup in H. Therefore, since *1)* Φ is Zariski dense in V_C, *2)* φ_V is an algebraic homomorphism, *3)* the restriction of T to Φ coincides with the restriction of φ_V to Φ (for φ_V is the extension to V_C of the restriction of T to Φ), we have

$$\varphi_V(V_C) = A(T(\Phi)), \tag{9}$$

where $A(T(\Phi))$ denotes the Zariski closure of $T(\Phi)$ in H. On the other hand, since we have by assumption *1)* Φ is a subgroup of finite index in $\Omega \cap L$, *2)* $T(\Phi)$ is unipotent, in view of Lemmas 1.3.3 and 1.3.1 $A(T(\Phi))$ is an algebraically connected algebraic subgroup of finite index in $\tilde{V}$. From this and (9) we obtain (8). From (8) it follows immediately that

$$\tilde{\varphi}_V(\mathfrak{V}_C) = \tilde{\mathfrak{V}}^0, \tag{10}$$

where $\tilde{\mathfrak{V}}^0$ denotes the Lie algebra of the group $\tilde{V}^0$.

As we have already mentioned, $\Omega \cap V$ is a normal subgroup in Ω. Therefore, $T(\Omega \cap V)$ is a normal subgroup in $T(\Omega)$. This implies that $T(\Omega)$ normalizes $\tilde{V}$, and consequently also $\tilde{V}^0$. Therefore, in view of Lemma 1.2.8 $\tilde{L}$ normalizes $\tilde{V}^0$. Thus, we can define a representation ψ_1 of the group $\tilde{L}$ into the complex algebraic group $GL(\tilde{\mathfrak{V}}^0)$, ($GL(\tilde{\mathfrak{V}}^0)$ is the group of non-singular linear transformations of the complex linear space $\tilde{\mathfrak{V}}^0$) by the following formula:

$$\psi_1(l)(\mathfrak{v}) = (\mathrm{Ad}\,l)\mathfrak{v}, \tag{11}$$

where $\mathfrak{v} \in \tilde{\mathfrak{V}}^0$. In view of Lemma 1.2.1 ψ_1 is an algebraic representation of the complex algebraic group $\tilde{L}^0$ into the complex algebraic group $GL(\tilde{\mathfrak{V}}^0)$.

Since V is a normal subgroup in F, V_C is a normal subgroup in F_C. Therefore, one can define a representation ψ_2 of the complexification L_C of the group L into the group $GL(\mathfrak{V}_C)$ by the formula

$$(\psi_2(l))(\mathfrak{v}) = (\mathrm{Ad}\,l)\mathfrak{v}, \tag{12}$$

where $\mathfrak{v} \in \mathfrak{V}_C$. In view of Lemma 1.2.1 ψ_2 is an algebraic representation of the complex algebraic group L_C into the complex algebraic group $GL(\mathfrak{V}_C)$.

From (7) and the definition of the mapping ψ_2 we see that for any $\omega \in \Omega \cap L$, $\psi_2(\omega)$ preserves the kernel of the mapping $\tilde{\varphi}_V$. Therefore, since *1)* $\Omega \cap L$ is Zariski dense in L by assumption, *2)* L is Zariski dense in L_C (by the definition of complexification), and *3)* ψ_2 is an algebraic representation, for any $l \in L_C$, $\psi_2(l)$ preserves the kernel of the mapping $\tilde{\varphi}_V$. Therefore, the representation ψ_2 induces, in a natural way, a representation ψ_3 of the group L_C on the quotient space $\mathfrak{B}_C/\text{Ker}\,\tilde{\varphi}_V$. Since the representation ψ_2 is algebraic, ψ_3 is an algebraic representation of the complex algebraic group L_C. In view of (10) we can identify $\mathfrak{B}_C/\text{Ker}\,\tilde{\varphi}_V$ with $\tilde{\mathfrak{B}}^0$ and assume that ψ_3 is a representation of L_C in the complex linear space $\tilde{\mathfrak{B}}^0$. Then it follows from (7) and the definition of ψ_3 and ψ_1 that for any $\omega \in \Omega \cap L$ we have

$$\psi_3(\omega) = \psi_1(T(\omega)). \tag{13}$$

Since $\tilde{L} \cap C(\tilde{V}^0) = e$ (by assumption), and $\tilde{V}^0$ is connected (by assumption), thus the representation ψ_1 is faithful. Then ψ_1 is a faithful algebraic representation of the complex algebraic group $\tilde{L}$. Therefore, in view of Lemmas 8.1 and 1.2.17, $\psi_1(\tilde{L})$ is a complex algebraic group and the inverse homomorphism ψ_1^{-1} of ψ from the group $\psi_1(\tilde{L})$ into the group $\tilde{L}$, is well defined, and is a regular algebraic homomorphism.

Since for any $\omega \in \Omega \cap L$ (13) holds, we have

$$\psi_3(\Omega \cap L) = \psi_1(T(\Omega \cap L)). \tag{14}$$

Now $\Omega \cap L$ is Zariski dense in L (by assumption), and the representation ψ_3 is algebraic. Therefore

$$A(\psi_3(\Omega \cap L)) = \psi_3(L), \tag{15}$$

where $A(\psi_3(\Omega \cap L))$ denotes the algebraic closure of the subgroup $\psi_3(\Omega \cap L)$. On the other hand, since $T(\Omega \cap L)$ is Zariski dense in $\tilde{L}$ (by the definition of $\tilde{L}$), and the representation ψ_1 is algebraic, we have

$$A(\psi_1(T(\Omega \cap L))) = \psi_1(\tilde{L}), \tag{16}$$

where $A(\psi_1(T(\Omega \cap L)))$ denotes the algebraic closure of the subgroup $\psi_1(T(\Omega \cap L))$. Combining (14), (15) and (16) we get

$$\psi_3(L) = \psi_1(\tilde{L}). \tag{17}$$

In view of (17) and the fact that ψ_1^{-1} is well defined, and is a regular algebraic homomorphism, $\psi_1^{-1}\psi_3$ is well defined, and is a regular algebraic homomorphism of the complex algebraic group L_C into the complex algebraic group $\tilde{L}$. Now, since for all $\omega \in \Omega \cap L$ (13) holds, the restriction of $\psi_1^{-1}\psi_3$ to $\Omega \cap L$ coincides with the restriction of T to $\Omega \cap L$, i.e. $\psi_1^{-1}\psi_3$ is the extension to L_C of the restriction of the representation T to $\Omega \cap L$.

Since F is the semi-direct product of L and V in the sense of § 1, the complex algebraic group F_C is the semi-direct product (in the sense of § 1) of the algebraic subgroup L_C and the algebraic normal subgroup V_C. This implies that the mappings l and v associating to every $f \in F_C$ the factors $l(f) \in L_C$ and $v(f) \in V_C$ in the decomposition $f = l(f) \cdot v(f)$ (which is unique, since F_C is the semi-direct product of $L_{\mathbf{C}}$ and $V_{\mathbf{C}}$), are regular algebraic mappings. We define now the mapping φ of the group $F_{\mathbf{C}}$ into the group H by the following formula:

$$\varphi(f) = \psi_1^{-1}\psi_3(l(f))\cdot\varphi_V(v(f)) \tag{18}$$

for any $f \in F_{\mathbf{C}}$ (as always, the dot stands for group multiplication). Since $\psi_1^{-1}\psi_3$, l and v are regular algebraic mappings of complex algebraic manifolds, φ is a regular algebraic mapping of the complex algebraic group $F_{\mathbf{C}}$ into the complex algebraic group H.

Since for any $\omega \in \Omega$ and any $v \in V_{\mathbf{C}}$ (6) holds and the restriction of φ_V to Φ coincides with the restriction of T to Φ (by the definition of φ_V), the restriction of φ_V to $\bigcup\limits_{\omega\in\Omega} \omega\Phi\omega^{-1}$ coincides with the restriction of T to $\bigcup\limits_{\omega\in\Omega} \omega\Phi\omega^{-1}$. Therefore, since φ_V and T are homomorphisms, the restriction of φ_V to the subgroup $\hat{\Phi}$ generated by $\bigcup\limits_{\omega\in\Omega} \omega\Phi\omega^{-1}$, coincides with the restriction of T to $\hat{\Phi}$. It is clear, on the other hand, that Φ is a normal subgroup in Ω.

Put now $\Omega_\varphi = (\Omega \cap L)\cdot\hat{\Phi}$. Ω_φ is a subgroup, since $\hat{\Phi}$ is a normal subgroup in Ω. Since *1)* the restriction of $\psi_1^{-1}\psi_3$ to $\Omega \cap L$ coincides with the restriction of T to $\Omega\cap L$, *2)* the restriction of φ_V to $\hat{\Phi}$ coincides with the restriction of φ_V to $\hat{\Phi}$ *3)* T is a representation, in view of (18) the restriction of φ to Ω_φ coincides with the restriction of T to Ω_φ.

As mentioned already Φ is Zariski dense in V. Therefore, since $\hat{\Phi} \supset \Phi$, $\hat{\Phi}$ is Zariski dense in V. From this and Lemma 5.4 it follows that the quotient space $V/\hat{\Phi}$ is compact. Therefore, since the volume of the quotient space $L/\Omega\cap L$ is finite, in view of Lemma 1.1.2 Ω_φ is a lattice in F (here we applied the fact that L can be identified with F/V in a natural way). From this and the discreteness of Ω we get that Ω_φ is a subgroup of finite index in Ω. Therefore, since *1)* Ω is Zariski dense in F (by assumption), *2)* F is Zariski dense in $F_{\mathbf{C}}$ (by the definition of complexification), *3)* F and consequently $F_{\mathbf{C}}$ too, are algebraically connected, in view of Lemma 8.8 Ω_φ is Zariski dense in $F_{\mathbf{C}}$. Thus, we have that the restriction of the regular algebraic mapping φ to the Zariski dense subgroup Ω_φ coincides with the restriction of the homomorphism T to Ω_φ. Therefore, in view of Lemma 1.2.6, φ is a homomorphism. Thus we conclude that φ is an algebraic homomorphism of the complex algebraic group $F_{\mathbf{C}}$, coinciding with T on the subgroup Ω_φ of finite index in Ω, i.e., φ is the desired representation.

Now, in order to complete the proof of the lemma, it remains to observe that any subgroup of finite index of the group Ω is Zariski dense in $F_{\mathbf{C}}$ (since $F_{\mathbf{C}}$ is connected, and Ω is Zariski dense in $F_{\mathbf{C}}$) and consequently, the homomorphism φ having the required properties, is unique (for, any two algebraic homomorphisms coinciding on a Zariski dense subgroup coincide everywhere).

§ 16. Rationality of the group Γ

In this section we will use the notations of §§ 7, 11, 12 and 13. Recall that G and Γ satisfy the conditions formulated at the beginning of §§ 7 and 13.

Now we prove some simple auxiliary algebraic statements. We recall that for any real algebraic group H, $H_{\mathbf{C}}$ stands for its complexification.

Lemma 16.1. If V is a non-trivial algebraic subgroup of G, contained in no proper algebraic normal subgroup of G, then $V_{\mathbf{C}}$ is contained in no proper algebraic normal subgroup of $G_{\mathbf{C}}$.

Proof. Since $V \neq e$, it follows easily from the conditions of the lemma that the intersection $N \cap G$ of any algebraic normal subgroup N of the group $G_{\mathbf{C}}$ containing $V_{\mathbf{C}}$ coincides with G. Since G is Zariski dense in $G_{\mathbf{C}}$ (because $G_{\mathbf{C}}$ is the complexification of G), we have $N = G_{\mathbf{C}}$. The lemma is proved.

Lemma 16.2. If V is an orispheric subgroup of G, then $V_{\mathbf{C}}$ is an orispheric subgroup of $G_{\mathbf{C}}$.

Proof. Since V is Zariski dense in $V_{\mathbf{C}}$, $V_{\mathbf{C}}$ is defined over $\mathbf{R}$, and consequently, in view of Lemma 1.2.13, the normalizer $N(V_{\mathbf{C}})$ is defined over $\mathbf{R}$. Therefore, according to Lemma 1.3.5, the unipotent radical W of the group $N(V_{\mathbf{C}})$ is defined over $\mathbf{R}$. Now W is connected (in view of Lemma 1.3.2). Therefore, in view of Lemma 1.2.11, $W \cap G$ is Zariski dense in G. On the other hand, $W \cap G$ is clearly a unipotent normal subgroup in $N(V)$. Since V is orispheric, we have

$$W \cap G \subset V. \tag{1}$$

Since $W \cap G$ is algebraically dense in W, (1) implies

$$W \subset V_{\mathbf{C}}. \tag{2}$$

On the other hand, it is clear that

$$W \supset V_{\mathbf{C}}. \tag{3}$$

The assertion of the lemma follows from (2) and (3).

Lemma 16.3. Let U be a non-trivial unipotent subgroup of the group Γ. If $\overline{U}$ is an orispheric subgroup of the group G, then $C(\overline{U}_{\mathbf{C}}) \subset \overline{U}_{\mathbf{C}}$ (recall that G and Γ satisfy the conditions formulated at the beginning of § 7).

Proof. Since G and Γ satisfy the conditions formulated at the beginning of § 7, and $U \neq e$, in view of Lemma 4.8 U is not contained in any proper algebraic normal subgroup of the group G, and, consequently, neither is $\overline{U}$. Therefore, according to Lemma 16.1 neither is $\overline{U}_{\mathbf{C}}$ contained in any proper algebraic normal subgroup of the group $G_{\mathbf{C}}$. On the other hand, since $\overline{U}$ is orispheric, in view of Lemma 16.2 $\overline{U}_{\mathbf{C}}$ is also orispheric. Thus, $\overline{U}_{\mathbf{C}}$ is an orispheric subgroup in the semisimple complex algebraic group G_C with trivial centre (the centre of G_C is trivial in view of the conditions formulated at the beginning of § 7), and $\overline{U}_C$ is not contained in any proper algebraic normal subgroup of the group G_C. Thus, to complete the proof of the lemma, it is enough to refer to Lemma 11.16.

Recall now that $B(\Gamma)$ denotes the set of regular unipotent subgroups of the group Γ (i.e. the set of all subgroups $U \subset \Gamma$, for which $U = \overline{U} \cap \Gamma$).

Theorem 16.4. *There are in Γ subgroups $U_1 \in B(\Gamma)$ and $U_2 \in B(\Gamma)$ different from e such that* 1) *$\overline{U}_1$ and $\overline{U}_2$ are opposite orispheric subgroups of the group G,* 2) $\dim \tilde{A}(U_1) > \dim \overline{U}_1$ *and* $\dim \tilde{A}(U_2) > \dim \overline{U}_2$.

Proof. We have two possibilities:

a) all maximal unipotent subgroups of Γ are ultramaximal,

b) there is in Γ a maximal, but not ultramaximal unipotent subgroup.

Case a). According to Theorem 13.20, there are in Γ subgroups $U_1 \in B(\Gamma)$ and $U_2 \in B(\Gamma)$ different from the identity such that *1)* $\overline{U}_1$ and $\overline{U}_2$ are not maximal unipotent subgroups of G, *2)* $\overline{U}_1$ and $\overline{U}_2$ are opposite orispheric subgroups of G. Therefore it remains to show that

$$\dim \tilde{A}(U_1) > \dim \overline{U}_1 \quad \text{and} \quad \dim \tilde{A}(U_2) > \dim \overline{U}_2. \tag{4}$$

Since in the analysed case all maximal unipotent subgroups of Γ are ultramaximal, and $\overline{U}_1$ is not a maximal unipotent subgroup in G, in view of Lemma 13.4 U_1 is not a maximal unipotent subgroup of Γ. Let U_1' be a maximal unipotent subgroup of Γ, in which U_1 is a proper subgroup. We showed at the beginning of § 13 that any proper subgroup of an arbitrary unipotent group is a proper subgroup of its normalizer, thus there is an element $u_0 \in U_1'$ such that *1)* $u_0 \notin U_1$, *2)* u_0 normalizes U_1. Since $U_1 \in B(\Gamma)$, $u_0 \in \Gamma$ and $u_0 \notin U_1$, we have $u_0 \notin \overline{U}_1$. Now, since u_0 is unipotent, and $\overline{U}_1$ is an algebraic unipotent subgroup, we have

$$u_0^n \notin \overline{U}_1 \quad \text{for any} \quad n \neq 0. \tag{5}$$

Further, since u_0 normalizes U_1, the element u_0 normalizes $\overline{U}_1$ also, i.e.,

$$u_0 \in N(\overline{U}_1). \tag{6}$$

Since *1)* u_0 is unipotent, *2)* $u_0 \in \Gamma$, it follows from (6) that

$$u_0 \in \overline{S}N(\overline{U}_1) \cap \Gamma. \tag{7}$$

From (5) and (7) we obtain that $\overline{U}_1$ is a subgroup of infinite index in $\tilde{A}(U_1)$. Now, since $\overline{U}_1$ and $\tilde{A}(U_1)$ are algebraic groups, we have the first inequality of (4). The second inequality of (4) can be proved analogously. Thus, in Case a) the theorem is proved.

Case b). Let U_1 be an arbitrary maximal, but not ultramaximal unipotent subgroup of the group Γ. We note first that U_1 is non-trivial in view of Lemma 7.5.

As mentioned several times, Γ is Zariski dense in G. Thus, since $\overline{U}_1$ is unipotent and G is semisimple, there exists an element $\gamma_0 \in \Gamma$ such that

$$\gamma_0 \overline{U}_1 \gamma_0^{-1} \cap \overline{U}_1 = e. \tag{8}$$

Put $U_2 = \gamma_0 U_1 \gamma_1^{-1}$ and note that $\overline{U}_2 = \gamma_0 \overline{U}_1 \gamma_0^{-1}$. Then, in consequence of (8) and Lemma 12.6 we have

$$\overline{U}_1 \cap N(\overline{U}_2) = \overline{U}_2 \cap N(\overline{U}_1) = e. \tag{9}$$

Since U_1 is a maximal unipotent subgroup of the group Γ, according to Theorem 11.13 $\overline{U}_1$, and consequently also $\overline{U}_2$ are orispheric subgroups of the group G. From this, (9) and Lemmas 10.9 and 10.14 we get that $\overline{U}_1$ and $\overline{U}_2$ are opposite orispheric subgroups. On the other hand, since U_1 is not ultramaximal, $\dim \tilde{A}(U_1) > \dim \overline{U}_1$, and consequently $\dim \tilde{A}(U_2) > \dim \overline{U}_2$. The last two statements imply the assertion of the theorem in Case b). The proof is complete.

From Lemmas 11.7, 11.16 and 11.17 we have

Lemma 16.5. Let $U \neq e$ be a unipotent subgroup of the group Γ. If $\overline{U}$ is an orispheric subgroup of the group G, then $v(\tilde{A}(U)/\tilde{A}(U) \cap \Gamma) < \infty$.

Lemma 16.6. Let $U_1 \in B(\Gamma)$ and $U_2 \in B(\Gamma)$ be such that $\overline{U}_1$ and $\overline{U}_2$ are opposite orispheric subgroups. Then $\tilde{\tilde{A}}(U_1) \cap \tilde{\tilde{A}}(U_2) \cap \Gamma$ is a lattice in $\tilde{\tilde{A}}(U_1) \cap \tilde{\tilde{A}}(U_2)$.

Proof. It follows from Lemmas 7.3 and 6.9 that $\tilde{A}(U_1)$ and $\tilde{A}(U_2)$ have property P. Therefore, since $\tilde{\tilde{A}}(U_1)$ and $\tilde{\tilde{A}}(U_2)$ are of finite index in $\tilde{A}(U_1)$ and $\tilde{A}(U_2)$, $\tilde{\tilde{A}}(U_1)$ and $\tilde{\tilde{A}}(U_2)$ have property P. Now, as was shown in § 6 (see Lemma 6.5), the intersection of two subgroups having property P also has property P. On the other hand, $\overline{U}_1$ and $\overline{U}_2$ are opposite orispheric subgroups, and so in view of Lemma 12.5, $\tilde{\tilde{A}}(U_1)$ is the semi-direct product (in the sense of § 1) of $\tilde{\tilde{A}}(U_1) \cap \cap \tilde{\tilde{A}}(U_2)$ and $\overline{U}_1$. Further, the subgroup $\overline{U}_1$ is integer (according to Lemma 5.4). The above facts and Lemma 6.10 imply that $\tilde{\tilde{A}}(U_1) \cap \tilde{\tilde{A}}(U_2) \cap \Gamma$ is a subgroup of finite index in $\varrho(\tilde{\tilde{A}}(U_1) \cap \Gamma)$, where ϱ is the natural mapping $\varrho: \tilde{\tilde{A}}(U_1) \to \to \tilde{\tilde{A}}(U_1)/\overline{U}_1 = \tilde{\tilde{A}}(U_1) \cap \tilde{\tilde{A}}(U_2)$. On the other hand, since $\tilde{\tilde{A}}(U_1)$ is a subgroup of finite index in $\tilde{A}(U_1)$, Lemma 16.5 shows that $\tilde{\tilde{A}}(U_1) \cap \Gamma$ is a lattice in $\tilde{\tilde{A}}(U_1)$. From the last two statements and the fact $\overline{U}_1$ is integer, the assertion of the lemma easily follows.

Lemma 16.7. Let V^1 and V^2 be algebraic unipotent subgroups of the group G, $V^1_{\mathbf{C}}$ and $V^2_{\mathbf{C}}$ the complexifications of the groups V^1 and V^2, $N(V^2)$ the normalizer of V^2 in G, and $N(V^2_{\mathbf{C}})$ the normalizer of $V^2_{\mathbf{C}}$ in $G_{\mathbf{C}}$. If $V^1 \cap N(V^2) = e$, then $V^1_{\mathbf{C}} \cap N(V^2_{\mathbf{C}}) = e$.

Proof. It is easy to see that the algebraic subgroup $V^1_{\mathbf{C}} \cap N(V^2_{\mathbf{C}})$ is defined over $\mathbf{R}$. On the other hand, it is unipotent, and consequently, in view of Lemma

1.3.1, it is connected. Since *1)* in any connected complex algebraic group which is defined over **R**, the set of real points is algebraically dense, *2)* it is easy to see that the set of real points of the group $V^1_{\mathbf{C}} \cap N(V^2_{\mathbf{C}})$ coincides with $V^1 \cap N(V^2)$, we conclude that if $V^1_{\mathbf{C}} \cap N(V^2_{\mathbf{C}}) \neq e$, then $V_1 \cap N(V_2) \neq e$. The lemma is proved.

Lemma 16.8. Let $U_1 \in B(\Gamma)$ and $U_2 \in B(\Gamma)$ be such that $\overline{U}_1$ and $\overline{U}_2$ are opposite orispheric subgroups. Then $\tilde{\tilde{A}}(U_1) \cap \tilde{\tilde{A}}(U_2) \cap \Gamma$ is Zariski dense in $\tilde{\tilde{A}}(U_1) \cap \tilde{\tilde{A}}(U_2)$.

Proof. As shown in the course of the proof of Lemma 16.6, *1)* $\tilde{\tilde{A}}(U_1)$ is the semi-direct product (in the sense of § 1) of $\tilde{\tilde{A}}(U_1) \cap \tilde{\tilde{A}}(U_2)$ and $\overline{U}_1$, *2)* $\tilde{\tilde{A}}(U_1) \cap \cap \tilde{\tilde{A}}(U_2) \cap \Gamma$ is a subgroup of finite index in $\varrho(\tilde{\tilde{A}}(U_1) \cap \Gamma)$, where ϱ is the natural mapping

$$\varrho: \tilde{\tilde{A}}(U_1) \to \tilde{\tilde{A}}(U_1)/\overline{U}_1 = \tilde{\tilde{A}}(U_1) \cap \tilde{\tilde{A}}(U_2).$$

Let us note now that *1)* ϱ, as is easy to see, is an algebraic homomorphism, *2)* $\tilde{\tilde{A}}(U_1) \cap \Gamma$ is Zariski dense in $\tilde{\tilde{A}}(U_1)$ (for $\tilde{A}(U_1) = A(\overline{S}N(\overline{U}_1) \cap \Gamma)$, and $\tilde{\tilde{A}}(U_1)$ is the algebraically connected component of the identity in the group $\tilde{A}(U_1)$). This implies that $\varrho(\tilde{\tilde{A}}(U_1) \cap \Gamma)$ is Zariski dense in $\tilde{\tilde{A}}(U_1) \cap \tilde{\tilde{A}}(U_2)$. Therefore, since *1)* $\tilde{\tilde{A}}(U_1) \cap \tilde{\tilde{A}}(U_2)$ is algebraically connected (because $\tilde{\tilde{A}}(U_1) \cap \tilde{\tilde{A}}(U_2) = \varrho(\tilde{\tilde{A}}(U_1))$, and $\tilde{\tilde{A}}(U_1)$ is algebraically connected), *2)* $\tilde{\tilde{A}}(U_1) \cap \tilde{\tilde{A}}(U_2) \cap \Gamma$ is a subgroup of finite index in $\varrho(\tilde{\tilde{A}}(U_1) \cap \Gamma)$, we find that $\tilde{\tilde{A}}(U_1) \cap \tilde{\tilde{A}}(U_2) \cap \Gamma$ is Zariski dense in $\tilde{\tilde{A}}(U_1)$. The lemma is proved.

Let $G_{\mathbf{C}}$ be the complexification of the group G. Since G satisfies the conditions formulated at the beginning of § 7, $G_{\mathbf{C}}$ is a connected semisimple complex algebraic group with trivial center. Therefore, in view of classical theorems of Chevalley [22], we may assume without loss that $G_{\mathbf{C}}$ is defined over **Q**, i.e. the complex algebraic group $G_{\mathbf{C}}$ can be realized as an algebraic subgroup of $GL(n, \mathbf{C})$ (n is a suitable integer) defined by polynomials with rational coefficients.

Since $G_{\mathbf{C}}$ is defined over **Q**, one can assign an automorphism ψ_G of the group $G_{\mathbf{C}}$ to every automorphism ψ of the field of complex numbers (by applying ψ to the matrix coefficients of elements in $G_{\mathbf{C}} \subset GL(n, \mathbf{C})$. Observe that, unless ψ is the identity transformation, ψ_G is not an algebraic automorphism of the complex algebraic group $G_{\mathbf{C}}$, i.e. ψ_G is an automorphism in the sense of abstract group theory.

One easily gets the following

Lemma 16.9. Let ψ be an automorphism of the field **C**, and A an arbitrary algebraic subgroup of $G_{\mathbf{C}}$. Then $\psi_G(A)$ is an algebraic subgroup of $G_{\mathbf{C}}$.

Remark. Lemma 16.9 can be proved by applying ψ to the coefficients of the polynomials defining A in $G_{\mathbf{C}}$.

For any regular unipotent subgroup $U \in B(\Gamma)$ let $\tilde{\tilde{A}}_{\mathbf{C}}(U)$ denote the complexification of the group $\tilde{\tilde{A}}(U)$.

Lemma 16.10. Let $U_1 \in B(\Gamma)$ and $U_2 \in B(\Gamma)$ be such that $\overline{U}_1$ and $\overline{U}_2$ are opposite orispheric subgroups, and ψ an arbitrary automorphism of the field $\mathbf{C}$. Then, there exist in $\tilde{\tilde{A}}(U_1)\cap\Gamma$ and $\tilde{\tilde{A}}(U_2)\cap\Gamma$ subgroups $\Omega^1_\psi\subset\tilde{\tilde{A}}(U_1)\cap\Gamma$ and $\Omega^2_\psi\subset\tilde{\tilde{A}}(U_2)\cap\Gamma$ of finite index so that the restrictions of ψ_G to Ω^1_ψ and Ω^2_ψ can be extended to algebraic homomorphisms of the complex algebraic groups $\tilde{\tilde{A}}_{\mathbf{C}}(U_1)$ and $\tilde{\tilde{A}}_{\mathbf{C}}(U_2)$ into the complex algebraic group $G_{\mathbf{C}}$.

Proof. Put $F=\tilde{\tilde{A}}(U_1)$, $V=\overline{U}_1$, $L=\tilde{\tilde{A}}(U_1)\cap\tilde{\tilde{A}}(U_2)$, $\Omega=\tilde{\tilde{A}}(U_1)\cap\Gamma$, $T=(\psi_G)_\Omega$, where $(\psi_G)_\Omega$ denotes the restriction of ψ_G to Ω, and show that F, V, L, Ω and T satisfy the conditions of Lemma 15.3. To this purpose we make the following remarks:

Remark 1. Since $L\cap\Omega=(\tilde{\tilde{A}}(U_1)\cap\tilde{\tilde{A}}(U_2))\cap(\tilde{\tilde{A}}(U_1)\cap\Gamma)=\tilde{\tilde{A}}(U_1)\cap\tilde{\tilde{A}}(U_2)\cap\Gamma=$ $=L\cap\Gamma$, in view of Lemma 16.6 the quotient space $L/L\cap\Gamma$ is of finite volume.

Remark 2. Since $\tilde{A}(U_1)=A(\bar{S}N(U_1)\cap\Gamma)$, and $\tilde{\tilde{A}}(U_1)$ is the algebraically connected component of the identity in the group $\tilde{A}(U_1)$, we get that $\tilde{\tilde{A}}(U_1)\cap\Gamma$ is Zariski dense in $\tilde{\tilde{A}}(U_1)$, and consequently $\Omega=F\cap\Gamma$ is Zariski dense in F.

Remark 3. Since $\Omega=\tilde{\tilde{A}}(U_1)\cap\Gamma\supset\overline{U}_1\cap\Gamma\supset U_1$, we have $\Omega\cap N\supset U_1$, and consequently $\Omega\cap V$ is Zariski dense in $V=\overline{U}_1$.

Remark 4. In view of Lemma 16.8 $L\cap\Omega=L\cap\Gamma$ is Zariski dense in L.

Remark 5. Since ψ_G is induced by an automorphism of the field of complex numbers, $\psi_G(\overline{U}_1)$ is unipotent. Therefore $T(\Omega\cap V)\subset\psi_G(\overline{U}_1)$ is unipotent.

Remark 6. Denote by $\tilde{V}$ the algebraic closure of $T(\Omega\cap V)$ in $G_{\mathbf{C}}$, and by $\tilde{V}^0$ the algebraically connected component of the identity of the group $\tilde{V}$. Since $T(\Omega\cap V)$ is unipotent (by the previous remark), $\tilde{V}$ is an algebraic unipotent group. Therefore, in view of Lemma 1.3.1 $\tilde{V}$ is connected, i.e.

$$\tilde{V}^0=\tilde{V}. \tag{10}$$

Remark 7. Denote by $\tilde{L}$ the algebraic closure of $T(\Omega\cap L)$ in $G_{\mathbf{C}}$. As an easy consequence of Lemma 16.9, we have

$$\tilde{L}\subset\psi_G(\tilde{\tilde{A}}_{\mathbf{C}}(U_1)\cap\tilde{\tilde{A}}_{\mathbf{C}}(U_2))\subset\psi_G(\tilde{\tilde{A}}_{\mathbf{C}}(U_2)). \tag{11}$$

Since $\Omega\cap V$ is Zariski dense in V (according to Remark 3), and so is in $V_{\mathbf{C}}$, as is easy to see, Lemma 16.9. implies

$$\tilde{V}=\psi_G(V_{\mathbf{C}})=\psi_G((\overline{U}_1)_{\mathbf{C}}). \tag{12}$$

Further, since $\overline{U}_1\cap N(\overline{U}_2)=e$ (in view of Lemma 10.14 and the fact that $\overline{U}_1$ and $\overline{U}_2$ are opposite orispheric subgroups), according to Lemma 16.7, we have

$$(\overline{U}_1)_{\mathbf{C}}\cap N((\overline{U}_2)_{\mathbf{C}})=e. \tag{13}$$

On the other hand, it is easy to see that

$$\tilde{\tilde{A}}_{\mathbf{C}}(U_2) \subset ((N(\overline{U}_2))_{\mathbf{C}}) \subset N((\overline{U}_2)_{\mathbf{C}}), \tag{14}$$

where $(N(\overline{U}_2))_{\mathbf{C}}$ denotes the complexification of the group $N(\overline{U}_2)$. (13), (14) and Lemma 16.3 imply

$$\tilde{\tilde{A}}_{\mathbf{C}}(U_2) \cap C((\overline{U}_1)_{\mathbf{C}}) = e, \tag{15}$$

where $C((\overline{U}_1)_{\mathbf{C}})$ is the centralizer of $(\overline{U}_1)_{\mathbf{C}}$ in $G_{\mathbf{C}}$. Since ψ_G is an automorphism, and under an automorphism the centralizer of the image of an arbitrary subgroup equals the image of the centralizer of this subgroup, (10), (11), (12) and (15) imply

$$\tilde{L} \cap C(\tilde{V}^0) = e, \tag{16}$$

where $C(\tilde{V}^0)$ is the centralizer of $\tilde{V}^0$ in $G_{\mathbf{C}}$.

It follows from the above remarks that F, V, L, Ω and T satisfy the conditions of Lemma 15.3. Therefore, there is in $\tilde{\tilde{A}}(U_1) \cap \Gamma$ a subgroup Ω^1_ψ of finite index such that the restriction of ψ_Ω to Ω^1_ψ can be extended to an algebraic homomorphism of the complex algebraic group $\tilde{\tilde{A}}_{\mathbf{C}}(U_1)$ into the complex algebraic group $G_{\mathbf{C}}$. Further it is clear that one can analogously prove the existence of Ω^2_ψ with the required properties. The proof of the lemma is complete.

Theorem 16.11. *Let ψ be an arbitrary automorphism of the complex number field. Then the restriction of ψ_G to Γ can be extended to an algebraic automorphism of the complex algebraic group $G_{\mathbf{C}}$.*

Proof. According to Theorem 16.4 there exist in Γ subgroups $U_1 \in B(\Gamma)$ and $U_2 \in B(\Gamma)$ different from the identity such that 1) $\overline{U}_1$ and $\overline{U}_2$ are opposite orispheric subgroups of the group G, 2) $\dim \tilde{A}(U_1) > \dim \overline{U}_1$ and $\dim \tilde{A}(U_2) > \dim \overline{U}_2$. Denote by W the subset of G consisting of all $g \in G$ for which $\overline{U}_1$ and $g\overline{U}_2 g^{-1}$ are opposite orispheric subgroups. Since $\overline{U}_1$ and $\overline{U}_2$ are opposite, in view of Lemmas 10.9 and 10.13 W is a non-empty Zariski open subset in G. Therefore, since $G_{\mathbf{C}}$ is the complexification of G, there exists in the complex algebraic manifold $G_{\mathbf{C}}$ a non-empty Zariski open subset $W_{\mathbf{C}}$, for which $W_{\mathbf{C}} \cap G = W$.

Recall that $\tilde{\tilde{A}}_{\mathbf{C}}(U_1)$ and $\tilde{\tilde{A}}_{\mathbf{C}}(U_2)$ denote the complexifications of the groups $\tilde{\tilde{A}}(U_1)$ and $\tilde{\tilde{A}}(U_2)$. Put now $H = G_{\mathbf{C}}$, $\Lambda = \Gamma$, $R = \tilde{\tilde{A}}(U_1)$, $S = \tilde{\tilde{A}}(U_2)$. As in § 14 (see the section preceding Lemma 14.15), $V(R, S)$ is the set of all $h_0 \in H = G_{\mathbf{C}}$ such that $\dim((h_0 S h_0^{-1}) \cap R) = \min_{h \in H} \dim((hSh^{-1}) \cap R)$. Put $\tilde{V} = V(R, S) \cap W_{\mathbf{C}}$, $D = G_{\mathbf{C}}$ and write $\tilde{T}$ for the restriction of ψ_G to $\Gamma = \Lambda$. We show that H, Λ, R, S, $\tilde{T}$, $\tilde{V}$ and D satisfy the conditions of Theorem 14.15. To this end, we make the following remarks:

Remark 1. We have remarked several times that Γ is Zariski dense in G, thus, since $G_{\mathbf{C}}$ is the complexification of G, Γ is Zariski dense in $G_{\mathbf{C}}$.

Remark 2. Since $\tilde{A}(U_1)=A(\bar{S}N(\bar{U}_1)\cap\Gamma)$, and $\tilde{\tilde{A}}(U_1)$ is the algebraically connected component of the identity of the group $\tilde{A}(U_1)$, $\tilde{\tilde{A}}(U_1)\cap\Gamma$ is Zariski dense in $\tilde{\tilde{A}}(U_1)$. Therefore, $R\cap\Lambda=\tilde{\tilde{A}}_{\mathbf{C}}(U_1)\cap\Gamma=\tilde{\tilde{A}}(U_1)\cap\Gamma$ is Zariski dense in $R=\tilde{\tilde{A}}_{\mathbf{C}}(U_1)$. Similarly, $S\cap\Lambda$ is Zariski dense in S.

Remark 3. As in § 14, for any $\lambda\in\Lambda\cap V(R, S)$ we write $(R, S)_\lambda$ for the connected component of the identity of the intersection $(\lambda S\lambda^{-1})\cap R$. Since R and S are the complexifications of the groups $\tilde{\tilde{A}}(U_1)$ and $\tilde{\tilde{A}}(U_2)$, R and S are defined over $\mathbf{R}$. Therefore, for any $\lambda\in\Lambda\cap V(R, S)$, the connected group $(R, S)_\lambda$ is defined over $\mathbf{R}$, and consequently, the set F_λ of all real points of the group $(R, S)_\lambda$ is Zariski dense in $(R, S)_\lambda$. On the other hand, it is easy to see that $F_\lambda=(\tilde{\tilde{A}}(U_1)\cap\lambda\tilde{\tilde{A}}(U_2)\lambda^{-1})^0$, where $(\tilde{\tilde{A}}(U_1)\cap\lambda\tilde{\tilde{A}}(U_1)\lambda^{-1})^0$ denotes the algebraically connected component of the identity of the group $\tilde{\tilde{A}}(U_1)\cap\lambda\tilde{\tilde{A}}(U_1)\lambda^{-1}$.

Remark 4. Let λ_0 be an arbitrary element in $\Lambda\cap\bar{V}$. Since $\bar{V}\supset W_{\mathbf{C}}$ and $W_{\mathbf{C}}\cap G=W$, $\lambda_0\in W\cap\Gamma=W\cap\Lambda$. Therefore *1)* $\lambda_0 U_2\lambda_0^{-1}\in B(\Gamma)$, *2)* $\lambda_0\bar{U}_2\lambda_0^{-1}$ and $\bar{U}_1$ are opposite orispheric subgroups of the group G (in view of the definition of set W). We also note that the following equality obviously holds:

$$\overline{\lambda_0 U_2\lambda_0^{-1}}=\lambda_0\bar{U}_2\lambda_0^{-1},\quad\text{where}\quad\overline{\lambda_0 U_2\lambda_0^{-1}}$$

denotes the algebraic closure of $\lambda_0 U_2\lambda_0^{-1}$ in G. The above facts and Lemma 16.8 imply that $\tilde{\tilde{A}}(U_1)\cap\lambda_0\tilde{\tilde{A}}(U_2)\lambda_0^{-1}\cap\Gamma$ is Zariski dense in $\tilde{\tilde{A}}(U_1)\cap\lambda_0\tilde{\tilde{A}}(U_2)\lambda_0^{-1}$. From this, Remark 3 and Lemmas 1.2.7 and 8.8 it follows easily that $\Lambda\cap(R, S)_{\lambda_0}$ is Zariski dense in $(R, S)_{\lambda_0}$.

Remark 5. Since $\bar{U}_1$ and $\bar{U}_2$ satisfy the conditions of Lemma 16.10, $\tilde{T}$ is the restriction of ψ_G to $\Lambda=\Gamma$, $R=\tilde{\tilde{A}}_{\mathbf{C}}(U_1)$ and $S=\tilde{\tilde{A}}_{\mathbf{C}}(U_2)$, thus there are in R and S subgroups Λ_R and Λ_S of finite index such that the restrictions of $\tilde{T}$ to Λ_R and Λ_S can be extended to algebraic representations T^R and T^S of the groups R and S.

Remark 6. Since *1)* Λ is Zariski dense in $H=G_{\mathbf{C}}$ (see Remark 1), *2)* $D=G_{\mathbf{C}}$, *3)* $\psi_G(G_{\mathbf{C}})=G_{\mathbf{C}}$ (for ψ_G is an automorphism), in view of Lemma 16.9, $\tilde{T}(\Lambda)$ is Zariski dense in D.

Remark 7. Since *1)* $W_{\mathbf{C}}$ is non-empty and Zariski open in $H=G_{\mathbf{C}}$, *2)* $V(R, S)\neq\emptyset$ (by definition), *3)* $V(R, S)$ is easily seen to be Zariski open in H, *4)* $H=G_{\mathbf{C}}$ is algebraically connected (for G is algebraically connected), we conclude that $V=V(R, S)\cap W_{\mathbf{C}}$ is non-empty and Zariski open in $H=G_{\mathbf{C}}$.

Remark 8. It follows from Remarks 1 and 7 that $\Lambda\cap\bar{V}$ is non-empty. Let λ_0 be an arbitrary element of the set $\Lambda\cap\bar{V}$. Repeating the arguments used in Remark 4 and applying Lemma 12.5 we obtain that $\tilde{\tilde{A}}(U_1)$ is the semi-direct product (in the sense of § 1) of $\tilde{\tilde{A}}(U_1)\cap\lambda_0\tilde{\tilde{A}}(U_2)\lambda_0^{-1}$ and $\overline{U}_1$. Therefore, since *1)* dim $\tilde{A}(U_1)>$ $>$dim $\overline{U}_1$ (see the beginning of the proof of the theorem), *2)* dim $\tilde{\tilde{A}}(U_1)=$ $=$dim $\tilde{A}(U_1)$, (for $\tilde{\tilde{A}}(U_1)$ is an algebraically connected component of the group $\tilde{A}(U_1)$), we have

$$\dim\left(\tilde{\tilde{A}}(U_1)\cap\lambda_0\,\tilde{\tilde{A}}(U_2)\lambda_0^{-1}\right) > 0. \tag{17}$$

Since $R=\tilde{\tilde{A}}_{\mathrm{C}}(U_1)$, and $S=\tilde{\tilde{A}}_{\mathrm{C}}(U_2)$, from (17) we get

$$\dim\,(R,\,S)_{\lambda_0} > 0. \tag{18}$$

On the other hand, as shown in Remark 4, $\Lambda\cap(R,\,S)_{\lambda_0}$ is Zariski dense in $(R,\,S)_{\lambda_0}$. From this and (18) we conclude that

$$\Lambda\cap(R,\,S)_{\lambda_0} \neq e. \tag{19}$$

Since $\Lambda=\Gamma$, and G and Γ satisfy the conditions formulated at the beginning of § 7, in view of Lemma 4.8 $\Lambda\cap(R,\,S)_{\lambda_0}$ is contained in no proper algebraic normal subgroup of the group G. Therefore, in view of Lemma 16.1, neither is $\Lambda\cap(R,\,S)_{\lambda_0}$ contained in any proper algebraic subgroup of the group G_{C}. This and Lemma 16.9 imply that $\tilde{T}(\Lambda\cap(R,\,S)_{\lambda_0})=\psi_G(\Lambda\cap(R,\,S)_{\lambda_0})$ is contained in no proper algebraic normal subgroup of the group $\psi_G(G_{\mathrm{C}})=G_{\mathrm{C}}=D$ ($D=G_{\mathrm{C}}$ by definition). We have

$$\tilde{T}(\Lambda\cap(R,\,S)_{\lambda_0})\subset T^R(R,\,S)_{\lambda_0}\subset\bigcup_{\lambda\in\Lambda\cap\overline{V}}(T^R(R,\,S)_{\lambda}),$$

thus the minimal algebraic normal subgroup of the group D containing $\bigcup_{\lambda\in\Lambda\cap\overline{V}}(T^R(R,\,S_{\lambda}))$, coincides with D.

Remark 9. Since $D=G_{\mathrm{C}}$ and G satisfies the conditions formulated at the beginning of § 7, the centre of D is trivial.

Remark 10. The groups $R=\tilde{\tilde{A}}_{\mathrm{C}}(U_1)$ and $S=\tilde{\tilde{A}}_{\mathrm{C}}(U_2)$ are algebraically connected, since $\tilde{\tilde{A}}(U_1)$ and $\tilde{\tilde{A}}(U_2)$ are algebraically connected.

It follows from Remarks 1—10 that H, Λ, R, S, $\tilde{T}$, $\overline{V}$ and D satisfy the conditions of Theorem 14.15, and consequently, $\tilde{T}$ extends to an algebraic homomorphism $\tilde{T}^H$ of the group $H=G_{\mathrm{C}}$ into the group $D=G_{\mathrm{C}}$. We show that $\tilde{T}^H$ is an isomorphism. To do this, we observe that *1)* since $\tilde{T}(\Lambda)$ is Zariski dense in H (by Remark 6), $\tilde{T}^H(H)$ is Zariski dense in D, *2)* since $\tilde{T}^H$ is an algebraic homomorphism of complex algebraic groups, in view of Lemma 8.1 $\tilde{T}^H(H)$ is an algebraic subgroup in D. Therefore

$$\tilde{T}^H(H) = D. \tag{20}$$

Since $H=G_{\mathrm{C}}$ and $D=G_{\mathrm{C}}$, in view of (20), the kernel Ker $\tilde{T}^H$ of the homomorphism $\tilde{T}^H$ is discrete. From this, connectivity of G_{C} (§ 7) and the triviality of the centre of the group $H=G_{\mathbf{C}}$ we get

$$\operatorname{Ker}\tilde{T}^H = e. \tag{21}$$

As a consequence of (20), (21) and Lemma 1.2.17 we have $\tilde{T}^H$ is an isomorphism. Thus we have shown, that the restriction $\tilde{T}$ of the automorphism ψ_G to $\Lambda=\Gamma$ extends to an algebraic homomorphism $\tilde{T}^H$ of the group $H=G_{\mathbf{C}}$ into the group $D=G_{\mathbf{C}}$, further $\tilde{T}^H$ is an isomorphism. The theorem is proved.

The following lemma is well known.

Lemma 16.12. If an element d of the field $\mathbf{C}$ is invariant under all automorphisms of the field $\mathbf{C}$, then $d\in\mathbf{Q}$ (as above, $\mathbf{C}$ and $\mathbf{Q}$ denote the fields of complex numbers and rational numbers, respectively).

Remark. For the proof of Lemma 16.12 one should *1)* choose an arbitrary transcendence basis L of the field $\mathbf{C}$ over the field $\mathbf{Q}$, *2)* take the field $K\subset\mathbf{C}$ generated by L and *3)* remark that $\mathbf{C}$ is the algebraically closed algebraic extension of the field K (for the definition of transcendence basis and proof of its existence, see [45] § 12).

It is easy to see that we have the following

Lemma 16.13. Let H be an arbitrary complex (real) algebraic group, and φ an arbitrary algebraic automorphism of the group H. Then, for any $h\in H$, we have $\operatorname{Tr}(\operatorname{Ad}\varphi(h))=\operatorname{Tr}(\operatorname{Ad}h)$, where Ad denotes the adjoint representation of the group H.

Recall again that G and Γ satisfy the conditions formulated at the beginning of §§ 7 and 13.

Let $G_{\mathbf{C}}$ be, as above, the complexification of the group G. Write $\operatorname{Ad}_{\mathbf{C}}$ for the adjoint representation of the group $G_{\mathbf{C}}$.

Theorem 16.14. *For any* $\gamma\in\Gamma$, $\operatorname{Tr}(\operatorname{Ad}_{\mathbf{C}}\gamma)\in\mathbf{Q}$.

Proof. Let ψ be an arbitrary automorphism of the field $\mathbf{C}$. According to Theorem 16.11, the restriction of ψ_G to Γ can be extended to an algebraic automorphism of the group $G_{\mathbf{C}}$. From this and Lemma 16.13 we get

$$\operatorname{Tr}(\operatorname{Ad}\psi_G(\gamma)) = \operatorname{Tr}(\operatorname{Ad}\gamma) \tag{22}$$

for any $\gamma\in\Gamma$. From (22) and Lemma 16.12 the assertion of the theorem follows.

The following lemma can be easily seen:

Lemma 16.15. Let F and H be two complex algebraic groups defined over some field K ($K\subset\mathbf{C}$), and let f be an algebraic isomorphism between F and H also defined over K. Denote by F_K and H_K the set of all K-points in F and H. Then $f(F_K)=H_K$.

Remark 1. The assumption F and H are groups, is, of course, not important.

Remark 2. For the proof of Lemma 16.15 it is sufficient to note that the inverse mapping f^{-1} of the mapping f is easily seen to be defined over K.

Now we make the following

Remark 16.16. Let Ad denote the adjoint representation of the group G. Since the centre of the complexification $G_{\mathbf{C}}$ of G is trivial, Lemma 16.15 implies that Ad G is a real algebraic group and Ad presents an algebraic isomorphism between the real algebraic groups G and Ad G. Therefore, using Ad, we can identify the real algebraic group G with the real algebraic group Ad G.

Recall now that a **Q**-form of an algebraic **R**-group H is a pair $F=(\tilde{H},f)$, of an algebraic **R**-group $\tilde{H}$ defined over **Q** and an algebraic isomorphism $f:\tilde{H}\to H$, and that for any ring A of real numbers ($A\subset\mathbf{R}$) the image $f(\tilde{H}_A)$ is denoted by F_A. (As above **Q** denotes the field of rational numbers.)

Theorem 16.17. *There exists a* **Q**-*form* F *of the algebraic* **R**-*group* G *such that* $\Gamma\subset F_{\mathbf{Q}}$.

Before starting the proof of Theorem 16.17 we present some results of E. B. Vinberg on rings of definition of dense subgroups in semisimple linear groups (see [23]). To do this, first we give the necessary definitions.

Definition 16.18. Let V be a finite-dimensional vector space over the field D, and $A\subset D$ some subring. A set $L\subset V$ is called an *A-lattice*, if it is a finitely generated A-submodule and if the natural linear mapping $F\otimes_A L\to V$ is an isomorphism. If $K\subset D$ is a field, then a K-lattice is the same as a K-form of the space V.

Definition 16.19. Let Δ be a family of linear transformations of the space V. An integrally closed Noetherian ring $A\subset F$ is called a *ring of definition* of Δ if there is in V an A-lattice invariant under Δ. In this case we will also say that Δ can be defined over A.

Lemma 16.20. Let H be a semisimple algebraic **C**-group, and Λ a Zariski dense subgroup. An integrally closed Noetherian ring $A\subset\mathbf{C}$ is a ring of definition of Ad Λ if and only if $A\supset\operatorname{Tr}(\operatorname{Ad}\lambda)$ for all $\lambda\in\Lambda$ (Ad is the adjoint representation of the group H).

Lemma 16.21. Let Δ be a family of linear transformations of the complex linear space V, P a subfield of **C**, and $A\subset P$ an integrally closed Noetherian ring. Assume that Δ is fully reducible and can be defined over A. Then for any P-form U of the space V invariant under Δ, there exists an A-lattice invariant under Δ and contained in U.

Remark. Lemmas 16.20 and 16.21 correspond to Theorem 1 and Lemma 5 in [23]. We note that in [23] the results just introduced are proved for arbitrary algebraically closed fields of characteristic 0, and not only for the complex number field **C**.

The proof of Theorem 16.17. As mentioned several times, Γ is Zariski dense in G, and consequently also in $G_{\mathbf{C}}$. Therefore, it follows from Theorem 16.14 and Lemma 16.20 that $\mathrm{Ad}_{\mathbf{C}}\,\Gamma$ is defined over **Q**, where as in Theorem 16.14, $\mathrm{Ad}_{\mathbf{C}}$ denotes the adjoint representation of the group $G_{\mathbf{C}}$.

Denote by $\mathfrak{G}$ and $\mathfrak{G}_{\mathbf{C}}$ the Lie algebras of the groups G and $G_{\mathbf{C}}$. It is clear that $\mathfrak{G}$ is an **R**-form of the complex linear space $\mathfrak{G}_{\mathbf{C}}$, and, since $\Gamma\subset G$, the **R**-form $\mathfrak{G}$ is invariant under $\mathrm{Ad}\,\Gamma$.

Since Γ is Zariski dense in $G_{\mathbf{C}}$ and $G_{\mathbf{C}}$ is semisimple, $\mathrm{Ad}\,\Gamma$ is fully reducible. From this, the last two paragraphs and Lemma 16.21 we find that there exists a **Q**-form L of the space $\mathfrak{G}_{\mathbf{C}}$, contained in the **R**-form $\mathfrak{G}$. Now this is equivalent to the existence of a basis $e_1, \ldots, e_n$ in $\mathfrak{G}$, in which all elements of $\mathrm{Ad}\,\Gamma$ can be written as matrices with rational coefficients. Thus, since Γ is Zariski dense in G, in view of Lemma 8.7 $\mathrm{Ad}\,G$ is defined over **Q** in the basis $e_1, \ldots, e_n$. Put now $F=(\tilde{G},f)$, where $\tilde{G}=\mathrm{Ad}\,G$, and $f=\mathrm{Ad}^{-1}$, and note that according to Remark 16.16 f is an algebraic isomorphism. The above facts imply that F is the desired **Q**-form. The proof of Theorem 16.17 is complete.

§ 17. Intersections of the subgroup Γ with parabolic subgroups

In this section we will use the notations of §§ 7, 11, 12, 13 and 16. Recall that G and Γ satisfy the conditions formulated at the beginning of §§ 7 and 13, and that π is the natural projection $\pi: G\to G/\Gamma$.

According to Theorem 16.17 there exists a **Q**-form F of the group G such that $\Gamma\subset F_{\mathbf{Q}}$. Therefore, we can assume that the algebraic **R**-group G is defined over **Q**, so that

$$\Gamma\subset G_{\mathbf{Q}}. \tag{1}$$

The following simple lemma is well known.

Lemma 17.1. Let H be an arbitrary algebraic **R**-group defined over **Q**, and u a unipotent element contained in $H_{\mathbf{Q}}$. Then there exists a non-zero n (depending on u), such that $u^n\in H_{\mathbf{Z}}$ (**Z** is the ring of integers).

From (1) and Lemma 17.1 we obtain

Lemma 17.2. Let u be an arbitrary unipotent element contained in Γ. Then there exists a non-zero n (depending on u), such that $u^n\in G_{\mathbf{Z}}$.

For any subgroup H, defined over **Q**, of the group G, and any ring $K\subset\mathbf{R}$, H_K denotes the intersection $H\cap G_K$ (for the definition of G_K see § 1).

Lemma 17.3. Let U be an arbitrary unipotent subgroup of the group Γ. Then 1) $\overline{U}_{\mathbf{Z}}$ is Zariski dense in $\overline{U}$, 2) the quotient space $\overline{U}/\overline{U}_{\mathbf{Z}}$ is compact, 3) the subgroups U and $\overline{U}_{\mathbf{Z}}$ are commensurable.

Proof. As mentioned in § 5 (before Lemma 5.4), ln U generates the Lie algebra $\mathfrak{U}$ of the group $\bar{U}$ as a linear space. From this and Lemma 17.2, it follows that $\ln(U \cap \bar{U}_Z)$ generates $\mathfrak{U}$ as a linear space. Therefore, in view of Lemma 5.3 the quotient space $\bar{U}/U \cap \bar{U}_Z$ is compact. From this and the discreteness of U and U_Z we see that U and $\bar{U}_Z$ are commensurable.

The compactness of the quotient space $\bar{U}/\bar{U}_Z$ is a consequence of that of the quotient space $\bar{U}/U \cap \bar{U}_Z$. In view of Lemma 5.6 it follows from this that $\bar{U}_Z$ is Zariski dense in $\bar{U}$. The proof of the lemma is complete.

Lemmas 17.3 and 8.7 imply the following

Lemma 17.4. Let U be an arbitrary unipotent subgroup of the group Γ. Then the algebraic subgroup $\bar{U}$ of G is defined over $\mathbf{Q}$.

In consequence of part 3) of Lemma 17.3 we have

Lemma 17.5. Let U be an arbitrary unipotent subgroup of the group Γ. Then the set $\mathrm{Ad}\,(G_Z \cap N(\bar{U}))\Delta_U$ is finite (as above, Δ_U denotes the quasi-lattice corresponding to U, and $N(\bar{U})$ the normalizer of $\bar{U}$ in G).

Lemma 17.6. Let U be an arbitrary unipotent subgroup of the group Γ. Then we have $N(\bar{U}) \cap G_Z = \bar{S}N(\bar{U}) \cap G_Z$.

Proof. According to Lemma 17.3 $\bar{U}_Z = \bar{U} \cap G_Z$ is Zariski dense in $\bar{U}$. Now it remains to observe that *1)* G_Z is a discrete subgroup in G, *2)* $\bar{U}_Z$ is a unipotent subgroup in G_Z, *3)* as it can be seen from the proof of Lemma 7.13, the assertion of Lemma 7.13 holds for all discrete subgroups of the group G, as well as for Γ.

The following lemma can be easily seen:

Lemma 17.7. Let D and F be complex algebraic groups defined over $\mathbf{Q}$, and f an algebraic homomorphism $f: D \to F$ defined over $\mathbf{Q}$. Then $f(D)$ is an algebraic subgroup of the group F defined over $\mathbf{Q}$, and the homomorphism $f^{-1}: f(D) \to D$ is defined over Q.

Recall (see § 1) that for any complex (real) algebraic group F and any ring $K \subset \mathbf{C}$ $(K \subset \mathbf{R})$, F_K denotes the group of all points of F defined over K, whose determinant is an invertible element of the ring K.

From Lemma 17.7 one can easily deduce the following

Lemma 17.8. Let D and F be complex algebraic groups defined over $\mathbf{Q}$, and f an algebraic homomorphism $f: D \to F$ defined over Q. Then the subgroups D_Z and $f^{-1}(F_Z)$ are commensurable.

Definition 17.9. Let D be an algebraic $\mathbf{C}$-group ($\mathbf{R}$-group) defined over $\mathbf{Q}$, V a complex (real) linear space, $e_1, \ldots, e_n$ a basis in the space V, and T an algebraic representation of the group D in the space V. We say that T is defined over $\mathbf{Q}$ in the basis $e_1, \ldots, e_n$, if the matrix coefficients of T corresponding to the basis $e_1, \ldots, e_n$, can be given as polynomials with coefficients in $\mathbf{Q}$.

From Lemma 17.8 one gets easily the following

Lemma 17.10. Let D be an algebraic **C**-group, defined over **Q**, V a complex linear space, $e_1, \dots, e_n$ a basis in V, and T a faithful algebraic representation of the group D in the space V, defined over **Q** in the basis $e_1, \dots, e_n$. Denote by $L_{\mathbf{Z}}$ the group of all linear transformations of the space V which can be written in the basis $e_1, \dots, e_n$ as integer coefficient matrices of determinant ± 1. Then the subgroups $D_{\mathbf{Z}}$ and $T^{-1}(L_{\mathbf{Z}})$ are commensurable.

Recall that *1)* $B(\Gamma)$ denotes the set of all regular unipotent subgroups of the group Γ (see § 7), *2)* for any algebraic subgroup $H \subset G$, $H_{\mathbf{C}}$ denotes its complexification, *3)* for any subgroup $F \subset G$, F^u stands for the set of all unipotent elements of F.

Lemma 17.11. Let $U \in B(\Gamma)$ satisfy $(C(\bar{U}) \cap \Gamma)^u \subset U$. If H is an algebraic subgroup of G defined over **Q**, such that *1)* $H \subset \bar{S}N(\bar{U})$, *2)* H has property P, *3)* $H_{\mathbf{C}} \cap C(\bar{U}_{\mathbf{C}}) = e$ ($C(\bar{U}_{\mathbf{C}})$ is the centralizer of $\bar{U}_{\mathbf{C}}$ in $G_{\mathbf{C}}$), then the subgroups $H_{\mathbf{Z}} = H \cap G_{\mathbf{Z}}$ and $H \cap \Gamma$ are commensurable.

Proof. Let $\mathfrak{U}$ and $\mathfrak{U}_{\mathbf{C}}$ denote the Lie algebras of the groups $\bar{U}$ and $\bar{U}_{\mathbf{C}}$, and $e_1, \dots, e_n$ an arbitrary basis of the quasi-lattice Δ_U. As emphasized several times (see § 5 and Lemma 7.1), Δ_U is a quasi-lattice of full dimension in $\mathfrak{U}$. Therefore $e_1, \dots, e_n$ form a basis in the real linear space $\mathfrak{U}$, and consequently in the complex linear space $\mathfrak{U}_{\mathbf{C}}$ too. On the other hand, it follows from Lemma 17.2 and the definition of the quasi-lattice Δ_U that all e_i $(1 \leqq i \leqq n)$ are contained in $\ln(\bar{U}_{\mathbf{Q}})$.

Denote by T the representation of the group $H_{\mathbf{C}}$ in the space $\mathfrak{U}_{\mathbf{C}}$ defined by the formula $T_h(\mathfrak{u}) = (\mathrm{Ad}\, h)\mathfrak{u}$ for any $h \in H_{\mathbf{C}}$ and $\mathfrak{u} \in \mathfrak{U}_{\mathbf{C}}$. From Lemma 17.4 and what we have said in the previous paragraph it follows easily that the algebraic representation T is defined over **Q** in the basis $e_1, \dots, e_n$. On the other hand, in view of *3)* in the condition of the lemma and by connectivity of $\bar{U}_{\mathbf{C}}$, the representation T is exact. Therefore, in view of Lemma 17.10 the subgroups $H_{\mathbf{Z}}$ and $T^{-1}(L_{\mathbf{Z}})$ are commensurable, where $L_{\mathbf{Z}}$ denotes the group of all linear transformations of the space $\mathfrak{U}_{\mathbf{C}}$ that can be represented in the basis $e_1, \dots, e_n$ with integer matrices of determinant ± 1.

Now *1)* $U = \bar{U} \cap \Gamma$ (for, $U \in B(\Gamma)$), *2)* H normalizes $\bar{U}$ (for, $H \subset \bar{S}N(\bar{U})$), thus $H \cap \Gamma$ normalizes U. Therefore, since Δ_U is uniquely determined by U, we have

$$T(H \cap \Gamma) \subset L_{\mathbf{Z}}. \tag{2}$$

From (2) and the commensurability of $H_{\mathbf{Z}}$ and $T^{-1}(L_{\mathbf{Z}})$ we get that $(H \cap \Gamma) \cap H_{\mathbf{Z}}$ is a subgroup of finite index in $H \cap \Gamma$.

Since $H \subset \bar{S}N(\bar{U})$, in view of Lemma 17.5 the set $(\mathrm{Ad}\, H_{\mathbf{Z}})\Delta_U$ is finite. Therefore, since $U \in B(\Gamma)$ and $(C(\bar{U}) \cap \Gamma)^U \subset U$ (by assumption), in view of Lemma 7.6 the set $\pi(H_{\mathbf{Z}})$ is relatively compact in G/Γ. Now H has property P. These last two statements and Lemma 6.13 imply that $\pi_H(H_Z)$ is relatively compact in $H/H\cap\Gamma$, where π_H is the natural mapping $\pi_H: H\to H/H\cap\Gamma$. Therefore, since $(H\cap\Gamma)\cap H_Z$ is, as shown earlier, a subgroup of finite index in $H\cap\Gamma$, we get that $\tilde{\pi}(H_Z)$ is relatively compact in $H/H\cap\Gamma\cap H_Z$, where $\tilde{\pi}$ is the natural mapping $\tilde{\pi}: H\to H/H\cap\Gamma\cap H_Z$. From this, Lemma 9.9 and the fact that $H_Z\supset H\cap\Gamma\cap H_Z$ it follows that $H\cap\Gamma\cap H_Z$ is a subgroup of finite index in H_Z. The proof of the lemma is complete (because it has been shown in the previous paragraph that $H\cap\Gamma\cap H_Z$ is of finite index in $H\cap\Gamma$).

Lemma 17.12. Let $U_1\in B(\Gamma)$ and $U_2\in B(\Gamma)$ be such that *1)* $U_1\neq e$ and $U_2\neq e$, *2)* $\overline{U}_1$ and $\overline{U}_2$ are opposite orispheric subgroups. Then the subgroups $N(\overline{U}_1)\cap N(\overline{U}_2)\cap G_Z$ and $N(\overline{U}_1)\cap N(\overline{U}_2)\cap\Gamma$ are commensurable.

Proof. In consequence of Lemmas 17.6 and 7.13 we have

$$N(\overline{U}_1)\cap N(\overline{U}_2)\cap G_Z = N(\overline{U}_1)\cap \bar{S}N(\overline{U}_2)\cap G_Z \tag{3}$$

and

$$N(\overline{U}_1)\cap N(\overline{U}_2)\cap\Gamma = N(\overline{U}_1)\cap \bar{S}N(\overline{U}_2)\cap\Gamma. \tag{4}$$

Since $\overline{U}_1$ and $\overline{U}_2$ are opposite orispheric subgroups, in view of Lemma 10.33 we have

$$N(\overline{U}_1)\cap \bar{S}N(\overline{U}_2) = \bar{S}N(\overline{U}_1)\cap \bar{S}N(\overline{U}_2). \tag{5}$$

From (3), (4) and (5) we have

$$N(\overline{U}_1)\cap N(\overline{U}_2)\cap G_Z = \bar{S}N(\overline{U}_1)\cap \bar{S}N(\overline{U}_2)\cap G_Z \tag{6}$$

and

$$N(\overline{U}_1)\cap N(\overline{U}_2)\cap\Gamma = \bar{S}N(\overline{U}_1)\cap \bar{S}N(\overline{U}_2)\cap\Gamma. \tag{7}$$

Since $U_1\neq e$, and $\overline{U}_1$ is an orispheric subgroup in G, according to Lemma 16.3 it follows that

$$C((\overline{U}_1)_C)\subset(\overline{U}_1)_C. \tag{8}$$

Further, since $\overline{U}_1$ and $\overline{U}_2$ are opposite orispheric subgroups, in view of Lemma 10.14 we have

$$N(\overline{U}_2)\cap\overline{U}_1 \neq e. \tag{9}$$

From (9) and Lemma 16.7 we get

$$N((\overline{U}_2)_C)\cap(\overline{U}_1)_C = e. \tag{10}$$

On the other hand, it is easy to see that

$$N((\overline{U}_2)_C)\supset(N(\overline{U}_2))_C, \tag{11}$$

where $(N(\overline{U}_2))_C$ denotes the complexification of the group $N(\overline{U}_2)$. It follows from (8), (10) and (11) that

$$(N(\overline{U}_2))_C\cap C((\overline{U}_1)_C) = e. \tag{12}$$

Denote $\bar{S}N(\bar{U}_1) \cap \bar{S}N(\bar{U}_2)$ by H. Then, by virtue of (12)

$$H_{\mathbf{C}} \cap C((\bar{U}_1)_{\mathbf{C}}) = e. \tag{13}$$

Now we make the following remarks:

Remark 1. In view of Lemma 7.3, $\bar{S}N(\bar{U}_1)$ and $\bar{S}N(\bar{U}_2)$ have property P. Therefore, according to Lemma 6.5 $H = \bar{S}N(\bar{U}_1) \cap \bar{S}N(\bar{U}_2)$ has property P.

Remark 2. Since $\bar{U}_1$ and $\bar{U}_2$ are defined over $\mathbf{Q}$ (in view of Lemma 7.4), $\bar{S}N(\bar{U}_1)$ and $\bar{S}N(\bar{U}_2)$ are easily seen to be defined over $\mathbf{Q}$. Therefore, the subgroup $H = \bar{S}N(\bar{U}_1) \cap \bar{S}N(\bar{U}_2)$ is defined over $\mathbf{Q}$.

As a consequence of (13), Remarks 1 and 2, and Lemma 17.11 we get that the subgroups

$$H_{\mathbf{Z}} = H \cap G_{\mathbf{Z}} = \bar{S}N(\bar{U}_1) \cap \bar{S}N(\bar{U}_2) \cap G_{\mathbf{Z}}$$

and

$$H \cap \Gamma = \bar{S}N(\bar{U}_1) \cap \bar{S}N(\bar{U}_2) \cap \Gamma$$

are commensurable, whence, in view of (7) the assertion of the lemma follows.

The following lemma is well known (see Corollary 6.4 in [4]):

Lemma 17.13. Let F be an algebraic R-group defined over $\mathbf{Q}$, $H \subset F$ an algebraic subgroup, and N an algebraic normal subgroup. If H and N are defined over $\mathbf{Q}$, and F is the semi-direct product (in the sense of § 1) of H and N, then the group $H_{\mathbf{Z}} \cdot N_{\mathbf{Z}}$ is of finite index in $F_{\mathbf{Z}}$.

Lemma 17.14. Let $U_1 \in B(\Gamma)$ and $U_2 \in B(\Gamma)$ be such that *1)* $U_1 \neq e$ and $U_2 \neq e$ *2)* $\bar{U}_1$ and $\bar{U}_2$ are opposite orispheric subgroups. Then *1)* the subgroups $N(\bar{U}_1) \cap G_{\mathbf{Z}}$ and $N(\bar{U}_1) \cap \Gamma$ are commensurable, *2)* the subgroups $N(\bar{U}_2) \cap G_{\mathbf{Z}}$ and $N(\bar{U}_2) \cap \Gamma$ are commensurable.

Proof. Since $\bar{U}_1$ and $\bar{U}_2$ are opposite orispheric subgroups, $N(\bar{U}_1) \cap N(\bar{U}_2)$ is a Levi subgroup in $N(\bar{U}_1)$. Therefore, since $\bar{S}N(\bar{U}_1) \supset \bar{U}_1$, and $\bar{U}_1$ is the unipotent radical in $N(\bar{U}_1)$ (for $\bar{U}_1$ is orispheric), we have

$$\bar{S}N(\bar{U}_1) = (N(\bar{U}_2) \cap \bar{S}N(\bar{U}_1)) \tilde{\times} \bar{U}_1, \tag{14}$$

where $\tilde{\times}$ stands for the semi-direct product in the sense of § 1. Since $\bar{U}_1$ and $\bar{U}_2$ are opposite orispheric subgroups, (14) and Lemma 10.33 imply that

$$\bar{S}N(\bar{U}_1) = (\bar{S}N(\bar{U}_1) \cap \bar{S}N(\bar{U}_2)) \tilde{\times} \bar{U}_1. \tag{15}$$

Now we make the following remarks:

Remark 1. $\bar{S}N(\bar{U}_1)$ has property P (in view of Lemma 7.3).

Remark 2. The subgroup $\bar{S}N(\bar{U}_1) \cap \bar{S}N(\bar{U}_2)$ has property P (see Remark 1 at the end of the proof of Lemma 17.12).

Remark 3. The quotient space $\bar{U}_1/U_1$ is compact (in view of Lemma 5.4).

Remark 4. The subgroup $\bar{S}N(\bar{U}_1)\cap\bar{S}N(\bar{U}_2)$ is defined over $\mathbf{Q}$ (see Remark 2 at the end of the proof of Lemma 17.12).

Remark 5. The subgroup $\bar{U}_1$ is defined over $\mathbf{Q}$ (in view of Lemma 17.4).

Remark 6. $\bar{U}_1\cap\Gamma=U_1$ (since $U_1\in B(\Gamma)$).

Remark 7. Put $H=\bar{S}N(\bar{U}_1)\cap\bar{S}N(U_2)$. It follows from (15), Remarks 1, 2, 3, 6 and Lemma 6.11 that the subgroup $(H\cap\Gamma)\cdot U_1$ is of finite index in $\bar{S}N(\bar{U}_1)\cap\Gamma$.

Remark 8. From (15), Remarks 4 and 5 and Lemma 17.13 we obtain that the subgroup $(H\cap G_{\mathbf{Z}})\cdot(\bar{U}_1)_{\mathbf{Z}}$ is of finite index in $\bar{S}N(\bar{U}_1)\cap G_{\mathbf{Z}}$.

Remark 9. It follows from (15) and Lemmas 17.3 and 17.11 that the subgroups $(H\cap\Gamma)\cdot U_1$ and $(H\cap G_{\mathbf{Z}})\cdot(\bar{U}_1)_{\mathbf{Z}}$ are commensurable.

In consequence of Remarks 7, 8 and 9, the subgroups $\bar{S}N(\bar{U}_1)\cap\Gamma$ and $\bar{S}N(U_1)\cap G_{\mathbf{Z}}$ are commensurable. Therefore, in view of Lemmas 7.13 and 17.6, the subgroups $N(\bar{U}_1)\cap\Gamma$ and $N(\bar{U}_1)\cap G_{\mathbf{Z}}$ are commensurable, which proves Part *1)* of the lemma. Part *2)* can be established in a similar way. The lemma is proved.

Lemma 17.15. Let U_1 be an arbitrary maximal unipotent subgroup of the group Γ. Then there is a maximal unipotent subgroup U_2 of Γ such that $\bar{U}_1$ and $\bar{U}_2$ are opposite orispheric subgroups of the group G.

Proof. Just as in the analysis of Case b) of Theorem 16.4, one can prove that there exists a $\gamma_0\in\Gamma$ for which $\gamma_0\bar{U}_1\gamma_0^{-1}\cap U_1=e$. Put $U_2=\gamma_0 U_1\gamma_0^{-1}$. Then, the same way as in Case b) of Theorem 16.4 one finds that $\bar{U}_1$ and $\bar{U}_2$ are opposite orispheric subgroups of the group G. Now, to complete the proof of the lemma it remains to observe that U_2 is a maximal unipotent subgroup of the group Γ, since $U_2=\gamma_0 U_1\gamma_0^{-1}$ and U_1 is maximal.

Theorem 17.16. *Let U be an arbitrary maximal unipotent subgroup of the group Γ. Then the subgroups $N(\bar{U})\cap G_{\mathbf{Z}}$ and $N(\bar{U})\cap\Gamma$ are commensurable.*

Proof. By Lemma 17.15 there exists a maximal unipotent subgroup U' of the group Γ, such that $\bar{U}$ and U' are opposite orispheric subgroups of the group G. Note that, since U and U' are maximal, we have *1)* $U_1\in B(\Gamma)$ and $U_2\in B(\Gamma)$ (in view of Lemma 7.7), *2)* $U_1\neq e$ and $U_2\neq e$ (in view of Lemma 7.5). This and Lemma 17.14 imply the assertion of the theorem.

Theorem 17.17. *Let U be an arbitrary maximal unipotent subgroup of Γ. Then $\bar{U}_{\mathbf{Z}}=\bar{U}\cap G_{\mathbf{Z}}$ is a maximal unipotent subgroup of the group $G_{\mathbf{Z}}$.*

Proof. Suppose $\bar{U}_{\mathbf{Z}}=\bar{U}\cap G_{\mathbf{Z}}$ is not a maximal unipotent subgroup of the group $G_{\mathbf{Z}}$. Then $\bar{U}_{\mathbf{Z}}$ is a proper subgroup in some unipotent subgroup $\hat{U}$ of the group $G_{\mathbf{Z}}$. Now $\hat{U}$ is nilpotent according to Lemma 1.3.1. Therefore, in view of Lemma

13.1 there is a $u_0 \in \hat{U} \subset G_{\mathbf{Z}}$ such that *1)* $u_0 \notin \overline{U}_{\mathbf{Z}}$, *2)* $u_0 \in N(\overline{U}_{\mathbf{Z}})$ where $N(\overline{U}_{\mathbf{Z}})$ denotes the normalizer of $\overline{U}_{\mathbf{Z}}$ in G. Since $u_0 \in N(\overline{U}_{\mathbf{Z}})$, and $\overline{U}_{\mathbf{Z}}$ is Zariski dense in $\overline{U}$ (according to Lemma 17.3), $u_0 \in N(\overline{U})$. Therefore, since $u_0 \in G_{\mathbf{Z}}$, we have

$$u_0 \in N(\overline{U}) \cap G_{\mathbf{Z}}. \tag{16}$$

In consequence of (16) and the commensurability of $N(\overline{U}) \cap G_{\mathbf{Z}}$ and $N(\overline{U}) \cap \Gamma$ (see Theorem 17.16), there exists a $n \neq 0$ for which

$$u_0^n \in N(\overline{U}) \cap \Gamma. \tag{17}$$

Since the element u_0^n is unipotent, and U is a maximal unipotent subgroup of Γ, in view of (17) and Lemmas 1.3.1 and 7.4 we have $u_0^n \in U$ and consequently

$$u_0^n \in \overline{U}. \tag{18}$$

Since $n \neq 0$, u_0 is unipotent and $\overline{U}$ is an algebraic unipotent group, in view of Lemmas 1.3.1 and 5.1 we find that $u_0 \in \overline{U}$. Therefore, $u_0 \in G_{\mathbf{Z}}$, implies $u_0 \in \overline{U} \cap G_{\mathbf{Z}} = \overline{U}_{\mathbf{Z}}$, which is a contradiction. The proof of the theorem is complete.

An algebraic subgroup P of the group G is called a *parabolic* **Q**-*subgroup*, if *1)* P is a parabolic subgroup of the group G, *2)* P is defined over **Q**. A parabolic **Q**-subgroup P is called a *minimal* parabolic **Q**-subgroup if P does not contain, as proper subgroup, any other parabolic **Q**-subgroup of the group G. One can define the analogous notions: *unipotent* **Q**-subgroup, *maximal unipotent* **Q**-subgroup etc.

Since all results quoted in § 10 have been proved by their authors for arbitrary field of characteristic 0, we have

Remark 17.18. Lemmas 10.8—10.33 and the paragraph following Lemma 10.25 remain valid, if we replace the parabolic subgroups, maximal **R**-split torus, maximal unipotent subgroups etc. by parabolic **Q**-subgroups, maximal **Q**-decomposable torus, maximal unipotent **Q**-subgroups etc. Further, an analogous classification can be introduced for parabolic **Q**-subgroups and orispheric **Q**-subgroups normalized by an arbitrary maximal **Q**-decomposable torus of the group, as that introduced in § 10 for parabolic and orispheric subgroups normalized by some maximal **R**-split torus. In the sequel when considering **Q**-subgroups of G, we shall directly refer to Lemmas 10.8—10.33 meaning the analogous lemmas for **Q**-subgroups.

Theorem 17.19. *Let U be an arbitrary maximal unipotent subgroup of the group Γ. Then 1) $\overline{U}$ is a maximal unipotent* **Q**-*subgroup of G 2) $N(\overline{U})$ is a minimal parabolic* **Q**-*subgroup of G.*

Proof. According to Lemma 17.4, the algebraic unipotent subgroup $\overline{U}$ is defined over **Q**, i.e. $\overline{U}$ is a unipotent **Q**-subgroup. Let V be an arbitrary maximal unipotent **Q**-subgroup, containing $\overline{U}$. Since V is unipotent and defined over **Q**, the quotient space $V/V_{\mathbf{Z}}$ is compact ($V_{\mathbf{Z}} = V \cap G_{\mathbf{Z}}$), (see 6.10 in [4]). Now, since

$\overline{U}_Z = \overline{U} \cap G_Z$ is, according to Theorem 17.17, a maximal unipotent subgroup in G_Z, and $V \supset \overline{U}$, we have $V_Z = U_Z$. The last two assertions and Lemma 5.6 imply $V = \overline{U}$. Thus, part *1)* is proved. Now part *1)* and Lemma 10.17 imply part *2)* (recall Remark 17.18).

Theorem 17.20. *Let U be an arbitrary maximal unipotent subgroup of the group Γ, and P a parabolic* **Q***-subgroup of G such that 1) $P \supset N(\overline{U})$, 2) $P \neq G$. Then the subgroups $P \cap G_Z$ and $P \cap \Gamma$ are commensurable.*

Proof. By Lemma 17.15 there is a maximal unipotent subgroup U' of the group Γ, such that $\overline{U}$ and $\overline{U}'$ are opposite orispheric subgroups of the group G. Then, in view of Theorem 17.19, $N(\overline{U})$ and $N(\overline{U}')$ are opposite minimal parabolic **Q**-subgroups of the group G. Therefore, in view of Lemma 10.16 the intersection $N(\overline{U}) \cap N(\overline{U}')$ contains some maximal **Q**-decomposable torus T of the group G (recall Remark 17.18). Since $P \supset N(\overline{U})$, $P \supset T$ holds and consequently T normalizes P. Write Σ for the root system of the group G with respect to the torus T, Σ^+ for the set of positive roots corresponding to $\overline{U}$, and Σ_0 for the set of simple roots in Σ^+. Then, with the notations introduced in § 10 before Lemma 10.25, we have

$$\overline{U} = H_{\Sigma_0}, \quad \overline{U}' = H_{\Sigma_0^-} \tag{19}$$

and

$$P = P_D, \quad \text{where} \quad D \subset \Sigma_0 \tag{20}$$

((20) holds in view of Lemma 10.25, Remark 17.18 and the facts that a) T normalizes P, b) $P \supset N(\overline{U})$). Now put $P' = P_{D^-}$. Then, in view of Lemma 10.31, (19), (20), *1)* $P' \supset N(U')$ *2)* P and P' are opposite parabolic **Q**-subgroups of the group G.

Denote the unipotent radicals of the groups P and P' by W and W'. In view of Theorem 17.19 $\overline{U}$ is a maximal unipotent **Q**-subgroup in G, and consequently also in P. Therefore, since the unipotent radical of an arbitrary **Q**-group F is contained in all maximal unipotent **Q**-subgroups of the group F, we have

$$\overline{U} \supset W. \tag{21}$$

Analogously

$$\overline{U}' \supset W'. \tag{22}$$

Since W is a unipotent **Q**-subgroup, thus (see 6.10 in [4]), the quotient space W/W_Z is compact ($W_Z = W \cap G_Z$). On the other hand, in view of Lemma 17.3, the subgroups $\overline{U} \cap \Gamma$ and $\overline{U}_Z = \overline{U} \cap G_Z$ are commensurable. (21) and the last two statements imply that the quotient space $W/W \cap \Gamma$ is compact. Therefore, in view of Lemma 5.6 the subgroup $U_1 = W \cap \Gamma$ is Zariski dense in W, i.e.

$$W = \overline{U}_1. \tag{23}$$

Analogously, the subgroup $U_2 = W' \cap \Gamma$ is Zariski dense in W', i.e.

$$W' = \overline{U}_2. \tag{24}$$

Since W and W' are the unipotent radicals of the opposite parabolic subgroups P and P', they are opposite orispheric subgroups of the group G. On the other hand, since $P \neq G$, we have $W \neq e$, and consequently $W' \neq e$. The last two statements and equalities (23) and (24) imply that U_1 and U_2 satisfy the conditions of Lemma 17.14 (relations $U_1 \in B(\Gamma)$ and $U_2 \in B(\Gamma)$ are consequences of the equalities $U_1 = W \cap \Gamma$ and $U_2 = W' \cap \Gamma$). Therefore, the subgroups $N(\overline{U}_1) \cap G_{\mathbf{Z}}$ and $N(\overline{U}_1) \cap \Gamma$ are commensurable. On the other hand, since P is parabolic and W is the unipotent radical of P, in view of (23) and Lemma 10.21 we have $P = N(\overline{U}_1)$. The last two statements imply the assertion of the theorem.

Theorem 17.21. *The group G is simple over* $\mathbf{Q}$ (i.e., every proper algebraic normal subgroup of G defined over $\mathbf{Q}$ reduces to e).

Proof. Let H be a proper algebraic normal subgroup of G defined over $\mathbf{Q}$. Since G is algebraically connected, semisimple and the centre of the complexification of G is trivial, in view of Lemma 1.4.2 we find that

$$G = H \times H', \tag{25}$$

where H' is an algebraic normal subgroup of the group G. Remark that H' in (25) is uniquely determined by H. Therefore, since H is defined over $\mathbf{Q}$, H' is also defined over $\mathbf{Q}$.

Let now U be an arbitrary maximal unipotent subgroup of the group Γ. Then, according to Theorem 17.17 $\overline{U}_{\mathbf{Z}} = \overline{U} \cap G_{\mathbf{Z}}$ is a maximal unipotent subgroup of the group $G_{\mathbf{Z}}$. Therefore, since H and H' are subgroups defined over $\mathbf{Q}$, it is easy to see that $(H \cap \overline{U}_{\mathbf{Z}}) \times (H' \cap \overline{U}_{\mathbf{Z}})$ is of finite index in $\overline{U}_{\mathbf{Z}}$. From this and Lemma 17.3 we get that $(H \cap U) \times (H' \cap U)$ is of finite index in U. On the other hand, *1)* $U \neq e$ (in view of Lemma 7.5), *2)* U is unipotent, thus U is infinite. The last two statements imply that either $H \cap U \neq e$, or $H' \cap U \neq e$. Therefore, since *1)* G and Γ satisfy the conditions formulated at the beginning of § 7, *2)* H is a proper algebraic normal subgroup of the group G, an application of Lemma 4.8 immediately gives that $H = e$. The theorem is proved.

Lemma 17.22. Let F be a real algebraic group, $V \subset F$ a unipotent algebraic normal subgroup, and $L \subset F$ a reductive algebraic subgroup. Denote by $C(L)$ the centralizer of L in F, and by $(\overline{L, V})$ the algebraic commutator of the subgroups L and V (for the definition of algebraic commutator see § 15). If $V \cap C(L) = e$, then $(\overline{L, V}) = V$.

Proof. Denote by $\mathfrak{B}$ the Lie algebra of the group V, and by $(L, \mathfrak{B})$ the commutator of the subgroup L and the Lie algebra $\mathfrak{B}$ (i.e. the subalgebra generated by the elements of the form $\mathfrak{v} - (\mathrm{Ad}\, l)\mathfrak{v}$, where $l \in L$ and $\mathfrak{v} \in \mathfrak{B}$). Since V is algebraically connected (in view of Lemma 1.3.1), to prove the lemma, it is sufficient to show that

$$(L, \mathfrak{B}) = \mathfrak{B}. \tag{26}$$

Denote $(L, \mathfrak{B})$ by $\mathfrak{B}_0$. It is clear that $\mathfrak{B}_0$ is invariant under Ad L. Therefore, since L is reductive, in view of Lemmas 1.2.1 and 1.4.4 there is a subspace $\mathfrak{B}'\subset\mathfrak{B}$, invariant under Ad L such that $\mathfrak{B}=\mathfrak{B}_0\times\mathfrak{B}'$. Thus, it follows from the definition of $\mathfrak{B}_0$ that $\mathfrak{v}-(\mathrm{Ad}\, l)\mathfrak{v}=0$ for any $l\in L$ and $\mathfrak{v}\in\mathfrak{B}'$, i.e.

$$\mathfrak{B}'\subset\mathfrak{C}(L), \tag{27}$$

where $\mathfrak{C}(L)$ is the centralizer of L in the Lie algebra $\mathfrak{B}$. From (27) and the relation $V\cap C(L)=e$ (by assumption) we conclude $\mathfrak{B}'=e$, i.e. $\mathfrak{B}=\mathfrak{B}_0$. Consequently (26) holds, which as mentioned already, proves the lemma.

Recall that for any parabolic subgroup P, $R_u(P)$ stands for its unipotent radical.

Lemma 17.23. Let P and P' be parabolic subgroups of the group G. If $P'\supset P$, then $R_u(P')\subset R_u(P)$.

Proof. Since P is a parabolic subgroup of G, there exists in G a maximal unipotent subgroup V, contained in P. Now, since $P'\subset G$ and $P\subset P'$, we get that V is a maximal unipotent subgroup in P'. Therefore, in view of Lemma 10.8

$$V\supset R_u(P'). \tag{28}$$

From (28) and the inclusion $P'\supset P$ we get that $R_u(P')$ is a unipotent algebraic normal subgroup in P. Now this immediately implies the assertion of the lemma.

Lemma 17.24. Let F and H be algebraic subgroups of the group G, and let $H\subset F$. Write $C(H)$ for the centralizer of H in G, D for the minimal algebraic subgroup in G containing $F\cup C(H)$, H_F for the minimal algebraic normal subgroup of F containing H, and H_D for the minimal algebraic normal subgroup of D containing H. Then $H_F=H_D$.

Proof. Denote by D_0 the subgroups of the group G generated (in the sense of abstract group theory) by the union $F\cup C(H)$. Let $d\in D_0$. Then $d=c_1\cdot f_1\ldots c_i\cdot f_i$ for some i and some $c_1, \ldots, c_i\in C(H)$ and $f_1, \ldots, f_i\in F$. Since every c_j $(1\leqq j\leqq i)$ commutes with H (for they belong to $C(H)$), we have $dHd^{-1}=fHf^{-1}$ where $f=f_1\ldots f_i$. From this it follows

$$\bigcup_{d\in D_0} dH\,d^{-1}=\bigcup_{f\in F} fHf^{-1}. \tag{29}$$

From (29) and the fact that D_0 is easily seen to be Zariski dense in D, the assertion of the lemma follows.

Lemma 17.25. Let P be an arbitrary parabolic subgroup of the group G, L any Levi subgroup in P, and N an arbitrary algebraic normal subgroup of the group L. Then the algebraic commutator $\overline{(N, R_u(P))}$ of the subgroups N and $R_u(P)$, is an orispheric subgroup of the group G.

Proof. Write H for the minimal algebraic normal subgroup of the group P containing N, and V for the unipotent radical of the group H. We show that

$$V = H \cap R_u(P) = \left(\overline{N, R_u(P)}\right). \tag{30}$$

Indeed, since L is a Levi subgroup in P, we get $P=L\cdot R_u(P)$. Therefore, since N is a normal subgroup in L,

$$\bigcup_{p\in P} pNp^{-1} = \bigcup_{u\in R_u(P)} uNu^{-1}. \tag{31}$$

Further, since $R_u(P)$ is normalized by the subgroup N (because $R_u(P)$ is a normal divisor in P, and $N\subset P$), we obtain that N normalizes $(\overline{N, R_u(P)})$. Therefore, since *1)* N belongs to a Levi subgroup of the group P, *2)* $(\overline{N, R_u(P)})$ belongs to the unipotent radical of the group P, *3)* N and $(\overline{N, R_u(P)})$ are algebraic subgroups, we get that $N\cdot(\overline{N, R_u(P)})$ is an algebraic subgroup and

$$N\cdot(\overline{N, R_u(P)}) = N \tilde{\times} (\overline{N, R_u(P)}), \tag{32}$$

where $\tilde{\times}$ stands for semi-direct product in the sense of section IV of § 1. From this and (31) we get

$$H = N\tilde{\times}(\overline{N, R_u(P)}). \tag{33}$$

Now, since N is an algebraic normal subgroup in the reductive group L, it is reductive. Therefore it is easy to see that (33) implies (30) (we observe that $(\overline{N, R_u(P)})\subset R_u(P)$).

Denote by $C(N)$ the centralizer of N in G, by P' the minimal algebraic subgroup of the group G containing $C(N)\cup P$, and by H' the minimal algebraic normal subgroup of the group P' containing N. According to Lemma 17.24, we have

$$H' = H. \tag{34}$$

Furthermore, since $P'\supset P$, and P is a parabolic subgroup, according to Lemma 17.23 we have

$$R_u(P')\subset R_u(P). \tag{35}$$

Since N is reductive, according to Lemma 10.2 $C(N)$ is also reductive. Therefore, since $P'\supset C(N)$, we have

$$R_u(P')\cap C(N) = e. \tag{36}$$

From (36) and Lemma 17.22 we obtain

$$(\overline{N, R_u(P')}) = R_u(P'). \tag{37}$$

Since H is a normal subgroup in P' (in view of (34)), and the unipotent radical is uniquely determined, V is a unipotent algebraic normal subgroup in P'. Therefore we have

$$V\subset R_u(P'). \tag{38}$$

On the other hand, by (30), (35) and (37) we have

$$V = \overline{(N, R_u(P))} \supset \overline{(N, R_u(P'))} = R_u(P'). \tag{39}$$

Combining (38) and (39) we obtain

$$V = R_u(P'). \tag{40}$$

From (30), (40) and Lemma 10.9 the assertion of the lemma follows (it should be observed that P' is parabolic, since $P \subset P'$, and P is parabolic).

Remark Lemma 17.24 is only used in the proof of Lemma 17.25. One can give another proof of Lemma 17.25 using Lemmas 10.25 and 10.29 and the following facts: *1)* P normalizes $\overline{(N, R_u(P))}$, *2)* if (in the notation of § 10) $P = P_D$, $L = P_D \cap P_{D^-}$, $\alpha \in \Sigma_0$ is a simple root, $U_\alpha \subset R_u(P)$ and $U_\alpha \not\subset \overline{(N, R_u(P))}$, then $U_\alpha \subset C(N)$.

As a consequence of Lemma 7.5 and Theorem 17.19, we have the following

Lemma 17.26. In the group G there exist proper parabolic $\mathbf{Q}$-subgroups (i.e. parabolic $\mathbf{Q}$-subgroups not coinciding with G).

Lemma 17.27. Let P be a parabolic $\mathbf{Q}$-subgroup of the group G, maximal in the class of all proper parabolic $\mathbf{Q}$-subgroups of G. Denote by $\bar{P}_{\mathbf{Z}}$ the algebraic closure of the subgroup $P_{\mathbf{Z}} = P \cap G_{\mathbf{Z}}$, and by $\bar{P}^0_{\mathbf{Z}}$ the algebraically connected component of the identity in the group $\bar{P}_{\mathbf{Z}}$. Then $\bar{P}^0_{\mathbf{Z}} \supset R_u(P)$, but $\bar{P}^0_{\mathbf{Z}} \neq R_u(P)$.

Proof. Since the unipotent algebraic subgroup $R_u(P)$ is defined over $\mathbf{Q}$, (for P is defined over $\mathbf{Q}$), thus, in view of section 6.10 of [4], the quotient space $R_u(P)/R_u(P) \cap G_{\mathbf{Z}}$ is compact. Therefore, in view of Lemma 5.6, $R_u(P) \cap G_{\mathbf{Z}}$ is Zariski dense in $R_u(P)$. From this and the algebraic connectedness of the group $R_u(P)$ (see Lemma 1.3.2), it follows immediately that

$$\bar{P}^0_{\mathbf{Z}} \supset R_u(P). \tag{41}$$

P is a parabolic $\mathbf{Q}$-subgroup of the group G, and so in view of Remark 17.28 and Lemma 10.16, one can easily show that there exist T and S such that *1)* T is a maximal $\mathbf{R}$-split torus of the group G, *2)* S is a maximal $\mathbf{Q}$-decomposable torus of the group G, *3)* T is defined over $\mathbf{Q}$, *4)* $P \supset T \supset S$. We distinguish two cases: a) $T = S$, b) $T \neq S$.

Case a) Since the real rank of G is greater than or equal to two, we have $\dim T \geqq 2$, and consequently, since in the analysed case $S = T$, we have

$$\dim S \geqq 2. \tag{42}$$

Note that, since $S \subset P$, S normalizes P. Therefore, according to the classification of parabolic $\mathbf{Q}$-subgroups normalized by a maximal $\mathbf{Q}$-decomposable torus (see Remark 17.18 and § 10), we have

$$P = P_D, \tag{43}$$

where $D\subset\Sigma$, and Σ is the root system of the group G corresponding to the torus S (here we have used the notation introduced in § 10 before Lemma 10.25). Since P is a maximal parabolic $\mathbf{Q}$-subgroup of the group G, it is easy to see that D consists of a single element. Therefore, in view of (42), Remark 17.18 and Lemma 10.26, $R_u(P)$ is not a maximal unipotent $\mathbf{Q}$-subgroup of the group G. On the other hand, since P is a parabolic $\mathbf{Q}$-subgroup of the group G, P contains some maximal unipotent $\mathbf{Q}$-subgroup V of the group G. The last two statements, Remark 17.18 and Lemma 10.8 imply that

$$P\supset V\supset R_u(P) \quad \text{but} \quad V\neq R_u(P). \tag{44}$$

Since V is a unipotent algebraic group defined over $\mathbf{Q}$, in view of section 6.10 in [4] and Lemma 5.6, $V_{\mathbf{Z}}=V\cap G_{\mathbf{Z}}$ is Zariski dense in V. Therefore, since V is algebraically connected (in view of Lemma 1.3.1), applying (44) and (41) we get that

$$\bar{P}^0_{\mathbf{Z}}\supset V\neq R_u(P) \quad \text{but} \quad \bar{P}^0_{\mathbf{Z}}\supset R_u(P). \tag{45}$$

Thus, the lemma is proved in Case a).

Case b) For any torus A defined over $\mathbf{Q}$, $X_{\mathbf{Q}}(A)$ denotes the group of all rational characters, defined over $\mathbf{Q}$, of the torus A. Denote by $\tilde{T}$ the intersection of the kernels of the elements of $X_{\mathbf{Q}}(T)$. Then (see section 1.4 in [5]) *1)* $\tilde{T}$ is the almost direct product of $\tilde{T}$ and S, *2)* the torus $\tilde{T}$ is anisotropic over $\mathbf{Q}$ (i.e. $\tilde{T}$ is defined over $\mathbf{Q}$ and $X_{\mathbf{Q}}(\tilde{T})=1$). On the other hand, since in the analysed case $T\neq S$, and T is algebraically connected (for T is a torus), we have $\dim T>\dim S$. From the last two statements and Theorem 11.8 in [4] we obtain

$$\dim \tilde{T}\geqq 1. \tag{46}$$

and

$$\text{the quotient space } \tilde{T}/\tilde{T}\cap G_{\mathbf{Z}} \text{ is compact.} \tag{47}$$

Since $\tilde{T}$ is $\mathbf{R}$-split (because $\tilde{T}\subset T$, and T is $\mathbf{R}$-split), in view of (46) $\tilde{T}$ is not compact. From this and (47) it follows that $\tilde{T}_{\mathbf{Z}}=\tilde{T}\cap G_{\mathbf{Z}}$ is infinite. Therefore, since *1)* $\tilde{T}\subset P$, *2)* $\tilde{T}\cap R_u(P)=e$ (for $\tilde{T}$ is a torus, and $R_u(P)$ is a unipotent subgroup), we get easily that $\bar{P}^0_{\mathbf{Z}}\neq R_u(P)$, whence, in view of (41), the assertion of the lemma follows in Case b). The proof of the lemma is complete.

In order to avoid misunderstanding, we emphasize that in the sequel the parabolic $\mathbf{Q}$-subgroups, maximal in the class of all proper parabolic $\mathbf{Q}$-subgroups, will be called maximal parabolic $\mathbf{Q}$-subgroups.

Lemma 17.28. Let P be a maximal parabolic $\mathbf{Q}$-subgroup of the group G, and L an arbitrary Levi subgroup in P defined over $\mathbf{Q}$. Write $\bar{L}_{\mathbf{Z}}$ for the algebraic closure of the subgroup $L_{\mathbf{Z}}=L\cap G_{\mathbf{Z}}$, and $\bar{L}^0_{\mathbf{Z}}$ for the algebraically connected component of the identity of the group $\bar{L}_{\mathbf{Z}}$. Then the algebraic commutator $\overline{(\bar{L}^0_{\mathbf{Z}}, R_u(P))}$ coincides with $R_u(P)$.

Proof. For the proof of the lemma we make some remarks:

Remark 1. As an easy consequence of Lemmas 17.27 and 17.13, we get $\dim \bar{L}_{\mathrm{Z}}^0 > 0$.

Remark 2. Let l be an arbitrary element in $L_{\mathbf{Q}} = L \cap G_{\mathbf{Q}}$. Then, it is easy to see that the subgroup $(lL_{\mathrm{Z}}l^{-1}) \cap L_{\mathrm{Z}}$ is of finite index in L_{Z}. Therefore, in view of Lemma 8.8

$$l\bar{L}_{\mathrm{Z}}^0 l^{-1} = \bar{L}_{\mathrm{Z}}^0. \tag{48}$$

Further, as P is a parabolic subgroup of G, (see Corollary 4.5 in [4]), P is algebraically connected. Therefore, since L is a Levi subgroup in P, L is algebraically connected. Now, since L is defined over $\mathbf{Q}$ (by assumption), the above facts imply, in view of Lemma 1.2.11, that $L_{\mathbf{Q}}$ is Zariski dense in L, whence, in view of (48), it follows that $\bar{L}_{\mathrm{Z}}^0$ is a normal subgroup in L.

Remark 3. Since P is defined over $\mathbf{Q}$, in view of Lemma 1.3.5 $R_u(P)$ is defined over $\mathbf{Q}$. Therefore, the minimal algebraic normal subgroup of the group G containing $R_u(P)$, is also defined over $\mathbf{Q}$. Now this implies, since G is simple over $\mathbf{Q}$ according to Theorem 17.21, that the minimal algebraic normal subgroup of G containing $R_u(P)$, coincides with G.

Remark 4. Since P is a parabolic subgroup, in view of Lemma 10.9 $R_u(P)$ is an orispheric subgroup of the group G. From this, Remark 3 and Lemma 11.16 we get $C(R_u(P)) \subset R_u(P)$, where $C(R_u(P))$ denotes the centralizer of $R_u(P)$ in G.

Remark 5. Since L is a Levi subgroup in P, we have $L \cap R_u(P) = e$. Therefore, in view of Remark 4, $L \cap C(R_u(P)) = e$. From this and Remark 1 we obtain

$$\overline{(\bar{L}_{\mathrm{Z}}^0, R_u(P))} \neq e. \tag{49}$$

Remark 6. Remark 2 and Lemma 17.25 imply that $\overline{(\bar{L}_{\mathrm{Z}}^0, R_u(P))}$ is an orispheric subgroup of the group G. On the other hand, since *1)* $\bar{L}_{\mathrm{Z}}^0$ is defined over $\mathbf{Q}$ (in view of Lemmas 8.7 and 1.2.12), *2)* $R_u(P)$ is defined over $\mathbf{Q}$ (in view of Lemma 1.3.5), the algebraic commutator $\overline{(\bar{L}_{\mathrm{Z}}^0, R_u(P))}$ is also defined over $\mathbf{Q}$. Thus $\overline{(\bar{L}_{\mathrm{Z}}^0, R_u(P))}$ is an orispheric subgroup of $\mathbf{Q}$ defined over G.

Remark 7. Since P is a maximal parabolic $\mathbf{Q}$-subgroup of the group G, applying 10.9 we get that $R_u(P)$ is a minimal orispheric $\mathbf{Q}$-group of the group G.

Since $\overline{(\bar{L}_{\mathrm{Z}}^0, R_u(P))} \subset R_u(P)$ (for $R_u(P)$ is an algebraic normal subgroup in P), it follows from Remarks 5, 6, and 7 that $\overline{(\bar{L}_{\mathrm{Z}}^0, R_u(P))} = R_u(P)$. The lemma is proved.

From Lemma 17.13 and the inclusion relation (41) we can get easily the following

Lemma 17.29. Let P be a maximal parabolic **Q**-subgroup in G, L a Levi subgroup of P defined over **Q**, let $\bar{P}^0_{\mathbf{Z}}$ denote the same as in Lemma 17.27, and $\bar{L}^0_{\mathbf{Z}}$ the same as in Lemma 17.28. Then $\bar{P}^0_{\mathbf{Z}}$ is the semi-direct product (in the sense of § 1) of $\bar{L}^0_{\mathbf{Z}}$ and $R_u(P)$ (in this lemma the maximality of P is not essential).

Recall that for any two subgroups $A, B \subset G$, $\{A, B\}$ denotes the set of all commutators of the form $aba^{-1}b^{-1}$, where $a \in A$, and $b \in B$.

Lemma 17.30. Let P be a maximal parabolic **Q**-subgroup of the group G, L a Levi subgroup of P defined over **Q**, and $A \subset (G_{\mathbf{Q}} \cap R_u(P))$ a subset containing the identity of the group G. For every $a \in A$ we choose an arbitrary subgroup D of finite index in the group $(aLa^{-1}) \cap G_{\mathbf{Z}}$. If the minimal algebraic subgroup $\tilde{A}$ containing $\{A, \bar{L}^0_{\mathbf{Z}}\}$ (for the definition of $\bar{L}^0_{\mathbf{Z}}$ see Lemma 17.28) coincides with $R_u(P)$, then the subgroup F generated by the union $\bigcup_{a \in A} D_a$ is of finite index in $P_{\mathbf{Z}}$ (here, as well as in the previous lemma, the maximality of P is not essential).

Proof. Denote by B the minimal algebraic subgroup containing the union $\bigcup_{a \in A} (a\bar{L}^0_{\mathbf{Z}}a^{-1})$. Since A contains the identity, it is easy to see that $B \supset \{A, \bar{L}^0_{\mathbf{Z}}\}$. Therefore, since B is an algebraic subgroup, and $\tilde{A} = R_u(P)$ (by assumption), we have

$$B \supset R_u(P). \tag{50}$$

Further, since $A \subset R_u(P)$, and $\bar{L}^0_{\mathbf{Z}}$ normalizes $R_u(P)$ (for, $\bar{L}^0_{\mathbf{Z}} \subset P$), we obtain $\bigcup_{a \in A} (a\bar{L}^0_{\mathbf{Z}}a^{-1}) \subset \bar{L}^0_{\mathbf{Z}} \cdot R_u(P)$. Therefore, since $\bar{L}^0_{\mathbf{Z}} \cdot R_u(P) = \bar{P}^0_{\mathbf{Z}}$ (in view of Lemma 17.29), and $\bar{P}^0_{\mathbf{Z}}$ is an algebraic subgroup, we get

$$B \subset \bar{P}^0_{\mathbf{Z}} = \bar{L}^0_{\mathbf{Z}} \cdot R_u(P). \tag{51}$$

On the other hand, since $e \in A$, we have $B \supset \bar{P}^0_{\mathbf{Z}}$. Therefore, in consequence of (50) and (51) we have

$$B = \bar{P}^0_{\mathbf{Z}}. \tag{52}$$

Denote by $\bar{F}$ and $\bar{D}_a$ ($a \in A$) the Zariski closures of the subgroups F and D_a. Now $A \subset G_{\mathbf{Q}}$, thus for any $a \in A$ the subgroups $a(L \cap G_{\mathbf{Z}})a^{-1}$ and $(aLa^{-1}) \cap G_{\mathbf{Z}}$ are commensurable. Therefore, since D_a is of finite index in $(aLa^{-1}) \cap G_{\mathbf{Z}}$, the subgroups $a(L \cap G_{\mathbf{Z}})a^{-1}$ and D_a are commensurable. From this, the definition of $\bar{L}^0_{\mathbf{Z}}$, and Lemma 8.8 we see that for any $a \in A$, $\bar{D}_a \supset a\bar{L}^0_{\mathbf{Z}}a^{-1}$. Therefore, since $F \supset \bigcup_{a \in A} D_a$, and F is an algebraic subgroup we have $\bar{F} \supset B$. This and (52) imply

$$\bar{F} \supset \bar{P}^0_{\mathbf{Z}}. \tag{53}$$

On the other hand, since it is easy to see that $F \subset P_{\mathbf{Z}}$ holds, we have

$$\bar{F} \subset \bar{P}_{\mathbf{Z}} \tag{54}$$

(for the definition of $\bar{P}_{\mathbf{Z}}$ see Lemma 17.27).

Since *1)* D_e (e is the identity of the group G) is of finite index in $L_Z = L \cap G_Z$ (by assumption), *2)* $L_Z \cdot (G_Z \cap R_u(P))$ is of finite index in P_Z (in view of Lemma 17.13), *3)* it is easy to see that $F \subset P_Z$, *4)* $F \supset D_e$ (by the definition of F), it is easy to show that $D_e \cdot (F \cap R_u(P))$ is of finite index in F. From this, Lemma 8.8, inclusion formulae (50) and (41), and the fact that $D_e \subset L$ we obtain

$$\overline{F \cap R_u(P)} = R_u(P), \tag{55}$$

where $\overline{F \cap R_u(P)}$ denotes the algebraic closure of the subgroup $F \cap R_u(P)$. Since $F \subset P_Z$, we see that F is discrete. Therefore (55) and Lemma 5.4 imply that the quotient space $R_u(P)/F \cap R_u(P)$ is compact, whence it follows that $F \cap R_u(P)$ is a subgroup of finite index in $R_u(P) \cap G_Z$. Therefore, since *1)* D_e is of finite index in L_Z *2)* $L_Z(G_Z \cap R_u(P))$ is of finite index in P_Z we get that $F \supset D_e \cdot (F \cap R_u(P))$ is of finite index in P_Z. The lemma is proved.

Lemma 17.31. Let V be an arbitrary real unipotent algebraic group of dimension n. Denote by V^n the direct product $\overbrace{V \times \ldots \times V}^{n}$ and by $W_n \subset V^n$ the set of all $(v_1, \ldots, v_n) \in V^n$ for which the minimal algebraic subgroup containing $\bigcup_{1 \leq i \leq n} v_i$ coincides with V. Then W_n is a non-empty Zariski open subset in V^n.

Proof. Denote by $\mathfrak{V}$ the Lie algebra of the group V, and by ln the logarithmic mapping $\ln : V \to \mathfrak{V}$ (ln is well defined in view of Lemmas 1.3.1 and 5.1). Let $v = (v_1, \ldots, v_n) \in V^n$. Write V_v for the subgroup generated by $\bigcup_{1 \leq i \leq n} v_i$, $\overline{V}_v$ for the algebraic closure of the subgroup V_v, and $\mathfrak{V}_v \subset \mathfrak{V}$ for the subalgebra generated by $\bigcup_{1 \leq i \leq n} \ln v_i$. The Campbell—Hausdorff formula implies

$$\ln V_v \subset \mathfrak{V}_v. \tag{56}$$

As shown in § 5, the logarithmic mapping is polynomial in case of unipotent algebraic **R**-groups. Therefore, (56) yields

$$\ln \overline{V}_v \subset \mathfrak{V}_v. \tag{57}$$

Furthermore, V_v is a connected, simply connected nilpotent Lie group (in view of Lemma 1.3.1) and in view of Lemma 5.1, $\ln \overline{V}_v$ coincides with the Lie algebra of the group $\overline{V}_v$. From this and the relation $\ln \overline{V}_v \supset \bigcup_{1 \leq i \leq n} \ln v_i$ we get the inclusion

$$\ln \overline{V}_v \supset \mathfrak{V}_v. \tag{58}$$

Combining (57) and (58) we get

$$\ln \overline{V}_v = \mathfrak{V}_v. \tag{59}$$

Since V is a nilpotent Lie group (in view of Lemma 1.3.1), the Lie algebra $\mathfrak{V}$ of the group V is nilpotent. Denote by $l(\mathfrak{V})$ the nilpotency class of the Lie algebra $\mathfrak{V}$ (i.e. the length of the lower central series of the Lie algebra $\mathfrak{V}$). Then it is easy to see that $\mathfrak{V}_v$ is generated, as a linear space, by the union of all commutators of $\ln v_1, \ldots, \ln v_n$ whose weight does not exceed $l(\mathfrak{V})$ (for definition of the weight of a commutator see § 5). This, equality (59) and the fact that ln is a polynomial mapping, easily imply that W_n is Zariski open in V^n. Now, in order to complete the proof of the lemma, we note that in view of (59) for any basis $e_1, \ldots, e_n$ of the linear space $\mathfrak{V}$, we have $(\exp e_1, \ldots, \exp e_n) \in W_n$, and consequently W_n is non-empty. The lemma is proved.

Lemma 17.32. Let F be a real algebraic group, $V \subset F$ a unipotent algebraic normal subgroup in F, A a subset contained in V and B an arbitrary subset in F. Assume that the minimal algebraic subgroup $\tilde{A}$ containing $\{A, B\}$, coincides with V, and denote $\dim V$ by n. Then there exist elements $a_1, \ldots, a_n \in A$ and $b_1, \ldots, b_n \in B$ such that the minimal algebraic subgroup containing $\bigcup_{1 \leqq i \leqq n} (a_i b_i a_i^{-1} b_i^{-1})$ coincides with V.

Proof. Since $\tilde{A} = V$, it is easy to see that there are $a_1, \ldots, a_n \in A$ and $b_1, \ldots, b_n \in B$ such that *1)* $a_1 b_1 a_1^{-1} b_1^{-1} \neq e$ *2)* for any j $(2 \leqq j \leqq n)$ either the minimal algebraic subgroup $\tilde{A}_j$ containing $\bigcup_{1 \leqq i \leqq j} (a_i b_i a_i^{-1} b_i^{-1})$ coincides with V, or $a_{j+1} b_{j+1} a_{j+1}^{-1} b_{j+1}^{-1}$ does not belong to $\tilde{A}_j$. Now, for any j $(1 \leqq j \leqq n)$, $\tilde{A}_j$ is algebraically connected (in view of Lemma 1.3.1), thus *1)* $\dim \tilde{A}_1 \geqq 1$ *2)* for any j $(2 \leqq j \leqq n)$ we have either $\tilde{A}_j = V$, or $\dim \tilde{A}_{j+1} > \dim \tilde{A}_j$. Now this easily implies that $a_1, \ldots, a_n \in A$ and $b_1, \ldots, b_n \in B$ are the desired elements. The lemma is proved.

Lemma 17.33. Let P be a maximal parabolic $\mathbf{Q}$-subgroup of the group G, and L a Levi subgroup in P defined over $\mathbf{Q}$. Denote $R_u(P)$ by V, the dimension of the group V by n, and the direct product $\overbrace{V \times \ldots \times V}^{n}$ by V^n. Let $\bar{L}^0_{\mathbf{Z}}$ denote the same as in Lemma 17.28. Write $X \subset V^n$ for the set of all $v = (v_1, \ldots, v_n) \in V^n$ for which the minimal algebraic subgroup containing $\{v, \bar{L}^0_{\mathbf{Z}}\}$ coincides with V ($\{v, \bar{L}^0_{\mathbf{Z}}\}$ denotes the union $\bigcup_{1 \leqq i \leqq n,\, l \in \bar{L}^0_{\mathbf{Z}}} (v_i l v_i^{-1} l^{-1})$). Then X contains some non-empty Zariski open subset $\tilde{X}$ of the real algebraic manifold V^n.

Proof. Since P is a maximal parabolic $\mathbf{Q}$-subgroup in G, L is a Levi subgroup in P defined over $\mathbf{Q}$, and $V = R_u(P)$, according to Lemma 17.28, the algebraic commutator $(\bar{L}^0_{\mathbf{Z}}, V)$ coincides with V. Therefore, in view of Lemma 17.32 and the definition of algebraic commutators, there are $l_1, \ldots, l_n \in \bar{L}^0_{\mathbf{Z}}$ and $\bar{v}_1, \ldots, \bar{v}_n$ for which the minimal algebraic subgroup containing $\bigcup_{1 \leqq i \leqq n} (\bar{v}_i l_i \bar{v}_i^{-1} l_i^{-1})$ coincides with V. On the other hand, from Lemma 17.31 it follows easily that the set $\tilde{X} \subset V^n$ of all $(v_1, \ldots, v_n) \in V^n$, for which the minimal algebraic subgroup containing $\bigcup_{1 \leqq i \leqq n} (v_i l_i v_i^{-1} l_i^{-1})$ coincides with V, is Zariski open in V^n (here we apply the fact that in any algebraic group the commutator $h_1 h_2 h_1^{-1} h_2^{-1}$ algebraically depends on h_1 and h_2). The last two statements imply the assertion of the lemma.

Lemma 17.34. Let P be a parabolic **Q**-subgroup of the group G for which the subgroups $P_{\mathbf{Z}}=P\cap G_{\mathbf{Z}}$ and $P\cap\Gamma$ are commensurable. Then, for any $\gamma\in\Gamma$ the subgroups $(\gamma P\gamma^{-1})\cap G_{\mathbf{Z}}$ and $(\gamma P\gamma^{-1})\cap\Gamma$ are commensurable.

Proof. Since $\Gamma\subset G_{\mathbf{Q}}$ (see (1)), for any $\gamma\in\Gamma$ the subgroups $\gamma P_{\mathbf{Z}}\gamma^{-1}$ and $(\gamma P\gamma^{-1})\cap G_{\mathbf{Z}}$ are commensurable. Therefore, since *1)* the subgroups $P_{\mathbf{Z}}$ and $P\cap\Gamma$ are commensurable (by assumption), *2)* it is easy to see that for any $\gamma\in\Gamma$, $\gamma(P\cap\Gamma)\gamma^{-1}=(\gamma P\gamma^{-1})\cap\Gamma$, we get that for any $\gamma\in\Gamma$ the subgroups $(\gamma P\gamma^{-1})\cap G_{\mathbf{Z}}$ and $(\gamma P\gamma^{-1})\cap\Gamma$ are commensurable. The lemma is proved.

Lemma 17.35. Let P and P' be opposite parabolic subgroups of the group G. Then the mapping $f: R_u(P)\times P'\to G$ assigning to $(u, p)\in R_u(P)\times P'$ the product $u\cdot p\in G$ is an isomorphism of the algebraic manifold $R_u(P)\times P'$ onto a Zariski open subset of G.

Proof. As above, for any algebraic subgroup $H\subset G$, $H_{\mathbf{C}}$ denotes the complexification. It is easy to see that $P_{\mathbf{C}}$ and $P'_{\mathbf{C}}$ are opposite parabolic subgroups of the group $G_{\mathbf{C}}$. Therefore, since the field **C** is algebraically closed, in view of part *2)* of Proposition 4.10 in [5], the mapping $f_{\mathbf{C}}: R_u(P_{\mathbf{C}})\times P'_{\mathbf{C}}\to G_{\mathbf{C}}$ assigning to $(u, p)\in R_u(P_{\mathbf{C}})\times P'_{\mathbf{C}}$ the product $u\cdot p\in G_{\mathbf{C}}$ is an isomorphism of the algebraic manifold $R_u(P_{\mathbf{C}})\times P'_{\mathbf{C}}$ onto some Zariski open subset of $G_{\mathbf{C}}$ (here we apply the fact, mentioned in Section 0.3 in [5], that for any algebraic k-group G, $G_{\bar{k}}$ and G can be identified, where $\bar{k}$ is the algebraic closure of the field k). On the other hand, it is easy to see that the subgroups $R_u(P_{\mathbf{C}})$ and $P'_{\mathbf{C}}$ and the mapping f are defined over **R**. Therefore, since $G=(G_{\mathbf{C}})_{\mathbf{R}}$, f is an isomorphism, and **C** is a separable extension of the field **R** (because the characteristic of the field **R** is equal to 0), we have the following consequence: if $f(u, p)\in G$, then $u\in(R_u(P_{\mathbf{C}}))_{\mathbf{R}}$, and $p\in(P'_{\mathbf{C}})_{\mathbf{R}}$. Now, in order to complete the proof of the lemma, it remains to observe that *1)* $P'=(P'_{\mathbf{C}})_{\mathbf{R}}$ (by the definition of complexification), *2)* in view of Lemma 1.3.6 $R_u(P)=(R_u(P_{\mathbf{C}}))_{\mathbf{R}}$. The lemma is proved.

Lemma 17.36. Let P and P' be opposite parabolic subgroups of the group G. Denote by $W\subset G$ the Zariski open subset consisting of all $g\in G$, for which P and $gP'g^{-1}$ are opposite parabolic subgroups (W is open in view of Lemma 10.13). Denote by f the mapping $f: W\to R_u(P)$ assigning to every $w\in W$ the unique (in view of Lemma 10.22) element $f(w)=u\in R_u(P)$ for which $wP'w^{-1}=uP'u^{-1}$. Then *1)* f is a regular algebraic mapping, *2)* $f(W)=R_u(P)$.

Proof. As an easy consequence of Lemmas 10.11 and 10.22, we have $W=R_u(P)\cdot P'$. Therefore, in view of Lemma 17.35, any $w\in W$ can be uniquely represented in the form $w=u(w)\cdot p(w)$, where $u(w)\in R_u(P)$ and $p(w)\in P'$, further $u(w)$ and $p(w)$ depend on w regularly and algebraically. Further, in view of Lemma 10.22, it is easy to see that $f(w)=u(w)$. The last two statements imply part *1)*.

Since $P \cap P'$ is a Levi subgroup in P (because P and P' are opposite), for any $u \in R\ (P)$, $u(P \cap P')u^{-1}$ is also a Levi subgroup in P. Now for any $u \in R_u(P)$ we have

$$(uP'u^{-1}) \cap P = (uP'u^{-1}) \cap (uPu^{-1}) = u(P' \cap P)u^{-1}.$$

The last two statements imply that for any $u \in R_u(P)$, $uP'u^{-1}$ and P are opposite parabolic subgroups. Therefore *1)* $R_u(P) \subset W$ *2)* the restriction of f to $R_u(P)$ is the identity mapping. This implies part *2)*. The proof of the lemma is complete.

Lemma 17.37. Let P be a maximal parabolic $\mathbf{Q}$-subgroup of the group G. Then the subgroup $P \cap G_{\mathbf{Z}} \cap \Gamma$ is of finite index in $P \cap G_{\mathbf{Z}}$.

Proof. In view of Remark 17.18 and Lemma 10.10 there is a parabolic $\mathbf{Q}$-subgroup P' opposite to P. Note that, since $P \neq G$, we have $P' \neq G$. Therefore, in view of Theorems 17.19 and 17.20 and the classification of parabolic $\mathbf{Q}$-subgroups (see Remark 17.18 and § 10), there is a $g_0 \in G_{\mathbf{Q}}$ such that the subgroups $(g_0 P' g_0^{-1}) \cap G_{\mathbf{Z}}$ and $(g_0 P' g_0^{-1}) \cap \Gamma$ are commensurable. Denote $g_0 P' g_0^{-1}$ by P_0. Since *1)* P and P' are opposite, *2)* $P_0 = g_0 P' g_0^{-1}$, *3)* as mentioned several times, Γ is Zariski dense in G, applying Lemma 10.13, we can easily obtain that there is a $\gamma_0 \in \Gamma$ such that P and $\gamma_0 P_0 \gamma_0^{-1}$ are opposite parabolic subgroups.

Denote $\gamma_0 P_0 \gamma_0^{-1}$ by $\hat{P}$, and note that *1)* since $g_0 \in G_{\mathbf{Q}}$ and $\gamma_0 \in G_{\mathbf{Q}}$ (in view of (1)), and P' is a parabolic $\mathbf{Q}$-subgroup, thus $\hat{P}$ is a parabolic $\mathbf{Q}$-subgroup, *2)* since the subgroups $(g_0 P' g_0^{-1}) \cap G_{\mathbf{Z}}$ and $(g_0 P' g_0^{-1}) \cap \Gamma$ are commensurable, and $\gamma_0 \in G_{\mathbf{Q}}$, in view of Lemma 17.34, the subgroups $\hat{P} \cap G_{\mathbf{Z}}$ and $\hat{P} \cap \Gamma$ are commensurable.

Denote by $W \subset G$ the set of all $g \in G$ for which P and $g\hat{P}g^{-1}$ are opposite parabolic subgroups. Since P and $\hat{P}$ are opposite, in view of Lemma 10.13, W is a non-empty Zariski open subset of the group G. Denote by f the mapping $f: W \to R_u(P)$ assigning to every $w \in W$ the unique (in view of Lemma 10.22) element $f(w) = u \in R_u(P)$ for which $w\hat{P}w^{-1} = u\hat{P}u^{-1}$.

Denote $\dim R_u(P)$ by n, $R_u(P)$ by V, and the direct products $\overbrace{W \times \ldots \times W}^{n}$ and $\overbrace{V \times \ldots \times V}^{n}$ by W^n and V^n. Then the mapping $f: W \to R_u(P) = V$ introduced above, induces in a natural way a mapping $f_n: W^n \to V^n$. Since in view of Lemma 17.36 f is a regular algebraic mapping and $f(W) = V^n$, it is easy to see that f_n is a regular algebraic mapping and $f_n(W^n) = V$ Therefore, for any non-empty Zariski open subset $B \subset V^n$, $f_n^{-1}(B)$ is a non-empty Zariski open subset in W^n.

Denote $P \cap \hat{P}$ by L. Since *1)* P is a parabolic $\mathbf{Q}$-subgroup (by assumption) *2)* as mentioned already, $\hat{P}$ is a parabolic $\mathbf{Q}$-subgroup, *3)* P and $\hat{P}$ are opposite (by the definition of $\hat{P}$), L is a Levi subgroup in P defined over $\mathbf{Q}$. Let, as above (see Lemmas 17.28 and 17.33), $\bar{L}_{\mathbf{Z}}^0$ denote the algebraically connected component of the identity in the algebraic closure $\bar{L}_{\mathbf{Z}}$ of the subgroup $L_{\mathbf{Z}} = L \cap G_{\mathbf{Z}}$, and $X \subset V^n$ the set of all $v = (v_1, \ldots, v_n) \in V^n$, for which the minimal algebraic subgroup containing $\{v, \bar{L}_{\mathbf{Z}}^0\}$, coincides with V ($\{v, \bar{L}_{\mathbf{Z}}^0\}$ denotes the union $\bigcup_{1 \leq i \leq n,\, l \in \bar{L}_{\mathbf{Z}}^0} (v_i l v_i^{-1} l^{-1})$).

Then, according to Lemma 10.33, X contains a non-empty Zariski open subset $\tilde{X}$ of the real algebraic manifold V^n. Therefore, according to the last sentence of the previous paragraph $f_n^{-1}(X)$ contains a non-empty Zariski open (in W^n) subset $f_n^{-1}(\tilde{X})$. Now, as we have mentioned earlier, W is Zariski open in G, and consequently, W^n is Zariski open in G^n, where G^n denotes the direct product $\overbrace{G\times\ldots\times G}^{n}$. Therefore, since $f_n^{-1}(\tilde{X})$ is non-empty and Zariski open in W^n and $f_n^{-1}(X)\supset f_n^{-1}(\tilde{X})$, $f_n^{-1}(X)$ contains a non-empty Zariski open (in G^n) subset $f_0^{-1}(\tilde{X})$.

Denote the direct product $\overbrace{\Gamma\times\ldots\times\Gamma}^{n}$ by Γ^n. As mentioned several times, Γ is Zariski dense in G, thus Γ^n is Zariski dense in G^n. Therefore, in view of the last sentence of the previous paragraph we have

$$\Gamma^n\cap f_n^{-1}(X)\neq\emptyset. \tag{60}$$

Let $\gamma=(\gamma_1, \ldots, \gamma_n)$ be an arbitrary element of the set $\Gamma^n\cap f_n^{-1}(X)$. For any i $(1\leqq i\leqq n)$ $f(\gamma_i)$ is denoted by v_i. Then, in view of the definition of mapping f_n, $v=(v_1, \ldots, v_n)$ is an element of the set X. Therefore, according to the definition of the set X, the minimal algebraic subgroup containing $\{v, \bar{L}_{\mathbf{Z}}^0\}$, coincides with $V=R_u(P)$.

For any i $(1\leqq i\leqq n)$, D_i denotes the intersection $(v_iLv_i^{-1})\cap G_{\mathbf{Z}}\cap\Gamma$. Since *1)* $L=P\cap\hat{P}$, *2)* for any i $(1\leqq i\leqq n)$, $v_i\in R_u(P)$ and $v_i=f(\gamma_i)$, according to the definition of mapping f, we have

$$\begin{aligned} v_iLv_i^{-1} &= v_i(P\cap\hat{P})v_i^{-1} = (v_iPv_i^{-1})\cap(v_i\hat{P}v_i^{-1}) = \\ &= P\cap(v_i\hat{P}v_i^{-1}) = P\cap(\gamma_i\hat{P}\gamma_i^{-1}). \end{aligned} \tag{61}$$

As mentioned at the beginning of the proof, the subgroups $\hat{P}\cap G_{\mathbf{Z}}$ and $\hat{P}\cap\Gamma$ are commensurable, thus, in view of Lemma 17.34 for any i $(1\leqq i\leqq n)$, the subgroups $(\gamma_i\hat{P}\gamma_i^{-1})\cap G_{\mathbf{Z}}$ and $(\gamma_i\hat{P}\gamma_i^{-1})\cap\Gamma$ are commensurable. Therefore, since for any i $(1\leqq i\leqq n)$ in view of (61) $v_iLv_i^{-1}\subset\gamma_i\hat{P}\gamma_i^{-1}$ holds, for any i $(1\leqq i\leqq n)$ the subgroups $(v_iLv_i^{-1})\cap G_{\mathbf{Z}}$ and $(v_iLv_i^{-1})\cap\Gamma$ are commensurable. From this we see that for any i $(1\leqq i\leqq n)$ the subgroup D_i is of finite index in $(v_iLv_i^{-1})\cap G_{\mathbf{Z}}$. Note that, since *1)* $L\subset\hat{P}$ (for, $L=P\cap\hat{P}$), *2)* as mentioned already, the subgroups $\hat{P}\cap G_{\mathbf{Z}}$ and $\hat{P}\cap\Gamma$ are commensurable, the subgroups $L\cap G_{\mathbf{Z}}$ and $L\cap\Gamma$ are commensurable, and consequently, the subgroup $D_0=L\cap G_{\mathbf{Z}}\cap\Gamma$ is of finite index in $L_{\mathbf{Z}}$. Denote the union $e\cup v$ by A, where e is the identity of the group G, and note that, since the minimal algebraic subgroup containing $\{v, \bar{L}_{\mathbf{Z}}^0\}$, coincides with $V=R_u(P)$ (see the previous section), the minimal algebraic subgroup containing $\{A, \bar{L}_{\mathbf{Z}}^0\}$ coincides with $V=R_u(P)$. The last three statements and Lemma 17.30 imply that the subgroups F generated by $\bigcup_{0\leqq i\leqq n} D_i$ is of finite index in $P_{\mathbf{Z}}=P\cap G_{\mathbf{Z}}$. Since for any i $(1\leqq i\leqq n)$, $D_i\subset\Gamma$ (for, $D_0=L\cap G_{\mathbf{Z}}\cap\Gamma$), and for $1\leqq i\leqq n$, $D_i=v_iLv_i^{-1}\cap G_{\mathbf{Z}}\cap\Gamma$, we have $F\subset\Gamma$. The last two assertions imply that the subgroup $P\cap\Gamma\cap G_{\mathbf{Z}}$ is of finite index in $P_{\mathbf{Z}}$. The lemma is proved.

From the classification of parabolic **Q**-subgroups and orispheric **Q**-subgroups (see Remark 17.18 and § 10) we easily get the following

Lemma 17.38. Let P be an arbitrary parabolic **Q**-subgroup of the group G. Then there is a maximal unipotent **Q**-subgroup V of G such that $P \supset N(V)$, where $N(V)$ is the normalizer of V in G.

Lemma 17.39. Let P be a maximal parabolic **Q**-subgroup of the group G. Then the subgroups $P \cap G_{\mathbf{Z}}$ and $P \cap \Gamma$ are commensurable.

Proof. According to Lemma 17.38 there is a maximal unipotent **Q**-subgroup V of the group G such that

$$P \supset N(V), \tag{62}$$

where $N(V)$ is the normalizer of V in G. Now, since V is a unipotent algebraic **R**-group defined over **Q**, the quotient space $V/V_{\mathbf{Z}}$ is compact (see 6.10 in [4]). On the other hand, since *1)* $P \cap G_{\mathbf{Z}} \cap \Gamma$ is a subgroup of finite index in $P \cap G_{\mathbf{Z}}$ (according to Lemma 17.37), *2)* $V \subset P$ (in view of (62)), we see that $V \cap G_{\mathbf{Z}} \cap \Gamma$ is a subgroup of finite index in $V_{\mathbf{Z}} = V \cap G_{\mathbf{Z}}$. The last two statements imply that the quotient space $V/V \cap \Gamma$ is compact. Therefore, since V is a unipotent algebraic **R**-group, in view of Lemma 5.6 the subgroup $V \cap \Gamma$ is Zariski dense in V.

Put $U = V \cap \Gamma$ and note that the last sentence of the previous paragraph is equivalent to the fact that

$$\overline{U} = V. \tag{63}$$

We show that U is a maximal unipotent subgroup of the group Γ. Assume, on the contrary, that in Γ there is a unipotent subgroup U', in which U is a proper subgroup. By $U' \subset \Gamma$ and $\Gamma \subset G_{\mathbf{Q}}$ (see inclusion (1)), we have $U' \subset G_{\mathbf{Q}}$. On the other hand, since U' is unipotent, in view of Lemma 1.3.3 the algebraic closure $\overline{U}'$ of the subgroup U is unipotent. The last two statements and Lemma 8.7 imply that $\overline{U}'$ is a unipotent **Q**-subgroup of the group G. Therefore, since *1)* $\overline{U}' \supset \overline{U}$ (for, $U' \supset U$), *2)* V is a maximal unipotent **Q**-subgroup of the group G, in view of (63) we have

$$\overline{U}' = V. \tag{64}$$

Since $U' = \Gamma$ and $U' \subset \overline{U}'$, (64) leads to $U' \subset \overline{U}' \cap \Gamma = V \cap \Gamma = U$. Consequently, U cannot be a proper subgroup in U', which is a contradiction. Thus U is a maximal unipotent subgroup of the group Γ. Now this, (62), (63) and Theorem 17.20 imply the assertion of the lemma.

Since every proper (i.e. different from G) parabolic **Q**-subgroup is contained in some maximal parabolic **Q**-subgroup, Lemma 17.39 implies the following

Theorem 17.40. *Let P be an arbitrary proper parabolic* **Q**-*subgroup of the group G. Then the subgroups $P \cap G_{\mathbf{Z}}$ and $P \cap \Gamma$ are commensurable.*

Remark 17.41. The assumption of the properness of P in Theorem 17.40 was made only because in case $P=G$ the author was not able to prove that the subgroups $P \cap G_{\mathbf{Z}}$ and $P \cap \Gamma$ are commensurable.

Remark 17.42. Strictly speaking, in all assertions of this section we should write $F_{\mathbf{Q}}, F_{\mathbf{Z}}, F_{\mathbf{C}}$ etc. rather than $G_{\mathbf{Q}}, G_{\mathbf{Z}}, G_{\mathbf{C}}$ etc., where F is the **Q**-form appearing in Theorem 16.17, mentioned also at the beginning of this section.

As a consequence of Remark 17.42 and Theorems 16.17, 17.21 and 17.40, we have the following

Theorem 17.43 *(Main Theorem).*

Let G be a semisimple algebraically connected real algebraic group such that 1) the centre of the complexification of G is trivial, 2) G has no compact factors, 3) the real rank of G is greater than or equal to two, and let Γ be a discrete subgroup of G such that 1) the quotient space G/Γ of the group G by the subgroup Γ is of finite volume, but is not compact, (i.e. Γ is a non-uniform lattice in G), *2) the pair* (G, Γ) *is irreducible. Then there is a* **Q***-form F of the group G, for which*

1) $\Gamma \subset F_{\mathbf{Q}}$

2) G is simple over **Q** *(with respect to the* **Q***-form F),*

3) for any proper (i.e. different from G) *parabolic subgroup P of G, defined over* **Q** *(with respect to the* **Q***-form F), the subgroups* $P \cap F_{\mathbf{Z}}$ *and* $P \cap \Gamma$ *are commensurable,*

4) the **Q***-rank* of G (with respect to the* **Q***-form F) is greater than zero.*

Remark 1. Part *4)* follows from Lemma 17.26 and Remark 17.18 and the section after Lemma 10.25.

Remark 2. The conditions made on G and Γ in the formulation of the Main Theorem, coincide with those made at the beginning of §§ 7 and 13.

§ 18. Appendix to §§ 10 and 13

I. Let $F \subset GL(n, \mathbf{R})$ $(GL(n, \mathbf{C}))$ be a real (complex) algebraic group of linear transformations of the space $\mathbf{R}^n (\mathbf{C}^n)$. The group F is said to be *fully reducible* if for every subspace $V \subset \mathbf{R}^n (\mathbf{C}^n)$, invariant with respect to F, there exists a subspace $V' \subset \mathbf{R}^n (\mathbf{C}^n)$, invariant with respect to F such that $\mathbf{R}^n = V \times V'$ $(\mathbf{C}^n = V \times V')$.

The following lemma is an easy consequence of 3.1 and 3.2 of [4].

Lemma 18.1. Let F be an algebraic subgroup of the group $GL(n, \mathbf{R})$ $(GL(n, \mathbf{C}))$. The following conditions are equivalent:

1) F is reductive;

2) F is fully reducible.

* i.e., the dimension of the maximal **Q**-decomposable torus of G.

Let now a positive-definite Hermitian form ψ be given on $\mathbf{R}^n(\mathbf{C}^n)$. We agree to say that the subgroup $F \subset GL(n, \mathbf{R})$ $(GL(n, \mathbf{C}))$ is self-adjoint with respect to ψ if for every $f \in F$, the subgroup F contains f^* too, where f^* denotes the linear transformation of $\mathbf{R}^n(\mathbf{C}^n)$ adjoint to f with respect to ψ. Let us formulate now as lemmas, certain results of Mostow on the relation between fully reducible subgroups and self-adjoint ones.

Lemma 18.2. Let F be an algebraic group of automorphisms of the rea (complex) space. F is self-adjoint with respect to some positive-definite Hermitian form if and only if F is fully reducible.

Lemma 18.3. Let $F_1 \supset \ldots \supset F_n$ be a decreasing sequence of fully reducible algebraic groups of automorphisms of the real (complex) linear space $\mathbf{R}^n(\mathbf{C}^n)$. Then there exists on $\mathbf{R}^n(\mathbf{C}^n)$ a positive-definite Hermitian form with respect to which $F_1, \ldots, F_n$ are simultaneously self-adjoint.

Remark. Lemma 18.2 combines Theorems 7.1 and 7.2 of [16], Lemma 18.3 is in turn an easy consequence of Theorems 7.1, 7.2 and 7.3 of [16].

II. In this section we are going to make some remarks concerning § 10.

Remark 18.4. Lemma 10.2 is a simple consequence of Lemmas 18.1, 18.2 and 18.3.

Remark 18.5. Let H be a semisimple real (complex) algebraic group over some field $K \subset \mathbf{R}$ ($K \subset \mathbf{C}$). Furthermore, let $T \subset H$ be any maximal K-decomposable torus and let Σ be the set of roots of H with respect to the torus T. In § 10 and the subsequent sections we quite often used implicitly a theorem of A. Borel and J. Tits (see Theorem 2, § II of [21] or Theorem 3.3 of [20]) which says that Σ is a root system (as for the definition of root systems, see § 7 of [19] or § 11 of [21]).

Remark 18.6. Statement *2)* of Lemma 10.5 follows from the fact that for every $\tilde{\alpha} \in \hat{\Sigma}$, $2\tilde{\alpha}$ is not a root. Statement 3 of Lemma 10.5 is a consequence of the fact that for simple H the root system Σ is irreducible (i.e. the Dynkin schema of H is connected) and of some well-known properties of root systems (see § 7 of [19]).

Remark 18.7. Lemma 10.4 easily follows from Corollary 5.9 of [5] and from the fact that the normalizer of any maximal unipotent subgroup of H is a minimal parabolic subgroup of H.

Remark 18.8. As for Lemmas 10.10 and 10.22, see 4.8 of [5]; for Lemma 10.11 see 4.3 of [5]; for Lemmas 10.12, 10.18 and 10.19, see 4.13, 4.14 and 4.16 of [5]. Lemma 10.13 follows from Proposition 4.10 and Lemma 4.12 of [5]. As for Lemma 10.14 see Proposition 4.10 of [5]; for Lemma 10.16, see Corollary 4.18 of [5]. Lemma 10.17 follows from Proposition 5.14 of [5] and Lemma 10.9.

Lemmas 10.9 and 10.20 easily follow from Lemmas 10.17 and 10.19 (and from the fact F is semi-direct product of a reductive and a unipotent group). Lemma 10.23 easily follows from Lemma 10.20.

Remark 18.9. Lemma 10.8 easily follows from Lemma 10.17 of the present paper and from 4.13, 4.14 and 4.16 of [5].

Remark 18.10. It is not hard to prove that the classification of parabolic subgroups introduced in § 10 coincides with that introduced in [5]. For this see 5.12 and 5.14 of [5].

Remark 18.11. Lemma 10.26 easily follows from Theorem 3.13 of [5].

Remark 18.12. Let us draw attention to one of the differences between the terminology of [5] and that of the present paper.

In the present paper by algebraic **R**-group an algebraic subgroup of the group $GL(n, \mathbf{R})$ is understood. In the terminology of [5] this is a group of the real points of some algebraic **C**-group defined over **R**. Nevertheless, it is not hard to prove that the results of [5] cited in the above remarks are valid for all algebraic **R**-groups as defined here.

III. In this section and the subsequent ones Lemma 10.5 and 10.29 will be proved. The notations will be those introduced in § 10 when formulating Lemma 10.4 and preparing Lemma 10.29. It is easy to see that Lemmas 10.5 and 10.29 are to be proved only if *1)* H is algebraically connected *2)* $Z(H_{\mathbf{C}})=e$. (In addition we note that the statements of § 10 will be used in the following in case $H=G$. In this case, however, $H=G$ is algebraically connected and $Z(H_{\mathbf{C}})=Z(G_{\mathbf{C}})=e$ (see conditions formulated at the beginning of § 7)). Then, on account of Lemma 1.4.2, H decomposes into the direct sum of algebraically connected simple algebraic **R**-groups whose centre of complexification is trivial. Therefore, as is easily seen, it is enough to prove Lemmas 10.5 and 10.29 in case *1)* H is simple, *2)* $Z(H_{\mathbf{C}})=e$, *3)* H is algebraically connected (this will be assumed in the following sections III and IV).

Proof of 1 *of Lemma* 10.5. Let $N(V)$ denote the normalizer V in G. Since the centre of any group is invariant with respect to all the automorphisms of the group, $Z(V)$ is a normal subgroup of $N(V)$, i.e.

$$N(V)\subset N(\mathfrak{Z}(V)), \tag{1}$$

where $N(\mathfrak{Z}(V))$ denotes the normalizer of $Z(V)$ in H. Denote by $\mathfrak{z}(V)$ the Lie algebra of the group $\mathfrak{Z}(V)$. Then from (1) it follows easily that

$$N(V)\subset N(\mathfrak{z}(V)), \tag{2}$$

where $N(\mathfrak{z}(V))$ denotes the normalizer of $\mathfrak{z}(V)$ in the group H.

Recall that for a real algebraic group F we denote by $F_{\mathbf{C}}$ its complexification. Let S be an arbitrary maximal torus of the group $(N(V))_{\mathbf{C}}$ which contains $T_{\mathbf{C}}$. Since V is a maximal unipotent subgroup of H, according to Lemma 10.17* $N(V)$ is a parabolic subgroup in H and consequently $(N(V))_{\mathbf{C}}$ is a parabolic subgroup in $H_{\mathbf{C}}$. Therefore (see 2.3 and 4.1 of [5]), $(N(V))_{\mathbf{C}}$ contains some maximal torus S' of the group $H_{\mathbf{C}}$. But then, taking into account the fact (see Theorem 4.21 of [5]) that in a complex algebraic group any two maximal tori are conjugate, we find that S and S' are conjugate and thus S is a maximal torus of $H_{\mathbf{C}}$.

Denote now by D a minimal algebraic subgroup of $H_{\mathbf{C}}$, containing the union $S \cup V_{\mathbf{C}}$. Since *1)* $(N(V))_{\mathbf{C}}$ normalizes $V_{\mathbf{C}}$ because $N(V)$ normalizes V, *2)* $S \subset (N(V))_{\mathbf{C}}$, *3)* $V_{\mathbf{C}}$ is nilpotent (see Lemma 1.3.1), *4)* S is connected (because it is a torus), *5)* $V_{\mathbf{C}}$ is connected, we obtain that D is a connected solvable algebraic subgroup of $H_{\mathbf{C}}$. Therefore D is contained in some Borel subgroup B of $H_{\mathbf{C}}$. Since (see the previous paragraph) the torus S is maximal in $H_{\mathbf{C}}$ and so is in B, and B is a connected solvable algebraic **C**-group, we get that (see [1] and [3])

$$B = S \tilde{\times} W, \tag{3}$$

where W is the unipotent radical of B and $\tilde{\times}$ denotes the semidirect product (in the sense of § 1). Since $V_{\mathbf{C}}$ is unipotent and belongs to B, (3) implies that

$$V_{\mathbf{C}} \subset W. \tag{4}$$

Denote by $\Sigma_{\mathbf{C}}$ the root system of $H_{\mathbf{C}}$ corresponding to the torus S. Since B is a Borel subgroup of $H_{\mathbf{C}}$ and S is a maximal torus of $H_{\mathbf{C}}$ contained by B, we obtain that (see 2.3 of [5]) it is possible to introduce an order ψ on $\Sigma_{\mathbf{C}}$ such that $\alpha \in \Sigma_{\mathbf{C}}^{+}$ ($\Sigma_{\mathbf{C}}^{+}$ denotes the set of all roots positive with respect to the order ψ) if and only if $N_\alpha \subset W$, where $N_\alpha = \exp(\mathfrak{N}_\alpha)$ and $\mathfrak{N}_\alpha$ is the subspace corresponding to the root α. Denote by $\hat{\Sigma}_{\mathbf{C}}$ the set of the highest roots in $\Sigma_{\mathbf{C}}^{+}$, by $Z(W)$ the centre of W and by $\mathfrak{Z}(W)$ the Lie algebra of $Z(W)$. As S is a maximal torus in the complex algebraic group $H_{\mathbf{C}}$, for any two $\alpha, \beta \in \Sigma_{\mathbf{C}}^{+}$ such that $\alpha + \beta$ is a root we have

$$[\mathfrak{N}_\alpha, \mathfrak{N}_\beta] = \mathfrak{N}_{\alpha+\beta}, \tag{5}$$

where $[\mathfrak{N}_\alpha, \mathfrak{N}_\beta]$ denotes the commutator of $\mathfrak{N}_\alpha$ and $\mathfrak{N}_\beta$ (regarding (5), see [22] and 2.5 of [5]). From (5), it is easy to deduce that

$$\mathfrak{Z}(W) = \bigoplus_{\alpha \in \hat{\Sigma}_{\mathbf{C}}} \mathfrak{N}_\alpha. \tag{6}$$

Now we are going to prove that for any $\alpha' \in \hat{\Sigma}_{\mathbf{C}}$ and $\alpha'' \in \hat{\Sigma}_{\mathbf{C}}$ we have

$$\alpha'_T = \alpha''_T, \tag{7}$$

* Although Lemmas 10.8—10.23 are formulated after Lemma 10.5, we still may use them since (see II of the present §) for their proof Lemmas 10.5—10.7 are not used.

where α'_T and α''_T denote the restrictions to T of α' and α'', respectively. In proving this we distinguish two cases: *1)* the group H is absolutely simple (i.e. $H_{\mathbf{C}}$ is almost simple), *2)* H is not absolutely simple.

Case 1. Since $H_{\mathbf{C}}$ is almost simple, the system $\Sigma_{\mathbf{C}}$ is irreducible and thus (see § 7 of [19]) $\hat{\Sigma}_{\mathbf{C}}$ consists of one root alone from which (7) is immediate.

Case 2. Since *1)* $H_{\mathbf{C}}$ is semisimple because H is simple, *2)* $Z(H_{\mathbf{C}})=e$ as assumed at the end of the first paragraph of this section, *3)* $H_{\mathbf{C}}$ is not almost simple, by making use of Lemma 1.4.1 we easily obtain the decomposition

$$H_{\mathbf{C}} = H_1 \times \ldots \times H_i \quad (i \geqq z), \tag{8}$$

where $H_1, \ldots, H_i$ are simple algebraic **C**-groups of positive dimension. Now H is a simple algebraic **R**-group. Therefore, any (not necessarily algebraic) proper normal subgroup of H reduces to e. Thus (8) implies that we have

$$\text{either } H \cap H_j = e \quad \text{or} \quad H \cap H_j = H. \tag{9}$$

Denote by φ the unique nontrivial automorphism of **C** which leaves the elements of the field **R** invariant and associate with φ an automorphism φ_H of H in the same way as in Lemma 16.9 in case ψ and ψ_G. As the set of the fixed points of φ coincides with **R**, we conclude that any algebraic subgroup of H which is invariant with respect to φ_H is defined over **R**. Moreover, as is easily seen, (8) is the unique decomposition of $H_{\mathbf{C}}$ into a product of simple (in abstract group-theoretical sense) groups, thus for every j we have

$$\varphi_H(H_j) = H_{k(j)} \tag{10}$$

for some $k(j)$ $(1 \leqq k(j) \leqq i)$. From the last two statements and the fact that $\varphi_H^2 = E$ (because $\varphi^2 = E$) it follows that $H_1 \times H_{k(1)}$ is an algebraic subgroup of $H_{\mathbf{C}}$ defined over **R**. $(k(1) \neq 1$ because, in the contrary case, in virtue of (8) and the algebraic connectedness of H, H_1 would be a proper algebraic normal subgroup over **R** of positive dimension which is impossible as H is simple.) On the other hand, since $H_{\mathbf{C}}$ is connected (because H is algebraically connected), each H_j $(1 \leqq j \leqq i)$ is connected, too. The last two statements and Lemma 1.2.11 imply that the group $(H_1 \times H_{k(1)})_{\mathbf{R}}$ of the real points of $H_1 \times H_{k(1)}$ is Zariski dense in $H_1 \times H_{k(1)}$. Thus, $H \cap (H_1 \times H_{k(1)}) \neq \emptyset$. On the other hand, (8) implies that $H \cap (H \times H_{k(1)})$ is a normal subgroup of H. The last two statements and the simplicity of H give

$$H \cap (H_1 \times H_{k(1)}) = H. \tag{11}$$

Since $H_{\mathbf{C}}$ is the complexification of H, we can conclude from (11)

$$H_{\mathbf{C}} = H' \times H'', \tag{12}$$

where $H' = H_1$ and $H'' = H_{k(1)}$. Since *1)* $\varphi_H(H') \cap H' = e$ (because $\varphi_H(H') = \varphi_H(H_1) = H_{k(1)}$ and $k(1) \neq 1$), *2)* the restriction to H of φ_H is the identity transformation, we have

$$H \cap H' = e. \tag{13}$$

Similarly we obtain

$$H \cap H'' = e. \tag{14}$$

As the set of the fixed points of φ_H is identical with H and $\varphi_H^2 = E$, from (12), (13) and (14) it follows easily that

$$H = \{h \in H_C : h = h'\varphi(h'), h' \in H'\}, \tag{15}$$

which means that H consists of exactly those points of H_C which admit a representation $h = h'\varphi(h')$ with $h' \in H'$.

Denote by π_1 and π_2 the unique projections of H_C onto H' and H'', respectively, and by $\tilde{\pi}_1$ and $\tilde{\pi}_2$ their restrictions to H. From (15) it follows easily that if H' and H'' are considered as real algebraic groups, then $\tilde{\pi}_1$ and $\tilde{\pi}_2$ are algebraic isomorphisms. Therefore, since V is a maximal unipotent subgroup of H, we get that $\tilde{\pi}_1(V)$ and $\tilde{\pi}_2(V)$ are maximal unipotent subgroups of H' and H'', respectively (here we have made use of the simple fact that if an element u of a complex algebraic group F is unipotent if F is considered as a real algebraic group, then u is also a unipotent element of the complex algebraic group F). From this it is easy to conclude that $\tilde{\pi}_1(V) \times \tilde{\pi}_2(V)$ is a maximal unipotent subgroup of H_C and consequently, the complex dimension of a maximal unipotent subgroup of H_C equals the real dimension of the maximal unipotent subgroup V of H.

Therefore, as is easily seen, V_C is a maximal unipotent subgroup of H_C. From this and (4) it follows that

$$V_C = W. \tag{16}$$

(16) and the fact that $\varphi_H(V) = V$ (because $V \subset H$) imply

$$\varphi_H(W) = W. \tag{17}$$

Let $W' = W \cap H'$ and $W'' = W \cap H''$ and let $Z(W')$ (resp. $Z(W'')$) denote the centre of W' (resp. W''). It is easy to see that

$$W = W' \times W''. \tag{18}$$

This implies that

$$Z(W) = Z(W') \times Z(W''). \tag{19}$$

Since $\varphi_H(H') = H''$, from (17) and (19) we can conclude

$$\varphi_H(Z(W')) = Z(W'). \tag{20}$$

We note now that *1)* as W is a maximal unipotent subgroup of H_C, W' and W'' are maximal unipotent subgroups of H' and H'', *2)* S being a maximal torus in H_C, $S' = S \cap H'$ and $S'' = S \cap H''$ are maximal tori in H' and H'', *3)* since S normalizes W (see (3)), S' normalizes W' and S'' normalizes W''. This fact, the simplicity of H' and H'' and the complex analogue of *3)* in Lemma 10.5 and (5) imply that

$$Z(W') = N_{\alpha'}, \quad Z(W'') = N_{\alpha''}, \tag{21}$$

where

$$\alpha', \alpha'' \in \hat{\Sigma}_{\mathbf{C}}.$$

From (20) and (21) we can conclude that

$$\varphi_H(N_{\alpha'}) = N_{\alpha''}. \tag{22}$$

Since the elements of the torus T are invariant with respect to the automorphism ψ, we get from (22) that

$$\alpha'_T = \alpha''_T \tag{23}$$

holds true, where α'_T (resp. α''_T) denotes the restriction of α' (resp. α'') to T. On the other hand, (6), (19) and (21) imply

$$\hat{\Sigma}_{\mathbf{C}} = \{\alpha', \alpha''\}. \tag{24}$$

From (23) and (24) we get (7). Thus (7) is proved in both cases.

Since $\tilde{\alpha}$ is the highest root in Σ^+,

$$\mathfrak{Z}(V) \supset \mathfrak{A}_{\tilde{\alpha}} \tag{25}$$

easily follows. Suppose that

$$\mathfrak{Z}(V) \neq \mathfrak{A}_{\tilde{\alpha}}. \tag{26}$$

Then, on account of $T \subset N(V)$, from (2), (25) and (26) we can conclude that for some $\alpha_0 \in \Sigma^+$ different from $\tilde{\alpha}$ we have

$$\mathfrak{Z}(V) \cap \mathfrak{A}_{\alpha_0} \neq 0. \tag{27}$$

Since V is a maximal unipotent subgroup of H, Lemmas 10.17 and 10.18 imply

$$N(V) = C(T) \rtimes V. \tag{28}$$

Furthermore, *1)* $C(T) \subset N(\mathfrak{A}_{\alpha_0})$ (for $C(T)$ commutes with T), *2)* $C(T) \subset N(\mathfrak{Z}(V))$ on account of (28) and (2). Therefore we have

$$C(T) \subset N(\mathfrak{Z}(V) \cap \mathfrak{A}_{\alpha_0}). \tag{29}$$

On the other hand, the fact that V commutes with $Z(V)$ shows that

$$V \subset N(\mathfrak{Z}(V) \cap \mathfrak{A}_{\alpha_0}). \tag{30}$$

From (28), (29) and (30) it follows that

$$N(V) \subset N(\mathfrak{Z}(V) \cap \mathfrak{A}_{\alpha_0}). \tag{31}$$

Denote by L the complexification of the linear space $\mathfrak{Z}(V) \cap \mathfrak{A}_{\alpha_0}$. In virtue of the fact that $W \subset (N(V))_{\mathbf{C}}$, from (31) we get that

$$W \subset N(L). \tag{32}$$

(32) and the unipotence of W imply that there is in L a nonzero vector invariant under W. Then, taking into account that L belongs to the Lie algebra $\mathfrak{B}_{\mathbf{C}}$ of the group $V_{\mathbf{C}}$ (because of $Z(V) \subset V$), we get from (16)

$$\mathfrak{l} \in \mathfrak{Z}(W) \cap L, \quad \mathfrak{l} \neq 0. \tag{33}$$

From (6) and (33) one can easily deduce the existence of an $\alpha_1 \in \hat{\Sigma}_C$ for which $(\alpha_1)_T = \alpha_0$. By making use of (25), we can prove the existence of an $\alpha_2 \in \hat{\Sigma}_C$ with $(\alpha_2)_T = \tilde{\alpha}$ in a similar way. But the last two statements contradict (7). Thus (26) led to a contradiction. The proof of Lemma 10.5 is complete.

IV. If L_1 and L_2 are subspaces of a linear space L, $L_1 + L_2$ will stand for their sum (i.e. $L_1 + L_2$ consists of all elements of the form $l_1 + l_2$ with $l_1 \in L_1$, $l_2 \in L_2$). For the proof of Lemma 10.29 we need

Lemma 18.13. Let $\mathfrak{G}$ be a Lie algebra over a field K, $\mathfrak{A}$ a subalgebra of $\mathfrak{G}$ and $\mathfrak{N}$ a linear subspace of $\mathfrak{G}$. Denote by $\overline{\mathfrak{N}}$ the subalgebra of $\mathfrak{G}$ generated by $\mathfrak{N}$. If $\mathfrak{A} + \mathfrak{N}$ is invariant with respect to ad $\mathfrak{A}$, then $\mathfrak{A} + \overline{\mathfrak{N}}$ is a subalgebra.

Proof. If L is a linear subspace of $\mathfrak{G}$, denote by L' the smallest linear subspace of $\mathfrak{G}$ which contains L and all the commutators of form $[l_1, l_2]$ with $l_1, l_2 \in L$. Set $\mathfrak{N}_1 = \mathfrak{N}'$ and $\mathfrak{N}_{i+1} = \mathfrak{N}_i'$ for $i \geqq 1$. Let $x, y \in \mathfrak{N}$ and $\mathfrak{a} \in \mathfrak{A}$. Then the Jacobi identity shows that

$$[\mathfrak{a}, [x, y]] = -[y, [\mathfrak{a}, x]] - [x, [y, \mathfrak{a}]]. \tag{34}$$

Since $\mathfrak{A} + \mathfrak{N}$ is supposed invariant with respect to ad $\mathfrak{A}$, we get

1) $[\mathfrak{a}, x] = \mathfrak{a}_1 + x_1$, $[\mathfrak{a}, y] = \mathfrak{a}_2 + y_1$ with $\mathfrak{a}_1, \mathfrak{a}_2 \in \mathfrak{A}$, $x_1 y_1 \in N$;

2) $[y, [\mathfrak{a}, x]] = [y, \mathfrak{a}_1 + x_1] = [y, \mathfrak{a}_1] + [y, x_1] \in \mathfrak{A} + \mathfrak{N}'$,

and analogously $[x, [y, \mathfrak{a}]] \in \mathfrak{A} + \mathfrak{N}'$. These together imply $[\mathfrak{a}, [x, y]] \in \mathfrak{A} + \mathfrak{N}'$. From this the invariance of $\mathfrak{A} + \mathfrak{N}' = \mathfrak{A} + \mathfrak{N}_1$ with respect to ad $\mathfrak{A}$ immediately follows. An induction argument will show the invariance of all other subspaces $\mathfrak{A} + \mathfrak{N}_i$ for any i.

Consequently, as $\overline{\mathfrak{N}} = \bigcup \mathfrak{N}_i$, we see that $\mathfrak{A} + \overline{\mathfrak{N}}$ is invariant with respect to ad $\mathfrak{A}$. This means that for every $x \in \overline{\mathfrak{N}}$ and $\mathfrak{a} \in \mathfrak{A}$ we have

$$[x, \mathfrak{a}] \in \overline{\mathfrak{N}} + \mathfrak{A}. \tag{35}$$

From (35) and the fact that $\overline{\mathfrak{N}}$ and $\mathfrak{A}$ are subalgebras we obtain the assertion of the lemma.

Proof of Lemma 10.29. We are going to use the notations introduced in § 10 just before Lemma 10.25. Furthermore, denote by $\mathfrak{H}$, $\mathfrak{B}^+$, $\mathfrak{B}^-$ and $\mathfrak{C}$ the Lie algebras of the groups H, V^+, V^- and $C(T)$, respectively, and denote by $\mathfrak{B}_{\Sigma_0}$ the linear space generated by $\bigcup_{\alpha \in \Sigma_0} \mathfrak{A}_\alpha$, and by $\overline{\mathfrak{B}}_{\Sigma_0}$ the subalgebra generated by $\mathfrak{B}_{\Sigma_0}$. Let us prove first that

$$\overline{\mathfrak{B}}_{\Sigma_0} = \mathfrak{B}^+. \tag{36}$$

Indeed, *1)* it is easy to see that $\mathfrak{C}$ normalizes $\mathfrak{B}_{\Sigma_0}$, *2)* since Σ_0 is the set of the simple roots in Σ^+, we obtain that for any $\alpha \in \Sigma_0$ and $\beta \in \Sigma^-$ the commutator $[\mathfrak{A}_\alpha, \mathfrak{A}_\beta]$ belongs to $\mathfrak{C} + \mathfrak{B}^-$, and so for any $v_1 \in \mathfrak{B}^-$ and $v_2 \in \mathfrak{B}_{\Sigma_0}$ we have $[v_1, v_2] \in \mathfrak{C} + \mathfrak{B}^-$. Therefore for any $\mathfrak{b} \in \mathfrak{C} + \mathfrak{B}^-$ the subspace $\mathfrak{C} + \mathfrak{B}^- + \mathfrak{B}_{\Sigma_0}$ is invariant with respect to ad $\mathfrak{b}$. On the other hand, the definition of $\mathfrak{C}$ and $\mathfrak{B}^-$ shows that $\mathfrak{A}=\mathfrak{C}+\mathfrak{B}^-$ is a subalgebra. From the last two statements and Lemma 18.13 it follows that $\mathfrak{B}=\mathfrak{A}+\overline{\mathfrak{B}}_{\Sigma_0}$ is also a subalgebra.

Denote by N the normalizer of the subalgebra $\mathfrak{B}$ in the group H.

The adjoint representation of H being algebraic, N is an algebraic subgroup of H. We note now that *1)* $C(T)$ normalizes $\mathfrak{B}^-$, *2)* $C(T)$ normalizes $\mathfrak{C}$, *3)* $C(T)$ normalizes V_{Σ_0} and consequently $\overline{V}_{\Sigma_0}$, too. Therefore $C(T)$ normalizes $\mathfrak{B}$. On the other hand, since V^- is algebraically connected (because V^- is unipotent) and the Lie algebra $\mathfrak{B}^-$ of the group V^- belongs to $\mathfrak{B}$, we see that V^- normalizes $\mathfrak{B}$. The last two statements imply $N\supset C(T)\cup V^-$, and consequently, taking into account the fact that *1)* V^- is a maximal unipotent subgroup of H, *2)* $C(T)$ is the centralizer of a maximal torus splitting over $\mathbf{R}$, *3)* $C(T)$ normalizes V^-, in virtue of Lemmas 10.17 and 10.18, they also imply that N contains a parabolic subgroup of H. Being an algebraic subgroup of H, N is parabolic. Now *1)* $\mathfrak{B}_{\Sigma_0}\subset\mathfrak{B}$ implies $N\supset \bigcup_{\alpha\in\Sigma_0} U_\alpha$, *2)* $N\supset V^-$ (as mentioned already) implies $N\supset \bigcup_{\alpha\in\Sigma^-} U_\alpha$, *3)* as $N\supset T$ (because of $N\supset C(T)$) we get that N is normalized by T. From the last two statements and Lemma 10.25 it follows easily that $N\supset \bigcup_{\alpha\in\Sigma} U_\alpha \cup C(T)$. Therefore, the Lie algebra of N contains $\mathfrak{H}$. The algebraic connectedness of H and the fact that N is an algebraic subgroup of H imply $N=H$. Therefore, since H is semisimple and N is the normalizer of $\mathfrak{B}$ in H, we obtain $\mathfrak{B}=H$. But $\mathfrak{B}=\mathfrak{A}+\overline{\mathfrak{B}}_{\Sigma_0}=\mathfrak{C}+\mathfrak{B}^-+\mathfrak{B}^-_{\Sigma_0}$. From this (36) easily follows.

For any $D\subset\Sigma_0$, we denote by $\mathfrak{B}_D$ the linear subspace generated by $\bigcup_{\alpha\in D}\mathfrak{A}_\alpha$, by $\overline{\mathfrak{B}}_D$ the ideal generated by $\mathfrak{B}_D$, of the Lie algebra $\mathfrak{B}^+$, by $\hat{\mathfrak{B}}_D$ the subalgebra generated by $\mathfrak{B}_D$, and by $\mathfrak{H}_D$ the Lie algebra of the group H_D (let us recall that we are using the notations of § 10). Then, as is easily seen, we have

$$\mathfrak{B}_D\subset\mathfrak{H}_D \tag{37}$$

and

$$\hat{\mathfrak{B}}_{\Sigma_0-D}\cap\mathfrak{H}_D=\emptyset. \tag{38}$$

On the other hand, since $\overline{\mathfrak{B}}_D$ is an ideal of $\mathfrak{B}^+$, we have

$$\overline{\mathfrak{B}}_{\Sigma_0}=\overline{\mathfrak{B}}_D+\hat{\mathfrak{B}}_{\Sigma_0-D}. \tag{39}$$

From (36), (37), (38) and (39) it follows easily that

$$\overline{\mathfrak{B}}_D=\mathfrak{H}_D. \tag{40}$$

From (40) and the algebraic connectedness of H_D (see Lemma 1.3.1) the statement of Lemma 10.29 easily follows.

V. *Proof of Lemma* 13.14. Since H is an algebraic $\mathbf{R}$-group, for some n, the group H is an algebraic subgroup of $GL(n,\mathbf{R})$. Now it is easy to see that H and

T are reductive. Therefore, on account of Lemmas 18.1, 18.2 and 18.3, there exists on $\mathbf{R}^n$ a positive-definite Hermitian form ψ with respect to which H and T are simultaneously self-adjoint. Denote by I_T the restriction to H of the operation of taking adjoint with respect to ψ. Since H and T are self-adjoint with respect to ψ, we get that I_T is an involutive automorphism of H, transforming T into itself. Furthermore, since the torus T is self-adjoint with respect to ψ, we obtain that all elements of T are of diagonal form in the same orthonormal (with respect to ψ) basis of $\mathbf{R}^n$. From this and the fact that all eigenvalues of any element of T are real, it follows easily that for any $t \in T$ we have $I_T(t)=t$. All these facts imply that I_T is the desired automorphism. The lemma is proved.

§ 19. Rigidity of nonuniform lattices

I. Let us begin with some well-known definitions.

Definition 19.1. Let H be a complex (real) Lie group. Let Λ be a countable group and T^t $(0 \leqq t \leqq 1)$ a family of representations of Λ in H. We agree to say that the family T^t is *continuous at a point* x $(0 \leqq x \leqq 1)$ if for every $\lambda \in \Lambda$ we have $\lim_{t \to x} T^t(\lambda) = T^x(\lambda)$.

Definition 19.2. Let H, Λ and T^t be the same as above. The family T^t is *trivial at* x $(0 \leqq x \leqq 1)$ if there exist an $\varepsilon > 0$ and a family h_t $(x-\varepsilon \leqq t \leqq x+\varepsilon)$ of elements of H such that *1)* for $t \in [x-\varepsilon, x+\varepsilon]$ we have $T^t(\Lambda) = h_t T^x(\Lambda) h_t^{-1}$, *2)* $\lim_{t \to x} h_t = e$.

Definition 19.3. Let H be a complex (real) Lie group and Λ a countable subgroup of H. A *deformation* of the subgroup Λ in H is a family of representations T^t $(0 \leqq t \leqq 1)$ of Λ in H such that *1)* for $t=0$ and for any $\lambda \in \Lambda$ we have $T^t(\lambda)=\lambda$, *2)* the family T^t is continuous at the point 0.

Definition 19.4. Let H and Λ be the same as in Definition 19.3. Λ is said to be a *rigid* subgroup of H if any deformation T^t $(0 \leqq t \leqq 1)$ of Λ in H is trivial at $t=0$.

II. Let us recall that G and Γ satisfy the conditions formulated at the beginning of §§ 7 and 13.

Theorem 19.5. *Γ is a rigid subgroup of G.*

Theorem 19.6. *Let G' be an algebraically connected semisimple real algebraic group such that 1) G' has no compact factors, 2) the centre of the complexification of G' is trivial, and let Γ be a lattice in G'. Then any isomorphism φ (in the abstract group-theoretical sense) between Γ and Γ' can be extended to an algebraic isomorphism between G and G'.*

The proof of these theorems will be presented in sections IV and V of this §. Both of these theorems will be deduced from assertions presented in III (actually from Theorem 19.10 and Lemma 19.7).

III. As in § 17, from now on we suppose that G is defined over $\mathbf{Q}$ in such a way that (1) of § 17 is satisfied.

Recall that for any parabolic subgroup P of G we denote by $R_u(P)$ its unipotent radical. Also recall that a maximal parabolic $\mathbf{Q}$-subgroup of G is a parabolic $\mathbf{Q}$-subgroup which is maximal in the class of nontrivial (different from G) parabolic $\mathbf{Q}$-subgroups of G.

Lemma 19.7. Let P_1 and P_2 be opposite parabolic $\mathbf{Q}$-subgroups of G. Denote $P_1 \cap P_2$ by L, the algebraic closure of the subgroup $L_{\mathbf{Z}} = L \cap G_{\mathbf{Z}}$ in G by $\bar{L}_{\mathbf{Z}}$, and the algebraically connected component of the identity of $\bar{L}_{\mathbf{Z}}$ by $\bar{L}^0_{\mathbf{Z}}$. Let T be a representation of Γ in a complex algebraic group H. Denote the algebraic closure in H of the subgroup $T(R_u(P_1) \cap \Gamma)$ (resp. $T(R_u(P_2) \cap \Gamma)$ by V_1 (resp. V_2), the algebraically connected component of the identity of the group V_1 (resp. V_2) by V_1^0 (resp. V_2^0), and the centralizer of the subgroup V_1^0 (resp. V_2^0) in H by $C(V_1^0)$ (resp. $C(V_2^0)$). Suppose that *1)* P_1 is a maximal parabolic $\mathbf{Q}$-subgroup of G, *2)* $C(V_1^0) \cap C(V_2^0) = e$. Then

1) there can be found in $R_u(P_1) \cap \Gamma$ (resp. $R_u(P_2) \cap \Gamma$) a subgroup U_1 (resp. U_2) of finite index such that the restriction of T to U_1 (resp. U_2) can be extended to an algebraic representation in H of the complexification $(R_u(P_1))_{\mathbf{C}}$ (resp. $(R_u(P_2))_{\mathbf{C}}$) of $R_u(P_1)$ (resp. $R_u(P_2)$):

2) the restriction of T to $\bar{L}^0_{\mathbf{Z}} \cap \Gamma$ can be extended to a representation in H of the complexification $L' = (\bar{L}^0_{\mathbf{Z}})_{\mathbf{C}}$ of $\bar{L}^0_{\mathbf{Z}}$.

Proof. First of all we note that, since P_1 is a maximal parabolic $\mathbf{Q}$-subgroup of G and P_2 is a parabolic $\mathbf{Q}$-subgroup opposite to P_1, the classification of parabolic $\mathbf{Q}$-subgroups (see Remark 17.18 and § 10) easily gives that P_2 is also a maximal parabolic $\mathbf{Q}$-subgroup of G. Let us present now some remarks.

Remark 1. Since P_1 and P_2 are opposite parabolic $\mathbf{Q}$-subgroups of G, L is a Levi subgroup defined over $\mathbf{Q}$ both in P_1 and P_2. Therefore, by making use of the fact that P_1 and P_2 are maximal parabolic $\mathbf{Q}$-subgroups of the group G, we obtain that (see Lemma 17.28) the algebraic commutators $\overline{(\bar{L}^0_{\mathbf{Z}}, R_u(P_1))}$ and $\overline{(\bar{L}^0_{\mathbf{Z}}, R_u(P_2))}$ coincide with $R_u(P_1)$ and $R_u(P_2)$, respectively.

Remark 2. Since P_1 and P_2 are defined over $\mathbf{Q}$, in virtue of Lemma 1.3.5 the unipotent algebraic $\mathbf{R}$-groups $R_u(P_1)$ and $R_u(P_2)$ are also defined over $\mathbf{Q}$. Therefore, taking into account 6.10, we get that the quotient spaces $R_u(P_1)/(R_u(P_1) \cap G_{\mathbf{Z}})$ and $R_u(P_2)/(R_u(P_2) \cap G_{\mathbf{Z}})$ are compact. From this and 3 of Theorem 17.43 the compactness of the quotient spaces $R_u(P_1)/R_u(P_1) \cap \Gamma$ and $R_u(P_2)/R_u(P_2) \cap \Gamma$ follows. Therefore, the fact that $R_u(P_1)$ and $R_u(P_2)$ are unipotent algebraic $\mathbf{R}$-groups implies, on account of Lemma 5.6, that the subgroups $R_u(P_1)\cap\Gamma$ and $R_u(P_2)\cap\Gamma$ are Zariski dense in $R_u(P_1)$ and $R_u(P_2)$, respectively.

Remark 3. Denote the algebraic closures in the complexification G_C of G of the subgroups $P_1\cap G_Z$ and $P_2\cap G_Z$ by S_1 and S_2, and the algebraically connected components of the identities of S_1 and S_2 by S_1^0 and S_2^0, respectively. Then, in virtue of Lemmas 1.2.7 and 8.8, $S_1^0\cap G_Z$ (resp. $S_2^0\cap G_Z$) is Zariski dense in S_1^0 (resp. S_2^0). Therefore, since *1)* $S_1^0\subset(P_1)_C$, *2)* $S_2^0\subset(P_2)_C$ ($(P_1)_C$ and $(P_2)_C$ are complexifications of P_1 and P_2, respectively), *3)* as it can be seen easily, $(P_1)_C\cap G_Z=P_1\cap G_Z$ and $(P_2)_C\cap G_Z=P_2\cap G_Z$, we obtain, in virtue of 3 of Theorem 17.43 and Lemma 8.8, that $S_1^0\cap\Gamma$ (resp. $S_2^0\cap\Gamma$) is Zariski dense in S_1^0 (resp. S_2^0).

Remark 4. Since the quotient spaces $R_u(P_1)/(R_u(P_1)\cap G_Z)$ and $R_u(P_2)/(R_u(P_2)\cap G_Z)$ are compact (see Remark 2), taking into account Lemma 5.6 and the definition of complexification we obtain that the subgroup $R_u(P_1)\cap G_Z$ and $R_u(P_2)\cap G_Z$ are Zariski dense in $(R_u(P_1))_C$ and $(R_u(P_2))_C$, respectively. This fact and the algebraic connectedness of the groups $R_u(P_1)$ and $R_u(P_2)$ (see Lemma 1.3.2) imply that $S_1^0\supset(R_u(P_1))_C$ and $S_2^0\supset(R_u(P_2))_C$. Moreover, it is easy to see that $(R_u(P_1))_C$ and $(R_u(P_2))_C$ are unipotent algebraic normal subgroups of S_1^0 and S_2^0 respectively.

Remark 5. Remarks 1—4, the connectedness of the groups S_1^0 and S_0^2 and Lemma 15.2 imply that there exists a subgroup U_1 (resp. U_2) in $R_u(P_1)\cap\Gamma$ (resp. $R_u(P_2)\cap\Gamma$) such that *1)* $T(U_1)$ and $T(U_2)$ are unipotent, *2)* U_1 (resp. U_2) is algebraically dense in $R_u(P_1)$ (resp. $R_u(P_2)$). Then *1)* on account of Lemma 5.11, the restrictions of T to U_1 and U_2 can be extended to algebraic representations of the complex algebraic groups $(R_u(P_1))_C$ and $(R_u(P_2))_C$ in the complex algebraic group H, *2)* on account of Lemma 5.6, the quotient spaces $R_u(P_1)/U_1$ and $R_u(P_2)/U_2$ are compact, and so U_1 and U_2 are subgroups of finite index in $R_u(P_1)\cap\Gamma$ and $R_u(P_2)\cap\Gamma$. From this it follows that U_1 and U_2 satisfy the requirements of *1)* of the lemma.

Remark 6. The restrictions of T to U_1 and U_2 have algebraic extensions to $(R_u(P_1))_C$ and $(R_u(P_2))_C$. Denote these representations by F_1 and F_2 and their images $F_1((R_u(P_1))_C)$ and $F_2((R_u(P_2))_C)$ by E_1 and E_2. Then, since *1)* U_1 and U_2 are Zariski dense in $R_u(P_1)$ and $R_u(P_2)$ respectively (see Remark 5), *2)* $R_u(P_1)$ and $R_u(P_2)$ are algebraically connected, we have $E_1=V_1^0$ and $E_2=V_2^0$.

Remark 7. Denote the Lie algebras of the groups $(R_u(P_1))_C$, $(R_u(P_2))_C$, V_1^0 and V_2^0 by $\mathfrak{R}_1$, $\mathfrak{R}_2$, $\mathfrak{V}_1$ and $\mathfrak{V}_2$, respectively. Then, since the homomorphisms F_1 and F_2 are algebraic, they canonically induce the following homomorphisms φ_1 and φ_2 of the Lie algebras: $\mathfrak{R}_1\xrightarrow{\varphi_1}\mathfrak{V}_1$ and $\mathfrak{R}_2\xrightarrow{\varphi_2}\mathfrak{V}_2$. Moreover, because of $E_1=V_1^0$ and $E_2=V_2^0$ (see Remark 6), φ_1 and φ_2 are endomorphisms. Then, φ_1 and φ_2 canonically induce the endomorphism $\varphi:\mathfrak{R}_1\times\mathfrak{R}_2\xrightarrow{\varphi}\mathfrak{V}_1\times\mathfrak{V}_2$.

Remark 8. Since $\bar{L}_Z^0$ normalizes both $R_u(P_1)$ and $R_u(P_2)$, we get that $L'=(\bar{L}_Z^0)_C$ normalizes both $(R_u(P_1))_C$ and $(R_u(P_2))_C$. Therefore it is possible to define representations ψ_1 and ψ_2 of the group L' in the complex linear spaces $\mathfrak{R}_1$ and $\mathfrak{R}_2$, respectively, by the formulae

$$(\psi_1(l))(r) = (\mathrm{Ad}\, l)r \quad (l \in L', r \in \mathfrak{R}_1) \tag{1}$$

and

$$(\psi_2(l))(r) = (\mathrm{Ad}\, l)r \quad (l \in L', r \in \mathfrak{R}_2). \tag{2}$$

By Lemma 1.2.1, ψ_1 and ψ_2 are algebraic representations of the complex algebraic group L'. Then ψ_1 and ψ_2 canonically induce an algebraic representation ψ of the group L' in the complex linear space $\mathfrak{R}_1 \times \mathfrak{R}_2$.

Remark 9. Denote by N_1 (resp. N_2) the normalizer of the subgroup V_1^0 (resp. V_2^0) in H. Since $N_1 \cap N_2$ normalizes both V_1^0 and V_2^0, we can define representations α_1 and α_2 of the group $N_1 \cap N_2$ in the complex linear spaces $\mathfrak{V}_1$ and $\mathfrak{V}_2$ by formulae

$$(\alpha_1(n))(\mathfrak{v}) = (\mathrm{Ad}\, n)\mathfrak{v} \quad (n \in N_1 \cap N_2, \mathfrak{v} \in \mathfrak{V}_1) \tag{3}$$

and

$$(\alpha_2(n))(\mathfrak{v}) = (\mathrm{Ad}\, n)\mathfrak{v} \quad (n \in N_1 \cap N_2, \mathfrak{v} \in \mathfrak{V}_2) \tag{4}$$

respectively. On account of Lemma 1.2.1, α_1 and α_2 are algebraic representations of the complex algebraic group $N_1 \cap N_2$ ($N_1 \cap N_2$ is indeed an algebraic group in virtue of Lemma 1.2.8). Then α_1 and α_2 canonically induce an algebraic representation α of the group $N_1 \cap N_2$ in the complex linear space $V_1 \times V_2$. Now, since $C(V_1^0) \cap C(V_2^0) = e$ by assumption, and V_1^0 and V_2^0 are connected, the representation α is faithful.

Remark 10. Since $\bar{L}_Z^0$ normalizes both $R_u(P_1)$ and $R_u(P_2)$, we get that $\bar{L}_Z^0 \cap \Gamma$ normalizes both $R_u(P_1) \cap \Gamma$ and $R_u(P_2) \cap \Gamma$. Therefore $T(\bar{L}_Z^0 \cap \Gamma)$ normalizes both $T(R_u(P_1) \cap \Gamma)$ and $T(R_u(P_2) \cap \Gamma)$. From this it follows that $T(L_Z^0 \cap \Gamma)$ normalizes both V_1 and V_2, consequently both V_1^0 and V_2^0, thus

$$T(\bar{L}_Z^0 \cap \Gamma) \subset N_1 \cap N_2 \tag{5}$$

holds.

Remark 11. Since U_1 and U_2 are Zariski dense in $R_u(P_1)$ and $R_u(P_2)$, respectively (see Remark 5), the argument in Lemma 15.3 (see the paragraph of § 15 containing (6) and (7)) shows that

$$\varphi_1(\mathrm{Ad}\,(l)r) = (\mathrm{Ad}\,(T(l)))\varphi_1(r) \quad (l \in \bar{L}_Z^0 \cap \Gamma, r \in \mathfrak{R}_1) \tag{6}$$

and

$$\varphi_2(\mathrm{Ad}\,(l)r) = (\mathrm{Ad}\,(T(l)))\varphi_2(r) \quad (l \in \bar{L}_Z^0 \cap \Gamma, r \in \mathfrak{R}_2) \tag{7}$$

are also valid (for the definition of φ_1 and φ_2 see Remark 7).

By the formulae (1), (2), (3), (4), (5), from (6) and (7) it follows that for any $l \in \bar{L}_Z^0 \cap \Gamma$ the equalities

$$\varphi_1 \circ \psi_1(l) = \alpha_1(T(l)) \circ \varphi_1 \tag{8}$$

and

$$\varphi_2 \circ \psi_2(l) = \alpha_2(T(l)) \circ \varphi_2 \tag{9}$$

hold, where the symbol $\circ$ denotes composition of mappings. (8), (9) and the definition of the mapping φ and that of the representations φ and α (see Remarks 7, 8 and 9) imply that for any $l \in \bar{L}^0_{\mathbf{Z}} \cap \Gamma$ we have

$$\varphi \circ \psi(l) = \alpha(T(l)) \circ \varphi. \tag{10}$$

Remark 12. By (10), we can easily see that for any $l \in \bar{L}^0_{\mathbf{Z}} \cap \Gamma$ the linear transformation $\psi(l)$ sends the kernel of φ into itself. Therefore, since φ is an endomorphism of $\mathfrak{V}_1 \times \mathfrak{V}_2$ (see Remark 7), the formula

$$\varrho(l) = \varphi \circ \psi(l) \circ \varphi^{-1} \quad (l \in \bar{L}^0_{\mathbf{Z}} \cap \Gamma) \tag{11}$$

defines a representation ϱ of the group $\bar{L}^0_{\mathbf{Z}} \cap \Gamma$ in the complex linear space $\mathfrak{V}_1 \times \mathfrak{V}_2$ (the fact that ϱ is a representation can be easily proved). Moreover, on account of (10) we have

$$\varrho(l) = \alpha(T(l)) \quad (l \in \bar{L}^0_{\mathbf{Z}} \cap \Gamma). \tag{12}$$

We also note that, since ψ is an algebraic representation of the complex algebraic group L' (see Remark 8) and φ is a linear mapping, it is easy to see that ϱ is an algebraic representation of the group $L' \cap \Gamma = \bar{L}^0_{\mathbf{Z}} \cap \Gamma$ (the fact that ϱ is an algebraic representation is understood in the sense that the matrix coefficients of the representation of ϱ are restrictions to $L' \cap \Gamma$ of regular rational functions on L'). Therefore, taking into account Lemma 1.2.6, ϱ can be extended to an algebraic representation ϱ' of the complex algebraic group L' in the complex linear space $\mathfrak{V}_1 \times \mathfrak{V}_2$. But then, since $L' \cap \Gamma = \bar{L}^0_{\mathbf{Z}} \cap \Gamma$ is Zariski dense in $\bar{L}^0_{\mathbf{Z}}$ and consequently in L', (5) and (12) imply

$$\varrho'(L') \subset \alpha(N_1 \cap N_2). \tag{13}$$

Remark 13. Since α is a faithful algebraic representation of the complex algebraic group $N_1 \cap N_2$ (see Remark 9), in virtue of Lemmas 1.2.17 and 8.1, $\alpha(N_1 \cap N_2)$ is an algebraic subgroup of the complex algebraic group $GL(\mathfrak{V}_1 \times \mathfrak{V}_2)$ and the mapping $\alpha^{-1}: \alpha(N_1 \cap N_2) \xrightarrow{\alpha^{-1}} N_1 \cap N_2$ is an algebraic homomorphism of complex algebraic groups. Therefore, on account of (13) and since ϱ' is algebraic we have that the homomorphism $\alpha^{-1} \circ \varrho'$ of the complex algebraic group L into the complex algebraic group $N_1 \cap N_2$ is defined and algebraic. On the other hand, ϱ' is an extension of ϱ. Therefore by (12), $\alpha^{-1} \circ \varrho'$ is an extension to L' of the restriction to $\bar{L}^0_{\mathbf{Z}} \cap \Gamma$ of the representation T. From the last two statements, the assertion of *2)* of the lemma follows.

Combining Remarks 5 and 13 we get the lemma.

Lemma 19.8. Let F be an algebraically connected complex (real) algebraic group, $L \subset F$ an algebraic subgroup, $N \subset F$ an algebraic normal subgroup, $\Lambda \subset F$ a Zariski dense subgroup and T a representation of Λ in the complex algebraic group H. Now if *1)* $F = L \tilde{\times} N$ ($\tilde{\times}$ denotes the semidirect product in the sense of

§ 1), *2)* $L \cap \Lambda$ (resp. $N \cap \Lambda$) is Zariski dense in L (resp. N), *3)* the restriction to $L \cap \Lambda$ of T extends to an algebraic representation T_L of the group L, *4)* there exists in $N \cap \Lambda$ a subgroup Δ of finite index such that the restriction to Δ of T extends to an algebraic representation T_N of the group N, *5)* the subgroup generated by $\Lambda \cap (L \cap N)$ is of finite index in Λ, then there exists in $F \cap \Lambda$ a subgroup Ω of finite index such that

1) the restriction to Ω of T extends to an algebraic representation T_F of F,

2) $\Omega \supset L \cap \Lambda$.

Proof. Since N is a normal subgroup in F, we have that $N \cap \Lambda$ is a normal subgroup in Λ. Therefore, on account of the fact that Δ is a subgroup of finite index of $N \cap \Lambda$, for every $\lambda \in \Lambda$ the intersection $\Delta \cap (\lambda \Delta \lambda^{-1})$ is a subgroup of finite index in N. On the other hand F is algebraically connected and $F = L \tilde{\times} N$. Thus N is algebraically connected. From the last two statements and the fact that $N \cap \Lambda$ is Zariski dense in N, by Lemma 8.8 it follows that for every $\lambda \in \Lambda$ the intersection $\Delta \cap (\lambda \Delta \lambda^{-1})$ is Zariski dense in N. Then, the argument followed in Lemma 15.3 (see the paragraph of § 15 containing (6)) gives that for any $\lambda \in \Lambda$ and $n \in N$ we have

$$T_N(\lambda n \lambda^{-1}) = T_\lambda \cdot T_N(n) T_\lambda^{-1}, \tag{14}$$

where $T_\lambda = T(\lambda)$ and $T_\lambda^{-1} = T(\lambda^{-1})$. (14) implies that the restriction of T to $\bigcup_{\lambda \in \Lambda} (\lambda \Delta \lambda^{-1})$ coincides with that of T_N to $\bigcup_{\lambda \in \Lambda} (\lambda \Delta \lambda^{-1})$. Therefore, since T and T_N are homomorphisms, the restriction of T to the subgroup $\hat{\Delta}$ generated by $\bigcup_{\lambda \in \Lambda} (\lambda \Delta \lambda^{-1})$ coincides with that of T_N to $\hat{\Delta}$. On the other hand, it is evident that $\hat{\Delta}$ is a normal subgroup in Λ. Thus $\Omega = (L \cap \Lambda) \cdot \hat{\Delta}$ is a subgroup. Moreover, since $N \cap \Lambda$ is a normal subgroup in Λ, we get that $\Lambda' = (L \cap \Lambda).(N \cap \Lambda)$ is a subgroup and as is easily seen, Λ' coincides with the subgroup generated by $\Lambda \cap (L \cap N)$. Therefore, on account of assumption *5)* of the lemma, the index of Λ' in Λ is finite. On the other hand, since $\hat{\Delta} \supset N$ and Δ has a finite index in $N \cap \Lambda$, the index of $\Omega = (L \cap \Lambda) \cdot \hat{\Delta}$ in $\Lambda' = (L \cap \Lambda) \cdot (N \cap \Lambda)$ is also finite. These imply that Ω is of finite index in Λ.

Since $F = L \tilde{\times} N$, every element f can be uniquely written in the form $f = l(f) \cdot n(f)$, where $l(f) \in L$, $n(f) \in N$, and the mapping $l\colon F \to L$ (resp $n\colon F \to N$) associating $l(f)$ (resp. $n(f)$) to f is regular algebraic. Now for every $f \in F$ set

$$T_F(f) = T_L(l(f)) \cdot T_N(n(f)). \tag{15}$$

Since each of the mappings l, n, T_L and T_N is regular algebraic, T_F is also a regular algebraic mapping of F into H. Furthermore, since *1)* T_L is an extension of the restriction to $L \cap \Lambda$ of T, *2)* the restriction of T_N to $\hat{\Delta}$ coincides with that of T to $\hat{\Delta}$ (see the previous paragraph), it follows easily from (15) that the restriction of T_F to $\Omega = (L \cap \Lambda) \cdot \hat{\Delta}$ coincides with that of T to Ω. On the other hand, since *1)* the index of Ω in Λ is finite (see the end of the previous paragraph), *2)* Λ is

Zariski dense in F, *3)* F is algebraically connected, we get, by Lemma 8.8, that Ω is Zariski dense in F. The last three statements and Lemma 1.2.6 imply that T_F is a regular algebraic homomorphism of F into H whose restriction to Ω coincides with that of T to Ω. Now to complete the proof of the lemma, it is enough to note that *1)* Ω has a finite index in Λ, *2)* $\Omega=(L\cap\Lambda)\cdot\hat{\Delta}$ implies that $\Omega\supset L\cap\Lambda$. The lemma is proved.

Lemma 19.9. Let P_1 and P_2 be opposite parabolic **Q**-subgroups of the group G, and let T be a representation of the group Γ in the complex algebraic group H. Let $L, L_Z, \bar{L}_Z, \bar{L}_Z^0, V_1, V_2, V_1^0, V_2^0, C(V_1^0)$ and $C(V_2^0)$ stand for the same as in Lemma 19.7. Denote by S_1 and S_2 the algebraic closures of the subgroups $P_1\cap G_Z$ and $P_2\cap G_Z$ in the complexification G_C of G, and by S_1^0 and S_2^0 the algebraically connected components of the identity in S_1 and S_2, respectively. Then, if P_1 is a maximal parabolic **Q**-subgroup of G and $C(V_1^0)\cap C(V_2^0)=e$, there exist in $S_1^0\cap\Gamma$ and $S_2^0\cap\Gamma$ subgroups Ω_1 and Ω_2 of finite index, such that

1) the restrictions of T to Ω_1 and Ω_2 extend to algebraic representations of the complex algebraic groups S_1^0 and S_2^0 in the complex algebraic group H,

2) $\Omega_1\supset\bar{L}_Z^0\cap H$ and $\Omega_2\supset\bar{L}_Z^0\cap\Gamma$.

Proof. We make the following remarks.

Remark 1. Since P_1 and P_2 are opposite parabolic **Q**-subgroups of the group G, we see that L is a Levi subgroup of P_1 defined over **Q**.

Remark 2. From Remark 1 and Lemma 17.13 it follows that

$$(L\cap G_Z)\cdot(R_u(P_1)\cap G_Z)$$

is of finite index in $P_1\cap G_Z$.

Remark 3. In virtue of *3)* of Theorem 17.43, the subgroups $P_1\cap G_Z$ and $P_1\cap\Gamma$ are commensurable.

Remark 4. Remark 3 implies that *1)* $L\cap G_Z$ and $L\cap\Gamma$ as well as *2)* $R_u(P_1)\cap G_Z$ and $R_u(P_1)\cap\Gamma$ are commensurable.

Remark 5. The previous two remarks imply that $(L\cap\Gamma)\cdot(R_u(P_1)\cap\Gamma$ is of finite index in $P_1\cap\Gamma$.

Remark 6. Let L' denote the complexification of the group $\bar{L}_Z^0$. Since $\bar{L}_Z^0$ is by definition algebraically connected, L' is connected. Therefore, on account of Remark 4, the definition of the subgroup $\bar{L}_Z^0$ and Lemma 8.8, we obtain that the subgroup $L'\cap\Gamma=\bar{L}_Z^0\cap\Gamma$ is Zariski dense in L'.

Remark 7. The fact that $R_u(P_1)\cap\Gamma$ is Zariski dense in $R_u(P_1)$ (see Remark 2 in the proof of Lemma 19.7) implies that $R_u(P_1)\cap\Gamma$ is Zariski dense in the complexification $(R_u(P_1))_C$.

Remark 8. The conditions in Lemma 19.9 are the same as in Lemma 19.7.

Remark 9. From Remarks 7 and 8 and from *1)* of Lemma 19.7 it follows that there exists in $R_u(P_1)\cap\Gamma$ a subgroup U_1 of finite index such that the restriction to U_1 of T extends to an algebraic representation of the complex algebraic group $(R_u(P_1))_{\mathrm{C}}$ in the complex algebraic group H.

Remark 10. Remarks 6 and 8 and *2)* of Lemma 19.7 imply that the restriction to $\bar{L}^0_{\mathrm{Z}}\cap\Gamma$ of T extends to an algebraic representation of L' in H.

Remark 11. The complex algebraic group S_1^0 is by definition connected.

Remark 12. Denote by $(\bar{P}_1)_{\mathrm{Z}}$ the Zariski closure of the subgroup $P_1\cap G_{\mathrm{Z}}$ in G and by $(\bar{P}_1)^0_{\mathrm{Z}}$ the algebraically connected component of the identity of the group $(\bar{P}_1)_{\mathrm{Z}}$. Then, by virtue of Lemma 17.29, we have $(\bar{P}_1)^0_{\mathrm{Z}}=\bar{L}^0_{\mathrm{Z}}\tilde{\times}R_u(P_1)$. Therefore, since, as is easy to see, S_1^0 is the complexification of the group $(\bar{P}_1)_{\mathrm{Z}}$, we get that $S_1^0=L'\tilde{\times}(R_u(P_1))_{\mathrm{C}}$, where $L'=(\bar{L}^0_{\mathrm{Z}})_{\mathrm{C}}$ as above.

From Remarks 5, 6, 7, 9, 10, 11, 12 and Lemma 19.8 it follows that in $S_1^0\cap\Gamma=(\bar{P}_1)^0_{\mathrm{Z}}$ there exists a subgroup Ω_1 of finite index such that *1)* the restriction of T to $S_1^0\cap\Gamma$ extends to an algebraic representation of S_1^0 in the group H, *2)* $\Omega_1\supset\bar{L}^0_{\mathrm{Z}}\cap\Gamma$. It can be proved in the same way that in $S_2^0\cap\Gamma$ there exists a subgroup Ω_2 of finite index such that *1)* the restriction to Ω_2 of T extends to an algebraic representation of S_2 in H, *2)* $\Omega_2\supset\bar{L}^0_{\mathrm{Z}}\cap\Gamma$. From these the lemma follows.

Remark. The proof of Lemmas 19.7, 19.8 and 19.9 shows much analogy at several points with that of Lemma 15.3.

Theorem 19.10. *Let P_1 and P_2 be opposite parabolic* **Q**-*subgroups of G and T a representation of Γ in a connected complex algebraic group E whose centre is trivial. Let $L, L_{\mathrm{Z}}, \bar{L}_{\mathrm{Z}}, \bar{L}^0_{\mathrm{Z}}, V_1, V_2, V_1^0, V_2^0, C(V_1^0), C(V_2^0)$ stand for the same as in Lemma* 19.7. *Now if 1) P_1 is a maximal parabolic* **Q**-*subgroup of the group G, 2) $C(V_1^0)\cap C(V_2^0)=e$, 3) $T(\Gamma)$ is Zariski dense in E, 4) the subgroup $T(\bar{L}^0_{\mathrm{Z}}\cap\Gamma)$ is different from the identity and is not contained in any proper algebraic normal subgroup of E, then T extends to an algebraic representation T^G of the complex algebraic group G_{C} in the complex algebraic group E.*

Proof. Let us make the following remarks.

Remark 1. Conditions *1)* and *2)* of the theorem imply that P_1, P_2 and T satisfy the conditions of Lemmas 19.7 and 19.9.

Remark 2. Let S_1^0 and S_2^0 be the same as in Lemma 19.9. Then, Remark 1 and Lemma 19.9 imply that in $S_1^0\cap\Gamma$ and $S_2^0\cap\Gamma$ there exist subgroups Ω_1 and Ω_2 of finite index, respectively, such that

1) the restriction to Ω_1 (resp. Ω_2) of T extends to an algebraic representation T_1 (resp. T_2) of S_1^0 (resp. S_2^0) in E,

2) $\Omega_1 \supset \bar{L}_{\mathbf{Z}}^0 \cap \Gamma$ and $\Omega_2 \supset \bar{L}_{\mathbf{Z}}^0 \cap \Gamma$.

Remark 3. Since P_1 and P_2 are opposite parabolic **Q**-subgroups, we get that $L = P_1 \cap P_2$ is a Levi subgroup defined over **Q** both in P_1 and P_2. Therefore we have

$$P_1 = L \tilde{\times} R_u(P_1) \tag{16}$$

and

$$P_2 = L \tilde{\times} R_u(P_2). \tag{17}$$

Remark 4. Denote by D_1 (resp. D_2) the algebraic closure of $P_1 \cap G_{\mathbf{Z}}$ (resp. $P_2 \cap G_{\mathbf{Z}}$) in G, and by D_1^0 (resp. D_2^0) the algebraically connected component of the unity of D_1 (resp. D_2). Then, since L is a Levi subgroup defined over **Q** both in P_1 and P_2 (see Remark 3), we get, by virtue of Lemma 17.29, that

$$D_1^0 = \bar{L}_{\mathbf{Z}}^0 \tilde{\times} R_u(P_1) \tag{18}$$

and

$$D_2^0 = \bar{L}_{\mathbf{Z}}^0 \tilde{\times} R_u(P_2). \tag{19}$$

Remark 5. By Remark 3, L is a Levi subgroup defined over **Q** both in P_1 and P_2. Therefore, Remark 2 in the proof of Lemma 17.28 gives that $\bar{L}_{\mathbf{Z}}^0$ is a normal subgroup of L. From this and equalities (16), (17), (18) and (19) it follows easily that D_1^0 and D_2^0 are normal subgroups of P_1 and P_2, respectively.

Remark 6. Let $g \in G$ be such an element for which the parabolic subgroups P_1 and gP_1g^{-1} are opposite. Then, since P_1 and P_2 are opposite, in virtue of Lemma 10.22 one can find an element $u_g \in R_u(P_1)$ such that

$$gP_2g^{-1} = u_g P_2 u_g^{-1}. \tag{20}$$

From (20) we see that $u_g^{-1} \cdot g$ normalizes the subgroup P_2. On the other hand, being a parabolic subgroup, P_2 coincides with its normalizer. The last two statements imply that $u_g^{-1} g \in P_2$. Therefore, since $D_{\mathbf{Z}}^0$ is a normal subgroup of P_2 (see Remark 5), one has $u_g^{-1} g D_2^0 g^{-1} u_g = D_2^0$. From this it follows that

$$gD_2^0 g^{-1} = u_g D_2^0 u_g^{-1}. \tag{21}$$

Furthermore, since P_1 and P_2 are opposite, on account of Lemma 10.14, we have

$$R_u(P_1) \cap P_2 = P_1 \cap R_u(P_2) = e. \tag{22}$$

Now $L = P_1 \cap P_2$ and $\bar{L}_{\mathbf{Z}}^0 \subset L$. Therefore, (18), (19) and (22) imply

$$D_1^0 \cap D_2^0 = \bar{L}_{\mathbf{Z}}^0. \tag{23}$$

Since $R_u(P_1) \subset D_1^0$ (see (18)) and $u_g \in R_u(P_1)$, we obtain that u_g normalizes D_1^0. From this and (21)—(22) we get that

$$D_1^0 \cap (gD_2^0 g^{-1}) = D_1^0 \cap (u_g D_2^0 u_g^{-1}) = (u_g D_1^0 u_g^{-1}) \cap (u_g D_2^0 u_g^{-1}) =$$

$$= u_g(D_1^0 \cap D_2^0) u_g^{-1} = u_g \bar{L}_{\mathbf{Z}}^0 u_g^{-1}. \tag{24}$$

Remark 7. Denote by $W \subset G$ the set of the elements $g \in G$ for which the parabolic subgroups P_1 and $gP_2 g^{-1}$ are opposite. Using the fact that P_1 and P_2 are opposite, in virtue of Lemma 10.13 we obtain that W is Zariski dense in G.

Remark 8. As is easy to see, S_1^0 and S_2^0 are complexifications of the groups D_1^0 and D_2^0.

Remark 9. Since, on account of Remark 6, (24) holds for every $g \in W$, from Remarks 7 and 8 and Lemma 14.2 it is not hard to deduce that for any $g_0 \in W$ we have

$$\dim (S_1^0 \cap g_0 S_2^0 g_0^{-1}) = \min_{g \in G} \dim (S_1^0 \cap g S_2^0 g^{-1}). \tag{25}$$

Remark 10. Let γ be an arbitrary element of $\Gamma \cap W$. Then $\Gamma \subset G_{\mathbf{Q}}$ (see (1) in § 17), thus $\gamma \in G_{\mathbf{Q}}$. Therefore, as P_2 is defined over $\mathbf{Q}$, we get that $\gamma P_2 \gamma^{-1}$ is also defined over $\mathbf{Q}$. On the other hand, because of $\gamma \in W$, the parabolic subgroups P_1 and P_2 are opposite. These last two statements and 4.8 of [5] imply the existence of an element $u_\gamma \in G_{\mathbf{Q}}$ satisfying

$$P_1 \cap (\gamma P_2 \gamma^{-1}) = P_1 \cap (u_\gamma P_2 u_\gamma^{-1}). \tag{26}$$

Then, by (24), we have

$$D_1^0 \cap (\gamma D_2^0 \gamma^{-1}) = u_\gamma \bar{L}_{\mathbf{Z}}^0 u_\gamma^{-1}. \tag{27}$$

Taking into account the fact that $u_\lambda \in G_{\mathbf{Q}}$, from (27) it follows easily that the subgroup $G_{\mathbf{Z}} \cap D_1^0 \cap \gamma D_2^0 \gamma^{-1}$ is commensurable with $u_\gamma (G_{\mathbf{Z}} \cap \bar{L}_{\mathbf{Z}}^0) u_\gamma^{-1}$. Therefore, since *1)* $G_{\mathbf{Z}} \cap \bar{L}_{\mathbf{Z}}^0$ is Zariski dense in $\bar{L}_{\mathbf{Z}}^0$ (see the definition of $\bar{L}_{\mathbf{Z}}^0$ and Lemma 8.8), *2)* $\bar{L}_{\mathbf{Z}}^0$ is algebraically connected, by Lemma 8.8 we get that $G_{\mathbf{Z}} \cap D_1^0 \cap \gamma D_2^0 \gamma^{-1}$ is Zariski dense in $D_1^0 \cap \gamma D_2^0 \gamma^{-1}$. On the other hand, since $D_1^0 \subset P_1$, assertion *3)* of Theorem 17.43 implies that the subgroups $G_{\mathbf{Z}} \cap D_1^0 \cap \gamma D_2^0 \gamma^{-1}$ and $\Gamma \cap D_1^0 \cap \cap \gamma D_2^0 \gamma^{-1}$ are commensurable. The last two statements, the algebraic connectedness of the group $D_1 \cap (\gamma D_2^0 \gamma^{-1})$ (this latter fact is evident by (27)) and Lemma 8.8 imply that $\Gamma \cap D_1^0 \cap \gamma D_2^0 \gamma^{-1}$ is Zariski dense in $D_1^0 \cap \gamma D_2^0 \gamma^{-1}$.

Remark 11. Since $\Gamma \subset G$, from Remark 8 it follows that $S_1^0 \cap \Gamma = D_1^0 \cap \Gamma$ and $S_2^0 \cap \Gamma = D_2^0 \cap \Gamma$. From this, Remarks 8 and 10 and Lemma 8.8 it is easy to conclude that for any $\gamma \in \Gamma \cap W$ the subgroup $\Gamma \cap S_\gamma$ is Zariski dense in S_γ where S_γ denotes the algebraically connected component of the identity of the group $S_1^0 \cap (\gamma S_2^0 \gamma^{-1})$.

Remark 12. Since W is Zariski dense and open in G (see Remark 7), it is easy to see that there exists in $G_{\mathbf{C}}$ a Zariski dense and open subset W such that $W' \cap G = W$.

Remark 13. Let $X \subset G_{\mathbf{C}}$ denote the set of elements $g_0 \in G_{\mathbf{C}}$ for which $\dim (S_1^0 \cap g_0 S_2^0 g_0^{-1}) = \min_{g \in G_{\mathbf{C}}} \dim (S_1^0 \cap g S_2^0 g^{-1})$. Lemma 14.2 implies that X is Zariski open in $G_{\mathbf{C}}$. We note that X is evidently not empty.

Remark 14. From Remarks 12 and 13, it follows that $W' \cap X$ is a non-empty Zariski open subset of $G_{\mathbf{C}}$.

Remark 15. We want to refer to Theorem 14.15 and therefore modify the notations. Put *1)* $H = G_{\mathbf{C}}$, $\Lambda = \Gamma$, *2)* $R = S_1^0$, $S = S_2^0$ (for the definition of S_1^0 and S_2^0, see Lemma 19.9), *3)* $\tilde{T} = T$, $D = E$, *4)* $\Lambda_R = \Omega_1$, $\Lambda_S = \Omega_1$ (for the definition of Ω_1 and Ω_2 see Remark 2), *5)* $\bar{V} = W' \cap X$ (for the definition of W' and X see Remarks 12 and 13).

We will show in the subsequent remarks that H, Λ, R, S, $\tilde{T}$, D, Λ_R, Λ_S and $\bar{V}$ satisfy the conditions of Theorem 14.15.

Remark 16. By the density theorem of Borel, $\Lambda = \Gamma$ is Zariski dense in $H = G_{\mathbf{C}}$.

Remark 17. The algebraic subgroups $R = S_1^0$ and $S = S_2^0$ are algebraically connected by their definition.

Remark 18. According to Remark 3 in the proof of Lemma 19.7, the subgroups $R \cap \Lambda = S_1^0 \cap \Gamma$ and $S \cap \Lambda = S_2^0 \cap \Gamma$ are Zariski dense in $R = S_1^0$ and $S = S_2^0$, respectively.

Remark 19. By the assumption of the theorem, the group $D = E$ is a connectedi complex algebraic group with a trivial centre.

Remark 20. Since $R = S_1^0$ and $S = S_2^0$, from the definitions of X (see Remark 13) and $V(R, S)$ (see the paragraph preceding Theorem 14.15) it follows that $X = V(R, S)$.

Remark 21. Remark 14 and 20 imply that *1)* $\bar{V} = W' \cap X$ is a nonempty Zariski open subset of $H = G_{\mathbf{C}}$, *2)* $\bar{V} \subset V(R, S)$.

Remark 22. Remark 11 implies that *1)* in Theorem 14.15 is satisfied.

Remark 23. From Remark 2, it follows that conditions *2)* and *3)* in Theorem 14.15 are also satisfied (put $T^R = T_1$ and $T^S = T_2$, where T_1 and T_2 are defined in Remark 2).

Remark 24. $\tilde{T}(\Lambda) = T(\Gamma)$ is Zariski dense in $D = E$ by the assumption of the present theorem. Therefore condition *4)* in Theorem 14.15 is satisfied.

Remark 25. Since P_1 and P_2 are opposite, we have $e \in W$, where e is the identity of G (W has been defined in Remark 7). Therefore, *1)* since $W' \cap G = W$ (see Remark 12), we get $e \in W'$, *2)* it follows from Remark 9 and the definition of X (see Remark 13) that $e \in X$, *3)* $e \in \Gamma$ evidently. Thus $e \in W' \cap X \cap \Gamma = \bar{V} \cap \Lambda$.

Remark 26. The relation

$$S_1^0 \cap S_2^0 \supset \bar{L}_Z^0 \tag{28}$$

follows from the definition of $\bar{L}_Z^0$, S_1^0, S_2^0 (see Lemmas 19.7 and 19.9) and from the fact that $L = P_1 \cap P_2$. Since $\bar{L}_Z^0$ is by definition algebraically connected, it follows from (28) that the algebraically connected component of the identity of the group $S_1^0 \cap S_2^0$ contains $\bar{L}_Z^0$. Therefore, since *1)* $e \in W' \cap X \cap \Gamma = \bar{V} \cap \Lambda$ (see Remark 25), *2)* $S_1^0 = R$, and *3)* $S_2^0 = S$, we have

$$\bar{L}_Z^0 \cap \Gamma \subset \bigcup_{\lambda \in \bar{V} \cap \Lambda} (R, S)_\lambda, \tag{29}$$

where, in accordance with the paragraph preceding Theorem 14.15, $(R, S)_\lambda$ denotes the connected component of the identity of $(\lambda S \lambda^{-1}) \cap R$. Since *1)* $\Omega_1 \supset \bar{L}_Z^0 \cap \Gamma$ (see Remark 2), *2)* T_1 is an extension to S_1^0 of the restriction of T to $\Omega_1 \subset S_1^0 \cap \Gamma$ (see Remark 2), *3)* $T^R = T_1$ (see Remark 23), we get from (25) that

$$T(\bar{L}_Z^0 \cap \Gamma) = T_1(\bar{L}_Z^0 \cap \Gamma) = T^R(\bar{L}_Z^0 \cap \Gamma) \subset \bigcup_{\lambda \in \bar{V} \cap \Lambda} (T^R(R, S)_\lambda). \tag{30}$$

(30) and assumption *4)* of the present theorem imply that *5)* of Theorem 15.15 is also satisfied.

Remarks 16, 17, 18, 19, 21, 22, 23, 24 and 26 tell us that H, Λ, R, S, $\tilde{T}$, D, Λ_R, Λ_S and $\bar{V}$ satisfy the conditions of Theorem 14.15. Therefore, by Remark 15 and Theorem 14.15, we obtain that $T = \tilde{T}$ extends to an algebraic representation $T^G = \tilde{T}^H$ of $G_C = H$ in the group $E = D$. The theorem is proved.

Remark. The above proof is very similar to that of Theorem 16.11.

IV. Now we are going to prove Theorem 19.5.

Lemma 19.11. Let U_1 and U_2 be opposite orispherical subgroups of G. Denote by $\mathfrak{G}$, $\mathfrak{U}_1$ and $\mathfrak{U}_2$ the Lie algebras of the groups G, U_1, and U_2, respectively, by $\mathfrak{A} \subset \mathfrak{G}$ the subalgebra generated by $\mathfrak{U}_1 \cup \mathfrak{U}_2$. Now, if $\mathfrak{A}_1$ does not reduce to the identity and is not contained in any nontrivial algebraic normal subgroup of G, then $\mathfrak{A} = \mathfrak{G}$.

Proof. From the assertion 4.11 of [5] it follows that any minimal algebraic subgroup A of G containing $U_1 \cup U_2$, is a normal subgroup of G. Therefore, since $U_1 \neq e$ and U_1 is not contained in any nontrivial algebraic normal subgroup of G we get that $A = G$. On the other hand, it is easy to see that the Lie algebra of the group A contains $\mathfrak{A}$. The last two statements prove the lemma.

Lemma 19.12. Let X be a connected topological space, f a continuous mapping of X into the real (complex) algebraic group H, and x_0 a point of X.

Now, if *1)* $f(x_0)$ is a unipotent element of H, *2)* for every $x \in X$ there exists an integer $a(x)$ such that $(f(x))^{a(x)}$ is a unipotent element of H, then $f(x)$ is unipotent for every $x \in X$.

Proof. Obviously it is enough to prove the lemma in case $H = GL(n, \mathbf{R})(GL(n, \mathbf{C}))$. Then for every $h \in H$, let $\alpha_i(h)$ $(1 \leqq i \leqq n)$ denote the coefficients of the characteristic polynomial of the matrix h. It is plain that for every i $(1 \leqq i \leqq n)$ the function α_i is continuous on H. On the other hand from condition *2)* of the lemma we get that for any $x \in X$ every eigenvalue of the matrix $f(x)$ is a root of unity, consequently, for any i $(1 \leqq i \leqq n)$ the set $\alpha_i(f(x))$ is countable. From the last two statements and the continuity of f we can conclude that for any i $(1 \leqq i \leqq n)$ the set $\alpha_i(f(x))$ is a connected countable subset of $\mathbf{R}(\mathbf{C})$. But such a subset cannot be anything else but a point, therefore for any i $(1 \leqq i \leqq n)$ the function $\alpha_i(f(x))$ is a constant on X. On the other hand from condition *1)* of the lemma we get that for any i $(1 \leqq i \leqq n)$, $\alpha_i(f(x_0)) = 1$. The last two statements imply the validity of the lemma.

Let Λ be a countable group and H a topological group. Denote by $T(\Lambda, H)$ the set of the representations of Λ in H. Let us introduce a topology in $T(\Lambda, H)$ in the usual way: $W \subset T(\Lambda, H)$ is open if and only if for every $T \in W$ there exist an integer n, elements $\lambda_1, \ldots, \lambda_n$ of Λ and neighbourhoods $U_1, \ldots, U_n$ of the points $T(\lambda_1), \ldots, T(\lambda_n)$ in H such that W contains all $T' \in T(\Lambda, H)$ for which $T'(\lambda_i) \in U_i$ for every i $(1 \leqq i \leqq n)$. It is easy to see that the family T^t $(0 \leqq t \leqq 1)$ of representations of Λ in H is continuous at the point $x \in [0, 1]$ (see Definition 19.1) if and only if the mapping of $[0, 1]$ into $T(\Lambda, H)$ given by T^t, is continuous.

It is also easy to see that the following well-known lemma is valid.

Lemma 19.13. Let Λ be a finitely generated (and hence countable) group and H a real (complex) algebraic group. Then the space $T(\Lambda, H)$ equipped with the topology just introduced is a real (complex) algebraic manifold and so the topological space $T(\Lambda, H)$ is locally connected.

Remark 19.14. In the course of Remarks 1—5 in the proof of Lemma 19.7 we have not used the fact that $C(V_1^0) \cap C(V_2^0) = e$. Therefore, since the proof of statement *1)* of Lemma 19.7 ends with Remark 5, it is valid without the assumption $C(V_1^0) \cap C(V_2^0) = e$.

Lemma 19.15. Let P be a maximal parabolic $\mathbf{Q}$-subgroup of G and let T^t $(0 \leqq t \leqq 1)$ be a Deformation (see Definition 19.3) of the subgroup Γ in G. Then there exists an $\varepsilon > 0$ such that for any $t \in [0, \varepsilon]$ the subgroup $T^t(R_u(P) \cap \Gamma)$ is unipotent.

The proof will be carried out in a series of remarks. Denote by $T(\Gamma, G)$ the space of representations of Γ in G.

Remark 1. On account of Remark 17.18 and Lemma 10.10, in G there exists a parabolic $\mathbf{Q}$-subgroup P' opposite to P. Therefore by virtue of Remark 19.14, Lemma 19.7 and Remark 5 in the proof of Lemma 19.7, for every $T\in T(\Gamma, G)$ there exists a subgroup of finite index U_T in $R_u(P)\cap\Gamma$ such that $T(U_T)$ is unipotent.

Remark 2. Let γ be an arbitrary element of $R_u(P)\cap\Gamma$. Then, by the fact that for every $T\in T(\Gamma, G)$ the subgroup U_T has finite index in $R_u(P)\cap\Gamma$, we see that for every $T\in T(\Gamma, G)$ there exists an integer $a(T)$ such that $(T(\gamma))^{a(T)}$ is a unipotent element of G.

Remark 3. Denote by $T(R_u(P)\cap\Gamma, G)$ the space of all representations of the group $R_u(P)\cap\Gamma$ in G, and by k the canonical mapping $T(\Gamma, G)\xrightarrow{k}T(R_u(P)\cap\Gamma, G)$ which assigns to every $T\in T(\Gamma, G)$ its restriction to $R_u(P)\cap\Gamma$. Then the mapping k is evidently continuous.

Remark 4. Since $R_u(P)\cap\Gamma$ is a discrete unipotent subgroup, by Lemmas 1.3.1, 5.4 and 5.1 we obtain that $R_u(P)\cap\Gamma$ is finitely generated. Therefore, on account of Lemma 19.13, the space $T(R_u(P)\cap\Gamma, G)$ is locally connected.

Remark 5. Denote by $T_0(R_u(P)\cap\Gamma, G)$ the connected component of the space $T(R_u(P)\cap\Gamma, G)$ which contains the identical representation T^0 of the group Γ. Then, in virtue of Remark 2 and Lemma 19.12, for every $T\in T_0(R_u(P)\cap\Gamma, G)$ and for every $\gamma\in R_u(P)\cap\Gamma$ the element $T(\gamma)$ is unipotent.

Remark 6. Since T^t is a deformation, T^t is continuous at 0. Therefore, by Remarks 3 and 4 there exists an $\varepsilon>0$ such that for every $t\in[0, \varepsilon]$ we have $k(T^t)\in T_0(R_u(P)\cap\Gamma, G)$.

The last two remarks prove the lemma.

Lemma 19.16. Let W be a discrete unipotent subgroup of G and let T^t $(0\leqq t\leqq 1)$ be a deformation (see Definition 19.3) of W in G. For any t $(0\leqq t\leqq 1)$ denote by W_t the Zariski closure of $T^t(W)$ in G, by $\overline{W}$ the Zariski closure of W in G and by $\mathfrak{W}_t$ and $\overline{\mathfrak{W}}$ the Lie algebras of W_t and W respectively. Now, if for every t $(0\leqq t\leqq 1)$ the representation T^t is unipotent (i.e. the group $T^t(W)$ is unipotent), then $\lim_{t\to 0}\mathfrak{W}_t=\overline{\mathfrak{W}}$, where $\mathfrak{W}_t$ and $\overline{\mathfrak{W}}$ are considered as elements of the Grassmann manifold $G(\mathfrak{G})$ of all linear subspaces of the Lie algebra $\mathfrak{G}$ of the group G.

Proof. First of all, let us recall that for any unipotent element u of G, we denote by $\ln u$ the unique element of the Lie algebra $\mathfrak{G}$ for which $\exp(\ln u)=u$. With the discrete unipotent subgroup W associate now the quasi-lattice Δ_W in the same way as we did in § 5 (see the paragraph preceding Lemma 5.7). Let us make the following observations.

Remark 1. $\Delta_W\subset\ln W$ (see the paragraph preceding Lemma 5.7).

Remark 2. Δ_W is a quasi-lattice of full dimension in $\overline{\mathfrak{W}}$ (see Lemma 7.1).

Remark 3. Let $e_1, \ldots, e_n$ be a basis of the quasi-lattice Δ_W ($n=\dim W$). Then, since *1)* T^t is a deformation of W into G (by assumption), *2)* for every t ($0 \leqq t \leqq 1$) the representation T^t is unipotent (by assumption again), *3)* for every i ($1 \leqq i \leqq n$) we have $\exp e_i \in W$ (see Remark 1), we get that *1)* the element $T^t(\exp e_i)$ is defined and unipotent, *2)* the mapping f_i of $[0, 1]$ into G, sending $t \in [0, 1]$ into $\ln T^t(\exp e_i)$, is continuous at 0.

Remark 4. For every t ($0 \leqq t \leqq 1$), T^t is a unipotent representation of the discrete unipotent subgroup W, thus, in virtue of Lemma 5.9, for every t the representation T^t extends to an algebraic unipotent representation $\tilde{T}^t$ of the group $\overline{W}$. For any t ($0 \leqq t \leqq 1$), denote by S^t the representation of the Lie algebra $\overline{\mathfrak{W}}$ in the Lie algebra $\mathfrak{G}$ which corresponds to the algebraic representation $\tilde{T}^t$. Then, since $\exp \overline{\mathfrak{W}} = \overline{W}$ (see Lemma 5.1) and $\tilde{T}^t$ is an algebraic extension of the unipotent representation T^t, we get that for every i ($1 \leqq i \leqq n$) and t ($0 \leqq t \leqq 1$) the following relation holds:

$$S^t(e_i) = f_i(e_i), \tag{31}$$

where f_i denotes the same mapping as in Remark 3.

Remark 5. Since *1)* $e_1, \ldots, e_n$ is a basis of Δ_W, *2)* for every t ($0 \leqq t \leqq 1$) the mapping S^t is linear, *3)* the facts that $\tilde{T}^t$ is an algebraic representation and it is an extension of T^t imply that $S^t(\overline{\mathfrak{W}}) = \mathfrak{W}_t$ for every t ($0 \leqq t \leqq 1$), we get that from (31) and Remark 3 the lemma follows.

Lemma 19.17. Let T^t ($0 \leqq t \leqq 1$) be an arbitrary deformation of Γ in G. Then for any $\gamma \in \Gamma$ different from the identity there exists an $\varepsilon(\gamma) > 0$ such that for any $t \in [0, \varepsilon(\gamma)]$ the element $T^t(\gamma)$ is not contained in any proper algebraic normal subgroup of the complexification G_C of G.

Proof. Since G and Γ satisfy the conditions formulated at the beginning of § 7, and $\gamma \neq e$, on account of Lemma 4.8, γ does not belong to any proper algebraic normal subgroup of G. On the other hand, since G is semisimple, the number of the algebraic normal subgroups of G is finite. From these last two statements it is easy to deduce the existence of an $\varepsilon(\gamma) > 0$ such that for any $t \in [0, \varepsilon(\gamma)]$ the element $T^t(\gamma)$ is not contained in any proper algebraic normal subgroup of G. From this and the fact that the intersection of any algebraic normal subgroup of G_C with G is an algebraic normal subgroup of G, the lemma follows.

Lemma 19.18. Let $\mathfrak{A}$ be a finite-dimensional real (complex) Lie algebra, and let L_1 and L_2 be two linear subspaces of $\mathfrak{A}$. Denote by $G(\mathfrak{A})$ the Grassmann manifold of all linear subspaces of $\mathfrak{A}$. Let f_1 and f_2 be two mappings of $[0, 1]$ into $G(\mathfrak{A})$. Now if *1)* $L_1 \cup L_2$ generates $\mathfrak{A}$ as a Lie algebra, *2)* $f_1(0) = L_1$, $f_2(0) = L_2$, *3)* f_1 and f_2 are continuous at 0, then there exists an $\varepsilon > 0$ such that for every $t \in [0, \varepsilon]$, $f_1(t) \cup f_2(t)$ generates $\mathfrak{A}$, as a Lie algebra.

Proof. The conditions of the lemma imply that there exists in $\mathfrak{A}$ a basis $e_1, \ldots, e_n$ ($n = \dim \mathfrak{A}$) such that for every i ($1 \leqq i \leqq n$), e_i is either an element of one of the spaces L_1 and L_2 or it is a commutator of some weight of elements from $L_1 \cup L_2$. Thus, since $\lim_{t\to 0} f_1(t) = L_1$ and $\lim_{t\to 0} f_2(t) = L_2$ (by assumption), we get that for any i, in $f_1(t) \cup f_2(t) \cup K_t$ there exists an element $e_i(t)$ such that $\lim_{t\to 0} e_i(t) = e_i$ ($1 \leqq i \leqq n$). Here K_t denotes the set of commutators of all possible weights of elements of $f_1(t) \cup f_2(t)$. Then, $e_1(t), \ldots, e_n(t)$ form a basis of $\mathfrak{A}$, and the lemma is proved.

Let us turn now to the

Proof of Theorem 19.5. According to Lemma 17.26, there exists a proper parabolic **Q**-subgroup in G. Therefore, there exists in G a parabolic **Q**-subgroup P_1 which is maximal in the class of the proper parabolic **Q**-subgroups, i.e. there exists in G a maximal parabolic **Q**-subgroup P_1. On account of Remark 7.18 and Lemma 10.10, there exists in G a parabolic **Q**-subgroup P_2 opposing to P_1. Denote by U_1 and U_2 the Zariski closures of the subgroups $R_u(P_1) \cap \Gamma$ and $R_u(P_2) \cap \Gamma$, by $U_{1,t}$ and $U_{2,t}$ ($0 \leqq t \leqq 1$) the Zariski closures in G of the subgroups $T^t(R_u(P_1) \cap \Gamma)$ and $T^t(R_u(P_2) \cap \Gamma)$, by $(U_1)_{\mathbf{C}}$, $(U_2)_{\mathbf{C}}$, $(U_{1,t})_{\mathbf{C}}$ and $(U_{2,t})_{\mathbf{C}}$ ($0 \leqq t \leqq 1$) the complexifications of the groups U_1, U_2, $U_{1,t}$ and $U_{2,t}$, by $\mathfrak{U}_1$, $\mathfrak{U}_2$, $\mathfrak{U}_{1,t}$, $\mathfrak{U}_{2,t}$, $(\mathfrak{U}_1)_{\mathbf{C}}$, $(\mathfrak{U}_2)_{\mathbf{C}}$, $(\mathfrak{U}_{1,t})_{\mathbf{C}}$, $(\mathfrak{U}_{2,t})_{\mathbf{C}}$ and U_1^0, U_2^0, $U_{1,t}^0$, $U_{2,t}^0$, $(U_1)_{\mathbf{C}}^0$, $(U_2)_{\mathbf{C}}^0$, $(U_{1,t})_{\mathbf{C}}^0$, $(U_{2,t})_{\mathbf{C}}^0$ the corresponding Lie algebras and the algebraic connected components of the identities of the groups U_1, U_2, $U_{1,t}$, $U_{2,t}$, $(U_1)_{\mathbf{C}}$, $(U_2)_{\mathbf{C}}$, $(U_{1,t})_{\mathbf{C}}$ and $(U_{2,t})_{\mathbf{C}}$, respectively. We proceed now along a series of remarks.

Remark 1. Since P_1 is defined over **Q**, the unipotent radical is defined over **Q** (see Lemma 1.35). Therefore, in virtue of 6.10 of [4] and *3)* of Theorem 17.43, the quotient space $R_u(P_1)/(R_u(P_1) \cap \Gamma)$ is compact. Consequently, by Lemma 5.6, the subgroup $R_u(P_1) \cap \Gamma$ is Zariski dense in $R_u(P_1)$, i.e. we have

$$U_1 = R_u(P_1). \tag{32}$$

Similarly,

$$U_2 = R_u(P_2). \tag{33}$$

Remark 2. Since P_1 is parabolic and $\neq G$ (by the definition of maximal parabolic **Q**-subgroups), we have

$$R_u(P_1) \neq e \tag{34}$$

(see the paragraphs following Lemmas 10.25 and 10.26). (32) and (34) imply

$$R_u(P_1) \cap \Gamma \neq e. \tag{35}$$

Remark 3. By the conditions formulated at the beginning of § 7 and by Lemma 4.8, we obtain that the intersection of Γ with any proper algebraic normal subgroup of G equals $\{e\}$. From this and (35) it follows that $R_u(P_1) \cap \Gamma$ and consequently $R_u(P_1)$ are not included in any proper algebraic normal subgroup of G.

Remark 4. Since P_1 and P_2 are opposite parabolic subgroups, therefore $R_u(P_1)$ and $R_u(P_2)$ are opposite orispherical subgroups. Thus, (32), (33), (34), Remark 3 and Lemma 19.11 imply that $\mathfrak{U}_1 \cup \mathfrak{U}_2$ generates $\mathfrak{G}$ as a Lie algebra ($\mathfrak{G}$ is the Lie algebra of G).

Remark 5. From Lemmas 19.15 and 19.16 we obtain that

$$\lim_{t\to 0} \mathfrak{U}_{1,t} = \mathfrak{U}_1 \quad \text{and} \quad \lim_{t\to 0} \mathfrak{U}_{2,t} = \mathfrak{U}_2. \tag{36}$$

From Remark 4, (36) and Lemma 19.18 it follows that there exists an $\varepsilon_0>0$ such that for every $t\in[0, \varepsilon_0]$ the union $\mathfrak{U}_{1,t}\cup\mathfrak{U}_{2,t}$ generates $\mathfrak{G}$ as a Lie algebra.

Remark 6. Since $(U_{1,t})_{\mathbf{C}}$ and $(U_{2,t})_{\mathbf{C}}$ are the complexifications of $U_{1,t}$ and $U_{2,t}$, it follows from Remark 5 that for every $t\in[0, \varepsilon_0]$ the union $(\mathfrak{U}_{1,t})_{\mathbf{C}}\cup(U_{2,t})_{\mathbf{C}}$ generates $\mathfrak{G}_{\mathbf{C}}$ as a Lie algebra, where $\mathfrak{G}_{\mathbf{C}}$ denotes the Lie algebra of the complexification $\mathfrak{G}_{\mathbf{C}}$ of G.

Remark 7. For every subgroup H of $G_{\mathbf{C}}$ or subalgebra $\mathfrak{H}$ of the Lie algebra $\mathfrak{G}_{\mathbf{C}}$, $C(H)$ (resp. $C(\mathfrak{H})$) will stand for the centralizer of H (resp. of $\mathfrak{H}$) in $G_{\mathbf{C}}$. Since *1)* G and consequently $G_{\mathbf{C}}$ are semisimple, *2)* $G_{\mathbf{C}}$ is connected and the center of it is trivial (see the conditions formulated at the beginning of § 7), it follows from Remark 6 that for every $t\in[0, \varepsilon_0]$

$$C((\mathfrak{U}_{1,t})_{\mathbf{C}})\cap C((\mathfrak{U}_{2,t})_{\mathbf{C}}) = e. \tag{37}$$

This implies that for any $t\in[0, \varepsilon_0]$

$$C((U_{1,t})^0_{\mathbf{C}})\cap C((U_{2,t})^0_{\mathbf{C}}) = e \tag{38}$$

is also valid.

Remark 8. Remark 6 and the connectedness of the group $G_{\mathbf{C}}$ imply that for every $t\in[0, \varepsilon_0]$ the minimal algebraic subgroup of $G_{\mathbf{C}}$ containing $(U_{1,t})^0_{\mathbf{C}}\cup(U_{2,t})^0_{\mathbf{C}}$ is identical with $G_{\mathbf{C}}$. Therefore, by Lemma 1.2.18, for any $t\in[0, \varepsilon_0]$ the subgroup generated by $T^t(R_u(P_1)\cap\Gamma)\cup T^t(R_u(P_2)\cap\Gamma)$ is Zariski dense in $G_{\mathbf{C}}$. Consequently, for any $t\in[0, \varepsilon_0]$ the subgroup $T^t(\Gamma)$ is Zariski dense in $G_{\mathbf{C}}$.

Remark 9. Put $L=P_1\cap P_2$ and denote by $\bar{L}_{\mathbf{Z}}$ the Zariski closure of the subgroup $L_{\mathbf{Z}}=L\cap G_{\mathbf{Z}}$ in G, and by $\bar{L}^0_{\mathbf{Z}}$ the algebraically connected component of the identity of $\bar{L}_{\mathbf{Z}}$. Since P_1 and P_2 are opposite $\mathbf{Q}$-subgroups of G, we see that L is a Levi subgroup in P_1 defined over $\mathbf{Q}$. Since P_1 is maximal, by Remark 1 the proof of Lemma 17.28 yields

$$\dim \bar{L}^0_{\mathbf{Z}} > 0. \tag{39}$$

Furthermore, since *1)* $L_{\mathbf{Z}}$ is Zariski dense in $\bar{L}_{\mathbf{Z}}$ (by definition), *2)* the subgroups $L_{\mathbf{Z}}$ and $\bar{L}_{\mathbf{Z}}\cap\Gamma$ are commensurable (in virtue of 3) of Theorem 17.43 and the fact that $L_{\mathbf{Z}}$ is contained in the proper parabolic $\mathbf{Q}$-subgroup P_1, by Lemmas 1.2.7 and 8.8 we have that $\bar{L}^0_{\mathbf{Z}}\cap\Gamma$ is Zariski dense in $\bar{L}^0_{\mathbf{Z}}$. From this and (39) it follows

$$\bar{L}^0_{\mathbf{Z}} \cap \Gamma \neq e. \tag{40}$$

Remark 10. (40) and Lemma 19.17 imply the existence of an $\varepsilon'>0$ such that for all $t\in[0, \varepsilon']$ the set $T^t(\bar{L}^0_{\mathbf{Z}}\cap\Gamma)$ is not contained in any proper algebraic normal subgroup of G.

Remark 11. Set $\tilde{\varepsilon}=\min(\varepsilon_0, \varepsilon')$ (ε_0 and ε' come from Remark 5 and 10, respectively). Then, since P_1 is a maximal parabolic **Q**-subgroup of G and $G_{\mathbf{C}}$ is a complex connected algebraic group with a trivial centre, it follows from Remarks 7, 8 and 10 and Theorem 19.10 that for any $t\in[0, \tilde{\varepsilon}]$ the representation T^t extends to an algebraic representation $\tilde{T}^t$ of $G_{\mathbf{C}}$ into itself.

Remark 12. From Remarks 8, 11 and Lemma 8.1 we obtain that for every $t\in[0, \tilde{\varepsilon}]$ the representation $\hat{T}^t$ is an epimorphism, i.e. $\hat{T}^t(G_{\mathbf{C}})=G_{\mathbf{C}}$.

Remark 13. Remark 12 shows that for $t\in[0, \tilde{\varepsilon}]$ we have $\dim(\operatorname{Ker}\hat{T}^t)=0$, i.e. for $t\in[0, \tilde{\varepsilon}]$ the kernel $\operatorname{Ker}\hat{T}^t$ is discrete. Therefore, since *1)* in any connected Lie group any discrete normal subgroup belongs to the centre of the group, *2)* $G_{\mathbf{C}}$ is connected and has no centre (see the beginning of § 7), for every $t\in[0, \tilde{\varepsilon}]$ the representation $\hat{T}^t$ is a monomorphism.

Remark 14. From Remarks 13 and 14 it follows that for any $t\in[0, \tilde{\varepsilon}]$ the representation $\hat{T}^t$ is an isomorphism. Thus $\hat{T}^t$ is an algebraic automorphism of the group $G_{\mathbf{C}}$ for every $t\in[0, \tilde{\varepsilon}]$. Note that $\varepsilon_0>0$ and $\varepsilon'>0$ imply $\tilde{\varepsilon}>0$.

Remark 15. Let us associate to the discrete unipotent subgroups $W_1=R_u(P_1)\cap\Gamma$ and $W_2=R_u(P_2)\cap\Gamma$ the quasi-lattices Δ_{W_1} and Δ_{W_2}, respectively, the same way as in the definition preceding Lemma 5.7 in § 5. From (32), (33) and Lemma 7.1 it follows that Δ_{W_1} and Δ_{W_2} are quasi-lattices of full dimension in $\mathfrak{U}_1$ and $\mathfrak{U}_2$, respectively.

Remark 16. Choose bases $e_{1,1}, \dots, e_{1,n}$ and $e_{2,1}, \dots, e_{2,m}$ in the quasi-lattices Δ_{W_1} and Δ_{W_2}, ($n=\dim\Delta_{W_1}, m=\dim\Delta_{W_2}$).* Then from Remarks 4 and 15 we get that $(\bigcup_{1\leq i\leq n} e_{1,i})\cup(\bigcup_{1\leq j\leq m} e_{2,j})$ generates $\mathfrak{G}$ as a Lie algebra.

Remark 17. For $t\in[0, \varepsilon]$, denote by S^t the automorphism of the Lie algebra $\mathfrak{G}_{\mathbf{C}}$ corresponding to the algebraic automorphism $\hat{T}^t$ of $\mathfrak{G}_{\mathbf{C}}$. Then, as is easy to see from the proof of Lemma 19.16 (see Remark 3 in the proof of Lemma 19.16 and (31)), we have

$$\lim_{t\to 0} S^t(e_{1,i}) = e_i \quad (1\leq i\leq n) \tag{41}$$

and

$$\lim_{t\to 0} S^t(e_{2,j}) = e_j \quad (1\leq j\leq n). \tag{42}$$

* In fact, $n=m$.

Remark 18. (41), (42) and Remark 16 imply that $\lim_{t\to 0} S^t = E$, where E is the identity operator of the space $\mathfrak{G}_C$ and the limit is taken in the natural topology of $GL(\mathfrak{G}_C)$.

Remark 19. It is well known (see [26]) that for any semisimple complex Lie group H, Ad H is a closed subgroup of finite index in the Lie group of all automorphisms of the Lie algebra $\mathfrak{H}$ of H.

From Remarks 18.19 and the connectedness of the group G_C the assertion of Theorem 19.5 follows.

V. In this section Theorem 19.6 will be proved.

Lemma 19.19. Let Γ_0 be a subgroup of finite index of the group Γ. Then Γ_0 cannot be represented as the direct product of proper subgroups.

Proof. Suppose the contrary holds, i.e. $\Gamma_0 = \Gamma_1 \times \Gamma_2$, where $\Gamma_1 \neq e$, $\Gamma_2 \neq e$. Denote by $\bar{\Gamma}_1$ and $\bar{\Gamma}_2$ the Zariski closure in G of the groups Γ_1 and Γ_2 respectively. Since $\Gamma_0 = \Gamma_1 \times \Gamma_2$, Γ_0 normalizes Γ_2, and consequently $\bar{\Gamma}_2$. On the other hand, since *1)* Γ is Zariski dense in G (in virtue of conditions formulated at the beginning of § 7 and the density theorem of Borel), *2)* Γ_0 has a finite index in Γ, *3)* G is algebraically connected (see the beginning of § 7), by Lemma 8.8, we obtain that Γ_0 is Zariski dense in G. The last two statements and Lemma 1.2.8 imply that G normalizes $\bar{\Gamma}_2$. Therefore, $\bar{\Gamma}_2$ is an algebraic normal subgroup in G. Therefore, since *1)* from Lemma 4.8 and conditions formulated at the beginning of § 7, it follows that the intersection of Γ with any proper algebraic normal subgroup of G equals $\{e\}$, *2)* $\bar{\Gamma}_2 \supset \Gamma_2$, *3)* $\Gamma_2 \neq e$, we get that

$$\bar{\Gamma}_2 = G. \tag{43}$$

Analogously,

$$\bar{\Gamma}_1 = G. \tag{44}$$

On the other hand, since $\Gamma_0 = \Gamma_1 \times \Gamma_2$, we find that Γ_1 commutes with Γ_2. Therefore, on account of Lemma 1.2.9, $\bar{\Gamma}_1$ commutes with $\bar{\Gamma}_2$, and consequently (43) and (44) give that G is commutative. But this contradicts the semisimplicity of G. The proof is complete.

Lemma 19.20. Let N be a nilpotent Lie group, $\mathfrak{N}$ the Lie algebra of N and Ad the adjoint representation of N. Then Ad N is a unipotent subgroup of $GL(\mathfrak{N})$.

Proof. Let $e = N_0 \subset N_1 \subset \ldots \subset N_i = N$ be an upper central series of N (see 10.2 of [8]). Induction on j $(1 \leqq j \leqq i)$ shows that for every j $(1 \leqq j \leqq i)$ the subgroup N_j is closed in N. Since every closed subgroup of a Lie group is a Lie subgroup, we get that for every j $(1 \leqq j \leqq i)$, N_j is a Lie subgroup of N. Let $\mathfrak{N}_j$ denote the Lie algebra of N_j. Since the subgroups N_j form a central series (see [8]), for every j $(1 \leqq j \leqq i)$, the commutator $[N, N_j]$ belongs to N_{j-1}. This implies that for any j

$(1 \leq j \leq i)$ and $n \in N$ we have $(E - \mathrm{Ad}\, n)\mathfrak{N}_j \subset \mathfrak{N}_{j-1}$, where E is the identity operator. Therefore, for any $n \in N$, $(E - \mathrm{Ad}\, n)^i = E$ holds, which completes the proof.

Lemma 19.21. Let U be a maximal unipotent subgroup of Γ, U' a subgroup of finite index in U and γ an element of Γ. If the subgroup generated by $\gamma \cup U'$ is nilpotent then $\gamma \in U$.

Proof. Denote by $\overline{U}$ (resp. $\overline{U}'$) the Zariski closure in G of U (resp. $\overline{U}'$) and by $\mathfrak{U}$ the Lie algebra of $\overline{U}$. Then, since U is unipotent, by Lemmas 1.3.3 and 1.3.1 we get that $\overline{U}$ is algebraically connected. Therefore, U' being of finite index in U, Lemma 8.8 gives

$$\overline{U}' = \overline{U}. \tag{45}$$

Since U is a maximal unipotent subgroup of Γ, on account of Lemma 11.18, we have

$$C(\overline{U}) \subset \overline{U}, \tag{46}$$

where $C(\overline{U})$ denotes the centralizer of $\overline{U}$ in G. As we have already noted, $\overline{U}$ is algebraically connected. Therefore, it follows from (46) that

$$C(\mathfrak{U}) \subset \overline{U}, \tag{47}$$

where $C(\mathfrak{U})$ denotes the centralizer of the Lie algebra $\mathfrak{U}$ in G.

Suppose now

$$\gamma \notin U, \tag{48}$$

and denote by $\overline{M}$ the Zariski closure of M in G and make the following observations.

Remark 1. Because of $M \supset U'$, by (45) we have $\overline{M} \supset \overline{U}$.

Remark 2. Since M is nilpotent, in virtue of Lemma 1.2.3, $\overline{M}$ is nilpotent too. Therefore, because of $\gamma \in M$, Remark 1 implies that the subgroup F generated by $\gamma \cup \overline{U}$ is nilpotent.

Remark 3. Denote by Q the subgroup generated by $\gamma \cup U$. Then, on account of $U \subset \overline{U}$ and Remark 2, the subgroup Q is nilpotent.

Remark 4. By (48), U is a proper subgroup of Q. Therefore, Q being nilpotent (see Remark 3), by Lemma 13.1 there exists an element q of Q such that q normalizes U and at the same time

$$q \notin U. \tag{49}$$

Remark 5. Since *1)* q normalizes U, *2)* $q \notin U$, *3)* $q \in \Gamma$ (because $\gamma \cup U \subset \Gamma$ and Q is generated by $\gamma \cup U$), *4)* U is a maximal unipotent subgroup of Γ, *5)* U is nilpotent (by Lemma 1.3.1), in virtue of Lemma 7.4, the element q is not unipotent.

Remark 6. Since U is a maximal unipotent subgroup of Γ, Lemma 7.7 gives $U=\bar{U}\cap\Gamma$. Therefore, as $q\in\Gamma$ but $q\notin U$ (see (49)), we have $q\notin\bar{U}$.

Remark 7. Denote by $N(\bar{U})$ the normalizer of U in G, and by T the restriction of the adjoint representation of $N(\bar{U})$ to the Lie algebra $\mathfrak{U}$ of the group U. Then, by Lemmas 1.2.1 and 1.28, $N(\bar{U})$ is an algebraic subgroup of G and T is an algebraic representation of $N(\bar{U})$.

Remark 8. Since q normalizes U (see Remark 4), q normalizes $\bar{U}$ too, i.e. $q\in N(\bar{U})$ and thus T is defined on q. Furthermore, $\bar{U}$ is unipotent in (view of Lemma 1.3.3 and the fact that U is unipotent). Therefore, from Remarks 5, 6, 7, the inclusion (46) and Lemma 1.3.8, it follows that $T(q)$ is not a unipotent linear representation.

Remark 9. Denote by $\bar{F}$ the Zariski closure in G of F (for the definition of F see Remark 2). Then, F being nilpotent (see Remark 2), in virtue of Lemma 1.2.3, $\bar{F}$ is nilpotent, too. Therefore, by Lemma 19.20, $\tilde{\mathrm{A}}\mathrm{d}\, F$ is unipotent, where $\tilde{\mathrm{A}}\mathrm{d}$ denotes the adjoint representation of the group $\bar{F}$.

Remark 10. Since *1)* $F\subset\bar{F}$, *2)* $Q\subset F$ (because of $U\subset\bar{U}$), *3)* $q\in Q$, we get $q\in\bar{F}$. On the other hand, $\bar{F}\supset F$ and $F\supset\bar{U}$ (see Remark 2) imply $\bar{F}\supset\bar{U}$. These and Remark 9 imply that $T(q)$ is unipotent (for the definition of T see Remark 7).

Remark 8 and 10 show that the assumption (48) leads to a contradiction, and this proves the lemma.

Lemma 19.22. Let H be an algebraically connected semisimple real algebraic group, V an orispherical subgroup of H and Λ a discrete subgroup of H. If *1)* the real rank of H (i.e. the dimension of a maximal **R**-split torus of H) equals 1, *2)* $V\neq e$, *3)* $V\cap\Lambda$ is Zariski dense in V, then $V\cap\Lambda$ is a subgroup of finite index in $N(V)\cap\Lambda$, where $N(V)$ denotes the normalizer of V in H.

Proof. Similarly to the end of § 10, denote by $\bar{S}N(V)$ the subgroup consisting of the elements of $N(V)$ with determinant ± 1 under the natural action of $N(V)$ over the Lie algebra $\mathfrak{B}$ of the group V. The proof will be carried out in a series of remarks.

Remark 1. Since *1)* the real rank of H is equal to 1, *2)* V is an orispheric subgroup of H, *3)* $V\neq e$, we can apply the classification of orispheric subgroups introduced in § 10 (see Lemma 10.9, the paragraph following Lemma 10.25, and Lemma 10.26) and we get that V is a maximal unipotent subgroup of the group H.

Remark 2. From Lemmas 1.2.1 and 1.2.8 it follows that $\bar{S}N(V)$ is an algebraic subgroup of the group H.

Remark 3. Let T be any maximal **R**-split torus of $\bar{S}N(V)$. Suppose $\dim T>0$. Then, since the real rank of H equals 1, T is a maximal **R**-split torus of H. Therefore, since *1)* T normalizes V (because of $T \subset \bar{S}N(V)$), *2)* V is a maximal unipotent subgroup of H (by Remark 1), by Lemma 13.5 we get

$$T \cap N(V) \neq T \cap \bar{S}N(V). \tag{50}$$

But (50) contradicts the fact that $T \subset \bar{S}N(V)$ and $\bar{S}N(V) \subset N(V)$. Thus the assumption dim $T > 0$ leads to a contradiction. Consequently, the real rank of the group S is zero.

Remark 4. It is well known that if the real rank of a reductive algebraic **R**-group F is zero, then it is compact in topological sense. This follows easily for instance from the existence of the Iwasawa decomposition for reductive groups (see 1.11 of [4] or from Lemmas 18.1 and 18.3 of this paper). From this it follows easily that if the real rank of an algebraic **R**-groups D equals zero, then the quotient group of D with respect to its unipotent radical $R_u(D)$ is compact.

Remark 5. Note that, by Lemmas 1.2.1 and 1.3.4, $V \subset \bar{S}N(V)$. Therefore, since *1)* V is a maximal unipotent subgroup in H (see Remark 1) and consequently, also in $\bar{S}N(V)$, *2)* $\bar{S}N(V)$ normalizes V, we obtain that V is the unipotent radical of the group $\bar{S}N(V)$.

Remark 6. Remarks 3, 4 and 5 imply that the quotient group $\bar{S}N(V)/V$ is compact.

Remark 7. Since, by assumption, $V \cap \Lambda$ is Zariski dense in V, on account of Lemma 5.4, the quotient space $V/(V \cap \Lambda)$ is compact. From this and Remark 6 it follows that the quotient space $\bar{S}N(V)/(V \cap \Lambda)$ is also compact. Since Λ is discrete, $V \cap \Lambda$ has a finite index in $\bar{S}N(V) \cap \Lambda$.

Remark 8. The proof of Lemma 7.13 shows that the assertion of Lemma 7.13 is valid not only for G and Γ, but for any algebraic **R**-group F and for any discrete subgroup ψ of F. Therefore, since $V \cap \Lambda$ is Zariski dense in V, we get that $N(V) \cap \Lambda = \bar{S}N(V) \cap \Lambda$.

Now the assertion of the Lemma is a consequence of Remarks 7 and 8.

Lemma 19.23. Let U be an arbitrary unipotent subgroup of the group Γ. Denote by $\bar{U}$ the Zariski closure of U in G and by $N(\bar{U})$ the normalizer of $\hat{U}$ in G. If U is a maximal unipotent normal subgroup of $N(\bar{U}) \cap \Gamma$ then $\hat{U}$ is an orispheric subgroup of G.

Proof. Denote by V the unipotent radical of the group $N(\bar{U})$ and suppose the assertion of the lemma does not hold. Then, by the definition of the orispheric subgroups, $\bar{U}$ is a proper subgroup in V, i.e.

$$\bar{U} \subset V, \quad \text{but} \quad \bar{U} \neq V. \tag{51}$$

We shall make a number of remarks.

Remark 1. By Lemma 17.4, the algebraic subgroup $\bar{U}$ is defined over **Q**.

Remark 2. Remark 1 and Lemmas 1.2.13 and 1.3.5 imply that the unipotent algebraic subgroup V is defined over **Q**. Therefore, there exists a maximal unipotent **Q**-subgroup $\tilde{V}$ in G which contains V. On account of Remark 17.18 and Lemma 10.17, $N(\tilde{U})$ is a parabolic **Q**-subgroup of G, where $N(\tilde{U})$ denotes the normalizer of $\tilde{V}$ in G.

Remark 3. By Lemma 7.5 and *1)* of Lemma 17.3, $(G_Z)^u \neq e$, where $(G_Z)^u$ denotes the set of unipotent elements of G_Z.

Remark 4. It follows from Remark 3 that $\tilde{V} \neq e$. Thus, since *1)* $\tilde{V}$ is algebraically and topologically connected (see Lemma 1.3.1), *2)* G is semisimple, we get that $N(\tilde{V}) \neq G$.

Remark 5. By Remarks 2 and 4, we obtain that $N(\tilde{V})$ is a nontrivial parabolic **Q**-subgroup of G. Therefore, in view of *3)* of Theorem 17.13, the subgroups $N(\tilde{V}) \cap G_Z$ and $N(\tilde{V}) \cap \Gamma$ are commensurable, and consequently, since $V \subset N(\tilde{V})$ (in fact $V \subset \tilde{V}$ and $\tilde{V} \subset N(\tilde{V})$), the subgroups $V \cap G_Z$ and $V \cap \Gamma$ are commensurable, too.

Remark 6. Now the unipotent algebraic **R**-group V is defined over **Q** (see Remark 2). Thus, on account of 6.10 of [4], the quotient space $V(V \cap G_Z)$ is compact. Therefore, by Remark 5, the quotient space $V/V \cap \Gamma$ is also compact, and consequently, by Lemma 5.6, the subgroup $V \cap \Gamma$ is Zariski dense in V. From this and (51) it follows that $U \subset \bar{U} \cap \Gamma$ is a proper subgroup of $V \cap \Gamma$.

Remark 7. Since V is a unipotent normal subgroup in $N(\bar{U})$ (in fact V is the unipotent radical of the group $N(\bar{U})$), therefore $V \cap \Gamma$ is a normal subgroup in $N(\bar{U}) \cap \Gamma$.

From Remarks 6 and 7 we can conclude that U is not a maximal unipotent normal subgroup in $N(\bar{U}) \cap \Gamma$. Thus, the negation of the validity of the lemma leads to a contradiction and that completes the proof.

Lemma 19.24. Let V be an arbitrary orispheric **Q**-subgroup of G, different from the identity. Denote by N the normalizer of V in G. Then $V \cap \Gamma$ is a maximal unipotent normal subgroup of $N \cap \Gamma$.

Proof. Suppose the contrary holds. Then $V \cap \Gamma$ is a proper subgroup of some unipotent normal subgroup U of $N \cap \Gamma$. Denote by $\bar{U}$ (resp. $\bar{N}_Z$) the Zariski closure of the subgroup U (resp. $N_Z = N \cap G_Z$), and by $\bar{N}_Z^0$ the algebraic connected component of the identity of $\bar{N}_Z$. Now the proof will be completed in several steps.

Remark 1. Since the orispheric **Q**-subgroup V is different from the identity, by Lemmas 1.2.13 and 10.9, $N(V)$ is a proper (i.e. different from G) parabolic **Q**-subgroup of G. Therefore, in virtue of *3)* of Theorem 17.43, the subgroups $N_Z = N \cap G_Z$ and $N \cap \Gamma$ are commensurable.

Remark 2. The facts that N is an algebraic subgroup of G (see Lemma 1.2.8) and $U \subset N$, imply $\overline{U} \subset N$.

Remark 3. Remarks 1 and 2 imply that the subgroups $\overline{U} \cap \Gamma$ and $\overline{U}_{\mathbf{Z}} = \overline{U} \cap G_{\mathbf{Z}}$ are commensurable.

Remark 4. Since $V \subset N$, by Remark 1 we get that the subgroups $V \cap \Gamma$ and $V_{\mathbf{Z}} = V \cap G_{\mathbf{Z}}$ are commensurable.

Remark 5. Since U is unipotent, by Lemma 1.3.3 so is $\overline{U}$. Therefore, by Lemma 1.3.1, $\overline{U}$ is algebraically connected.

Remark 6. Since U is Zariski dense in $\overline{U}$ (by the definition of $\overline{U}$), from Remarks 3 and 5 and Lemma 8.8 it follows that the subgroup $\overline{U} = \overline{U} \cap G_{\mathbf{Z}}$ is Zariski dense in $\overline{U}$.

Remark 7. The subgroups $N_{\mathbf{Z}} = N \cap G_{\mathbf{Z}}$ and $N \cap \Gamma$ are commensurable (see Remark 1). Thus, by Lemmas 1.2.6 and 8.8, the subgroup $\overline{N}_{\mathbf{Z}} \cap \Gamma$ is Zariski dense in $\overline{N}_{\mathbf{Z}}^{0}$.

Remark 8. $\overline{N}_{\mathbf{Z}}^{0} \cap \Gamma$ normalizes U (in fact, $\overline{N}_{\mathbf{Z}}^{0} \subset N$ and $N \cap \Gamma$ normalizes U). Thus $\overline{N}_{\mathbf{Z}}^{0} \cap \Gamma$ normalizes $\overline{U}$ too. Therefore, in view of Remark 7 and Lemma 1.2.8, $\overline{N}_{\mathbf{Z}}^{0}$ normalizes $\overline{U}$. On the other hand, it follows from Remarks 5 and 6 that $\overline{U} \subset \overline{N}_{\mathbf{Z}}^{0}$. The last two statements imply that $\overline{U}$ is a unipotent normal subgroup of $\overline{N}_{\mathbf{Z}}^{0}$.

Remark 9. Since V is an orispheric subgroup of G, V is the unipotent radical of N, according to the definition of orispheric subgroups.

Remark 10. Let h be an arbitrary element of the subgroup $N_{\mathbf{Q}} = N \cap G_{\mathbf{Q}}$. Then, as is easy to see, $(hN_{\mathbf{Z}}h^{-1}) \cap N_{\mathbf{Z}}$ is a subgroup of finite index both in $N_{\mathbf{Z}}$ and $hN_{\mathbf{Z}}h^{-1}$. Therefore, by Lemmas 1.2.7 and 8.8, $h\overline{N}_{\mathbf{Z}}^{0}h^{-1} = \overline{N}_{\mathbf{Z}}^{0}$. Thus, h being an arbitrary element of $N_{\mathbf{Q}}$, it follows that $N_{\mathbf{Q}}$ normalizes $\overline{N}_{\mathbf{Z}}^{0}$.

Remark 11. Since V is an orispheric subgroup in G, by Lemma 10.9, the subgroup N is parabolic. Consequently (see 4.3 of [5]), the subgroup N is algebraically connected. On the other hand, V is defined over $\mathbf{Q}$, thus, by Lemma 1.2.13, N is also defined over $\mathbf{Q}$. The last two statements and Lemma 1.2.11 imply that $N_{\mathbf{Q}}$ is Zariski dense in N.

Remark 12. It follows from Remarks 11, 12, and Lemma 1.2.8 that N normalizes $\overline{N}_{\mathbf{Z}}^{0}$, i.e. $\overline{N}_{\mathbf{Z}}^{0}$ is a normal subgroup of N.

Remark 13. Denote by W the unipotent radical of $\overline{N}_{\mathbf{Z}}^{0}$. Then, since *1)* the unipotent radical is uniquely defined, *2)* by Remark 12, $\overline{N}_{\mathbf{Z}}^{0}$ is a normal subgroup of N, we get that W is a unipotent normal subgroup of N. Therefore, since V is the unipotent radical of the group N, we have $W \subset V$.

Remark 14. Since $\bar{U}$ is a unipotent normal subgroup in N_Z^0, we have $\bar{U} \subset W$. Therefore $W \cap \Gamma \supset \bar{U} \cap \Gamma \supset U$.

Remark 15. Since $V \cap \Gamma$ is a proper subgroup in U (see the beginning of the proof of the lemma) and $U \subset \Gamma$, we get that $U \not\subset V$. Therefore by Remark 14, $W \cap \Gamma \not\subset V$, and consequently $W \not\subset V$.

Remark 13 and 15 show a contradiction, thus the lemma is proved.

Lemma 19.25. Let V be a maximal unipotent **Q**-subgroup of G. Then there exist maximal parabolic **Q**-subgroups $P_1, \ldots, P_n$ (n is the **Q**-rank of G) such that *1)* for every i $(1 \leqq i \leqq n)$ we have $R_u(P_i) \subset V$, *2)* the minimal algebraic subgroup containing $\bigcup_{1 \leqq i \leqq n} R_u(P_i)$ coincides with V.

Proof. In view of Remark 17.18 and Lemma 10.17, the normalizer $N(V)$ of V in G is a minimal parabolic **Q**-subgroup of G. On the other hand, since V is a maximal unipotent **Q**-subgroup of G and by Lemmas 1.2.13 and 1.3.5 $R_u(N(V))$ is defined over **Q**, we obtain that $V = R_u(N(V))$ $(R_u(N(V))$ is the unipotent radical of $(N(V))$. Therefore, by the classification of parabolic **Q**-subgroups (see Remark 17.18 and § 10), there is in G a maximal **Q**-decomposable torus and an order ψ of the root system Σ of G corresponding to T, such that $V = H_{\Sigma^+}$, where Σ^+ is the set of roots in Σ positive with respect to ψ, and H_{Σ^+} denotes the same as in § 10 (see (29) in § 10, and Remark 17.18).

Let us recall that for $\alpha \in \Sigma$, $\mathfrak{U}_\alpha$ denotes the root subspace corresponding to α, and U_α stands for $\exp(\mathfrak{U}_\alpha)$. Then it follows from the classification of the parabolic **Q**-subgroups normalized by T (see Remark 17.18 and § 10) that for any $\alpha \in \Sigma^+$ there exists a root $d \in \Sigma_0$ (Σ_0 is the set of simple roots in Σ^+) such that $U_\alpha \subset H_d$ (for the definition of H_d see (29) in § 10). On the other hand, by Remark 17.18, Lemmas 10.25 and 10.26 it is clear that for every $d \in \Sigma_0$, H_d is the unipotent radical of the maximal parabolic **Q**-subgroup P_d. The last two statements and the equality $V = H_{\Sigma^+}$ (see the end of the preceding paragraph) prove the lemma.

Definition 19.26. Let H be a group and $H_1, \ldots, H_n$ a finite family of its subgroups. We say that $\bigcup_{1 \leqq i \leqq n} H_i$ *almost generates* H if for any family $H_1', \ldots, H_n'$ of subgroups of finite index in $H_1, \ldots, H_n$, the subgroup generated by $\bigcup_{1 \leqq i \leqq n} H_i'$ is of finite index in H.

Lemma 19.27. Let U be a maximal unipotent subgroup of Γ. Denote by $\bar{U}$ the Zariski closure of U in G. Then there exist maximal parabolic **Q**-subgroups $P_1, \ldots, P_n$ (n is the **Q**-rank of G) in G such that *1)* $\bigcup_{1 \leqq i \leqq n} R_u(P_i) \subset \bar{U}$, *2)* $\bigcup_{1 \leqq i \leqq n} R_u(P_i) \cap \Gamma \cap U$ almost generates U.

Proof. By Theorem 17.19, $\bar{U}$ is a maximal unipotent **Q**-subgroup of G. Therefore, by Lemma 19.25, there exist maximal parabolic **Q**-subgroups $P_1, \ldots, P_u$

in G such that *1)* for any i $(1 \leqq i \leqq n)$, $R_u(P_i) \subset V$, *2)* the minimal algebraic subgroup containing $\bigcup_{1 \leqq i \leqq n} R_u(P_i)$ coincides with V. Then, by Lemma 1.3.5, the unipotent algebraic subgroups $R_u(P_i)$ $(1 \leqq i \leqq n)$ are defined over $\mathbf{Q}$. Therefore, on account of 6.10 of [4], the quotient spaces $R_u(P_i)/R_u(P_i) \cap G_{\mathbf{Z}}$ are compact. On the other hand, since $R_u(P_i) \subset \overline{U}$ for any i $(1 \leqq i \leqq n)$, statement 3 of Lemma 17.3 implies that the subgroups $R_u(P_i) \cap \Gamma$ and $R_u(P_i) \cap G_{\mathbf{Z}}$ are commensurable for every i $(1 \leqq i \leqq n)$. From the last two statements it follows that for any i $(1 \leqq i \leqq n)$ the quotient space $R_u(P_i)/R_u(P_i) \cap \Gamma$ is also compact. Now, note that *1)* the maximality of U implies $U = \overline{U} \cap \Gamma$ (Lemma 7.7), *2)* since $U = \overline{U} \cap \Gamma$ and $R_u(P_i) \subset \overline{U}$, for every i $(1 \leqq i \leqq n)$ $R_u(P_i) \cap \Gamma \cap U = R_u(P_i) \cap \Gamma$. From these last two statements and Lemma 5.6 we conclude that if U_i' $(1 \leqq i \leqq n)$ is any subgroup of finite index in $R_u(P_i) \cap \Gamma \cap U$ then U_i' is Zariski dense in $R_u(P_i)$. Therefore, since *1)* $\bigcup_{1 \leqq i \leqq n} R_u(P_i) \subset \overline{U}$, *2)* the minimal algebraic subgroup containing $\bigcup_{1 \leqq i \leqq n} R_u(P_i)$ coincides with $\overline{U}$, by Lemmas 1.2.17 and 5.4 we see that $P_1, \ldots, P_n$ are the desired subgroups. The lemma is proved.

From now on, we will use the notations introduced when formulating Theorem 19.6, i.e. G' is an algebraically connected semisimple real algebraic group without compact factors whose complexification has no centre, Γ' is a lattice in G' and φ is an isomorphism (in abstract group-theoretical sense) between Γ and Γ'.

Lemma 19.28. Let U be an arbitrary maximal unipotent subgroup of Γ. Then there exists a subgroup U' of finite index in U such that $\varphi(U')$ is a unipotent subgroup of G'.

Proof. According to Lemma 19.27, there exist maximal parabolic $\mathbf{Q}$-subgroups $P_1, \ldots, P_n$ in G such that *1)* $\bigcup_{1 \leqq i \leqq n} R_u(P_i) \subset \overline{U}$, *2)* $\bigcup_{1 \leqq i \leqq n} (R_u(P_i) \cap \Gamma \cap U)$ almost generates U. Then, since *1)* for any i $(1 \leqq i \leqq n)$, P_i is a maximal parabolic $\mathbf{Q}$-subgroup, *2)* by Remark 17.18 and Lemma 10.10, for every i there exists a parabolic $\mathbf{Q}$-subgroup P_i' opposing to P_i, by Remark 19.14 and Lemma 19.7 for every i $(1 \leqq i \leqq n)$ there exists a subgroup U_i' of finite index in $R_u(P_i) \cap \Gamma \cap U$ such that the subgroup $\varphi(U_i')$ is unipotent. Therefore, denoting by U' the subgroup generated by $\bigcup_{1 \leqq i \leqq n} U_i'$, we obtain that since *1)* $\bigcup_{1 \leqq i \leqq n} (R_u(P_i) \cap \Gamma \cap U)$ almost generates U, *2)* by Lemma 1.3.1 U and consequently U' and $\varphi(U')$ are nilpotent, by Lemma 7.4* we conclude that *1)* $\varphi(U')$ is unipotent, *2)* U' is of finite index in U. The lemma is proved.

Lemma 19.29. The quotient space G'/Γ' is not compact.

Proof. By virtue of Lemma 7.5, $\Gamma^u \neq e$ (Γ^u is the set of unipotent elements of Γ). Let u be a unipotent element of Γ different from the identity. Then, since any non-

* In the proof of Lemma 7.4 we did not use the conditions imposed upon G at the beginning of § 7.

zero power of any unipotent element different from the identity, is not equal to the identity, the subgroup U_0 generated by u is infinite. Let U be an arbitrary maximal unipotent subgroup of Γ containing U_0. Since U_0 is infinite, so is U. Therefore, by Lemma 19.28 there exists an element $u' \in U$ different from the identity such that $\varphi(u')$ is unipotent. Since φ is an isomorphism, $\varphi(u') \neq e$. Therefore, Γ contains unipotent elements different from the identity. Thus, since *1)* as is well known (see e.g. [25]) any uniform subgroup of a semisimple real algebraic group consists only of semisimple elements, *2)* G' is semisimple, we get that G'/Γ' is not compact. The lemma is proved.

Lemma 19.30. The pair (G', Γ') is irreducible.

Proof. Since φ is an isomorphism between Γ and Γ', by Lemma 19.19 we get that none of the subgroups of finite index of Γ' can be represented as a direct product of proper subgroups, whence the lemma follows (here we used the facts that *1)* G' has no compact factors, *2)* Γ' is a lattice in G', *3)* the Haar measure of any noncompact Lie group is infinite).

Remark 19.31. Since *1)* G' is an algebraically connected semisimple real algebraic group, *2)* G' has no compact factors, *3)* the centre of the complexification of G is trivial, *4)* Γ' is a lattice in G', *5)* the quotient space G'/Γ' is not compact (see Lemma 17.29), *6)* the pair (G', Γ') is irreducible, we get that G' and Γ' satisfy the conditions formulated at the beginning of § 7.

Remark 19.32. Since the real rank of G is greater than 1, in virtue of Remark 19.31 all the assertions proved in §§ 7, 11 and 12 remain valid if we replace G by G' and Γ by Γ'.

From Remark 19.32, Theorem 11.13 and Lemma 7.5 we can conclude

Lemma 19.33. Let U' be an arbitrary maximal unipotent subgroup of Γ'. Denote by $\overline{U}'$ the Zariski closure of U' in G. Then $\overline{U}'$ is an orispheric subgroup of G' different from the identity.

Lemma 19.34. Let U be an arbitrary maximal unipotent subgroup of Γ. Denote by U_φ the set of all elements u of U for which $\varphi(u)$ is unipotent. Then $\varphi(U_\varphi)$ is a maximal unipotent subgroup of Γ'.

Proof. On account of Lemma 1.3.1 the group U is nilpotent. Therefore the group U is nilpotent too. In view of Lemma 19.28 this implies that U_φ is a subgroup. Therefore, in virtue of Lemma 19.28 U_φ is a subgroup of finite index in U.

Suppose now that $\varphi(U_\varphi)$ is not a maximal unipotent subgroup of Γ', i.e. $\varphi(U_\varphi)$ is a proper subgroup of some unipotent subgroup U of the group Γ'. Then, *1)* since φ is an isomorphism, U_φ is a proper subgroup of $\varphi^{-1}(U')$, *2)* since a) U_φ consists of those $x \in U$ for which $\varphi(U_\varphi)$ is unipotent, b) the subgroup U' is unipotent, we get $U \cap \varphi^{-1}(U') = U_\varphi$, *3)* since U' is nilpotent (see Lemma 1.3.1(and φ is an isomorphism, the group $\varphi^{-1}(U')$ is nilpotent. Therefore, on account of Lemma 13.1, there exists in $\varphi^{-1}(U')$ an element γ_0 such that *1)* γ_0 normalizes U_φ, *2)* $\gamma_0 \notin U$. Then, since $\varphi^{-1}(U')$ is nilpotent and $\gamma_0 \cup U \subset \varphi^{-1}(U')$, the subgroup N generated by $\gamma_0 \cup U_\varphi$ is nilpotent. Therefore, since U_φ is a subgroup of finite index in U, and U is a maximal unipotent subgroup of Γ, in virtue of Lemma 19.21, we get that $\gamma_0 \in U$, which is a contradiction. Thus the assumption that $\varphi(U_\varphi)$ is not a maximal unipotent subgroup of Γ' led us to a contradiction. The lemma is proved.

Lemma 19.35. The real rank of G' is greater than 1.

Proof. Suppose the contrary holds true, i.e. the **R**-rank of G' is equal to 1, and recall (see § 13) that a maximal unipotent subgroup U of Γ is said to be ultramaximal if $\dim \tilde{A}(U) = \dim \overline{U}$ (for the definition of $A(U)$ and $\overline{U}$ see § 7 and 11).

Now we distinguish two cases:

1) all the maximal unipotent subgroups of Γ are ultramaximal.

2) Γ possesses a maximal unipotent subgroup which is not ultramaximal.

Case 1). In virtue of Lemma 13.6, there exist in Γ maximal unipotent subgroups U' and U'' such that *1)* $\overline{U}' \cap \overline{U}'' \neq e$, *2)* $\overline{U}' \neq \overline{U}''$ (as always, for a unipotent subgroup U of Γ, $\overline{U}$ stands for its Zariski closure in G). Since U' and U'' are maximal, in virtue of Lemma 19.34, $\varphi(U'_\varphi)$ and $\varphi(U''_\varphi)$ are maximal unipotent subgroups of Γ', where U'_φ and U''_φ are defined in the same manner as U_φ in Lemma 19.34. Therefore, in view of Lemma 19.33, V' and V'' are orispheric subgroups of G' different from the identity, where V' and V'' denote the Zariski closures in G' of the subgroups $\varphi(U'_\varphi)$ and $\varphi(U''_\varphi)$, respectively. Observe now that *1)* $\overline{U}' \cap \overline{U}'' \neq e$, *2)* on account of Lemma 5.4, $\overline{U}'$ and $\overline{U}''$ are integer subgroups of G (for the definition of integer subgroups see § 7), *3)* since $\overline{U}'$ and $\overline{U}''$ are integer, in virtue of Lemmas 6.7 and 6.8, $\overline{U}' \cap \overline{U}''$ is integer, too, *4)* on account of Lemma 1.3.1, the subgroups $\overline{U}'$ and U'' are unipotent and, consequently, $\overline{U}' \cap \overline{U}''$ is unipotent, too, *5)* since $U' = \overline{U}' \cap \Gamma$ and $U'' = \overline{U}'' \cap \Gamma$ (because of the maximality of U' and U''), we have $\overline{U}' \cap \overline{U}'' \cap \Gamma = U' \cap U''$, *6)* since U'_φ and U''_φ have finite indices in U' and U'' respectively, $U'_\varphi \cap U''_\varphi$ has a finite index in $U' \cap U''$. *3)*, *5)* and *6)* imply that the quotient space $\overline{U}' \cap \overline{U}'' / U'_\varphi \cap U''_\varphi$ is compact, thus by *1)*, *4)* and Lemmas 1.3.1 and 5.1 it follows $U'_\varphi \cap U''_\varphi \neq e$. Therefore, since φ is an isomorphism, the intersection of the orispheric subgroups V' and V'' differs from $\{e\}$. But then, taking into account the assumption on the real rank of G', from Lemmas 10.9 and 10.16 and the classification of orispheric subgroups (see § 10) it is easy to deduce that $V' = V''$. The unipotent subgroup $V' \cap \Gamma' = V'' \cap \Gamma'$ contains, however, both $\varphi(U'_\varphi)$ and $\varphi(U''_\varphi)$, which implies, by the maximality of $\varphi(U'_\varphi)$ and $\varphi(U''_\varphi)$ that $\varphi(U'_\varphi) = \varphi(U''_\varphi)$.

Therefore, φ being an isomorphism, $U'_\varphi = U''_\varphi$ follows. On the other hand, since *1)* U'_φ and U''_φ have finite indices in U' and U'', respectively, *2)* $\bar{U}'$ and $\bar{U}''$ are algebraically connected (see Lemmas 1.3.3 and 1.3.1), we can deduce from Lemma 8.8 that U'_φ and U''_φ are Zariski dense in $\bar{U}'$ and $\bar{U}''$, respectively. The last two statements imply $\bar{U}' = \bar{U}''$. Thus we have a contradiction to $\bar{U}' \neq \bar{U}''$. Case 1) is proved.

Case 2). Let U be a maximal but not ultramaximal unipotent subgroup of Γ, and let U_φ stand for the same set as in the formulation of Lemma 19.34. Then, on account of Lemma 19.34, $\varphi(U_\varphi)$ is a maximal unipotent subgroup of Γ'. Therefore, in view of Lemma 19.33, V is an orispheric subgroup of G different from the identity, where V denotes the Zariski closure of the subgroup $\varphi(U_\varphi)$ in G'. Then, since *1)* we have supposed that the real rank of G' equals 1, *2)* $\varphi(U_\varphi) = V \cap \Gamma'$ (because $\varphi(U_\varphi)$ is a maximal unipotent subgroup of Γ'), in virtue of Lemma 19.22 we get that the subgroup $\varphi(U_\varphi)$ has a finite index in $N(V) \cap \Gamma'$, where $N(V)$ denotes the normalizer of V in G'. On the other hand, since $\varphi(U_\varphi) = V \cap \Gamma'$, we have $N(V) \cap \Gamma' \supset N_{\Gamma'}(\varphi(U_\varphi))$, where $N_{\Gamma'}(\varphi(U_\varphi))$ denotes the normalizer of $\varphi(U_\varphi)$ in Γ'. The last two statements give that the subgroup $\varphi(U_\varphi)$ has a finite index in $N_{\Gamma'}(\varphi(U_\varphi))$. Therefore, since φ is an isomorphism, U_φ has finite index in $N_\varphi(U_\varphi)$, where $N_\Gamma(U_\varphi)$ denotes the normalizer of U_φ in Γ. On the other hand, since *1)* U_φ has finite index in U, *2)* $\bar{U}$ is algebraically connected (see Lemmas 1.3.3 and 1.3.1), we get that U_φ is Zariski dense in $\bar{U}$, and it follows easily that $N_\Gamma(U_\varphi) \supset N(\bar{U}) \cap \Gamma$. These last two statements imply that the index of U_φ in $N(\bar{U}) \cap \Gamma$ is finite. Consequently, the quotient group $\tilde{A}(U)/\bar{U}$ is finite (as usually (see § 11 and after) $\tilde{A}(U) = A(\bar{S}N(\bar{U}) \cap \Gamma))$. From this it follows that dim $\tilde{A}(U) =$ dim $(\bar{U})$ which contradicts the fact that U is not ultramaximal. Thus the lemma is proved in Case 2) as well.

Remark 19.36. From Remark 19.32 and Lemma 19.35 it follows that G' and Γ' satisfy the conditions formulated at the beginning of §§ 7 and 13. Therefore, since φ is an isomorphism, all the assertions proved for G, Γ and φ remain valid if G, Γ and φ are replaced by G', Γ' and φ^{-1} respectively.

Lemma 19.37. Denote by Γ^u (resp. $(\Gamma')^u$) the set of all unipotent elements of Γ (resp. Γ'). Then $\varphi(\Gamma^u) = (\Gamma')^u$.

Proof. Let U be an arbitrary maximal unipotent subgroup of Γ and let U_φ denote the set of those $u \in U$ for which $\varphi(u)$ is unipotent. Then, in view of Lemma 19.34, $\varphi(U_\varphi)$ is a maximal unipotent subgroup of Γ'. Therefore, since *1)* $U_\varphi = \varphi^{-1}(\varphi(U_\varphi))$, *2)* U_φ is unipotent (because $U_\varphi \subset U$), by Remark 19.36 and Lemma 19.34 we have that U_φ is a maximal unipotent subgroup of Γ. Thus $U_\varphi \subset U$ implies $U_\varphi = U$. Hence we have proved that for any maximal unipotent subgroup U of Γ, the subgroup $\varphi(U)$ is unipotent. Furthermore, since any unipotent element of Γ is contained in some maximal unipotent subgroup of Γ, we have

$$\varphi(\Gamma^u) \subset (\Gamma')^u. \tag{52}$$

On the other hand, by making use of Remark 19.36, we can prove similarly that

$$\varphi^{-1}((\Gamma')^u) \subset \Gamma^u. \tag{53}$$

These last two inclusions prove the lemma.

Lemma 19.38. Let V be an arbitrary orispheric **Q**-subgroup different from the identity. Denote by W the Zariski closure of the subgroup $\varphi(V \cap \Gamma)$ in G'. Then *1)* W is an orispherical subgroup in G', *2)* $W \cap \Gamma' \neq e$.

Proof. Since $V \neq e$ and V is connected (see Lemma 1.3.1), $\dim V > 0$. Therefore, by the semisimplicity of G and by Lemmas 1.2.1 and 10.9, $N(V)$ is a proper parabolic **Q**-subgroup of G. Consequently, on account of *3)* of Theorem 19.43, the subgroups $N(V) \cap G_{\mathbf{Z}}$ and $N(V) \cap \Gamma$ are commensurable. Therefore $V_{\mathbf{Z}} = V \cap G_{\mathbf{Z}}$ and $V \cap \Gamma$ are commensurable, too. Thus, the quotient space $V/V_{\mathbf{Z}}$ being compact (in view of 6.10 of [4]), by Lemma 5.6 the subgroup $V \cap \Gamma$ is Zariski dense in V.

Denote by $N(V)$ the normalizer of V in G, by $N_\Gamma(V \cap \Gamma)$ that of $V \cap \Gamma$ in Γ and by $N_{\Gamma'}(\varphi(V \cap \Gamma))$ that of $\varphi(V \cap \Gamma)$ in Γ'. Let us make the following remarks:

Remark 1. Since, as it has just been proved, $V \cap \Gamma$ is Zariski dense in V, we get that $N(V) \cap \Gamma = N_\Gamma(V \cap \Gamma)$.

Remark 2. From Remark 1 and Lemma 19.24 it follows that $V \cap \Gamma$ is a maximal unipotent normal subgroup of $N_\Gamma(V \cap \Gamma)$. Therefore, since φ is an isomorphism, by Lemma 19.37 $\varphi(V \cap \Gamma)$ is a maximal unipotent normal subgroup of $N_{\Gamma'}(\varphi(V \cap \Gamma))$.

Remark 3. Denote by M the normalizer of the subgroup $\varphi(V \cap \Gamma)$ in $W \cap \Gamma'$ and note that *1)* since $\varphi(V \cap \Gamma)$ is Zariski dense in W, $N_{\Gamma'}(\varphi(V \cap \Gamma))$ normalizes $W \cap \Gamma'$, *2)* since $N_{\Gamma'}(\varphi(V \cap \Gamma))$ normalizes both $\varphi(V \cap \Gamma)$ and $W \cap \Gamma'$, therefore $N_{\Gamma'}(\varphi(V \cap \Gamma))$ normalizes M, *3)* since $\varphi(V \cap \Gamma)$ is unipotent (in view of Lemma 19.37), by Lemma 1.3.1 W and consequently $W \cap \Gamma'$ and M are unipotent, *4)* since M normalizes $\varphi(V \cap \Gamma)$ and is contained in Γ', we have $M \subset N_{\Gamma'}(\varphi(V \cap \Gamma))$. Thus, M is a unipotent normal subgroup of $N_{\Gamma'}(\varphi(V \cap \Gamma))$. Therefore, on account of Remark 2, $M \supset \varphi(V \cap \Gamma)$ implies

$$M = \varphi(V \cap \Gamma). \tag{54}$$

Remark 4. The fact that the subgroup $W \cap \Gamma'$ is unipotent (see Remark 3), implies by Lemma 1.3.1 that $W \cap \Gamma'$ is nilpotent. From this, Lemma 13.1 and (54) we conclude $W \cap \Gamma' = \varphi(V \cap \Gamma)$.

Remark 5. Since *1)* $W\cap\Gamma'=\varphi(V\cap\Gamma)$, *2)* the subgroup $\varphi(V\cap\Gamma)$ is Zariski dense in W (by the definition of W), it is easy to see that $N_{\Gamma'}(\varphi(V\cap\Gamma))=N(W)\cap\Gamma$, where $N(W)$ denotes the normalizer of W in G'. Thus, $\varphi(V\cap\Gamma)$, being Zariski dense, by Remark 2, Remark 19.36 and Lemma 19.23 W is orispherical.

Remark 6. Since, as was proved at the beginning of the lemma, *1)* $V\cap\Gamma$ is Zariski dense in G, *2)* $V\neq e$ (by assumption), we have $V\cap\Gamma\neq e$. Therefore the fact that φ is an isomorphism between Γ and Γ' implies $W\cap\Gamma'\supset\varphi(V\cap\Gamma)\neq e$.

Now the assertion of the lemma follows from Remarks 5 and 6.

Lemma 19.39. Let V_1 and V_2 be opposite orispheric **Q**-subgroups of G. Denote by W_1 and W_2 the Zariski closures of the subgroups $\varphi(V_1\cap\Gamma)$ and $\varphi(V_2\cap\Gamma)$ in G', by W_1^0 and W_2^0 the algebraically connected components of the identities of the complexifications $(W_1)_{\mathbf{C}}$ and $(W_2)_{\mathbf{C}}$ of W_1 and W_2 and by $C(W_1^0)$ and $C(W_2^0)$ the centralizers of W_1^0 and W_2^0 in the complexification $G_{\mathbf{C}}$ of G', respectively. If $V_1\neq e$, then $C(W_1^0)\cap C(W_2^0)\neq e$.

The proof will be carried out in a series of remarks.

Remark 1. Since V_1 and V_2 are orispheric **Q**-subgroups of G, by Lemma 19.38 W_1 and W_2 are orispheric subgroups of G'.

Remark 2. Since $(W_1)_{\mathbf{C}}$ and $(W_2)_{\mathbf{C}}$ are unipotent algebraic subgroups of $G_{\mathbf{C}}$ (in view of Lemmas 19.37 and 1.3.3), by Lemma 1.3.1 $W_1^0=(W_1)_{\mathbf{C}}$ and $W_2^0=(W_2)_{\mathbf{C}}$.

Remark 3. Since V_1 and V_2 are opposite orispheric subgroups of G, we have $V_1\cap V_2=e$. Consequently $V_1\cap V_2\cap\Gamma=(V_1\cap\Gamma)\cap(V_2\cap\Gamma)=e$.

Remark 4. Since φ is an isomorphism, from Remark 3 it follows that $\varphi(V_1\cap\Gamma)\cap\varphi(V_2\cap\Gamma)=e$.

Remark 5. Since $\varphi(V_1\cap\Gamma)$ and $\varphi(V_2\cap\Gamma)$ are Zariski dense in W_1 and W_2 (by the definition of W_1 and W_2), respectively, by Remark 1 and Lemma 5.4 the quotient spaces $W_1/\varphi(V_1\cap\Gamma)$ and $W_2/\varphi(V_2\cap\Gamma)$ are compact.

Remark 6. From Remark 5 it follows that $\varphi(V_1\cap\Gamma)$ and $\varphi(V_2\cap\Gamma)$ have finite indices in $W_1\cap\Gamma'$ and $W_2\cap\Gamma'$, respectively.

Remark 7. Since $\varphi(V_1\cap\Gamma)\subset\Gamma'$ and $\varphi(V_2\cap\Gamma)\subset\Gamma'$, from Remark 6 it follows that the quotient spaces $W_1/(W_1\cap\Gamma')$ and $W_2/(W_2\cap\Gamma')$ are compact. Therefore, on account of Remark 19.36 and Lemmas 6.7 and 6.8, the quotient space $(W_1\cap W_2)/(W_1\cap W_2\cap\Gamma')$ is compact.

Remark 8. Since W_1 and W_2 are unipotent algebraic subgroups of G' (see Remark 2), $W_1\cap W_2$ is also a unipotent subgroup of G'.

Remark 9. From Remarks 7, 8 and Lemma 5.6 we get that the subgroup $W_1\cap W_2\cap\Gamma'$ is Zariski dense in $W_1\cap W_2$.

Remark 10. Since (see Remark 8) the algebraic subgroup $W_1 \cap W_2$ is unipotent, by Lemma 1.3.1 $W_1 \cap W_2$ is algebraically connected.

Remark 11. From Remark 6 it follows that the subgroup $\varphi(V_1 \cap \Gamma) \cap \varphi(V_2 \cap \Gamma)$ has finite index in $W_1 \cap W_2 \cap \Gamma'$. Therefore, by virtue of Remarks 9 and 10 and Lemma 8.8, the subgroup $\varphi(V_1 \cap \Gamma) \cap \varphi(V_2 \cap \Gamma)$ is Zariski dense in $W_1 \cap W_2$.

Remark 12. From Remarks 4 and 11 it follows that $W_1 \cap W_2 = e$.

Remark 13. Since $V_1 \neq e$, and V_2 is an orispheric subgroup opposing to V_1, it is easy to see that $V_2 \neq e$. Consequently, both V_1 and V_2 are orispheric **Q**-subgroups different from the identity.

Remark 14. Remark 13 and *2)* of Lemma 19.38 imply that $W_1 \cap \Gamma' \neq e$ and $W_2 \cap \Gamma' \neq e$. On the other hand, since $\varphi(V_1 \cap \Gamma)$ and $\varphi(V_2 \cap \Gamma)$ are Zariski dense in W_1 and W_2 respectively, and φ is an isomorphism between Γ and Γ', we get that $W_1 \cap \Gamma'$ and $W_2 \cap \Gamma'$ are Zariski dense in W_1 and W_2.

Remark 15. From Remarks 19.36, 13, 14 and Lemma 16.3 it follows that $C((W_1)_{\mathbf{C}}) \subset (W_1)_{\mathbf{C}}$ and $C((W_2)_{\mathbf{C}}) \subset (W_2)_{\mathbf{C}}$, where $C((W_1)_{\mathbf{C}})$ and $C((W_2)_{\mathbf{C}})$ denote the centralizers of the subgroups $(W_1)_{\mathbf{C}}$ and $(W_2)_{\mathbf{C}}$ in $G_{\mathbf{C}}$.

Remark 16. Since $(W_1)_{\mathbf{C}}$ and $(W_2)_{\mathbf{C}}$ as complexifications of W_1 and W_2, they are defined over **R**. Therefore the algebraic subgroups $(W_1)_{\mathbf{C}}$ and $(W_2)_{\mathbf{C}}$ are also defined over **R**. On the other hand, since (see Remark 2) $(W_1)_{\mathbf{C}}$ is unipotent, by Lemma 1.3.1 $(W_1)_{\mathbf{C}}$ is connected. The last two statements and Lemma 1.2.11 imply that $W_1 \cap W_2$ is Zariski dense in $(W_1)_{\mathbf{C}} \cap (W_2)_{\mathbf{C}}$. Therefore, Remark 12 gives $(W_1)_{\mathbf{C}} \cap (W_2)_{\mathbf{C}} = e$.

Remarks 2, 15 and 16 imply the lemma.

Now we turn to the

Proof of Theorem 19.6. By Lemma 17.26, there exists in G a proper parabolic **Q**-subgroup. Therefore there exists in G a parabolic **Q**-subgroup maximal in the class of all proper parabolic **Q**-subgroups. Let P_1 be such a maximal parabolic **Q**-subgroup. Then, in view of Remark 17.18 and Lemma 10.10, there exists in G a parabolic **Q**-subgroup P_2 opposite to P_1. Now as usual, the remainder part of the proof will be carried out in a series of remarks.

Remark 1. Since P_1 and P_2 are parabolic **Q**-subgroups, in virtue of Lemma 10.21 and 1.3.5 $R_u(P_1)$ and $R_u(P_2)$ are opposite orispheric **Q**-subgroups.

Remark 2. Since P_1 is maximal and hence a proper parabolic **Q**-subgroup in G, in view of Lemma 10.21 $R_u(P_1) \neq e$.

Remark 3. Denote by W_1 and W_2 the Zariski closure of the subgroups $\varphi(R_u(P_1) \cap \Gamma)$ and $\varphi(R_u(P_2) \cap \Gamma)$ in G, and by W_1^0 and W_2^0 the algebraically connected components of the identities in the complexifications $(W_1)_{\mathbf{C}}$ and $(W_2)_{\mathbf{C}}$ of

W_1 and W_2 respectively. From Remarks 1 and 2 and Lemma 19.39 it follows that $C(W_1^0) \cap C(W_2^0) = e$, where $C(W_1^0)$ and $C(W_2^0)$ stand for the centralizers of the subgroups W_1^0 and W_2^0 in $G'_{\mathbf{C}}$.

Remark 4. Since G' is an algebraically connected semisimple real algebraic group without compact factors, and Γ' is a lattice in G', on account of the density theorem of Borel ([2]), the subgroup Γ' is Zariski dense in G'. Therefore, since φ is an isomorphism between Γ and Γ', we get that $\varphi(\Gamma)$ is Zariski dense in G'.

Remark 5. Put $L = P_1 \cap P_2$ and denote by $\bar{L}_{\mathbf{Z}}$ the Zariski closure of $L_{\mathbf{Z}}$ in G. Furthermore, let $\bar{L}_{\mathbf{Z}}^0$ be the algebraic connected component of the identity in the group $\bar{L}_{\mathbf{Z}}$. Then *1)* from Lemmas 17.27 and 17.13 it follows easily that $\dim \bar{L}_{\mathbf{Z}}^0 > 0$, *2)* since L belongs to the maximal parabolic $\mathbf{Q}$-subgroup P_1, by Lemma 17.39 the subgroups $\bar{L}_{\mathbf{Z}}^0 \cap G_{\mathbf{Z}}$ and $\bar{L}_{\mathbf{Z}}^0 \cap \Gamma$ are commensurable, *3)* in view of Lemma 1.2.7, $\bar{L}_{\mathbf{Z}}^0 \cap G_{\mathbf{Z}}$ has a finite index in $L_{\mathbf{Z}}$, *4)* $\bar{L}_{\mathbf{Z}}^0$ is algebraically connected (by definition). This and Lemma 8.8 imply that $\bar{L}_{\mathbf{Z}}^0 \cap \Gamma \neq e$. Thus, since φ is an isomorphism, it follows from Remark 19.36 and Lemma 4.8 that $\varphi(\bar{L}_{\mathbf{Z}}^0 \cap \Gamma)$ is not contained in any proper algebraic normal subgroup of G'. Therefore, since $G'_{\mathbf{C}}$ is the complexification of G', neither is $\varphi(\bar{L}_{\mathbf{Z}}^0 \cap \Gamma)$ contained in any proper algebraic normal subgroup of $G'_{\mathbf{C}}$.

Remark 6. From Remarks 3, 4, 5 and the fact that P_1 is a maximal parabolic $\mathbf{Q}$-subgroup, and from Theorem 19.10, it follows that φ extends to an algebraic homomorphism φ_G of $G_{\mathbf{C}}$ into $G'_{\mathbf{C}}$.

Remark 7. By making use of Remark 19.36 and Remark 6 we obtain that φ^{-1} extends to an algebraic homomorphism $\tilde{\varphi}$ of $G'_{\mathbf{C}}$ into $G_{\mathbf{C}}$. Then $\varphi_G \circ \varphi$ is an algebraic homomorphism of $G'_{\mathbf{C}}$ into $G'_{\mathbf{C}}$ and restriction of it to Γ' is the identity transformation. Therefore, in view of Remark 4, $\varphi_G \circ \varphi$ is the identical isomorphism of the group $G'_{\mathbf{C}}$. Analogously, by making use of Remark 19.36, we obtain that $\varphi \circ \varphi_G$ is the identical isomorphism of $G_{\mathbf{C}}$. Since φ_G is an algebraic homomorphism of $G_{\mathbf{C}}$ into $G'_{\mathbf{C}}$, the last two statements imply that φ_G is an algebraic isomorphism between $G_{\mathbf{C}}$ and $G'_{\mathbf{C}}$.

Remark 8. Since *1)* Γ is Zariski dense in G (in virtue of the density theorem of Borel), *2)* the homomorphism φ_G is algebraic, *3)* φ_G is an extension of φ, *4)* Γ' is Zariski dense in G' (see Remark 4), we get that $\varphi_G(G) = G'$.

Remarks 6, 7 and 8 imply the assertion of Theorem 19.6.

REFERENCES

[1] Borel, A., Groupes linéaires algébriques, *Ann. of Math.*, **64** (1956), 20—82.

[2] ———, Density properties of certain subgroups of semisimple groups without compact components, *Ann. of Math.*, **72** (1960), 179—188.

[3] ———, *Linear algebraic groups,* (Notes by H. Bass), Benjamin, New York—Amsterdam, 1969.

[4] ———, and Harish-Chandra, Arithmetic subgroups of algebraic groups, *Ann. of Math.*, **75** (1962), 485—535.

[5] ———, and Tits, J., Groupes réductifs, *Publ. Math. I.H.E.S.*, No. **27** (1965), 55—150.

[6] Garland, H. and Raghunathan, M. S., Fundamental domains for lattices in *R*-rank 1 semi-simple groups, *Ann. of Math.*, **92** (1970), 279—326.

[7] Weissfeiler, B. Ju. (Вейсфейлер Б. Ю.), Об одном классе унипотентных подгрупп полупростых алгебраических групп, *УМН,* **21**:2 (1966), 222—223.
(On a class of unipotent subgroups of semisimple algebraic groups.)

[8] Hall, M., *The theory of groups,* Macmillan, New York, 1959.

[9] Zassenhaus, H., Beweis eines Satzes über diskrete Gruppen, *Abh. Math. Sem. Hansischer Univ.*, **19** (1938), 289—312.

[10] Malcev, A. I. (Мальцев А. И.), Дискретные подгруппы нильпотентных групп Ли, *Изв. АН СССР,* **13**:1 (1949), 9—32.
(Discrete subgroups of nilpotent Lie groups.)

[11] Siegel, C. L., Some remarks on discontinuous groups, *Ann. of Math.*, **46** (1945), 708—718.

[12] Pontrjagin, L. S. (Понтрягин Л. С.), *Непрерывные группы,* Гостехиздат, М., 1954. (*Topological groups*, Princeton Univ. Press, 1939.)

[13] Auslander, L., On radicals of discrete subgroups of Lie groups, *Amer. J. Math.*, **85** (1963), 145—150.

[14] Wang, Hsien-Chung, On the deformations of lattice in a Lie group, *Amer. J. Math.*, **85** (1963), 189—212.

[15] Mostow, G. D., Factor spaces of solvable groups, *Ann. of Math.*, **60** (1954), 1—27.

[16] ———, Self-adjoint groups, *Ann. of Math.*, **62** (1955), 44—55.

[17] ———, Fully reducible subgroups of algebraic groups, *Amer. J. Math.*, **78** (1956), 200—221.

[18] ———, Homogeneous spaces with finite invariant measures, *Ann. of Math.*, **75** (1962), 17—38.

[19] Bourbaki, N., *Groupes et algèbres de Lie, Chap. II, Systèmes de racines,* Hermann, Paris.

[20] Tits, J., Groupes semi-simple isotropes, *in Colloque sur la théorie des groupes algébriques Bruxelles,* 1962, pp. 137—147, Gauthier-Villars, 1962.

[21] Borel, A., *Ensembles fondamentaux pour les groupes arithmétiques et formes automorphes,* Faculté des sciences de Paris, 1967.

[22] Chevalley, C., Sur certains groupes simples, *Tohoku Math, J.*, **7** (1955), 14—66.

[23] Vinberg, E. B. (Винберг Э. Б.), Кольца определения плотных подгрупп полупростых линейных групп, *Изв. АН СССР,* **35** (1971), 45—55.
(The ring of definition of a dense subgroup of a semisimple linear group.)

[24] Pjateckij-Šapiro, I. I. (Пятецкий-Шапиро И. И.), Дикспетные подгруппы Ли, *Труды Московского матем. об-ва,* **18** (1968), 3—18.
(Discrete Lie subgroups.)

[25] Kajdan, D. A. and Margulis, G. A. (Каждан Д. А., Маргулис Г. А.), Доказательство гипотезы Сельберга, *Матем. сб.,* **75** (117):1 (1968), 163—168.
(Proof of Selberg's conjecture.)

[26] *Séminaire Sophus Lie. Théorie des algèbres de Lie. Topologie de groupes de Lie,* Secrétariat Mathématique, Paris, 1955. (*Теория алгебр Ли. Топология групп Ли,* ИЛ, М., 1962).

[27] Chevalley, C., *Theory of Lie groups*, Princeton, 1946.
[28] ———, On the topological structure of solvable groups, *Ann. of Math.*, **42** (1941).
[29] Rosenlicht, M., Some basic theorems on algebraic groups, *Amer. J. Math.*, **78** (1956), 401—443.
[30] ———, Some rationality questions of algebraic groups, *Annali di Mat. IV*, **43** (1957), 25—50.
[31] [32] Raghunathan, M. S., Cohomology of arithmetic subgroups of algebraic groups, I—II, *Ann. of Math.*, **86** (1967), 409—424; **87** (1968), 279—304.
[33] Cassels, J. W. S., *An introduction to the geometry of numbers*, Springer-Verlag, 1959.
[34] Selberg, A., On discontinuous groups in higher dimensional symmetric spaces, *in Contributions to function theory* (Internat. Colloq. function theory, Bombay, 1960), pp. 147—164, Tata Institute, Bombay, 1960.
[35] Weil, A., On discrete subgroups of Lie groups, *Ann. of Math.*, **75** (1962), 578—602.
[36] Pjateckij-Šapiro, I. I. (Пятецкий-Шапиро И. И.), Автоморфные функции и арифметические группы, in *Proceedings of International Congress of Mathematicians (Moscow 1966)*, pp. 232—247, ,,Mir", M., 1968.
(Automorphic functions and arithmetic groups.)
[37] Vinberg, E. B. (Винберг Э. Б.), Дискретные группы, порожденные отражениями в пространстве Лобачевского, *Матем. сб.*, **80** *(122)* (1969), 600—615.
(Discrete groups generated by reflections in the Lobačevskij space.)
[38] Makarov, V. S. (Макаров В. С.), Об одном классе дискретных групп пространства Лобачевского, имеющих бесконечную фундаментальную область конечной меры *ДАН СССР,* **167** (1966), 30—33
(On a class of discrete groups of the Lobačevskij space, having an infinite fundamental domain of finite measure.)
[39] Prasad, M. and Raghunathan, M. S., Some structural properties of lattices in semisimple groups, *preprint*.
[40] Margulis, G. A. (Маргулис Г. А.), К проблеме арифметичности дискретных групп, *ДАН СССР,* **187** (1969), 518—520.
(On the problem of arithmeticity of discrete groups.)
[41] ———, Дискретные подгруппы вещественных полупростых групп Ли, *Матем. сб.,* **80** (122) (1969), 600—615.
(Discrete subgroups of real semisimple Lie groups.)
[42] ———, On the action of unipotent groups in the space of lattices, *this volume*.
[43] Matsushima, Y., On the discrete subgroups and homogeneous spaces of nilpotent Lie groups, *Nagoya Math. J.*, **2** (1951), 95—110.
[44] Mostow, G. D., The rigidity of locally symmetric spaces, *in Actes, Congrès intern. Math., 1970,* Vol. **2**, pp. 187—197, Gauthier-Villars, Paris, 1971.
[45] Zariski, O. and Samuel, P., *Commutative algebra*, D. van Nostrand, 1958.

Received December 15, 1971.

SPHERICAL FUNCTIONS AND HECKE OPERATORS*

FRIEDRICH I. MAUTNER

PART I

EIGENFUNCTIONS

§ 1. Let G be a topological group and $\mathscr{G}$ a closed subgroup. Let us suppose that on the homogeneous space $\mathscr{G}\diagdown G$ there exists an invariant measure and let us consider the space $L^2(\mathscr{G}\diagdown G)$ of **C**-valued square integrable functions with respect to this measure. Let $\text{end}_G(L^2)$ be the algebra of all continuous linear transformations T of this L^2 which commute with right translations R_g by the elements g of G. If the homogeneous space $\mathscr{G}\diagdown G$ is a real or p-adic variety for which the theory of distributions of Schwartz [12] or its generalization by Bruhat [3] holds, one knows that T is given by a distribution $t \in \mathscr{D}(\mathscr{G}\diagdown G)$ which is invariant under right multiplication by the elements of the subgroup $\mathscr{G}$:

$$Tf = t \hat{*} f, \; R_g t = t \quad \text{for} \quad g \in \mathscr{G} \tag{1.1}$$

where $\hat{*}$ is the convolution on $\mathscr{G}\diagdown G$.

If the subgroup $\mathscr{G}$ is compact, we write K instead of $\mathscr{G}$ and observe that our distributions are bi-invariant by the elements of K: Thus we obtain "spherical distributions" (which are not functions in general) which form the algebra $\text{end}_G\big(L^2(K\diagdown G)\big)$.

If the subgroup $\mathscr{G}$ is discrete we write Γ instead of $\mathscr{G}$. Let us suppose that there exists a double coset $\Gamma g \Gamma$, which is a *finite* union of cosets Γ_{g_j}:

$$\Gamma g \Gamma = \bigcup_{j=1}^{h} \Gamma_{g_j} \quad \text{with} \quad h < \infty. \tag{1.2}$$

Let $\delta_{\Gamma_{g_j}}$ be the Dirac-delta measure of the point Γ_{g_j} of the homogeneous space $\Gamma\diagdown G$ and t the finite sum

$$t = \sum_{j=1}^{h} \delta_{\Gamma_{g_j}}. \tag{1.3}$$

Clearly (1.3) defines a distribution (measure) t on $\Gamma\diagdown G$ which is right-invariant by the elements γ of $\Gamma: R_\gamma t = t$ for $\gamma \in \Gamma$.

Example: Let Ω be any field of characteristic zero and $G = \mathbf{G}_\Omega$ the projective group $PG(\Omega, N)$. As subgroup Γ we take the modular group $PG(\mathbf{Z}, N)$. Let us

* Some of this was also presented at the University of Paris in several seminars; see references [8] and [9].

consider on the homogeneous space $\Gamma\backslash G$ the following distribution t_n: In the group $GL(\Omega, N)$ we take the set of all matrices with integral coefficients and fixed determinant $n>0$. Its natural image (mod. scalars) in the projective group $PG(\Omega, N)$ is clearly bi-invariant by the elements of the modular group $\Gamma=PG(\mathbf{Z}, N)$ and is a finite union of cosets Γ_{g_j}; let t be the corresponding distribution (1.3). Thus we obtain for each positive integer n a linear transformation T_n, defined by

$$(T_n f)(g) = (t \hat{*} f)(g) = \sum_{j=1}^{h} f(g_j^{-1} g); \tag{1.4}$$

as the values of the functions f one can take here any (additive) commutative (semi-) group. These are the "Hecke operators" which were already considered by Kronecker and Hurwitz. Let us remark that in our definition (1.4) the field Ω (of characteristic zero) is arbitrary; in particular T_n is already defined over the field $\mathbf{Q}$ of rational numbers, as linear transformation over the homogeneous space $\mathbf{G_Z}\backslash\mathbf{G_Q}$. Let us also remark that definition (1.4) is easily adapted to other groups, also to fields of characteristic $\neq 0$ (for example to algebraic function fields in one variable over a finite constant field).

§ 2. Until recently T_n had been considered mainly as a linear transformation of $L^2(\mathbf{G_Z}\backslash\mathbf{G_R})$, in fact usually as a linear transformation of the subspace of holomorphic cusp forms of this Hilbert space. Considered as linear transformations of $L^2(\mathbf{G_Z}\backslash\mathbf{G_R})$ the Hecke operators T_n do not generate the algebra of all $\mathbf{G_R}$-endomorphisms of this Hilbert space. This algebra is non-commutative already in the case of $L^2(SL(\mathbf{Z}, 2)\backslash SL(\mathbf{R}, 2))$, because the discrete series of irreducible unitary representations of $SL(\mathbf{R}, 2)$ [1] occurs in the decomposition of this Hilbert space with multiplicity >1 in general, since this multiplicity is equal to the dimension$_{\mathbf{C}}$ of the space of holomorphic modular cusp forms. The $\mathbf{G_R}$-endomorphisms of this Hilbert space $L^2(SL(\mathbf{Z}, 2)\backslash SL(\mathbf{R}, 2))$ which are not generated by the Hecke operators T_n are still quite mysterious: they are apparently given by distributions which seem to come from "badly behaved orbits" (in the sense of Mackey) under right multiplication R_γ $(\gamma\in\Gamma)$ on the homogeneous space $\Gamma\backslash G$. This phenomenon of "badly behaved orbits" is known in the theory of dual objects of groups and C^*-algebras, where one has the impression that this still quite mysterious phenomenon may imply the existence of irreducible representations which can in some sense not be calculated. The same phenomenon may perhaps be responsible for the difficulties which one has found if one wishes to decompose such spaces as $L^2(SL(\mathbf{Z}, 2)\backslash SL(\mathbf{R}, 2))$ explicitly. It seems that the problem of this explicit decomposition had already been considered by Hilbert: Let ∇ be the hyperbolic Laplacian $y^2\left(\frac{\partial^2}{\partial x^2}+\frac{\partial^2}{\partial y^2}\right)$ acting in the upper half plane $z=x+iy$, $y>0$. Let us consider the space $L^2(\mathbf{D}; y^{-2}dx\,dy)$ of all modular functions $\psi(z)$ which are square integrable over a fundamental domain **D** of $SL(\mathbf{Z}, 2)$ with respect to the measure $y^{-2}dxdy$ and the eigenvalue problem

$$\nabla\psi = \lambda\psi \in L^2(\mathbf{D}; y^{-2}\,dx\,dy). \tag{2.1}$$

It seems that Hilbert already asked if there are relations between this point spectrum $\{\lambda\}$ and the zeros of the Riemann zeta function $\zeta(s)$. This question has been studied above all by Selberg [13] and C. L. Siegel. The Dirichlet series $\sum a_n n^{-s}$ which Selberg has associated with certain homogeneous spaces $\Gamma\backslash G$ (e.g. $SL(\mathbf{Z}, 2)\backslash SL(\mathbf{R}, 2)$) are still far from the Riemann zeta function, because in the series $\sum a_n n^{-s}$ constructed by Selberg, n varies over an infinite compositum of quadratic fields, not only over **Z**.

The eigenvalue problem (2.1) is still very difficult, even for the modular group. For example one still does not know whether the spectrum in (2.1) is simple for the modular group $SL(\mathbf{Z}, 2)$, nor whether there exist non-trivial expressions for the square integrable eigenfunctions, for example Dirichlet series similar to Selberg's Dirichlet series mentioned above.

One makes little progress if one observes that $T_n\nabla = \nabla T_n$ and if one replaces (2.1) by

$$T_n\psi = \lambda_n\psi \tag{2.2}$$

and considers this eigenvalue problem over the field **R** of real numbers. This does not diminish the difficulties which are perhaps caused by the badly behaved orbits of the right multiplications R_γ on $\Gamma\backslash G$. One makes much progress if one considers the eigenvalue problem (2.2) over the field **Q** of rational numbers. A solution of this problem will be outlined below: All solutions of (2.2) will be obtained with ψ varying over all eigenfunctions defined on $\mathbf{G_Z}\backslash\mathbf{G_Q}$.

§ 3. Let $\mathbf{G_Q} = PG(\mathbf{Q}, 2)$ denote the group of linear fractional transformations $x \to g(x) = (g_{11}x + g_{12})(g_{21}x + g_{22})^{-1}$, with $g_{ij} \in \mathbf{Q}$, $g_{11}g_{22} - g_{12}g_{21} \neq 0$ and $\mathbf{G_Z} = PG(\mathbf{Z}, 2)$ the modular subgroup defined by $g_{ij} \in \mathbf{Z}$ and $g_{11}g_{22} - g_{12}g_{21} = \pm 1$. We shall study the homogeneous space $\mathbf{G_Z}\backslash\mathbf{G_Q}$ as a discrete space, the space $L(\mathbf{G_Z}\backslash\mathbf{G_Q})$ of all complex valued functions defined on $\mathbf{G_Z}\backslash\mathbf{G_Q}$ and the action of the Hecke operators T_n considered as linear transformations of the linear space $L(\mathbf{G_Z}\backslash\mathbf{G_Q})$.

Let us observe that

$$\mathbf{G_Z}\backslash\mathbf{G_Q} \cong \prod_{p\neq\infty} (\mathbf{G_{Z_p}}\backslash\mathbf{G_{Q_p}}) \cong \prod_{p\neq\infty} \mathbf{G_{Z_p}}\backslash \prod_{p\neq\infty} \mathbf{G_{Q_p}} \tag{3.1}$$

where the products $\prod_{p\neq\infty}$ are restricted direct products, except $\prod \mathbf{G_{Z_p}}$ which is a Cartesian direct product. These isomorphisms preserve the Hecke operators T_n (by left multiplication) and right multiplication R_g by the elements g of $\mathbf{G_Q}$. For general algebraic groups (3.1) is equivalent to the class number of **G** (in the adelic sense) being one. In the case of the groups of 2×2 matrices considered here, (3.1) is an easy consequence of the theory of elementary divisors.

Using (3.1), the "global" problem

$$T_n \psi = \lambda_n \psi \in L(\mathbf{G_Z} \backslash \mathbf{G_Q}) \tag{3.2}$$

of finding all "global" eigenfunctions can thus be reduced to the "local" problem

$$T_{p^\nu} \psi = \lambda_{p^\nu} \psi \in L(\mathbf{G}_{\mathbf{Z}_p} \backslash \mathbf{G}_{\mathbf{Q}_p}). \tag{3.3}$$

Let us consider the local case. Let us write $\mathfrak{G}$ instead of $PG(\Omega, 2)$ where Ω is now a local non-Archimedean field, $\mathfrak{O}$ its valuation ring, $\mathfrak{U}$ the group of units of $\mathfrak{O}$ (i.e. $\mathfrak{U}=\mathfrak{O}^*$), $\mathfrak{P}$ the maximal ideal of $\mathfrak{O}$ and $N\mathfrak{P}$ the norm of $\mathfrak{P}$. The group $\mathfrak{K}=PG(\mathfrak{O}, 2)$ is a maximal compact subgroup of $\mathfrak{G}$ and the structure of the algebra $L^0(\mathfrak{K}\backslash\mathfrak{G}/\mathfrak{K})$ of all spherical functions of compact support is well known [10]. Let us remark that this algebra is *-isomorphic to the subalgebra of the algebra of Hecke operators generated by T_p if $\Omega=\mathbf{Q}_p$.

The problem (3.3) is thus replaced by the problem of finding all solutions $\psi \in L(\mathfrak{K}\backslash\mathfrak{G})$ of

$$f * \psi = \lambda_f \psi \tag{3.4}$$

with $f \in L^0(\mathfrak{K}\backslash\mathfrak{G}/\mathfrak{K})$; *in particular the Hecke operator* T_{p^ν} *is replaced by convolution by a spherical function of compact support,* which enables one to solve the problem (3.3) completely:

Theorem: *All complex valued solutions* ψ *of* (3.4) *are* (suitable) *matrix coefficients of the representations of* $\mathfrak{G}$ *in the principal series.*

One can take the representations of the "principal series" as explicit multiplier representations acting on the projective line over the given local field Ω: Let $v(x)$ be a complex valued function on Ω and α a character of the multiplicative group Ω^* into $\mathbf{C}^*$. One has

$$[M_\alpha(g)v](x) = \alpha[(g_{12}x+g_{22})^{-2} \det g]\, v\left(\frac{g_{11}x+g_{21}}{g_{12}x+g_{22}}\right) \tag{3.5}$$

if $g_{12}x+g_{22} \neq 0$. One verifies easily that for $\alpha(a)=|a|^s$ (where $|a|$ is the normalized $\mathfrak{P}$-adic absolute value of $a \in \Omega^*$) i.e. $\alpha(\varepsilon)=1$ for $\varepsilon \in \mathfrak{U}$, the solutions ξ_α of the equations

$$M_\alpha(k)\xi_\alpha = \xi_\alpha \quad \text{for all} \quad k \in \mathfrak{K} \tag{3.6}$$

form a one-dimensional space over $\mathbf{C}$ and that the function

$$\psi(g) = \langle M_\alpha(g^{-1})\xi_\alpha, \eta \rangle \tag{3.7}$$

is a local eigenfunction of the Hecke operators, i.e. is a solution of (3.4), if η is any distribution on the projective line over Ω. This distribution η is essentially arbitrary, it suffices that the inner product $\langle\ ,\ \rangle$ in (3.4) converges in some sense, which will become clear later. Considered as a function of the parameter s, the eigenfunction $\psi(g)=\psi(g, s)$ in (3.7) may have zeros; in this case the partial derivative with respect to s is also an eigenfunction.

One shows that all local eigenfunctions of the Hecke operators are obtained in this way. This result does not remain correct if one replaces $\mathfrak{K}$ by a congruence subgroup, because in this case representations of the discrete series enter in general. However one can consider ind $\mathfrak{u}$ for any irreducible unitary representation $\mathfrak{u}$ of $\mathfrak{K}$ instead of the 1-representation of $\mathfrak{K}$ considered above. In general the matrix coefficients of the principal series are not the only eigensolution, matrix coefficients of the discrete series will also occur. If u belongs to the "principal series mod p^ν" [12] corresponding to a primitive congruence character, the principal series M_α of $\mathfrak{G}$ suffices.

§ 4. In order to prove this theorem, it is useful to diagonalize the T_{p^ν}-module $L^0(\mathfrak{K}\backslash\mathfrak{G})$ of complex-valued functions of compact support on $\mathfrak{K}\backslash\mathfrak{G}$, which is clearly also an R_g-module ($g\in\mathfrak{G}$). Let us consider at first only right multiplications R_k by the elements k of $\mathfrak{K}$. This gives a Fourier development

$$L^2(\mathfrak{K}\backslash\mathfrak{G}) = \sum_u L^2(\mathfrak{K}\backslash\mathfrak{G}) * u, \tag{4.1}$$

where $*$ is here convolution on $\mathfrak{K}$, and u varies over a (suitable) set of matrix coefficients of irreducible unitary representations of $\mathfrak{K}$. Let us apply the Fourier transform

$$f \to \int_{\mathfrak{G}} f(g)\, M_\alpha(g)\, dg \tag{4.2}$$

to the submodule $L^0(\mathfrak{K}\backslash\mathfrak{G}) * u$. One shows that the image consists of *all* polynomials $F(s)$ in $N\mathfrak{P}^{\pm s}$ having certain zeros and satisfying a functional equation $F(s)\to F(1-s)$. Hence every $L^0(\mathfrak{K}\backslash\mathfrak{G}/\mathfrak{K})$-homomorphism of this submodule into $\mathbf{C}$ is given by a complex number c and is of the form

$$F \to F^{(j)}(c) \tag{4.3}$$

where

$$F^{(j)}(c) = \frac{\partial^j F}{\partial s^j}(c), \quad \text{if} \quad \frac{\partial^l F}{\partial s^l}(c) = 0$$

for all F and $0 \leqq l < j$. From this one obtains all eigenfunctions ψ as inner products (3.7).

Instead of the Fourier series development (4.1) it is also useful and interesting to consider the Fourier series development with respect to a compact Cartan subgroup $\mathfrak{R}$ of $\mathfrak{G}$: Let δ be an element of the local field Ω such that $\sqrt{\delta}\notin\Omega$ and let $\mathfrak{R}$ be the image of the group of matrices $\begin{pmatrix} r_1 & r_2 \\ r_2\delta & r_1 \end{pmatrix}$ with $r_1, r_2 \in \Omega$, $r_1^2 - r_2^2\delta \neq 0$, by the natural homomorphism of $GL(\Omega, 2)$ on $PG(\Omega, 2)$. Let $\hat{\mathfrak{R}}$ be the group of characters ϱ of this compact group $\mathfrak{R}$ and let us consider the Fourier series development

$$L^2(\mathfrak{K}\backslash\mathfrak{G}) = \sum_{\varrho\in\hat{\mathfrak{R}}} L^2(\mathfrak{K}\backslash\mathfrak{G}) * \varrho \tag{4.4}$$

where $*$ is here convolution on $\mathfrak{R}$.

Let us apply the Fourier transform (4.2) to the submodule $L^0(\mathfrak{K}\backslash\mathfrak{G})*\varrho$. One shows that the image consists again of all polynomials $F(s)$ in $N\varrho^{\pm s}$ which have certain zeros and satisfy a functional equation $s\to 1-s$. That shows again that all $L^0(\mathfrak{K}\backslash\mathfrak{G}/\mathfrak{K})$-homomorphisms of the submodule $L^0(\mathfrak{K}\backslash\mathfrak{G})*\varrho$ are of the form (4.3). As a consequence one obtains all eigenfunctions of the algebra $L^0(\mathfrak{K}\backslash\mathfrak{G}/\mathfrak{K})$, in $L(\mathfrak{K}\backslash\mathfrak{G})*\varrho$: They are all of the form

$$\psi_{s,\varrho}(g) = \langle M_\alpha(g^{-1})\xi_\alpha, \eta_{\alpha,\varrho}\rangle \xi_\varrho(s) \tag{4.5}$$

where $\alpha(a)=|a|^s$ and $\eta_{\alpha,\varrho}$ is determined (up to a multiplicative constant c) by

$$M_\alpha(r)\eta_{\alpha,\varrho} = \varrho(r)\eta_{\alpha,\varrho} \quad \text{for} \quad r\in\mathfrak{R} \tag{4.6}$$

which implies

$$\eta_{\alpha,\varrho}(x) = c\cdot\alpha(1-x^2\delta)\varrho(r) \tag{4.7}$$

where $x=r(0)=r_2\cdot r_1^{-1}$, if r is given by the matrix $\begin{pmatrix} r_1 & r_2 \\ r_2\delta & r_1 \end{pmatrix}$ (mod. scalars). The factor $\zeta_\varrho(s)$ is a polynomial in $N\mathfrak{R}^{\pm s}$ which has its zeros on $Rs=\frac{1}{2}$.

Let us observe that each orbit of $\mathfrak{K}$ and of $\mathfrak{R}$ (by right multiplication) on $\mathfrak{K}\backslash\mathfrak{G}$ is finite, since $\mathfrak{K}$ and $\mathfrak{R}$ are compact and $\mathfrak{K}\backslash\mathfrak{G}$ discrete. Hence the developments (4.1) and (4.4) are trivially convergent. Hence one sees that if one takes for each $\varrho\in\hat{\mathfrak{R}}$ an eigenfunction $\psi_\varrho\in L(\mathfrak{K}\backslash\mathfrak{G})*\varrho$, all with the same eigenvalue, and arbitrary complex numbers c_ϱ, the series $\sum_{\varrho\in\hat{\mathfrak{R}}} c_\varrho\psi_\varrho$ converges to an eigenfunction, and the distribution η in (3.7) is given by $\eta=\sum c_\varrho\eta_\varrho$, the convergence being trivial. The same observations hold for u instead of ϱ: One obtains again an eigenfunction $\sum c_u\psi_u$ if $T_p\psi_u=\lambda\psi_u$ with λ independent of u and c_u arbitrary complex numbers. If f is an arbitrary element of $L^0(\mathfrak{K}\backslash\mathfrak{G})$, its Fourier transform (4.2) is thus a sequence of polynomials $F_u(s)$ or $F_\varrho(s)$.

Let us consider a little more closely the case of a compact Cartan subgroup $\mathfrak{R}$ of $\mathfrak{G}$. Let $\mathfrak{p}$ be a generator of the maximal ideal $\mathfrak{P}$ and p the element of $\mathfrak{G}$ given by the matrix $\begin{pmatrix} \mathfrak{p} & 0 \\ 0 & 1 \end{pmatrix}$. One shows that

$$\mathfrak{G} = \bigcup_{n=0}^{\infty} \mathfrak{K}p^n\mathfrak{R}. \tag{4.8}$$

Thus it suffices to know the function $\psi_{s,\varrho}(g)$ for $g=p^n$ with $n\geqq 0$.

Let us suppose now that the quadratic extension $\Omega(\sqrt{\delta})$ of Ω which defines $\mathfrak{R}$ is unramified and $2\nmid N\mathfrak{P}$, i.e. that $\mathfrak{R}$ is the group given by the matrices $\begin{pmatrix} r_1 & r_2 \\ r_2\delta & r_1 \end{pmatrix}$ (mod $\Omega^*\cdot I$) with $r_j\in\Omega$, $r_1^2-r_2^2\delta\neq 0$, $\delta\in\mathfrak{U}$, $\sqrt{\delta}\notin\Omega$ and $2\nmid N\mathfrak{P}$. Given a $\mathbf{C}$-valued function $f(g)$ defined on $\mathfrak{G}$ satisfying

$$f(kgr) = f(g)\varrho(r) \tag{4.9}$$

for all $k\in\mathfrak{K}$, $g\in\mathfrak{G}$ and $r\in\mathfrak{R}$, and a fixed character ϱ of $\mathfrak{R}$, $\varrho\neq 1$. Let $\mathfrak{R}^{(m)}$ be the congruence subgroup of those elements of $\mathfrak{R}$ for which $r_2\in\mathfrak{P}^m$ and μ the minimal value of m for which $\varrho(\mathfrak{N}^m)=1$. One shows that (4.9) implies

$$f(\mathfrak{p}^n) = 0 \quad \text{for} \quad 0 \leqq n < \mu. \tag{4.10}$$

For $n \geqq \mu$ a calculation which will not be reproduced here gives the following result for the eigenfunction $\psi_{s,\varrho}(\mathfrak{p}^n)$ defined by (4.5):

$$\varphi_{s,\varrho}(\mathfrak{p}^n) = \langle M_\alpha(\mathfrak{p}^n)\xi_\alpha{}^{\text{‘}}\, \eta_{\alpha,\varrho}\rangle =$$

$$= \xi_\alpha(0)\bar{\eta}_{\alpha,\varrho}(0)\alpha(\mathfrak{p}^n)[\delta_{0,\mu}(1+q^{-1})-\varrho(w)(I_1+I_2+I_3-\delta_{1,\mu}q^{-1})] \tag{4.11}$$

where $q=N\mathfrak{P}$, w is the element of $\mathfrak{N}$ given by the matrix $\begin{pmatrix}0 & 1\\ \delta & 0\end{pmatrix}$ and $\alpha(x)=|x|^s= =q^{-s\,\mathrm{ord}\,(x)}$ and the expressions I_1, I_2, I_3 are given by

$$I_1 = \frac{q^{(1-2s)\max(1,\mu)}-q^{-1+(1-2s)\max(1,\mu-1)}}{1-q^{1-2s}}$$

$$I_2 = \frac{q^{-1+(1-2s)\max(n+1,\mu-1)}-q^{-1+(1-2s)\max(1,\mu-1)}}{1-q^{1-2s}}$$

$$I_3 = \alpha(\mathfrak{p}^{-2n})\frac{q^{-\max(n+1,\mu)}-q^{-1\max(n+1,\mu-1)}}{1-q^{-1}} \tag{4.12}$$

If $\mu \geqq 2$ and $n \geqq \mu$ one obtains

$$\frac{q^{\frac{n}{2}}\bar{\varrho}(w)}{\xi_\alpha(0)\eta_{\alpha,\varrho}(0)}\cdot\varphi_{s,\varrho}(\mathfrak{p}^n)\cdot\left(q^{s-\frac{1}{2}}-q^{\frac{1}{2}-s}\right) = q^{n\left(s-\frac{1}{2}\right)}(q^{s-1}+1)(1-q^{s-1})q^{(1-2s)\mu}+$$

$$+q^{-n\left(s-\frac{1}{2}\right)}q^{s-\frac{3}{2}}(1+q^{s-1})(1-q^{s-1}).$$

This gives the following functional equation

$$\frac{\varphi_{s,\varrho}(\mathfrak{p}^n)}{\varphi_{1-s,\varrho}(\mathfrak{p}^n)} = \frac{(1+q^{s-1})(1-q^{s-1})}{(1+q^{-s})(1-q^{-s})}\cdot q^{(1-2s)\mu} \quad \text{for} \quad \mu \geqq n. \tag{4.13}$$

Let us note that an elementary computation shows that $\varphi_{s,\varrho}(g)=0$ for all $g\in\mathfrak{G}$ only on points of the line $Rs={}^1/_2$. Hence one has to multiply $\varphi_{s,\varrho}$ by a polynomial $\zeta_\varrho(s)=\prod(q^s-q^{s_j})$ with $Rs_j={}^1/_2$, in order to obtain the most general eigenfunction in $L(\mathfrak{K}\diagdown\mathfrak{G})*\varrho$.

$$\psi_{s,\varrho}(\mathfrak{p}^n) = \zeta_\varrho(s)\varphi_{s,\varrho}(\mathfrak{p}^n) \quad \text{with} \quad \zeta_\varrho(s) = \prod(q^s-q^{s_j}) \tag{4.14}$$

where s_j varies over those values of the parameter s for which $\varphi_{s,\varrho}(g)$ vanishes identically in g.

Suppose now that f is a function defined on $\mathfrak{K}\diagdown\mathfrak{G}$ whose support is contained in a single double-coset $\mathfrak{K}p^m\mathfrak{K}=\mathfrak{K}p^m\mathfrak{N}$ (the quadratic extension corresponding to $\mathfrak{N}$ being unramified). Let us consider the Fourier transform (4.2) of f and let us put again $\langle M_\alpha(g^{-1})\xi_\alpha, \eta_{\alpha,\varrho}\rangle=\varphi_{s,\varrho}(g)$. We have

$$\int_{\mathfrak{G}} f(g)\,\varphi_{s,\varrho}(g)\,dg = F_\varrho(s). \tag{4.15}$$

As $f(g) = 0$ if $g \notin \mathfrak{K} p^m \mathfrak{R}$ we have

$$F_\varrho(s) = \int_{\mathfrak{K} p^m \mathfrak{R}} f(g) \varphi_{s,\varrho}(g)\, dg = \int_{\mathfrak{R}^{(m)} \setminus \mathfrak{R}} f(p^m r) \varphi_{s,\varrho}(p^m r)\, dr \tag{4.16}$$

where $\mathfrak{R}^{(m)}$ is the subgroup of $\mathfrak{R}$ which leaves the point $\mathfrak{K} p^m$ of the homogeneous space $\mathfrak{K} \setminus \mathfrak{G}$ fixed (by right multiplication). Thus $\mathfrak{R}^{(m)} = (p^{-m} \mathfrak{R} p^m) \cap \mathfrak{R}$. Hence $\mathfrak{R}^{(m)}$ is the congruence subgroup of $\mathfrak{R}$ given by the matrices $\begin{pmatrix} r_1 & r_2 \\ r_2\delta & r_1 \end{pmatrix}$ (mod $\Omega^* \cdot I$) for which $r_2 \in \mathfrak{P}^m$. Hence the integral $\int_{\mathfrak{R}^{(m)} \setminus \mathfrak{R}}$ is a finite sum, the index of $\mathfrak{R}^{(m)}$ in $\mathfrak{R}$ being finite. If $f(kgr) = f(g)\varrho(r)$ for $k \in \mathfrak{K}$, $g \in \mathfrak{G}$ and $r \in \mathfrak{R}$, one obtains for the fixed character ϱ of $\mathfrak{R}$ the very explicit result

$$F_\varrho(s) = f(p^n) \int_{\mathfrak{R}^{(n)} \setminus \mathfrak{R}} \varrho(r) \varphi_{s,\varrho}(p^n r)\, dr =$$

$$= f(p^n) \varphi_{s,\varrho}(p^n) \int_{\mathfrak{R}^{(n)} \setminus \mathfrak{R}} |\varrho(r)|^2\, dr = f(p^n) \varphi_{s,\varrho}(p^n) \mu_n \tag{4.17}$$

where μ_n is the measure of $\mathfrak{R}^{(n)} \setminus \mathfrak{R}$ (=measure of the double coset $\mathfrak{K} p^n \mathfrak{R}$). If f is arbitrary, but still $f(g) = 0$ for $g \notin \mathfrak{K} p^n \mathfrak{R}$, one obtains

$$F_\varrho(s) = \varphi_{s,\varrho}(p^n) \int_{\mathfrak{R}^{(n)} \setminus \mathfrak{R}} f(p^n r)\, \bar{\varrho}(r)\, dr. \tag{4.18}$$

The group $\mathfrak{R}^{(m)} \setminus \mathfrak{R}$ is the Galois group of a finite Abelian extension of the local field $\Omega(\sqrt{\delta})$, which is well known in local class field theory. The last integral (=finite sum) is the Fourier transform of f with respect to this finite Abelian group $\mathfrak{R}^{(m)} \setminus \mathfrak{R}$.

Let us observe also that the image of the Hecke operator T_p under the Fourier transform (4.2) is multiplication by $(N\mathfrak{P})^s + (N\mathfrak{P})^{1-s}$, i.e.

$$\text{if} \quad F_\varrho(s) = \int f(g) \varphi_{s,\varrho}(g)\, dg$$

$$\text{then} \quad \int (T_p f)(g) \varphi_{s,\varrho}(g)\, dg = [(N\mathfrak{P})^s + (N\mathfrak{P})^{1-s}]\, F_\varrho(s) \tag{4.19}$$

where $\varphi_{\varrho,s}$ is given by (4.11) and (4.12). Hence we have obtained a completely explicit "diagonalisation" of the double module $L^0(\mathfrak{K} \setminus \mathfrak{G})$. This suggests the possibility of "local complex multiplication".

§ 5. Let us consider now the "global" eigenfunctions of the Hecke operators. So Ω denotes now an algebraic number field or an algebraic function field of transcendence degree 1 over a finite constant field. For each non-Archimedean place p of Ω let Ω_p be the p-adic completion of Ω, $\mathfrak{O}_p$ the valuation ring of Ω_p and

$$G_p = PG(\Omega_p, 2) \supset K_p = PG(\mathfrak{O}_p, 2).$$

For an arbitrary given set P of non-Archimedean places p of Ω, let G_P be the restricted direct product of the groups G_p with respect to the compact open subgroups K_p and K_P the direct product of the compact groups K_p ($p \in P$).

Let us consider the global analogue of the local Fourier series development (4.1):

$$L^2(K_P \backslash G_P) = \sum_u L^2(K_P \backslash G_P) * u \tag{5.1}$$

where u varies over a suitable (orthogonal) set of matrix-coefficients of irreducible unitary representations of the compact group K_P and $*$ is convolution on K_P. For any $f \in L(K_P \backslash G_P)$ we can compute the convolution

$$(f*u)(g) = \int_{K_P} f(gk^{-1})u(k)\,dk$$

as follows. The orbit under K_P of the point $K_P g$ (by right multiplication) in the space $K_P \backslash G_P$ is finite because K_P is compact and $K_P \backslash G_P$ discrete; hence there exists a finite number h_g of elements k_j of K_P such that

$$K_P g K_P = \bigcup_{j=1}^{h_g} K_P g k_j; \tag{5.2}$$

we can assume that this union is disjoint. Clearly h_g and k_j vary with g. Supposing that the measure of K_P is 1, we obtain

$$(f*u)(g) = \frac{1}{h_g} \sum_{j=1}^{h_g} f(gk_j^{-1})u(k_j). \tag{5.3}$$

Let us assume now that ψ is an eigenfunction of the Hecke operators, defined on $K_P \backslash G_P$. Then the convolution $\psi * u$ is again an eigenfunction and is contained in the subspace $L(K_P \backslash G_P) * u$. We can suppose that

$$u(k) = \prod_{p \in P} u_p(k_p) \quad \text{where} \quad k = \prod_{p \in P} k_p$$

where $u_p(k_p) = 1$ for almost all $p \in P$. Now the eigenfunction of the Hecke operators in the space $L(K_P \backslash G_P) * u$ is unique (up to a constant multiple), hence one finds

$$\sum_{j=1}^{h_g} \psi(gk_j^{-1})u(k_j) = c_u \prod_{p \in P} \psi_{s,u_p}(g_p) \tag{5.4}$$

where $g = \prod_p g_p$, ψ_{s,u_p} is the eigenfunction contained in $L(K_p \backslash G_p) * u_p$ described in § 4 and c_u a complex constant. We have

$$T_p \psi_{s,u_p} = [(N\mathfrak{P})^s + (N\mathfrak{P})^{1-s}]\psi_{s,u_p}$$

where $\mathfrak{P}$ is the prime ideal of $\mathfrak{O}_p$. Let us also note that one can take g to be given by a diagonal matrix, i.e. $g = \prod_p p^{\nu_p}$ and that ψ_{s,u_p} is the elementary spherical function [11] if $u_p = 1$. Hence the product $\prod_p \psi_{s,u_p}(g_p)$ is finite for each element $g = \prod_p g_p$ of G_P, in the sense that all but a finite factors in the product $\prod_p \psi_{s,u_p}(g_p)$ are 1.

36*

Formulas of the type (5.4) seem to be new for the classical modular forms, even for such classical eigenfunctions of the Hecke operators as the Eisenstein series $E(z,s)=y^s\sum|mz+n|^{-2s}$, or the holomorphic modular eigen-forms. Yet (5.4) is quite explicit: One knows the matrix coefficients u quite explicitly, so that the left side of (5.4) is an explicit finite sum, like a Hecke operator with "coefficients $u(k_j)$" (acting on the right, whereas the Hecke operators T_n "without coefficients" act on the left). Apparently *the restriction of a modular function or form to the upper half plane of an imaginary quadratic field* $\mathbf{Q}(\sqrt{d})$ *has very special properties, which are still badly known.*

Hence let us consider somewhat more closely the "rational points" of a fundamental domain of the modular group: Let $\mathbf{Q}(\sqrt{d})$ be an imaginary quadratic extension of $\mathbf{Q}$ and $R_{\mathbf{Q}}$ the subgroup of $\mathbf{G}_{\mathbf{Q}}=PG(\mathbf{Q},2)$ given by the matrices $\begin{pmatrix} a & b \\ bd & a \end{pmatrix}$ (mod $\mathbf{Q}^*\cdot I$) with $a,b\in\mathbf{Q}$ and $a^2-b^2d\neq 0$. The space $\mathbf{G}_{\mathbf{Z}}\backslash\mathbf{G}_{\mathbf{Q}}/R_{\mathbf{Q}}$ of double cosets $\mathbf{G}_{\mathbf{Z}}gR_{\mathbf{Q}}$ $(g\in\mathbf{G}_{\mathbf{Q}})$ can be identified in a natural manner with $\mathbf{G}_{\mathbf{Z}}\backslash\{\mathbf{Q}(\sqrt{d})-\mathbf{Q}\}$, because the homogeneous space $\mathbf{G}_{\mathbf{Q}}/R_{\mathbf{Q}}$ can be identified with the set $\mathbf{Q}(\sqrt{d})-\mathbf{Q}$ of elements $z=x+y\sqrt{d}$ for which $y\neq 0$.

As the isomorphism (3.1) preserves right multiplication by the elements of $\mathbf{G}_{\mathbf{Q}}$, hence in particular by the elements of $R_{\mathbf{Q}}$, one obtains a natural and explicit isomorphism

$$\mathbf{G}_{\mathbf{Z}}\backslash\mathbf{G}_{\mathbf{Q}}/R_{\mathbf{Q}} \cong K_P\backslash G_P/R_{\mathbf{Q}} \tag{5.5}$$

where P is here the set of all rational prime numbers.

This space of double cosets is no longer a product $\prod_p$ of "local" spaces. In order to obtain such products one has to "adelise" $R_{\mathbf{Q}}$: Let $R_{\mathbf{Q}_p}$ be the p-adic completion of $R_{\mathbf{Q}}$ in $\mathbf{G}_{\mathbf{Q}_p}$ and R_P the restricted direct product of the groups $R_{\mathbf{Q}_p}$ with respect to the compact open subgroups $R_{\mathbf{Q}_p}\cap K_p$, $p\in P$. Right multiplication by the elements $\mathfrak{r}$ of the compact quotient group $R_{\mathbf{Q}}\backslash R_P$ is well defined in the space $K_P\backslash G_P/R_{\mathbf{Q}}$. The quotient group $R_{\mathbf{Q}}\backslash R_P$ is a (compact Abelian) Galois group of an Abelian extension of $\mathbf{Q}(\sqrt{d})$ which is well known in class field theory and in the theory of complex multiplication. Let us note that $R_{\mathbf{Q}}$ is a discrete subgroup of R_P, because the number of units of the *imaginary* quadratic field $\mathbf{Q}(\sqrt{d})$ is finite.

Thus one sees that the space $L(\mathbf{G}_{\mathbf{Z}}\backslash\mathbf{G}_{\mathbf{Q}}/R_{\mathbf{Q}})$ is a double module. It admits the Hecke operators (on the left) and right multiplication by the elements $\mathfrak{r}$ of the compact Abelian Galois group $R_{\mathbf{Q}}\backslash R_P$. This double module appears to be a fundamental building stone of the theory of complex multiplication and the preceding results give a complete and quite explicit description of this double module as a direct sum of polynomial modules. Let ϱ now denote an arbitrary character of the compact Abelian group $R_{\mathbf{Q}}\backslash R_P$ and let us consider the Fourier series development

$$L(\mathbf{G}_{\mathbf{Z}}\backslash\mathbf{G}_{\mathbf{Q}}/R_{\mathbf{Q}}) = \sum_{\varrho\in\widehat{R_{\mathbf{Q}}\backslash R_P}} L(K_P\backslash G_P/R_{\mathbf{Q}})*\varrho \tag{5.6}$$

where $*$ is here convolution (on the right) over the compact group $R_{\mathbf{Q}} \backslash R_P$. Let us note that each orbit of the compact group $R_{\mathbf{Q}} \backslash R_P$ by right multiplications on the discrete space $K_P \backslash G_P / R_{\mathbf{Q}}$ is finite, hence the Fourier series (5.6) are always convergent. For the submodule $L^0(\mathbf{G_Z} \backslash \mathbf{G_Q} / R_{\mathbf{Q}}) * \varrho$ one obtains an explicit and complete description as a polynomial module. Its image under the Fourier transform given by the "global" principal series (i.e. the global analogue of (4.2)) consists of all polynomials $F(\ldots, s_p, \ldots)$ in the infinitely many variables $p^{\pm s_p}$ which have certain zeros and satisfy an infinite number of functional equations $s_p \to 1 - s_p$.

If $f \in L(\mathbf{G_Z} \backslash \mathbf{G_Q} / R_{\mathbf{Q}})$ one has

$$(f * \varrho)(g) = \int_{R_{\mathbf{Q}} \backslash R_P} f(g\mathfrak{r}^{-1}) \varrho(\mathfrak{r})\, d\mathfrak{r} = \frac{1}{h'_g} \sum_{j=1}^{h'_g} f(g\mathfrak{r}_j^{-1}) \varrho(\mathfrak{r}_j) \tag{5.7}$$

as analogue of (5.3). In particular, if f is an eigenfunction of the Hecke operators, so is $f * \varrho$. But $f * \varrho$ is a product of local eigenfunctions, which are polynomials. If $\sqrt{d} \notin \mathbf{Q}_p$, $p \neq 2$ and $\mathbf{Q}_p(\sqrt{d})$ is unramified, these polynomials are given by (4.11). If $\sqrt{d} \in \mathbf{Q}_p$, the local eigenfunction is still of the form

$$\langle M_{\alpha_p}(g)\, \xi_{\alpha_p}, \eta_{\alpha_p, \varrho_p} \rangle \quad \text{where} \quad M_{\alpha_p}(\mathfrak{r}_p) \eta_{\alpha_p, \varrho_p} = \varrho(\mathfrak{r}_p)\, \eta_{\alpha_p, \varrho_p}$$

for $\mathfrak{r}_p \in R_{\mathbf{Q}_p}$ and $|x|_p^{s_p} = \alpha_p(x)$ for $x \in \mathbf{Q}_p^*$. For $\sqrt{d} \in \mathbf{Q}_p$ the subgroup $R_{\mathbf{Q}_p}$ is a split torus of $\mathbf{G}_{\mathbf{Q}_p}$, hence conjugate to the subgroup given by diagonal matrices, and the above inner product can again be computed quite explicitly as polynomial in $p^{\pm s_p}$.

The Fourier development (5.6) to which one is led in this context, in fact forced to introduce, is clearly arithmetically significant and natural. It leads, as we have just seen to quite explicit product formulas which seem to be new even for the classical modular functions and forms and which are in terms of finite expressions like (5.7) which are like "Hecke operators with characters". For each imaginary quadratic extension $\mathbf{Q}(\sqrt{d})$ of $\mathbf{Q}$ one is thus led to a new kind of Fourier series development (5.6), (5.7) which is significant in the class field theory of $\mathbf{Q}(\sqrt{d})$ and in the theory of complex multiplication. If one applies this to real-continuous or complex holomorphic modular functions or forms $f(z)$ one gets new kinds of linear relations for these Fourier coefficients. These relations seem to be quite new even if one takes for $f(z)$ such classical modular functions as the Eisenstein series $E(z, s) = y^s \sum |mz+n|^{-2s}$, the elliptic invariant $j(z)$, etc.

If ψ is an eigenfunction of the Hecke operators $\in L(\mathbf{G_Z} \backslash \mathbf{G_Q} / R_{\mathbf{Q}})$ one obtains in the above manner the natural Fourier development $\psi = \sum \psi * \varrho$ where the functions $\psi * \varrho$ are the eigenfunctions explicitly described above; in particular they are naturally parametrized by their eigenvalues. Hence the Ramanujan hypothesis (order of magnitude of cusp forms) becomes a special case of a kind of problem, which is well known in Fourier analysis and in spectral analysis: What is the spectrum of a given commutative *-algebra (in this case the algebra generated by the Hecke operators) in a given subspace? What are the special properties of the developments $\psi=\sum\psi*\varrho$ and $\psi=\sum\psi*u$ when ψ is the restriction of a real-continuous or complex-holomorphic function to $\mathbf{G_Q}$? Can one find an arithmetic characterisation of modular functions or forms, which is useful in this context? These problems seem to have analogies in other parts of mathematics, especially with the quite fundamental and general problem of constructing global solutions from local solutions (for example in the theory of quadratic forms) or of deducing global properties from local properties.

Let us note that instead of considering the space $L(\mathbf{G_Z}\backslash\mathbf{G_Q}/R_\mathbf{Q})$ of all complex valued functions of $\mathbf{G_Z}\backslash\mathbf{G_Q}$ which are right invariant by the elements of $R_\mathbf{Q}$, we could have studied by the same methods the space of all complex-valued functions on $\mathbf{G_Z}\backslash\mathbf{G_Q}$ which transform under right-multiplication by the elements of $R_\mathbf{Q}$ according to a fixed character of $R_\mathbf{Q}$; if one takes $e^{2\pi il\theta}$ as character of $R_\mathbf{Q}$, the modular forms are contained in our function space.

§ 6. Let us consider a little more the "Fourier transform" (4.2) given by the "local principal series" $M_\alpha=M_{\alpha_p}$ defined for example by (3.5). As M_α is the induced representation

$$\operatorname*{ind}_{\mathfrak{A}\mathfrak{B}\uparrow\mathfrak{G}}\alpha,\ \alpha\in\hat{\mathfrak{A}},\ \alpha(ab)=\alpha(a) \tag{6.1}$$

where $\mathfrak{A}$ is the subgroup of $\mathfrak{G}$ given by the diagonal matrices, $\mathfrak{B}$ the subgroup of translations, $M_\alpha(g)$ is right multiplication by g in a space of (measurable, continuous or square integrable) functions $v(g)$ defined on $\mathfrak{G}$ which satisfy

$$v(abg)=\alpha(a)v(g)\quad\text{for all}\quad a\in\mathfrak{A},\, b\in\mathfrak{B}\quad\text{and}\quad g\in\mathfrak{G}. \tag{6.2}$$

Hence $\mathfrak{G}=\mathfrak{A}^+\mathfrak{B}\mathfrak{K}$ where $\mathfrak{A}^+=\{p^n\}_{n\in\mathbf{Z}}$ and (6.2) imply that v is determined by its restriction to $\mathfrak{K}$. Hence for a (somewhat smooth and decreasing) function $f(g)$ defined on $\mathfrak{G}$, the operator $\mathbf{F}(\alpha)=\int_{\mathfrak{G}}f(g)M_\alpha(g)\,dg$ is for each $\alpha\in\hat{\mathfrak{A}}^+$ for which this integral converges, given by an integral kernel $\check{f}(k,\varkappa)$

$$[\mathbf{F}(\alpha)v](k)=\int_{\mathfrak{K}}\check{f}(k,\varkappa)v(\varkappa)\,d\varkappa \tag{6.3}$$

as is well known, where $\check{f}$ is given by the now well-known expression

$$\check{f}(k,\varkappa)=\int_{\mathfrak{A}}\Big(\int_{\mathfrak{B}}f(kab\varkappa^{-1})\,db\Big)\alpha(a)\,da \tag{6.4}$$

with k and $\varkappa\in\mathfrak{K}$.

Let us suppose for the sake of convergence that f is of compact support, i.e. $f\in L^0(\mathfrak{K}\backslash\mathfrak{G})$ and let us introduce in the Hilbert space $L^2(\mathfrak{A}\mathfrak{B}\cap\mathfrak{K}\backslash\mathfrak{K})$ of the restrictions to $\mathfrak{K}$ of functions $v(abg)=v(g)$ an $\mathfrak{K}$-adapted orthonormal base. In this base the operator $\mathbf{F}(\alpha)$ is given for each $\alpha\in\hat{\mathfrak{A}}^+$ by a matrix which has only one non-vanishing column; in this column the matrix coefficients are given by

$$F_u(\alpha) = \int_{\mathfrak{A}} \int_{\mathfrak{B}} \int_{\mathfrak{K}} f(ab\varkappa^{-1}) u(\varkappa)\, db\alpha(a)\, da \tag{6.5}$$

where u varies again over a (suitable) set of matrix coefficients of irreducible unitary representations of $\mathfrak{K}$. Let us suppose now that $f \in L^0(\mathfrak{K}\backslash\mathfrak{G}) * u$ and we obtain

$$F_u(\alpha) = \int_{\mathfrak{A}} \int_{\mathfrak{B}} f(ab)\, db\, \alpha(a)\, da. \tag{6.6}$$

Evidently $f(kg) = f(g)$ implies that the function $f(ab)$ is determined by its restriction to the elements $\mathfrak{p}^n b$ (with $b \in \mathfrak{B}$ and $n = 1, 2, \ldots$), hence

$$F_u(\alpha) = \sum_{n=0}^{\infty} \tilde{f}(\mathfrak{p}^n)\, \alpha(\mathfrak{p}^n) \tag{6.7}$$

provided the measure of $\mathfrak{A}$ is normalized so that the total measure of $\mathfrak{A} \cap \mathfrak{K}$ is 1 and $\tilde{f}$ is given by

$$\tilde{f}(\mathfrak{p}^n) = \int_{\mathfrak{B}} f(\mathfrak{p}^n b)\, db. \tag{6.8}$$

As $\alpha(\mathfrak{p}^n) = \alpha_p(\mathfrak{p}^n) = N\mathfrak{P}^{-ns}$ and f is of compact support, hence $\tilde{f}(\mathfrak{p}^n)$ of finite support, one sees that (6.7) is a Fourier polynomial given by

$$F_u(\alpha) = \sum_n \tilde{f}(\mathfrak{p}^n)\, N\mathfrak{P}^{-ns}. \tag{6.9}$$

Essentially the same observations apply if one considers the Fourier series development (4.4) relative to a compact Cartan subgroup $\mathfrak{R}$ of $\mathfrak{G}$ instead of (4.1), $F_\varrho(\alpha)$ instead of $F_u(\alpha)$ and $\mathfrak{G} = \mathfrak{A}^+\mathfrak{B}\mathfrak{R}$. If f is of compact support, one obtains again finite sums of the form

$$F_\varrho(\alpha) = \sum_n \tilde{f}(\mathfrak{p}^n)\, N\mathfrak{P}^{-ns} \tag{6.10}$$

where ϱ varies now over the characters of $\mathfrak{R}$. If one assumes $f \in L^1(\mathfrak{K}\backslash\mathfrak{G}) * u$ (or $\in L^1(\mathfrak{K}\backslash\mathfrak{G}) * \varrho$) the sums (6.9) and (6.10) are not finite in general, but converge uniformly in the "critical strip" $0 \leqq Rs \leqq 1$, such that $F_u(s)$ is (for each u) a holomorphic function of s in $0 < Rs < 1$, continuous in $0 \leqq Rs \leqq 1$ and so is $F_\varrho(s)$ for each $\varrho \in \hat{\mathfrak{R}}$.

If one takes instead of the local group $\mathfrak{G}$ a restricted direct product G_P (as in § 5 above) the elements a (mod K_P) are of the form $a \equiv \prod_{p \in P} \mathfrak{p}^{n_p}$ (mod K_P), their characters α are parametrized by the variables s_p $(p \in P)$ if one puts $\alpha(a) = \prod (N\mathfrak{P})^{-n_p s_p}$. Hence for each u we obtain now as the global analogue of (6.9)

$$F_u(\ldots, s_p, \ldots) = \sum \tilde{f}(\prod \mathfrak{P}^{n_p}) \prod (N\mathfrak{P})^{-n_p s_p} \tag{6.11}$$

where $\tilde{f}$ is the global analogue of (6.8) and u varies over a suitable set of matrix coefficients of the compact group K_P. For each such u we obtain thus a function of infinitely many complex variables $(N\mathfrak{P})^{\pm s_p}$. If f is of compact support the sum in (6.11) is finite and hence $F_u(\dots, s_p, \dots)$ is for each u a polynomial in the (infinitely many, if P is infinite) variables $N\mathfrak{P}^{\pm s_p}$. If f is summable, $F_u(\dots, s_p, \dots)$ is no longer a polynomial in general, but a uniform limit of polynomials in each "strip" $(0<)\varepsilon \leqq Rs_p \leqq 1-\varepsilon$, $p \in P$, hence a holomorphic function of the variables $s_p (p \in P)$ given by (6.11). If the set P is infinite one obtains thus a very interesting class of holomorphic functions $F_u(\dots, s_p, \dots)$ of infinitely many complex variables $s_p (p \in P)$. If f is summable then F_u is given by the series (6.11) which is in this case absolutely convergent for $0 \leqq Rs_p \leqq 1$. We have thus obtained a rather remarkable T_n-module of holomorphic functions of the variables s_p $(p \in P)$ which satisfy one functional equation $s_p \to 1-s_p$ for each $p \in P$. The action of $T_{\mathfrak{P}}$ on $F_u(\dots, s_p, \dots)$ is multiplication by $(N\mathfrak{P})^{s_p} + (N\mathfrak{P})^{1-s_p}$.

Each function $F_u(\dots, s_p, \dots)$ is determined by its restriction $F_u(s)$ to the diagonal $s = \dots = s_p = \dots = s_{p'} = \dots$. This yields an alternative method of calculating the Fourier inversion formula

$$F_u(\dots, s_p, \dots) \to F_u(s) \to f. \tag{6.12}$$

The transformation $f \to \tilde{f}$ has been studied by me [10] in the case where P consists of one non-Archimedean place p and $u=1$. For this case I have obtained the result

$$\tilde{f}(\mathfrak{p}^n) = N\mathfrak{P}^n \left[\frac{I-\partial^2}{I-N\mathfrak{P}\cdot\partial^2} f \right] (\mathfrak{p}^n) \quad \text{for} \quad n \geqq 0 \tag{6.13}$$

where $(\partial f)(\mathfrak{p}^n) = f(\mathfrak{p}^{n+1})$. One obtains the result that for an arbitrary set P of non-Archimedean places p the transformation $f \to \tilde{f}$ is a product of two transformations: One is multiplication by the function $\prod (N\mathfrak{P})^{n_p}$, the other is determined by the "Mellin transform" of multiplication by $\zeta_{P,u}(2s-1)/\zeta_{P,u}(2s)$ where $\zeta_{P,u}$ is the "zeta function of the pair P, u". If $u=1$ one has

$$\zeta_{P,1}(s) = \prod_{p \in P} (1 - N\mathfrak{P}^{-s})^{-1},$$

in particular for the module $L^0(G_{\mathbf{Z}} \backslash G_{\mathbf{Q}})$ one obtains the Riemann zeta function if $u=1$, since P is the set of all rational prime numbers in this case.

Thus one sees that the structure of T_n-modules like $L^0(K_P \backslash G_P)$, $L^0(K_P \backslash G_P) * u$, etc. is essentially determined by the functions $\zeta_{P,u}(s)$. As $u_p = 1$ for all but a finite number of places $p \in P$, one sees that $\zeta_{P,u}/\zeta_{P,1}$ is a polynomial in a finite number of variables $(N\mathfrak{P})^{\pm s_p}$, namely in those for which $u_p \neq 1$. These results (and other considerations) lead to the following *conjecture:*

Properties of the Riemann zeta function (or more generally of $\zeta_{P,u}$) are equivalent to polynomial properties, for example to properties of the polynomial modules introduced and studied above.

Turán [17—19] has obtained relations between the zeros of the Riemann zeta function and the zeros of the partial sums of the Dirichlet series $\sum n^{-s}$. The "polynomial properties" conjectured here are different, because the non-commutative structure of the group $PG(2)$ enters essentially.

In any case the zeta function enters here with vengeance. Clearly the Euler product $\prod_{p \in P} (1 - N\mathfrak{P}^{-s_p})^{-1}$ converges in every region $Rs_p > 1+\varepsilon$ ($p \in P$, $\varepsilon > 0$) and is a holomorphic function of the variables s_p ($p \in P$) in each region $Rs_p > 1+\varepsilon$, $\varepsilon > 0$. It is not difficult to see that this is not a meromorphic function of the variables s_p ($p \in P$) in a "compact region" like $-1 \leqq Rs_p \leqq 2$, $0 \leqq \mathrm{Im}\, s_p \leqq 2\pi/\log N\mathfrak{P}$, if P is infinite. Hence the "infinite Weyl group" generated by the transformations $s_p \to 1 - s_p$ remains well hidden in the Euler product, in fact in the whole of arithmetic so far!

On the other hand this infinite Weyl group is completely explicit in such algebras of spherical functions as $L^0(K_P \backslash G_P / K_P)$ and $L^1(K_P \backslash G_P / K_P)$. Here one has the impression that one has finally succeeded in avoiding the zeta function. However, it is the transformation $f \to \tilde{f}$ which determines the Fourier transform of these algebras. Hence the zeta function reenters here with vengeance, as in an ancient Greek tragedy!

If one replaces $\mathbf{G_Z}$ by a congruence subgroup $\mathbf{G_Z^{(m)}}$, the results presented above do not remain correct. The principal series M_α is no longer sufficient for the decomposition of $L^2(\mathbf{G_Z^{(m)}} \backslash \mathbf{G_Z})$, representations of the discrete series of G_p enter for $p|m$. This is quite an abrupt and qualitative difference between the modular group $\mathbf{G_Z}$ and its congruence subgroups $\mathbf{G_Z^{(m)}}$, which does not seem to have been anticipated in the classical theory of modular functions and forms.

Let us also observe that one must not omit $\mathbf{G_Z}$ completely in $L^2(\mathbf{G_Z} \backslash \mathbf{G_Q})$. The regular representation of $PG(2, \mathbf{Q})$ is of type II_1 in the sense of the theory of operator algebras of Murray and von Neumann, in which case the irreducible unitary representations, which enter into a direct integral decomposition of $L^2(\mathbf{G_Q})$ are by no means unique, apparently not even all "calculable". Thus one sees that the modular group plays an essential role in the problems considered here, even from the point of view of operator algebras and the general theory of unitary representations.

PART II

EXPLICIT TRACE FORMULAS

§ 7. From now on $\mathbf{G}$ denotes an algebraic matrix group defined over an algebraic number field Ω or an algebraic function field Ω of transcendence degree 1 over a finite constant field. We shall study the action of the Hecke operators on $L^2(\mathbf{G_A}/\mathbf{G}_\Omega)$ where $\mathbf{A}$ denotes the ring of adeles of Ω. Instead of $\mathbf{G}_{\Omega_p}$ we write G_p, instead of $\mathbf{G}_{\mathfrak{O}_p}$ we shall write K_p, so that K_p is the subgroup of those elements of G_p whose matrix coefficients are in the ring $\mathfrak{O}_p$ and whose determinant is in $\mathfrak{U}_p$, the group of p-adic units (for every non-Archimedean place p of Ω).

We shall also write $G_\infty = \prod_{p \in \infty} G_p$, $G_0 = \prod_{p \notin \infty} G_p$, G instead of $\mathbf{G_A}$, so that $G = G_\infty \times G_0$, and $K_0 = \prod_{p \notin \infty} K_p$; hence K_0 is a compact open subgroup of G_0. As $\mathbf{G}_\Omega$ is a discrete subgroup of $\mathbf{G_A}$, considered as the group of principal ideles, we shall write often Γ instead of $\mathbf{G}_\Omega$.

Let $L^0(K_0 \diagdown G_0/K_0)$ be the space of all complex-valued functions on G_0, which are of compact support and left- and right-invariant by the elements of K_0. For every (sufficiently decreasing) measure (distribution), μ we define an operator T_μ by

$$(T_\mu \psi)(g) = (\mu * \psi)(g) = \int \psi(\mathbf{g}^{-1} g)\, d\mu(\mathbf{g}) \tag{7.1}$$

where ψ varies over $L^2(G/\Gamma)$. In particular, if f is a (sufficiently smooth and decreasing, e.g. integrable) function defined on a closed subgroup $\mathscr{G}$ of G, we write T_f for the operator

$$(T_f \psi)(g) = \int_{\mathscr{G}} f(\mathbf{g}) \psi(\mathbf{g}^{-1} g)\, d\mathbf{g} \tag{7.2}$$

where $d\mathbf{g}$ refers to a Haar measure on $\mathscr{G}$. We shall be particularly interested in the case where $f_0 \in L^0(K_0 \diagdown G_0/K_0)$:

$$(T_{f_0} \psi)(g) = \int_{G_0} f(g_0) \psi(g_0^{-1} g)\, dg_0 \tag{7.3}$$

where dg_0 refers to Haar measure on G_0. One verifies that if $\mathbf{G} = PG(2)$ and $\Omega = \mathbf{Q}$ the operators T_{f_0} defined by (7.3) are exactly the finite $\mathbf{C}$-linear combinations of the Hecke operators T_n acting on $L^2(PG(2, \mathbf{R})/PG(2, \mathbf{Z}))$, provided f_0 varies over $L^0(K_0 \diagdown G_0/K_0)$. To see this one identifies the space of double cosets $K_0 \diagdown G/\Gamma$ with $PG(2, \mathbf{R})/PG(2, \mathbf{Z})$ and uses $G = K_0 G_\infty \Gamma$.

In general one knows [2] that this space of double cosets is finite:

$$G = \bigcup_{j=1}^{h} K_0 G_\infty g_j \Gamma \quad \text{with} \quad g_j \in G \quad \text{and} \quad h < \infty. \tag{7.4}$$

If $h > 1$, the "adelic" definition (7.3) of Hecke operators is not equivalent to the definition given in § 1, even for the homogeneous space $\mathbf{G_R}/\mathbf{G_Z}$. It seems that the "adelic" definition (7.3) of Hecke operators is the most useful for arriving at certain explicit and concrete results. For example, I have shown [5] that one can express the trace of T_{f_0} as finite sum of Tamagawa numbers multiplied by Minkowski—Siegel—Tamagawa—Weil "alphas", provided things converge. If **G** is semisimple and G/Γ compact, all the expressions converge which enter into Selberg's trace formula

$$\sum_{W_\gamma} m\left(\frac{G(\gamma)}{\Gamma(\gamma)}\right) f^\natural(\gamma) \quad \text{where } f^\natural(\gamma) = \int_{G(\gamma)\setminus G} f(x^{-1}\gamma x)\, dx \tag{7.5}$$

and where $G(\gamma)$ is the centralizer of γ in G, $\Gamma(\gamma)$ the centralizer of γ in Γ and all Haar measures are normalized canonically (in the sense of Tamagawa—Weil; cf. [21]). The calculation of $f^\natural$ gives surprisingly explicit results: Let W_γ be the conjugacy class of γ in Γ, considered as affine variety (in the space of matrices over Ω). If f_0 is the characteristic function of the double coset $K_0 g_0 K_0$ and

$$K_0 g_0 K_0 = \bigcup_{j=1}^{r} K_0 m_j \quad \text{with} \quad r < \infty \quad \text{and} \quad m_j \in G_0 \tag{7.6}$$

(K_0 being compact and open in G_0) I have shown [5] that

$$f_0^\natural(\gamma) = \sum_{i=1}^{r'} \alpha_0(W_\gamma, M_i) \quad r' \leqq r \tag{7.7}$$

where M_i are lattices in the space of matrices over Ω and

$$\alpha_0(W_\gamma, M_i) = \prod_{p \notin \infty} \alpha_p(W_\gamma, M_i); \tag{7.8}$$

the α_p are Minkowski—Siegel—Tamagawa—Weil p-adic representation numbers of the variety W_γ, obtained as canonical measures of the p-adic completions of $W_\gamma \cap M_i$. If G_∞ is compact one can take f_∞=constant 1 on G_∞, define a function f on $G = G_\infty \times G_0$ by $f = f_\infty \cdot f_0$. One then obtains

$$f^\natural(\gamma) = \sum_{i=1}^{r'} \alpha(W_\gamma, M_i) \tag{7.9}$$

where

$$\alpha(W_\gamma, M_i) = \prod_{\text{all } p} \alpha_p(W_\gamma, M_i) \tag{7.10}$$

are the Minkowski—Siegel—Tamagawa—Weil representation numbers of the M_i-lattice points of the variety W_γ.

If the class number h in (7.4) is 1 and the group G_∞ compact, I have obtained [5] the identity

$$\sum_{W_\gamma} m\left(\frac{G(\gamma)}{\Gamma(\gamma)}\right) \sum_{i=1}^{r'} \alpha(W_\gamma, M_i) = \int_G f(g)\, dg. \tag{7.11}$$

This is a very explicit generalization of certain of the classical class number relations to higher dimension. If $h=1$, the main theorem in the analytic theory of quadratic forms of Minkowski and Siegel, generalized by Ono, Tamagawa, Weil et al., implies

$$\alpha(W_\gamma, M_i) = \text{number of points of } W_\gamma \cap M_i. \tag{7.12}$$

If $\mathbf{G}$ is the orthogonal group of the quadratic form $x_1^2+\ldots+x_N^2$ one knows that if $\Omega=\mathbf{Q}$, the lattice in Ω^N is $\mathbf{Z}^N$ and $3 \leqq N \leqq 8$, then the number of classes in the genus is one. The centralizers $G(\gamma)$ are direct products of adelized orthogonal subgroups, so that their Tamagawa numbers $m(G(\gamma)/\Gamma(\gamma))$ are known. Thus (7.11) and (7.12) are in this case very explicit linear identities which were not known for $N \geqq 5$. For $N=3$ one can embed W_γ in Ω^3, so that one obtains Minkowski—Siegel "alphas" of ternary quadratic forms, hence some of the classical class number relations.

There is an essential difference between these identities, some of which were for $N=3$ already known to Kronecker, and the general identity (7.11) in higher dimensions: *The general identity* (7.11) *is in higher dimensions no longer a relation between class numbers of algebraic number fields, but a linear identity where the centralizers* $\mathbf{G}(\gamma)$ *occur, which are in general non-commutative algebraic groups.*

It seems to be too early to try to compute the integral $f_0^\natural(g) = \int_{G(g)\setminus G} f_0(x^{-1}gx)\,dx$ if f_0 is a function on G_0 which is not bi-invariant by the elements of K_0, but is a matrix-valued function satisfying

$$f_0(k_0 g_0 k_0') = u(k_0) f_0(g_0) u(k_0') \tag{7.13}$$

where $k_0, k_0' \in K_0$, $g_0 \in G_0$ and u is an arbitrary (irreducible unitary) representation of K_0, because one knows very few of the representations of the compact groups K_p if p is non-Archimedean and $\dim K_p > 3$. For the group $PG(2, \mathfrak{O}_p)$ the irreducible unitary representations are known (see [15] and [16]) and one can calculate $f_0^\natural$ for the corresponding functions (7.13). In particular if u belongs to the "principal series mod $\mathfrak{P}^m$" of representations of $PG(2, \mathfrak{O}_p)$, which is parametrized by a congruence character mod $\mathfrak{P}^m$ [11], we write $u = u_\chi$. In this case the expression for $f_0^\natural$ where $f_0^\natural$ satisfies (7.12) with $u=u_\chi$ is obtained from the expression for $f_0^\natural$ with $f_0 \in L^0(K_0 \setminus G_0 / K_0)$ by multiplying each term by a certain value of χ. The analogy with the Riemann—von Mangoldt formula generalized by Weil [20] is quite remarkable, in particular the analogy with passage in the Riemann—von Mangoldt—Weil formula for the zeta function to the corresponding formula for an L-series.

§ 8. The analogy between the trace formula (7.5) for $G = SL(2, \mathbf{R})$ and Γ a discrete (uniform) subgroup, and the Riemann—von Mangoldt—Weil formula was discovered and used by Selberg [13] and this led him to the construction of certain Dirichlet series for which he succeeded in showing that the non-trivial zeros are always on the line $Rs = \frac{1}{2}$ or on the real interval (0, 1).

One can generalize the methods and results of Selberg, if one takes for G the group of ideles of a semisimple linear group (as in § 7 above) instead of the group $G = SL(2, \mathbf{R})$ considered by Selberg and for Γ the subgroup of principal ideles. If one assumes the space G/Γ of idele classes to be compact and the functions sufficiently smooth and decreasing one obtains the identity

$$\sum_{W_\gamma} m\left(\frac{G(\gamma)}{\Gamma(\gamma)}\right) f^\natural(\gamma) = \sum \lambda_j \tag{8.1}$$

where the λ_j are the eigenvalues of the convolution operator T_f in the space $L^2(G/\Gamma)$. It is too early to try to study this formula in detail for an arbitrary semisimple group $\mathbf{G}$. We are still too far from a sufficient knowledge of the irreducible unitary representations of the groups G_p and even of the groups K_p if p is non-Archimedean.

Even in the case where $\mathbf{G}$ is the orthogonal group of a ternary quadratic form over Ω, which we assume to be non-zero over Ω, formula (8.1) leads to certain results and relations which are still quite difficult to understand. In fact I have the impression that the concepts which exist now in the theory of numbers and functional analysis are not sufficient to find interpretations for many of the consequences of (8.1).

For example if f is an integrable spherical function (i.e. bi-invariant under a maximal compact subgroup K) defined on the adelized orthogonal group G of a ternary quadratic non-zero form over Ω, i.e. $f \in L^1(K \backslash G/K)$ one knows that its Fourier transform on G is a holomorphic function $F(\dots, s_p, \dots)$ of infinitely many complex variables s_p in each strip $\varepsilon < Rs_p < 1 - \varepsilon$, $\varepsilon > 0$. Let us write $\mathfrak{s} = (\dots, s_p, \dots)$ and let us observe that the mapping $f \to \lambda_j$ in (8.1) is for each j a *-homomorphism of the algebra $L^1(K \backslash G/K)$ into $\mathbf{C}$. Hence there exists for each j a point $\mathfrak{s}_j$ such that $\lambda_j = F(\mathfrak{s}_j)$ for all $f \in L^1(K \backslash G/K)$.

Thus formula (8.1) *characterizes in this case an ideal in a ring of holomorphic functions of infinitely many complex variables s_p which is not determined by a single Dirichlet series, even if one knows its zeros!*

Even if one supposes that G_∞ is compact and f_∞ an irreducible character of $G_\infty = K_\infty$ which implies that (with $f = f_\infty \cdot f_0$ and $f_0 \in L^1(K_0 \backslash G_0/K_0)$ the rank of the linear transformation T_f is finite, hence the set of points $\mathfrak{s}_j$ finite, formula

(8.1) still characterizes an ideal in a ring of holomorphic functions of infinitely many complex variables, which one cannot yet use and interpret completely. The analogy with the Riemann—von Mangoldt—Weil formula (cf. [20]) for the zeta function of an algebraic curve (or variety) over a finite constant field is still remarkable, but the relationship between $f^{\natural}$ and $F(\mathfrak{s})$ is no longer given by a Fourier series in one single variable: $F(\mathfrak{s})$ is still a holomorphic function of infinitely many variables s_p. Formula (8.1) has thus a global significance which one has to find in the arithmetic of global fields. Let us remark here that this search for an interpretation of (8.1) in global arithmetic becomes even more exciting and provoking if one replaces biinvariance of f_0 under K_0 by the condition that f_0 satisfy (7.12) with $u=u_\chi$ in the principal series mod $\mathfrak{P}^m$ of representations of K_0: The left-side of (8.1) is for such $u=u_\chi$ quite close to the left-side of the Riemann—von Mangoldt—Weil formula for the Dirichlet L-series or for certain L-polynomials [11]. It is therefore natural to ask whether there exist interpretations of (8.1) in the arithmetic of certain global fields like $\mathbf{Q}(\sqrt[m]{1})$ and whether (8.1) implies results about arithmetic progressions.

There are also local to global and global to local relations which occur naturally in this context. Certain of these relations are obtained by a specialization of almost all the variables s_p: Let us suppose that for almost all places l except for a finite number of places p which are unramified in the quaternion algebra corresponding to the given ternary quadratic form, the function f_l is a fixed self-adjoint idempotent of compact support and f_p still an arbitrary (spherical, integrable) function on G_p. If $f=\prod_p f_p \prod_l f_l$, the Fourier transform F of f is a function of a finite number of variables s_p only. In particular if one leaves only one unramified place $\mathfrak{p}$ where $f_{\mathfrak{p}}$ is not a fixed self-adjoint idempotent, but arbitrary (spherical), then F becomes a function of one single complex variable s_p, and formula (8.1) is in this case quite close to (and in some cases identical with) the Riemann—von Mangoldt—Weil formula for the zeta function of an algebraic curve over a finite constant field if $\mathfrak{p}$ is non-Archimedean, to the corresponding formula for the zeta function which Selberg [13] associated with the homogeneous space $\mathbf{G_R}/\mathbf{G_Z}$, if $\mathfrak{p}$ is Archimedean.

In this case where F is a function of a single variable s_p, formula (8.1) characterizes one single Dirichlet series in the variable s_p, whose non-trivial zeros are always on $Rs_p=\frac{1}{2}$ (or on the real interval (0, 1) if representations of the exceptional series enter into the spectrum of the given algebra of convolution operators T_{f_p}). As $f^{\natural}$ can be expressed in terms of "alphas" of the given ternary quadratic form and values of congruence characters χ, the coefficients of these Dirichlet series can be expressed in terms of these alphas and values of χ. *Thus one obtains a new kind of relationship between arithmetic and analytic functions. A class of Dirichlet series whose non-trivial zeros are on* $Rs=\frac{1}{2}$ or (0, 1) *and whose coefficients have expressions in terms of alphas and congruence characters* χ.

These methods give new results even for the "zeta function" of Selberg [13] in the case of an arithmetic subgroup of $PG(2, \mathbf{R})$ whose number of classes in the genus is 1, if one takes $\Omega = \mathbf{Q}$, $\mathbf{G}$ again the orthogonal group of a ternary (non-zero) quadratic form, $f = f_\infty \cdot f_0$, with f_0 the characteristic function of K_0 (considered as a function on G_0) and f_∞ a (sufficiently smooth and decreasing) spherical function on $G_\infty \cong PG(2, \mathbf{R})$. As $f_\infty^\natural$ can be expressed by "alphas" of the given ternary quadratic form (assumed to be non-zero over $\mathbf{Q}$, but indefinite now) the coefficients of the corresponding zeta function of Selberg [13] can be expressed in terms of these "ternary alphas". For the proof one has to use only the methods and results described above, in particular (7.7) and $G = K_0 G_\infty \Gamma$. One can vary Selberg's [13] zeta functions and obtain a family of functions which depend on congruence characters, but still have their non-trivial zeros on $Rs = \frac{1}{2}$ or $(0, 1)$, if one takes at a finite number of finite places p instead of the characteristic function of K_p a minimal self-adjoint idempotent e_p of compact support; the existence of such idempotents e_p i.e. elementary positive definite functions of compact support, was found by myself [11]. According to the thesis of J. Shalika [15] the corresponding irreducible unitary representations of the group $G_p \cong PG(2, \Omega_p)$ are parametrized in a natural manner by certain congruence characters χ which belong to quadratic extensions of Ω_p and the integrals $e_p^\natural$ can be expressed in terms of these χ's. The corresponding convolution operator T_f is in this case convolution by the function $f = f_\infty \cdot f_0$ on the subspace of $L^2(G/\Gamma)$ of functions

$$\psi(k_0 g) = u(k_0)\psi(g) \quad (k_0 \in K_0,\ g \in G)$$

which is naturally isomorphic to the space of functions

$$\psi'(g_\infty \varepsilon) = u(\varepsilon)\psi'(g_\infty) \quad (g_\infty \in G_\infty,\ \varepsilon \in \Gamma \cap K_0)$$

which are square integrable over $G_\infty / \Gamma \cap K_0$.

One obtains thus many analogies of certain zeta functions and L-series and can construct a large class of Dirichlet series which satisfy the Riemann hypothesis, if one takes in formula (8.1) an algebra of functions f for which F depends on *one* complex variable only. The resulting formulas (8.1) generalize Selberg's [13] identities in a manner which justifies one assuming the existence of important relations with arithmetic, which are still unknown, even if one remains within the class of Dirichlet series of one single complex variable s_p. The full information contained in (8.1) when F depends on all the variables s_p will undoubtedly remain a deep source for a long time to come.

We have assumed throughout § 7 and § 8 that the space $\mathbf{G}_\mathbf{A}/\mathbf{G}_\Omega$ is compact, in order to insure convergence of the trace formula. It is natural to conjecture that some of the results outlined above remain correct for an arbitrary semisimple linear group, even if the space of idele-classes is not compact. In particular I conjecture that the trace of the restriction of the Hecke operators defined by (7.3) to the space of cusp forms can be expressed in terms of Tamagawa numbers and Minkowski—Siegel—Tamagawa—Weil alphas for an arbitrary semi-simple algebraic linear group **G** defined over an algebraic number field or function field of transcendence degree one over a finite constant field, even if the space of ideleclasses is not compact.

REFERENCES

[1] Bargmann, V., Unitary representations of the Lorentz group, *Annals of Math.*, **48** (1947), 568—640.

[2] Borel, A. and Harish-Chandra, Arithmetic subgroups of algebraic groups, *Annals of Math.* **75** (1962), 485—535.

[3] Bruhat, F., Distributions sur un groupe localement compact et applications, *Bull. Soc. Mat. France,* **89** (1961), 43—57.

[4] *Actes, Congrès intern. math., 1970,* Gauthier-Villars, Paris, 1971.

[5] Mautner, F. I., The trace of Hecke operators, *Monatsh. für Math.,* **72** (1968), 137—143.

[6] ———, On the zeros of certain *L*-polynomials, *Monatsh. für Math.,* **73** (1969), 238—249.

[7] ———, Fonctions propres des opérateurs de Hecke, *C. R. Ac. Sci. Paris,* Série A, **269** (1969), 940—943; **270** (1970), 89—92.

[8] ———, Fonctions sphériques et opérateurs de Hecke, *in Séminaire Delange—Pisot—Poitou,* Paris, Feb. 1970.

[9] ———, Quelques remarques sur la trace des opérateurs de Hecke, *Colloquium Université Paris* IV, Jan. 1971.

[10] ———, Spherical functions over $\mathfrak{p}$-adic fields I, *Amer. J. Math.,* **80** (1958), 441—457.

[11] ———, Spherical functions over $\mathfrak{p}$-adic fields II, *Amer. J. Math.,* **86** (1964), 171—200.

[12] Schwartz, L., *Théorie des distributions*, Hermann et Cie., Paris, 1950—51.

[13] Selberg, A., Harmonic analysis and discontinuous groups on weakly symmetric Riemann spaces, *J. Indian Math. Soc.,* **20** (1956), 47—87.

[14] ———, Discontinuous groups and harmonic analysis, *in Proc. Int. Congr. Stockholm (1962).*

[15] Shalika, A., *Thesis,* John Hopkins University, 1966.

[16] Silberger, A. J., *PG2 over the p-adics, its representations, spherical functions and Fourier analysis,* Springer Lecture Notes, Vol. **166** (1970).

[17] Turán, P., On some approximative Dirichlet polynomials in the theory of the zeta function of Riemann, *Det. Kgl. Danske Vid. Selsk. Mat. Fys. Medd.,* **24** (1948), 2—36.

[18] ———, On some approximative Dirichlet polynomials in the theory of zeta function of Riemann, *Acta Math. Acad. Sci. Hung.,* **10** (1959), 277—298.

[19] ———, A theorem on Diophantine approximation with application to Riemann zeta-function, *Acta Szeged,* **XXI** (1960), 311—318.

[20] Weil, A., Sur les formules explicites de la théorie des nombres premiers, *Comm. Sem. Math. Univ. Lund,* vol. suppl. 1952, 252—265.

[21] ———, *Adeles and algebraic groups,* Lecture Notes, Institute for Advanced Study, Princeton 1961.

References added in proof:

Mautner, F. I., *C. R. Ac. Sci. Paris, Série A,* **274** (1972) 1092—1095; **275** (1972), 353—356; 739—742; **278** (1974), 207—209.

DISCRETE SUBGROUPS FROM A GEOMETRIC VIEWPOINT

G. D. MOSTOW

§ 1. Introduction

Let G_i be a semisimple analytic group having no compact normal subgroup other than (1), and let Γ_i be a discrete subgroup such that G_i/Γ_i has finite invariant measure ($i=1, 2$). We are concerned with the problem:

Does Γ_1 determine the pair (G_1, Γ_1)? That is, given an isomorphism $\theta:\Gamma_1\to\Gamma_2$, does θ extend to an analytic isomorphism of G_1 onto G_2?

Example. $G=G_1=G_2=PSL(2, \mathbf{R})$. The group G operates on the upper half-plane $X=\{z;\ \mathrm{Im}\, z>0\}$ transitively and is the group of all analytic transformations of X onto X. By uniformization theory, X is the simply-connected covering space of all Riemann surfaces of genus greater than one. Let S_i be a compact Riemann surface of genus $p>1$, and let $\Gamma_i=\pi_1(S_i)$ $(i=1, 2)$. Then $\Gamma_i\subset G$ and G/Γ_i is compact. Since S_1 and S_2 are homeomorphic, there is an isomorphism $\theta:\Gamma_1\to\Gamma_2$. It is easy to see that θ extends to an analytic automorphism of G if and only if the S_i are analytically equivalent, which, of course, is not the case in general.

But this counter-example, looked at from a certain point of view, provides a key for solving our problem.

We adopt the viewpoint of *transformation groups*. Let $\psi:S_1\to S_2$ be a diffeomorphism and let $\varphi:X\to X$ be a lift of φ, and let $\theta:\Gamma_1\to\Gamma_2$ be the isomorphism induced by ψ. Then $\varphi(\lambda x)=\theta(\lambda)\varphi(x)$ for all $\lambda\in\Gamma_1$, $x\in X$. Thus, regarding Γ_i as transformation groups on X, φ is a Γ-space isomorphism. In other words the Γ-space structure of X is not refined enough to distinguish the analytic type of $\Gamma\diagdown X$.

Performing the Cayley transform, we can take $X=\{z;\ |z|<1\}$.

Consider now Γ_1 acting on $X\cup X_0$, where $X_0=\partial X$. If $\Gamma_1\diagdown X$ and $\Gamma_2\diagdown X$ are analytically equivalent, then there is a *Möbius* transformation φ of $X\cup X_0$ onto $X\cup X_0$ with $\varphi(\lambda x)=\theta(\lambda)\varphi(x)$ for all $\lambda\in\Gamma_1$ and $x\in X\cup X_0$. Our approach is motivated by the converse observation:

Theorem 1. *Suppose there exists a (Γ-space) morphism $\varphi:X\cup X_0\to X\cup X_0$ of class C^1. Then θ extends to an analytic isomorphism of G.*

This result generalizes to all semisimple group G without compact normal subgroups.

Indeed, let K be a maximal compact subgroup of G, and set $X=G/K$. We define X_0 as the *maximal boundary* of X in the sense of Furstenberg. More explicitly, let ϱ denote the adjoint representation of G. We can select coordinates in the underlying linear space so that

$$\varrho(G) = \varrho(G) \cap P(n, \mathbf{R}) \cdot \varrho(G) \cap O(n, \mathbf{R})$$

where

$$P(n, \mathbf{R}) = \{x \in GL(n, \mathbf{R}); {}^t x = x, x > 0\}$$

$$O(n, \mathbf{R}) = \{x \in GL(n, \mathbf{R}); {}^t x = x^{-1}\}.$$

Let

$$S(n, \mathbf{R}) = \{x \in GL(n, \mathbf{R}); {}^t x = x\}.$$

Let μ denote the map $gK \to g^t g$, and let π denote the canonical map of $S(n, \mathbf{R})-0$ onto the projective space $[S]$ of lines through 0 in the linear space $S(n, \mathbf{R})$. Then $\pi \circ \mu$ is an injective G-space morphism of X into $[S]$, the action on $[S]$ being induced by the linear transformations $s \to gs^t g$. We identify X with its image in $[S]$ and let $\overline{X}$ denote its closure. Satake has proved that

(i) $\overline{X}$ is a finite union of G-orbits,

(ii) $\overline{X}$ contains a unique compact G-orbit X_0; X_0 is also the G-orbit of lowest dimension and lies in the closure of all G-orbits in $\overline{X}$.

(iii) The stabilizer of a point in X_0 is a minimal parabolic subgroup of G. This compact G-orbit is Furstenberg's maximal boundary.

The proof of Theorem 1 depends heavily on the structure theory of restricted roots on real semisimple Lie algebras (cf. Proc. of Symposia in Pure Math. v. 9 (1966), p. 413).

Given the known facts on the non-existence of deformations of Γ, it was natural to conjecture that the hypothesis on the existence of a C^1-morphism $\varphi : X \cup X_0 \to X \cup X_0$ is superfluous.

Indeed, the existence of a continuous Γ-space morphism φ of X_1 into X_2 is elementary — at least in case Γ_1 has no torsion. For the space $X_i = G_i/K_i$ is topologically Euclidean space and Γ_i operates freely on X_i. Therefore X_i is a universal Γ_i-bundle and the existence of a morphism $\varphi : X_1 \to X_2$ inducing $\theta : \Gamma_1 \to \Gamma_2$ is a well-known fact (cf. Steenrod, *Fiber Bundles*, Princeton University Press). However, the task of proving that the Γ-space morphism $\varphi : X_1 \to X_2$ has *boundary* values is more difficult.

§ 2. Real hyperbolic space

The first case in which I could execute the foregoing idea was in the case $G_1 = G_2 = SO(1, n)/(\pm 1)$. Here the space X is real hyperbolic n-space of constant negative curvature. To start with, I assumed that $\Gamma_1 \backslash X$ and $\Gamma_2 \backslash X$ were diffeomorphic and compact. Any diffeomorphism ψ of compact manifolds is quasi-conformal (we say ψ is k-quasi-conformal if

$$\overline{\lim_{r\to 0}} \frac{a(r,p)}{b(r,p)} < k \quad \text{for almost all } p$$

and finite for all p, where $a(r,p)$ [resp. $b(r,p)$] is the radius of the circumscribed [resp. inscribed] ball of $\psi(B(r,p))$, where $B(r,p)$ is the ball of center p and radius r). On the other hand, the G invariant metric on X can be realized as the Poincaré metric

$$\frac{|dx|}{1-|x|^2}$$

on $\{x \in \mathbf{R}^n;\ |x|<1\}$, and therefore G operates conformally on X. Consequently, the lift $\varphi: X \to X$ of ψ is a quasi-conformal morphism. The fact that the φ has boundary values on $X_0 = \partial X$ is a basic theorem on quasi-conformal maps (cf. Mostow, Rigidity of Hyperbolic Space Forms, Publ. IHES, v. 34 (1968)). Indeed, the theory of quasi-conformal maps tells us that φ induces a boundary homeomorphism φ_0 of X_0 which is quasi-conformal. But we cannot yet apply our Theorem 1 above because a quasi-conformal map need not be of class C^1. Indeed a quasi-conformal map must be differentiable only *almost* everywhere.

However, using the fact the Γ operates ergodically on X_0, one can show that φ_0 is conformal. (One shows that

$$F(g) = [{}^t\varphi(\dot{g}(0))\varphi(\dot{g}(0))]\,[\det(\varphi(\dot{g}(0)))]^{-2/n}$$

is constant as on G. Thus φ_0 is 1-quasi-conformal on X_0 and therefore φ_0 is conformal on X_0 if $n>2$).

Therefore $\theta(\lambda) = \varphi_0 \lambda \varphi_0^{-1}$ with $\varphi_0 G \varphi_0^{-1} = G$. Thus one finds (cf. Publ. IHES. v. 34).

Theorem 2. *Let Γ_i be a discrete subgroup of $G = SO(1,n)$ such that G/Γ_1 and G/Γ_2 have finite measure. Let $\varphi: X \to X$ be a Γ-space morphism which is quasi-conformal. Then φ induces a boundary map $\varphi_0: X_0 \to X_0$ which is conformal, provided $n>2$, and*

$$\theta(\lambda) = \varphi_0 \lambda \varphi_0^{-1} \quad \textit{for all} \quad \lambda \in \Gamma, \quad \textit{with} \quad \varphi_0 G \varphi_0^{-1} = G.$$

§ 3. Symmetric spaces of rank greater than one

Theorem 3. *Let G_i be a semisimple analytic group with no compact normal subgroup, and let Γ_i be a discrete subgroup ($i=1,2$). Let $\theta: \Gamma_1 \to \Gamma_2$ be an isomorphism. Assume G_i/Γ_i is compact. Then θ extends to an analytic isomorphism, provided that G has no factors of* **R**-*rank one.*

The extension of θ can be described as follows: Let X_i be the symmetric space G_i/K_i associated to $G_i\{i=1,2\}$. Let $X_{i,0}$ be the Furstenberg maximal boundary of X_i. Let $\varphi: X_1 \to X_2$ be a Γ-space morphism.

The key point is to prove that φ induces a *continuous* boundary map $\varphi_0 : X_{1,0} \to X_{2,0}$. We do this with the help of the following notion.

Definition. Let Y_1 and Y_2 be metric spaces, and let $\varphi : Y_1 \to Y_2$ be a continuous map. We call φ a (k, b) *pseudo-isometry* if

1. $d(\varphi(p), \varphi(q)) \leqq kd(p, q)$ for all p, q in Y_1
2. $d(\varphi(p), \varphi(q)) \geqq k^{-1}d(p, q)$ for all p, q in Y_1

such that $d(p, q) \geqq b$. Here, k and b are positive real numbers.

It is not hard to see that the Γ-space morphism $\varphi : X_1 \to X_2$ can be selected to be a (k, b) pseudo-isometry for some k and b. One then establishes the following facts.

Lemma 3.1. *Let X be a symmetric Riemannian space of rank r. Let F be an r-dimensional flat geodesic subspace in X. Let $\pi : X \to F$ denote the orthogonal projection of X onto F. Let $p \in X$ and let V be an r-surface in X containing p, and let $J(p)$ denote the Jacobian at p of the restriction of π to V. Then*

$$J(p) \geqq cd(p, F)^{-1/2}$$

where c is a constant depending only on X.

Lemma 3.2. *Let $\varphi : X_1 \to X_2$ be a Γ-space morphism which is a pseudo-isometry. Let F_1 be a maximal flat subspace of X_1. Assume $\Gamma_1 \backslash \Gamma_1 F$ is compact. Then there is a constant c and a unique flat subspace F_2 in X_2 such that*

$$hd(\varphi(F_1), F_2) < c$$

where hd denotes Hausdorff distance and moreover, the constant c is independent of the particular F_1 such that $\Gamma_1 \backslash \Gamma_1 F_1$ is compact.

Lemma 3.3. *The set of maximal flat subspaces F_1 such that $\Gamma_1 \backslash \Gamma_1 F_1$ is compact is dense in the set of all maximal flat subspaces of X_1.*

Lemma 3.4. *Let φ be as in Lemma* 3.1. *Then for any maximal flat subspace F_1 of X_1 there is a unique flat subspace F_2 in X_2 such that*

$$hd(\varphi(F_1), F_2) < c^1.$$

Definition. Let X be a symmetric Riemannian space of non-positive curvature. A flat (geodesic) subspace S is called *singular* if it lies in more than one flat subspaces. Let F be a maximal flat subspace of X. Let $p \in F$ and let ${}^xS$ denote the union of all singular geodesics containing x. A connected component of $F - {}^xS$ is called a *chamber* of origin x.

Any chamber is bounded by chamber *walls*, just as in the case of Cartan subalgebras.

Lemma 3.5. *Let F and F^1 be maximal flat subspaces in the symmetric Riemannian space X of non-positive curvature. Set $T_r(F)=\{x\in X;\ d(x, F)<r\}$. Then $T_r(F^1)\cap F$ lies within finite Hausdorff distance of a convex set made up of chambers or chamber walls.*

Lemma 3.6. *Let $\varphi: X_1 \to X_2$ be as in Lemma* 3.2, *and let ${}^{<}F_1$ be a chamber in X. Then there is a chamber ${}^{<}F_2$ in X_2 such that*

$$hd(\varphi({}^{<}F_1),\ {}^{<}F_2) < \infty.$$

It may be of interest to observe that the proof of Lemma 3.6 requires the following analogue of Brouwer's Invariance of Domain Theorem:

Let $\varphi: \mathbf{R}^r \to \mathbf{R}^s$ be a pseudo-isometry. If $s \leqq r$, then $s=r$ and φ is surjective.

To complete the proof that φ induces a boundary map $X_{1,0} \to X_{2,0}$, we utilize the following alternative definition of Furstenberg's maximal boundary.

Let $\mathscr{X}_0$ denote the set of all chambers in a symmetric Riemannian space X. We define two chambers ${}^{<}F$ and ${}^{<}F^1$ to be equivalent if and only if

$$hd({}^{<}F,\ {}^{<}F^1) < \infty.$$

Let X_0 denote the set of equivalence classes in $\mathscr{X}_0$, topologized in the natural way. Then X_0 is the Furstenberg maximal boundary of X.

With this definition, we deduce directly from Lemma 3.6

Lemma 3.7. *φ induces a continuous map $\varphi_0: X_{1,0} \to X_{2,0}$. Moreover, φ_0 is a homeomorphism.*

Definition. Let ${}^{<}S$ be a chamber or chamber wall in X. Set

$$P({}^{<}S) = \{g\in G;\ hd(g\,{}^{<}S,\ {}^{<}S) < \infty\}.$$

Lemma 3.8. *$P({}^{<}S)$ is a parabolic subgroup of G and conversely every parabolic subgroup of G is $P({}^{<}S)$ for some chamber or chamber wall ${}^{<}S$.*

Definition. Let P be a parabolic subgroup of G, and let X_0 be the maximal boundary of the symmetric space X associated to G. Set

$$|P| = \{x\in X_0;\ G_x \subset P\}$$

where G_x denotes the isotropy subgroup.

It follows from Lemma 3.6 that

Lemma 3.9. *Given any parabolic subgroup P_1 in G_1, there is a unique parabolic subgroup P_2 in G_2 such that*

$$\varphi_0(|P_1|) = |P_2|.$$

Lemma 3.9 has a remarkable interpretation in terms of Tits geometries. It asserts that φ_0 induces a continuous incidence-preserving isomorphism of the

Tits geometry of G_1 onto the Tits geometry of G_2. By the generalized fundamental theorem of projective geometry for Tits geometries, φ_0 is induced by an isomorphism α of G_1 onto G_2, provided G_1 has no factors of **R**-rank $\leqq 1$. It follows that

$$\theta(\lambda) = \alpha(\lambda) \quad \text{for all} \quad \lambda \in \Gamma$$

and α is the desired extension!

§ 4. Concluding remarks

1. We cannot conclude the existence of $\alpha: G_1 \to G_2$ if G_1 is of rank 1 because in that case, the Tits geometry of G_1 consists only of points and there are no incidence relations. Nevertheless, the theorem should remain true except for the case $G = SL(2, \mathbf{R})$ where it is known to be false.

2. In case G_1/Γ_1 has finite measure and is not compact, the result should also be valid. What has to be proved is that the Γ-space morphism $\varphi: X_1 \to X_2$ can be selected to be a pseudo-isometry. That in turn should follow from the announced results of Margulis on the arithmeticity of such subgroups Γ.

ON THE WEIL–JACQUET–LANGLANDS THEOREM

I. I. PJATECKIJ-ŠAPIRO

We shall use the usual notations of adelic theory. By $\mathfrak{k}$ we mean an arithmetic field. The letter $\mathfrak{p}$ denotes a place of $\mathfrak{k}$; $\mathfrak{k}_{\mathbf{A}}$ and $I_{\mathfrak{k}}$ are the adele ring and the idele group, respectively. $\mathfrak{k}_{\mathfrak{p}}$ denotes the completion of $\mathfrak{k}$ at $\mathfrak{p}$; $\mathfrak{r}_{\mathfrak{p}}$ the ring of integers of $\mathfrak{k}_{\mathfrak{p}}$, and $\mathfrak{r}_{\mathfrak{p}}^*$ the group of units of $\mathfrak{r}_{\mathfrak{p}}$.

By G we mean $GL(2)$ considered as an algebraic group over $\mathfrak{k}$.

H. Jacquet and R. P. Langlands have proved the following remarkable result [1]:

The irreducible admissible representation π of $G_{\mathbf{A}}$ belongs to $L_0^2(G_{\mathfrak{k}}\backslash G_{\mathbf{A}})=L_0^2$ if and only if for all characters ω of $\mathfrak{k}^*\backslash I_{\mathfrak{k}}$, $L(s, \omega\otimes\pi)$ and $L(s, \omega^{-1}\otimes\tilde{\pi})$ are entire functions which are bounded in every vertical strip of finite width and satisfy the functional equation

$$L(s, \omega\otimes\pi) = \varepsilon(s, \omega\otimes\pi)\, L(1-s, \omega^{-1}\otimes\tilde{\pi}) \tag{1}$$

where $\tilde{\pi}$ is the contragredient representation, L and ε are the functions introduced in [1].

A. Weil was the first who obtained results of this kind, but he used a different language. To explain consequences of Weil's results we will introduce the following definition. An irreducible admissible representation π of $G_{\mathbf{A}}$ satisfies the condition A_ω, where ω is a character of $\mathfrak{k}^*\backslash I_{\mathfrak{k}}$, if $L(s, \omega\otimes\pi)$ and $L(s, \omega^{-1}\otimes\tilde{\pi})$ are entire functions bounded in each vertical strip of finite width, and satisfy the functional equation (1). From Weil's results it follows:

1) if π satisfies A_ω for all ω ramified only outside S (where S is a finite or a rare infinite set of finite places at which π is unramified), then $\pi\in L_0^2$ [2], *2) if $\mathfrak{k}=\mathbf{Q}$, then the fulfilment A_ω for a well-chosen finite set of characters of $\mathfrak{k}^*\backslash I_{\mathfrak{k}}$ guarantees $\pi\in L_0^2$* [4].

Remark. In fact, Weil's results are essentially stronger. Weil gave his results in terms of Dirichlet series. Considering representations of $G_{\mathbf{A}}$ is more or less the same as considering Dirichlet series possessing an Euler product everywhere with Euler factors of degree 2 at almost all $\mathfrak{p}$. Weil did not require that the Dirichlet series has an Euler product everywhere.

A. Weil conjectured that for any arithmetic field $\mathfrak{k}$ the fulfilment of A_ω for all ω which are ramified only at a large enough finite set of places of $\mathfrak{k}$ should imply that $\pi \in L_0^2$ [3].

The aim of this paper is to prove this conjecture. (It is easy to show that the fulfilment of A_ω only for a finite set of characters does not guarantee that $\pi \in L_0^2$, except the case $\mathfrak{k}=\mathbf{Q}$.)

In fact, we will prove that if π is an irreducible admissible representation and A_ω is valid for all characters of $\mathfrak{k}\backslash I_\mathfrak{k}$ ramified only at a properly chosen finite set of places, then there exists an irreducible representation π' of $G_\mathbf{A}$ such that π is almost isomorphic to π', i.e. $\pi_\mathfrak{p} \cong \pi'_\mathfrak{p}$ for almost all $\mathfrak{p}$. The uniqueness of such π' follows from [1] and [9].

I am very grateful to D. Kajdan, R. Godement, G. Harder and A. Weil for helpful discussions.

Now we introduce some notations:

Throughout this paper by S we denote a finite nonempty set of places of $\mathfrak{k}$ containing all infinite places. We denote by O_s the ring of elements of $\mathfrak{k}$ which are integers outside S. We put $\Gamma=G_{O_s}$. For any positive divisor $m=\sum m_\mathfrak{p}\,\mathfrak{p}$ we introduce three congruence subgroups:

$$\Gamma(m) = \left\{\gamma\in\Gamma \,\middle|\, \gamma\equiv\begin{pmatrix}1 & 0\\ 0 & 1\end{pmatrix}(m)\right\}$$

$$\Gamma_0(m) = \left\{\gamma\in\Gamma \,\middle|\, \gamma\equiv\begin{pmatrix}* & *\\ 0 & *\end{pmatrix}(m)\right\}$$

$$\Gamma_1(m) = \left\{\gamma\in\Gamma \,\middle|\, \gamma\equiv\begin{pmatrix}* & *\\ 0 & 1\end{pmatrix}(m)\right\}$$

where the stars mean elements of O_s.

We shall call the first one the *principal congruence subgroup* and the others the *standard congruence subgroups.*

Definition 1. Let P be a set of finite places, $W\cap S=\emptyset$, r a positive number. The set Δ of generators of $\Gamma_\nu(m)$, $\nu=0, 1$ is a *set of type* (P, r) if for each $\gamma=(\gamma_{ij})\in\Delta$ with $\gamma_{22}\neq 0$

$$v_\mathfrak{p}\left(\frac{1}{\gamma_{22}}\right) \geqq -re(P, \mathfrak{p}) \tag{2}$$

where $e(P, \mathfrak{p})=1$ if $\mathfrak{p}\in P$, $e(P, \mathfrak{p})=0$ if $\mathfrak{p}\notin P$, $v_\mathfrak{p}(x)$ denotes, as usually, a normalized discrete valuation.

Under the assumption that m is a principal divisor of the ring O_s: $m=(\alpha)$, $\alpha\in O_s$, it is easy to verify that the inequality (2) is equivalent to: each $\gamma=(\gamma_{ij})\in\Delta$ with $\gamma_{22}\neq 0$ can be written in the form

$$\gamma = b\tau^{-1}z\tau,\ b\in B_\mathfrak{k},\ z = \begin{pmatrix}1 & \hat{z}\\ 0 & 1\end{pmatrix},\ \tau = \begin{pmatrix}0 & 1\\ \alpha & 0\end{pmatrix} \tag{3}$$

where $v_\mathfrak{p}(\hat{z}) \geqq -re(P, \mathfrak{p})$.

For each finite place $\mathfrak{p}$ we introduce

$$U_n = \left\{ g \in GL(2, r_{\mathfrak{p}}) \,\middle|\, g \equiv \begin{pmatrix} * & * \\ 0 & * \end{pmatrix} (\mathfrak{p}^n) \right\}. \tag{4}$$

We will call U_n the standard compact subgroup of level n. Let $\pi_{\mathfrak{p}}$ be an irreducible admissible representation of $G_{\mathfrak{p}}$. We will call $a = a(\pi_{\mathfrak{p}})$ the degree of ramification of $\pi_{\mathfrak{p}}$ if a is the smallest nonnegative number such that in $\pi_{\mathfrak{p}}$ there exists a vector stabilized by the standard compact subgroup of level $a = a(\pi_{\mathfrak{p}})$. Novodvorskij [7] and Casselman [9] have proved the uniqueness (up to a scalar factor) of this vector. We will call it the NC vector.

Let π be an irreducible admissible representation of $G_{\mathbf{A}}$. It is well known that π can be written as a tensor product $\pi = \otimes \pi_{\mathfrak{p}}$ where $\pi_{\mathfrak{p}}$ are irreducible admissible representations of $G_{\mathfrak{p}}$ [1], [5], [6]. (For the case of Archimedean places under an admissible representation of $G_{\mathfrak{p}}$ we mean a minimal $G_{\mathfrak{p}}$-invariant subspace which contains an admissible representation of the corresponding Hecke algebra [1], [6]. The slight difference between the definitions of Hecke algebras in [1] and [6] is not essential for us.) For each finite place $\mathfrak{p}$ we will call $a(\pi, \mathfrak{p}) = a(\pi_{\mathfrak{p}})$ the degree of ramification at $\mathfrak{p}$. If P is a set of finite places of $\mathfrak{k}$, we put

$$G_P = \prod_{\mathfrak{p} \in P} G_{\mathfrak{p}},\ U_0(m) = \prod_{\mathfrak{p} \in P} U_{m_{\mathfrak{p}}},\ m = \sum_{\mathfrak{p} \in P} m_{\mathfrak{p}} \mathfrak{p}.$$

We will call $U_0(m)$ the standard compact subgroup of $G_{\mathfrak{p}}$. We put $U'_n = \left\{ g \in U_n \,\middle|\, g \equiv \begin{pmatrix} * & * \\ 0 & 1 \end{pmatrix} (\mathfrak{p}^n) \right\}$ and $U_1(m) = \prod U'_{m_{\mathfrak{p}}}$. We will use the term standard compact group for $U_1(m)$ as well.

Theorem. Let $\pi = \otimes \pi_{\mathfrak{p}}$ *be an admissible irreducible preunitary* representation of* $G_{\mathbf{A}}$ *such that the restriction of* π *to the center is trivial on* $\mathfrak{k}^*$ *and let* S *be a nonempty set of places, containing all infinite ones, such that* O_s *is a principal ideal ring. Assume that*

I) *T and P are sets of finite places of $\mathfrak{k}$ with the following properties:*

1) *for each* $\mathfrak{p} \in T$, $a(\pi, \mathfrak{p}) > 0$,

2) *for each* $\mathfrak{p} \in P$, $a(\pi, \mathfrak{p}) = 0$,

3) *for a suitable* v *the group* $\Gamma_v(m), m = \sum_{\mathfrak{p} \notin S+T} a(\pi, \mathfrak{p})\mathfrak{p}$ *contains a set of generators of type* (P, r).

II) A_{ω} *is satisfied for all characters* ω *such that 1)* ω *is nonramified outside* $S+T+P$, *2) the degree of ramification* $d(\omega)$ *satisfies the inequality*

$$d(\omega) \leqq \begin{matrix} 2r, & \text{if} \quad \mathfrak{p} \in P \\ a(\pi, \mathfrak{p}) & \text{if} \quad \mathfrak{p} \in T \end{matrix} \tag{5}$$

* In the sense of [6].

III) *there exists no pair of quasi-characters* μ *and* ν of $\mathfrak{k}^*\diagdown I_\mathfrak{k}$ *such that* $\pi_\mathfrak{p}=\pi(\mu_\mathfrak{p}, \nu_\mathfrak{p})$ *for all* $\mathfrak{p}\in S$ *or* $\mathfrak{p}\notin S$ *and* $a(\pi, \mathfrak{p})=0$. (For the definition $\pi(\mu_\mathfrak{p}, \nu_\mathfrak{p})$ see [1] p. 92.)

Then there exists a unique irreducible representation $\pi'=\otimes\pi'_\mathfrak{p}$ *of* $G_\mathbf{A}$ *such that 1)* $\pi'\in L_0^2(G_\mathfrak{k}\diagdown G_\mathbf{A})$, *2)* $\pi'_\mathfrak{p}\cong\pi_\mathfrak{p}$ *if* $\mathfrak{p}\in S$ *or* $a(\pi, \mathfrak{p})=0$.

The proof of the theorem is based on 7 lemmas which will be proved later. Now we sketch how the theorem follows from them.

A continuous function $f(g)$ on $G_\mathbf{A}$ will be called an *automorphic form* if

1) $f(\gamma g)=f(g)$ for all $\gamma\in G_\mathfrak{k}$,

2) $f(g)$ is U-finite on the right (U denotes the maximal compact subgroup of $G_\mathbf{A}$),

3) for each elementary idempotent ξ in the Hecke algebra H the space $\{\varrho(\xi h)f|h\in H\}$ is of finite dimension,

4) if $\mathfrak{k}$ is a number field $f(g)$ is slowly increasing in the sense of [1] p. 330, i.e. for any compact subset $M\subset G_\mathbf{A}$ there exist C_1 and C_2 such that

$$\left|f\left(\begin{pmatrix} a & 0\\ 0 & 1\end{pmatrix} g\right)\right| \leqq C_1|a|^{C_2} \quad \text{for all} \quad g\in M.$$

Let $\mathscr{A}$ be the $G_\mathbf{A}$-invariant linear space generated by all automorphic forms. (This definition of $\mathscr{A}$ is slightly different from that in [1].) For a compact subgroup U we denote by $\mathscr{A}^U$ the subspace fixed by U.

We put

$$\pi_S = \bigotimes_{\mathfrak{p}\in S} \pi_\mathfrak{p},\ \pi^S = \bigotimes_{\mathfrak{p}\notin S} \pi_\mathfrak{p}.$$

By Lemmas 1, 2, 4, 5 there exists a canonical embedding of π_S into $\mathscr{A}$.

Denote by $\pi'\subset\mathscr{A}$ the minimal representation of $G_\mathbf{A}$ which contains π_S. We prove that by virtue of Lemmas 3, 6, 7, π' contains an irreducible component, which is almost isomorphic to π (i.e. $\pi'_\mathfrak{p}\cong\pi_\mathfrak{p}$ for almost all $\mathfrak{p}$). From III follows $\pi\in L_0^2$ (cf. [1] p. 340.)

Many ideas are taken from [1] and [4]. For instance, the consideration of the Whittaker model is an essential point. Lemma 4 is proved by the same considerations as the corresponding statement in [1]. The linear functional l_0 plays the role of the modular form in [4]. (For the definition of l_0 see the lemmas.)

Remarks 1) It is known that if $\mathfrak{k}$ is a number field or card $S\geqq 2$ then the group $\Gamma_\nu(m)$ has a finite system of generators [10]. Hence it is of type (P, r) for some finite P. This implies that if A_ω is satisfied for every ω which is ramified at a properly chosen finite set of places of $\mathfrak{k}$ then $\pi\in L_0^2$. Moreover, we can choose a P which does not intersect a given set of places (by multiplying γ by an elementary matrix on the right). If S contains a non-Archimedean or real place and Card $S\geqq 2$, then $\Gamma_1(S)$ is generated by its elementary matrices and hence $P=\emptyset$ (cf. L. N. Wasserstein [11]).

2) If $\mathfrak{k}=\mathbf{Q}$, $S=\{\infty\}$ then the group $\Gamma_0(m)$ has a finite set of generators. By making use of the system of generators which was given in [4] one obtains that if A_ω holds for a properly chosen finite set of ω then $\pi\in L_0^2$ (4).

3) Take $T=\{\mathfrak{p}\in S|a(\pi,\mathfrak{p})>0\}$ then $m=(1)$.

Under the assumption that O_s is a principal ideal ring, the group $\Gamma=G_{O_s}$ is generated by $w=\begin{pmatrix}0&1\\-1&0\end{pmatrix}$ and the upper triangular subgroup [10]. This implies that all assumptions of our theorem are satisfied in the case $P=\emptyset$. Hence, if A_ω is valid for all ω ramified only outside S then $\pi\in L_0^2$. For instance, take $\mathfrak{k}=\mathbf{Q}$. One can show that if A_ω holds for every ω with ramification degree at $\mathfrak{p}$ not exceeding $a(\pi,\mathfrak{p})$, then $\pi\in L_0^2$.

Before the proof we have to introduce some notations. For a set S of places of $\mathfrak{k}$, we shall write

$$G_S=\prod_{\mathfrak{p}\in S}G_{\mathfrak{p}},\quad G^S=\prod_{\mathfrak{p}\notin S}G_{\mathfrak{p}}=G_{\bar{S}}$$

where $\bar{S}$ means the complement to S. (Of course, $G_{\bar{S}}$ is considered as a restricted product.) It is obvious, that $G_{\mathbf{A}}=G_SG^S$ for any S. For each algebraic subgroup $\mathscr{G}$ of G one can define $\mathscr{G}_S$ and $\mathscr{G}^S$ respectively.

We denote by i_S the natural projection of $G_{\mathbf{A}}$ to G_S and by i^S the natural projection from $G_{\mathbf{A}}$ to G^S. It is clear that for each $g\in G_{\mathbf{A}}$ we have $g=i_S(g)i^S(g)$, if G_S and G^S are considered as subgroups of $G_{\mathbf{A}}$.

Let Γ be a subgroup of finite index in G_{O_s}. Similarly to the definition of $\mathscr{A}$ one can introduce $\mathscr{A}_\Gamma$. For instance, $f(g)$, $g\in G_S$ is called *slowly increasing* if for a compact subset $M\subset G_S$ there exist c_1, c_2 such that

$$\left|f\left(\begin{pmatrix}a&0\\0&1\end{pmatrix}g\right)\right|\leqq c_1|a|^{c_2}\quad\text{for all}\quad g\in M \tag{6}$$

where $a\in I_S=\prod_{\mathfrak{p}\in S}\mathfrak{k}_{\mathfrak{p}}^*$.

Throughout this paper S means a nonempty finite set of places of $\mathfrak{k}$, containing all Archimedean places and such that O_S is a principal ideal ring. Sometimes this last assumption can be dropped (for instance, in Lemma 1).

Lemma 1. *Let π_S be an admissible irreducible representation of G_S. Denote by $g\xi$ the result of the application of $g\in G_S$ to $\xi\in\pi_S$. Assume that there exists a linear functional l_0 on π_s such that*

1) $l_0(\gamma\xi)=l_0(\xi)$ for all $\gamma\in\Gamma$ and $\xi\in\pi_s$ where Γ is a congruence subgroup of G_{O_s};

2) for each $\xi\in\pi_s$ the function $l_0\left(\begin{pmatrix}a&0\\0&1\end{pmatrix}g\xi\right)$ is slowly increasing.

Then the mapping

$$\xi\to\varphi_\xi(g)=l_0(g\xi)$$

defines an imbedding of π_s in $\mathscr{A}_\Gamma$.

Proof. The first assumption implies $\varphi_\xi(\gamma g)=\varphi_\xi(g)$ for all $\gamma\in\Gamma$. From *2)* it follows that $\varphi\in\mathscr{A}_\Gamma$.

Lemma 2. *If U is an open compact subgroup of* G^s *and*

$$i^s(G_{\mathfrak{k}})U = G^s \tag{7}$$

then

$$\mathscr{A}^u = \mathscr{A}_{\Gamma_u}, \tag{8}$$

where

$$\Gamma_U = \{\gamma\in G_{\mathfrak{k}} | i^s(\gamma)\in U\}. \tag{9}$$

Proof. It follows from (7) and $G_{\mathbf{A}}=G_sG^s$ that

$$G_{\mathfrak{k}}\diagdown G_{\mathbf{A}}/U \cong \Gamma_U\diagdown G_s. \tag{10}$$

Hence, a function f on $\Gamma_U\diagdown G_s$ can be considered as a function f on $G_{\mathfrak{k}}\diagdown G_{\mathbf{A}}$, U-invariant on the right. It is easy to verify that if f is slowly increasing on G_s then $\bar{f}$ is slowly increasing on $G_{\mathbf{A}}$.

Let $U(m)$ be the standard compact subgroup of G^s. It is easy to see that $\Gamma_{U(m)}=$ $=\Gamma_0(m)$, where $\Gamma_0(m)$ is the standard congruence subgroup of $\Gamma=G_{O_s}$, $m=\sum_{\mathfrak{p}\notin S} m_{\mathfrak{p}}\mathfrak{p}$.

Hence, by virtue of Lemma 2 one can define the imbedding

$$\mathscr{A}_{\Gamma_0(m)}\to\mathscr{A}. \tag{11}$$

We will call it the canonical imbedding.

Denote by $\bar{G}_{\mathfrak{k}}^s$ the completion of $G_{\mathfrak{k}}$ with respect to the topology defined by all congruence subgroups of G_{O_s}. It is obvious that $\bar{G}_{\mathfrak{k}}^s$ is isomorphic to the closure of $i_s(G_{\mathfrak{k}})$ in G^s.

Lemma 3. *Let* π_s *be a representation of* G_s *satisfying all conditions of Lemma* 1 *and let* $\hat{\pi}_s$ *be the space of all linear functions on* π_s *endowed with the action of* G_s, $(gl)(\xi)=l(g^{-1}\xi)$.

Denote by $\Omega\subset\hat{\pi}_s$ *the subspace spanned by* $G_{\mathfrak{k}}$ *translates of* l_0. (For the definition of l_0, see Lemma 1.)

Denote by π' *the minimal* $G_{\mathbf{A}}$ *invariant subspace of* $\mathscr{A}$ *containing the image of* π_s *under the canonical imbedding* (11).

Then 1) the representation of $G_{\mathfrak{k}}$ *in* Ω *can be extended to an admissible representation of* $G_{\mathfrak{k}}^s$ *2)* π' *is of the form* $\pi_s\otimes\pi'^s$, *where* π'^s *is a representation of* G^s, *3) there exists a linear mapping:* $\Omega\to\pi'^s$ *which commutes with the* $\bar{G}_{\mathfrak{k}}^s$ *and sends* l_0 *to a vector which is invariant with respect to the corresponding standard compact subgroup of* G^s.

Proof. The first assertion follows from the fact that any vector in Ω is fixed by a congruence subgroup of G_{O_s}.

To simplify the notations throughout the proof of this lemma we will write π and π^s instead of π' and π'^s. We put

$$\pi^s = \mathrm{Hom}(\pi_s, \pi) \tag{12}$$

where π_s and π are considered as representations of G_s. Since π is a representation of $G_{\mathbf{A}}=G_sG^s$ one can naturally define the action of G^s on π^s. Under our assumptions π_s is irreducible and π is fully reducible (since $\pi\in\mathscr{A}$), therefore

$$\pi \cong \pi_s \otimes \pi^s \tag{14}$$

as a representation of $G_{\mathbf{A}}$. The second assertion is proved.

Let us agree to denote by ξ vectors in π_s and by η vectors in π^s. For each $\xi\in\pi_s$, $\eta\in\pi^s$ denote by $\varphi_{\xi\otimes\eta}(g)$ the corresponding function on $G_{\mathfrak{k}}\diagdown G_{\mathbf{A}}$. It is easy to see that

$$\varphi_{\xi\otimes\eta}(\gamma g) = \varphi_{\xi\otimes\eta}(g) \quad \text{for} \quad \gamma\in G_{\mathfrak{k}} \tag{15}$$

$$\varphi_{g_0(\xi\otimes\eta)}(g) = \varphi_{\xi\otimes\eta}(gg_0) \quad \text{for} \quad g_0\in G_{\mathbf{A}}. \tag{16}$$

Associate with each $\eta\in\pi^s$ the following linear functional on π_s

$$l_\eta(\xi) = \varphi_{\xi\oplus\eta}(e), \quad e = \begin{pmatrix}1 & 0\\ 0 & 1\end{pmatrix}. \tag{17}$$

It gives the mapping:

$$\alpha: \pi^s \to \hat{\pi}_s. \tag{18}$$

For each $\gamma\in G_{\mathfrak{k}}$ we have

$$\begin{gathered}\gamma_s(l_\eta(\xi)) = l_\eta(\gamma_s^{-1}\xi) \overset{(17)}{=} \varphi_{\gamma_s^{-1}\xi\otimes\eta}(e) \overset{(16)}{=} \\ \varphi_{\xi\otimes\eta}(\gamma_s^{-1}) \overset{(15)}{=} \varphi_{\xi\otimes\eta}(\gamma^s) \overset{(16)}{=} \varphi_{\xi\otimes\gamma^s\eta}(e)\end{gathered} \tag{19}$$

where the numbers show that equality follows from the corresponding formula.

Hence α commutes with the action of $G_{\mathfrak{k}}$. First we will prove that the kernel π_0 of α is trivial. It is obvious that π_0 is invariant with respect to $\bar{G}^s_{\mathfrak{k}}$.

Novodvorskij has shown that the restriction of an irreducible representation of G^s to $\bar{G}^s_{\mathfrak{k}}$ is irreducible [8].

Then π_0 is invariant with respect to G^s. From the definition of π and π^s it follows that π_0 is trivial.

From the description of the canonical imbedding of $\mathscr{A}_{\Gamma_v(m)}$ into $\mathscr{A}$ it follows that there exists a $\eta_0\in\pi^s$ such that

$$l_0(\xi) = \varphi_{\xi\otimes\eta_0}(e)$$

and η_0 is stable under the action of $U(m)$. Since π is a minimal $G_{\mathbf{A}}$-invariant subspace in L^2_0 containing π_s, the representation π^s is generated by the G^s-translates of η_0. From $G^s=i^s(G_{\mathfrak{k}})U(m)$ one obtains that the $\bar{G}^s_{\mathfrak{k}}$-translates of η_0 generate the whole π^s. Thus the image of π^s (under α) coincides with Ω. The lemma is proved.

Now we recall some notions from [1]. From now on we will denote by B the upper triangular subgroup of G and by Z the unipotent subgroup of B. Let

χ be a character of A, trivial on $\mathfrak{k}$. Denote by W_χ the space of all functions on $G_{\mathbf{A}}$ satisfying

$$f(zg) = \chi(z) f(g)$$

for all $z \in Z_A$.

For each irreducible admissible infinite-dimensional representation of π of $G_{\mathbf{A}}$ there exists a unique mapping

$$\xi \to f_\xi(g),\ f_\xi(g) \in W_\chi$$

such that

$$f_{g_0 \xi} = f(gg_0).$$

Denote by $W_\chi(\pi)$ the set of all functions of the form $f_\xi(g)$, where $\xi \in \pi$.

Denote by $W(\pi)$ the set of all functions of the form

$$\varphi_\xi(g) = \sum_{\alpha \in \mathfrak{k}^*} f_\xi\left(\begin{pmatrix} \alpha & 0 \\ 0 & 1 \end{pmatrix} g\right). \tag{20}$$

It is clear that *1)* $W(\pi)$ is invariant under a right translation by $g \in G_{\mathbf{A}}$ and *2)* the representation of $G_{\mathbf{A}}$ on $W(\pi)$ is isomorphic to the representation on π.

For each $\xi \in \pi$ we have

$$\varphi_\xi(bg) = \varphi_\xi(g), \quad b \in B_{\mathfrak{k}}, \quad g \in G_{\mathbf{A}} \tag{21}$$

$$\varphi_{g_0 \xi}(g) = \varphi_\xi(gg_0), \quad g_0 \in G_{\mathbf{A}}. \tag{22}$$

Lemma 4. *Let X be a compact subgroup of $I_{\mathfrak{k}}$. Assume that M is a linear subspace of $W(\pi)$ and*

$$\varphi\begin{pmatrix} ax & 0 \\ 0 & 1 \end{pmatrix} = \varphi\begin{pmatrix} a & 0 \\ 0 & 1 \end{pmatrix} \quad \text{for all} \quad \varphi \in M \quad \text{and any} \quad a \in I_{\mathfrak{k}},\ x \in X.$$

If A_ω is satisfied for all characters of $\mathfrak{k}^* \setminus I_{\mathfrak{k}}$ which are trivial on X then for any $\varphi \in M$

$$\varphi_\xi(w) = \varphi_\xi(e), \quad w = \begin{pmatrix} 0 & 1 \\ -1 & 0 \end{pmatrix}, \quad e = \begin{pmatrix} 1 & 0 \\ 0 & 1 \end{pmatrix}.$$

The proof is similar to the proof for the case $M = W(\pi)$, $X = 1$. (See [1] pp. 363—368.) We put

$$\psi_\pi(\xi, s) = \int_{I_{\mathfrak{k}}} f_\xi\begin{pmatrix} a & 0 \\ 0 & 1 \end{pmatrix} |a|^{s-\frac{1}{2}} d^* a = \int_{\mathfrak{k}^* \setminus I_{\mathfrak{k}}} \varphi_\xi\begin{pmatrix} a & 0 \\ 0 & 1 \end{pmatrix} |a|^{s-\frac{1}{2}} d^* a,$$

$$\tilde{\psi}_\pi(\xi, s) = \int f_\xi\begin{pmatrix} a & 0 \\ 0 & 1 \end{pmatrix} |a|^{s-\frac{1}{2}} \eta^{-1}(a)\, d^* a = \int_{\mathfrak{k}^* \setminus I_{\mathfrak{k}}} \varphi_\xi\begin{pmatrix} a & 0 \\ 0 & 1 \end{pmatrix} |a|^{s-\frac{1}{2}} \eta^{-1}(a)\, d^* a$$

where $\eta(a)$ is defined by the relation

$$\begin{pmatrix} a & 0 \\ 0 & a \end{pmatrix} \xi = \eta(a)\, \xi, \quad \xi \in \pi.$$

(These functions are similar to the functions $\psi(g, s, w)$ and $\tilde{\psi}(g, s, w)$ introduced in [1].)

We put

$$\Phi_\pi(\xi, s) = L^{-1}(s, \pi)\psi_\pi(\xi, s)$$
$$\tilde{\Phi}_\pi(\xi, s) = L^{-1}(s, \tilde{\pi})\psi_{\tilde{\pi}}(\xi, s).$$

As is shown in [1]

$$\tilde{\Phi}_\pi(w\xi, 1-s) = \varepsilon(s, \pi)\Phi_\pi(\xi, s).$$

From the assumption that $L(s, \pi)$ and $L(s, \tilde{\pi})$ are entire functions satisfying the functional equation:

$L(s, \pi) = \varepsilon(s, \pi)L(1-s, \tilde{\pi})$ it follows that $\psi_\pi(\xi, s) = \psi_\pi(W\xi, 1-s)$.

From this and the assumption that $L(s, \pi)$ and $L(s, \tilde{\pi})$, are entire functions bounded in vertical strips and by making use of the lemmas on pp. 368—369 of [1] one obtains that

$$\int_{r^*}\left(\varphi\begin{pmatrix} ay & 0 \\ 0 & 1\end{pmatrix} - \varphi\left(w\begin{pmatrix} ay & 0 \\ 0 & 1\end{pmatrix}\right)\right)dy = 0$$

where $r^* = \prod r_\mathfrak{p}^*$ ($r_\mathfrak{p}^*$ is the group of units of $\mathfrak{k}_\mathfrak{p}$),

Using in fact that each $\varphi(g) \in W(\pi\otimes\omega)$ is of the form $\varphi_1(g)\omega(\det g)$, where $\varphi_1 \in \in W(\pi)$ one obtains that if A_ω is satisfied then

$$\int_{r^*}\left(\varphi\begin{pmatrix} ay & 0 \\ 0 & 1\end{pmatrix} - \varphi\left(w\begin{pmatrix} ay & 0 \\ 0 & 1\end{pmatrix}\right)\right)\omega(y)\,dy = 0.$$

From this formula our lemma can be obtained trivially.

Lemma 5. *Let π be an admissible irreducible representation of $G_\mathbf{A}$ satisfying all assumptions of the theorem. We put* $\pi_S = \bigotimes_{\mathfrak{p}\in S} \pi_\mathfrak{p}$, $\pi^S = \bigotimes_{\mathfrak{p}\notin S} \pi_\mathfrak{p}$ *and introduce*

$$l_0(\xi) = \varphi_{\xi\otimes\eta_0}(e), \quad \xi\in\pi_S. \tag{23}$$

where $\varphi_{\xi\otimes\eta}(g)$ is defined in (20) *and η_0 is a* NC *vector in* π^s (i.e. $\bigotimes_{\mathfrak{p}\notin s}\eta_\mathfrak{p}$ where $\eta_\mathfrak{p}$ *are local* NC-*vectors*).

Then l_0 satisfies all assumptions of Lemma 1 *for the standard congruence subgroup* $\Gamma_\nu(\tilde{m})$, $\tilde{m} = \sum_{\mathfrak{p}\notin S} a(\pi, \mathfrak{p})\mathfrak{p}$.

Proof. In order to make the main idea clear we consider first the simplest case. Assume that $P=T=\emptyset$ and $a(\pi, \mathfrak{p})=0$ for $\mathfrak{p}\notin S$. Then η_0 is stabilized by the maximal compact subgroup U of G and $\Gamma=G_{O_s}$ is generated by elements which are either of the form

$$bwzw^{-1},\ w = \begin{pmatrix} 0 & 1 \\ -1 & 0\end{pmatrix},\quad b\in B_k,\quad z\in Z_{O_s} \tag{24}$$

or of

$$\begin{pmatrix} * & * \\ * & 0\end{pmatrix}. \tag{25}$$

From Lemma 4 (take $X=\prod_{\mathfrak{p}\notin s} r_{\mathfrak{p}}^*$) it follows that for each $\xi\in\pi_s$

$$\varphi_{\xi\otimes\eta_0}(e) = \varphi_{\xi\otimes\eta_0}(w). \tag{26}$$

For each $\gamma\in\Gamma$ we have $\gamma^s\eta_0=\eta_0$ and hence by (21) and (22)

$$l_0(\gamma\xi) = \varphi_{\gamma_s\xi\otimes\eta_0}(e) = \varphi_{\gamma_s\xi\otimes\gamma^s\eta_0}(e) = \varphi_{\gamma(\xi\oplus\eta_0)}(e) = \varphi_{\xi\otimes\eta_0}(\gamma). \tag{27}$$

Take a γ which is of the form (24). We have

$$\varphi_{\xi\otimes\eta_0}(bwzw^{-1}) \overset{(21)}{=} \varphi_{\xi\otimes\eta_0}(wzw^{-1}) \overset{(22)}{=} \varphi_{zw^{-1}(\xi\otimes\eta_0)}(w) \hat{=} \varphi_{(zw^{-1})_s\xi\otimes\eta_0}(w) \overset{(26)}{=}$$

$$= \varphi_{(zw^{-1})_s\xi\otimes\eta_0}(e) \hat{=} \varphi_{zw^{-1}(\xi\otimes\eta_0)}(e) \overset{(22)}{=} \varphi_{\xi\otimes\eta_0}(zw^{-1}) \overset{(21)}{=} \varphi_{\xi\otimes\eta_0}(w^{-1}) \overset{(26)}{=} \varphi_{\xi\otimes\eta_0}(e), \tag{28}$$

the number or sign over a passage shows that equality follows from the corresponding formula and ^ means: $(zw^{-1})^s\eta_0=\eta_0$.

Each γ from (25) can be represented in the form $\gamma=bw$, $b\in B_t$. Therefore

$$\varphi_{\xi\otimes\eta_0}(bw) \overset{(21)}{=} \varphi_{\xi\otimes\eta_0}(w) \overset{(26)}{=} \varphi_{\xi\oplus\eta_0}(e). \tag{29}$$

From (27), (28) it follows that for a γ which is of the form (24) or (25) we have

$$l_0(\gamma\xi) = l_0(\xi), \quad \xi\in\pi_s \tag{30}$$

It follows easily from the definition of l_0 that if $\gamma_1, \gamma_2\in\Gamma$ satisfy (30) then so does $\gamma_1\gamma_2$.

According to our assumption the elements of the forms (24) and (25) generate the whole of Γ. Hence (30) is satisfied for all $\gamma\in\Gamma$. We have verified the first assumption of Lemma 1. The second assumption is proved in [1] and [6] pp. 329—332.

Now let us consider the general case. Denote by M the set of all functions having the following form

$$\varphi_{\xi\otimes\xi_T\otimes\eta_1}(g) \tag{31}$$

where

$$\xi\in\pi_S, \quad \xi_T\in\pi_T = \bigotimes_{\mathfrak{p}\in T}\pi_{\mathfrak{p}}, \quad \eta_1\in\pi^{S+T} = \bigotimes_{\mathfrak{p}\otimes S+T}\pi_{\mathfrak{p}}$$

is an NC-vector and ξ_T is stabilized by the following compact subgroup of G_T:

$$U_T = \left\{g\in GL(2, r_T) \middle| g\equiv\begin{pmatrix}1 & 0\\ 0 & 1\end{pmatrix}\left(\sum_{\mathfrak{p}\in T}(\pi, \mathfrak{p})\mathfrak{p}\right)\right\}.$$

Let $\overline{M}$ be the set of all functions of the form

$$\varphi_{\xi\otimes\xi_T\otimes\xi_w\eta_2}(g) \tag{32}$$

where ξ and ξ_T are the same as in (31), $\xi_w\in\pi_w$, $\eta_2\in\pi^{S+T+W}$, η_2 is an NC-vector, and ξ_w is stabilized by

$$U_w = \left\{g\in GL(2, r_w) \middle| g\equiv\begin{pmatrix}1 & 0\\ 0 & 1\end{pmatrix}\left(2r\sum_{\mathfrak{p}\in W}\mathfrak{p}\right)\right\}.$$

As a consequence of our assumption and Lemma 4 one obtains

$$\varphi(w) = \varphi(e) \tag{33}$$

for all $\varphi \in \overline{M}$.

Using (22) one obtains for any $\varphi \in M$ and z of form (3)

$$\varphi_1(g) = \varphi(gz) \in \overline{M}. \tag{34}$$

It is clear that τ (defined in (3)) belongs to the normalizer of the standard compact subgroup of level $a(\pi, \mathfrak{p})$ of $G_\mathfrak{p}$ for $\mathfrak{p} \notin S+T$. This and the uniqueness (up to a scalar factor) of $\eta_\mathfrak{p}$ yield

$$\tau\eta_\mathfrak{p} = C_\mathfrak{p}\eta_\mathfrak{p}, \quad C_\mathfrak{p} \in \mathbf{C}, \quad \mathfrak{p} \notin S+T. \tag{35}$$

Hence if $\varphi(g) \in \overline{M}$ then $\varphi_2(g) = \varphi(g\tau^{-1}) \in \overline{M}$. Putting it together with (34) one obtains that

$$\hat{\varphi}(g) = \varphi(gz\tau) \in \overline{M} \tag{36}$$

for all $\varphi(g) \in M$.

For each $\varphi \in M$ we have

$$\varphi(\tau^{-1}z\tau) = \varphi\left(\begin{pmatrix}\alpha^{-1} & 0\\ 0 & -1\end{pmatrix} wz\tau\right) \overset{(21)}{=} \varphi(wz\tau) = \hat{\varphi}(w) \overset{(33)}{=} \hat{\varphi}(e) =$$

$$= \varphi(z\tau) \overset{(21)}{=} \varphi(\tau) = \varphi\left(\begin{pmatrix}1 & 0\\ 0 & -\alpha\end{pmatrix} w\right) \overset{(21)}{=} \varphi(w) \overset{(33)}{=} \varphi(e) \tag{37}$$

where z satisfies the condition (3).

By assumption there exist a system Δ of generators of $\Gamma_\mathfrak{v}(m)$ which are of the following form

$$\gamma = b\tau^{-1}z\tau \tag{38}$$

$$\gamma = \begin{pmatrix} * & * \\ * & 0\end{pmatrix} \tag{39}$$

where τ and z are the same as in (3).

We have

$$\varphi(b\tau^{-1}z\tau) \overset{(21)}{=} \varphi(\tau^{-1}z\tau) \overset{(37)}{=} \varphi(e). \tag{40}$$

Those γ of the form (39) can be considered as in (29).

We have proved that

$$\varphi(\gamma) = \varphi(e), \quad \varphi \in M \tag{41}$$

for all $\gamma \in \Delta$. In order to prove (41) for all $\gamma \in \Gamma_\mathfrak{v}(m)$ it is enough to verify that if (41) is valid for $\gamma_i \in \Gamma_0(m)$, $i=1, 2$ then (41) is valid for $\gamma_1\gamma_2$. We put $\tilde{\varphi}(g) = \varphi(g\gamma_2)$ for $\varphi \in M$. From (31) and (22) it follows that $\tilde{\varphi}(g) \in M$. We have $\varphi(\gamma_1\gamma_2) = \tilde{\varphi}(\gamma_1) = \tilde{\varphi}(e) = \varphi(\gamma_2) = \varphi(e)$.

Hence we have proved that (41) is valid for all $\gamma \in \Gamma_\mathfrak{v}(m)$. Similarly to (27) one can obtain the statement of the lemma (do not forget that $\Gamma_\mathfrak{v}(\tilde{m}) \subset \Gamma_\mathfrak{v}(m)$).

Lemma 6. *Let* $\pi^s = \otimes \pi_\mathfrak{p}$ *and* $\tilde{\pi}^s = \otimes \tilde{\pi}_\mathfrak{p}$ *be two admissible irreducible representations of* G^s. *Assume that there exists a linear bijective mapping* $\alpha : \pi^s \to \tilde{\pi}^s$ *which commutes with the action of* $B_\mathfrak{t}$ *(*$B_\mathfrak{t}$ *is considered naturally imbedded into* G^s*) and* *sends the* NC-*vector of* π^s *to a vector of* $\tilde{\pi}^s$ *which is stabilized by the corresponding standard compact subgroup.*

Then for all $\mathfrak{p} \notin S$ *where* $\pi_\mathfrak{p}$ *is unramified there is an isomorphism*

$$\pi_\mathfrak{p} = \tilde{\pi}_\mathfrak{p}. \tag{42}$$

Proof. Denote by N the one-dimensional subspace π^s generated by the NC vector and put $\tilde{N} = \alpha(N)$.

Denote by G_π the subgroup of G^s which is the restricted product of the following subgroups

$$\begin{aligned} &U_{a(\pi,\mathfrak{p})} \quad \text{if} \quad a(\pi, \mathfrak{p}) > 0, \\ &G_\mathfrak{p} \qquad\quad \text{if} \quad a(\pi, \mathfrak{p}) = 0. \end{aligned} \tag{43}$$

We put $U_\pi = \prod_{\mathfrak{p} \notin S} U_{a(\pi,\mathfrak{p})}$.

Denote by H the Hecke algebra consisting of double cosets $U_\pi g U_\pi$, $g \in G_\pi$. It is easy to see that in each coset gU_π, $g \in G_\pi$ there is a $b \in B_\mathfrak{k}$. Therefore the representation of H on N and $\tilde{N}$ are isomorphic. It is well known that if $\pi_\mathfrak{p}$ and $\tilde{\pi}_\mathfrak{p}$ are unramified irreducible representations and if the representations of the Hecke algebra on the spaces $N_\mathfrak{p}$ and $\tilde{N}_\mathfrak{p}$ are isomorphic, then $\pi_\mathfrak{p}$ and $\tilde{\pi}_\mathfrak{p}$ are isomorphic. (Here $N_\mathfrak{p}$ and $\tilde{N}_\mathfrak{p}$ means $GL(2, r_\mathfrak{p})$ stable subspaces.) From this the required statement follows.

The case when $\tilde{N}$ is invariant only with respect to $U_1(m)$ can be considered analogously.

Lemma 7. *Let* π^s *be an admissible irreducible representation of* G^s *and* η_0 *the corresponding* NC-*vector. Then each vector* $\eta \in \pi^s$ *can be represented in the form:*

$$\eta = \sum c_i b_i \eta_0, \quad b_i \in B_\mathfrak{k}, \quad C_i \in \mathbf{C}. \tag{44}$$

Proof. It is easy to verify that the subspace of π^s consisting of all vectors of the form (44) is invariant with respect to the action of B^s. Hence in order to prove the lemma it is enough to verify that any B^s-invariant subspace which contains η_0 coincides with π^s. Obviously it is enough to prove the corresponding local statement, i.e. that the $B_\mathfrak{p}$-translates of $\eta_\mathfrak{p}$ generate the whole $\pi_\mathfrak{p}$. If $\pi_\mathfrak{p}$ is supercuspidal the required statement is trivial, since $\pi_\mathfrak{p}$ is $B_\mathfrak{p}$-irreducible [1], [6]. If $\pi_\mathfrak{p}$ is not supercuspidal, then using the explicit description of the $B_\mathfrak{p}$-invariant subspaces ([6] pp. 1—36) it is easy to complete the proof of the lemma.

The proof of the theorem. We consider the linear functional $l_0(\xi)$ on π^s which is defined in (23). By virtue of Lemmas 5 and 1 we obtain the imbedding of π_s into $\mathscr{A}_{r_v}(\tilde{m})$. To each $\eta \in \pi^s$ we assign the following linear functional on π_s

$$l_\eta(\xi) = \varphi_{\xi \otimes \eta}(e). \tag{45}$$

For each $b \in B_\mathfrak{k}$, we have

$$l_\eta(b^{-1}\xi) = \varphi_{b_s^{-1}\xi\otimes\eta}(e) \overset{(22)}{=} \varphi_{\xi\otimes\eta}(b_s^{-1}) \overset{(21)}{=} \varphi_{\xi\otimes\eta}(b^s) \overset{(22)}{=} \varphi_{\xi\otimes b^s\eta}(e) \tag{46}$$

where $b_s = i_s(b)$, $b^s = i^s(b)$.

From (46) by virtue of Lemma 7 it follows that $l_\eta \in \Omega$ (for the definition of Ω, see Lemma 3).

Denote by π' the minimal $G_{\mathbf{A}}$-invariant subspace in $\mathscr{A}$, containing the image of π_s under the canonical imbedding. According to Lemma 3 we have $\pi' = \pi_s \otimes \pi'^s$. The composition of the two mappings $\pi^s \to \Omega$ and $\Omega \to \pi^s$ (defined in Lemma 3) yields a mapping $\alpha: \pi^s \to \pi'^s$ such that $\alpha(\eta_0)$ is a vector fixed by the corresponding standard compact group. From (46) and Lemma 3 it follows that α commutes with the action of $B_{\mathfrak{k}}$. Let $\tilde{\pi}$ be an irreducible component of π'. By Lemma 6 one obtains that $\tilde{\pi}_{\mathfrak{p}} \cong \pi_{\mathfrak{p}}$ for all $\mathfrak{p}$, where $\pi_{\mathfrak{p}}$ are unramified. The theorem is proved.

REFERENCES

[1] Jacquet, H. and Langlands, R. P., *Automorphic forms on GL(2)*, Springer Lecture Notes, Vol. **114**, 1970.

[2] Weil, A., *Dirichlet series and automorphic forms,* Springer Lecture Notes, Vol. **189,** 1970.

[3] Weil, A., Zeta-functions and Mellin transforms, *in Colloquium on algebraic geometry, Bombay, 1968*, pp. 409—426, Tata Institute, Bombay, 1969.

[4] ———, Über die Bestimmung Dirichletscher Reihen durch Funktionalgleichungen, *Math. Ann.,* **168** (1967), 149—156.

[5] Gelfand, I. M., Graev, M. I. and Pjateckij-Šapiro, I. I. (Гельфанд И. М., Граев М. И., Пятецкий-Шапиро И. И.), *Теория представлений и автоморфные функции,* „Наука", М., 1966 (*Representation theory and automorphic functions,* W. B. Saunders Company, 1969).

[6] Godement, R., *Notes on Jacquet—Langlands' theory,* Princeton, 1970.

[7] Novodvorskij, M. E. (Новодворский М. Е.), О некоторых неподвижных векторах в бесконечномерных представлениях групп Шевалле, *Функциональный анализ* **5**: 1 (1971), 87—88.
(On certain fixed vectors in infinite-dimensional representations of the Chevalley groups.)

[8] ———, О представлениях некоторых адельных групп, *УМН* **27**: 3 (1972), 206.
(On the representations of certain adelic groups.)

[9] Casselman, R., On some results of Atkin and Lehner, *preprint.*

[10] O'Meara, O. T., On the finite generation of linear groups over Hasse domains, *J. Reine Angew. Math.* **217** (1965), 79—108.

[11] Wasserstein, L. N. (Васерштейн Л. Н.), О группе SL_2 над дедекиндовыми кольцами арифметического типа, *Мат, сб.*, **89** (1972), 313—322.
(On the group SL_2 over Dedekind rings of arithmetic type.)

EULER SUBGROUPS

I. I. PJATECKIJ-ŠAPIRO

Let $\mathcal{G}$ be a reductive algebraic group over a global field $\mathfrak{k}$ and G be an algebraic subgroup of $\mathcal{G}$ which is also defined over $\mathfrak{k}$. Suppose that there exists a nontrivial morphism $\mu: G \to \mathfrak{k}^*$.

The integral

$$\int_{G_{\mathfrak{k}} \backslash G_A} f(g)\,|\mu(g)|^s\,dg \tag{1}$$

where $f(g)$ is an automorphic form on $\mathcal{G}_A$, may be used for defining the zeta function only if it may be transformed into a product taken over all places of $\mathfrak{k}$. For instance, we are in such situation in the case $\mathcal{G} = GL(2, \mathfrak{k})$, $G = \left\{\begin{pmatrix}\alpha & 0\\ 0 & 1\end{pmatrix}, \alpha \in \mathfrak{k}^*\right\}$.

More precisely, if $f(g)$ is a parabolic form on $GL(2, \mathfrak{k}) \backslash GL(2, A)$ we have an identity

$$\int_{\mathfrak{k}^* \backslash I_{\mathfrak{k}}} f\begin{pmatrix}\alpha & 0\\ 0 & 1\end{pmatrix} |\alpha|^s\,d\alpha = \int_{I_k} f_\varkappa \begin{pmatrix}\alpha & 0\\ 0 & 1\end{pmatrix} |\alpha|^s\,d\alpha \tag{2}$$

where

$$f_\chi(g) = \int_{Z_{\mathfrak{k}} \backslash Z_A} f(zg)\,\overline{\chi(z)}\,dz \tag{3}$$

$Z = \left\{\begin{pmatrix}1 & *\\ 0 & 1\end{pmatrix}\right\}$ and χ is a nontrivial character of $Z_{\mathfrak{k}} \backslash Z_A$.

The aim of this paper is to give examples of similar identities for groups different from $GL(2)$.

For any place $\mathfrak{p}$ of $\mathfrak{k}$ we denote by $W_\chi^{\mathfrak{p}}$ the space of all functions $f: \mathcal{G}_{\mathfrak{p}} \to \mathbf{C}$ such that $f(zg) = \chi(z) f(g)$, where Z is an algebraic subgroup of $\mathcal{G}$ defined over $\mathfrak{k}$, χ is a character $Z_{\mathfrak{p}} \to \mathbf{C}$. The group $\mathcal{G}_{\mathfrak{p}}$ acts on $W_\chi^{\mathfrak{p}}$ by right translations.

We shall introduce the following property of $W_\chi^{\mathfrak{p}} \cdot (U^{\mathfrak{p}})$. For any irreducible admissible representation π of $\mathcal{G}_{\mathfrak{p}}$ $\dim \operatorname{Hom}_{\mathcal{G}_{\mathfrak{p}}} (\pi, W_\chi^{\mathfrak{p}}) \leq 1$.

The fact that the right-hand side of (2) can be represented as an Euler product is based on the property $(U^{\mathfrak{p}})$ [1].

The groups $\mathcal{G} = GL(2, \mathfrak{k})$, $Z = \left\{\begin{pmatrix}1 & *\\ 0 & 1\end{pmatrix}\right\}$ give us an example where $(U^{\mathfrak{p}})$ is satisfied for all $\mathfrak{p}$ [1].

The results of Gelfand and Kajdan show that $(U^{\mathfrak{p}})$ holds if $\mathcal{G}$ is a Chevalley group, Z is its maximal unipotent subgroup, χ is a nondegenerate character of Z and $\mathfrak{p}$ is a finite place of $\mathfrak{k}$ [2].

The well-known method of description of the representations of principal series of Chevalley groups shows that if $(U^{\mathfrak{p}})$ holds for all $\mathfrak{p}$ then

$$\dim Z \geqq \dim_{\mathbf{C}} \text{ (maximal unipotent subgroup of } \mathcal{G}_{\mathbf{C}}). \tag{4}$$

An algebraic subgroup G of $\mathcal{G}$ defined over $\mathfrak{k}$ is called an *Euler subgroup* if
1) there exists a nontrivial morphism $\mu: G \to \mathfrak{k}^*$
2)

$$\int_{G_{\mathfrak{k}} \backslash G_A} f(g)\,|\mu(g)|^s\,dg = \int_{B_A} f_\chi(b)\,|\mu(b)|^s\,db \tag{5}$$

where B is a homogeneous space corresponding to a subgroup of G, Z is an algebraic subgroup of $\mathcal{G}$ defined over $\mathfrak{k}$ and satisfying (4), χ is a character of Z_A which is trivial on $Z_{\mathfrak{k}}$ and $f_\chi(g)$ is defined by a formula similar to (3).

In Section 1 I show that for $\mathcal{G}=GL(n)$ the subgroup $G=GL(n-1)$ is an Euler subgroup. In section 2 I give examples for Euler subgroups of $\mathcal{G}$ different from $GL(n)$. I give a sufficient condition for a subgroup G to be an Euler subgroup. It generalizes some results of Rankin [4], Selberg [5], [6], Ogg [7], and Andrianov [9], and gives a clue for consideration of further examples. I want to mention that the language used by these authors was quite different from that of this paper.

Let $\mathfrak{k}$ be a global field of positive characteristic. Using the results of [2] I prove in Section 1 the following.

Multiplicity-one theorem. The multiplicity of an irreducible representation of $GL(n, A)$ in $L_0^2(GL(n, \mathfrak{k}) \backslash GL(n, A))$ is not greater than one.

It seems very plausible that a statement similar to the Gelfand—Kajdan theorem is true in the Archimedean case. If it were proved we should have a multiplicity-one theorem for $GL(n)$ over a number field.*

I do not know whether the multiplicity-one theorem is valid for groups different from $GL(n)$.

Section 3 is devoted to computation of local factors in this case: $\mathcal{G}=\mathrm{Sp}(4, \mathfrak{k})$. The main idea of this computation is contained in Andrianov's paper [9].

Now, from the point of view of the theory of zeta-functions, only in case $GL(2)$ is the theory more or less complete. However even in this case many important problems are unsolved. I want to mention one of them.

Let $\mathfrak{k}$ be a global field and K a Galois extension of $\mathfrak{k}$. Let $\pi = \otimes \pi_{\mathfrak{p}}$ be an irreducible representation of $GL(2, A_{\mathfrak{k}})$ belonging to $L_0^2(GL(2, \mathfrak{k}) \backslash GL(2, A_{\mathfrak{k}}))$.

Does there exist an irreducible representation

$$\tilde{\pi} = \otimes \tilde{\pi}_{\mathfrak{p}} \in L_0^2(GL(2, K) \backslash GL(2, A_K)) \quad \text{of} \quad GL(2, A_K)$$

* Now it has been proved by Shalika, *Bull. Amer. Math. Soc.* 79:2 (1973).

such that for each place $\mathfrak{p}$ of $\mathfrak{k}$ and each $\tilde{\mathfrak{p}}$ of K lying over $\mathfrak{p}$ the representations π and $\tilde{\pi}_{\mathfrak{p}}$ naturally correspond to each other, i.e. the representation of the Weil—Šafarevič group corresponding to $\tilde{\pi}_{\tilde{\mathfrak{p}}}$ is the restriction of the representation of the Weil—Šafarevič group corresponding to $\pi_{\mathfrak{p}}$? A positive answer to this question implies that zeta functions of arithmetic curves satisfy the Hasse—Weil conjecture. Here by arithmetic curve I mean a curve obtained as a factor of the upper half plane by an arithmetic group. Hitherto the Hasse—Weil conjecture was proved by Shimura only for some well-chosen models of arithmetic curves. Of course it is not the only example for relations between zeta functions of infinite-dimensional representations of adele groups.

The remarkable results of A. Weil, G. Shimura and P. Deligne give the hope that zeta functions of algebraic varieties may be expressed in terms of zeta functions of infinite-dimensional representations of adele groups of reductive groups.

Now I am going to say a few words about two problems. The first is the problem of the analytic continuation of integral (1) to the whole complex plane. In section 2 I give two methods of analytic continuation of integrals of this kind. The first method I call the Hecke method and the second one I call the Rankin—Selberg method. Moreover there exist examples of Euler subgroups (for instance Example 3 of § 2) where both of these methods fail. I think that integral (1) can always be extended to the whole complex plane as a meromorphic function. I want to mention that there are examples where the analytic continuation of the integral (1) is a meromorphic function with an infinite set of poles. More precisely this is the case in any application of the Rankin—Selberg method.

The second problem refers to finite-dimensional representations. Let $\mathscr{G}$ be a Chevalley group over a finite field. We call an irreducible representation of $\mathscr{G}$ in a finite-dimensional space V over **C** *hyperanalytic* if there exist two proper subgroups G and Z and an irreducible representation of Z in V_0 such that

$$V = \sum_{g \in G} gV_0. \tag{6}$$

It is known that every analytic representation of $GL(n)$ is hyperanalytic (with $G=GL(n-1)$ and Z maximal unipotent subgroup of $GL(n)$) [12]. This is closely related to the fact that $GL(n-1)$ is Euler subgroup of $GL(n)$.

The Andrianov theory for Sp (4) corresponds to the fact that Sp (4) has a hyperanalytic representation. This is the analytic representation of Sp (4) of minimal dimension. Gelfand proved (not published) that this representation is hyperanalytic in our sense.

Apparently there is a close connection between the examples of Euler subgroups and the examples of decomposition (6). I expect that to each example of a hyperanalytic representation of $\mathscr{G}$ over a finite field there corresponds an example of an Euler subgroup of $\mathscr{G}$ over a global field.

I am very thankful to I. M. Gelfand, D. A. Kajdan, R. Godement, A. N. Andrianov and A. A. Kirillov for many pleasant and fruitful discussions. I am especially obliged to D. A. Kajdan who has read the manuscript.

I also thank very much G. van Dijk and F. Zak for the linguistic correction of this paper.

§ 1. Representations of $GL(n)$

We denote by $\mathfrak{k}$ any global field (i.e. a number field or a field of algebraic functions of one variable over a finite field). If $\mathfrak{p}$ is a place of $\mathfrak{k}$ we denote by $\mathfrak{k}_\mathfrak{p}$ the corresponding completion of $\mathfrak{k}$.

Denote by $\mathscr{G}$ a reductive group over $\mathfrak{k}$.

The group $\mathscr{G}_A$ operates by right translations on the Hilbert space $L^2(\mathscr{G}_\mathfrak{k}\diagdown\mathscr{G}_A)$ of all measurable functions $\varphi(g)$ such that

$$\varphi(\gamma g) = \varphi(g) \quad \text{for every} \quad \gamma \in \mathscr{G}_\mathfrak{k} \tag{1}$$

and

$$\int_{\mathscr{G}_\mathfrak{k}\diagdown\mathscr{G}_A} |\varphi(g)|^2 \, dg < \infty. \tag{2}$$

Denote by $L_0^2 = L_0^2(\mathscr{G}_\mathfrak{k}\diagdown\mathscr{G}_A)$ the closed invariant subspace of $L^2(\mathscr{G}_\mathfrak{k}\diagdown\mathscr{G}_A)$ consisting of all functions φ such that

$$\int_{R_\mathfrak{k}\diagdown R_A} \varphi(rg) \, dr = 0 \quad (\text{almost all } g \in \mathscr{G}_A) \tag{3}$$

for any cuspidal subgroup R of G (a cuspidal subgroup is the unipotent radical of a parabolic subgroup).

Let C be the centre of $\mathscr{G}$ and let ψ be a character of $\mathscr{G}_A$ which is trivial on $\mathscr{G}_\mathfrak{k}$. Denote by $L_{0,\psi}^2 = L_{0,\psi}^2(\mathscr{G}_\mathfrak{k}\diagdown\mathscr{G}_A)$ the Hilbert space which consists of all functions on $\mathscr{G}_\mathfrak{k}\diagdown\mathscr{G}_A$ which satisfy (3) and the following conditions

$$\varphi(cg) = \psi(c)\varphi(g) \quad \text{for every} \quad c \in C_A \tag{4}$$

$$\int_{\mathscr{G}_\mathfrak{k} C_A\diagdown\mathscr{G}_A} |\varphi(g)|^2 \, dg < \infty. \tag{5}$$

We can (by means of Fourier transforms) decompose L_0^2 into a continuous sum of unitary irreducible representations of $L_{0,\psi}^2$. It is well known that $L_{0,\psi}^2$ is a discrete sum of unitary representations of $\mathscr{G}_A$ each occurring with finite multiplicity [3]. We call a function $\varphi(g)$ on $\mathscr{G}_\mathfrak{k}\diagdown\mathscr{G}_A$ elementary if *1)* the space generated by the right translates of $\varphi(g)$ by the elements of $\mathscr{G}_A$, is isomorphic to an admissible irreducible representation $H = \otimes H_\mathfrak{p}$ of $\mathscr{G}_A$ *2)* under this isomorphism φ corresponds to a decomposable vector $\xi = \otimes \xi_\mathfrak{p} \in H$ *3)* there exists a maximal compact subgroup $U \subset \mathscr{G}_A$ such that the space generated by the right translates of $\varphi(g)$ by elements of U is finite-dimensional.

We shall denote by $A(\mathscr{G})$ the set of all elementary functions. Of course, $A(\mathscr{G})$ is not a linear space over **C**. Denote by $S(\mathscr{G})$ the linear space consisting of all C^∞ slowly increasing functions on $\mathscr{G}_\mathfrak{k} \backslash \mathscr{G}_A$. Of course, the linear span $\tilde{A}(\mathscr{G})$ of $A(\mathscr{G})$ belongs to $S(\mathscr{G})$.

From now on we shall consider in this section only the case $\mathscr{G}=GL(n)$, $G=GL(n-1)$. Denote by Z a maximal unipotent subgroup of $\mathscr{G}$. Let θ be a character of Z_A which is trivial on $Z_\mathfrak{k}$. We shall suppose that θ is nondegenerate, i.e. θ is nontrivial on every subgroup corresponding to a simple root. For instance, if Z is the upper triangle subgroup we can put

$$\theta(z) = v\left(\sum_{k=1}^{n-1} z_{kk+1}\right), \tag{6}$$

where v is a character of $\mathfrak{k} \backslash A$.

For any $f(g) \in S(\mathscr{G})$ we can define

$$f_\theta(g) = \int_{Z_\mathfrak{k} \backslash Z_A} \overline{\theta(z)} f(zg)\, dz. \tag{7}$$

For $GL(2)$, Jacquet and Langlands have found a very simple explicit formula which shows that any $f(g) \in \tilde{A}(\mathscr{G})$ can be uniquely defined by means of $f_\theta(g)$. It says

$$f(g) = \sum f_\theta(\delta g), \tag{8}$$

where δ runs over the set of all matrices of the form

$$\begin{pmatrix} \delta & 0 \\ 0 & 1 \end{pmatrix}, \quad \delta \in \mathfrak{k}^*. \tag{9}$$

They also proved two important corollaries to (8).

Corollary 1. *The multiplicity-one theorem. The multiplicity of an irreducible representation H of $GL(2, A)$ in L_0^2 is not greater than one.*

Remark. An important particular case of this theorem was considered by Hecke. He proved that $SL(2, \mathbf{Z})$-modular forms are defined uniquely up to a scalar factor by the eigenvalues of all Hecke operators. It is equivalent to the Jacquet—Langlands theorem under the additional assumption that all non-Archimedean components $H_\mathfrak{p}$ of H are of class one.

Corollary 2. If $f(g) \in A(\mathscr{G})$, then the integral

$$\mathscr{I}(s) = \int_{\mathfrak{k}^* \backslash I_\mathfrak{k}} f\begin{pmatrix} a & 0 \\ 0 & 1 \end{pmatrix} |a|^s\, da \tag{10}$$

can be decomposed into an Euler product.

Remark. It is very simple to prove that *1)* the integral (10) converges absolutely in the half plane $\operatorname{Re} s > c$ for sufficiently large c, *2)* $\mathscr{I}(s)$ can be extended to the whole complex plane, as an entire function, *3)* $\mathscr{I}(s)$ has a "nice" functional equation.

The purpose of this section is to generalize formula (8) and its corollaries to $GL(n)$, for arbitrary n.

Theorem I.* *For* $f(g) \in \tilde{A}(\mathscr{G})$ *we have*

$$f(g) = \sum_{\delta \in (Z \cap G)_{\mathfrak{k}} \backslash G_{\mathfrak{k}}} f_\theta\left(\begin{pmatrix} \delta & 0 \\ 0 & 1 \end{pmatrix} g\right) \tag{11}$$

where the series converges absolutely and uniformly on any compact subset of $\mathscr{G}_A$.

Proof. For the sake of simplicity we suppose that Z is the group of upper triangular matrices and $\theta = \nu(\sum_{k=1}^{n-1} Z_{k,k+1})$ where ν is a character of $\mathfrak{k} \backslash A$.

Denote by P^n the subgroup of $GL(n)$, consisting of all matrices of the form

$$\begin{pmatrix} * & * \dots & * \\ \vdots & & \\ * & * \dots & * \\ 0 & 0 \dots 0 & 1 \end{pmatrix}.$$

By a standard cuspidal subgroup of P^n we mean a subgroup of P^n consisting of all matrices of the form

$$\begin{pmatrix} E_{k_1} & \dots & * \\ & \ddots & \vdots \\ 0 & & E_{k_r} \end{pmatrix}, \quad k_1 + \dots + k_r = n.$$

Denote by S_n the set of all C^∞-functions (i.e. locally constant with respect to non-Archimedean variables and infinitely differentiable with respect to Archimedean variables) on P_A^n such that

$$\varphi(\gamma p) = \varphi(p) \quad \text{for every} \quad \gamma \in P_{\mathfrak{k}}^n, \tag{12}$$

$\int_{R_{\mathfrak{k}} \backslash R_A} \varphi(rp)\, dr = 0$ for every standard cuspidal subgroup R.

It is easy to verify that $f(g) \in \tilde{A}(\mathscr{G})$ implies that $\varphi(p) = f(pg_0)$ belongs to S_n for every g_0. This shows that it is enough to prove the following

Lemma 1. For any $\varphi(p) \in S_n$ we have

$$\varphi(p) = \sum_{\delta \in Z_{\mathfrak{k}} \backslash P_{\mathfrak{k}}^n} \varphi_\theta(\delta p), \quad \varphi_\theta(p) = \int_{Z_{\mathfrak{k}} \backslash Z_A} \varphi(zp) \overline{\theta(z)}\, dz \tag{13}$$

where the series converges absolutely and uniformly on any compact subset of P_A^n.

* This formula was independently obtained by Shalika.

Proof. First we introduce the subgroup $Y \subset P^n$ which consists of all matrices of the form

$$\begin{pmatrix} E & y \\ 0 & 1 \end{pmatrix}, \quad y = \begin{pmatrix} y_1 \\ \vdots \\ y_{n-1} \end{pmatrix}$$

We define

$$\varphi_1(p_1) = \int_{Y_\mathfrak{k} \backslash Y_A} \varphi(yp)\, v(y_{n-1})\, dy. \tag{14}$$

It is not difficult to verify that

$$\varphi_1\left(\begin{pmatrix} \gamma & \beta \\ 0 & 1 \end{pmatrix} p\right) = \varphi_1(p) \quad \text{for any} \quad \gamma \in P_\mathfrak{k}^{n-1}, \quad \beta \subset \mathfrak{k}^{n-1}. \tag{15}$$

Denote by P^n the stationary subgroup of $\varphi_1(p)$. If χ is a nontrivial character of Y_A which is trivial on $Y_\mathfrak{k}$ we can find $\delta \in G_\mathfrak{k} = GL(n-1, \mathfrak{k})$ such that

$$\varphi_1\left(\begin{pmatrix} \delta & 0 \\ 0 & 1 \end{pmatrix} p\right) = \int_{Y_\mathfrak{k} \backslash Y_A} \varphi(yp)\, \chi(y)\, dy. \tag{16}$$

Moreover Y is a standard cuspidal subgroup and hence

$$\int_{Y_\mathfrak{k} \backslash Y_A} \varphi(yp)\, dy = 0. \tag{17}$$

By making use of the ordinary Fourier expansion, we obtain from (15), (16) and (17) that

$$\varphi(p) = \sum_{\delta \in \tilde{P}_\mathfrak{k} \backslash P_\mathfrak{k}^n} \varphi_1(\delta p) \tag{18}$$

implies that the series (18) converges absolutely and uniformly on any compact subset of P_A^n.

Consider the function ψ on P_A^{n-1}, defined by

$$\psi(p) = \varphi_1\left(\begin{pmatrix} p & 0 \\ 0 & 1 \end{pmatrix} p_0\right), \quad \varphi(p) \in S_n \tag{19}$$

where p_0 is a fixed element of P_A^n.

It is clear that $\psi(p)$ is a C^∞-function on P_A^{n-1} and $\psi(\gamma p) = \psi(p)$ for any $\gamma \in P_\mathfrak{k}^{n-1}$. Let R be a standard cuspidal subgroup of P^{n-1}. We introduce the subgroup $R(k)$, $1 \leqq k \leqq n-2$ of P^n such that

$$R(k) = \left\{ \begin{pmatrix} r & y^k \\ 0 & 1 \end{pmatrix} \middle| y = \begin{pmatrix} y_1 \\ \vdots \\ y_k \\ 0 \\ \vdots \\ 0 \end{pmatrix} \right\}.$$

It is obvious that there exists K_0, $1 \leqq K_0 \leqq n-2$ such that $R^0 = R(K_0)$ is a standard cuspidal subgroup of P^n. We put $\tilde{R} = RY$. Since $R \subset P^{n-1} \subset \tilde{P}^n$, we have $\chi(ryr^{-1}) =$

$=\chi(y)$ for $r \in R_A$. Thus there exists an extension $\tilde{\chi}(r)$ of $\chi(y)=\nu(y_{n-1})$ to $\tilde{R}_A$ such that $\tilde{\chi}(r)=1$ for $r \in R_A$.

It is clear that R^0 is a normal subgroup of $\tilde{R}$ and $\tilde{\chi}(r)=1$ for $r \in R_A^0$. This shows that

$$\int_{R_{\mathfrak{k}}\setminus R_A} \psi(rp)\,dr = \int_{\tilde{R}_{\mathfrak{k}}\setminus R_A} \varphi(rp)\chi(r)\,dr = \int_{\tilde{R}_{\mathfrak{k}}\tilde{R}_A^0\setminus \tilde{R}_A} \chi(r_1)\,dr_1 \int_{\tilde{R}_{\mathfrak{k}}^0\setminus R_A^0} \varphi(r_0 r_1 p)\,dr = 0 \quad (20)$$

so we have proved that $\psi(p) \subset S_{n-1}$.

Now we shall prove our lemma by induction on n. For $n=2$ formula (18) is equivalent to the lemma. Suppose $n>2$ and suppose our lemma is true for $n-1$. Then (18) and the expansion for $\psi(p)$ (which follows from the inductive assumption) yield the statement of the lemma for n.

Now we can prove the theorem. Let $f(g) \in \tilde{A}(\mathscr{G})$. For $p \in P_A^n$ and g_0 fixed, we put $\varphi(p)=f(pg_0)$.

It is obvious that $\varphi(p) \in S_n$ and hence the expansion (11) holds for $f(g)$.

From this the statement of the theorem follows for $f(g)$.

Now I have to give some definitions. Let θ be a character of $Z_{\mathfrak{p}}$. Consider the space $S_\theta(\mathscr{G}_{\mathfrak{p}})$ of all measurable locally integrable functions on $\mathscr{G}_{\mathfrak{p}}$ for which

$$f(zg) = \theta(z) f(g) \quad \text{for any} \quad z \subset Z_{\mathfrak{p}}, \quad g \in \mathscr{G}_{\mathfrak{p}}. \quad (21)$$

Let H be an irreducible admissible representation of $\mathscr{G}_{\mathfrak{p}}$. Denote by $\mathrm{Hom}\,(H, S_\theta(\mathscr{G}_{\mathfrak{p}}))$ the set of all mappings $\delta: H \to S_\theta(\mathscr{G}_{\mathfrak{p}})$ which satisfy the following conditions:

1) If $h_n \to h_0$, $f_n=\delta(hn)$, then

$$\int_K f_n(g)\,dg \to \int_K f_0(g)\,dg \quad (22)$$

for every compact $K \subset \mathscr{G}_{\mathfrak{p}}$;

2) If $\delta(h)=f(g)$, then for any $g_0 \in \mathscr{G}_{\mathfrak{p}}$ we have $\delta(T_{g_0}h)=f(gg_0)$ where T_{g_0} is the operator on H corresponding to g_0.

It is obvious that $\mathrm{Hom}\,(H, S_\theta(\mathscr{G}_{\mathfrak{p}}))$ is a linear space over $\mathbf{C}$.

Gelfand and Kajdan obtained that $\dim_{\mathbf{C}} \mathrm{Hom}\,(H, S_\theta(\mathscr{G}_{\mathfrak{p}})) \leqq 1$ for any non-Archimedean local field $\mathfrak{k}_{\mathfrak{p}}$ [2]. A similar statement is likely to hold in the Archimedean case as well.* This is the reason why we have to formulate our theorem in the following weak form.

Theorem 2.** *For every irreducible unitary representation H of $\mathscr{G}_A$ we have*

$$\dim \mathrm{Hom}\,(H, L^2_{0,\psi}) \leqq \prod_{\mathfrak{p}} \dim\,(H_{\mathfrak{p}}, S_\theta(\mathscr{G}_{\mathfrak{p}})), \quad (23)$$

where $\mathfrak{p}$ runs over all Archimedean places of $\mathfrak{k}$.

* Now it has been proved by Shalika.

** This theorem has been independently proved by Shalika.

Proof. Let θ be a character of Z_A which is equal to 1 on $Z_{\mathfrak{k}}$. We introduce the space $S_\theta(\mathscr{G}_A)$ consisting of all measurable locally integrable functions on $\mathscr{G}_A$ such that

$$f(zg) = \theta(z)f(g) \quad \text{for any} \quad z \in Z_A, \quad g \in \mathscr{G}_A. \tag{24}$$

As in the local case we can define $\operatorname{Hom}(H, S_\theta(\mathscr{G}_A))$. Consider the mapping: $\mu: L^2_{0,\psi} \to S_\theta(\mathscr{G}_A)$

$$\mu: f(g) \to f_\theta(g) = \int_{Z_{\mathfrak{k}} \setminus Z_A} f(zg)\theta(z)\,dz. \tag{25}$$

Theorem 1 shows that the kernel of μ is zero. Hence for any unitary irreducible representation H, we have

$$\dim_{\mathbf{C}} \operatorname{Hom}(H, L^2_{0,\psi}) \leqq \dim_{\mathbf{C}} \operatorname{Hom}(H, S_\theta(\mathscr{G}_A)). \tag{26}$$

Following the pattern given in [1] we obtain from (26) the statement of our theorem.

I. M. Gelfand and D. A. Kajdan introduced a very important function which depends on two representations: one of $GL(n, \mathfrak{k}_p)$ and one of $GL(n-1, \mathfrak{k}_p)$. (This function also depends on a character of $\mathfrak{k}^+$ which will be fixed and will not appear in our notation.)

I. M. Gelfand and D. A. Kajdan conjectured [2] that there exists a zeta function $\zeta(\varrho, \sigma)$ which depends on two irreducible representations (ϱ, σ): one of $GL(n, A)$ and one of $GL(n-1, A)$ where $\varrho \in L^2_0(GL(n, \mathfrak{k}) \setminus GL(n, A))$, $\sigma \in S(GL(n-1))$. (I should like to point out that this conjecture has stimulated me very much.) More precisely, they have suggested:

1) the local Γ-function can be represented in the form

$$\Gamma(\varrho_p, \sigma_p) = \frac{\zeta(\varrho_p, \sigma_p)}{\zeta(\hat{\varrho}_p, \hat{\sigma}_p)} \varepsilon(\varrho_p, \sigma_p)$$

where ^ means the contragredient representation.

2) $\zeta(\varrho, \sigma) = \prod \zeta(\varrho_p, \sigma_p)$ is a function holomorphic on the set of pairs of representations (ϱ, σ) which has been described above and satisfies the functional equation

$$\zeta(\hat{\varrho}_p, \hat{\sigma}_p) = \zeta(\varrho_p, \sigma_p)\varepsilon(\varrho_p, \sigma_p). \tag{27}$$

They have also suggested that (27) implies

$$\varrho \in L^2_0(GL(n, \mathfrak{k}) \setminus GL(n, A)).$$

The following does not contain the definition of the desired zeta function. We are going to consider some integral which might be useful for it. As for this integral, we prove that it has properties analogous to those of the similar integral for $GL(2)$.

Let $f \in \tilde{A}(\mathscr{G})$, $\varphi \in S(G)$. We put

$$\mathscr{I}(f, \varphi, s) = \int_{G_{\mathfrak{k}} \diagdown G_A} f\begin{pmatrix} h & 0 \\ 0 & 1 \end{pmatrix} \varphi(h) |h|^s \, dh \tag{28}$$

where $|h| = |\det h|$.

Theorem 3. *For any* $f \in A(\mathscr{G})$, $\varphi \in S(\mathscr{G})$

1. $\mathscr{I}(f, \varphi, s)$ *converges absolutely in the half plane* $\operatorname{Re} s > c_1$ *for some* c_1*;*

2. $\mathscr{I}(f, \varphi, s)$ *is an entire function bounded in vertical strips;*

3. *if the right translations of* φ *by* G_A *generate an irreducible representation then* $\mathscr{I}(f, \varphi, s)$ *decomposes into an Euler product.*

Proof. First we have to prove the following

Lemma. If $f(g) \in A(\mathscr{G})$ then $\tilde{f}(g) = f(g'^{-1}) \in A(\mathscr{G})$.

Proof. First for any $\gamma \in \mathscr{G}_{\mathfrak{k}}$, $g \in \mathscr{G}_A$ we have

$$\tilde{f}(\gamma g) = f(\gamma'^{-1} g'^{-1}) = f(g'^{-1}) = \tilde{f}(g)$$

By making use of the equality $dg'^{-1} = dg$, we obtain

$$\int_{C_A \mathscr{G}_{\mathfrak{k}} \diagdown \mathscr{G}_A} |\tilde{f}(g)|^2 \, dg = \int_{C_A \mathscr{G}_{\mathfrak{k}} \diagdown \mathscr{G}_A} |f(g)|^2 \, dg. \tag{29}$$

If Z is a cuspidal subgroup the Z'^{-1} is also cuspidal. Consequently

$$\int_{Z_{\mathfrak{k}} \diagdown Z_A} \tilde{f}(zg) \, dz = \int_{Z_{\mathfrak{k}}'^{-1} \diagdown Z_A'^{-1}} f(zg'^{-1}) \, dz = 0. \tag{30}$$

Now the mapping $\delta: f \to \tilde{f}$ transforms an admissible irreducible representation into one with the same properties. That completes our proof.

Now we can prove our theorem.

It is well known that if $f(g) \in \tilde{A}(\mathscr{G})$ then f rapidly decreases at ∞. That shows that $\mathscr{I}(f, \varphi, s)$ converges absolutely in the half plane $\operatorname{Re} s > c_0$ for some c_0.

In order to prove that $\mathscr{I}(f, \varphi, s)$ is an entire function we have to use the usual trick.

We write

$$\mathscr{I} = \int_{|g| \geqq 1} + \int_{|g| < 1}.$$

It is obvious that the first integral is an entire function. Consider the second integral

$$\int_{|g|<1} f\begin{pmatrix} g & 0 \\ 0 & 1 \end{pmatrix} \varphi(g) |g|^s \, dg = \int_{|g|>1} \tilde{f}\begin{pmatrix} g & 0 \\ 0 & 1 \end{pmatrix} \tilde{\varphi}(g) |g|^{-s} \, dg$$

Using the lemma we obtain that $\tilde{f} \in A(\mathscr{G})$ and hence this integral is also an entire function. To prove the existence of the Euler product for $\mathscr{I}(f, \varphi, s)$ we have to prove the following formula:

$$\mathscr{I}(f, \varphi, s) = \int_{Z'_A \backslash G_A} f_\theta \begin{pmatrix} h & 0 \\ 0 & 1 \end{pmatrix} \varphi_{\theta^{-1}}(h)\,|h|^s\,dh \tag{31}$$

where Z' is the upper triangle subgroup of G. Using (11), we have

$$\mathscr{I}(f, \varphi, s) = \int_{Z'_\mathfrak{k} \backslash G_A} f_\theta \begin{pmatrix} h & 0 \\ 0 & 1 \end{pmatrix} \varphi(h)\,|h|^s\,dh = \int_{Z_A \backslash G_A} f_\theta \begin{pmatrix} h & 0 \\ 0 & 1 \end{pmatrix} \varphi_{\theta^{-1}}(h)\,|h|^s\,dh$$

where

$$\varphi_{\theta^{-1}}(g) = \int_{Z'_\mathfrak{k} \backslash Z'_A} \theta(z)\,\varphi(zg)\,dz.$$

By making use of the Gelfand—Kajdan theorem we obtain for $\varphi \in S(G), f \in A(\mathscr{G})$ satisfying assumptions 3) the decompositions

$$f_\theta(g) = f^\infty(g^\infty) \prod_\mathfrak{p} f_{\theta_\mathfrak{p}}(g_\mathfrak{p})$$

$$\varphi_{\theta^{-1}}(g) = \varphi^\infty(g^\infty) \prod \varphi^\mathfrak{p}_{\theta_\mathfrak{p}^{-1}}(g_\mathfrak{p})$$

where the products are taken over all non-Archimedean places.

This gives

$$\mathscr{I}(f, \varphi, s) = \mathscr{I}_\infty \prod_{\mathfrak{p}n} \mathscr{I}_\mathfrak{p}$$

where $\mathfrak{p}$ runs over all finite places and

$$\mathscr{I}_\infty = \int_{Z_\infty \backslash G_\infty} f^\infty(g^\infty)\,\varphi^\infty(g^\infty)\,dg^\infty$$

$$\mathscr{I}_\mathfrak{p} = \int_{Z_\mathfrak{p} \backslash G_\mathfrak{p}} f^\mathfrak{p}_{\theta_\mathfrak{p}}\,\varphi_{\theta_\mathfrak{p}^{-1}}\,dg_\mathfrak{p}.$$

Here $G_\infty = \prod_\mathfrak{p} G_\mathfrak{p}$ (the product is taken over all Archimedean $\mathfrak{p}$).

§ 2. Some examples of Euler subgroups

In this section Theorem 1 gives some sufficient conditions for a subgroup $G \subset \mathscr{G}$ to be an Euler subgroup and Theorems 2 and 3 give sufficient conditions for the existence of a meromorphic continuation of the integral $\mathscr{I}(f, G, S)$.

Some examples of Euler subgroups will also be given. Example 1 for the case $K = \mathfrak{k} \oplus \mathfrak{k}$ and $\mathfrak{k} = \mathbf{Q}$ was considered by Rankin [4], Selberg [5], Ogg [7]. The case of arbitrary $\mathfrak{k}$ was considered from the point of view of the theory of representations by H. Jacquet (R. Godement's oral communication).

Andrianov was the first who considered Example 4. It seems to me that Examples 2 and 3 were not studied by anyone yet. Theorems 2 and 3 guarantee the possibility of a meromorphic continuation of $\mathscr{I}(f, G, s)$ to whole complex plane in Examples 1, 2, 4.

I do not know if a meromorphic continuation of $\mathcal{I}(f, G, s)$ to the whole complex plane exists in Example 3.

Of course, it is not difficult to give many other applications of Theorem 1.

Before formulating Theorem 1 let us agree upon the following notation. Let G be an algebraic group over $\mathfrak{k}$ and X and Y two of its algebraic subgroups defined over $\mathfrak{k}$. We shall write $G=XY$, if each element g of $G_{\mathfrak{k}}(G_A)$ can be uniquely represented in the form $g=xy$, where $x\in X_{\mathfrak{k}}(X_A)$, $y\in Y_{\mathfrak{k}}(Y_A)$.

Theorem 1. *Let $\mathcal{G}$ be a reductive algebraic group over $\mathfrak{k}$ and $H, N, S, G=NH$, $P=GS$, $Z=NS$ algebraic subgroups of $\mathcal{G}$ defined over $\mathfrak{k}$ with the following properties:*

1. *N has no nontrivial algebraic characters: $N\to\mathfrak{k}^*$,*
2. *the left invariant measure on G_A can be represented in the form*

$$\theta(h)\,dn\,dh \tag{1}$$

where $\theta(h)$ is a character of H_A, dn and dh are left invariant measures on N_A and H_A, respectively.

3. *S is a commutative unipotent normal subgroup of P.*
4. *There exists a nontrivial character $\chi_0: S_A\to\mathbf{C}$ with the following properties:*

a) $\chi_0(x)=1$ *for* $x\in S_k$,

b) $\chi_0(nxn^{-1})=\chi_0(x)$ *for* $x\in S_A$, $n\in N_A$.

This shows that χ_0 can be extended to a character χ_0 of Z_A such that $\chi_0(n)=1$ for $n\in N_A$ (we use the same symbol for the character and its extension).

5. *$H_{\mathfrak{k}}$ acts on $S_{\mathfrak{k}}-0$ simply transitively. Hence each nontrivial character χ of $S_{\mathfrak{k}}\setminus S_A$ can be uniquely represented as $\chi_0(hxh^{-1})$ with $h\in H_{\mathfrak{k}}$.*
6. *P has a normal subgroup R which is a cuspidal subgroup of $\mathcal{G}$.*

For every function $f\in A(\mathcal{G})$ and any algebraic character $\mu: G\to\mathfrak{k}^*$ the identity

$$\int_{G_{\mathfrak{k}}\setminus G_A} f(g)\,\nu(\mu(g))\,dg = \int_{H_A} f_{\chi_0}(h)\,\nu(\mu(h))\,\theta(h)\,dh \tag{2}$$

holds, where

$$f_{\chi_0}(g) = \int_{Z_{\mathfrak{k}}\setminus Z_A} \chi_0(z) f(zg)\,dz, \tag{3}$$

ν is any character of $\mathfrak{k}^*\setminus I_{\mathfrak{k}}$ with sufficiently large real part.

Proof. First of all, Z has no nontrivial algebraic character and hence $Z_{\mathfrak{k}}\setminus Z_A$ has a finite volume [10], [11]. It shows that the integral (3) is correctly defined.

We assume that both integrals in (2) converge absolutely.

We put

$$\psi_0(g) = \int_{S_{\mathfrak{k}}\setminus S_A} \chi_0(s) f(sg)\,ds, \quad \psi_1(g) = \int_{S_{\mathfrak{k}}\setminus S_A} f(sg)\,ds. \tag{4}$$

The group S is commutative and hence the usual theory of Fourier series and assumption *5)* give

$$f(g) = \psi_1(g) + \sum_{h \in H_{\mathfrak{k}}} \psi_0(hg) \quad \text{for any} \quad f(g) \in A(\mathcal{G}). \tag{5}$$

Hence

$$\int\limits_{G_{\mathfrak{k}} \diagdown G_A} f(g)\, \nu(\mu(g))\, dg = \int\limits_{G_{\mathfrak{k}} \diagdown G_A} \psi_1(g)\, \nu(\mu(g))\, dg + \int\limits_{G_{\mathfrak{k}} \diagdown G_A} \Big(\sum_h \psi_0(hg)\, \nu(\mu(g))\Big)\, dg. \tag{6}$$

We shall show that the first term in the right-hand side of (6) equals zero. It follows from *3)* that μ can be extended to $\tilde{\mu}: P \to \mathfrak{k}^*$ such that $\tilde{\mu}(s) = 1$. Obviously, any unipotent subgroup lies in the kernel of $\tilde{\mu}$ and hence $\tilde{\mu}(R) = 1$, where R is defined in *6)*. Using this and the definition of ψ_1 in (4), we get

$$\int\limits_{G_{\mathfrak{k}} \diagdown G_A} \psi_1(g)\, \nu(\mu(g))\, dg = \int\limits_{P_{\mathfrak{k}} \diagdown P_A} f(g)\, \nu(\tilde{\mu}(g))\, dg = 0 \tag{7}$$

because the cuspidal subgroup R is a normal subgroup of P, $f \in A(\mathcal{G})$ and hence for any $p \in P_A$

$$\int\limits_{R_{\mathfrak{k}} \diagdown R_A} f(rp)\, \nu(\tilde{\mu}(rp))\, dr = \nu(\tilde{\mu}(p)) \int\limits_{R_{\mathfrak{k}} \diagdown R_A} f(rp)\, dr = 0.$$

Now we shall consider the second term of (6). Denote by F a fundamental domain in G_A with respect to the left translations by $G_{\mathfrak{k}}$. From $G = NH$ it follows that $F_1 = \bigcup_{h \in H_{\mathfrak{k}}} hF$ is a fundamental domain in G_A with respect to the left translations by $N_{\mathfrak{k}}$. Using this fact, we obtain

$$\int\limits_F \Big(\sum_{h \in H_{\mathfrak{k}}} \psi_0(hg)\Big)\, \nu(\mu(g))\, dg = \sum_{h \in H_{\mathfrak{k}}} \int\limits_F \psi_0(hg)\, \nu(\mu(g))\, dg =$$

$$= \sum_{h \in H_{\mathfrak{k}}} \nu(\mu(h^{-1})) \int\limits_{hF} \psi_0(g)\, \nu(\mu(g))\, dg = \int\limits_{N_{\mathfrak{k}} \diagdown G_A} \psi_0(g)\, \nu(\mu(g))\, dg \tag{8}$$

because ν is trivial on $\mathfrak{k}^*$ and hence $\nu(\mu(\gamma)) = 1$ for $\gamma \in G_{\mathfrak{k}}$.

Applying $G = NH$ we get that $N_{\mathfrak{k}} \diagdown G_A = N_{\mathfrak{k}} \diagdown N_A H_A$. By making use of $\mu(N) = 1$ and of *1)* and *2)* we obtain

$$\int\limits_{N_{\mathfrak{k}} \diagdown G_A} \psi_0(g)\, \nu(\mu(g))\, dg = \int\limits_{H_A} f_{\chi_0}(h)\, \nu(\mu(h))\, \theta(h)\, dh \tag{9}$$

where

$$f_{\chi_0}(g) = \int\limits_{Z_{\mathfrak{k}} \diagdown Z_A} \chi_0(z) f(zg)\, dz = \int\limits_{N_{\mathfrak{k}} \diagdown N_A} \psi_0(ng)\, dn. \tag{10}$$

Remark. It is possible to prove a similar theorem under more general assumptions. For instance, it is not necessary to require the simple transitivity in *5)*.

Before considering some examples of applications of our theorem we shall discuss the problem of the analytic continuation of the integral (2). We have

$\nu(x)=\nu_0(x)|x|^s$, $x\in I_{\mathfrak{k}}$. Let s be a complex variable and denote by $\mathscr{I}(s)=\mathscr{I}(f, G, s)$ the integral defined in the left-hand side of (2). It is clear that $\mathscr{I}(s)$ is an analytic function in the half-plane $\operatorname{Re} s>c$ for sufficiently large c.

Of course, there does not necessarily exist an analytic continuation of $\mathscr{I}(s)$ to the whole complex plane as an entire function or as a function with a finite number of poles. But it is plausible that there exists a meromorphic continuation to the whole complex domain. From now on we assume that $G\cap C=\{1\}$, where C is the center of $\mathscr{G}$ and $G^0=\operatorname{Ker}\mu$ has no nontrivial algebraic character.

Now we describe two methods of analytic continuation. I shall call the first one the Hecke method and the second one the Rankin—Selberg method.

Theorem 2. *(The Hecke method.) If there exists an automorphism* $\sigma: g\to g^\sigma$ *of* $\mathscr{G}$ *such that*

(i) $\sigma(G)=G$,

(ii) $\mu(g^\sigma)=\mu^{-1}(g)$ *for every* $g\in G$,

then for any $f\in A(\mathscr{G})$ *the integral* $\mathscr{I}(f, G, s)$ *is an entire function.*

Proof. First we shall prove that if $f(g)\in A(\mathscr{G})$ then $\tilde{f}(g)=f(g^\sigma)\in A(\mathscr{G})$. It is clear that *1)* if $f(\gamma g)=f(g)$ for every $\gamma\in\mathscr{G}_{\mathfrak{k}}$ then

$$\tilde{f}(\gamma g)=f(\gamma^\sigma g^\sigma)=f(g^\sigma)=\tilde{f}(g);$$

2) if R is a cuspidal subgroup then $\sigma(R)$ is also a cuspidal subgroup; *3)* if the space generated by the right translates of $f(g)$ by $\mathscr{G}_A$ is irreducible and admissible then the same is true for $f(g)$. Hence $\tilde{f}(g)\in A(\mathscr{G})$.

To prove our theorem we use the usual trick. Denote by G^0, G^1, G^{-1} the subsets in G_A where $|\mu(g)|=1$. $|\mu(g)|>1$, $|\mu(g)|<1$ respectively. We have

$$\mathscr{I}(s)=\int_{G_{\mathfrak{k}}\backslash G^{-1}}+\int_{G_{\mathfrak{k}}\backslash G^{1}}+\int_{G_{\mathfrak{k}}\backslash G^{0}}. \tag{11}$$

By making use of the fact that under our assumptions $f(g)$ rapidly decreases at ∞, we obtain that the first and the second terms are holomorphic in the whole complex plane. The substitution $g\to g^\sigma$ shows that the third term is also holomorphic.

Theorem 3. *(The Rankin—Selberg method.) Let* $\tilde{G}$ *be a reductive group over* $\mathfrak{k}$ *and* G *its parabolic subgroup. For any* $f\in S(\tilde{G})$ *the integral* $\mathscr{I}(f, G, s)$ *can be represented in the form*

$$\mathscr{I}(f, G, s)=\int_{\tilde{G}_{\mathfrak{k}}\backslash\tilde{G}_A} f_0(g)\,E(g, s)\,dg \tag{12}$$

where $f_0(g)$ *is contained in the space generated by the right translates of* $f(g)$ *by the elements of a maximal compact subgroup* $U\subset\tilde{G}_A$; $E(g, s)$ *is an Eisenstein series.*

Proof. Let us fix a maximal compact subgroup $U \subset \tilde{G}_A$. We introduce

$$f_0(g) = \int_W f(gw)\, v_0(\mu(w)), \quad W = U \cap G_A \tag{14}$$

where the measure dw on W is chosen so that

$$\int_W dw = 1.$$

It is clear that

$$\mathscr{I}(f, G, s) = \mathscr{I}(f_0, G, s) \tag{15}$$

and

$$f_0(gw) = v_0^{-1}(\mu(w)) f_0(g) \quad \text{for} \quad w \in W. \tag{16}$$

It is obvious that the space L_0 generated by the right translates $f_0(gu)$, $u \in U$ is finite-dimensional. Let us consider a basis

$$f_1(g), f_2(g), \dots, f_n(g) \quad \text{of} \quad L_0$$

such that

$$f_1(g) = f_0(g),\ f_k(g) = f_0(g^{u_k}) \quad \text{for some} \quad u_k \in U, \quad 2 \leqq k \leqq n. \tag{17}$$

Denote by $F(g)$ the row vector the components of which are taken from (17) (in the natural order).

We have

$$F(gu) = F(g) A(u), \quad g \in \tilde{G}_A, \quad u \in U \tag{18}$$

where $A(u)$ is a matrix function on U. From (16), (18) and the definition of F it follows that

$$a_{11}(w) = v_0^{-1}(\mu(w)),\ a_{1i}(w) = 0, \quad 2 \leqq i \leqq k. \tag{19}$$

We introduce the column-vector

$$\Phi(g) = \begin{pmatrix} v(\mu(g)) \\ 0 \\ \vdots \\ 0 \end{pmatrix}, \quad g \in G_A. \tag{20}$$

For $\tilde{g} \in \tilde{G}_A$ we define

$$\Phi(\tilde{g}) = \Phi(gu) = A^{-1}(u)\Phi(g), \quad g \in G_A, \quad u \in U. \tag{21}$$

(19) shows that $\Phi(\tilde{g})$ is correctly defined. From (18) and (21) it follows that $F(g)\Phi(g)$ is stable under the action of U by right translations. We note that for almost all g, one has $g^{-1}Ug \cap \tilde{G}_{\mathfrak{t}} = U \cap C_{\mathfrak{t}}$, where C is the center of $\tilde{G}$. From this we get

$$\int_{C_A G_{\mathfrak{t}} \setminus \tilde{G}_A} \psi(g)\, dg = c \int_{C_A G_{\mathfrak{t}} \setminus G_A} \psi(g)\, dg \tag{22}$$

where c is a scalar and $\psi(g)$ is any function on $C_A G_{\mathfrak{t}} \setminus \tilde{G}_A$ satisfying the following assumptions: *1)* $\psi(g)$ is stable under the right action of U, *2)* the integral on the right-hand side of (22) converges absolutely.

Applying (22) to $\psi(g)=F(g)\Phi(g)$ we get

$$\int\limits_{C_A G_{\mathfrak{k}}\backslash G_A} F(g)\Phi(g)\,dg = \int\limits_{C_A G_{\mathfrak{k}}\backslash G_A} F(g)\Phi(g)\,dg = \int\limits_{C_A G_{\mathfrak{k}}\backslash G_A} f_0(g)\nu(\mu(g))\,dg. \quad (23)$$

Using the definition of $F(g)$ we obtain a function $\theta(g,s)$ on $\tilde{G}_A$ such that

$$\int\limits_{C_A G_{\mathfrak{k}}/\tilde{G}_A} F(g)\Phi(g)\,dg = \int\limits_{C_A G_{\mathfrak{k}}\backslash \tilde{G}_A} f_0(g)\theta(g,s)\,dg \qquad (24)$$

and $\theta(g,s)$ is a holomorphic function of s stable under the left action of $G_{\mathfrak{k}}$.

We introduce the Eisenstein series

$$E(g,s) = \sum \theta(\gamma g, s). \qquad (25)$$

We have

$$\int\limits_{C_A G_{\mathfrak{k}}\backslash \tilde{G}_A} f_0(g)\theta(g,s)\,dg = \int\limits_{C_A \tilde{G}_{\mathfrak{k}}\backslash \tilde{G}_A} f_0(g)E(g,s)\,dg. \qquad (26)$$

Finally, combining (15), (23), (24) and (26) we get (12).

Now we are going to give some examples.

1) Let K be a two-dimensional algebra over $\mathfrak{k}$ which is isomorphic to $\mathfrak{k}\oplus\mathfrak{k}$ or a quadratic extension of $\mathfrak{k}$. Denote by i an imbedding: $\mathfrak{k}\to K$ which is the diagonal imbedding if $K=\mathfrak{k}\oplus\mathfrak{k}$ and the natural imbedding if K is a quadratic extension of $\mathfrak{k}$. Denote by σ the nontrivial automorphism of K which is trivial on $\mathfrak{k}$. The imbedding $i:\mathfrak{k}\to K$ defines an imbedding: $GL(2,\mathfrak{k})\to GL(2,K)$. We put $\mathcal{G}=GL(2,K)$, and set

$$H=\left\{\begin{pmatrix}\alpha & 0\\ 0 & 1\end{pmatrix},\ \alpha\in\mathfrak{k}\right\}$$

$$N=\left\{\begin{pmatrix}1 & x\\ 0 & 1\end{pmatrix},\ x\in\mathfrak{k}\right\}$$

$$S=\left\{\begin{pmatrix}1 & x\\ 0 & 1\end{pmatrix},\ x\in K,\ \sigma(x)=-x\right\}.$$

Furthermore, let $G=NH$, $P=GS$ and

$$R=\left\{\begin{pmatrix}1 & x\\ 0 & 1\end{pmatrix},\ x\in K\right\},\quad \chi_0(x)=\nu\ (\operatorname{tr}\alpha x)$$

where $\alpha\in K$, $\alpha^\sigma=-\alpha$ and ν is a nontrivial character on $\mathfrak{k}\backslash A_{\mathfrak{k}}$.

It is easy to show that assumptions (1—8) are satisfied for these subgroups.

2) Let $\mathfrak{A}$ be a division algebra over $\mathfrak{k}$. Denote by $n(x)$, $x\in\mathfrak{A}$, the norm on $\mathfrak{A}$. Let $\mathcal{G}=GL(2,\mathfrak{A})$.

Put

$$H=\left\{\begin{pmatrix}\alpha & 0\\ 0 & e\end{pmatrix},\ \alpha\in\mathfrak{A}^*\right\}$$

$$N=\left\{\begin{pmatrix}\beta & 0\\ 0 & \beta\end{pmatrix},\ \beta\in\mathfrak{A}^*,\ n(\beta)=1\right\}$$

$$S=\left\{\begin{pmatrix}1 & x\\ 0 & 1\end{pmatrix},\ x\in\mathfrak{A}\right\}.$$

Now, let $G=NH$, $P=GS$, $R=S$ and $\chi_0=\nu(\mathrm{tr}x)$ where ν is a nontrivial character of $\mathfrak{k}\setminus A_{\mathfrak{k}}$. It is not too difficult to verify that the assumptions (1—8) are satisfied.

3) We shall give an example in which no intermediate reductive subgroup exists, in other words there is no reductive subgroup $\tilde{G}$ such that $\mathscr{G}\supset\tilde{G}\supset G$.

Let $\mathscr{G}=GL(2,\mathfrak{k})\times\ldots\times GL(2,\mathfrak{k})$ (n times), and let

$$H=\left\{\begin{pmatrix}\alpha & 0\\ 0 & 1\end{pmatrix},\ldots\begin{pmatrix}\alpha & 0\\ 0 & 1\end{pmatrix}\right\}$$

$$N=\left\{\begin{bmatrix}1 & x_1\\ 0 & 1\end{bmatrix},\ldots\begin{pmatrix}1 & x_n\\ 0 & 1\end{pmatrix},\ \sum_{i=1}^{n}x_i=0\right\}$$

$$S=\left\{\begin{pmatrix}1 & x\\ 0 & 1\end{pmatrix},\ldots\begin{pmatrix}1 & x\\ 0 & 1\end{pmatrix}\right\}.$$

We put $\chi_0(x)=\nu(x)$ for $x\in S$ (ν is a nontrivial character of $\mathfrak{k}\setminus A_{\mathfrak{k}}$), and $R=NS$. It is obvious that if $n>2$, then no intermediate reductive group exists.

4) Let K be a quadratic extension of $\mathfrak{k}$. Denote by $\mathrm{End}_{\mathfrak{k}}\,K$ the ring of all $\mathfrak{k}$-linear endomorphisms of K which is considered as a two-dimensional space over $\mathfrak{k}$. Denote by $r(x)$ the result of the application $r\in\mathrm{End}_{\mathfrak{k}}\,K$ to $x\in K$. Denote by $\mathrm{tr}\,(r)$ and $n(r)$ the trace and the determinant of r, respectively. By means of the natural embedding $K\hookrightarrow\mathrm{End}_{\mathfrak{k}}\,K$ one can define $\mathrm{tr}\,(x)$ and $n(x)$ for $x\in K$. It is obvious that $\mathrm{tr}\,x=x+x^{\sigma}$ and $n(x)=xx^{\sigma}$, where σ is a nontrivial automorphism of K, coinciding with the identity on $\mathfrak{k}$.

Let $V=K^2$ and let $\alpha(v_1, v_2)$ be a nontrivial skew-form on V with values in K. Set $\varrho(v_1, v_2)=\mathrm{tr}\,\alpha(v_1, v_2)$. It is obvious that ϱ is a nondegenerate skew-form with values in $\mathfrak{k}$. Each element $g\in\mathrm{End}_{\mathfrak{k}}\,V$ can be written in the form

$$g=\begin{pmatrix}g_{11} & g_{12}\\ g_{21} & g_{22}\end{pmatrix}\tag{27}$$

where $g_{ij}\in\mathrm{End}_{\mathfrak{k}}\,K$.

In the following we shall assume that $\mathscr{G}$ is a factor group of the group $\tilde{\mathscr{G}}$ consisting of all $g\in\mathrm{End}_{\mathfrak{k}}\,V$ for which

$$\varrho(gx, gy)=\alpha\varrho(x, y),\quad \alpha\in\mathfrak{k}^*,\quad x, y\in\mathfrak{k}^4.\tag{28}$$

holds.

Now we introduce the subgroup R consisting of all $g\in\mathscr{G}$ having the form

$$g=\begin{pmatrix}e & r\\ 0 & e\end{pmatrix},\quad r\in\mathrm{End}_{\mathfrak{k}}\,K.\tag{29}$$

It is easy to verify that g of form (29) belongs to $\mathscr{G}$ if and only if $r\in\mathrm{End}_{\mathfrak{k}}\,K$ is symmetric, i.e.

$$\mathrm{tr}\,(r(x)y)=\mathrm{tr}\,(xr(y))\tag{30}$$

for all $x, y\in K$. Denote by $\mathrm{End}_{\mathfrak{k}}^{S}\,K$ the set of all $r\in\mathrm{End}_{\mathfrak{k}}\,K$ satisfying (30).

It is easy to verify that the image of K under the natural imbedding $K \hookrightarrow \mathrm{End}_{\mathfrak{k}} K$ belongs to $\mathrm{End}_{\mathfrak{k}}^S K$. Set

$$\tilde{H} = \left\{ \begin{pmatrix} \alpha & 0 \\ 0 & e \end{pmatrix}, \ \alpha \in \mathfrak{k}^* \right\}$$

$$\tilde{S} = \left\{ \begin{pmatrix} e & x \\ 0 & e \end{pmatrix}, \ x \in \mathrm{End}_{\mathfrak{k}}^S K, \ \mathrm{tr}\ xa = 0 \ \text{for all} \ a \in K \right\}$$

$$\tilde{N} = \left\{ \begin{pmatrix} \delta & x \\ 0 & \delta^\sigma \end{pmatrix}, \ \sigma \in K^*, \ x \in K \right\}.$$

We define H, S, N as the images of $\tilde{H}$, $\tilde{S}$, $\tilde{N}$ in $\mathscr{G}$ under the natural homomorphism: $\tilde{\mathscr{G}} \to \mathscr{G}$.

We put $\chi_0(x) = \nu(\mathrm{tr}\,(\hat{\sigma}x))$, where $\hat{\sigma} \in \mathrm{End}_{\mathfrak{k}} K$ is defined by the relation $\hat{\sigma}(x) = x^\sigma$ for all $x \in K$, and ν is a nontrivial character. It is not difficult to show that all assumptions (1—6) are satisfied.

§ 3. Computation of the local factors for a symplectic group of rank two

The idea of the proof of the key lemma (Lemma 1) is taken from Andrianov's paper [9].

Let $\mathfrak{k}$ be a local non-Archimedean field; denote by $\mathcal{O}$ the ring of integers, by $\mathfrak{p}$ the prime ideal of $\mathcal{O}$, by ω a generator of $\mathfrak{p}$ and by v the discrete valuation of $\mathfrak{k}$ such that $v(\omega) = 1$. We assume that the characteristic of the residue field of $\mathfrak{k}$ is not equal to 2.

We shall introduce the following skew symmetric matrix

$$\mathscr{I} = \begin{pmatrix} 0 & e \\ -e & 0 \end{pmatrix}, \quad e = \begin{pmatrix} 1 & 0 \\ 0 & 1 \end{pmatrix}. \tag{1}$$

Denote by $\mathscr{G}$ the set of all $g \in M(4, \mathfrak{k})$ such that

$$g' \mathscr{I} g = \alpha \mathscr{I}, \quad \alpha \in \mathfrak{k}^*. \tag{2}$$

Here and in the following g' means the transpose of g.

Let T denote the subgroup consisting of all $g \in \mathscr{G}$ which are of the form

$$\begin{pmatrix} e & t \\ 0 & e \end{pmatrix}, \quad t' = t. \tag{3}$$

The characters of T can be written in form

$$\chi(t) = \nu\ (\mathrm{tr}\,(\beta t)), \quad \beta \in M(2, \mathfrak{k}), \ \beta' = \beta \tag{4}$$

where ν is the standard character on $\mathfrak{k}^+$.

The normalizer $N(T)$ of T equals PT, where $P \cong GL(2, \mathfrak{k}) \times \mathfrak{k}^*$ consists of the following matrices

$$P = \begin{pmatrix} r & 0 \\ 0 & \alpha r'^{-1} \end{pmatrix} \quad r \in GL(2, \mathfrak{k}), \ \alpha \in \mathfrak{k}^*. \tag{5}$$

Let $\chi(ptp^{-1})=v(\text{tr}(\beta_1 t))$ where x is of form (4). Then

$$\beta_1 = \alpha^{-1}r'\beta r. \tag{6}$$

We call the character (4) nondegenerate if $\det \beta \neq 0$. Let χ be a nondegenerate character of T. Denote by P^χ the stabilizer of χ in P. Let $Z^\chi = P^\chi T$. It is clear that χ can be extended to the character of Z^χ such that $\chi(p)=1$ for each $p \in P$. It is not difficult to prove that $P \cong \hat{K}$, where K is a semisimple two-dimensional algebra over $\mathfrak{k}$, σ its canonical automorphism and $\hat{K}$ is the semidirect product of K^* and $\{1, \sigma\}$.

The following types of K can occur: $K=\mathfrak{k}\oplus\mathfrak{k}$ or $K=\mathfrak{k}(\sqrt{\delta})$, where $\sqrt{\delta}\notin\mathfrak{k}$.

To a representation of K one can attach a character $\chi(t)=v(\text{tr}(\hat{\sigma}t))$ where $\hat{\sigma}$ satisfies the relation $\hat{\sigma}^{-1}a'\hat{\sigma}=a^\sigma$. It is easy to check that P^χ contains a subgroup of index two

$$\begin{pmatrix} a & 0 \\ 0 & \alpha a'^{-1} \end{pmatrix}, \quad a \in K, \ \alpha = \det a. \tag{7}$$

Denote by W_χ the set of all functions $f(g)$ on $\mathscr{G}$ such that

$$f(zg) = \chi(z)f(g) \tag{8}$$

for every $z \in Z^\chi$.

Let π be an irreducible representation of $\mathscr{G}$ containing a vector stable under the action of a maximal compact subgroup of $\mathscr{G}$. It is well known that such a vector is unique up to a scalar factor. Denote this vector by ξ_0.

Set

$$V_n = \{M \in M(4, \mathcal{O}) \mid M'\mathscr{I}M = \alpha\mathscr{I}, v(\alpha) = n\}. \tag{9}$$

It is well known that

$$V_n = \sum_{i=1}^{r} M_i \mathscr{G}_{\mathcal{O}}.$$

It follows that $\sum_{i=1}^{r} T_{M_i}\xi_0$ is $\mathscr{G}_{\mathcal{O}}$-stable and hence

$$\sum_{i=1}^{r} T_{M_i}\xi_0 = \lambda_n(\pi)\xi_0 \tag{11}$$

where $\lambda_n(\pi) \in \mathbf{C}$.

For any irreducible admissible representation π of $\mathscr{G}$ denote by $W_\chi(\pi)$ some subrepresentation of W_χ which is isomorphic to π. Assume that π contains the vector ξ_0 defined above. Denote by $f_0(g)$ the corresponding function in $W_\chi(\pi)$.

Collecting together all we know about $f_0(g)$ we have

$$f_0(zg) = \chi(z)f_0(g), \quad z \in Z^\chi \tag{12}$$

$$f_0(gu) = f_0(g), \quad u \in \mathscr{G}_{\mathcal{O}} \tag{13}$$

$$\sum_{i=1}^{r} f_0(gM_i) = \lambda_n(\omega)f_0(g). \tag{14}$$

Applying the Iwasawa decomposition we see that M_i can be taken to the form

$$\begin{pmatrix} a & b \\ 0 & d \end{pmatrix}, \quad a, b, d \in M(2, \mathcal{O}), \; ab' = ba', \; ad' = \omega^n e. \tag{15}$$

Denote by $V_n(a)$ the collection of all $M \subset V_n$ which are of the form (15) with a given a.

Denote by A_n the set of all $a \in M(2, \mathcal{O})$ such that $\omega^n a^{-1} \in M(2, \mathcal{O})$. We shall call $a_1, a_2 \in A_n$ *equivalent* if $a_1 a_2^{-1} \in GL(2, \mathcal{O}_p)$. Denote by $B(a)$ the set of all $b \in M(2, \mathcal{O})$ such that $a^{-1}b = b'a'^{-1}$. We shall call $b_1, b_2 \in B(a)$ *congruent* if $a^{-1}(b_1 - b_2) \in M(2, \mathcal{O})$.

We put

$$\psi_\chi(a) = \sum \chi(\omega^{-n} ba') \tag{16}$$

where b runs over all incongruent elements of $B(a)$. (14) and (15) show that

$$\sum f_0 \begin{pmatrix} a & 0 \\ 0 & \omega^n a'^{-1} \end{pmatrix} \psi_\chi(a) = \lambda_n(\pi) f_0(E) \tag{17}$$

where a runs over all inequivalent elements of A_n.

Lemma 1. If $\psi_\chi(a) \neq 0$, $a \in A_n$, then a is equivalent to $\tilde{a} \in K$. More precisely, let $\chi = v(\operatorname{tr} \hat{\sigma} t)$, where v is the standard character on $\mathfrak{k}^*$.

We have *1)* if $K = \mathfrak{k} \oplus \mathfrak{k}$

$$K = \left\{ \begin{pmatrix} x_1 & 0 \\ 0 & x_2 \end{pmatrix}, \; x_i \in \mathfrak{k} \right\}, \; \hat{\sigma} = \begin{pmatrix} 0 & 1 \\ 1 & 0 \end{pmatrix}, \tag{18}$$

then

$$\tilde{a} = \begin{pmatrix} \omega^k & 0 \\ 0 & \omega^r \end{pmatrix}, \; \max(k, r) = n, \; \psi_\chi(a) = q^{n + 2\min(k, r)}, \tag{19}$$

2) if $\mathfrak{k}(\sqrt{\delta}) = K$ is a quadratic extension of k,

$$K = \left\{ \begin{pmatrix} x_0 & x_1 \\ x_1 \delta & x_0 \end{pmatrix}, \; x_0, x_1 \in \mathfrak{k} \right\}, \; \hat{\sigma} = \begin{pmatrix} \delta & 0 \\ 0 & -1 \end{pmatrix}, \tag{20}$$

then

$$\hat{a} = \begin{cases} \omega^n e, & v(\delta) = 0 \\ \omega^n e \quad \text{or} \quad \begin{pmatrix} 0 & \omega^{n-1} \\ \omega^{n-1}\delta & 0 \end{pmatrix}, & v(\delta) = 1, \end{cases} \tag{21}$$

$$\psi_\chi(\tilde{a}) = \begin{cases} q^{3n}, & v(\delta) = 0 \\ q^{3n} \quad \text{or} \quad q^{3n-2}, & v(\delta) = 1. \end{cases} \tag{22}$$

Proof. It is clear that if $\psi_\chi(a) \neq 0$, then $\chi(\omega^{-n} ba') = 1$ for all $b \in B(a)$, i.e.

$$\operatorname{tr}(\omega^{-n} \hat{\sigma} ba') \in \mathcal{O}. \tag{23}$$

Each $a \in A_n$ is equivalent to $\bar{a}$ which is of the form

$$\bar{a} = \begin{pmatrix} \omega^k & c \\ 0 & \omega^r \end{pmatrix}. \tag{24}$$

It is easy to show that $\bar{a} \in A_n$ if and only if

$$n \geqq k \geqq 0, n \geqq r \geqq 0, k > v(c) \geqq \max(0, k+r-n) \quad \text{or} \quad c = 0. \tag{25}$$

Moreover $b=(b_{ij}) \in B(\bar{a})$ if and only if

$$b_{ij} \in \mathcal{O} \tag{26}$$

$$b_{12}\omega^r = b_{21}\omega^k + b_{22}c \tag{27}$$

First we consider the case $K=\mathfrak{k}\oplus\mathfrak{k}$. We have

$$\operatorname{tr}(\omega^{-n}\hat{\sigma}ba') = 2b_{12}\omega^{r-n}. \tag{28}$$

(27), (23) and (28) show that $\max(k, r)=n$.

If $r=n$ it follows from (25) that $c=0$ and hence $a \in K$. If $r<n$, then $k=n$. (23), (27) and (28) imply that $v(c) \geqq 0$ and hence, according to (25), $c=0$. This means $\bar{a} \in K$. It is clear that $\psi_\chi(a)$ equals the number of incongruent elements in $B(a)$. We get $\psi_\chi(\bar{a})=q^{n+2\min(k,r)}$.

Now we consider the case $K=\mathfrak{k}(\sqrt{\delta})$ and $\sqrt{\delta} \notin \mathfrak{k}$. Suppose that the representation K is of the form (20).

It follows from (23) that

$$\delta(b_{11}\omega^k + b_{12}c) - b_{22}\omega^r \in \omega^n\mathcal{O}. \tag{29}$$

First we take b which is of the form

$$\begin{pmatrix} b_{11} & 0 \\ 0 & 0 \end{pmatrix}. \tag{30}$$

From (29) it follows that $k+v(\delta) \geqq n$. If $v(\delta)=0$ then $k=n$. Now we take

$$b = \begin{pmatrix} 0 & b_{12} \\ 0 & b_{22} \end{pmatrix}. \tag{31}$$

(25) shows that $v(c) \geqq r$ and hence (27) gives

$$b_{12} = c\omega^{-r}b_{22}. \tag{32}$$

(32) and (29) show that

$$v(\delta c^2 - 1) \geqq n-r. \tag{33}$$

Using the fact that $\sqrt{\delta} \notin \mathfrak{k}$, we obtain from (33) that $n=r$. Thus it follows from (25) that $c=0$, $\bar{a}=\omega^n e$. Now we consider the case $v(\delta)=1$. Suppose first that $k=n$. As in the previous case, it follows from (25) that $v(c) \geqq r$. If b is of the form (31) then (33) holds again and shows that $n=r$. Suppose now that $k=n-1$. Let

$$b = \begin{pmatrix} 0 & b_{12} \\ b_{21} & 0 \end{pmatrix}. \tag{34}$$

The substitution of b into (27) gives

$$b_{12}\omega^r = b_{21}\omega^{n-1} \tag{35}$$

We want to show that $r=n$ again. Suppose that $r<n$, then $v(b_{12})\geqq n-1-r$ and there exists $\hat{b}\in B(a)$ of the form (34) for which $v(b_{12})=n-1-r$.

Substituting $\hat{b}$ into (29) shows $v(c)\geqq r$. For b of form (31) it follows from (27) that

$$b_{12} = b_{22}c\omega^{-r}.$$

If we substitute b of form (31) into (29) we obtain

$$v(\delta c^2\omega^{-2r}-1) \geqq n-r,$$

and this is a contradiction since $v(\delta c^2\omega^{-2r})>0$. So the equality $r=n$ is proved. Application of (25) shows that $c=0$. Finally, we obtain

$$\bar{a} = \begin{pmatrix}\omega^{n-1} & 0\\ 0 & \omega^n\end{pmatrix} = \begin{pmatrix}0 & \omega^{n-1}\\ \omega^{n-1}\delta & 0\end{pmatrix}u, \quad u\in GL(2, \mathcal{O}).$$

$\psi_\chi(a)$ can be computed in the same way. Hence our lemma is proved.

We define a formal power series as

$$\Phi(t) = \sum_{n=0}^{\infty} f_0\begin{pmatrix}\omega^n e & 0\\ 0 & e\end{pmatrix}t^n$$

$$\Lambda(t) = \sum_{n=0}^{\infty} \lambda_n(\pi)\,t^n.$$

Lemma 2.

$$\Phi(tq^3) = f_0(E)\,\varrho_K(t)\,\Lambda(t), \tag{36}$$

where

$$\varrho_K(t) = \begin{cases} \dfrac{1-qt}{1+qt} & \text{if } K=\mathfrak{k}\oplus\mathfrak{k},\\ (1+qt)^{-1} & \text{if } K \text{ is a ramified extension of } \mathfrak{k},\\ 1 & \text{if } K \text{ is an unramified extension of } \mathfrak{k}.\end{cases}$$

Proof. First we consider the case $K=\mathfrak{k}\oplus\mathfrak{k}$. Formula (17) and Lemma 1 imply

$$\sum_{r=0}^{n-1}\left(f_0\begin{pmatrix}a_r & 0\\ 0 & \omega^n a_r^{-1}\end{pmatrix}+f_0\begin{pmatrix}b_r & 0\\ 0 & \omega^n b_r^{-1}\end{pmatrix}\right)q^{n+2r}+f_0\begin{pmatrix}a_n & 0\\ 0 & \omega^n a_n^{-1}\end{pmatrix}q^{3n} = \lambda_n(\pi)f_0(E) \tag{37}$$

where

$$a_r = \begin{pmatrix}\omega^r & 0\\ 0 & \omega^n\end{pmatrix},\; b_r = \begin{pmatrix}\omega^n & 0\\ 0 & \omega^r\end{pmatrix}. \tag{38}$$

Moreover

$$\begin{pmatrix}a_r & 0\\ 0 & \omega^n a_r^{-1}\end{pmatrix}\begin{pmatrix}c_r & 0\\ 0 & c_r^\sigma\end{pmatrix} = \begin{pmatrix}\omega^n e & 0\\ 0 & e\end{pmatrix} \tag{39}$$

where

$$c_r = \begin{pmatrix}1 & 0\\ 0 & \omega^{r-n}\end{pmatrix}\in K$$

and hence

$$\begin{pmatrix} c_r & 0 \\ 0 & c_r^\sigma \end{pmatrix} \in P^\varkappa$$

(P^χ is the stabilizer of χ_σ).

It shows that

$$f_0\begin{pmatrix} a_r & 0 \\ 0 & \omega^n a_r^{-1} \end{pmatrix} = f_0\begin{pmatrix} \omega^n e & 0 \\ 0 & e \end{pmatrix}. \tag{40}$$

Analogously, we have

$$f_0\begin{pmatrix} b_r & 0 \\ 0 & \omega^n b_r^{-1} \end{pmatrix} = f_0\begin{pmatrix} \omega^r e & 0 \\ 0 & e \end{pmatrix}. \tag{41}$$

A substitution of (40), (41) in (37) gives

$$2\sum_{r=0}^{n-1} f_0\begin{pmatrix} \omega^r e & 0 \\ 0 & e \end{pmatrix} q^{n+2r} + f_0\begin{pmatrix} \omega^n e & 0 \\ 0 & e \end{pmatrix} q^{3n} = \lambda_n(\pi) f_0(E). \tag{42}$$

Analogously, we can prove that

$$f_0\begin{pmatrix} \omega^{n-1} e & 0 \\ 0 & e \end{pmatrix} q^{3n-2} + f_0\begin{pmatrix} \omega^n e & 0 \\ 0 & e \end{pmatrix} q^{3n} = \lambda_n(\pi) f_0(E) \tag{43}$$

if K is a ramified extension of $\mathfrak{k}$,

$$f_0\begin{pmatrix} \omega^n e & 0 \\ 0 & e \end{pmatrix} q^{3n} = f_0(E)\, \lambda_n(\pi) \tag{44}$$

if K is an unramified extension of $\mathfrak{k}$.

By formal computation the statement of our lemma can be easily deduced from (42), (43) and (44).

Denote by H the following subgroup

$$H = \left\{ \begin{pmatrix} \alpha & 0 \\ 0 & 1 \end{pmatrix},\ \alpha \in \mathfrak{k} \right\}. \tag{45}$$

Lemma 3.

$$\int_{H_\mathfrak{p}} f_0(h) |h|^s \, d^* h = f_0(E)\, \varrho_K(q^{-s-3})\, \Lambda(q^{-s-3})\, \mu(H_{E_\mathfrak{p}}) \tag{46}$$

where $E_\mathfrak{p}$ is a group of units of $\mathfrak{k}$.

Proof. $f_0(h)$ is invariant under the action of $H_{E_\mathfrak{p}}$. Thus it is enough to find $f_0(h_n)$, where

$$h_n = \begin{pmatrix} \omega^n e & 0 \\ 0 & e \end{pmatrix}. \tag{47}$$

By making use of (12) it is not too difficult to prove that $f_0(h_n)=0$ for $n<0$. Combining this fact with the result of Lemma 4 we get the desired equality.

Remark. Lemma 2 shows that the restriction of $f_0(g)$ to H is uniquely defined by the representation π. Of course it does not guarantee the uniqueness of the space $W_\chi(\pi)$. Perhaps, it can be proved by using the same method or that of paper [2].

REFERENCES

[1] Jacquet, H., Langlands, R. P., *Automorphic Forms on* $GL(2)$, Springer Lecture Notes, Vol. **114**, 1970.

[2] Gelfand, I. M. and Kajdan, D. A., Representations of the group $GL(n, K)$ where K is a local field, *this volume*.

[3] Gelfand, I. M., Graev, M. I. and Pjateckij-Šapiro, I. I. (Гельфанд И. М., Граев М. И., Пятецкий-Шапиро И. И.), *Теория представлений и автоморфные функции, М.* „Наука", 1966. (English translation: *Representation theory and automorphic functions*, W. B. Saunders Company, Phil.—London—Toronto, 1969).

[4] Rankin, R. A., Contributions to the theory of Ramanujan's function $\tau(n)$ and similar arithmetical functions, *Proc. Cambridge Philos. Soc.*, **35** (1939), 357—372.

[5] Selberg, A., Bemerkungen über eine Dirichletsche Reihe, die mit der Theorie der Modulformen nahe verbunden ist, *Arch. Math. Naturvid.*, **43** (1940), 47—50.

[6] Selberg, A., On the estimation of Fourier coefficients of modular forms, *Proc. Symp. in Pure Math.*, Vol. **VIII**, (1965).

[7] Ogg, A. P., On a convolution of L-series, *Invent. math.*, **7** (1969), 297—312.

[8] Langlands, R. P., Eisenstein series, *Proc. Symp. in Pure Math.*, Vol. **IX**, 1966.

[9] Andrianov, A. N., Zeta functions and Siegel modular forms, (*this volume*).

[10] Borel, A., *Introduction aux groupes arithmétiques*, Hermann, Paris, 1969.

[11] Harder, G., Minkowskische Relationstheorie über Funktionenkörpern, *Inventiones math.*, **7** (1969), 33—54.

[12] Gelfand, S. I. (Гельфанд С. И.), Представления полной линейной группы над конечным полем, *Мат. сб.*, **83** (1970), 15—41.
(Representations of the general linear group over a finite field.)

ON THE CHARACTERS OF CERTAIN FINITE GROUPS

T. A. SPRINGER

Introduction

The groups to be considered here are the finite analogues of the semisimple Lie groups, that is the Chevalley groups and their twisted analogues. A uniform description of these finite groups has been given by Steinberg in [15]; they can all be realized as the group of fixed points G_σ of a suitable endomorphism σ of a semisimple group G, defined over the algebraic closure of a finite field.

Section 1 reviews known material about the groups G_σ. Section 2 discusses facts about parabolic subgroups of reductive linear algebraic groups. Most of these facts (but not all of them) are contained in the standard reference [3]. In section 3 we mention some results about characters of finite groups. This section also contains a version of Brauer's characterization of generalized characters, for the case of the groups G_σ. Sections 4 and 5 deal with the notion of cusp form and its application to the groups G_σ. The main results are due to Harish-Chandra, their applications to character theory were discussed (in a slightly less general setting) in [11]. The present treatment is somewhat simpler than that given in [11]. Section 6 contains material about tori and about regular elements and regular characters of the finite groups determined by the σ-stable maximal tori of G. As an application we deduce an asymptotic formula for the number of conjugacy classes of G_σ (see 6.19).

In section 7 the results of the preceding sections are applied to prove that, in favorable circumstances, a σ-stable maximal torus T of G and a regular character φ of the corresponding finite group T_σ determine an irreducible character $\chi_{T,\varphi}$ of G_σ, which is a cusp form if T is minisotropic.

The construction uses ideas from the theory of exceptional characters of finite groups. The results are still very incomplete.

I believe that a further study of the character theory of the groups G_σ should begin with a closer investigation of the characters $\chi_{T,\varphi}$ of section 6. In particular, one should try to find the explicit values of these characters. For the values on the regular semisimple elements see the conjecture of [11, p. 117]. A conjecture about the values of $\chi_{T,\varphi}$ on unipotent elements (in certain cases) is made in [13, p. 152]. Another interesting problem is that of constructing, in some explicit way, the irreducible representation with character $\chi_{T,\varphi}$. Perhaps the representation can be realized in a cohomology space of the Tits building of G_σ, for a suitable system of coefficients.

§ 1. The groups

1.1. *Notations and recollections*

Let G be a group. If S is a subset of G, we denote the normalizer and centralizer of S in G by $N_G(S)$ and $Z_G(S)$, respectively. If $x, y \in G$ we write ${}^x y = xyx^{-1}$, $y^x = x^{-1}yx$. If G is a linear algebraic group its identity component is denoted by G^0 and its unipotent radical by $R_u(G)$.

Let G be a connected linear algebraic group over an algebraically closed field. Let T be a maximal torus of G. If H is a closed subgroup of G which is normalized by T we denote by $\Phi(T, H)$ the set of roots of T in H, i.e. the nontrivial rational characters of T which occur in the adjoint representation of T in the Lie algebra of H. The character group X of T is a free Abelian group of rank $r = \dim T$.

Let now G be a connected reductive algebraic group; put $N = N_G(T)$. The Weyl group $W = N/T$ acts on $V = X \otimes \mathbf{R}$ as a reflection group, with root system $\Phi = \Phi(T, G)$. If $a \in \Phi$ denote by U_a the one-parameter unipotent subgroup defined by a. The group U_a is normalized by T.

Fix an order on Φ. Let U be the subgroup of G generated by the U_a with $a > 0$ and let $B = T \cdot U$. Then B is a Borel subgroup of G, with unipotent radical U and all Borel subgroups of G can be obtained in this manner (from a maximal torus T and an order of $\Phi(T, G)$).

Bruhat's lemma asserts that (G, B, N) is a Tits system in the sense of [5, p. 28, Remarque 1].

If G is a connected reductive linear algebraic group we have a decomposition $G = G' \cdot S$, where G' is the derived group of G (which is semisimple), whereas S is a central torus in G such that $G' \cap S$ is finite. The rank of G' is called the semi-simple rank of G.

1.2. *The σ-setup*

Let G be a linear algebraic group. Assume that σ is a surjective endomorphism of G (in the sense of algebraic groups). Denote by G_σ its group of fixed points and assume that G_σ is *finite*. The finite groups to be considered are those obtained in this manner from a connected reductive G.

1.3. *Examples*

(a) Let G be a connected semisimple linear algebraic group defined over a finite field k, let σ be the Frobenius endomorphism of G. Then $G_\sigma = G(k)$, the finite group of k-rational points of G. If G is k-split and adjoint then G_σ is a Chevalley group. The so-called Steinberg groups are obtained from a G which is not k-split.

(b) The groups of Suzuki and Ree can also be obtained as a G_σ (but do not fall under the heading (a)).

1.4. The groups G_σ obtained from a semisimple G are discussed at length in [15], some complementary results are given in [14, Part E].

It is also shown in [15, p. 75, 11.6] that the finite groups of the previous example are the only groups G_σ coming from a quasi-simple group G.

We next recall some results which we need later. The basic fact for the study of the finite groups G_σ is the following generalization of a theorem of Lang.

1.5. *Theorem. Let G be a connected linear algebraic group and σ a surjective endomorphism of G such that G_σ is finite. Then the map $x \to x(\sigma x)^{-1}$ of G into G is surjective.*

The proof is given in [15, § 10].

1.6. *Lemma.* (i) *Let σ be a surjective endomorphism of the linear algebraic group G. Let H be a closed subgroup of G such that $\sigma H \subset H$ and $H \cap \operatorname{Ker} \sigma \subset H^0$. Then $\sigma H = H$.*

(ii) *Let C be the centre of G. Then* $\operatorname{Ker} \sigma \subset C^0$.

This is proved in [14, p. 172, 2.5].

If A is a group and σ an endomorphism of A we denote by $H^1(\sigma, A)$ the set of classes in A for the equivalence relation: $a \sim b$ if $a = cb\sigma(c)^{-1}$ for some $c \in A$.

If S is a set and σ a permutation of A we denote by S_σ the set of fixed points of σ.

1.7. *Proposition. Let G, σ be as in 1.5, let M be a left homogeneous space for G on which σ acts such that*

$$\sigma(g \cdot m) = \sigma(g) \cdot \sigma(m) \quad \textit{for} \quad g \in G,\ m \in M.$$

(i) *M contains a point fixed by σ;*

(ii) *Fix $m_0 \in M_\sigma$, let H be the isotropy subgroup of m_0 in G. Assume that H is a closed subgroup of G. Then the elements of the orbit space $G_\sigma \backslash M_\sigma$ are in one-one correspondence with the elements of $H^1(\sigma, H/H^0)$.*

The proof (which is based on 1.5) is given in [14, p. 174, 2.7]. A direct consequence of 1.7 is the following result of [14, p. 175, 2.9].

1.8. *Proposition. Let G, σ be as in 1.5. Then σ fixes a Borel subgroup B and a maximal torus $T \subset B$. Any two such couples are conjugate by an element of G_σ.*

1.9. Assume now that G is a connected reductive linear algebraic group and that σ is as before. Let T be as in 1.8. Put $N = N_G(T)$, this is a σ-stable subgroup. Then $(G_\sigma, B_\sigma, N_\sigma)$ is a Tits system. This is proved in [15, p. 72, 11.1]. The Weyl group of the Tits system is $W_\sigma = N_\sigma / T_\sigma$ (which acts as a reflection group in a suitable vector space). It should perhaps be remarked that in [15] only the case of a semisimple G is treated. The extension to a reductive group is immediate.

Let Φ be the root system of G with respect to T, let σ^* be the adjoint action on the character group X, given by $\sigma^* a(t)=a(\sigma t)$ $(a \in X, t \in T)$. Since σ fixes T, the U_a's are permuted by σ, hence σ induces a permutation of Φ. An analysis leads to the following result [14, p. 187, 1.4]. We fix a positive-definite inner product on $V=X \otimes \mathbf{R}$ which is invariant under the Weyl group $W=N/T$.

1.10. *Proposition.* (i) *There exists a permutation π of Φ, which keeps positive roots positive and positive integers q_a such that $\sigma^* a=q_a \pi a$ for all $a \in \Phi$. The q_a are powers of the characteristic exponent of the underlying field. Further $\prod q_a>1$ for every π-orbit in Φ;*

(ii) *Assume that Φ is irreducible. Then there exists an isometry τ of V and a positive real number q such that $\sigma^*=q\tau$. If θ is a π-orbit then $\prod_{a \in \theta} q_a=q^{|\theta|}$ (where $|\theta|$ denotes the number of elements of θ). q depends only on σ, not on T. Finally $q>1$.*

1.11. *Examples.* (a) Let G be a quasi-simple linear algebraic group defined over the finite field k, assume that G is k-split. In that case we have $\pi=id$, $\tau=id$, $q=|k|$ (the number of elements of k).

(b) G is quasi-simple of type B_2, in characteristic 2. There is an ordering of Φ such that the set Φ^+ of positive roots consists of 4 elements a, b, $a+b$, $a+2b$. There exists an endomorphism σ of G such that (with the previous notations) $\sigma^* a=2^{n+1} b$, $\sigma^* b=2^n a$, where n is an integer $\geqq 1$. We have $\pi a=b$, $\pi b=a$; $\tau a=\sqrt{2} \cdot b$, $\tau b=(\sqrt{2})^{-1} a$, $q=2^{n+\frac{1}{2}}$. G_σ is a Suzuki-group. For more details see [15, § 11].

1.12. Let G be quasi-simple. Using the notations of 1.9, we shall say that the triple (Φ, X, τ) of a root system in V, a lattice X in V and an isometry τ of V, is the *type* of (G, σ) and that q is the *parameter* of (G, σ).

In sections 6 and 7 we shall be interested in "asymptotic" properties, in the sense that we fix the type of (G, σ) and consider large values of the parameter.

§ 2. Parabolic groups

2.1. Let G be a connected linear algebraic group, over an algebraically closed field. Let $P \subset G$ be a parabolic subgroup of G, i.e. a closed subgroup containing a Borel subgroup. It is known that P is the normalizer of its unipotent radical $R_u(P)$. Let B be a Borel subgroup in P, let $T \subset B$ be a maximal torus of G. Put $\Phi=\Phi(T, G)$. There exists an order on Φ such that $\Phi(T, B)$ is the set Φ^+ of positive roots. Let Δ be the basis of Φ(=the system of simple roots) defined by the order. If $r \in \Phi$ then we have an expression

$$r=\sum_{a \in \Delta} m_a(r) a,$$

where the integers $m_a(r)$ are either all $\geqq 0$ or all $\leqq 0$. We have $\Phi(T, P) \subset \Phi$.

2.2. *Lemma.* (i) *There exists a subset* Δ_1 *of* Δ *such that* $r \in \Phi$ *lies in* $\Phi(T, P)$ *if and only if* $m_a(r) \geqq 0$ *for all* $a \in \Delta_1$;

(ii) $r \in \Phi(T, R_u(P))$ *if and only if* $m_a(r) > 0$ *for at least one* $a \in \Delta_1$, *or equivalently:* $r \in \Phi(T, R_u(P))$ *if and only if* $r \in \Phi(T, P)$, $-r \notin \Phi(T, P)$.

2.2 is a familiar result (see [3, p. 86, 4.3]).

2.3. Recall that a closed connected subgroup L of P is called a Levi subgroup of P if P is the semi-direct product $P = L \cdot R_u(P)$ (in the sense of algebraic groups). We then have:

2.4. *Proposition. Let* $T \subset P$ *be a maximal torus of* G.

(i) *There exists a unique Levi subgroup* L *of* P *which contains* T, *we have* $\Phi(T, L) = \{r \in \Phi(T, P) | -r \in \Phi(T, P)\}$;

(ii) *Two Levi subgroups of* P *are conjugate by a unique element of* $R_u(P)$.

For the proof see [3, p. 78—80, 3.13 and 3.14].

The next results are also contained in [3, p. 86, 4.4]. In view of their importance for what follows we give a proof.

2.5. *Lemma. Let* B *be a Borel subgroup of* G, *let* P *be a parabolic subgroup. Then* $C = (B \cap P)$. $R_u(P)$ *is a Borel subgroup contained in* P.

It follows from Bruhat's lemma that $B \cap P$ contains a maximal torus T of G. C^0 then contains T, also C^0 is clearly solvable. Hence C^0 is contained in a Borel subgroup of G. Now $\Phi(T, C) = (\Phi(T, B) \cap \Phi(T, P)) \cup \Phi(T, R_u(P))$. It follows from 2.2 that if $r \notin \Phi(T, R_u(P))$, $-r \notin \Phi(T, R_u(P))$ we have $r, -r \in \Phi(T, P)$. This implies that for each $r \in \Phi$ we have $r \in \Phi(T, C)$ or $-r \in \Phi(T, C)$. This shows that $\Phi(T, C)$ contains at least half of the roots of Φ, whence $\dim C^0 \geqq \dim B$. It follows that C^0 must be a Borel subgroup. Since C^0 is then its own normalizer, we have $C = C^0$, which proves the lemma.

2.6. *Proposition. Let* P, P' *be two parabolic subgroups of* G, *let*

$$Q = (P \cap P') \cdot R_u(P).$$

(i) Q *is a parabolic subgroup contained in* P, *we have* $R_u(Q) = (P \cap R_u(P')) \cdot R_u(P)$;

(ii) *the following conditions are equivalent:*

(a) $Q = P$, (b) $P \cap R_u(P') \subset R_u(P)$, (c) P' *contains a Levi subgroup of* P.

That Q is parabolic follows from 2.5. Let $T \subset P \cap P'$ be a maximal torus. Then $\Phi(T, Q) = (\Phi(T, P) \cap \Phi(T, P')) \cup \Phi(T, R_u(P))$. The second part of (i) now follows from 2.2 (ii).

To prove (ii) first observe that (a) implies (b). If (b) holds then $R_u(Q) = R_u(P)$, whence $Q = P$, so that (a) holds. That (c) implies (a) is clear. Let L, L' be the Levi subgroups of P, P' containing T (see 2.4). If $Q = P$ then 2.4 (i) implies that we must have $L \cap L' = L$, whence $L \subset L'$. Hence (a) implies (c).

2.7. We next discuss the action of an endomorphism σ. So assume now, moreover, that we are given a surjective endomorphism σ of G with G_σ finite. We want to discuss σ-stable parabolic subgroups P of G. i.e. with $\sigma P = P$.

2.8. *Proposition. Let P be a parabolic subgroup with $\sigma P \subset P$.*

(i) *We have $\sigma P = P$, $\sigma R_u(P) = R_u(P)$;*

(ii) *There exists a σ-stable Levi subgroup of G, and any two of them are conjugate by an element of $R_u(P)_\sigma$.*

(i) follows from 1.6 and (ii) follows by applying 1.7 to the space M of all Levi subgroups of P. The group H of 1.7 (ii) is now a connected linear algebraic group (a Levi subgroup of P by 2.4 (ii)), so that G_σ acts transitively on M_σ by 1.7 (ii).

2.9. It is known that if G is a connected reductive algebraic group defined over a field k and if P, P' are two parabolic k-subgroups of G we have $P = P'$ if and only if their groups $P(k)$, $P'(k)$ of k-rational points are equal. We need similar results in the σ-setup, which we discuss now.

2.10. *Lemma Let P, P' be σ-stable parabolic subgroups of G with $P \subset P'$. If $P_\sigma = P'_\sigma$ then $P = P'$.*

Let $U = R_u(P')$. Then U is σ-stable and $U \subset R_u(P)$. Replace P by P/U and P' by P'/U. Since $(P/U)_\sigma \cong P_\sigma/U_\sigma$, $(P'/U)_\sigma \cong P'_\sigma/U_\sigma$ (as a consequence of 1.5, see for example [14, p. 175, 2.11]), we see that it suffices to deal with the case that G is reductive and that $P' = G$. Let $B \subset P$ be a σ-stable Borel subgroup, let $T \subset B$ be a σ-stable maximal torus (these exist by 1.8). Let $N = N_G(T)$, $W = N/T$. Then σ induces an automorphism of W (as a consequence of 1.6). B defines an order of $\Phi = \Phi(T, G)$. Let $w_0 \in W$ be the element which sends positive roots into negative ones. Since σ fixes B, we have that $\sigma w_0 = w_0$. Then there exists $n_0 \in N_\sigma$ representing w_0 by [14, p. 175, 2.11]. If $P_\sigma = G_\sigma$, we have $n_0 \in P$. But then P contains the Borel subgroup $B^- = n_0 B n_0^{-1}$. Since B and B^- generate G, we have $P = G$.

2.11. *Proposition. Let P, P' be two σ-stable parabolic subgroup of G. If $P_\sigma = P'_\sigma$ then $P = P'$.*

Put $Q = (P \cap P') \cdot R_u(P)$. Q is parabolic by 2.6, moreover $\sigma Q \subset Q$. Hence Q is σ-stable (2.8 (i)). We have $P_\sigma = (P_\sigma \cap P'_\sigma) \cdot R_u(P)_\sigma \subset Q_\sigma$. On the other hand we know that $Q \subset P$. Hence $Q_\sigma = P_\sigma$, whence $Q = P$ by 2.10. Consequently (2.6 (ii)) P' contains a Levi subgroup of P. Let $T \subset P \cap P'$ be a maximal torus, let L and L' be the Levi groups of P, P', respectively, which contain T.

We then have $L' \supset L$. Interchanging P and P' we find $L \supset L'$, hence $L = L'$. Then $P \cap P' = L \cdot V$, where $V = R_u(P) \cap R_u(P')$, $P_\sigma = (P \cap P')_\sigma = L_\sigma \cdot V_\sigma$. Hence we have $V_\sigma = R_u(P)_\sigma$. Now if V is any subgroup of $R_u(P)$ which is normalized by T and σ-stable then the order of V_σ equals $\prod_{a \in \Phi(T,V)} q_a$, where q_a is as in 1.10

(i) (see [15, p. 76, 11.8]). 1.10 (ii) now shows that $V_\sigma = R_u(P)_\sigma$ implies $V = R_u(P)$. This concludes the proof of 2.11.

We continue with a result of the same nature.

2.12. *Proposition. Let P, P' be two σ-stable parabolic subgroup of G. Put $Q = (P \cap P') \cdot R_u(P)$. Then P is a σ-stable parabolic subgroup. The following conditions are equivalent:*

(a) $Q = P$, (b) $Q_\sigma = P_\sigma$, (c) $P_\sigma \cap R_u(P')_\sigma \subset R_u(P)_\sigma$, (d) *$P'$ contains a σ-stable Levi subgroup of P.*

The equivalence of (a) and (b) follows by 2.10. If (a) holds then $R_u(Q) = R_u(P)$, whence (c). Next let (c) hold. Let $T \subset P \cap P'$ be a σ-stable maximal torus of G. $P \cap R_u(P')$ and $P \cap R_u(P') \cap R_u(P)$ are σ-stable subgroups of $R_u(P')$ which are normalized by T. A computation of the orders of the corresponding groups of fixed points of σ (using [15, p. 76, 11.8], as in the proof of 2.11) then shows that $P_\sigma \cap R_u(P')_\sigma \cap R_u(P)_\sigma = P_\sigma \cap R_u(P')_\sigma$ implies $P \cap R_u(P') \subset R_u(P)$. By 2.6 (ii) we now conclude that (c) implies (a). The equivalence of (d) and the other conditions follows by considering the Levi subgroups of P, P', Q containing T.

2.13. *Corollary. Assumptions as in* 2.12. *The following conditions are equivalent:*

(a) *$P_\sigma \cap R_u(P')_\sigma \subset R_u(P)_\sigma$ and $P'_\sigma \cap R_u(P)_\sigma \subset R_u(P')_\sigma$,*

(b) *P and P' have a common σ-stable Levi subgroup.*

2.14. *Corollary. Same assumptions. If $R_u(P)_\sigma = R_u(P')_\sigma$ then $P = P'$.*

If $R_u(P)_\sigma = R_u(P')_\sigma$ the conditions of 2.13 (a) are satisfied. Hence P and P' have a common σ-stable Levi subgroup. It then follows that $P_\sigma = P'_\sigma$. Hence $P = P'$ by 2.11.

2.15. Let $\mathscr{P}$ be the set of σ-stable parabolic subgroups of G. We say that $P, P' \in \mathscr{P}$ are *associated* if there exists $x \in G_\sigma$ such that ${}^xP$ and P' have a common σ-stable Levi subgroup. This defines an equivalence relation on $\mathscr{P}$, let $\mathscr{C}$ be the corresponding set of equivalence classes. We define an order on $\mathscr{C}$ as follows: let $c, c' \in \mathscr{C}$, then $c \leqq c'$ if there exists $P \in c$ and $P' \in c'$ such that P' contains a σ-stable Levi subgroup of P. Let $c \leqq c'$, let $P \in c$, $P' \in c'$. We then also write $P \leqq P'$. Clearly $P \leqq P'$, $P' \leqq P$ implies that P and P' are associated.

2.16. *Example.* Let $G = GL_n$, over a finite field k, let σ be the corresponding Frobenius endomorphism. Then there is a bijection of $\mathscr{C}$ onto the set of partitions of n, i.e. sequences of integers $\lambda = (n_1, \ldots, n_r)$ with $n_1 \geqq \ldots \geqq n_r > 0$, $n_1 + n_2 + \ldots + n_r = n$. The order of $\mathscr{C}$ corresponds to the following order of partitions: $(n_1, \ldots, n_r) \leqq (m_1, \ldots, m_s)$ if the first partition is a "refinement" of the second one, i.e. the m_i are each replaced by a partition λ_i and the n_j are obtained by putting the parts of all λ_i together, in the right order. All this follows readily from the well-known description of the parabolic subgroups of GL_n (see e.g. [12, p. 128—129, 2.7]).

2.17. *Proposition. Let P be a σ-stable parabolic subgroup of G, let L be a σ-stable Levi subgroup of P. The σ-stable parabolic subgroups of G contained in P are the semi-direct products of the σ-stable parabolic subgroups of L with $R_u(P)$. Two such parabolic subgroups are conjugate by an element of G_σ if and only if their intersections with L are conjugate by an element of L_σ.*

The proof is like that of the corresponding result without σ in [3, p. 86—87, 4.4] and is omitted.

2.18. Let $P=L\cdot R_u(P)$ be a σ-stable parabolic subgroup G, L being a σ-stable Levi subgroup. Put $W_L=N_G(L)/L$. If L' is another σ-stable Levi subgroup of P, then there exists a unique $x\in R_u(P)_\sigma$ such that $L'={}^xL$. This implies that there exists a canonical isomorphism $W_L \simeq W_{L'}$. Therefore we may write W_P instead of W_L.

2.19. *Lemma. Let T be a σ-stable maximal torus in L.*

(i) *W_L is isomorphic to $N_G(T)\cap N_G(L)/N_L(T)$;*

(ii) *σ induces an automorphism of W_L, every element w of $W_{L,\sigma}=(W_L)_\sigma$ is represented by an element n_w in $N_G(T)_\sigma\cap N_G(L)_\sigma$.*

If $x\in N_G(L)$ then ${}^xT$ is a maximal torus of L, hence conjugate to T by an element of L. This remark readily implies (i). The first point of (ii) follows from 1.6, the rest is a consequence of [14, p. 175, 2.11].

2.20. *Proposition. The map $W_{L,\sigma}\to P_\sigma\backslash G_\sigma/P_\sigma$ given by $w\to P_\sigma n_w P_\sigma$ is injective.*

Here n_w is as in 2.19 (ii). Let $W=N_G(T)/T$ be the Weyl group of G, relative to T, let $W'=N_L(T)/T$ be the Weyl group of L'. The theory of Tits systems shows that the double cosets $P_\sigma xP_\sigma$ are in a one-one correspondence with the double cosets $W'_\sigma wW'_\sigma$ ($w\in W_\sigma$) [5, p. 28, Remarque 2]. If $n_w\in N_G(T)_\sigma$ represents $w\in W_\sigma$ then $W'_\sigma wW'_\sigma$ corresponds to $P_\sigma n_w P_\sigma$. 2.20 now follows from the remark that two distinct elements w, w' of $W_{L,\sigma}$ determine distinct double cosets $W'_\sigma wW'_\sigma$, $W'_\sigma w'W'_\sigma$ in W_σ.

We terminate this section with a remark about the orders of the groups P_σ.

2.21. *Proposition. Let P, P′ be two associated σ-stable parabolic subgroups of G. Then P_σ and P'_σ have the same order.*

We may assume that P and P' have a common σ-stable Levi subgroup L. It then suffices to prove that $R_u(P)_\sigma$ and $R_u(P')_\sigma$ have the same order. Let T be a maximal torus in L, which is σ-stable. Using again [15, p. 76, 11.8] we see that the order $|R_u(P)_\sigma|$ equals $\prod_{a\in\Phi(T,R_u(P))} q_a$ (where q_a is as in 1.10 (i)). Since $\Phi(T,G)-\Phi(T,L)=\Phi(T,R_u(P))\cup(-\Phi(T,R_u(P)))$ (by 2.2 (ii)) we conclude that

$$|R_u(P)_\sigma|^2=\prod_{a\in\Phi(T,G)-\Phi(T,L)} q_a.$$

The right-hand side depends only on L, which proves the assertion.

§ 3. Some results on finite groups

3.1. If S is a finite set we denote by $|S|$ the number of its elements and by $C(S)$ the vector space of complex-valued functions on S. The hermitian form $\langle\, ,\rangle_S$ on $C(S)$ is defined by

$$\langle f, g\rangle_S = |S|^{-1} \sum_{s \in S} f(s)\overline{g(s)} \quad (f, g \in C(S)).$$

Let A be a finite group. We make $C(A)$ into an algebra by the convolution product: if $f, g \in C(A)$ then

$$f * g(x) = \sum_{y \in A} f(y) g(y^{-1}x).$$

$C(A)$, with this product, is isomorphic to the group algebra of A over the complex field. A operates on $C(A)$ by left (right) translations; the left (right) ideals of $C(A)$ are the subspaces of $C(A)$ which are stable by all left (right) translations.

$f \in C(A)$ is called invariant if $f(x) = f(yxy^{-1})$ $(x, y \in A)$. The invariant functions form a commutative subalgebra $I(A)$ of $C(A)$ (which is the centre of $C(A)$).

Let ϱ be a (complex) representation of A, let ω be its equivalence class. We denote by χ_ϱ or χ_ω the character of ϱ. The invariant functions χ_ϱ span $I(A)$. An $f \in C(A)$ is called a generalized character if it is an integral linear combination of χ_ϱ.

3.2. *Induced representations*

Let A be a finite group, let B be a subgroup of A. Let σ be a right representation of B in a complex vector space W. Let V be the space of functions $f: A \to W$ such that

$$f(xy) = f(x)\sigma(y) \quad (x \in A, y \in B).$$

Put $\varrho(x)f(y) = f(x^{-1}y)$ $(x, y \in A)$. Then ϱ is a (left) representation of A in V, the representation of A induced by σ. Define a linear map $i_{B\to A}$ of $I(B)$ into $I(A)$ by

$$i_{B\to A}f(x) = |B|^{-1} \sum_{\substack{y \in A \\ yxy^{-1} \in B}} f(yxy^{-1}), \tag{1}$$

then we have

$$\chi_\varrho = i_{B\to A}\bar{\chi}_\sigma$$

(see [8, p. 46, 9.1]).

We next recall some well-known formulae. We denote by $r_{A\to B}$ the restriction map $C(A) \to C(B)$. For $x \in A$ denote by $f \to {}^x f$ the isomorphism $I(B) \to I({}^x B)$ given by

$$({}^x f)({}^x y) = f(y) \quad (y \in B).$$

3.3. *(Frobenius duality.) If $f \in I(A)$, $g \in I(B)$ then*

$$\langle f, i_{B\to A} g\rangle_A = \langle r_{A\to B} f, g\rangle_B.$$

3.4. *(Mackey's formula.) Let B and C be subgroups of A, let* $f \in I(B)$, $g \in I(C)$. *Then*

$$\langle i_{B\to A}f, i_{C\to A}g\rangle_A = \sum_{B\backslash A/C} \langle r_{B\to B\cap {}^xC}f, r_{{}^xC\to B\cap {}^xC}{}^xg\rangle_{B\cap {}^xC}$$

(see [8, p. 47, 9.4 and p. 51, 9.8]).

We next recall a lemma of Brauer—Suzuki, to be used in section 7. The proof is given in [8, p. 124—125, 23.3].

3.5. *Lemma. Let* $n \geqq 2$, *let* f_{ij} $(1 \leqq i, j \leqq n)$ *be a family of generalized characters of A such that* $f_{ij}(1) = 0$ *and*

$$\langle f_{ij}, f_{kl}\rangle_A = \delta_{ik} - \delta_{il} - \delta_{jk} + \delta_{jl}.$$

Then there exists a sign $\varepsilon = \pm 1$ *and irreducible characters* $\chi_1, \ldots, \chi_n$ *of A such that*

$$f_{ij} = \varepsilon(\chi_i - \chi_j).$$

ε *is uniquely determined if* $n > 2$.

3.6. *A remark on Brauer's theorem.*

We shall give now a reformulation of Brauer's well-known characterization of generalized characters of finite groups (see [8, p. 79, 15.3]), for the case of the finite groups G_σ discussed in section 1. This result will not be used in the sequel.

Let G be a connected reductive linear algebraic group, let σ be a surjective endomorphism of G such that its group of fixed points G_σ is finite.

3.7. *Proposition.* $f \in I(G_\sigma)$ *is a generalized character of* G_σ *if and only if the restriction of f to the following subgroups A of* G_σ *is a generalized character of A:*

(a) $A = P_\sigma$, *where P is a* σ*-stable proper parabolic subgroup of G,*

(b) $A = N_G(T)_\sigma$, *where T is a* σ*-stable maximal torus of G.*

Brauer's theorem asserts that this is so if one lets A run through the elementary subgroups of G_σ. Here an elementary subgroup of G_σ is a direct product of a cyclic group and an l-group (where l is some prime number). In order to establish 3.7 it therefore suffices to show that any elementary subgroup of G_σ is contained in a group of the form (a) or (b). Let A be elementary. Let p be the characteristic of the underlying field.

(a) $|A|$ is divisible by p. Then there is $x \in A$ such that x is unipotent (i.e. of p-power order) and that $A \subset Z_{G_\sigma}(x)$. We now use an argument of Borel—Tits (see [4, p. 95 and 2.3, p. 99]) to find a σ-stable proper parabolic subgroup P of G such that $A \subset P$.

Let U be the subgroup of G generated by x, let $N_1 = N_G(U)$, $U_1 = U \cdot R_u N_1$, $N_i = N_G(U_{i-1})$, $U_i = U_{i-1} \cdot R_u N_i$ $(i \geqq 2)$. The N_i form an increasing sequence of closed subgroups of G, they are σ-stable and they contain A. We have that N_i is a parabolic subgroup of G for large i, which must be a proper one.

(b) $|A|$ is prime to p. In this case A is a nilpotent group of semisimple automorphisms of G. That A normalizes a σ-stable maximal torus is a known result, [14, p. 210—211, 5.16 (b)].

§ 4. Cusp forms

4.1. Let G be a connected linear algebraic group with an endomorphism σ as in 1.5. If H is a closed σ-stable subgroup we write $i_{H\to G}$ and $r_{G\to H}$ for $i_{H_\sigma\to G_\sigma}$ and $r_{G_\sigma\to H_\sigma}$ (see 3.2). We also write $\langle\ ,\ \rangle_G$ for $\langle\ ,\ \rangle_{G_\sigma}$ (see 3.1). Since we will consider only one σ, no confusion will arise.

We denote by $C'(G_\sigma)$ the subspace of $C(G_\sigma)$ formed by the functions which are right invariant by $R_u(G)_\sigma$. Put $I'(G_\sigma)=I(G_\sigma)\cap C'(G_\sigma)$ (notations as in 3.1).

Let P be a σ-stable parabolic subgroup of G. If $f\in C(G_\sigma)$ we put

$$f_P(x) = |R_u(P)_\sigma|^{-1} \sum_{y\in R_u(P)_\sigma} f(xy).$$

f_P is a function on G_σ which is right invariant by $R_u(P)_\sigma$.

4.2. *Lemma.* (i) *Let L be a σ-stable Levi subgroup of P, let $f\in I'(P_\sigma)$, $g\in I(G_\sigma)$. Then*

$$\langle i_{P\to G}f, g\rangle_G = \langle r_{P\to L}f, r_{G\to L}g_P\rangle_L;$$

(ii) *Let $Q\subset P$ be a σ-stable parabolic subgroup, let $f\in C(G_\sigma)$. Then*

$$(r_{G\to P}f_P)_Q = r_{G\to P}f_Q.$$

(i) follows by Frobenius duality (3.3) and (ii) is proved by a trivial calculation.

4.3. *Definition.* $f\in C(G_\sigma)$ *is called a cusp form or a cuspidal function if*

(a) *f is right invariant by $R_u(G)_\sigma$,*

(b) *$f_P=0$ for all proper σ-stable parabolic subgroups P of G.*

The space ${}^0C(G_\sigma)$ of cusp forms in $C(G_\sigma)$ is invariant by right and left translations, hence is a twosided ideal in $C(G_\sigma)$. We put ${}^0I(G_\sigma)=I(G_\sigma)\cap{}^0C(G_\sigma)$.

A (complex) representation of G_σ is called cuspidal if all its matrix elements are cusp forms. Its character is then called a cuspidal character.

4.4. *Lemma. Let G be reductive. Let ϱ be a representation of G_σ, with character χ. The following conditions are equivalent:*

(i) *ϱ is cuspidal;*

(ii) *for any σ-stable proper parabolic subgroup P of G the restriction of ϱ to $R_u(P)_\sigma$ does not contain the trivial representation of $R_u(P)_\sigma$;*

(iii) *For any P of* (ii) *we have* $\sum_{x\in R_u(P)_\sigma} \chi(x)=0.$

If ϱ is cuspidal then $\sum_{x \in R_u(P)_\sigma} \varrho(x)=0$ for the P of (ii). This implies (ii). Also (ii) and (iii) are equivalent by familiar properties. If (ii) holds then $\sum_{y \in R_u(P)_\sigma} \varrho(xy) = \varrho(x) \sum \varrho(y)=0$ for all $x \in G_\sigma$, from which one infers that (i) holds.

Warning: it is not true that $f \in I(G_\sigma)$ is a cusp form if $\sum_{x \in R_u(P)_\sigma} f(x)=0$ for all proper σ-stable parabolics P. This is only so for characters of representations (a counter-example is easily given, for example with $G=SL_2$).

4.5. *Proposition. Let G be reductive. Let χ be an irreducible character of G_σ. There exists a σ-stable parabolic subgroup P of G and an irreducible cuspidal character φ of P_σ such that $\langle \chi, i_{P \to G} \varphi \rangle_G \neq 0$.*

Let ϱ be an irreducible representation of G_σ in a complex vector space V such that $\chi=\chi_\varrho$. Let P be a σ-stable parabolic such that $\sum_{x \in R_u(P)_\sigma} \chi(x) \neq 0$ (such P exists, for example $P=G$) and choose a P which is minimal for this property. The restriction of ϱ to $R_u(P)_\sigma$ contains the trivial representation, let W be the subspace of V whose elements are fixed by $R_u(P)_\sigma$. Then $W \neq 0$. If $x \in P_\sigma$ we have $\varrho(x)W=W$, which gives us a representation τ of P on W. The character of τ is $\theta=r_{G \to P} \chi_P$. It follows from the choice of P (using 4.2 (ii) and 4.4 (iii)) that θ is a cuspidal character of P_σ. It follows from 4.2 (i) that $\langle i_{P \to G} \theta, \chi \rangle_G \neq 0$. 4.4 (iii) implies that all irreducible characters of P_σ occurring in θ are also cuspidal. There must be one of them (call it φ) which has the required property.

For the next corollary we introduce a notation. Let $f \in C(G_\sigma)$, let f_P be as in 4.1. We write $f_P \sim 0$ if

$$\sum_{y \in P_\sigma} f_P(xy) g(y) = 0$$

for all $x \in G_\sigma$ and all $g \in {}^0C(P_\sigma)$.

4.6. *Corollary. Let G be reductive. Let $f \in C(G_\sigma)$ be such that $f_P \sim 0$ for all σ-stable parabolic subgroups P of G (including $P=G$). Then $f=0$.*

The set of all such f is a two-sided ideal in $C(G_\sigma)$. It follows from 4.5 that this ideal cannot contain any irreducible character of G_σ, hence it must be the zero ideal. Another proof of the corollary may be given by the method of proof of [11, p. 102, 3.2].

4.7. It follows from 4.5 that the irreducible representations of G_σ (G reductive) can be found as follows. One takes an irreducible cuspidal representation of some P_σ, induces it up to G_σ and decomposes the induced representation. Any irreducible representation of G_σ appears in one of these induced representations. In the next section we shall give some results about the decomposition of the induced representations.

§ 5. Induced representations

We keep the notations of the previous section. G is assumed to be reductive.

5.1. Let P, P' be two σ-stable parabolic subgroups of G. By 2.6 (i) we know that $Q=(P\cap P')\cdot R_u(P)$ and $Q'=(P\cap P')\cdot R_u(P')$ are parabolic subgroups of P and P', respectively. Let T be a σ-stable maximal torus in $P\cap P'$, let L and L' be the Levi subgroups of P and P' containing T. They are σ-stable. We have $Q=(L\cap L')\cdot(L\cap R_u(P'))\cdot R_u(P)$, and similarly for Q'. Moreover, $L\cap L'$ is a common Levi subgroup of Q and Q'. With these notations we have the following result.

5.2. *Lemma. Let $f\in I'(P_\sigma)$, $f'\in I'(P'_\sigma)$. Denote by φ, φ' the restrictions of f and f' to $P_\sigma\cap P'_\sigma$.*

(i) $\langle\varphi,\varphi'\rangle_{P\cap P'}=\langle r_{P\to L\cap L'}f_Q, r_{P'\to L\cap L'}f'_{Q'}\rangle_{L\cap L'}$;

(ii) *If $f\in{}^0I(P_\sigma)$ then $\langle\varphi,\varphi'\rangle_{P\cap P'}=0$ unless P' contains a σ-stable Levi subgroup of P;*

(iii) *If $f\in{}^0I(P_\sigma)$, $f'\in{}^0I(P'_\sigma)$ then $\langle\varphi,\varphi'\rangle_{P\cap P'}=0$ unless P and P' have a common σ-stable Levi subgroup.*

Put $U=R_u(P)$, $U'=R_u(P')$. We have $P_\sigma\cap P'_\sigma=(L_\sigma\cap L'_\sigma)\cdot(L_\sigma\cap U'_\sigma)\cdot(U_\sigma\cap L'_\sigma)\cdot\cdot(U_\sigma\cap U'_\sigma)$. It follows that, up to a scalar factor, $\langle\varphi,\varphi'\rangle_{P\cap P'}$ equals

$$\sum f(xyz)\overline{f'}(xyz) = \sum f(xz\cdot z^{-1}yz)\overline{f'}(xy) = \sum f(xz)\overline{f'}(xy),$$

where x, y, z run through $L_\sigma\cap L'_\sigma$, $L'_\sigma\cap U_\sigma$, $L_\sigma\cap U'_\sigma$, respectively. Using 2.6 (i) we find the first assertion, up to a scalar factor, which is easily seen to be 1 (take f and f' constant). (ii) then follows from 2.12 and (iii) is a consequence of (ii).

5.3. We keep the same notations. If L is a σ-stable Levi subgroup of P, we write $W_L=N_G(L)/L$, as in 2.18. Let $w\in W_{L,\sigma}=(W_L)_\sigma$, let $n\in N_G(L)_\sigma$ represent w. For $f\in I'(P_\sigma)$ we define $w\cdot f$ as follows. Let $x\in P_\sigma$ and write $x=yz$, with $y\in L_\sigma$, $z\in R_u(P)_\sigma$. Then

$$w\cdot f(x) = f(n^{-1}yn).$$

This defines an action of $W_{L,\sigma}$ on $I'(P_\sigma)$. Clearly ${}^0I(P_\sigma)$ is a stable subspace of $I'(P_\sigma)$ for this action.

If L' is another σ-stable Levi subgroup there exists a unique $u\in R(P)_\sigma$ such that $L'={}^uL$, hence u induces an isomorphism $\varphi: W_L \xrightarrow{\sim} W_{L'}$. It is easily checked that

$$\varphi(w)\cdot f = w\cdot f \quad (f\in I'(P_\sigma)).$$

5.4. *Proposition.* (i) *Let $f\in{}^0I(P_\sigma)$, $f'\in{}^0I(P'_\sigma)$. Then $\langle i_{P\to G}f, i_{P'\to G}f'\rangle_G=0$ unless P and P' are associated;*

(ii) *Assume that P and P' have a common σ-stable Levi subgroup L. If f and*

f' *are as before, we have*

$$\langle i_{P\to G}f, i_{P'\to G}f'\rangle_G = \sum_{w\in W_{L,\sigma}} \langle r_{P\to L}f, r_{P'\to L}(w\cdot f')\rangle_L;$$

(iii) *If, moreover,* f *and* f' *have the same restriction to* L *then* $i_{P\to G}f=i_{P'\to G}f$.

By Mackey's formula (3.4) we have

$$\langle i_{P\to G}f, i_{P'\to G}f'\rangle_G = \sum_{P_\sigma\backslash G_\sigma/P'_\sigma} \langle r_{P\to P\cap{}^xP'}f, r_{{}^xP'\to P\cap{}^xP'}{}^xf'\rangle_{P\cap{}^xP'} \tag{1}$$

(i) now follows from 5.2 (iii). It also follows that the only nonzero terms in the right-hand side of (1) are those for which $P\cap{}^xP'$ contains a σ-stable Levi subgroup of P, say L'. If L is as in (ii), then there exists $a\in P_\sigma$ such that ${}^aL'=L$ and there exists $b\in P_\sigma$ such that $L'={}^{xb}L$. Hence ${}^{axb}L=L$, so that $axb\in N_G(L)$. Hence the only double cosets of $P_\sigma\backslash G_\sigma/P'_\sigma$ which give a contribution to the right-hand side of (1) are those containing an element of $N_G(L)_\sigma$. Now (ii) follows from 2.20 and 5.2 (i).

It follows from (ii) that we have in the situation of (iii) that

$$\langle i_{P\to G}f, i_{P\to G}f\rangle_G = \langle i_{P\to G}f, i_{P'\to G}f'\rangle_G = \langle i_{P'\to G}f', i_{P'\to G}f'\rangle_{G'}$$

from which one concludes that $i_{P\to G}f=i_{P'\to G}f'$.

With the notations of 5.4 we have the following corollary.

5.5. *Corollary. Let* φ *be an irreducible cuspidal character of* P_σ. *Then* $\langle i_{P\to G}\varphi, i_{P\to G}\varphi\rangle_G$ *equals the order of the isotropy group of* φ *in* $W_{L,\sigma}$. *Hence* $i_{P\to G}\varphi$ *is an irreducible character of* G_σ *if no nontrivial element of* $W_{L,\sigma}$ *fixes* φ.

If φ is as in 5.5, then $\langle r_{P\to L}\varphi, r_{P\to L}w\cdot\varphi\rangle_L$ equals 1 or 0 if $w\cdot\varphi=\varphi$ or $w\cdot\varphi\neq\varphi$, respectively. The assertions of the corollary now follow from 5.4 (ii).

5.6. Let $\mathscr{E}$ be the set of irreducible characters of G_σ. Let $\mathscr{C}$ be the set of classes of associated σ-stable parabolic subgroups (2.15). Let $c\in\mathscr{C}$, let $P\in c$. We denote by $\mathscr{E}(c)$ the set of irreducible characters of G_σ which occur in an induced character $i_{P\to G}\varphi$, φ being an irreducible cuspidal character of P_σ. If follows from 5.4 (iii) that $\mathscr{E}(c)$ depends indeed on c only. Moreover, 5.4 (i) shows that $\mathscr{E}(c)\cap\mathscr{E}(c')=\emptyset$ if $c\neq c'$. We then obtain a partition $\mathscr{E}=\bigcup_{c\in\mathscr{C}}\mathscr{E}(c)$ of $\mathscr{E}$ into disjoint nonempty subsets.

Equivalently, there is a decomposition of the group algebra

$$C(G_\sigma) = \bigotimes_{c\in\mathscr{C}} \mathscr{J}(c),$$

where $\mathscr{J}(c)$ is a nonzero twosided ideal, namely the one determined by the $\chi\in\mathscr{E}(c)$. If c is the class of G, then $\mathscr{J}(c)={}^0C(G_\sigma)$.

5.7. We shall now consider in more detail the representation of G_σ induced by a cuspidal irreducible representation of P_σ. So let τ be a cuspidal irreducible right representation of P_σ in a complex vector space W. Let V be the space of functions $f: G_\sigma \to W$ such that

$$f(xy) = f(x)\tau(y) \quad (x \in G_\sigma,\ y \in P_\sigma).$$

The representation of G_σ by left translation in V is the induced representation ϱ (see 3.2). If φ is the character of τ then $\chi_\varrho = i_{P\to G}\bar{\varphi}$.

Let $\mathscr{A}$ be the commuting algebra of ϱ, i.e. the algebra of all linear transformations of V which commute with the $\varrho(x)$ $(x \in G_\sigma)$. It is know that $\mathscr{A}$ is a semisimple algebra over $\mathbf{C}$ of dimension $\langle i_{P\to G}\varphi, i_{P\to G}\varphi\rangle_G$. Let L and W_L be as in 5.3, let Γ be the subgroup of $W_{L,\sigma}$ whose elements fix φ. By 5.5 we then have $\dim \mathscr{A} = |\Gamma|$.

If $\gamma \in \Gamma$ let $n_\gamma \in N_G(L)_\sigma$ be a representative of γ. There exists a nonsingular linear transformation t_γ of W such that

$$\tau(n_\gamma^{-1} x n_\gamma) = t_\gamma^{-1}\tau(x)\, t_\gamma \quad (\gamma \in \Gamma,\ x \in L_\sigma).$$

If $\gamma, \gamma' \in \Gamma$ then we have $t_{\gamma\gamma'} = c(\gamma, \gamma')t_\gamma t_{\gamma'}$, where $c(\gamma, \gamma') \in \mathbf{C}^*$. Then $(\gamma, \gamma') \to$ $\to c(\gamma, \gamma')$ is a 2-cocycle of Γ in $\mathbf{C}^*$, its cohomology class $\alpha \in H^2(\Gamma, \mathbf{C}^*)$ is independent of the choice of the t_γ.

We now put for $f \in V$, $\gamma \in \Gamma$

$$(T_\gamma f)(x) = |R_u(P)_\sigma|^{-1} \sum_{y \in R_u(P)_\sigma} f(xyn_\gamma)\, t_\gamma^{-1}.$$

5.8. *Proposition.* (i) *T_γ is a linear transformation of V, which commutes with all $\varrho(x)$ $(x \in G_\sigma)$;*

(ii) *The T_γ form a basis of $\mathscr{A}$ $(\gamma \in \Gamma)$.*

It is clear that $(T_\gamma f)(xz) = (T_\gamma f)(x)$ if $z \in R_u(P)_\sigma$. Now let $z \in L_\sigma$. Then using that

$$f(xzyn_\gamma) = f(x \cdot {}^z y n_\gamma)\,\tau(n_\gamma^{-1} z n_\gamma) = f(x \cdot {}^z y n_\gamma)\, t_\gamma^{-1}\tau(z)\, t_\gamma,$$

if $y \in R_u(P)_\sigma$, $\gamma \in \Gamma$ we see that

$$T_\gamma(xz) = T_\gamma f(x) \cdot \tau(z).$$

It follows that $T_\gamma f \in V$. That T_γ commutes with left translations is obvious. This establishes (i).

Since, as we saw above, $\dim \mathscr{A} = |\Gamma|$ it suffices to prove that the T_γ are linearly independent. So let the $c_\gamma (\gamma \in \Gamma)$ be a set of complex numbers with

$$\sum_{\gamma \in \Gamma} c_\gamma \sum_{y \in R_u(P)_\sigma} f(xyn_\gamma)\, t_\gamma^{-1} = 0, \tag{2}$$

for all $x \in G_\sigma$ and all $f \in V$.

Fix a nonzero vector $w \in W$ and define a function $f_0 \in V$ by

$$f_0(x) = 0 \quad \text{if} \quad x \notin P_\sigma,\ f_0(x) = w \cdot \tau(x) \quad \text{if} \quad x \in P_\sigma.$$

Fix $\delta \in \Gamma$ and apply (2) with $x=n_\delta^{-1}, f=f_0$. We obtain

$$\sum_{\gamma \in \Gamma} c_\gamma \sum_{\substack{y \in R_u(P)_\sigma \\ n_\delta^{-1} y n_\gamma \in P_\sigma}} w \cdot \tau(n_\delta^{-1} y n_\gamma)\, t_\gamma^{-1} = 0. \tag{3}$$

Now if $n_\delta^{-1} y n_\gamma \in P_\sigma$ for some $y \in R_u(P)_\sigma$, we must have $\gamma=\delta$ by 2.20. Hence (3) becomes

$$c_\delta w \cdot \sum_{\substack{y \in R_u(P)_\sigma \\ n_\delta^{-1} y n_\delta \in P_\sigma}} \tau(n_\delta^{-1} y n_\delta)\, t_\delta^{-1} = 0.$$

But if $\gamma \in \Gamma$ all terms in the last sum are equal.

It follows that $c_\delta=0$, for all $\delta \in \Gamma$. This establishes (ii).

5.9. Using 5.8 one can give an explicit description of the algebra $\mathscr{A}$. We shall not write down the formulas here. In the case that P is a Borel subgroup B and τ the trivial representation these formulas show that $\mathscr{A}$ is isomorphic to the Hecke algebra $H_{\mathbf{C}}(G_\sigma, B_\sigma)$ (see [5, p. 54, ex. 22]), which is a well-known fact. It is known that then $\mathscr{A}$ is isomorphic to the group algebra of the Weyl group $W_{L,\sigma}$ (see [5, p. 56, ex. 27]). Notice that L is now a maximal torus.

$i_{B\to G}\,1$ contains the important Steinberg character. For its properties and those of the corresponding irreducible Steinberg representation we refer to [15, p. 99, 15.5], [7] and [10]. Recent work of Benson, Curtis, Iwahori and Kilmoyer on the decomposition of $i_{G\to B}\,1$ is discussed in [1] and [9]. In [11, p. 108—109] some conjectures on the structure of the general $\mathscr{A}$ are discussed, for the case that σ is a Frobenius endomorphism. This discussion carries over to the more general case of an arbitrary σ.

§ 6. Maximal tori, regular semisimple elements

G is again assumed to be reductive, σ is as before. Let T be a σ-stable maximal torus of G. We use the notations of 1.9. $\hat{T}_\sigma$ denotes the character group of T_σ.

6.1. *Proposition.* (i) *N_σ/T_σ is isomorphic to W_σ;*
(ii) *$\hat{T}_\sigma$ is isomorphic to $X/(\sigma^*-1)X$;*
(iii) *The order of T_σ is* $|\det(\sigma^*-1)|$.

(i) follows from 2.19 (ii) and (ii) is proved in [14, p. 188, 1.7.]. (iii) is a consequence of (ii).

Let $\mathscr{G}$ be the set of all G_σ-conjugacy classes of σ-stable maximal tori of G. Define $H^1(\sigma, W)$ as just before 1.7.

6.2. *Proposition. There is a bijection φ of $\mathscr{G}$ onto $H^1(\sigma, W)$.*

This is proved in [14, p. 186, 1.2].

6.3. Let T be a σ-stable maximal torus, let $t \in \mathcal{G}$ be its class. Let $w \in W$ represent $c = \varphi(t)$, according to 6.2. We put

$$Z_\sigma(w) = \{s \in W | sw(\sigma s)^{-1} = w\}.$$

If w is replaced by an equivalent element (for the relation used in defining $H^1(\sigma, W)$) then $Z_\sigma(w)$ is replaced by a conjugate subgroup. Hence the order $|Z_\sigma(w)|$ depends only on the class c of w, we write $z_c = |Z_\sigma(w)|$.

6.4. *Lemma.* $\sum_{c \in H^1(\sigma, W)} z_c^{-1} = 1$.

This is easily proved; it is another formulation of the fact that the number of elements in W is the sum of the number of elements in the classes c.

Let T be as before.

6.5. *Proposition.* (i) *$N_G(T)_\sigma / T_\sigma$ is isomorphic to $Z_\sigma(w)$.*

(ii) *Assume moreover G to be quasi-simple and let q and τ be as in* 1.10 (ii), *let f_τ be the characteristic polynomial of τ. Then $|T_\sigma| = |f_\tau(q)|$.*

(i) follows from the discussion in [14, p. 186—187, 1.3]. (ii) is proved in [14, p. 188, 1.7 (b)].

6.6. *Examples.* (a) Let G be a connected semisimple linear algebraic group defined over the finite field k, which is k-split, let σ be the corresponding Frobenius endomorphism. Take for T a maximal k-split torus. Then σ acts trivially on W and $H^1(\sigma, W)$ is the set of conjugacy classes of W. 6.2 now show that the G_σ-conjugacy classes of σ-stable maximal tori correspond to the conjugacy classes in the Weyl group W.

(b) In the case of the example 1.11 (b) it turns out that there are three classes of σ-stable maximal tori T of G. An application of 6.5 (ii) shows that the order T_σ has the values q^2-1, $q^2-\sqrt{2}q+1$, $q^2+\sqrt{2}q+1$, respectively. See [14, p. 190—191] for a fuller discussion.

6.7. T is called *minisotropic* if T is not contained in a σ-stable proper parabolic subgroup of G. If G is a semisimple linear algebraic group over a finite field and σ its Frobenius endomorphism then the minisotropic tori are just the anisotropic ones (in the sense of [2, p. 217]).

With the notations of 1.9, let $V = X \otimes \mathbf{R}$, let V_1 be the subspace of V spanned by the roots of G with respect to T.

6.8. *Lemma.* (i) *T is minisotropic if and only if σ^* fixes no ray in V_1;*

(ii) *Assume moreover G to be quasi-simple and let τ be as in* 1.10 (ii). *Then T is minisotropic if and only if τ has no eigenvalue* 1 *in V_1.*

See [14, p. 191—192, 1.12].

6.9. *Example.* In the situation of 6.6 (a), the classes of anisotropic maximal k-tori correspond to the classes of elements of W which have no eigenvalue

1 in V. Such elements exist, a "Coxeter element" is an example (see [5, Ch. V, § 6]).

6.10. *Regular semisimple elements.*

We keep the same notations. W_σ acts on T_σ and $\hat{T}_\sigma$. An element $x \in T_\sigma$ (resp. $\varphi \in \hat{T}_\sigma$) is called *regular* if the isotropy group of x (resp. φ) in W_σ is reduced to the identity.

6.11. *Lemma. Let* $x \in T_\sigma$ *be regular.*

(i) $Z_G(x)^0 = T$.

(ii) *If* U *is a* σ*-stable maximal torus such that* $x \in U_\sigma$ *then* $U = T$.

$H = Z_G(x)^0$ is a connected reductive σ-stable subgroup of G, containing T. We have $N_H(T) \subset N_G(T)$. By 6.1 (i) every element of $(N_H(T)/T)_\sigma$ is represented by an element of $N_H(T)_\sigma$. But since all elements of H fix x, we must have $N_H(T)_\sigma = T_\sigma$. Using 6.5 (i) and 6.4 one then sees that H has only one class of σ-stable maximal tori. This can only be if $H = T$ (as follows by the discussion of [14, p. 189—191]). This establishes (i).

Let U be as in (ii). Then U is a maximal torus of $Z_G(T)^0 = T$, whence $U = T$.

6.12. *Remark.* The above notion of regularity, which is the most convenien one for the finite groups G_σ, is unfortunately not the same as the notion of regularity discussed in [14, p. 216, 1.7]. 6.11 shows that our regularity implies regularity in the sense of loc. cit. The converse need not be true, but is true if G is a semisimple and simply connected group, as follows from the connectedness theorem of [14, p. 197, 3.9].

6.13. We next discuss an asymptotic result about regular elements, and related questions. To have a simple formulation we assume henceforth G to be quasi-simple. Let q be as in 1.10 (ii).

Let S be a subset of G_σ (or a set related to G_σ in some other way). Let f, g be functions of a positive real variable. We shall write

$$|S| = f(q) + O(g(q))$$

if there exists a constant a, depending only on the type of (G, σ) (see 1.12) such that

$$||S| - f(q)| \leqq a g(q).$$

6.14. Let T be as in 6.1, put $r = \dim T$. If $w \in W$ denote by T_w the subgroup of T of elements fixed by w. If $w \in W_\sigma$ then T_w is σ-stable. Any nonregular element of T_σ is contained in some $(T_w)_\sigma$ $(w \in W_\sigma, w \neq e)$. If $w \in W$ denote by $n(w)$ the dimension of the space of vectors fixed by w in the space V of 1.9.

6.15. *Proposition.* (i) *If* $w \in W_\sigma$ *then* $|(T_w)_\sigma| = O(q^{n(w)})$;

(ii) *the number of regular elements of* T_σ *equals* $q^r + O(q^{r-1})$;

(iii) *the same is true for the number of regular elements of* $\hat{T}_\sigma$.

Let again X be the character group of T. The character group of the diagonalizable algebraic group T_w (in the sense of [2, p. 199—200]) is $Y=X/(w-1)X$. Let Y_t be the torsion subgroup of Y. Then Y/Y_t is the character group of the torus T_w^0 and it follows that $\dim T_w^0=n(w)$. It then follows from 6.1 (iii) and 6.5 (ii) that $|(T_w^0)_\sigma|=q^{n(w)}+O(q^{n(w)-1})$. We also have $|(T_w)_\sigma|\leq|(T_w^0)_\sigma|(T_w:T_w^0)$ and $(T_w:T_w^0)=|Y_t|$ has an order bounded by a constant depending only on the type of (G,σ). This establishes (i).

(ii) is a direct consequence of (i). Let $\alpha_w x=wx\cdot x^{-1}$ $(x\in T)$. If $w\in W_\sigma$, α_w is a homomorphism $T\to T$, which commutes with σ. A character $\varphi\in\hat{T}_\sigma$ is nonregular if and only if it is trivial on a subgroup $\alpha_w(T_\sigma)$, for some $w\neq 1$ in W_σ. Now the exact sequence of algebraic groups

$$1\to T_w\to T\to \alpha_w T\to 1.$$

leads to an exact sequence

$$1\to (T_w)_\sigma\to T_\sigma\to(\alpha_w T)_\sigma\to H^1(\sigma,T_w)\to 1$$

from which one infers that the order of the subgroup of $\hat{T}_\sigma$ whose elements are trivial on $\alpha_w T_\sigma$ equals

$$|(T_w)_\sigma|\cdot|H^1(\sigma,T_w)|^{-1},$$

which is at most $|(T_w)_\sigma|$. (iii) then follows from (i).

6.16. *Corollary. Let moreover G be simply connected. Then the number of regular semisimple conjugacy classes of G_σ equals $q^r+O(q^{r-1})$.*

Fix a maximal σ-stable torus T of G. If $x\in G_\sigma$ is regular semisimple, its centralizer $Z_G(x)$ is connected by the result recalled in 6.12, hence is a maximal torus U by 6.11 (i). If $y\in U_\sigma$ is conjugate to x in G, then it is known that there exists $n\in N_G(U)$ with $y=nxn^{-1}$. Then 1.7 implies that there then also exists $n\in N_G(U)_\sigma$ with this property. Now the G_σ-conjugacy class of U is described by an element $c\in H^1(\sigma, W)$ (where $W=N_G(T)/T$, see 6.2) and we conclude from 6.5 (i) and 6.15 (ii) that there are $z_c^{-1}q^r+O(q^{r-1})$ regular semisimple conjugacy classes of G_σ which are represented by elements of U_σ (where z_c is as in 6.4). The assertion now follows from 6.4.

6.17. We want to give a similar statement for all conjugacy classes of G_σ. For this we have to make a restrictive assumption on the characteristic p of the underlying field. We recall that p is said to be good (bad) for the group G (quasi-simple) if p does not (resp. does) divide a highest root coefficient of the root system of G. The bad primes are: $p=2$ if G is of type B_n, C_n, D_n, $p=2,3$ if G is of type E_6, E_7, F_4, G_2, $p=2,3,5$ if G is of type E_8. See [14, p. 178] for a fuller discussion.

If p is good then the cases of the Ree and Suzuki groups cannot occur. So then G is a quasi-simple linear algebraic group over a finite field, and σ is its Frobenius endomorphism.

6.18. *Lemma. Assume p to be good. Then the number of unipotent conjugacy classes of* G_σ *is bounded by a constant (depending on the type and possibly the characteristic p).*

Since p is good, the number of unipotent classes of G is finite by [14, p. 185, 5.4]. Let $x \in G_\sigma$ be unipotent. One knows that the number of distinct conjugacy classes of G_σ which lie in the same conjugacy class of G as x is $|H^1(\sigma, Z_G(x)/Z_G(x)^0)|$, see [14, p. 176—177, 3.4]. It then follows that the number of unipotent conjugacy classes of G_σ is bounded by

$$\sum (Z_G(x) : Z_G(x)^0),$$

x running through a set of representatives of the unipotent classes of G.

6.19. *Theorem. Let G be a quasi-simple and simply connected, assume p to be good. Then the number of conjugacy classes of* G_σ *equals* $q^r + O(q^{r-1})$ *(the constant involved in the term* $O(q^{r-1})$ *depending possibly also on p).*

By 6.16 it suffices to show that the number of classes of elements which are not regular semisimple is $O(q^{r-1})$. This we shall do now. If x is semisimple but not regular, x is a nonregular element in some σ-stable maximal torus. It follows from 6.5 (ii) and 6.15 (ii) that the number of nonregular semisimple classes is $O(q^{r-1})$.

Now if x is semisimple, the number of conjugacy classes of G_σ represented by elements of the form $x \cdot y$ with $y \in G_\sigma$ unipotent and $xy = yx$, equals the number of unipotent conjugacy classes in $H = Z_G(x)$ (see [14, p. 176, 3.1]). H is connected since G is simply connected, and is reductive. Write $H = H'. S$, where H' is the derived group and S a central torus (both σ-stable). The number of unipotent classes in H_σ equals the same number in H'_σ. Now H' is semisimple, moreover one knows that p is good for H' if p is good for H (this follows from [14, p. 178, 4.1 and 4.3]). The assertion then follows by using 6.18.

§ 7. Construction of characters

G is still assumed to be quasi-simple of rank r. We denote by q the parameter of (G, σ).

7.1. Let T be a σ-stable maximal torus of G. Put $W_T = W = N_G(T)/T$. Let S be a subset of T_σ which satisfies the following condition: for all $x \in G_\sigma - N_G(T)_\sigma$ the intersection $T_\sigma \cap {}^x T_\sigma$ lies in S. Using the fact that two elements of T which are conjugate by an element of G are so by an element of $N_G(T)$, it follows from 6.11 (ii) that all such intersections consist of nonregular elements of T_σ.

We say that two characters $\varphi, \varphi' \in \hat{T}_\sigma$ are *equivalent*, if they have the same restriction to S. We denote the equivalence class of φ by $c(\varphi)$.

Let $\hat{T}_{\sigma,r}$ denote the set of regular characters of T_σ.

7.2. *Theorem.* (i) *Let* $\varphi \in \hat{T}_{\sigma,r}$ *be such that* $c = c(\varphi)$ *meets at least* 3 *orbits of* W_σ *in* $\hat{T}_\sigma$. *There exists a unique sign* $\varepsilon_c = \pm 1$ *and unique irreducible characters* $\chi_{T,\varphi'} (\varphi' \in c)$ *of* G_σ *such that*

$$i_{T \to G}(\varphi - \varphi') = \varepsilon_c (\chi_{T,\varphi} - \chi_{T,\varphi'}),$$

the characters $\chi_{T,\varphi'}$ *are independent of the choice of* S *(subject to the conditions imposed).*

(ii) *Let* $\varphi, \varphi' \in T_{\sigma,r}$ *both satisfy the condition of* (i). *Then* $\chi_{T,\varphi} = \chi_{T,\varphi'}$ *if and only if* φ' *lies in the* W_σ*-orbit of* φ.

Let φ, φ' and φ_1, φ_1' two pairs of equivalent regular characters of T_σ. We shall compute $\langle i_{T \to G}(\varphi - \varphi'), i_{T \to G}(\varphi_1 - \varphi_1') \rangle_G$. Using Mackey's formula (3.4) the computation reduces to that of

$$\langle r_{T \to T \cap {}^xT}(\varphi - \varphi'),\ r_{{}^xT \to T \cap {}^xT}({}^x\varphi_1 - {}^x\varphi_1') \rangle_{T \cap {}^xT},$$

with $x \in G_\sigma$.

If ${}^xT \neq T$ then the intersection $T_\sigma \cap {}^xT_\sigma$ lies in S, hence $\varphi - \varphi'$ vanishes on $T_\sigma \cap {}^xT_\sigma$. It follows that we only have to consider the x normalizing T. Mackey's formula now gives

$$\langle i_{T \to G}(\varphi - \varphi'),\ i_{T \to G}(\varphi_1 - \varphi_1') \rangle = \eta_{\varphi,\varphi_1} - \eta_{\varphi,\varphi_1'} - \eta_{\varphi',\varphi_1} + \eta_{\varphi',\varphi_1'} \tag{1}$$

where $\eta_{\varphi,\varphi'} = 1$ if φ and φ' are in the same W_σ-orbit and $\eta_{\varphi,\varphi'} = 0$ otherwise.

Let c be as in (i). Let $\varphi_i \in c$ $(1 \leqq i \leqq n)$ be in distinct W_σ-orbits. Put $f_{ij} = i_{T \to G}(\varphi_i - \varphi_j)$. Then (1) shows that we are in the situation of 3.5. The existence and uniqueness of ε_c and the $\chi_{T,\varphi}$ now follow from 3.5. It is clear that $\chi_{T,\varphi}$ is independent of the choice of S. This establishes (i).

If c and c_1 are distinct equivalence classes as in (i) and if $\varphi, \varphi' \in c$, $\varphi_1, \varphi_1' \in c_1$ represent 4 distinct W_σ-orbits of regular characters, then we conclude from (1) and (i) that

$$\langle \chi_{T,\varphi} - \chi_{T,\varphi'}, \chi_{T,\varphi_1} - \chi_{T,\varphi_1'} \rangle_G = 0,$$

which is readily seen to imply that $\chi_{T,\varphi} \neq \chi_{T,\varphi_1}$. It follows that $\chi_{T,\varphi} = \chi_{T,\varphi'}$ implies that φ and φ' are in the same W_σ-orbit. The converse follows from the uniqueness statement of (i), using that $i_{T \to G}\varphi = i_{T \to G}\varphi'$ if $\varphi' \in W_\sigma \cdot \varphi$.

The next result shows that in certain cases 7.2 yields cuspidal characters of G_σ.

7.3. *Theorem. Assume moreover* T_σ *to be minisotropic. Let* S *now be the set of all nonregular elements of* T_σ. *Let* $\varphi \in \hat{T}_{\sigma,r}$ *be such that* $c(\varphi)$ *meets more than* $|W|$ *orbits of* W_σ *in* T_σ. *Then the irreducible character* $\chi_{T,\varphi}$ *of* 7.2 *is cuspidal.*

Let $\varphi' \in c(\varphi)$. Let P be a σ-stable proper parabolic subgroup of G. If $x \in T_\sigma \cap P_\sigma$ is a regular element, then $Z_P(x)^0$ must be a maximal torus of P, which by 6.11 (i) has to be T. But then $T \subset P$, contradicting the assumption that T is minisotropic.

Hence $T_\sigma \cap P_\sigma$ consists of nonregular elements, and the same is true for $T_\sigma \cap {}^x P_\sigma$ $(x \in G_\sigma)$. An argument as used in the proof of 7.2 now shows that we have for any invariant function ψ on P_σ

$$\langle i_{T\to G}(\varphi - \varphi'), i_{P\to G}\psi \rangle_G = 0. \tag{2}$$

Take for ψ a cuspidal irreducible character of P_σ. It follows from 5.5 that the number of irreducible characters of G_σ occurring in $i_{P\to G}\psi$ is at most $|W|$. Now 7.2 and (2) show that

$$\langle \chi_{T,\varphi}, i_{P\to G}\psi \rangle_G = \langle \chi_{T,\varphi'}, i_{P\to G}\psi \rangle_G.$$

We conclude that $\langle \chi_{T,\varphi}, i_{P\to G}\psi \rangle_G \neq 0$ can only be if $c(\varphi)$ has at most $|W|$ elements. This establishes 7.3.

7.4. Let T and S be as in 7.1, let $x \in G_\sigma$. The inner automorphism of G_σ induced by x defines isomorphisms of T_σ and $\hat{T}_\sigma$ onto ${}^x T_\sigma$ and ${}^x \hat{T}_\sigma$, respectively. It is clear that ${}^x T$ and ${}^x S$ satisfy the conditions of 7.1. Hence if $\varphi \in \hat{T}_{\sigma,r}$ is as in 7.2 (i), then so is ${}^x\varphi$ (with respect to ${}^x T$) and we have that

$$\chi_{T,\varphi} = \chi_{{}^x T, {}^x \varphi}.$$

Next let T and T_1 be two σ-stable maximal tori which are not conjugate by an element of G_σ. Let $S_1 \subset T_1$ be a set as in 7.1. Assume that S and S_1 contain all intersections $T \cap {}^x T_1$, ${}^x T \cap T_1$, respectively $(x \in G_\sigma)$.

7.5. *Proposition. Let φ and φ_1 be regular characters of T_σ and $(T_1)_\sigma$, with the property of* 7.2 (i). *Then* $\chi_{T,\varphi} \neq \chi_{T_1,\varphi_1}$.

It follows as before by using Mackey's formula that

$$\langle i_{T\to G}(\varphi - \varphi'), i_{T_1}(\varphi_1 - \varphi_1') \rangle_G = 0$$

if φ' is equivalent to φ, and φ_1' to φ_1 (with respect to S and S_1). Hence

$$\langle \chi_{T,\varphi} - \chi_{T,\varphi'}, \chi_{T_1,\varphi_1} - \chi_{T_1,\varphi_1'} \rangle_G = 0,$$

which implies the assertion.

7.6. In order for the results stated in 7.2 and 7.3 to have any sense, it must be established that there are cases where they can be applied (which is not obvious). We shall deal with this point now. First we establish some consequences of 7.2 and 7.3 of an "asymptotic" nature. After that a few examples will be given where these results apply. We fix the type of (G, σ) (recall that G is quasi-simple). By 6.2 the G_σ-conjugacy classes of σ-stable maximal tori are described by entities which depend on the type only and not on the parameter q. We can therefore fix the type of T and consider the situation for large q. This is what we shall do in the next result, in which 7.2 and 7.3 are applied. The subset S of T_σ is the set of all nonregular elements.

7.7. *Proposition. Assume that the nonregular elements of T_σ are contained in a subgroup of order $O(q^{r-1})$.*

(i) *There are $q^r + O(q^{r-1})$ regular characters of T_σ for which $\chi_{T,\varphi}$ is defined, the number of distinct irreducible characters of G_σ so obtained is $(W_T)^{-1} q^r + O(q^{r-1})$.*

(ii) *If moreover T is minisotropic then all except $O(q^{r-1})$ of the characters $\chi_{T,\varphi}$ are cuspidal.*

Let H be a subgroup of T_σ of order $O(q^{r-1})$, which contains all nonregular elements. We may then take $S=H$. Let $a>0$. It is clear from 6.15 (iii) that for all regular characters φ of T_σ, except at most $O(q^{r-1})$ of them, we have that $|c(\varphi)|>a$. Now 7.7 quickly follows from 7.2 and 7.3.

7.8. *Examples.* (a) Let G be a quasi-simple split group over a finite field k with q elements, let σ be the Frobenius endomorphism. Denote by T_0 a k-split maximal torus of G, let W_0 be its Weyl group. Let T be a Coxeter torus in G, i.e. a maximal torus in G, defined over k, whose G_σ-conjugacy class corresponds to the Coxeter class of W_0, by the map of 6.2 (notice that $H^1(\sigma, W_0)$ is the set of conjugacy classes of W_0). Let $W=N_G(T)/T$. By 6.5 (i) we have that W_σ is isomorphic to the centralizer of a Coxeter element of W_0, which is a cyclic group generated by that element (see [6, Proposition 30]). Let w be a corresponding generator of W_σ. Let h be the order of w, this is the Coxeter number of W_0, see [5, p. 117]. The nonregular elements of T_σ are those of the groups $(T_{w^d})_\sigma$ of 6.14, where d divides h. Let $n(w^d)$ be, as in 6.14, the dimension of the space of vectors fixed by w^d. It then follows from 6.15 (i) that the condition stated in the opening sentence of 7.7 is satisfied if we have

$$\sum_{\substack{d \mid h \\ d<h}} n(w^d) < r. \tag{3}$$

Now the $n(w^d)$ can be found from the eigenvalues of a Coxeter element, which are well known (see [5, p. 118] and the tables of [5, p. 250—275]). One checks without trouble that, for example, (3) holds whenever G is an exceptional quasi-simple group (i.e. of type E_6, E_7, E_8, F_4, G_2).

Another case where (3) holds in that where G is of type A_l, with $l+1$ a power of a prime p. However, in that case it is easily seen that there is a maximal $T_{w^d}(\neq T)$, of dimension $<r$, from which one also sees that the condition of 7.7 holds. Since a Coxeter torus is minisotropic (see 6.8), all $\chi_{T,\varphi}$ except $O(q^{r-1})$ of them obtained are cuspidal.

(b) Let G be a simple adjoint group defined over the finite field k. σ is again the Frobenius endomorphism. Let T be a maximal k-torus which is not contained in a proper reductive k-subgroup of G of dimension $>r$. Let $x \in T_\sigma$. Then $Z_G(x)^0$ must be either G or T, from which one concludes that all elements of $T_\sigma - \{e\}$ are regular. In this case we can apply 7.2 with $S=\{e\}$. We have that T_σ is a trivial intersection set in the sense of [8, p. 123]. Example of this situation: Let T be such that T has no proper k-subgroups. This means that the map σ^* of 1.9 defines a linear map of $X\otimes\mathbf{Q}$ which acts irreducibly in that $\mathbf{Q}$-vectorspace (which is the case, for example, if T is a Coxeter torus in a split group of type G_2, F_4, E_8). Let $H\neq T$ be a minimal connected reductive k-subgroup which contains T. The assumption on T implies that H is again quasi-simple, and that T has the required property relative to the adjoint group of H.

Another example where T_σ is a trivial intersection set is given by the Coxeter torus in PGL_p, where p is a prime number.

REFERENCES

[1] Benson, C. T. and Curtis, C. W., On the degrees and rationality of certain characters of finite Chevalley groups, *in Proc. Symp. Pure Math.,* Vol. **21**, pp. 1—5, Amer. Math. Soc., Providence, R. I., 1971.

[2] Borel, A., *Linear algebraic groups,* Benjamin, New York, 1969.

[3] Borel, A. and Tits, J., Groupes réductifs., *Publ. math. I.H.E.S.* **27** (1965), 55—151.

[4] Borel, A. and Tits, J., Eléments unipotents et sous-groupes paraboliques de groupes réductifs I, *Inv. math.,* **12** (1971), 95—104.

[5] Bourbaki, N., *Groupes et algèbres de Lie,* Chap. IV, V, VI, Hermann, Paris, 1968.

[6] Carter, R. W., Conjugacy classes in the Weyl groups, *Comp. Math.* **25** (1972), 1—59.

[7] Curtis, C. W., The Steinberg character of a finite group with a (B, N)-pair, *J. Algebra,* **4** (1966), 433—441.

[8] Feit, W., *Characters of finite groups,* Benjamin, New York, 1967.

[9] Kilmoyer, R., The reflection character of a finite group with a (B, N)-pair, *in Proc. Symp. Pure Math.,* Vol. **21**, pp. 91—94, Amer. Math. Soc., Providence, R. I., 1971.

[10] Solomon, L., The Steinberg character of a finite group with BN-pair, *in Proc. Sympos. Theory of finite groups,* pp. 213—221, Benjamin, New York, 1969.

[11] Springer, T. A., Cusp forms for finite groups, *in Seminar on algebraic groups and related finite groups,* pp. 97—120, Springer Lecture Notes, Vol. **131**, 1969.

[12] Springer, T. A., Characters of special groups, *ibid.,* pp. 121—166.

[13] Springer, T. A., Generalization of Green's polynomials, *in Proc. Symp. Pure Math.,* vol. **21**, pp. 149—153, Amer. Math. Soc., Providence, R. I., 1971.

[14] Springer, T. A. and Steinberg, R., Conjugacy classes, *in Seminar on algebraic groups and related finite groups,* pp. 167—266, Springer Lecture Notes, Vol **131**, 1969.

[15] Steinberg, R., Endomorphisms of linear algebraic groups, *Amer. Math. Soc. Memoirs,* no. **80** (1968).

Additional references (added April 1974):

[16] Benson, C. T. and Curtis, C. W., On the degree and rationality of certain characters of finite Chevalley groups, *Trans. Amer. Math. Soc.*, **165** (1972), 251—274.

[17] Curtis, C. W., Iwahori, N. and Kilmoyer, R., Hecke algebras and characters of parabolic type of finite groups with (B, N)-pairs, *Publ. Math. I. H. E. S.*, **40** (1971), 81—116.

[18] Srinivasan, B., Isometries in finite groups of Lie type, *in Finite groups '72 (Proc. Gainesville Conference in Finite Groups)*, pp. 142—146, North-Holland Publ. Corp., Amsterdam, 1973.

THE ZETA FUNCTION OF A CUSPIDAL REPRESENTATION OF A FINITE GROUP $GL_n(k)$

T. A. SPRINGER

This note contains some fairly trivial results about the zeta function of the title. A functional equation of the usual type is established. For the determination of the constant occurring in it we have to use the character theory of $GL_n(k)$. A possible connection with recent work of S. I. Gelfand and D. A. Kajdan is discussed at the end.

I am grateful to D. A. Kajdan for an interesting discussion.

1. Let k be a finite field with q elements. We fix a nontrivial character χ of the additive group of k.

Let $G=G_n=GL_n(k)$, $X=M_n(k)$ (the space of $n\times n$-matrices over k), we assume that $n>1$. We denote by $C(X)$ the space of complex valued functions on X. If $x, y\in X$ we write

$$\langle x, y\rangle = \chi(\mathrm{Tr}\,(xy)).$$

$\langle\,,\rangle$ is a bicharacter on $X\times X$, which we use to define a Fourier transform on $C(X)$: If $f\in C(X)$ then its Fourier transform $\hat{f}$ is given by

$$\hat{f}(x) = q^{-\frac{n^2}{2}} \sum_{y\in X}\langle x, y\rangle f(y). \tag{1}$$

We then have

$$\hat{\hat{f}}(x) = f(-x). \tag{2}$$

Let V be a complex vector space, let φ be a function on G with values in V. If $f\in C(X)$ we write

$$Z(f, \varphi) = \sum_{x\in G} f(x)\,\varphi(x);$$

this is the "zeta function" of φ (which is a function on $C(X)$ with values in V).

Moreover let

$$\theta(\varphi, x) = q^{-\frac{n^2}{2}} \sum_{y\in G}\langle x, y\rangle\,\varphi(y) \quad (x\in X). \tag{3}$$

A simple computation then shows that

$$Z(f, \varphi) = \sum_{x\in X}\theta(\varphi, -x)\hat{f}(x). \tag{4}$$

Now let π be a representation of G in a finite-dimensional vector space. It is then immediate that we have, if $x \in X$, $y \in G$

$$\begin{cases} \theta(\pi, xy) = \pi(y)^{-1}\theta(\pi, x), \\ \theta(\pi, yx) = \theta(\pi, x)\,\pi(y)^{-1}. \end{cases} \tag{5}$$

Lemma. Let π be a cuspidal representation of G. Then $\theta(\pi, x)=0$ if x is a singular element of X.

For the notion of cuspidal representation see [5, section 4]. If $x=0$ then the assertion comes from the fact that $\sum \pi(x)=0$, which follows from the observation that a cuspidal representation of G cannot contain the trivial representation. Next let $x \neq 0$ be singular, of rank $r>0$. Let $e_r=(x_{ij})$ be the matrix with $x_{ij}=0$ unless $1 \leqq i=j \leqq r$, and $x_{ii}=1$ for $1 \leqq i \leqq r$. There then exist $y, z \in G$ such that $x=ye_rz$. Using (5) one sees that it suffices to consider the case that $x=e_r$. Let U be the subgroup of G whose elements (x_{ij}) are such that $x_{ii}=1$, $x_{ij}=0$ unless $1 \leqq i \leqq r$, $r+1 \leqq j \leqq n$. U is the set of k-rational points of the unipotent radical of a proper parabolic subgroup of GL_n, moreover we have $ue_r=e_r$ for all $u \in U$. Hence

$$q^{r(n-r)}\theta(\pi, e_r) = \sum_{u \in U} \theta(\pi, ue_r) = \theta(\pi, e_r) \sum_{u \in U} \pi(u) = 0,$$

by (5) and the cuspidality of π. This proves the lemma.

Proposition. Let π be an irreducible cuspidal representation of G. There exists a constant $\varepsilon(\pi)$ such that

$$Z(f, \pi) = \pi(-1)\,\varepsilon(\pi)\,Z(\hat{f}, \pi^{-1}). \tag{6}$$

By (5) we have that $\theta(\pi, 1)$ commutes with all $\pi(x)$, hence is a scalar multiple $\varepsilon(\pi)$ of 1, by Schur's lemma. The proposition now follows from (4) and (5), using the lemma.

(6) is the "functional equation of the zeta function".

2. We keep the notation of the Proposition. We shall now give an explicit determination of the constant $\varepsilon(\pi)$. For this we need detailed information about the character theory of G, for which we refer to [4]. Let ψ be the character of π. Then

$$\psi(1)\,\varepsilon(\pi) = q^{-\frac{n^2}{2}} \sum_{x \in G} \chi(\operatorname{Tr} x)\,\psi(x). \tag{7}$$

To compute this sum we use [4]. It follows from *loc. cit.* that the cuspidal (or discrete series) character ψ can be given as follows. Let k_n be the extension of degree n of k. Call a character φ of k_n^* regular if φ is not fixed by any nontrivial power of the Frobenius automorphism F of k_n/k. Then the cuspidal characters ψ of G are parametrized by the orbits of the Galois group of k_n/k in the set of regular characters of k_n^* (see [4, p. 149, 7.12 and p. 155, 8.6]). If φ is such a character, let ψ_φ be the corresponding cuspidal character (so that $\psi_{F\varphi}=\psi_\varphi$) and π_φ a corresponding representation. Now observe that the function $x\to\chi(\mathrm{Tr}\,x)$ on G depends only on the semisimple part of x. Its principal parts (see [4, p. 143]) can then be determined by [4, p. 143—144, 6.6]. The principal parts and the degree of ψ_φ are given in [loc. cit., p. 149, 7.12]. Using [loc. cit., 6.3 (i)] one then easily computes the sum occurring in (7).

The result is

$$\varepsilon(\pi_\varphi) = (-1)^{n-1} q^{-\frac{n}{2}} \sum_{x\in k_n^*} \chi(\mathrm{Tr}_{k_n/k}\, x)\,\varphi(x), \tag{8}$$

a Gaussian sum.

3. *Remarks.* (a) One can ask whether a functional equation like (6) exists for any irreducible representation of G. This is not so, as one sees in the case of the trivial representation 1. Write $Z(f)$ for $Z(f, 1)$. It follows from (5) that $\theta(1, x)$ depends only on the rank of x.

Put $\alpha_r=\theta(1, x)$ if $\operatorname{rank} x=r$ and let

$$Z_r(f) = \sum_{\operatorname{rank} x=r} f(x),$$

so that $Z_n(f)=Z(f)$.

We then have

$$Z(f) = \alpha_n Z(\hat{f}) + \sum_{0\leqq r<n} \alpha_r Z_r(\hat{f}). \tag{9}$$

One can show without much difficulty that

$$\alpha_r = (-1)^r q^{-\frac{n}{2}} (q-1)\dots(q^{n-r}-1) \quad (0\leqq r<n),$$

from which one sees that $\alpha_r\neq 0$.

If we had a functional equation $Z(f)=\alpha Z(\hat{f})$, then (9) would imply a relation

$$\beta Z_n(\hat{f}) + \sum_{0\leqq r<n} \alpha_r Z_r(\hat{f}) = 0,$$

which is impossible because the $Z_r(f)$ $(0\leqq r\leqq n)$ are linearly independent functions on $C(X)$ (as one can easily check).

(b) The Gaussian sum (8) also occurs in connection with some recent work [3] of S. I. Gelfand and D. A. Kajdan, who have studied for the finite group G a question discussed by I. M. Gelfand and Kajdan in their paper [1] for the case of the general group over a local field. We write G_n instead of G, since the dependence on n is now of importance. Let $\pi(\pi')$ be an irreducible representation of G_n (resp. G_{n-1}). One wishes to associate with this pair a complex number $\Gamma(\pi, \pi')$, in the same way as was done in [1] for the case of local fields. S. I. Gelfand and Kajdan have done this explicitly for $n=3$, using the detailed information about G_3 contained in [2]. The number $\Gamma(\pi, \pi')$ turns out in this case to be something like the Gaussian sum (8).

Now the representation theory of G_n suggests another way of defining $\Gamma(\pi, \pi')$, for *cuspidal* representations π and π'. More generally, let π and π' be cuspidal representations of G_m and G_n, with m and n relatively prime. Let ψ and ψ' be the characters of π, π'. We know that ψ and ψ' come from regular characters φ, φ' of k_m and k_n, respectively (see no. 2). Since m and n are relatively prime, we have $k_{mn}=k_m \otimes_k k_n$. Define a character Φ of k^*_{mn} by

$$\Phi(x) = \varphi(N_{k_{mn}/k_m}x)\,\varphi'(N_{k_{mn}/k_n}x).$$

One checks that Φ is a regular character of k^*_{mn}, which then determines a cuspidal character ψ of G_{mn}. Let Π be a corresponding irreducible representation. The number $\varepsilon(\Pi)$, defined in no. 2, is then given by

$$\varepsilon(\Pi) = (-1)^{mn-1} q^{-\frac{1}{2}mn} \sum_{x \in k^*_{mn}} \chi(\mathrm{Tr}_{k_{mn}/k}\, x)\,\varphi(N_{k_{mn}/k_m}x)\,\varphi'(N_{k_{mn}/k_n}x).$$

If $m=3$, $n=2$ the number $\Gamma(\pi, \pi')$ defined by S. I. Gelfand and Kajdan turns out to be equal to $\varepsilon(\Pi)$ (up to a scalar factor, independent of π and π'). It would be interesting to know if a relation exists between their method of obtaining $\Gamma(\pi, \pi')$ and the zeta function of this note.

REFERENCES

[1] Gelfand, I. M. and Kajdan, D. A., Representations of the group $GL(n, K)$ where K is a local field, *this volume*.

[2] Gelfand, S. I. (Гельфанд С. И.) Представления полной линейной группы над конечным полем, *Матем. сборник*, **83** *(125)*, (1970), 15—41.
(Representations of the general linear group over a finite field.)

[3] Gelfand, S. I. and Kajdan, D. A., private communication.

[4] Springer, T. A., Characters of special groups, *in Seminar on algebraic groups and related finite groups,* Springer Lecture Notes, Vol. **131**, 1970.

[5] Springer, T. A., On the character of certain finite groups, *this volume.*

ON THE FOURIER TRANSFORM OF SECOND DEGREE CHARACTERS

SHUNICHI TANAKA

1. This note is an exposition of two theorems connected with the Fourier transform of second degree characters on a locally compact Abelian group. One of them (Theorem 1) asserts that the Fourier transform of a second degree character is again a second degree character up to a constant of absolute value one and the other (Theorem 2) gives a sufficient condition such that this constant is one. These results were obtained by Weil [3] as corollaries of his general theory of certain group of unitary operators. Cartier [1] also has given a direct and elementary proof of these results.

Here we give another proof of Theorem 1. Our proof is based on the well-known lemma which characterizes a bounded operator commuting with the multiplications of all characters. The proof of Theorem 2 is not essentially different from that of Weil or Cartier and is included here only for completeness.

2. Let G be a locally compact Abelian group and G^* be its dual group. We denote by $\langle x, x^*\rangle$ the value of $x^* \in G^*$ at $x \in G$. Let $S(G)$ be the space of Schwartz—Bruhat functions on G. For φ in $S(G)$, its Fourier transform φ^* is defined by

$$\varphi^*(x^*) = \int_G \varphi(x)\langle x, x^*\rangle\, dx \quad x^* \in G^*,$$

where dx is a Haar measure on G. φ^* is in $S(G^*)$ and the mapping $W: \varphi \to \varphi^*$ is an isomorphism of $S(G)$ onto $S(G^*)$. If we normalize dx^* suitably, we can extend W to a unitary transformation of $L^2(G, dx)$ onto $L^2(G^*, dx^*)$. If ϱ is an isomorphism of G onto G^*, we can define the operator W_ϱ by

$$W_\varrho \varphi(x) = |\varrho|^{\frac{1}{2}}(W\varphi)(x\varrho) = |\varrho|^{\frac{1}{2}} \int_G \varphi(y)\langle y, x\varrho\rangle\, dy,$$

where $|\varrho|$ is the module of the isomorphism ϱ.

For $h \in G$, let T_h be a unitary operator on $L^2(G, dx)$ defined by

$$(T_h\varphi)(x) = \varphi(x+h).$$

Let S_h denote the unitary operator on $L^2(G^*, dx^*)$ defined by

$$(S_h\varphi)(x^*) = \varphi(x^*)\overline{\langle h, x^*\rangle}.$$

It is easy to see $S_h W = W T_h$. T_h and S_h induce automorphisms of $S(G)$ and $S(G^*)$ respectively.

For a bounded measurable function $a(x)$ on G, let M_a be a operator defined by the multiplication of $a(x)$:

$$(M_a \varphi)(x) = a(x)\varphi(x).$$

The operators of the type M_a are characterized by the following

Lemma. Let A be a bounded operator on $L^2(G^*, dx^*)$ commuting with S_h ($h \in G$). Then there exists a bounded measurable function $a(x^*)$ on G^* such that $A = M_a$.

For the proof, see the Appendix in [2].

3. We call a continuous function f on G a second degree character if $|f(x)| = 1$ for $x \in G$ and there exists a homomorphism $\varrho = \varrho_f$ from G to G^* such that

$$f(x+y) = f(x)f(y)\langle x, y\varrho\rangle$$

for $x, y \in G$. For a second degree character f, M_f is a unitary operator and it induces an automorphism of $S(G)$ (see Section 12 in [3]).

From now on we only consider a second degree character f whose associated homomorphism ϱ_f is an onto isomorphism. The convolution $f * \varphi$ of f with φ in $S(G)$ is rewritten as follows:

$$(f*\varphi)(x) = \int_G f(x-y)\varphi(y)\,dy = f(x)\int_G \langle y, -x\varrho\rangle f(-y)\varphi(y)\,dy.$$

In other words,

$$|\varrho|^{\frac{1}{2}}(f*\varphi) = M_f W_{-\varrho} M_{\check{f}}\varphi, \tag{1}$$

where $\check{f}(x) = f(-x)$. So $\varphi \to |\varrho|^{\frac{1}{2}}(f*\varphi)$ is an automorphism of $S(G)$ and can be extended to a unitary operator of $L^2(G, dx)$.

Theorem 1. *f being as above, there exists a constant $\gamma = \gamma(f)$ of absolute value one such that for all φ in $S(G)$, we have*

$$|\varrho|^{\frac{1}{2}}(f*\varphi)^*(x^*) = \gamma\varphi^*(x^*)f(x^*\varrho^{-1})^{-1} \quad x^* \in G^*. \tag{2}$$

Remark. If we put $x^* = 0$ in (2), we have

$$|\varrho|^{\frac{1}{2}}\int_G (f*\varphi)(x)\,dx = \gamma\int_G \varphi(x)\,dx. \tag{3}$$

Conversely if we put $\varphi(x) = \psi(x)\langle x, x^*\rangle$ in (3), we have (2) for the function ψ. So (2) and (3) are equivalent.

Proof. Let U be an automorphism of $S(G^*)$ defined by

$$U\varphi = W(|\varrho|^{\frac{1}{2}} f*(W^{-1}\varphi)).$$

U is extended to a unitary operator on $L^2(G^*, dx^*)$. As U commutes with S_h $(h \in G)$, there exists a bounded measurable function $a(x^*)$ on G^* such that $U = M_a$. By the unitarity of U, $|a(x^*)| = 1$ almost everywhere. $a(x^*)$ is continuous because U is an automorphism of $S(G^*)$. So we have

$$(|\varrho|^{\frac{1}{2}} f * \varphi)^*(x^*) = a(x^*)\varphi^*(x^*)$$

for $x^* \in G^*$ and $\varphi \in S(G)$. Putting $x^* = 0$ here, we have the identity (3) which is equivalent to (2).

4. Let Γ be a closed subgroup of G and Γ_* be the closed subgroup of G^* associated with Γ by duality, i.e.

$$\Gamma_* = \{\xi^* \in G^*; \langle \xi, \xi^* \rangle = 1 \quad \text{for} \quad \xi \in \Gamma\}.$$

Let $d\xi$ and $d\xi^*$ be Haar measures on Γ and Γ_* respectively. If we normalize $d\xi^*$ suitably, we have the following formula, which one may call the (generalized) Poisson summation formula (see Sections 16, 17, 18 in [3]).

$$\int_\Gamma \varphi(\xi)\, d\xi = \int_{\Gamma_*} \varphi^*(\xi^*)\, d\xi^*, \quad \varphi \in S(G). \tag{4}$$

Theorem. Let f be a second degree character on G whose associated homomorphism $\varrho = \varrho_f$ is an isomorphism of G onto G^*. If, moreover, $f(\xi) = 1$ for $\xi \in \Gamma$ and ϱ gives an isomorphism Γ onto Γ_*, we have $\gamma(f) = 1$.

Proof. For φ in $S(G)$, we consider the integral

$$I = \int_\Gamma (f * \varphi)(\xi)\, d\xi.$$

We have

$$I = \int_\Gamma f(\xi)(\check{f}\varphi)^*(-\xi\varrho)\, d\xi,$$

using the identity (1). Then

$$I = \int_\Gamma (\check{f}\varphi)^*(-\xi\varrho)\, d\xi = c \int_{\Gamma_*} (\check{f}\varphi)^*(\xi^*)\, d\xi^*,$$

where c is a positive constant. By (4), we have

$$I = c\int_\Gamma (\check{f}\varphi)(\xi)\, d\xi = c\int_\Gamma \varphi(\xi)\, d\xi.$$

We go back to the definition of the integral I and have

$$I = \int_{\Gamma_*} (f * \varphi)^*(\xi^*)\, d\xi^*$$

by (4). Then

$$I = |\varrho|^{-\frac{1}{2}}\gamma(f) \int_{\Gamma_*} \varphi^*(\xi^*) f(\xi^* \varrho^{-1})^{-1}\, d\xi^*$$

by Theorem 1. We have

$$I = |\varrho|^{-\frac{1}{2}} \gamma(f) \int_{\Gamma_*} \varphi^*(\xi^*)\, d\xi^* = |\varrho|^{-\frac{1}{2}} \gamma(f) \int_{\Gamma} \varphi(\xi)\, d\xi.$$

So we have obtained the identity

$$c \int_{\Gamma} \varphi(\xi)\, d\xi = |\varrho|^{-\frac{1}{2}} \gamma(f) \int_{\Gamma} \varphi(\xi)\, d\xi$$

for all φ in $S(G)$ and $\gamma(f)=1$ follows.

REFERENCES

[1] Cartier, P., Über einige Integralformeln in der Theorie der quadratischen Formen, *Math. Z.*, **84** (1964), 93—100.

[2] Saito, M., Représentations unitaires du groupe de déplacements dans un plan *p*-adique, *J. Math. Soc. Japan,* **19** (1967), 411—425.

[3] Weil, A., Sur certains groupes d'opérateurs unitaires, *Acta Math.,* **111** (1964), 143—211.

THE RÔLE OF AFFINE WEYL GROUPS IN THE REPRESENTATION THEORY OF ALGEBRAIC CHEVALLEY GROUPS AND THEIR LIE ALGEBRAS*

DAYA-NAND VERMA

Summary

This is the first of a series of papers on the rational representations of a connected semisimple affine algebraic group (in short, an algebraic Chevalley group) G over a field of characteristic $p \neq 0$, and the representations of the related modular Lie algebra as well as the p-modular representations of the finite Chevalley (sub)groups. This paper has two parts, the first of which is purely elementary in that it deals only with root-systems. In this part is formulated a "weighted mean-value conjecture" on the famous dimension polynomial of H. Weyl, after giving preliminaries on affine Weyl groups. The object of Part II is two-fold: to give a quick account (§§ 4, 5 and 6) of some of the known facts in the representation theory of G with special reference to an earlier conjecture of the author (which has been proved true, by J. E. Humphreys, when p exceeds the Coxeter number) that defines the relevance of affine Weyl groups, and to put forward several new conjectures (§§ 5, 7 and 8) on certain decomposition numbers. The binding force of the two parts is provided by Conjecture IV in § 7. The Introduction (§ 0) describes the contents in greater detail, and the Epilogue (§ 8) points out a seemingly *metamathematical* aspect of our study when the characteristic p is regarded as a variable.

Added in proof: Special attention must be drawn to the various Footnotes added in proof, where we remark on the recent developments.

List of notations

§ 1.1 Δ, l, W, $\langle , \rangle$, E, X, X'; Δ_+, X^+; $T_\lambda (\lambda \in X)$, $\tilde{W}$, $\tilde{W}'$; α^v(for $\alpha \in \Delta$) [NOTE: The reader is urged to identify α^v with the notation $\alpha^\vee$ standardized by Bourbaki [3]; here α^v has been used throughout for printer's convenience, as there is no chance of confusion]; α_i, λ_i, n_i, R_i $(1 \leq i \leq l)$; α_0,

* This is a revised and expanded version of part of the material presented at the Budapest Summer School on Group Representations, 1971, under the title "On rational representations of algebraic Chevalley groups in non-zero characteristics". The rest of the material will appear in a series of subsequent papers. The computations summarized in § 3, concerning the verification of Conjecture I for types G_2 and A_3, were done subsequent to the lectures, with the assistance of the Computer Section of the Tata Institute of Fundamental Research. See also the acknowledgement contained in the Footnote (**) in § 5.

§ 1.2 λ_0, n_0, R_0, J_0; $E^+ \supset E^\# \supset E^\star$; $\sigma_0 \in W$; $\theta_j (j \in J_0)$; $\Omega = \{\omega_j | j \in J_0\}$, $\Omega_0 = \{\gamma_j | j \in J_0\}$;

§ 1.3 $l(\ )$; $\delta_\sigma, I_\sigma, w_\sigma (\sigma \in W)$; $W^\#$; $r = |\Delta_+|$, $h_\Delta = \frac{2r}{l}$;

§ 1.4 $\tilde{W}^+ \supset \tilde{W}^\#$; $\delta \in X$; $E^\star_{\text{top}}$; $j_0 \in J_0$, $w_0 \in \tilde{W}_0^\#$, $\omega_{j_0} \in \Omega$;

§ 1.5 E_J, $\tilde{W}^J$, $\tilde{W}_J \supset \tilde{W}_J^+ \supset \tilde{W}_J^{++} \supset \tilde{W}_J^\#$ $(J \subsetneqq \{0, 1, 2, \ldots, l\})$; $(E^\star)^w$ is called w-simplex for $w \in \tilde{W}'$;

§ 1.6 partial-orders $\overset{\mathbf{A}}{\leqq}$ and $\overset{\mathbf{B}}{\leqq}$ on $\tilde{W}'$, with restriction to $\tilde{W}^+$ denoted by $\leqq$;

§ 2.1 $D(\ )$; $b_\sigma \in \mathbf{Z}$ (for $\sigma \in W$) defined by Conjecture I;

§ 4.1 $\mathfrak{g}_\mathbf{C}$; G, B, T; $\mathfrak{g}_K, \mathfrak{b}_K, \mathfrak{h}_K$; $U_\mathbf{Z}$; irreducible G-module $M_\lambda (\lambda \in X^+)$; partial-order $\leqq$ on X; char M,

§ 4.2 M^{Fr} (for arbitrary G-module M); $X^\#$;

§ 4.4—4.5 "Weyl module" $\bar{V}_\lambda$, $c_{\lambda\mu}$, $\gamma_{\lambda\mu} (\lambda, \mu \in X^+)$; $c_{\lambda\mu} (\lambda, \mu \in X^\#)$; partial-order $\leqq^\#$ on $X^\#$; Y_0;

§ 6.1—6.4 $\mathbf{u}$; Λ; bijection $\pi : X^\# \to \Lambda$; $\mathbf{u}^-, \mathbf{b}'$; $Z_\lambda, Q_\lambda, d_\lambda, a_\lambda$ (for $\lambda \in \Lambda$);

§ 5.1 $X^\star$; $X_J^\star$ $(J \subsetneqq \{0, 1, 2, \ldots, l\}$ here and in the notations below);

§ 5.5 $c^J_{w'w} \in \mathbf{Z}$ (for $w', w \in \tilde{W}_J^{++}$) defined by Conjecture III in § 5.4;

§ 7.3 $d^J_w \in \mathbf{Z}$ (for $w \in \tilde{W}_J^\#$) defined by Conjecture III′ in § 7.1.

CONTENTS

§ 0. Introduction

Given a complex semisimple Lie algebra $\mathfrak{g}_\mathbf{C}$ of rank l, there is a homogeneous polynomial D (the Weyl polynomial) in l variables with rational coefficients, which yields the dimensions of the irreducible $\mathfrak{g}_\mathbf{C}$-modules when the variables assume positive integer values (cf. the opening paragraph of § 2; it is well known that the irreducible $\mathfrak{g}_\mathbf{C}$-modules are parametrized, up to isomorphism, by l-tuples of positive integers). In Part I (§ 2) we present a new conjecture on this polynomial which amounts to a weighted mean-value assertion when the variables undergo certain affine transformations. In § 1 we give preliminaries on the affine Weyl group, in terms of which it is easier and more natural to introduce these affine transformations denoted by w_σ for σ running in the Weyl group W of $\mathfrak{g}_{\mathrm{C}}$. However, most of the discussion in § 1 is required only in Part II from § 5 onwards; thus, a reader who is familiar with standard facts on affine Weyl groups and is not interested in the representation-theoretic discussion in Part II may proceed directly to § 2 after noting the definition of w_σ in § 1.3. Denoting by D_σ the (inhomogeneous) polynomial obtained by transforming D under w_σ, Conjecture I says that there should exist a unique integer linear combination $\sum_\sigma b_\sigma \cdot D_\sigma$ which equals the constant polynomial 1. The linear independence of $\{D_\sigma | \sigma \in W\}$ means simply that these span the ($|W|$-dimensional) space of W-harmonic polynomials, and may be much easier to prove; that would force b_σ to be rational numbers, but the main import of the Conjecture lies in the assertion that b_σ must be integers. (*See Footnote* (²) *to Conjecture* I.)

We have verified the conjecture in four cases of low rank (§ 3) when $\mathfrak{g}_{\mathrm{C}}$ is simple of type A_2, B_2, C_2, or A_3 (the case of A_1 being trivial since $|W|=2$ and the two D_σ's are simply x and $1-x$). But we are unable to evolve any reasonable guess on the values of the "weighting function" b_σ. As a matter of fact the values are extremely mysterious in the case of type G_2; and though I knew of the values in case of types A_2 and B_2 for quite a while, the computation for types G_2 and A_3 could only be made possible through computer print-outs of the explicit expansions of all the D_σ's. Thus, it must be admitted that we have no "reason" to believe in this conjecture short of these verifications and the beauty of the related Conjecture IV in Part II (§ 7); see the argument in the second paragraph of § 7 deducing Conjecture I from Conjecture IV. *(See Footnotes* (²) *and* (⁶) *to these two Conjectures.)*

Part II of this paper concerns the representation theory of algebraic Chevalley groups defined over a field of characteristic $p \neq 0$ and their Lie algebras. The basic unsolved problem here is the computation of characters of the irreducible representations (which are known to be parametrized by l-tuples of non-negative integers in exactly the same manner as in characteristic 0). Even the dimensions of these representations are not known. Our guess, however, is that the problem of giving explicitly a p-analogue of the Weyl's dimension formula would turn out to be no simpler than that of obtaining an explicit p-analogue of the Weyl's character formula for irreducible $\mathfrak{g}_{\mathrm{C}}$-modules. Though we may be far from providing a satisfactory solution to this problem, one can easily reduce the problem to that of determining a finite set of integers $c_{\lambda\mu}$; this reduction is described in detail in § 4 (our exposition in § 4 and § 6.1—6.4 can be read independently of Part I, in contrast with the rest of Part II which uses § 1 heavily) since it appears here for the first time. The numbers $c_{\lambda\mu}$ describe the composition factors of certain indecomposable modules whose characters are given by Weyl's formula for irreducible $\mathfrak{g}_{\mathbf{C}}$-modules. Determination of these fundamental constants is a different (and indeed, a difficult) matter with which we do not concern ourselves here. Nevertheless, it suits us to present here two conjectures (in § 5) on these integers which are couched in terms of the affine Weyl group of § 1; *we draw special attention of the reader to the third paragraph of* § 5, where we have introduced a modification in the action of the affine Weyl group (denoted $\tilde{W}'$) and related objects on the real vector space E spanned by the root-system, this modification being dependent on the prime p (we thought this to be more convenient than introducing from the start an action of $\tilde{W}'$ on E dependent on a parameter t, such that $t=1$ gives the action in § 1 and $t=p$ gives that for § 5 onwards). The contraption $\tilde{W}'$ enters the representation theory purely by virtue of a conjecture of this author dating back to 1966 (see the Footnote** in § 5), which has been called the "Harish-Chandra principle for characteristic $p\neq 0$", and which has been proved to be true by Humphreys for large p.

In § 6 we come to the representations of the Lie algebras associated to Chevalley groups. In this expository section the emphasis is neither on irreducible representations alone (which can all be lifted to the algebraic group uniquely) nor on all representations (which are hardly understood at present), but is geared specifically towards the results of Humphreys on the principal indecomposable modules (PIM's, in short) of the restricted enveloping algebra **u** of the Lie algebra. (The rôle of these PIM's in the representation theory of Chevalley groups has recently assumed added significance; while it is being shown in [19] that these can be lifted, perhaps uniquely, to the algebraic group, in [8] way is paved for interpreting the PIM's of the p-modular group-ring of a finite Chevalley group in terms of the PIM's of **u**.) Thus, the highlight of § 6 is the theorem of Humphreys [7] which gives a certain decomposition of the PIM's Q_λ of **u** (and hence the dimensions of the Q_λ) in terms of the decomposition of the "induced modules" Z_λ which are the analogues for **u** of the infinite-dimensional induced $\mathfrak{g}_{\mathbf{C}}$-modules studied initially in [15] (and later by the school of Gelfand [6]); here the "decomposition" of Q_λ stands for that in the Grothendieck-ring of the semisimple category of projective modules for a "Borel subalgebra" of **u**, whereas that of Z_λ refers to the determination of the multiplicities of its composition factors as a **u**-module, i.e. to the decomposition of Z_λ in the Grothendieck-ring of all **u**-modules. It is this decomposition of Z_λ (given by certain numbers d_λ) that is the subject of the main conjecture in § 7. For concluding remarks on the behaviour of the numbers $c_{\lambda\mu}$ and d_μ when p varies, we refer the reader to Epilogue (§ 8), particularly the opening words there.

For purely technical reasons, we shall assume $\mathfrak{g}_{\mathbf{C}}$ (and hence the algebraic Chevalley group of type $\mathfrak{g}_{\mathbf{C}}$) to be simple, i.e. the root-system of $\mathfrak{g}_{\mathbf{C}}$ is assumed indecomposable.

PART I

A CONJECTURE ON WEYL'S DIMENSION POLYNOMIAL

§ 1. Affine Weyl groups

1.1. Let Δ be an indecomposable reduced root-system [3, p. 151] in a real vector space E with positive-definite inner product $\langle , \rangle$ which is unique up to a scalar multiple. Let W be the Weyl group of Δ generated by the reflections $R_\alpha = R_{-\alpha}$ on E for $\alpha \in \Delta$, where R_α sends x to $x - \langle x, \alpha^\vee \rangle \alpha$ with $\alpha^\vee$ defined to mean $2\alpha / \langle \alpha, \alpha \rangle$. Let X' denote the root-lattice $\mathbf{Z} \cdot \Delta$ (the $\mathbf{Z}$-span of Δ), and X be the weight-lattice consisting of all $\lambda \in E$ such that $\langle \lambda, \alpha^\vee \rangle \in \mathbf{Z}$ for every $\alpha \in \Delta$. Let $\{\alpha_1, \alpha_2, \ldots, \alpha_l\}$ be a choice of fundamental (=simple) roots in Δ, and Δ_+ (resp. $\Delta_- = -\Delta_+$) the corresponding set of positive (resp. negative) roots with $\Delta = \Delta_+ \cup \Delta_-$ (disjoint union). Define the fundamental weights $\lambda_1, \lambda_2, \ldots, \lambda_l$ by $\langle \lambda_i, \alpha_j^\vee \rangle = \delta_{ij}$ (=the Kronecker delta), so that

$$\sum_i \mathbf{Z} \lambda_i = X \supseteq X' = \sum_i \mathbf{Z} \alpha_i .$$

The action of W on the group X maps each coset of X' therein onto itself, i.e. the natural action of W on X/X' is trivial. The elements of $X^+ = \sum_i \mathbf{Z}^+ \lambda_i$ (the set of non-negative integer linear combinations of $\lambda_1, \lambda_2, \ldots, \lambda_l$) are called dominant weights; each W-orbit in X contains a unique dominant weight. The set $X^+ \cap \Delta$ of dominant roots has cardinality equal to the number of W-orbits in Δ, which is 1 or 2 according as all elements of Δ have same length or not. The element(s) of the set $X^+ \cap \Delta$ are therefore unambigously named "long dominant root" (also called highest root) and "short dominant root", which do coincide if all roots have the same length. *We shall denote the short dominant root by* $-\alpha_0$. The set $\Delta^\vee = \{\alpha^\vee \mid \alpha \in \Delta\}$ of "coroots" is again a root-system (called the root-system inverse to Δ) such that the set $\{\alpha_1^\vee, \alpha_2^\vee, \ldots, \alpha_l^\vee\}$ affords a choice of fundamental (co)roots in it; relative to this choice $-\alpha_0^\vee$ is the long dominant (co)root of $\Delta^\vee$ (and if β were to denote the long dominant root of Δ, $\beta^\vee$ would be the short dominant (co)root in $\Delta^\vee$). The set $\{\alpha_0, \alpha_1, \ldots, \alpha_l\}$ has the property that $\langle \alpha_i, \alpha_j^\vee \rangle$ is a non-positive integer for distinct i, j in $\{0, 1, \ldots, l\}$. The subset $\{\alpha_0, \alpha_1, \ldots, \alpha_{j-1}, \alpha_{j+1}, \ldots, \alpha_l\}$ obtained by deleting α_j is a fundamental system for a subroot-system of Δ which is all of Δ if and only if $n_j = 1$ in the expansion

$$-\alpha_0^\vee = \sum_{i=1}^{l} n_i \alpha_i^\vee$$

of the long dominant coroot in the fundamental coroots. By the *extended Dynkin diagram* we shall mean the graph with $l+1$ nodes labelled 0, 1, ..., l, in which the nodes labelled i and j (for $i \neq j$) are joined by $\langle \alpha_i, \alpha_j^\vee \rangle \cdot \langle \alpha_j, \alpha_i^\vee \rangle (=0, 1, 2,$ or $3)$ -fold lines, omitting the arrows that are sometimes placed to distinguish the role of i and j in case $\langle \alpha_i, \alpha_j^\vee \rangle \neq \langle \alpha_j, \alpha_i^\vee \rangle$. It is known that this diagram determines Δ up to isomorphism. We remark that the common usage is to call it the "extended Dynkin diagram of $\Delta^\vee$".

Let D (resp. D') denote the group consisting of all translations T_λ operating on E or X, for $\lambda \in X$ (resp. $\lambda \in X'$), where T_λ sends x to $x+\lambda$. We denote by $\tilde{W}$ (resp. $\tilde{W}'$) the group $D \cdot W$ (resp. $D' \cdot W$) of affine transformations of E or X, generated by D (resp. D') and W. It is clear that $\tilde{W}$ (resp. $\tilde{W}'$) is the semidirect extension of W by the normal subgroup $D = X$ (resp. $D' = X'$) on which the action of W is known. From the fact that $\lambda - \lambda^\sigma \in X'$ for $\lambda \in X$ and $\sigma \in W$, it is easily seen that $\tilde{W}'$ is a normal subgroup of $\tilde{W}$.

For linear and affine transformations *we shall denote the operation on the right, and shall compose them accordingly;* thus for $\lambda \in E$ its image under $w \in \tilde{W}$ will be written λ^w, or occasionally λw for typographical reasons, and $(\lambda^w)^{w'} = \lambda^{ww'}$. With this convention we introduce the abbreviations

$$R_0 = R_{\alpha_0} \cdot T_{-\alpha_0}, \quad R_i = R_{\alpha_i} \quad \text{for} \quad 1 \leqq i \leqq l.$$

It is known that $\tilde{W}'$ (resp. W) is a Coxeter group on generators $R_0, R_1, \ldots, R_l$ (resp. $R_1, R_2, \ldots, R_l$). The order of the product $R_i R_j$, for distinct i, j in $\{0, 1, \ldots, l\}$, is 2, 3, 4, or 6 according as the nodes labelled i and j in our extended Dynkin diagram are joined by 0, 1, 2, or 3-fold lines. An element of W whose fixpoint-space has codimension 1 in E is called a reflection, and two reflections are called parallel if their fixed hyperplanes are parallel. Every reflection is of the form $T_{n\alpha} R_\alpha = R_\alpha T_{-n\alpha}$ for unique $\alpha \in \Delta_+$ and $n \in \mathbf{Z}$; one also knows that every reflection lies in $\tilde{W}'$ and is in fact conjugate to one of the generators $\{R_i | 0 \leqq i \leqq l\}$ under an element of $\tilde{W}'$. (More generally, it can be shown that every finite subgroup of $\tilde{W}$ is isomorphic to a subgroup of W, and every finite subgroup of $\tilde{W}'$ is conjugate under an element of $\tilde{W}'$ to a subgroup of one of the $l+1$ finite subgroups generated by $\{R_i | 0 \leqq i \leqq l, i \neq j\}$ for $0 \leqq j \leqq l$.) We shall call a point of E, W-regular (resp. $\tilde{W}'$-regular, $\tilde{W}$-regular) if its stabilizer in W (resp. $\tilde{W}'$, $\tilde{W}$) is trivial. The group $\tilde{W}'$ will be called the *affine Weyl group*. We remark that the common usage is to call it the "affine Weyl group of the root-system $\Delta^\vee$". The reader may refer to [3, (Chapter VI, § 2)] or [9] for details of some of the standard facts on affine Weyl groups which are stated here without proof.

1.2. Consider the *fundamental cone* (also called the dominant Weyl chamber)

$$E^+ = \sum_i \mathbf{R}^+ \lambda_i = \{\lambda \in E | \langle \lambda, \alpha_i^\vee \rangle \geqq 0 \quad \text{for} \quad 1 \leqq i \leqq l\},$$

the *fundamental parallelepiped*

$$E^{\#} = \{\lambda \in E | 0 \lneqq \langle \lambda, \alpha_i^{\vee} \rangle \leqq 1 \text{ for } 1 \leqq i \leqq l\},$$

and the *fundamental simplex*

$$E^{\star} = \{\lambda \in E^{+} | \langle \lambda, -\alpha_0^{\vee} \rangle \leqq 1\}.$$

These are the fundamental regions for the actions of W, D, and $\tilde{W}'$ respectively on E, in the strict sense that every point of E has a unique image in each of these under the action of the respective group. Note that $E^{\star}$ lies in the closure of $E^{\#}$ but not in $E^{\#}$; the reason for our *ad hoc* choice of omitting all walls through 0 (from the closure of $E^{\#}$) in the definition of $E^{\#}$, would be clear only in Part II. The interior of $E^{\star}$ is one of the connected components of the complement in E of the union of the fixed hyperplanes of all the reflections in $\tilde{W}'$, and the affine Weyl group operates simply-transitively on the set of all connected components in this complement. Thus the collection $\{(E^{\star})^w | w \in \tilde{W}'\}$ of the corresponding closed simplices forms a tessellation (=covering) of E, which is parametrized by $\tilde{W}'$ itself; we shall call $(E^{\star})^w$ the w-simplex. The walls of $E^{\star}$ are precisely the reflecting hyperplanes for the Coxeter generators $\{R_0, R_1, \dots, R_l\}$ of $\tilde{W}'$. (In the more general situation when Δ is the union of k mutually orthogonal indecomposable reduced root-systems, $E^{\star}$ would be the Cartesian product of k simplices and thus would have $l+k$ faces and $\geqq l+k$ vertices. It is merely to avoid the cumbersomeness resulting from such a situation that we have assumed Δ indecomposable.) It is easy to see that $n_i^{-1}\lambda_i$ is the vertex of $E^{\star}$ opposite the face fixed by R_i, where n_0 is the integer 1 and λ_0 is the zero element of X (so that the n_i's satisfy $\sum_{i=0}^{l} n_i \alpha_i^{\vee} = 0$). Thus

$$E^{\star} \cap X = \{\lambda_j | j \in J_0\} \quad \text{where} \quad J_0 = \{j \in \{0, 1, \dots, l\} | n_j = 1\}.$$

In view of the fact that the $\tilde{W}'$-orbit of $\lambda \in X$ is precisely the set $\lambda + X'$ one finds that each coset of X' in X meets $E^{\star}$ in a unique element of $E^{\star} \cap X$. Thus

$$X/X' = \{\lambda_j + X' | j \in J_0\}$$

is an enumeration of the group X/X', with $[X:X'] = |J_0|$. (We are using the standard notations for index and cardinality.) It is known that $X' = X$ (i.e. all the n_i's in the expansion of $-\alpha_0^{\vee}$ in the fundamental coroots are greater than 1) if and only if Δ is of type G_2, F_4, or E_8.

The action of $\tilde{W}$ on E permutes the simplices $(E^{\star})^w$ in our tessellation parametrized by $w \in \tilde{W}'$, which shows that $\tilde{W}$ is the semidirect extension of the stabilizer Ω of $E^{\star}$ by the normal subgroup $\tilde{W}'$. Let γ_j denote (as a temporary notation) the unique element of $\tilde{W}'$ which takes $E^{\star}$ to its image $E^{\star} - \lambda_j$ under the translation $T_{-\lambda_j} \in \tilde{W}$. Since $0 \in E^{\star} - \lambda_j$ for $j \in J_0$, and since each simplex contains a unique point of X', we find that γ_j must lie in the stabilizer W of the zero element of E in $\tilde{W}$. On the other hand, since the map $X/X' \to \tilde{W}/\tilde{W}'$ sending the coset $\lambda + X'$ to the coset $\tilde{W}'T_\lambda$ is clearly (well-defined and is) an isomorphism of groups, considering its composite with the canonical isomorphism $\tilde{W}/\tilde{W}' \to \Omega$, we see that for every $j \in J_0$ there is a unique element of Ω lying in the coset $\tilde{W}'T_{\lambda_j}$ (viz. the image of $\lambda_j + X'$ under this composite); it is clear that this element is simply $\gamma_j T_{\lambda_j}$ (which sends $E^\star$ to $(E^\star - \lambda_j) + \lambda_j$, and hence lies in Ω). Setting $\omega_j = \gamma_j T_{\lambda_j}$, we have the enumeration

$$\Omega = \{\omega_j | j \in J_0\}, \quad \text{with} \quad |\Omega| = [X:X'] = |J_0|.$$

For $j \in J_0$, the action of $\omega_j \in \Omega$ on $\tilde{W}'$ (through conjugation) permutes the Coxeter generators $\{R_i | 0 \leqq i \leqq l\}$ in the same manner in which it permutes the vertices $\{n_i^{-1}\lambda_i | 0 \leqq i \leqq l\}$ of $E^\star$. Thus, if θ_j denotes the permutation of $\{0, 1, \ldots, l\}$ such that $\omega_j^{-1} R_i \omega_j = R_{\theta_j(i)}$, then

$$\left(\frac{\lambda_i}{n_i}\right)^{\omega_j} = \frac{\lambda_{\theta_j(i)}}{n_{\theta_j(i)}}, \quad \text{i.e.} \quad \lambda_i^{\gamma_j} = \lambda_{\theta_j(i)} - n_i \lambda_j = \lambda_{\theta_j(i)} - n_{\theta_j(i)} \lambda_j.$$

Verifying directly the fact that the ordered basis

$$\alpha_0^\vee, \alpha_1^\vee, \ldots, \alpha_{j-1}^\vee, \alpha_{j+1}^\vee, \ldots, \alpha_l^\vee$$

is the dual, with respect to the W-invariant pairing $\langle\, , \rangle$, of the ordered basis

$$\lambda_0 - n_0\lambda_j, \quad \lambda_1 - n_1\lambda_j, \ldots, \lambda_{j-1} - n_{j-1}\lambda_j, \quad \lambda_{j+1} - n_{j+1}\lambda_j, \ldots, \lambda_l - n_l\lambda_j,$$

one concludes immediately that $\alpha_i^{\gamma_j} = \alpha_{\theta_j(i)}$. Since $\lambda_0^{\gamma_j} = \lambda_{\theta_j(0)} - \lambda_j$, we have $\theta_j(0) = j$. This shows that γ_j^{-1} sends α_j to $\alpha_0 \in \Delta_-$, and sends every other fundamental root into Δ_+. Thus γ_j^{-1} sends all positive roots in the subroot-system Δ_j of rank $l-1$ spanned by $\{\alpha_i | 1 \leqq i \leqq l,\ i \neq j\}$ into Δ_+; conversely, if α is a positive root not in Δ_j, say $\alpha = \sum_{i=1}^{l} a_i \alpha_i$ with $a_j \neq 0$, then looking at

$$\beta = (\alpha)^{\gamma_j^{-1}} = \sum_{i=1}^{l} a_i \alpha_{\theta_j^{-1}(i)} = a_j \alpha_0 + \sum_{\substack{i=1 \\ i \neq j}}^{l} a_i \alpha_{\theta_j^{-1}(i)}$$

we find at least one coefficient (in the expansion of β in fundamental roots) negative, which gives $\alpha^{\gamma_j^{-1}} \in \Delta_-$. If σ_0 (resp. σ_j) denotes the unique element of maximal length in W (resp. in the Weyl group $W(\Delta_j)$ of Δ_j), verifying that $\sigma_j \sigma_0$ is the unique element of W which makes precisely those positive roots negative which do not lie in Δ_j (here uniqueness follows from the standard theory of Weyl groups), we are forced to the conclusion $\gamma_j^{-1} = \sigma_j \sigma_0$. Thus we have the formula

$$\omega_j = \sigma_0 \sigma_j T_{\lambda_j} \quad (\text{for } j \in J_0)$$

for the elements of Ω; note that every element of $W(\Delta_j)$, and hence σ_j, fixes λ_j, which gives also $\omega_j = \sigma_0 T_{\lambda_j} \sigma_j$. One calls σ_0 the "apposition element" of W.

Note that the permutation θ_j of $\{0, 1, \ldots, l\}$ is an automorphism of the extended Dynkin diagram, and that its action is completely determined by its action on the subset J_0. (The latter claim follows from the fact that J_0 is a proper subset of $\{0, 1, \ldots, l\}$ in all cases except those of type A_l, and in such cases the extended Dynkin diagram is a tree and the nodes corresponding to J_0 are precisely the "end-nodes" of the diagram. Incidentally, this also shows that γ_j is the unique element of W sending α_0 to α_j and preserving the set $\{\alpha_0, \alpha_1, \ldots, \alpha_l\}$.) The action of θ_j on J_0 itself can be read-off from the group-structure of $\{\lambda_j+X' \mid j\in J_0\}$ (which is quite easy to write down in each given case), in view of the following straightforward equivalences on $i, j, k\in J_0$:

$$\theta_j(i) = k \Leftrightarrow \lambda_i+\lambda_j \equiv \lambda_k \,(\mathrm{mod}\, X') \Leftrightarrow \lambda_i^{\omega_j} = \lambda_k \Leftrightarrow \alpha_i^{\gamma_j} = \alpha_k \Leftrightarrow$$

$$\Leftrightarrow \theta_i\theta_j = \theta_k \Leftrightarrow \omega_i\omega_j = \omega_k \Leftrightarrow \gamma_i\gamma_j = \gamma_k.$$

The isomorphism $\omega_j \to \gamma_j$ (for $j\in J_0$) of the group Ω onto the subgroup $\Omega_0=\{\gamma_j \mid j\in J_0\}$ of W, given by the last equivalence above, can also realized as follows: Elements of Ω leave fixed the barycentre $\beta_0=|J_0|^{-1} \sum_{j\in J_0} \lambda_j$ of the sub-simplex of $E^\star$ having vertices $E^\star \cap X$; note that β_0 does not lie in X except in the three extreme cases (types G_2, F_4, E_8) when $X'=X$, $J_0=\{\lambda_0\}$ and $\beta_0=\lambda_0=0$. Thus we have $\beta_0^{\gamma_j}+\lambda_j=\beta_0$, which gives

$$\gamma_j^{-1}T_{\beta_0}\gamma_j = T_{\beta_0-\lambda_j} = T_{\beta_0}T_{\lambda_j}^{-1}, \quad \text{i.e.} \quad \gamma_j = T_{\beta_0}\gamma_j T_{\lambda_j}T_{-\beta_0} = T_{\beta_0}\omega_j T_{-\beta_0}.$$

Hence conjugation by $T_{-\beta_0}$ gives our isomorphism $\Omega \to \Omega_0$.

1.3. Now let $l(\,)$ denote the usual length-function on $\tilde{W}'$ in terms of its Coxeter generators $\{R_0, R_1, \ldots, R_l\}$ as in the standard theory of Coxeter groups. Thus for $w\in \tilde{W}'$, $l(w)$ is the smallest integer k such that $w=R_{i_1}R_{i_2}\ldots R_{i_k}$ for $0\leqq i_q\leqq l$, $1\leqq q\leqq k$; such an expression for w (with $k=l(w)$) is called a reduced expression. One knows that $l(w)$ equals the number of reflecting hyperplanes (i.e. fixed hyperplanes of reflections in $\tilde{W}'$) that separate $E^\star$ from the w-simplex $(E^\star)^w$. The restriction of $l(\,)$ to W coincides with the length-function on W defined in terms of its Coxeter generators $\{R_1, R_2, \ldots, R_l\}$; for $\sigma\in W$, $l(\sigma)$ also equals the number of elements of Δ_+ that are carried into Δ_- under σ.

For $\lambda\in E$ and $\sigma\in W$, let $\lambda^{(\sigma)}$ denote the unique translate under X (i.e. image under D) of λ^σ such that $\lambda^{(\sigma)}\in E^\#$. In general $\lambda^{(\sigma)}$ depends very much on both λ and σ. But when λ lies in the interior of $E^\star$ the difference $\lambda^{(\sigma)}-\lambda^\sigma$ is independent of λ, and can be seen to be equal to

$$\delta_\sigma = \sum_{i\in I_\sigma} \lambda_i, \quad \text{where} \quad I_\sigma = \{i|1 \leqq i \leqq l,\ l(\sigma R_i) < l(\sigma)\}.$$

Thus $\lambda^{(\sigma)}=\lambda w_\sigma$ where $w_\sigma=\sigma T_{\delta_\sigma}\in \tilde{W}$. *It is this set $\{w_\sigma|\sigma\in W\}$ of affine transformations of E that is basic to the formulation of our conjecture in Part I* (§ 2).

Let λ be an interior of $E^\star$; thus λ is $\tilde{W}'$-regular. Setting

$$W^\# = \{\sigma\in W|w_\sigma\in \tilde{W}'\} = \{\sigma\in W|\delta_\sigma\in X'\},$$

it is clear that the $\tilde{W}'$-orbit of λ meets $E^{\#}$ precisely in the set $\{\lambda^{(\sigma)}|\sigma\in W^{\#}\}$ of cardinality $|W^{\#}|=[W:\Omega_0]$. The last equality follows from $(\lambda^{(\gamma)})^{(\sigma)}=\lambda^{(\gamma_\sigma)}$ for $\gamma\in\Omega_0$ and $\sigma\in W$, the map $\gamma\to w_\gamma$ being simply the isomorphism $\Omega_0\to\Omega$ already encountered (with $\gamma_j\to\omega_j=w_{\gamma_j}$). This shows that $W^{\#}$ is a set of distinguished right-coset representatives for the right-coset space

$$\Omega_0\backslash W=\{\Omega_0\cdot\sigma|\sigma\in W\}.$$

Incidentally, this also gives the famous formula for the order of the Weyl group:

$$|W| = l!\,n_1 n_2\dots n_l[X:X'],$$

as one sees on making an elementary calculation verifying that the volume of $E^{\#}$ is $l!\,n_1 n_2\dots n_l$ times the volume of $E^{\star}$. Let us digress slightly at this point to mention another important formula involving the n_i's (this time their sum), which we shall require later:

$$\sum_{i=1}^{l} n_i = h_\Delta - 1, \quad \text{where} \quad h_\Delta(=\text{the Coxeter no. of }\Delta) = \frac{2r}{l},\ r=|\Delta_+|.$$

An alternative interpretation of this formula is that the "height" $\sum_{i=1}^{l} n_i$ of the long dominant coroot $-\alpha_0^\vee$ equals the largest "exponent" of W; this is a special case of a deep observation of Kostant and Steinberg on the "duality" between the two partitions of the number r given by the exponents (which are known to add up to r) on one hand, and by the number of elements of Δ_+ of various heights on the other hand.

1.4. It is well known, and easily seen, that the set $\tilde{W}^+$ defined by

$$\tilde{W}^+ = \{w\in\tilde{W}'|l(wR_i) > l(w) \quad \text{for} \quad 1\leq i\leq l\}$$

is a set of distinguished left-coset representatives for the left-coset space $\tilde{W}'/W=\{w\cdot W\,|\,w\in W'\}$. The cone E^+ is tessellated by the images of $E^{\star}$ under $\tilde{W}^+$, just as the closure of $E^{\#}$ is tessellated by the images under

$$\tilde{W}^{\#} = \{w_\sigma|\sigma\in W^{\#}\} = \{w_\sigma|\sigma\in W\}\cap\tilde{W}'.$$

In particular, $\tilde{W}^{\#}$ is contained in $\tilde{W}^+$.

Let us note that in the tessellation $\{(E^{\star})^w\,|\,w\in\tilde{W}^{\#}\}$ of the closure of $E^{\#}$ there is a unique simplex which contains $\delta=\sum_{i=1}^{l}\lambda_i$. We call this the "top simplex" in $E^{\#}$, and the corresponding element w_0 of $\tilde{W}^{\#}$ the "top element" of $\tilde{W}^{\#}$. Writing $E^{\star}_{\text{top}}$ for the top simplex we have

$$(E^{\star})^{w_0} = E^{\star}_{\text{top}} = -E^{\star}+\delta = (E^{\star})^{\sigma_0 T_\delta},$$

which shows that $w_0 T_{-\delta}\sigma_0\in\Omega$. Denoting by j_0 the unique index in J_0 such that $\delta\equiv\lambda_{j_0}$ (mod X'), we see that the zero element in E goes under w_0 to $\delta-\lambda_{j_0}$ (this being the only point of $E^{\star}_{\text{top}}$ in the root-lattice) and hence under $w_0T_{-\delta}\sigma_0$ to

$(-\lambda_{j_0})^{\sigma_0}=\lambda_{j_0}$ where this last equality is immediate from $\delta^{\sigma_0}=-\delta$. Since ω_{j_0} is the only element of Ω which sends $\lambda_0=0$ to λ_{j_0}, we have $w_0=\omega_{j_0}\sigma_0 T_\delta$. From this one easily gets $w_0=\sigma_{j_0}T_{\delta-\lambda_{j_0}}$, on substituting $\omega_{j_0}^{-1}=\sigma_0 T_{\lambda_{j_0}}\sigma_{j_0}$ proved earlier. We call ω_{j_0} the "apposition element" of Ω.

1.5. We shall now establish certain relativizations of the concepts and notations $\tilde{W}^+$ and $\tilde{W}^\#$, relative to a proper subset J of $\{0, 1, \ldots, l\}$ (empty subset permitted). Let us write

$$E_J^\star = \{\lambda \in E^\star | \lambda^{R_j} = \lambda \Leftrightarrow j \in J\}.$$

This is an open subset in its own affine-span (which passes through the origin if and only if $0 \in J$). When J is empty this is precisely the interior of $E^\star$, whose boundary is the disjoint union of $E_J^\star$ as J ranges over the $2^{l+1}-2$ other subsets. The stabilizer of $\lambda \in E_J^\star$ in $\tilde{W}'$ is precisely the subgroup $\tilde{W}^J$ generated by $\{R_j | j \in J\}$. In the right-coset space $\tilde{W}^J \backslash \tilde{W}'$ each coset $\tilde{W}^J \cdot w$ has a unique element of minimal length, all such elements forming the set $\tilde{W}_J$ of distinguished right-coset representatives, where

$$\tilde{W}_J = \{w \in \tilde{W}' | l(R_j w) > l(w) \quad \text{for} \quad j \in J\};$$

and the double-coset space

$$\tilde{W}^J \backslash \tilde{W}'/W = \{\tilde{W}^J \cdot w \cdot W | w \in \tilde{W}'\}$$

has $\tilde{W}_J^+ = \tilde{W}_J \cap \tilde{W}^+$ as a set of distinguished double-coset representatives. The $\tilde{W}'$-orbit of $\lambda \in E_J^\star$ is parametrized by $\tilde{W}_J$ (i.e. $w \to \lambda^w$ is a bijection from $\tilde{W}_J$ to this orbit), and the intersection of this orbit with E^+ is precisely $\{\lambda^w | w \in \tilde{W}_J^+\}$ parametrized by $\tilde{W}_J^+$. Note that this intersection lies in the interior of E^+ if and only if $E_J^\star$ itself does, which happens precisely when J is either empty or the singleton $\{0\}$. Setting

$$\tilde{W}_J^{++} = \{w \in \tilde{W}_J | (E_J^\star)^w \text{ lies in the interior of } E^+\},$$

and

$$\tilde{W}_J^\# = \{w \in \tilde{W}_J | (E_J^\star)^w \text{ lies in } E^\#\},$$

it is clear that $\tilde{W}_J^{++}$ (resp. $\tilde{W}_J^\#$) is the subset of $\tilde{W}_J$ which parametrizes the intersection of the $\tilde{W}'$-orbit of $\lambda \in E_J^\star$ with the interior of E^+ (resp. with $E^\#$). Thus $\tilde{W}_J^\# \subset \tilde{W}_J^{++} \subseteq \tilde{W}_J^+ \subset \tilde{W}_J$, where the middle inclusion is also strict whenever J is nonempty and contains an index other than 0. It is also clear that $\tilde{W}_J^\# = \tilde{W}^\#$ if and only if J is empty. We claim that $\tilde{W}_J^\# \subseteq \tilde{W}^\#$ in general. For, suppose w lies in $\tilde{W}_J^\#$ but not in $\tilde{W}^\#$. Then w sends the interior of $E^\star$ outside $E^\#$; but since $\lambda^w \in E^\#$ for $\lambda \in E_J^\star$ there is a wall of $E^\#$ on which λ^w lies and which separates the fundamental simplex from the w-simplex $(E^\star)^w$. Let $w' = wR$ where R is the reflection in this wall. Then $\lambda^{w'} = \lambda^w$ implies $w' \in \tilde{W}^J \cdot w$; and since the said wall of $E^\#$ does not separate the fundamental simplex and the w'-simplex we have

$l(w') < l(w)$, which contradicts the fact that $w \in \tilde{W}_J$ is the unique element of minimal length in the coset $\tilde{W}^J \cdot w$. Thus we have

$$\tilde{W}_J^{\#} \subseteq \tilde{W}^{\#} \cap \tilde{W}_J^{+} \quad \text{and} \quad \tilde{W}_J^{\#} = \tilde{W}^{\#} \cap \tilde{W}_J^{++}.$$

We remark that $\tilde{W}_J^{\#}$ could be empty, and that this situation does occur for some J if and only if Δ is not of type G_2, F_4, or E_8. This follows from the more precise claim that *$\tilde{W}_J^{\#}$ is empty if and only if the complement of J in* $\{0, 1, \ldots, l\}$ *consists of a single index j with* $j \in J_0$, $j \neq j_0$, and that *whenever $\tilde{W}_J^{\#}$ is nonempty the top element w_0 of $\tilde{W}^{\#}$ lies in* ($\tilde{W}_J^{++}$ and therefore in) $\tilde{W}_J^{\#}$. For this first consider the situation when the complement of J is the singleton $\{i\}$; then $E_J^{\star}$ consists of just one point, viz. the vertex $n_i^{-1}\lambda_i$. If $i \in J_0$ then $n_i = 1$ and the $\tilde{W}'$-orbit of $E_J^{\star}$ is contained in the set of 2^l vertices of the parallelepiped $E^{\#}$ (the intersection of the closure of $E^{\#}$ with the weight-lattice); but none of these lie in $\tilde{W}'$-orbit of λ_i unless $\delta - \lambda_i \in X'$, i.e. $i = j_0$. This proves the "if" part of the claim. Now note that the image of $n_i^{-1}\lambda_i$ under $w_0 = \omega_{j_0}\sigma_0 T_\delta$ is of the form $\delta - n_k^{-1}\lambda_k$ with $n_k = n_i$ (for $n_i^{-1}\lambda_i$ goes to $n_j^{-1}\lambda_j$ under ω_{j_0} if $j = \theta_{j_0}(i)$, and so to $n_k^{-1}\lambda_k$ under $\omega_{j_0}\tau_0 = \omega_{j_0} \cdot (-\sigma_0)$ if R_k is the image of R_j under the Dynkin outer automorphism of W which is known to be given by conjugation under the endomorphism $\tau_0 = -\sigma_0$ of E sending α to $-\alpha^{\sigma_0}$). It is clear that $\delta - n_k^{-1}\lambda_k \in E^{\#}$ if $n_k > 1$, which takes care of our assertion when J has cardinality l. In other cases, the closure of $E_J^{\star}$ has at least two vertices $n_i^{-1}\lambda_i$ and $n_{i'}^{-1}\lambda_{i'}$ and its image under ω_0 contains a point of the form $\delta - a(n_k^{-1}\lambda_k) - (1-a)(n_{k'}^{-1}\lambda_{k'})$ with $0 < a < 1$, which clearly lies in $E^{\#}$ (all the coefficients in the expansion of this point in the basis $\lambda_1, \lambda_2, \ldots, \lambda_l$ of E being > 0 and $\leqq 1$); this shows that $(E_J^{\star})^{w_0}$ is contained in $E^{\#}$, i.e. $w_0 \in W_J^{\#}$ as claimed.

1.6. We shall now introduce a canonical partial-order $\leqq$ on the set $\tilde{W}^{+}$, which would be needed in § 5 for streamlining Conjecture II and its applications. It is defined as the restriction to $\tilde{W}^{+}$ of the "Bruhat partial-order" $\overset{\mathrm{B}}{\leqq}$ on $\tilde{W}'$ (see the next paragraph) which was introduced in [16] for arbitrary Coxeter groups in terms of given Coxeter generators (and earlier in [15] and [15A] for finite Weyl groups; the name "Bruhat partial-order" in that case was motivated by its relation with the "Bruhat decomposition" parametrized by a Weyl group). In these references $\overset{\mathrm{B}}{\leqq}$ was denoted simply $\leqq$, but here we have made this variation because there is another partial-order $\overset{\mathrm{A}}{\leqq}$ on $\tilde{W}'$ which we shall call the "affine partial-order" ("A" for "affine" and "B" for "Bruhat") for the definition of which one needs the affine Weyl group structure of $\tilde{W}'$. This $\overset{\mathrm{A}}{\leqq}$ is a more natural partial-order for the representation-theoretic purposes we have in mind in Part II, but on the set $\tilde{W}^{+}$ the two orderings coincide.

$\overset{\mathrm{B}}{\leqq}$ is defined as the weakest partial-order on $\tilde{W}'$ such that

$$l(wR) < l(w) \quad \text{implies} \quad wR \overset{\mathrm{B}}{\leqq} w,$$

for arbitrary reflection R and element w in $\tilde{W}'$. Thus, for w, w' in $\tilde{W}'$, if $w^{-1}w'$ is a reflection one has the dichotomy that one and exactly one of $w \overset{\mathrm{B}}{\leqq} w'$ and $w' \overset{\mathrm{B}}{\leqq} w$ holds. It has been shown in [17, Proposition 2] that $\overset{\mathrm{B}}{\leqq}$ coincides with the (weakest) partial-order $\overset{\mathrm{B}'}{\leqq}$ obtained by demanding $wR \overset{\mathrm{B}'}{\leqq} w$ for only those pairs (w, R) for which $l(wR)=l(w)-1$. It is also known [17, Corollary 1] that if $w=R_{i_1}R_{i_2}\dots R_{i_m}$ is a reduced expression then $w' \overset{\mathrm{B}}{\leqq} w$ if and only if w' is of the form $R_{j_1}R_{j_2}\dots R_{j_n}$ for some subsequence $(j_1, j_2, \dots, j_n)$ of the sequence of indices $(i_1, i_2, \dots, i_m)$ (all indices ranging in $\{0, 1, \dots, l\}$), where this expression for w' can also be assumed reduced by passing to a further subsequence.

The partial-order $\overset{\mathrm{A}}{\leqq}$ on $\tilde{W}'$ is defined to be the weakest partial-order with the following property: Let $w \in \tilde{W}'$ and R be a reflection parallel to R_α for $\alpha \in \Delta_+$; for an interior point λ of $(E^\star)^w$ the sign of the (nonzero) real number $c=c_\lambda$ defined by $\lambda-\lambda^R=c\alpha$ is clearly independent of λ (depends only on w and R); then we require that

$$wR \overset{\mathrm{A}}{\leqq} w \text{ whenever } c>0 \text{ (for some, and hence all, } \lambda).$$

Once again we have the dichotomy "$w \overset{\mathrm{A}}{\leqq} w'$ versus $w' \overset{\mathrm{A}}{\leqq} w$" in case $w^{-1}w'$ is a reflection. Let us observe that for arbitrary (w, R) there is an odd number of reflections parallel to R which separate the w-simplex from the wR-simplex, so that if $R', R'', R''', \dots$ is the enumeration of these reflections in the order of their closeness to the w-simplex then $wR=wR'R''R'''\dots$; and for an interior point λ in the w-simplex $(E^\star)^w$, denoting its successive mirror-images by $\lambda'=\lambda R'$, $\lambda''=\lambda'R''$, $\lambda'''=\lambda''R'''$, ..., one finds the consecutive differences $\lambda-\lambda'$, $\lambda'-\lambda''$, $\lambda''-\lambda'''$, ... to be of the form $c\alpha$, $(1-c)\alpha$, $c\alpha$, ... alternately, where $0<|c|<1$. It follows that $\overset{\mathrm{A}}{\leqq}$ coincides with the (weakest) partial-order $\overset{\mathrm{A}'}{\leqq}$ obtained by demanding $wR \overset{\mathrm{A}'}{\leqq} w$ whenever $0<c<1$ (where c is defined as before by $\lambda-\lambda^R=c\alpha$, and $0<c<1$ holds either for all or for none $\lambda \in$ *interior* $(E^\star)^w$).

It is easy to see that the condition $c>0$ on the pair (w, R) is equivalent to the following:

$$\begin{cases} l(wR) > l(w) & \text{if the } R\text{-plane } \textit{does not meet} \text{ interior } E^+, \\ l(wR) < l(w) & \text{if the } R\text{-plane } \textit{does meet} \text{ interior } E^+, \end{cases}$$

where R-plane means the fixed hyperplane of R. It follows that for w, w' in the Weyl group W (resp. for w, w' in the set $\tilde{W}^+$ of distinguished coset representatives for $\tilde{W}'/W$), if $w^{-1}w'$ is a reflection then

$$w \overset{\mathrm{A}}{\leqq} w' \text{ is equivalent to } w' \overset{\mathrm{B}}{\leqq} w \text{ (resp. } w \overset{\mathrm{B}}{\leqq} w').$$

This statement is true even without the restriction that $w^{-1}w'$ be a reflection; to deduce this generalization it suffices to show that for $w, w' \in W$ (resp. $\tilde{W}^+$), if $w \overset{\mathrm{A}}{\leqq} w'$ then one can find reflections $S_1, S_2, \dots, S_k$ with

$$w \overset{\mathrm{A}}{\leqq} wS_1 \overset{\mathrm{A}}{\leqq} wS_1S_2 \overset{\mathrm{A}}{\leqq} \dots \overset{\mathrm{A}}{\leqq} wS_1S_2\dots S_k = w' \qquad (1.1)_{\mathrm{A}}$$

such that all the $wS_1S_2 \dots S_i$ lie in W (resp. $\tilde{W}^+$), and also that the similar statement holds with $\overset{B}{\leqq}$ in place of $\overset{A}{\leqq}$ throughout. Of course, the existence of $S_1, S_2, \dots, S_k$ such that $(1.1)_A$ (resp. $(1.1)_B$, obtained by replacing A by B in $(1.1)_A$) is guaranteed from the definition of $\overset{A}{\leqq}$ (resp. $\overset{B}{\leqq}$). When w, w' are in W one can prove that in both cases all the S_i, and hence the $wS_1S_2 \dots S_i$, lie in W; for the case of $\overset{A}{\leqq}$ this follows from the fact that the union of all the σ-simplices for σ running in W lies inside the "strip" $\{\lambda \in E \mid -1 \leqq \langle \lambda, \alpha^\vee \rangle \leqq 1\}$ for every $\alpha \in \Delta_+$ (in fact, it is precisely the intersection of all such strips for α running over the short roots only); and for the case of $\overset{B}{\leqq}$ this follows from the fact that if $\sigma \in W$, $l(\sigma S) < l(\sigma)$ for a reflection S implies $S \in W$. When w, w' are in $\tilde{W}^+$, the existence of S_i's satisfying $(1.1)_B$ with $wS_1S_2 \dots S_i \in \tilde{W}^+$ is proved in [17, Proposition 3], which gives $w \overset{B}{\leqq} w' \Rightarrow w \overset{A}{\leqq} w'$. We omit the proof of the converse, i.e. of the existence of S_i's satisfying $(1.1)_A$ with $wS_1S_2 \dots S_i \in \tilde{W}^+$, as our proof is unduly involved and this result is not required here. This is why we consider the partial-order $\leqq$ on $\tilde{W}^+$ as the restriction of $\overset{B}{\leqq}$.

1.7. The identity is the unique smallest element of $\tilde{W}'$ under $\overset{B}{\leqq}$, whereas there is no smallest or largest element under $\overset{A}{\leqq}$. The "atoms" of $(\tilde{W}', \overset{B}{\leqq})$ are precisely the Coxeter generators, and it can be shown that when $l > 1$ the only automorphisms of this partial-order are the Dynkin outer automorphisms of the group $\tilde{W}'$ (coming from the automorphisms of the extended Dynkin diagram) permuting the Coxeter generators. (See Footnote [1].) On the other hand, $(\tilde{W}', \overset{A}{\leqq})$ has lots of automorphisms: for $\lambda \in X'$ the right-multiplication by $T_\lambda \in \tilde{W}'$ preserves $\overset{A}{\leqq}$; and more generally, if $\lambda \in \lambda_j + X'$ (for $j \in J_0$) then the map sending $w \in \tilde{W}'$ to the unique element $\omega_j w T_\lambda$ lying in $\tilde{W}'$ as well as the right-coset $\Omega w T_\lambda$, is an automorphism of this partial-order. Also, the Dynkin outer automorphism of the group W permuting the generators $\{R_i \mid 1 \leqq i \leqq l\}$ which lifts canonically to an automorphism of $\tilde{W}'$ (given by the unique automorphism of the extended Dynkin diagram fixing the node labelled 0), gives a distinguished involution of the affine partial-order; it sends $w \in \tilde{W}'$ to its conjugate $\tau_0 w \tau_0$ where $\tau_0 = -\sigma_0$ is the endomorphism of E sending λ to $-\lambda^{\sigma_0}$ and hence permuting the fundamental roots; thus this involution is nontrivial if and only if $-id.$ does not lie in W. All

[1] *(Added in Proof.)* It has been shown in a recent preprint of Mr. A. van den Hombergh (Katholieke Universiteit, Nijmegen) through a somewhat complicated analysis, that if Γ is an arbitrary indecomposable Coxeter group of rank >2 and ϱ an automorphism of the Bruhat partial order structure on Γ (which obviously permutes the atoms), then ϱ is determined by its restriction to the set of Coxeter generators of Γ; in contrast, when Γ has rank 2 (which means it is finite or infinite dihedral), say with $|\Gamma| = 2m$ $(3 \leqq m \leqq \infty)$, the structure of this partial-order is easily described upon noting that for $0 < i < m$ there are exactly 2 elements of "height" i (viz. the elements of length i), which shows that the automorphism group is the direct product of $m-1$ copies of the cyclic group of order 2.

the automorphisms of $(\tilde{W}', \overset{A}{\leqq})$ are composed of the ones just described; this is seen on observing that the preceding involution is the only one fixing the identity element.

One gets a better understanding of the affine partial-order on looking at it as a partial-order on the tessellation of E, i.e. on the set of w-simplices as w runs in $\tilde{W}'$. Then the automorphism $w \to \omega_j w T_\lambda$ (for $\lambda \in \lambda_j + X'$) corresponds merely to the translation sending the w-simplex $(E^\star)^w$ to $(E^\star)^w + \lambda$. There is also an interesting anti-involution (order-reversing bijection of period 2) of $(\tilde{W}', \overset{A}{\leqq})$, sending $(E^\star)^w$ to $-(E^\star)^w$, which commutes with the involution described above (which takes $(E^\star)^w$ to $(E^\star)^{\tau_0 w \tau_0} = (E^\star)^{w\tau_0}$), the composite being the anti-involution $w \to w\sigma_0$. When we compose this latter anti-involution with the "translation by δ", viz. $w \to \omega_{j_0} w T_\delta$ (ω_{j_0} being the apposition element of Ω), we get a distinguished anti-involution of the affine partial-order on $\tilde{W}'$ which maps $\tilde{W}^\#$ onto itself; it is given by $w \to \omega_{j_0} w \sigma_0 T_\delta = \omega_{j_0} w \omega_{j_0} w_0$, so that its effect on the tessellation is $(E^\star)^w \to (E^\star)^{w\omega_{j_0} w_0}$. It seems likely that all the automorphisms of the partial-ordering $(\tilde{W}^\#, \overset{A}{\leqq}) = (\tilde{W}^\#, \overset{B}{\leqq})$ are composed of this anti-involution and the involution $w \to \tau_0 w \tau_0$ of the last paragraph.

It is clear from the preceding that

$$\tilde{W}^\# \subseteq \{w \in \tilde{W}^+ | w \leqq w_0\} = \{w \in \tilde{W}' | w \overset{B}{\leqq} w_0\}.$$

It can be seen that equality holds here if and only if Δ is of type A_1, A_2 or B_2. The distinguishing feature of these types responsible for this phenomenon is that every positive root is either fundamental or dominant; for other types one can find even a short root α which is neither. Since α is short, $\langle \alpha, \beta^\vee \rangle$ lies between -1 and 1 for all roots β, in particular for $\beta = \alpha_i$ which gives $\delta - \alpha \in X^+$ (with $\alpha \notin X^+$); one sees that if R is the reflection parallel to R_α sending δ to $\delta - \alpha$, then the image of the top simplex $(E^\star)^{w_0}$ under R lies in E^+ but not in $E^\#$. Thus one has $w_0 R \leqq w_0$ with $w_0 R \notin W^\#$. For type G_2, see the Tessellation Diagram in § 3, which indicates the four members of $\{w \in \tilde{W}^+ | w \leqq w_0\}$ not lying in $\tilde{W}^\#$.

§ 2. Harmonic polynomials and Conjecture I

2.1. The polynomial referred to in the title of Part I is the following function on E:

$$D(\lambda) = \frac{\prod_{\alpha \in \Delta_+} \langle \lambda, \alpha^\vee \rangle}{\prod_{\alpha \in \Delta_+} \langle \delta, \alpha^\vee \rangle} \quad \text{for} \quad \lambda \in E,$$

where $\delta = \sum_i \lambda_i = \frac{1}{2} \sum_{\alpha \in \Delta_+} \alpha$, as usual. Because of homogeneity one could replace $\alpha^\vee$ by α both in the numerator and the denominator. But the form given here emphasizes the fact that if we consider D as a polynomial in the coefficients $x_1, x_2, \ldots, x_l$ in the expansion of λ in the fundamental weights, $\lambda = \sum_i x_i \lambda_i$, then the denominator is an integer and the numerator is a product of linear forms in $x_1, x_2, \ldots, x_l$ (called root-forms) having non-negative integer coefficients; these forms $\langle \lambda, \alpha^\vee \rangle$ can be easily written down in any given case once the Dynkin diagram of Δ is given. The value of $D(\lambda)$ is therefore $\geqq 0$ for $\lambda \in X^+$, and that of $D(\lambda+\delta)$ is strictly positive; in fact, the latter number is an integer, being the dimension of the "irreducible $\mathfrak{g}_{\mathbf{C}}$-module with highest weight λ", where $\mathfrak{g}_{\mathbf{C}}$ is the simple Lie algebra over $\mathbf{C}$ possessing Δ as its root system. (It should be noted that when the "coordinates" $x_1, x_2, \ldots, x_l$ of $\lambda = \sum_i x_i \lambda_i$ are non-negative integers, i.e. $\lambda \in X^+$, those of $\lambda+\delta$ are strictly positive integers, and conversely; that justifies the parenthetical statement in the opening sentence of the Introduction.)

The polynomial $D(\lambda)$ is determined by Δ and the choice of positive roots therein. But a different choice of Δ_+ (equivalently, of the fundamental roots) affects $D(\lambda)$ only by a multiplicative factor of -1. This is because all choices of Δ_+ are conjugate under W, and D is skew (or alternating) with respect to W in the sense that $D(\lambda^\sigma) = \det \sigma \cdot D(\lambda)$ for $\sigma \in W$. As a matter of fact, it is known that every skew-invariant polynomial on E is of the form $D(\lambda) \cdot P(\lambda)$ where P is W-invariant. There is another property that distinguishes D, viz. that up to scalar multiples it is the unique W-harmonic homogeneous polynomial of maximal degree. A polynomial on E is called W-harmonic if it is annihilated by all W-invariant homogeneous differential operators of positive degree. (We consider only operators with constant coefficients. The algebra of such operators is canonically isomorphic to the symmetric algebra $S(E)$ of E; for $u \in S(E)$ one denotes the corresponding differential operator by ∂_u. The action of ∂_u on the algebra of polynomial functions on E, which is simply the algebra $S(E^*)$, is obtained by requiring that for $p \in S(E^*)$ the value of $\partial_u p$ at $x \in E$ be equal to the inner-product $\langle u, p^x \rangle$ where p^x is the translated polynomial $p^x(\lambda) = p(\lambda + x)$ and $\langle \, , \rangle$ is the canonical pairing between $S(E)$ and $S(E^*)$ extending that between E and E^*. When we consider a polynomial function on E as a formal polynomial in the coordinates $x_1, x_2, \ldots, x_l$ of $\lambda = \sum_i x_i \lambda_i$, we can identify the first-order homogeneous differential operator ∂_{λ_i} in the above notation with the formal partial derivation $\frac{\partial}{\partial x_i}$ in the standard sense.)

The fundamental fact about W-harmonic polynomials (cf. Steinberg [11]) is that the space of all such polynomials has dimension $|W|$ and is spanned by D and all its (partial) derivatives. This generalizes, in fact, to the situation when W is an arbitrary finite pseudoreflection group (in the accepted Bourbaki usage), which includes the noncrystallographic (=non-Weyl) finite Coxeter groups; only the definition of D has to be slightly modified to take into account the orders of the generating pseudoreflections which can be greater than 2.

2.2. We can now state our conjecture on D in terms of the affine transformations $w_\sigma \in \tilde{W}$ (for $\sigma \in W$) introduced in § 1; we remark that though D depends only on

W this conjecture depends on the extension $\tilde{W}$ (as isomorphic Weyl groups can give rise to distinct extensions), and does not make sense for even the noncrystallographic finite Coxeter groups.

Conjecture I. *There exist unique scalars* b_σ, $\sigma \in W$, *such that*

$$\sum_{\sigma \in W} b_\sigma \cdot D(\lambda^{w_\sigma}) = 1 \quad \textit{for all} \quad \lambda \in E; \tag{2.1}$$

these "weightings" b_σ *are necessarily integers.* (See Footnote[2].)

Let us write $D_\sigma(\lambda) = D(\lambda^{w_\sigma}) = \det \sigma \cdot D(\lambda + (\delta_\sigma)^{\sigma^{-1}})$, and note that D_σ is a harmonic polynomial for all σ. In fact for arbitrary $\mu \in E$ the translate H^μ of a harmonic polynomial H is again harmonic, as is clear from Taylor's theorem and the fact that the derivatives of H are harmonic. It follows that the first half of the conjecture simply asserts the linear independence of the set $\{D_\sigma | \sigma \in W\}$ of inhomogeneous polynomials. Since $(\delta_\sigma)^{\sigma^{-1}} \in X$, all the D_σ belong to the **Q**-span of D and its successive derivatives with respect to ∂_λ for $\lambda \in X$; since this **Q**-span is a **Q**-form of the space of harmonic polynomials, it is clear that all b_σ are rational numbers. Thus the crux of the conjecture lies in the assertion $b_\sigma \in \mathbf{Z}$. However, it should be emphasized that our main interest is not in the existence and uniqueness of b_σ's, but in finding an explicit method of computing their values. Unfortunately, even the examination of several low rank cases in the next section sheds little light on this problem.

[2] *(Added in Proof.)* In the original (preprint) version of this paper it was hoped that b_σ's would be necessarily non-negative (as is the case for rank 2 and for type A_3 cf. § 3). But shortly afterwards (in Autumn 1972) S. G. Hulsurkar found methods for hand-computation of the numbers b_σ (more efficient than our brutal technique in § 3), and carrying it through for types B_3 and C_3 discovered that in both these cases some of the b_σ's are negative. Later, refining on his methods, he proved this conjecture in a form stronger than stated (and required) by us, by showing that the set $\{D_\sigma | \sigma \in W\}$ is a **Z**-basis of the "lattice" of those harmonic polynomials that take integer values on the weight-lattice X. The mystery shrouding the values of b_σ has also disappeared, as it turns out that $b_\sigma = \sum_\tau b_{\sigma,\tau}$ where $[b_{\sigma,\tau}]$ is the integer matrix inverse to the $|W| \times |W|$ matrix $[D(-\delta_\sigma^{\sigma^{-1}\tau} + \delta_\tau)]$; the cruical step in Hulsurkar's work rests in the discovery that the latter matrix is unimodular.

In a forthcoming paper I shall show that the integers b_σ arise naturally as the degrees of certain (unique) virtual T-modules given by $\beta_\sigma \in \mathbf{Z} \cdot e^X$ (for notations see § 4) such that

$$e^\lambda = \sum_\sigma \operatorname{ch}(\lambda^\sigma + \delta_\sigma) \cdot \beta_\sigma \quad \text{for all} \quad \lambda \in X, \quad \text{where} \quad \operatorname{ch} \mu = \frac{\sum \det \tau \cdot e^{\mu^\tau}}{\sum \det \tau \cdot e^{\delta^\tau}}$$

This formula refines Conjecture I on one hand (taking "trace" on both sides one gets (2.1)), and implies on the other hand the fact the ring $\mathbf{Z} \cdot e^X$ is free as a module over its subring of W-invariants; it seems that this fundamental fact of "semisimple Lie Theory" remained unnoticed until topologists discovered it (it was proved by Harsh Pittie in 1970 who attributes it as a conjecture of Luke Hodgkin). *See also Footnote* [6]. In subsequent paper(s) I hope to show that these virtual T-modules β_σ enter even the complex representation theory of finite Chevalley groups in a fundamental manner.

2.3. Our "inhomogeneous weighted mean-value conjecture" above has a known homogeneous analogue which asserts

$$|W|^{-1} \sum_{\sigma \in W} D(\lambda+\mu^{\sigma}) = D(\lambda) \quad \text{for} \quad \lambda, \mu \in E. \tag{2.2}$$

This is in fact true for an arbitrary finite linear group W with D an arbitrary W-harmonic polynomial, and is quite easy to prove: Since the translation operator T_μ sending $D(\lambda)$ to $D^\mu(\lambda)=D(\lambda+\mu)$ is a differential operator (Taylor's theorem), with constant term in it equal to 1 (corresponding to the fact that the highest degree homogeneous term in the expansion of D^μ is D itself), the W-invariant operator

$$\left(\sum_{\sigma \in W} T_{\mu^\sigma}\right) - |W|$$

has zero constant term, and hence annihilates every harmonic polynomial, giving (2.2). This argument is due to Steinberg.

It is conceivable that an appropriate generalization of (2.1) would render it easier to prove. Venturing in one possible direction of generalization, we may ask if the following is true:

$$\sum_{\sigma \in W} b_\sigma \cdot D(\lambda+\mu^{w_\sigma}) \quad \text{is independent of} \quad \mu \in E, \tag{2.3}$$

where b_σ are as in the conjecture. We have verified this only for the case of type A_2, apart from the trivial case of type A_1.

2.4. Note that there is exactly one value of $\sigma \in W$, viz. $\sigma=\sigma_0$, such that $D_\sigma(\lambda)=D(\lambda^{w_\sigma})$ has a nonzero constant term, i.e. does not vanish for $\lambda=0$; this is clear from the fact that $D_\sigma(0)=D(\delta_\sigma)\neq 0$ if and only if δ_σ is W-regular. Therefore, for $\lambda=0$ (2.1) specializes to $\sum_{\sigma \in W} b_\sigma \cdot D(\delta_\sigma)=1$, which gives $b_{\sigma_0}=1$.

Let us also observe that the weighting function $\sigma \to b_\sigma$ on W is constant on each right-coset with respect to the subgroup Ω_0 introduced in § 1, i.e. $b_{\gamma\sigma}=b_\sigma$ for $\gamma \in \Omega_0$ and $\sigma \in W$. This is clear from the fact that for λ in the interior of $E^\star$ one has

$$\sum_\sigma b_{\gamma\sigma} D_\sigma(\lambda) = \sum_\sigma b_\sigma D_{\gamma^{-1}\sigma}(\lambda) = \sum_\sigma b_\sigma D(\lambda^{(\gamma^{-1}\sigma)}) = \sum_\sigma b_\sigma D((\lambda^{(\gamma^{-1})})^{(\sigma)}) = 1.$$

Thus the weighting function is determined by its restriction to the set $W^{\#}$ of distinguished coset representatives for $\Omega_0 \backslash W$, and (2.1) becomes

$$\sum_{\sigma \in W^{\#}} b_\sigma \left(\sum_{j \in J_0} D(\lambda^{\omega_j w_\sigma}) \right) = 1.$$

§ 3. Evidence in low ranks

Given below are the explicit expressions for the polynomial $D(\lambda)=$ $=D(x_1, x_2, \ldots, x_l)$ in $\lambda=x_1\lambda_1+x_2\lambda_2+\ldots+x_l\lambda_l \in E$ in the four cases of rank 2 and 3 in which Conjecture I has been verified, along with the Dynkin diagrams which reveal our convention for the labelling of fundamental roots; the arrows in the diagrams go from a long root to a short one:

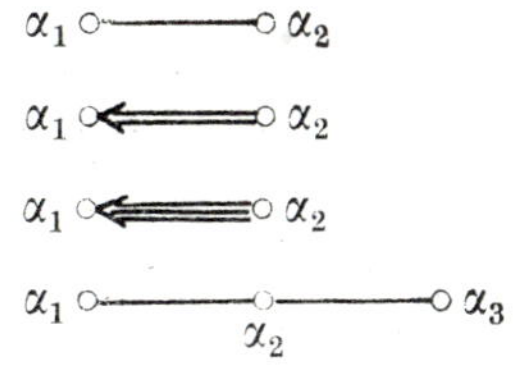

A_2:	$2.D(x, y) = xy(x+y)$,	$\lvert W\rvert = 6$;
B_2:	$6.D(x, y) = xy(x+y)(x+2y)$,	$\lvert W\rvert = 8$;
G_2:	$120.D(x, y) = xy(x+y)(x+2y)(x+3y)(2x+3y)$,	$\lvert W\rvert = 12$;
A_3:	$12.D(x, y, z) = xyz(x+y)(y+z)(x+y+z)$,	$\lvert W\rvert = 24$;

where $x=x_1$, $y=x_2$, $z=x_3$. The integer coefficient in front of $D(x, \ldots)$ comes from the normalization $D(1, \ldots, 1)=1$. The Coxeter diagrams of $\tilde{W}'$ are then as follows:

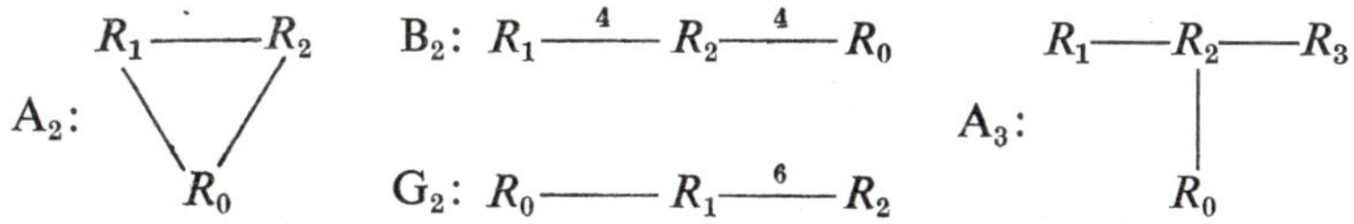

We shall first give our verification of Conjecture I in the three rank-two cases. This is done in the three tables below, for the respective cases, where the value of D_σ for $\sigma=R_{i_1}R_{i_2}\ldots R_{i_k}\in W$ can be read off explicitly from the column headed by the juxtaposition $i_1 i_2 \ldots i_k$ according to the following rule: In the first column $D_{x\ldots x}$ (resp. $D_{y\ldots y}$), with the subscripts appearing i times, stands for $\left(\frac{\partial}{\partial x}\right)^i D$ $\left(\text{resp. } \left(\frac{\partial}{\partial y}\right)^i D\right)$ as in elementary calculus, and the 6, 8 or 12 polynomials given in the first column of the respective table form a basis of the space of W-harmonic polynomials. (The mixed partial derivative $D_{x\ldots xy\ldots y}$ can be computed in terms of this from the differential equation

$$\frac{\partial^2 D}{\partial x \partial y} = \frac{\partial^2 D}{\partial x^2} + \frac{1}{c}\frac{\partial^2 D}{\partial y^2}$$

satisfied by the polynomial D in these cases, where c equals 1, 2 or 3 according as Δ is of type A_2, B_2 or G_2 (the number of lines in the Dynkin diagram); the invariant differential (Casimir) operator $(\partial_x)^2+c^{-1}(\partial_y)^2-\partial_x\partial_y$ is precisely the one corresponding to the second-degree Killing invariant.) Now the column $i_1 i_2 \ldots i_k$ (which stands for $\sigma = R_{i_1} R_{i_2} \ldots R_{i_k}$) simply gives the expansion of the inhomogeneous polynomial D_σ in terms of the basis of W-harmonics given in the first column (which has been so normalized, through suitable denominators, that all the coefficients in these expansions are integers). Thus the value of D_σ is simply the inner product of the column under σ with the left-end column (the part appearing between the two horizontal lines). All blank places stand for the entry "zero".

The first part of Conjecture I asserts merely the non-singularity of the $|W| \times |W|$ matrices showing the expansion of the D_σ's in these tables; and the values of b_σ's given in the bottom rows verify the second part. Our verifications were done manually, except that the 12×12 matrix in the case of type G_2 was obtained from a computer print-out of the D_σ's.

TABLE FOR TYPE A_2

	$\Omega_0 R_1 R_2 R_1$			Ω_0		
$\sigma =$	121	1	2	12	21	id.
D	−1	−1	−1	1	1	1
D_x	1	1	−1	−1		
D_y	1	−1	1		−1	
$\frac{1}{2} D_{xx} = \frac{1}{2} y$	−3	1	1		1	
$\frac{1}{2} D_{yy} = \frac{1}{2} x$	−3	1	1	1		
1	1					
$b_\sigma =$	1	1	1	1	1	1

TABLE FOR TYPE B_2

	$\Omega_0 \cdot R_2$		$\Omega_0 \cdot R_1 R_2$		$\Omega_0 \cdot R_2 R_1 R_2$		Ω_0	
$\sigma =$	1212	2	1	12	21	212	121	id.
D	1	−1	−1	1	1	−1	−1	1
D_x	−1	−2	1	−2	1		1	
D_y	−1	1	−1	1	−1	1		
$\frac{1}{2} D_{xx}$	3		1		−1		−1	
$\frac{1}{2} D_{yy}$	2	1		−1		−1		
$\frac{1}{6} D_{xxx} = \frac{1}{6} y$	−13	4	1	4	1		1	
$\frac{1}{6} D_{yyy} = \frac{1}{6} x$	−11	2	1	2	1	2		
1	1							
$b_\sigma =$	1	1	1	1	2	2	2	2

$\sigma =$	121212	2	12	212	1212	21212	1	21	121	2121	12121	id.
D	1	−1	1	−1	1	−1	−1	1	−1	1	−1	1
D_x	−1	−3	−3	−3	−3		1	2	2	1	1	
D_y	−1	1	2	2	1	1	−1	−1	−1	−1		
$\frac{1}{2} D_{xx}$	3	−3	−3	3	3		1			−1	−1	
$\frac{1}{6} D_{yy}$	5	3			−3	−3	−1	−1	1	1		
$\frac{1}{6} D_{yyy} = \frac{1}{6} x(x+3y)(2x+3y)$	−3	1	−1	−1	1	1						
$\frac{1}{6} D_{xxx} = \frac{1}{6} y(x+y)(x+2y)$	−10						1	−1	−1	1	1	
$\frac{1}{24} D_{xxxx} = \frac{1}{24} y(2x+3y)$	39	9	9	−9	−9		1			−1	−1	
$\frac{1}{216} D_{yyyy} = \frac{1}{24} x(x+2y)$	55	9			−9	−9	1	1	−1	−1		
$\frac{1}{080} D_{yyyyy} = \frac{1}{60} x$	−149	9	18	18	9	9	1	1	1	1		
$\frac{1}{120} D_{xxxxx} = \frac{1}{60} y$	−211	27	27	27	27		1	2	2	1	1	
1	1											
$b_\sigma =$	1	1	1	3	2	4	1	1	3	9	4	2

At the top of the tables for types A_2 and B_2, are indicated the right-cosets of Ω_0 in W, on each of which the weighting function is constant. In the adjoining Tessellation Diagrams (Figs 1—3) are also shown the numbers b_σ for $\sigma \in W$ in bold numerals inside the corresponding w_σ-simplices (which fill up $E^{\#}$), as also are shown the points $\lambda^{(\sigma)}$ for $\lambda \in interior\ (E^{\star})$ and $\sigma \in W$. The points $\lambda^{(\sigma)} = \lambda^{w_\sigma}$

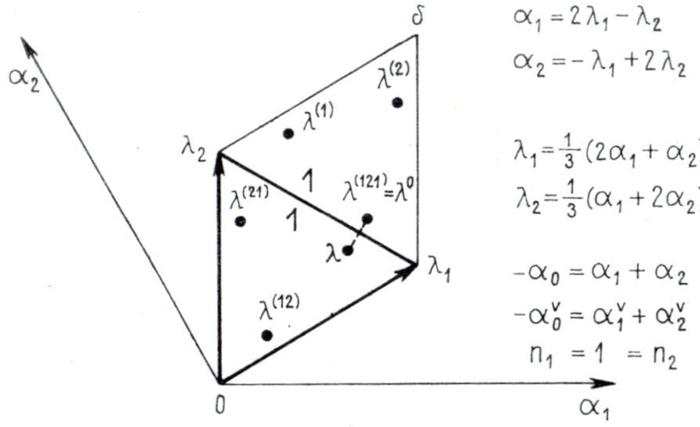

Fig. 1. Tessellation diagram for type A_2

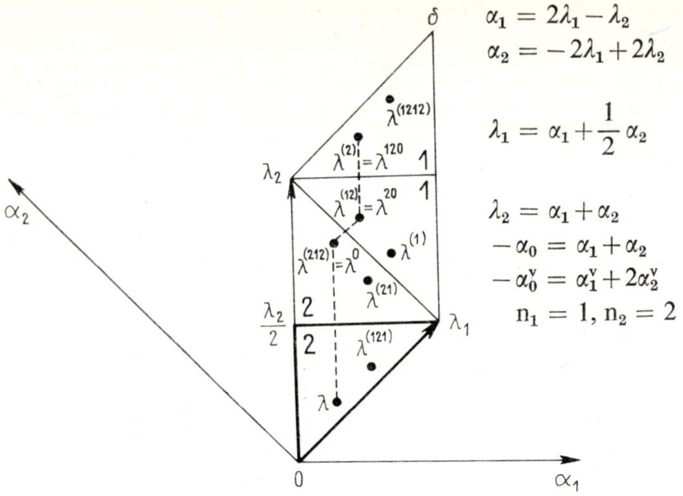

Fig. 2. Tessellation diagram for type B_2

for $\sigma \in W$, comprising the intersection of the $\tilde{W}'$-orbit of λ with $E^{\#}$, are shown linked through dotted lines joining pairs of elements that are adjacent with respect to the partial-order $\leqq$. As in the tables we have followed again the juxtaposition convention in these Diagrams, so that $\lambda^{(\sigma)}$ (resp. λ^{w}) for $\sigma \in W$ (resp. $w \in \tilde{W}^{\#}$) given by a reduced expression $R_{i_1} R_{i_2} \dots R_{i_k}$ with the indices $i_1, i_2, \dots, i_k$ ranging in $\{1, 2\}$ (resp. $\{0, 1, 2\}$), has been abbreviated to

$$\lambda^{(i_1 i_2 \cdots i_k)} \quad (\text{resp. } \lambda^{i_1 i_2 \cdots i_k}).$$

Lastly, we come to the case of type A_3. Here we omit the explicit values of the 24 polynomials D_σ because of the enormity of the size of the corresponding table (once again obtained through a computer print-out of these inhomogeneous polynomials), nor do we give a diagrammatic visualization of the tessellation of $E^{\#}$ (it is not very hard to construct a 3-dimensional model of this tessellation, as also a plane projection of this model). We give only the final listing of the weighting function b_σ for $\sigma \in W^{\#}$, in the table on p. 676, showing also (schematically) the "linkage diagram" for the partial-order on $\tilde{W}$ (analogous to the linkage formed by $\{\lambda^{w} \mid w \in W^{\#}\}$ in the rank-two diagrams above, shown by the dotted lines there).

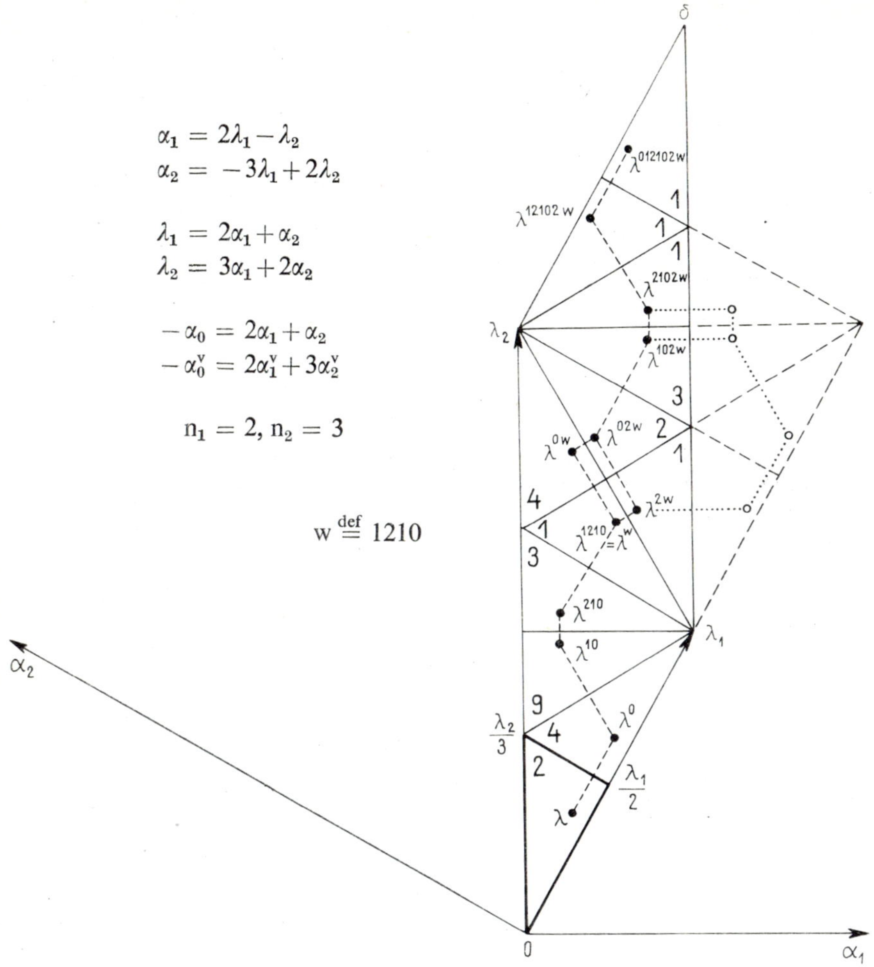

σ	121212	2	12	212	1212	21212	1	21	121	2121	12121	id
w_σ	012102w	12102w	2102w	102w	02w	0w	2w	1210=w	210	10	0	id

Fig. 3. Tessellation diagram for type G_2

In the Tessellation Diagram for type G_2 we also exhibit, incidentally, all the w'-simplices for $w' \in \tilde{W}^+$ satisfying $w' \leqq w_0$. The four triangles shown by broken lines correspond to the four such values of w' not lying in $\tilde{W}^{\#}$, viz.

$$R_1 R_2 w, \quad R_0 R_1 R_2 w, \quad R_1 R_0 R_1 R_2 w, \quad \text{and} \quad R_2 R_1 R_0 R_1 R_2 w,$$

where w stands for $R_1 R_2 R_1 R_0$ given by the juxtaposition $w = 1210$ which has been used throughout this diagram for the sake of clarity in printing.

TABLE FOR TYPE A_3

$\sigma \in W^{\#}$	b_σ	$\lambda^{(\sigma)}$	w_σ	linkage
$R_1 R_3$	1	$(1-x, x+y+z, 1-z)$	$w_0 = R_2 R_1 R_3 R_0$	
$R_2 R_1 R_3$	1	$(1-x-y, x+y+z, 1-y-z)$	$R_1 R_3 R_0$	
$R_1 R_2 R_1 R_3$	1	$(1-y, y+z, 1-x-y-z)$	$R_3 R_0$	
$R_3 R_2 R_1 R_3$	1	$(1-x-y-z, x+y, 1-y)$	$R_1 R_0$	
$R_1 R_3 R_2 R_1 R_3$	2	$(1-y-z, y, 1-x-y)$	R_0	
id.	0	(x, y, z)	id.	

In this case there are two elements of $\tilde{W}^+$ that lie below w_0 with respect to $\leqq$ but do not belong to $\tilde{W}^{\#}$, viz. $R_2 R_1 R_0$ and $R_2 R_3 R_0$.

PART II

REPRESENTATIONS AND DECOMPOSITIONS IN CHARACTERISTIC $p \neq 0$

§ 4. Fundamental constants $c_{\lambda\mu}$ and the character formula

As in Part I, let Δ be an indecomposable reduced root-system, and $\mathfrak{g}_{\mathbf{C}}$ be the simple Lie algebra over $\mathbf{C}$ (determined up to isomorphism) possessing Δ as its root-system.

4.1. Let G be an algebraic Chevalley group, i.e. a connected semisimple (affine) algebraic group, of type $\mathfrak{g}_{\mathbf{C}}$ (or of type Δ) defined over a field K of characteristic $p \neq 0$; one denotes by G_L the group of L-rational points of G for $L \supseteqq K$. Thus, depending on one's approach, G itself may be identified either with $G_{\bar{K}}$ for the algebraic closure $\bar{K}$ of K, or with G_L for a very large algebraically closed (universal) field L, or else with the functor $L \to G_L$; G is characterised by its affine-ring $A = A_K$ (being $K[G]$ in the notation of [2], our canonical reference for standard facts summarized here too briefly), which is a Hopf algebra over K, such that G_L is identified with the maximal-ideal spectrum of $A_L = A \underset{K}{\otimes} L$ for every field $L \supseteqq K$, the group-operation on G_L coming from the coproduct in A_L in the standard manner. By a G-module we shall always mean an algebraic G-module defined over K, using the words "module" and "representation" interchangeably. Thus K is the underlying field of every representation space V, and concretely the representation affords a homomorphism of G_L into $\mathrm{Aut}_L\, V_L$ for every overfield L of K, where $V_L = V \underset{K}{\otimes} L$.

Let us assume that G is simply-connected, and split over K. The former assumption means that every projective representation (i.e. algebraic homomorphism into $PGL(n)$) of G comes from a G-module (i.e. factors through $GL(n)$); this is particularly suitable for representation-theoretic purposes, and there is no loss in this assumption because of the existence and uniqueness of simply-connected (universal) coverings. The latter assumption means that G contains a maximal torus T which is defined and split over K, i.e. is isomorphic over K to $(\mathbf{G}_m)^l$ where $\mathbf{G}_m = GL(1)$ is the "multiplicative group" (which, as a group-functor, assigns to $L \supseteqq K$ the multiplicative group of L). One knows that under these assumptions G is uniquely determined (up to isomorphism) by K and Δ, and that K may be taken to be the prime field $\mathbf{F}_p$; this means existence of an $\mathbf{F}_p$-form of the affine-ring, which was described explicitly by Kostant purely in terms of the root-system Δ (via the $\mathbf{Z}$-form $U_{\mathbf{Z}}$ of the enveloping algebra $U_{\mathbf{C}}$ of $\mathfrak{g}_{\mathbf{C}}$ which we shall describe below in some detail, without discussing the affine-ring further as it is not required by us here); cf. [2, § 4]. We find it convenient not to specify K, but treat it as a floating (working) field. Thus the Lie algebra $\mathfrak{g}_K$ of G is a vector space over K of dimension $\dim G = \dim \mathfrak{g}_{\mathbf{C}} = |\Delta| + l$. The elements of $\mathfrak{g}_K$ may be identified with either the right or the left-invariant derivations of A_K. It may be remarked that unlike the situation in characteristic 0, the object $\mathfrak{g}_K$ is not entirely "local"; in particular, the Lie algebra of two isogenous groups (in the sense of algebraic groups) may not be isomorphic. In fact, one knows that this pathology arises if and only if p divides the "connexion number" $[X:X']$ of Δ where X, X' are as in § 1; in this case $\mathfrak{g}_K$ is nonsimple, and the only time that happens again is for types F_4 with $p=2$ and G_2 with $p=3$; but we shall not require this fact. An explicit description of $\mathfrak{g}_K$ in terms of Δ, essentially due to Chevalley, is as follows.

Let $\mathfrak{g}_{\mathbf{Z}} = \mathfrak{h}_{\mathbf{Z}} + \sum_{\alpha \in \Delta} \mathfrak{g}_{\mathbf{Z}}^{\alpha}$ be a Chevalley lattice in $\mathfrak{g}_{\mathbf{C}}$. This means the following: (i) $\mathfrak{h}_{\mathbf{C}} = \mathfrak{h}_{\mathbf{Z}} \otimes_{\mathbf{Z}} \mathbf{C}$ is a Cartan subalgebra of $\mathfrak{g}_{\mathbf{C}}$ on which elements of Δ are considered as linear functions and $\mathfrak{h}_{\mathbf{Z}}$ consists of all $h \in \mathfrak{h}_{\mathbf{C}}$ such that $\lambda(h) \in \mathbf{Z}$ for all λ in the weight-lattice X of Δ (as defined in § 1, with $\Delta \subset X \subset \mathfrak{h}_{\mathbf{C}}^* =$ vector space dual of $\mathfrak{h}_{\mathbf{C}}$); thus if $h_\alpha \in \mathfrak{h}_{\mathbf{C}}$ is defined by $\lambda(h_\alpha) = \langle \lambda, \alpha^\vee \rangle$ for all $\lambda \in \mathfrak{h}_{\mathbf{C}}^*$, then putting $h_i = h_{\alpha_i}$ one has $\mathfrak{h}_{\mathbf{Z}} = \sum_{i=1}^{l} \mathbf{Z} h_i$; (ii) $\mathfrak{g}_{\mathbf{C}}^{\alpha} = \mathfrak{g}_{\mathbf{Z}}^{\alpha} \otimes_{\mathbf{Z}} \mathbf{C}$ is the α-rootspace of $\mathfrak{g}_{\mathbf{C}}$ with respect to $\mathfrak{h}_{\mathbf{C}}$, and if e_α denotes a generator of the (additive) infinite-cyclic group $\mathfrak{g}_{\mathbf{Z}}^{\alpha}$, then $[e_\alpha, e_{-\alpha}] = \pm 2h_\alpha$; and (iii) if $U_{\mathbf{Z}}$ denotes the subring of the universal enveloping algebra $U_{\mathbf{C}}$ of $\mathfrak{g}_{\mathbf{C}}$ generated by $\left\{ \frac{e_\alpha^n}{n!} \middle| \alpha \in \Delta, n \in \mathbf{Z}^+ \right\}$, then $\mathfrak{g}_{\mathbf{Z}}$ is stable under the adjoint action of $U_{\mathbf{Z}}$ on $\mathfrak{g}_{\mathbf{C}}$. It is customary to choose e_α's such that $[e_\alpha, e_{-\alpha}] = 2h_\alpha$, and we write e_i for e_{α_i} and f_i for $e_{-\alpha_i}$, for $1 \leqq i \leqq l$. It may be remarked that the set

$$\left\{ \frac{e_i^n}{n!}, \frac{f_i^n}{n!} \middle| 1 \leqq i \leqq l, n \in \mathbf{Z}^+ \right\}$$

generates $U_{\mathbf{Z}}$ as a ring [18].

The Lie algebra $\mathfrak{g}_K$ of G can be identified with $\mathfrak{g}_{\mathbf{Z}} \otimes_{\mathbf{Z}} K$ in such a way that $\mathfrak{h}_K = \mathfrak{h}_{\mathbf{Z}} \otimes_{\mathbf{Z}} K$ is the Lie algebra of the maximal torus T, and identifying X with the character group $X(T)$ of T (i.e. the group of algebraic homomorphisms of T into $\mathbf{G}_m$ with the group-operation given by tensoring these 1-dimensional T-modules, which is written additively), the subspace $\mathfrak{g}_K = \mathfrak{g}_{\mathbf{Z}}^{\alpha} \otimes_{\mathbf{Z}} K$ of $\mathfrak{g}_K$ becomes its α-rootspace with respect to the adjoint action of T. Thus, we are identifying Δ with the root-system of G with respect to T, as is common practice, and this should lead to no confusion. In general, for any (algebraic) T-module M and for $\lambda \in X(T) = X$ we denote by $M_{(\lambda)}$ its λ-weight-space consisting of all vectors in M on which T acts by λ, so that one has $M = \sum_{\lambda \in X} M_{(\lambda)}$ in view of the complete reducibility of T-modules; thus $\mathfrak{g}_{K(\alpha)}$ (for the adjoint action of T on $\mathfrak{g}_K$) is the same as $\mathfrak{g}_K^\alpha$. The structure of a T-module is determined by the dimensions of the various weight-spaces, i.e. by the element *char M* of the group-ring of X over $\mathbf{Z}$, called the formal character of M, which assigns $\lambda \to \dim M_{(\lambda)}$. More precisely, denoting by $\mathbf{Z} \cdot e^X$ the set of all $\mathbf{Z}$-linear combinations $\sum_{\lambda \in X} n_\lambda \cdot e^\lambda$ of formal symbols $\{e^\lambda \mid \lambda \in X\}$ where all but a finite number of n_λ's are zero, and making it into a ring under the usual coordinatewise addition and convolution multiplication with $e^\lambda \cdot e^\mu = e^{\lambda+\mu}$, we have

$$\operatorname{char} M = \sum_{\lambda \in X} (\dim M_{(\lambda)}) \cdot e^\lambda \in \mathbf{Z} \cdot e^X.$$

The formal character of a G-module M shall mean simply that of M as a T-module upon restriction. From the standard fact that the Weyl group W of Δ can be identified with the quotient N/T in a canonical manner, where N is the normalizer of T in G, it is immediate that the formal character of a G-module is a W-invariant element of $\mathbf{Z} \cdot e^X$ under the natural action of W.

Let B be the Borel subgroup of G having as its Lie algebra $\mathfrak{b}_K = \mathfrak{h}_K + \sum_{\alpha \in \Delta_+} \mathfrak{g}_K^\alpha$, where Δ_+ is a given choice of positive roots in Δ. One knows that every irreducible G-module contains a unique 1-dimensional highest-weight-space, in the sense that it is a weight-space for T whose elements remain fixed under the unipotent radical $[B, B]$ of B. If this weightspace corresponds to the character $\lambda \in X$, we say λ is the highest weight of the given module. Necessarily $\lambda \in X^+$ (the dominant elements of X with respect to the fixed choice of positive roots), and if $\mu \in X$ is a "weight" of the given G-module, in the sense that the μ-weight-space is nonzero, then $\lambda - \mu \in \mathbf{Z} \cdot \Delta_+$, which we rewrite as $\mu \leqq \lambda$ by setting up the canonical partial-order $\leqq$ on X compatible with the addition under which $\mathbf{Z}^+ \cdot \Delta_+$ is the set of all elements $\geqq 0$. Conversely, to every $\lambda \in X^+$ there corresponds an irreducible G-module M_λ with highest weight λ, which is unique up to isomorphism. As stated in the Introduction, one of the long-standing unsolved problems in the study of G-modules is to give a formula for *char* M_λ in terms of $\lambda \in X^+$.

4.2. A famous theorem of Steinberg reduces the study of all irreducible G-modules M_λ for $\lambda \in X^+$, to that of the p^l basic ones with λ lying in the set

$$X^\# = \{\sum_i m_i \lambda_i \in X \mid 0 \leqq m_i \leqq p-1 \text{ for } 1 \leqq i \leqq l\};$$

the elements of $X^\#$ will be called the "restricted dominant characters". The formulation of this theorem hinges on the fact that G is defined over $\mathbf{F}_p$ (as stated earlier) *and* that all the M_λ's are definable over $\mathbf{F}_p$; we will be able to see the latter fact from the construction of M_λ by the standard technique of "reduction *mod p*", which we shall need to describe presently in some detail. Now if M is any G-module defined over the prime field, one defines its Frobenius-twist

M^{Fr} to be the G-module whose underlying space is M itself, such that the new action of $g \in G_L$ on $M_L = M \underset{K}{\otimes} L$ is obtained as follows: take a basis of the $\mathbf{F}_p$-form $M_{\mathbf{F}_p}$ of $M = M_K$ and write the matrix of (the old action of) g with respect to this basis of M_L; then the new matrix obtained by replacing each entry by its pth power gives the new (Frobenius-twisted) action of g on M_L with respect to the same basis. It is immediate that the μ-weight-space of M coincides with the $p\mu$-weight-space of M^{Fr}, which gives *char* $M^{\mathrm{Fr}} = (\textit{char } M)^{\mathrm{Fr}}$, where

$$\textit{for} \quad \chi = \sum_{\lambda} n_\lambda e^\lambda \in \mathbf{Z}e^X, \quad \textit{we define} \quad \chi^{\mathrm{Fr}} = \sum_{\lambda} n_\lambda e^{p\lambda}.$$

Steinberg's tensor-product theorem then essentially says that for $\lambda \in X^+$ and $\mu \in X^\#$ the module $M_\lambda^{\mathrm{Fr}} \otimes M_\mu$ is irreducible and hence isomorphic to $M_{p\lambda+\mu}$; in particular, $M_\lambda^{\mathrm{Fr}} \cong M_{p\lambda}$. It follows that if $\lambda \in X^+$ has the p-expansion

$$\lambda = \mu_0 + \mu_1 p + \ldots + \mu_k p^k, \quad \text{with} \quad \mu_j \in X^\# \quad \text{for} \quad 0 \leqq j \leqq k,$$

then

$$M_\lambda \cong M_{\mu_0} \otimes M_{\mu_1}^{\mathrm{Fr}} \otimes \ldots \otimes M_{\mu_k}^{\mathrm{Fr}^k},$$

where Fr^j denotes the j-fold iteration of Fr, and consequently,

$$\text{char } M_\lambda = \prod_{j=0}^{k} (\text{char } M_{\mu_j})^{\mathrm{Fr}^j}. \tag{4.1}$$

4.3. In order to define the constants $c_{\lambda\mu}$ in the title of this section, we need to quickly introduce certain G-modules $V_{\lambda,K}$ which are modelled after the irreducible $\mathfrak{g}_{\mathbf{C}}$-module $V_{\lambda,\mathbf{C}}$ with highest weight $\lambda \in X^+$, so that the formal character of $V_{\lambda,K}$ coincides with that of $V_{\lambda,\mathbf{C}}$. We caution in advance that $V_{\lambda,K}$ is not uniquely determined by λ alone; however, there are only a finite number of choices (up to isomorphism). The formal character of $V_{\lambda,K}$ is determined by λ, and is given by Weyl's character formula as in characteristic 0; thus *char* $V_{\lambda,K}$ is independent of p. Also (by Proposition 2 below) the composition factors of $V_{\lambda,K}$ are determined by λ alone; these are far from being independent of p though, and in fact for every λ one can find a bound for p beyond which $V_{\lambda,K}$ is irreducible. This is not a new result, and follows from the "discriminant criterion" described below.

Let $V_{\lambda,\mathbf{Z}}$ be a $U_{\mathbf{Z}}$-stable lattice in $V_{\lambda,\mathbf{C}}$ ("admissible **Z**-form" in the terminology of [2]). One sees that up to dilatation, i.e. multiplication by a nonzero scalar, there are only a finite number of choices for $V_{\lambda,\mathbf{Z}}$; this follows from the fact that if an arbitrary nonzero element v_0 in the highest-weightspace of $V_{\lambda,\mathbf{C}}$ is required to be a primitive (=nondivisible) vector in $V_{\lambda,\mathbf{Z}}$ (as can always be ensured upon dilatation, as the highest-weight-space is 1-dimensional), then there exist two uniquely determined extremes for $V_{\lambda,\mathbf{Z}}$:

$$V_{\lambda,\mathbf{Z}}^{\min} \leqq V_{\lambda,\mathbf{Z}} \leqq V_{\lambda,\mathbf{Z}}^{\max}.$$

It is clear that the minimal lattice $V_{\lambda,\mathbf{Z}}^{\min}$ has to be simply the image $v_0 U_{\mathbf{Z}}$

of v_0 under the action of $U_{\mathbf{Z}}$ on $V_{\lambda,\mathbf{C}}$. Let $v_0^* \in (V_{\lambda,\mathbf{C}})^*$ = the vector space dual to $V_{\lambda,\mathbf{C}}$, be the linear function having value 1 at v_0 and vanishing at all the weight-spaces of $V_{\lambda,\mathbf{C}}$ except the highest one. Then v_0^* is a lowest-weight-vector of the contragredient $\mathfrak{g}_{\mathbf{C}}$-module $(V_{\lambda,\mathbf{C}})^* = V_{\lambda^*,\mathbf{C}}$ with highest weight $\lambda^* = \lambda^{\tau_0} = -\lambda^{\sigma_0}$. Then, setting $V_{\lambda^*,\mathbf{Z}}^{\min} = v_0^* U_{\mathbf{Z}}$, one has

$$V_{\lambda,\mathbf{Z}}^{\max} = \{x \in V_{\lambda,\mathbf{C}} \mid y(x) \in \mathbf{Z} \text{ for all } y \in V_{\lambda^*,\mathbf{Z}}^{\min}\}.$$

An alternate way of describing this is in terms of the "contragredient bilinear form" introduced by W. J. Wong [20]:

Let θ denote the (unique) anti-automorphism of $U_{\mathbf{C}}$ which interchanges e_i and f_i for $1 \leqq i \leqq l$ (and hence leaves the h_i's fixed); the existence of θ is classical and follows, for example, from the now well-known Harish-Chandra—Jacobson—Serre presentation of the Lie algebra $\mathfrak{g}_{\mathbf{C}}$ (equivalently, of the associative algebra $U_{\mathbf{C}}$) in terms of the generators $\{e_1, e_2, \ldots, e_l, f_1, f_2, \ldots, f_l\}$. Then $V_{\lambda,\mathbf{C}}$ possesses a unique symmetric bilinear form $(\,,\,)$ such that $(v_0, v_0) = 1$ and $(x \cdot e_i, y) = (x, y \cdot f_i)$ for $1 \leqq i \leqq l$ and $x, y \in V_{\lambda,\mathbf{C}}$; more generally $(x \cdot u, y) = (x, y \cdot u^\theta)$ for $u \in U_{\mathbf{C}}$, and $(\,,\,)$ is determined by the fact that $(v_0 \cdot u, v_0 \cdot u') = c$ if the projection of $v_0 u' u^\theta$ on the highest-weight-space along the sum of all other weight-spaces is cv_0. The restriction of $(\,,\,)$ to $V_{\lambda,\mathbf{Z}}^{\min}$ takes integer values, and one has

$$V_{\lambda,\mathbf{Z}}^{\max} = \{x \in V_{\lambda,\mathbf{C}} \mid (x, y) \in \mathbf{Z} \text{ for all } y \in V_{\lambda,\mathbf{Z}}^{\min}\}.$$

The "reduction mod p" $V_{\lambda,K} = V_{\lambda,\mathbf{Z}} \otimes_{\mathbf{Z}} K$ of any $U_{\mathbf{Z}}$-stable lattice $V_{\lambda,\mathbf{Z}}$ in $V_{\lambda,\mathbf{C}}$, becomes a G-module in a natural manner [2] (here simply-connectedness of G is required, just as in characteristic 0), such that the associated infinitesimal action of $\mathfrak{g}_K$ coincides with that of $\mathfrak{g}_{\mathbf{Z}} \otimes_{\mathbf{Z}} K$ (which acts naturally on $V_{\lambda,\mathbf{Z}} \otimes_{\mathbf{Z}} K$, since $\mathfrak{g}_{\mathbf{Z}} \subset U_{\mathbf{Z}}$), with which $\mathfrak{g}_K$ has already been identified. The module $V_{\lambda,K}$ need not be irreducible as we shall see presently, and its internal structure may vary with the choice of $V_{\lambda,\mathbf{Z}}$. Perhaps $V_{\lambda,K}$ is always indecomposable; this is certainly true of the two extreme choices of $V_{\lambda,\mathbf{Z}}$. In fact $V_{\lambda,K}^{\min} = V_{\lambda,\mathbf{Z}}^{\min} \otimes_{\mathbf{Z}} K$ has a unique maximal proper G-submodule (contained in the sum of all the weight-spaces except the highest one), the quotient by which is the irreducible module M_λ; and on the other extreme

$$V_{\lambda,\mathbf{Z}}^{\max} \otimes_{\mathbf{Z}} K = V_{\lambda,K}^{\max} = (V_{\lambda^*,K}^{\min})^*$$

has M_λ as its unique irreducible submodule, on account of the duality given by the last equality. The injection

$$i_{\lambda,\mathbf{Z}} : V_{\lambda,\mathbf{Z}}^{\min} \hookrightarrow V_{\lambda,\mathbf{Z}}^{\max}$$

gives, on reduction mod p, only a homomorphism

$$i_{\lambda,K}: V_{\lambda,K}^{\min} \to V_{\lambda,K}^{\max},$$

whose kernel is the maximal proper submodule of $V_{\lambda,\mathbf{Z}}^{\min}$ and image is the unique copy of M_λ in $V_{\lambda,K}^{\max}$; in other words, $i_{\lambda,K}$ factors through M_λ. Hence we have the "discriminant criterion": $V_{\lambda,K}$ is irreducible if and only if p does not divide the index $[V_{\lambda,\mathbf{Z}}^{\max}: V_{\lambda,\mathbf{Z}}^{\min}]$; this index equals the discriminant of Wong's contragredient form $(\ ,\ )$ on $V_{\lambda,\mathbf{Z}}^{\min}$ described above (cf. [2, Theorem 2]). It does not seem easy to give a formula for this number. (See Footnote[3].) It would be particularly interesting to determine all "small" prime divisors of this index for each λ; prime divisors not exceeding the Coxeter number h_Δ of Δ are of special interest, since the ideas of the next section provide an alternate method towards deciding the irreducibility of $V_{\lambda,K}$ particularly when $p > h_\Delta$ (cf. Proposition 3).

It may be mentioned at this point that for the case of type A_1, i.e. for $G \cong SL_2$, all $V_{\lambda,K}$ for $\lambda \in X^{\#}$ are irreducible. But this is an exceptional situation; it will be shown in a later paper that when Δ is not of type A_1, there are values of $\lambda \in X^{\#}$ (in fact, for large p, a sizeable percentage of values exist) such that $V_{\lambda,K}$ is not irreducible. The first example of a non-irreducible $V_{\lambda,K}$ with $\lambda \in X^{\#}$ was given by Curtis [5, p. 859], for type A_2 ($G \cong SL_3$).

4.4 Let us write $\overline{V}_\lambda = V_{\lambda,K}^{\min}$, and for $\mu \in X^+$ define $c_{\lambda\mu}$ to be the multiplicity $[\overline{V}_\lambda : M_\mu]_G$, of the occurrence of M_μ as a composition factor of the G-module $\overline{V}_\lambda$.

[3] *(Added in Proof.)* This problem does not seem intractable now. First note that the required index $N_\lambda = [V_{\lambda,\mathbf{Z}}^{\max}: V_{\lambda,\mathbf{Z}}^{\min}]$ for $\lambda \in X^+$, is the product of the partial indices $N_{\lambda,\mu} = [V_{\lambda,\mathbf{Z}(\mu)}^{\max} : V_{\lambda,\mathbf{Z}(\mu)}^{\min}]$ as μ ranges over all the weights (since every $U_{\mathbf{Z}}$-stable $\mathbf{Z}$-form is compatible with the weight-space decomposition); clearly it suffices to determine $N_{\lambda,\mu}$ for $\mu \in X^+$. I have solved this problem only when μ is "sufficiently close" to λ in this special sense: the dimension of the μ-weightspace of $V_{\lambda,\mathbf{C}}$ equals $P(\lambda-\mu)$ (which is the leading term in Kostant's multiplicity formula) where $P(\,)$ is Kostant's partition function. Based on that result it seems plausible that the value of $N_{\lambda,\mu}$ is given in general by

$$N_{\lambda,\mu} = \prod_{\alpha \in \Delta^+} \prod_{j \in \mathbf{Z}^+} \binom{(\lambda+\delta)(h_\alpha)-1}{j}^{m_j(\alpha)} \quad \left(\binom{\cdot}{\cdot} \text{ being the usual binomial symbol}\right)$$

where $m_j(\alpha) = m_j^{\lambda;\mu}(\alpha)$ is the dimension of the subspace of $V_{\lambda,\mathbf{C}(\mu)}$ annihilated by e_α^{j+1} but not by e_α^j; when μ is sufficiently close to λ it is easily seen that $m_j(\alpha) = P(\lambda-\mu-j\alpha) - P(\lambda-\mu-(j+1)\alpha)$. The suggested expression for $N_{\lambda,\mu}$ is of particular interest for $\mu=0$, since it appears in a crucial manner in the work of K. R. Parthasarathy, R. Ranga Rao and V. S. Varadarajan (*Ann. Math.* 85 (1967), Theorem 4.2 on page 424) on the principal series representations of complex semi-simple Lie groups; cf. § 8.5.10(b) in Dixmier's book cited in Footnote [5]. My proof of this formula (for μ sufficiently close to λ) is obtained by refining certain ideas of N. N. Šapovalov (cf. § 7.8.23 in Dixmier's book) that are not unrelated to the computations of the 3 authors quoted above; it will be published in due course.

In this connection, reader's attention may also be drawn to N. Burgoyne's ideas in § 1 of "Modular representations of some finite groups" in *Proc. of Symposia in Pure Mathematics,* Vol. XXI, AMS, Providence, 1971.

(See Footnote[4].) It is clear that $c_{\lambda\lambda}=1$, and that $c_{\lambda\mu}\neq 0$ implies (μ is a weight of $\bar{V}_\lambda$ and therefore) $\mu\leqq\lambda$; in particular, $\lambda-\mu\in X'$ is a necessary condition for $c_{\lambda\mu}\neq 0$. We have $\bar{V}_\lambda \underset{G}{\sim} \sum_{\mu\in X^+} c_{\lambda\mu}M_\mu$, where $\underset{G}{\sim}$ denotes equivalence in the Grothendieck ring of the category of G-modules, and thus

$$\sum_{\mu} c_{\lambda\mu}\cdot \mathrm{char}\, M_\mu = \mathrm{char}\, \bar{V}_\lambda = \frac{\sum_{\sigma\in W} \det\sigma\cdot e^{((\lambda+\delta)^\sigma)}}{\sum_{\sigma\in W} \det\sigma\cdot e^{(\lambda^\sigma)}} \tag{4.2}$$

where the second equality comes from the Weyl's character formula for $V_{\lambda,\mathbf{C}}$.

The linear equations (4.2) for the "unknowns" $\chi_\mu=\mathrm{char}\, M_\mu$ can be solved in terms of the "known" quantities $\bar{\chi}_\mu=\mathrm{char}\,\bar{V}_\mu$, in the form $\chi_\lambda=\sum_\mu \gamma_{\lambda\mu}\bar{\chi}_\mu$, where the matrix $[\gamma_{\lambda\mu}]_{X^+}$ is the inverse of the infinite matrix $[c_{\lambda\mu}]_{X^+}$, the outer subscript X^+ denoting the set over which the matrix indices λ, μ run. With respect to any enumeration of the set X^+ which is stronger than the partial order $\leqq$ (definition at the end of § 4.1), the matrix $[c_{\lambda\mu}]_{X^+}$ is unipotent in the sense that it is triangular with diagonal entries all 1. As a matter of fact, in order to compute χ_{μ_0} for given $\mu_0\in X^+$, one needs to consider (4.2) only for λ ranging over the finite set $Y=Y(\mu_0)$ consisting of those $\lambda\in X^+$ such that there exists a sequence $\mu_1, \mu_2, \ldots, \mu_{k-1}, \mu_k=\lambda$ with $c_{\mu_{j-1}\mu_j}\neq 0$ for $1\leqq j\leqq k$. Since for $\lambda\in Y$, $\mu\in X^+$,

$$c_{\lambda\mu}\neq 0 \quad \text{implies} \quad \mu\in Y(\lambda)\subseteqq Y(\mu_0)=Y,$$

we have for $\lambda\in Y$,

$$\sum_{\mu\in Y} c_{\lambda\mu}\chi_\mu = \bar{\chi}_\lambda,$$

i.e. only the finite segment $[c_{\lambda\mu}]_Y$ of $[c_{\lambda\mu}]_{X^+}$ needs to be inverted.

Proposition 1 *(Character formula). There exist unique integers $\gamma_{\lambda\mu}$ for λ, $\mu\in X^+$, such that for given λ only finitely many $\gamma_{\lambda\mu}$ are nonzero and*

$$\mathrm{char}\, M_\lambda = \frac{\sum_\mu \left(\sum_{\sigma\in W} \det\sigma\cdot\gamma_{\lambda\mu}\cdot e^{((\mu+\delta)^\sigma)}\right)}{\sum_{\sigma\in W}\det\sigma\cdot e^{(\delta^\sigma)}}. \tag{4.3}$$

By virtue of the standard fact that the set $\{\bar{\chi}_\lambda \mid \lambda\in X^+\}$ is a $\mathbf{Z}$-free basis of the W-invariants in the group-ring $\mathbf{Z}\cdot e^X$ (see [3]: VI. 3.4, Proposition 3 and Example 2, on p. 187-188), and the fact that the matrix $[\gamma_{\lambda\mu}]_{X^+}$ in $\sum_\mu \gamma_{\lambda\mu}\bar{\chi}_\mu=\chi_\lambda$ is uni-

[4] *(Added in Proof.)* In a recent preprint (titled 'On the modular representations of the general linear and symmetric groups') R. W. Carter and G. Lusztig have introduced the imaginative term "Weyl Module" for $\bar{V}_\lambda$, albeit in the special situation of the general (or the special) linear groups. The allusion to Weyl is to emphasize the fact that the dimension is given by Weyl's formula as in characteristic zero: dim $\bar{V}_\lambda=D(\lambda+\delta)$; in the special situation of Carter and Lusztig it has an additional association with the treatment given by H. Weyl (in his classic monograph "Classical groups") to the problem of decomposing the tensor spaces in accordance with certain symmetries. Our notation $\bar{V}_\lambda$ conforms that in [7].

potent, one finds that $\{\chi_\lambda | \lambda \in X^+\}$ is also a **Z**-free basis of the formal characters of G-modules. Hence we have

Proposition 2. *The composition factors of a G-module are determined by its formal character.*

Because of (4.1) one is primarily interested in computing the formal characters of the p^l basic irreducible G-modules with highest weight in $X^{\#}$. We claim that these are completely determined by our knowledge of the "decomposition"

$$\bar{V}_\lambda \underset{G}{\sim} \sum_{\mu \in X^+} c_{\lambda\mu} M_\mu$$

for λ ranging in $X^{\#}$ alone. To see this, let us impose the new partial-order $\leqq^{\#}$ on $X^{\#}$, defined to be the weakest one such that whenever $\mu_0 + \mu_1 p + \ldots + \mu_k p^k \leqq \lambda$ with $\lambda, \mu_0, \mu_1, \ldots, \mu_k$ all in $X^{\#}$, one has $\mu_i \leqq^{\#} \lambda$ for $0 \leqq i \leqq k$; and let $\leqq'$ denote any linear order on $X^{\#}$ stronger than $\leqq^{\#}$. Then χ_λ can be computed, recursively with respect to $\leqq'$, i.e. assuming that χ_μ is known for all $\mu \leqq' \lambda$, simply by substitution in

$$\chi_\lambda = \bar{\chi}_\lambda - \sum_{\substack{\mu \in X^+ \\ \mu \neq \lambda}} c_{\lambda\mu} \chi_\mu,$$

where the χ_μ's appearing on the right-hand side are known by virtue of (4.1) and the definition of $\leqq^{\#}$. Our claim is proved; and *as a corollary we note that a knowledge of* $[c_{\lambda\mu}]_{\lambda \in X^{\#}, \mu \in X^+}$ *determines,* by virtue of Proposition 2, *the entire matrix* $[c_{\lambda\mu}]_{X^+}$. It would be very interesting to devise a formula (short of a computer algorithm) which enables one to write down, efficiently and explicitly, the decompositions of $\bar{V}_\lambda$ for all $\lambda \in X^+$ in terms of those for $\lambda \in X^{\#}$ alone. (Here and elsewhere in this paper, by the "decomposition" of a module we mean that in the Grothendieck group of the category of modules at hand.)

4.5 When we substitute $\chi_\mu = \prod_j (\chi_{\mu_j})^{\mathrm{Fr}^j}$ for $\mu = \sum_j \mu_j p^j$, with all μ_j's in $X^{\#}$, in $\sum_\mu c_{\lambda\mu}\chi_\mu = \bar{\chi}_\lambda$, what we get is a set of *nonlinear* equations in $\{\chi_\nu | \nu \in X^{\#}\}$. However, under certain circumstances, it is possible to replace these by a set of linear equations

$$\sum_{\mu \in X^{\#}} C_{\lambda\mu} \chi_\mu = \chi_\lambda \quad \text{for} \quad \lambda \in X^{\#}, \tag{4.4}$$

where the coefficients $C_{\lambda\mu}$ are no longer integers, but instead lie in the group-ring $\mathbf{Z} \cdot e^X$ itself; in fact $C_{\lambda\mu}$ is a W-invariant element of the subring $\mathbf{Z} \cdot e^{pX}$ consisting of $\{\chi^{\mathrm{Fr}} | \chi \in e^X\}$. To this end let us define

$$Y_0 = \{\nu \in X^+ | c_{\lambda, \mu + \nu p} \neq 0 \quad \text{for some} \quad \lambda, \mu \in X^{\#}\};$$

then

$$\begin{aligned} \chi_\lambda &= \sum_{\substack{\mu \in X^{\#} \\ \nu \in Y_0}} c_{\lambda, \mu + \nu p} \cdot \chi_{\mu + \nu p} \\ &= \sum_{\mu \in X^{\#}} \Big(\sum_{\nu \in Y_0} c_{\lambda, \mu + \nu p} \cdot \chi_\nu^{\mathrm{Fr}} \Big) \chi_\mu, \end{aligned}$$

which gives (4.4) with

$$C_{\lambda\mu} = \Big(\sum_{\nu\in Y_0} c_{\lambda,\mu+\nu p}\cdot\chi_\nu\Big)^{\mathrm{Fr}}. \tag{4.5}$$

Our "coefficients" in (4.4) till now are themselves polynomials in the unknowns $\{\chi_\nu \mid \nu\in X\}$. If for a given value of p it can be shown that $\bar{V}_\nu$ is irreducible for $\nu\in Y_0$, one can replace χ_ν in (4.5) by the known quantity $\bar{\chi}_\nu$; in Corollary to Proposition 3 in § 5, we shall see that such is indeed the case at least for $p\geqq 2h_\Delta-3$, using the fact that

$$Y_0 \subseteqq \{\sum_i m_i\lambda_i\in X^+ \mid \sum_{i=1}^{l} m_i n_i \leqq h_\Delta-2\}, \tag{4.6}$$

where $h_\Delta=(\sum_i n_i)+1$ is the Coxeter number as in § 1.

To see (4.6) we note that

$$\begin{aligned} Y_0 &\subseteqq \{\nu\in X^+ \mid \mu+p\nu \leqq \lambda \quad \text{for some} \quad \lambda,\mu\in X^{\#}\}\\ &\subseteqq \{\nu\in X^+ \mid \langle\mu+p\nu, -\alpha_0^\vee\rangle \leqq \langle\lambda, -\alpha_0^\vee\rangle \quad \text{for some} \quad \lambda,\mu\in X^{\#}\}\\ &\subseteqq \{\nu\in X^+ \mid \langle p\nu, -\alpha_0^\vee\rangle \leqq \langle (p-1)\delta, -\alpha_0^\vee\rangle\}\\ &= \left\{\nu\in X^+ \mid \langle \nu, -\alpha_0^\vee\rangle \leqq \frac{p-1}{p}(h_\Delta-1)\right\}, \end{aligned}$$

where the second (resp. the third) inclusion follows in view of the fact that $-\alpha_0$ has non-negative inner product with all the fundamental roots (resp. all the fundamental weights), and hence also with all elements of $\mathbf{Z}_+\cdot\Delta_+$ (resp. all elements of $X^{\#}\subset X^+$, and so also all elements of the form $(p-1)\delta-\lambda$ for $\lambda\in X^{\#}$). The inclusion (4.6) follows in view of the fact that $h_\Delta=\frac{2r}{l}$ is an integer. This also shows that $Y_0\subseteqq X^{\#}$ if $p\geqq h_\Delta-2$. It is shown in § 5.4 (Example 2) that $Y_0=\{0\}$ for types A_2 and B_2 (and of course also for type A_1). It seems likely that for all other types $Y_0 \supsetneqq \{0\}$.

Once again, the linear equations (4.4) can be solved (in those cases where $C_{\lambda\mu}$'s are known) in the same way as (4.2) earlier, because the matrix $[C_{\lambda\mu}]_{X^{\#}}$ is unipotent with respect to $\leqq'$ (any linear-order on $X^{\#}$ stronger than $\leqq^{\#}$):

$$\begin{cases} C_{\lambda\lambda} = 1 = \text{the identity element of } \mathbf{Z}\cdot e^X, \text{ and}\\ C_{\lambda\mu} \neq 0 \text{ implies } (\mu+\nu p \leqq \lambda \text{ for some } \nu\in X^+ \text{ and hence}) \ \mu \leqq^{\#} \lambda. \end{cases}$$

The fact that $C_{\lambda\mu}$'s are themselves formal characters is no obstacle to inverting the unipotent matrix as the inverse can be explicitly written down as a polynomial in the $C_{\lambda\mu}$'s. In fact there is no need to handle the entire matrix of size as large as $|X^{\#}|=p^l$; in order to compute χ_{μ_0} for $\mu_0\in X^{\#}$ one need consider (4.4) simply for λ ranging in the set $Y'=Y'(\mu_0)$ consisting of $\lambda\in X^{\#}$ such that there exists a sequence $\mu_1, \mu_2, \ldots, \mu_k$ with $C_{\mu_{j-1}\mu_j} \neq 0$ for $1 \leqq j \leqq k$. The validity of the "Harish-Chandra principle" to be discussed in § 5 implies that the cardinality of Y' never exceeds $|W|$, which is quite comfortable for large p. Note that $Y(\mu) \subseteqq Y'(\mu_0) + pY_0$; the set Y_0 can also be "relativized", i.e. replaced by a subset $Y_0(\mu_0)$ of cardinality bounded by a number independent of p, which would show that $|Y|$ is also bounded by a number independent of p.

§ 5. The "Harish-Chandra principle" and two conjectures on $c_{\lambda\mu}$

5.1 It was conjectured by this author in 1967, in letters to C. W. Curtis and T. A. Springer, that if the irreducible G-modules M_λ and M_μ (for $\lambda, \mu \in X^+$) occur as composition factors of some indecomposable G-module then $\lambda + \delta \equiv \equiv (\mu + \delta)^\sigma \pmod{pX}$ must hold for some $\sigma \in W$. This conjecture was preceded by an infinitesimal version (to be discussed in § 6) which was announced to Curtis orally in 1966 (see Footnote**). It has been dubbed "Harish-Chandra principle in nonzero characteristic", because of its close analogy with the theorem of Harish-Chandra on the infinitesimal characters of $\mathfrak{g}_{\mathbf{C}}$ and because of a central role of this theorem in the study of $\mathfrak{g}_{\mathbf{C}}$-modules (infinite-dimensional in general), particularly in the study of the universal $\mathfrak{g}_{\mathbf{C}}$-modules generated by highest-weight-vectors investigated in [15] and [1]. This analogy is clear from the detailed discussion in [7 (§ 3)], where our conjectures have been proved true for $p > h_\Delta$ (Theorems 3.1, 4.1 and 5.1).

It is believed that the Harish-Chandra principle (along with its infinitesimal version) holds for all p, including the bad cases when the Lie algebra $\mathfrak{g}_K$ of G is not simple. The decompositions of $\bar{V}_\lambda$ (for λ ranging in a set larger than $X^\#$) for the case of type G_2 with $p=3$, given by Springer in [10, p. 103], as well as a few such decompositions for the case of type F_4 with $p=2$ that are implied by [14, Tables I and II], are all consistent with our view.

As in Part I (§ 1), let $\tilde{W}$ be the semidirect extension of W by a group D isomorphic to X. *However, we find it convenient to modify the action of D on X* slightly now: for $\lambda \in X$ we define the corresponding element of D to be the translation on X (or on $E = X \otimes_{\mathbf{Z}} \mathbf{R}$) by $p\lambda$ and not λ as before. *Accordingly* D', $\tilde{W}$, $\tilde{W}'$, Ω, *and the elements* w_σ *of* $\tilde{W}^\#$, *all have modified action on X or E which is dependent on p.* This does not affect the abstract structure of any of these groups, nor does it affect the partial orders on $\tilde{W}^+$, $\tilde{W}^\#$ ect. Also, *we find it convenient not to*

** The genesis of this conjecture lies in the author's attempt to stream-line the results of Braden in his Ph. D. thesis, written under the guidance of Curtis [4], determining the infinitesimal decomposition of $\bar{V}_\lambda$ for $\lambda \in X^\#$ in the case of A_2. I learnt of these results during the part of the summer of 1966 that I spent at the University of Oregon under an invitation from Professor Curtis, who initiated me at that time to the p-modular representation theory of finite Chevalley groups in characteristic p and its connection with the representations of the related Lie algebras.

change the meaning of $E^{\#}$ *and* $E^{\star}$, so that it is $pE^{\#}$ (resp. $pE^{\star}$) and not $E^{\#}$ (resp. $E^{\star}$) which is now the fundamental region in E for the action of D (resp. $\tilde{W}'$).

With this convention, the Harish-Chandra principle states that if M_λ and M_μ are composition factors of an indecomposable G-module then $\lambda+\delta$ and $\mu+\delta$ must lie in the same $\tilde{W}$-orbit. Since all weights of an indecomposable G-module lie in one single coset of X', clearly $\lambda-\mu\in X'$ is another necessary condition. We claim that when $p\nmid[X:X']$ these two conditions together imply (and therefore are equivalent to) the condition that $\lambda+\delta$ and $\mu+\delta$ be conjugate under $\tilde{W}'$. In view of $\tilde{W}=\tilde{W}'\Omega$, it suffices to consider the situation when $\lambda+\delta$ and $\mu+\delta$ are conjugate under $\Omega=\{\omega_j \mid j\in J_0\}$ (cf. § 1.2); then

$$\mu+\delta = (\lambda+\delta)^{\omega_j} = (\lambda+\delta)^{\sigma_0\sigma_j}+p\lambda_j \equiv (\lambda+\delta)+p\lambda_j \pmod{X'}$$

(as $\sigma_0\sigma_j\in W$), gives $p\lambda_j\in X'$; because of $p\nmid[X:X']$ we have $\lambda_j\in X'$, which gives $\lambda_j=0$ and $\omega_j=id.$ as desired.

Let us call $\lambda, \mu\in X$, $\tilde{W}'$*-linked if* $\lambda+\delta$ *and* $\mu+\delta$ *are conjugate under* $\tilde{W}'$. It is clear that every element of X is $\tilde{W}'$-linked to a unique element of the set

$$X^{\star} = (X\cap pE^{\star})-\delta = \{\lambda\in X \mid p^{-1}(\lambda+\delta)\in E^{\star}\}$$
$$= \{\lambda\in X \mid \langle\lambda+\delta, \alpha_i^\vee\rangle \geqq 0 \quad \text{for} \quad 1\leqq i\leqq l, \quad \text{and} \quad \langle\lambda+\delta, -\alpha_0^\vee\rangle \leqq p\},$$

which is partitioned into disjoint subsets $X_J^{\star}$ for $J\subsetneqq\{0, 1, \ldots, l\}$, where

$$X_J^{\star} = (X\cap pE_J^{\star})-\delta = \{\lambda\in X^{\star} \mid (\lambda+\delta)^{R_j} = \lambda+\delta \Leftrightarrow j\in J\}.$$

Then $\mu\in X$ can be written in the form $\mu=(\lambda+\delta)^w-\delta$ with $\lambda\in X_J^{\star}$, $w\in\tilde{W}_J$, where J, λ, w are all unique; and $\mu\in X^+$ (resp. $X^{\#}$) if and only if $w\in\tilde{W}_J^{++}$ (resp. $w\in\tilde{W}_J^{\#}$). We recall that $\tilde{W}_J^{\#}$ is empty precisely when the complement of J in $\{0, 1, \ldots, l\}$ is a singleton $\{j\}$ with $j\in J_0$, $j\neq j_0$. Note that $X^{\star}$ does not lie (entirely) in X^+, and that

$$X^{\star}\cap X^+ = X^{\star}\cap X^{\#} = X_{\emptyset}^{\star}\cup X_{\{0\}}^{\star},$$

where this intersection (resp. the subset $X_{\emptyset}^{\star}$) is nonempty if and only if

$$\langle\delta, -\alpha_0^\vee\rangle \leqq p \quad (\text{resp.}\ \langle\delta, -\alpha_0^\vee\rangle \lneqq p),$$

i.e. $p\geqq h_\Delta-1$ (resp. $p\geqq h_\Delta$). Let it be noted (from the classification of root-systems) that the Coxeter number h_Δ is always a composite number except in the case of type A_{p-1} (when $h_\Delta=p$), and in this case $X_{\emptyset}^{\star}$ consists of the point 0 alone. This shows that for $p<h_\Delta$ every point of X has a nontrivial stabilizer in $\tilde{W}'$ (and hence in $\tilde{W}$), and for $p\leqq h_\Delta$ every point has a nontrivial stabilizer in $\tilde{W}$; for the latter statement one needs to observe that if $p=h_\Delta$ (type A_l, $l=p-1$), then the set $X\cap pE_{\emptyset}^{\star}=X_{\emptyset}^{\star}+\delta$ consists of only one point δ which turns out to be the barycentre of the p vertices $\{p\lambda_j \mid j\in J_0\}$ of the fundamental simplex $pE^{\star}$ and is therefore fixed by Ω (cf. § 1). We shall call $\lambda\in X$ *p-regular* if its stabilizer in $\tilde{W}'$ is the identity; for such λ the stabilizer in $\tilde{W}$ is also the identity provided $p\nmid[X:X']$.

For $J=\{1, 2, \ldots, l\}$ the set $X_J^\star$ has $-\delta$ as its solitary element, and the only element of $X^\#$, $\tilde{W}'$-linked to it is $(p-1)\delta$. Since no dominant weight of $\bar{V}_{(p-1)\delta}$ is $\tilde{W}'$-linked to the highest weight, the Harish-Chandra principle implies that this G-module is irreducible which gives

$$\dim M_{(p-1)\delta} = p^r \quad (\text{where } r = |\Delta_+|). \tag{5.1}$$

This may be regarded as the first evidence in favour of our conjecture, because (5.1) is known to be true for all p. Since

$$\dim M_\lambda \leqq \dim \bar{V}_\lambda < p^r \quad \text{for} \quad \lambda \in X^\#, \lambda \neq (p-1)\delta,$$

(5.1) follows from the fact that at least one member of $\{M_\lambda | \lambda \in X^\#\}$ has dimension p^r; this fact is a consequence of Steinberg's theorem [2, Theorem 7.5] asserting that upon restriction to $G_{\mathbf{F}_p}$ (for the form of G defined and split over $\mathbf{F}_p$) the set $\{M_\lambda | \lambda \in X^\#\}$ furnishes all the irreducible $KG_{\mathbf{F}_p}$-modules, and another celebrated discovery of Steinberg that there is an irreducible representation of $\mathbf{C}G_{\mathbf{F}_p}$ of degree p^r (the p-Sylow order of $\mathbf{F}_p$) which remains irreducible upon "reduction mod p in Brauer's sense", i.e. upon passing to $KG_{\mathbf{F}_p}$. The module $M_{(p-1)\delta}$ is called the Steinberg module (for the p-modular group-ring $KG_{\mathbf{F}_p}$, or the algebraic Chevalley group G, or for its Lie algebra $\mathfrak{g}_K$, for each of which it is irreducible; see also § 6), and has the property that no indecomposable module other than itself can possess it as a composition factor. Curtis has generalized the existence and uniqueness of the Steinberg module to a class of Tits systems that generalizes the class of finite Chevalley groups.

5.2. The situation (5.1) above is a special case of the following more general consequence of the Harish-Chandra principle.

For a given dominant weight λ, if the set

$$\{\mu = (\lambda+\delta)^w - \delta | w \in \tilde{W}, \ \mu \leqq \lambda, \ \mu \in X^+\} \tag{5.2}$$

contains no element besides λ, then the G-module $\bar{V}_\lambda$ is irreducible.

From this we shall deduce the following Proposition, using the fact that the hypothesis of (5.2) is satisfied when $\lambda \in X^\star \cap X^+$, i.e. when $\langle \lambda, -\alpha_0^\vee \rangle \leqq p - \langle \delta, -\alpha_0^\vee \rangle$; it would suffice for us to use this fact for $p > h_\Delta$ in which case the result of [7] on the validity of the Harish-Chandra principle is applicable.

Proposition 3. If $\lambda = \sum_i m_i \lambda_i \in X^+$ *satisfies* $\sum_{i=1}^{l} m_i n_i \leqq p - h_\Delta + 1$ *then the G-module $\bar{V}_\lambda$ is irreducible.*

The earlier known results on the irreducibility of $\bar{V}_\lambda$ are to be found in [5, Theorem 1] and [10, Corollary 4.3], and are mostly contained in the above. Let us note that the Proposition is vacuous when $p < h_\Delta - 1$. Postponing the values $h_\Delta - 1$ and h_Δ of p, let us assume $p > h_\Delta$, which implies $p \nmid [X:X']$ as $[X:X']$ never exceeds h_Δ. Then in the definition of the set involved in the hypothesis of (5.2) $\tilde{W}$ can be replaced by $\tilde{W}'$. This hypothesis is fulfilled because the set of elements of X^+ in the $\tilde{W}'$-orbit of $\lambda+\delta$ has $\lambda+\delta$ as the smallest element (with respect to the partial order $\leqq$); this latter statement is clear from the observation that the identity is the smallest element of $\tilde{W}^+$ (with respect to $\leqq$, and hence with respect to the affine partial order, introduced in § 1). This proves the Proposition in the major case $p>h_\Delta$. For the remaining values of p, we find that the hypothesis of (5.2) is satisfied in a most trivial manner, viz. that $\bar{V}_\lambda$ has no dominant weight at all except λ (which guarantees the irreducibility of $\bar{V}_\lambda$ irrespective of the characteristic), so that (5.2) and the Harish-Chandra principle need not be invoked. This is clear from the fact that for $p=h_\Delta-1$ (resp. $p=h_\Delta$, which can hold only when Δ is of type A so that $n_i=1$ for all i) the only value(s) of λ for which the hypothesis of the Proposition holds is $\lambda_0=0$ (resp. are $\{\lambda_i \mid 0\leqq i\leqq l\}$).

Note that in the last part of the proof above we have essentially appealed to the standard fact that when $G\cong SL_{l+1}$ the fundamental representations $\bar{V}_{\lambda_i}$ are irreducible (irrespective of p), being precisely the exterior powers of the natural $(l+1)$-dimensional representation. In general, it is not true that all the $\bar{V}_{\lambda_i}$'s are irreducible irrespective of p; the correct generalization of the preceding phenomenon for type A is the following: *For $j\in J_0$ the G-module $\bar{V}_{\lambda_j}$ is irreducible for every p.* This follows from the fact that $\bar{V}_\lambda$ has no dominant weight except λ if and only if $\lambda\in\{\lambda_j \mid j\in J_0\}$, which in turn is an immediate consequence of the observation that if $\lambda\in X^+$ lies in the coset $\lambda_i+X'\in X/X'$ (with $i\in J_0$) then $\lambda\geqq\lambda_i$.

Combining (4.6) and Proposition 3 we have

Corollary to Proposition 3. *If $p\geqq 2h_\Delta-3$ then $\bar{V}_\lambda$ is irreducible for λ lying in the set Y_0 defined in* § 4.5.

5.3. For the rest of this section we assume that $p\nmid[X:X']$ and the Harish-Chandra principle is valid for the prime p at hand (unless otherwise stated). Thus, whenever M_λ and M_μ are "linked" in the sense that there exists a chain of indecomposable G-modules in which each consecutive pair has a composition factor in common and M_λ (resp. M_μ) occurs as a composition factor of the first (resp. the last) in this chain, the highest weights λ and μ are $\tilde{W}'$-linked, i.e. $\lambda+\delta$ and $\mu+\delta$ are $\tilde{W}'$-conjugate.

With this assumption the solving of (4.2) is greatly simplified, because now one has

$$Y(\mu_0)\subseteqq\{\mu\in X^+ \mid \mu\leqq\mu_0 \quad\text{and}\quad \mu=(\mu_0+\delta)^w-\delta \quad\text{for}\quad w\in\tilde{W}'\}.$$

The conjecture below puts an even stricter limit on $Y(\mu_0)$ which could turn out to be optimal in most cases.

Conjecture II. *For $\mu_0, \mu\in X^+$, the nonvanishing of the decomposition number $c_{\mu_0\mu}$ defined in* § 4 *implies that the following condition* (A) *holds:*

(A) "*there exist reflections* $S_1, S_2, \ldots, S_k$ *such that defining* $\mu_i = (\mu_{i-1}+\delta)^{S_i} - \delta$ *successively for* $1 \leqq i \leqq k$, *one has* $\mu_i + \delta \in X^+$ *and*

$$\mu_0 \geqq \mu_1 \geqq \ldots \geqq \mu_{k-1} \geqq \mu_k = \mu\text{''}.$$

The reader who is familiar with the author's work in [15] on the universal $\mathfrak{g}_{\mathbf{C}}$-modules generated by highest-weight-vectors (see Footnote[5] in § 5.4), would recognize some formal similarity between the above and the Conjecture 1 of [15A] which has recently been proved true by I. M. Gelfand and collaborators [1] through an elegant tensor product trick. The following recasting is immediate, and provides our main justification for introducing the partial-order $\leqq$ on $\tilde{W}^+$.

Conjecture II *(Restatement). Suppose* $\mu', \mu \in X^+$ *are* $\tilde{W}'$*-linked, so that*

$$\mu' = (\lambda+\delta)^{w'} - \delta, \ \mu = (\lambda+\delta)^{w} - \delta \quad \text{for} \quad \lambda \in X_J^{\star}, \ w', \ w \in \tilde{W}_J^{++},$$

where J, λ, w', w *are all unique. Then*

$$c_{\mu'\mu} \neq 0 \quad \textit{implies} \quad w \leqq w'.$$

It is possible to give substantial confirmation of this Conjecture in case of type A_2 i.e. $G \cong SL_3$, but we relegate it to a later paper. Below we give an example in this case which shows that *the* μ_i's *in the original statement of Conjecture* II *cannot always be chosen inside* X^+. With this evidence it becomes reasonable to consider the G-modules $\bar{V}_\mu$ and M_μ not just for $\mu \in X^+$, but also, somewhat artificially, for $\mu \in X^+ - \delta$, defining these to be 0-dimensional in case $\mu + \delta$ is W-singular i.e. lies on a wall of E^+, in accordance with dim $\bar{V}_\mu = D(\mu+\delta)$.

Example 1. In the adjoining diagram (Fig. 4) consider an arbitrary point λ in $X_{\{0\}}^{\star}$, i.e. a point of X lying on the open line-segment joining $-\delta+p\lambda_1$ and $-\delta+p\lambda_2$; say $\lambda+\delta = i\lambda_1 + (p-i)\lambda_2$ for $1 \leqq i \leqq p-1$. Setting

$$\mu' + \delta = (\lambda+\delta)R_1R_2R_0 = 2p\lambda_1 + i\lambda_2,$$

one immediately sees that the set

$$\{\mu = (\lambda+\delta)^w - \delta \,|\, w \in \tilde{W}_{\{0\}}^{+}, \ \mu \leqq \mu'\}$$

of weights of $\bar{V}_{\mu'}$ which lie in $X^+ - \delta$ and are linked to μ', consists of 4 elements μ', μ'', μ''' and λ where

$$\mu'' + \delta = (\lambda+\delta)R_1R_0 = (p+i)\lambda_2 \notin X^+ + \delta,$$

and

$$\mu''' + \delta = (\lambda+\delta)R_2R_0 = (2p-i)\lambda_1 \notin X^+ + \delta.$$

In this case we can give the decomposition $\bar{V}_{\mu'} \underset{G}{\sim} \sum_{\mu} c_{\mu'\mu} M_\mu$ on pure dimensionality considerations: on one hand we have

$$\dim \bar{V}_{\mu'} = pi(2p+i), \quad \dim \bar{V}_\lambda = \frac{1}{2}pi(p-i) = \dim M_\lambda,$$

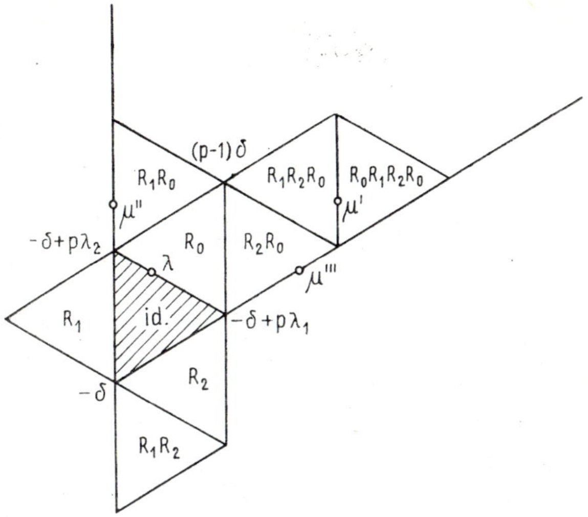

Fig. 4. Diagram for Example 1

and on the other hand

$$M_{\mu'} = M_{\lambda_1}^{Fr} \otimes M_{(p-1)\lambda_1+(i-1)\lambda_2}$$

gives $\dim M_{\mu'}=3\cdot\frac{1}{2}pi(p+i)$; here we have used the irreducibility of $\overline{V}_\lambda$ and also of $\overline{V}_\mu$ for $\mu+\delta=p\lambda_1+i\lambda_2$, which can be seen either from (5.2) or by appealing to Braden [4] as long as $p\geqq 5$. (Actually Braden gives the infinitesimal decomposition numbers $c'_{\lambda\mu}$ to be defined in § 6, but in this case it happens that $c'_{\lambda\mu}=c_{\lambda\mu}$, as we remark at the end of § 6.) Now verifying

$$\dim M_{\mu'} = \dim \overline{V}_{\mu'} - \dim \overline{V}_\lambda,$$

we are forced to the exact sequence

$$0 \to M_\lambda \to \overline{V}_{\mu'} \to M_{\mu'} \to 0,$$

which gives $c_{\mu'\lambda}=1$. The Conjecture is clearly verified with $k=2$ $(\mu_0=\mu')$, and μ_1 can be taken to be either μ'' or μ''' neither of which lies in X^+.

5.4. Our next conjecture says that the value of $c_{\mu'\mu}$ in the restatement of Conjecture II depends only on w and w' (and not on λ).

Conjecture III. *Let* $J \subsetneqq \{0, 1, \ldots, l\}$, $\nu_1, \nu_2 \in X_J^{\star}$ *and* $w', w \in \tilde{W}_J^{++}$ *be all given, and write*

$$\mu_i = (\nu_i+\delta)^w - \delta, \quad \mu'_i = (\nu_i+\delta)^{w'} - \delta \quad \text{for} \quad 1 \leqq i \leqq 2.$$

Then we must have $c_{\mu'_1\mu_1}=c_{\mu'_2\mu_2}$.

This is analogous to a phenomenon for the composition factors of the universal $\mathfrak{g}_{\mathbb{C}}$-modules generated by highest-weight-vectors studied in [15] and [1], the appearance of this latter phenomenon being indicated by certain information that was communicated to me orally by I. N. Bernstein during the Budapest Summer School on Group Representations, 1971; it may be mentioned at this point that, contrary to Theorems 1 and 7 of [15A] and a similar announcement in the original version of [6], the composition factors of the said $\mathfrak{g}_{\mathbf{C}}$-modules can have multiplicity greater than one, as was shown in [1, p. 9]. (See Footnote[5].)

Example 2. Here we shall show that for the case of type B_2, each of the two conjectures II and III yields that

$$\text{for} \quad \mu' \in X^{\#}, \; c_{\mu'\mu} \neq 0 \quad \text{implies} \quad \mu \in X^{\#}, \tag{5.3}$$

i.e. the set Y_0 defined in § 4 reduces to $\{0\}$. (The same conclusion is valid for type A_2, and is easily obtained either from the Harish-Chandra principle or from Braden's determination of $c_{\lambda\mu}$ in this case. For the case of type G_2 it is suspected that this statement is false; this is based on the fact that $\{w \in \tilde{W}^+ \mid w \leqq w_0\}$ properly contains $\tilde{W}^{\#}$, as was noted in § 1 and § 3; the existence of this possibility is also stated in [7, p. 76].) We first consider the general case $p \geqq 5$, under which the Harish-Chandra principle is valid in view of [7], taking up $p=2$ and $p=3$ later. That then Conjecture II implies (5.3) is straightforward from the fact that if w, w' are defined as in the restatement of Conjecture II (then because of $w' \in \tilde{W}^{\#}$, which is equivalent to $\mu' \in X^{\#}$) one has $w \leqq w' \leqq w_0$ and hence $w \in \tilde{W}^{\#}$.

[5] *(Added in Proof.)* The recently published book "Algèbres Enveloppantes" (Cahiers Scientifiques XXXVII, Gauthier-Villars, 1974) by J. Dixmier contains an exhaustive account of these $\mathfrak{g}_{\mathbf{C}}$-modules which have been termed Verma modules. But the problem of determination of their composition multiplicities remains open (on account of the said counterexample in [1]); no one seems to have even a plausible guess at these numbers. (And, therefore, with extreme apology for a rather arbitrary and vague guess, I should like to suggest that these multiplicities may not be unrelated to the numbers $b_{\sigma,\tau}$ mentioned in Footnote [2].) There is an obvious parallel between that problem and the present problem of determination of the numbers $c_{\lambda\mu}$, but we think that this (latter) problem should better be compared with the problem of constructing "resolution" of $\mathfrak{g}_{\mathbf{C}}$-irreducibles in terms of the sums of (i.e. the category generated by) Verma modules which was at once proposed and solved by I. N. Bernstein, I. M. Gelfand and S. I. Gelfand in the sequel to [1] ("Differential operators on the base affine space and a study of $\mathfrak{g}$-modules", in the present volume, already made famous by Dixmier in his reports in Séminaire Bourbaki (Exposé 425) and the above-mentioned book). More precisely, replacing **C** by the p-adic field and resorting to Kostant's **Z**-form $U_{\mathbf{Z}}$ of the enveloping algebra, one can construct certain "p-adic integral forms of Verma modules"; it may be possible to obtain the long-cherished "p-analog of Weyl's character formula" (i.e. to obtain the numbers $\gamma_{\lambda\mu}$ directly without having to get $c_{\lambda\mu}$ first) by a suitable resolution involving these forms of Verma modules in a manner parallel to that of Bernstein—Gelfand—Gelfand.

At this point it is hard to resist the temptation of musing on the appearance of affine Weyl groups in the Bruhat—Tits theory of $\mathfrak{p}$-adic semisimple groups (though those are "dual" to affine Weyl groups here; cf. end of § 1.1). The natural question is whether this "coincidence" could be related in some manner. Could it be that such a relationship is revealed precisely through the said p-adic integral forms?

To see that Conjecture III also implies (5.3), we look at the various possibilities for the location of μ outside $X^{\#}$ (yet satisfying $c_{\mu'\mu} \neq 0$ for some $\mu' \in X^{\#}$) in the various triangles (or on their edges) in Fig. 5. The triangles shown there are

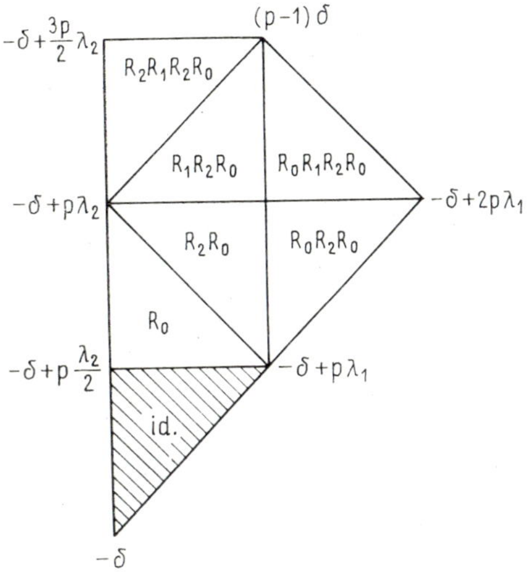

Fig. 5. Diagram for Example 2

part of the tessellation of E by $(pE^{\star})^w - \delta$ where w ranges over $\widetilde{W}'$, the element w being shown inside the corresponding triangle. It is clear that μ must lie inside the quadrilateral with vertices

$$-\delta, \quad -\delta + \frac{3}{2}p\lambda_2, \quad (p-1)\delta, \text{ and } -\delta + 2p\lambda_1,$$

because no $\mu \in X^+$ outside this region satisfies $\mu \leqq \mu'$ for $\mu' \in X^{\#}$. From the irreducibility of $\overline{V}_{(p-1)\delta}$ (the Steinberg module) it is also clear that μ has to be an interior point of this quadrilateral, which is decomposed into seven triangles three of which lie outside $p \cdot E^{\#}$. If μ lies in either of the triangles marked $R_2 R_1 R_2 R_0$ and $R_0 R_1 R_2 R_0$, including the edge common to the triangles marked $R_0 R_1 R_2 R_0$ and $R_0 R_2 R_0$, then every $\mu' \in X^{\#}$ which is $\widetilde{W}'$-linked to μ satisfies $\mu' \leqq \mu$; so we must try μ in the interior of the $R_0 R_2 R_0$-triangle. Take $\mu + \delta = (p+i)\lambda_1 + j\lambda_2$, with i, j positive integers such that $i + 2j < p$. Then the only $\mu' \in X^{\#}$ which is $\widetilde{W}'$-linked to μ and does not satisfy $\mu' \leqq \mu$ lies in the $R_1 R_2 R_0$-triangle, and is given by

$$\mu' + \delta = (p-i)\lambda_1 + (p-j)\lambda_2;$$

since

$$\mu' - \mu = (p - 2i - 2j)\alpha_1 + (p - 2i - j)\alpha_2,$$

one has $\mu \leqq \mu'$ (a necessary condition for $c_{\mu'\mu} = 0$) if and only if $i + j < \frac{1}{2}p$. Thus $c_{\mu'\mu} = 0$, in case $i + j < \frac{1}{2}p$, and hence by Conjecture III for all i, j (satisfying $i + 2j < p$). This completes the two "proofs" of (5.3) under $p \geqq 5$.

For $p=2$ there is nothing to prove because the aforementioned quadrilateral has no interior point lying in X but not in $X^{\#}$. For $p=3$ there is only one such point, viz. $\mu=3\lambda_1$; and the only value of $\mu' \in X^{\#}$, other than the Steinberg highest weight $(p-1)\delta$, satisfying $\mu \leqq \mu'$ is $\mu'=\lambda_1+2\lambda_2$, with $\mu'-\mu=\alpha_2$. The points μ, μ' are not in the same $\tilde{W}$-orbit; however, we do not require to invoke the Harish-Chandra principle in this case to conclude $c_{\mu'\mu}=0$. We have only to show that no nonzero μ-weight-vector of $\bar{V}_{\mu'}$ is annihilated by $\mathfrak{g}_K^+ = \sum_{\alpha \in \Delta_+} \mathfrak{g}_K^\alpha$. We claim that the 1-dimensional μ-weight-space of the 40-dimensional G-module $\bar{V}_{\mu'}$ is certainly not annihilated by $\mathfrak{g}_K^{\alpha_2}$. This is immediate from the irreducibility of the 3-dimensional SL_2-module spanned by the μ', μ, and μ''-weight-spaces, where $\mu''=\mu'-2\alpha_2$, which is stable under the action of the subgroup of G corresponding to the 3-dimensional simple Lie subalgebra generated by $\mathfrak{g}_K^{\alpha_2}$ and $\mathfrak{g}_K^{-\alpha_2}$. This completes our discussion of (5.3).

5.5. Now assuming Conjecture III, the decomposition $\bar{V}_{\mu'} \underset{G}{\sim} \sum_{\mu} c_{\mu'\mu} M_\mu$ for $\mu'=(\lambda+\delta)^{w'}-\delta$ with $\lambda \in X_J^\star$ and $w' \in \tilde{W}_J^{++}$, assumes the form

$$\bar{V}_{\mu'} \underset{G}{\sim} \sum_{w \in \tilde{W}_J^{++}} c^J_{w'w} M_{(\lambda+\delta)^w-\delta},$$

where the (well-determined) constants $c^J_{w'w}$ do not depend on λ. Writing $\chi^J_w(\lambda)$ (resp. $\bar{\chi}^J_w(\lambda)$) for $\chi_{(\lambda+\delta)^w-\delta}$ (resp. $\bar{\chi}_{(\lambda+\delta)^w-\delta}$), for $\lambda \in X_J^\star$, we have

$$\bar{\chi}^J_{w'}(\lambda) = \sum_{w \in \tilde{W}_J^{++}} c^J_{w'w} \chi^J_w(\lambda). \tag{5.4}$$

This simplifies the problem in § 4 of solving the linear equations (4.2). By virtue of Conjecture II, the matrix $[c^J_{w'w}]$ for w', w ranging in $\tilde{W}_J^{++}$, is again unipotent with respect to (any linear order stronger than) the partial order $\leqq$, since $c^J_{w'w} \neq 0$ implies $w \leqq w'$, and $c^J_{ww}=1$. As one is primarily interested in evaluating χ_μ for $\mu \in X^{\#}$, i.e. $\chi^J_w(\lambda)$ for $w \in \tilde{W}_J^{\#}$, it suffices to invert the submatrix $[c^J_{w'w}]$ for w', w running over the finite subset $Y_J=\{w \in \tilde{W}_J \mid w \leqq w_0\}$. This is particularly convenient for large p, since the number and size of the matrices $[c^J_{..}]_{Y_J}$ to be inverted are independent of p; in the Epilogue (§ 8) we shall comment on how the matrix-entries themselves are expected to be largely independent of p. Denoting the inverse by $[\gamma^J_{w'w}]$ we have

$$\dim M_{(\lambda+\delta)^{w'}-\delta} = \sum_{w \in \tilde{W}_J^{++}} \gamma^J_{w'w} D\big((\lambda+\delta)^w\big). \tag{5.5}$$

§ 6. PIM's of the u-algebra and Humphreys' numbers d_λ

6.1. The Lie algebra of an algebraic group in characteristic $p \neq 0$ possesses a 1-ary operation, called p-power operation commonly denoted $x \to x^{[p]}$, under which it becomes a restricted Lie algebra (also called p-Lie algebra), introduced and studied initially by Jacobson and Zassenhaus. Treating the elements of the Lie algebra as the right- or left-invariant derivations of the affine-ring, this operation arises directly from the fact that the pth power of a derivation is again a derivation. This operation is given explicitly for the basis elements

$$\bar{h}_i = h_i \otimes 1_K \in \mathfrak{h}_{\mathbf{Z}} \underset{\mathbf{Z}}{\otimes} K = \mathfrak{h}_K \quad \text{and} \quad \bar{e}_\alpha = e_\alpha \otimes 1_K \in \mathfrak{g}_{\mathbf{Z}}^\alpha \underset{\mathbf{Z}}{\otimes} K = \mathfrak{g}_K^\alpha$$

(for $1 \leqq i \leqq l$ and $\alpha \in \Delta$) of the Lie algebra $\mathfrak{g}_K$ of our Chevalley group G by

$$\bar{h}_i^p = \bar{h}_i, \; \bar{e}_\alpha^p = 0.$$

The quotient $\mathbf{u}$ of the universal enveloping algebra of $\mathfrak{g}_K$ by the ideal generated by the elements of the form $x^p - x^{[p]}$ for $x \in \mathfrak{g}_K$ is called the *restricted* enveloping algebra (also called the *u*-algebra) of $\mathfrak{g}_K$; it is the universal object with respect to the *restricted* $\mathfrak{g}_K$-modules, viz. those representations for which the action of $x^{[p]}$ coincides with the pth power of the endomorphism representing $x \in \mathfrak{g}_K$. Thus restricted $\mathfrak{g}_K$-modules and $\mathbf{u}$-modules are synonymous. We treat $\mathfrak{g}_K$ as a subspace of $\mathbf{u}$. By virtue of a Birkhoff—Witt type basis of $\mathbf{u}$ (with exponents in the monomials ranging from 0 to $p-1$ only), one has $\dim_K \mathbf{u} = p^{\dim G}$.

A most remarkable fact about $\mathbf{u}$ (which does not seem to have been noticed earlier) is that, in contrast with the above construction of $\mathbf{u}$ as a *quotient* of the ordinary enveloping algebra, it can also be identified with the *sub*algebra of $U_K = U_{\mathbf{Z}} \underset{\mathbf{Z}}{\otimes} K$ generated by $\mathfrak{g}_K$; this can be argued either by noticing that

$$h_i^p - h_i \in p \cdot U_{\mathbf{Z}} \quad \text{and} \quad e_\alpha^p \in p \cdot U_{\mathbf{Z}} \quad \text{for} \quad 1 \leqq i \leqq l \quad \text{and} \quad \alpha \in \Delta$$

(of which the latter is trivial), or else through more sophisticated arguments which involve identifying U_K with the algebra of all (right- or) left-invariant differential operators of the affine-ring A_K of G. We shall investigate this new object U_K elsewhere (as we do not require it here), calling it the "*hyper*enveloping algebra *for* G", since it is the noncommutative analog of the hyperalgebra introduced by Dieudonné for a commutative algebraic group in nonzero characteristic (which is again the algebra of invariant differential operators). Suffice it to say here that U_K is a Hopf algebra in some sense dual to A_K, and that though it is not finitely generated as an algebra, for representation-theoretic purposes it behaves infinitely better than either the ordinary or the restricted enveloping algebra of $\mathfrak{g}_K$, in that every finite-dimensional U_K-module can be canonically lifted to a G-module (just as a finite-dimensional $U_{\mathbf{C}}$-module lifts to the connected simply-connected complex Lie group with Lie algebra $\mathfrak{g}_{\mathbf{C}}$).

At any rate, the infinitesimal of a G-module is naturally defined as a restricted representation of $\mathfrak{g}_K$. But this correspondence from the isomorphism classes of G-modules to those of (finite-dimensional) $\mathbf{u}$-modules is neither one-one nor onto. Nevertheless, there are good reasons (even for someone interested purely in group representations, — for algebraic, or finite, Chevalley groups) to study the category of all $\mathbf{u}$-modules; apart from the classical result of Curtis and Steinberg [2, § 7] giving a bijection between the isomorphism classes of the irreducible modules for algebras $\mathbf{u}$ and $KG_{\mathbf{F}_p}$, there have recently arisen some new indications in support of this view: in [19] and the forthcoming joint work [8] we provide certain structural relationships between the projective modules for the two associative algebras, whereas Humphreys has already demonstrated [7] how some of the structure of the indecomposable projective modules for $\mathbf{u}$ can be understood by analysing certain $\mathbf{u}$-modules Z_λ which cannot normally be lifted to the group G, which are the analogues of the subject-matter of [15] and [1]. (*Added in Proof:* That is, in the terminology of Dixmier's book quoted in Footnote (5) in § 5.4, Z_λ are the u-algebra analogues of Verma modules.) We shall presently describe the modules Z_λ and the said results of Humphreys in detail, one of the main objectives of this paper (Parts I and II combined) being the formulation of certain conjectures on the decomposition of these modules Z_λ.

It is easy to see that the infinitesimal action of $\mathfrak{g}_K$ on the G-module $M_{p\lambda} \cong M_\lambda^{\mathrm{Fr}}$ is trivial for all $\lambda \in X^+$; consequently, by Steinberg's theorem, for $\lambda \in X^+$ and $\mu \in X^\#$ one has

$$M_{p\lambda+\mu} \underset{\mathbf{u}}{\cong} (\dim M_\lambda) \cdot M_\mu. \tag{6.1}$$

It is a result of Curtis [2, 6.4-6.6] that M_μ is $\mathbf{u}$-irreducible for every $\mu \in X^\#$, and that every irreducible $\mathbf{u}$-module is isomorphic to a member of $\{M_\mu \mid \mu \in X^\#\}$ which are mutually nonisomorphic (as we shall see presently). Thus every irreducible $\mathbf{u}$-module lifts uniquely to a G-module. It is known that there exist indecomposable $\mathbf{u}$-modules which do not arise as infinitesimal of any G-module; for the case of type A_1 ($G \cong SL_2$) it is not difficult to show that the indecomposable $\mathbf{u}$-module Z_λ to be defined presently for $\lambda \in X^\#$, has this property with one exception when (for $\lambda = (p-1)\delta$) it is irreducible (giving the Steinberg module), and it is believed that the same is true for all types. In [19] we shall show that every projective $\mathbf{u}$-module lifts to a G-module, generalizing a phenomenon for the case of SL_2 that was very recently observed to follow from certain computations of A. Vincent Jeyakumar on p-modular representations of the group $SL_2(\mathbf{F}_p)$. However, we do not know the answer to the crucial question: Do there exist nonisomorphic G-modules such that the associated actions of $\mathbf{u}$ are indecomposable and isomorphic? Clearly, indecomposability is essential, for the left and right sides in (6.1) are nonisomorphic G-modules.

6.2. There is a natural homomorphism from $X = X(T)$ into the dual $\mathfrak{h}_K^*$ of the Lie algebra $\mathfrak{h}_K$ of T obtained by taking the infinitesimals of the irreducible representations (characters) of T. Its kernel is $p \cdot X$ and the image is the set of "restricted infinitesimal weights", viz. the set of all those elements of $\mathfrak{h}_K^*$ which (considered as 1-dimensional representations of the Abelian Lie algebra $\mathfrak{h}_K$) afford restricted representations of $\mathfrak{h}_K$. Thus if $\mathfrak{h}_{\mathbf{F}_p} = \sum_i \mathbf{F}_p \cdot \bar{h}_i$ denotes the canonical $\mathbf{F}_p$-form of $\mathfrak{h}_K$, this image is identified with the $\mathbf{F}_p$-dual $\mathfrak{h}_{\mathbf{F}_p}^*$; following Humphreys we shall use the notation Λ in place of $\mathfrak{h}_{\mathbf{F}_p}^*$. The map $X \to \Lambda$, which induces an isomorphism from X/pX to Λ, gives a bijection $\pi: X^{\#} \to \Lambda$ on restriction to the subset $X^{\#}$. We shall use letters λ, μ to denote typical elements of Λ as well as $X^{\#}$; and for $\lambda \in \Lambda$ the **u**-modules $M_{\pi^{-1}(\lambda)}$ and $\bar{V}_{\pi^{-1}(\lambda)}$ shall be written simply M_λ and $\bar{V}_\lambda$ (and in the same vein, for $\lambda \in X^{\#}$, the module $Z_{\pi(\lambda)}$ defined in the following paragraph may also be denoted by Z_λ); in spite of this notational abuse (which can hardly lead to confusion) it is important to maintain a conceptual distinction between the group $\Lambda \cong X/pX$ and the set $X^{\#}$.

Let us note that the action of the Borel subalgebra $\mathfrak{b}_K = \mathfrak{h}_K + \sum_{\alpha \in \Delta_+} \mathfrak{g}_K^\alpha$ on the **u**-irreducible module M_λ for $\lambda \in \Lambda$, stabilizes a unique 1-dimensional subspace, viz. the highest-weight-space of this G-module (stabilized by B), on which the action of $\mathfrak{h}_K$ is given by the linear function λ; this shows that the irreducible **u**-modules M_λ are mutually nonisomorphic for $\lambda \in \Lambda$. Further, since each G-module M_λ is defined over $\mathbf{F}_p$, these **u**-irreducibles possess $\mathbf{F}_p$-forms with respect to the canonical $\mathbf{F}_p$-form of **u**, viz. the u-algebra of the "form" $\mathfrak{g}_{\mathbf{F}_p} = \mathfrak{g}_{\mathbf{Z}} \otimes_{\mathbf{Z}} \mathbf{F}_p$ of $\mathfrak{g}_K$. It follows that the **u**-modules M_λ are absolutely irreducible for $\lambda \in \Lambda$ (or $X^{\#}$).

In accordance with [7], for $\lambda \in \Lambda$ let Z_λ denote the (right) **u**-module $1_\lambda \otimes_{\mathbf{b}} \mathbf{u}$ obtained by inducing from the 1-dimensional **b**-module 1_λ for the "Borel subalgebra" **b** of **u** generated by $\mathfrak{b}_K$, where on 1_λ the action of $\mathfrak{h}_K$ is given by the restricted weight λ and that of the radical

$$\mathfrak{g}_K^+ = [\mathfrak{b}_K, \mathfrak{b}_K] = \sum_{\alpha \in \Delta_+} \mathfrak{g}_K^\alpha$$

of $\mathfrak{b}_K$ is trivial. If $\mathbf{u}^-$ denotes the subalgebra of **u** generated by $\mathfrak{g}_K^- = \sum_{\alpha \in \Delta_-} \mathfrak{g}_K^\alpha$, one finds from the vector space isomorphism $\mathbf{u} \cong \mathbf{b} \otimes \mathbf{u}^-$ (Birkhoff—Witt theorem for **u**) that Z_λ is a free $\mathbf{u}^-$ module of rank 1, and thus has dimension p^r for $r = |\Delta_+|$. The module Z_λ is generated by a unique 1-dimensional subspace of the λ-weight-space which is annihilated by $\mathfrak{g}_K^+$ (is "standard cyclic with highest weight λ" in the terminology of [7]), and is universal with this property, in that every **u**-module with this property is a quotient of Z_λ. It is proved in [7, Proposition 1.1] that for $\lambda \in \Lambda$ the **u**-module $\bar{V}_\lambda$ has this property, so that one has the surjections $Z_\lambda \to \bar{V}_\lambda \to M_\lambda$. (For $\lambda \in X^+$, $\lambda \notin X^{\#}$, the dimension of $\bar{V}_\lambda$ generally exceeds p^r; however, it is easy to see that for no such λ can the **u**-module $\bar{V}_\lambda$ be "standard cyclic".) It should be noted that though the G-module M_λ has the λ-weight-space 1-dimensional for all $\lambda \in X^+$, in the infinitesimal situation normally the λ-weight-space of the **u**-irreducible M_λ for $\lambda \in \Lambda$ has dimension greater than one (and hence the same holds for $\bar{V}_\lambda$ and Z_λ), on account of the merging of weights when one passes from the decomposition for the torus T to that for the Lie algebra $\mathfrak{g}_K$. In fact, every weight-space of every Z_λ has dimension p^{r-l}. Considering the decomposition of $\mathbf{u}^-$ as a direct sum of the subspace of scalars and the augmentation ideal (generated by $\mathfrak{g}_K^-$), one finds that every proper submodule of Z_λ is contained in the image of the 1-dimensional generating subspace under this augmentation ideal. This shows that Z_λ has a unique maximal proper submodule (the kernel of $Z_\lambda \to M_\lambda$), and that Z_λ has no irreducible module other than M_λ as its quotient, and thus is indecomposable.

6.3. On account of the standard facts from the theory of Artinian rings (which applies to **u** because of its finite dimensionality), we know that (i) up to isomorphism there exists a unique projective **u**-module Q_λ (called the projective *cover* of M_λ) such that M_λ is its only irreducible quotient, (ii) every indecomposable **u**-projective is isomorphic to some Q_λ for $\lambda \in \Lambda$, which are clearly all indecomposable, (iii) the category of projective **u**-modules is semisimple, i.e. every projective module is isomorphic to a direct sum $\sum_{\lambda \in \Lambda} n_\lambda Q_\lambda$, and (iv) the (right-regular) representation of **u** on **u** (which is free, and therefore projective, over **u**) is equivalent to the direct sum

$$\sum_{\lambda \in \Lambda} (\dim M_\lambda) \cdot Q_\lambda ,$$

so that every Q_λ does occur in this representation. The modules Q_λ are called PIM's (principal indecomposable modules). Two PIM's Q_λ and Q_μ are called *linked* if there exists a chain of PIM's beginning at Q_λ and ending at Q_μ such that the consecutive pairs have a common composition factor; we write $Q_\lambda \sim Q_\mu$ in that case. Further, one knows (from the theory of Artinian rings) that the sum of all PIM's belonging to one linkage class (i.e. equivalence class under $\sim$) forms a 2-sided ideal of **u** generated by a primitive central idempotent, and thus is a minimal direct summand in the associative algebra **u**, called a "block" of **u**. Since both M_λ and Q_λ are defined over the prime field, one finds that the semisimple part of each block, i.e. the quotient by its radical, is "split" in the sense that it is a direct sum of matrix algebras, — exactly one of dimension $(\dim M_\lambda)^2$ for each λ corresponding to Q_λ's appearing in the linkage class. From the decomposition of **u** as a direct sum (as K-algebra) of blocks, it is clear that if Q_λ and Q_μ are not linked then M_λ and M_μ cannot be "linked" either (as **u**-modules, in a sense similar to that for G-modules: see the beginning of § 5.3), the converse being trivial. It must be noted that if M_λ and M_μ are linked as G-modules (for $\lambda, \mu \in X^{\#}$) then these are also linked as **u**-modules, but not conversely: for in the former case one necessarily has $\lambda - \mu \in X'$, whereas this is far from necessary in the case of **u**-modules as we shall see presently.

The infinitesimal version of the Harish-Chandra principle (cf. § 5) conjectures that for $\lambda, \mu \in \Lambda$ if M_λ and M_μ occur as a composition factor of an indecomposable **u**-module (or more generally, if M_λ and M_μ are linked as **u**-modules) then $\lambda + \bar{\delta}$ and $\mu + \bar{\delta}$ must be conjugate under the Weyl group W (which acts canonically on $\Lambda \cong X/pX$), where $\bar{\delta}$ stands for the image of $\delta \in X^{\#}$ under π. *When this conclusion on λ, μ holds, let us introduce the terminology that λ and μ are W-linked;* in the notation of [7] this is expressed as $\lambda \sim \mu$. It must be observed that $\lambda, \mu \in \Lambda$ are W-linked if and only if the restricted dominant (global) weights $\pi^{-1}(\lambda), \pi^{-1}(\mu) \in X^{\#}$

are $\tilde{W}$-linked (and *not* just $\tilde{W}'$-linked), in the sense that $\pi^{-1}(\lambda)+\delta$ and $\pi^{-1}(\mu)+\delta$ are $\tilde{W}$-conjugate. (The term "$\tilde{W}'$-linked" was introduced in § 5.1.)

The conjecture on the linkage of **u**-modules that we just mentioned was proved to be true in [7], provided that $p>h_\Delta$. In this (infinitesimal) case the converse is also easy, and follows from the fact that if a fundamental reflection $R_i\in W$ sends $\lambda+\bar{\delta}$ to $\mu+\bar{\delta}=\lambda+\bar{\delta}-k\bar{\alpha}_i$ with $1\leqq k\leqq p-1$ (where $\bar{\alpha}_i$ stands for the projection of the global root $\alpha_i\in\Delta\subset X$ under the natural projection $X\to\Lambda$), then there is a nonzero homomorphism from Z_μ to Z_λ (if v and v' are nonzero elements in the 1-dimensional generating subspaces of the highest-weight-spaces of Z_λ and Z_μ respectively, then this homomorphism is determined by $v'\to v\bar{f}_i^k$), which shows that the indecomposable **u**-module Z_λ has both M_λ and M_μ as composition factors. In fact, pursuing this argument further, Humphreys has found [7, Theorem 2.2] (without assuming the Harish-Chandra principle) that *if λ and μ are W-linked elements of Λ then $Z_\lambda \underset{\mathbf{u}}{\sim} Z_\mu$*, where $\underset{\mathbf{u}}{\sim}$ denotes equivalence in the Grothendieck ring of the category of **u**-modules, i.e. Z_λ and Z_μ have identical composition factors (multiplicity included). It follows that if $\lambda,\mu\in X^\#$ are $\tilde{W}$-linked then M_λ and M_μ are linked as **u**-modules, which corroborates the statement immediately preceding the last paragraph.

6.4. Now let $\mathbf{b}'$ be the subalgebra of **u** generated by the Borel subalgebra $\mathfrak{b}'_K=\mathfrak{h}_K+$ $+\mathfrak{g}_K^-$ opposite to $\mathfrak{b}_K$. It is readily seen, by looking at the regular representation of $\mathbf{b}'$ on $\mathbf{b}'$, that the modules $\{Z_\lambda\,|\,\lambda\in\Lambda\}$ exhaust the (isomorphism classes of) projective $\mathbf{b}'$-modules (upon restriction to $\mathbf{b}'$), the unique irreducible quotient of Z_λ being the 1-dimensional irreducible $\mathbf{b}'$-module $1'_\lambda$ on which $\mathfrak{g}_K^-$ acts trivially and $\mathfrak{h}_K$ acts through the restricted weight λ. Since **u** is $\mathbf{b}'$-free (under right-multiplication, as is clear from the Birkhoff—Witt theorem for u-algebras) and Q_λ is a direct summand in **u**, we see that the restriction of Q_λ to $\mathbf{b}'$ is $\mathbf{b}'$-projective. Hence there is a decomposition in the semisimple category of projective $\mathbf{b}'$-modules:

$$Q_\lambda \underset{\mathbf{b}'}{\cong} \sum_\mu \oplus [Q_\lambda : Z_\mu]_{\mathbf{b}'\text{-proj}} \cdot Z_\mu,$$

where the meaning of "$\mathbf{b}'$-proj" in subscript is clear.

We are now ready to state the most interesting consequence of the (infinitesimal) Harish-Chandra principle that Humphreys has derived [7, Theorem 4.5]:

If we define $d_\mu=[Z_\mu:M_\mu]_{\mathbf{u}}(=[Z_\lambda:M_\mu]_{\mathbf{u}}$ for every $\lambda\in\Lambda$ which is W-linked to μ), then

$$Q_\lambda \underset{\mathbf{b}'}{\sim} d_\lambda\cdot\Big(\sum_{\mu\sim\lambda}\oplus Z_\mu\Big),$$

i.e. the multiplicity $[Q_\lambda : Z_\mu]_{\mathbf{b}'\text{-proj}}$ is d_λ or 0 according as Q_λ is linked to Q_μ or not.

It follows that if a_λ is the cardinality of the W-linkage class of λ in Λ then

$$\dim Q_\lambda = a_\lambda d_\lambda p^r.$$

In order to compute a_λ one has only to know the stabilizer $W^{\lambda+\delta}$ of $\lambda+\delta$ in W, for then one has $|W|=a_\lambda \cdot |W^{\lambda+\delta}|$. For $p \nmid [X:X']$ we claim that $W^{\lambda+\delta}$ is again a reflection subgroup. Since Λ is the set of "integral points" on the torus E/pX, this fact can be obtained from the study of Steinberg of the action of Weyl groups on tori [12, Lemma 4.1], but the following observation not only gives this fact directly but also shows how to compute the order of this stabilizer. Let $\mu=\pi^{-1}(\lambda) \in X^{\#}$, and denote by $\tilde{W}^{\mu+\delta}$ the stabilizer of $\mu+\delta$ in $\tilde{W}$. Then $p \nmid [X:X']$ gives $\tilde{W}^{\mu+\delta} \subset \tilde{W}'$ by exactly the same type of argument as in the fourth paragraph of § 5. The subgroup $\tilde{W}^{\mu+\delta}$ is then generated by reflections; in fact, if $\nu \in X_J^{\star}$ is the unique element of $X^{\star}$ which is $\tilde{W}'$-linked to μ then $\tilde{W}^{\mu+\delta}$ is conjugate to the stabilizer $\tilde{W}^J$ in $\tilde{W}'$ of $\nu+\delta$ ($\tilde{W}^J$ is defined in § 1.4 as the subgroup by the reflections $\{R_j \mid j \in J\}$). Since $\tilde{W}^{\mu+\delta}$ contains no translations, the projection $\tilde{W}' \to W$ gives an isomorphism of $\tilde{W}^{\mu+\delta}$ onto $W^{\lambda+\delta}$. Thus to calculate a_λ (for $p \nmid [X:X']$) one needs only determine $J \subsetneqq \{0, 1, \ldots, l\}$, for the Coxeter relations on $\{R_j \mid j \in J\}$ can be then read off from the extended Dynkin diagram and the order of the group generated by them readily computed by standard methods. This effectively solves the problem of computing a_λ in any given case. (Things could be accounted for even in the more general situation when p is allowed to divide $[X:X']$, through a slightly more detailed analysis; see § 7.2.)

6.5. In the next section we offer a conjecture on the decomposition of the **u**-module Z_λ which implies a neat formula ((7.2)) for Humphreys' numbers d_λ. This formula involves the constants b_σ of Conjecture I and the infinitesimal decomposition numbers $c'_{\lambda\mu}=[\bar{V}_\lambda : M_\mu]_{\mathbf{u}}$ analogous to the constants $c_{\lambda\mu}$, and thus forms the *raison d'être* for the two parts of this paper. The numbers $c'_{\lambda\mu}$ for $\lambda, \mu \in \Lambda$ can be readily derived from a knowledge of $c_{\lambda\mu}$ for $\lambda \in X^{\#}$ and $\mu \in X^{+}$; we recall that the latter set of numbers were shown to be sufficient, in § 4, for determining the formal characters of all the irreducible G-modules and their dimensions. Explicitly,

$$\begin{aligned} \bar{V}_\lambda &\underset{\mathrm{G}}{\sim} \sum_{\mu \in X^+} c_{\lambda\mu} M_\mu \\ &\underset{\mathrm{G}}{\sim} \sum_{\mu'' \in X^{\#}} \Big(\sum_{\mu' \in Y_0} c_{\lambda, \mu''+p\mu'} M_{\mu''} \otimes M_{\mu'}^{\mathrm{Fr}} \Big) \\ &\underset{\mathbf{u}}{\sim} \sum_{\mu \in X^{\#}} \Big(\sum_{\mu' \in Y_0} c_{\lambda, \mu+p\mu'} \cdot \dim M_{\mu'} \Big) \cdot M_\mu , \end{aligned}$$

which gives

$$c'_{\lambda\mu} = \sum_{\mu' \in Y_0} c_{\lambda, \mu+p\mu'} \cdot \dim M_{\mu'} = \operatorname{trace} C_{\lambda\mu}, \tag{6.2}$$

where $C_{\lambda\mu}$ is as in (4.5), and the trace of an element $\sum_\lambda n_\lambda e^\lambda$ of $\mathbf{Z} \cdot e^X$ means the sum $\sum_\lambda n_\lambda$ of all the coefficients (multiplicities). Here we have repeated the earlier abuse of identifying indices in $X^{\#}$ and Λ via the bijection π. In case p is large enough to guarantee the irreducibility of $\bar{V}_{\mu'}$ for $\mu' \in Y_0$ (cf. Corollary to Proposition 3 in § 5), one substitutes $\dim M_{\mu'}=D(\mu'+\delta)$ in (6.2) and obtains a cleaner formula for $c'_{\lambda\mu}$. Note that on has $c'_{\lambda\mu}=c_{\lambda\mu}$ whenever $Y_0=\{0\}$ (as is the case for types A_2 and B_2; cf. Example 2, § 5).

§ 7. Conjecture on the decomposition of induced modules Z_λ

7.1. The following is independent of the conjectures in § 5.

Conjecture IV. *Let* $\lambda \in X^\star$ (i.e. $\lambda \in X$ and $p^{-1}(\lambda+\delta) \in E^\star$), *and for* $\sigma \in W$ *let us denote* $(\lambda+\delta)^{w_\sigma}-\delta$ by $\lambda_{(\sigma)}$. *Then for every* $\nu \in \Lambda$ *which is W-linked* (definition in § 6.3) *to the image of* λ *in* Λ (under the canonical projection $X \to \Lambda$) *one must have*

$$Z_\nu \underset{\mathbf{u}}{\sim} \sum_{\sigma \in W} b_\sigma \cdot \bar{V}_{\lambda_{(\sigma)}}. \tag{7.1}$$

It should be noted that when $\lambda+\delta$ is not $\tilde{W}$-regular (in particular, when it is not $\tilde{W}'$-regular, i.e. when $\lambda \in X_J^\star$ for J nonempty), there may be several values of σ which give rise to same $\lambda_{(\sigma)}$, and that some of the values of $\lambda_{(\sigma)}+\delta$ could even be W-singular (since the image of $X^\star+\delta$ under $w_\sigma \in \tilde{W}^\# \subset \tilde{W}^+$ lies only in X^+ and not always in $X^+ + \delta$) in which case we use the convention of treating $\bar{V}_{\lambda_{(\sigma)}}$ as the zero-module in consonance with the formula dim $\bar{V}_{\lambda_{(\sigma)}} = D(\lambda_{(\sigma)}+\delta)$. (See Footnote[6].)

Note that the equality of the dimensions of the left and right-hand sides of (7.1) is already asserted by Conjecture I; one sees this by recalling the modification in the action of $\tilde{W}$ on X that has been introduced in Part II (third paragraph

[6] *(Added in Proof.)* I have been able to prove Conjecture IV rather recently by obtaining an explicit formula for the formal character of the G-module Q_μ for μ of the form $\mu=(p-1)\delta - -\lambda^* \in p \cdot E^\star$ (* denoting the Dynkin outer automorphism) with $\lambda \in X_J^\star+\delta$ ($J \subsetneqq \{0, 1, \ldots, l\}$) which guarantees

$$|W| \cdot Z_{\lambda-\delta} \sim \mathbf{u} |\tilde{W}^J| \cdot Q_\mu \underset{G}{\approx} \sum_{\sigma \in W} \bar{V}_{(p-1)\delta+\lambda^\sigma},$$

where the second equivalence was known for a while. For years I was unable to figure out exactly how to replace (in the Grothendieck ring of G-modules) an arbitrary Weyl module (e.g. one of the $|W|$ terms in the last sum) in terms of Weyl modules whose highest weights are restricted (i.e. lie in $X^\#$), though this was evident from the intuition gained by various computations (this means that a suitable analogue of Steinberg's tensor-product theorem holds with Weyl modules in place of irreducible modules). My first breakthrough came on seeing the recursive argument needed to justify this; though explicit formulae were lacking, it was enough to yield the existence of virtual $\mathfrak{g}_{\mathbf{C}}$-modules with formal characters $B_\sigma \in (\mathbf{Z} \cdot e^X)^W$ (independent of λ as well as p) such that

$$|\tilde{W}^J| \cdot \text{char } Q_\mu = \sum_{\sigma \in W} (\bar{B}_\sigma)^{\text{Fr}} \cdot \text{char } \bar{V}_{\lambda^\sigma + p\delta_\sigma - \delta}$$

which gives $Z_{\lambda-\delta} \underset{\mathbf{u}}{\sim} \sum_{\sigma \in W} (|W|^{-1} \cdot \deg B_\sigma) \cdot \bar{V}_{\lambda^{(\sigma)}-\delta}$ where $\lambda^{(\sigma)} = \lambda^{w_\sigma} = (\lambda-\delta)_{(\sigma)}+\delta$. From this Conjectures IV and I follow, except for the integrality of $b_\sigma = |W|^{-1} \cdot \deg B_\sigma$. (Here "deg" means the virtual degree, and is the same as "trace" used in formula (6.2)). The next breakthrough came only a few weaks later, when I realized that the ideas at the end of Footnote [2] can be used for a quick proof of the preceding as well as a closed expression for B_σ in terms of the virtual T-modules β_σ (explicitly, if β_σ is written as $\sum s_\lambda \cdot e^\lambda$, one has $B_\sigma = \sum s_\lambda \cdot (\sum_{\tau \in W} e^{\lambda^\tau})$, in consonance with $|W|^{-1} \cdot \deg B_\sigma = \deg \beta_\sigma = b_\sigma$). The details will appear in due course.

of § 5.1) under which (2.1) becomes

$$\sum_{\sigma \in W} b_\sigma \cdot D\big(p^{-1}(\lambda+\delta)^{w_\sigma}\big) = 1,$$

whence

$$\sum_{\sigma \in W} b_\sigma \cdot D(\lambda_{(\sigma)}+\delta) = p^r = \dim Z_\nu,$$

as $D(\ )$ is homogeneous of degree r. As a matter of fact, the validity of Conjecture IV for infinitely many values of p implies that of Conjecture I itself, for if (2.1) holds for $\lambda \in E^\star \cap p^{-1}X$ for the given values of p then it must hold for all λ, as the union of $E^\star \cap p^{-1}X$ over those values is (topologically dense and hence) Zariski-dense.

Substituting $\bar{V}_{\lambda_{(\sigma)}} = \sum_{\mu \in X^\#} c'_{\lambda_{(\sigma)}\mu} M_\mu$ in (7.1), we have by (6.2) the following equivalent formulation of (7.1):

$$d_\mu = \sum_{\mu' \in Y_0} \Big(\sum_{\sigma \in W} b_\sigma \cdot c_{\lambda_{(\sigma)}, \mu+p\mu'} \Big) \dim M_{\mu'} \tag{7.2}$$

for all $\mu \in X^\#$ that are $\tilde{W}$-linked to $\lambda \in X^\star$, where we write d_μ for $d_{\pi(\mu)}$ $(\mu \in X^\#)$ in keeping with the similar notational abuses earlier. (Once again we recall that for λ, μ in $X^\#$, the infinitesimal weights $\pi(\lambda)$ and $\pi(\mu)$ are W-linked if and only if λ and μ are $\tilde{W}$-linked.)

The following statement for d_μ is analogous to the statement in Conjecture III for the fundamental constant $c_{\lambda\mu}$.

Conjecture III′ *(Corollary to Conjectures* III *and* IV*). Let* $J \subsetneqq \{0, 1, \ldots, l\}$, $\nu_1, \nu_2 \in X_J^\star$, *and* $w \in \Omega$. $\tilde{W}^\# = \{w_\sigma \mid \sigma \in W\}$ *be all given, and write* $\mu_i = (\nu_i+\delta)^w - \delta$ *for* $1 \leqq i \leqq 2$. *Then for* $p \nmid [X:X']$ *one must have* $d_{\mu_1} = d_{\mu_2}$.

Due to (7.2), to prove this corollary it would suffice to show that for $\lambda \in X_J^\star$, $w \in \tilde{W}_J^{++}$, $\mu = (\lambda+\delta)^w - \delta$, the sum $\sum_{\sigma \in W} b_\sigma \cdot c_{\lambda_{(\sigma)}\mu}$ depends only on J and w (and not on λ). But this is immediate from Conjecture III which gives $c_{\lambda_{(\sigma)}\mu} = c^J_{w_\sigma w}$ in the notation introduced towards the end of § 5. The rôle of the hypothesis $p \nmid [X:X']$ is to guarantee that (for $i=1, 2$) ν_i is the unique element of $X_J^\star$ which is $\tilde{W}$-linked to μ_i and satisfies $\nu_i \equiv \mu_i \pmod{X'}$; cf. the fourth paragraph of § 5.

7.2. In this paragraph we briefly comment on the modification one needs to introduce in order to remove the assumption $p \nmid [X:X']$ in the preceding. For $\lambda \in X^\star$, let Ω^λ denote the stabilizer of λ in Ω (defined in § 1), and for an element μ of $X^\#$ which is $\tilde{W}$-linked to λ define the "virtual Humphreys number" d'_μ to be $d_\mu \cdot |\Omega^{\lambda+\delta}|^{-1}$. With this definition, we might strengthen Conjecture III′ above, to assert that the value of d'_μ for $\mu = (\lambda+\delta)^w - \delta$, $\lambda \in X_J^\star$, depends only on J and w. This would have a very pleasant repercussion on the formula in § 6.4 for the dimension of $Q_\mu = Q_{\pi(\mu)}$ for $\mu \in X^\#$. Setting

$$a'_\mu = |W| \cdot |\tilde{W}'^{\mu+\delta}|^{-1}$$

where $\tilde{W}'^{\mu+\delta}$ is the stabilizer of $\mu+\delta$ in $\tilde{W}'$, one finds

$$a_\mu = |W| \cdot |\tilde{W}^{\mu+\delta}|^{-1} = a'_\mu \cdot |\Omega^{\lambda+\delta}|^{-1}$$

where λ is any element of $X^\star$ that is $\tilde{W}$-linked to μ; and one has

$$\dim Q_\mu = a'_\mu d'_\mu p^r.$$

Since the value of a'_μ for $\mu=(\lambda+\delta)^w-\delta, \lambda \in X_J^\star$, depends only on J and w (the order of the stabilizer of $\mu+\delta$ or $\lambda+\delta$ in $\tilde{W}'$ being simply the order of the Coxeter group $\tilde{W}^J$ generated by $\{R_j \mid j \in J\}$), it follows (from the preceding unlabelled extension of Conjecture III′) that the dimension of Q_μ depends only on J and w, even when p divides $[X:X']$. We shall prove this for the case of type A_2 in a later paper, calling the preceding (suggested) phenomenon, the "ramification" phenomenon (term suggested by Humphreys) for the constants a_μ and d_μ, the number $|\Omega^{\lambda+\delta}|$ being the measure of this ramification.

7.3. To get ready for the concluding remarks in § 8, let us make the following notation: If $\mu \in X^\#$ is $\tilde{W}'$-linked to $\lambda \in X_J^\star$ by $\mu=(\lambda+\delta)^w-\delta$ for $w \in \tilde{W}_J^\#$, then assuming the truth of Conjecture III′, we write d_w^J in place of d_μ. Since an element ω of Ω sends $X_J^\star$ to some $X_{J'}^\star$, (J' being precisely the image of J under the permutation θ_j of § 1.2, in case $\omega=\omega_j$ for $j \in J_0$), it follows from Conjecture III′ that $d_w^J=d_{w'}^{J'}$, where $w'=\omega^{-1}w\omega \in \tilde{W}_{J'}^\#$.

§ 8. Epilogue

A metamathematical (?) *conjecture*

The philosophy underlying the relationship between Parts I and II of this paper is deeper than the superficial dimensionality connection between Conjectures I and IV we exhibited in the second paragraph of § 7. It rests in the suggestion that in some inexplicable ways several phenomena concerning the representation theory of algebraic Chevalley groups (and their Lie algebras) *of a given type* and of characteristic $p \neq 0$, happen to behave extremely smoothly with respect to the characteristic p (when p is treated as an indeterminate, or as a real variable if one prefers, particularly in case of the numerical invariants of this theory when these invariants are treated as functions of p), perhaps with a handful of exceptions. The most classical example of a "characteristic p phenomenon" behaving smoothly with respect to p is, of course, the formula for the order of the finite Chevalley group $G_{\mathbf{F}_q}$ (over the finite field $\mathbf{F}_q$ of cardinality q) for the semisimple algebraic group G, which is known to be a polynomial in q (this polynomial being dependent only on the type, or root-system, of G).

Concretely speaking we like to offer the following (assuming that the earlier Conjectures are true).

Conjecture V. *For* $p \nmid [X:X']$, *the fundamental constants* $c^J_{w'w}$ (for w', w lying in $\tilde{W}_J^{++}$) *and the Humphreys numbers* d^J_w (for w in $\tilde{W}_J^{\#}$), *do not depend on p.*

The remarks in § 7.2 indicate that the restriction $p \nmid [X:X']$ could be lifted for the case of the Humphreys numbers d^J_w if we replace them by the virtual Humphreys number defined there.

It follows from (7.2) that when p is large enough to guarantee that Y_0 is independent of p as well as dim M_λ is independent of p for each $\lambda \in Y_0$ (cf. Corollary to Proposition 3 in § 5.2), the second half of Conjecture V follows from the first half.

It follows from Conjecture V that for $\mu + \delta \in (X_J^+ + \delta)^w$, $w \in \tilde{W}^{\#}$, the number $p^{-r} \cdot \dim Q_\mu$ must be independent not only of μ but also of p (depends only on J and w).

Yet another dimensionality consequence of Conjecture V is the following (cf. formula (5.5)):

Corollary to Conjecture V. Let $J \subsetneqq \{0, 1, \ldots, l\}$, *and for* $\lambda \in X_J^+$ let $\lambda + \delta = \sum_{i=0}^{l} m_i \lambda_i$ *be the expansion in the "homogeneous coordinates"* m_i *satisfying* $\sum_{i=0} m_i n_i = p$, *where* $\lambda_0 = 0$ *(as in* § 1*); thus* m_i *is zero or (strictly) positive according as* $i \in J$ *or* $i \notin J$. *Then for each* $w \in \tilde{W}_J^{++}$ *there exists a polynomial in* $l+1-|J|$ *variables (depending on J and w, but not on p), such that the value of this polynomial obtained on substituting the above homogeneous coordinates* m_i *of* $\lambda + \delta$, *equals the dimesion of the irreducible G-module* $M_{(\lambda+\delta)^w - \delta}$ *for the Chevalley group G (of the given type) in characteristic* $p \neq 0$.

We like to wind up by asking the following question: Is there some way of obtaining the matrix $[c^J_{w'w}]$ for all J knowing it only for the case J empty?

REFERENCES

[1] Bernstein, I. N., Gelfand, I. M. and Gelfand, S. I., (Бернштейн И. Н., Гелфанд И. М. Гелфанд С. И.), Структура представлений, порожденных векторами старшего веса, *Функционалный анализ и его приложения*, **5**:(1971), 1—9.
(The structure of representations generated by vectors of highest weight.)

[2] Borel, A., Properties and linear representations of Chevalley groups, *in Seminar on algebraic groups and related finite groups*, Springer Lecture Notes, Vol. **131**, 1970.

[3] Bourbaki, N., *Groupes et algébres de Lie,* Chap. IV—VI, Hermann, Paris, 1969.

[4] Braden, B., Restricted representations of classical Lie algebras of type A_2 and B_2, *Bull. Amer. Math. Soc.*, **73** (1967), 482—486.

[5] Curtis, C. W., On projective representations of certain finite groups, *Proc. Amer. Math. Soc.*, **11** (1960), 852—860.

[6] Gelfand, I. M., The cohomology of infinite dimensional Lie algebras; some questions in integral geometry, *in Actes, Congrès intern. math., 1970,* Vol. **1**, pp. 95—111, Gauthier-Villars, Paris, 1971.

[7] Humphreys, J. E., Modular representations of classical Lie algebras and semisimple groups, *J. Algebra,* **19** (1971), 51—79.

[8] Humphreys, J. E. and Verma, D.-N., Projective modules for finite Chevalley groups, *Bull. Amer. Math. Soc.*, **79** (1973), 467—468.
[9] Iwahori, N. and Matsumoto, H., On some Bruhat decomposition and the structure of the Hecke ring of *p*-adic Chevalley groups, *Publ. I.H.E.S.*, **25**, (1965), 237—280.
[10] Springer, T. A., Weyl's character formula for algebraic groups, *Inventiones Math.*, **5** (1968), 85—105.
[11] Steinberg, R., Differential equations invariant under finite reflection groups, *Trans. Amer. Math. Soc.,* **112** (1964), 392—400.
[12] ———, Endomorphisms of linear algebraic groups, *Memoirs of the AMS,* no. **80** (1968).
[13] ———, *Lectures on Chevalley groups* (mimeograph), Yale lecture notes, New Haven, 1968.
[14] Veldkamp, F. D., Representations of algebraic groups of type F_4 in characteristic 2, *J. Algebra,* **16** (1970), 326—339.
[15] Verma, D.-N., Structure of certain induced representations of complex semisimple Lie algebras, *Dissertation,* Yale University, New Haven, 1966.
[15A] ———, Structure of certain induced representations of complex semisimple Lie algebras, (announcement of [15]), *Bull. Amer. Math. Soc.,* **74** (1968), 160—166.
[16] ———, Möbius inversion for the Bruhat ordering on a Weyl group, *Ann. Sci. École Norm. Sup.* 4e-série, **4** (1971), 393—398.
[17] ———, A strengthening of the exchange property of Coxeter groups *(to appear).*
[18] ———, A note on Chevalley lattice and its Kostant envelope *(in preparation).*
[19] ———, Exponentiating the projective modules of the *u*-algebra associated to an algebraic Chevalley group *(in preparation).*
[20] Wong, W. J., Representations of Chevalley groups in characteristic *p*, *in Proc. of Symposia in Pure Mathematics,* Vol. XXI, AMS, Providence, 1971.

Received October 20, 1972.

ELEMENTARY REPRESENTATIONS OF A SEMISIMPLE COMPLEX LIE GROUP

D. P. ŽELOBENKO

Let G be a semisimple complex connected Lie group, B a Borel subgroup, H the Cartan subgroup of B, and A the set of all (continuous) characters of the group B with can be identified with the set of all characters of the group H in a natural way. For each $\alpha \in A$ we put[1].

$$e(\alpha) = \mathrm{Ind}_B^G \left(\alpha, C^\infty(G)\right). \tag{1}$$

The representation $e(\alpha)$ is called the *elementary representation* of the group G with *signature* α [20]. Each of these representations can be extended to a representation of G on a Hilbert space, and the family of representations obtained in this way contains the Gelfand—Neumark "principal series" [7] as a "real section" (by the parameters α).

The study of the representations $e(\alpha)$ is of interest e.g. from the point of view of harmonic analysis in the class of finitary or rapidly decreasing (fundamental or generalized) functions on the group G [5], [24]. First of all, questions arise about the irreducibility and mutual equivalence of the representations $e(\alpha)$ as well as about the description of the family $P(\alpha, \beta)$ of all (continuous) intertwining operators $e(\alpha) \to e(\beta)$.

By now, the first question (on irreducibility) has been solved completely [21], but the remaining two in "general position" only. In what follows we shall list the known results and formulate several unsolved problems.

1. *The case of rank 1.* The group $G = SL(2, \mathbf{C})$ of rank 1 is well studied [6], [23]. In this case $\dim P(\alpha, \beta) = 0, 1$ $\left(P(\alpha, \beta)\right.$ is a linear space$\left.\right)$, and all possible homomorphisms $e(\alpha) \to e(\beta)$ are determined by the following schemes:

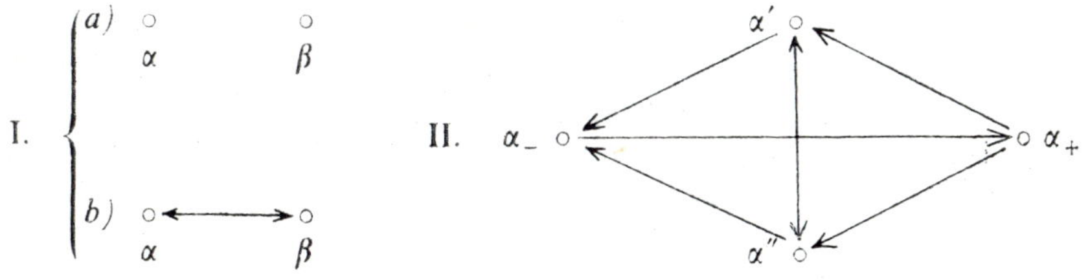

[1] The representation (1) acts by right translation on the subspace of all functions $f(x) \in C_0^\infty(G)$ such that $f(bx) = \alpha(b) f(x)$, $b \in B$, $x \in G$.

Here each circle corresponds to one of the signatures α, β and each arrow to a homomorphism (intertwining operator) $e(\alpha)\to e(\beta)$. Scheme I concerns signatures α, β in "general position", when the representations $e(\alpha)$ are irreducible, and the existence (lack) of arrows means the equivalence (inequivalence) of $e(\alpha)$, $e(\beta)$. Scheme II connects the pair of "degenerate" signatures α_-, α_+ with the pair α', α'' of "nondegenerate" ones. Here the diagrams consisting of diagonal and vertical arrows are commutative, while the diagrams of any two arrows one of which is horizontal are exact.

Scheme II illustrates the structure of all submodules of the G-module $e(\alpha)$. $e(\alpha_-)$ contains the submodule $e(\alpha')\sim e(\alpha'')$, and the factor module by this is irreducible and finite-dimensional. The horizontal arrow maps this factor module onto a submodule of $e(\alpha_+)$, the factor by which can be identified with $e(\alpha')\sim\sim e(\alpha'')$.

2. *The differential of* $e(\alpha)$. There is the following remarkable way to algebraize the study of the G-modules $e(\alpha)$. Let K be the maximal compact subgroup in G so that $G=BK$ (the Iwasawa decomposition), and let $V(\alpha)$ be the set of all K-finite vectors in the space $E(\alpha)$ of the representation $e(\alpha)$. The elements of $V(\alpha)$ are real analytic functions on G and the subspace $V(\alpha)$ is invariant under infinitely small right translations on G. Put

$$\varepsilon(\alpha) = de(\alpha)|V(\alpha), \tag{2}$$

where $de(\alpha)$ is the differential of the representation $e(\alpha)$. The representation $\varepsilon(\alpha)$ so obtained is called an *elementary representation (module) of the Lie algebra* of the group G. We shall denote by $\mathfrak{g}$ the real Lie algebra of the group G and by $\mathfrak{G}$ its associative hull over the field of complex numbers. Let us note that every representation $\varepsilon(\alpha)$ is also a $\mathfrak{G}$-module.

Theorem 1. *There exists a one-to-one correspondence (determined by restriction) between the submodules of the G-module* $e(\alpha)$ *and the submodules of the* $\mathfrak{g}$*-module (*$\mathfrak{G}$*-module)* $\varepsilon(\alpha)$.

Here, of course, we view $e(\alpha)$ as a topological module (with the C^∞ topology) and restrict ourselves to consider submodules associated with closed invariant subspaces. However, the C^∞ topology could be replaced by any other locally convex topology of the space $V(\alpha)$ of the representation $\varepsilon(\alpha)$, provided that G has a representation on the completion of $V(\alpha)$ with respect to this topology,[2] whose differential coincides with $\varepsilon(\alpha)$ on $V(\alpha)$.

A factor module of an arbitrary submodule of $e(\alpha)$ $(\varepsilon(\alpha))$ will be called a *component* of $e(\alpha)$ $(\varepsilon(\alpha))$. Theorem 1 can be improved as follows:

[2] By a representation of the group G in the topological vector space E we mean any homomorphism $g\to T_g$, $g\in G$, into the group of invertible linear operators of the space E, for which the vector function $T_g\xi$, $\xi\in E$, is separately continuous on $G\times E$.

Theorem 2. *To every irreducible component of* $\varepsilon(\alpha)$ *there corresponds a completely irreducible*[3] *component of* $e(\alpha)$.

We remark that the modules $e(\alpha)$ are "semi-irreducible" (for infinite-dimensional representations of the group G, H. Weyl's theorem on complete irreducibility ceases to be valid). For $SL(2, \mathbf{C})$ this fact was remarked by the author in 1958, cf. [6], [23].

3. *The theorem on irreducibility*. Let us write every character of the group H in the form $\exp[(\lambda, x)+(\mu, \bar{x})]$, where x is the inverse image of the element $h \in H$ under the canonical mapping $\mathfrak{h} \to H$ ($\mathfrak{h}$ is the Lie algebra of the group H), $x \to \bar{x}$ is the complex conjugation in the basis of simple roots, and (x, y) is the Cartan—Killing-form. We put $\lambda = p - \delta$, $\mu = q - \delta$, where δ is one half of the sum of the positive roots in the algebra $\mathfrak{h}$. Remark that $p - q \in \Lambda$ where Λ is the set of weights of the subgroup K (uniqueness condition of the character on the torus $H \cap K$). We shall identify the signature α with the pair of vectors $p|q$, $p-q \in \Lambda$. The difference $\nu = p - q$ will be called the *index* of the signature α.

For each $x \in \mathfrak{h}$ and every root ω we put $x_\omega = 2(x, \omega)/(\omega, \omega)$. We shall say that the signature α is *natural* for ω if both p_ω and q_ω are natural numbers. The signature α will be called *degenerate* if it is natural for at least one root ω. In the opposite case α is called *non-degenerate*.

Theorem 3. *The representation* $e(\alpha)$ *is completely irreducible if and only if the signature* α *is nondegenerate*.

The proof of Theorem 3 can be found in [21]. Particular cases of this theorem have been considered by a number of authors. Thus for the classical matrix groups the irreducibility of the representations of the principal series was proved by I. M. Gelfand and M. A. Neumark [7], for the representations of the principal series in general position by F. Bruhat [4], for representations "of class 0" in [15]. The condition of complete irreducibility in Theorem 3 can be replaced by that of topological irreducibility. In particular, Theorem 3 implies the following

Corollary. The representations of the principal (nondegenerate) Gelfand—Neumark series are irreducible.

In fact, it is not difficult to distinguish a "strip" in the neighbourhood of the signatures α corresponding to the representations of the fundamental series, where the representations $e(\alpha)$ remain (completely) irreducible.

4. *The theorem on cyclicality*. In order to obtain a more detailed description of the modules $e(\alpha)$, it is convenient to notice the following facts. Every module

[3] A representation T_g is called *completely irreducible* (by Godement) [12], if the weakly closed linear hull of the operators T_g, $g \in G$ contains every continuous linear operator of the space E.

$e(\alpha)$ is K-finite, i.e. it contains any irreducible representation of the subgroup K with finite multiplicity. The representation δ_0 with minimal highest weight is of special interest among these representations, it always occurs with multiplicity one. Let $f(\alpha)$ be the closed cyclic hull of δ_0 with respect to the group G, and let $\varphi(\alpha)=df(\alpha)|V(\alpha)$. The submodules $f(\alpha)\subset e(\alpha)$ and $\varphi(\alpha)\subset\varepsilon(\alpha)$ will be called the *principal* submodules in $e(\alpha)$, $\varepsilon(\alpha)$ respectively.

In order to describe the principal submodules, in the class A of all signatures we distinguish two subclasses A_-, A_+ as follows. The set $A_-(A_+)$ consists of all signatures which are not natural for any root $\omega>0$ ($\omega<0$).

Theorem 4. *If* $\alpha\in A_-$ *then* $f(\alpha)=e(\alpha)$. *If* $\alpha\in A_+$ *then the principal submodule* $f(\alpha)$ *is completely irreducible.*

Analogous statements are valid for the modules $\varepsilon(\alpha)$. Cf. [26], for the particular case of representations of "class 0" [22]. Theorem 4 will be referred to as the *theorem on cyclicality.*

Corollary 1. *If* $\alpha\in A_-$ *then the representation* $e(\alpha)$ *is cyclic and its cyclic vectors belong to* δ_0.

Corollary 2. *If* $\alpha\in A_+$ *then the only irreducible submodule of* $e(\alpha)$ *is* $f(\alpha)$.

If $\alpha\in A_+$ we call $f(\alpha)$ the minimal representation (module) of the group G with signature α [22]. Analogously, $\varphi(\alpha)$, $\alpha\in A_+$, will be called the *minimal module of the algebra* $\mathfrak{g}$ *(or* $\mathfrak{G}$*) with signature* α.

5. *The symmetry operators* $W_s(\alpha)$. Now we shall describe a certain class of intertwining operators between the representations $e(\alpha)$ $(\varepsilon(\alpha))$. Let $\mathbf{W}$ be the Weyl group associated with $\mathfrak{g}$ (as a complex Lie algebra). For each $s\in\mathbf{W}$ put $s\alpha=sp|sq$ if $\alpha=p|q$.

Theorem 5. *For each* $s\in\mathbf{W}$ *there exists an operator-valued function* $W_s(\alpha)$, *which is meromorphic in* $V(\alpha)$, *taking its values in* $P(\alpha, s\alpha)$ *and satisfying the conditions*

$$W_{s\sigma}(\alpha) = W_s(\sigma\alpha)W_\sigma(\alpha), \quad s, \sigma\in\mathbf{W}. \tag{3}$$

The function $W_s(\alpha)$ *can be normed in such a way that it becomes a constant* (not depending on α) *non-degenerate operator on the finite-dimensional subspace of the representation* δ_0 *of the subgroup* K.

The formulation of Theorem 5 needs to be made precise. Since the operators from $P(\alpha, \beta)$ commute with the right translations on K, they take $V(\alpha)$ into $V(\beta)$ (and intertwine $\varepsilon(\alpha)$, $\varepsilon(\beta)$).

Consider the projection $x(g)\to x(k)$, $k\in K$, mapping $V(\alpha)$ into the algebra $S(K)$ of all spherical functions on K, and remark that the image of $V(\alpha)$ under this mapping is the subspace $L(v)$ belonging to the left regular representation

$$\lambda(k_0)x(k) = x(k_0'k), \quad k_0, k\in K, \tag{4}$$

where $k' = \bar{k}^{-1}$ (the bar denotes the conjugation in G induced by the complex conjugation in $\mathfrak{g}$ with respect to the Cartan—Weyl basis). Since the functions $x(g) \in V(\alpha)$ are completely determined by their restrictions to K (by virtue of the Iwasawa decomposition), the mapping $V(\alpha) \to L(v)$ is an isomorphism. By being meromorphic in $V(\alpha)$ we mean being meromorphic in $L(v)$ for each fixed v with respect to the complex parameters α, for instance with respect to the vector $\tau = p + q$ which we call the *exponent* of the signature α.

Having agreed to write (3) in the form $W_{s\sigma} = W_s W_\sigma$, we obtain that the mapping $s \to W_s$ is a homomorphism of the Weyl group. From Theorem 5 we obtain the following

Corollary. The family of operators $W_s(\alpha)$ is generated by the r generators $W_i(\alpha) = W_{s_i}(\alpha)$, $i = 1, 2, \ldots, r$; $r = \operatorname{rank} G$, where s_i denotes the reflection with respect to the simple root ω_i, $i = 1, 2, \ldots, r$.

The operators $W_i(\alpha)$ will be called *symmetry operators* with respect to the Weyl group, and the operators $W_i(\alpha)$ *simple reflections* (with respect to the roots ω_i).

The generators $W_s(\alpha)$ can be written explicitly as meromorphic functions in $L(v)$. Let us consider in $S(K)$ the differential of the representation (4) and restrict it to the three-dimensional subalgebra ϑ generated by the simple root ω_i. We remark that every irreducible representation of a three-dimensional simple Lie algebra is equal to a symmetric power ϑ^n of the two-dimensional (fundamental) representation ϑ, $n = 0, 1, \ldots$. Let L^n be the maximal subspace in $S(K)$ where the representation coincides with a multiple of ϑ^n, $L^n(v)$ its intersection with $L(v)$ and p_n the projector of $L(v)$ onto $L^n(v)$. Then the operator $W_i(\alpha)$ has the form

$$W_i(\alpha) = \lambda(s_i) \sum_{n=n_0}^{\infty} (c_n(\alpha) p_n) \tag{5}$$

where

$$c_n(\alpha) = (-1)^{n-n_0} \pi_n(s_i\alpha)/\pi_n(\alpha), \quad \pi_n(\alpha) = (\tau_i - n_0 - 2)(\tau_i - n_0 - 4) \ldots (\tau_i - n),$$

$x_i = x_{\omega_i}$ for every vector $x \in \mathfrak{h}$, v is the index and τ is the exponent of the signature α, $\lambda(s)$ is the operator of left translation (4) corresponding to the element $s \in W$ which is canonically embedded in K (the operator $\lambda(s)$ is uniquely defined in the space $L(v)$ up to a scalar multiplier). Formula (5) enables us to compute all zeros and poles of the function $W_i(\alpha)$ directly.

Let us note that if $r = 1$ the family $W_s(\alpha)$ reduces to the single function $W_1(\alpha)$ whose action is represented in diagrams I and II by the horizontal and vertical arrows. A proof of Theorem 5 can be obtained in the following steps: *1)* verify formula (5) for the case $r = 1$; *2)* use the induction rule that for every intertwining operator in the class of the elementary representations of the three-dimensional subgroup $G_i \subset G$, generated by the simple root ω_i, one can construct an intertwining operator for the whole group G; *3)* verify the equality (3) which follows from Schur's lemma and Theorem 3 on the irreducibility.

The symmetry operators $W_s(\alpha)$ have been studied by a number of authors [5], [20], [13][4]. The method presented here, which enables one to study all zeros and poles of these operators, was suggested by the author in [20]. This is based on the case $r=1$ which had been worked out earlier in [23]. We remark that formula (5) for $r=1$ was proved in [23] by means of harmonic analysis on $SL(2, \mathbf{C})$, however this can be replaced by a significantly simpler (so far unpublished) algebraic method.

6. *The zeros and poles of the operators* $W_s(\alpha)$. We recall that the algebra $S(K)$ is the set of all finite linear combinations of matrix elements of the group K. Let us consider the gradation in $S(K)$ according to the main weights of the matrix elements and the corresponding gradation in $L(\nu)$:

$$L(\nu) = \bigotimes_{\lambda \geqq \lambda_0} V^\lambda, \quad \lambda_0 = |\nu|,$$

where $|\nu|$ denotes the dominating vector in $\mathfrak{h}$ lying on the same orbit as ν with respect to the Weyl group ($\lambda \geqq \mu$ is the lexicographic ordering of the weights). Since the operators $W_s(\alpha)$ commute with the right translations on K, they admit the corresponding direct decomposition

$$W_s(\alpha) = \bigotimes_{\lambda \geqq \lambda_0} W_s^\lambda(\alpha),$$

where $W_s^\lambda(\alpha)$ sends V^μ to zero if $\mu \neq \lambda$ and maps V^λ into the corresponding subspace of the linear space $L(s\nu)$. Since V^λ is finite-dimensional, every operator $W_s^\lambda(\alpha)$ can be determined by a $d_\lambda \times d_\lambda$ square matrix, where $d_\lambda = n_\lambda(\nu)$ is the multiplicity of the weight ν in the irreducible representation π^λ of main weight λ. We put

$$d_s^\lambda(\alpha) = \det W_s^\lambda(\alpha).$$

Applying Theorem 5 and formula (5) we easily deduce the equality

$$d_s^\lambda(\alpha) = \pm \pi^\lambda(s\alpha)/\pi^\lambda(\alpha),$$

where

$$\pi^\lambda(\alpha) = \prod_{\omega > 0} \prod_{\varkappa} (\tau_\omega - 2\varkappa - 2)^{n_\varkappa}.$$

Here ω is a root, $\varkappa$ is a natural number and the (integer) exponents $n_\varkappa \geqq 0$ are different from zero for a finite number of the values of $\varkappa$. We also recall that $\tau = p + q$ is the exponent of the signature α. A more detailed description of the numbers $n_\varkappa$ can be found in [22].

[4] For semisimple real Lie groups the symmetry operators were constructed in Schiffmann's work [16]. Cf. also [21].

Theorem 6. *The operator* $W_s(\alpha)$ *is defined and continuous if and only if* $d_s^\lambda(\alpha)\neq\infty$ *for* $\lambda\geqq\lambda_0$. *The operator* $W_s(\alpha)$ *is continuously invertible if and only if* $d_s^\lambda(\alpha)\neq 0$ *for* $\lambda\geqq\lambda_0$, *moreover in the latter case*

$$W_s(\alpha)^{-1} = W_{s^{-1}}(s\alpha).$$

As a particular case of Theorem 6 we obtain

Corollary 1. *If* $\alpha\in A_-$, *then every operator* $W_s(\alpha)$ *is defined and continuous.*

Using Theorem 4 and the operators $W_s(\alpha)$ we can give an effective description of the space $F(\alpha)$ belonging to the minimal representation $f(\alpha)$, $\alpha\in A_+$. Let s_0 be the representation of the Weyl group which takes all negative roots into positive ones.

Corollary 2. *The space* $F(\alpha)$ *of the minimal representation* $f(\alpha)$, $\alpha\in A_+$ *coincides with the image*[5] *of the symmetry operator* $C_0(\alpha)=W_{s_0}(s_0\alpha)$:

$$F(\alpha) = \operatorname{im} C_0(\alpha). \tag{6}$$

7. *The discrete symmetry operators.* Using the above mentioned induction rule we can describe in the general case the family of intertwining operators generated by the diagonal arrows in diagram II.

Consider the differential of the representation (4) extended to the complex algebra $\mathfrak{g}$ and put

$$\chi_{-i} = \lambda(e_{-\omega_i}),\ \chi_{+i} = \lambda(e_{\omega_i}),\quad i = 1, 2, \ldots, r.$$

We recall that $x_i = x_{\omega_i}$, $i=1, 2, \ldots, r$ for every vector $x\in\mathfrak{h}$.

Theorem 7. *If at least one of the projections* p_i, q_i *of the signature* α *is natural then the corresponding operator*

$$S_i(\alpha) = X_{-i}^{p_i},\quad \bar{S}_i(\alpha) = X_{+i}^{q_i}$$

is an intertwining operator of class $P(\alpha, \beta)$, *where* $\beta=s_ip|q$ *in the first case and* $\beta=p|s_iq$ *in the second.*

If both projections p_i and q_i are natural then the quadruple of signatures

$$\alpha_- = s_ip|s_iq,\ \alpha' = s_ip|q,\ \alpha'' = p|s_iq,\quad \alpha = p|q$$

satisfies diagram II generated by the operators

$$W_i(\alpha),\quad \alpha = \alpha_-, \alpha', \alpha'';\quad S_i(\alpha), \alpha = \alpha'', \alpha_+,\quad \bar{S}_i(\alpha), \alpha = \alpha', \alpha_+.$$

Here, as above, diagrams consisting of diagonal arrows are commutative and a diagram generated by $W_i(\alpha_-)$ and one of the operators S_i, $\bar{S}_i$ is exact.

The operators S_i, $\bar{S}_i$ will be called *discrete symmetry operators* [20].

[5] The operator $C_0(\alpha)$ is defined on the whole space $E(\beta)$ belonging to the elementary representation $e(\beta)$, $\beta=s_0\alpha$. Replacing $e(\alpha)$ by $\varepsilon(\alpha)$ we obtain an analogous description of the space $\Phi(\alpha)$ of the representation $\varphi(\alpha)$.

8. *Elementary representations in general position.* Let A_{reg} be the set of all nondegenerate signatures. Since $A_{\text{reg}} = A_- \cap A_+$, for every $\alpha \in A$ both the operator $W_s(\alpha)$ and its inverse are defined and continuous, consequently $e(\alpha)$ is equivalent to $e(s\alpha)$. It turns out that the converse of this statement is also true.

Theorem 8. *If* $\alpha, \beta \in A_{\text{reg}}$ *then the elementary representations* $e(\alpha)$, $e(\beta)$ *are equivalent if and only if* $\beta = s\alpha$ *for some* $s \in \mathbf{W}$.

Corollary. If $\alpha, \beta \in A_{\text{reg}}$ *then* $\dim P(\alpha, \beta) = 0$ *or* $\dim P(\alpha, \beta) = 1$ *according to whether* $e(\alpha)$ *and* $e(\beta)$ *are equivalent or not.*

The proof of Theorem 8 is based upon the results of the "operational calculus" for the algebra $\mathfrak{G}$ [22]. One can also apply the trace formula [1], [7] for the representations $e(\alpha)$.

9. *Elementary representation in degenerate points.* If the signature α is degenerate then the module $e(\alpha)$ is reducible (Theorem 3). In order to describe the structure of submodules of $e(\alpha)$ we agree to say that the module γ is a semi-direct sum of the modules δ_i, $i = 1, 2, \ldots, n$ if γ admits a linear filtration by submodules γ_i, $i = 1, 2, \ldots, n$, $\gamma_1 = \gamma$, $\gamma_{n+1} = (0)$ such that $\gamma_i / \gamma_{i+1} \sim \delta_i$. In this case we write

$$\gamma = \vec{\sum_i} \delta_i = \delta_1 \rightarrow \delta_2 \rightarrow \ldots \rightarrow \delta_n.$$

Theorem 9. *Every module* $e(\alpha)$ *where* $\alpha = p \,|\, q$ *is the semi-direct sum of minimal modules* $f(\beta)$, $\beta = sp \,|\, \sigma q$ *for certain* $s, \sigma \in \mathbf{W}$:

$$e(\alpha) = f_1 \rightarrow f_2 \rightarrow \ldots \rightarrow f_n, \tag{7}$$

where each f_i *is isomorphic to* $f(\alpha_i)$. *Here* $e(\alpha)$ *is the semi-direct sum of the same modules* f_i *arranged in another order.*

The proof of Theorem 9 is based upon formula (6) and on Theorem 7, which enables us to reduce the study of the submodule $e(\beta)/\ker C_0(\alpha)$, $\beta = s_0 \alpha \in A_-$ to the consideration of the representations $e(\gamma)$ $\gamma = sp \,|\, \sigma q$, $s, \sigma \in \mathbf{W}$. Here we apply finite induction on $|sp - \sigma q|$ (with respect to the lexicographic ordering).

Among the signatures $sp \,|\, \sigma q$ occurring in the decomposition (7) there are two "extreme" points, one of which is nondegenerate (in which case $f(\beta) = e(\beta)$), while the other is of the form $s\alpha$, $s \in \mathbf{W}$, i.e. it lies on the same orbit as α does with respect to the Weyl group. Since $f(s\alpha)$ contains δ_0, this component occurs in (7) with multiplicity 1.

10. *Minimal representations of the group G.* Let us introduce the notation $\mu(\alpha)$ for the component from (7) containing δ_0. Consequently $\mu(\alpha)$ is equivalent to $\mu(\beta)$ if $\beta = t\alpha$, $t \in \mathbf{W}$. The following proposition generalizes Theorem 8:

Theorem 10. *The minimal representations* $\mu(\alpha)$, $\mu(\beta)$ *are equivalent if and only if* $\beta = s\alpha$, $s \in \mathbf{W}$.

The proof of this theorem relies on the "operational calculus" for the algebra $\mathfrak{G}$[6].

As a result we obtain that formula (7) yields a decomposition of elementary representations into a semi-direct sum of irreducible (completely irreducible) representations.

11. *Operational calculus for the algebra* $\mathfrak{G}$. Let us consider the gradation of $S(K)$ according to main weights (i.e. the equivalence classes of the irreducible representations of the group K) and let p^λ be the projector of highest weight λ with respect to this gradation. Let us fix two highest weights λ, μ and for every $x \in \mathfrak{G}$ put

$$x(\alpha) = p^\lambda e(\alpha, x) p^\mu, \tag{8}$$

where $e(\alpha, x)$ is the operator of the representation $\varepsilon(\alpha)$ corresponding to the element $x \in \mathfrak{G}$. By an *operational calculus* for $\mathfrak{G}$ we mean a description of the family of functions of type (8) for all pairs of highest weights λ, μ. Let $\Pi^{\lambda,\mu}$ be the set of all functions $x(\alpha) = p^\lambda x(\alpha) = x(\alpha) p^\mu$ defined on $L(\nu)$, where ν is the index of α, which depend on $\tau = p+q$ polynomially and satisfy the symmetry conditions

$$C^\lambda(\alpha) x(\alpha) = x(\alpha) C^\lambda(\alpha), \tag{9}$$

where

$$C^\lambda(\alpha) = p^\lambda C(\alpha) = C(\alpha) p^\lambda.$$

Here $C(\alpha)$ is one of the symmetry operators

$$W_i(\alpha),\ S_i(\alpha),\ \bar{S}_i(\alpha),\quad i = 1, 2, \ldots, r.$$

We note that the functions of type (8) belong to $\Pi^{\lambda\mu}$. The fundamental result of the operational calculus is

Theorem 11. *Every element* $x(\alpha) \in \Pi^{\lambda\mu}$ *can be written in the form* (8) *for some* $x \in \mathfrak{G}$.

In fact we could consider an even finer gradation in $S(K)$ by decomposing each projector p^λ into the direct sum of projectors p_j^λ, $j = 1, \ldots, n$, with respect to a fixed basis e_j in the space of the irreducible representation with highest weight λ. (This decomposition is generated by the right translations on K.) Replacing p^λ by p_i^λ and p^μ by p_j^μ we obtain, instead of $\Pi^{\lambda\mu}$, the family $\Pi_{ij}^{\lambda\mu}$ which can be described similarly.

12. *Theorems of Paley—Wiener type.* The family of the representations $g \to e(\alpha, g)$, $g \in G$ enables us to build up a harmonic analysis of operators in classes of functions on the group G. Following I. M. Gelfand and M. A. Neumark [8], we put for every function $x(g)$ on G

$$e(\alpha, x) = \int x(g) e(\alpha, g)\, dg. \tag{10}$$

The results of the L^2-theory for the transforms of the type (10) are well known [7], [8]. However, it is especially interesting to study function algebras with respect

[6] An analogous method was worked out by B. Kostant [12] for the study of representations "of class 0" of a semisimple real Lie group.

to convolution, for instance $X=C_0^\infty(G)$. Let us fix the highest weights λ, μ and put for each $x\in X$

$$x(\alpha) = p^\lambda e(\alpha, x)p^\mu. \tag{11}$$

We remark that the right-hand side of (11) can be written in the form $e(\alpha, e^\lambda x e^\mu)$, where e^λ, e^μ are the projectors in X generated by the left and right translations on the members of the subgroup K, respectively. Let us put $X^{\lambda\mu}=e^\lambda X e^\mu$ and denote by $A^{\lambda\mu}$ the set of all operator functions

$$x(\alpha) = p^\lambda x(\alpha) = x(\alpha)p^\mu\colon\ L(\nu)\to L(\nu)$$

(ν is the index of α), which satisfy the symmetry conditions (9) and which are entire functions of exponential growth of $\tau=p+q$, with estimates of form

$$\|\alpha^n x(\alpha)\| \leqq C_n e^{a|\mathrm{Re}\,\varrho|}, \quad n = 0, 1, 2, \dots, \tag{12}$$

where $\alpha^n=\|p\|^n\|q\|^n$, $n=0, 1, 2, \dots$, and $\|X\|$ denotes the norm of the finite-dimensional matrix function $x=p^\lambda x=xp^\mu$. Notice that the Fourier transformation (10) maps $X^{\lambda\mu}$ into $A^{\lambda\mu}$.

Theorem 12. *The Fourier transformation* (10) *sets up an isomorphism between the families* $X^{\lambda\mu}$ *and* $A^{\lambda\mu}$ *for every pair of highest weights* λ, μ.

Moreover, it is easy to construct compact subsets $K(a)\subset G$ such that every function $x(g)$ vanishes outside them if and only if the estimate (12) holds for the corresponding Fourier transform [24]. Since the algebra X is "well-approximated" by the families $X^{\lambda\mu}$, Theorem 12 also implies a global theorem of Paley—Wiener type for the whole algebra X [24]. Similar results can be obtained for other classes of finite or rapidly decreasing (fundamental or generalized) functions on G[7]. Rapidly decreasing smooth functions on G were also studied in the work [5] of I. M. Gelfand and M. I. Graev. The case $G=SL(2, \mathbf{C})$ had been studied earlier in the works [6], [23].

13. *Classification of the irreducible representations.* Let $M=(T, E, \tau)$ be a locally convex G-module, i.e. a representation $g\to T(g)$ of the group G on a sequentially complete vector space E with separated locally convex topology τ.[8] We shall say that the module $M_0=(T_0, E_0, \tau_0)$ is a *reduction* of the module M, if $E_0\subset E$, $\tau_0\geqq\tau$, and $T_0(g)=T(g)|E_0$. A reduction is called a *condensation* if E_0 is dense in E. The module M is said to be *variationally irreducible* if the lattice of its reductions consists of (0) and M only. Moreover, M is called *extremally irreducible* if it is variationally irreducible and every continuous operator $A\colon E\to E$ which commutes with all the operators $T(g)$ is a multiple of the identity.

There exists a one-to-one correspondence between the variationally irreducible

[7] Since our survey is devoted to the study of elementary representations, we will not go into the results of harmonic analysis in more detail. To this end, see [13], [15].

[8] See the footnote on p. 708.

G-modules and the algebraically irreducible X-modules, $X=C_0^\infty(G)$, [25]. If M is a completely irreducible G-module, then there exists a unique condensation $M_0 \subset M$ (i.e., unique upto topology) which is extremally irreducible. Therefore, the study of completely irreducible G-modules is reducible to that of the extremally irreducible ones. Two G-modules are defined to be equivalent if there exists a topological isomorphism between them intertwining the representation operators.

Theorem 13. *Every extremally irreducible G-module is equivalent to a condensation $\mu_0(\alpha)$ of one of the modules $\mu(\alpha)$.*

Corollary 1. *Every completely irreducible G-module contains a condensation which is equivalent to $\mu_0(\alpha)$ for a suitable α.*

Corollary 2. *There exists a one-to-one correspondence between $A/\mathbf{W}$ and the set of all equivalence classes of the extremally irreducible modules of the group G.*

Here $A/\mathbf{W}$ denotes the set of all orbits of the Weyl group in the manifold A of all signatures (i.e. all characters of the Borel subgroup B). The proof of Theorem 13 is based on Theorem 12, i.e. on harmonic analysis in the class X [25]. Corollary 1, in a somewhat different formulation, was obtained earlier in the work [14] of M. A. Neumark and the present author.

14. *Leading vectors, δ-duality.* Let $e[\alpha]$ be the extension of $e(\alpha)$ to the class of generalized functions. The representation $e[\alpha]$ can be defined as the representation of the group G conjugate to $e(\hat{\alpha})$, or the Hermitian conjugate to $e(\dot{\alpha})$, where $\hat{\alpha}=-p|-q$ and $\dot{\alpha}=-\bar{q}|-\bar{p}$ provided $\alpha=p|q$. Correspondingly, let $\mu[\alpha]$ be the extension of $\mu(\alpha)$ for $\alpha\in A_+$.

Concerning the representations $e[\alpha]$ we have a remarkable analogue of the Cartan—Weyl theory [25], [26] which is connected with the existence of weighted vectors for the Borel subgroups. In particular, let B be the Borel subgroup inducing $e(\alpha)$, B^+ be the Hermitian conjugate subgroup. Let δ_a be the δ-function on G, concentrated in the point $a\in G$; we consider δ_a as a functional on $e(\hat{\alpha})$, i.e., as an element of $e[\alpha]$. Put $\delta=\delta_e$, $\delta^*=\delta_G$ where e is the identity and σ a Weyl conjugation in G. It is easy to verify that δ and δ^* are weighted vectors of B and B^*, respectively.

The weighted vectors of B and B^* are called the smallest and highest vectors of $e[\alpha]$, respectively.

Furthermore, consider as above the $\mathfrak{G}$-module $\varepsilon(\alpha)$ in $V(\alpha)$ and put $\varepsilon^+(\alpha)= =de[\hat{\alpha}]\,|\,\Lambda(\alpha)$, where $\Lambda(\alpha)$ is the cyclic hull of the highest vector δ^*. The canonical bilinear form between $e(\alpha)$ and $e(\hat{\alpha})$ induces a duality between $\varepsilon(\alpha)$ and $\varepsilon(\hat{\alpha})$ which we call δ-duality[9]. It can be shown that the submodules $\varepsilon(\alpha)$, $\varepsilon^+(\alpha)$ are

[9] Another way to construct δ-duality was suggested by D.-N. Verma (oral communication).

not necessarily closed in the dual topologies of the pair $V(\alpha)$, $\Lambda(\alpha)$. However, if the signature α satisfies certain conditions then we can claim that the submodule $\varphi(\alpha)\subset\varepsilon(\alpha)$ corresponding to the minimal representation $\mu(\alpha)$ is closed and its annihilator $\varphi_0(\alpha)=ann\,\varphi(\alpha)$ is maximal in $\varepsilon^+(\alpha)$. This implies that $\varphi(\alpha)= =ann\,\varphi_0(\alpha)$, moreover the space of the representation $\mu(\alpha)$ is determined in $e(\alpha)$ by a system of differential equations

$$\mathscr{D}x(g) = 0, \quad \mathscr{D}\in I(\alpha),$$

where $I(\alpha)$ is a certain family of infinitesimal operators of left translation on G generated by a nilpotent subgroup $N^+\subset B^+$. The system $I(\alpha)$ is called the *indicator system* of $\mu(\alpha)$ [26]. The explicit description of this system is connected with the structure of the Harish-Chandra modules $\chi^+(\alpha)=\mathfrak{G}/\mathfrak{G}^+(\alpha)$, where $\mathfrak{G}^+(\alpha)$ denotes the left ideal generated by the elements

$$n^+,\ x-\alpha(x),\ \alpha(x) = (\lambda, x)+(\mu, \bar{x}),\ x\in\mathfrak{g},\ \lambda = p-\delta,\ \mu = q-\delta.$$

The structure of the modules $\chi^+(\alpha)$ was studied in [2] and [17]; however, no complete description is known so far.

If $\alpha\in A_+$, the module $e[\alpha]$ contains as irreducible component $\mu[\alpha]=C_0(\alpha)e[s_0\alpha]$, which is an extension of $\mu(\alpha)$ to the class of generalized functions. The module $\mu[\alpha]$ contains the cyclic (highest) vector $\zeta_\alpha=C_0(\alpha)\eta_{s_0\alpha}$, which can be given by the function $\alpha(g)=\alpha(\zeta hz)=\alpha(h)$, for $\zeta\in N$, $h\in H$, $z\in N^+$, where N and N^+ stand for the nilpotent commutators of B and B^+, respectively. In view of the equivalence relations among the $\mu[\alpha]$, in general there exist several linearly independent highest vectors of $\mu[\alpha]$ (with different weights). If all projections p_i, q_i are positive integers, then ζ_α is the classical highest vector of the finite-dimensional representation $\mu[\alpha]=\mu(\alpha)$.

More details about the highest vectors of $e[\alpha]$ can be found in [25], [26]. Highest vectors of representations of "class 0" of semisimple real Lie groups have also been considered by S. Helgason [9].

15. *The structure of the submodules of* $e(\alpha)$. The description of all submodules of $e(\alpha)$ (in degenerate points α) is still unknown. Therefore we shall only consider here several scattered particular results.

We remark that on the orbit of each signature with respect to the Weyl group there always exists a point $\alpha=s_0(p|\sigma q)$ with the following properties: *1)* the projections p_i, q_i are non-negative integers if $\omega_i\in\Pi_0$, where Π_0 is a certain subsystem of simple roots; *2)* $\sigma\in W_0$ where W_0 is the subgroup of the Weyl group generated by Π_0; *3)* the projections p_ω, q_ω are integers for $\omega\in\Delta_0$ only, where Δ_0 denotes the system of all roots generated by Π_0. The set of all such signatures α will be denoted by $N_\sigma(\Pi_0)$. We remark that $N_\sigma(\Pi_0)\subset A_-$. By means of certain induction considerations on the rank of the group G we may restrict our attention to the single case in which $\Pi_0=\Pi$ (in this case all projections p_ω, q_ω are integers). We put $N_\sigma=N_\sigma(\Pi)$.

Now we shall introduce a partial ordering in the Weyl group as follows. We say that $\sigma < s$ if the element s can be written as a reduced chain $s = s_{i_1} s_{i_2} \dots s_{i_n}$ (s_i denotes the reflection with respect to ω_i) in such a way that σ is obtained from s by omitting some of the factors s_i. Let us note that the immediate successors of the element s are those of the form $s_\omega s$, $l(s_\omega s) > l(s)$, where s_ω is the reflection with respect to the root ω and $l(s)$ denotes the length of the element s, i.e. the number of factors s_i in the reduced chain of s. This implies that our ordering coincides with that introduced in [2], [18].

Theorem 14. *If* $\alpha \in N_\sigma$ *the representation* $e(\alpha)$ *can be written as the semi-direct sum*

$$e(\alpha) = \sum_{s \geqq \sigma} k_s \mu(p/sq), \tag{13}$$

where $k_s \geqq 1$ *for* $s \geqq \sigma$. *The cyclic component of this representation is* $\mu(p\,|\,\sigma q)$.

We also remark that $\mu(p\,|\,s_0 q)$ is an irreducible submodule of $e(\alpha)$. Formula (13) remains valid for all signatures $t\alpha$, $t \in \mathbf{W}$ (however in this case another system of ordering of the submodules $\mu(\alpha)$ is to be chosen). The proof of Theorem 15 is not complicated and is based on Corollary 2 of Theorem 6.

Example. $G = SL(3, \mathbf{C})$, $\alpha = s_0(\delta\,|\,\delta)$, where δ denotes one half of the sum of the positive roots. In this case $k_s = 1$ for all $s \in \mathbf{W}$.

16. *Intertwining operators.* Write $P(\alpha, \beta)$ for the set of all (continuous) homomorphisms $e(\alpha) \to e(\beta)$ and $\Pi(\alpha, \beta)$ for the set of all (algebraic) homomorphisms $\varepsilon(\alpha) \to \varepsilon(\beta)$. The correspondence between $e(\alpha)$ and $\varepsilon(\alpha)$ yields a natural imbedding of $P(\alpha, \beta)$ into $\Pi(\alpha, \beta)$.

Conjecture 1. $P(\alpha, \beta) = \Pi(\alpha, \beta)$.

The validity of Conjecture 1 can be verified easily in case $G = SL(2, \mathbf{C})$. Let $A \in P(\alpha, \beta)$, then A commutes with right translations from K and the kernel function theorem of L. Schwartz guarantees that A is an operator of convolution over the group K

$$A\varphi = a * \varphi, \quad \varphi \in E(\alpha)$$

by some generalized function a, where $E(\alpha)$ stands for the space of representation $e(\alpha)$ realized on the group K. Applying the functional $\langle \cdot, \delta \rangle$ to both sides of the equality (δ denotes the δ-function on K) we get $\langle \varphi, a \rangle = \langle A\varphi, \delta \rangle = \langle \varphi, A^* \delta \rangle$, hence $a = A^* \delta$, where A^* stands for the adjoint of A. Now δ is a smallest vector of $e[\hat{\alpha}]$, so a is a smallest vector of $e[\hat{\beta}]$. It can be shown on the other hand that to every smallest vector of $e[\hat{\beta}]$ there exists a corresponding operator in $P(\alpha, \beta)$ (with suitable α)[10].

17. *Unsolved problems.* We shall make a list of several unsolved problems connected with the modules $e(\alpha)$.

[10] See e.g. [5]. The function $k(s, \sigma) = \det(s\sigma)$ is called the Möbius function of the group $\mathbf{W}$.

I. *Description of the structure of all submodules of* $e(\alpha)$.

In particular, how to decide when $e(\beta)$ is imbeddable in $e(\alpha)$.

A solution to the second question was obtained in the work [2] for the case of the cyclic modules $\chi^+(\alpha)$. In view of the considerations of δ-duality it can be expected that a solution for our case can be obtained in analogous terms.

II. *Description of the family* $P(\alpha, \beta)$ *of all intertwining operators (homomorphisms) between the modules* $e(\alpha)$ *(and the same for* $\varepsilon(\alpha)$*)*.

The solution of the analogous problems for $\mu(\alpha)$ is known [22]: the only intertwining operator $\mu(\alpha) \to \mu(\beta)$, $\alpha \in A_-$, $\beta \in A_+$ is $W_s(\alpha)$, $s\alpha > \beta$. Especially interesting is the case $\beta = \dot{\alpha} = -\bar{q}| - \bar{p}$, in which $W_s(\alpha)$ determines a Hermitian form on $L(v)$ which is invariant with respect to $\mu(\alpha)$ and $\mu(\dot{\alpha})$ (but is not necessarily positive-definite). However, the following problem is still unsolved.

III. *When is* $W_s(\alpha)$ *positive-definite for* $s\alpha = \dot{\alpha}$, $\alpha \in A_-$?

According to what has been said above, a solution to this problem would enable us to determine all signatures α such that the representation $\mu(\alpha)$ is unitary. In view of a theorem of M. A. Neumark and the present author [14], this would further make it possible to enumerate all irreducible unitary representations of the group G, up to unitary equivalence.

IV. *Study the embedding* $\mu(\alpha) \subset e(\alpha)$ *for* $\alpha \in A_+$.

In particular, it is interesting to decide when the space $F(\alpha)$ of the representation $\mu(\alpha)$ coincides with the null-space of the system of the symmetry operators (in this case $\mu(\alpha) = \pi(\alpha)$), as well as when $\mu(\alpha)$ is induced by a finite-dimensional representation of a parabolic subgroup. In [27] this author made the conjecture that the latter property is fulfilled by every $\mu(\alpha)$; however, it turns out to be not the case. A counterexample has been communicated to the author by D. A. Kajdan and I. N. Bernstein.

V. *Is* $\mu(\alpha)$ *extremally irreducible?*

The author takes this occasion to thank Professors B. Kostant, D.-N. Verma, I. N. Bernstein and D. A. Kajdan for discussing this paper and giving valuable remarks.

REFERENCES

[1] Berezin, F. A. (Березин Ф. А.), Операторы Лапласа на полупростых группах Ли, *Труды Моск. матем. о-ва,* **6** (1957), 371—463.
(Laplace operators on semisimple Lie groups).

[2] Bernstein, I. N., Gelfand, I. M. and Gelfand, S. I. (Бернштейн И. Н., Гельфанд И. М., Гельфанд С. И.), Структура представлений, порожденных векторами старшего веса, *Функциональный анализ и его приложения,* **5**:1 (1971), 1—9.
(The strcture of representations generated by a vector of highest weight.)

[3] ———, ———, ——— Differential operators on the base affine space and a study of g-modules, in *this volume*.

[4] Bruhat, F., Sur les représentations induites des groupes de Lie, *Bull. Soc. Math. France* **84** (1956), 97—205.

[5] Gelfand, I. M. and Graev, M. I. (Гельфанд И. М., Граев М. И.), Преобразования Фурье быстро убывающих функций на комплексных полупростых группах, *Докл. АН СССР*, **131** (1960), 496—499.
(Fourier transforms of rapidly decreasing functions on complex semisimple groups.)

[6] ———, ——— and Vilenkin, N. Ja. (Виленкин Н. Я.), *Интегральная геометрия и связанные с ней вопросы теории представлений*, „Обобщенные функции", вып. **5**, Москва, Физматгиз, 1962.
(Integral geometry and its connections with representations theory.)

[7] ——— and Neumark, M. A. (Наймарк М. А.), Унитарные представленя классических групп, *Труды матем. ин-та им. В. А. Стеклова АН СССР*, 36, (1950). (*Unitäre Darstellungen der klassischen Gruppen*, Akademie Verlag, Berlin, 1957.)

[8] Harish-Chandra, The Plancherel formula for complex semisimple Lie groups, *Trans. Amer. Math. Soc.*, **76** (1954), 485—528.

[9] Helgason, S., Applications of the Radon transform to representations of semisimple Lie groups, *Proc. Nat. Acad. Sci. USA*, **60** (1969), 3—6.

[10] Jacobson, N., *Lie algebras*. (Джекобсон Н., *Алгебры Ли*, „Мир", Москва, 1964.)

[11] Knapp, A. W. and Stein, E. M., Intertwining operators for semisimple groups, *Ann. of Math.*, **93** (1971), 489—578.

[12] Kostant, B., The existence and irreducibility of a certain series of representations, *Bull. Amer. Math. Soc.*, **76** (1969), 68—93.

[13] Kunze, R. A. and Stein, E. M., Uniformly bounded representations, I, III, *Amer. J. Math.*, **83** (1961), 723—786; **89** (1967), 385—442.

[14] Neumark, M. A. and Želobenko, D. P. (Наймарк М. А., Желобенко Д. П.) Описание вполне неприводимых представлений полупростой комплексной группы Ли, *Изв. АН СССР*, сер. матем., **34** (1970), 57—85.
(Description of completely reducible representations of semisimple complex Lie groups.)

[15] Parthasarathy, K. R., Ranga Rao and Varadarajan, V. S., Representations of complex semisimple Lie groups and Lie algebras, *Ann. of Math.*, **85** (1967), 383—429.

[16] Schiffmann, G., Intégrales d'entrelacement, *C. R. Acad. Sci. Paris*, **266** (1968), 47—49.

[17] Verma, D. N., Structure of certain induced representations of complex semisimple Lie algebras, *Bull. Amer. Math. Soc.*, **74** (1968), 160—166.

[18] ———, Möbius inversion for a Bruhat ordering of a Weyl group, *preprint*, 1971.

[19] Želobenko, D. P. (Желобенко Д. П.), К теории линейных представлений комплексных и вещественных групп Ли, *Труды Моск. матем. о-ва*, **12** (1963), 53—98.
(On the theory of linear representations of complex and real Lie groups.)

[20] ———, Симметрия в классе элементарных представлений полупростой комплексной группы Ли, *Функциональный анализ и его приложения*, **1**:2 (1967), 15—38.
(Symmetry in the class of elementary representations of semisimple complex Lie groups.)

[21] ———, Анализ неприводимости в калссе элементарных представлений полупростой комплексной группы Ли, *Изв. АН СССР*, сер. матем., **32** (1968), 108—133.
(Analysis of irreducibility in the class of elementary representations of a complex Lie group.)

[22] ———, Операционное исчисление на полупростой комплексной группе Ли, *Изв. АН СССР*, сер. матем., **33** (1969), 931—973.
(Operational calculus on semisimple complex Lie groups.)

[23] ———, О гармоническом анализе функций на полупростых группах Ли, I. *Изв. АН СССР*, сер. матем., **27**:6 (1963), 1343—1396.
(On harmonic analysis of functions on semisimple Lie groups, I.)

[24] ———, О гармоническом анализе функций на полупростых группах Ли, II. *Изв. АН СССР,* сер. матем., **33**: 6 (1969), 1255—1295.
(On harmonic analysis of functions on semisimple Lie groups, II.)

[25] ———, Классификация экстремально неприводимых и нормально неприводимых представлений полупростой комплексной группы Ли, *Изв. АН СССР,* сер. матем. (The classification of extremally irreducible and normally irreducible representations of semisimple complex Lie groups.)

[26] ———, О циклических модулях для полупростой комплексной группы, Ли, *Изв. АН СССР,* сер. матем. (to appear)
(On cyclic modules for semisimple complex Lie groups.)

[27] ———, О представлениях класса 0 полупростой комплексной группы Ли, *Функциональный анализ и его приложения,* **4**: 2 (1970), 85—86.
(On representations of class 0 of semisimple complex Lie groups.)

PROGRAMME OF THE SUMMER SCHOOL

SECTION A(=ADVANCED)

August 16

Gelfand, I. M.: Introductory talk

August 17

Tanaka, S.: Cartier—Weil formula
Bernstein, I. N.: Differential operators on the base affine space
Duflo, M.: Induced representations of Lie algebra and applications to the study of the center of the universal enveloping algebra

August 18

Gelfand, S. I.: Representations of the general linear group over finite fields
Labesse, J. P.: Arithmetic equivalence and trace formula for $SL(2)$

August 19

Gelfand, I. M., Kajdan, D. A.: Noncommutative reciprocity law
Pjateckij-Šapiro, I. I.: The representations in the space $GL(n, A)\setminus GL(n, k)$

August 23

Godement, R.: Zeta functions and parabolic forms
Van Dijk, G.: Admissible representations of $\mathfrak{p}$-adic groups

August 24

Andrianov, A. N.: Zeta functions associated with Siegel modular forms of genus 2
Kirillov, A. A.: The method of orbits in representation theory
Tanaka, S.: Cusp forms connected with the quadratic extensions

August 25

Mostow, G. D.: Discrete subgroups from a geometrical view-point
Harder, G.: On the cohomology of $SL(2, O)$
Springer, T. A.: Asymptotic construction of characters of finite groups

August 26

Harder, G. (cont.)
Springer, T. A.: The Steinberg function of finite Lie algebras

August 27

Kostant, B.: Polarizations and a generalized Fourier transform
Želobenko, D. P.: Elementary representations of semisimple Lie algebras
Mackey, G. W.: Some analogies between semi-direct products and semisimple Lie groups

August 30

Parthasarathy, R.: On the realization of the discrete series for a semisimple Lie group
Kostant, B.: Existence and uniqueness of certain series of representations
Goto, M.: Non-discrete uniform subgroups of semisimple Lie groups

August 31

Kostant, B.: (cont.)
Verma, D.-N.: On rational representations of Chevalley groups in nonzero characteristic
Parthasarathy, R.: (cont.)

September 1

Verma, D.-N.: (cont.)
Mautner, F. I.: Spherical functions and Hecke operators

September 2

Bhanu Murthy, T. S.: The Plancherel formula for $SU(p, q)$

SPECIAL LECTURES FOR PHYSICISTS

August 25—26—27 (4 p.m.)

Mackey, G. W.: Induced representations and their applications to quantum mechanics

SECTION B(=BEGINNERS)

August 16

Gelfand, I. M.: Introductory talk

August 17

Tanaka, S.: Cartier—Weil formula
Andrianov, A. N.: Local fields
Kirillov, A. A.: Representation theory of finite and compact groups

August 18

Andrianov (cont.)
Kirillov (cont.)

August 19

Kirillov (cont.)
Bernstein, I. N.: Representations of semisimple Lie algebras

August 23

Bernstein (cont.)
Želobenko, D. P.: Gelfand—Cetlin formulas of the representations of $GL(n, \mathbf{C})$

August 24

Želobenko (cont.)

August 25

Želobenko (cont.)

August 26

Kajdan, D. A.: Lecture on number theory

August 27

Kajdan (cont.)

August 31

Andrianov, A. N.: Modular forms, zeta functions and elliptic curves

September 1

Andrianov (cont.)

September 2

Andrianov (cont.)
Pjateckij-Šapiro, I.I.: The Hasse—Weil conjecture for arithmetic curves
Tanaka, S.: Representations of $SL(2, F_q)$

September 3

Bhanu Murthy, T. S.: Homogeneous complex manifolds and group representations

LIST OF CONTRIBUTORS
AND FOREIGN PARTICIPANTS OF THE SUMMER SCHOOL

Andrianov, A. N. — Steklov Institute of Mathematics, Leningrad.

Bernstein, I. N. — Moscow State University.

Bhanu Murthy, T. S. — Ramanujan Institute, Madras.

van Dijk, Gerrit — State University of Utrecht.

Duflo, Michel — University of Paris VII.

Gelfand, I. M. — Institute of Applied Mathematics, Moscow.

Gelfand, S. I., — Institute of Information Transmission of the Academy of Sciences of the USSR, Moscow.

Godement, Roger — University of Paris VII.

Goto, Morikuni — University of Pennsylvania.

Harder, Günter — University of Bonn.

Kajdan, D. A. — Moscow State University.

Kirillov, A. A. — Moscow State University.

Kostant, Bertram — Massachusetts Institute of Technology.

Labesse, Jean-Pierre — University of Paris VII.

Lachaud, Gilles — University of Paris VII.

Mackey, George W. — Harvard University.

Margulis, G. A. — Institute of Information Transmission of the Academy of Sciences of the USSR, Moscow.

Mautner, Friedrich I.

Mostow, G. D. — Yale University.

Parthasarathy, R. — Tata Institute, Bombay.

Pjateckij-Šapiro, I. I. — Institute of Applied Mathematics, Moscow.

Rodier, François — École Normale Supérieure, Paris.

Springer, T. A. — State University of Utrecht.

Tanaka, Shunichi — Osaka University.

Verma, Daya-Nand — Tata Institute, Bombay.

Želobenko, D. P. — People's Friendship University, Moscow.

And many other participants from Hungary.